SPIE Milestone Series
Volume MS 7

Selected Papers on
Scattering in the Atmosphere

Craig F. Bohren
Editor

Brian J. Thompson
General Editor, SPIE Milestone Series

SPIE OPTICAL ENGINEERING PRESS

A Publication of SPIE—The International Society for Optical Engineering
Bellingham, Washington USA

Library of Congress Cataloging-in-Publication Data

Selected papers on scattering in the atmosphere / Craig F. Bohren, editor.
    p.   cm. — (SPIE milestone series) (SPIE ; MS 7)
    1. Light—Scattering. 2. Meteorological optics.
    3. Atmospheric physics.   I. Bohren, Craig F., 1940–
    II. Series.
    QC976.S3S45   1989
    551.5′65—dc20   89-24059
                                                 CIP

ISBN 0-8194-0283-4 (hardbound)
ISBN 0-8194-0240-0 (softbound)

Published by
SPIE—The International Society for Optical Engineering
P.O. Box 10, Bellingham, Washington 98227-0010 USA
Telephone 206/676-3290 (Pacific Time) • Telex 46-7053

Introduction to the Series

There is no substitute for reading the original papers on any subject even if that subject is mature enough to be critically written up in a textbook or a monograph. Reading a well-written book only serves as a further stimulus to drive the reader to seek the original publications. The problems are, which papers and in what order?

As a serious student of a field, do you really have to search through all the material for yourself, and read the good with the not-so-good, the important with the not-so-important, and the milestone papers with the merely pedestrian offerings? The answer to all these questions is usually yes, unless the authors of the textbooks or monographs that you study have been very selective in their choices of references and bibliographic listings. Even in that all-too-rare circumstance, the reader is then faced with finding the original publications, many of which may be in obscure or not widely held journals.

From time to time and in many disparate fields, volumes appear that are collections of reprints that represent the milestone papers in the particular field. Some of these volumes have been produced for specific topics in optical science, but as yet no systematic set of volumes has been produced that covers connected areas of optical science and engineering.

The editors of each individual volume in this series have been chosen for their deep knowledge of the world literature in their fields; hence, the selection of reprints chosen for each book has been made with authority and care.

On behalf of SPIE, I thank the individual editors for their diligence, and we all hope that you, the reader, will find these volumes invaluable additions to your own working library.

Brian J. Thompson
University of Rochester
September 1989

 *Other Books in SPIE's Milestone Series*

Selected Papers on:

Optical Computing
H. John Caulfield, Gregory Gheen, *Editors*

Light Scattering
Milton Kerker, *Editor*

Optomechanical Design
Donald C. O'Shea, *Editor*

Infrared Design
R. Barry Johnson and William L. Wolfe, *Editors*

Laser Scanning and Recording
Leo Beiser, *Editor*

Forthcoming Milestones:

Deposition of Optical Coatings
Michael R. Jacobson, *Editor*

Fiber Optic Gyroscopes
Robert B. Smith, *Editor*

Surface-Enhanced Raman Scattering
Milton Kerker, *Editor*

Infrared Fiber Optics
James A. Harrington, *Editor*

Optical Fibers in Medicine
Abraham Katzir, *Editor*

Radiometry
Irving J. Spiro, *Editor*

Visual Communication: Technology and Applications
T. Russell Hsing, Andrew G. Tescher, *Editors*

Selected Papers on
Scattering in the Atmosphere

Contents

iii Introduction to the Series, Brian J. Thompson
ix Preface, Craig F. Bohren

Section One
Historical Keynote Paper

3 **Random reflections on the history of atmospheric optics**
W. E. K. Middleton (*Journal of the OSA* 1960)

Section Two
Molecular Scattering and Skylight

9 **On the light from the sky, its polarization and colour**
Lord Rayleigh (J. W. Strutt) (*Philosophical Magazine* 1871)

26 **On the transmission of light through an atmosphere containing small particles in suspension, and on the origin of the blue of the sky**
Lord Rayleigh (J. W. Strutt) (*Philosophical Magazine* 1899)

35 **Theories of the color of the sky**
E. L. Nichols (*Proceedings of the American Physical Society* 1908)

50 **On the complex anisotropic molecule in relation to the dispersion and scattering of light**
L. V. King (*Proceedings of the Royal Society of London* 1923)

75 **Explanation of the brightness and color of the sky, particularly the twilight sky**
E. O. Hulburt (*Journal of the OSA* 1953)

81 **Tables of the refractive index for standard air and the Rayleigh scattering coefficients for the spectral region between 0.2 and 20.0 L and their application to atmospheric optics**
R. Penndorf (*Journal of the OSA* 1957)

88 **Light scattering in the earth's atmosphere**
G. V. Rozenberg (*Soviet Physics Uspekhi* 1960)

115 **On the Rayleigh-scattering optical depth of the atmosphere**
A. T. Young (*Journal of Applied Meteorology* 1981)

118 **Rayleigh scattering**
A. T. Young (*Physics Today* 1982)

Section Three
Multiple Scattering

127   **Radiation through a foggy atmosphere**
A. Schuster (*Astrophysical Journal* 1905)

149   **Matrix methods for multiple-scattering problems**
S. Twomey, H. Jacobowitz, H. B. Howell (*Journal of the Atmospheric Sciences* 1966)

157   **The delta-Eddington approximation for radiative flux transfer**
J. H. Joseph, W. J. Wiscombe, J. A. Weinman (*Journal of the Atmospheric Sciences* 1976)

165   **Multiple scattering of light and some of its observable consequences**
C. F. Bohren (*American Journal of Physics* 1987)

175   **Numerically stable algorithm for discrete-ordinate-method radiative transfer in multiple scattering and emitting layered media**
K. Stamnes, S.-C. Tsay, W. J. Wiscombe, K. Jayaweera (*Applied Optics* 1988)

Section Four
Clouds

185   **Noctilucent clouds**
F. H. Ludlam (*Tellus* 1957)

210   **Scattering and polarization properties of water clouds and hazes in the visible and infrared**
D. Deirmendjian (*Applied Optics* 1964)

220   **The influence of pollution on the shortwave albedo of clouds**
S. Twomey (*Journal of the Atmospheric Sciences* 1977)

224   **Solar radiative transfer in cirrus clouds. part I: single-scattering and optical properties of hexagonal ice crystals**
Y. Takano, K.-N. Liou (*Journal of the Atmospheric Sciences* 1989)

241   **Solar radiative transfer in cirrus clouds. part II: theory and computation of multiple scattering in an anisotropic medium**
Y. Takano, K.-N. Liou (*Journal of the Atmospheric Sciences* 1989)

Section Five
Polarization

261   **The illumination and polarization of the sunlit sky on Rayleigh scattering**
S. Chandrasekhar, D. D. Elbert (*Transactions of the American Philosophical Society* 1954)

273   **Brightness and polarization of the daylight sky at various altitudes above sea level**
R. Tousey, E. O. Hulburt (*Journal of the OSA* 1947)

288   **Interpretation of the polarization of Venus**
J. E. Hansen, J. W. Hovenier (*Journal of the Atmospheric Sciences* 1974)

312 **Light scattering in the atmosphere and the polarization of sky light**
Z. Sekera (*Journal of the OSA* 1957)

319 **Skylight polarization during a total solar eclipse: a quantitative model**
G. P. Können (*Journal of the OSA A* 1987)

Section Six
Rainbows, Coronas, and Glories

329 **On the intensity of light in the neighbourhood of a caustic** and **Supplement**
G. B. Airy (*Transactions of the Cambridge Philosophical Society* 1838, 1849)

358 **Coronae and iridescent clouds**
G. C. Simpson (*Quarterly Journal of the Royal Meteorological Society* 1912)

367 **A theory of the anti-coronae**
H. C. van de Hulst (*Journal of the OSA* 1947)

374 **Mie theory and the glory**
H. C. Bryant, A. J. Cox (*Journal of the OSA* 1966)

378 **Infrared rainbow**
R. G. Greenler (*Science* 1971)

380 **Complex angular momentum theory of the rainbow and the glory**
H. M. Nussenzveig (*Journal of the OSA* 1979)

393 **Why can the supernumerary bows be seen in a rain shower?**
A. B. Fraser (*Journal of the OSA* 1983)

Section Seven
Ice Crystal Phenomena

399 **A general theory of halos**
C. S. Hastings (*Monthly Weather Review* 1920)

408 **Circumscribed halos**
R. G. Greenler, A. J. Mallmann (*Science* 1972)

412 **A simple theory of certain heliacal and anthelic halo arcs: the long hexagonal ice prism as a kaleidoscope**
R. A. R. Tricker (*Quarterly Journal of the Royal Meteorological Society* 1973)

420 **What size of ice crystals causes the halos?**
A. B. Fraser (*Journal of the OSA* 1979)

427 **Some ice crystals that made halos**
W. Tape (*Journal of the OSA* 1983)

433 **Multiple-scattering effects in halo phenomena**
E. Tränkle, R. G. Greenler (*Journal of the OSA A* 1987)

Section Eight
Mirages and the Green Flash

445 **Elementare theorie der atmosphärischen spiegelungen**
A. Wegener (*Annalen der Physik* 1918)

473 **Normal atmospheric dispersion as the cause of the "green flash" at sunset, with illustrative experiments**
Lord Rayleigh (R. J. Strutt) (*Proceedings of the Royal Society A* 1930)

481 **Observations and theoretical reconstruction of the green flash**
G. E. Shaw (*Pure and Applied Geophysics* 1973)

494 **The green flash and clear air turbulence**
A. B. Fraser (*Atmosphere* 1975)

Section Nine
Extinction

505 **On the atmospheric transmission of sun radiation and on dust in the air**
A. Ångström (*Geografiska Annaler* 1929)

516 **On the atmospheric transmission of sun radiation II**
A. Ångström (*Geografiska Annaler* 1930)

546 **The blue sun of 1950 September**
R. Wilson (*Monthly Notices of the Royal Astronomical Society* 1951)

Section Ten
Visibility

561 **Theorie der horizontalen sichtweite**
H. Koschmieder (*Beiträge zur Physik der freien Atmosphäre* 1924)

584 **Theorie der horizontalen sichtweite II: kontrast und sichtweite**
H. Koschmieder (*Beiträge zur Physik der freien Atmosphäre* 1925)

595 **The reduction of apparent contrast by the atmosphere**
S. Q. Duntley (*Journal of the OSA* 1948)

608 **Optics of sunbeams**
D. K. Lynch (*Journal of the OSA A* 1987)

Section Eleven
Backscattering

613 **Errors inherent in the radar measurement of rainfall at attenuating wavelengths**
W. Hitschfeld, J. Bordan (*Journal of Meteorology* 1954)

623 **Determination of aerosol height distributions by lidar**
F. G. Fernald, B. M. Herman, J. A. Reagan (*Journal of Applied Meteorology* 1972)

631 **Analysis of atmospheric lidar observations: some comments**
F. G. Fernald (*Applied Optics* 1984)

633 **Lidar inversion with variable backscatter/extinction ratios**
J. D. Klett (*Applied Optics* 1985)

639 Author Index
641 Subject Index

Preface

When I lived in Wales, BBC Radio broadcast a weekly program entitled "Desert Island Discs." It is almost unnecessary to describe its format since the name alone gives it away. Every week celebrities of various stripes were invited to chat about works of music they would want with them if stranded on a desert island. Inspired by this format, as I selected the following papers on atmospheric scattering I asked myself which ones might accompany me into island exile. Surely I would want those that would help me understand the sky since I would have plenty of time to watch it.

Marooned on my imaginary desert island, I would observe the sky in the light of Rayleigh's papers, and in my mind's eye return to a time when its blueness was dimly understood. Yet papers by Nichols and by Hulburt would remind me that Rayleigh's work, although an important step forward, did not give a full account of the color and brightness of the sky, especially at twilight.

With polarizing sunglasses I would explore the patterns of skylight polarization elucidated by Chandrasekhar and Elbert, whose work was the culmination of that begun by Rayleigh more than eighty years before.

On an island shore I would have an unobstructed horizon for observing the distorted images of the setting or rising sun, which were first related to the temperature structure of the atmosphere by Wegener, who met his end in Greenland, a very large and cold desert island. And I would surely see the green flash, despite its undeserved reputation as a rare phenomenon. Rayleigh and Shaw would help me understand its color, whereas Fraser would remind me that the crenellations of the sun seen when it is above the horizon result from horizontal rather than vertical inhomogeneities in the atmosphere.

Although ice is not commonly associated with desert islands, from time to time small ice crystals would inhabit the sky high overhead, their presence manifested by arcs and halos of stunning beauty. I would certainly want papers such as those by Hastings, Greenler and Mallmann, and Tricker to enhance my enjoyment of these displays.

Atmospheric visibility would be of great concern to me as I scanned the horizon each day for a rescue ship. Depending on my distance from the shipping lanes, such ships might be seen as infrequently as the blue moon and sun that Wilson was so fortunate to observe.

These topics in atmospheric scattering and the authors of papers about them give a clue to the primary criterion that guided me in selecting papers for this volume: I gave pride of place to those that interpret phenomena readily observable by the human eye, unaided except for possibly binoculars or polarizing filters. About two-thirds of the papers selected are related to directly observable phenomena resulting from scattering in the atmosphere, whether by molecules or by particles.

Anyone who essays to gather together under common cover a collection of papers on any subject and pronounce them important knows that among his customers are many who would have chosen otherwise. Indeed, except for Rayleigh's two papers and possibly a few others, it is unlikely that anyone else faced with my task would have produced the same list. One who compiles does so uneasily, knowing full well that all tastes cannot be served, all egos cannot be soothed, and all heroes cannot be eulogized.

No doubt I will be cursed roundly in some quarters for having omitted papers of pathbreaking importance or of having included ones judged to be mere footnotes to sacred texts. But, gentle reader, before you vent your rage, reflect on the constraints that hobbled me.

Space, of course, was the major constraint. I was limited to perhaps fifty papers, certainly no more than seventy-five. And the aggregate number of pages could not exceed 700. Thus an average paper could not be much longer than ten pages. This immediately ruled out many review papers, no matter how enlightening they might be. I have included only one or two papers that might be called reviews, and the reason for these exceptions is discussed below. Length also barred many important papers published in an era before publication charges, when scientists enjoyed the luxury of unveiling granitic monoliths brought to light after years of gestation instead of having to swell their resumes with a steady flow of gravel. I excluded from consideration chapters of books and encyclopedias, both because of their length and to avoid difficulties obtaining reproduction rights. Finally, I excluded work not published in refereed journals.

As I stated previously, the primary criterion that guided me was that a chosen paper should shed some light on the observable consequences of atmospheric scattering. But as I further narrowed my choices, I was guided by subsidiary criteria. One of them was the desire to recognize the major contributors to the field. Thus I chose names as much as topics. I searched my thoughts for salient names, and then tried to find papers to fit them.

Contributors to atmospheric scattering can be grouped into three categories: the founding fathers, the immediate past generation, and the present generation. Among the founding fathers I include Airy, Rayleigh, Schuster, and Wegener. The immediate past generation, the members of which are either deceased or officially retired, would include Middleton, Hulburt, Deirmendjian, Sekera, van de Hulst, and Ångström. Finally, the present generation is represented by Fraser, Greenler, Tape, Können, and Lynch, who can be counted on to have something more to say before the lights go out.

Another subsidiary criterion was the desire to make some papers more widely accessible. Only the world's greatest libraries have under one roof all the papers in this collection.

The advent of quantum mechanics and discoveries in nuclear and particle physics have engendered the expectation of frequent revolutions in science, of momentous, sudden, and discrete events. This expectation has been brought to a fever pitch by the incessant and ubiquitous press-agentry that is now as much a part of science as test tubes, spectrometers, and Bunsen burners. Yet viewed in the cold light of day, much science is more evolutionary than revolutionary. Before Chadwick discovered the neutron, for all intents and purposes, it did not exist. But one morning it suddenly popped, so to speak, out of the vacuum, as did the positron and all the other so-called "elementary" particles, the number and variety of which are beginning to pall. This was not so with atmospheric scattering.

Rayleigh did not discover the hitherto unobserved blue sky; he enunciated a law of scattering that helped to explain its origin. He had predecessors and successors, all forming part of a continuum. Rayleigh's contributions were noteworthy pinnacles on a ridge rising to a summit that has not quite been reached, just as the summit of Kangchenjunga remains untrodden. And indeed, all the papers in this collection could be so described. In the following paragraphs, I comment on my reasons for selecting some of them, although I shall not bore you by justifying every one.

I have taken Middleton's "Random reflections on the history of atmospheric optics" as the keynote paper. This serves the dual purpose of honoring one whose contributions to atmospheric science, especially its history, occupy volumes, and of setting the stage for what follows. Middleton's paper supplies an especially good historical perspective on the papers by Koschmieder and by Duntley.

The Russians have a reputation for asserting that everything in science was accomplished by them first. To a certain extent, they are correct: much original work by them took time to diffuse west, especially if first published in Russian. Yet Western reaction to this nationalistic braggadocio often has been to ignore Russian work, even up to the present day. To strike a balance between wild claims and neglect, I included Rozenberg's review paper, which gives the Russian perspective on our subject.

Young's paper on Rayleigh scattering is also a review, one of the two I included in violation of my precepts. Without his clear-headed unraveling of the many meanings of "Rayleigh scattering," this term would have fast been on its way to meaning nothing.

When selecting a few papers on multiple scattering to include, I was faced with the problem that there are hundreds if not thousands of papers on this subject strewn throughout journals in astrophysics, atmospheric science, and engineering. A thick volume could be devoted to the bewildering array of methods of solution of the equation of radiative transfer. As a compromise, I chose Schuster's pioneering paper and then a few papers on the methods likely to be most familiar to atmospheric scientists. Readers who are ardent partisans of other methods will have to wait for another more comprehensive—and possibly mammoth—volume.

I chose the paper by Twomey et al. knowing that their doubling method had been outlined previously by van de Hulst in a report, "A new look at multiple scattering," published by the Goddard Institute for Space Studies in 1963. To have included this report would have veered from two of my guidelines.

With only a twinge of embarrassment, I took the editorial privilege of including one of my own papers. Although not a seminal paper in the scientific sense, it does, I believe, represent a milestone of expository form. It is to my knowledge the first paper on multiple scattering aimed at physics teachers and their students. Multiple scattering is rarely mentioned even briefly in textbooks on optics despite its relevance to observable phenomena.

Just as with multiple scattering in general, there are scores of papers on scattering by clouds. I could include only a few, so again I chose papers that help elucidate observations. I wanted to include a paper on noctilucent clouds if for no other reason than to call attention to atmospheric clouds quite unlike those of our everyday experience. But in choosing Ludlam's paper, I was deterred somewhat by his advocacy of the now discarded notion that noctilucent cloud particles are of volcanic or extraterrestrial origin. Because of this, one expert reacted with horror to my suggestion that this paper be included. Yet he could not suggest a better one, so the paper stayed. This choice reflects a criterion that did *not* guide me: I did not exclude papers on the grounds that they contain errors or run contrary to the current conventional wisdom.

The paper by Twomey on pollution and clouds is interesting in its own right, but especially in light of the present hand-wringing over the role of clouds in climate change. Do clouds make the earth warmer or cooler? Tune in tomorrow for the latest episode in this exciting adventure series.

I chose the two-part paper by Takano and Liou because it is the latest in a long series on cirrus clouds by Liou and his collaborators and because of its tie to observations of ice crystal phenomena.

Sekera was not, by today's standards, a prolific publisher of papers in scientific journals. Much of his work was published as research reports, and one of his best known publications is a review chapter on polarization of skylight, the length of which and its publication in Volume 48 of *Handbuch der Physik* precluded its inclusion in this collection. To call attention to his work, I included a journal article that is a shortened version of his review chapter and was published in the same year.

I chose Greenler's paper on the infrared rainbow in part because he told me that much to his surprise it has been the paper of his that has attracted the most attention and in part because of the following incident. Just before Thanksgiving last year, one of my former students called me to ask, "Do you know about the infrared rainbow? I just heard something about it on the radio." My response was, "Of course, Greenler did something on it, but that was several years ago." Within a few hours of this puzzling conversation, I went upstairs to take a bath. I brought my shortwave radio with me, tuned to the Voice of America Spanish-language broadcast, and turned on the water. As the water was flowing into the tub, the words *"arco iris infrarojo"* and *"el profesor Robert Greenler de la Universidad de Wisconsin en Milwaukee"* suddenly filtered through the noise of splashing water. I grabbed frantically for the taps, but by the time I got the water turned off the broadcaster had moved on to another topic. This nearly simultaneous transmission of news of the infrared rainbow to a gringo in Washington, D.C., and Spanish-speaking people from Mexico to Tierra del Fuego was an omen I could not ignore.

The papers by Greenler and Mallmann and by Tränkle and Greenler were chosen to bracket the work of Greenler and his collaborators: these papers are their first and their latest—but not, I trust, their last—on ice crystal phenomena. I should note that the first paper is not original in an absolute sense since, unbeknownst to its authors at the time, the circumscribed halo had been treated analytically much earlier by Wegener ["Theorie der Haupthalos," *Archiv der Deutschen Seewarte* 43, 1–32 (1926)]. Nevertheless, the approach of Greenler and Mallmann was different from that of their predecessors and enabled them subsequently to solve problems that did not have analytical solutions.

You may be surprised to find papers on atmospheric refraction included in a collection of papers on scattering. If so, it is because you have been misled by the scores of textbook writers who have been shockingly slow to learn that refraction *is* scattering. The time has come—indeed, is long overdue—to dispel the widespread misconception that there is some fundamental difference between scattering and refraction. The work of Ewald, Oseen, and others showing how the laws of (specular) reflection and refraction can be obtained from a scattering point of view is now hoary with age. The refractive index of a gas is one plus a term proportional to the molecular number density and the molecular polarizability; the scattering coefficient of a gas is proportional to the number density and the polarizability squared, signaling that refraction and scattering are brothers under the skin. Refractive indices are associated with electric fields, scattering coefficients with the squares of such fields. Refraction may be considered as forward scattering in which the coherence properties of the scatterers determine what is observed, whereas scattering without qualification refers to all those other directions for which coherence may be ignored.

Another reason for including a section on refraction was to display one of Wegener's contributions to meteorology, which have been eclipsed by his formerly controversial but now accepted ideas about continental drift.

One of the advantages afforded to a compiler of milestone papers in a scientific field is that of righting historical wrongs. The empirical coefficients in the expression for the wavelength dependence of extinction by atmospheric particles are usually called the Ångström coefficients. Yet if one reads Ångström's original papers, one learns that he acknowledges Lundholm as the originator of these coefficients. So let us hear more in the future about the Lundholm instead of the Ångström coefficients. We do no honor to Ångström by crediting him with something to which he laid no claim.

Since the advent of lidar, many papers have been published on atmospheric backscattering. I had to content myself with including only a few of them. Again, the subject of backscattering probably merits its own volume.

Before I began compiling this collection, I wrote to several colleagues telling them what I had in mind and asking for advice. Most of them responded, but some did not—the ones who might disagree most with my choice of papers. To those who responded to my call for help I am grateful: Tom Ackerman, Ed Andreas, Kinsel Coulson, Chris Fairall, Alistair Fraser, Gunther Können, Bob Greenler, John Howard, John Olivero, John Reagan, Gary Thomas, Henk van de Hulst, Steve Warren, and Warren Wiscombe. Although I heeded their suggestions, it was not possible to follow all of them. Indeed, some of my correspondents had none of their suggestions followed. One such correspondent who deserves special mention is Ed Andreas. I originally had intended to include a few papers on scattering in a turbulent atmosphere, which caused me to seek Ed's advice. He responded with a detailed list of possible papers. But as my collection began to swell even without papers on turbulence—a subject about which I am remarkably ignorant—I decided to do without them. Scattering in a turbulent atmosphere deserves its own separate volume, and I know just the man to undertake it.

As always, the colleague to whom I owe the most is Alistair Fraser, three of whose papers are included not to reward him nor even because he suggested them (he did not) but rather because they are as likely to be of interest to others as they have been to me.

Although Emilie McWilliams, the head librarian of the Earth and Mineral Science Library, eventually came to dread my repeated forays into her domain in search of recondite journals, she nevertheless always cheerfully helped me shake the dust from what I sought.

Lorretta Palagi at SPIE was a marvel of efficiency who took from my shoulders much of the burden that would have crushed me. I thank her for this and for her patience with my tardiness and vacillation.

Brian Thompson, the general editor of this series, responded immediately with offers of help and suggestions when I was faced with problems of finding authors and journals. When Brian asked me to compile this collection, I sensed the unseen hand of Milton Kerker, to whom I am grateful for his vote of confidence in suggesting me as someone to follow in his footsteps as an editor in the Milestone Series.

Although collecting papers does not qualify as research, it is an adjunct to it and I would therefore be remiss if I did not acknowledge the National Science Foundation and NASA, two agencies that supported my research during the time I was collecting papers. I am also grateful to Gary Salzman, who with partial support from the National Institutes of Health, provided me with a summer home at Los Alamos National Laboratory

Craig F. Bohren
Department of Meteorology
The Pennsylvania State University

September 1989

Section One
Keynote Historical Paper

Random Reflections on the History of Atmospheric Optics

W. E. KNOWLES MIDDLETON

National Research Council, Ottawa, Ontario, Canada

The history of ideas on the blue of the sky, the rainbow, and the visibility of distant objects is briefly sketched. In particular, it is shown that the theory of the horizontal visual range was largely rediscovered after about 175 years.

YOUR Ives medalist for 1959 suddenly finds himself in the company of the great, which is a rather frightening experience. He feels a little as though he had absentmindedly walked into a room which he had no business to enter, a room full of important people—some, alas, ghosts!—doing important things. It is a little difficult to get used to.

It seems to be the custom for an Ives medalist to talk in a general way about his favorite optical subject. I do not propose to depart from tradition. I want to talk briefly about atmospheric optics, about the history of atmospheric optics. This has the advantage of permitting me to indulge my taste for the history of science, and to deplore the lack of papers on optical history at meetings of the Society.

Atmospheric optics probably started when it occurred to some early specimen of *Homo sapiens* to wonder why the sky was such a beautiful blue after a night of pouring rain. A great deal was written about this by the ancient Greeks, notably by Aristotle. My respect for the ancient Greeks cannot conceal the fact that practically all of this was wild speculation, much of it very hard for us in the twentieth century to understand, even in the best translations. Let us consider for a moment the subsequent ideas regarding the blue of the sky.

The great Leonardo da Vinci[1] had one, a painter's idea. A translucent layer of white paint over black looks bluish; hence the air against the outer darkness. Not an explanation, but in the light of our present sophistication not a bad analogy. The idea that the blue of the sky was due to the clear air itself was held by Mariotte, Bouguer, Euler, Leslie, and de Saussure,[2] the last-named being of the very sound opinion that the air molecules scatter more blue light than yellow or red. On page 414 of reference 2 occurs a passage which I have translated as follows, "But the air is not perfectly transparent; its elements always reflect some rays of light, especially blue rays. It is these reflected rays which produce the blue color of the sky. The purer the air, and the greater the extent of this pure air, the darker the blue color."

But others had other ideas. Chappuis,[3] who first measured the absorption spectrum of ozone in the visible, thought the blue of the sky might possibly be due to ozone. This was before anyone knew how little ozone there is in the sky; but it ought to be of interest to the members of this Society that in 1953 an earlier recipient of this medal, Dr. E. O. Hulburt, found[4] that the color of the *twilight* sky is greatly affected by the absorption by ozone, but not that of the daytime sky.

Naturally many people have thought that the blue light of the sky must be reflected by particles larger than the molecules of the atmospheric gases; but what kind of particles? One way of explaining the phenomenon is to invent some suitable particles, and Clausius, a great physicist in other fields and a man of imagination, seriously suggested in 1849[5] that the color of the clear blue sky was due to innumerable tiny bubbles of water, like soap bubbles without the soap!

The great Newton postulated particles too. He thought that the color of the sky was produced by exceedingly small water droplets acting as thin plates, and producing the blue of the first order of what we now call interference.[6] Tyndall who made many beautiful experiments on the scattering of light by aerosols,[7] realized that scattering was involved, but continued to think in terms of small particles or droplets. One particularly interesting observation was made by Forbes.[8]

He happened to see the sun through the steam escaping from the funnel of a locomotive. It looked bright red through a certain part of the jet. He then borrowed a small steam boiler from an engineering firm and experimented in the laboratory, finding that a white light always looked red through the portion of the jet just prior to the visible cloud of steam. In a later paper[9] he concluded that the transmission of red light through just-condensing water vapor provides the main explanation for red sunsets. Presumably the blue light must go somewhere else but Forbes did not pay as much attention to this.

Obviously what was wanted was more mathematics and less speculation. This was abundantly provided by Lord Rayleigh,[10] who explained not only the blue color of skylight but also its polarization, and calculated the extinction of a beam of light passing through pure air. The much more difficult question of the effect of multiple scattering had to wait another seventy years for

[1] L. da Vinci, *Trattato della pittura*, quoted by Pernter and Exner, *Meteorologische Optik* (1922), second edition, p. 614.
[2] H. B. de Saussure, Mem. Acad. Turin **4**, 409–424 (1789).
[3] J. Chappuis, Compt. rend. **91**, 985–986 (1880).
[4] E. O. Hulburt, J. Opt. Soc. Am. **43**, 113–118 (1953).
[5] R. Clausius, Ann. Physik **16**, 161–188 (1849).
[6] I. Newton, *Opticks* (1704), Book II, Pt. 3, Prop. 7.
[7] J. Tyndall, Proc. Roy. Soc. London **17**, 223–233 (1869).
[8] J. D. Forbes, Trans. Roy. Soc. Edinburgh **14**, 371–374 (1840).
[9] J. D. Forbes, Trans. Roy. Soc. Edinburgh **14**, 375–391 (1840).
[10] Lord Rayleigh, Phil. Mag. **41**, 107–120, 274–279 (1871), and many other papers. See Rayleigh, *Collected Works*.

its resolution in a brilliant series of papers by Chandrasekhar, collected and extended in his book, *Radiative Transfer*.[11]

Colorimetrists will be interested in early attempts to *measure* the color of the sky. The first was probably that of de Saussure,[2] who prepared a one-dimensional blue scale by pigment mixture, and held it so that it could be compared with the sky while the sun illuminated it. A hundred and forty years later Ostwald and Linke[12] did the same thing, apparently with no reference to their predecessor. In between these dates there had been a fair amount of crude spectroradiometry, the best being probably due to Lord Rayleigh.

As the theory gives an inverse-fourth-power law for the dependence of scattering on wavelength, we should expect the color of the bluest blue sky to be that of a Planckian radiator at a very high temperature. It actually turns out that skies of lesser purity still lie very near the Planckian locus. Quite the best measurements were made in the early 1920's by Priest with a remarkable colorimeter depending on the rotary dispersion of quartz,[13] and this fact was made evident.

Let us now turn to another well-known phenomenon, the rainbow, the symbol of Iris, Juno's handmaid, according to the Greeks, or the sign of God's covenant with Noah, according to the Hebrews.[14]

Aristotle had a great deal to say about it. We must remember that in his day the laws of refraction were unknown, the nature of light equally mysterious. With all the goodwill towards the ancient Greeks which seems to be a habit with modern man, the best that Sayeli[15] was able to write at the end of seventeen pages of commentary was that, "All in all, considering the general level of scientific knowledge at the time, the Aristotelian explanation of the rainbow was a great contribution to scientific knowledge, and no real progress was made on the subject until the end of the thirteenth and the beginning of the fourteenth centuries." It is difficult for a twentieth-century scientist to feel that Aristotle's confused speculations are an explanation in any sense of the word. Let us look in on the thirteenth century, where Witelo, according to J. F. Scott,[16] was the first to realize that the rainbow is due to both reflection and refraction. Roger Bacon (*ca* 1214–1292) wrote,[17] "Wherefore the rainbow must be produced by infinite reflections or refractions in numberless drops of water falling without an interval." He also observed

the geometry of the rainbow correctly,[18] but apart from this he does not seem to have made any experiments, and it cannot be said that he provided an explanation.

But at the very beginning of the 14th century a great advance was made in the theory of the rainbow. Theodoric (or Dietrich) of Freiberg wrote a work *De Iride et Radialibus Impressionibus* which according to Sarton[19] was not completely published until 1914.

Theodoric came close to being three hundred years ahead of his time. After he had realized that light must be refracted and reflected in raindrops, he actually made experiments with a glass globe filled with water and a ray of sunlight in a darkened room. He found that one internal reflection and two refractions could produce a bow, two reflections and two refractions a larger bow. He found that if one looked at the illuminated sphere, the color one saw depended on the position of the eye, and he reasoned correctly that the various colors of the rainbow came from drops which happened to be at various angles in relation to the position of the sun.

All this anticipated Descartes. Unfortunately Theodoric's thinking was hampered by his use of the classical "meteorological sphere," which puts the observer at the center; puts the sun, the raindrop, and anything else the observer may be looking at, on the circumference, all at the same distance from the observer. Theodoric's numbers came out quite wrong.[20]

After this the subject sank back into Peripatetic commentary for about three hundred years. It was rescued in 1637 by Descartes.

Descartes experimented[21] with a globe full of water, as had Theodoric three hundred years before. But he did more. Using the newly discovered law of refraction he calculated the paths of many rays through a sphere of water. "I found that after one reflection and two refractions there are many more rays which can be seen at an angle of from forty-one to forty-two degrees than at any smaller angle, and that there are none which can be seen at a larger angle." He found also that two reflections and two refractions produce a concentration at fifty-one to fifty-two degrees, thus explaining the secondary rainbow.

But he did not give a satisfactory explanation of the colors. This remained for Newton,[22] who had discovered the dispersion of light by making the great experiment pictured on the Society's emblem. He was able to calculate the distribution of colors in the bow.

What remained was the explanation of the so-called supernumerary bows—the narrow bows which are sometimes seen inside the primary rainbow. This had to wait until the beginning of the 19th century, when

[11] S. Chandrasekhar, *Radiative Transfer* (Clarendon Press, Oxford, 1950).

[12] W. Ostwald and F. Linke, Meteorol. Z. **45**, 367–370 (1928).

[13] I. G. Priest, J. Opt. Soc. Am. **7**, 1175–1209 (1923).

[14] For a very extensive history of man's ideas about the rainbow, see C. B. Boyer, *The Rainbow from Myth to Mathematics* (Thomas Yoseleff, New York and London, 1959).

[15] A. M. Sayeli, Isis **30**, 65–83 (1939).

[16] J. F. Scott, *The Scientific Work of René Descartes* (Taylor and Francis, London), n.d.

[17] Roger Bacon, *Opus Majus*, translated by R. B. Burke (University of Pennsylvania Press, Philadelphia, 1928), Vol. 1, p. 235.

[18] See reference 17, Vol. 2, p. 592 ff.

[19] G. Sarton, *Introduction to the History of Science*, Vol. III. Pt. i, p. 706.

[20] For a good account see reference 14, p. 110 ff.

[21] Descartes, *Les Météores*, in *Oeuvres de Descartes, VI*, edited by Adam and Tannery, (Paris, 1897–1913), 12 Vols. and Suppl.

[22] See reference 6, Book I, Pt. 2, Prop. 9.

Thomas Young[23] discovered the interference of light and immediately saw how the supernumerary arcs could be explained. The detailed mathematical analysis was finally done by Airy.[24] By calculating the shape of the wave front after it had passed through the drop, the amplitude of the light vibrations could be found exactly. The mathematics involve an integral which cannot be evaluated in terms of elementary functions. Careful experiment[25] has confirmed Airy's theory, which seems to be adequate.

For a number of years in the 1930's I used to give a graduate course in atmospheric optics, and of course I would work through Airy's theory. I always felt that it was a really elegant piece of useless knowledge. Now this summer (1959) I was in Professor A. C. S. van Heel's laboratory at Delft in Holland. Imagine my delight when I found that Professor van Heel and Mr. Walter, a student, had made use of Airy's theory to measure the index of refraction of glass with very great accuracy. This they have done by making a sphere from the glass, illuminating it by collimated monochromatic light, and measuring the difference between the angles of the primary and secondary rainbows with a theodolite. The method, while simple, is so sensitive that only the most homogeneous glass can make full use of it. Professor van Heel plans to publish this elegant method very soon, but graciously permitted me to mention in this address this interesting application of the theory of atmospheric optics.

Finally let us turn to the transmission of light through the atmosphere, and the problem of the visual range.

Obviously the history of this subject could not begin before that of photometry. It is interesting that the history of photometry as a serious discipline began with Bouguer,[26] and that of atmospheric transmission also. No one who has not read the *Traité* (an extremely rare book) can have any idea of the fertility of invention that Bouguer displayed. The fundamental principles of photometry, even the less elementary ones, emerged suddenly and almost complete.

One of these is the exponential law of attenuation of a collimated beam in a homogeneous medium, the law which frequently is, and always should be referred to as Bouguer's law,[27] though it is often wrongly ascribed to Lambert and even to Beer. Now it is not quite clear what led Bouguer to this law, but it is quite clear that one of the first uses he made of it was to calculate from the apparent intensity of the moon at various altitudes above the horizon, the transmissivity of the air.[28] He

also dealt with the transmission of light coming from a body at a finite distance,[29] deriving the law which is usually ascribed to Allard[30] of the French Lighthouse Service.

But Bouguer went much further. I must confess that I have never read entirely through the 368 pages of the *Traité*. In preparing this address I discovered that the very end of the book is of most interest from the standpoint of atmospheric optics; in particular from page 337 to the end. Bouguer considered an object seen near the horizon at a distance of a few miles with a high sun. The apparent brightness of such an object, he realized, will be the sum of the light reflected from the object and attenuated by the air, and the light from the sun scattered by the air into the eye of the observer. He integrated the contributions of many elementary layers (dx) by a geometrical construction, and showed that (in modern notation) the apparent brightness of the object at a distance x is

$$B(x) = ae^{-\sigma x} + b(1 - e^{-\sigma x}).$$

At this stage a and b are not well determined; indeed they suffer from a highly speculative theory of diffuse reflection which was one of Bouguer's rare flights of fancy. But near the very end of the book he makes a very significant observation. He has just completed the calculation of a table[31] showing the apparent brightness of the air light (*couleur aérienne*) at various distances in very clear air. These are values of $b(1 - e^{-\sigma x})$. Now suppose, he says, we look at a very large mountain covered with woods, which reflects very little light toward the spectator. At about 30 leagues[32] this will have a relative brightness of 0.2525 according to his table, while that *of the whole atmosphere, or of the sky which we shall see beside the mountain*,[33] will be 0.2575. Since the difference is about 1/51 of the whole, the mountain will be visible if we look carefully.

Later on he distinguishes between negative contrast, as of a dark mountain against the sky, and positive, as of a snow-covered peak, and shows why the bright object is visible at a greater distance. There can be no doubt that Bouguer was quite clear about the main factors determining the horizontal visual range, and that he effectively stated the law which has lately been called Koschmieder's law.[34]

The same theory was given in algebraic terms (and

[23] T. Young, Phil. Trans. Roy. Soc. London **92**, 387–397 (1802).

[24] G. B. Airy, Trans. Cambridge Phil. Soc. **6**, 379–403 (1838).

[25] For example, W. H. Miller, Trans. Cambridge Phil. Soc. **70**, 277–286 (1842).

[26] P. Bouguer, (a) *Essai d'Optique sur la gradation de la lumière* (Paris, 1729). (b) *Traité d'Optique sur la gradation de la lumière,* posthumous, edited by N. L. de la Caille (Paris, 1760).

[27] See F. Perrin, J. Opt. Soc. Am. **38**, 72–74 (1948).

[28] See reference 26 (b), p. 82.

[29] See reference 26(b), p. 265 *ff.*

[30] E. Allard, *Memoire sur l'intensité et la portée des phares* (Dunod, Paris, 1876).

[31] Reference 26(b), p. 360.

[32] There were 20 leagues to the degree of latitude.

[33] The italics are mine. In view of the importance of this point, I give the original, "Mais si l'atmosphère est bien serein, la montagne sera couverte de toute la lumière aérienne renvoyée par la masse d'air comprise entre l'objet et l'oeil de l'observateur; et cette couleur sera exprimée par (2525/10000)q, pendant que celle de toute l'atmosphère, ou celle du ciel qu'on verra à côté, le sera par (2575/10000)q."

[34] H. Koschmieder, Beitr. Phys. freien Atmos. **12**, 33–53 and 171–181 (1924).

in almost modern notation) by J. H. Lambert.[35] His derivation of the fundamental equation is not quite rigorous, but he saw quite clearly what was involved. He made no acknowledgments in this paper, although it is most unlikely that he had not read the *Traité* by 1774. In addition to the main idea, he had the useful idea of "optical equilibrium," rediscovered twice[36] in more recent times.

For a long time after Bouguer and Lambert the history of the visual range was an anticlimax. In 1789 de Saussure, the inventor of the hair hygrometer and an excellent experimenter, made an attempt[37] to measure the transparency of the air by a method which depended on the relation between visual acuity and contrast. The precision of this would probably be very poor. The succeeding century produced nothing very sound in this field. The stimulus provided by aviation in the 20th century was perhaps necessary to revive interest in the subject.

[35] J. H. Lambert, Mem. Acad. Berlin, 74–80 (1774).

[36] M. Hugon, Sci. et ind. phot. **1**, 161–168, 201–212 (1930). S. Q. Duntley, J. Opt. Soc. Am. **38**, 179–191 (1948).

[37] H. B. de Saussure, Mem. Acad. Turin **4**, 425–440 (1789).

Section Two
Molecular Scattering and Skylight

ON THE LIGHT FROM THE SKY, ITS POLARIZATION AND COLOUR.

[*Phil. Mag.* XLI. pp. 107—120, 274—279; 1871.]

IT is now, I believe, generally admitted that the light which we receive from the clear sky is due in one way or another to small suspended particles which divert the light from its regular course. On this point the experiments of Tyndall with precipitated clouds seem quite decisive. Whenever the particles of the foreign matter are sufficiently fine, the light emitted laterally is blue in colour, and, in a direction perpendicular to that of the incident beam, is *completely polarized.*

About the colour there is no *prima facie* difficulty; for as soon as the question is raised, it is seen that the standard of linear dimension, with reference to which the particles are called small, is the wave-length of light, and that a given set of particles would (on any conceivable view as to their mode of action) produce a continually increasing disturbance as we pass along the spectrum towards the more refrangible end; and there seems no reason why the colour of the compound light thus scattered laterally should not agree with that of the sky.

On the other hand, the direction of polarization (perpendicular to the path of the primary light) seems to have been felt as a difficulty. Tyndall says, "......the polarization of the beam by the incipient cloud has thus far proved itself to be *absolutely independent of the polarizing-angle.* The law of Brewster does not apply to matter in this condition; and it rests with the undulatory theory to explain why. Whenever the precipitated particles are sufficiently fine, no matter what the substance forming the particles may be, the direction of maximum polarization is at right angles to the illuminating beam, the polarizing angle for matter in this condition being invariably 45°. This I consider to be a point of capital importance with reference to the present question"*. As to the importance there will not be two opinions; but I venture to think that the difficulty is imaginary, and is caused mainly by misuse of the word reflection. Of course there is nothing in the etymology of reflection or refraction to forbid their application in this sense; but the words have acquired technical meanings, and become associated with certain well-known laws called after them. Now a moment's consideration of the principles according to which reflection and refraction are explained in the wave theory is sufficient to show that they have no application unless the surface of the disturbing body is larger than many square wave-lengths; whereas the particles to which the sky is supposed to owe its illumination must be *smaller* than the wave-length, or else the explanation of the colour breaks down. The idea of polarization by reflection is therefore out of place; and that "the law of Brewster does not apply to matter in this condition" (of extreme fineness) is only what might have been inferred from the principles of the wave theory.

Nor is there any difficulty in foreseeing what, according to the wave theory, the direction of polarization ought to be. Conceive a beam of plane-polarized light to move among a number of particles, all small compared with any of the wave-lengths. The foreign matter, if optically denser than air, may be supposed to *load* the æther so as to increase its *inertia* without altering its resistance to distortion, provided that we agree to neglect effects analogous to chromatic dispersion. If the particles were away, the wave would pass on unbroken and no light would be emitted laterally. Even with the particles retarding the motion of the æther, the same will be true if, to counterbalance the increased inertia, suitable forces are caused to act on the æther at all points where the inertia is altered. These forces have the same period and *direction* as the undisturbed luminous vibrations themselves. The light actually emitted laterally is thus the same as would be caused by forces exactly the opposite of these acting on the medium otherwise free from disturbance; and it only remains to see what the effect of such forces would be.

On account of the smallness of the particles, the forces acting throughout the volume of any one are all of the same intensity and direction, and may be considered as a whole. The determination of the motion in the æther, due to the action of a periodic force at a given point, requires, of course, the aid of mathematical analysis; but very simple considerations will lead us to a conclusion on the particular point now under discussion. In the first place there is a complete symmetry round the direction of the force. The disturbance, consisting of transverse vibrations, is propagated outwards in all directions from the centre; and in consequence of the symmetry, the

<hr>

* *Phil. Mag.* S. 4, vol. XXXVII. p. 388.

direction of vibration in any ray lies in the plane containing the ray and the axis; that is to say, the direction of vibration in the scattered or diffracted ray makes with the direction of vibration in the incident or primary ray the least possible angle. The symmetry also requires that the intensity of the scattered light should vanish for the ray which would be propagated along the axis; for there is nothing to distinguish one direction transverse to the ray from another. We have now got what we want. Suppose, for distinctness of statement, that the primary ray is vertical, and that the plane of vibration is that of the meridian. The intensity of the light scattered by a small particle is constant, and a maximum for rays which lie in the vertical plane running east and west, *while there is no scattered ray along the north and south line.* If the primary ray is unpolarized, the light scattered north and south is entirely due to that component which vibrates east and west, and is therefore *perfectly polarized,* the direction of its vibration being also east and west. Similarly any other ray scattered horizontally is perfectly polarized, and the vibration is performed in the horizontal plane. In other directions the polarization becomes less and less complete as we approach the vertical, and in the vertical direction itself altogether disappears.

So far, then, as disturbance by very small particles is concerned, theory appears to be in complete accordance with the experiments of Tyndall and others. At the same time, if the above reasoning be valid, the question as to the direction of the vibrations in polarized light is decided in accordance with the view of Fresnel. Indeed the observation on the plane of polarization of the scattered light is virtually only another form of Professor Stokes's original test with the diffraction-grating. In its present shape, however, it is free from certain difficulties both of theory and experiment, which have led different physicists who have used the other method to contradictory conclusions. I confess I cannot see any room for doubt as to the result it leads to*.

The argument used is apparently open to a serious objection, which I ought to notice. It seems to prove too much. For if one disturbing particle is unable to send out a scattered ray in the direction of original vibration, it would appear that no combination of them (such as a small body may be supposed to be) could do so, at least at such a distance that the

---

* I only mean that *if* light, as is generally supposed, consists of transversal vibrations similar to those which take place in an elastic solid, the vibration must be normal to the plane of polarization. There is unquestionably a formal analogy between the two sets of phenomena extending over a very wide range ; but it is another thing to assert that the vibrations of light are really and truly to-and-fro motions of a medium having mechanical properties (with reference to small vibrations) like those of ordinary solids. The fact that the theory of elastic solids led Green to Fresnel's formulæ for the reflection and refraction of polarized light seems amply sufficient to warrant its employment here, while the question whether the analogy is more than formal is still left open.

body subtends only a small solid angle. Now we know that when light vibrating in the plane of incidence falls on a reflecting surface at an angle of 45°, light *is* sent out according to the law of ordinary reflection, whose direction of vibration is perpendicular to that in the incident ray. And not only is this so in experiment, but it has been proved by Green* to be a consequence of the very same view as to the nature of the difference between media of various refrangibilities as has been adopted in this paper. The apparent contradiction, however, is easily explained. It is true that the disturbance due to a foreign body of any size is the same as would be caused by forces acting through the space it fills in a direction parallel to that in which the primary light vibrates; *but these forces must be supposed to act on the medium as it actually is—that is, with the variable density.* Only on the supposition of complete uniformity would it follow that no ray could be emitted parallel to the line in which the forces act. When, however, the sphere of disturbance is small compared with the wave-length, the want of uniformity is of little account, and cannot alter the law regulating the intensity of the vibration propagated in different directions.

Having disposed of the polarization, let us now consider how the intensity of the scattered light varies from one part of the spectrum to another, still supposing that all the particles are many times smaller than the wave-length even of violet light. The whole question admits of analytical treatment; but before entering upon that, it may be worth while to show how the principal result may be anticipated from a consideration of the *dimensions* of the quantities concerned.

The object is to compare the intensities of the incident and scattered rays; for these will clearly be proportional. The number (i) expressing the ratio of the two amplitudes is a function of the following quantities:—T, the volume of the disturbing particle; r, the distance of the point under consideration from it; λ, the wave-length; b, the velocity of propagation of light; D and D', the original and altered densities: of which the first three depend only on space, the fourth on space and time, while the fifth and sixth introduce the consideration of mass. Other elements of the problem there are none, except mere numbers and angles, which do not depend on the fundamental measurements of space, time, and mass. Since the ratio i, whose expression we seek, is of no dimensions in mass, it follows at once that D and D' only occur under the form $D : D'$, which is a simple number and may therefore be omitted. It remains to find how i varies with T, r, λ, b.

Now, of these quantities, b is the only one depending on time; and therefore, as i is of no dimensions in time, b cannot occur in its expression. We are left, then, with T, r, and λ; and from what we know of the dynamics

* *Camb. Phil. Trans.* vol. VII. 1837.

of the question, we may be sure that i varies directly as T and inversely as r, and must therefore be proportional to $T \div \lambda^2 r$, T being of three dimensions in space. In passing from one part of the spectrum to another λ is the only quantity which varies, and we have the important law :—

When light is scattered by particles which are very small compared with any of the wave-lengths, the ratio of the amplitudes of the vibrations of the scattered and incident light varies inversely as the square of the wave-length, and the intensity of the lights themselves as the inverse fourth power.

I will now investigate the mathematical expression for the disturbance propagated in any direction from a small particle which a beam of light strikes.

Let the vibration corresponding to the incident light be expressed by $A \cos (2\pi bt/\lambda)$. The acceleration is

$$- A \left(\frac{2\pi}{\lambda}\, b\right)^2 \cos \frac{2\pi}{\lambda}\, bt;$$

so that the force which would have to be applied to the parts where the density is D', in order that the wave might pass on undisturbed, is, per unit of volume,

$$- (D' - D)\, A \left(\frac{2\pi b}{\lambda}\right)^2 \cos \frac{2\pi}{\lambda}\, bt.$$

To obtain the total force which must be supposed to act over the space occupied by the particle, the factor T must be introduced. The opposite of this conceived to act at O (the position of the particle) gives the same disturbance in the medium as is actually caused by the presence of the particle. Suppose, now, that the ray is incident along OY, and that the direction of vibration makes an angle α with the axis of x, which is the line of the scattered ray under consideration—a supposition which involves no loss of generality, because of the symmetry which we have shown to exist round the line of action of the force. The question is now entirely reduced to the discovery of the disturbance produced in the æther by a given periodic force acting at a fixed point in it. In his valuable paper " On the Dynamical Theory of Diffraction " [*], Professor Stokes has given a complete investigation of this problem ; and I might assume the result at once. The method there used is, however, for this particular purpose very indirect, and accordingly I have thought it advisable to give a comparatively short cut to the result, which will be found at the end of the present paper. It is proved that if the total force acting at O in the manner supposed be $F \cos (2\pi bt/\lambda)$, the resulting disturbance in the ray propagated along OX is

$$\zeta = \frac{F \sin \alpha}{4\pi b^2 Dr}\ \cos \frac{2\pi}{\lambda}\, (bt - r).$$

[*] *Camb. Phil. Trans.* vol. IX. p. 1, 1849.

Substituting for F its value, we have

$$\zeta = A\,\frac{D'-D}{D}\,\frac{\pi T}{r\lambda^2}\,\sin\alpha\,\cos\frac{2\pi}{\lambda}(bt - r)^*,$$

an equation which includes all our previous results and more.

One reservation, however, must not be omitted. Since we have supposed the medium uniform throughout, whereas it really has a different density at the place where the force acts, our investigation does not absolutely correspond to the actual circumstances of the case. As before remarked, no error is on that account to be feared in the law determining the intensity of the vibration in different directions; but it is probable that the coefficient, so far as it depends on $D : D'$, may be changed†, and there may be a change in the phase comparable with $(2\pi/\lambda) \times$ the linear dimension of the particle, which is of importance when the scattered and primary waves have to be compounded.

So much for a single particle. In actual experiments, as, for instance, with Professor Tyndall's " clouds," we have to deal with an immense number of such particles; and the question now is to deduce what their effect must be from the results already obtained. Were the particles absolutely motionless, the partial waves sent out in any direction from them would have permanent relations as to phase, and the total disturbance would have to be found by compounding the *vibrations* due to all the particles. Such a supposition, however, would be very wide of the mark; for, in consequence of the extreme smallness of λ, the slightest motion of any particle will cause an alteration of phase passing through many periods in a less time than the eye could appreciate. Our particles are, then, to be treated as so many *unconnected* sources of light; and instead of adding the *vibrations*, we must take the *intensities* represented by their squares. Only in one direction is a different treatment necessary, namely along the course of the primary light. I mention this because it would not otherwise appear how the reduction in the intensity of the transmitted light is effected; but we do not require to follow the details of the process, because, when once we know the intensity of the light emitted laterally, the principle of energy will tell us what the primary wave has lost.

The intensity of the light scattered from a cloud is thus equal to

$$A^2\,\frac{(D'-D)^2}{D^2}\,\sin^2\alpha\,\frac{\pi^2 . \Sigma T^2}{\lambda^4 r^2},$$

* [1898. The factor π was omitted in the original paper.]

† I find that no alteration of any kind is needed.—Jan. 20. [1899. See *Phil. Mag.* XLI. p. 452, 1871 ; This Collection Art. 9 below.]

where ΣT^2 is the sum of all the squares of T. If T^2 be understood to denote the mean square of T (*not* the square of the mean value of T), and m be the number of particles,

$$\Sigma T^2 = m \cdot T^2.$$

If the primary light be unpolarized, the intensity in a direction making an angle β with its course becomes

$$A^2 \frac{(D' - D)^2}{D^2} (1 + \cos^2 \beta) \frac{m \pi^2 T^2}{\lambda^4 r^2}.$$

Backwards from the cloud the light is thus twice as bright as normally. To the light scattered nearly in the direction of the primary ray our expression does not apply.

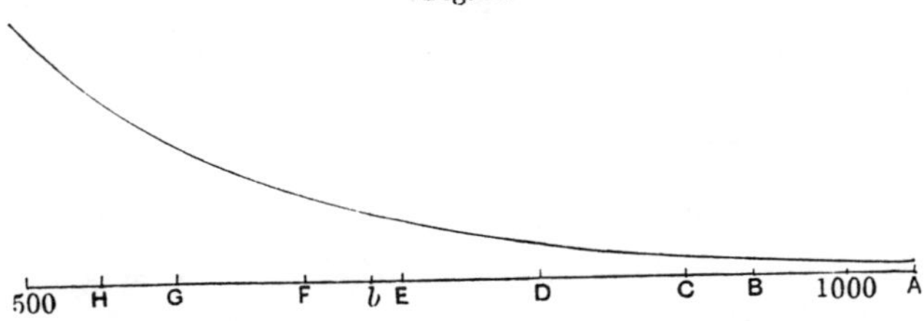

Fig. 1.

Fig. 1 shows the curve representing the intensity of the scattered light for each part of the spectrum, referred to the intensity in the primary light as a standard. The abscissa being proportional to λ, the base line represents the diffraction-spectrum with the principal fixed lines. Over the brighter portion of the spectrum from B to G the curve differs but little from a straight line, while the small curvature is turned downwards, indicating a deficiency in the green and yellow.

Before making out the theory, I had endeavoured to ascertain by observation the actual prismatic composition of the blue of the sky, and had obtained preliminary results. The experimental method (the description of which I must reserve for another opportunity) was fully adequate to the comparison of two given lights; but the difficulty was to find something to compare the blue light with. In the only complete set of observations that I have hitherto been able to make, the blue of the sky (apparently a very good one) taken from the neighbourhood of the zenith was compared with sunlight diffused through white paper. About thirty consistent comparisons were made, ranging over the spectrum from C to beyond F, and a curve drawn on the plan of fig. 1. I do not give the complete curve, because I hope before long to complete and confirm the observations; but the following numbers will give an idea of the results :—

$C.$	$D.$	$b_3.$	$F.$	
25	40	63	80	from fig. 1.
25	41	71	90	observed.

The upper line gives the theoretical intensities for the fixed lines C, D, b_3, F, while the lower gives the observed ratios between the lights (sky and diffused sunlight), the two sets of numbers being made to agree at C. Considering the difficulties and uncertainties of the case, the two curves agree very well; and it should be noticed that the sky compared with diffused light was even bluer than theory makes it, on the supposition that the diffused light through the paper may be taken as similar to that whose scattering illuminates the sky. It is possible that the paper was slightly yellow; or the cause may lie in the yellowness of sunlight as it reaches us compared with the colour it possesses in the upper regions of the atmosphere. It would be a mistake to lay any great stress on the observations in their present incomplete form; but at any rate they show that a colour more or less like that of the sky would result from taking the elements of white light in quantities proportional to λ^{-4}. I do not know how it may strike others; but individually I was not prepared for so great a difference as the observations show, the ratio for F being more than three times as great as for C.

There is one point in which our calculations do not exactly meet the case of the sky. In the experiments with precipitated clouds the total quantity of light scattered is quite insignificant compared with the incident beam; but it is by no means so clear that the same is the case with the sky. Each particle is thus struck, not only by the direct light of the sun, but also by that scattered from others. It does not seem that the chromatic effects would be much affected by this consideration; but it is worth notice that the conclusion as to complete polarization perpendicular to the incident ray would have to be modified. To see this, imagine, as before, the light (unpolarized) incident along OY upon a particle O; we have seen that the ray diffracted along OX contains no vibration parallel to OY. By the aid, however, of another particle P in the xy plane such a vibration may be communicated to it; for in the ray diffracted from P to O there is a component vibration in the xy plane perpendicular to PO, which, when again diffracted along OX, will give a component parallel to OY. This is perhaps the explanation of the incomplete polarization of sky-light at right angles to the solar beams; but it must be remembered that an insufficient fineness in some of the particles of foreign matter would have a like result.

By many physicists, from Newton downwards, the light of the sky has been supposed to be reflected from thin plates, and the colour to be the blue of the first order in Newton's scale. Such a view is fundamentally different from that adopted in this paper, though it might not at first seem so. In support of this assertion, it may be sufficient to notice that the two theories are at variance as to the law connecting the intensity with wave-length. By an argument from dimensions similar to that already used, it is easy to find how the intensity of the light reflected from a thin plate (thin, that is, compared with *any* of the wave-lengths) varies with λ. Instead of our former quantities, T, r, λ, we now have merely λ, and δ the thickness of the plate. Since the reflected vibration necessarily varies as δ, it must also be proportional to λ^{-1}, and so the *intensity* of the reflected light $\propto \lambda^{-2}$ instead of λ^{-4}. The ordinary analytical expression for the reflected light leads readily to the same conclusion (Airy's *Tracts*, p. 297). There can, I think, be no question that the composition of the light of the sky agrees more nearly with the latter than with the former law.

The principle of energy makes it clear that the light emitted laterally is not a new creation, but only diverted from the main stream. If I represent the intensity of the primary light after traversing a thickness x of the turbid medium, we have

$$dI = -\, kI\lambda^{-4}\, dx,$$

where k is a constant independent of λ. On integration,

$$I = I_0 \epsilon^{-k\lambda^{-4}x},$$

if I_0 correspond to $x = 0$,—a law altogether similar to that of absorption, and showing how the light tends to become yellow and finally red as the thickness of the medium increases. Fig. 2 shows a series of curves representing the composition of the originally white light after passing through thicknesses in the ratio of 1, 2, 4, 8, 16, 32. The reader will observe how little of the violet light remains when the red is still in nearly its original force. I cannot but think that this rapid diversion of the rays of short

Fig. 2.

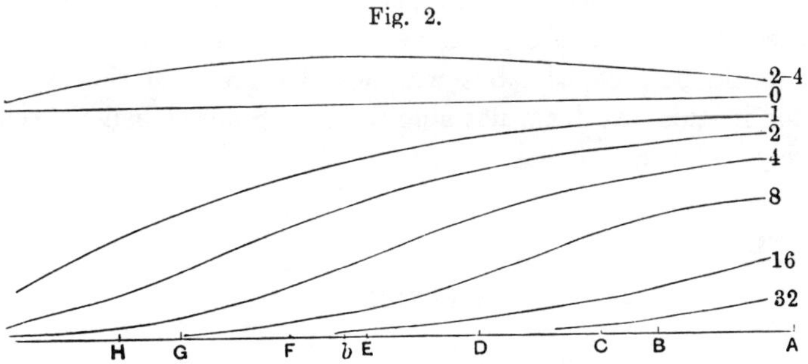

wave-length has a good deal to do with the absence of light of the highest refrangibility from the direct rays of the sun. For the line A at the extreme red and R near the upper limit of the photographic spectrum the wave-lengths are 7617 and 3108. The ratio of the fourth powers is about 36 : 1 ; so that, whatever the fraction representing the transmission of A may be, its 36th power will give the transmission of R. To take an instance, if ·9 of the ray A gets through, only ·018 of R would be able to penetrate. For the rays of still higher refrangibility, which Professor Stokes found abundant in the electric light but missing in the solar rays, the fraction would be smaller still; but I am not aware of any measurements of smaller wave-length on which to found a calculation.

We have hitherto supposed that the light scattered by the finely divided matter reaches the eye without modification, and we have taken no account of any change in the composition of the primary light before diffraction. If x be the total length of the path of the ray through the turbid medium, we may express the quality of the light in terms of x; for it makes no difference whether the lateral leakage takes place before diffraction or after. In fact

$$I \propto \lambda^{-4} \epsilon^{-k\lambda^{-4}x},$$

an expression which shows that I vanishes for very small as well as for very large values of λ, while for some definite value (say Λ) it rises to a maximum (I_0). Expressing I in terms of I_0 and Λ, we have

$$\frac{I}{I_0} = \frac{\Lambda^4}{\lambda^4} \epsilon^{1-\Lambda^2/\lambda^2};$$

from which we may fall back on our original law by supposing Λ indefinitely small, and replacing $\Lambda^4 I_0$ by a finite constant. An approximate idea of the character of these lights may be obtained by subtracting the successive curves of Fig. 2. Thus the difference of the curves marked 2 and 4 represents a light having its maximum brightness (of course relatively to the primary light) in the blue-green portion of the spectrum. I find by calculation that, if the maximum intensity be at b and be taken as unity, the intensities at G and C are given by the numbers ·713, ·710 respectively. The colour would be greenish; but whether the green of the sky is to be accounted for in this way I am not able to say. Some, I believe, consider it to be entirely a contrast effect.

APPENDIX.

Within a space T, small in all its dimensions against λ, and situated at the origin of coordinates, let a force parallel to OZ, and, so far as it depends upon the time, expressed by a simple circular function, act on the medium. If ξ, η, ζ denote the displacements parallel to the axes at the point x, y, z, and

$$\delta = \frac{d\xi}{dx} + \frac{d\eta}{dy} + \frac{d\zeta}{dz},$$

$$\frac{d^2\xi}{dt^2} = b^2\nabla^2\xi + (a^2 - b^2)\frac{d\delta}{dx},$$

$$\frac{d^2\eta}{dt^2} = b^2\nabla^2\eta + (a^2 - b^2)\frac{d\delta}{dy},$$

$$\frac{d^2\zeta}{dt^2} = b^2\nabla^2\zeta + (a^2 - b^2)\frac{d\delta}{dz} + Z,$$

$$\quad\quad\quad\quad\quad\quad\quad\quad\quad\quad\quad\quad\quad\quad\quad\quad\quad\quad\quad\quad(\text{A})*$$

where ∇^2 stands for $d^2/dx^2 + d^2/dy^2 + d^2/dz^2$; a^2 and b^2 are constants depending on the nature of the medium supposed to be isotropic. For the luminiferous æther, Green has shown that a is to be regarded as indefinitely great[†].

To represent the periodic force, write for Z, $Z\epsilon^{int}$. Similar transformations will then apply to ξ, η, ζ, and δ; so that on substitution in (A) and dividing out the common factor ϵ^{int}, there results

$$(b^2\nabla^2 + n^2)\,\xi + (a^2 - b^2)\,d\delta/dx = 0,$$

$$(b^2\nabla^2 + n^2)\,\eta + (a^2 - b^2)\,d\delta/dy = 0,$$

$$(b^2\nabla^2 + n^2)\,\zeta + (a^2 - b^2)\,d\delta/dz = -Z.$$

$$\quad\quad\quad\quad\quad\quad\quad\quad\quad\quad\quad\quad\quad\quad\quad\quad\quad\quad(\text{B})$$

Writing

$$\frac{d\xi}{dy} - \frac{d\eta}{dx} = \varpi_3 \;\text{ &c.,}$$

we obtain from (B) by differentiation and subtraction,

$$(b^2\nabla^2 + n^2)\,\varpi_3 = 0,$$

$$(b^2\nabla^2 + n^2)\,\varpi_1 = dZ/dy,$$

$$(b^2\nabla^2 + n^2)\,\varpi_2 = -dZ/dx.$$

$$\quad\quad\quad\quad\quad\quad\quad\quad\quad\quad\quad\quad\quad\quad\quad\quad\quad(\text{C})$$

ϖ_1, ϖ_2, ϖ_3 are the *rotations* of the elements of the medium round axes parallel to those of coordinates.

The disturbance which we are investigating is that caused and maintained by the force Z acting within the space T. Accordingly[‡]

$$\varpi_3 = 0,$$

$$\varpi_1 = -\frac{1}{4\pi b^2}\iiint\frac{dZ}{dy}\frac{\epsilon^{\pm ikr}}{r}\,dx\,dy\,dz,$$

r being the distance between the element $dx\,dy\,dz$ and the point where ϖ_1 is estimated, and

$$k = 2\pi/\lambda = n/b. \quad\quad\quad\quad\quad\quad\quad\quad\quad\quad(\text{D})$$

---

* Thomson and Tait's *Natural Philosophy*, § 698.

† *Camb. Phil. Trans.* vol. VII.

‡ Helmholtz, *Crelle's Journal*, 1860.

Since $\epsilon^{\pm ikr}$ will be finally multiplied by ϵ^{int}, and the disturbance which we are dealing with is propagated *outwards* from T, it is evident that the *lower* sign is to be employed. Now

$$\int \frac{\epsilon^{-ikr}}{r} \frac{dZ}{dy}\, dy = \left[Z\, \frac{\epsilon^{-ikr}}{r} \right] - \int Z\, \frac{d}{dy} \left(\frac{\epsilon^{-ikr}}{r} \right) dy,$$

of which the term within brackets vanishes, because the value of Z is only finite within the space T. Thus

$$\varpi_1 = \frac{1}{4\pi b^2} \iiint Z\, \frac{d}{dy} \left(\frac{\epsilon^{-ikr}}{r} \right) dx\,dy\,dz.$$

The factor $\dfrac{d}{dy}\left(\dfrac{\epsilon^{-ikr}}{r} \right)$ within the space T is sensibly constant, so that, if Z stand for the mean value of Z over the volume T,

$$\varpi_1 = \frac{TZ}{4\pi b^2} \frac{d}{dy} \left(\frac{\epsilon^{-ikr}}{r} \right), \qquad \varpi_2 = -\frac{TZ}{4\pi b^2} \frac{d}{dx} \left(\frac{\epsilon^{-ikr}}{r} \right),$$

$$x\varpi_1 + y\varpi_2 \propto \left(x\frac{d}{dy} - y\frac{d}{dx} \right) \cdot \frac{\epsilon^{-ikr}}{r} = 0.$$

And if $R = \sqrt{x^2 + y^2}$,

$$\varpi = \frac{x\varpi_2 - y\varpi_1}{R} = -\frac{TZ}{4\pi b^2 R} \left(x\frac{d}{dx} + y\frac{d}{dy} \right) \cdot \frac{\epsilon^{-ikr}}{r}$$

$$= -\frac{TZ}{4\pi b^2} \frac{d}{dR} \cdot \frac{\epsilon^{-ikr}}{r} = \frac{TZ \sin\alpha}{4\pi b^2} \frac{d}{dr} \cdot \frac{\epsilon^{-ikr}}{r},$$

where α denotes the angle between r and z.

The resultant rotation at any point is thus about an axis perpendicular to the plane passing through the point and the axis of Z; and its magnitude is given by ϖ. In differentiating $(r^{-1} e^{-ikr})$ with respect to r, we may neglect the term divided by r^2 as altogether insensible, kr being an exceedingly great quantity at any ordinary distance from the origin of disturbance. Thus

$$\varpi = \frac{-ik \cdot TZ \sin\alpha}{4\pi b^2} \cdot \frac{\epsilon^{-ikr}}{r}, \quad \dotsi \dotsi \dotsi (\mathrm{E})$$

which completely determines the rotation at any point. For a given disturbance it is seen to be everywhere about an axis perpendicular to r and the direction of the force, and in magnitude dependent only on the angle between these two directions (α) and on the distance (r).

The intensity of the light, however, is more usually expressed in terms of the actual displacement in the plane of the wave. In order to find the connexion between the two quantities, it will be more convenient to suppose the scattered ray parallel to x, and that the force F (for Z is no longer appropriate) acts in the plane of zx at an angle α with Ox. ϖ becomes identical with ϖ_2; that is, with $d\zeta/dx$; for ξ as well as η is zero; so that

$$\zeta = \int \varpi\, dr = \frac{TF \sin \alpha}{4\pi b^2} \cdot \frac{\epsilon^{-ikr}}{r}.$$

Restoring the factor ϵ^{int}, we have

$$\zeta = \frac{TF \sin \alpha}{4\pi b^2} \cdot \frac{\epsilon^{i(nt-kr)}}{r};$$

or throwing away the imaginary part,

$$\zeta = \frac{TF \sin \alpha}{4\pi b^2 r} \cdot \cos \frac{2\pi}{\lambda} (bt - r). \quad \ldots\ldots\ldots\ldots\ldots(F)$$

This corresponds to a total accelerating force equal to $FT \cos (2\pi bt/\lambda)$; a result which agrees with that of Professor Stokes's more complete investigation, with the exception of a slight difference of notation.

In the February Number of the *Philosophical Magazine* I have propounded a theory of the scattering of light by particles which are small in *all* their dimensions compared with the wave-length of light, and have applied the results to explain the phenomena presented by the sky. Another theory has been given by Clausius, who attributes the light of the sky to reflection from water-bubbles, and has developed his views at length in a series of papers in Poggendorff's *Annalen* and Crelle's *Journal**.

Starting from the ordinary laws of reflection and refraction, he has no difficulty in showing that, were the atmosphere charged with globes of water in sufficient quantity to send us the light which we actually receive, a star instead of appearing as a point would be dilated into a disk of considerable magnitude. But the requirements of the case are satisfied if we suppose the spheres hollow, like bubbles; for then, on account of the parallelism of the surfaces, but little effect is produced by refraction on a wave of light. At the same time, if the film be sufficiently thin, the light reflected from it will be the blue of the first order, and so the colour of the sky is apparently accounted for.

Apart from the difficulty of seeing how such bubbles could be formed, there is a formidable objection to this theory, mentioned by Brücke (Pogg. *Ann.* vol. LXXXVIII. p. 363)—that the blue of the sky is a much better colour than the blue of the first order. That it is so appears clearly from the

* Pogg. *Ann.* vols. LXXII. LXXVI. LXXXVIII. *Crelle*, vols. XXXIV. XXXVI.

measurements quoted in the February Number, and from the theoretical composition of the blue of the first order*. Nor can we escape from this difficulty by supposing, with Brücke, that the greater part of the light from the sky has been reflected more than once.

Brücke also brings forward an experiment of great importance when he shows that mastic precipitated from an alcoholic solution scatters light of a blue tint. He remarks that it is impossible to suppose that the particles of mastic are in the form of bubbles.

In his last utterance on this subject†, Clausius replies to the objections urged by Brücke and others against his theory, and shows that, if the illumination of the sky is due to thin plates at all, those thin plates *must* be in the form of bubbles. While admitting that if the particles are very small the ordinary laws of reflection and refraction no longer apply‡, and that therefore this case is not excluded by his argument, he still holds to his original view as to the nature of the reflecting matter in the sky, considering that the polarization of the light indicates that it has undergone *regular* reflection. His concluding paragraph so well sums up the case that I cannot do better than quote it. "Das Resultat der vorstehenden Betrachtungen kann ich hiernach kurz so zusammenfassen. Soweit man die gewöhnlichen Brechungs- und Reflexionsgesetze als gültig anerkennt, glaube ich auch meine früheren Schlüsse festhalten zu müssen, nämlich, dass in der Atmosphäre Dampfbläschen vorhanden seyen, und dass sie die Hauptursache der in ihr stattfindenden Lichtreflexion und ihrer Farben bilden. Nimmt man aber an, die in der Atmosphäre wirksamen Körperchen seyen so klein, dass jene Gesetze auf sie keine Anwendung mehr finden, dann sind auch diese Schlüsse ungültig. Auf diesen Fall ist aber auch die Theorie der Farben dünner Blättchen nicht mehr anwendbar, und er bedarf vielmehr einer neuen Entwickelung, bei welcher noch besonders berücksichtigt werden muss, in wiefern diese Annahme mit der Polarisation des vom Himmel kommenden Lichtes und mit der angenähert bekannten Grösse der in den Wolken vorhandenen Wassertheilchen vereinbar ist."

* I find that I omitted to explain why it is that the light dispersed from small particles is of so much richer a hue than that reflected from very thin films. In the latter case the reflected wave may be regarded as the sum of the disturbances originating in the elementary parts of the film, and these elementary parts may be assimilated to the small particles of the former supposition. The integration is best effected by dividing the surface into the *zones of Huyghens*; and it is proved in works on physical optics that the total effect is just half of that due to the first zone. Now the zones of Huyghens vary as the wave-length; and thus it appears that in the integration the long waves gain an advantage which diminishes the original preponderance of their quicker-timed rivals.

† Pogg. *Ann.* vol. LXXXVIII. p. 543.

‡ In many departments of science a tendency may be observed to extend the field of familiar laws beyond their proper limits. Thus the properties of gross matter are often assumed to hold equally good for molecules. An example more analogous to that which suggests this remark is to be found in the common explanation of the mode of action of the speaking-trumpet.

Clausius does not seem to have followed up the line of research here indicated. My investigation (written, it so happens, before seeing Clausius's papers) shows in the clearest manner the connexion between the smallness of the particles and the polarization of the light scattered from them. Indeed I must remark that in this respect there is an advantage over the theory of thin plates, according to which the direction of complete polarization would be about 76° from the sun. It would be a singular coincidence if the action of secondary causes were to augment this angle to 90°—its observed magnitude. It seems, therefore, not too much to say that, if the illumination of the sky were due to suspended water-bubbles, neither its colour nor its polarization would agree with what is actually observed.

In his celebrated paper on Fluorescence*, Professor Stokes makes the following significant remark :—" Now this result appears to me to have no remote bearing on the question of the direction of the vibrations in polarized light. So long as the suspended particles are large compared with the waves of light, reflection takes place as it would from a portion of the surface of a large solid immersed in the fluid, and no conclusion can be drawn either way. But if the diameter of the particles be small compared with the length of a wave of light, it seems plain that the vibrations in a reflected ray cannot be perpendicular to the vibrations in the incident ray." This is the only passage that I have met with in which the theory of the reflection of light from very small particles is touched upon.

If it be assumed, as in the theories of Green and Cauchy of reflection at plane surfaces, that the effect of dense matter is merely to load the ether, it follows rigorously that the direction of vibration cannot be turned through a right angle when light is scattered from small particles. But all we know in the first instance is that the velocity of propagation of luminous waves is less in ordinary transparent matter than in vacuum ; and this may be accounted for as well by a diminished rigidity as by an increased density. In the first case a scattered ray *might* be composed of vibrations perpendicular to those of the incident beam ; so that the matter is not quite so clear as it would seem from the argument of Professor Stokes. I believe, however, that good reasons may be given for rejecting the view that the difference between media of varying refrangibility is one of rigidity. The point is an important one, and I propose to recur to it later.

The experiments of Professor Tyndall† with precipitated clouds exhibit more clearly than had been done by Brücke the relation between the size of the particles and the nature of the dispersed light. The observation that the polarization is complete perpendicular to the track of the incident light is in itself sufficient to disprove the theory of bubbles. As the particles increase

----

* *Phil. Trans.* 1852, p. 526.
† *Phil. Mag.* vol. XXXVII. p. 385. *Phil. Trans.* 1870.

in magnitude, the azure and polarization are gradually lost. During the transition a different and more complicated set of phenomena present themselves, which will furnish a test for the theory when it is extended so as to include the consideration of particles which are no longer very small in comparison with the waves of light.

All who have written on this subject seem to have taken for granted that the foreign matter in the atmosphere is water or ice. Even Tyndall, who expressly says that any particles, if small enough, will do, still believes in the presence of water-particles. But this view is encumbered with considerable difficulty; for even if, in virtue of its transparency to radiant heat, the air in the higher regions of our atmosphere is at a very low temperature, it would still be capable of absorbing the very small quantity of water which is sufficient to explain the blue of the sky. At any rate it is difficult to imagine particles of water smaller than the wave-length endowed with any stability. These difficulties might perhaps be got over if there were any strong argument in favour of the water-particles; but of the existence of such I am not aware. Every one knows that a blue haze evidently akin to the azure of the sky obliterates the details and modifies the colour of a distant mountain; and this, when it occurs on a hot day, cannot possibly be attributed to aqueous particles. On the face of it, there is no reason for supposing that near the earth's surface the foreign matter is of one kind and at a great altitude another. If it were at all probable that the particles are all of one kind, it seems to me that a strong case might be made out for common salt. Be this as it may, the optical phenomena can give us no clue.

The apparatus by means of which the comparison was made between sky light and that of the sun diffused through white paper, was originally arranged for measurements of the absolute absorption of coloured fluids for the various rays of the spectrum, and had been applied rather extensively in experiments having that object. In the shutter of a darkened room were placed two slits in the same vertical line, each about three inches long, and a foot apart. At the other end of the room was an arrangement of prisms and lenses for producing a pure spectrum on a screen in the ordinary way. At first only one prism was used; but I soon introduced another, and the number might probably be further increased with advantage. It is even more important to have a great dispersion in these experiments than in the ordinary spectroscope. Two spectra would thus be thrown on the screen one over the other, but by means of a very obtuse-angled prism situated in front of the dispersion-prisms they are brought together so as exactly to overlap. The double spectrum thus formed passes through a horizontal slit in the screen placed so as to receive it. Close behind is an opaque card carrying a small vertical slit, which can be slid along so as to allow any desired part of the spectrum to pass through. At the beginning and end of a set of experiments the card is removed, and the principal fixed lines are observed through an eyepiece and referred to a scale situated just over the horizontal aperture.

When the experimenter looks through the eye-slit in the direction of the lens, he sees the two parts of the obtuse prism illuminated with light, in each case homogeneous, and, if the adjustments are properly made, belonging to the same part of the spectrum. By varying the breadths of the original slits, the two parts of the field may be made equally bright; and when the match is attained, the breadths are inversely proportional to the richness of the lights behind them in the homogeneous ray under consideration. But if the object be to make a complete comparison between two lights, it is often more convenient to leave the widths of the slits arbitrary, and then, by sliding the card, to seek that part of the spectrum which allows a match. It was in this way that the observations on the light of the sky were made. To give an idea of the degree of accuracy to which the comparisons may be made, I may mention that in my experiments on absorption, the means of six observations were usually correct to about one in 50 or 60. In the less-luminous parts of the spectrum the error might be somewhat greater.

The difficulty, however, of getting a satisfactory result with the blue of the sky does not lie in the inaccuracy of the measurements, but in the arbitrary character of the light with which it is compared. In order to test the theory in a strict manner, the second light ought to be similar in composition to that which lights up the sky. Now the sky is lit not only by the direct rays of the sun, but also by itself and by the bright surface of the earth. It is evident, therefore, that the requirements of the case are very imperfectly met by taking as the second light that of the sun as received by us, even if the translucent material through which we diffuse it effects no change in the quality. A nearer approximation to what we want would probably be found in the diffused light of a thoroughly cloudy day. But here we meet with an experimental difficulty; for the method described is only available to compare two lights both given at once. A suitable artificial light might no doubt be used as a middle term to be afterwards eliminated; but a candle or a lamp would hardly be available, on account of the yellowness of their light. On the other hand, the bluer radiation from burning magnesium would probably be inconvenient, and difficult to keep constant in quality from day to day. I am, however, in hopes that, by a method founded on a different principle, I may be able to compare the blue of one day's clear sky with the white light from the clouds on another.

Reprinted from *Scientific Papers*, Vol. III, 397–405, Cambridge University Press (1902).

ON THE TRANSMISSION OF LIGHT THROUGH AN ATMOSPHERE CONTAINING SMALL PARTICLES IN SUSPENSION, AND ON THE ORIGIN OF THE BLUE OF THE SKY.

[Philosophical Magazine, XLVII. pp. 375—384, 1899.]

THIS subject has been treated in papers published many years ago*. I resume it in order to examine more closely than hitherto the attenuation undergone by the primary light on its passage through a medium containing small particles, as dependent upon the number and size of the particles. Closely connected with this is the interesting question whether the light from the sky can be explained by diffraction from the molecules of air themselves, or whether it is necessary to appeal to suspended particles composed of foreign matter, solid or liquid. It will appear, I think, that even in the absence of foreign particles we should still have a blue sky†.

The calculations of the present paper are not needed in order to explain the general character of the effects produced. In the earliest of those above

---

* *Phil. Mag.* XLI. pp. 107, 274, 447 (1871); XII. p. 81 (1881). [Vol. I. pp. 87, 104, 518.]

† My attention was specially directed to this question a long while ago by Maxwell in a letter which I may be pardoned for reproducing here. Under date Aug. 28, 1873, he wrote :—

"I have left your papers on the light of the sky, &c. at Cambridge, and it would take me, even if I had them, some time to get them assimilated sufficiently to answer the following question, which I think will involve less expense to the energy of the race if you stick the data into your formula and send me the result....

"Suppose that there are N spheres of density ρ and diameter s in unit of volume of the medium. Find the index of refraction of the compound medium and the coefficient of extinction of light passing through it.

"The object of the enquiry is, of course, to obtain data about the size of the molecules of air. Perhaps it may lead also to data involving the density of the æther. The following quantities are known, being combinations of the three unknowns,

M = mass of molecule of hydrogen ;

N = number of molecules of any gas in a cubic centimetre at 0° C. and 760 B.

s = diameter of molecule in any gas :—

referred to I illustrated by curves the gradual reddening of the transmitted light by which we see the sun a little before sunset. The same reasoning proved, of course, that the spectrum of even a vertical sun is modified by the atmosphere in the direction of favouring the waves of greater length.

For such a purpose as the present it makes little difference whether we speak in terms of the electromagnetic theory or of the elastic solid theory of light; but to facilitate comparison with former papers on the light from the sky, it will be convenient to follow the latter course. The small particle of volume T is supposed to be small in all its dimensions in comparison with the wave-length (λ), and to be of optical density D' differing from that (D) of the surrounding medium. Then, if the incident vibration be taken as unity, the expression for the vibration scattered from the particle in a direction making an angle θ with that of *primary vibration* is

$$\frac{D'-D}{D}\,\frac{\pi T}{r\lambda^2}\sin\theta\,\cos\frac{2\pi}{\lambda}(bt-r)^{*}, \quad \ldots\ldots\ldots\ldots\ldots(1)$$

r being the distance from T of any point along the secondary ray.

In order to find the whole emission of energy from T we have to integrate the square of (1) over the surface of a sphere of radius r. The element of area being $2\pi r^2\sin\theta\,d\theta$, we have

$$\int_0^{\pi}\frac{\sin^2\theta}{r^2}\,2\pi r^2\sin\theta\,d\theta = 4\pi\int_0^{\frac{1}{2}\pi}\sin^3\theta\,d\theta = \frac{8\pi}{3}\,;$$

so that the energy emitted from T is represented by

$$\frac{8\pi^3}{3}\,\frac{(D'-D)^2}{D^2}\,\frac{T^2}{\lambda^4}\,, \quad \ldots\ldots\ldots\ldots\ldots\ldots(2)$$

Known Combinations.

$MN=$ density.

Ms^2 from diffusion or viscosity.

Conjectural Combination.

$\dfrac{6M}{\pi s^3}=$ density of molecule.

" If you can give us (i) the quantity of light scattered in a given direction by a stratum of a certain density and thickness ; (ii) the quantity cut out of the direct ray ; and (iii) the effect of the molecules on the index of refraction, which I think ought to come out easily, we might get a little more information about these little bodies.

" You will see by *Nature*, Aug. 14, 1873, that I make the diameter of molecules about $\frac{1}{10000}$ of a wave-length.

" The enquiry into scattering must begin by accounting for the great observed transparency of air. I suppose we have no numerical data about its absorption.

" But the index of refraction can be numerically determined, though the observation is of a delicate kind, and a comparison of the result with the dynamical theory may lead to some new information."

Subsequently he wrote, " Your letter of Nov. 17 quite accounts for the observed transparency of any gas." So far as I remember, my argument was of a general character only.

* The factor π was inadvertently omitted in the original memoir.

on such a scale that the energy of the primary wave is unity per unit of wave-front area.

The above relates to a single particle. If there be n similar particles per unit volume, the energy emitted from a stratum of thickness dx and of unit area is found from (2) by introduction of the factor $n\,dx$. Since there is no waste of energy on the whole, this represents the loss of energy in the primary wave. Accordingly, if E be the energy of the primary wave,

$$\frac{1}{E}\frac{dE}{dx} = -\frac{8\pi^3 n}{3}\frac{(D'-D)^2}{D^2}\frac{T^2}{\lambda^4}; \quad \dots\dots\dots\dots\dots(3)$$

whence

$$E = E_0 e^{-hx}, \quad \dots\dots\dots\dots\dots\dots\dots(4)$$

where

$$h = \frac{8\pi^3 n}{3}\frac{(D'-D)^2}{D^2}\frac{T^2}{\lambda^4}. \quad \dots\dots\dots\dots\dots\dots(5)$$

If we had a sufficiently complete expression for the scattered light, we might investigate (5) somewhat more directly by considering the resultant of the primary vibration and of the secondary vibrations which travel in the same direction. If, however, we apply this process to (1), we find that it fails to lead us to (5), though it furnishes another result of interest. The combination of the secondary waves which travel in the direction in question has this peculiarity, that the phases are no more distributed at random. The intensity of the secondary light is no longer to be arrived at by addition of individual intensities, but must be calculated with consideration of the particular phases involved. If we consider a number of particles which all lie upon a primary ray, we see that the phases of the secondary vibrations which issue along this line are all the same.

The actual calculation follows a similar course to that by which Huygens' conception of the resolution of a wave into components corresponding to the various parts of the wave-front is usually verified. [See for example Vol. III. p. 74.] Consider the particles which occupy a thin stratum dx perpendicular to the primary ray x. Let AP (Fig. 1) be this stratum and O the point where the vibration is to be estimated. If $AP = \rho$, the element of volume is $dx \cdot 2\pi\rho\,d\rho$, and the number of particles to be found in it is deduced by introduction of the factor n. Moreover, if $OP = r$, $AO = x$, $r^2 = x^2 + \rho^2$, and $\rho\,d\rho = r\,dr$. The resultant at O of all the secondary vibrations which issue from the stratum dx is by (1), with $\sin\theta$ equal to unity,

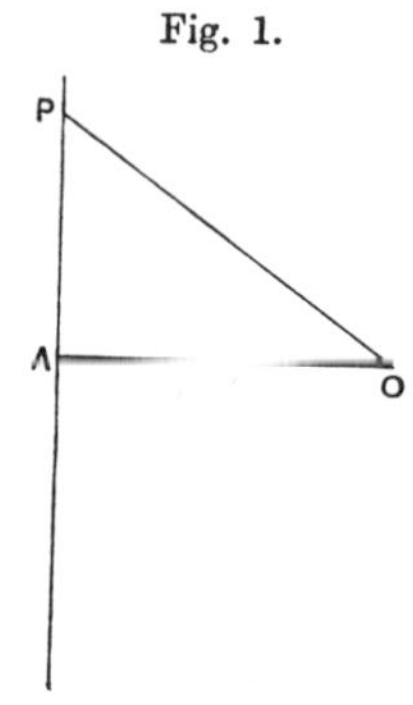

Fig. 1.

$$n\,dx \cdot \int_x^\infty \frac{D'-D}{D}\frac{\pi T}{r\lambda^2}\cos\frac{2\pi}{\lambda}(bt-r)\,2\pi r\,dr,$$

or

$$n\,dx \cdot \frac{D'-D}{D}\frac{\pi T}{\lambda}\sin\frac{2\pi}{\lambda}(bt-x). \quad \dots\dots\dots\dots\dots(6)$$

To this is to be added the expression for the primary wave itself, supposed to advance undisturbed, viz., $\cos \frac{2\pi}{\lambda}(bt - x)$, and the resultant will then represent the whole actual disturbance at O as modified by the particles in the stratum dx.

It appears, therefore, that to the order of approximation afforded by (1) the effect of the particles in dx is to modify the phase, *but not the intensity*, of the light which passes them. If this be represented by

$$\cos \frac{2\pi}{\lambda}(bt - x - \delta), \quad \dots\dots\dots\dots\dots\dots\dots(7)$$

δ is the *retardation* due to the particles, and we have

$$\delta = nT dx\,(D' - D)/2D. \quad \dots\dots\dots\dots\dots\dots(8)$$

If μ be the refractive index of the medium as modified by the particles, that of the original medium being taken as unity, $\delta = (\mu - 1)\,dx$, and

$$\mu - 1 = nT\,(D' - D)/2D. \quad \dots\dots\dots\dots\dots\dots(9)$$

If μ' denote the refractive index of the material composing the particles regarded as continuous, $D'/D = \mu'^2$, and

$$\mu - 1 = \tfrac{1}{2}nT\,(\mu'^2 - 1), \quad \dots\dots\dots\dots\dots\dots(10)$$

reducing to

$$\mu - 1 = nT\,(\mu' - 1) \quad \dots\dots\dots\dots\dots\dots(11)$$

in the case where $\mu' - 1$ can be regarded as small.

It is only in the latter case that the formulæ of the elastic-solid theory are applicable to light. In the electric theory, to be preferred on every ground except that of easy intelligibility, the results are more complicated in that when $(\mu' - 1)$ is not small, the scattered ray depends upon the shape and not merely upon the volume of the small obstacle. In the case of *spheres* we are to replace $(D' - D)/D$ by $3\,(K' - K)/(K' + 2K)$, where K, K' are the dielectric constants proper to the medium and to the obstacle respectively*: so that instead of (10)

$$\mu - 1 = \frac{3nT}{2}\,\frac{\mu'^2 - 1}{\mu'^2 + 2}. \quad \dots\dots\dots\dots\dots\dots(12)$$

On the same suppositions (5) is replaced by

$$h = 24\pi^3 n\,\frac{(\mu'^2 - 1)^2}{(\mu'^2 + 2)^2}\,\frac{T^2}{\lambda^4}. \quad \dots\dots\dots\dots\dots(13)$$

On either theory

$$h = \frac{32\pi^3\,(\mu - 1)^2}{3n\lambda^4}, \quad \dots\dots\dots\dots\dots\dots(14)$$

---

* *Phil. Mag.* XII. p. 98 (1881). [Vol. I. p. 533.] For the corresponding theory in the case of an *ellipsoidal* obstacle, see *Phil. Mag.* Vol. XLIV. p. 48 (1897). [Vol. IV. p. 305.]

a formula giving the coefficient of transmission in terms of the refraction, and of the *number of particles per unit volume.*

We have seen that when we attempt to find directly from (1) the effect of the particles upon the transmitted primary wave, we succeed only so far as regards the retardation. In order to determine the attenuation by this process it would be necessary to supplement (1) by a term involving

$$\sin 2\pi\,(bt - r)/\lambda;$$

but this is of higher order of smallness. We could, however, reverse the process and determine the small term in question *à posteriori* by means of the value of the attenuation obtained indirectly from (1), at least as far as concerns the secondary light emitted in the direction of the primary ray.

The theory of these effects may be illustrated by a completely worked out case, such as that of a small rigid and fixed spherical obstacle (radius c) upon which plane waves of sound impinge*. It would take too much space to give full details here, but a few indications may be of use to a reader desirous of pursuing the matter further.

The expressions for the terms of orders 0 and 1 in spherical harmonics of the velocity-potential of the secondary disturbance are given in equations (16), (17), § 334. With introduction of approximate values of γ_0 and γ_1, viz.

$$\gamma_0 + kc = \tfrac{1}{3}k^3c^3, \qquad \gamma_1 + kc = \tfrac{1}{2}\pi + \tfrac{1}{6}k^3c^3,$$

we get

$$[\psi_0] + [\psi_1] = -\frac{k^2c^3}{3r}\left(1 + \frac{3\mu}{2}\right)\cos k\,(at - r) + \frac{k^5c^6}{9r}\left(1 - \frac{3\mu}{4}\right)\sin k\,(at - r), \dots(15)\dagger$$

in which c is the radius of the sphere, and $k = 2\pi/\lambda$. This corresponds to the primary wave

$$[\phi] = \cos k\,(at + x), \dots\dots\dots\dots\dots\dots\dots(16)$$

and includes the most important terms from all sources in the multipliers of $\cos k\,(at - r)$, $\sin k\,(at - r)$. Along the course of the primary ray $(\mu = -1)$ it reduces to

$$[\psi_0] + [\psi_1] = \frac{k^2c^3}{6r}\cos k\,(at - r) + \frac{7k^5c^6}{36r}\sin k\,(at - r). \quad\dots\dots(17)$$

We have now to calculate by the method of Fresnel's zones the effect of a distribution of n spheres per unit volume. We find, corresponding to (6), for the effect of a layer of thickness dx,

$$2\pi n\,dx\,\{\tfrac{1}{6}kc^3\sin k\,(at + x) - \tfrac{7}{36}k^4c^6\cos k\,(at + x)\}. \dots\dots\dots(18)$$

---

* *Theory of Sound,* 2nd ed. § 334.

† [1902. μ here denotes the sine of the latitude.]

To this is to be added the expression (16) for the primary wave. The coefficient of $\cos k (at + x)$ is thus altered by the particles in the layer dx from unity to $(1 - \frac{7}{18} k^4 c^6 \pi n \, dx)$, and the coefficient of $\sin k (at + x)$ from 0 to $\frac{1}{3} k c^3 \pi n \, dx$. Thus, if E be the energy of the primary wave,

$$dE/E = -\tfrac{7}{9} k^4 c^6 \pi n \, dx;$$

so that if, as in (4), $E = E_0 e^{-hx}$,

$$h = \tfrac{7}{9} \pi n k^4 c^6. \quad\quad\dots\dots\dots\dots\dots\dots\dots\dots\dots\dots(19)$$

The same result may be obtained indirectly from the *first* term of (15). For the whole energy emitted from one sphere may be reckoned as

$$\frac{k^4 c^6}{9 r^2} \int_{-1}^{+1} 2\pi r^2 (1 + \tfrac{3}{2}\mu)^2 \, d\mu = \frac{7 \pi k^4 c^6}{9}, \quad\quad \dots\dots\dots\dots(20)$$

unity representing the energy of the primary wave per unit area of wave-front. From (20) we deduce the same value of h as in (19).

The first term of (18) gives the refractivity of the medium. If δ be the retardation due to the spheres of the stratum dx,

$$\sin k\delta = \tfrac{1}{3} k c^3 \pi n \, dx,$$

or

$$\delta = \tfrac{1}{3} \pi n c^3 \, dx. \quad\quad \dots\dots\dots\dots\dots\dots\dots\dots(21)$$

Thus, if μ be the refractive index as modified by the spheres, that of the original medium being unity,

$$\mu - 1 = \tfrac{1}{3} \pi n c^3 = \tfrac{1}{4} p, \quad\quad \dots\dots\dots\dots\dots\dots\dots\dots(22)$$

where p denotes the (small) ratio of the volume occupied by the spheres to the whole volume. This result agrees with equations formerly obtained for the refractivity of a medium containing spherical obstacles disposed in cubic order*.

Let us now inquire what degree of transparency of air is admitted by its molecular constitution, *i.e.*, in the absence of all foreign matter. We may take $\lambda = 6 \times 10^{-5}$ centim., $\mu - 1 = \cdot0003$; whence from (14) we obtain as the distance x, equal to $1/h$, which light must travel in order to undergo attenuation in the ratio $e : 1$,

$$x = 4\cdot4 \times 10^{-13} \times n. \quad\quad \dots\dots\dots\dots\dots\dots(23)$$

The completion of the calculation requires the value of n. Unfortunately this number—according to Avogadro's law the same for all gases—can hardly be regarded as known. Maxwell† estimates the number of molecules under standard conditions as 19×10^{18} per cub. centim. If we use this value of n, we find

$$x = 8\cdot3 \times 10^6 \text{ cm.} = 83 \text{ kilometres,}$$

<hr>

* *Phil. Mag.* Vol. xxxiv. p. 499 (1892). [Vol. iv. p. 35.] Suppose $m = \infty$, $\sigma = \infty$.

† "Molecules," *Nature*, viii. p. 440 (1873).

as the distance through which light must pass in air at atmospheric pressure before its intensity is reduced in the ratio of $2\cdot7 : 1$.

Although Mount Everest appears fairly bright at 100 miles distance as seen from the neighbourhood of Darjeeling, we cannot suppose that the atmosphere is as transparent as is implied in the above numbers; and of course this is not to be expected, since there is certainly suspended matter to be reckoned with. Perhaps the best data for a comparison are those afforded by the varying brightness of stars at various altitudes. Bouguer and others estimate about $\cdot8$ for the transmission of light through the entire atmosphere from a star in the zenith. This corresponds to $8\cdot3$ kilometres of air at standard pressure. At this rate the transmission through 83 kilometres would be $(\cdot8)^{10}$, or $\cdot11$, instead of $1/e$ or $\cdot37$. It appears then that the actual transmission through 83 kilometres is only about 3 times less than that calculated (with the above value of n) from molecular diffraction without any allowance for foreign matter at all. And we may conclude that the light scattered from the molecules would suffice to give us a blue sky, not so very greatly darker than that actually enjoyed.

If n be regarded as altogether unknown, we may reverse our argument, and we then arrive at the conclusion that n cannot be greatly less than was estimated by Maxwell. A lower limit for n, say 7×10^{18} per cubic centimetre, is somewhat sharply indicated. For a still smaller value, or rather the increased individual efficacy which according to the observed refraction would be its accompaniment, must lead to a less degree of transparency than is actually found. When we take into account the known presence of foreign matter, we shall probably see no ground for any reduction of Maxwell's number.

The results which we have obtained are based upon (14), and are as true as the theories from which that equation was derived. In the electromagnetic theory we have treated the molecules as spherical continuous bodies differing from the rest of the medium merely in the value of their dielectric constant. If we abandon the restriction as to sphericity, the results will be modified in a manner that cannot be precisely defined until the shape is specified. On the whole, however, it does not appear probable that this consideration would greatly affect the calculation as to transparency, since the particles must be supposed to be oriented in all directions indifferently. But the theoretical conclusion that the light diffracted in a direction perpendicular to the primary rays should be *completely* polarized may well be seriously disturbed. If the view, suggested in the present paper, that a large part of the light from the sky is diffracted from the molecules themselves, be correct, the observed incomplete polarization at $90°$ from the Sun may be partly due to the molecules behaving rather as elongated bodies with indifferent orientation than as spheres of homogeneous material.

Again, the suppositions upon which we have proceeded give no account of *dispersion*. That the refraction of gases increases as the wave-length diminishes is an observed fact; and it is probable that the relation between refraction and transparency expressed in (14) holds good for each wave-length. If so, the falling off of transparency at the blue end of the spectrum will be even more marked than according to the inverse fourth power of the wave-length.

An interesting question arises as to whether (14) can be applied to highly compressed gases and to liquids or solids. Since approximately $(\mu - 1)$ is proportional to n, so also is h according to (14). We have no reason to suppose that the purest water is any more transparent than (14) would indicate; but it is more than doubtful whether the calculations are applicable to such a case, where the fundamental supposition, that the phases are entirely at random, is violated. When the volume occupied by the molecules is no longer very small compared with the whole volume, the fact that two molecules cannot occupy the same space detracts from the random character of the distribution. And when, as in liquids and solids, there is some approach to a regular spacing, the scattered light must be much less than upon a theory of random distribution.

Hitherto we have considered the case of obstacles small compared to the wave-length. In conclusion it may not be inappropriate to make a few remarks upon the opposite extreme case and to consider briefly the obstruction presented, for example, by a shower of rain, where the diameters of the drops are large multiples of the wave-length of light.

The full solution of the problem presented by spherical drops of water would include the theory of the rainbow, and if practicable at all would be a very complicated matter. But so far as the direct light is concerned, it would seem to make little difference whether we have to do with a spherical refracting drop, or with an opaque disk of the same diameter. Let us suppose then that a large number of small disks are distributed at random over a plane parallel to a wave-front, and let us consider their effect upon the direct light at a great distance behind. The plane of the disks may be divided into a system of Fresnel's zones, each of which will by hypothesis include a large number of disks. If α be the area of each disk, and ν the number distributed per unit of area of the plane, the efficiency of each zone is diminished in the ratio $1 : 1 - \nu\alpha$, and, so far as the direct wave is concerned, this is the only effect. The *amplitude* of the direct wave is accordingly reduced in the ratio $1 : 1 - \nu\alpha$, or, if we denote the relative opaque area by m, in the ratio $1 : 1 - m$*. A second operation of the same kind will reduce the

* The *intensity* of the direct wave is $1 - 2m$, and that of the scattered light m, making altogether $1 - m$.

amplitude to $(1-m)^2$, and so on. After x passages the amplitude is $(1-m)^x$, which if m be very small may be equated to e^{-mx}. Here mx denotes the whole opaque area passed, reckoned per unit area of wave-front; and it would seem that the result is applicable to any sufficiently sparse random distribution of obstacles.

It may be of interest to give a numerical example. If the unit of length be the centimetre and x the distance travelled, m will denote the projected area of the drops situated in one cubic centimetre. Suppose now that a is the radius of a drop, and n the number of drops per cubic centimetre, then $m = n\pi a^2$. The distance required to reduce the *amplitude* in the ratio $e : 1$ is given by

$$x = 1/n\pi a^2.$$

Suppose that $a = \frac{1}{20}$ centim., then the above-named reduction will occur in a distance of one kilometre $(x = 10^5)$ when n is about 10^{-3}, *i.e.* when there is about one drop of one millimetre diameter per litre.

It should be noticed that according to this theory a distant point of light seen through a shower of rain ultimately becomes invisible, not by failure of definition, but by loss of intensity either absolutely or relatively to the scattered light.

Reprinted from *Proceedings of the American Physical Society,* Vol. XXVI(6), 497–511 (1908).

Theories of the Color of the Sky.[1]

By Edward L. Nichols.

IN asking your attention to-day, even briefly, to the consideration of the present state of our knowledge concerning the causes of the color of the sky it may be truly said that I am inviting you to leave the thronged thoroughfares of our science for some quiet side street where there is little going on and you may even suspect that I am coaxing you into some blind alley, the inhabitants of which belong to the dead past. Nevertheless, the problem is not a simple one. It cannot be considered as settled or likely to be soon completely solved.

Review of Existing Theories.

It is true that the theory suggested by Brücke,[2] appears to many minds conclusive and all sufficient. The famous demonstrations of Tyndall[3] and the masterly theoretical work of Rayleigh[4] may be said to have established Brücke's view that the color of the sky is that of a turbid medium. More recent experimental work has been in a general way confirmatory of this theory.

At the same time it should not be forgotten that other explanations of the color of the sky have been offered. Clausius[5] in a classical paper, has shown that the color of the sky is deducible on theoretical grounds from the assumption of the existence of minute vesicles or water bubbles suspended in the air. Hagenbach[6] has pointed to the reflection from surfaces of moving bodies of air as the probable cause of the blueness of the sky. Spring[7] has called attention to the fact that oxygen, ozone, water and hydro-

[1] Presidential address delivered at the New York meeting of the Physical Society, February 29, 1908.

[2] Brücke, Poggendorff's Annalen, LXXXVIII., p. 363, 1852.

[3] Tyndall, Proc. Royal Society, XVII., p. 223, 1868 ; also Phil. Mag. (4), XXXVII., p. 388.

[4] Rayleigh, Phil. Mag. (4), XLI., pp. 107 and 447, 1871.

[5] Clausius, Pogg. Annalen, LXXVI., p. 161 ; LXXXIV., p. 449, 1849.

[6] Hagenbach, Pogg. Annalen, CXLVII., p. 77, 1872.

[7] Spring, Bulletins de l'Academie Belge (3), XXXVI., p. 504.

gen peroxide, all of which are constituents of our atmosphere are blue, and that they are present in sufficient quantity to amply account for the preponderance of the shorter wave-lengths in the light which reaches us from the atmosphere. Lallemand[1] and also Hartley[2] have sought for an explanation in the fluorescence of the ozone of the upper atmosphere, and finally subjective color[3] has been advanced as a possible explanation.

Of these various theories only that of Clausius can be discarded without further consideration. That the atmosphere is a turbid medium we have overwhelming experimental evidence but that this turbidity is in itself the sole source of color does not necessarily follow. It has been a characteristic of nearly all the literature of this subject that each author has emphasized some one explanation to the exclusion of all others, although Pernter[4] who is one of the most strenuous advocates of the theory of turbidity, is inclined to consider the fluorescence of ozone as a possible additional factor. Jansen[5] alone, so far as I know, in an interesting summary of the literature of the subject treats the color of the sky as due to the combination of a number of factors.

We may state the case as follows:

1. The turbidity of the atmosphere would of itself give us a blue sky, but the ideal medium of Rayleigh would afford a distribution of intensities to which the actual sky rarely if ever corresponds.

2. Even were the atmosphere free from particles of dust, condensed water vapor or other extraneous matter it would not, according to Rayleigh's latest paper, be optically empty, to use the term employed by Tyndall, but would be blue by virtue of reflections from the molecules of the air itself.

3. If there were no other source of blueness, the color of the air, according to Spring, would give us a blue sky by virtue of the selective absorption-color of various of its constituents. The objections to the adoption of this as a factor are obvious and are regarded by many writers as insuperable but their arguments are not in my opinion conclusive.

4. Reflection from surfaces in a troubled atmosphere as pointed out by Hagenbach, would give us light from the sky increasing in intensity relatively to sunlight in proportion to the square of the wave-length. This is quite sufficient to account for the average blueness of the sky but not for the intenser blueness frequently observed. It cannot therefore be regarded as the sole or most important factor.

[1] Lallemand, Comptes Rendus, LXXV., p. 707, 1872.

[2] Hartley, Nature, XXXIX., p. 474, 1889.

[3] Nichols, Phil. Mag. (5), VIII., p. 425, 1879; also Trans. Kansas Acad. of Science X., p. 111.

[4] Pernter, Denkschriften d. Akad. d. Wissenschaften zu Wien, LXXIII., p. 301, 1901.

[5] Jansen, Das Weltall, V., 1904–5, p. 37 et seq.

5. Fluorescence as a factor of blueness of the sky cannot.be definitely considered at the present time for lack of experimental data concerning it.

6. As regards the subjective or physiological factor it may be said that were there no other cause the sky would undoubtedly appear blue ; for we still see it blue where measurements with the spectrophotometer indicate a composition relatively much weaker in the shorter wave-lengths of the spectrum than the *average* composition of sunlight.   In the present paper I shall, however, consider only the objective factors.

The problem of the color of the sky resolves itself then into a determination of the relative importance of these various factors, the existence of all of which with the possible exception of fluorescence must be regarded as experimentally established.   Such an analysis is unfortunately a difficult matter on account of the complexity of the phenomena and because the observer cannot create his conditions nor hold them constant.

The phenomena of aerial polarization indicate beyond any doubt the turbidity of the air as one source of the blueness of the sky.   In the case of the ideal turbid medium of Rayleigh, in which all the particles are small as compared with the wave-length of light, the composition of the diffused light, as is well known, is such that the ratio of the intensity of the reflected ray to that of the incident ray varies inversely as the fourth power of the wave-length.   This would give us a sky about twelve times as bright in the extreme violet as in the red, as compared with sunlight.   Rayleigh's analysis may be said to have found complete verification in the experimental study of artificial turbid media, particularly in the studies of the colors of the steam jet by Bock.[1]   Spectrophotometric measurements of the sky itself have, however, led to widely varying results.   Zettwuch,[2] who made many measurements of the sky at Rome, calls especial attention to its extreme variability.   Crova,[3] who has made more measurements of this sort than any other investigator with whose work I am acquainted, obtained, in the course of measurements at Montpellier extending over more than two years, relations for the variation of intensity with wave-length in which the exponential constant ranged from 1.61 to 6.44.   Crova's measurements unfortunately extend only through the brighter parts of the spectrum from $.635\,\mu$ to $.510\,\mu$; the extreme red and the violet being omitted.   If, however, we are to accept his exponential coefficients as even approximately correct, we shall have to admit other sources for the blue of the sky than that of turbidity. That there is further evidence to the same effect will be shown later.

[1] Bock, Wiedemann's Annalen, LXVIII., p. 674, 1899.

[2] Zettwuch, Philos. Magazine (6), IV., p. 199, 1902.

[3] Crova, C. R., CIX., p. 493; CXII., p. 1178; also Annalen de Chimie et de Physique (6), XX., p. 480.

Certain measurements with a portable spectrophotometer, which I have undertaken in various localities, during a recent vacation may be of interest as indicating the nature of the phenomena with which we have to deal even if they afford insufficient data for the solution of the general question.

Comparisons of the light from the sky with direct sunlight by means of the spectrophotometer are chiefly interesting as a means of determining the relation between actual skies and the ideal sky imagined by Rayleigh. For other purposes such measurements are of little value on account of the variable quality of sunlight after passage through the earth's atmosphere, and because the sky receives a portion of its illumination from light reflected from clouds and from the surface of the earth ; and further because of absorption which the blue light from the sky suffers from the turbid medium itself. To indicate the variations in the quality of sunlight which are encountered in making direct comparisons of sky and sun I may say that the range in the relative intensity of sunlight when compared with the acetylene flame employed as a standard source in my measurements was between 9.9 at Brienz, Switzerland, on August 19, 1907, and 10.2 at Andermatt on August 9 of that year, on the one hand, and 3.57 at Brienz on August 13, where these quantities express the intensity of the sun's spectrum in the violet ($.418\,\mu$) as compared with the spectrum of the standard flame. Both spectra were taken of equal intensity in the red at $.725\,\mu$.

Measurements of artificial turbid media are not open to this difficulty. Bock, whose steam jet, made turbid by the action of hydrochloric acid gas, seems to have afforded the most favorable conditions for investigation, found the following relative intensities, the intensity in the red being taken as unity :

Red	1.00
Yellow	1.52
Green	2.89
Blue	4.35
Violet	9.81

Rayleigh's formula for the ideal medium gives for the corresponding regions :

Red	1.00
Yellow	1.98
Green	2.82
Blue	4.92
Violet	8.50

Bock also made comparisons between sky and sun and found on five successive days the relative intensities of the violet, as compared with red, to be as 5.7, 8.7, 6.5, 5.2, 4.3. Of these only one shows an approach to the ideal value.

In comparing the spectrum of the skylight with that of an artificial standard, the flame sources hitherto employed, namely, the petroleum flame, by Vogel; the Carcel flame, by Crova; and the Hefner flame, by Frl. Köttgen; are too weak in the violet to admit of satisfactory measurements in that region. The acetylene flame which I have employed in my recent experiments is better in this respect, being relatively about three times as bright in the violet as the sources previously employed, and I have attempted to cover a much wider range than earlier observers, namely, from $.725\,\mu$ in the red to $.390\,\mu$ in the extreme violet. It is in these regions at the extreme ends of the visible spectrum, hitherto uncovered by spectrophotometric observations, that some of the most interesting and striking variations of the light from the sky occur.

The instrument employed was planned with special reference to portability. It consists essentially of a simplified Lummer-Brodhun spectrophotometer with a vertical collimator V, containing a nicol N, and a horizontal collimator H, as shown in Fig. 1. For the single prism of the

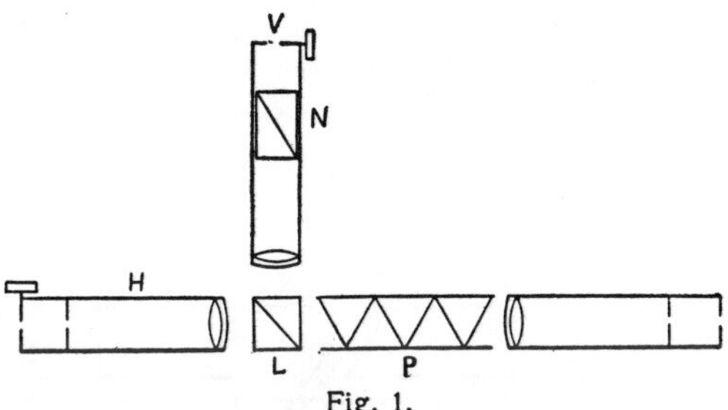

Fig. 1.

original instrument a set of direct vision prisms P is substituted. The comparison source is a small flat acetylene flame surrounded with a hooded chimney of black metal. In front of the flame is a diaphragm with an aperture of 2 mm. diameter so that only the central portions of the flame send light to the slit. About 3 cm. in front of this diaphragm is a circular screen of milk glass which within the limits of the visible spectrum exhibits no marked selective absorption. This lamp is mounted in front of the horizontal slit of the spectrophotometer and access of all light from other sources is excluded by screens.

The acetylene lamp is supplied from a small portable generator and a light rubber bag is used as a storage reservoir and to equalize the pressure of the gas supplied to the flame. Such a comparison source does not realize the conditions as to uniformity of intensity which can be obtained in the laboratory but the variations were rarely troublesome and indeed the degree of constancy was quite sufficient for the purposes of studying so variable a spectrum as that which we obtain from the sky.

Numerous measurements of the spectrum of the sky in various localities by means of this apparatus showed, in general, far greater relative intensities of the longer wave-lengths than one would expect from the theory of Rayleigh. Direct comparisons of skylight and sunlight made by substituting for the standard flame a screen smoked with magnesium oxide and exposed to the sun's direct rays, with adequate precautions to prevent

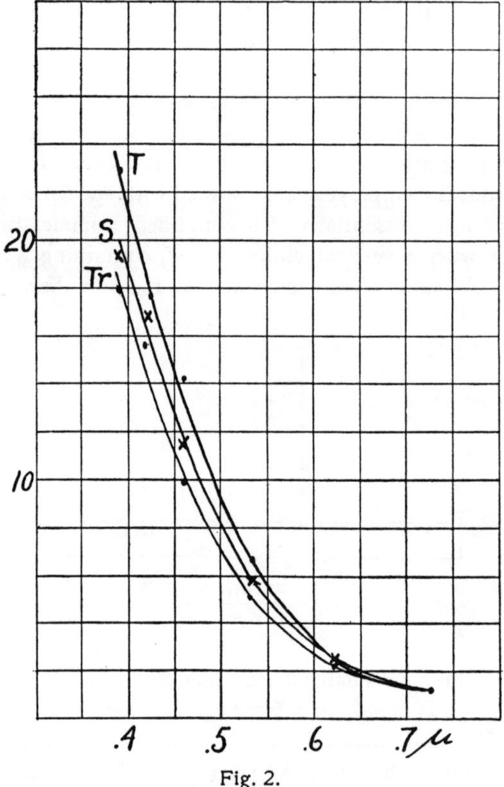

Fig. 2.

illumination from other sources, showed that the skies under consideration were as a rule very far from corresponding to the light which would be obtained from an ideal turbid medium.

Among the most obvious causes of such discrepancies are:

(*a*) The presence of larger reflecting particles in the atmosphere, sometimes invisible and sometimes forming masses of mist or cloud.

(*b*) Absorption by transmission through the turbid medium itself.

(*c*) Illumination of the atmosphere by light reflected from the surface of the earth.

To determine definitely and quantitatively the relative importance of these and other possible factors is a matter of great difficulty; but the study of numerous curves obtained under different conditions throws some light upon the question. It was found that curves taken at dawn and in the twilight after sunset showed a much greater relative preponderance of blue than those taken during the day. These curves moreover were all of the same type and in close agreement with each other as regards the distribution of intensities. In Fig. 2 a group of such curves is given, all reduced for comparison to equal intensity in the extreme red.

While it is impossible to make a direct comparison between such spectra observed at dawn or twilight with the spectrum of the rays by means of which the sky overhead is illuminated, an indirect approximate estimate may be made by means of comparisons made during the day immediately following or immediately preceding, as the case may be, of sky and lamp and then of sky and sun. From several such computations it appears that the skylight curves at dawn and twilight not only show far greater constancy of type than do mid-day measurements, but also a fair approximation to the *ideal* sky. I found for example from measurements made at Sterzing (Tyrol) on July 18 at 4:55 A. M. (before sunrise) and the comparison of these with measurements on the same day between 8 and 9 A. M., in which ratios between sky and lamp on the one hand and sky and sun on the other were obtained, that the probable early morning sky/sun ratio was 8.9 at wave-length $.418\,\mu$ both being reduced to unity at $.725\,\mu$. The ratio for the ideal sky after Rayleigh's formula for these wave-lengths is 9.3.

Since these dawn-twilight skies by fair weather show but litttle variation as to composition I have taken the average of the data at my disposal and have called the resulting curve the *typical dawn curve*. For convenience this is arbitrarily given an intensity of unity in the extreme red ($.725\,\mu$) and other ordinates express the ratio of the skylight to the light from the flame, wave-length for wave-length, throughout the spectrum. The characteristic difference between the typical dawn curve, or any individual dawn curve by fine weather, since these are found to agree closely in composition, and the spectrum of the sky in the middle of the day referred to the same comparison source is shown in Fig. 3. It will be seen that while the skylight taken as a whole is greatly increased in intensity the actual intensity of the blue and the violet is much less affected than are the longer wave-lengths.

The gradual change in the quantity and character of the light from the sky as the sun rises in the heavens is shown in Fig. 4, which contains curves obtained at Sterzing (Tyrol) at 4:45 A. M., before sunrise (*a*), at 5:40, soon after sunrise (*b*), at 7:00 (*c*) and at 8:25 A. M. respectively, on a cloudless and unusually clear morning.

This relation of curves taken after sunrise, to the typical dawn curve is
more obvious if we plot the ratios of the curves to be compared for each
wave-length of the spectrum.   The results for curves *b*, *c* and *d* of Fig. 4
are given in Fig. 5.

These curves of ratios are significant because they repeat themselves
persistently as to type, although with certain systematic variations, in the
case of numerous measurements taken in various localities.

In fine weather, especially in summer, the moisture of the atmosphere
tends to condense and to form masses of sunlit cloud which gather in the

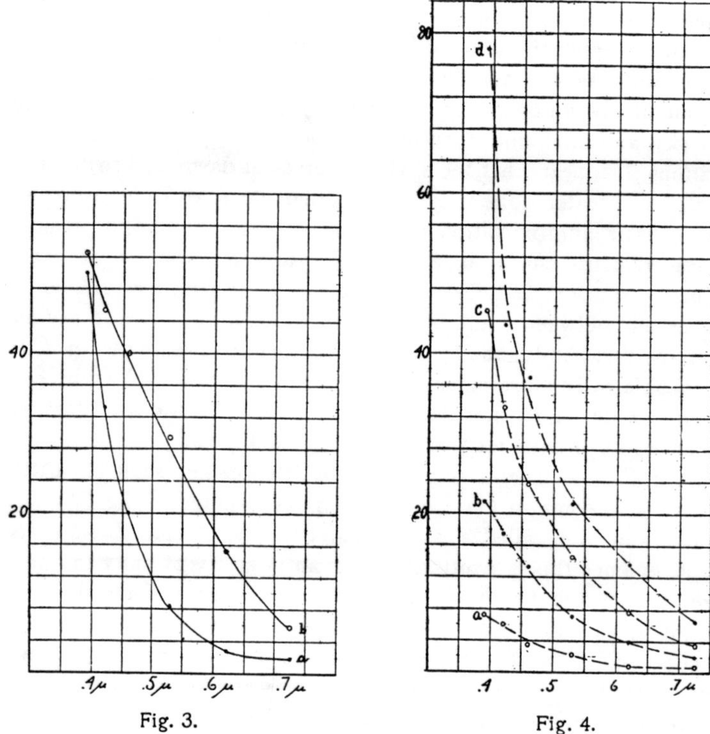

Fig. 3.   Fig. 4.

middle of the day and disappear towards sunset.   The sky/acetylene
curve then takes on another distinctive form well illustrated in Fig. 6.
The two curves there shown were taken at Sterzing about two hours after
the completion of the set of morning measurements plotted in Fig. 4.

Fig. 7 shows the culmination of this effect in the curve for 1:30 P. M.
obtained from a very bright sky with masses of cumuli, its persistence
with much reduced intensities at 5:50 P. M. and its almost entire disappearance at 6:45 P. M.   The two curves for 7:20 and 7:40 P. M. —

about sunset — show complete return to the early morning type. This form of curve is especially characteristic of summer weather in the mountains and the droop towards the ultra-violet was most pronounced in high localities, as at Andermatt, Samaden (Engadine), on the Rhone Glacier and on the Brienzer Rothhorn. It occurred occasionally, but in less fully developed form in the winter skies of Algeria and Sicily.

The study of the sky, on foggy mornings when the air was filled with sunlit mist through which the sun and the blue sky were dimly visible yielded other and slightly different curves, always similar in form and characteristic of this state of the atmosphere (Fig. 8, *a*). Overcast skies gave distinctive forms (Fig. 8, *b*), notable for their general resemblance to certain curves obtained from the clear sky. The relations of these various types will be considered in a forth-coming paper. They are mentioned here only because of a certain bearing upon the theory.

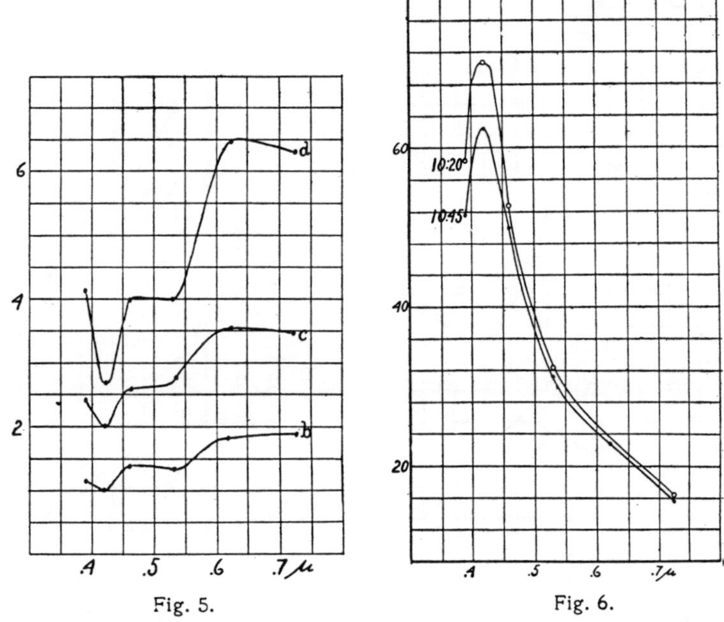

Fig. 5.Fig. 6.

There is good reason to think certain features of the curves of ratios illustrated in Fig. 5 are due to the illumination of the sky by light reflected from the earth. The maximum at $.625\,\mu$ and the minimum at $.525\,\mu$ occur in all curves taken when the landscape is under direct sunlight. They are most marked in measurements made at stations in the open country, as at Taormina, Trafoi and Sterzing, less prominent in the winter skies at Biskra and Palermo and least of all in observations at Vienna where the sky was always seen through a pall of smoke. Mist and over-cast skies, of which the curves in Fig. 8 (*a* and *b*) are typical, are devoid of these two selective features, as will be seen from the curves of ratios given in Fig. 9. Two of these examples, *B* and

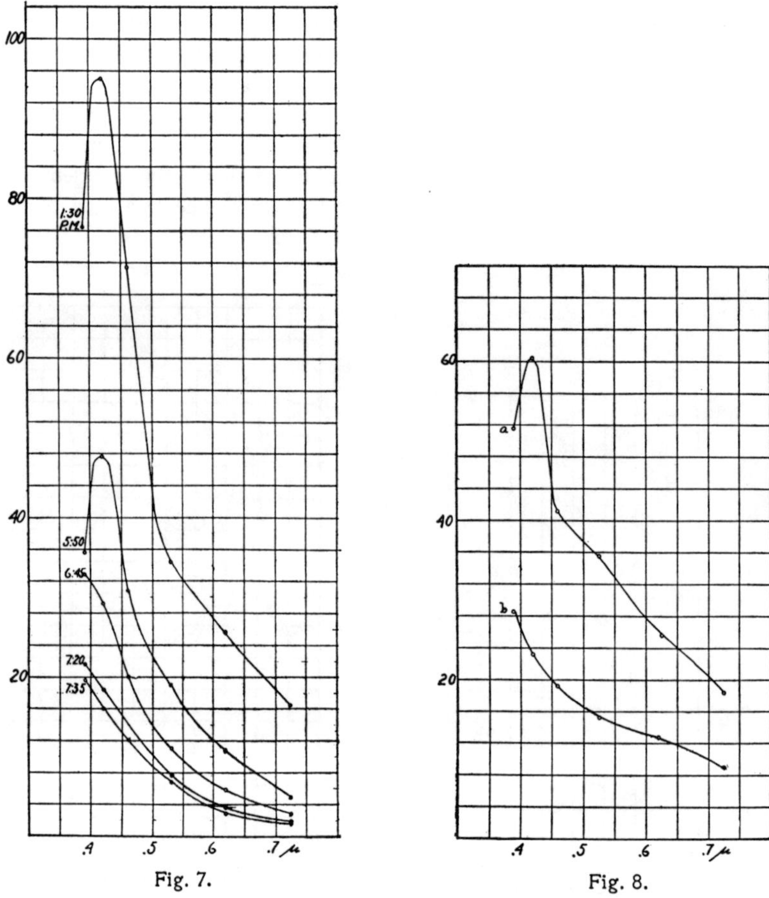

Fig. 7.   Fig. 8.

Z, are for fog, measured at Bebek on the Bosphorus and at Zell am See in Austria, and two, *V* and *T*, for clouded skies at Vienna and Taormina respectively when the landscape was untouched by sunlight. The absence of the features under consideration, and which are so obvious in Fig. 5, will be noted.   These features are also missing in measurements taken at sea during bright weather.   The fact that the region $.525\,\mu$ corresponds to one of the strongest absorption bands of the spectrum of light transmitted by chlorophyl and $.625\,\mu$ to the brighest portion of

that spectrum may be taken to indicate that presence of foliage modifies
to a measurable extent the character of the light from the sky.   In other
words we have, when the landscape is
sunlit, a weak chlorophyl spectrum
superimposed upon the true spectrum
of the light from the sky !

We have, therefore, in light reflected
from the earth one of the causes of the
deviation of many skies from the distribution of intensities to be expected
from an ideal turbid medium illuminated only by direct sunlight.   Reflection from cloud masses is often another
source of illumination and by means of
the spectrophotometer the selective
character of such light, which closely
resembles that from fog or mist, can
readily be determined.

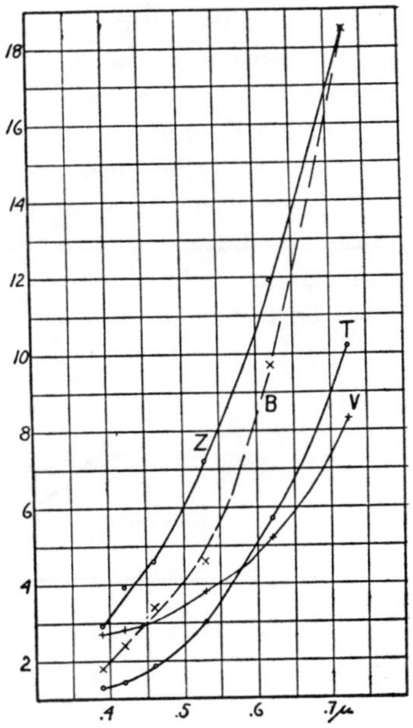

Fig. 9.

Inferences from the Polarization
of the Sky.

Some of the strongest arguments for
regarding the atmosphere as a turbid
medium are derived from the study
of the character of the polarization of skylight.

The phenomena described by Tyndall as the result of his study of
artificial turbid media and his application of the same to the sky are
confirmed and in a manner explained by observations with the spectrophotometer.   Pernter[1] who made extended observations upon the light
diffused by alcoholic solutions of mastic gum found the degree of polarization of the red, green and blue to vary in a remarkable manner
according to the color of the emulsion.   He found in general the light
emitted at right angles to the illuminating beam to be less polarized the
whiter the appearance of the medium — which is also true of the sky.
For blue emulsions the green showed the greatest polarization, then the
blue and then the red.   For whiter emulsions the red was most strongly
polarized and the effect diminished toward the violet.   He found the
same to be true of blue and whitish skies respectively.

In comparisons of the distribution of polarization in the spectrum of
skylight from the zenith, taking the ratio of the brightness of the component polarized in the sun's vertical plane to that at right angles to the
same, I obtained the results shown graphically in Fig. 10.

[1] Pernter, l. c.

It will be seen from the figure that the polarization of the sky is sometimes :

(*a*) Greatest in the violet.

(*b*) Greater in the green, yellow or blue than in the red or violet.

(*c*) Greatest in the red.

(*d*) Least in the middle of the spectrum.

(*e*) Nearly uniform throughout the visible spectrum.

The ideal turbid medium of Rayleigh would give complete polarization of all wave-lengths at an angle of 90°.   If, with Pernter,[1] we consider that a medium may have this character for longer waves and yet reflect unpolarized light of shorter wave-length, the various cases noted above may be accounted for as follows :

(*a*) When the polarization increases towards the violet as in the ob-

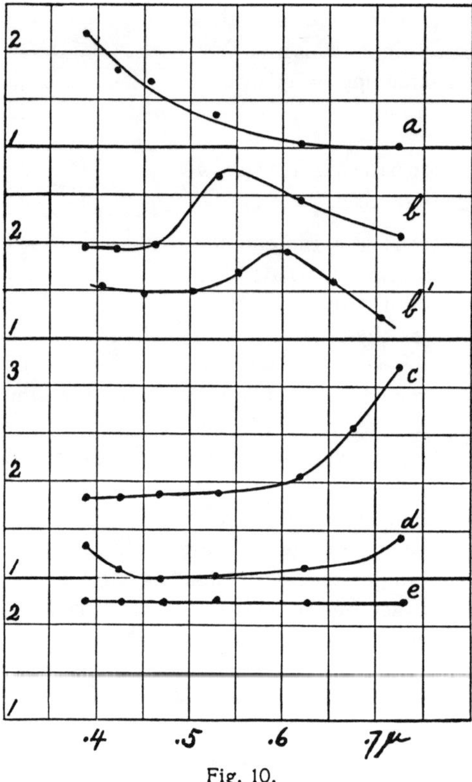

Fig. 10.

servations made at Biskra, February 23 (Fig. 10, *a*), we may suppose the light to come from an *ideal* medium with an admixture of coarser reflect-

[1] Pernter, l. c.

ing particles. Since light from the latter will be relatively weaker in the violet the percentage of polarized light will be greatest in that region and will diminish as we pass toward the red. The distribution of intensities in such a sky will obviously show greater relative strength at the red end of the spectrum than would the ideal sky.

(*b*) Observations made at Salzburg, July 5, 1907 (Fig. 10, *b*), also at sea on February 9 (Fig. 10, *b'*). The medium from which the light is derived may be a mixture of the *ideal* medium, with a medium ideal for green but not for violet, together with particles reflecting unpolarized light throughout the spectrum.

(*c*) Observations made at Chur on August 4 (Fig. 10, *c*). The *ideal* medium may be mixed with a medium ideal for red but not for green and the shorter wave-lengths, and also with coarser particles.

(*d*) Observations made at Biskra on February 24 (Fig. 10, *d*). It is possible that we have in this example a combination of the conditions producing *a* and *c*. Light from a region or layer of the atmosphere giving preponderance of polarization in the red may be considered as mingled with light from regions giving polarization chiefly in the violet.

(*e*) Case *e* offers greater difficulties. It might of course indicate an ideal sky but none of the skies for which this condition as to polarization was observed approached the ideal in relative intensity in the violet. The skies for which a uniform distribution of polarization was actually obtained were, indeed, very far from the ideal as to distribution of intensities. We may imagine as one of several combinations of circumstances which would produce this condition, a sky consisting of the ideal medium mingled with a coarser medium reflecting unpolarized light as in case (*a*) and that the preponderance of the polarized violet is removed by absorption of the light diffused by the ideal medium in passing through extended regions filled with the same. The change of composition would be analogous to that observed in fluorescent solutions of considerable concentration, where there is always a shift of the fluorescence band toward the infra-red. One may also imagine other combinations where the effects described under *b*, *c* and *d* happen to balance each other giving uniformity throughout the spectrum. Skies giving polarization curves of the types *b*, *b'* and *c* might obviously be expected to show the phenomenon designated by Tyndall as *residual blue*.

EVIDENCE CONCERNING OTHER FACTORS IN THE
PRODUCTION OF SKYLIGHT.

As has already been pointed out, the blue color of the air itself and a blue or violet due to fluorescence of ozone or other components of the atmosphere are to be regarded as possible factors in the production of the color of the sky. Data upon this subject are very incomplete and not

47

altogether conclusive.   It is significant however that Crova in certain of
his measurements, referred to in a previous paragraph, obtained an
exponent much higher than that demanded by Rayleigh's theory.   I find
furthermore that all available measurements which give a near approach
to the proper relation between the extreme red and the violet of the
spectrum, namely, an increase of intensity of the violet which is inversely

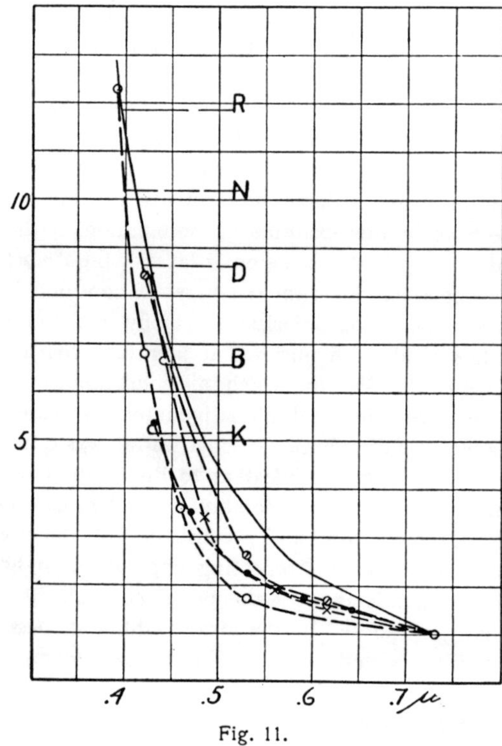

Fig. 11.

proportional to the fourth power of the wave-length, give values relatively
much too small in the middle of the spectrum.   This is true of the
twilight and dawn curves obtained from my own measurements and likewise of the only set of measurements made by Bock in which the intensity of the violet approached the value corresponding to that of the ideal
sky.   While Miss Köttgen made no direct comparisons of sky and
sun she measured both skylight and sunlight in terms of the Hefner lamp.
Making use of her data one can compute the sky/sun ratios under the assumption that the sunlight at the time that the sun curve was taken was
of the same quality as that during the comparison of skylight with the
standard flame.   I have made such a computation and find the resulting

curve to have the same peculiarity, that is to say, like the curve plotted from Bock's data and like my own curves there is marked relative weakness throughout the middle of the spectrum. This peculiarity could be readily explained by assuming a turbid medium containing particles too large to give the ideal sky but having a distinct blue color of its own due either to absorption, in accordance with Spring's theory, or it might be due to fluorescence in the violet end of the spectrum or to absorption and fluorescence combined. The nature of the departure of these curves from the ideal form, which is illustrated·in Fig. 11, is quite consistent with and indeed explanatory of the occasional high values of the exponential factor found by Crova. In Fig. 11, R is the curve for the ideal turbid medium, N is one of the very few cases in which I obtained a relation between red and violet in accord with Rayleigh's theory. D is from my typical dawn curve, B and K are from the data of Bock and Köttgen respectively.

Summary.

It has been my endeavor in this paper to show :

1. That while there is good reason for regarding the sky as a turbid medium, the experimental study of the spectrum of skylight affords evidence of a distribution of intensities which cannot be altogether accounted for by the assumption of an atmosphere conforming to Rayleigh's formula nor of a turbid medium containing coarser particles.

2. That the illumination of the atmosphere by selectively reflected light from the surface of the earth and from cloud masses and mist modifies the character of the light from the sky to an extent which while perhaps not readily discernible with the unaided eye is definite and unmistakable when the sky is studied with the spectrophotometer.

3. That the deviation of the observed distribution of intensities recorded by several investigators indicates a blue absorption color of the air or, since the preponderance in the violet appears to be variable in amount, the existence of fluorescence of some unstable factor of the atmosphere such as ozone or both.

Physical Laboratory of Cornell University,
February, 1908.

Reprinted with permission from *Proceedings of the Royal Society of London,*
Vol. A104, 333–357 (1923).

On the Complex Anisotropic Molecule in Relation to the Dispersion and Scattering of Light.

By Louis V. King, D.Sc., Macdonald Professor of Physics, McGill University, Montreal.

(Communicated by Prof. A. S. Eve, F.R.S. Received January 1, 1923.)

§ 1. *Introduction.*

It has now been known for several years that the molecules of a gas consisting of positive and negative charges set into forced vibrations by the electric field of an incident light-wave are responsible, not only for the refractive properties of the medium, but also for lateral scattering and extinction as exemplified on a large scale by the blue of the sky and the colour of the setting sun.*   In fact, until comparatively recently, observations on the extinction of solar radiation of various wave-lengths by the earth's atmosphere have provided the only data† by means of which the theory of molecular scattering and extinction could be tested, this by satisfactory evaluations of the number of molecules per cubic centimetre of air under standard conditions of temperature and pressure,‡ making use for the purpose of Rayleigh's well-known extinction formula based on the idea of the *symmetrical molecule, i.e.,* one in which the

---

* Lord Rayleigh, ' Phil. Mag.,' vol. 37, pp. 375–384 (1899).

† ' Annals of the Smithsonian Astrophysical Observatory,' Washington, vols. II (1908) and III (1913).

‡ Schuster, A., ' Nature,' vol. 81, p. 97 (1909).   Natanson, L., ' Bull. Int. de l'Acad. des Sciences de Cracovie' (1910).   King, L.V., ' Phil. Trans,' vol. 212, A (1912).   A comprehensive survey of the entire field is given by Cabannes, J., Thèse, " Sur la diffusion de la lumière par les molécules des gaz transparents," ' Annales de Physique,' vol. 15 (1921). Further references given in this section are, for the most part, supplementary.

dispersion electrons move in the direction of the electric vector in the light-wave.*

As far as observations were then available, theory gave a tolerably good account of measurements of sky-intensity, both as regards quality and polarization.† The difficulty in this case is to take into account the illumination of the atmosphere by itself, a problem capable of reasonably simple solution in terms of integral equations only if the curvature of the earth is disregarded.‡ From the experimental point of view, satisfactory observations are made difficult by the omnipresent and ever-varying dust content of the atmosphere at ordinary levels, to say nothing of the " haziness " and extinction due to the presence of water-vapour.§

For these reasons, observations of atmospheric extinction and polarization were not sufficiently precise to detect effects due to possible molecular anisotropy, until recent laboratory investigations on light scattering in gases revealed a marked departure as regards polarization from theoretical requirements based on symmetrical molecules.

§ 2. *The Anisotropic Molecule.*

That the molecules of a dust-free gas are able to scatter light seems to have been first demonstrated by Cabannes‖ in 1913, although the unexpected result of *depolarization* or incomplete polarization was first established by Lord Rayleigh,¶ and substantially verified by the later observations of Cabannes,** in France, and of Gans, in Argentina.††

On rotating the analysing nicol complete extinction is not obtained, and the ratio of the minimum to the maximum is termed the " depolarization," a

* King, L. V., ' Nature,' vol. 93, p. 557 (1914). Fowle, F. E., ' Astrophys. Journ.,' vol. 40, p. 435 (1914). Dember, H., ' Ann. der Physik,' vol. 49, pp. 599–610 (1916). Fowle, ' Smithsonian Miscellaneous Collections,' vol. 69, No. 3, May, 1918).

† In addition to observations by the Smithsonian Astrophysical Observatory (' Annals,' vol. III) we may mention those by Bauer and Moulin, ' Comptes Rendus,' vol. 151, p. 864 (1910), and by Pacini, ' Nuovo Cimento,' vol. 10, pp. 131–167 (1915).

‡ King, L. V., ' Phil. Trans.,' vol. 212, A, p. 380 (1912).

§ In this connection recent observations cited by Raman are of interest (' Nature,' p. 75, Jan. 19, 1922), also " Molecular Diffraction of Light " (Univ. of Calcutta Press, 1922).

‖ Cabannes, J., " Sur la diffusion de la lumière par l'air," ' C.R.,' vol. 160, pp. 62–63 (1915).

¶ Strutt, R. J. (Lord Rayleigh), ' Roy. Soc Proc.,' vol. 94, A, p. 453 (1918) ; vol. 95, A, pp. 155–176 (1919) ; vol. 95, A, pp. 476–479 (1919).

** Cabannes, J., ' Annales de Physique,' vol. 15 (1921).

†† Gans, R., " Asymmetrie von Gasmolekeln," ' Ann. der Physik,' vol. 65, pp. 97–123 (1921).

quantity characteristic of the gas, determinable with fair accuracy, and indicative of the asymmetry of the molecule.

The simple " asymmetrical " or " anisotropic " molecule seems to have been first discussed in a fundamental paper by Langevin.*   In this paper by means also of the statistical theory of fluctuations of " molecular orientations " in electric and magnetic fields he has accounted successfully for the outstanding features of electric and magnetic double refraction.

The simple anisotropic molecule will, in general, have three natural vibration periods, corresponding to the free oscillations along each of the principal axes, while an aggregate of such molecules will possess, in general, three *molecular absorption* bands.   But most substances have a much more complicated spectrum, so that the simple anisotropic molecule cannot be expected to account in detail for all the optical properties of transparent media.   When, however, it is employed in the classical electromagnetic theory of radiation, in place of the familiar dispersion electron assumed to move in the direction of the electric vector in the light-wave, many optical phenomena are satisfactorily accounted for.

In the following sections there is a systematic discussion of the associated theories of dispersion and scattering based on a general type of complex anisotropic molecule, developed in such a way that the formulae may be applied to scattering in dust-free transparent liquids, recent observations of this phenomena now being available.†

§ 3. *Dispersion Theory in Terms of the Simple Anistropic Molecule.*

The refractive index of a transparent medium is due to the reaction of the positive and negative charges, constituting atoms and molecules, on the propagation of electro-magnetic disturbances.

We will first consider a simple anisotropic molecule, *i.e.*, one containing a

* Langevin, P., " Sur les biréfringences électrique et magnétique," ' Le Radium,' vol. 7, pp. 249–60 (1910).

† Martin, W. H., " The Tyndall Effect in Liquids," ' Roy. Soc. Proc.,' Canada, vol. 7, p. 219 (1913) ; ' Chem. Abs.,' vol. 8, p. 3739 (1914).   ·" The Scattering of Light by Dust-Free Liquids," ' Journ. Phys. Chem.,' vol. 24, p. 478 (June, 1920) ; vol. 26, p. 75 (Jan., 1922) ; " The Relation between Light Absorption and Light Scattering for Liquids," ' Journ. Phys. Chem.,' vol 26, p. 471 (May, 1922).   Bibliography, ' Trans. Roy. Soc., Canada,' vol 16, p. 276 (1922).

Kenrick, F. B., ' Journ. Phys. Chem.,' vol 26, p. 72 (Jan., 1922).

Ramanathan, K. R., " The Molecular Scattering of Light in Vapours and in Liquids, etc.," ' Proc. Roy. Soc.,' vol. 102, A, p. 151 (Nov., 1922).

single dispersion electron of mass m, and charge e. This will have three equations of motion of the type—

$$m\,(\ddot{\xi}+p_1{}^2\xi) = e\,(E_1+P_1),\qquad(1)$$

where E is the electric vector, and P the polarization vector due to all other vibrating electrons in the medium, and ξ, η, ζ are displacements along its three principal axes of vibration (*see* fig. 1).

For periodic displacements due to the electric force in a light wave proportional to e^{ipt}, we have—

$$\xi = (A/e)\,(E_1+P_1),\quad \eta = (B/e)\,(E_2+P_2),\quad \zeta = (C/e)\,(E_3+P_3),\qquad(2)$$

where

$$A = \frac{e^2}{m}\cdot\frac{1}{p_1{}^2-p^2},\quad B = \frac{e^2}{m}\cdot\frac{1}{p_2{}^2-p^2},\quad C = \frac{e^2}{m}\cdot\frac{1}{p_3{}^2-p^2}.\qquad(3)$$

Corresponding to the three natural periods of vibration $2\pi/p_1$, etc., we have the

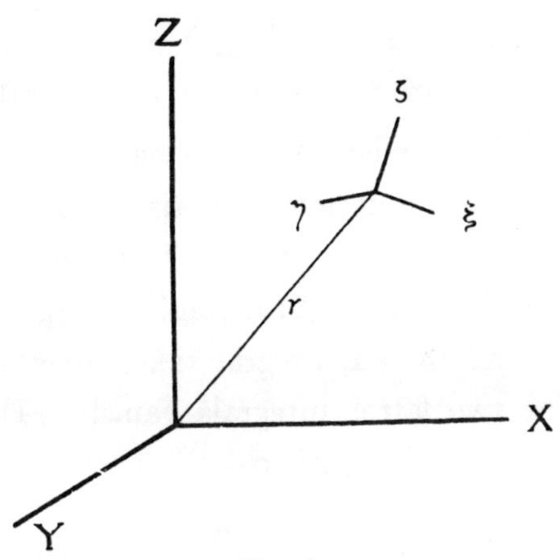

FIG. 1.

wave lengths λ_1, λ_2, λ_3, given by the relations $\lambda_1 = 2\pi c/p_1$, ..., where c is the velocity of light in vacuo.

Referred to axes (X, Y, Z) fixed in the medium, we denote by—

$$Z\,(x,\,t) = |\,Z\,|\cos p\,(t-x/c) = |\,Z\,|\cos\,(pt-2\pi x/\lambda)\qquad(4)$$

the electric force parallel to the axis of Z due to a plane-polarized light-wave acting on the molecule at P, whose co-ordinates are $(x,\,y,\,z)$. This electric force displaces all the electrons of the molecules giving rise to a polarized field $Z_p\,(x,\,t)$ which in isotropic media is assumed parallel to $Z\,(x,\,t)$, and depends only on the molecules in the immediate neighbourhood of P so that $Z_p\,(x,\,t)$ and $Z\,(x,\,t)$ do not differ appreciably in phase.

From (8) and (9) we obtain—

$$w = \frac{\mu^2}{4\pi}\frac{dZ}{dt} \tag{10}$$

where

$$\frac{3\,(\mu^2-1)}{\mu^2+2} = \frac{4\pi n_0 \bar{k}}{1-s} \tag{11}$$

and

$$\frac{Z+Z_p}{Z} = \frac{\mu^2+2}{3}\cdot\frac{1}{1-s} = \frac{\mu^2-1}{4\pi n_0 \bar{k}}. \tag{12}$$

Equation (11) leads to the well-known dispersion formula of the Lorentz type, and Havelock* has shown that (with $s = 0$, and $\lambda_2 = \lambda_3$, *i.e.* $p_2 = p_3$) Kirn's† observations on dispersion in Hydrogen are thereby accurately interpreted, while the calculated value of the depolarization agrees closely with Lord Rayleigh's‡ experimental determination.

§ 4. *Scattering of Light by a Simple Anisotropic Molecule.*

We proceed, now, to investigate the light scattered by an anisotropic molecule situated at the origin O (fig. 2).   As before, we suppose that the incident light

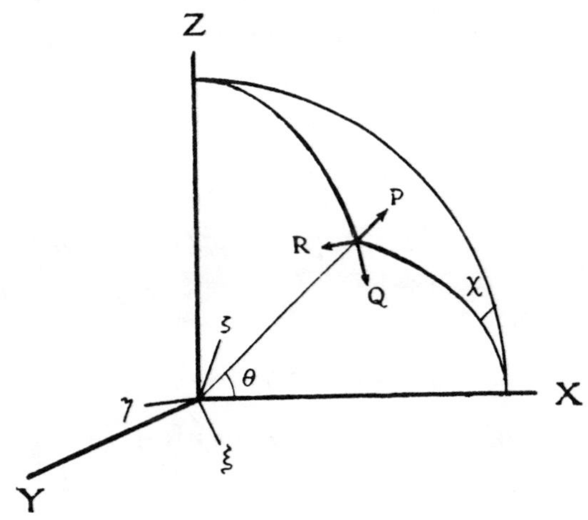

Fɪɢ. 2.

is propagated along the X-axis and is plane-polarized, the electric vector being parallel to the Z-axis.   It will be remembered that in a plane-polarized electro-magnetic wave, the electric force (light vector) is perpendicular to the plane of polarization.

* Havelock, T. H., ' Roy. Soc. Proc.,' vol. 101, A, pp. 154–164 (1922).   A formula of this type is implied in a paper by Langevin on electric and magnetic double refraction (' Le Radium,' vol 7, p. 255 (1910), equation (36)), although no comparison is made with experiment.

† Kirn, M., ' Ann. de Phys.,' vol. 64, pp. 556–576 (1921).

‡ Lord Rayleigh, ' Roy. Soc. Proc.,' vol. 97, A, p. 449 (1920).

The principal axes (ξ, η, ζ) of the molecule are defined as before in section 3. In examining the light scattered to a distant point P, we take POX as the plane of reference. The scattered wave at P is completely specified by the electric vector P along OP, Q perpendicular to the plane POX, and R perpendicular to both P and Q. The direction cosines of the mutually perpendicular vectors P, Q, R are given by the table—

	P	Q	R
X	$\cos\theta$	0	$-\sin\theta$
Y	$\sin\theta\sin\chi$	$\cos\chi$	$\cos\theta\sin\chi$
Z	$\sin\theta\cos\chi$	$-\sin\chi$	$\cos\theta\cos\chi$

$$(13)$$

θ and χ being the angles indicated in fig. 2.

If $(\alpha_p, \alpha_q, \alpha_r)$ be the components of acceleration parallel to (P, Q, R) of a moving charge at O, the electrical intensities in the resulting spherical wave are given by*

$$P = 0, \quad Q = -\frac{e}{c^2 r}\, a_q, \quad R = -\frac{e}{c^2 r}\, a_r, \tag{14}$$

provided velocities and accelerations remain within the limits of *quasi*-stationary motion.

From equation (5) it is seen that the components of acceleration of the charge e along the axes (ξ, η, ζ) are—

$$\ddot\xi = -p^2\,(A/e)\,n_1\,(Z + Z_p), \quad \ddot\eta = -p^2\,(B/e)\,n_2\,(Z + Z_p),$$
$$\ddot\zeta = -p^2\,(C/e)\,n_3\,(Z + Z_p). \tag{15}$$

Since
$$a_q = \ddot\xi \cos\widehat{\xi Q} + \ddot\eta \cos\widehat{\eta Q} + \ddot\zeta \cos\widehat{\zeta Q}$$
$$a_r = \ddot\xi \cos\widehat{\xi R} + \ddot\eta \cos\widehat{\eta R} + \ddot\zeta \cos\widehat{\zeta R} \tag{16}$$

and
$$\cos\widehat{\xi Q} = \cos\widehat{\xi X}\,.\,\cos\widehat{X Q} + \cos\widehat{\xi Y}\,.\,\cos\widehat{Y Q} + \cos\widehat{\xi Z}\,.\,\cos\widehat{Z Q},$$

with similar equations for the remaining cosines in (16), we find from (13)—

$$P = 0, \quad Q = \frac{p^2}{c^2 r}(Z + Z_p)(j\cos\chi - k\sin\chi)$$
$$R = \frac{p^2}{c^2 r}(Z + Z_p)(-h\sin\theta + j\cos\theta\sin\chi + k\cos\theta\cos\chi) \tag{17}$$

where
$$h = An_1 l_1 + Bn_2 l_2 + Cn_3 l_3$$
$$j = An_1 m_1 + Bn_2 m_2 + Cn_3 m_3$$
$$k = An_1^2 + Bn_2^2 + Cn_3^2 \tag{18}$$

* Lorentz, H. A., 'The Theory of Electrons' (Teubner's, 1909), art. 39, p. 51.

Squaring and adding, we have the relation,

$$h^2 + j^2 + k^2 = A^2 n_1^2 + B^2 n_2^2 + C^2 n_3^2. \tag{19}$$

It is easily seen that on averaging for a law of distribution of molecular orientations unaltered by changes in sign of the direction cosines (*e.g.*, depending on their squares), we have—

$$\overline{jk} = 0, \quad \overline{kh} = 0, \quad \overline{hj} = 0. \tag{20}$$

Furthermore, if the law of distribution depends only on the angles which the principal axes of the molecule make with the Z-axis, *i.e.*, on (n_1, n_2, n_3) (as in problems of electric and magnetic double refraction), the average values $\overline{h^2}$ and $\overline{j^2}$ are unaltered by interchanging the axes X and Y, or (l_1, l_2, l_3) and (m_1, m_2, m_3) so that $\overline{h^2} = \overline{j^2}$. In these circumstances we have from (19) ·

where, from (18),
$$\left.\begin{aligned}
2\overline{h^2} + \overline{k^2} &= A^2\overline{n_1^2} + B^2\overline{n_2^2} + C^2\overline{n_3^2} \\
\overline{k^2} &= A^2\overline{n_1^4} + B^2\overline{n_2^4} + C^2\overline{n_3^4} + 2BC\overline{n_2^2 n_3^2} \\
&\quad + 2CA\overline{n_3^2 n_1^2} + 2AB\overline{n_1^2 n_2^2}
\end{aligned}\right\} . \tag{21}$$

Since the axes (ξ, η, ζ) may, in an *isotropic distribution* of molecular orientations, be mutually interchanged without affecting the final result of averaging, we have, since

$$\begin{aligned}
\overline{n_1^2} + \overline{n_2^2} + \overline{n_3^2} &= 1, \\
\overline{n_1^2} = \overline{n_2^2} = \overline{n_3^2} &= \tfrac{1}{3}.
\end{aligned} \tag{22}$$

Furthermore,

$$\overline{n_1^4} = \overline{n_2^4} = \overline{n_3^4} = \tfrac{1}{2} \int_0^\pi \cos^4 \theta \sin \theta \, d\theta = \tfrac{1}{5}, \tag{23}$$

and, since

$$\overline{n_1^4} + \overline{n_2^4} + \overline{n_3^4} + 2\overline{n_2^2 n_3^2} + 2\overline{n_3^2 n_1^2} + 2\overline{n_1^2 n_2^2} = 1$$
$$\overline{n_1^2 n_2^2} = \overline{n_2^2 n_3^2} = \overline{n_3^2 n_1^2} = \tfrac{1}{15}. \tag{24}$$

Thus, for *equally probable molecular orientations*,[*] it follows from (12) that

when
$$\left.\begin{aligned}
2\overline{h^2} + \overline{k^2} &= \tfrac{1}{3}\Sigma A^2, \quad \overline{k^2} = \tfrac{1}{15}(3\Sigma A^2 + 2\Sigma BC) \\
\Sigma A^2 &= A^2 + B^2 + C^2 \quad \text{and} \quad \Sigma BC = BC + CA + AB.
\end{aligned}\right\} \tag{25}$$

[*] The results of averaging for equally probable molecular orientations obtained in this section are in agreement with those of the late Lord Rayleigh, obtained by direct integrations in terms of the Eulerian angles (Lord Rayleigh, ' Phil. Mag.,' vol 25, pp. 373–381 (1918). ' Scientific Papers,' vol VI, pp. 540–548).

It is convenient to introduce the ratio ρ defined by the relation

$$\rho = \frac{2\overline{h^2}}{\overline{h^2} + \overline{k^2}} = \frac{2\,(\Sigma A^2 - \Sigma BC)}{4\Sigma A^2 + \Sigma BC}\,. \tag{26}$$

From the last equation of (18) and (22) we have

$$\overline{k} = \tfrac{1}{3}\,(A + B + C). \tag{27}$$

Eliminating ΣA^2 and ΣBC between (25), (26) and (27) we obtain

$$\overline{h^2} = \overline{j^2} = 3\,(\overline{k})^2 \cdot \frac{\rho}{6 - 7\rho}, \quad \overline{k^2} = 3\,(\overline{k})^2 \cdot \frac{2 - \rho}{6 - 7\rho} \tag{28}$$

and

$$3\overline{k^2} - 4\overline{h^2} = 3\,(\overline{k})^2. \tag{29}$$

§5. *Dispersion Theory in terms of the Complex Anisotropic Molecule.*

(i) *Monomolecular Media.*—It is hardly to be supposed that the simple anisotropic molecule dealt with in the preceding sections can represent the actual state of affairs in any physical medium. We proceed, therefore, to consider a more general type of molecule made up of any number of positive and negative charges (referred to as dispersion charges) in static equilibrium under the mutual action of intermolecular forces.* Without specifying the exact nature of these forces about which we are still in ignorance, we will suppose that the system is stable for small displacements, and that each charge is capable of vibration about its equilibrium position. For each dispersion charge three axes may be found with respect to which the equations of motion are given by equation (1). These *principal directions* will not, however, be parallel for all the dispersion charges affected by the incident light-wave.

We denote, by the following table, the direction cosines of the principal directions of the ith dispersion charge with respect to the axes (X, Y, Z) fixed in the medium.

	ξ_i	η_i	ζ_i	
X	$^i l_1$	$^i l_2$	$^i l_3$	
Y	$^i m_1$	$^i m_2$	$^i m_3$	(30)
Z	$^i n_1$	$^i n_2$	$^i n_3$	

* The writer has thought it desirable in the sequel to use the term *dispersion charge* in preference to *dispersion electron*, to denote the positive and negative parts of atomic systems which, by their vibrations in the periodic electric field of a light-wave, give rise to the phenomenon of dispersion.

Since the results of §3 apply to the ith dispersion charge in each molecule of an aggregate, its contribution to the polarization field at O is, by (8)—

$$^iZ_p = (\tfrac{4}{3}\pi n_0 \bar{k}_i + s_i)\,[Z + Z_p], \qquad (31)$$

where

$$\bar{k}_i = \mathrm{A}_i{}^i\bar{n}_1{}^2 + \mathrm{B}_i{}^i\bar{n}_2{}^2 + \mathrm{C}_i{}^i\bar{n}_3{}^2, \qquad (32)$$

and s_i is defined by (7) with respect to the ith dispersion electron.

The total polarization field due to the i dispersion charges in each molecule of which there are n_0 per unit volume is given by—

$$Z_p = \sum_i Z_p = (Z + Z_p)\sum_i (\tfrac{4}{3}\pi n_0 \bar{k}_i + s_i). \qquad (33)$$

Similarly, the total current density along the Z-axis is, as in (17),

$$w = \frac{1}{4\pi}\frac{dz}{dt} + n_0\,\{\textstyle\sum_i \bar{k}_i\}\,\frac{d}{dt}[Z + Z_p]. \qquad (34)$$

Writing

$$\bar{\mathrm{K}} = \sum_i \bar{k}_i \quad \text{and} \quad \mathrm{S} = \sum_i s_i, \qquad (35)$$

we finally obtain the formulae—

$$\frac{3\,(\mu^2 - 1)}{\mu^2 + 2} = \frac{4\pi n_0 \sum_i \bar{k}_i}{1 - \sum_i s_i} = \frac{4\pi n_0 \bar{\mathrm{K}}}{1 - \mathrm{S}}, \qquad (36)$$

and

$$\frac{Z + Z_p}{Z} = \frac{\mu^2 + 2}{3}\cdot\frac{1}{1 - \sum_i s_i} = \frac{\mu^2 - 1}{4\pi n_0 \sum_i \bar{k}_i} = \frac{\mu^2 - 1}{4\pi n_0 \bar{\mathrm{K}}}, \qquad (37)$$

corresponding to (11) and (12) for the simple anisotropic molecule.

For an isotropic distribution of molecular orientations, we have[*]

$$\overline{{}^i n_1{}^2} = \overline{{}^i n_2{}^2} = \overline{{}^i n_3{}^2} = \tfrac{1}{3}$$

for all values of i, so that (36) leads to a general dispersion formula of the Lorentz type, applicable to a medium made up of complex anisotropic molecules having $3i$ absorption bands—

$$\frac{3\,(\mu^2 - 1)}{\mu^2 + 2} = \frac{e^2 n_0}{3\pi c^2 m\,(1 - \sum_i s_i)}\sum_i\left[\frac{1}{{}^i\lambda_1{}^{-2} - \lambda^{-2}} + \frac{1}{{}^i\lambda_2{}^{-2} - \lambda^{-2}} + \frac{1}{{}^i\lambda_3{}^{-2} - \lambda^{-2}}\right]. \quad (38)$$

If the values of $e_i{}^2/m$ for the dispersion charges are not all equal, it is readily seen that we have only to write the terms $e_i{}^2/m_i$ under the summation sign in (38).

(ii) *Mixtures and Compounds.*—The summations in (33) and (34) may be extended to include the dispersion charges belonging to different molecules of

[*] *See* equation (52) *infra.*

a mixture, or different atoms of a compound. We obtain for (36) and (37) the equations—

$$\frac{3\,(\mu^2-1)}{(\mu^2+2)} = \frac{4\pi\Sigma_r n_r \overline{K}_r}{1-\underset{r}{\Sigma} S_r} \quad \text{and} \quad \frac{Z+Z_p}{Z} = \frac{\mu^2+2}{3}\cdot\frac{1}{1-\underset{r}{\Sigma} S_r} = \frac{\mu^2-1}{4\pi\,\Sigma_r n_r \overline{K}_r}, \quad (39)$$

the summations $\underset{r}{\Sigma}$ being extended to include the number of different kinds of molecules concerned, there being n_r per unit volume of each. We denote by μ_r the refractive index of the constituent r alone, when the molecular density is N_r, the molecular weight m_r and the mass density D_r. Then, provided that mixture or chemical combination has no appreciable effect on the (A, B, C)'s of equation (3) owing to molecular (or atomic) interaction, we have for that constituent—

$$\frac{3\,(\mu_r^2-1)}{\mu_r^2+2} = \frac{4\pi N_r \overline{K}_r}{1-S_r}. \qquad (40)$$

The molecular refractivity of the constituent r is defined by the relation—

$$M_r = \frac{m_r}{D_r}\cdot\frac{\mu_r^2-1}{\mu_r^2+2}. \qquad (41)$$

Since $D_r = m_r N_r$, we see from (40) that

$$\tfrac{4}{3}\pi\overline{K}_r = (1-S_r)\,M_r. \qquad (42)$$

If w_r is the weight of the constituent r per unit volume of the mixture or compound of density D, we have—

$$w_r = m_r n_r, \quad D = \underset{r}{\Sigma} w_r, \quad n_r/N_r = w_r/D_r. \qquad (43)$$

Furthermore, if τ_r denotes the proportion of each constituent by weight, we have

$$\tau_r = w_r / (\underset{r}{\Sigma} w_r) = w_r/D. \qquad (44)$$

From (39) we have on making use of (41), (43), (44)—

$$\frac{1}{D}\frac{\mu^2-1}{\mu^2+2} = \frac{\underset{r}{\Sigma}\left[\dfrac{\tau_r\,(1-S_r)}{D_r}\dfrac{\mu_r^2-1}{\mu_r^2+2}\right]}{1-\underset{r}{\Sigma} S_r}. \qquad (45)$$

For a large number of substances it is found experimentally that the additive law of Lorentz

$$\frac{1}{D}\frac{\mu^2-1}{\mu^2+2} = \underset{r}{\Sigma}\frac{\tau_r}{D_r}\frac{\mu_r^2-1}{\mu_r^2+2} \qquad (46)$$

holds to a high degree of accuracy.*

* Chéneveau, ' Les propriétés optiques des solutions ' (Gauthier-Villars, 1913).

On comparing (45) and (46) it would appear that for each of the constituents the term S_r is negligibly small if not actually zero, a result which is difficult to justify theoretically and which must have some physical significance connected with the geometrical configuration of dispersion electrons in molecules of media in the liquid state.

The formula (39), (45) and (46) also apply when the different molecules (or atoms) form a compound, provided that the inter-atomic forces representing chemical combination have a negligible effect on the (A, B, C)'s of equation (3). If there are c_r atoms of type r for which the *atomic refractivity* is specified by M_r of (42), we see that (39) leads to the result for the molecular refractivity of the compound M,

$$\frac{m}{D}\frac{\mu^2-1}{\mu^2+2} = M = \frac{\sum_r c_r\,(1-S_r)\,M_r}{1-\sum_r S_r} \tag{47}$$

where the molecular weight of the compound m is given by—

$$m = \sum_r c_r m_r, \qquad \text{and} \qquad \tau_r = c_r m_r/m.$$

It is found experimentally that the terms S_r are negligible for a large number of substances, when (47) takes the simple form—

$$M = \sum_r c_r M_r.$$

The atomic refractivities M_r cannot be measured directly, but are indirectly calculated from the known refractive indices of compounds. The subject is fully discussed in an interesting paper by Silberstein.*

§ 6. *Scattering of Light by a Complex Anisotropic Molecule.*

If we have several dispersion charges in a complex molecule, each will contribute to the components (P, Q, R) of the scattered radiation amounts given by equations (17) and (18). Hence taking a summation for all the dispersion charges the components of electric force in the radiation scattered by a complex molecule is given by—

$$\left. \begin{aligned} P &= 0, \quad Q = \frac{p^2}{c^2 r}\,(Z+Z_p)\,(J\cos\chi - K\sin\chi) \\[2ex] R &= \frac{p^2}{c^2 r}\,(Z+Z_p)\,(-H\sin\theta + J\cos\theta\sin\chi + K\cos\theta\cos\chi) \end{aligned} \right\} \tag{48}$$

* Silberstein, L., " Molecular Refractivity and Atomic Interaction," ' Phil. Mag.,' vol. 33, pp. 92–128 (Jan., 1917).

where

$$H = \sum_i h_i = \sum_i (A_i{}^i n_1{}^i l_1 + B_i{}^i n_2{}^i l_2 + C_i{}^i n_3{}^i l_3)$$
$$J = \sum_i j_i = \sum_i (A_i{}^i n_1{}^i m_1 + B_i{}^i n_2{}^i m_2 + C_i{}^i n_3{}^i m_3) \Bigg\}.$$
$$K = \sum_i k_i = \sum_i (A_i{}^i n_1{}^2 + B_i{}^i n_2{}^2 + C_i{}^i n_3{}^2) \qquad (49)$$

The principal directions of each dispersion charge will not, in general, be parallel. We refer them to *molecular axes* (ξ, η, ζ) fixed in the molecule, whose direction cosines with respect to X, Y, Z) and (ξ_i, η_i, ζ_i) respectively are given the tables—

	ξ	η	ζ			ξ	η	ζ
Y	l_1	l_2	l_3	ξ_1		$^i a_1$	$^i b_1$	$^i c_1$
X	m_1	m_2	m_3	η_1		$^i a_2$	$^i b_2$	$^i c_2$
Z	n_1	n_2	n_3	ζ_1		$^i a_3$	$^i b_3$	$^i c_3$

$$(50)$$

The direction cosines in (49) are connected with these of the axes (ξ, η, ζ) by the linear relations—

$$^i l_r = {}^i a_r l_1 + {}^i b_r l_2 + {}^i c_r l_3$$
$$^i m_r = {}^i a_r m_1 + {}^i b_r m_2 + {}^i c_r m_3 \Bigg\} \ r = 1, 2, 3. \qquad (51)$$
$$^i n_r = {}^i a_r n_1 + {}^i b_r n_2 + {}^i c_r n_3$$

In order to obtain an expression for the intensity of radiation scattered in a particular direction, we have to compute $H^2 + J^2 + K^2$ from (49) and average for all possible orientations of the molecular axes (ξ, η, ζ).

We find as a typical term

$$H^2 = \sum_i [A_i{}^2 \, {}^i n_1{}^2 \, {}^i l_1{}^2 + B_i{}^2 \, {}^i n_2{}^2 \, {}^i l_2{}^2 + C_i{}^2 \, {}^i n_3{}^2 \, {}^i l_3{}^2 + 2 B_i C_i{}^i n_2{}^i n_3{}^i l_2{}^i l_3 + \ldots + \ldots]$$
$$+ 2 \sum_i \sum_s [A_i{}^i n_1{}^i l_1 + B_i{}^i n_2{}^i l_2 + C_i{}^i n_3{}^i l_3][A_s{}^s n_1{}^s l_1 + B_s{}^s n_2{}^s l_2 + C_s{}^s n_3{}^s l_3],$$

where in the double summation $i = 1, 2, 3, \ldots m$ and $s = 1, 2, 3 \ldots m$, the values $r = s$ being excluded.

Expressions for J^2 and K^2 may be written down from the above by replacing the l's successively by m's and n's.

From (30) we have

$$\cos \widehat{\xi_i \eta_s} = {}^i l_1{}^s l_2 + {}^i m_1{}^s m_2 + {}^i n_1{}^s n_2 \quad (i \neq s)$$

with similar expressions obtained by rotating $(\xi_i \eta_i \zeta_i)$ with the suffixes (1, 2, 3) of the direction cosines $({}^i l_r, {}^i m_r, {}^i n_r)$, and independently (ξ_s, η_s, ζ_s) with the suffixes (1, 2, 3) of $({}^s l_r, {}^s m_r, {}^s n_r)$, where $r = 1, 2, 3$.

Making use of the well-known relations between direction cosines in (30) we find

$$H^2 + J^2 + K^2 = \sum_i [A_i^2 \, {}^i n_1^2 + B_i^2 \, {}^i n_2^2 + C_i^2 \, {}^i n_3^2]$$

$$+ 2 \sum_i \sum_s [A_i A_s \, {}^i n_1 \, {}^s n_1 \cos \widehat{\xi_i \xi_s} + B_i B_s \, {}^i n_2 \, {}^s n_2 \cos \widehat{\eta_i \eta_s} + C_i C_s \, {}^i n_3 \, {}^s n_3 \cos \widehat{\zeta_i \zeta_s}$$

$$+ B_i C_s \, {}^i n_2 \, {}^s n_3 \cos \widehat{\eta_i \zeta_s} + \text{two similar terms}$$

$$+ B_s C_i \, {}^s n_2 \, {}^i n_3 \cos \widehat{\eta_s \zeta_i} + \text{two similar terms}].$$

On proceeding to average for an isotropic distribution of molecular axes, we easily see from (51), making use of (22), that

$$\overline{{}^i n_1^2} = \overline{{}^i n_2^2} = \overline{{}^i n_3^2} = \tfrac{1}{3}$$

and
$$\overline{{}^i n_2 \, {}^s n_3} = \tfrac{1}{3} ({}^i a_2 \, {}^s a_3 + {}^i b_2 \, {}^s b_3 + {}^i c_2 \, {}^s c_3) = \tfrac{1}{3} \cos \widehat{\eta_i \zeta_s}, \&\text{c.} \qquad (52)$$

Making use of the notation

$$\sum_i \sum_s (A_i B_i C_i)(A_s B_s C_s) \cos^2 (\xi_i \eta_i \zeta_i)(\xi_s \eta_s \zeta_s)$$

$$= \sum_i \sum_s [A_i A_s \cos^2 \widehat{\xi_i \xi_s} + B_i B_s \cos^2 \widehat{\eta_i \eta_s} + C_i C_s \cos^2 \widehat{\zeta_i \zeta_s}$$

$$+ B_i C_s \cos^2 \widehat{\eta_i \zeta_s} + C_i A_s \cos^2 \widehat{\zeta_i \xi_s} + A_i B_s \cos^2 \widehat{\xi_i \eta_s}$$

$$+ B_s C_i \cos^2 \widehat{\eta_s \xi_i} + C_s A_i \cos^2 \widehat{\zeta_s \xi_i} + A_s B_i \cos^2 \widehat{\xi_s \eta_i}], \qquad (53)$$

we find after some reductions

$$\overline{H^2} + \overline{J^2} + \overline{K^2} = \tfrac{1}{3} \sum_i [A_i^2 + B_i^2 + C_i^2]$$

$$+ \tfrac{2}{3} \sum_i \sum_s (A_i B_i C_i)(A_s B_s C_s) \cos^2 (\xi_i \eta_i \zeta_i)(\xi_s \eta_s \zeta_s). \qquad (54)$$

From (49) and (51) we notice that the averaged product terms vanish, *i.e.*

$$\overline{HJ} = 0, \quad \overline{JK} = 0, \quad \overline{KH} = 0. \qquad (55)$$

Furthermore, since H is the same function of (l_1, l_2, l_3) as J is of (m_1, m_2, m_3), it follows as in § 4 that—

$$\overline{H^2} = \overline{J^2}. \qquad (56)$$

From equations (32), (35) and (52) we see that

$$\overline{K} = \tfrac{1}{3} \sum_i (A_i + B_i \quad C). \qquad (57)$$

Squaring the expression for K in (49), we find

$$K^2 = \sum_i (A_i^2 \, {}^i n_1^4 + B_i^2 \, {}^i n_2^4 + C_i^2 \, {}^i n_3^4)$$

$$+ 2 \sum_i [B_i C_i \, {}^i n_2^2 \, {}^i n_3^2 + C_i A_i \, {}^i n_3^2 \, {}^i n_1^2 + A_i B_i \, {}^i n_1^2 \, {}^i n_2^2]$$

$$+ 2 \sum_i \sum_s [A_i \, {}^i n_1^2 + B_i \, {}^i n_2^2 + C_s \, {}^i n_3^2][A_s \, {}^s n_1^2 + B_s \, {}^s n_2^2 + C_s \, {}^s n_3^2], \qquad (58)$$

where $i = 1, 2, 3 \cdots m$, $s = 1, 2, 3 \cdots m$, and $i \neq s$ in the double summation.

From (60) we have as a typical squared product term

$${}^{i}n_1{}^{2\,s}n_2{}^2 = [({}^{i}a_1 n_1{}^2 + {}^{i}b_1 n_2{}^2 + {}^{i}c_1 n_3{}^2) + 2\,({}^{i}b_1{}^{i}c_1 n_2 n_3 + {}^{i}c_1{}^{i}a_1 n_3 n_1 + {}^{i}a_1{}^{i}b_1 n_1 n_2)]$$
$$\times [({}^{s}a_2 n_1{}^2 + {}^{s}b_2 n_2{}^2 + {}^{s}c_2 n_3{}^2) + 2\,({}^{s}b_2{}^{s}c_2 n_2 n_3 + {}^{s}c_2{}^{s}a_2 n_3 n_1 + {}^{s}a_2{}^{s}b_2 n_1 n_2)].$$

Remembering, from (23) and (24), that

$$\overline{n_1{}^4} = \overline{n_2{}^4} = \overline{n_3{}^4} = \tfrac{1}{5} \quad \text{and} \quad \overline{n_2{}^2 n_3{}^2} = \overline{n_3{}^2 n_1{}^2} = \overline{n_1{}^2 n_3{}^2} = \tfrac{1}{15}, \tag{59}$$

we find on averaging the above expression for an isotropic distribution of molecular axes

$$\overline{{}^{i}n_1{}^{2\,s}n_2{}^2} = ({}^{i}a_1{}^{2\,s}a_2{}^2 + {}^{i}b_1{}^{2\,s}b_2{}^2 + {}^{i}c_1{}^{2\,s}c_2{}^2)\,\overline{n_1{}^4}$$
$$[{}^{i}a_1{}^2\,({}^{s}b_2{}^2 + {}^{s}c_2{}^2) + {}^{i}b_1{}^2\,({}^{s}c_2{}^2 + {}^{s}a_2{}^2) + {}^{i}c_1{}^2\,({}^{s}a_2{}^2 + {}^{s}b_2{}^2)]\,\overline{n_2{}^2 n_3{}^2}$$
$$+ 4\,[{}^{i}b_1{}^{i}c_1{}^{s}b_2{}^{s}c_2 + {}^{i}c_1{}^{i}a_1{}^{s}c_2{}^{s}a_2 + {}^{i}a_1{}^{i}b_1{}^{s}a_2{}^{s}b_2]\,\overline{n_2{}^2 n_3{}^2}.$$

Noting that ${}^{s}a_2{}^2 + {}^{s}b_2{}^2 + {}^{s}c_2{}^2 = 1$, and making use of (52), we find

$$\overline{{}^{i}n_1{}^{2\,s}n_2{}^2} = ({}^{i}a_1{}^{2\,s}a_2{}^2 + {}^{i}b_1{}^{2\,s}b_2{}^2 + {}^{i}c_1{}^{2\,s}c_2{}^2)\,(\overline{n_1{}^4} - 3\overline{n_2{}^2 n_3{}^2}) + (1 + 2\cos^2 \widehat{\xi_i \eta_s})\,\overline{n_2{}^2 n_3{}^2}.$$

Inserting numerical values from (59), the first term of the above expression vanishes, and we are left with

$$\overline{{}^{i}n_1{}^{2\,s}n_2{}^2} = \tfrac{1}{15}\,\{1 + 2\cos^2 \widehat{\xi_i \eta_s}\}, \tag{60}$$

with similar expressions obtained by rotating the suffixes (1, 2, 3) as already explained.

It is obvious that for the principal directions of the ith dispersion charge the result of averaging for an isotropic distribution of molecular orientations is to give for all values of i

$$\overline{{}^{i}n_1{}^4} = \overline{{}^{i}n_2{}^4} = \overline{{}^{i}n_3{}^4} = \tfrac{1}{5} \quad \text{and} \quad \overline{{}^{i}n_2{}^2\,{}^{i}n_3{}^2} = \overline{{}^{i}n_3{}^2\,{}^{i}n_1{}^2} = \overline{{}^{i}n_1{}^2\,{}^{i}n_2{}^2} = \tfrac{1}{15} \tag{61}$$

as may easily be verified by making use of (51). It is easily seen that with these formulæ (60) is in agreement as a particular case, since on identifying i and s we leave $\widehat{\xi_i \eta_i} = \tfrac{1}{2}\pi$, and on identifying the suffixes 1 and 2 and at the same time i and s, we have $\widehat{\xi_i \zeta_i} = 0$.

Making use of (60) and (61) we obtain finally as a result of averaging (58),

$$\overline{\mathrm{K}^2} = \tfrac{1}{5} \sum_i (\mathrm{A}_i{}^2 + \mathrm{B}_i{}^2 + \mathrm{C}_i{}^2) + \tfrac{2}{15} \sum_i (\mathrm{B}_i \mathrm{C}_i + \mathrm{C}_i \mathrm{A}_i + \mathrm{A}_i \mathrm{B}_i)$$
$$+ \tfrac{2}{15} \sum_i \sum_s [\mathrm{A}_i \mathrm{A}_s + \mathrm{B}_i \mathrm{B}_s + \mathrm{C}_i \mathrm{C}_s + \mathrm{B}_i \mathrm{C}_s + \mathrm{C}_i \mathrm{A}_s + \mathrm{A}_i \mathrm{B}_s + \mathrm{B}_s \mathrm{C}_i + \mathrm{C}_s \mathrm{A}_i + \mathrm{A}_s \mathrm{B}_i]$$
$$+ \tfrac{4}{15} \sum_i \sum_s (\mathrm{A}_i \mathrm{B}_i \mathrm{C}_i)(\mathrm{A}_s \mathrm{B}_s \mathrm{C}_s) \cos^2 (\xi_i \eta_i \zeta_i)(\xi_s \eta_s \zeta_s),$$

where the abbreviated notation of (53) is employed in the last term.

Introducing the squared value of (57) and substituting from (54) we find

$$\overline{K^2} = \tfrac{3}{5}\,(\overline{K})^2 + \tfrac{2}{5}\,(\overline{H^2} + \overline{J^2} + \overline{K^2}),$$

or remembering from (56) that $\overline{H^2} = \overline{J^2}$, we have

$$3\overline{K^2} - 4\overline{H^2} = 3\,(\overline{K})^2 \qquad (62)$$

If, as in (26), we define ρ by the equation

$$\rho = 2\overline{H^2}/(\overline{H^2} + \overline{K^2}) \qquad (63)$$

we find from (62) that

$$\overline{H^2} = \overline{J^2} = 3\,(\overline{K})^2 \cdot \frac{\rho}{6 - 7\rho} \quad \text{and} \quad \overline{K^2} = 3\,(\overline{K})^2 \cdot \frac{2 - \rho}{6 - 7\rho}. \qquad (64)$$

§ 7. *Scattering and Extinction of Light in Gases : Invariant Formulæ.*

In a gas not too near the critical point, we suppose the molecules to be sufficiently far apart that the distribution of molecular orientations is unaffected by intermolecular forces.* Owing to the rapidly changing positions of the gaseous molecules due to their high velocities, it is usually assumed that each molecule contributes to the *intensity* of the scattered radiation.† Adding the squares of the amplitudes of the vectors (P, Q, R) given in (48) for

$$\nu = n_0 \overline{V} \qquad (65)$$

molecules contained in a finite volume $\overline{V}$, we have

$$\left.\begin{array}{ll}
\Sigma\,\mathrm{P}^2 = 0 & \Sigma\,\mathrm{Q}^2 = \dfrac{p^4}{c^4 r^2}\;Z + Z_p{}^2 \displaystyle\sum_1^\nu (\mathrm{J}\cos\chi - \mathrm{K}\sin\chi)^2 \\[3ex]
\Sigma\,\mathrm{R}^2 = \dfrac{p^4}{c^4 r^2}\;Z + Z_p{}^2 \displaystyle\sum_1^\nu (-\mathrm{H}\sin\theta + \mathrm{J}\cos\theta\sin\chi + \mathrm{K}\cos\theta\cos\chi)^2
\end{array}\right\} (66)$$

where H, J, K have the values defined by (49).

Owing to the existence of molecular rotations, we may expect an isotropic distribution of molecular orientations. We may thus write

$$\sum_1^\nu \mathrm{J}^2 = \nu\overline{J^2}, \qquad \sum_1^\nu \mathrm{K}^2 = \nu\overline{K^2}, \qquad \sum_1^\nu \mathrm{H}^2 = \nu\overline{H^2},$$

* Except, possibly, for a short interval of time at the " instant " of collision, an effect which may explain deviations from the law $(\mu - 1)/\rho = $ constant in gases at low pressures. Posejpal, V., " Sur la variation de la réfraction des gaz avec la pression, etc.," ' Jour. de Physique et le Radium,' vol. II, p. 85 (March, 1921).

† This point is discussed at length by Sir J. J. Thomson, ' Phil. Mag.,' vol. 40, p. 400 (1920).

and remembering that by (55) the averaged product-terms of (H, J, K) vanish,

$$\Sigma\;Q^{2} = \frac{p^{4}}{c^{4}r^{2}}\;Z + Z_{p}^{2}\nu\,(\overline{J^{2}}\cos^{2}\chi + \overline{K^{2}}\sin^{2}\chi)$$

$$\Sigma\;R^{2} = \frac{p^{4}}{c^{4}r^{2}}\;Z + Z_{p}^{2}\nu\,(\overline{H^{2}}\sin^{2}\theta + \overline{J^{2}}\cos^{2}\theta\sin^{2}\chi + \overline{K^{2}}\cos^{2}\theta\cos^{2}\chi) \tag{67}$$

We proceed to consider two cases, depending on whether the incident light is polarized or unpolarized.

Case I. Incident light plane-polarized.—If we denote by I the intensity of the incident light, and by $I_{s}(\theta, \chi)$ that of the scattered light in a direction (θ, χ) with reference to the incident beam (fig. 2), we have, for the ratio of the scattered to the incident radiation,

$$\frac{I_{s}(\theta, \chi)}{I} = \frac{\Sigma\;Q^{2} + \Sigma\;R^{2}}{Z^{2}}. \tag{68}$$

Remembering that $\overline{H^{2}} = \overline{J^{2}}$, and making use of (65), we find after some reductions that (67) and (68) give

$$\frac{r^{2}I_{s}(\theta, \chi)}{VI} = \frac{Z + Z_{p}^{2}}{Z^{2}}\frac{p^{4}}{c^{4}}\,n_{0}\,[\overline{H^{2}} + \overline{K^{2}} + (\overline{H^{2}} - \overline{K^{2}})\sin^{2}\theta\cos^{2}\chi]. \tag{69}$$

Substituting for the ratio $Z + Z_{p}\,/\,Z$ from (37), and making use of (64), we may write (69) in the form, remembering that $p = 2\pi c/\lambda$,

$$\frac{r^{2}I_{s}(\theta, \chi)}{VI} = \frac{\pi^{2}}{\lambda^{4}}\frac{(\mu^{2}-1)^{2}}{n_{0}}\cdot\frac{6}{6 - 7\rho}\,[1 - (1 - \rho)\cos^{2}\chi\sin^{2}\theta]. \tag{70}$$

If we examine the scattered light in the direction (θ, χ) by means of a Nicol prism, we find for the *depolarization*, or ratio of minimum to maximum intensities, on rotating the prism

$$\rho(\theta, \chi) = \frac{\Sigma\;R^{2}}{\Sigma\;Q^{2}} = \left[\frac{\overline{H^{2}} + \overline{K^{2}}\cos^{2}\theta}{\overline{H^{2}}\cos^{2}\chi + \overline{K^{2}}\sin^{2}\chi} - \cos^{2}\theta\right]$$

$$= \left[\frac{\rho + (2 - \rho)\cos^{2}\theta}{\rho\cos^{2}\chi + (2 - \rho)\sin^{2}\chi} - \cos^{2}\theta\right]. \tag{71}$$

This formula is greatly simplified when the incident beam is polarized so that $\chi = \tfrac{1}{2}\pi$, when $\rho(\theta, \tfrac{1}{2}\pi) = 2/(2 - \rho)$ which is independent of θ, while for

$$\chi = 0, \qquad \rho(\theta, 0) = \sin^{2}\theta + (2/\rho - 1)\cos^{2}\theta.$$

Case II. Incident light unpolarized.—In almost all experimental arrangements an unpolarized incident beam is employed. By averaging the intensities given in (66) for all possible orientations χ of the plane of polarization, the amplitude Z , and therefore $Z + Z_p$ remaining constant, we obtain, on re-placing $\sin^2 \chi$ and $\cos^2 \chi$ by their mean values $\frac{1}{2}$,

$$\frac{r^2\,\mathrm{I}_s\,(\theta)}{\mathrm{VI}} = \tfrac{1}{2}\frac{\pi^2}{\lambda^4}\frac{(\mu^2-1)^2}{n_0} \cdot \frac{6\,(1+\rho)}{6-7\rho}\left\{1+\frac{1-\rho}{1+\rho}\cos^2\theta\right\}, \qquad (72)$$

while the depolarization in a direction θ with the incident beam is given by

$$\rho\,(\theta) = \rho + (1-\rho)\cos^2\theta. \qquad (73)$$

From this formula we notice that $\rho\,(\tfrac{1}{2}\pi) = \rho$ is the depolarization when the scattered light is examined by a Nicol or double-image prism at right angles to the unpolarized incident beam.

For light scattered at right angles to the incident beam ($\theta = \tfrac{1}{2}\pi$) we have

$$\frac{r^2\mathrm{I}_s(\tfrac{1}{2}\pi)}{\mathrm{VI}} = \tfrac{1}{2}\frac{\pi^2}{\lambda^4}\frac{(\mu^2-1)^2}{n_0} \cdot \frac{6\,(1+\rho)}{6-7\rho}, \qquad (74)$$

which agrees with Cabannes' modification of Rayleigh's well-known formula by the introduction of the factor $6\,(1+\rho)/(6-7\rho)$ depending on molecular anisotropy, as a result of which observed and theoretical values of relative scattering in various dust-free gases are brought into much closer agreement.*

It does not appear to have been pointed out that molecular anisotropy results in a deviation from the well-known $(1 + \cos^2 \theta)$ law for relative intensities of scattered light in various directions with the incident beam. From (72) and (74) we have

$$\mathrm{I}_s\,(\theta) = \mathrm{I}_s\,(\tfrac{1}{2}\pi)\left\{1 + \frac{1-\rho}{1+\rho}\cos^2\theta\right\}. \qquad (75)$$

Owing to molecular anisotropy, Rayleigh's well-known formula for extinction of light in gaseous media as a result of scattering is sensibly modified, as, in fact, was foreseen by that writer.†

The rate of flow of energy from an element of volume $\mathrm{V} = \mathrm{S}dx$, of cross-section S and thickness dx, across an element of surface $r^2d\omega$ subtending a small solid angle $d\omega = \sin\theta\,d\theta\,d\phi$ at the origin in a direction θ with the incident beam is, on integrating with respect to the azimuth ϕ,

$$c\int \mathrm{I}_s(\theta)\,.\,r^2d\omega = c\,r^2\mathrm{I}_s(\tfrac{1}{2}\pi)\,.\,2\pi\int_0^\pi\left\{1 + \frac{1-\rho}{1+\rho}\cos^2\theta\right\}\sin\theta\,d\theta, \qquad (76)$$

c being the velocity of light.

Since the present theory does not take into account any dissipation of energy by individual molecules, the rate of flow of energy expressed in (76) is equal to

* Cabannes, J., ' Annales de Physique,' vol. 15, p. 130 (1921).

† Lord Rayleigh, ' Scientific Papers,' vol. VI., p. 546.

the rate of loss of energy from the incident beam of cross-section S on travelling a distance dx, *i.e.*

$$- c\,S dI = c S \kappa I dx \tag{77}$$

where κ is the coefficient of extinction defined by the relation

$$d\,I/dx = -\kappa\,I.$$

On equating (76) and (77), integrating the latter and making use of (74) with $V = S dx$, we find for the coefficient of attenuation the expression—

$$\kappa = \frac{8\pi^3}{3\lambda^4} \frac{(\mu^2 - 1)^2}{n_0} \cdot \frac{6 + 3\rho}{6 - 7\rho}. \tag{78}$$

It is interesting to note that the formulae (71), (72), (74) and (78) have been derived for a very general type of molecule, and that these results, involving experimentally observed quantities are *invariant* with respect to such details of molecular constitution as number and magnitude of dispersion charges, their position and the mutual orientation of their principal directions.

Although the results thus far obtained have been based on a general type of "static" molecule, the theory is by no means opposed to the modern conceptions of the "dynamic" atom. For wave-lengths long compared to molecular dimensions, we may suppose those perturbations which contribute principally to dispersion to consist of forced oscillations of each atomic system of electrons with respect to the corresponding positive system.

§ 8. *Scattering and Extinction of Light in Gaseous Mixtures.*

(i) *Theory.*—Consider a mixture of r different kinds of molecules, there being n_r per unit volume of each. As in § 5 (ii), we denote for each constituent, the molecular density by N_r, the molecular weight by m_r and the mass-density by D_r. If w_r is the weight of each constituent per unit volume of mixture of density D, we have

$$w_r = m_r n_r, \qquad D = \sum_r w_r, \qquad n_r/N_r = w_r/D_r, \tag{79}$$

and the proportion τ_r of each constituent by weight is given by

$$\tau_r = w_r/D. \tag{80}$$

As in the preceding section, we shall suppose that each molecule contributes independently to the *intensity* of the scattered light. Limiting ourselves to the case of *unpolarized incident light*, for which $\overline{\cos^2 \chi} = \overline{\sin^2 \chi} = \frac{1}{2}$, we have for the light scattered by $\sum n_r V$ molecules contained in a volume V, from equation (67)

$$\Sigma \ Q \ ^2 = \tfrac{1}{2}\frac{p^4 V}{c^4 r^2}\ Z + Z_p \ ^2 \Sigma_r n_r [\overline{H_r{}^2} + \overline{K_r{}^2}]$$

$$\Sigma \ R \ ^2 = \tfrac{1}{2}\frac{p^4 V}{c^4 r^2}\ Z + Z_p \ ^2 \Sigma_r n_r [2\overline{H_r{}^2} + (\overline{K_r{}^2} - \overline{H_r{}^2})\cos^2 \theta], \qquad (81)$$

in which $\overline{H_r{}^2}$ and $\overline{K_r{}^2}$ refer to expressions of the form (49) for the rth type of complex anisotropic molecule averaged for all possible independent molecular orientations.

Z_p refers to the polarization field of the medium made up of the r types of molecules. If it has been ascertained experimentally that Lorentz's additive law (46)

$$\frac{1}{D}\frac{\mu^2 - 1}{\mu^2 + 2} = \Sigma_r \frac{\tau_r}{D_r}\frac{\mu_r{}^2 - 1}{\mu_r{}^2 + 2} \qquad (82)$$

is obeyed to a sufficient degree of accuracy, we conclude that S_r is negligibly small for each constituent, and therefore from (39) we have—

$$Z + Z_p \ / \ Z \ = \tfrac{1}{3}(\mu^2 + 2). \qquad (83)$$

It follows from (81) that the intensity scattered in the direction θ is given by

$$\frac{r^2 I_s(\theta)}{VI} = \frac{r^2}{V}\cdot\frac{\Sigma \ Q \ ^2 + \Sigma \ R \ ^2}{Z^2}$$

$$= \tfrac{1}{2}\frac{p^4}{c^4}\left(\frac{\mu^2 + 2}{3}\right)^2 \Sigma_r n_r [(3\overline{H_r{}^2} + \overline{K_r{}^2}) + (K_r{}^2 - H_r{}^2)\cos^2 \theta]. \qquad (84)$$

If we consider that the rth constituent alone is present in the volume V, we write $n_r = 0$ for all types of molecules except the rth, when $n_r = N_r$. If examined under the same conditions as the mixture (r, θ and I being the same), we have

$$\frac{r^2 \ ^r I_s(\theta)}{VI} = \tfrac{1}{2}\frac{p^4}{c^4}\left(\frac{\mu_r{}^2 + 2}{3}\right)^2 . N_r [(3\overline{H_r{}^2} + \overline{K_r{}^2}) + (\overline{K_r{}^2} - \overline{H_r{}^2})\cos^2 \theta].$$

Substituting in (84), and making use of (79) and (80), we obtain an additive law for the intensity of scattered light in the form

$$\frac{1}{D}\frac{I_s(\theta)}{(\mu^2 + 2)^2} = \Sigma_r \frac{\tau_r}{D_r}\frac{^r I_s(\theta)}{(\mu_r{}^2 + 2)^2}. \qquad (85)$$

If k is the coefficient of extinction by scattering for the mixture, and k_r that of the rth constituent for density D_r, we see on integrating (85) over a sphere of large radius to obtain the rate of flow of energy from an element of volume of unit cross-section and thickness dx, that the corresponding additive law is

$$\frac{1}{D}\frac{\kappa}{(\mu^2 + 2)^2} = \Sigma_r \frac{\tau_r}{D_r}\frac{k_r}{(\mu_r + 2)^2}. \qquad (86)$$

The *depolarization* of the mixture at right angles to the incident beam is given by writing $\theta = \frac{1}{2}\pi$ in (81). We then have

$$\rho = \frac{\Sigma \ R^2}{\Sigma \ Q^2} = \frac{2\Sigma_r n_r \overline{H_r^2}}{\Sigma_r n_r (\overline{H_r^2} + \overline{K_r^2})}$$

Making use of (64) for each constituent and (42), we easily derive the result given below for mixtures for which Lorentz's additive law (82) holds,

$$\frac{\rho}{6-7\rho} = \frac{\Sigma_r n_r M_r^2 \rho_r/(6-7\rho_r)}{\Sigma_r n_r M_r^2} = \frac{\Sigma_r \dfrac{v_r}{N_r} \cdot \left(\dfrac{\mu_r^2-1}{\mu_r^2+2}\right)^2 \cdot \dfrac{\rho_r}{6-7\rho_r}}{\Sigma_r \dfrac{v_r}{N_r} \cdot \left(\dfrac{\mu_r^2-1}{\mu_r^2+2}\right)^2}, \quad (87)$$

where M_r is the molecular refractivity of the constituent r as defined by (41) and v_r the volume of the constituent r per unit volume of mixture given by

$$v_r = w_r/D_r = n_r/N_r. \quad (88)$$

These formulae have been derived without approximation, so that the hypothesis of molecular independence, as regards orientation and contribution to intensity of the scattered light, may be tested out by comparison with experiments on gases under high pressures, on saturated vapours and on liquids.

(ii) *Application to atmospheric extinction.*—In view of the fact that the extinction coefficient of the earth's atmosphere is accurately known for various wave-lengths over the range of the visible spectrum,* and provides an accurate means of computing the molecular density of gases at 0° and 760 mm. pressure,† it is of considerable importance to investigate the effect of molecular anisotropy which enters into the expression for the extinction coefficient as shown by (78). Referring refractive indices and extinction coefficients to 0° and 760 mm., we may without sensible error replace (μ_r^2+2) by 3 in (86). In these circumstances, the molecular density of each constituent by itself is constant for all gases, *i.e.* $N_r = n_0$, and by (80) and (88) we may write—

$$\kappa = \Sigma_r \frac{n_r}{n_0} \kappa_r = \Sigma_r v_r \kappa_r, \quad (89)$$

where v_r is the proportion by volume of each constituent.

For gaseous media we have, approximately,

$$\kappa_r = \frac{32}{3} \frac{\pi^3}{\lambda^4} \frac{(\mu_r-1)^2}{n_0} \cdot \frac{6+3\rho_r}{6-7\rho_r}. \quad (90)$$

* 'Annals of the Smithsonian Astrophysical Observatory,' Washington, vols. III (1913) and IV (1922).

† King, L. V., 'Nature,' vol. 93, p. 557 (1914). Fowle, F. E., 'Astrophysical Jour.,' vol. 40, p. 435 (1914).

If air be considered as an aggregate of symmetrical molecules of refractive index μ_a, the molecular density n_a has hitherto been calculated from the uncorrected formula—

$$k = \frac{32}{3} \frac{\pi^3}{\lambda^4 n_a} (\mu_a - 1)^2.$$

Comparing this with (98) we see that—

$$n_0 / n_a = \sum_r v_r \left(\frac{\mu_r - 1}{\mu_a - 1}\right)^2 \cdot \frac{6 + 3\rho_r}{6 - 7\rho_r}, \tag{91}$$

the right hand side representing the factor by which the uncorrected value n_a must be multiplied to give the correct value of n_0.

Considering the three principal components of the atmosphere, we have the following data :—

Table I.

	Oxygen.	Nitrogen.	Argon.	Air.	Remarks.
r_r	0·210	0·781	0·009	1·000	$\lambda = 4600$ Å.
$(\mu_r - 1)$	27·2 × 10⁻⁵	30·0 × 10⁻⁵	28·4 × 10⁻⁵	29·5 × 10⁻⁵	Rayleigh,* white light.
ρ_r	0·094	0·0406	0·032	0·050	

From this data we obtain $n_0 = n_a \times 1 \cdot 084$. If we make use of Cabannes' determinations of depolarization, *i.e.* $0 \cdot 054$ for oxygen, $0 \cdot 028$ for nitrogen, and $0 \cdot 00$ for argon, formula (91) gives $1 \cdot 049$ for this correction factor.†

A further correction arises from the fact that in the outer atmosphere, in isothermal equilibrium, the relative molecular density of the rth constituent depends on the molecular weight according to the formula‡

$$\log (n_r / n_{r0}) = -m_r g z / (\text{RT}) \tag{92}$$

in which the curvature of the earth is neglected. In this equation, n_{r0} is the molecular density at the lower boundary of the outer atmosphere from which the height z is also measured.

In computing the extinction of light by the earth's atmosphere, it is necessary

* Lord Rayleigh, ' Roy. Soc. Proc.,' vol. 94, A, p. 453 (1918).

† Cabannes (' Annales de Physique,' vol. 15, p. 30 (1921)) inadvertently assumes that $I_s(\theta) = (1 + \cos^2 \theta) I_s (\tfrac{1}{2}\pi)$ instead of (84) and derives $(6 + 6\rho)/(6 - 7\rho)$ as the factor by which the coefficient of extinction is affected by molecular anisotropy, instead of the correct value given in (87). Using his own values for the depolarization quoted above, he obtains $1 \cdot 07$ for the correction factor, and from independent observations on relative scattering, the value $1 \cdot 10$. He concludes, finally, that Fowle's estimate of $n = 2 \cdot 69 \times 10^{19}$ should be increased to $2 \cdot 92 \times 10^{19}$.

‡ Jeans, ' The Dynamical Theory of Gases,' 2nd Edition, p. 351 (1916).

to take into account the variable proportion of oxygen and nitrogen in the formulæ (89) and (90) for mixtures as far as the stratosphere $(p = \frac{1}{4}p_0)$. Within the stratosphere, to the level considered, the relative proportions of the constituent gases remain the same owing to convection. The further consideration of this interesting question must, however, be left to a future paper.

With regard to experimental evidence on the constitution of the extreme limits of the upper atmosphere, it may be pointed out that observations on the depolarization of the zenith sky after sunset may, on comparison with values for gaseous mixtures computed from laboratory data, enable us to test the isothermal theory.*

(iii) *Test of Theory of Scattering in Gaseous Mixtures.*—When pressure and temperature are far removed from the critical values of any constituent, we may write $N_r = n_0$, and replace $\mu_r^2 + 2$ by 3. In these circumstances, the additive law (85) becomes—

$$I_s(\theta) = \Sigma v_r{}^r I_s(\theta), \tag{93}$$

while the depolarization of the mixture, given by (87), becomes—

$$\frac{\rho}{6-7\rho} = \frac{\sum_r v_r \dfrac{\rho_r(\mu_r-1)^2}{6-7\rho_r}}{\sum_r v_r(\mu_r-1)^2}. \tag{94}$$

As far as the writer is aware, no observations have as yet been carried out for the special purpose of testing the laws of scattering and depolarization in gaseous mixtures, and such are greatly to be desired.

We may, however, test formula (94) by computing the depolarization of air in terms of that of its two most important constituents, oxygen and nitrogen, making use of the data of Table I. We easily find $\rho = 0 \cdot 0505$ (calculated), in good agreement with $\rho = 0 \cdot 050$ (observed). This satisfactory agreement goes far to justify the hypothesis of molecular independence assumed by the theory.†

Further observations over a wide range of pressures and temperatures are greatly to be desired, especially in gases capable of chemical combination, and in gases near the critical point. The depolarization is very sensitive to intermolecular influence as affecting complete independence of molecular orientations.

---

* Fabry, C., " Remarques sur la diffusion de la lumière par les gaz," ' Jour. de Physique,' vol. VII, pp. 89–102 (1917). Raman, C. V., ' Molecular Diffraction of Light ' (University of Calcutta Press, 1922).

† According to Cabanne's determinations, $\rho_1 = 0 \cdot 054$ for oxygen and $\rho_2 = 0 \cdot 028$ for nitrogen. Formula (94) gives for air $\rho = 0 \cdot 034$ (calculated) to be compared with $\rho = 0 \cdot 040$ (observed). The agreement is not so satisfactory in this case.

The same remark applies to the effect on depolarization of molecular aggregation, which, even in the case of a single gas, may be expected to give rise to an observable effect as the critical point is approached. Observations on light scattering in vapours in contact with their liquids would give some indications of molecular aggregation, and of circumstances of molecular collisions which determine peculiarities in the experimentally determined equation of state not interpretable in terms of the dynamical theory of gases.

Summary and Conclusions.

Observations by Cabannes, the present Lord Rayleigh, Gans and others have shown that the light scattered by various gases in a direction at right angles to the incident beam is not completely polarized. This is accounted for by Cabannes on the theory of the *simple anisotropic molecule* as developed by Langevin in 1910 to account for electric and magnetic double refraction. Such a molecule contains a single dispersion electron acted on by unequal quasi-elastic restoring forces along *three principal directions* and capable of vibrating with three different frequencies.

The present writer has extended the theory to gaseous and liquid media composed of *complex anisotropic molecules*, in which there are any number of dispersion charges, not necessarily equal, whose principal directions of vibration are not parallel.

For an isotropic medium in which all molecular orientations are equally probable, a general dispersion formula of the Lorentz type is derived.

In gaseous media, owing to continual changes of position, each molecule contributes independently to the *intensity* of the scattered radiation. For unpolarized incident light of intensity I, the depolarization ρ is the ratio of the minimum to maximum intensity when the light scattered at right angles is examined by a Nicol prism, and, as observed in gases, is a quantity characteristic of the molecule. The intensity $I_s(\theta)$ of light of wave-length λ scattered in a direction θ with the incident beam to a distance r from a volume V is given by the formula—

$$\frac{r^2 I_s(\theta)}{VI} = \tfrac{1}{2}\,\frac{\pi^2}{\lambda^4}\frac{(\mu^2-1)^2}{n}\cdot\frac{6(1+\rho)}{6-7\rho}\left\{1 + \frac{1-\rho}{1+\rho}\cos^2\theta\right\}, \qquad (95)$$

where μ is the refractive index of the medium corresponding to molecular density n.

The corresponding expression for the coefficient of extinction by scattering is—

$$\kappa = \frac{8\pi^3}{3\lambda^4}\frac{(\mu^2-1)^2}{n}\cdot\frac{6+3\rho}{6-7\rho}. \qquad (96)$$

A remarkable feature of these formulæ is their *invariance* with respect to such details of molecular structure as number and magnitude of dispersion charges, their position, and the mutual orientation of their principal directions.

Formulæ of the same invariant character are derived for scattering, depolarization and extinction of light in gaseous mixtures. In particular, the depolarization of air calculated in terms of that of its principal constituents, oxygen and nitrogen, is in good agreement with observation, making use of the present Lord Rayleigh's data. An important application of these formulae lies in the re-determination of Avogadro's constant from observations on the extinction of solar radiation by the earth's atmosphere, correcting for molecular anisotropy and the probable variation in the proportions of its constituent gases above the stratosphere.

An invariant formula analogous to (95), based on Smoluchouski's theory of fluctuations of density, has been obtained for the scattering of light by dust-free liquids, and is in satisfactory agreement with Martin's observations. This phase of the subject the writer hopes to deal with in a future paper.

Explanation of the Brightness and Color of the Sky, Particularly the Twilight Sky

E. O. HULBURT

Naval Research Laboratory, Washington, D. C.

(Received August 15, 1952)

By use of the Rayleigh scattering theory for pure air, primary scattering and known upper air densities from rockets, the brightness of the zenith twilight sky was calculated and values were obtained two to four times greater than those observed at Sacramento Peak, New Mexico (J. Opt. Soc. Am. 42, 353 (1952)). The attenuation of the atmosphere there was observed to be twenty percent above that of the Rayleigh theory, and the ozone thickness was measured to be about 2.2 mm. When the absorption of the Chappuis band of ozone in the visible spectrum was added to the Rayleigh theory, the calculated sky brightness came into agreement with observation for solar depression angles below the horizon from about 0° to 6°. For the sun below 7° the calculated zenith sky brightness fell rapidly below the observed brightness, showing that primary scattering in the atmosphere above about 60 km does not contribute appreciably to the brightness, as has long been known (J. Opt. Soc. Am. 28, 227, 1938).

Calculation showed that during the day the clear sky is blue according to Rayleigh, and that ozone has little effect on the color of the daylight sky. But near sunset and throughout twilight ozone affects the sky color profoundly. For example, in the absence of ozone the zenith sky would be a grayish green-blue at sunset becoming yellowish in twilight, but with ozone the zenith sky is blue at sunset and throughout twilight (as is observed), the blue at sunset being due about $\frac{1}{3}$ to Rayleigh and $\frac{2}{3}$ to ozone, and during twilight wholly to ozone.

INTRODUCTION

ONE group of observers from this Laboratory measured[1] the brightness of the twilight sky from Sacramento Peak, New Mexico, altitude 2800 meters, and another group measured[2] the optical transmission of the air above Mount Lemon, Arizona, altitude 2500 meters and about 500 km from Sacramento Peak. The ozone thicknesses were also measured at the two mountains. When corrected for the absorption of ozone, the transmission of the Mount Lemon atmosphere was found to be only about 20 percent above that of the Rayleigh scattering theory for pure air. Therefore, as described in this paper, the Rayleigh theory was used to calculate the brightness of the twilight sky at Sacramento Peak. When this was done for the zenith sky, using first-order scattering and known upper air densities from rockets, theoretical values 2 to 4 times the observed values were obtained for solar depression angles below the horizon from 0° to 7°. However, the discrepancy was removed when the absorption of the Chappuis band of ozone from about 4500 to 7000A was introduced in addition to the Rayleigh attenuation; the theoretical brightness values then came into agreement with observation. For the sun below 7° below the horizon the calculated sky brightness fell rapidly below the observed brightness showing that the zenith sky brightness in this case was due to multiple scattering, as has long been known, and was not influenced perceptibly by the air above about 60 km.

When the color of the sky was calculated, it was found, for example, that for pure air and no ozone the color of the zenith sky was the Rayleigh blue (i.e., approximately of an intensity distribution inversely with the fourth power of the wavelength) during the day, but as the sun neared the horizon it became much less blue, and when the sun was below 3° below the horizon it became slightly yellowish; these last two statements are contrary to observation. Again, when the absorption of the Chappuis band of ozone was introduced, the color of the daylight sky was only imperceptibly changed from Rayleigh blue and the colors of the zenith sunset and twilight sky became blue, the blue of the sunset sky being due about $\frac{1}{3}$ to Rayleigh and $\frac{2}{3}$ to ozone and of the twilight sky wholly to ozone; all of these blues were about the same color to the eye. This is in accord with observation.

That the Chappuis band of ozone exerts such noticeable effects on the brightness and color of the sky, particularly the twilight sky, has not previously been realized, except in one case. In two interesting papers Dubois[3] has obtained spectrophotometric evidence that the "earth shadow," which is visible in the sky opposite the sun during the early part of twilight, is caused by the selective absorption of the Chappuis band of ozone.

DATA OF SKY BRIGHTNESS, AIR DENSITY, RAYLEIGH SCATTERING, AND OZONE ABSORPTION

In Table I, reference 1, are given the complete data of the twilight sky brightness B ca ft^{-2} obtained during May and June, 1951, at Sacramento Peak, New Mexico, altitude 2800 meters, latitude 32° 47′ north, longitude 105° 49′ west. B is the photopic brightness, or the brightness seen with the light adapted eye. Some of the observed values of B are in Figs. 1 and 2. The photopic transmission of the Sacramento Peak atmosphere was measured by means of the sun and a visual photometer and found to be but little above that of the Rayleigh

[1] Koomen, Lock, Packer, Scolnik, Tousey, and Hulburt, J. Opt. Soc. Am. 42, 353 (1952).

[2] L. Dunkelman and R. Scolnik, J. Opt. Soc. Am., to be published.

[3] J. Dubois, Ann. géophys. 7, 103, 145 (1951).

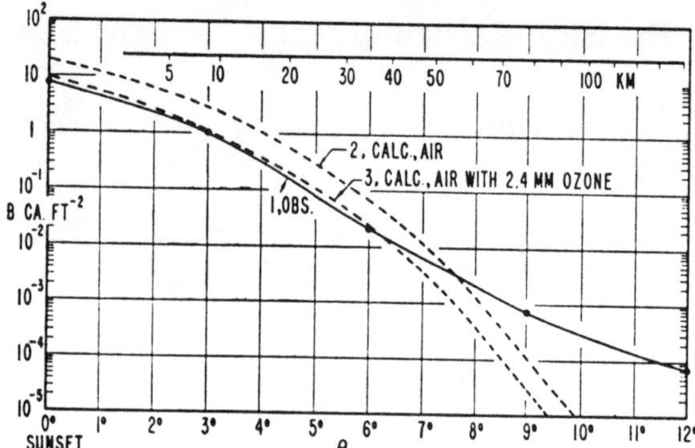

FIG. 1. Zenith sky brightness at Sacramento Peak; full line, observed (see reference 1); dotted lines, calculated.

scattering theory for pure air, showing that there was very little haze present. In October, 1951, spectral measurements[2] of the atmosphere above Mount Lemon, Arizona, altitude 2500 meters, showed that throughout the spectrum from 3000 to 7000A, when correction was made for the ozone absorption, the transmission was only about 20 percent above the Rayleigh theory. Mount Lemon is about 500 km from Sacramento Peak, and because of the similarity of the climate at the two mountains it was assumed in the following pages that the transmission of the Sacramento Peak air was close to that of Rayleigh for pure air.

Measurements of the ozone thickness at Sacramento Peak with a new Dobson photoelectric spectrophotometer gave a value 2.2 mm at STP. The value was a little below rocket values[4] in the area and in the following calculations a value 2.4 mm was used.

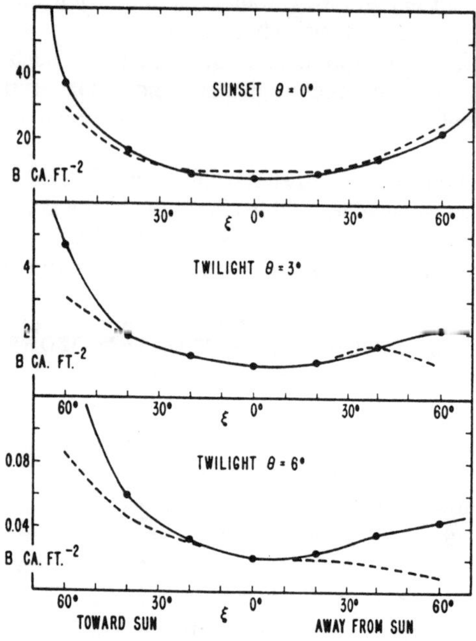

FIG. 2. Brightness of sky on meridian through sun; full lines observed (see reference 1); dotted lines, calculated.

[4] Johnson, Purcell, and Tousey, J. Geophys. Research **56**, 583 (1951).

The average air density d gram cm^{-3} for various heights z above sea level observed by means of rockets[5] is given in Table I. Since the composition of the main constituents of the atmosphere is approximately constant with altitude up to about 100 km, the molecular density n of column 3, Table I was obtained by dividing d by 4.82×10^{-23} gram, the mass of the average air particle. When log n was plotted against z, the points fell on a slightly wavering line of average slope $-1/H$. Hence

$$n = n_0 e^{-z/H}, \qquad (1)$$

where n_0 is the value of n at the surface and H is termed the "scale height." The average value of H was 7 km. Actually H was 9.0, 6.6, 6.3, and 6.9 km for the 10 km intervals from $z=0$ to 40 km. The value $H=7$ km was used throughout this paper; no conclusions were reached which depended critically on the exact value of H. The temperature[6] at various altitudes is given in column 4, Table I.

Some well-known relations which are derived from (1) for a curved earth are as follows. In a direction of

TABLE I. Atmospheric density and temperature from rocket data.[a]

Altitude z km	d	n	Temp. °C
0	1.2×10^{-3}	2.54×10^{19}	17
10	4.2×10^{-4}	8.72×10^{18}	-43
20	9.2×10^{-5}	1.91×10^{18}	-63
30	1.9×10^{-5}	3.85×10^{17}	-38
40	4.3×10^{-6}	8.93×10^{16}	-13
50	1.3×10^{-6}	2.70×10^{16}	-3
60	3.8×10^{-7}	7.90×10^{15}	-13
70	1.2×10^{-7}	2.49×10^{15}	-63
80	2.5×10^{-8}	5.20×10^{14}	-83
90	4.0×10^{-9}	8.32×10^{13}	-63
100	8×10^{-10}	1.66×10^{13}	-33

[a] See reference 5.

zenith angle ξ the total number of molecules N between altitudes z and z_1 is

$$N = H(n - n_1) \sec\xi, \qquad (2)$$

where n and n_1 are the molecular densities at the resepective altitudes. Equation (2) is accurate for a flat earth and begins to be in appreciable error for a curved earth when $\xi > 80°$. When $z_1 = \infty$, $n_1 = 0$; in this case N is the total number of molecules from n to the outside of the atmosphere, and (2) becomes

$$N = CHn \sec\xi, \qquad (3)$$

where C is the correction for a curved earth. C has been evaluated[6] in gamma-functions which are incomplete for $\xi \neq 90°$. C is a slowly varying function of H, and for $H = 7$ km, $C = 1$ for $\xi = 0°$ to 75°; $C = 0.98, 0.94, 0.87, 0.72$, and 0.46 for $\xi = 80°, 83°, 86°, 88°$, and 89°, respectively. For $\xi = 90°$ the gamma-function is complete

[5] Havens, Koll, and LaGow, J. Geophys. Research **57**, 59 (1952).
[6] E. O. Hulburt, Phys. Rev. **55**, 639 (1939).

and

$$N = Kn, \qquad (4)$$

where $K = (\pi r H/2)^{\frac{1}{2}}$ and $r = 6.3 \times 10^8$ cm is the radius of the earth. Equations (3) and (4) contain approximations based on the fact that n falls to small values at altitudes z which are small with respect to r.

The attenuation coefficient β_λ for light of wavelength λ passing through a medium is defined by

$$i_\lambda = i_{0\lambda} e^{-\beta_\lambda x}, \qquad (5)$$

where $i_{0\lambda}$ and i_λ are the intensities entering and leaving a thickness x of the medium. For air the attenuation[7] resulting from Rayleigh scattering is

$$\beta_\lambda = \frac{8\pi^3}{3} \frac{(\mu_\lambda - 1)^2}{n^2 \lambda^4} \frac{6(1+\delta)}{6-7\delta} \left(3 + \frac{1-\delta}{1+\delta} \right), \qquad (6)$$

where μ_λ is the refractive index of air and n is the number of molecules cm^{-3} at STP. The polarization defect is $\delta = 0.04$. In (6) β_λ is the attenuation per air molecule and when used in (5), x is the number of air molecules.

Table II. Attenuation coefficients and sunlight.

λ	β_λ per air molecule	β_λ' per cm of ozone at $-60°C$	$\psi_\lambda i_{0\lambda}$ visibility $\times$ sunlight
4000A	16.7×10^{-27}	$\cdots$	0.0003
4500	10.3	0.0034	0.0417
5000	6.68	0.053	0.358
5500	4.54	0.120	1.000
6000	3.17	0.175	0.615
6500	2.28	0.084	0.0936
7000	1.71	0.030	0.0032
7500	1.29	$\cdots$	0.0000

Values of β_λ calculated from (6) for the visible spectrum are listed in column 2, Table II.

If ϕ is the angle between the incident and scattered rays, the scattering coefficient $\sigma_{\phi\lambda}$ per air molecule per steradian is[7]

$$\sigma_{\phi\lambda}/\beta_\lambda = \frac{3}{4\pi} \left(1 + \frac{1-\delta}{1+\delta} \cos^2\phi \right) \Big/ \left(3 + \frac{1-\delta}{1+\delta} \right). \qquad (7)$$

The absorption coefficient β_λ' of the Chappuis band of ozone, measured by Vassy[8] at a temperature of 18°C, is plotted in Fig. 1. β_λ' is defined by (5) in which x cm is the ozone thickness reduced to STP. Data[9] for β_λ' at temperatures $-42°$, $-80°$, and $-105°$C are plotted as dots in Fig. 3; the dots were joined by the dotted lines. The temperature of the atmosphere in the main ozone region from 20 to 30 km was, from Table I, about $-60°$C. For the purpose of the present calculations β_λ' was determined for $-60°$ by interpolation and extrapolation in Fig. 3, and is listed in column 3, Table II.

[7] R. Tousey and E. O. Hulburt, J. Opt. Soc. Am. 37, 78 (1947).
[8] A. Vassy, Ann. phys. 11 ser. 16, 145 (1941).
[9] A. Vassy and E. Vassy, J. Chem. Phys. 16, 1163 (1948).

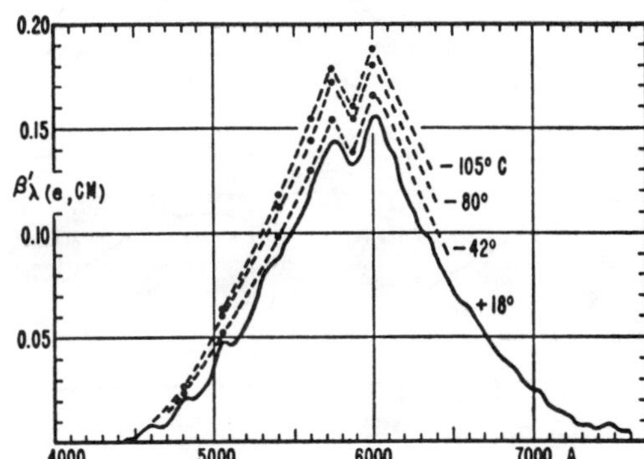

Fig. 3. Absorption coefficient of the Chappuis band of ozone and its temperature dependence, after A. Vassy and E. Vassy (see references 8 and 9).

In column 4, Table II, are given in relative units the values of the intensity of sunlight $i_{0\lambda}$ outside the atmosphere multiplied by the visibility function ψ_λ of the light adapted eye. The value of the solar illumination i_0 outside the atmosphere is 12 000 foot candles, the value being uncertain to at least 10 percent.

BRIGHTNESS AND COLOR OF A TWILIGHT SKY OF PURE AIR AND NO OZONE

An atmosphere was assumed distributed according to Table I composed of pure air with no ozone. The brightness B of the zenith sky at sunset was calculated by referring to Fig. 4. A solar ray of intensity $i_{0\lambda}$ for wavelength λ outside the terrestrial atmosphere passes through N air molecules to point P in the observer's zenith at sunset (or sunrise) where the molecular density is n and the altitude is z. At n the solar intensity is $i_{0\lambda} e^{-\beta_\lambda N}$. An amount $\sigma_{90°\lambda} n dz$ of this is scattered vertically downward, and is further attenuated by an amount $e^{-\beta_\lambda H (n_0 - n)}$ in reaching the surface; n_0 is the value of n at the surface. Then β_λ, the brightness at wavelength λ, is

$$B_\lambda = \int_0^\infty \sigma_{90°\lambda} i_{0\lambda} e^{-\beta_\lambda N} e^{-\beta_\lambda H (n_0 - n)} n \, dz. \qquad (8)$$

Substituting (4) and $-H dn$ for $n dz$ in (8) leads to

$$B_\lambda = i_{0\lambda} \sigma_{90°\lambda} H e^{-\beta_\lambda H n_0} \int_0^{n_0} e^{-\beta_\lambda (K-H) n} \, dn. \qquad (9)$$

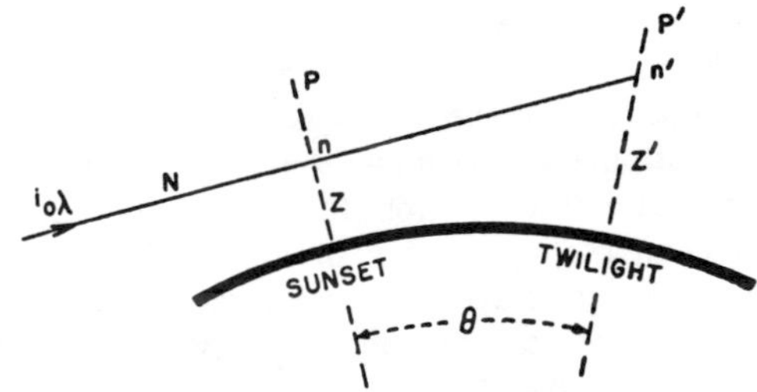

Fig. 4. Geometry of twilight scene.

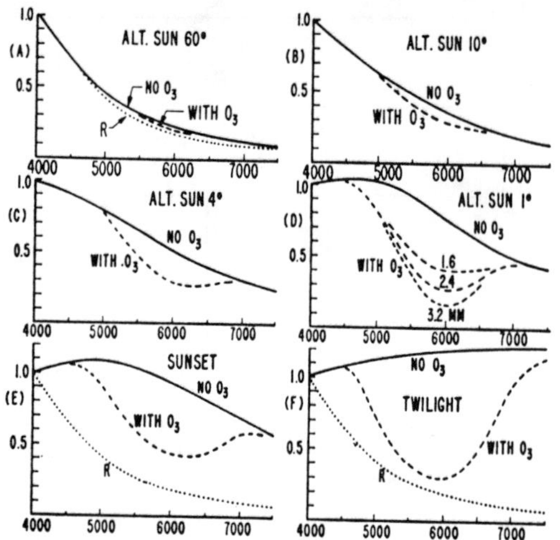

FIG. 5. Zenith sky color calculated for various altitudes of the sun for pure air with and without 2.4 mm of ozone. Curve R is Rayleigh blue.

Integrating (9) yields

$$B_\lambda = i_{0\lambda} \frac{\sigma_{90°\lambda}}{\beta_\lambda} \frac{H}{K-H} e^{-\beta_\lambda H n_0}\left[1 - e^{-\beta_\lambda(K-H)n_0}\right]. \quad (10)$$

With $n_0 = 1.8 \times 10^{19}$ for the altitude of Sacramento Peak and with $i_{0\lambda} = 12\,000$ ft-ca in (10), one finds $B_\lambda = 16.6$ ca ft^{-2}, for wavelength 5500A.

Now the photopic brightness B is

$$B = \int B_\lambda \psi_\lambda i_{0\lambda} d\lambda \Big/ \int \psi_\lambda i_{0\lambda} d\lambda, \quad (11)$$

where the integrals are taken over the visible spectrum, and $\int \psi_\lambda i_{0\lambda} d\lambda = i_0 = 12\,000$ foot candles. The values of B_λ were calculated from (10) for a number of wavelengths in the visible spectrum and are plotted in the full line curve of Fig. 5 (E), all being multiplied by a constant factor to make B_λ unity at 4000A. With the values of B_λ and with $\phi_\lambda{}'i_{0\lambda}$ from Table II, B was calculated from (11). It came out to be 16.2 ca ft^{-2}, which was about twice the observed value 8 ca ft^{-2}. It is plotted in Fig. 1 at the sunset point of the dotted curve 2, Fig. 1. We note in passing that B_λ from (10) for 5500A is approximately the same as photopic B from (11); as is of course the case when dealing with the transmission of sunlight through optical filters which are not too highly colored, because, as seen in Table II, $\psi_\lambda i_{0\lambda}$ peaks strongly at 5500A.

The full line curve of Fig. 5 (E) brings out the fact that the color of the zenith sky at sunset of an atmosphere of pure air, with no ozone, is a pale blue-greenish gray, quite different from the blue of the daylight sky. For comparison, curve R of Fig. 5 (E) is the plot of the Rayleigh scattering coefficients from column 2, Table II, which was observed by Rayleigh[10] to be

[10] See Table 777, *Smithsonian Physical Tables* (1932), eighth revised edition.

approximately the spectral distribution of the daylight sky, and may be termed "Rayleigh blue."

In order to calculate the brightness B of the zenith sky during twilight we refer to Fig. 4 in which the observer is at an angular distance θ from the sunset point and his zenith is P'; θ is the depression angle of the sun below his horizon. The solar ray which passes the sunset meridian at altitude z crosses his zenith at an altitude z' where the molecular density is n'. Then

$$z' - z \sec\theta = r(\sec\theta - 1), \quad (12)$$

where r is the radius of the earth. We shall be concerned with $\theta < 10°$. Then $\sec\theta$ is within 2 percent of unity and (12) becomes approximately

$$z' - z = r\theta^2/2. \quad (13)$$

From (1) and (13)

$$n' = n/A, \quad (14)$$

where

$$A = \exp(r\theta^2/2H). \quad (15)$$

We take θ greater than about 4°; hence n' is about 1/10 of n or less and the solar ray which reaches n' passes through nearly $2N$ air molecules and is attenuated by an amount $e^{-2\beta_\lambda N}$. An amount $\sigma_{\theta\lambda} n' dz'$ is scattered downward and is further attenuated by $e^{-\beta_\lambda H(n_0 - n')}$ in reaching the surface. $\sigma_{\theta\lambda}$ is nearly the same as $\sigma_{90°\lambda}$. Then for the zenith twilight sky

$$B_\lambda = \int_0^\infty \sigma_{\theta\lambda} i_{0\lambda} e^{-2\beta_\lambda N} e^{-\beta_\lambda H(n_0 - n')} n' dz'. \quad (16)$$

Introducing (4), (14), and $-H dn' = n' dz'$ into (16) and integrating leads to

$$B_\lambda = i_{0\lambda} \frac{\sigma_{\theta\lambda}}{\beta_\lambda} \frac{H}{2AK-H} e^{-\beta_\lambda H n_0}\left[1 - e^{-\beta_\lambda(2K - H/A)n_0}\right]. \quad (17)$$

The values of B_λ were calculated from (17) for wavelengths of the visible spectrum and from these and (11), B was calculated for θ from 4° to 10° and is plotted in the dotted curve 2, Fig. 1. Again, B was nearly the same as B_λ for 5500A. It is seen that for $\theta = 4°$ to 6°

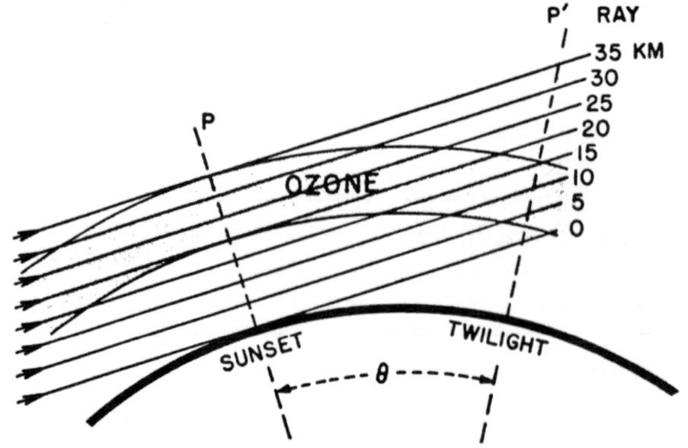

FIG. 6. Geometry of twilight scene with ozone.

the calculated values of B were about 3 times the observed values, and that for $\theta > 7.5°$ the calculated values fell rapidly below the observed values. Between $\theta = 0°$ and $4°$ curve 2, Fig. 1, was merely drawn in by eye; it could be calculated more precisely by evaluating the incomplete gamma-functions[6] in this interval, but such precision was uncalled for.

It turned out that the exponent in the bracket of (17) was much larger than 1 and hence the bracket was close to 1. Therefore the wavelength dependence of B_λ was controlled by the term $\exp(-\beta_\lambda H n_0)$ and hence was the same for all values of $\theta > 4°$. The variation of B_λ with wavelength is shown by the full line curve of Fig. 5 (F), in which B_λ for 4000A was adjusted to unity. The curve shows that the zenith sky during twilight of an atmosphere of pure air with no ozone would be faintly yellowish, which is of course contrary to the fact that it is blue or purple-blue.

BRIGHTNESS AND COLOR OF A SKY OF PURE AIR AND 2.4 MM OF OZONE

In the rocket flights over New Mexico ozone was observed[4] to be distributed mainly between 18 and 35 km, and of vertical thickness 2.38 mm at STP in October, the epoch of maximum ozone. As a fair approximation and one lending itself to ease of calculation, an ozone bank was here assumed entirely between 20 and 35 km, of ozone density constant with height between these levels, and of total vertical thickness 2.4 mm at STP. The brightness and color of the sky were then calculated for the same atmosphere of pure air of Table I together with the ozone. In this case one did not have the luxury of the algebra of Eqs. (10) and (17), because the integrals became unmanagable. Recourse was had to numerical integration; the method is indicated in Fig. 6. Parallel rays of sunlight traversing the sunset meridian horizontally were taken at each 5 km increase of altitude; in Fig. 6 the ray is labeled according to its altitude at the sunset point. Refraction of the rays in the atmosphere increased B_λ by less than 10 percent and was neglected. The thicknesses of air molecules and of ozone traversed by each ray from the sun to the observer's zenith P' were calculated, the first by the methods of the previous section, and the second by methods of geometry which are too simple to need description here. The attenuation of each ray was calculated from the coefficients of Table II. The fractional amount of each ray for each wavelength scattered downward and its attenuation in reaching the surface were calculated; these were added, multiplied by $\psi_\lambda i_{0\lambda}$ of Table II, and B_λ and B were determined.

The calculated values of B thus obtained were plotted in curve 3, Fig. 1, and although slightly greater than the observed values of curve 1, Fig. 1, probably agree with them as well as expected for θ from 0° to 6°. For example, at sunset the observed and calculated values of B were 8 and 10 ca ft^{-2}; and for $\theta = 6°$, were 0.022 and 0.026 ca ft^{-2}. Two small effects have been omitted

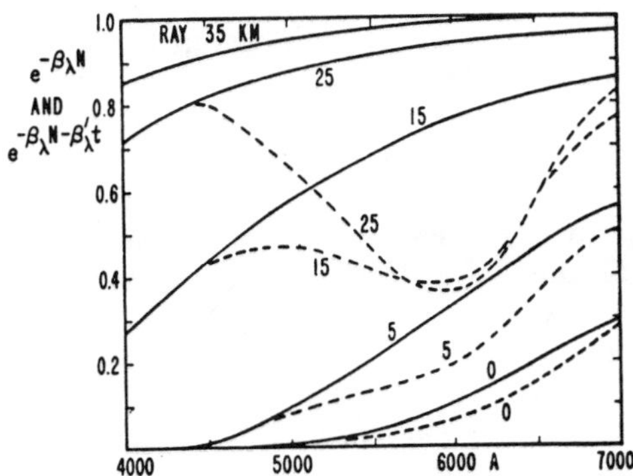

FIG. 7. Attenuation of sunset rays reaching the zenith at altitude marked on each curve; full lines, air; dotted lines, air with 2.4 mm of ozone.

in the calculation, one which increases B and one which decreases B. The first is multiple scattering. Since B was calculated from primary scattering only, multiple scattering should also be included, but probably contributes less than 20 percent for $\theta < 6°$ (see Table II, reference 11). The second is the spectral line and band absorption in the visible spectrum of other atmospheric gases, as oxygen and water vapor; but there are no quantitative data to enable the effect to be assessed. Finally, there is probably always some atmospheric dust or haze present which may either increase or decrease B depending on its distribution.

The calculation of B was repeated using 2.0 mm and 3.0 mm of ozone. With 2.0 mm the values of B were about 8 and 17 percent greater than B of curve 3, Fig. 1, at sunset and during twilight, respectively; and with 3.0 mm were 14 and 29 percent less, respectively.

Some details of the calculations of the various rays of Fig. 6 are given in Fig. 7 for sunset; all of the rays are not included because the figure would become confused. The full lines of Fig. 7 are the attenuations of air only, i.e., $\exp{-\beta_\lambda N}$, N being the number of air molecules along the ray, and the dotted lines are the additional absorption resulting from the Chappuis band of ozone, i.e., $\exp{-\beta_\lambda' t}$, t being the thickness of ozone along the ray. The curves bring out the strong effect of the Chappuis absorption in reducing the intensity of some of the rays and in changing their yellow color to blue or blue-purple. The curve for ray 0 shows that the setting (or rising) sun is yellow to an observer on the surface, as is observed when the air is very clear. The colors of the zenith sky at sunset and during twilight of the atmosphere of pure air with 2.4 mm of ozone are shown by the dotted curves of Fig. 5 (E) and (F), and are seen to be blue or blue-purple, perhaps somewhat unsaturated. This agrees with qualitative observation, but no quantitative measurements of the colors of the twilight sky are available with which to make comparison.

Since ozone modifies the color of the zenith twilight sky it was of interest to determine its effect on the color of the daylight sky. This was readily done from Eq. (23), reference 7, and the results are shown in Fig. 4 (A) to (D) for the zenith sky with 2.4 mm of ozone. It is seen that for the sun above 10° ozone has little effect and that the zenith sky color is close to Rayleigh blue, as is observed.[10] When the sun nears the horizon a sky without ozone would become gray-blue, but with sufficient ozone it remains fairly blue. The influence of several thicknesses of ozone is shown in Fig. 5 (D). By similar calculations it was found that a sky of pure air and no ozone becomes whiter toward the horizon no matter what the altitude of the sun, and that ozone gives an appreciable blue coloration to the horizon sky only when the sun is fairly low. However, ordinarily there is haze present in even a cloudless sky and the color of the sky near the horizon is usually confused by the whitening effects of haze, and by multiple scattering effects which have never been calculated.

In order to see to what extent primary scattering accounts for the brightness of other places than the zenith in the twilight sky, B was calculated for pure air and 2.4 mm of ozone for places in the sky on the meridian through the sun for various zenith angles for depressions of the sun below the horizon $\theta = 0°$, 3°, and 6°. The methods underlying Eqs. (10) and (17) and the geometry of Fig. 5 were used in the calculation. The calculated and observed curves, given in Fig. 2, agree near the zenith moderately well. For ξ greater than 30° or 40° the calculated values are below the observed values, which is the result of the neglect of multiple scattering in the calculated curves.

Referring to Fig. 1 the agreement within perhaps 30 percent of the calculated curve 3 with the observed curve 1 for θ from 0° to 6.5° shows that the observed zenith sky brightness was accounted for almost entirely by primary scattering. Now, primary scattering occurs in the zenith air which is in direct sunlight above the earth's shadow. The position overhead of the boundary of the earth's shadow as a function of θ, calculated from (12) with $z = 0$, is given in the upper scale of abscissas of Fig. 1. The scale shows, for example, that when the sun is 6.5° below the horizon the boundary of the earth's shadow is 41 km overhead, the air above 41 km being

in direct sunlight and the air below 41 km being in shadow. Therefore, working backwards, if one measures zenith B during twilight with an accuracy of better than, say, 5 percent, and if one knows that the air is very pure, one can by means of the Rayleigh theory, with due correction for ozone absorption, determine the atmospheric pressure up to 41 km, and possibly but not probably to 50 km, with a confidence of hardly better than 30 percent, because further corrections arising from absorption of water vapor, oxygen, and the slight haze which may be present in even the purest of surface air may be appreciable.

As θ increases above 6.5° the calculated curve 3, Fig. 1, descends rapidly below the observed curve 1, showing that there is not enough air in the upper levels to produce by primary scattering the observed values of zenith B for $\theta > 6.5°$. Hence, measurements of B can yield no information about the density of the air above about 50 km. This conclusion was reached from similar experiments in 1938,[11] which went one step further and showed that B for $\theta > 7°$ was due almost entirely to multiply scattered light from air in shadow in the lower levels below 50 km. However, the idea has persisted that one can learn about the density of the air in the upper levels from measurements of B. For example, Van de Hulst[12] stated in italics that "twilight photometry forms potentially the most powerful method for observing scattering in the high atmosphere." Such a statement is erroneous for altitudes above 50 km. Many people, the most recent being Ashburn,[13] have calculated atmospheric densities up to as high as 150 km from measurements of zenith B during twilight in relative units. The densities which they obtained differed among themselves by several orders of magnitude (see Fig. 3, reference 13), did not agree with the rocket results, and above about 50 km were without meaning.

[11] E. O. Hulburt, J. Opt. Soc. Am. **28**, 227 (1938).
[12] H. C. Van de Hulst, edited by G. P. Kuiper, *The Atmospheres of the Earth and Planets* (University of Chicago Press, Chicago, 1952), revised edition, Chap. III, p. 84.
[13] E. V. Ashburn, J. Geophys. Research **57**, 85–93 (1952), and references therein.

Tables of the Refractive Index for Standard Air and the Rayleigh Scattering Coefficient for the Spectral Region between 0.2 and 20.0 μ and Their Application to Atmospheric Optics

RUDOLF PENNDORF*

Geophysics Research Directorate, Air Force Cambridge Research Center, Air Research and Development Command, Cambridge, Massachusetts

(Received June 8, 1955)

The refractive index of standard air has been tabulated for the range 0.2 to 20.0 μ, namely, 0.2(0.01)0.8; 0.8(0.1)2.0; 2.0(0.5)10.0; 10(1)20 μ. The values are based on Edlén's formula. For computations of Rayleigh scattering, the scattering cross sections, the mass and volume scattering coefficients have been tabulated for the same values of wavelength as for the refractive index. The optical thickness and the transmissivity of the standard atmosphere is given next, and finally, the mean scattering coefficient for visual observations which is $\beta_v = 1.23 \times 10^{-7}$ (cm^{-1}).

I. INTRODUCTION

THE refractive indexes of standard air and the Rayleigh scattering coefficients are frequently needed for problems in atmospheric optics. The standard textbooks and tables are not too helpful, because they list the values either in an unusable form, or in very wide steps, or not accurately enough. Furthermore, the textbook data are based on old experiments leading to values which do not reflect the present status of accuracy.

II. REFRACTIVE INDEX OF STANDARD AIR

Edlén[1] discussed the dispersion of air and found a new dispersion formula based on the available experimental results. His formula has already been adopted by the Joint Commission for Spectroscopy in 1952. *Standard air* is defined as dry air containing 0.03% CO_2 by volume at normal pressure 760 mm Hg ($= 1013.250$ mb) and having an air temperature of 15°C.

The refractive index n_s for such a standard air at air temperature 15°C is given by Edlén as

$$(n_s - 1)10^{-8} = 6432.8 + \frac{2949810}{146 - \nu^2} + \frac{25540}{41 - \nu^2}, \quad (1)$$

where $\nu = 1/\lambda$ and the wavelength λ is measured in μ. Thus the dimension of ν is μ^{-1}.

This formula has been evaluated for the range 0.2 to 20 μ in steps of 0.01 μ between 0.2 and 0.8 μ, in steps of 0.1 μ from there on up to 2 μ, in steps of 0.5 μ between 2 and 10 μ, and finally, in steps of 1.0 μ. These steps are sufficient because changes of n_s become smaller the larger the wavelength.

For most practical purposes, the *influence of pressure and temperature* can be written as

$$(n - 1) = (n_s - 1)\left(\frac{1 + \alpha t_s}{1 + \alpha t}\right)\frac{p}{p_s}, \quad (2)$$

where $t_s = 15$°C, $t =$ the actual air temperature in °C, $p =$ the actual pressure, and $p_s = 760$ mm Hg ($= 1013.25$ mb). Formula (2) has been evaluated for air temperatures $t = -30$°C, $t = -15$°C, $t = \pm 0$°C, $t = 15$°C, $t = 30$°C, and $p = p_s$ for the same wavelengths as formula (1). The results of the computations are given in Table I. *It is seen that the influence of the air temperature is considerable and it cannot be neglected.*

Water vapor influences the refractive index only slightly as compared to the temperature influence. An example may illustrate the magnitude of the correction. For $\lambda = 0.55$ μ and $n_s = 1.00027773$, the addition of an equivalent of 2 mm Hg water vapor will reduce the refractive index to $n = 1.00027762$; 5 mm Hg water vapor will reduce it to $n = 1.00027746$ and 10 mm Hg water vapor will reduce it to $n = 1.00027718$. A glance at Table I shows that in comparison with the temperature influence *such corrections are unnecessary for most practical applications.* It may, however, be added that the water vapor correction cannot be neglected in the cm-wavelength spectrum (radar).

The extrapolation to the infrared is a short and relatively safe extrapolation. Several checks can be made with experimental measurements. For $\lambda = 1.65$ μ Rank et al.[2] measured in the laboratory $n_s = 1.00027323$ for standard air at air temperature $t_0 = 15$°C. This result agrees well within the experimental error with the value obtained from Edlén's formula, namely $n_s = 1.00027317$. Another set of measurements by Rank and Shearer[3] for $\lambda = 1.5300$ μ, 1.5349 μ, and 1.6541 μ gave results for $n_s - 1$ which are persistently higher (about 0.1 to 0.2%) than by applying Edlén's formula.[1] Essen[4] measured the refractive index for air at 1.25 cm and 3.26 cm. His values for $n_s - 1$ are also about 0.2% higher than Edlén's; the values $(28815 \pm 10) \times 10^{-8}$ at 1.25 cm and $(28810 \pm 10) \times 10^{-8}$ at 3.26 cm are higher than Edlén's value of 28757×10^{-8} for $t = 0$°C. Rank and Shearer[3] conclude that Edlén's formula gives values

---

* Now at Electronics Research Laboratory, Advanced Research and Development Division, AVCO Manufacturing Company, Boston, Massachusetts.

[1] B. Edlén, J. Opt. Soc. Am. **43**, 339 (1953).

[2] Rank, Shull, Bennett, and Wiggins, J. Opt. Soc. Am. **43**, 952 (1953).

[3] D. H. Rank and J. N. Shearer, J. Opt. Soc. Am. **44**, 575 (1954).

[4] L. Essen, Proc. Phys. Soc. (London) **66B**, 189 (1953).

TABLE I. Refractive indexes n of standard air for
selected air temperatures t.

Wavelength in μ	(n-1) x 10^8 for				
	t = -30°C	t = -15°C	t = ±0°C	t = +15°C	t = +30°C
0.20	38406	36174	34187	32408	30802
0.21	37617	35430	33484	31742	30169
0.22	36945	34797	32886	31175	29621
0.23	36526	34403	32514	30821	29294
0.24	36079	33982	32116	30444	28936
0.25	35725	33648	31800	30145	28652
0.26	35422	33363	31530	29889	28408
0.27	35160	33116	31297	29668	28198
0.28	34931	32900	31093	29475	28015
0.29	34730	32711	30915	29306	27845
0.30	34552	32544	30756	29156	27711
0.31	34392	32395	30616	29022	27584
0.32	34253	32262	30490	28903	27471
0.33	34126	32142	30377	28796	27369
0.34	34011	32034	30275	28699	27277
0.35	33907	31936	30182	28611	27194
0.36	33813	31847	30098	28532	27118
0.37	33727	31766	30022	28459	27049
0.38	33648	31692	29951	28393	26986
0.39	33576	31624	29887	28332	26928
0.40	33509	31561	29828	28276	26875
0.41	33448	31504	29773	28224	26825
0.42	33392	31450	29723	28176	26780
0.43	33339	31401	29676	28132	26737
0.44	33290	31355	29633	28091	26699
0.45	33245	31312	29592	28052	26662
0.46	33203	31273	29555	28017	26629
0.47	33164	31235	29520	27984	26597
0.48	33127	31201	29487	27953	26568
0.49	33092	31168	29457	27924	26540
0.50	33060	31138	29428	27896	26514
0.51	33030	31109	29401	27871	26490
0.52	33001	31083	29376	27847	26467
0.53	32974	31057	29352	27824	26445
0.54	32949	31034	29329	27803	26425
0.55	32925	31011	29308	27783	26406
0.56	32903	30990	29288	27764	26388
0.57	32881	30970	29269	27746	26371
0.58	32861	30951	29251	27729	26355
0.59	32842	30933	29234	27712	26339
0.60	32824	30915	29218	27697	26325
0.61	32806	30899	29202	27682	26311
0.62	32790	30884	29188	27669	26298
0.63	32775	30869	29174	27656	26285
0.64	32760	30855	29161	27643	26273
0.65	32746	30842	29148	27631	26262
0.66	32732	30829	29136	27620	26251
0.67	32719	30817	29124	27609	26241
0.68	32707	30805	29113	27598	26231
0.69	32695	30794	29103	27589	26222
0.70	32684	30784	29093	27579	26213
0.71	32673	30774	29084	27570	26204
0.72	32663	30764	29074	27561	26196
0.73	32653	30754	29065	27553	26188
0.74	32644	30746	29057	27545	26180
0.75	32634	30737	29049	27537	26173
0.76	32626	30729	29041	27530	26166
0.77	32617	30721	29034	27523	26159
0.78	32609	30713	29027	27516	26153
0.79	32601	30706	29020	27509	26146
0.80	32594	30699	29013	27503	26140
0.90	32533	30641	28959	27451	26091
1.00	32489	30600	28920	27415	26056
1.10	32457	30570	28891	27388	26031
1.20	32433	30547	28870	27367	26011
1.30	32414	30529	28853	27351	25996
1.40	32399	30515	28840	27339	25984
1.50	32387	30504	28829	27329	25974
1.60	32377	30495	28820	27320	25966
1.70	32369	30487	28813	27313	25960
1.80	32362	30481	28807	27308	25954
1.90	32356	30475	28802	27303	25950
2.0	32351	30470	28797	27298	25946
2.5	32335	30455	28782	27285	25933
3.0	32326	30447	28775	27277	25925
3.5	32321	30442	28770	27272	25921
4.0	32317	30438	28767	27270	25918
4.5	32315	30436	28765	27268	25916

TABLE I.—(continued).

Wavelength in μ	(n-1) x 10^8 for				
	t = -30°C	t = -15°C	t = ±0°C	t = +15°C	t = +30°C
5.0	32314	30435	28763	27267	25915
5.5	32312	30433	28762	27265	25914
6.0	32311	30432	28761	27264	25913
6.5	32310	30432	28760	27264	25913
7.0	32309	30431	28760	27263	25912
7.5	32309	30431	28759	27263	25912
8.0	32309	30430	28759	27262	25912
8.5	32308	30430	28759	27262	25911
9.0	32308	30430	28759	27262	25911
9.5	32308	30429	28758	27262	25911
10.0	32308	30429	28758	27262	25911
11.0	32307	30429	28758	27261	25910
12.0	32307	30429	28758	27261	25910
13.0	32307	30429	28757	27261	25910
14.0	32307	30428	28757	27261	25910
15.0	32307	30428	28757	27261	25910
16.0	32306	30428	28757	27261	25910
17.0	32306	30428	28757	27261	25910
18.0	32306	30428	28757	27260	25910
19.0	32306	30428	28757	27260	25910
20.0	32306	30428	28757	27260	25910
∞	32305.7	30427.5	28756.5	27259.9	25909.2

in the infrared which are about 0.1 to 0.2% too low, but I think the agreement is good enough for most of the problems if one bears in mind that it is an excellent formula to describe the refractive index from the UV to the IR. Based on the above experimental data, *the computed n values are considered as very reliable for the spectral range from 0.2 to 20 μ.*

III. RAYLEIGH SCATTERING COEFFICIENT FOR STANDARD AIR

Another parameter of importance in atmospheric optics are the various scattering coefficients of standard air. For such standard air the Rayleigh *scattering cross sections*, σ, per molecule for incident unpolarized (natural) light is given by

$$\sigma = \frac{8\pi^3(n_s^2-1)^2}{3\lambda^4 N_s^2}\left(\frac{6+3\rho_n}{6-7\rho_n}\right), \qquad (3)$$

where N_s is the number density (number of molecules/cm³) and ρ_n is the depolarization factor (see Table II). The other terms are defined in Eq. (1). The first term $8\pi^3(n_s^2-1)^2/3\lambda^4 N_s^2$ is the classical Rayleigh term for optically isotropic molecules. The second term $(6+3\rho_n)/(6-7\rho_n)$ is the depolarization term which expresses the influence of the optically anisotropic molecules on scattering. Since (3) describes the scattering cross section of one molecule, it is independent of the temperature of the gas.

The following numerical values have been selected for the computation of σ: $N_s = 2.54743 \times 10^{19}$ (cm⁻³) for $t = +15°C$ and $\rho_n = 0.035$ (see Table II). This choice leads to the value 1.061 for the depolarization term $(6+3\rho_n)/(6-7\rho_n)$. The results are given in Table III.

For most applications in atmospheric optics, however, two other definitions are more convenient. The

TABLE II. Depolarization factor ρ_n of atmospheric gases for incident unpolarized light.

Gas	Lord Rayleigh 1919 [a]	Gans 1921 [b]	Cabannes Granier 1923 [c]	Rao 1927 [d]	Anantakrishnan 1935 [e]	Volkmann 1935 [f]	Vaucouleur 1949 [g]	Parthasarathy 1951 [h]	Gucker Basu 1953 [i]
air (experimental)	0.042		0.041	0.0415			0.0310 ± 0.0005	0.0365	0.0330
air (computed)	0.036		0.038	0.038				0.0364	0.035
O_2	0.060	0.0665	0.0645	0.0642				0.0600	0.0540
N_2	0.030	0.030	0.0375	0.0357				0.0305	0.0305
A	0.032				0.0042		0.00	0.0006	
CO_2	0.080	0.073	0.098	0.097		0.0724		0.0922	0.0805
N_2O	0.14	0.12	0.122	0.12	0.1295	0.102		0.1197	
NO			0.026					0.0218	
CO	0.032		0.017					0.080	
H_2	0.017		0.022	0.0274	0.009 ± 0.0009			0.0221	
He	0.42							0.025	

[a] Lord Rayleigh (R. J. Strutt), Proc. Roy. Soc. (London) **A95**, 155 (1919).
[b] R. Gans, Ann. Physik **65**, 97 (1921).
[c] J. Cabannes and J. Granier, J. phys. radium **4**, 429 (1923).
[d] S. R. Rao, Indian J. Phys. **2**, 61 (1927).
[e] R. Anantakrishnan, Proc. Indian Acad. Sci. **2**, 153 (1935).
[f] H. Volkmann, Ann. Physik **24**, 457 (1935).
[g] G. de Vaucouleur, Compt. rend. **288**, 1485 (1949).
[h] S. Parthasarathy, Indian J. Phys. **25**, 21 (1951).
[i] F. T. Gucker and S. Basu, Sci. Rept. No. 1, AF19(122)-400 (University of Indiana, 1953).

Rayleigh mass scattering coefficient, κ, is

$$\kappa = \frac{8\pi^3}{3} \frac{(n_s^2-1)^2 N}{\lambda^4 N_s^2 \rho} \left(\frac{6+3\rho_n}{6-7\rho_n}\right), \qquad (4)$$

where ρ=density of air at temperature t and the *Rayleigh volume scattering coefficient, β,*

$$\beta = \frac{8\pi^3}{3} \frac{(n_s^2-1)^2 N}{\lambda^4 N_s^2} \left(\frac{6+3\rho_n}{6-7\rho_n}\right), \qquad (5)$$

where N is the number density at any atmospheric temperature t and at any pressure p. For standard pressure $p=1013.25$ mb and a normal variety of atmospheric temperatures the Rayleigh volume scattering coefficient can be written as

$$\beta = \frac{32\pi^3(n-1)^2}{3\lambda^4 N} \left(\frac{6+3\rho_n}{6-7\rho_n}\right), \qquad (6)$$

which is correct to about $\pm 10\%$. Since (6) is the form normally given in textbooks, one has to keep in mind that the accuracy of this approximation is not too high. The dimensions of the coefficients are: $\sigma = (\text{cm}^2)$, $\kappa = (\text{cm}^2/\text{g})$, $\beta = (\text{cm}^{-1})$.

The volume scattering coefficient, β, is most widely used because it is convenient to use the volume as a parameter when dealing with the free atmosphere. The absorption coefficients are also defined in such a way and have the same dimension as β. β and κ are computed for air with $N = N_0 = 2.68731 \; 10^{19}$ cm^{-3} at $T_0 = 273.16°$K and $\rho_0 = 1.2925 \times 10^{-3}$ g/cm^3. Because of the approximation made and the fact that the basic constants are correct to 5 figures, the results for σ, κ, β can be considered correct to 4 figures only. Machine computations have been carried out to 5 or 6 figures to minimize round-off errors. The results are given in Table III.

For considerations of the intensity scattered in a particular direction θ, *the angular Rayleigh scattering cross section, $\sigma_{R\theta}$,* has to be computed from

$$\sigma_{R\theta} = \frac{\pi^2(n_s^2-1)^2 2(2+\rho_n)}{\lambda^4 N_s^2(6-7\rho_n)} p(\cos\theta)$$

or

$$\sigma_{R\theta} = 1.07558 \times 10^{-38} p(\cos\theta)(n_s^2-1)^2/\lambda^4 \qquad (7)$$

where $p(\cos\theta)$ is the Rayleigh phase function.

$$p(\cos\theta) = 0.7629(1+0.932 \cos^2\theta) \qquad (8)$$

which is tabulated in Table IV; $(n_s^2-1)^2$ is tabulated in Table V.

For visual observations we can use the mean angular scattering cross section, which corresponds to the wavelength $\lambda = 0.55 \; \mu$ (see Chap. 6, where it is proven), and obtain

$$\sigma_{R\theta}(\lambda = 0.55 \; \mu) = 3.6301 \times 10^{-29} p(\cos\theta). \qquad (9)$$

This function is tabulated in Table IV for all scattering angles θ from 0° to 180° in steps of 1°, because it is a very important function for atmospheric visibility problems.

Since the simplification of isotropic air molecules is usually introduced and scattering intensities are computed for this case, it is necessary to evaluate the error caused by this simplified assumption. The angular scattering cross section for isotropic molecules is

TABLE III. Rayleigh scattering cross section σ and Rayleigh scattering coefficients κ and β for $t_0 = 0°C$ and $p = 1013.25$ mb.

Wavelength in μ	σ (cm^2)		$\kappa\left(\dfrac{cm^2}{g}\right)$		$\beta \times 10^8$ (cm^{-1})
0.20	3.551	10^{-25}	7.382	10^{-3}	954.2
0.21	2.802		5.826		753.1
0.22	2.244		4.666		603.1
0.23	1.836		3.818		493.4
0.24	1.511		3.142		406.1
0.25	1.258		2.616		338.2
0.26	1.057		2.199		284.2
0.27	8.959	10^{-26}	1.863		240.8
0.28	7.645		1.590		205.5
0.29	6.568		1.367		176.5
0.30	5.676		1.180		152.5
0.31	4.933		1.026		132.6
0.32	4.309		8.959	10^{-4}	115.8
0.33	3.782		7.863		101.6
0.34	3.334		6.931		89.59
0.35	2.951		6.135		79.29
0.36	2.622		5.450		70.45
0.37	2.337		4.860		62.82
0.38	2.091		4.348		56.20
0.39	1.877		3.902		50.43
0.40	1.689		3.512		45.40
0.41	1.525		3.170		40.98
0.42	1.380		2.869		37.08
0.43	1.252		2.603		33.65
0.44	1.139		2.368		30.60
0.45	1.038		2.158		27.89
0.46	9.482	10^{-27}	1.972		25.48
0.47	8.680		1.805		23.33
0.48	7.961		1.655		21.40
0.49	7.316		1.521		19.66
0.50	6.735		1.400		18.10
0.51	6.211		1.291		16.69
0.52	5.736		1.193		15.42
0.53	5.307		1.103		14.26
0.54	4.917		1.022		13.21
0.55	4.563	10^{-27}	9.486	10^{-5}	12.26
0.56	4.239		8.814		11.39
0.57	3.945		8.201		10.60
0.58	3.675		7.641		9.876
0.59	3.428		7.127		9.212
0.60	3.202		6.657		8.604
0.61	2.994		6.224		8.045
0.62	2.802		5.826		7.531
0.63	2.626		5.460		7.057
0.64	2.464		5.122		6.620
0.65	2.313		4.810		6.217
0.66	2.175		4.521		5.844
0.67	2.046		4.254		5.498
0.68	1.927		4.006		5.178
0.69	1.816		3.776		4.881
0.70	1.713		3.562		4.605
0.71	1.618		3.364		4.348
0.72	1.529		3.179		4.109
0.73	1.446		3.006		3.886
0.74	1.369		2.845		3.678
0.75	1.296		2.695		3.484
0.76	1.229		2.555		3.302
0.77	1.166		2.423		3.132
0.78	1.106		2.300		2.973
0.79	1.051		2.185		2.824
0.80	9.989	10^{-28}	2.077		2.684
0.90	6.212		1.292		1.670
1.00	4.065		8.452	10^{-6}	1.092
1.10	2.771		5.761		0.7447
1.20	1.954		4.062		0.5250
1.30	1.417		2.946		0.3807
1.40	1.052		2.188		0.2828
1.50	7.979	10^{-29}	1.659		0.2144
1.60	6.160		1.281		0.1655
1.70	4.831		1.005		0.1298
1.80	3.842		7.988	10^{-7}	0.1033
1.90	3.094		6.432		0.08314
2.0	2.519		5.238		0.06770
2.5	1.031		2.143		0.02770
3.0	4.968	10^{-30}	1.033		0.01335
3.5	2.681		5.574	10^{-8}	0.007204
4.0	1.571		3.267		0.004222
4.5	9.807	10^{-31}	2.039		0.002636

TABLE III.—(continued).

Wavelength in μ	σ (cm^2)		$\kappa\left(\dfrac{cm^2}{g}\right)$		β (cm^{-1})	
5.0	6.434		1.338		1.729	10^{-11}
5.5	4.394		9.136	10^{-9}	1.181	
6.0	3.102		6.450		8.337	10^{-12}
6.5	2.252		4.683		6.053	
7.0	1.582		3.289		4.251	
7.5	1.271		2.642		3.414	
8.0	9.815	10^{-32}	2.041		2.637	
8.5	7.701		1.601		2.069	
9.0	6.127		1.274		1.647	
9.5	4.935		1.026		1.326	
10.0	4.020		8.358	10^{-10}	1.080	
11.0	2.746		5.708		7.378	10^{-13}
12.0	1.938		4.030		5.209	
13.0	1.407		2.926		3.782	
14.0	1.046		2.175		2.812	
15.0	7.940	10^{-33}	1.651		2.134	
16.0	6.133		1.275		1.648	
17.0	4.813		1.001		1.293	
18.0	3.829		7.961	10^{-11}	1.029	
19.0	3.084		6.413		8.288	10^{-14}
20.0	2.512		5.223		6.751	

given as

$$^{(1)}\sigma_{R\theta} = \frac{\pi^2}{2} \frac{(n_s^2 - 1)^2}{\lambda^4 N_s^2}(1 + \cos^2\theta)$$

or

$$^{(1)}\sigma_{R\theta} = 7.6044 \times 10^{-39}(n_s^2 - 1)^2(1 + \cos^2\theta)/\lambda^4. \qquad (10)$$

Numerical values are computed for visual observations, where $\lambda = 0.55$, namely,

$$^{(1)}\sigma_{R\theta}(\lambda = 0.55\ \mu) = 2.5665 \times 10^{-29}(1 + \cos^2\theta) \qquad (11)$$

and the results tabulated for $\theta = 0(1°)180°$ in Table IV. This simplification leads to values of $\sigma_{R\theta}$ which are 4% to 7% lower than the $\sigma_{R\theta}$ computed correctly (4% for $\theta = 0°$, 5% for $\theta = 45°$, 6% for $\theta = 60°$, and 7% for $\theta = 90°$). This error is quite large and cannot be neglected for precision measurement, in the same way as multiple scattering cannot be neglected. For very crude measurements, however, it does not play any role.

IV. TEMPERATURE DEPENDENCE OF THE SCATTERING COEFFICIENT

In the last section β has been written in such a way that $\beta = \sigma N$, which is very convenient for computational purposes. The computed values in Table III give $\beta_0 = \sigma N_0$ for $t = 0°C$. If β is wanted for any temperature t, we can write (if the pressure $p = $ const, and the index 0 refers to $t = 0°C$ or $T = 273.16°K$)

$$NT = N_0 T_0$$

or

$$\beta = \sigma N_0(T_0/T) = \beta_0(T_0/T). \qquad (12)$$

If β_0 is taken from Table III, formula (12) seems to be the simplest way as far as computation of β is concerned, but this is not the way it is found in textbooks.

From the Lorentz-Lorentz law follows, by setting $n^2 + 2 = 3$,

$$\frac{n_s^2 - 1}{4\pi N_s} = \frac{n^2 - 1}{4\pi N} = \text{const.} \qquad (13)$$

TABLE IV. Rayleigh's phase function $p(\cos\theta)$, the angular Rayleigh scattering cross section for air assuming isotropic and anisotropic molecules for 0(1°)180° for visual observations ($\lambda = 0.55\,\mu$).

Scattering angle θ	$p(\cos\theta)$	$\sigma_{R\theta} \times 10^{28}$	$^{(1)}\sigma_{R\theta} \times 10^{28}$	Scattering angle θ
0	1.4739	5.350	5.133	180
1	1.4737	5.350	5.132	179
2	1.4730	5.347	5.130	178
3	1.4719	5.343	5.126	177
4	1.4704	5.338	5.121	176
5	1.4685	5.331	5.114	175
6	1.4661	5.322	5.105	174
7	1.4634	5.312	5.095	173
8	1.4601	5.300	5.083	172
9	1.4565	5.287	5.070	171
10	1.4525	5.273	5.056	170
11	1.4481	5.257	5.040	169
12	1.4432	5.239	5.022	168
13	1.4379	5.220	5.003	167
14	1.4323	5.199	4.983	166
15	1.4263	5.178	4.961	165
16	1.4199	5.154	4.938	164
17	1.4131	5.130	4.914	163
18	1.4060	5.109	4.888	162
19	1.3985	5.077	4.861	161
20	1.3908	5.049	4.833	160
21	1.3826	5.019	4.803	159
22	1.3741	4.988	4.773	158
23	1.3654	4.957	4.741	157
24	1.3563	4.924	4.708	156
25	1.3469	4.889	4.675	155
26	1.3373	4.855	4.640	154
27	1.3274	4.819	4.604	153
28	1.3172	4.782	4.567	152
29	1.3068	4.744	4.530	151
30	1.2962	4.705	4.491	150
31	1.2853	4.666	4.452	149
32	1.2743	4.626	4.412	148
33	1.2630	4.585	4.372	147
34	1.2516	4.543	4.331	146
35	1.2400	4.501	4.289	145
36	1.2283	4.459	4.246	144
37	1.2164	4.416	4.204	143
38	1.2044	4.372	4.160	142
39	1.1923	4.328	4.117	141
40	1.1801	4.284	4.073	140
41	1.1679	4.240	4.028	139
42	1.1556	4.195	3.984	138
43	1.1432	4.150	3.939	137
44	1.1308	4.105	3.895	136
45	1.1184	4.060	3.850	135
46	1.1060	4.015	3.805	134
47	1.0936	3.970	3.760	133
48	1.0812	3.925	3.716	132
49	1.0689	3.880	3.671	131
50	1.0567	3.836	3.627	130
51	1.0445	3.792	3.583	129
52	1.0324	3.748	3.539	128
53	1.0204	3.704	3.496	127
54	1.0086	3.661	3.453	126
55	0.9968	3.619	3.411	125
56	0.9852	3.577	3.369	124
57	0.9738	3.535	3.328	123
58	0.9626	3.494	3.287	122
59	0.9515	3.454	3.247	121
60	0.9407	3.415	3.208	120
61	0.9300	3.376	3.170	119
62	0.9196	3.338	3.132	118
63	0.9095	3.301	3.096	117
64	0.8996	3.265	3.060	116
65	0.8899	3.230	3.025	115
66	0.8805	3.196	2.991	114
67	0.8715	3.164	2.958	113
68	0.8627	3.132	2.927	112
69	0.8542	3.101	2.896	111
70	0.8461	3.071	2.867	110
71	0.8383	3.043	2.838	109
72	0.8308	3.016	2.812	108
73	0.8237	2.990	2.786	107
74	0.8169	2.966	2.762	106
75	0.8105	2.942	2.738	105
76	0.8045	2.920	2.717	104
77	0.7989	2.900	2.696	103
78	0.7936	2.881	2.677	102
79	0.7888	2.863	2.660	101

TABLE IV.—(continued).

Scattering angle θ	$p(\cos\theta)$	$\sigma_{R\theta} \times 10^{28}$	$^{(1)}\sigma_{R\theta} \times 10^{28}$	Scattering angle θ
80	0.7843	2.847	2.644	100
81	0.7803	2.833	2.629	99
82	0.7767	2.819	2.616	98
83	0.7735	2.808	2.605	97
84	0.7707	2.798	2.595	96
85	0.7683	2.789	2.586	95
86	0.7664	2.782	2.579	94
87	0.7649	2.777	2.574	93
88	0.7638	2.773	2.570	92
89	0.7631	2.770	2.567	91
90	0.7629	2.769	2.567	90

By inserting (13) into (5) we obtain

$$\beta = \frac{8\pi^3}{3}\,\frac{(n^2-1)^2}{\lambda^4 N}\,\frac{(6+3\rho_n)}{(6-7\rho_n)}. \tag{14}$$

This is the usual way of writing Rayleigh's volume scattering coefficient, whereby it is understood, but often forgotten, that n as well as N have to be computed for a specified temperature t. We have seen above that β can easily be computed from formula (9) using β_0 from Table III. The factor (T_0/T) shows that, *under actual atmospheric conditions, β can vary by $\pm 10\%$ and it is not allowed to treat β as a constant.*

V. OPTICAL THICKNESS OF THE ATMOSPHERE

The optical thickness for a standard atmosphere is given by

$$u(s,s') = \int_{s'}^{s} \beta\,ds, \tag{15}$$

TABLE V. $(n_s^2-1)^2$ for standard air at air temperature $t = 15°C$.

Wavelength in μ	$(n_s^2-1)^2 \times 10^7$	Wavelength in μ	$(n_s^2-1)^2 \times 10^7$	Wavelength in μ	$(n_s^2-1)^2 \times 10^7$
0.20	4.2024	0.55	3.0884	1.50	2.9882
0.21	4.0315	0.56	3.0841	1.60	2.9864
0.22	3.8887	0.57	3.0801	1.70	2.9849
0.23	3.8010	0.58	3.0763	1.80	2.9836
0.24	3.7085	0.59	3.0728	1.90	2.9826
0.25	3.6360	0.60	3.0694	2.0	2.9816
0.26	3.5746	0.61	3.0661	2.5	2.9786
0.27	3.5219	0.62	3.0631	3.0	2.9770
0.28	3.4762	0.63	3.0602	3.5	2.9759
0.29	3.4363	0.64	3.0574	4.0	2.9753
				4.5	2.9749
0.30	3.4012	0.65	3.0547		
0.31	3.3702	0.66	3.0522	5.0	2.9747
0.32	3.3425	0.67	3.0498	5.5	2.9743
0.33	3.3177	0.68	3.0475	6.0	2.9741
0.34	3.2955	0.69	3.0453	6.5	2.9740
				7.0	2.9739
0.35	3.2754	0.70	3.0432	7.5	2.9738
0.36	3.2572	0.71	3.0412	8.0	2.9737
0.37	3.2406	0.72	3.0393	8.5	2.9737
0.38	3.2255	0.73	3.0375	9.0	2.9736
0.39	3.2116	0.74	3.0357	9.5	2.9736
0.40	3.1989	0.75	3.0340	10.0	2.9736
0.41	3.1873	0.76	3.0324	11.0	2.9735
0.42	3.1765	0.77	3.0309	12.0	2.9734
0.43	3.1665	0.78	3.0293	13.0	2.9734
0.44	3.1572	0.79	3.0279	14.0	2.9734
0.45	3.1486	0.80	3.0265	15.0	2.9734
0.46	3.1407	0.90	3.0151	16.0	2.9734
0.47	3.1332			17.0	2.9734
0.48	3.1263	1.00	3.0071	18.0	2.9733
0.49	3.1198	1.10	3.0012	19.0	2.9733
		1.20	2.9967		
0.50	3.1137	1.30	2.9932	20.0	2.9733
0.51	3.1080	1.40	2.9904		
0.52	3.1026				
0.53	3.0976				
0.54	3.0928				

TABLE VI. The optical thickness u and the transmissivity τ for a standard atmosphere (isothermal atmosphere).

Wavelength in μ	u	τ	Wavelength in μ	u	τ
0.20	7.630	0.0005	0.55	0.09805	0.9066
0.21	6.022	0.0024	0.56	0.09110	0.9129
0.22	4.823	0.00804	0.57	0.08477	0.9187
0.23	3.946	0.0193	0.58	0.07897	0.9241
0.24	3.247	0.0389	0.59	0.07367	0.9290
0.25	2.704	0.06694	0.60	0.06880	0.9335
0.26	2.272	0.1031	0.61	0.06433	0.9377
0.27	1.925	0.1459	0.62	0.06022	0.9415
0.28	1.643	0.1934	0.63	0.05643	0.9451
0.29	1.411	0.2439	0.64	0.05294	0.9484
0.30	1.220	0.2952	0.65	0.04971	0.9515
0.31	1.060	0.3465	0.66	0.04673	0.9543
0.32	0.9260	0.3961	0.67	0.04397	0.9570
0.33	0.8127	0.4392	0.68	0.04141	0.9594
0.34	0.7164	0.4885	0.69	0.03903	0.9617
0.35	0.6341	0.5304	0.70	0.03682	0.9638
0.36	0.5634	0.5693	0.71	0.03477	0.9658
0.37	0.5023	0.6050	0.72	0.03286	0.9677
0.38	0.4494	0.6380	0.73	0.03107	0.9694
0.39	0.4033	0.6681	0.74	0.02941	0.9710
0.40	0.3630	0.6956	0.75	0.02786	0.9725
0.41	0.3277	0.7206	0.76	0.02641	0.9739
0.42	0.2966	0.7433	0.77	0.02505	0.9753
0.43	0.2691	0.7641	0.78	0.02378	0.9765
0.44	0.2447	0.7829	0.79	0.02258	0.9777
0.45	0.2231	0.8000	0.80	2.147×10^{-2}	0.9788
0.46	0.2038	0.8156	0.90	1.335	0.9867
0.47	0.1865	0.8299			
0.48	0.1711	0.8427	1.00	8.736×10^{-3}	0.9913
0.49	0.1572	0.8545	1.10	5.955	0.9941
			1.20	4.198	0.9958
0.50	0.1447	0.8653	1.30	3.044	0.9970
0.51	0.1335	0.8750	1.40	2.261	0.9977
0.52	0.1233	0.8840			
0.53	0.1140	0.8923			
0.54	0.1057	0.8997			

TABLE VI.—(continued).

Wavelength in μ	u	τ
1.50	1.715×10^{-3}	0.9983
1.60	1.324	0.9987
1.70	1.038	0.9990
1.80	8.257×10^{-4}	0.9992
1.90	6.649	0.9994
2.0	5.414	0.9995
2.5	2.215	0.9998
3.0	1.067	0.9999
3.5	5.761×10^{-5}	0.9999
4.0	3.377	1.0000
4.5	2.107	1.0000
5.0	1.382	1.0000
5.5	9.442×10^{-6}	1.0000
6.0	6.666	1.0000
6.5	4.840	1.0000
7.0	3.399	1.0000
7.5	2.730	1.0000
8.0	2.109	1.0000
8.5	1.655	1.0000
9.0	1.317	1.0000
9.5	1.060	1.0000
10.0	8.637×10^{-7}	1.0000
11.0	5.899	1.0000
12.0	4.165	1.0000
13.0	3.024	1.0000
14.0	2.248	1.0000
15.0	1.706	1.0000
16.0	1.318	1.0000
17.0	1.034	1.0000
18.0	8.227×10^{-8}	1.0000
19.0	6.627	1.0000
20.0	5.398	1.0000

and in the case where s' is at the surface of the earth and s the top of the atmosphere, we can then write

$$u = \int_0^\infty \beta\, dh = \sigma \int_0^\infty N\, dh \qquad (16)$$

since the scattering cross section, σ, is a constant. In the atmosphere,

$$N = N_0 e^{-h/H} \qquad (17)$$

where the scale height, H, is defined in the usual way. Based on the standard values adopted by ICAO, $H = 7995.75$ (m). Hence, if the atmosphere is isothermal throughout $N\,dh = 2.149 \times 10^{25}$ (molecules/cm²) in a column.

For a model atmosphere, where T varies with height h, the Rand model[5] was chosen, because N is tabulated for $\Delta h = 1.524$ km. We integrated stepwise from 0 to 70 km applying Simpson's rule for every 15.24 km using 10 coordinates. The result is $\int N\,dh = 2.177 \times 10^{25}$ (molecules/cm²) in a column, i.e., only 1.3% more than for the isothermal atmosphere.

Table VI gives the result for the isothermal atmosphere ($T = 0°C$). It would be of interest to compute it for various model atmospheres, especially the "Standard Atmosphere" which is presently under consideration for general adoption. The differences in u for various model atmospheres will not exceed a few per cent as compared to the numbers given in Table VI.

Furthermore, the transmissivity is computed, which is defined as

$$\tau = e^{-u} \qquad (18)$$

and tabulated also in Table VI. The values u and τ in Table VI are valid only for the optical mass 1, for oblique paths through the atmosphere u has to be multiplied by the optical air mass m, tables for m can be found in the *Smithsonian Meteorological Tables*[6] and similar books.

VI. MEAN RAYLEIGH SCATTERING COEFFICIENT FOR VISUAL OBSERVATIONS

In all problems where eye observations are used, a single Rayleigh scattering coefficient, β_v, has to be used based on the luminosity function ψ_λ, of the light adapted eye of a "standard observer." In this case,

$$\beta_v = \int_0^\infty I_\lambda \psi_\lambda \beta_\lambda d\lambda \Big/ \int_0^\infty I_\lambda \psi_\lambda d\lambda, \qquad (19)$$

where I_λ is the intensity of the solar radiation outside the atmosphere. The absorption by atmospheric gases in the visible region is negligible.

It is again better to compute the mean scattering cross section

$$\sigma_v = \int_0^\infty I_\lambda \psi_\lambda \sigma_\lambda d\lambda \Big/ \int_0^\infty I_\lambda \psi_\lambda d\lambda. \qquad (20)$$

[5] G. Grimminger, Rand Rept. R-105 (Rand Corporation, Santa Monica, California, 1948).

[6] R. J. List, *Smithsonian Meteorological Tables* (Smithsonian Institute, Washington, D. C., 1951), 6th revised edition, pp. 442–443.

The result gives $\sigma_v = 4.578 \times 10^{-27}$ (cm²) if for I_λ the intensity as given in the *Smithsonian Meteorological Tables*[7] is used. This corresponds to the σ_v value for $\lambda = 0.55\,\mu$, i.e., the maximum sensitivity of the eye. *For β_v we obtain at 0°C and 1013.25* mb $\beta_v = 1.23\times 10^{-7}$ (cm⁻¹) *or 0.0123* (km⁻¹) NTP, a value which agrees very well with $\beta_v = 0.0126$ (km⁻¹) as given by Tousey and Hulbert[8] and which is widely used. The difference is insignificant for practical applications. It should be pointed out, however, that this value can be used only under NTP conditions, for other temperatures and pressures the value N has to be computed and applied to that at any condition

$$\beta_v = \sigma_v N = \sigma_v (p/kT). \tag{21}$$

The mean angular Rayleigh scattering cross section $\sigma_{R\theta}$ for visual observation is described in Sec. 3 and tabulated in Table IV.

[7] See reference 6, p. 415.

[8] R. Tousey and E. O. Hulbert, J. Opt. Soc. Am. **37**, 78 (1947).

LIGHT SCATTERING IN THE EARTH'S ATMOSPHERE

(An essay on the 150th anniversary of the discovery by Arago of the polarization of light in the daytime sky, and on the 100th anniversary of the discovery by Govi of the polarization of light in scattering)

G. V. ROZENBERG

Usp. Fiz. Nauk **71**, 173–213 (June, 1960)

CONTENTS

1. Introduction 346
2. The Puzzle of the Daytime Sky and the Discovery of Light Scattering 347
3. Description of Light and the Properties of the Medium in the Scattering Act 352
4. Atmospheric Transmission and the Aerosol 354
5. The Polarization Map of the Sky; the Anisotropy of Molecules and Multiple Scattering 358
6. The Brightness Map of the Sky and the Scattering Function 361
7. Optical Probing of the Atmosphere and the Problem of Interpretation of the Data 364
8. Propagation of Light in Clouds and Fogs and Similar Problems 368
9. Radiation Climatology and the Optics of the Aerosol 368

1. INTRODUCTION

ATMOSPHERIC optics belongs to the number of ancient sciences whose origins are lost in prehistoric time. The various light phenomena in the atmosphere — the varying blue of the sky, the rosy dawn, rainbows and the extraordinary halo, fantastic mirages — have long capitvated poetic imagination, converting them into objects of religious cults. But even in ancient memorials of material culture and in the evidence of the historians, there are found traces of searching thought, attempting to discover the real nature of phenomena behind mystic coverings. In every age, very promising minds pursued this field. Thanks to them, atmospheric optics invariably occupied a conspicuous place in the process of understanding nature, never being relegated to scientific obscurity. It is true that its development was frequently marked by great mistakes and curious happenings, but it was no exception in this respect. It is far more important that it is connected with many great discoveries, which had a decisive effect on the history of science. For example, it suffices to recall it was precisely from attempts to explain the blue color of the sky (see below) that there arose one of the most important and widest fields of contemporary physics — the study of the scattering of radiation by matter. In the same way, measurements of transmission in the earth's atmosphere, as is well known, led at the beginning of our century to one of the most decisive proofs of the existence of molecules and the validity of the kinetic theory of gases. Also widely known is the role played in the development of spectroscopy by the discovery of selective absorption of light by the atmosphere.

It is not difficult to recall a number of similar examples. It is also not difficult to see that a basic factor in determining this role of investigations in the region of atmospheric optics is the scale of the earth's atmosphere, which makes it possible to carry to completion much finer observations that can be obtained (with the same measuring techniques) under laboratory conditions. It is just this circumstance that defines what to our point of view is the basic mark of atmospheric optics as a science. Setting aside the numerous and often very important problems of an applied character, we see that the development of atmospheric physics has always been directly connected with the leading advances of theoretical concepts and experimental technology, and sometimes has anticipated them, and that the fundamental difficulties of this science have always been the difficulties of physics in general. In particular, we recall that, in connection with the studies of atmospheric optics and the optics of planetary and stellar atmospheres, the theory of propagation (transfer) of radiation in a scattering medium sprang up and received its development, acquiring today great value in applied nuclear physics and already becoming a fundamental branch of mathematical physics. It is also impossible not to recall the direct connection of the development of the study of turbulence with investigations on the twinkling of stars and other distant light sources. Finally, we shall quickly see that the fundamental problems of the study of the optical properties of the atmosphere are not only tightly bound with the most important problems of colloidal optics, but are very close (methodologically, at any rate) to the fundamental problems of contemporary nuclear physics. In particular, this leads to the result that the solution (and even the correct formulation) of a whole series of, it would appear, especially classical problems of atmospheric physics would be impossible without bringing in the most advanced ideas of contemporary theoretical physics.

The practical value of atmospheric optics is also much broader than may appear at first glance. The optical properties of the atmosphere determine to a significant degree its optical and thermal states, and indeed the optical and thermal states of the earth's surface for sowing, engineering construction, etc. Knowledge of these states and skill in forecasting their change are highly necessary, both for the development of methods of weather forecasting and for the solution of all problems connected with conditions of visibility, illumination and exposure, including urgent problems of transport, construction, medicine and agrobiology. Moreover, optical methods can serve in the study of the atmosphere itself (including the stratosphere, which is scarcely accessible by other methods) and for the study of processes taking place in it. This has direct meteorological importance and acquires fundamental importance in connection with the development of stratospheric means of transportation and with the increasing necessity of the investigation of planetary atmospheres from without, in order to guarantee the possibility of penetration through them (to the surface of the planets) in space ships.

However, if we turn to the development of atmospheric optics in the last half century, a rather unusual picture is revealed. On the one hand, during this time, especially during the last two decades, our knowledge of the atmosphere and, in particular of its upper layers, experienced a remarkable progress, requiring a radical revision of most of the representations that had only recently become widely adopted. At the same time there was a vigorous intrusion of contemporary physico-mathematical means of investigation into the region of atmospheric physics, thus greatly changing the aspects of this science. On the other hand, the traditional aims and directions of investigation scarcely experienced the spirit of the time. The concepts of the modern investigations essentially preserve the same features as at the beginning of the century, differing chiefly in the scale and technology of the accomplishments, and not in the formulation of the problems.

Such a conflict between old-fashioned tendencies and the extreme modernization of methods of their accomplishment has quite a regular foundation, which will be clear from what follows. However, this does not do away with the necessity of surmounting it by means of a serious review both of the aims and methods of atmospheric-optical investigations, the more so that this conflict finds its concrete expression in a whole series of clearly painful facts.

In particular, contemporary atmospheric physics is characterized by a complete break between theory, which is rarely very refined, and experiment, occasionally very fine and highly perfected. As a rule, they developed very vigorously, but practically independently of one another, missing the decisive stage of mutual control. In addition, a situation often came about in which the cost of the experimental or theoretical investigations significantly exceeded the value of the results obtained — in some cases, it was the material expression of the lack of correspondence of the general direction of investigation, its methodology and the nature of the object. Below we shall establish that this lack of correspondence was due in the first place to a lack of those special characteristics with which the transition from passive, chiefly qualitative observations of natural phenomena was made to purposeful, quantitative analysis of its physical nature under conditions of controlled variability of the very object of investigation — the atmosphere.

Thus, in the process of its very intensive development, atmospheric optics came in very definite fashion to a boundary which required serious reconsideration both as to program and to methods of investigation. Therefore, it is appropriate to discuss briefly, but in historical aspect, the modern position of this science, which is the purpose of the present paper. The necessity of a detailed review is heightened by the fact that in recent years the leading role in atmospheric-optical investigation has been more and more clearly assumed by Soviet scientists, and to the fact that, notwithstanding this, for some years the problems of atmospheric optics have remained practically untreated in domestic literature. However, we shall limit ourselves to only one rather large group of problems, namely, problems touching on the properties of the atmosphere as a scattering medium and leading directly to the two notable discoveries mentioned in the subtitle.

2. THE PUZZLE OF THE DAYTIME SKY AND THE DISCOVERY OF LIGHT SCATTERING

Studies of the color and polarization of the daytime, cloudless sky in connection with the discovery and explanation of the phenomenon of light scattering form one of the most interesting pages in the history of science. One can find a detailed description, although somewhat unclear and in much already archaic, of the early stages of this investigation — approximately up to the first decade of the present century — in well-known texts of atmospheric optics, which have long become bibliographical rarities.[1,2]

Therefore, we shall dwell only on certain more important moments, without the knowledge of which it is not possible to make a correct estimate of the present position of atmospheric optics as a science.

It is difficult to show when the idea first arose that the brightness and color of the daytime sky were brought about by "reflection" of sunlight from the air or from particles contained in it — droplets and dust.

We find it already clearly formulated by one of the pioneers of experimental physics Al-Hazen[3] (XI century) and later by Kepler[4] and Leonardo da Vinci.[5] Newton[6] held these views and, following him, Mariotte, Bouguer, Euler,[7] Goethe,[8] Clausius[9] and others.

Such a point of view has been the prevailing one but by no means the only one, and even in our time there have been attempts to oppose it, if only in part by other representations. Thus, several authors, among them Brewster and others, and at the end of the XIX century Chapuis[10] and Spring,[11] have suggested that the dark blue of the sky is brought about completely by selective absorption of the light by the air or by solid or gaseous impurities contained in it (for example, ozone). Other authors — Lalleman,[12] Hartley,[13] et al., and more recently Cohn[14] and, even in 1953, a group of American authors,[15] attempted to find the reason for the blueness of the daytime sky in various phenomena of photo- and cathode-luminescence. Finally, for a long time, right down to the eighties of the last century, i.e., much later than the completion of the first spectrophotometric measurements of the cloudless sky, there existed hypotheses that the reason for the blueness lay in the peculiarities of the observer's perception, and had no physical character. In particular, in 1885, Pickering[16] published a description of researches especially set up for the purpose of proving this hypothesis.

However, one must not consider these hypotheses, which are clearly incompatible with the contemporary viewpoint, only as historical curiosity. Their popularity and endurance before the critical objectors testifies conclusively to the serious difficulties which arise in path of explanation of the illumination of the daytime sky by scattering of light. The object of most attention in this case was the color of the sky.

In passing, we note that still greater difficulties confronted the explanation of the blueness of the sea and other water bodies. From the time of Bunsen, who turned his attention to the sharply expressed selectivity of the absorption capability of water, which causes its greenish blue color in transmitted light, the viewpoint was established that the color of the sea is explained by precisely this circumstance. Aufsess[17] and Pietenpol[18] adhered persistently to this concept at the threshold of the XX century, emphasizing the possibility of complementary coloring of the water by all sorts of impurities. After the researches of Schwartzchild, Schuster et al. on the theory of propagation of light in scattering media, it was not difficult to see that such an explanation was insupportable, because, in the absence of scattering, the depth of the tank would not have reflected the light and it would have appeared black, independent of the absorption spectrum of water. However, the contrary point of view, according to which the color of the sea would be explained by scattering (by molecular impurities or impurities contained in the water — bubbles, plankton, etc) even though advanced by a number of authors,

appeared to be even less viable. From these theoretical considerations, it followed that in the absence of absorption, any law of scattering leads to the independence of the reflection coefficient of the container on the wavelength of light, i.e., the ocean would be white like the clouds.

Actually, as is well known, the color of the sea and other bodies of water is determined by the combined action of both factors. This was first shown in 1921 by Raman[19] and Shuleĭkin,[20] independently, and a rigorous theoretical analysis was completed only in the forties by Ambartsumyan, Chandrasekhar, Sobolev, and others.[21] With this, the role of fluorescence in the coloring of bodies of water remains only partially clear at the present time.

We shall not dwell on the first attempt to penetrate into the secret of the explanation of the blue sky, advanced by Leonardo da Vinci. This attempt proceeded from an opinion, which was widespread in his day, that the color is determined by the proportion in which light and darkness are mixed, and is not of interest today. In the same way, we shall discuss the hypothesis according to which the color of the sky is determined by the actual coloring of the particles of the air or of impurities contained in it. This hypothesis, which was advanced especially by Euler, was refuted by the difference of coloring of the atmospheric air in transmitted and reflected light, and it played no significant role in the XIX century.

The viewpoint of Newton was more widespread in the first half of the XIX century; according to this the blue color of the sky arises as the result of the interference of light upon its reflection from tiny water droplets contained in the air (like the interference colors of thin film). In the middle of the XIX century, Clausius[9] advanced against the hypotheses of Newton objections which were very weighty for their time but, as we now understand, were not always justifiable. Fundamentally, they reduce to the following. If water is actually contained in the air in the form of droplets, thus causing the appearance of such intense interference coloring of the sky, then the diffraction phenomena on these droplets ought to lead to the formation of powerful coronas, i.e., to an appreciable washing out of the outlines of heavenly bodies, which is not observed in a clear sky. Moreover, it was shown by Clausius to be inexplicable how the droplets of water could float in the air and why, if these actually were droplets of water, a rainbow is not observed in clouds and fogs. This stimulated Clausius to make an attempt to save the hypothesis of Newton by replacing the water droplets in it by thin-walled bubbles with their characteristic interference coloring. Upon increase of humidity, the thickness of the walls of the bubbles would, according to Clausius, increase, bringing about a whitening of the light reflected by the bubbles. The discovery of light scattering soon compelled Clausius to give up his hypothesis, but for a long time it remained an object of discussion. In particular, as late as 1883, it was actively defended by Müller,[22] who made an attempt to improve the representation of the interference nature of the blue color of the sky by taking into account the effects of multiple reflection from the bubbles.

However, the imperfections of the idea that the atmosphere, with water droplets and bubbles constantly contained in it, was a turbid medium (colloid), and that the color of the sky was determined by conditions of "reflection" of light from particles producing atmospheric turbidity were not the only places to look for the reason for those difficulties which it encountered among the competing assumptions. A more important role in this was played on the one hand by the well known, and (at that time) sharp difference between the character of the phenomena observed in transmitted light through a turbid medium — clouds and fog — and the picture of the bright blue sky, and on the other hand, the total absence of laboratory experiments simulating in some degree the blueness of the sky. It must be noted that in the first quarter of the XIX century, not only were the most significant characteristics of rainbows and various halos well known and, fundamentally, correctly understood, but the diffraction nature of corona had been discovered by Jordan and considered in detail by Fraunhofer. We also note that not long before (in 1790), the first apparatus for quantitative measurement of the blue color of the sky was constructed by Saussure,[23] and the character of the change in color from the zenith to the horizon was made clear. In 1799, these measurements were extended by Humboldt, and were later frequently repeated in a number of variations by a whole series of authors right down to our own day. The differences discovered here between the optical phenomena in clouds and fogs and in a clear sky were so striking that there was almost no doubt remaining of the necessity of seeking explanations in the two cases.

The first attempt at a laboratory study of the optical properties of a turbid medium was undertaken by Goethe[8] and had as its purpose simulation of the blue color of the sky. However, the object of observation — the lower, non-luminous part of a spirit flame illuminated by the sun — was unsuccessfully chosen by him, and, in spite of the qualitative confirmation of the basic idea (on a white background a light blue was observed and on a dark background a dark blue color), the experiment of Goethe would not be considered as convincing, especially being "supported" by his mistaken ideas as to the nature of light.

Successful laboratory simulation of light blue color of the sky was first achieved by Brucke[24] in 1853. Irradiating a water emulsion of gum arabic and observing it against a background of a black screen. Brucke observed a bright blue color, while in transmitted light, the emulsion had a reddish-yellow color. By the same means, Brucke demonstrated experimentally that one could seek in the optical properties of the emulsion explanations of the coloring of the vault of the sky and the characteristic reddening of the sun upon its approach to the horizon.

In 1860, Govi[25] carried out investigations with smoke from alcohol and tobacco, in which it was shown that color effects are produced in the scattering of light in smoke similar to those observed in the scattering of light by clouds and fogs. Finally, in 1869, that is, sixteen years after the experiments of Brucke, Tyndall[26] completed his brilliant experiment (which rapidly achieved wide fame) with the so-called "actinic clouds." The substance of these experiments was that as a result of chemical decomposition of the vapors of certain compounds in an illuminated vessel, an aerosol cloud was obtained, the particles of which gradually increased in size. In particular, the growth of the particles was manifest in changes of the coloring of the light scattered by them — a gradual transition was observed from azure blue, which reproduced completely the coloring of the southern sky, to white, characteristic of clouds, smokes, and fogs.

Thus the experiment of Brucke, as Clausius[27] correctly noted in 1853, demonstrated that the reflection of light from small particles takes place according to other laws than that from massive bodies, i.e., that in the case of small particles there is no longer a reflection of light, but a certain phenomenon which has obtained the name of scattering. Furthermore, the experiments of Brucke, Govi, and Tyndall have shown that the character of the scattering depends significantly on the character of the scattering particles, especially on their dimensions. Thus strong arguments appeared against the theory of Newton and Clausius, which the latter had pointed out.[27] However, the experiment of Brucke did not yield decisive arguments in support of the explanation of the brightness and color of the daytime sky by scattering. The blue of the light scattered by a colloid merely gave evidence of the admissibility of a similar explanation, but nowhere did it prove its validity. One had to seek the proof in other directions, and one did not have long to wait.

Even in 1809, François Arago,[28] viewing the daytime sky through a Nicol prism, discovered that the light coming from the sky was strongly polarized. Not only was the light of the sky polarized in this case, but also the light of a haze separating an observer from distant objects such as mountains. Therefore, there was no doubt that the very illumination of the air (or of impurities contained in it) was polarized. During the next 150 years, this phenomenon has served constantly as the object of persistent and numerous investigations, the results of which will be given below. But all attempts at its explanation by starting out from representations of the reflection of the sun's rays by particles of air or of impurities contained in it, put forth by Arago, Babinet, Brewster, Clausius and others, proved to be unsuccessful. In particular,

this was in its time a strong argument against the Newton-Clausius picture. The situation was no better relative to competing hypotheses, of which we mention only the hypothesis of Brewster,[29] actively supported by Rubenson.[68] Starting out from the fact that the maximum of the polarization of light of the daytime sky, discovered by Arago and investigated in detail by Brewster and Rubenson, made an angle φ with the vertical to the sun of about 90° (this corresponds approximately to twice Brewster's angle at the boundary of two media with similar indices of refraction), These authors expressed the conviction that the reflecting substances were not foreign particles but the air molecules themselves.

The solution of the intriguing puzzle of the polarization of the illumination of the daytime sky required a full half-century and was first obtained in the researches of the Turin astronomer Professor Govi, to which reference has already been made.[25] The investigation, the results of which were given in the form of two short letters to the Proceedings of the French Academy of Sciences, was undertaken with the aim of showing that the illumination of the daytime sky was brought about by the scattering of light, and that the illumination of comet tails, the light of which is also polarized, could be explained by the same phenomenon. Govi filled a closed and carefully shaded room with different smokes and directed a beam of sunlight into it through a slit in the shutter. Observing the luminous shaft through a polariscope, Govi discovered that the scattered light was strongly polarized, and that the maximum of the polarization did not occur at a 90° scattering angle. By subsequent experiments, Govi established the fact that if the scattering angle is changed over a wide range, then directions are observed in which the scattered light is completely unpolarized (neutral points according to the terminology of Arago) and intervals of angles are observed where the plane of polarization is rotated by 90° relative to the plane of scattering (negative polarization according to the terminology of Arago). In Sec. 5 we shall see that both these circumstances are characteristic of the illumination of the daytime sky (as demonstrated by Govi) and were first observed by Arago. On the basis of his experiments, Govi definitely concluded that the scattering of light by small particles had nothing in common with the reflection of light and represented an independent phenomenon.

Ten years later, there appeared the famous work of Tyndall[26] under the title "On the Blue Color of the Sky, The Polarization of Sky Light, and On the Polarization of Light by Cloudy Matter in General." Together with the evidence already mentioned on the effect of the dimensions of scattering particles on the coloring of light scattered by them, Tyndall reported the results of his very careful polarization measurements. It was established by him that the character of polarization changes sharply with change in the dimensions of the particles. If the particles are very small (dark blue color of the scattered light), then the degree of polarization is large and its maximum is observed at a scattering angle of 90°. With increase in the dimensions of the particles (whitening of the scattered light), the degree of polarization decreases and the maximum is displaced to one side while, in accordance with Govi, "neutral points" are observed and also regions of "negative polarization." Now the light scattered by colloids under laboratory conditions and the light emitted from the sky displayed at least two common features — the character of color and the character of polarization. This gave Tyndall a sufficiently strong foundation for assuming the identity of these phenomena, which immediately attracted great attention to him.

Along with this, a discovery was made which was most important in its consequences — the discovery of the change of polarization (i.e., spin) of radiation in the act of its scattering by matter. As is seen from the above, this discovery was made in 1809 by Arago under field conditions, but it was understood a half-century later and only after the independent laboratory experiments of Govi. We add that history has been unfair to both scholars. Knowledge of the discovery of Arago, which forms essentially one of the cornerstones of the organized structure of modern molecular optics, escaped the attention of later generations. The brilliant experiments of Tyndall not only underscored the remarkable discoveries of Brucke and Govi, but even eclipsed them — the name "Tyndall effect" was undeservedly but firmly attached to the phenomenon of light scattering by colloids.

In addition, the experiments of Tyndall again awakened an active interest in the problem of the daytime sky. Scarcely two years had passed before Rayleigh (then still Strutt) began the publication of a series of papers devoted to a theoretical discussion of this problem. These papers formed the basis of the future science of the scattering of radiation by matter.[30] Here Rayleigh started out from the concept that the scattering of light took place (similarly to the experiment of Tyndall) on small particles suspended in the air (the particles were supposedly of spherical shape), and considered light waves as waves in an elastic ether. Later, in 1899, he revised his theory on the basis of the electromagnetic theory of light,[31] which, however, did not change the fundamental conclusions: 1) outside of absorption bands of scattering particles, the intensity of light scattered by particles whose dimensions are much smaller than the wavelength of the light λ is proportional to λ^{-4}, and 2) the degree of polarization p of the scattered light depends upon the angle of scattering φ, and is equal to

$$p = \frac{\sin^2 \varphi}{1 + \cos^2 \varphi} \, . \tag{1}$$

Naturally, an attempt was immediately made to compare these conclusions with the observation of the daytime sky. In part, so far as the coloring of the daytime sky is concerned, such measurements were first completed by Rayleigh himself, and, after him, by Vogel, Krov, Tsetvukha, Bock and, especially careful measurements in 1886 by Abney and Festing. The measurements showed that in very clear weather the spectral distribution of the brightness of the daytime sky corresponded approximately to $\sim \lambda^{-4}$, but that, even under the most favorable conditions, the departures from this relation were considerable. As far as polarization is concerned, the polarization picture of the sky, which was rather well studied at this time, as a whole corresponded approximately to Eq. (1), departing, however, from theoretical expectations in a number of important details. Finally, spectral measurements of the transmission in the atmosphere which were completed, in particular by Abney and Festing, and also by Becquerel in the eighties of the last century, again demonstrated that, for very high transmission in the atmosphere, the atmospheric attenuation of light changes with the wavelength also in approximate correspondence with the Rayleigh law $\sim \lambda^{-4}$ (in contrast to the proportionality λ^{-2}, following from the theory of Clausius), but with very appreciable and variable departures from it. Below we shall return to a discussion of the nature of all of these deviations; here we shall only note that such excellent, semi-quantitative agreement of the spectral and polarization characteristics of the light emitted from the sky with theoretical predictions could be regarded as convincing evidence in support of the validity of the explanation of the illumination of the sky as the scattering of sunlight from small particles suspended in the air.

However, Rayleigh in 1899 gave up these assumptions and went over to the hypothesis that the scattering particles were the air molecules themselves.[31] This step, which returned to the ideas of Brewster, was all the more daring since Tyndall directly confirmed that, according to his measurements, air, dust particles and water droplets do not scatter light at all.[26] As is well known, an important discussion between L. I. Mandel'shtam and M. Planck[32] quickly arose over this hypothesis, the result of which was the creation in $1908-1910$ of the fluctuation theory of light scattering.[33] In particular, Planck showed that, in spite of the error discovered by Mandel'shtam in the initial assumption of Rayleigh on the incoherence of waves scattered by the separate molecules, the connection established by Rayleigh between the scattering power of the medium and its index of refraction was valid. He also obtained a reliable foundation of the theory of light scattering in finely dispersed colloids put forward in 1904 by Maxwell-Garnett,[34] which qualitatively explained the color of colloidal solutions. The starting point in this theory was the assumption that, if the colloidal particles and the distances between them are small in comparison with the wavelength, then one can attribute an index of refraction to the medium which will be connected with the polarizability of the individual particles by the Lorentz-Lorenz formula. Thus the content of the theory of Maxwell-Garnett was essentially reduced to consideration of cooperative (namely, dispersive) effects in scattering on a set of colloidal particles and the region of its applicability was limited only to extremely finely dispersed and very concentrated colloids, as a consequence of which this theory played no role at all in atmospheric optics.

The fact that the Rayleigh theory gave a correct qualitative description of the fundamental features of the illumination of the bright daytime sky — its color and polarization, and also the spectral dependence of transmission in the atmosphere — by means of very simple and graphic formulas, immediately brought forth universal acknowledgment of this theory. However, its triumph, which not only confirmed the hypothesis of Rayleigh that the atmospheric scattering of light bore an essentially molecular character, but which also demonstrated its decisive effect on the process of establishment of the molecular-kinetic point of view generally, was the determination of the Loschmidt number from measurements of the transmission of light (see Sec. 4).

Nevertheless, there did not exist any basis for abandoning the idea that aerosol particles suspended in air have a significant effect on the optical properties of the atmosphere, the more so since not all facts were explained by the Rayleigh theory. In particular, the brightness picture of the daytime sky and also the existence in it of neutral points were by no means compatible with it.

In the same year (1899) when the Rayleigh theory apparently gave the keys for the solution of the puzzle of the daytime sky, the theory of light scattering on foreign particles was first set forth. Love[35] rigorously investigated the problem of the diffraction of electromagnetic waves on a sphere. Within nine years, Mie again solved the same problem. But, in contrast with Love, he did not limit himself to the mathematical side of the problem, but compared in detail the conclusions of the theory with the optical properties of the colloidal suspensions of metals, which immediately gave publicity to this theory among experimentalists.

Subsequently, the experimental and theoretical investigations of the phenomena of radiation scattering by matter were developed very intensively and rapidly became the object of one of the fundamental branches of contemporary science. But the connection with the problems of atmospheric optics was not lost, and the effect of atmospheric-optical problems on the development of the theory of scattering continued to be decisive in many connections. Discussion of this connection will be to a significant extent the theme of what follows. However, it is first necessary to consider what parameters are needed to characterize the scattering medium and the scattering act itself.

3. DESCRIPTION OF LIGHT AND THE PROPERTIES OF THE MEDIUM IN THE SCATTERING ACT

If we turn to the phenomena of light scattering, we must characterize the atmospheric air as well as each colloid by their ability to absorb and scatter light of different wavelengths. However, upon close examination, it is clear that this ability, generally speaking, depends not only on the properties of the air itself but also on the character of polarization of the light.[37-40] Therefore, it is first necessary to seek out such optical characteristics of the scattering medium (in particular, atmospheric air) which would reflect the properties of the latter independently of the character of the light field. It is shown that this is possible only in the case when a transition is made in the description of the absorbed and scattered light beams from the ordinary intensities of the electric and magnetic fields of the light waves to other, specialized parameters.

A basic feature which we encounter in considering the propagation of light in a scattering medium is the multiplicity of scattering acts which are mutually incoherent, as a result of which a constant mixing of light beams with very different previous histories takes place. Inasmuch as not only an angular redistribution of the intensity of the light waves takes place in each scattering act, but also a change in its polarization, the scattered light is an incoherent, statistical mixture of beams of very different intensities and with various states of polarization.

However, the character and the result of this scattering act depend strongly on the polarization of the scattered light. Therefore, in problems of the scattering of radiation, one must characterize the latter by parameters which are additive for incoherent light beams and which cover descriptions such as their energy and the state of their polarization. Such parameters were first suggested by Stokes[41] in 1852, i.e., even before the discovery of light scattering, and for a long time were disregarded "owing to their uselessness." Subsequently they appeared in different variants in the works of Rayleigh, Poincaré, Becquerel, Wiener, Soleil, and others in connection with certain "exotic" problems, but remained completely unknown to a wide circle of physicists. Interest in them was reborn only in the 1940's precisely in connection with problems of scattering and propagation of light in turbid media, and, most important of all, in the atmosphere (references 42 — 47 and others; for more detail see reference 38). It was quickly shown that the Stokes parameters are connected in most direct fashion with the quantum mechanical radiation density matrices and historically form their first formulation. In this connection, it is impossible not to note that even the correct statement of such a characteristically

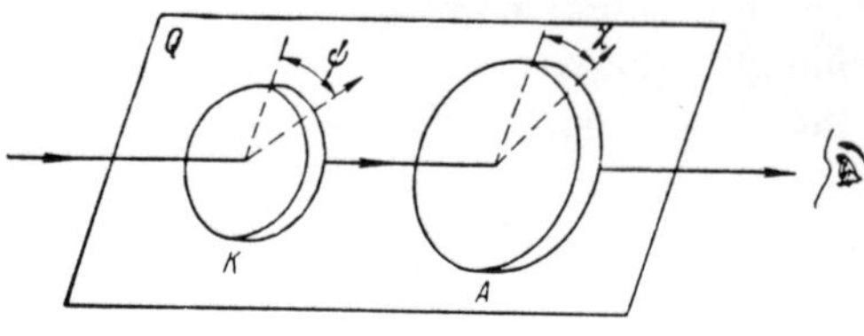

FIG. 1

classical problem as the multiple scattering of light in the atmosphere was completely impossible without taking on the most typical apparatus of quantum mechanics — the density matrix.

Bypassing the later generalization of the Stokes parameters (see, for example, reference 38), we can introduce them in their simplest form in the following way,[44,47] connecting them directly with one of the possible and most useful procedures of measurement. We shall assume that a compensator K, having a path length difference of a quarter-wavelength, and an analyzer A are placed successively in the path of the light beam (Fig. 1). We choose an arbitrary plane of reference Q, containing the direction of the ray, while the angles of rotation of the compensator ψ and of the analyzer χ about the direction of the light beam will be measured from the plane Q counterclockwise looking toward the ray. Then, the Stokes parameters of the light beam are, by definition,

$$\left.\begin{aligned}
S_1 &= I(\psi = 0, \ \chi = 0) + I(\psi = 90°, \ \chi = 90°), \\
S_2 &= I(\psi = 0, \ \chi = 0) - I(\psi = 90°, \ \chi = 90°), \\
S_3 &= 2I(\psi = 45°, \ \chi = 45°) - S_1, \\
S_4 &= S_1 - 2I(\psi = 0°, \ \chi = 45°),
\end{aligned}\right\} \qquad (2)$$

where $I(\psi, \chi)$ is the light intensity passing through the compensator and the analyzer for given values of the angles ψ and χ. It is easy to show[38] that

$$S_1 = I, \quad S_2 = Ip \cos 2\psi_0, \quad S_3 = Ip \sin 2\psi_0, \quad S_4 = Iq, \qquad (3)$$

where I is the total intensity of the light beam, p is its degree of polarization, q is the so-called degree of ellipticity of the polarization, and ψ_0 is the angle of rotation of the direction of maximum polarization relative to the reference plane Q (Figs. 1 and 2).

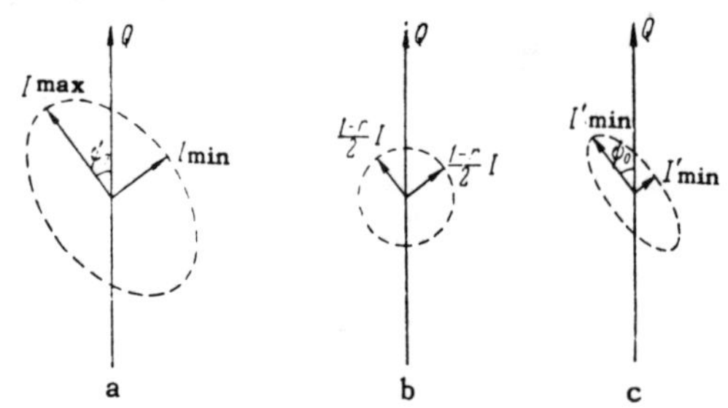

FIG. 2. a) Partially polarized beam

$$I = I_{max} + I_{min} = I' + I'', \qquad p = \frac{I_{max} - I_{min}}{I} = rp', \ q = rq'.$$

b) Depolarized components

$$I'' = (1 - r) I.$$

c) Completely polarized component

$$I' = I'_{max} + I'_{min} = rI, \qquad p' = \frac{I'_{max} - I'_{min}}{I'}, \qquad q' = \pm \frac{2\sqrt{I'_{max} - I'_{min}}}{I'}$$

Generally speaking, an arbitrary partially polarized light beam of intensity I can be represented as the sum of two incoherent beams — a completely (generally speaking, elliptically) polarized beam of intensity rI and a completely depolarized beam of intensity $(1-r)I$ (Fig. 2). The quantity $r \equiv \sqrt{p^2 + q^2}$ is the <u>value of the polarization</u>[48] or <u>degree of homogeneity</u>[44] of the light beam.

The four Stokes parameters S_i ($i = 1, 2, 3, 4$) can be regarded as the components of a single Stokes vector-parameter $\vec{S}$ in four-dimensional functional space,[44,38] which materially simplifies the writing down of the formulas. Therefore, the different letter symbols for the Stokes parameters usually employed in foreign literature (for example, in reference 40) appear to us to be irrational, the more so since a universally adopted system of notation does not exist.

If the plane of reference is turned by an angle ψ' in a counterclockwise direction (looking towards the ray) then the components of the Stokes parameter $\vec{S}$ change their values:

$$S'_i = \sum_j K_{ij}(\psi') S_j, \tag{4}$$

where the transformation matrix K_{ij} has the form

$$K = \begin{pmatrix} 1 & 0 & 0 & 0 \\ 0 & \cos 2\psi' & \sin 2\psi' & 0 \\ 0 & -\sin 2\psi' & \cos 2\psi' & 0 \\ 0 & 0 & 0 & 1 \end{pmatrix}, \tag{5}$$

and the quantities $S_1 = I$, p, q and r will be invariant relative to the transformation (5). For details on the properties of the Stokes vector-parameter, see reference 38.

The light ray undergoes attenuation in passage through the absorbing and scattering medium, and in the case of anisotropy of the medium for the scattering particles filling it, a change in the character of the polarization results. Both this and the other are recorded as the change of the vector-parameter of the light beam in its passage through an element of f path:[38]

$$dS_i = -\sum_j \varkappa_{ij} S_j \, dl, \tag{6}$$

where $\varkappa_{ij}$ ($i, j = 1, 2, 3, 4$) is the so-called <u>extinction matrix</u>. In the case of an isotropic medium, it degenerates into a scalar — the <u>attenuation (extinction) coefficient</u> k:

$$\varkappa_{ij} = k\delta_{ij}, \tag{7}$$

where δ_{ij} is the Kronecker symbol (for details see reference 39).

We now turn to the description of the event of light scattering. We shall assume that the element of volume dV is irradiated in the direction 1^0 by a light beam with Stokes vector-parameter $\vec{S}_0$, and we shall consider the light beam scattered by the element of volume dV in a certain direction 1 (Fig. 3). As a

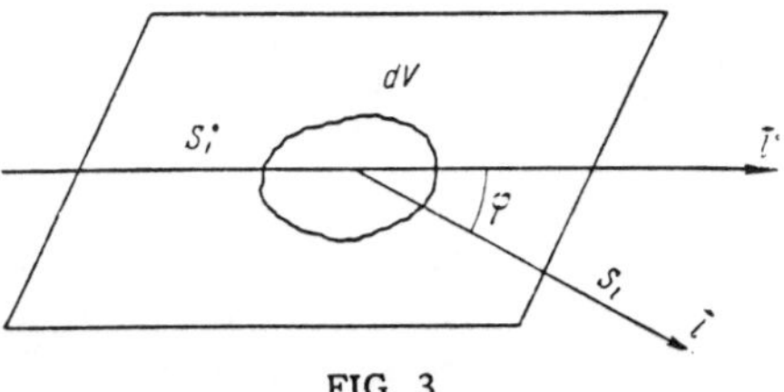

FIG. 3

reference plane Q both for the scattered and for the scattering beams, we shall take the plane of scattering which includes both the directions 1^0 and 1. Then, from the linearity of the equations of electrodynamics, and from the additivity of the Stokes vector-parameters for incoherent light beams, it follows[37,38,42-44] that the components of the Stokes vector-parameter of the scattering ($\vec{S}$) and radiating ($\vec{S}^0$) beams are connected by the relation

$$dS_i(1) = \frac{1}{r^2} \sum_j D_{ij}(1, 1^0) S_j^0(1^0) \, dV, \tag{8}$$

where r is the distance of the point of observation from the scattering element of volume, and D_{ij} are the components of a matrix of fourth rank which characterizes the scattering properties of the medium, independent of the state of polarization of the scattered light and referred to unit volume. We shall call it the <u>light scattering matrix</u> of the medium: its component D_{11} is called <u>coefficient of directed light scattering</u> or the differential transverse scattering cross section. If the scattering medium is isotropic (for example, as a result of a random distribution of orientation of anisotropic particles), then the components of the scattering matrix D_{ij} depend not on the directions 1^0 and 1, but only on the scattering angle φ between them (Fig. 3). In this (and only in this!) case is it possible[37-40] to introduce the concept of a <u>scattering coefficient of the medium</u> σ (or its total scattering cross section), independent of the state of polarization of the scattered light, by defining it as the fraction of the intensity of the light wave (referred to unit volume) incident on the scattering volume dV, which is scattered by the latter in all directions:

$$\sigma = \oint D_{11} \, d\omega, \tag{9}$$

where $d\omega$ is the element of solid angle of the scattered light beam. Then, in place of the matrix $D_{ij}(\varphi)$, we can introduce the <u>normalized scattering matrix</u> $f_{ij}(\varphi)$:

$$D_{ij}(\varphi) = \frac{\sigma}{4\pi} f_{ij}(\varphi), \tag{10}$$

where the component $f_{11}(\varphi)$, which satisfies the normalization condition

$$\frac{1}{4\pi} \oint f_{11}(\varphi) \, d\omega = 1, \tag{11}$$

in agreement with (9), is usually called the <u>scattering function or indicatrix</u>.

The energy carried away by the particles (or by the medium) from the wave irradiating it is not only scattered but is also partially absorbed, being transformed into other forms. It is evident that the fraction of the incident light intensity (referred to unit volume) absorbed by the medium, i.e., the coefficient (or cross section) of absorption α of the medium is determined as the difference of the attenuation and scattering coefficients

$$k = \alpha + \sigma, \qquad (12)$$

where α will be a scalar, generally speaking, only in an isotropic medium. We note that all three quantities in (12) have the dimensions of cm^{-1} or more conveniently, km^{-1} (the cross section referred to unit volume).

It is very important that the quantities (matrices) α, σ, and k be expressed in terms of the components of the scattering matrix $D_{ij}(\varphi)$ and the extinction matrix κ_{ij}, as a consequence of which the spectral and angular dependences of all $16 + 16 = 32$ components of these matrices exhaust all the information which can be obtained about the properties of the medium by studying the phenomena of scattering, absorption of light, or other radiation. Nevertheless, preference is usually given to direct measurements of the extinction coefficient k and, in certain cases, of the specific absorption coefficient $\beta = \alpha/\sigma$ (for further details, see reference 39).

Often it is convenient to consider the reduced scattering matrix in place of the matrices D_{ij} or f_{ij}; the components of this matrix, $\widetilde{f}_{ij}$, are given by the relation

$$\widetilde{f}_{ij}(\varphi) = \frac{D_{ij}(\varphi)}{D_{11}(\varphi)} = \frac{f_{ij}(\varphi)}{f_{11}(\varphi)}. \qquad (13)$$

The form of the scattering matrix $D_{ij}(\varphi)$ depends essentially on the properties of the scattering medium, in particular on the composition, dimensions, form and orientation of the particles suspended in it. The character of the anisotropy of the medium or the symmetry of the scattering particles is directly reflected in the number of independent and non-vanishing components of the scattering matrix.[40,43,44,49] However, the concrete form of the scattering matrix is known only for molecular scattering. For non-absorbing gases, with corrections for the anisotropy of the molecules in correspondence with the theory of Rayleigh-Cabanne (see Sec. 5), it has the form:[44,38]

$$f(\varphi) = \frac{3}{4+3d}\begin{pmatrix} 1+\cos^2\varphi+d & -\sin^2\varphi & 0 & 0 \\ -\sin^2\varphi & 1+\cos^2\varphi & 0 & 0 \\ 0 & 0 & 2\cos\varphi & 0 \\ 0 & 0 & 0 & 2\cos\varphi \end{pmatrix}, \qquad (14)$$

where the scattering coefficient (as the result of the absence of absorption, $k = \sigma$) is equal to*

$$\sigma = \frac{8\pi^3(n^2-1)^2}{\lambda^4 N}\frac{4+3d}{12-d}. \qquad (15)$$

Here φ is the scattering angle, N is the number of molecules per unit volume, n is the index of refraction of the medium, λ is the wavelength, $d = 4\Delta/(1-\Delta)$, and Δ is the depolarization of the scattered light at $\varphi = 90°$ and exposure of the scattering volume to linearly polarized light with an electric vector perpendicular to the plane of scattering.* It is essential that the plane of scattering is chosen as the plane of reference both for the incident and for the scattered beams.

In the case of scattering by spherical particles (Mie scattering), retaining the plane of scattering as the plane of reference Q for the incident and scattered beams, we have

$$f(\varphi) = \begin{pmatrix} f_1(\varphi) & f_2(\varphi) & 0 & 0 \\ f_2(\varphi) & f_1(\varphi) & 0 & 0 \\ 0 & 0 & f_3(\varphi) & f_4(\varphi) \\ 0 & 0 & -f_4(\varphi) & f_3(\varphi) \end{pmatrix}, \qquad (16)$$

i.e., the scattering matrix contains only four independent components. Both the scattering coefficient σ and the components of the scattering matrix are shown to be very sensitive to the wavelength, and the functions $f_i(\varphi)$ depend in complicated fashion on the scattering angle. The Mie theory makes it possible in principle to compute $\sigma(\lambda)$ and $f_i(\varphi, \lambda)$ for particles of a given dimension and with a given index of refraction. However, these calculations are actually so cumbersome that, in spite of the use of mathematical computers, they have been carried out only for a few cases, mostly in the absence of absorption and exclusively for σ (or k) and $f_1(\varphi)$ [in a few cases also for $f_2(\varphi)$]. As far as the functions $f_3(\varphi)$ and $f_4(\varphi)$ are concerned, they have never been computed up to the present time. It must also be kept in mind that qualitative discussions in this region, and also interpolation of computed data, must be used with extreme care in view of the very capricious character of the change of most of the computed quantities.

In addition to scattering by spherical particles, the problem of scattering by elliptical particles has been considered in principle. Other forms of particles have not yet been amenable to the theory. Further details on the character of the scattering of light on solid particles can be found in references 40, 50.

4. ATMOSPHERIC TRANSMISSION AND AEROSOLS

The transmission in the atmosphere in various parts of the spectrum is among its important optical characteristics. It determines the conditions of visibility of distant objects and, generally, the lighting conditions of our existence. On it also depend the radiative and thermal conditions both on the surface of the earth and in the atmosphere. It is not surpris-

---

*In reference 38, the factor 1/2 is incorrectly omitted in Eq. (57).

*Frequently the quantity $\rho = \dfrac{2\Delta}{1+\Delta}$ is introduced in place of Δ in the equations, i.e., depolarization at $\phi = 90°$ and irradiation by natural light, which is sometimes also denoted by Δ.

ing that, from the time of Bourguer and Lambert, who first definitely formulated the problem, and Saussure, who first attempted observations, a great deal of effort has been spent on measurements of atmospheric transmission. Setting aside the extensive and practically important group of operational and climatological problems, which go beyond the limits of this article, we consider the material accumulated during the course of two centuries from a single point of view only, namely, knowledge of the properties of the atmosphere itself.

The widest category of investigations is made up of the numerous observations and measurements carried out by meteorologists in the hope of getting some sort of connection between transmission and weather, and in this fashion to make it possible to forecast the latter. Lacking any sort of theoretical foundations and, as a rule, limited only to surface correlation estimates on the basis of comparatively short time observations, carried out only under favorable meteorological conditions, these measurements bore no fruit which would be of interest from our point of view. In addition to a certain quantity of curious observations, lost in a wide and varied collection of various cases not amenable to scientific analysis, one can extract from them only the conviction of the extremely and universally uncontrolled variability of the optical state of the atmosphere, even on completely clear days.

Such a paucity of results also exists for the repeated measurements of transmission, carried out by astronomers with the purpose of studying interferences which take place in their investigations of the earth's atmosphere. In spite of well-known facts, the variability of the atmosphere has generally been ignored in these measurements, and efforts have been directed toward the determination of some average (at best, seasonal transmission on clear days) somehow characteristic for a given observatory, i.e., quantities having no real importance, as we shall see below.

Against this background, which is most heterogeneous in purposes and methods of investigation, there stand out a comparatively small number of researches which are clearly directed toward the investigation of the optical properties of the atmosphere itself, and which are distinguished by their clearly thought out methodology and the perfection of the apparatus employed. The framework of this paper does not permit us to give a full account of each of these researches which left a notable mark in science. Below we shall only sketch rapidly contemporary information relating to transmission in the atmosphere and note some conclusions following from it. Here we shall put aside the broad and self-contained problem of the selective absorption of the gaseous phase of the atmosphere, including also its impurities, such as water vapor, ozone, and carbon dioxide.

We shall also not concern ourselves with the rather difficult and still far from satisfactory solution of the problems of methodology nnd technology of carrying out transmission measurements (see, for example, reference 51 and the literature cited there). We shall only remark that one usually measures either the attenuation coefficient k, averaged over a long or short distance (usually horizontal) or the vertical optical thickness of the atmosphere above the head of the observer (who is located at an altitude h)

$$\tau(h) = \int_h^\infty k(h)\, dh, \tag{17}$$

which is connected with the vertical transmission of the atmosphere at the same level by the relation

$$P(h) = e^{-\tau(h)} \tag{18}$$

and with the oblique transmission at an angle ζ to the zenith by the relation

$$T(\zeta, h) = P(h)^{m(\zeta,\, h)}, \tag{19}$$

where $m(\zeta, h)$ is the so-called air mass, which is accurately described (for not too large ζ) by the formula

$$m = \sec \zeta. \tag{20}$$

In place of τ, one frequently uses a quantity proportional to it, the vertical optical density of the atmosphere: $D = 0.43\,\tau$. We note that Eqs. (17) — (19) are valid only for monochromatic light or under conditions of the invariance of k within the measured wavelength interval.

The discovery of molecular light scattering raised the problem of determining the optical thickness of the earth's atmosphere, which is evaluated by means of this phenomenon. In fact, in accord with the Lorentz-Lorenz formula (see, for example, reference 52, page 42) for gases,

$$(n^2 - 1) = 4\pi N \left(\alpha + \frac{P_0^2}{3kT} \right), \tag{21}$$

where α is the polarizability of the molecules and P_0 is their constant dipole moment. Denoting by N_0 Loschmidt's number and by ρ_0 and n_0 the density and the index of refraction of air under normal conditions, and assuming that the composition of the air remains constant at least up through those altitudes where the attenuation of light (as a result of scattering) remains appreciable, we find that at any altitude

$$N = N_0 \frac{\varrho}{\varrho_0}, \quad (n^2 - 1) = (n_0^2 - 1) \frac{\varrho}{\varrho_0}. \tag{22}$$

Making use of Eq. (15), and taking it into account that the pressure at the point of observation is given by

$$p = g \int_h^\infty \rho\, dh,$$

we find, in accord with Eq. (17),

$$\tau(h) = \frac{8\pi^3 (n_0^2 - 1)^2}{\lambda^4 N_0 \varrho_0} \frac{p}{12 - d} \cdot \frac{4 + 3d}{12 - d}. \tag{23}$$

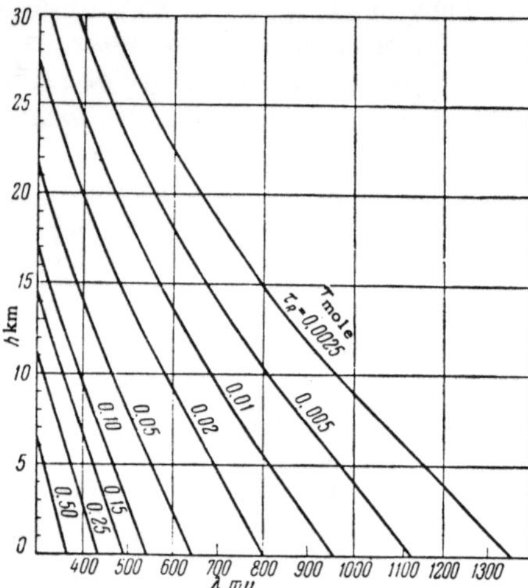

FIG. 4. Lines of equal optical thickness for an ideally pure atmosphere in altitude vs. wavelength coordinates.

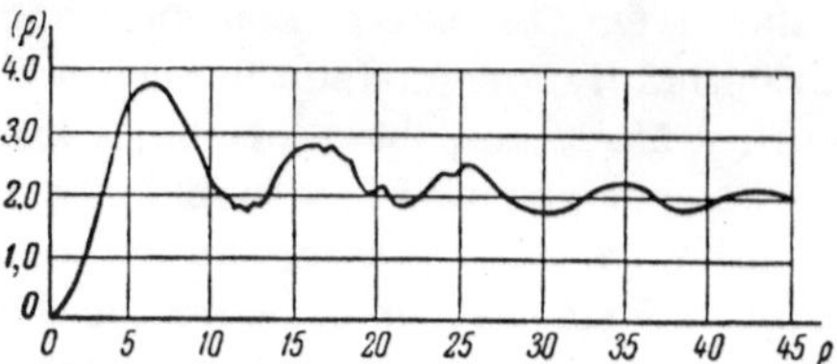

Corresponding values of τ as a function of the wavelength and altitude of the point of observation, according to the calculations of Penndorf,[53] are shown in Fig. 4, which takes into account contemporary data on the mean altitude variation of the atmospheric pressure. From (23) and from the absence of correlation between the value of the transmission and pressure at a fixed altitude of the point of observation, the conclusion immediately follows that aerosols, which are always present in some measure in the air (as we now know, at all altitudes at least up to 80 km), are almost completely responsible for the optical variation of the atmosphere (see below).

However, if we choose days with especially high transmission, then we can expect fulfilment of Eq. (23). In the early years of this century, the least accurate quantity in this formula was Loschmidt's number and, therefore, attempts were undertaken to determine it from this equation (without account of corrections then unknown for the anisotropy of molecules — see Sec. 5). Early calculations, both of Rayleigh and of Kelvin (1902), showed agreement that was completely satisfactory for that time (according to Kelvin, $N_0 = 2.47 \times 10^{19}$ instead of the actual value 2.67×10^{19}), which served not only as a first proof of the validity of the Rayleigh hypothesis on molecular light scattering, but was also a strong argument in favor of the molecular kinetic picture generally (see reference 54). However, a more careful analysis of the data (in particular, the excellent observations of the Smithsonian Institution over a period of many years) showed that atmospheric air was never free from aerosols, and that it was further necessary to introduce corrections for the selective absorption of the gas phase. Introduction of these corrections made it possible to improve the agreement appreciably — the error was lowered by Cabanne to ~ 8 per cent, by Kew to ~ 2 per cent, by Vassy to 1 per cent and recently by Toropova to ~ 0.5 per cent. At the same time it was shown that by com-

puting the part caused by molecular scattering and selective absorption of the gas phase from the total optical thickness (or the attenuation coefficient k) one could correctly determine the absorption due to the aerosol.

Such a separation of the aerosol components of k or τ, carried out by a number of authors, immediately brought to the forefront its variability both in absolute value and in spectral dependence. According to the Mie theory for spherical particles* $k = N\pi a^2 K(\rho)$, where N is the particle concentration and a is thin radius, $\rho = (2\pi a/\lambda) m$, and m is the relative index of refraction of the matter composing them. In particular, for water droplets (without consideration of dispersion) the function $K(\rho)$ has the form[55] shown in Fig. 5. It is at once evident that the spectral variability of atmospheric transmission is connected with the change in the character of the particles suspended in the air, particularly, the change in their dimensions. As the measurements of Vassy,[56] Rodionov[57] and others (for details, see reference 51) have shown, the aerosol attenuation of light often has a sharply expressed selective character. In the case of a semi-dispersed aerosol, which we usually encounter under ordinary conditions, the selectivity of the attenuation should be more or less smoothed out. If we average the data of measurements of different authors or of a single author for a series of days, then the aerosol attenuation of light is found to be independent of the wavelength. Since the aerosol is added to the air, which scatters in proportion to λ^{-4}, its effect is seen to be much greater in the long-wave than in the short-wave region of the spectrum. Therefore, under ordinary conditions, the spectral dependence of k or τ is weaker than follows from the Rayleigh law (the sky has a whitish light), and can sometimes be approximately expressed in the form $k \sim \lambda^{-n}$, where n varies from zero to 2 or 3.

On cloudless days, the attenuation coefficient of atmospheric air generally increases with increase in altitude. At low altitudes (up to $0.5 - 1$ km), this falling off takes place rapidly, chiefly as a result of the decrease in concentration of dust or water droplets, raised up by local air streams flowing over the earth. This decrease is then rapidly slowed and sometimes even changes to a weak increase up to an altitude of $3 - 5$ km, i.e., approximately to the upper boundary

____

*In the visible portion of the spectrum, cooperative effects (see reference 39) which can alter the formula given for k are not important for aerosols, and for molecular scattering play a role only at altitudes of around 100 km.

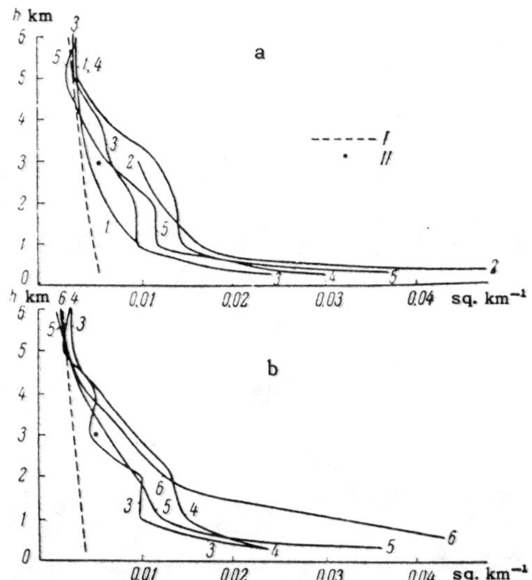

FIG. 6. Mean values of the attenuation coefficient k at different altitudes. 1 – Moscow oblast' (1956-57), 2 – Northern Kazakhstan (1956), 3 – Khar'kov oblast' (1957), 4 – Northern Caucasus (1957), 5 – Moscow oblast' (1958), 6 – West Germany (1944), I – Absolutely pure atmosphere, II – on the slopes of Elbrus (1957).

of the convective layer. Beyond that point, k decreases with altitude on the average, approximately according to the barometric formula, with frequent and very changeable disruptions as a result of aerosol (clouds) layers. As an example, we give the data obtained by Faraponova for measurements with an airplane[58] (Fig. 6). For measurements on mountain slopes, the altitude dependence is shown on the average to be more gradual than in the free atmosphere, without protruberances in the 3 — 5 region (Fig. 7), which makes it possible to explain the effect of the spreading out of the surface and mountain-valley circulations.[51] It remains to add that in the absence of clearly expressed cloud layers, the relative concentration of aerosols above 3 — 5 km varies but little with altitude and that on especially clear days the share of the total light attenuation of the atmosphere due to aerosols amounts to 30 — 50 per cent, which also explains the success of the first determinations of Loschmidt's number.

Aerosols, which include the liquid-drop phase of water, belong to the most variable component of the atmosphere both in a quantitative and in a qualitative

sense. Numerous measurements, both of the transmission and the brightness and polarization of the daytime sky (see Secs. 5 and 6) testify to the fact that the aerosol is distributed in the atmosphere not in homogeneous layers but in the form of separate flocculent accumulations, which are carried along by the wind and which undergo constant qualitative changes, including the results of condensation processes. Therefore, the picture of a horizontally homogeneous atmosphere does not correspond to reality even on very clear and calm days. This picture finds its clear expression, in particular, in the variability already noted in the transmission with respect to time, in the dependence on the direction (azimuth) of observation, and in the displacement along the earth's surface.[51] The absence of correlation between the variability of the transmission at different points in the spectrum is quite characteristic, and supports not only the quantitative but also the qualitative variability of the aerosol. These effects are exhibited in Fig. 8.[51] No less characteristic is the absence (in spite of the prevailing point of view) of correlation between horizontal and vertical transmissions of the atmosphere.[51]

If we now return to cases of strong turbidity of the air — clouds, fogs, precipitation, etc, then the given observations present a still more varied picture, which finds its natural explanation in the extreme variation of the microstructure of these formations. In general, there is now no doubt that all the effects observed here lie within the framework of the theory advanced by Mie (see, for example, references 40, 50, 59-61). However, as a result of the characteristic semi-dispersed character of the water aerosol and the varied nature of its size distribution, the observed phenomena lend themselves to quantitative analysis only with difficulty. In this case the chief role is played by barriers of two kinds. Firstly, the calculations of attenuation coefficients by the Mie theory far from cover all cases which one encounters in practice in water droplets. In particular, data referring to water absorption bands are almost completely absent, while very appreciable and peculiar anomalies are known to exist.[62] In the second place, we have no reliable data on the sub-microscopic fraction of water droplets, which is most active in optical behavior. Moreover, measurements under conditions of very high turbidity are complicated by effects

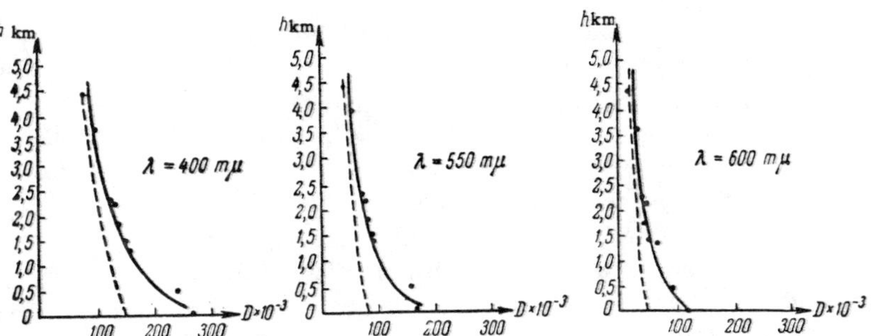

FIG. 7. Dependence of the mean vertical optical density of the atmosphere on altitude above sea level according to the measurements of various authors on the slopes of mountains. Dashed line — dust-free atmosphere.

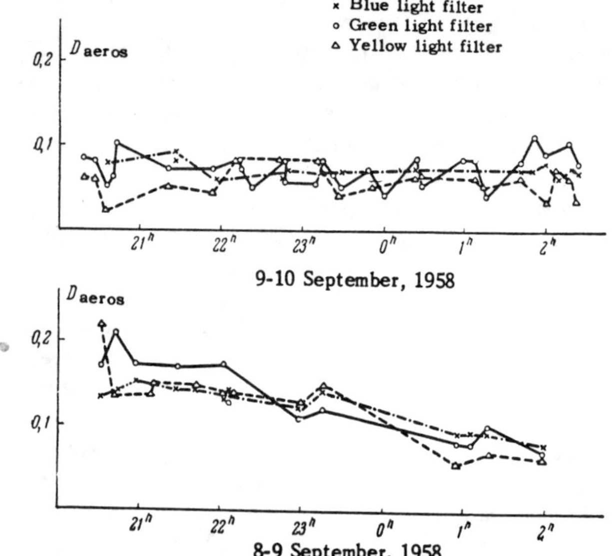

FIG. 8. Examples of variation of aerosol component of vertical optical density of the atmosphere in different parts of the spectrum.

of multiple light scattering, which are difficult to separate. Therefore, numerous efforts directed toward the use of data on transmission for the determination of the microstructure of clouds and fog or of atmospheric aerosols in general have not yet led to expected results, although they have created confidence in the significance of this problem.

Such, in general terms, are the results of 200 years of investigation of the transmission qualities of the atmosphere in the visible portion of the spectrum (for details, see references 40, 50, 51). We shall now consider what can be extracted from the data on the illumination of daytime sky.

5. THE POLARIZATION PICTURE OF THE SKY; THE ANISOTROPY OF MOLECULES AND MULTIPLE SCATTERING

The observation of Arago not only made possible the discovery of light scattering, but also had a permanent value in itself. At the same time, it raised questions whose solution required all the power of the physics of the twentieth century and directly stimulated the development of two of its important sections — molecular optics and the theory of transfer of radiation in turbid media.

Having discovered the polarization of the light received from the firmament, Arago went on to plot the polarization map of the sky and developed almost all its characteristic peculiarities. He established the fact that the maximum polarization is observed at an angle of about 90° to the direction of the sun's rays, and that (expressing it in our modern language) the plane of polarization coincides with the plane of scattering ("positive" polarization). He also discovered (in the vicinity of an anti-solar point) a region where

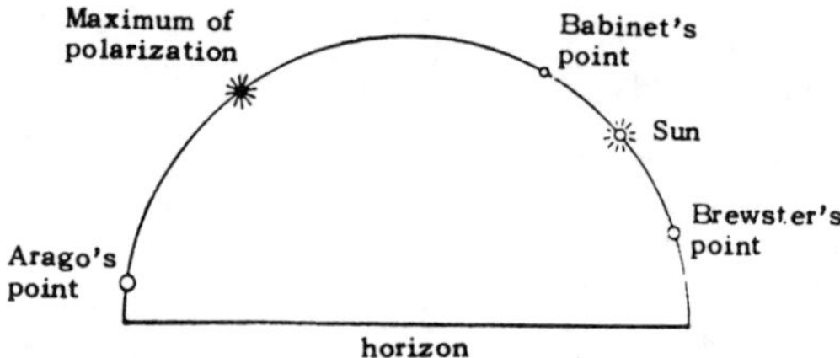

FIG. 9. Position of the neutral points and maximum polarization in the vertical of the sun.

the plane of polarization is perpendicular to the plane of scattering ("negative" polarization), while at the boundary of these two regions a "neutral point" is located in the vertical of the sun, from which completely depolarized light emerges. A similar picture is observed at night in moonlight, as was established by Arago, and confirmed 80 years later by Cornu and Pilchikov. The magnitude of the maximum polarization and the position of the neutral point, according to the observations of Arago, depend strongly on the state of the atmosphere, while sometimes the neutral point is shifted in the direction of the normal of the sun. In 1840, Babinet added still another neutral point to this picture, located not far from the sun, and later Brewster discovered a third neutral point located under the sun (Fig. 9).

Subsequent observations, pursued diligently up to our own day, have merely confirmed and made somewhat more accurate this qualitatively correct picture, and have added to it almost nothing new, although among the investigators we meet (in addition to the above) such names as Becquerel, Weber, Wild, Jensen, Forno, Süring, Pernter, Tikhanovskiĭ, and many others. Perhaps the only discoveries which contain points of importance were: 1) the observation by Pilchikov[63] (1892)(which is completely unnatural from the point of view of the Rayleigh theory) of the dependence

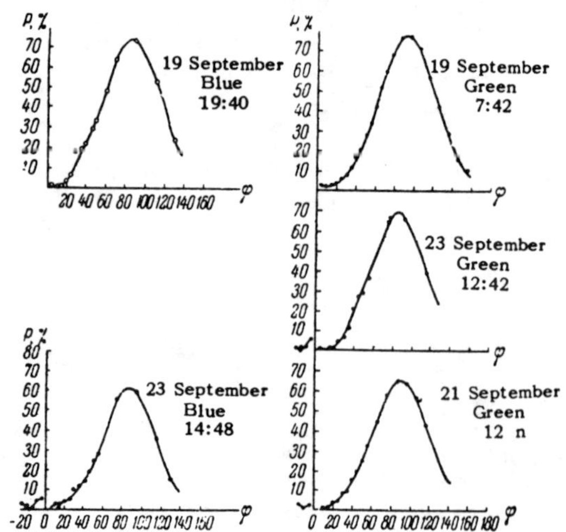

FIG. 10. Examples of the dependence of the degree of polarization of the illumination of the daytime sky on the angle of scattering on different days.

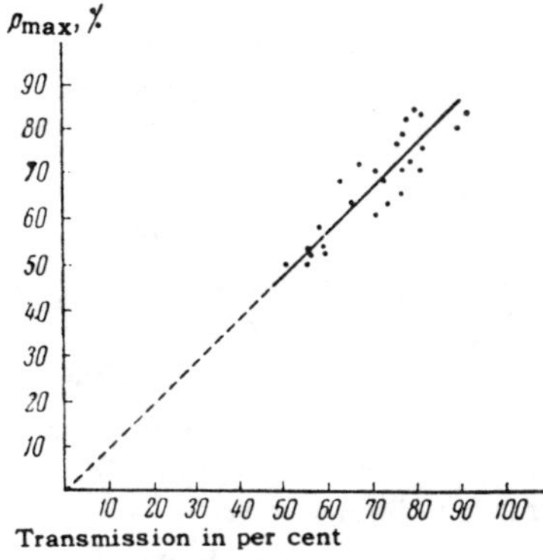

FIG. 11. Correlation between the maximum polarization of light of a daytime sky and the vertical transmission of the atmosphere.

of the degree of polarization on the wavelength ("dispersed polarization" in the terminology of Schirmann), which was subsequently confirmed by the observations of Tikhanovskiĭ and Pernter, and also 2) the discovery by Soret, Dorno, MacConnell and others of the effect of the albedo of the earth's surface on the polarization. One must also note the observation of Rubenson[68] (1864), confirmed later by Wild and Jensen, that the position of the maximum polarization, as a function of the weather, can be displaced somewhat along the vertical to the sun both in the direction of lower and in the direction of higher angles. As an illustration, a typical dependence of the degree of polarization on the scattering angle along the vertical to the sun, according to the data of Rozenberg and Turikov, is given in Fig. 10, while Fig. 11 gives the dependence obtained by the same authors of the degree of polarization at the maximum on the vertical transmission of the atmosphere.[51] The latter drawing shows that, in spite of the appreciable individual variations, there is a clearly expressed correlation, according to which the degree of polarization at the maximum is equal to the vertical transmission of the atmosphere, independently of the altitude of the sun, which is in good agreement with the observations both of Arago himself and also of all subsequent investigators of the sky. It is not possible for us to linger on subsequent details, and we must refer the readers to the excellent reviews of Jensen[64] and Dorno,[65,66] particularly because little has changed in this field since their time. Reviewing the

observational material, it remains to be said that in the main it was re-obtained in searches for empirical connections with the weather and, in spite of its immensity, it presents no interest for subsequent analysis. As a conspicuous example, we must introduce the researches of Roggenkamp,[67] who during his life carried out more than 15,000 sets of the position of the neutral points in the twilight hours (i.e., 15,000 evenings!) and carefully obtained from them . . . the arithmetic mean.

We saw above that the experiments of Govi and Tyndall on the scattering of light in colloids discovered practically all of the peculiarities noted by Arago in the polarization of the light of the daytime sky, and that precisely these peculiarities strengthened the conviction in that day of the identity of the two phenomena. On this point Rubenson in 1864 directly pointed out[68] that one must see the reason for all the variations of the polarization of the illumination of the daytime sky in the contamination of the air by dust and water droplets. However, with the appearance of the theory of Rayleigh, the general character of the picture was so well explained, and all other points of view being decisively relegated to second place, these peculiarities were regarded as surprising anomalies, requiring for their explanation the point of view of molecular scattering.

The pioneer in these searches was one of the most active students of the polarization of the illumination of the daytime sky, Soret,[69] who advanced in 1888 an idea that was completely new for his time on the possibility of multiple scattering of light (by analogy with the idea advanced by Muller[22] of its multiple reflection), and who attempted to apply it to an explanation of the existence of the neutral points. In essence this was first worked out in the region of the transfer of radiation into the scattering medium, showing, in spite of its extreme roughness of calculation, the serious effect on the subsequent form of all problems on the whole. The first equations of radiation transfer were quickly formulated by Khvol'son), and also by Schwartzchild and Schuster. They first started development of this branch of knowledge within the framework of

FIG. 12. Real and observed displacements of the position of the maximum polarization in different parts of the spectrum according to observations at different points and on different days. The solid line — theoretical expectation for the Rayleigh scattering and albedo of the earth's atmosphere 0.25.

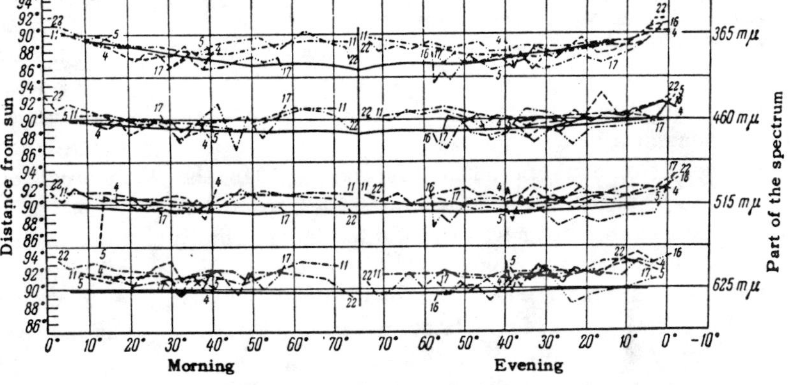

mathematical physics. However, in the initial formulation, there were only the equations of energy transfer, not taking into account polarization effects and of Soret's problem unsuitable for solution. Therefore it was again necessary for its solution to fall back on homemade methods (although more reliable than those used by Soret). This was done in 1914 by Ahlgrimm[70] and in 1927 by Tikhanovskiĭ,[71] who by means of long and complicated calculations showed that the idea of Soret was correct, and that actually secondary scattering of light in the atmosphere leads to the appearance in the sky of neutral points and to a certain τ-dependent decrease in the degree of polarization at the maximum. Here, the dispersion polarization discovered by Pilchikov found its explanation.

At that time an event took place which directed the thoughts of investigators along another path. In 1917-1918, Max Born advanced the idea of the anisotropy of molecules.[72] Simultaneously, Strutt (Rayleigh) the younger discovered[73] that in light scattering in vapors and gases the polarization is never complete in observations at a scattering angle $\varphi = 90°$. The latter phenomenon especially attracted attention in connection with the picture of the polarization of the sky, and in 1921 Cabanne[74] explained it by starting out from the idea of Born, and improved the Rayleigh theory by introducing in it an account of the anisotropy — see Eqs. (14) and (15) (see also references 52, 29, and 75). However, the correction of Cabanne did not solve the problem. For pure air it led to a value of the degree of polarization at $\varphi = 90°$ equal to 92 per cent, in place of the 84.5 per cent known from the data of Tikhanovskiĭ as the maximum observed in the atmosphere (see Fig. 14). Even account of secondary scattering with the Cabanne correction, carried out in 1927 by Tikhanovskiĭ, did not correct the situation. Even more incomprehensible was the effect which the weather had not only on the value of the polarization but also on the position of the neutral points. This stimulated Schirmann and Milch[76] to return to the hypothesis of Rubenson,[68] that one must seek the reason for polarization effects in the scattering of light by large particles, in correspondence with the theory of Mie, which was already well developed at this time. However, this attempt remained within the realm of abstract discussion, because concrete calculations of the polarization of light scattered by such particles are very few even today, and 20 years ago there were almost none at all.

The next step could be taken only when in 1946, the equation of radiation transfer, taking into account the polarization of the latter was formulated independently by Chandrasekhar[45] and Rozenberg[44] (see references 21 and 38). This matrix equation, now so widely used in atomic physics, was formulated by these authors, and also by V. V. Sobolev,[46] in connection with problems of polarization of light emitted from the sky. This formulation was shown to be possible only by use of the Stokes vector-parameter and the scattering matrix, and immediately made it possible to increase appreciably the number of problems that could be solved. In particular, it was clearly established that the polarization effects have an important influence not only on the polarization but on the intensity of multiply scattered light, even in the depth of the scattering medium.[49] At the same time, it was made clear that the calculations of Soret, Ahlgrimm, and Tikhanovskiĭ contained methodological errors. However, qualitative calculations on the rough scheme of Soret, carried out by Rozenberg,[77] showed that these errors did not change the general picture, and that the neutral points and their position in the sky are explained by secondary scattering. Finally, in 1954, Chandrasekhar and Elbert[78] published detailed tables of the brightness and polarization of the daytime sky at different altitudes of the sun under the assumption of purely molecular scattering. These tables, which were the result of careful solution of the rigorous equation of transfer for different assumptions on the albedo of the earth's surface essentially completed the solution of the problem of the dust-free atmosphere.

Sekera[79] quickly carried out careful investigations of the polarization of daylight sky by means of an automatic polarimeter, and established the fact that on average the picture drawn by Chandrasekhar and Elbert did not correspond at all badly with reality, but that the agreement again bore only a semi-quantitative character. In addition, systematic departures from this picture and also random variations, including short period ones are definitely observed. As an example of this we have the wandering of the maximum of the polarization for different parts of the spectrum, shown in Fig. 12. Recently Lipskiĭ,[80] measuring the polarization of the light of the daytime sky with very high angular and spectral resolution, also discovered fast and deep-seated vibrations of the degree of polarization, which were not correlated for different wavelengths. This leaves no doubt of the validity of the point of view already expressed by Rubenson, and also by Schirmann, that atmospheric aerosols are responsible for all variations of the polarization of the daytime sky light, that these aerosols are extremely variable, both quantitatively and qualitatively, and that their role in the scattering of the light by the atmosphere is never small. Inasmuch as the effects of secondary scattering are also not small, the division of the effects of molecular and aerosol scattering in the daytime sky is not possible, as Sekera[77] emphasized, because of the presence of nonlinear effects which cannot be neglected even at very high transmission. For the same reason, it is impossible to draw any reliable information from the data on the illumination of the daytime sky on the scattering properties of air in every case, inasmuch as one is dealing with such a fine-grained effect as polarization, i.e., with the components $f_{21}(\varphi)$ and $f_{31}(\varphi)$ of the scattering matrix — see Sec. 3. Such is the some-

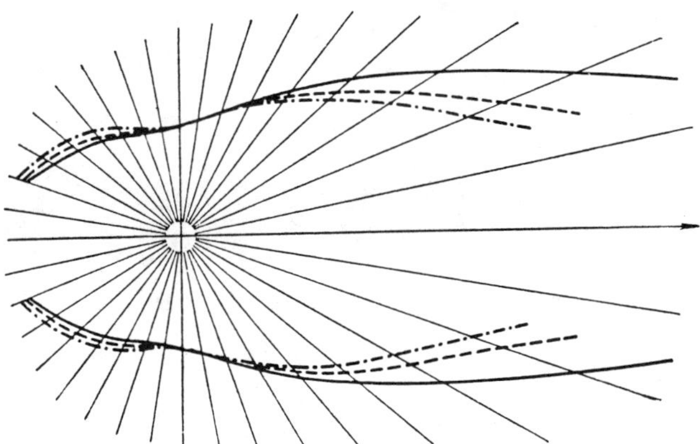

FIG. 13. Examples of the atmospheric indicatrices of scattering found from measurements on the brightness of the daytime sky.

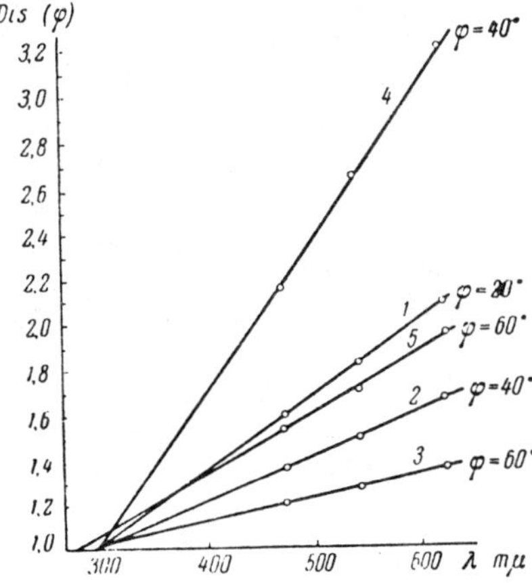

FIG. 14. Increase of the dissymmetry of the indicatrices of scattering with increase in the wavelength: 1, 2 and 3 — for atmospheric air; 4 and 5 — for aerosol component.

what unexpected result of the very intensive investigations over a century and a half, which as we have seen exerted an important effect on the development of optics and made possible a whole series of important discoveries.

6. THE BRIGHTNESS MAP OF THE SKY AND THE SCATTERING FUNCTION

Systematic measurements of the brightness of the daytime cloudless sky were begun in 1898 by Jensen[64] and continued to our own day by many authors. Especially broad and long-term observations were completed by Dorno[65] and in recent years by Pyaskovskaya-Fesenkova.[81] As a result of this, a very rich collection of brightness pictures of the sky has been accumulated under different meteorological and geographic conditions. However, the first comparisons of these pictures with the Rayleigh theory led to a remarkable result. If the polarization and the coloring of the sky corresponded approximately to theoretical expectations, then the angular distribution of the brightness had nothing in common with them. The idea then arose of using these data for experimental determination of the scattering function $f_{11}(\varphi)$ of atmospheric air, the more so since other ways for this measurement were not seen at that time. In broad outline (for details, see reference 81), the idea was the following. If one measures the brightness of the sky at different azimuths, but always along the almucantar of the sun (that is at points located below the horizon at the same angle as the sun is above it), then the attenuation of the sun's rays on their path to the scattering region and beyond it to the eye of the observer will be the same, and the difference in brightness will not be brought about by the angular dependence of the component $f_{11}(\varphi)$ of the scattering matrix. In this case, naturally, it is assumed that the secondary scattering does not greatly distort the picture, and that the atmosphere is homogeneous in the horizontal direction.

Furthermore, since the properties of the atmosphere change with altitude, the data obtained on the form of the scattering function $f_{11}(\varphi)$ will be certain weighted averages over the altitude.[82] A detailed analysis of data obtained in this fashion is contained in reference 81 (see also reference 51), and we shall limit ourselves to a certain amount of fundamental information.

First of all, it has been pointed out that the scattering indicatrices (functions) are extremely strongly elongated in the forward direction and terminate in a diffraction "nose," which is characteristic of the Mie theory, and which is responsible for the aureole around the sun. Examples of observed indicatrices[81] are shown in Fig. 13. It has been further shown that the form of the indicatrix is very sensitive to change of atmospheric conditions, not only in the region of small and large angles of scattering, especially in the region of the aureole, since this sensitivity is much less in the region of medium angles of scattering. Finally, it was discovered[81] that the prolateness of the indicatrix increases with increase in the wavelength, which is illustrated in Fig. 14,[81] where the so-called dissymmetry of the scattering function

$$Dis(\varphi) = \frac{f_{11}(\varphi)}{f_{11}(\pi - \varphi)}$$

is plotted along the ordinate. As Kastrov[83] has shown, this is the result of the decrease of the relative role of molecular scattering (or scattering by the submicroscopic fraction) as the wavelength increases and the scattering angle decreases.

Thus the brightness map of the sky testifies unambiguously to the fact that the scattering in the atmosphere bears an essentially aerosol character. It was very unexpected to learn that this character of the scattering is preserved up to very high frequencies, as follows from Fig. 15, which was obtained by measurement of the brightness of the sky with the aid of an "autostratostat."[84]

The methodology of the theoretical calculation of the brightness of the sky under different conditions on the basis of a solution of the equation of radiation transfer in a scattering medium was first developed by Kuznetsov,[85] and then by Chandrasekhar and Sobo-

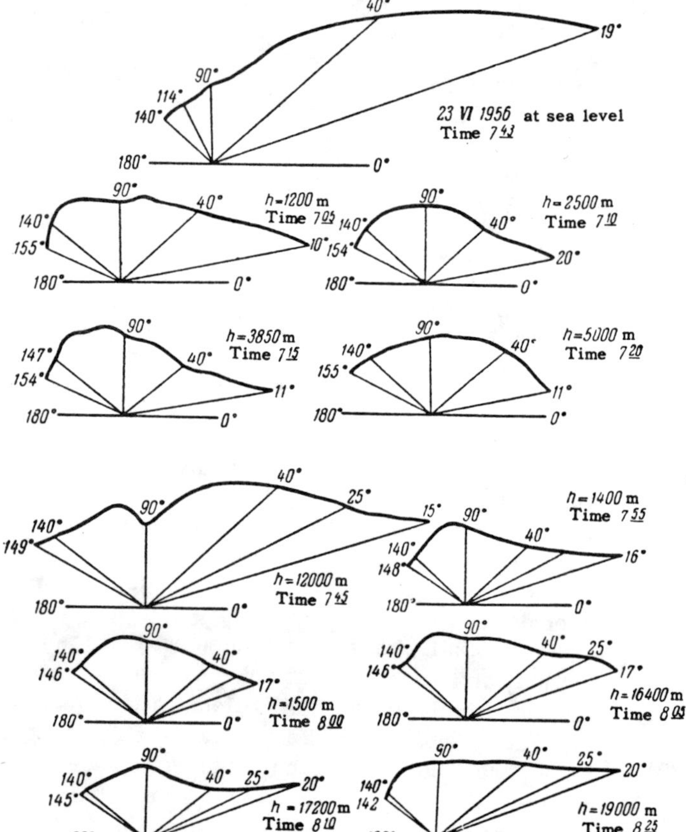

FIG. 15. Indicatrices of scattering of atmospheric air, measured by the brightness of the daytime sky at different altitudes.

lev,[21] unfortunately, without account of polarization effects. Making use of this method, a number of authors have carried out numerical calculations of the brightness map of the sky for different assumptions of the form of the indicatrix, and also the altitude dependence of the scattering coefficient, as a result of which we now have detailed tables covering a wide range of possible variations of these quantities.[86]

Analysis of these tables makes it possible, in particular, to estimate how much the effects of secondary scattering, and also reflections from the earth's surface, can distort the form of the scattering function drawn from data on the brightness of the daytime sky. It is shown[82] that these distortions are very significant as soon as the transmission of the entire thickness of the atmosphere becomes lower than approximately 85 percent, i.e., practically always, and do not decrease with increasing height of the observer. Therefore, it is not possible to draw anything except qualitative and highly tentative conclusions relative to $f_{11}(\varphi)$ from analysis of the light of the daytime sky. An exception is the case of extremely high transmission, which is less characteristic for the atmosphere. Besides, knowledge of the brightness and polarization maps of the sky and their variation has large importance by itself, both conceptually and in an applied sense, independent of the problem of the determination of the form of the scattering matrix of atmospheric air. For example, we recall that the polarization of the light of the daytime sky can be used for the purpose of orientation in space (bees, as is well known, orient themselves in just this fashion) and the illumination of our houses is determined by this brightness map.

Thus, we have shown that all the integral methods, based on the study of the illumination of the daytime sky, do not warrant placing any confidence in them and cannot supply us with trustworthy information on the form of the scattering matrix (we also add, and because this matrix is subjected to material and completely random changes with altitude and time). Therefore the naturally reliable method of study of the character of the scattering matrix and its variations is the method of local investigation of the scattering act by means of a directed light beam. However, the possibility of carrying out detailed measurements, which require the use of rather ideal lighting and observing means, has existed only in recent times, and measurements to date have been limited exclusively to the angular dependence of the component $f_{11}(\varphi)$, with no studies of either its spectral behavior or of the other components of the scattering matrix.

As was to be expected, there exists only a qualitative and very uncertain agreement among the (as yet) small amount of data of different authors. At sea level (and at the present time there is no basis for expecting [see Sec. 7] that the situation will be different elsewhere, at least up to altitudes of the order of $80 - 90$ km) the indicatrices of scattering have a typical aero-

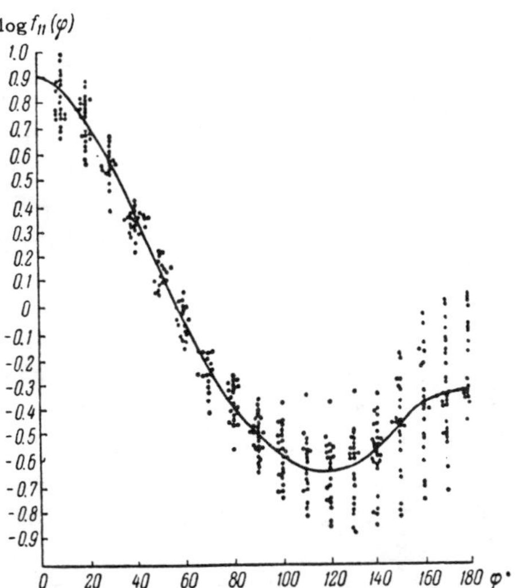

FIG. 16. Scattering function of atmospheric air at sea level (in logarithmic scale) from measurements over a series of nights.

sol character, with a sharp elongation forward and a more or less clearly marked aureole part. As an example, indicatrices measured at sea level by Chesterman and Stiles[87] are shown in Fig. 16. The protracted measurements of Barten'eva[88] have shown (in agreement with other authors) that the degree of elongation of the indicatrix corresponds well with the value of the scattering coefficient, increasing simultaneously with it, and the absence of any geographical dependence is clearly noted, except the frequency of repetition of this or that condition of transmission. Furthermore, it is observed that all the indicatrices intersect in a comparatively narrow range of scattering angles around $\varphi = 45°$, so that the coefficient of directed light scattering $D_{11}(45°)$ can be regarded with accu

racy as a measure of the transmission of air. (According to the data of Pyaskovskaya-Fesenkova,[81] the intersection of all indicatrices takes place at $\varphi = 60°$, which can be explained, in particular, by their distortion as a result of secondary scattering).

Any further judgments on the form of the scattering function for atmospheric air are impossible at the present time, because all measurements have been carried out without an accompanying control on the nature and microstructure of the aerosol component.

The other components of the scattering matrix have in general not been measured to date. An exception are the still preliminary experiments recently completed by Rozenberg and Rudometkina.[51] One can obtain a general picture of the character of polarization effects taking place in light scattering in the atmosphere from Fig. 17, in which a collection of photographs of a given part of a linearly polarized projected beam are mounted; these were carried out on a single night by means of a photocamera equipped with polaroids and filters ($\lambda_{eff} = 420 \pm 20\,\tau\mu$) directed at different angles φ to the beam (the visual angle of the camera is $\pm 15°$, and the ends of the picture of the ray are distorted by the vignetting effect). The arrows to the right show the direction (relative to the vertical) of the electric field vector of the wave in the projected beam (P) and passed by the analyzer on to the camera (A).

Furthermore, it was established experimentally by Rozenberg and Mikhaĭlin[89] that for known conditions (corresponding to the conclusions from the theory of Mie) the light scattered by the atmosphere (i.e., with particles of the aerosols impregnated in it) is elliptically polarized, while the degree of ellipticity q

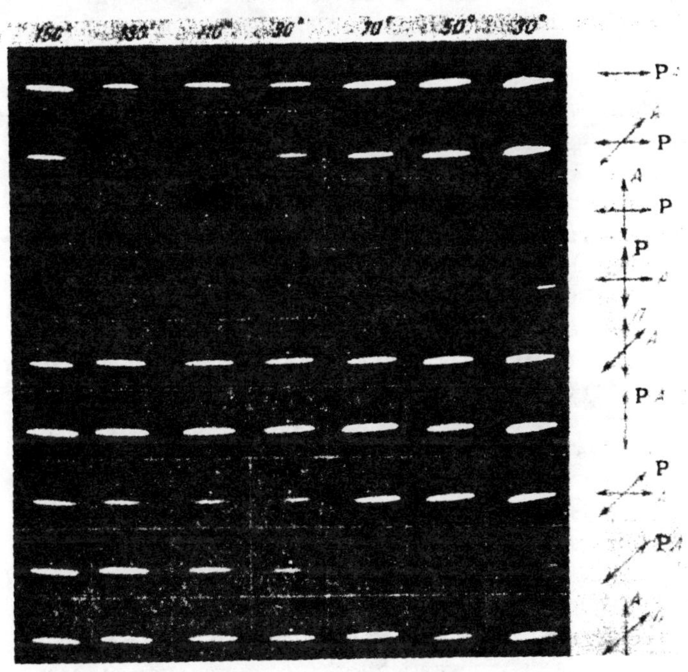

FIG. 17. Mosaic representation of the angular dependence o the brightness of a horizontal projector beam, on one of the night at different states of beam polarization (P), observed through di ferently oriented analyzers (A). Each picture subtends an angle interval of $\pm 15°$ relative to the value of φ indicated on top, and distorted at the edges by vignetting.

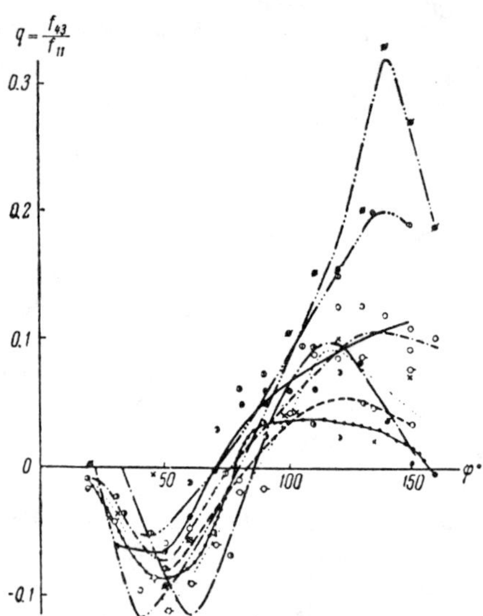

FIG. 18. Angular dependence of ellipticity of light scattered in the surface layer of the atmosphere, from measurements made in a series of successive nights.

$= f_{43}/f_{11}$ — see Eq. (16) — is never large, especially in the region of the rainbow (Fig. 18). It was shown here that the components of the matrix f_{41} and f_{42}, if they are not equal to zero, are in any event very small. But a more detailed investigation of the scattering matrix and its angular and spectral dependences is a thing of the future.

The problem of the investigation of the aureole is a special one. It follows from the Mie theory that the aureole appears only in the presence of sufficiently large particles and has a diffraction character, which allows one to describe it with a comparatively simple and general formula. In the case of a semi-dispersed aerosol, the angular structure of the aureole is obtained as a result of simple mutual superposition of the aureoles from all particles. As shown by Sliepcevich and later by Shifrin,[90] this makes it possible by comparatively simple means to calculate the distribution of the dimensionally large (more than several microns) droplet fraction of the aerosol which account for the formation of the aureole; this can be done from a knowledge of the angular structure of the aureole. There can be no doubt of the future of this method which attracts ever greater attention.

Finally, the problem considered by us of the form of the scattering matrix is directly related to a wide circle of various phenomena connected in their origin with the reflection, refraction and diffraction of light on water droplets and ice crystals — all sorts of rainbows, halos, coronas, and glories. For a long time they were the chief object of attention in atmospheric optics and many volumes have been devoted to their detailed description (see, for example, references 1, 2, and 91). In spite of the fact that over a period of 1000 years an uncountable number of investigators, frequently very experienced and notable, have taken part in their investigation, for the majority of them

there exists, besides the general description of the characteristic features and a detailed classification, only a qualitative explanation of the reasons on which they depend. Rigorous theory has been limited only to those phenomena which follow from the Mie theory, and principally (with the exception of rainbows) it touches only the general features of the phenomenon, omitting its details. So far as phenomena produced by ice crystals are concerned, there is no theory at all. Experimental data in this region are also very sketchy and there is no systematization, chiefly because of the absence of data on real conditions of the scattering medium to which they refer. But then this situation is met with in all regions of colloidal optics, and it provides the justification for the fact that the Mie theory remains today practically untested from a quantitative viewpoint after half a century.

7. OPTICAL PROBING OF THE ATMOSPHERE AND THE PROBLEM OF INTERPRETATION OF DATA

The idea of optical testing of the atmosphere is very simple. If we know how the laws of scattering are related to the properties of the scattering medium, then we can draw conclusions on the state of the latter (in the given case, the atmosphere) from the character of the light scattered by it. We have seen above that, as a consequence of the optical inhomogeneity of the atmosphere and of the obstacles presented by multiple scattering, use of integral effects (such as the illumination of the daytime sky or the transmission of the entire thickness of the atmosphere) does not provide reliable quantitative information on the optical state of the atmosphere at different altitudes at a given instant of time. We add that the decisive obstacle in this case is the complication of the theory of light propagation in scattering media which does not allow us to put into general form the solution of the inverse problem, i.e., the problem of the determination of the properties of a medium from its light field, including the angular dependence of all components of the scattering matrix.[39] Therefore, investigation of the optical proportion of the atmosphere must necessarily have a doubly local character, in which the effects reliably measured are only those of single scattering in a distinctly limited and comparatively small volume. Evidently, this can be done if the atmosphere is irradiated by a sharply bounded, directed light beam, the tracing and displacement of which makes it possible to test the different regions of the atmosphere.

This idea was first stated by Al-Hazen.[3] He noted that during twilight the shadow of the earth rose higher and higher and that night began when the entire atmosphere was in the shade. This permitted Al-Hazen, by measuring the duration of the twilight, to estimate the altitude of the atmosphere at 52,000 paces (later this calculation was improved by Kepler[4]), which was very high accuracy for the eleventh century.

In the period following, the problem of twilight, which embraces about 10 percent of the year at the equator and not less than 30 percent at the poles attracted the attention of J. and N. Bernoulli, Maupertuis, D'Alembert, Clausius and many others, remaining constantly in the field of view both of observers and of theoreticians. The modern explanation of twilight phenomena as caused by scattering and attenuation of the light of the sun by the earth's atmosphere was first advanced by Berzold[92] in 1863, i.e., immediately after the discovery of Govi, where the necessity of explaining the character of the polarization of the light of the twilight sky played no small role.

Subsequent development of the theory of twilight led along the line of explanation of the redness of daybreak. Only in 1923, did Fesenkov, for the first time after Al-Hazen and Kepler, again set up the inverse problem of using the twilight phenomena for optical testing of the high altitude layers of the atmosphere.[93] Careful analysis of the effect of the various factors on the brightness of the twilight sky permitted Fesenkov in 1930 to advance the very fruitful idea of the "twilight ray," later developed in detail by Staude and in a somewhat different form by Link. The substance of this idea briefly is that at each given moment the twilight light proceeds from the comparatively thin effective layer of the atmosphere, bounded above as a consequence of the rapid decrease in the air density with altitude and below because of the rapid increase (upon approach to earth) of the attenuation of the emitted rays of the sun in the lower levels of the atmosphere. During the twilight period this effective layer gradually rises (or falls), which makes it possible to examine systematically the scattering ability of the different layers of the atmosphere, beginning approximately at 20 km above sea level and higher, so long as we can generally distinguish against the background of the sky the light scattered at the high layers of the atmosphere.

During the twenties and forties, no doubts were raised by anyone that, beyond the limits of the troposphere, the atmospheric air is perfectly pure, and the problem of testing was regarded as the determination of the altitude dependence of the density and the temperature of the air at altitudes up to 100 or even 300 — 400 km, not attainable at that time by other means. The treatment of data of twilight observations, completed during this period by a number of authors, led to values that were completely reasonable in order of magnitude, in agreement with other indirect estimates, values which in their time played an important role as the stimulus toward reexamination of the traditional picture of an isothermal stratosphere.[95] The tendency to raise the accuracy of the estimate led in the middle of this period to the necessity of analyzing the role which secondary light scattering plays in the formation of the twilight. The difficulty of the problem is such that a rigorous solution of it has not been obtained to date. Approximate solutions, valid for different initial assumptions, have been shown to be

extremely contradictory, but in general prejudicial for the method as a whole, at least in its application to altitudes above 100 km. Such an uncertainty in the competence of the method on the one hand and the simultaneous development of rocket methods of investigation of the stratosphere on the other, has for a long time moved the twilight method to a back seat. However, the abundant observational material accumulated since that time makes it possible to draw the conclusion that for a suitable arrangement of observations and suitable treatment of observational data, one can, by studying the twilight, obtain rather trustworthy information on the scattering properties of the atmosphere in the range of altitudes approximately from 20 to 100 km.[96] Here attention[96] is turned to the extreme variability of the brightness, color,[97] and polarization[98] of the light of the twilight sky. The variability of this is so large and so characteristic that one can attribute it only to a single reason[44,96] — that, at all altitudes in the stratosphere, at least up to 80 — 90 km, i.e., up to the level of existence of the so-called silver clouds,[99] the basic substance for light scattering is not the air but aerosols with an inherent time and space variability both in quantity and in dimensions and nature of particles. In particular, if these aerosols have meteoric origin, then their appearance must be connected with the increase of ionic concentration, which makes understandable the connection noted by Khvostikov between the twilight polarization anomalies at the altitude of twilight light of around 80 km and the critical frequency of reflection of radio waves.[98] We add that an estimate of the concentration of aerosol particles which is necessary in order that scattering at altitudes of 50 — 80 km have an aerosol character[44] leads to quite reasonable numbers — of the order of 1 to 10^{-2} particles per cubic centimeter with dimensions of 0.1 to 0.3μ (see reference 99). Thus the problem of twilight testing of the stratosphere has changed sharply, and concerns now the observation of the transfer, sedimentation, and transformation of the aerosol in the mesosphere, i.e., investigations of the meteorology of extremely high altitudes, which have assumed a practical importance in recent years. The significance of the revival of the twilight method can be estimated by recalling the high expense of rocket investigations, which limits their widespread use, and the difficulty of using rocket technology for the investigation of aerosols.

In 1930, Synge (see reference 51) pointed out the possibility of testing the atmosphere by a ray from a projector. The progress of projector and measurement technology has made it possible to realize this possibility very quickly, while the "ceiling" of testing has steadily increased, although not reaching imposing figures in the 1950's — about 70 km. Here every American investigator, without exception, aimed at extracting from the probing data information on the density and temperature of the air at these altitudes, while in the Soviet Union, I. A. Khvostikov had already clearly

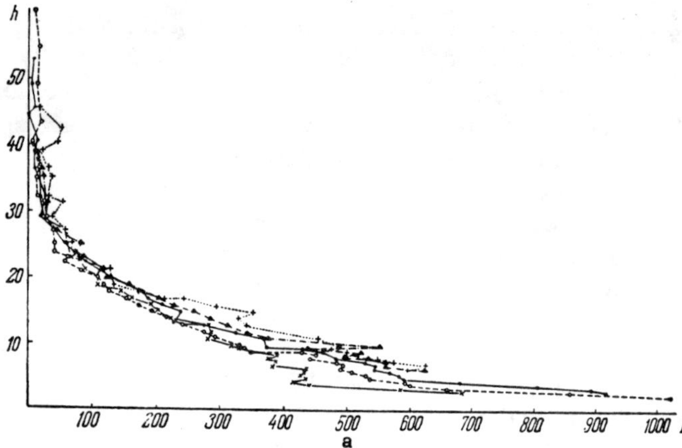

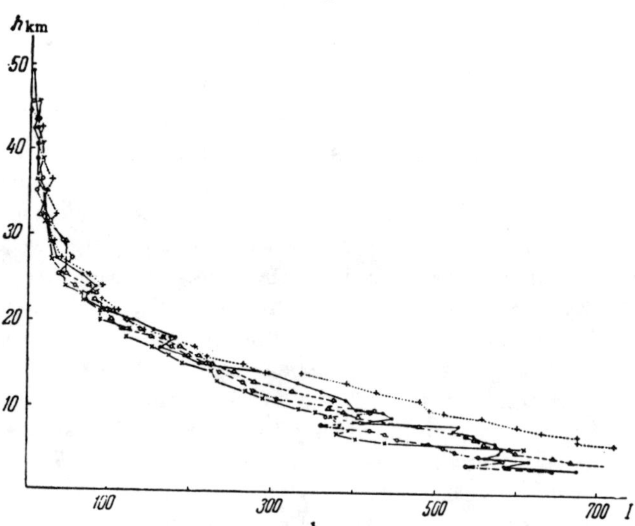

FIG. 19. Dependence of the brightness of the scattered light of the projector (in arbitrary units) on the altitude of the scattering volume according to measurements of a number of consecutive nights in August (a) and September (b), 1954.

aimed the investigation in 1944 in the direction of a study of the atmospheric aerosol, for which measurements of the brightness of the scattered light of the projector were from the outset accompanied by measurements of its polarization (for details, see reference 51).

In the period 1944-1958 many hundreds of optical profiles of the atmosphere were obtained in the laboratory of atmospheric optics of the Institute of Atmospheric Physics, Academy of Sciences, U.S.S.R., by means of projection probing by different methods and under different conditions; these made it possible to draw a number of completely unambiguous conclusions.[51]

It has been shown that the scattering ability of the atmosphere, even on very clear days and at all altitudes acceptable to projection probing, was subject to comparatively rapid and random changes, by a factor of at least 2 or 3. This is clearly seen in Fig. 19, where the intensities of the scattered light of a projector are plotted (in arbitrary units) as a function of the altitude of the scattering volume for two series of successive nights according to observations in Bakuryani (1954). Almost always, at one altitude or another, more or less clearly pronounced layers of aerosol were discovered, which frequently had an explicit cloud structure and which were made evident both by increase in the brightness of the scattered light of the projector, and also by the change in its polarization. An example of such a vertical optical profile of the earth's atmosphere can be seen in Fig. 20, in which data of measurements of the intensities of two linear components of the scattered light of the projector are shown for a single night (I_1 is the polarization in the plane of the scattering, I_2 is perpendicular to it). The measurements were carried out for the blue region of the spectrum and as a function of the altitude of the scattering volume. The degree of polarization of the scattered light almost never corresponds to the Rayleigh law, but falls off more slowly or more rapidly, while the polarization in the aerosol layers is frequently negative, as for example is the case in Fig. 21 at an altitude of around $22-23$ km (the altitudes are shown on the right along the ordinate, while the corresponding scattering angle is shown on the left of the ordinate). It is noted that the aerosol layers are most frequently observed in the region of the tropopause ($10-12$ km in our latitudes) and at the altitudes $22-25$ km, which corresponds to the usual altitude of the phenomenon of the so-called mother-of-pearl clouds. In the latter case, the data of projection testing made it possible for Driving to establish the fact that the clouds are formed by the supercooling of water droplets; he estimated their

FIG. 20. Dependence of two mutually perpendicular polarized components of the brightness of the scattered light of a projector on the altitude of the scattering volume for a single night.

dimensions and concentration.[51] Clearly expressed
aerosol layers were frequently observed even higher,
in every case up to 40 km. Sharp variations of the
ceiling of projection testing are characteristic; they
can be attributed only to strong variations of the scattering ability of the atmosphere at the corresponding
altitudes (40 — 70 km), i.e., the variability of the
aerosol located there,[51] which is in excellent agreement with the conclusions based on twilight observations.

Thus, the data of projection and twilight testing of
the atmosphere, together with the data of measurements of the brightness of the sky with "autostrato-
stats" (Sec. 6) leave no doubt that at all altitudes up
to 80 — 90 km aerosols are chiefly responsible for the
scattering of light in the atmosphere. Attempts at the
determination of densities and temperatures of the air
by optical means have led, as is well known, to a significant scatter of values, and to agreement with the
data by other methods only in order of magnitude. Now
this can be explained by the fact that, on an average,
the relative concentration of the aerosol varies slightly
with altitude, and that in the visible portion of the spectrum the scattering ability of the atmospheric aerosol
is, in the mean, close in order of magnitude to the scattering ability of pure air. Thus the only real result of
optical studies of the atmosphere at all altitudes can be
exclusively (if one puts aside the investigation of the
selective absorption of the gas phase) the study of op-

tical properties of the atmospheric aerosol and the
changes experienced by it —transfer, sedimentation,
crystallization, condensation, evaporation, etc. This
means that the use of optical methods for the investigation of such important meteorological processes as
those mentioned becomes the fundamental problem of
atmospheric optics, the more so since to date no
effective means of investigation of these processes
exist (especially if one is discussing processes in -
which submicroscopic particles take part).

However, we immediately come across the problem of the interpretation of sounding data. Essentially,
the problems facing us are identical with those of nuclear physics: it is necessary, by means of the scattering matrix, by its angular and spectral dependences, to make clear with the maximum completeness
obtainable the properties of the scattering medium
(the nature of the scattering particles, their size distribution etc.), i.e., one must solve the inverse problem of the theory of light scattering by the set of heterogeneous particles of the aerosol. After what has
been pointed out in the previous paragraph, it is
scarcely necessary to add that at the present time the
state of the theory makes it possible to carry out such
an analysis only in very few specially favorable cases,
and on a very limited scale, as a consequence of which
the effectiveness of the testing is negligible at the present time. Mostly, the observational data clearly reveal
certain processes, but interpretations are not supplied,
and the result is only an increase in the collection of
barren puzzles. Therefore, the central problem of
modern atmospheric optics is undoubtedly the detailed
explanation of the laws of light scattering by particles
of the aerosol, primarily by water droplets and ice
crystals, but also ensembles of these with different distribution laws.

This is the very general problem of colloidal optics, and it must be solved by employing the most effective means of contemporary experiment through a
direct combination of field and laboratory investigations with simultaneous treatment of theoretical representations. The greatest difficulty lies in the fact that
the submicroscopic fraction of the aerosol, which is
very active in its optical behavior, does not yield to
detailed study by other means at the present time.
This leads to the impossibility of direct comparison
of the optical characteristics of the medium with its
micro-physical parameters, which can be compensated
only by an abundance of optical information, i.e., by
the simultaneous operation of different optical means
of investigation in a joint study of the different aspects
of a single phenomenon. Only such a complex optical
experiment will be in a position to guarantee the representation and the possibility of analysis of the results, and at the same time the creation of bases for
the interpretation of the data of the optical sounding
of the atmosphere.

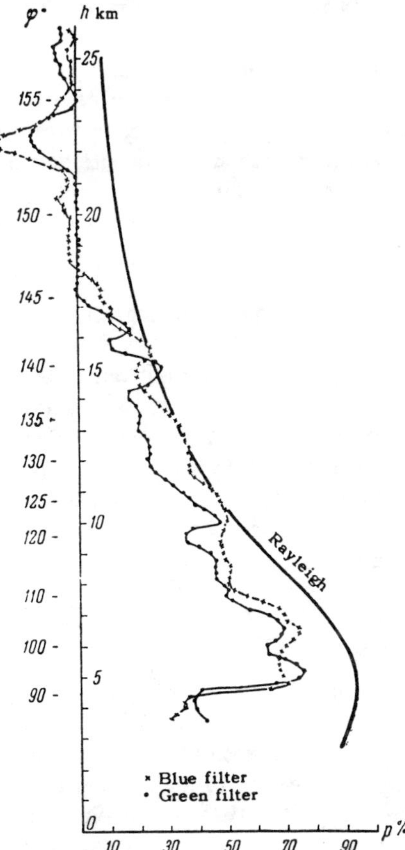

FIG. 21. Depend-
ence of the degree of
polarization of the
scattered light of a
projector on the alti-
tude of the scattering
volume in two parts
of the spectrum for a
single night.

8. PROPAGATION OF LIGHT IN CLOUDS AND FOGS AND SIMILAR PROBLEMS

In addition to the investigation of the dependence of the scattering matrix on the microphysical characteristics of the aerosol, the 20th century has put forth still another round of problems, which are very important for atmospheric optics both from an applied and from a theoretical point of view. The formulation of the equation of radiation transfer and the development of methods of its solution, in particular with the use of computer technology, has made possible the investigation of processes of propagation of radiation and radiation conditions in extremely turbid media — clouds, fogs, the sea, etc. The problems arising from the demands of atmospheric optics, among them the requirement of forecasting visibility, have long since pushed past its framework and achieved first-rate importance in astrophysics and nuclear physics, which makes it impossible to touch on them in any detail within the framework of this paper (see reference 21). Here we can only call to mind certain problems whose solution is most important from the point of view of an understanding of atmospheric-optical phenomena.

We immediately note that the biggest obstacle to the development of a modern theory of light propagation in strongly scattering media is again the absence of information on the form of the scattering matrix. As a result, the problem of calculating or estimating polarization effects, which are nowhere small, can neither be completely furnished nor solved with any certainty, with the exception of the trivial case of Rayleigh scattering, which has no relation to reality. This at once deprives the theory, no matter how inventive it might be, of practical significance, and sharply limits the possibility of its comparison with experimental data.

Even the scattering functions $f_{11}(\varphi)$ for real polydispersed systems remain today but little studied, either theoretically or experimentally. As a consequence, all the computations must be carried out relative to highly idealized systems, and their comparison with reality inevitably involves a highly qualitative, descriptive character, the more so since the experimental data on the light conditions in the scattering medium are very poor. Such a situation is emphasized by the absence of effective approximate methods of solution of the transfer equation, which would permit one to explain even qualitatively the general regularities, which forces theoreticians to direct their efforts to the collection of supposedly "rigorous" solutions of more or less typical, but, as a rule, barbarously stylized examples. Therefore, in spite of the serious progress in this region, with which the last 150 years have been marked, the situation remains a sad one, chiefly because of the complete separation between theory and experiment.

Turning to problems which await their solutions we must first of all point to the investigation of the laws of reflection of light from formations that strongly scatter it — clouds, snow, the sea, sand, soil, plant cover, etc., as a function of their microstructure, their absorbing ability and the angles of incidence and observation of the light beam. Here a large amount of observational material has been accumulated which finds to date no exhaustive theoretical explanation and generalization (see references 21, 39), consequences that are the more necessary in that they would permit a large step forward in a number of practically important problems.

Furthermore, detailed experimental and theoretical investigations of the light regime inside the scattering medium are necessary — from its boundaries to its depths, where the character of the light conditions is determined not from the properties of the illumination but from the properties of the medium itself.[21,39]

The whole region of three-dimensional problems of heat transfer remains today completely unstudied. To this is related the theory of twilight phenomena, and the theory of the so-called "ice" and "water" sky and, of first importance, the wide and practically important problem of the penetration of a projected ray through fog or other strongly scattering medium.[51] The last problem has a great methodological significance, because it is not clear at the present time to what measure the effects of multiple scattering are capable of preventing the interpretation of experiments on the determination of the coefficients and the scattering matrix in clouds and fogs, while the ways for the choice of the most rational measuring systems are very unclear.

Let us mention one of the most important (from our point of view) problems of modern optics — the creation of the foundations of the spectroscopy of dispersive media by the search for means of utilization of the effects of multiple scattering for experimental separation of the coefficients of absorption and scattering of dispersed phases.[39] In particular, in this way one should expect to obtain data currently absent on the absorptive ability of clouds and fogs in different portions of the spectrum.

Realization of the investigations mentioned requires the joint and directed efforts of theoreticians and experimentalists, while, as in other problems of light scattering, it is reasonable to join laboratory experiments with natural measurements on clouds, fogs, or oceanic water, subordinating these measurements to carefully thought-out general purposes.

9. RADIATION CLIMATOLOGY AND THE OPTICS OF THE AEROSOL

For centuries, and up to very recent times, the purposes of atmospheric optics as a science have been connected with hopes of using optical "objects" for weather forecasting. For example, evidence of this is clearly expressed by the subtitle of a book on atmospheric optics published in the Russian language as late as in 1924 by Brounov: "Optical Phenomena of the Sky in Connection with Weather Prediction." However, successes in this region were minimal and we now understand very well the reason why. The connection of the optical situation with the weather bears a doubly uncertain character. Meteorological processes in some fashion not yet clear to us in all the necessary detail bear on the fate and characteristics of the atmospheric aerosol. The latter in turn has a strong effect on the radiation and thermal conditions of the atmosphere, and at the same time on the behavior of meteorological processes. The optical state of the atmosphere reflects, again in a form not completely understandable to us, the instantaneous state of the atmospheric aerosol. Therefore, the way to the utilization of optical "objects" for weather forecasting lies, in the first place, in the discovery of relations between the nature of the aerosol and its optical properties and, in the second place, in clarification of the connection between the weather-producing processes and processes of transport and transformation of the aerosol. The lack of a future for these empirical searches of various "objects" is amply illustrated by their history.

Nevertheless, numerous and unremitting investigations in these directions have not been fruitless. They have led to the discovery and understanding of a great collection of atmospheric-optical phenomena, and if this period is regarded in retrospect, it is not difficult to establish the fact that precisely this qualitative description and explanation of all these phenomena constitutes the fundamental content of atmospheric optics. It is also not difficult to show that this stage of initial accumulation of facts and the creation of general qualitative representations on the nature of atmospheric-optical phenomena has already been exhausted. We shall set forth the general information on the basic optical characteristics of the atmosphere and on their variability in the measure in which they are attained by means of comparatively simple apparatus and uncomplicated theory. This information forms a reliable basis for general orientation on the behavior of the phenomena. An unusual epitaph of this purely observational stage is given by the book of M. Minnaert which has recently come into our hands: "Light and Color in Nature."[91] However, it is impossible to close one's eyes to the fact that this trend becomes archaic and by no means contains the fundamental tendencies of modern science. And these tendencies are connected with the transition from passive observation of natural phenomena to single-minded quantitative analysis of their physical nature; the analysis operates steadily on a broad invasion in this region of contemporary physico-mathematical means of investigation, both experimental and theoretical. This is most evident in the sharp change of problems and methods of investigation.

The observational aspect of atmospheric optics apparently has not lost its value, but it is now regarded in a completely different fashion. This is the development of radiation climatology on the basis of a statistical analysis of data, for example, on the transmission in the atmosphere in different parts of the spectrum, obtained on a far-flung grid of observation stations with the aid of mass-produced and inexpensive measuring apparatus. The aim of such a type of regular service must be to provide the national economy with operational and climatological information, which permits rational solutions of problems of transport, construction, illumination, agrotechnology, health science, etc. In addition there stands out sharply a completely independent group of problems, which requires a different approach in principle, which can conditionally be called the optics of the aerosol and which, in particular, should uncover the way for the investigation of meteorological problems by optical means.

The outstanding problems here have already been mentioned above. These are the study of the angular and spectral dependences of the scattering matrix as a way to a clarification of the microstructure of the aerosol, and the study of the laws of radiation propagation in strongly scattering media. However, in a real atmosphere we meet up with two circumstances which hinder the path of investigation. In the first place, there is the constant and uncontrolled variability of the atmosphere as an object of investigation, which does not permit one to reproduce the conditions of measurements. In the second place, there is the presence of the submicroscopic fraction, which is optically active, but which does not submit to identification by other methods. Therefore, the contemporary atmospheric-optical experiment takes on very specific marks. It must be a complex, many-sided optical investigation of isolated special cases, in which the volume of optical and other auxiliary information obtained is so large that it admits of an unambiguous theoretical analysis both in the sense of identification of the parameters of the scattering medium and in relation to comparing it with the different optical properties. Of course, such a rich, complex study can be carried out only by use of the most advanced means of contemporary measurement technology, and requires serious concentration of forces of highly skilled staffs of scientists on isolated, comparatively narrow, and clearly defined aspects of the problem. The complexity and purposefulness of the investigations are the most characteristic marks of the modern problems of optics and of the aerosol. At the same time, it is evident that isolated experiments in this direction, no matter how fine the apparatus with which they are carried out, lead only to additions to the collection of observed cases, but do not help in advancing the understanding of the physical laws.

It is perfectly natural that laboratory investigations on colloidal optics serve as a necessary complement to natural measurements, but, as at an earlier time, they cannot replace it, because the scale of the atmosphere permits observations, for modern measurement technology, of certain phenomena not yet accessible to laboratory conditions. Therefore, atmospheric optics in the correct understanding of its problems, as before, maintains its position as one of the leading advance posts in the study of the scattering of light and of its propagation in scattering media, and this circle of problems is undoubtedly related to a number of fundamental problems of modern optics generally.

[1] J. M. Pertner and F. M. Exner, Meteorologische Optik, Wien, 1910.

[2] P. I. Brounov, Атмосферная оптика (Atmospheric Optics) (Gostekhizdat, 1924).

[3] See, for example, Rozenberger, История физики, (History of Physics), vol. 1 (Gostekhizdat).

[4] J. Kepler, Ad Vittelionem paralipomena, Frankfurt, 1604.

[5] Leonardo da Vinci, Trattato della pittura, CXIII and CLI.

[6] I. Newton, Opticks, book II, part III, proposition VII, (Gostekhizdat, 1954).

[7] L. Euler, Letters to a German Princess, vol. 1.

[8] J. W. Goethe, Entwurf einer Farbenlehre, Didaktische Teil, 145-172, Goethes Werke, II Abteilung, Bd. I., Weimar.

[9] R. Clausius, Pogg. Ann. 72, 76, 84 (1847-51).

[10] Chapuis, Compt. rend. 91, 522 (1880).

[11] W. Spring, Bull, l'Acad. Bruxelles, 504 (1898).

[12] A. Lalleman, Compt. rend. 69, 18, 9, 282, 917, 1294 (1869); 75, 709 (1872); 79, 693 (1874).

[13] Hartley, Nature 39, 474 (1889).

[14] W. M. Cohn, Gerl. Beitr. z. Geophys. 37, 198 (1932).

[15] Miley, Cullington, and Bedinger, Trans. Amer. Geophys. Union 34, 680 (1953).

[16] W. H. Pickering, Metheorol. Z. 514 (1885).

[17] O. Aufsess, Die Farbe der Seen, München, 1903.

[18] W. B. Pietenpol, Trans. Wisconsin Acad. Sci. 19, 562.

[19] C. V. Raman, Proc. Roy. Soc. 101, 64 (1922).

[20] V. V. Shuleĭkin, Известия Ин-та физики и биофизики, (News of the Institute of Physics and Biophysics), 1922, p. 119.

[21] See, for example, V. V. Sobolev, Перенос лучистой энергии в атмосферах звезд и планет, (Transfer of Radiant Energy in the Atmospheres of Stars and Planets) (Gostekhizdat, 1956), S. Chandrasekhar, Radiative Transfer (Russian translation, IL, 1953).

[22] J. Müller, Lehrbuch d. Kosm. Physik 1883, p. 403.

[23] B. de-Saussure, Mem. de l'Acad. de Turin, 1790; J. de phys. 38, 199 (1791).

[24] E. Brücke, Pogg. Ann. 88, 363 (1852).

[25] G. Govi, Compt. rend. 51, 360, 669 (1860).

[26] D. Tindall, Proc. Roy. Soc. Lond. 17, 223 (1869).

[27] R. Clausius, Pogg. Ann. 88, 554 (1853).

[28] F. Arago, Oeuvres, Paris, 1858, t. 7, 10.

[29] D. Brewster, Phil. Mag. 33, 290 (1867).

[30] M. I. Strutt, Phil. Mag. 41(4), 107, 274, 447 (1871); Phil. Mag. 12(5), 81 (1881).

[31] Rayleigh, Phil. Mag. 47(5), 375 (1899).

[32] L. I. Mandel'shtam, Полное собрание трудов, (Complete collected works) (Academy of Sciences Press), vol. I, p. 104.

[33] M. Smoluchowski, Ann. Physik 25, 205 (1908); A. Einstein, Ann. Physik 33, 1275 (1910).

[34] Maxwell-Garnett, Philos. Trans. 203A, 385 (1904), 205A, 237 (1906).

[35] A. E. Love, Proc. London Math. Soc. 30, 308 (1899).

[36] G. Mie, Ann. Physik 25, 377 (1908).

[37] G. V. Rozenberg, Some Questions of the Propagation of Electromagnetic Waves in Turbid Media, Dissertation, Moscow, 1954.

[38] G. V. Rozenberg, Usp. Fiz. Nauk 56, 77 (1955).

[39] G. V. Rozenberg, Usp. Fiz. Nauk 69, 57 (1959), Soviet Physics-Uspekhi 2, 666 (1959).

[40] Van der Hulst, Light Scattering by Small Particles, London, 1957.

[41] G. Stokes, Trans. Cambr. Philos. Soc. 9, 339 (1852).

[42] G. Jones, J. Opt. Soc. Amer. 31, 488, 433, 500 (1941); 32, 486 (1942); 37, 107, 110 (1947).

[43] F. Perrin, J. Chem. Phys. 10, 415 (1942).

[44] G. V. Rozenberg, Peculiarities of the Polarization of Light Scattered by the Atmosphere under Conditions of Twilight Illumination, Dissertation, Moscow, 1946.

[45] S. Chandrasekhar, Astrophys. J. 105, 424 (1946).

[46] V. V. Sobolev, Уч. зап. ЛГУ, (Sci. Notes Leningrad State University) No. 16 (1949).

[47] Fano, J. Opt. Soc. Amer. 39, 859 (1949).

[48] L. D. Landau and E. M. Lifshitz, Квантовая механика (Quantum Mechanics) Gostekhizdat, 1948 [Engl. Transl. Pergamon, 1958].

[49] G. V. Rozenberg, Оптика и спектроскопия (Optics and Spectroscopy) 5, 440 (1948).

[50] K. S. Shifrin, Рассеяние света в мутной среде, (Scattering of Light in a Turbid Medium) Gostekhizdat, 1951.

[51] Georgievskiĭ, Driving, Zolotavina, Rozenberg, Feĭgel'son, and Khazanov, under the general editorship of Prof. G. V. Rozenberg, Прожекторный луч в атмосфере, (Projector Beams in the Atmosphere), Academy of Sciences Press, 1960.

[52] M. V. Vol'kenshteĭn, Молекулярная оптика, (Molecular Optics) Gostekhizdat, 1951.

[53] R. Penndorf, J. Opt. Soc. Amer. 47, No. 2 (1957).

[54] J. Perrin, Atoms, ONTI, 1932.

[55] J. Stratton and H. Houghton, Phys. Rev. 38, 159 (1931); H. Houghton and W. Chalker, J. Opt. Soc. Amer. 39, 955 (1949).

[56] A. and E. Vassy, J. phys. 10, 75, 403, 459 (1939).

[57] Rodionov, Pavlova, Rdutlovskaya, and Reĭnov, Izv. Akad. Nauk SSSR Ser. Geogr. i Geofiz. 135 (1952).

[58] G. P. Faraponova, Труды ЦАО, (Trans. Cent. Astron. Obs.) No. 23, 52 (1957); No. 32 (1959).

[59] I. A. Khvostikov, Usp. Fiz. Nauk 24, 165 (1940).

[60] L. M. Levin, Izv. Akad. Nauk SSSR, Ser. Geofiz., No. 10 (1958).

[61] K. S. Shifrin, Тр. Всес. заочн. Лесотехн. ин-та, (Trans. All-Union Correspondence Institute of Lumber Technology) No. 1 (1955).

[62] V. A. Malyshev, Труды X Совещания по спектроскопии, (Proc. Tenth Conference on Spectroscopy) L'vov, (1956).

[63] N. Pil'chikov, Compt. rend. **115**, 555 (1892).

[64] Ch. Jensen, Himmelstrahlung, Handb. Physik **19** (1928).

[65] C. Dorno, Himmelshelligkeit, Himmelspolarisation und Sonnenintensität in Davos 1911 bis 1918. Verötent. d. Preuss. Meteorol. Inst. Abh. **6**, No. 303 (1919).

[66] C. Dorno, Physik d. Sonnen und Himmelstrahlung, Braunschweig, 1919.

[67] F. Roggenkamp, Meteorol. Zs. **50**, 111 (1933).

[68] R. Rubenson, Memoire sur la polarisation de la lumiere atmospherique, Upsala, 1864.

[69] J. L. Soret, Ann. chim. et phys. 503 (1888); Arch. sci. phys. et natur. **20**, 439 (1888).

[70] F. Ahlgrimm, Jahrb. d. Hamburg. Wiss. Anstalten **32** (1914).

[71] I. I. Tikhanovskiĭ, Phys. Z. **28**, 252 (1927).

[72] M. Born, Verhandl. Phys. Ges. **19**, 43 (1917); **20**, 16 (1918).

[73] I. M. Strutt, Proc. Roy. Soc. **95**, 155 (1918).

[74] J. Cabannes, Ann. phys. **15**, 5 (1921); La diffusion moleculaire de la lumiere, Paris, 1929.

[75] M. Born, Optik (Russian translation), ONTI, 1933.

[76] M. A. Schirmann, Ann. Physik **59**, 493 (1919); **61**, 195 (1920).

[77] G. V. Rozenberg, Izv. Akad. Nauk SSSR Ser. Geograf. i Geofiz. **13**, 154 (1949).

[78] M. S. Chandrasekhar, and D. Elbert, Trans. Amer. Philos. Soc. **44**, 643 (1954).

[79] Z. Sekera, Handb. Phys. **48**, Geophys. **11**, 288 (1957).

[80] Yu. N. Lipskiĭ, Paper at the Conference of Actinometry and Atmospheric Optics, Leningrad, (1959).

[81] E. V. Pyaskovskaya-Fesenkova, Исследование рассеяния света в земной атмосфере (Investigation of Light Scattering in the Earth's Atmosphere) Acad. of Sciences Press, 1957.

[82] E. M. Feĭgel'son, Izv. Akad. Nauk SSSR, Ser. Geofiz., No. 10 (1958).

[83] V. G. Kastrov, loc. cit. ref. 58, No. 32 (1959).

[84] V. F. Belov, ibid., No. 23 (1947); B. A. Chayanov, ibid., No. 32 (1959).

[85] E. S. Kuznetsov, Izv. Akad. Nauk SSSR, Ser. Geogr. i Geofiz., No. 5, 247 (1943).

[86] E. S. Kuznetsov, Izv. Akad. Nauk SSSR, Ser. Geograf. i Geofiz. 9, No. 3, 204 (1945); Feĭgel'son, Malkevich, Kogan, Koronatov, Glazova, and Kuznetsov, Расчет яркости света в атмосфере при анизотропном рассеянии (Calculation of the Brightness of Light in Atmosphere for Anisotropic Scattering) Academy of Sciences Press, 1958; K. S. Shifrin and N. P. Pyatovskaya, Таблицы наклонной дальности видимости и яркости дневного неба (Tables of the Oblique Range of Visibility and the Brightness of the Daytime Sky), Gidrometizdat, 1959.

[87] W. D. Chesterman and W. S. Stiles, Symposium on Searchlights, Illum. Engng. Soc., London, 1948.

[88] O. D. Barten'eva, Paper at the Conference on Actinometry and Atmospheric Optics, Leningrad, 1959.

[89] G. V. Rozenberg and I. M. Mikhaĭlin, loc. cit. ref. 49, **5**, 671 (1958).

[90] R. O. Gumprecht and C. M. Sliepcevich, J. Phys. Chem. **57**, No. 1 (1953); J. H. Chin, C. M. Sliepcevich, and M. Tribus, J. Phys. Chem. **59**, No. 9 (1955); K. S. Shifrin, loc. cit. ref. 61, No. 2 (1956).

[91] M. Minnaert, Light and Color in Nature (Russ. Transl.), Fizmatgiz, 1959.

[92] Bezold, Pogg. Ann. **123**, 240 (1863).

[93] V. G. Fesenkov, Тр. Главн. Российск. астрофизич. обсерв. (Trans. of the Main Russian Astrophysical Observatory) **2**, 7 (1923).

[94] V. G. Fesenkov, Астрон. ж. (Astron. Journal) **7**, 100 (1932).

[95] For example, T. G. Megrelishvili and I. A. Khvostikov, Dokl. Akad. Nauk SSSR **59**, No. 7 (1948).

[96] G. V. Rozenberg, The Anatomy of Sunrise, Paper at the Conference on Actinometry and Atmospheric Optics, Leningrad, 1959.

[97] For example, T. G. Megrelishvili, Dokl. Akad. Nauk SSSR **52**, 127 (1946); **55**, 713 (1947); Paper at the Conference on Actinometry and Atmospheric Optics, Leningrad, (1959).

[98] See, for example, G. V. Rozenberg, Тр. Геофиз. ин-та АН СССР (Trans. Geophys. Inst. U.S.S.R. Acad. Sci.) No. 12, 35 (1949).

[99] F. H. Ludlam, Usp. Fiz. Nauk SSSR **65**, 407 (1958).

Translated by R. T. Beyer

On the Rayleigh-Scattering Optical Depth of the Atmosphere

A. T. YOUNG

Physics Department, Texas A&M University, College Station 77843

9 July 1980 and 4 December 1980

ABSTRACT

Hoyt's (1977) and Fröhlich and Shaw's (1980) attempts to revise Rayleigh-scattering optical depths, using Raman-free depolarization ratios, are incorrect (see Young, 1980). King's (1923) formula for the anisotropy correction requires that the rotational Raman light be *included* in the depolarization measurement. Recent depolarization data show that the scattering optical depths for dust-free air are only ~1% smaller than those used by Elterman (1968) and Penndorf (1957). Measured extinction values appreciably below standard Rayleigh values indicate systematic errors in measurement or reduction methods, not errors in the Rayleigh tables.

1. Introduction

Three years ago, Hoyt (1977) published revised Rayleigh-scattering optical depths, based on some early laser measurements of the depolarization of elastically scattered light (i.e., excluding rotational Raman lines). Such measurements are about four times smaller (Rowell *et al.*, 1971) than those that include the rotational scattering, so Hoyt's (1977) depolarization corrections were about four times smaller than those used by Elterman (1968)[1] and Penndorf (1957).

However, the classical formula (King, 1923) used by Hoyt (1977) requires that the rotational lines be included in the polarization measurement. To exclude them is to exclude the contribution of the Raman-shifted photons to the total molecular extinction. Hoyt's error was compounded by Fröhlich and Shaw (1980), who not only repeated it but confused the depolarization for natural light with that for (polarized) laser light (Young, 1980). Part of the confusion is due to different uses of the term Rayleigh scattering by different workers, and further confusion may be due to the variety of notations used by different authors. This note is intended to help atmospheric scientists find their way about in this jungle.

2. Origin of the depolarization correction

The scattering of light by anisotropic molecules was treated by Cabannes (1921), who showed that earlier work by Lord Rayleigh allowed the scattered intensity to be expressed in terms of the depolarization ratio ρ for randomly oriented molecules illuminated by natural (unpolarized) light. For 90° scattering, he found that anisotropic molecules scatter more light than isotropic molecules giving the same refractive index, by a factor

$$F_C = (6 + 6\rho)/(6 - 7\rho). \tag{1}$$

He assumed that the same factor applied to the total scattering. King (1923) showed that because the anisotropy alters the angular distribution of the scattered light, the total scattering is actually increased by

$$F_K = (6 + 3\rho)/(6 - 7\rho). \tag{2}$$

Kasten (1968) and many other authors call (2) the Cabannes anisotropy factor, despite its discovery by King. The factor exceeds unity for any nonzero anisotropy, because extinction is proportional to the mean-*square* polarizability (Rayleigh, 1918; Kasten, 1968), which is the squared mean that appears in Rayleigh's (1899) refractivity formula, plus an additional variance proportional to the square of the anisotropy.

As neither Cabannes nor King considered molecular rotation, their models produce no rotational Raman scattering (which was not discovered until 1928). Later, Cabannes and Rocard (1929) applied the theory to rotating molecules. In the static models, molecules with maximum polarizability parallel to the incident electric vector have larger individual cross sections than those with less favorable orientations. In the dynamical model, each individual molecule changes its cross section as it rotates. The mean cross section per molecule is the same in both cases, because all orientations are equally likely. But the rotating molecules radiate a wave train that is not purely sinusoidal, but is amplitude-modulated, due to the changing cross section. Such a wave is radiated by an AM radio station broadcasting a pure tone. As is well known, the spectrum of such a wave contains the original carrier frequency, plus two side bands at the sum and difference of the exciting and modulating frequencies. These side bands are the rotational Raman spectrum.

---

[1] Elterman, L., 1968: UV, visible and IR attenuation for altitudes to 50 km. AFCRL-68-0153. Air Force Geophysics Laboratory, L. G. Hanscom AFB, Bedford, MA, $$$ pp.

Clearly, molecular rotation does not alter the total scattering cross section, but does split off part of the classical Rayleigh scattering of stationary molecules into the Raman components (cf. Stuart, 1934; Fabelinskii, 1968). All the classical results are still valid when the language is changed to that of quantum mechanics, although the physical interpretation of the polarizabilities is slightly different (cf. Long, 1977). In particular, the King factor (2) is still valid, provided that the Raman scattering is included in the depolarization measurement.

In fact (Penney *et al.*, 1974), many classical relations are valid for any centered portion of the Rayleigh-Raman band; omitting the Raman lines gives less depolarization, which corresponds to the smaller cross section for scattering without change of frequency. But we cannot isolate this part of atmospheric extinction. We can only observe the attenuation of the incident beam by all processes together. Thus the full depolarization must be used to compute the entire optical depth for molecular scattering.

It is confusing that recent workers have used "Rayleigh scattering" for only the unshifted component. But changing the eponymy does not change the physics. Perhaps it would be safer to say simply "molecular scattering" and avoid the ambiguous use of Rayleigh's name.

3. Depolarization ratios

The early workers used the depolarization ratio for unpolarized incident light, because their strongest light sources were unpolarized. Similarly, the laser workers today generally use depolarizations for linearly polarized incident light. The former has been denoted by subscripts such as u or n (for unpolarized or natural); the latter, by p or l (for polarized or linear), or v or $\perp$, to indicate that the incident electric vector is perpendicular to the scattering plane. Some authors use primes; often one component has no subscript while the other does. It is often necessary to read the description of the experiment to determine just what kind of depolarization was measured. Still other depolarization ratios are occasionally used; Long (1977) gives a good discussion.

Many recent workers distinguish between depolarizations for the central component alone, and for the total scattering, including the Raman wings. Rowell *et al.* (1971) discuss the differences between these, but only for polarized incident light. Their notation, which uses superscripts of c for the central component, w for the Raman wings alone, and t for the total scattered radiation, is becoming fairly common.

The depolarization ρ_n^t required in (2) can be found from the other common forms:

$$\rho_n^t = 2\rho_v^t/(1 + \rho_v^t) = 8\rho_v^c/(1 + 8\rho_v^c). \quad (3)$$

Then (2) becomes

$$F_K = (6 + 3\rho_n^t)/(6 - 7\rho_n^t) = (3 + 6\rho_v^t)/$$
$$(3 - 4\rho_v^t) = (3 + 24\rho_v^c)/(3 - 4\rho_v^c). \quad (4)$$

I have examined several recent depolarization measurements, and find (Young, 1980) that the effective values for dry air are $\rho_n^t = 0.0279$ [which is close to Kasten's (1968) 0.0295, and not much below the value 0.0350 used by Penndorf (1957) and Elterman (1968)[1]], and $F_K = 1.048$ (Kasten: 1.051; Penndorf and Elterman: 1.061). Thus, modern data suggest about a 1.2% decrease from the older Rayleigh optical depths, but nothing like the 3.6% reduction proposed by Hoyt (1977), or the 4.5% of Fröhlich and Shaw (1980).

Remarkably, the best modern values differ inappreciably from King's (1923) $F_K = 1.049$, based on Cabannes' (1921) depolarization measurements. Hoyt's proposed corrections to Elterman and Penndorf are too large by a factor of 3.

4. Vibrational Raman effects

The molecular extinction estimated by including the rotational Raman lines in the depolarization will still be a slight underestimate, because it neglects the small contribution of vibrational Raman scattering to the total cross section. However, this involves the polarizability tensor for the excited vibrational states of the scattering molecules (Long, 1977), and cannot be simply related to the ground-state polarizabilities that contribute the bulk of the scattering. Fortunately, this error is only on the order of 0.1%, which is still somewhat smaller than other errors in estimating the scattering optical depth. But, as current uncertainties in F_K are only a few tenths of 1%, vibrational Raman scattering and other comparable effects, such as deviations from the ideal-gas equation of state, may soon have to be included.

5. Extinction measurements

Hoyt (1977) suggested that substantial discrepancies between published extinction data and tabulated molecular-scattering optical depths might be due to errors in the Rayleigh tables. We now see that this explanation is inadequate, so the fault must lie primarily with the extinction measurements.

Measurements that use only the Sun are particularly liable to systematic error, because solar heating causes convection that increases turbidity near midday (cf. Russell and Shaw, 1975). But temporal changes that are strongly correlated with solar air mass are not detectable in the Bouguer-Langley plot. Increased turbidity near noon produces lines of too low a slope, thus underestimating the extinction.

Other common sources of error are the use of $\sec z$ for air mass; or misuse of Bemporad's (1904) air mass tables, either by use of true instead of apparent (refracted) altitudes as arguments, or by failure to correct for local pressure and temperature. Since Bemporad's tables are for sea level, they require considerable modification for high-altitude sites. All these errors produce excessive air mass values near the horizon, again underestimating the extinction per air mass. Even if the errors seem negligible compared to the errors of single measurements, they persist in averaged data, and will exceed the much smaller random component of error in the mean.

These and other errors in extinction measurements are discussed at length by Young (1974). A very careful discussion, with an example of the proper use of Bemporad's tables, is given by Abbot *et al.* (1922).

Acknowledgment. This work was supported by NASA Planetary Atmospheres Grant NGR 44-001-117.

REFERENCES

Abbot, C. G., F. E. Fowle and L. B. Aldrich, 1922: *Annals of the Astrophysical Observatory of the Smithsonian Institution,* Vol. 4, 330–344.

Bemporad, A., 1904: Zur Theorie der Extinktion des Lichtes in der Eradatmosphäre. *Mitt. Grossherz. Sternwarte Heidelberg,* No. 4.

Cabannes, J., 1921: Sur la diffusion de la lumière par les molécules des gaz transparents. *Ann. Phys.,* **15,** 5–150.

——, and Y. Rocard, 1929: La théorie électromagnétique de Maxwell-Lorentz et la diffusion moléculaire de la lumière. *J. Phys. Rad.,* **10,** 52–71.

Fabelinskii, I. L., 1968: *Molecular Scattering of Light.* Plenum Press, 253–256.

Fröhlich, C., and G. E. Shaw, 1980: New determination of Rayleigh scattering in the terrestrial atmosphere. *Appl. Opt.,* **19,** 1773–1775.

Hoyt, D. V., 1977: A redetermination of the Rayleigh optical depth and its application to selected solar radiation problems. *J. Appl. Meteor.,* **16,** 432–436.

Kasten, F., 1968: Rayleigh-Cabannes Streuung in trockener Luft unter Berücksichtigung neuerer Depolarisations-Messungen. *Optik,* **27,** 155–166.

King, L. V., 1923: On the complex anisotropic molecule in relation to the dispersion and scattering of light. *Proc. Roy. Soc. London,* **A104,** 333–357.

Long, D. A., 1977: *Raman Spectroscopy.* McGraw-Hill.

Penndorf, R., 1957: Tables of the refractive index for standard air and the Rayleigh scattering coefficient for the spectral region between 0.2 and 20.0 μ and their application to atmospheric optics. *J. Opt. Soc. Amer.,* **47,** 176–182.

Penney, C. M., R. L. St. Peters, and M. Lapp, 1974: Absolute Raman cross sections for N_2, O_2, and CO_2. *J. Opt. Soc. Amer.,* **64,** 712–716.

Rayleigh, Lord (J. W. Strutt), 1899: On the transmission of light through an atmosphere containing small particles in suspension, and on the origin of the blue of the sky. *Phil. Mag.,* **47,** 375–384.

——, 1918: On the scattering of light by a cloud of similar small particles of any shape and oriented at random. *Phil. Mag.,* **35,** 373–381.

Rowell, R. L., G. M. Aval and J. J. Barrett, 1971: Rayleigh-Raman depolarization of laser light scattered by gases. *J. Chem. Phys.,* **54,** 1960–1964.

Russell, P. B., and G. E. Shaw, 1975: Comments on "The precision and accuracy of Volz sunphotometry." *J. Appl. Meteor.,* **14,** 1206–1209.

Stuart, H. A., 1934: *Molekülstruktur.* Verlag-Springer, 176–183.

Young, A. T., 1974: Atmospheric extinction. *Methods of Experimental Physics,* Vol. 12 A, *Astrophysics,* N. P. Carleton, Ed., Academic Press, 123–180.

——, 1980: Revised depolarization corrections for atmospheric extinction. *Appl. Opt.,* **19,** 3427–3428.

Rayleigh scattering

**Questions of terminology are resolved
during a historical excursion through the physics of
light-scattering by gas molecules**

Andrew T. Young

We all know that blue skies and red sunsets are due to the "Rayleigh scattering" of sunlight. But beyond that, Rayleigh scattering means different things to different scientists.

The "Rayleigh line" of Raman spectroscopists, who study the rotational and vibrational behavior of molecules by analyzing frequency shifts that occur when monochromatic light is scattered, is not the same as the "Rayleigh line" of Brillouin spectroscopists, who analyze light scattered by acoustic phonons, or density fluctuations. Their "Rayleigh line" is only the unshifted central component of the Raman spectroscopists' line, and contains less than 30% of the scattered energy; if we restricted "Rayleigh scattering" to it, we would have to say the blue sky is due chiefly to Brillouin scattering.

For many, Rayleigh scattering implies coherence or elastic scattering, or it brings to mind functions such as λ^{-4} and $(1 + \cos^2\theta)$. Yet the scattering of sunlight by air molecules or density fluctuations in the atmosphere is not strictly proportional either to λ^{-4} or to

$(1 + \cos^2\theta)$. And if we exclude inelastic scattering of light by rotating molecules, which accounts for about 3% of the sky light, only the remaining 96.6% of molecular scattering is without frequency shift. Furthermore, only 98.8% of *this* light is coherent in the forward direction.

The conflicting uses of Rayleigh's name, and the confusion about details of the phenomenon that bears his name, are not "just" matters of nomenclature. These problems recently caused an error in physics to be published in a reputable journal, to be amplified by other authors and cited approvingly by still others before being corrected[1]—in spite of having been foreseen and cautioned against[2] nearly fifty years ago! Also, a question of the accurate reflection of history is involved, as the conflicting uses of the term "Rayleigh line" illogically refer to phenomena discovered long after Lord Rayleigh's death.

Why do we honor Rayleigh with eponymy in molecular scattering? What phenomena did he actually explain? Is unchanged frequency a central feature of his work or not?

To see what use of Rayleigh's name is most appropriate, we will look at what

he did. Other, more suitable names will turn up to replace the conflicting and inappropriate uses of Rayleigh's.

The blue sky

To appreciate Rayleigh's contributions one must understand how the blue sky was explained in the mid-19th century. A number of observers, from Leonardo da Vinci to Johann Wolfgang von Goethe, had noticed the sky-blue color in smoke. Isaac Newton attributed the blue sky to first-order interference in "Vapors when they begin to condense and coalesce into small Parcels." Rudolf Clausius pointed out that refraction within macroscopic droplets would cause problems, which could be avoided if the droplets were hollow, so he argued for minute bubbles of water suspended in the air.

Ernst Wilhelm von Brücke objected that first-order interference blue is too unsaturated, and demonstrated the sky-blue color in hydrosols. In 1809, Dominique-François-Jean Arago discovered the strong polarization of the sky 90 degrees from the Sun (see figure 1), a finding inconsistent with Clausius's bubbles.

The color and polarization of sky light remained unexplained for another six decades. In 1869, John Tyndall[3] stated that "these questions constitute, in the opinion of our most eminent authorities, the two great standing enigmas of meteorology."

The authority most eminent in Tyndall's opinion was Sir John Herschel, who had stated the polarization problem in these terms:

The cause of the polarization is evidently a reflection of the sun's light upon *something*. The ques-

Andrew T. Young is adjunct associate professor of astronomy at San Diego State University, in San Diego, California.

tion is, On what? Were the angle of maximum polarization 76°, we should look to water or ice as the reflecting body, however inconceivable the existence in a cloudless atmosphere on a hot summer's day of unevaporated molecules (particles?) of water. But though we were once of this opinion, careful observation has satisfied us that 90°, or thereabouts, is a correct angle, and that therefore, whatever be the body on which the light has been reflected, *if polarized by a single reflection*, the polarizing angle must be 45°, and the index of refraction, which is the tangent of that angle, unity; in other words, the reflection would require to be made *in air upon air!*

George Stokes had argued, in his 1852 paper *On the Change of Refrangibility of Light*, that the transverse vibrations of light waves implied that particles smaller than the wavelength should not "diffract" in the direction of vibration any intensity from a polarized beam, and confirmed this experimentally with hydrosols. In 1868, Tyndall[4] noticed the sky-blue color in a beam of light that he was using to make photochemical smog. In his notebook, he wrote, "connect this blue with the colour of the sky," and he soon

> took a pleasure . . . in determining whether in all its bearings and phenomena the blue light was not identical with the light of the sky. This to the most minute detail appears to be the case. The incipient actinic clouds are to all intents and purposes pieces of artificial sky, and they furnish an experimental demonstration of the constitution of the real one.

He believed these clouds contained particles "whose diameters constitute but a very small fraction of the length of a wave of violet light."

Tyndall was puzzled that not only natural background aerosols but also those he made from "substances of widely different refractive indices, and therefore of very different polarizing-angles as ordinarily defined" by Brewster's law, all gave the same results, "*absolutely independent of the polarizing-angle.* The law of Brewster does not apply to matter in this condition; and it rests with the undulatory theory to explain why."

Rayleigh answered this challenge. In only his eighth published paper,[5] he pointed out that "the difficulty is imaginary, and is caused mainly by misuse of the word reflection," which has "no application unless the surface of the disturbing body is larger than many square wave-lengths . . . ."

By repeating Stokes's argument more clearly, he shows that the polarizing phenomena follow from the transverse-wave nature of light, adding, "I cannot see any room for doubt as to the result." He then uses dimensional analysis to show that

> the ratio of the amplitudes of the vibrations of the scattered and incident light varies inversely as the square of the wave-length, and the intensity of the lights themselves as the inverse fourth power.

His own measurements of the ratio of skylight to sunlight as a function of wavelength agree well with this "important law" and disagree with the λ^{-2} dependence of the first-order interference blue. Thus blue light of 450 nm wavelength, for example, is scattered more intensely than red light of 670 nm wavelength by a ratio of $(670/450)^4$ or about 5 to 1.

Furthermore, Rayleigh's detailed analysis produces the $(1 + \cos^2\theta)$ phase function; he suggests that the imperfect polarization of light from the sky is due to multiple scattering, "but it must be remembered that an insufficient fineness of some of the particles of foreign matter would have a like result"; and he shows, as Brücke had demonstrated with hydrosols, that the transmitted light is reddened by the "rapid diversion of the rays of short wave-length."

These substantial results are based entirely on the elastic-solid theory of the luminiferous ether. Ten years later,[6] he rederived them using Maxwell's electromagnetic theory. Here he also considered higher-order effects for spheres of finite size, for infinite cylinders and other possibilities.

Aerosol, or air itself?

Rayleigh's 1871 paper[5] begins,
It is now, I believe, generally ad-

mitted that the light which we receive from the clear sky is due in one way or another to small suspended particles which divert the light from its regular course. On this point the experiments of Tyndall with precipitated clouds seem quite decisive.

(The photograph in figure 2 was taken around the time Rayleigh wrote this.) Rayleigh accepts Tyndall's repeated assertions that clean gases are "optically empty" and incapable of scattering light. But he rejects the general assumption "that the foreign matter in the atmosphere is water or ice," as the "blue haze evidently akin to the azure of the sky...on a hot day, cannot possibly be attributed to aqueous particles." He suggests that "a strong case might be made out for common salt."

James Clerk Maxwell wrote to Rayleigh two years later, posing the question:

> Suppose that there are N spheres of density ρ and diameter s in unit of volume of the medium. Find the index of refraction of the compound medium and the coefficient of extinction of light passing through it.
>
> The object of the enquiry is, of course, to obtain data about the size of the molecules of air.

In 1899, Rayleigh succeeded in relating the molecular extinction to the refractive index of a gas and was able to show[7] that "even in the absence of foreign particles we should still have a blue sky." This relation also provided the first accurate method of estimating Avogadro's number. The replacement of aerosols by air molecules as the scatterers was itself as revolutionary as Rayleigh's earlier replacement of "reflection" by "scattering".

"To facilitate comparison with former papers," Rayleigh again develops the theory from the elastic-solid picture. "By considering the resultant of the primary vibration and of the secondary vibrations which travel in the same direction," he shows "that the phases are no more distributed at random." This *coherence* in the forward direction retards the phase of the resultant wave; the phase delay is related to the refractive index μ of the gas. If the energy of the incident beam falls with distance x, as $\exp(-hx)$, the extinction coefficient is

$$h = 32\pi^3(\mu - 1)^2/(3n\lambda^4) \qquad (1)$$

where n is the number of molecules per unit volume. Consequently, the "observed fact" that the index of refraction increases with decreasing wavelength means that "the falling off of transparency at the blue end of the spectrum will be even more marked than according to the inverse fourth power of the wave-length." (For air, the effective

Lord Rayleigh (John William Strutt), 1842–1919, in a photograph he took himself around 1870. (Photograph courtesy of John Arthur Strutt.)   Figure 2

power is about -4.08 in visible light.)

> Let us now inquire what degree of transparency of air is admitted by its molecular constitution, *i.e.*, in the absence of all foreign matter.... The calculation requires the value of n. Unfortunately this number—according to Avogadro's law the same for all gases—can hardly be regarded as known.

Rayleigh shows that Maxwell's crude estimate of 1.9×10^9 molecules per cubic centimeter agrees fairly well with values from Pierre Bouguer's stellar measures of atmospheric extinction.

> If n be regarded as altogether unknown, we may reverse our argument, and we then arrive at...a lower limit for n, say 7×10^{18} per cubic centimetre.... When we take into account the known presence of foreign matter, we shall probably see no ground for any reduction of Maxwell's number.

(The modern value is about 2.7×10^{19} per cm^3.)

It is remarkable that the relation between refractivity and extinction had been found in the 1880's by Ludwig Lorenz, during his development of the theory of refraction. Lorenz even used it to estimate the number density in air, and found 1.63×10^{19} molecules per cubic centimeter. But Lorenz published it in Danish, in a long paper on another topic, so like Stokes's explanation of the polarization, it was ignored.

Rayleigh's papers brought these matters to everyone's attention.

Subsequently, Sir Arthur Schuster showed that the best atmospheric extinction measurements closely agreed with improved laboratory values of n, a result that Rayleigh saw as "apparently justifying to the full the inference that the normal blue of the sky is due to molecular scattering."[8] A few years later, Charles Fabry's student Jean Cabannes measured molecular scattering in the laboratory,[9] thus finally refuting Tyndall's wrong notion.

Raleigh also pointed out that molecules in dense media cannot overlap, which "detracts from the random character of the distribution. And when, as in liquids and solids, there is some approach to a regular spacing, the scattered light must be much less...." We should note here that in our tenuous atmosphere, summing the scattering from individual molecules gives the same result as a calculation of scattering from density fluctuations in the gas. Thus these two ways of analyzing the scattering of sunlight in the atmosphere are equivalent.

Molecular anisotropy

So far, all is straightforward: Rayleigh has explained the blue sky with molecular scattering. But real molecules are not isotropic little spheres. Already in 1871, Rayleigh noted[10] that higher-order terms in the scattering would depend on "the *shape* of the disturbing particles." In 1881 he says,[6] "We must remember that our recent results are limited to particles of a spherical form," and adds, with a restraint rare among theoreticians a century later, "In the case of an ellipsoidal particle the problem is soluble; but it is perhaps premature to enter upon it, until experiment has indicated the existence of phenomena likely to be explained thereby."

In his great 1899 paper,[7] he observes that

> In the electric theory, to be preferred on every ground except that of easy intelligibility, the results are more complicated in that...the scattered ray depends upon the shape and not merely upon the volume of the small obstacle....
>
> If we abandon the restriction as to sphericity,...the theoretical conclusion that the light diffracted in a direction perpendicular to the primary rays should be *completely* polarized may well be seriously disturbed. If the view, suggested in the present paper, that a large part of the light from the sky is diffracted from the molecules themselves, be correct, the observed incomplete polarization at 90° from the Sun may be partly due to the

Table 1: Depolarization ratios (diatomic and linear molecules at 90°)

Spectral region	Unpolarized incident light	Vertically polarized incident light
Cabannes line	$\rho_o{}^c = \dfrac{6\epsilon}{180 + 7\epsilon}$	$\rho_v{}^c = \dfrac{3\epsilon}{180 + 4\epsilon}$
rotational wings	$\rho_o{}^w = 6/7$	$\rho_v{}^w = 3/4$
total scattering	$\rho_o{}^t = \dfrac{6\epsilon}{45 + 7\epsilon}$	$\rho_v{}^t = \dfrac{3\epsilon}{45 + 4\epsilon}$

A small depolarization ratio indicates a large degree of polarization in the scattered light. For air, ϵ has an effective value of approximately 0.22. The extinction relations given in the text require ρ_o, which was the depolarization usually measured before lasers were used. Now authors usually report $\rho_v{}^t$, and sometimes $\rho_v{}^c$, because they use laser light, which is polarized.

molecules behaving rather as elongated bodies with indifferent orientation than as spheres of homogeneous material.

In 1910, he again warns[8] that "If the shape be elongated, there would be incomplete polarization," and that "a molecule, especially a diatomic molecule, can hardly be supposed to behave as if it were the dielectric sphere of theory. Questions are here suggested for whose decision the time is perhaps not yet ripe."

Finally, in a 1918 paper[11] (his 430th publication) *On the Scattering of Light by a Cloud of Similar Small Particles of any Shape and Oriented at Random*, Rayleigh says, "My son's recent experiments upon light scattered by carefully filtered gases reveal a decided deficiency of polarization in the light emitted perpendicularly, and seem to call for a calculation of what is to be expected from particles of arbitrary shape."

He considers first an axially symmetric molecule, and then a general triaxial one, whose electric polarizabilities along its axes are A, B, and C. After averaging over all orientations, he calculates "the *ratio* of intensities of the two polarized components in the light scattered at right angles." This ratio, which had been measured by his son, Robert John Strutt,[12] was soon denoted by ρ in Cabannes's thesis,[13] and then was named the *depolarization* by Louis Vessot King.[14] We should note that a small depolarization ratio signifies a large degree of polarization in the scattered light.

Rayleigh treats two kinds of depolarizations. The first, which I shall call $\rho_v{}^t$ (where t and v denote total scattering and vertical polarization, respectively), occurs "when the primary light, propagated parallel to X, is completely polarized with vibrations parallel to Z, the direction of the secondary ray [the observed ray] being along OY," as shown in figure 3. The second, $\rho_o{}^t$ (which is almost twice as great), arises "if the primary light travelling in direction OX is unpolarized." Rayleigh finds that polarized light scatters into light that has a depolarization ratio of

$$\rho_v{}^t = \tag{2}$$
$$\frac{A^2 + B^2 + C^2 - AB - BC - CA}{3(A^2 + B^2 + C^2) + 2(AB + BC + CA)}$$

and that for unpolarized incident light, the scattered light has a depolarization ratio of

$$\rho_o{}^t = \tag{3}$$
$$\frac{2(A^2 + B^2 + C^2 - AB - BC - CA)}{4(A^2 + B^2 + C^2) + AB + BC + CA}$$

It is important not to confuse $\rho_o{}^t$ with $\rho_v{}^t$ as some recent authors have done.[1] The relation between the two depolarizations (given implicitly in table 1) is sometimes attributed to Kariamanikkam Krishnan or Cabannes, but was first published by King.[14]

The younger Strutt had problems with impurities (and probably stray light) in his first experiments,[12] and found spurious depolarizations of 3.2% for argon and 42% for helium. He later reported[15] much lower values, explaining that "the argon formerly used, which was, I believe, of French origin, had been put away in bottles. . . . A very definite smell of hydrogen sulphide was noticed in emptying the old bottles. The presence of a trace of this gas would quite account for a fog giving misleading results." As Cabannes[13] had obtained negligible depolarization from *his* argon, and had correctly concluded that the proper value to adopt was zero, while Rayleigh *fils* (figure 4) continued to claim a small residual depolarization for the monatomic gases, which "I do not think . . . can be explained away," the reader may judge on which side of the Channel something was rotten.

A more convenient parameter

Cabannes[13] noticed that only two functions of Rayleigh's A, B, and C appear in equations 2 and 3, $(A^2 + B^2 + C^2)$ and $(AB + BC + CA)$, which he denoted by ΣA^2 and ΣBC for short. These two independent quantities can also be expressed in many other ways.

For example, Rayleigh[11] inquired whether a want of equality among the coefficients A, B, C interferes with the relation between attenuation and refractive index, explained in my paper of 1899. The answer appears to be in the affirmative, since the attenuation depends upon $A^2 + B^2 + C^2$, while the refractive index depends upon $A + B + C$. . . .

This distinction between the mean-square polarizability and the square of the mean is fundamental. It is readily expressed in Cabannes's notation, as $\Sigma BC = [(\Sigma A)^2 - \Sigma A^2]/2$.

Cabannes[13] preferred to use the depolarization $\rho_o{}^t$ and the refractive index μ as the two independent parameters. In these terms, he found the intensity scattered at right angles to be larger by a factor

$$F_c = (6 + 6\rho_o')/(6 - 7\rho_o')$$

than for a gas of isotropic molecules with the same refractive index. He erroneously assumed that the extinction would increase by the same factor.

However, King[14] showed that because the molecular anisotropy makes the scattering more nearly isotropic, the extinction increases by the factor

$$F_k = (6 + 3\rho_o')/(6 - 7\rho_o')$$

instead. Thus, while the fraction of the incident unpolarized illumination that is scattered by a unit volume of gas into a unit solid angle at θ, is

$$\frac{\pi^2(\mu^2 - 1)^2}{2n\lambda^4}$$
$$\times \frac{6 + 6\rho_o'}{6 - 7\rho_o'} \left[1 + \frac{(1 - \rho_o')}{(1 + \rho_o')} \cdot \cos^2\theta \right]$$

which contains F_c, the extinction coefficient, found by integrating this expression over the sphere, contains F_k. (It is F_k times the extinction coefficient given in equation 1.)

Unfortunately, F_k is widely but mistakenly called "the Cabannes anisotropy factor." In fact, Cabannes[16] calls only F_c "*le facteur d'anisotropie;*" F_k does not appear explicitly in his work.

Later, George Placzek chose as parameters the trace α and the anisotropic part γ of the polarizability tensor, and these two invariants are widely used today. The refractivity is proportional to α, and the depolarizations, given in table 1, are more compactly expressed[17] by writing $\epsilon = (\gamma/\alpha)^2 = (\Sigma A^2 - \Sigma BC)/(\Sigma A/3)^2$. The parameter ϵ is a measure of the anisotropy in a molecule's polarizability. In this notation, the ratio of the mean-square polarizability to the square of the mean, α, is just $F_k = 1 + (2\epsilon/9)$. For air, ϵ has an effective value of about 0.22.

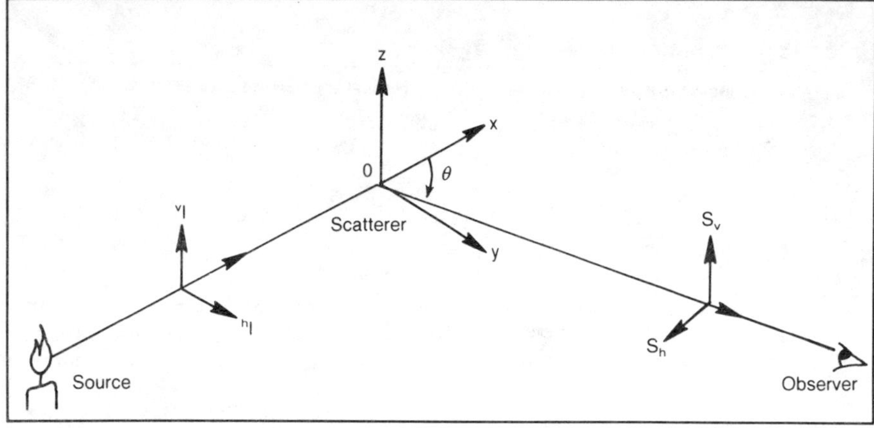

Scattering geometry and notation. Incident light (*I*) is scattered by a molecule at the origin (*O*) through an angle θ. The scattered light *S* is resolved into vertical and horizontal components S_v and S_h with electric vectors perpendicular and parallel to the scattering plane xy. Superscripts are used to denote the polarization of incident radiation; subscripts, the polarization of scattered radiation.   Figure 3

Moving molecules

The real confusion began with Chandrasekhara Raman's 1928 announcement of his observation of "a new radiation," due to *vibrational* combination scattering. Cabannes[18] recognized the new frequencies as the "optical beats" between the molecules' internal frequencies and that of the monochromatic illumination, an effect he had sought since 1924. He gave a classical analysis, showing that the changing polarizability presented by a rotating molecule also causes the scattered wave to be amplitude-modulated. The spectrum of scattered light then contains the incident "carrier" frequency and two side-bands shifted by twice the rotation frequency. Cabannes and Yves Rocard[18] gave the relative intensities of the shifted and unshifted parts in both polarization states and showed that they add up to Rayleigh's result for stationary molecules (see table 2).

Charles Manneback[19] treated the problem quantum-mechanically, and found that "in the limit $J \to \infty$ ... one gets full agreement with the results of Cabannes." At temperatures high enough to populate states with large rotational quantum number J, Cabannes's results are valid, "if one thinks of the Rayleigh line as split into just three parts," with all the rotationally-displaced lines on each side of the carrier frequency lumped together.

We will for brevity call the radiation displaced by vibrational fundamentals and overtones Raman radiation or lines, whereby we also think of the attendant fine structure as included. The undisplaced scattering, including its fine structure, we will call Rayleigh radiation or lines

says Manneback. Placzek and Edward Teller[20] likewise speak of the "rotational fine structure of the Rayleigh line."

Cabannes[21] said explicitly,

Let us suppose that there are no intramolecular vibrations and that the atomic nuclei remain fixed at their stable equilibrium positions; the coefficients A, B, C are constants.... One gets nothing in the scattered light but the exciting radiation (or, more exactly, the slightly-dispersed ensemble of the exciting line and the rotational lines.) This ensemble constitutes the radiation of Lord Rayleigh.

These papers are among the first to use the terms "Rayleigh scattering" and "Rayleigh radiation" for scattering by anisotropic molecules. Their usage is perfectly justified historically, as the younger Rayleigh discovered depolarization by anisotropic molecules and his father developed the basic theory. How, then, did the conflicting uses of the term "Rayleigh line" arise?

First, careless readers of Manneback's abstract[19] got the wrong idea, as he mentions "radiation scattered without change of frequency (Rayleigh lines) and with change of frequency (Raman lines)," and implies that only vibrational changes are meant, by referring to "rotational fine structure." Franco Rasetti had already called the rotational parts of Manneback's "Rayleigh" line "rotational Raman" lines. From here it was an easy step for S. Bhagavantam[22] to reproduce Cabannes's expression[18] for the middle of Manneback's "three parts" (^vC_v and ^vC_h in table 2), and to call this the "intensity of the true Rayleigh scattering," as opposed to the "pure rotational Raman lines" on either side of this "Rayleigh line." He finds that "the true *depolarisation* of the Rayleigh scattering ... should be only a *fraction* (about 1/4th) of the depolarization of *unresolved* scattering." (Compare ρ^c with ρ^t in table 1.)

Herbert Stuart[2] was aware of the dangers of this terminology:

... the broadening of the Rayleigh line, and thus the intensity of the unresolved fine structure, increases with the anisotropy of the scattering molecule and ... the edges, unlike the line center, are unpolarized. The intensity of all the rotational lines can thus make a marked contribution to the total radiation of strongly anisotropic molecules, and thereby appreciably influence the observed depolarization. Hence, one might at first fear that all previous depolarization measurements were counterfeit and worthless. But theoretical investigation immediately shows ... that the previously derived connection between the optical anisotropy and the usually observed depolarization remains true even when molecular rotation is taken into account.

In spite of this warning, the error he had foreseen has recently been committed,[1] as a direct result of restricting Rayleigh's name to the central line.

Because Cabannes[18] first predicted the intensity and polarization of this central line, I think it is appropriate to call it "the Cabannes line." Cabannes also deserves recognition for having first observed molecular scattering in the laboratory. Finally, as Cabannes correctly found no significant depolarization in the light scattered by argon —light that consists entirely of the central line—while the younger Rayleigh was claiming large spurious anisotropic effects, Cabannes deserves eponymy for his good experimental work on the line, but Rayleigh does not.

Thus, Rayleigh scattering is the sum of the Cabannes line and the rotational lines (see figure 5b).

Structure of Cabannes line

After the elder Rayleigh's death, but before the rotational structure of Rayleigh scattering had been thought of, Leon Brillouin and Leonid Isaakovich Mandel'shtam predicted a splitting of molecular scattering in dense media. Their doublet, due to the Doppler shifts caused by sound waves that create a Bragg reflection condition for light, can be though of as the *translational* "Raman" spectrum. (Doppler broadening due to thermal motions had been explicitly excluded from consideration by Cabannes and Rocard.[18])

Soon after the discovery of the Raman effect, Evgenii Feodorovich Gross,[23] in Leningrad, "attempted to find out whether in light scattered in various organic liquids the Raman lines, due to frequencies in the rotation spectrum, are present." He found, instead, a triplet structure "due to acoustic oscillations like those used by P.

Debye ... for explaining the variation of the specific heat of solids," which he verified by checking the angular dependence of the splitting. "Such an experiment was necessary because the presence of undisplaced and multiple components seemed to contradict the above interpretation." He called the components of the triplet "modified" and "unmodified."

Gross's "unmodified" line was finally explained by Lev Landau and Placzek[24] in a note, *Structure of the undisplaced scattering-line.* They say

If we consider the scattered light connected with the density fluctuations ... for fluids and gases that are not too dilute, the undisplaced line splits into a triplet. The two outer components of this form the familiar Brillouin–Mandelstam doublet with angle-dependent splitting $\Delta\nu = \pm \nu(v/c)\, 2\sin\theta/2$ (v the sound speed); but besides this there is still an undisplaced component, and the ratio of the two outer components to the intensity of the triplet is given to a good approximation by the ratio of specific heats c_v/c_p....

The widths of the 3 components can be given quantitatively; they are determined by viscosity and heat conductivity.

For gases, these results are valid so long as $l \ll \lambda/(2\sin\theta/2)$ (l the mean free path). For greater mean free paths the 3 components flow into one another and the structure of the line ultimately attains the Gaussian form, with angle-dependent width, as required by the Doppler effect.

(Notice that this width collapses to zero in the forward direction, as is necessary for this to be coherent.)

The central component of the Cabannes line, due to scattering from density fluctuations that do not propagate, could be called the "Gross line" after its discoverer. As Rayleigh had nothing to do with it, his name should clearly *not* be used. Likewise, the common term "Rayleigh–Brillouin scattering" is unsuitable for the resolved Cabannes line; "Landau–Placzek scattering" would be appropriate. Although Placzek contributed heavily to both the quantum-mechanical theory of the Cabannes line and the thermodynamical theory of its triplet fine structure, it seems historically inadequate to attach just one or two names to this triplet. Perhaps a descriptive term like "fluctuation scattering" is the most suitable way to refer to the triplet structure of the undisplaced central line.

Coherence

Even in low-density gases, at angles far enough from forward scattering that this Landau–Placzek triplet structure does not appear, the Cabannes line can be divided another way: into coherent and incoherent parts.

The incoherent part is the Q branch[19,20] of the rotational Raman band. (See figure 5.) It is due to transitions between degenerate states with different magnetic quantum numbers, but the same rotational and vibrational quantum numbers. This incoherent part, associated with the molecular anisotropy, has the same depolarization as the ordinary S-branch rotational lines that form the rotational "wings" (see table 2). In diatomic and linear molecules, such as oxygen, nitrogen and carbon dioxide, one-fourth of the rotational band intensity is in the Q branch, so the depolarization of the Cabannes line is about one-fourth that of the whole Rayleigh scattering, as Bhagavantam[22] found.

The coherent part (Placzek's "trace scattering") is associated with the mean polarizability and the refractive

Robert John Strutt, 1875–1947, son of John William. Sketch by Augustus John. (Courtesy of John Arthur Strutt.)   Figure 4

index. It corresponds to isotropic scatterers. It is spectrally superimposed on the rotational Q branch of the anisotropic scattering, which does not contribute to the refractive index because it is incoherent. The sum of the coherent or "trace" scattering and the incoherent rotational Q branch is the Cabannes line (see figure 5).

While the frequency displacement of the rotational S branches depends on molecular rotation, the incoherence of the rotational band (both Q and S branches) does not. The incoherence is due to the random orientations of the

---

Table 2: Relative intensities of scattered light

	Vertically-polarized-light input	Horizontally-polarized-light input	Natural-light input (the sum)
Cabannes line (no frequency shift)	$^vC_v = 180 + 4\epsilon$ $^vC_h = 3\epsilon$ $^vC_o = 180 + 7\epsilon$	$^hC_v = 3\epsilon$ $^hC_h = 3\epsilon + (180 + \epsilon)\cos^2\theta$ $^hC_o = 6\epsilon + (180 + \epsilon)\cos^2\theta$	$^oC_v = 180 + 7\epsilon$ $^oC_h = 6\epsilon + (180 + \epsilon)\cos^2\theta$ $^oC_o = (180 + 13\epsilon) + (180 + \epsilon)\cos^2\theta$ $= 12\epsilon + (180 + \epsilon)(1 + \cos^2\theta)$
Rotational Raman wing (frequency shifted)	$^vW_v = 12\epsilon$ $^vW_h = 9\epsilon$ $^vW_o = 21\epsilon$	$^hW_v = 9\epsilon$ $^hW_h = 9\epsilon + 3\epsilon\cos^2\theta$ $^hW_o = 18\epsilon + 3\epsilon\cos^2\theta$	$^oW_v = 21\epsilon$ $^oW_h = 18\epsilon + 3\epsilon\cos^2\theta$ $^oW_o = 39\epsilon + 3\epsilon\cos^2\theta$ $= 36\epsilon + 3\epsilon(1 + \cos^2\theta)$
Rayleigh scattering (total)	$^vT_v = 180 + 16\epsilon$ $^vT_h = 12\epsilon$ $^vT_o = 180 + 28\epsilon$	$^hT_v = 12\epsilon$ $^hT_h = 12\epsilon + (180 + 4\epsilon)\cos^2\theta$ $^hT_o = 24\epsilon + (180 + 4\epsilon)\cos^2\theta$	$^oT_v = 180 + 28\epsilon$ $^oT_h = 24\epsilon + (180 + 4\epsilon)\cos^2\theta$ $^oT_o = (180 + 52\epsilon) + (180 + 4\epsilon)\cos^2\theta$ $= 48\epsilon + (180 + 4\epsilon)(1 + \cos^2\theta)$

Superscripts give polarization of incident radiation; subscripts give polarization of scattered radiation; v, h, and o denote vertical, horizontal and natural (unpolarized), respectively. Placzek's coherent "trace scattering" is scaled to 180 here, to avoid fractions. Intensities are for linear and diatomic molecules.

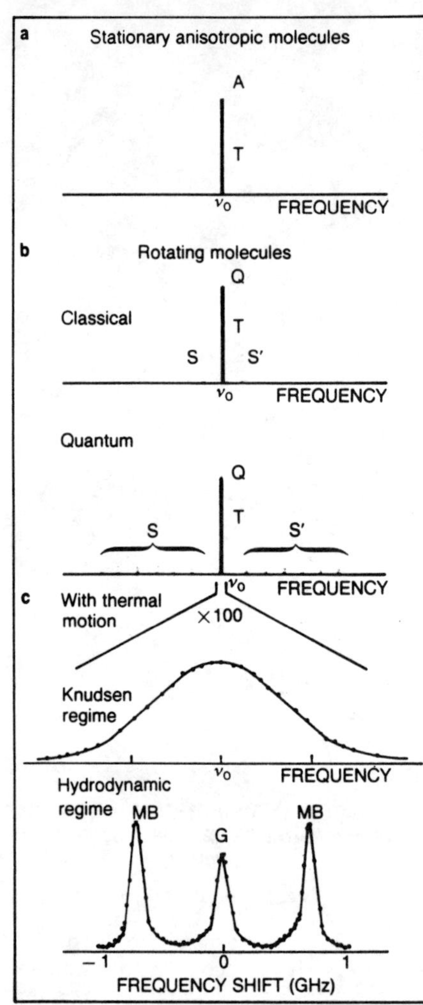

Spectral distribution of scattered monochromatic light. **a** The effect of molecular anisotropy alone gives two components in the scattered light: A depolarized incoherent anisotropy radiation (A), and a coherent part (T) due to the isotropic part of the polarizability. Both have the same frequency (ν_o) as the incident radiation. **b** Effects of molecular rotation. A splits classically into equal Stokes (S) and anti-Stokes (S') components, and an unshifted part, Q. All three components have the same depolarization ρ'', given in table 1. $S + S' = W$, the rotational Raman wing, and $Q + T = C$, the Cabannes line listed in table 2. Quantum mechanically, S and S' are unequal and split into rotational lines, forming the S branches of the rotational Raman band. The Q branch coincides with T, Placzek's "trace scattering" (scaled to 180 in table 2). **c** Effects of thermal motion become apparent when the frequency resolution is increased, here by a factor of 100. The profile of the Cabannes line depends on the density. At low densities molecules scatter independently, producing a Gaussian line-profile. At high densities the central Gross line (G, due to scattering from stationary density fluctuations) bisects the Mandel'shtam–Brillouin doublet (MB, due to scattering from moving density fluctuations, sound waves). In either case, the halfwidth of the whole pattern is proportional to $2\nu_o (v/c) \times \sin(\theta/2)$. Bottom curve is actual data for nitrogen, from Q. H. Lao, P. E. Schoen, B. Chu, J. Chem. Phys. **64**, 3547 (1976). Figure 5

molecules. Hence, the anisotropy radiation remains undiminished in liquids, where the coherent radiation, due to density fluctuations, is much reduced.

If we observe for a time short compared to the period of molecular rotation, our frequency resolution is so poor that we cannot separate the S-branch lines from the Cabannes line. Then we can regard the molecules as stationary, but randomly oriented. This is just the case Lord Rayleigh treated in his 1918 paper[11] (see figure 5a). The total Rayleigh scattering depends on the mean-square polarizability, while the refractive index and coherent scattering depend on the square of the mean.

Terminology

The whole of molecular scattering, the Cabannes line, and the Gross line have all been called "the Rayleigh line" by different workers. The latter two phenomena, unknown during Lord Rayleigh's lifetime, are clearly inappropriate uses of his name. The confusing term "Rayleigh line" should be avoided.

Although the Cabannes line is not monochromatic, and contains a mixture of coherent and incoherent scattering, one sometimes wants a single term to describe its cause. "Elastic scattering" may be appropriate, as the *internal* molecular energy remains unchanged. Brillouin spectroscopists, however, might want to restrict this term to Gross's "unmodified" line. The whole triplet structure might logically be attributed to "Landau–Placzek scattering" or "fluctuation scattering," composed of the Brillouin doublet and the Gross line. But this is inappropriate at low densities, where the Cabannes line has a Gaussian profile. Some consistent terminology is required to distinguish the Raman spectroscopists' "undisplaced line" from that of the Brillouin spectroscopists, to accommodate situations[17] that involve both Brillouin and rotational Raman scattering simultaneously.

Finally, what about the term "Tyndall scattering"? It is used for all particulate scattering by some authors, and applied to the Cabannes line by others. Scattering by turbid media (hydrosols) had been studied by Brücke and Stokes, among others, before Tyndall. The neutral points Tyndall found for large particles had been discovered earlier by Gilberto Govi. Historically, Tyndall's seminal contribution was to provoke Rayleigh's life-long interest in scattering, starting with *small spheres*. As large-sphere scattering is generally called "Mie scattering", it seems best to restrict "Tyndall scattering" to that by small spheres.

Molecular scattering should certainly be excluded, because of Tyndall's belief that clean gases were "optically empty." As Cabannes[16] says, it is in error that some authors apply "Tyndall phenomenon" to the molecular scattering of light, to which should remain attached the name of Lord Rayleigh.

* * *

I thank Professor G. V. Rozenberg, of the Institute of Atmospheric Physics in Moscow, for a stimulating discussion of the early history of light scattering. This work was supported by a NASA Planetary Atmospheres grant.

References

1. A. T. Young, Appl. Opt. **19**, 3427 (1980); A. T. Young, J. Appl. Meteorol. **20**, 328 (1981).
2. H. A. Stuart, *Molekülstruktur*, Springer, Berlin (1934), pages 167–183.
3. J. Tyndall, Philos. Mag. (4) **37**, 384 (1869).
4. J. Tyndall, Philos. Trans. Roy. Soc. London **160**, 333 (1870).
5. Lord Rayleigh (J. W. Strutt), Philos. Mag. (4) **41**, 107, 274 (1871); Sci. Papers I, 87 (1899).
6. Lord Rayleigh (J. W. Strutt), Philos. Mag. (5) **12**, 81 (1881); Sci. Papers I, 518 (1899).
7. Lord Rayleigh (J. W. Strutt), Philos. Mag. (5) **47**, 375 (1899); Sci. Papers IV, 397 (1903).
8. Lord Rayleigh (J. W. Strutt), Nature **83**, 48 (1910); Sci. Papers V, 540 (1912).
9. J. Cabannes, C. R. Acad. Sci. **160**, 62 (1915).
10. Lord Rayleigh (J. W. Strutt), Philos. Mag. (4) **41**, 447 (1891); Sci. Papers I, 104 (1899).
11. Lord Rayleigh (J. W. Strutt), Philos. Mag. (6) **35**, 373 (1918); Sci. Papers VI, 540 (1920).
12. R. J. Strutt, Proc. R. Soc. London **A 95**, 155, (1918).
13. J. Cabannes, Ann. Physique (9) **15**, 5 (1921).
14. L. V. King, Proc. R. Soc. London **A 104**, 333 (1923).
15. Lord Rayleigh (R. J. Strutt), Proc. R. Soc. London **A 98**, 57 (1921).
16. J. Cabannes, *La Diffusion Moléculaire de la Lumière*, Recueil des Conferences-Rapports de Documentation sur la Physique, Vol. 16, Presses Universitaires de France, Paris (1929).
17. G. W. Kattawar, A. T. Young, T. J. Humphreys, Astrophys. J. **243**, 1049 (1981).
18. J. Cabannes, C. R. Acad. Sci. **186**, 1201 (1928); J. Cabannes, Y. Rocard, J. Phys. Radium (6) **10**, 52 (1929).
19. C. Manneback, Z. Physik **62**, 224 and **65**, 574 (1930).
20. G. Placzek, E. Teller, Z. Physik **81**, 209 (1933).
21. J. Cabannes, Ann. Physique (10) **18**, 285 (1932).
22. S. Bhagavantam, Indian J. Phys. **6**, 331 (1931).
23. E. Gross, Nature **126**, 201 (1930).
24. L. Landau and G. Placzek, Phys. Z. der Sowjetunion **5**, 172 (1934). □

Section Three
Multiple Scattering

Reprinted from *The Astrophysical Journal,* Vol. XXI(1), 1–22 (1905).

RADIATION THROUGH A FOGGY ATMOSPHERE

By ARTHUR SCHUSTER

1. In discussing the transmission of light through a mass of gas, it is usual to consider only the effects of emission and absorption, and to neglect all effects of scattering. But when the absorbing mass holds fine particles of matter in suspension, the scattered light materially affects the character of the transmitted radiation. I propose to discuss the conditions under which "bright line" spectra or "dark line" spectra may be obtained from a radiating mass of gas, taking account of scattering. I call an atmosphere "foggy" when scattering takes place to an appreciable extent. The applications of the results of this investigation are, however, much wider than the title chosen would seem to imply, for there is some scattering even from the molecules of a homogeneous substance, and to that extent all bodies fall within the definition and may be called "foggy."

According to the investigations of Lord Rayleigh, the greater part of the light we receive from the sky is due to light scattered by the molecules of the air. This involves a diminution in the intensity of the direct rays amounting in our atmosphere to roughly 5 per cent. The effective thickness of stellar atmospheres may be great compared with that of the shell of air which surrounds our globe, and hence the effects of scattering may be of primary importance in interpreting the nature of stellar atmospheres.

2. The following notation will be used:

E = the total energy of radiation within a certain small range of wave-lengths sent out by unit surface of a completely black surface. E is a function of the temperature and wave-length.

S = the total energy of radiation incident within the same limit of wave-lengths on unit surface of a plane layer of the foggy gas.

R = the energy which leaves the plane layer per unit surface.

R_0 = the particular value of R for the case that there is no absorption.

R_c = the particular value of R for rays which are completely absorbed by an infinitely thin layer.

κ = the coefficient of absorption.

s = the coefficient of scattering.

The object of the investigation is to determine R in terms of S and E, if E refers to the temperature of the foggy gas. It will save needless repetition if it is understood once for all that our statements always refer to unit surface of the radiating or absorbing layer.

κ is a function of the wave-length which also depends on the density of the medium. If the medium is uniform, all molecules absorbing alike, κ would be proportional to the density. But in a mixture of different gases, κ must be considered proportional only to the quantity (measured per unit volume) of the particular substance which absorbs the wave-length in question. Similarly s depends on the number of the scattering particles, the scattering and absorbing particles not necessarily being of the same nature. If the scattering is of the nature of that which causes the blue color of the sky, the value of s varies inversely as the fourth power of λ, but in case of an ordinary cloud or mist, the dependence on wave-length is much less marked.

If S be the total intensity of radiation incident on a layer of small thickness dx, the radiation absorbed by the layer is $\kappa S dx$. The light emitted by the same layer in each direction is thus $\kappa E dx$. This follows from the law connecting absorption and radiation.

The light scattered by the layer is $sSdx$, of which one-half is sent forward and one-half returned backward.

The following variables will be introduced for convenience of expression:

$$\beta = \kappa/s ,$$
$$a = \sqrt{\kappa/(\kappa+s)} = \sqrt{\beta/(1+\beta)} ,$$
$$\therefore \beta = a^2/(1-a^2) .$$

β varies from zero to infinity, but a must lie between zero and unity.

$$\gamma = (1+a)/(1-a) \ ,$$
$$\therefore \gamma = 1 + 2\beta + \sqrt{\beta + \beta^2} \ .$$

3. Let (Fig. 1) $S_1 S_2$ be a surface sending out the radiation S, and let this radiation after passing through part of the foggy atmosphere be reduced to a value A and fall on a thin layer of thickness dx. The effect of the layer is to absorb energy amounting to $\kappa A\,dx$, and additionally to reduce the incident light by a quantity $sA\,dx$, which is not absorbed, but sent in equal amounts backward and forward as scattered light. If the stream of radiant energy in the opposite direction is B, we have similarly a diminution of energy equal to $(\kappa+s)\,B$, of which, however, $\tfrac{1}{2}sB$ is sent both forward and backward as scattered light.

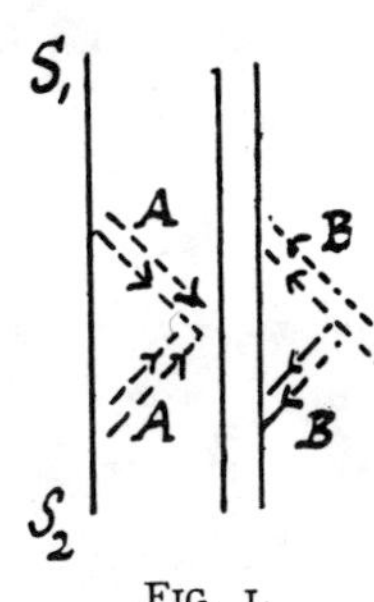

FIG. 1.

The layer also radiates energy in both directions equal to $\kappa E\,dx$. Collecting these effects, we obtain the equations:

$$\frac{dA}{dx} = \kappa(E-A) + \tfrac{1}{2}s(B-A) \tag{1}$$

$$\frac{dB}{dx} = \kappa(B-E) + \tfrac{1}{2}s(B-A) \ . \tag{2}$$

Combining (1) and (2) we find:

$$\frac{d(A+B)}{dx} = (\kappa+s)(B-A) \tag{3}$$

$$\frac{d(A-B)}{dx} = 2\kappa E - \kappa(A+B) \ . \tag{4}$$

Differentiating (3) and with the help of (4)

$$\frac{d^2(A+B)}{dx^2} = \kappa(\kappa+s)(A+B-2E) \tag{5}$$

If E is constant or varies uniformly with x, the last equation may be integrated, and we derive:

$$(A+B-2E) = Ke^{(\kappa+s)ax} + K_1 e^{-(\kappa+s)ax} \tag{6}$$

where K and K_1 are two constants and a has the value assigned to it in § 2.

If the temperature of the medium is constant, so that E has the same value throughout the scattering medium, differentiation gives

$$\frac{d(A+B)}{dx}=a(\kappa+s)(K_{\scriptscriptstyle\rm I}e^{(\kappa+s)ax}-K_{\scriptscriptstyle\rm I}e^{-(\kappa+s)ax})\qquad(7)$$

and hence by introducing (3)

$$B-A=a(Ke^{(\kappa+s)ax}-K_{\scriptscriptstyle\rm I}e^{-(\kappa+s)ax})\ .\qquad(8)$$

Equations (6) and (8) now allow us to obtain A and B separately, and we thus find:

$$\left.\begin{aligned}2A&=2E+(\mathrm{I}-a)Ke^{(\kappa+s)ax}+(\mathrm{I}+a)K_{\scriptscriptstyle\rm I}e^{-(\kappa+s)ax}\\2B&=2E+(\mathrm{I}+a)Ke^{(\kappa+s)ax}+(\mathrm{I}-a)K_{\scriptscriptstyle\rm I}e^{-(\kappa+s)ax}\end{aligned}\right\}\ .\qquad(9)$$

We consider x to be measured from the front surface of the foggy medium in the direction in which the radiation A proceeds. If no radiation enters the medium from the opposite direction, and if the radiation incident in the first absorbing layer be S, we have the conditions:

$$\left.\begin{aligned}\text{for } x&=0; & B&=0\\\text{for } x&=-t; & A&=S\end{aligned}\right\}\qquad(10)$$

the thickness of the medium being denoted by t.

We require to determine the emergent radiation which is equal to the value which A acquires when $x=0$. Denoting this by R, we have from the first of equations (9)

$$2R=2E+(\mathrm{I}-a)K+(\mathrm{I}+a)K_{\scriptscriptstyle\rm I}\ .\qquad(11)$$

Introducing (10) into (9) allows us to determine K and $K_{\scriptscriptstyle\rm I}$. We obtain in the first place the equations

$$0=2E+(\mathrm{I}+a)K+(\mathrm{I}-a)K_{\scriptscriptstyle\rm I}$$
$$2S=2E+(\mathrm{I}-a)Ke^{-a(\kappa+s)t}+(\mathrm{I}+a)K_{\scriptscriptstyle\rm I}e^{a(\kappa+s)t}\ ,$$

and these give

$$K=\frac{2\left[(\mathrm{I}-a)-(\mathrm{I}+a)e^{a(\kappa+s)t}\right]E-2(\mathrm{I}-a)S}{(\mathrm{I}+a)^2e^{a(\kappa+s)t}-(\mathrm{I}-a)^2e^{-a(\kappa+s)t}}\ ,$$

$$K_{\scriptscriptstyle\rm I}=\frac{2\left[(\mathrm{I}-a)e^{-a(\kappa+s)t}-(\mathrm{I}+a)\right]E+2(\mathrm{I}+a)S}{(\mathrm{I}+a)^2e^{a(\kappa+s)t}-(\mathrm{I}-a)^2e^{-a(\kappa+s)t}}\ .$$

Finally by substitution into (11)

$$R=2a\frac{\left[(\mathrm{I}+a)e^{a(\kappa+s)t}+(\mathrm{I}-a)e^{-a(\kappa+s)t}\right]E+2(S-E)}{(\mathrm{I}+a)^2e^{a(\kappa+s)t}-(\mathrm{I}-a)^2e^{-a(\kappa+s)t}}\ .\qquad(12)$$

Equation (12) contains the solution of our problem.

4. The equations of the last paragraph have been deduced under the assumption that the radiation throughout the absorbing mass is uniformly distributed in such a way that it does not depend on the angle between any direction considered and the normal drawn toward the same side. This supposition is obviously incorrect, for it appears that, even if it were to hold at any surface, e. g., the first surface of the layer dx (Fig. 1), absorption in that layer would destroy the uniformity owing to the greater absorption which the oblique rays suffer. To some extent the effect of scattering would act in the sense of partly restoring the equality of distribution; nevertheless serious errors might be introduced, if we attempted to obtain accurate values of κ and s by means of the application of equation (12). The complete investigation leads to equations of such complexity that a discussion becomes impossible, and I shall only use the solution obtained under the simplified conditions to deduce certain consequences which cannot be affected by the assumption made. The error committed might be allowed for by taking s and κ to be functions of the distance. When considered in this light, it is seen how useless the more complete calculation would be, because in the more important cases to which we have to apply our results, the coefficients of scattering and absorption vary in an unknown manner, and the error committed by the simplification of this problem becomes merged in other unavoidable uncertainties.

5. Before discussing the general results contained in equation (12) we may treat separately of some simple special cases. When the coefficient of absorption, and consequently a, is zero, we require to express the exponentials of (12) in a series, the first two terms being retained. But it is easier in this case to proceed directly. Equations (3) and (4) in this case become

$$\frac{d(A+B)}{dx}=s(B-A) \ ,$$

$$\frac{d(A-B)}{dx}=0 \ .$$

The second equation shows that $A-B$ is a constant which must be equal to R_0, the value of A at the front surface. The first equation may now be integrated, and gives

$$A+B=a-sR_0x \ .$$

As for $x=0$, $B=0$, and $A=R_o$, it follows that $a=R_o$; or replacing B by $A-R$,

$$2(A-R_o)=-sR_ox \ .$$

When $x=-t$, the value of A is equal to S, the incident radiation, hence

$$2(S-R_o)=sA_ot \ ;$$

or finally:

$$R_o=\frac{2}{2+st}S \ . \tag{13}$$

The equation shows that the emergent radiation diminishes with increasing thickness, but not so quickly as it would do if scattering acted in the same manner as absorption. If, for instance, we give to st the numerical value of ninety-eight, so that the emergent light is 2 per cent. of the incident light, doubling the layer would still give us 1 per cent. for the transmitted light, and with greater thicknesses the light would, roughly speaking, be inversely proportional to the thickness. But in the case of absorption, the double layer would only transmit 2 per cent. of 2 per cent., and the transmitted light would diminish in a geometric ratio, while the thickness increases in an arithmetic ratio.

6. When either st or κt is so large that practically no part of the original light is transmitted, we may neglect in (12) all terms except those containing an exponential with a positive argument, and this gives at once

$$R_c=\frac{2a}{1+a}E \tag{14}$$

When κ is large compared with s, a approaches unity and ultimately $R_c=E$. The radiation in that case becomes equal to that of a completely black surface, which agrees with the well-known law that absorption irrespective of scattering tends to make the radiation of all bodies equal to that of a black body when the thickness is increased.

But, as has been mentioned, scattering always exists, and has to be taken into account. It appears from the definition of a that it is always a fraction, and hence the factor of E in (14) is always smaller than one. It follows that the emergent radiation increases with the value of κ, and hence a luminous gas always gives a spectrum of bright lines, and does not approach with increasing thickness to the radiation from a black body, as it would do in the absence of scattering.

7. It is not possible to discuss equation (12) in its general form. In order to draw the appropriate conclusions in certain typical cases, we introduce other variables.

Put

$$e^{st}=r \; ; \qquad \beta=\kappa/s \; ,$$

and introduce a quantity γ defined by

$$\gamma=\frac{1+a}{1-a} \; .$$

We have then

$$a(\kappa+s)t=st(1+\beta)a \; ;$$

and also

$$a=\sqrt{\beta/(1+\beta)} \; .$$

Hence

$$\gamma=1+2\beta+2\sqrt{\beta+\beta^2} \tag{15}$$

$$\frac{2a}{1-a}=\gamma-1 \; ,$$

$$\frac{2a(1+a)}{(1-a)^2}=\frac{1+a}{1-a} \cdot \frac{2a}{1-a}=\gamma(\gamma-1) \; ,$$

$$\frac{4a}{(1-a)^2}=\gamma^2-1 \; .$$

Equation (12) now becomes

$$R=(\gamma-1)\frac{(\gamma r^{\sqrt{\beta+\beta^2}}+r^{-\sqrt{\beta+\beta^2}})E-(\gamma+1)(E-S)}{\gamma^2 r^{\sqrt{\beta+\beta^2}}-r^{-\sqrt{\beta+\beta^2}}}. \tag{16}$$

As (15) gives γ in terms of β, all factors of E and S are now expressed in terms of β and st. I have carried out the calculations for the three cases that st is equal to $\frac{1}{2}$, 1, and 2, respectively, and for a number of different values of β. If we calculate the coefficient in (16) and write it (16) in the form

$$R=aE+bS \; ,$$

Table I gives the coefficients of a and b.

TABLE I

		0.1	0.8	1	1.2	2	10
$st = 0.5$	a	0.0486	0.3258	0.3872	0.4428	0.6165	0.9762
	b	0.7604	0.5333	0.4820	0.4355	0.2911	0
$st = 1$	a	0.0944	0.5276	0.6019	0.6626	0.8142	0.9762
	b	0.6000	0.2902	0.2364	0.1926	0.0855	0
$st = 2$	a	0.1760	0.7147	0.7716	0.8111	0.8895	0.9762
	b	0.3971	0.0872	0.0574	0.0379	0.0074	0

The relative intensities of the dark and bright lines as they would
appear in the special cases considered will be most easily under-

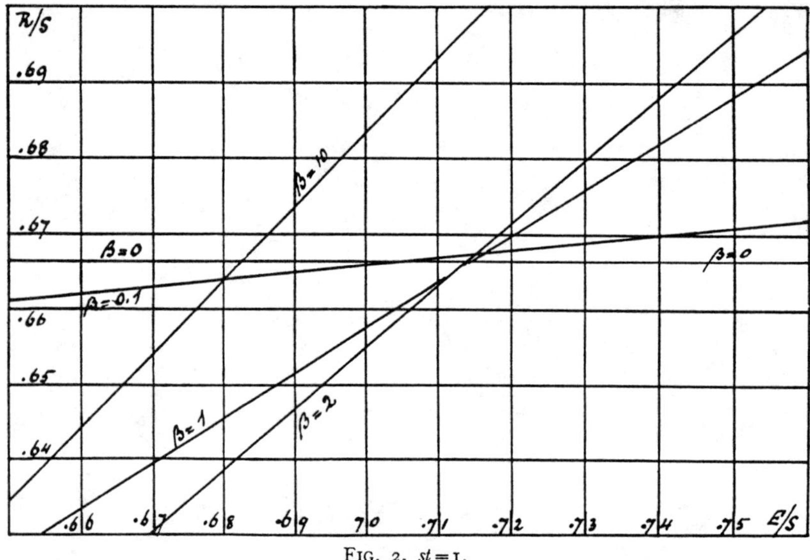

Fig. 2. $st = 1$.

stood if straight lines are drawn (Fig. 2) with E / S as abscissæ and
R / S as ordinates.   That figure refers to the case $st = 1$.   The horizontal line marked $\beta = 0$ gives R_0 / S, which defines the intensity of the
transmitted light when there is no absorption.   The value of R_0 / S is
obtained from (13), which shows that for $st = 1$, only two-thirds of
the incident light traverses the scattering medium.   If this medium
is capable of sending out any vibration defined as regards radiative
power by the fraction $\beta = \kappa / s$, the corresponding intensity may be

obtained from the curve by taking on the horizontal axis the magnitude E/S, which is the intensity of the black radiation at the temperature of the absorbing and scattering medium in terms of that of the incident light. The corresponding ordinate gives the transmitted light in terms of the same unit. If the point of the straight line corresponding to any particular value of β lies above the horizontal line marked $\beta = 0$, the appearance will be that of a bright line; if it lies below, a dark line would be observed. Starting with a comparatively low temperature of the foggy medium and gradually increasing it (i. e., gradually increasing the ordinates), it is seen that at first all homogeneous vibrations appear as dark lines, and if the temperature is sufficiently low (not shown in figure) the highest values of β give the greatest deficiency of light. This is in accordance with what takes place in the absence of scattering.

When the temperature is gradually raised, the most intense line represented in the figure ($\beta = 10$) ceases to be the darkest line, and ultimately when E/S is about 0.682, this line becomes brighter than the background. The next line to change from darkness to brightness is the line of lowest intensity $\beta = 0.1$, and when E/S is more than 0.715, all the lines are bright. The change from dark to bright lines takes place within a comparatively small range of temperature; nevertheless the possibility of the simultaneous appearance of dark and bright lines according to the intensity of absorption is shown by the figure. If, instead of a homogeneous line, we contemplate the case of narrow bands such as frequently occur, we must consider β to have a maximum value at the center of the band (e. g., $\beta = 10$) and to fall off on either side more or less rapidly to zero. At very low temperatures of the medium, the center of the band in this case would be darkest and at high temperatures brightest. But intermediate temperatures would give the appearance of a bright central line on a dark absorption band. Thus at a temperature of 0.69, the brightness of the center ($\beta = 10$) in terms of the intensity of the transmitted light is 0.674, which means that it is 1.2 per cent. brighter than the background, while towards both sides, where β has fallen to 2, the intensity is 0.637 or 4.5 per cent. darker than the background. The appearance is therefore that of an absorption band with a reversed line in the center.

Fig. 3 gives the diagram of radiation for $st = 0.5$. It is drawn to a somewhat larger scale than Fig. 2.

Fig. 4 represents the connection between transmitted light and coefficient of absorption when $st = 2$. In this case the unabsorbed radiation is scattered to the extent that only half the incident light is transmitted. The possibility of the simultaneous appearance of dark and bright lines, which carries with it the possibility of an absorption band with a reversed line at the center, is increased in

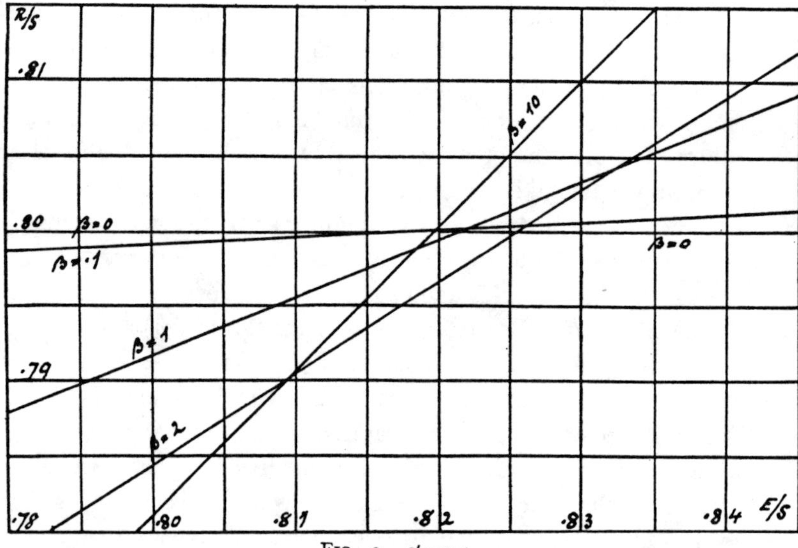

FIG. 3. $st = 0.5$.

this case. Thus when E/S is 0.54, a line defined by $\beta = 10$, shows an increase in brightness over the background $[(R-R_o)/R_o]$ of 5.4 per cent. and a weaker line ($\beta = 1$) gives a deficiency of light of 5.2 per cent.

Table II gives the values of E/S at which the transmitted radiation corresponding to different values of β is equal to that of the transmitted unabsorbed radiation ($\beta = 0$). The numbers given define the temperatures of the absorbing medium at which the transition from the dark to the bright lines takes place.

Table III gives in terms of R_o the intensities of the radiations when the temperature of the absorbing layer is the same as that of the background, the incident light S being in this case considered

to emanate from a black body.   The table shows the importance of the
effects of scattering on the production of bright line spectra; for,
neglecting this scattering, all the numbers would be equal to unity,
and we should only obtain the continuous spectrum of the background,
the medium not affecting the radiation at all.

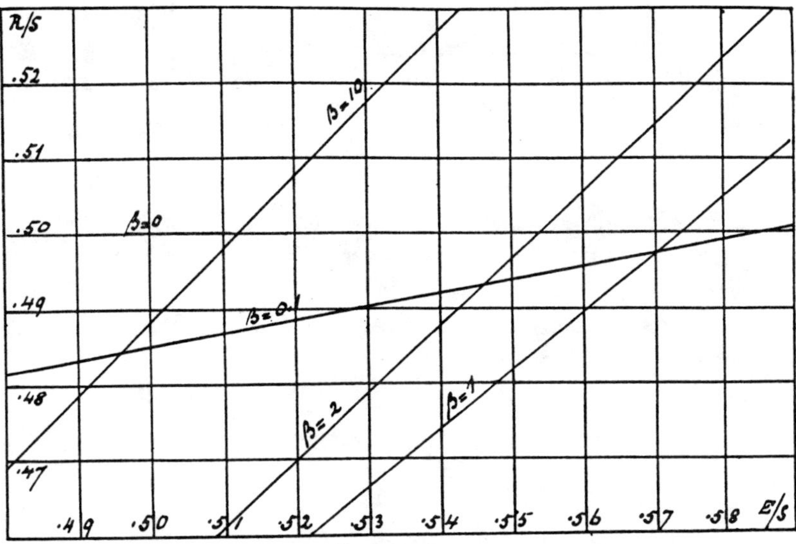

Fig. 4. $st = 2$.

TABLE II

$\beta =$		0.1	0.8	1	1.2	2	10
$st = 0.5\ldots$	$\dfrac{0.8 - a}{b}$	0.8149	0.8185	0.8213	0.8231	0.8254	0.8196
$st = 1\ldots$	$\dfrac{0.6667 - a}{b}$	0.7065	0.7137	0.7150	0.7155	0.7139	0.6831
$st = 2\ldots$	$\dfrac{0.5 - a}{b}$	0.5845	0.5776	0.5736	0.5697	0.5539	0.5123

TABLE III

$\beta =$		0.1	0.8	1	1.2	2	10
$st = 0.5\ldots$	$\dfrac{a + b}{0.8}$	1.011	1.074	1.086	1.098	1.134	1.220
$st = 1\ldots$	$\dfrac{3}{2}(a + b)$	1.042	1.227	1.257	1.283	1.350	1.464
$st = 2\ldots$	$2(a + b)$	1.146	1.604	1.658	1.698	1.794	1.952

8. In the problem as it is usually considered, absorption is introduced by having a cooler medium of uniform temperature in front of a hotter one, and the change of temperature is taken to be an abrupt one. But in considering the phenomena of radiation presented to us by celestial bodies, we must bear in mind that no such discontinuous variation in temperature is admissible. It seems therefore desirable to discuss, even if only in a very simple case, the radiation emitted by a gas of continuously varying temperature.

Equation (6) holds when the temperature variation of the medium is such that the radiation of a black body for the particular wavelength considered varies uniformly with x. We write now for the equation defining the temperature of the medium: $E=j-ux$, where j represents the radiation of a black body which is at the temperature of the external surface of the medium, and u being positive, the temperature increases toward the inside. By differentiation of (6) and the application of (3) we then obtain

$$(B-A)= -\frac{2u}{\kappa+s}+Kae^{(\kappa+s)ax}-K_{1}ae^{-(\kappa+s)ax}$$

Combining this with (6) we find

$$2A=2(j-ux)+\frac{2u}{\kappa+s}+K(1-a)e^{(\kappa+s)ax}+K_{1}(1+a)e^{-(\kappa+s)ax} \qquad (17)$$

At a certain distance, for which we may put $x=-t$, we may imagine a radiating black surface to be placed, having a temperature equal to that of the medium at that surface. The incident radiation A must therefore here coincide with the radiation E.
This gives

$$A=j+ut .$$

If t is very large, we find by applying (17) to this distant layer

$$0=\frac{2u}{\kappa+s}+K_{1}ae^{(\kappa+s)at} ,$$

which gives for K_{1} negligible values when t is sufficiently large.
Putting $x=0$ in (17), we now obtain

$$2R=2j+\frac{2u}{\kappa+s}+K(1-a) .$$

If we apply (6) to the case of $x=0$ when $B=0$, $A=R$, $E=j$, we also find, neglecting K_{1},

$$R=2j+K .$$

Hence, eliminating K,

$$(1+a)R = 2af + \frac{2u}{(\kappa+s)} \; . \tag{18}$$

The character of the radiation is seen from this equation to depend on the relative values of the radiation-gradient and the coefficient of scattering. In order to exemplify the conditions which regulate the nature of the spectrum, draw a plane through the medium at a distance t behind the front surface. Choose t to be such that, owing to scattering and independently of absorption, only 0.8 of the light incident on the layer passes through it. This gives $st=1$. Let $(1+m)f$ be the radiation passing through the plane at a distance t from the front surface. The temperature of the scattering medium varies therefore in such a manner that the radiation of a black body increases in the ratio $(1+m) : 1$, when $t=1/S$. As the radiation generally is $f-ux$, we have, when $st=1$,

$$f+ut = (1+m)f \; ,$$

or

$$ut = mf \; .$$

If in the second term of (18) we multiply numerator and denominator by t and substitute $st=1$, $\kappa t = \kappa / s = \beta$, we find

$$R = 2f\left(\frac{a}{1+a} + \frac{m}{(1+\beta)(1+a)}\right) ,$$

$$= 2f\left[\frac{a}{1+a} + m(1-a)\right] \tag{19}$$

It remains to discuss this equation. For $\beta=0$,

$$R_0 = 2mf$$

and this may be taken to be the intensity of the continuous background of the spectrum. If the radiative power of any homogeneous radiation is very large, so that β is very large,

$$R_c = f \; .$$

The lines with large radiative power appear therefore bright or dark according as m is less or greater than one-half.

Differentiating (19) with respect to a,

$$\frac{dR}{da} = 2f\left[\frac{1}{(1+a)^2} - m\right] ,$$

we see that as a increases from zero, the radiation increases or diminishes according as m is smaller or greater than one. A turning-point is reached when

$$m = \frac{1}{(1+a)^2} \, ,$$

or

$$a = \frac{1}{\sqrt{m}} - 1 \qquad\qquad (20)$$

As a is necessarily positive and smaller than one, this turning-point has a meaning applicable to our problem only when m is greater than one-quarter and smaller than unity. A maximum of radiation is reached in this case, and for values of the coefficient of absorption greater than those defined by (20) the radiation diminishes again. The radiation reaches the same value it has for $a = 0$, when

$$R = R_0 \, ,$$

or

$$m = \frac{a}{1+a} + m(1-a) \, ,$$

or

$$m = \frac{1}{1+a} \, .$$

As a is a positive fraction, it follows that in order that $R = R_0$ for a second value of a, it is necessary that m should be greater than one-half.

When there is a maximum, its value is easily found to be

$$R = 2f[m + (1 - \sqrt{m})^2] \, .$$

We may summarize our results thus:

Case I: $m < \frac{1}{4}$.—With increasing coefficient of absorption, the radiation increases. All homogeneous vibrations appear as bright lines. The brightness of the background ($\kappa = 0$) is given by

$$R_0 = 2mf \, ;$$

that of the brightest line

$$R_c = f \, .$$

Case II: $\frac{1}{4} < m < \frac{1}{2}$.—With increasing coefficient of absorption the radiation increases and reaches a maximum when

$$a = \frac{1}{\sqrt{m}} - 1 \, .$$

For greater values of a, the radiation diminishes. All lines are bright, but the lines with the greatest coefficient of absorption are not the brightest. Thus when $m=\frac{1}{2}$, the maximum radiation takes place when $a=0.414$ $(\beta=0.207)$, and gives an intensity of $1.172\,j$, while for infinite values of κ the intensity is j.

Case III: $\frac{1}{2}<m<1$.—The intensity rises to a maximum as in Case II, then diminishes until, when $a=\dfrac{1-m}{m}$, the radiation has the

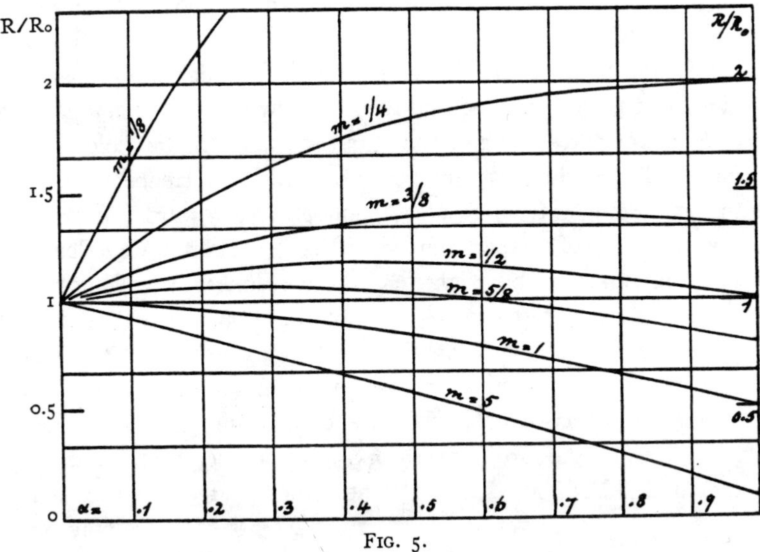

Fig. 5.

same value as when $a=0$. After this point it diminishes. Homogeneous radiations will in this case appear as bright lines when $a>\dfrac{1-m}{m}$ and as dark lines for greater values of a.

Case IV: $m>1$.—The intensity diminishes continuously with increasing coefficient of absorption. All homogeneous lines appear as dark lines. In Fig. 5 I have drawn the curves which give the intensity of radiation in terms of R_o. The abscissæ represent a and the ordinates:

$$\frac{a}{m(1+a)}+(1-a)\ .$$

Table IV gives the corresponding values of a and β.

α	β	α	β
0.1	0.0101	0.6	0.5625
0.2	0.0417	0.7	0.9608
0.3	0.0989	0.8	1.7778
0.4	0.1905	0.9	4.2632
0.5	0.3333	1.0	∞

In applying the results obtained, it should be remembered that m defines the "radiation-gradient," which depends not only on the "temperature-gradient," but also on the wave-length. The same increase in temperature will cause a greater radiation-gradient at the violet end than at the red end of the spectrum, and at comparatively low temperatures the radiation-gradient may in the violet be enormously larger than the temperature-gradient. Hence we conclude that with moderate increases of temperature toward the inside of a gas, the lines of a spectrum which have a shorter wave-length are much less likely to be bright than those of longer wave-length.

9. It may help the reader to draw the proper inferences from the preceding results if an elementary demonstration is given showing how bright line spectra are formed in an infinitely thick layer of incandescent gas, which, when the temperature is uniform, should, according to the ordinary theory, give a spectrum identical with that of a black body at the same temperature. In Fig. 6 let an observer look from E at a mass of gas giving out two homogeneous radiations λ_1 and λ_2, differing very largely in their emissive and absorptive powers κ_1, and κ_2 The wave-length λ_1 being that for which the emissive power is great, will chiefly be due to the radiation of a thin layer $L_1K_1K_2L_2$, because waves of the same wave-length coming from behind will be strongly absorbed by it. On the other hand, the vibrations of small emissive, and therefore small absorptive power will be due to the radiation of a much thicker layer. Neglecting in the first instance the loss of light due to scattering, we may draw H_1H_2 so that the layer $L_1H_1H_2L_2$ sends out a total intensity of waves, λ_2 representing the same fraction of the radiation of a black body of the same temperatures as does the layer $L_1K_1K_2L_2$ for

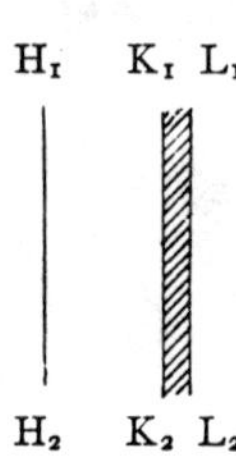

Fig. 6.

the wave-length λ_1. It is well known that whatever be the emissive powers, an indefinite increase in thickness will ultimately give the radiation of a black body. Now, introduce scattering in addition to the absorption. The wave-length λ_1 will not be much affected by scattering, as the light which leaves the gas has only traversed a small thickness of it. On the other hand, the wave-length for which the emissive and absorptive powers are small, being due to the radiation of a thick layer, will be much more weakened by the light scattered, and returned backward. Hence while, in spite of the scattering, λ_1 still shows an intensity nearly equal to that of the black body, the intensity of λ_2 is less. Consequently the radiation will be the stronger, the stronger the emissive power, and hence the gas gives a bright-line spectrum.

In this reasoning it has been supposed that the scattering is of the nature of that due to small bodies and is not a phenomenon primarily dependent on absorption. In the latter case it might be argued that the scattering might be stronger for the wave-length λ_1 in the same proportion as the emissive power is stronger. It is quite possible that a portion of the molecular scattering is selective in character, and, so far as this portion is concerned, our investigation does not apply, without more detailed consideration.

10. I may, in conclusion, briefly indicate the bearing of the results obtained on some problems of astrophysics. It has been shown that a spectrum of bright lines may be given by a mass of luminous gas, even if that gas is of great thickness. There is therefore no difficulty in explaining the existence of stars giving bright lines. The essential criterion which separates the bright-line emission from the dark-line absorption lies in the temperature-gradient of the luminous gas. If the increase of temperature toward the inside of a star is small, bright lines will appear, while the absorption spectra observed in the majority of cases accompany a more rapid variation of temperature. The temperature-gradient is chiefly regulated by the gravitational force, and a star in the early stages of condensation will therefore be in the condition most favorable for the bright-line emission. If the light is but feebly absorbed, so that we can look into considerable depths of the star, it may be possible that the outer regions contribute bright lines, while the hotter inner portions show absorption lines.

The possibility of the simultaneous presence of bright and dark lines of the same element, e. g., of hydrogen, is also strongly indicated by our theory. The matter has already been discussed in sections 7 and 8. The conditions under which the equations of sec. 8 have been deduced are more likely to apply to stars than the conditions of the problem as discussed in sec. 7, and as pointed out at the end of sec. 8, the lines of smaller frequency are those most liable to appear as bright lines. This agrees with the observed facts. The simultaneous appearance of bright lines of smaller and dark lines of greater frequency, has however, also been observed in cases where it is difficult to imagine that scattering plays an important part. I refer to Professor Hale's observations[1] on the spark-spectra observed in liquids. The proper explanation of these and similar observations suggests itself at once, if it is considered that the essential part of the effect of scattering lies in the diminution of the intensity of the continuous background. This diminution is not called for when the body giving the continuous spectrum has not infinitely great thickness and radiates with an intensity less than that belonging to a black body. Putting $s=0$ in (12), we obtain the ordinary equation for the radiation of a body sending out light of intensity S, which before reaching the observers traverses a body which is at a temperature for which the completely black radiation is E, viz.:

$$R=E+(S-E)e^{-\kappa l} .$$

For $\kappa = 0$, we have

$$R_0=S .$$

Hence the question of brightness or darkness for a particular wavelength depends on the sign of the quantity

$$R-R_0=(E-S)(1-e^{-\kappa l}) ,$$

and this depends entirely on the question as to whether $E-S$ is positive or negative. If S is the radiation due to a black body at a higher temperature than that of the absorbing body, $E-S$ is necessarily negative and an absorption line will appear. If the radiation S is not that of a black body, but, e. g., a radiation reduced by the same quantity in the red and blue, or even reduced in the same ratio, the peculiar dependence of the radiation curve on temperature and wave-

[1] *Astrophysical Journal*, **15**, 227, 1902.

length shows that $E-S$ which is now positive when the temperature
of the two bodies is equal, keeps positive longer with a diminishing
temperature of the absorbing layer when the wave-lengths are long
than when they are short. I need not enter more fully into this
question, because Professor Kayser[1] has fully discussed it in giving
what is practically an identical explanation. On applying Professor
Kayser's explanation to the case of stars, we meet, however, with
the very serious difficulty that we are obliged to consider the radiation
of the continuous spectrum which serves as background to be less
than that of a black body, which, on the views hitherto held, could
not be the case when the radiating body has a great thickness. The
consideration of the effect of scattering as explained by the present
investigation removes the difficulty. I differ from Kayser in so far
that he considers the existence of bright lines in stars to be an indication of high temperature. The small temperature-gradient seems,
on the contrary, to argue more in favor of relatively low temperatures.
Discussion on these and other connected matters is difficult, however,
owing to our ignorance of the relative values of the coefficient of
emission κ for different elements, and for different lines of the same
element. We do not even know whether in a series such as that
formed by the hydrogen lines κ increases or diminishes toward the
root of the series.

The appearance of bright hydrogen lines covered by the dark
calcium absorption, as presented by the spectrum of *Mira Ceti*,
presents no difficulty according to the views of the present investigation. It only implies that the interior of the star has a temperature-
gradient insufficient to reverse the hydrogen lines, and an outer
atmosphere containing cooler calcium vapor. I consider it indeed
as quite possible that, if we could remove the outer layers of the solar
atmosphere, we should obtain a spectrum of bright lines.

This brings me to the second consideration suggested by the previous investigation. The prominent part played by the H and K
lines of the solar spectrum in stellar atmospheres may be, to a great
extent, due to the high values of the coefficient κ. The experiments
of Sir William and Lady Huggins show conclusively that when
calcium gas is rendered luminous by the electric discharge under

[1] *Astrophysical Journal*, **14**, 313, 1901.

conditions under which the H and K lines can appear, they are most persistent and are seen even when only very minute quantities of the substance are present. We are justified in concluding from these experiments that the emissive power of H and K is very great. The same may be true of other lines characteristic of spark-spectra, and the appearance of these lines in the stellar spectra must therefore be treated with some caution. If a star in its process of condensation increases the temperature-gradient of its outer layers, those lines will first make their appearance as dark lines which have high values of κ. But I must defer the fuller discussion of this matter to another occasion.

The effect of scattering on the intensity of the continuous spectrum of what we call the photosphere of a star may be considerable. When the radiating layer of a gas is sufficiently cool to admit of the presence of particles of solid or liquid matter, of dimensions large compared with molecular dimensions, the reduction in luminosity would take place fairly equally throughout the range of the visible spectrum. There would consequently be no great alterations in the relative intensities of red or blue, and we could obtain a correct idea of the temperature of the radiating body by a thermal comparison of the intensity of radiation in different parts of the spectrum. But when the scattering is molecular, it is sixteen times as large for the extreme visible violet as for the extreme visible red. Consequently the radiation emitted by a mass of gas under these conditions would show the violet considerably weakened as compared with the red. This opens out the possibility that with increasing temperature the violet portion of the continuous spectrum of a star may diminish in intensity as compared with the red end. As will presently appear, we possess some independent evidence that the photosphere emits less violet light than it should do if it were a black surface, but a closer experimental investigation is necessary before this can be definitely established.

I consider that for this purpose the careful investigation of the distribution of intensity in the solar spectrum is a matter of urgent importance. It would be necessary, however, for a satisfactory solution of the problem to measure the intensities everywhere in the intervals between Fraunhofer lines, or, at any rate, to select portions of the spectrum where no prominent Fraunhofer lines are situated.

Unless this is done, we risk that the violet portion of the spectrum should show too small an intensity, merely because it contains a greater number of Fraunhofer lines.

The loss of light by scattering in the solar atmosphere renders it possible for bright lines to appear which are due to vibrations in the front layers, though the temperature at these layers may be less than that which supplies the continuous spectrum. If the scattering is insufficient, its effect may be to obliterate the Fraunhofer line without entirely converting it into a bright line. It is necessary, however, that some kind of law should exist as to which of the Fraunhofer lines are obliterated. The two striking facts to be explained are the absence of the ultra-violet portion of the hydrogen series and the absence of the helium lines. In both cases the lines appear with considerable brilliancy in the so-called chromosphere, and in the flash-spectrum observed at the beginning and end of total eclipses. With regard to the hydrogen series, observations on stellar spectra and laboratory experiments which have already been quoted, would have led us to expect that the ultra-violet lines would be more easily reversible than the less refrangible lines. If the Sun forms an exception, it seems to indicate that the violet part of the continuous spectrum is reduced in intensity relatively to the red portion, and that this reduction is not a mere temperature effect.

This consideration strengthens to some extent the idea that the comparative weakness of the ultra-violet radiation in solar stars is not due to a diminution of temperature. As already mentioned, molecular scattering in the photospheric region might account for the comparative poverty of the more rapid vibrations.

The behavior of helium cannot be due to the same cause, as none of its lines have been seen reversed in the solar spectrum. The correct explanation is, I believe, to be found in this case in the great height to which helium is found to rise above the photospheric layer. The previous investigation has shown that a bright line is more likely to appear when the product of the coefficient of scattering and the thickness of the absorbing layer is large. This may be caused either by a great coefficient of scattering or by a great thickness of the absorbing layer. It is true that this reasoning should apply equally to hydrogen and the metals which rise as high as helium, and I believe that it does apply. The absence of some of the hydrogen lines in the solar spectrum has already been noted. That the red and blue lines can be seen is no doubt a consequence of the fact that hydrogen exists in much greater quantities than helium, for it should be noted that the helium lines are not bright, but only insufficiently dark to be observed.

This comparative weakness of some Fraunhofer lines which are very prominent in the flash-spectrum, and are probably due to the high temperature of the portion of the solar disk emitting the correspondent radiation, has been commented upon by Mr. Evershed, whose explanation I consider in the main to be correct. Although a further discussion of some points of detail may be desirable, the matter is independent of scattering, and lies therefore outside the range of this communication.

Victoria University,
Manchester, Eng.

Matrix Methods for Multiple-Scattering Problems

S. Twomey

Naval Research Laboratory, Washington, D. C.

H. Jacobowitz and H. B. Howell

U. S. Weather Bureau, ESSA, Washington, D. C.

(Manuscript received 22 November 1965)

If the radiation field is approximated by a discrete distribution at points or latitude circles on the unit sphere, matrix relationships can be written between incident and reflected or transmitted radiation fields. The reflection and transmission matrices thus defined are shown to satisfy algebraic equations which can be used to compute the properties of thick layers by building up the thick layers from thinner sublayers, the starting point being a layer so thin that it is effectively a single scattering layer only.

---

1. Introduction

The problem of the transfer of radiation through a scattering medium is a classical one, and has many practical applications. There is an extensive literature ranging from purely mathematical investigations of some of the equations arising from the physical theory to physical observations of varying degrees of complexity.

For isotropic scattering and Rayleigh scattering, methods developed by Ambartsumian (1942) and Chandrasekhar (1944) and elaborated and extended by Chandrasekhar (1950), van de Hulst (1948), Kourganoff (1952) and others, have proved to be powerful and provided the mathematical bases for numerical calculations such as, for instance, the extended and detailed computations of intensities and polarizations of Rayleigh scattering in the atmosphere by Sekera and his co-workers (1960). In the context of asymmetric anisotropic scattering such as is produced by cloud layers and many atmospheric particulates, the theoretical situation is less satisfactory; Chandrasekhar's H-, X-, and Y-functions, for instance, cannot be applied to the case of the more complicated phase functions.

Since even the most elegant or elaborate mathematical solutions are likely to be reduced to a matrix formulation—i.e., addition—before numerical results are obtained, it seemed worthwhile to explore the problem from the beginning in a matrix description, which was in essence no more than an elaboration of the multiple-stream methods used by some early workers. It was found that a number of matrix relationships resulted, many of which were just the matrix equivalents of the integral equations of Chandrasekhar (1950). The matrix equations were such that reflection and transmission properties of finite layers could be obtained quickly for arbitrary scattering diagrams and without iteration or approximation (except for the

initial *physical* approximation in which a discrete distribution of radiant energy is used in place of a continuous one). The present paper will derive completely the fundamental matrix equations and will show how internal checking of results is provided by the existence of additional identities which are not used otherwise in the computations. A limited number of numerical examples will be presented, including some for isotropic and Rayleigh scattering, which are compared with available results from other sources.

After the first draft of this paper was completed, the authors read a monograph by van de Hulst (1963) in which an operator notation was developed and applied to the multiple scattering problem. The methods described in this paper and those developed by van de Hulst are essentially equivalent.

2. Terminology

Vector quantities, in the sense of one-dimensional arrays, will be designated by bold face, lower-case symbols; bold face capitals will be used for matrices and unless necessary for clarity the arguments of scattering and phase functions or matrices will be omitted, i.e., dependences such as $\mathbf{S}(\tau)$ will only be indicated explicitly when it seems necessary. Otherwise $\mathbf{S}$ alone will be used.

Directions will be described by an azimuth angle ϕ with respect to an arbitrary origin and by the cosine of the polar angle, which will be denoted by μ; $\mu = 1$ represents the vertically upwards direction, $\mu = 0$ the horizontal, and $\mu = -1$ vertically downwards. Optical depth τ is measured from an arbitrary level which may be envisaged for convenience as the top of the atmosphere; the energy remaining in a vertical beam of radiation is therefore a fraction $e^{-\tau}$ of its original value, while an oblique beam is attenuated to a fraction $e^{-\tau/|\mu|}$. The direct transmission of radiation in proportion to

$e^{-\tau/|\mu|}$ will always be distinguished from diffuse transmission during which one or more scatterings occur. This had the consequence that the *diffuse* transmission tends to *zero* (not to unity) when the optical depth τ tends to zero.

3. Scattering matrices

The multiple scattering of a continuous, distributed field of radiation can be approximated by the scattering of discrete streams in the directions (μ_1, φ_1), (μ_2, φ_2), $\cdots (\mu_N, \varphi_N)$, $(-\mu_1, \varphi_1)$, $(-\mu_2, \varphi_2)$, $\cdots (-\mu_N, \varphi_N)$, all scattering processes being considered as a redistribution of scattered energy among these $2N$ directions. Each "direction" can be visualized as a small cone containing the geometric vector. If the intensity in the direction (μ_i, φ_i) is $\mathbf{u}_i$ and the intensity in the direction $(-\mu_i, \varphi_i)$ is $\mathbf{v}_i$, the "upward" intensities $u_1, u_2, \cdots u_N$ can be grouped into a single vector $\mathbf{u}$ and the downward intensities $v_1, v_2, \cdots v_N$ into a vector $\mathbf{v}$. Then a reflection matrix $\mathbf{S} = \|s_{ij}\|$ and a transmission matrix $\mathbf{T} = \|t_{ij}\|$ can be defined for any layer such that an incident radiation field $\mathbf{u}_0$ in the downward hemisphere, for example, gives rise to a scattered field $\mathbf{Su}_0$ in the upward hemisphere, a diffusely transmitted field $\mathbf{Tu}_0$ in the downward hemisphere, and a directly transmitted field which is given by the vector $\mathbf{Eu}_0$, if $\mathbf{E}$ denotes a diagonal matrix with elements $e^{-\tau/\mu_i}$. The elements of $\mathbf{S}$ and $\mathbf{T}$ depend on the optical depth τ of the layer, on the choice of the set of directions (μ_i, φ_i), $i = 1, 2, \cdots N$, on the albedo, and on the scattering properties of the individual scattering elements. The latter can be described in a compatible manner by using matrices $\mathbf{P}$ and $\mathbf{B}$, defined as follows: when energy is removed from the jth direction, a fraction p_{ij} thereof reappears in the ith direction in the same sense, while a fraction b_{ij} reappears in the ith direction in the opposite sense to the original radiation. Thus $\mathbf{P}$ and $\mathbf{B}$ are analogous to, and derived from, the familiar phase function $p(\cos\theta)$, where θ is the angle between the two directions; p_{ij} corresponds to $p(\mu_i\mu_j + \sqrt{1-\mu_i^2}\sqrt{1-\mu_j^2}\cos[\varphi_i - \varphi_j])$ and b_{ij} corre-

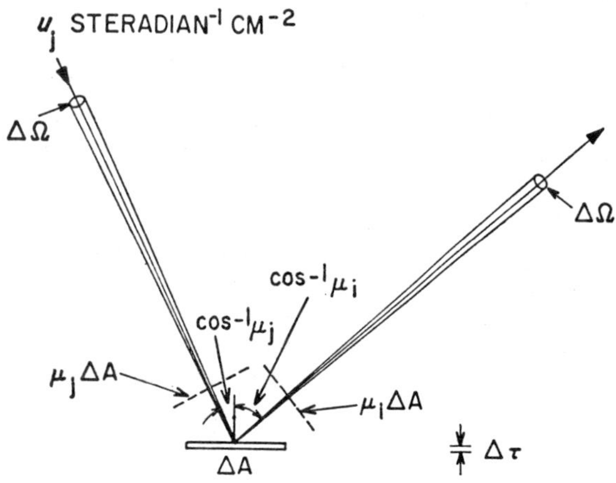

FIG. 1. Geometry of scattering by an infinitesimal layer.

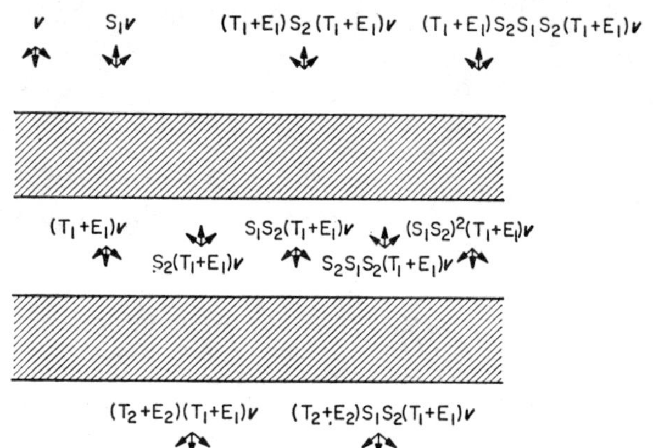

FIG. 2. Schematic chart of the individual contributions making up the whole reflected and transmitted intensity fields.

sponds to $p(-\mu_i\mu_j + \sqrt{1-\mu_i^2}\sqrt{1-\mu_j^2}\cos[\varphi_i - \varphi_j])$ if here μ_i, μ_j are defined to be positive quantities. Thus $\mathbf{B}$ relates scattering from an upward to a downward direction, or vice versa, whereas $\mathbf{P}$ relates to scattering from upward to upward or downward to downward directions.

For an infinitesimal layer only single scattering enters and the matrices $\mathbf{S}$ and $\mathbf{T}$ for such a layer can be related simply to $\mathbf{P}$ and $\mathbf{B}$. If, as in Fig. 1, a vector of intensities $\mathbf{u}_j$ is incident upon a layer of optical thickness $\Delta\tau$ in a direction (μ_j, φ_j), within a small solid angle $\Delta\Omega$, then within an element of area ΔA a fraction $\Delta\tau/\mu_j$ of the incident energy $\mu_j u_j \Delta A \Delta\Omega$ is removed; of this a fraction p_{ij} reappears (by definition) in the direction (μ_i, φ_i). In that direction the projected area of the element is $\mu_i\Delta A$, hence the *intensity* increment in the i-direction is obtained by dividing the energy increment by $\mu_i\Delta A \Delta\Omega$, i.e., it is $\mu_i^{-1}p_{ij}u_j\Delta\tau$. *Hence for an infinitesimal layer*, the scattering matrices are:

$$\mathbf{S}(\Delta\tau) = \mathbf{M}^{-1}\mathbf{B}\Delta\tau$$
$$\mathbf{T}(\Delta\tau) = \mathbf{M}^{-1}\mathbf{P}\Delta\tau, \tag{1}$$

where $\mathbf{M}$ is a diagonal matrix with elements μ_i.

4. Superposition of layers

If intensity defined by a vector $\mathbf{v}$ falls upon a layer of optical thickness τ_1 and characterized by scattering matrices $\mathbf{S}_1$ and $\mathbf{T}_1$, radiation $\mathbf{S}_1\mathbf{v}$ is scattered back and consists of radiation which did not penetrate through the layer; if a second layer characterized by τ_2, $\mathbf{S}_2$, and $\mathbf{T}_2$ is added below the first, the component $\mathbf{S}_1\mathbf{v}$ is unaffected, but the diffusely transmitted radiation field $\mathbf{T}_1\mathbf{v}$ and the directly transmitted field $\mathbf{E}_1\mathbf{v}$ are partly transmitted through the second layer and partly reflected by it to become an additional radiation field incident upon the first layer from below. This division continues as shown schematically in Fig. 2. Since the combination of the two layers must be equivalent to a single layer of thickness $\tau_1 + \tau_2$, and since $\mathbf{v}$ is arbitrary, one can write the relationship between the matrices S, T, and E for a layer of thickness $\tau_1+\tau_2$ and the matrices S_1, T_1, E_1, and S_2, T_2, E_2 as follows:

$$S=S_1+(T_1+E_1)S_2(I+S_1S_2+[S_1S_2]^2+\cdots)$$
$$\times(T_1+E_1), \quad (2a)$$

$$E+T=(T_2+E_2)(I+S_1S_2+[S_1S_2]^2+\cdots)$$
$$\times(T_1+E_1). \quad (2b)$$

For finite layers or non-conservative scattering, the matrix series converges and represents $(I-S_1S_2)^{-1}$, where I is the identity matrix.

Eqs. (2) are fundamental. They can be directly applied to build up scattering matrices for thick layers by superposing thinner layers upon each other. They also serve to provide matrix equations analogous to Chandrasekhar's non-linear integral equations. To obtain these all that is necessary is to let the layers referred to by Eqs. (2) become infinitesimal, in turn, and apply (1) for the infinitesimal layer.

5. Single-layer equations

If τ_2 is made infinitesimal, the relations (1) can be used for S_2 and T_2 in Eq. (2). Making these substitutions, and noting that for an infinitesimal layer the matrix E becomes $I-M^{-1}\Delta\tau$, one obtains

$$S+\frac{\partial S}{\partial\tau}\Delta\tau=S+(T+E)M^{-1}B\Delta\tau(I+\cdots)(T+E),$$

where the subscripts have been dropped and S, T, E, and M refer to a layer of optical thickness τ. Hence

$$M\frac{\partial S}{\partial\tau}=M(T+E)M^{-1}B(T+E). \quad (3a)$$

Similarly, from Eq. (2b),

$$M\frac{\partial T}{\partial\tau}=-T+(P+MSM^{-1}B)(T+E), \quad (3b)$$

and analogously, by letting $\tau_1\rightarrow\Delta\tau$ while $\tau_2=\tau$,

$$M\frac{\partial S}{\partial\tau}=B+PS+MSM^{-1}(P+BS)-S-MSM^{-1}, \quad (3c)$$

$$M\frac{\partial T}{\partial\tau}=-MTM^{-1}+(MTM^{-1}+E)(P+BS). \quad (3d)$$

These equations are equivalent to Chandrasekhar's non-linear integral equations. That a similar symmetry property holds for the scattering matrices as was proved by Chandrasekhar (1960, p. 171) for the scattering functions can be shown readily. For when τ is a sufficiently small thickness, $\Delta\tau$, S and T are merely $M^{-1}B\Delta\tau$ and $M^{-1}P\Delta\tau$. But B and P will be unaffected by transposition if b_{ij} and p_{ij} depend only on the angle between

the i- and j-directions. Hence, for an infinitesimal layer $[MS]=[MS]^*$ and $[MT]=[MT]^*$, where the asterisk signifies transposition. But if Eq. (3d) is transposed and these symmetry relations are inserted, there results

$$\frac{\partial}{\partial\tau}[MT]^*=-T+(P+MSM^{-1}B)(T+E),$$

in which the right hand side is identical to that of (3b), so that $\frac{\partial}{\partial\tau}[MT]^*=\frac{\partial}{\partial\tau}[MT]$; hence if the symmetry exists in T for any thickness, it will be maintained for all thicknesses. With respect to S, Eq. (3a) alone shows that $\frac{\partial}{\partial\tau}MS$ is unaffected by transposition provided only that $[MT]=[MT]^*$, i.e., that symmetry exists in T. Hence for all thicknesses the symmetry properties apply to S and T, i.e.,

$$[MS]^*=MS \qquad [MT]^*=MT$$
$$S^*=MSM^{-1} \qquad T^*=MTM^{-1}. \qquad (4)$$

For a semi-infinite layer $(\tau\rightarrow\infty)$, $\frac{\partial S}{\partial\tau}$ must vanish. Hence for such a layer

$$S+MSM^{-1}=B+PS+MSM^{-1}(P+BS). \quad (5)$$

The latter equation is the matrix equivalent of Chandrasekhar's equation[1] for a semi-infinite layer.

6. Harmonic scattering matrices

A considerable improvement in the possible fineness of subdivision of the specified directions can be obtained without much increase in complexity if a finite array of polar angles μ is used while variations of azimuth angle are treated by expanding the intensities and scattering functions as Fourier series in ϕ. Through the orthogonality of the trigonometric functions the equations separate into equations each of which involves only harmonics of the same order. The radiation fields can be described by harmonic vectors $\mathbf{u}^{(m)}$, $\mathbf{v}^{(m)}$, $m=0, 1, 2, \cdots$, such that the intensity in an (upward) direction $(+\mu_i,\phi)$, for example, is

$$u_i(\phi)=\sum_m [u_i^{(m)}\cos m\phi+u_i^{(-m)}\sin m\phi].$$

To describe multiple scattering by a given layer one could use a function continuous in the incident and emergent azimuth angles ϕ and ϕ' and discrete in μ, of the form $S_{ik}(\phi-\phi')$; this, in turn, can be expanded as a Fourier series in $(\phi-\phi')$, which by symmetry will contain only cosines. Writing this Fourier series as $\sum_m S_{ik}^{(m)}\cos m(\phi-\phi')$, one can derive the intensity increment in the direction (μ_i,ϕ) as a result of scattering from the latitude $\mu=\mu_k$ and the azimuth interval

---

[1] *Loc. cit.*, p. 94, Eq. (28).

$\phi \rightarrow \phi + \Delta\phi$ as

$$\sum_m S_{ik}{}^{(m)} \cos(\phi - \phi')$$
$$\times \sum_m [u_k{}^{(m)} \cos m\phi' + u_k{}^{(-m)} \sin m\phi'] \Delta\phi'.$$

Integration around each latitude belt, followed by summation over all belts gives the total scattered intensity increment in the direction (μ_i, ϕ) as

$$\Delta u_i = \Delta[\sum_m u_i{}^{(m)} \cos m\phi + \sum_m u_i{}^{(-m)} \sin m\phi]$$
$$= \sum_m f_m \sum_k S_{ik}{}^{(m)} [u_k{}^{(m)} \cos m\phi + u_k{}^{(-m)} \sin m\phi],$$

where

$$f_m = \int_0^{2\pi} \cos^2 m\phi \, d\phi = \pi \quad (m \neq 0)$$

and $f_0 = 2\pi$. Hence,

$$\Delta u_i{}^{(m)} = f_m \sum_k S_{ik}{}^{(m)} u_k{}^{(m)}.$$

Thus apart from the factor f_m, the harmonic scattering matrices $\mathbf{S}^{(m)}$ and $\mathbf{T}^{(m)}$ act on the corresponding vectors $\mathbf{u}^{(m)}$ and $\mathbf{v}^{(m)}$ exactly as $\mathbf{S}$ and $\mathbf{T}$ acted upon $\mathbf{u}$ and $\mathbf{v}$. It may be observed that the sine coefficients, for which negative superscripts were used, are identically zero in the scattered radiation as long as the incident radiation is symmetric in azimuth and that only positive values of m need be considered in practice.

Directly transmitted (unscattered) radiation is attenuated in proportion to the elements $e^{-\tau/|\mu_i|}$ of the diagonal matrix $\mathbf{E}$; the factor f_m does not enter, and hence the total (diffuse plus direct) transmission matrix to be applied to the mth harmonic is $f_m \mathbf{T}^{(m)} + \mathbf{E}$. One can proceed exactly as for Eq. (2) and so derive analogous equations for the harmonic matrices. Thus

$$f_m \mathbf{S}^{(m)} = f_m \mathbf{S}_1{}^{(m)} + (f_m \mathbf{T}_1{}^{(m)} + \mathbf{E}_1) \mathbf{S}_2{}^{(m)}$$
$$\times (\mathbf{I} + f_m{}^2 \mathbf{S}_1{}^{(m)} \mathbf{S}_2{}^{(m)} + \cdots)(f_m \mathbf{T}_1{}^{(m)} + \mathbf{E}_1), \quad (6)$$
$$\mathbf{E} + f_m \mathbf{T}^{(m)} = (f_m \mathbf{T}_2{}^{(m)} + \mathbf{E}_2)$$
$$\times (\mathbf{I} + f_m{}^2 \mathbf{S}_1{}^{(m)} \mathbf{S}_2{}^{(m)} + \cdots)(f_m \mathbf{T}_1{}^{(m)} + \mathbf{E}_1).$$

Obviously $\mathbf{E} = \mathbf{E}_1 \mathbf{E}_2$. For an infinitesimal layer it is evident that

$$\mathbf{S}^{(m)} \rightarrow \mathbf{M}^{-1} \mathbf{B}^{(m)} \Delta\tau,$$
$$\mathbf{T}^{(m)} \rightarrow \mathbf{M}^{-1} \mathbf{P}^{(m)} \Delta\tau. \quad (7)$$

Insertion of these infinitesimal matrices for $\mathbf{S}_1{}^{(m)}$ and $\mathbf{T}_1{}^{(m)}$ and for $\mathbf{S}_2{}^{(m)}$ and $\mathbf{T}_2{}^{(m)}$ give equations fully analogous to (3a)–(3d):

$$\mathbf{M}\frac{\partial}{\partial\tau}\mathbf{S}^{(m)} = \mathbf{M}[f_m \mathbf{T}^{(m)} + \mathbf{E}]\mathbf{M}^{-1}\mathbf{B}^{(m)}[f_m \mathbf{T}^{(m)} + \mathbf{E}], \quad (8a)$$

$$\mathbf{M}\frac{\partial}{\partial\tau}\mathbf{T}^{(m)} = -\mathbf{T}^{(m)} + [\mathbf{P}^{(m)} + f_m \mathbf{M}\mathbf{S}^{(m)}\mathbf{M}^{-1}\mathbf{B}^{(m)}]$$
$$\times [f_m \mathbf{T}^{(m)} + \mathbf{E}], \quad (8b)$$

$$\mathbf{M}\frac{\partial}{\partial\tau}\mathbf{S}^{(m)} = -\mathbf{S}^{(m)} - \mathbf{M}\mathbf{S}^{(m)}\mathbf{M}^{-1} + \mathbf{B}^{(m)} + f_m \mathbf{P}^{(m)}\mathbf{S}^{(m)}$$
$$+ f_m \mathbf{M}\mathbf{S}^{(m)}\mathbf{M}^{-1}[\mathbf{P}^{(m)} + f_m \mathbf{B}^{(m)}\mathbf{S}^{(m)}], \quad (8c)$$

$$\mathbf{M}\frac{\partial}{\partial\tau}\mathbf{T}^{(m)} = -\mathbf{M}\mathbf{T}^{(m)}\mathbf{M}^{-1} + [f_m \mathbf{M}\mathbf{T}^{(m)}\mathbf{M}^{-1} + \mathbf{E}]$$
$$\times [\mathbf{P}^{(m)} + f_m \mathbf{B}^{(m)}\mathbf{S}^{(m)}]. \quad (8d)$$

Again symmetry also exists, of the form

$$[\mathbf{S}^{(m)}]^* = \mathbf{M}\mathbf{S}^{(m)}\mathbf{M}^{-1}; \quad [\mathbf{T}^{(m)}]^* = \mathbf{M}\mathbf{T}^{(m)}\mathbf{M}^{-1},$$

while for semi-infinite layers $(\tau \rightarrow \infty, \mathbf{E} \rightarrow 0, \mathbf{T} \rightarrow 0)$,

$$\mathbf{S}^{(m)} + \mathbf{M}\mathbf{S}^{(m)}\mathbf{M}^{-1} = \mathbf{B} + f_m \mathbf{P}\mathbf{S}$$
$$+ f_m \mathbf{M}\mathbf{S}^{(m)}\mathbf{M}^{-1}[\mathbf{P}^{(m)} + f_m \mathbf{B}^{(m)}\mathbf{S}^{(m)}]. \quad (9)$$

Formally, therefore, these equations for the harmonic matrices can be obtained from Eqs. (2), (3), (4), and (5) for the entire matrices by adding the superscript to $\mathbf{S}$, $\mathbf{T}$, $\mathbf{B}$, and $\mathbf{P}$ and inserting in each term the factor f_m for each occurrence therein of any one of these matrices.

7. Normalization of the phase matrices P and B

A phase function $p(\theta)$ is normalized by a scalar multiplication; thus $\int_0^{2\pi} \int_{-1}^1 p \, d\mu \, d\phi$ represents the fraction of the incident energy scattered in all directions, and a scalar factor is all that is needed to make the integral to equal the albedo $\bar{\omega}$ and so normalize $p(\theta)$ to ensure that energy is conserved.

The normalization for the matrix case where a single scattering causes a fraction p_{ij} of incident energy in the direction associated with j to be scattered to the direction associated with i, and b_{ij} to the ith direction of opposite sense, is in the simplest case expressed by

$$\sum_i (p_{ij} + b_{ij}) = \bar{\omega}.$$

When harmonic scattering matrices are used one must integrate over azimuth and sum over latitude belts. The integration removes all but the zero-order harmonics and one obtains

$$f_0 \sum_i [p_{ij}{}^{(0)} + b_{ij}{}^{(0)}] = \bar{\omega}.$$

Since $\mathbf{P}$ and $\mathbf{B}$ must be symmetric, normalization is not a trivial matter. Even with isotropic scattering, orders of scattering of ten and higher must be included before even moderately good accuracy is realized for an optically deep layer, while asymmetric scattering (such as the Mie function represents) must involve considerably higher orders of scattering. Discrepancies of even 1 per cent cannot therefore be tolerated between the sums $\sum_i (p_{ij}+b_{ij})$ for different values of j, for the error would become grossly magnified at higher orders of scattering. A procedure had to be found, therefore, to change $\mathbf{P}$ and $\mathbf{B}$ slightly, while retaining symmetry, so that $\sum_i (p_{ij}+b_{ij})=\bar{\omega}$ was satisfied to considerable accuracy for every j.

Let $\mathbf{A}$ be a $2N\times 2N$ matrix containing as submatrices $\mathbf{P}$ in the upper left and lower right quarters and $\mathbf{B}$ in the upper right and lower left quarters. The normalization condition can be written

$$\sum_i a_{ij}=\bar{\omega} \quad (j=1, 2, \cdots 2N).$$

If now the vector $\mathbf{e}$ is defined such that the elements of $\mathbf{e}$ are all unity, the normalization condition can be written (for $\mathbf{A}$ is symmetric),

$$\mathbf{A}\mathbf{e}=\bar{\omega}\mathbf{e},$$

which defines $\mathbf{e}$ as an eigenvector of $\mathbf{A}$ with eigenvalue $\bar{\omega}$. (Since the remaining eigenvectors must be orthogonal with respect to $\mathbf{e}$, it follows that the sums of the elements of all other eigenvectors must be identically zero.) Hence, if $\mathbf{A}$ approximately satisfies the normalization requirement, it can be made to do so exactly if a rotation to principal axes is made to obtain the diagonal matrix $\mathbf{\Lambda}$ of the eigenvalues of $\mathbf{A}$ and the orthogonal matrix $\mathbf{U}$ of the eigenvectors. For then by inspection the column of $\mathbf{U}$ with all positive elements can be found; if these are not exactly equal they must be made so, by replacement by the average of the column elements; the remaining columns must be re-orthogonalized (e.g., by the Schmidt method) and then a modified and properly normalized $\mathbf{A}$ obtained by computation of $\mathbf{A}=\mathbf{U}\mathbf{\Lambda}\mathbf{U}^*$ using the adjusted $\mathbf{U}$.

8. Computational method

a) The computation of the single scattering matrices $\mathbf{B}$ and $\mathbf{P}$ and the harmonic single scattering matrices $\mathbf{B}^{(m)}$ and $\mathbf{P}^{(m)}$ was carried out by first computing a phase function $p(\theta)$ at close intervals in θ and assuming that $p(\theta)$ followed a straight line path between successive tabular points. (In the case of cloud scattering, the scattering functions were first integrated over a distribution of scatterers to obtain a properly averaged volume phase function). The tabular values of μ_i ($i=1, 2, \cdots N$) were then set up so that the μ_i were evenly spaced and the pairs $(+\mu_i, -\mu_i)$ symmetrically disposed. Thus for $N=10$, for example, the tabular μ_i were 0.05, 0.15, 0.25, 0.35, 0.45, 0.55, 0.65, 0.75, 0.85, 0.95. The (i,j) element of the phase matrices $\mathbf{P}$ and $\mathbf{B}$ accounted for energy scattered from the interval $\mu_j-\tfrac{1}{2}\Delta\mu\leqslant\mu'\leqslant\mu_j+\tfrac{1}{2}\Delta\mu$ into the interval $\mu_i-\tfrac{1}{2}\Delta\mu\leqslant\mu\leqslant\mu_i+\tfrac{1}{2}\Delta\mu$; the (i,j) elements $p_{ij}{}^{(m)}$ and $b_{ij}{}^{(m)}$ of the harmonic scattering matrices $\mathbf{P}^{(m)}$ and $\mathbf{B}^{(m)}$ are the Fourier coefficients in an expansion of the phase function as a function of ϕ, i.e.,

$$f_m p_{ij}{}^{(m)}=\int_0^{2\pi}\cos m\phi\int_{\mu_i-\frac{1}{2}\Delta\mu}^{\mu_i+\frac{1}{2}\Delta\mu}\int_{\mu_j-\frac{1}{2}\Delta\mu}^{\mu_j+\frac{1}{2}\Delta\mu}p(\mu\mu'+\sqrt{1-\mu^2}\sqrt{1-\mu'^2}\cos\phi)d\mu'd\mu d\phi,$$

$$f_m b_{ij}{}^{(m)}=\int_0^{2\pi}\cos m\phi\int_{\mu_i-\frac{1}{2}\Delta\mu}^{\mu_i+\frac{1}{2}\Delta\mu}\int_{\mu_j-\frac{1}{2}\Delta\mu}^{\mu_j+\frac{1}{2}\Delta\mu}p(-\mu\mu'+\sqrt{1-\mu^2}\sqrt{1-\mu'^2}\cos\phi)d\mu'd\mu d\phi.$$

b) Eqs. (3a) and (3b) for the entire scattering matrices, or Eqs. (8a) and (8b) for the harmonic matrices provide bases for numerical integration in steps of τ. For thin enough layers the single scattering relationships (1) provide a starting point and the equations for the first derivatives permit an integration in steps by repeated application of the latter. The procedure is easy to program and is fast, involving at each step only a small number of matrix multiplications.

In order to provide a continuous test of the numerical integration, formulae were also derived for the second derivatives. Eqs. (3), for instance, give the identities

$$\mathbf{M}\frac{\partial^2\mathbf{S}}{\partial\tau^2}=\left(\frac{\partial\mathbf{E}}{\partial\tau}+\frac{\partial\mathbf{T}}{\partial\tau}\right)^*\mathbf{B}(\mathbf{T}+\mathbf{E})$$
$$+\left[\left(\frac{\partial\mathbf{E}}{\partial\tau}+\frac{\partial\mathbf{T}}{\partial\tau}\right)^*\mathbf{B}(\mathbf{T}+\mathbf{E})\right]^*,$$

$$\mathbf{M}\frac{\partial^2\mathbf{T}}{\partial\tau^2}=-\frac{\partial\mathbf{T}}{\partial\tau}+(\mathbf{P}+\mathbf{S}^*\mathbf{B})\left(\frac{\partial\mathbf{T}}{\partial\tau}+\frac{\partial\mathbf{E}}{\partial\tau}\right)+\frac{\partial\mathbf{S}^*}{\partial\tau}\mathbf{B}(\mathbf{T}+\mathbf{E}),$$

while Eqs. (8) lead to a perfectly analogous pair of equations for

$$\mathbf{M}\frac{\partial^2\mathbf{S}^{(m)}}{\partial\tau^2} \quad\text{and}\quad \mathbf{M}\frac{\partial^2\mathbf{T}^{(m)}}{\partial\tau^2}.$$

Little additional computation is required to produce the second derivatives, for combinations such as $\mathbf{P}+\mathbf{S}^*\mathbf{B}$ and $\mathbf{T}+\mathbf{E}$ must be computed to obtain the first derivatives, while $\partial\mathbf{E}/\partial\tau$ is merely $-\mathbf{M}^{-1}\mathbf{E}$; also the second term in the equation for $\mathbf{M}\frac{\partial^2\mathbf{S}}{\partial\tau^2}$ is merely the transpose of the first. Once the second derivatives are known, the Taylor series

$$\mathbf{S}^{(m)}(\tau+\Delta\tau)=\mathbf{S}^{(m)}+\Delta\tau\frac{\partial\mathbf{S}^{(m)}}{\partial\tau}+\frac{(\Delta\tau)^2}{2!}\frac{\partial^2\mathbf{S}^{(m)}}{\partial\tau^2}+\cdots$$

$$\mathbf{T}^{(m)}(\tau+\Delta\tau)=\mathbf{T}^{(m)}+\Delta\tau\frac{\partial\mathbf{T}^{(m)}}{\partial\tau}+\frac{(\Delta\tau)^2}{2!}\frac{\partial^2\mathbf{T}^{(m)}}{\partial\tau^2}+\cdots$$

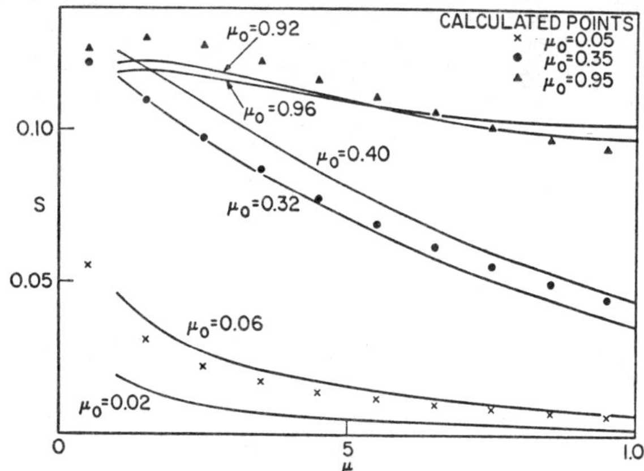

FIG. 3. Curves from Coulson *et al* (1960) for a Rayleigh atmosphere. The points superimposed were obtained by the matrix method described for $N=10$ with a Rayleigh phase function. $\tau=1$.

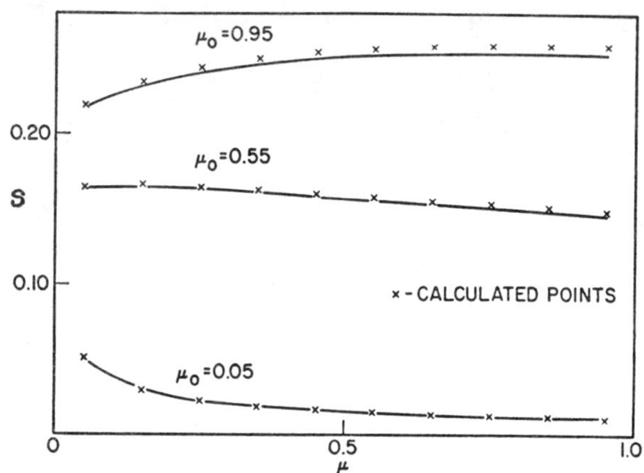

FIG. 4. Points obtained by the matrix method compared with curves by van de Hulst's (1964) asymptotic formula (isotropic scattering). $\tau=8.08$.

can be evaluated up to the quadratic term, so that a parabolic as well as a linear extrapolation becomes available to go from one value of τ to the next. The results of both procedures can be compared to ensure that the computations are accurate and the steps in τ not too coarse. An additional check is provided by the fact that there are equations which must be satisfied by $\mathbf{S}$ and $\mathbf{T}$ and which involve no derivatives. These equations are obtained by elimination of the derivatives; from (3a) and (3c), for example, it follows

that

$$\mathbf{S}+\mathbf{S}^* = \mathbf{B}+\mathbf{PS}+\mathbf{S}^*(\mathbf{P}+\mathbf{BS}) \\ -(\mathbf{T}+\mathbf{E})^*\mathbf{B}(\mathbf{T}+\mathbf{E}), \quad (10a)$$

while from (3b and (3d),

$$\mathbf{T}-\mathbf{T}^* = (\mathbf{P}+\mathbf{S}^*\mathbf{B})(\mathbf{T}+\mathbf{E})-(\mathbf{T}+\mathbf{E})^*(\mathbf{P}+\mathbf{BS}). \quad (10b)$$

From (8a) and (8c),

$$\mathbf{S}^{(m)}+[\mathbf{S}^{(m)}]^* = \mathbf{B}^{(m)}+f_m\mathbf{P}^{(m)}\mathbf{S}^{(m)} \\ +[\mathbf{S}^{(m)}]^*[\mathbf{P}^{(m)}+f_m\mathbf{B}^{(m)}\mathbf{S}^{(m)}] \\ -[f_m\mathbf{T}^{(m)}+\mathbf{E}]^*\mathbf{B}[f_m\mathbf{T}^{(m)}+\mathbf{E}], \quad (11a)$$

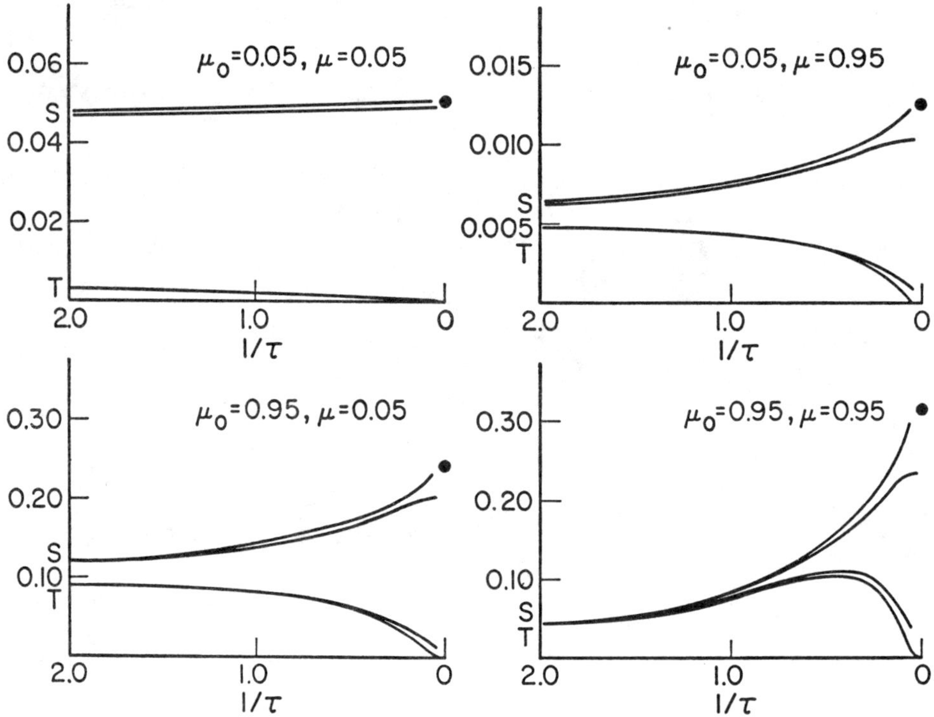

FIG. 5. The variation with optical thickness τ of selected reflection and transmission elements for isotropic scattering. Pairs of curves for $\bar{\omega}=1$ (upper) and 0.99 (lower).

and from (8b) and (8d),

$$\mathbf{T}^{(m)}-[\mathbf{T}^{(m)}]^*=[\mathbf{P}^{(m)}+f_m\mathbf{S}^{(m)*}\mathbf{B}^{(m)}][f_m\mathbf{T}^{(m)}+\mathbf{E}]$$
$$-[f_m\mathbf{T}^{(m)}+\mathbf{E}][\mathbf{P}^{(m)}+f_m\mathbf{B}^{(m)}\mathbf{S}^{(m)}]. \quad (11b)$$

These equations can be used to obtain an independent check at each or any stage of the calculation of a sequence of $\mathbf{S}$ and $\mathbf{T}$ matrices. These equations were programed for an IBM 7094 computer. Initial tests showed that only a few minutes were required to proceed through some hundreds of steps of the computation with the linear and parabolic formulae being both applied independently at each step. With steps of $\Delta\tau=0.005$, an accuracy of better than 1 per cent was indicated by both checking procedures.

c) Eqs. (2) or (6) permit one to take, for example, the scattering matrices for $\tau=0.100$ and build up directly scattering matrices, for $\tau=0.200, 0.300, 0.400\cdots$ up to any multiple of 0.100. The results of such a calculation can be checked in two ways: first, by arranging to have a range of values of τ within which scattering matrices have been computed by both superposition and integration methods, so that an intercomparison is possible; secondly, by inserting the $\mathbf{S}$ and $\mathbf{T}$ matrices or the $\mathbf{S}^{(m)}$ and $\mathbf{T}^{(m)}$ into the necessary relations (10) or (11), which must be satisfied identically. As experience was gained with the numerical computation it became apparent that a superior approach was to omit the Taylor expansion and to use the superposition formulae (2) and (6) from the start; the Taylor expansions are now used only for checking purposes.

Eqs. (11a) and (11b) which are explicit relationships between $\mathbf{S}^{(m)}$, $\mathbf{T}^{(m)}$ and known matrices [$\mathbf{E}$, $\mathbf{B}^{(m)}$ and $\mathbf{P}^{(m)}$], were used during the computational process to test the accuracy. They also could be used to improve the accuracy by direct iteration, and in the semi-infinite form (9), made it possible to proceed to a semi-infinite layer by iterating the equation using the S-matrix for a thick finite layer as a starting point.

9. Results

In the present paper, presentation and discussion of results will be limited to illustrative examples and comparison of results with those obtained by analytic methods for a few special phase functions (e.g., for isotropic or Rayleigh atmospheres). The primary goal of the work is to provide data on multiple Mie scattering by cloud layers, etc., but it was informative to apply the methods to those simpler phase functions for which analytic results exist.

At this point it is well to point out that the matrix equations are entirely rigorous. They are approximations only to the extent that the discrete scattering situations to which they apply can be considered as approximations to the physical situation where energy is directed in all directions and is not restricted to discrete latitude belts. Thus, if isotropic scattering matrices

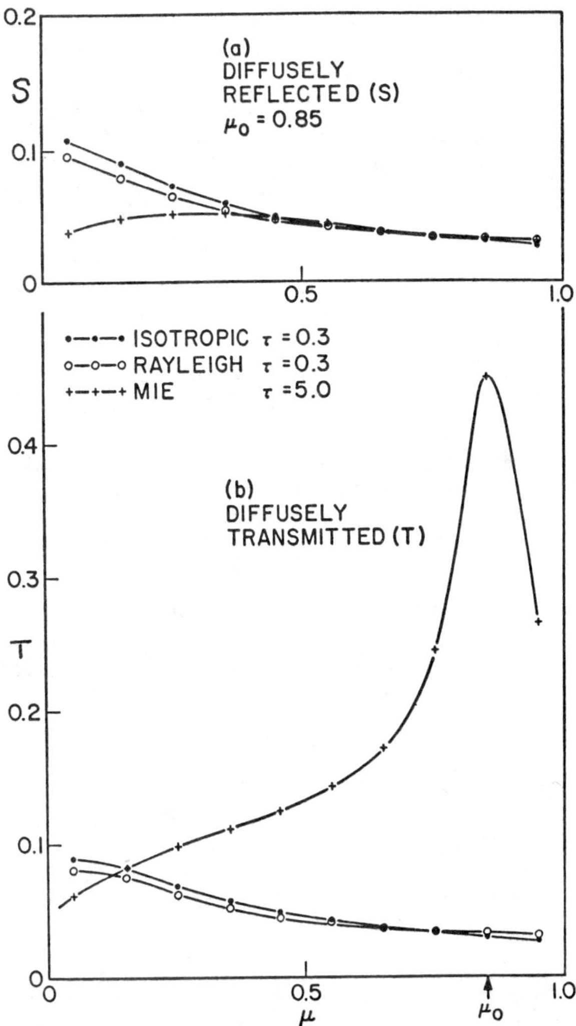

Fig. 6. Reflection and transmission for layers which reflect a fraction 0.15 of the incident intensity; the incident direction has $\mu_0=0.85$ (32° from the vertical).

for $N=10$ are compared with, say, Chandrasekhar's scattering functions by plotting points on a diagram in which the scattering functions describe a curve, the points would not lie on the curve even if both computations were completely free of error. But they will lie close to the curve, and if N were increased without limits the set of representative points would tend to the curve.

Fig. 3 shows a comparison of scattering elements of $\mathbf{S}^{(0)}$ for a Rayleigh phase function with values obtained analytically by Coulson *et al.* (1960). These data are not strictly comparable, since polarization was not included in our computations, whereas Coulson *et al.* included polarization effects fully.

Fig. 4 shows curves obtained by applying van de Hulst's (1964) asymptotic formula for thick isotropic atmospheres, superimposed upon which are the discrete points given by application of the matrix superposition equations.

Fig. 5 shows the variations of some selected reflection and transmission elements with optical depth, again for isotropic scattering. The values for a semi-infinite layer were obtained from Chandrasekhar's H-functions and are shown circled in the figure at $1/\tau = 0$.

After these and many similar comparisons had indicated that the method used was stable and reliable, it was applied to Mie scattering by a cloud of water droplets which corresponded roughly to stratocumulus at visible wavelengths. The scattering diagram was the total scattering from a Gaussian distibution with a mean radius of 13.16 wavelengths (e.g., $7.2\,\mu$ at $\lambda = 5500$ Å), and a coefficient of dispersion of 0.32. The results showed that the forward peak was evident in the diffuse transmission up to about $\tau = 10$, but became flattened out thereafter. For a given optical depth a Mie-scattering layer is necessarily a weaker reflector than a Rayleigh or isotropic layer, and comparisons at equal optical depths are not very informative. In Fig. 6 is shown a comparison of Rayleigh, isotropic and Mie layers which reflect back approximately equal amounts of energy. It will be noted that the forward peak is still very obvious in the transmission, whereas over much of the curves the reflected energy is distributed not too differently from the isotropic or Rayleigh distributions.

10. Conclusions

The matrix methods described herein have been successfully applied for the computation of the reflection and diffuse transmission of radiation through multiply-scattering layers. A considerable number of such computations is planned to give a reasonably detailed picture of the reflecting and transmitting properties of terrestrial clouds. In the method as described herein neither polarization nor asymmetric layers were included, but it is apparent that it can readily be modified to include these effects. Polarization can be treated by considering the four Stokes parameters rather than a single intensity, and modifying the definition of $\mathbf{B}$, $\mathbf{P}$, $\mathbf{S}$, $\mathbf{T}$, etc. accordingly; assymetric layers require differentiation between $\mathbf{S}$ and $\mathbf{T}$ for illumination entering from below, but with this differentiation, equations analogous to (6)–(9) can be written down without difficulty.

REFERENCES

Ambartsumian, V. A., 1942: On the light scattering in planetary atmospheres. *Russian Astronomical J.*, **19**, No. 5.

Chandrasekhar, S., 1944: On the radiative equilibrium of a stellar atmosphere. *Astro. J.*, **99**, 180–190.

——, 1950: *Radiative Transfer*. Oxford, England, Clarendon Press, 393 pp.

Coulson, K. L., J. V. Dave and Z. Sekera, 1960: *Tables Relating to Radiation Emerging from a Planetary Atmosphere with Rayleigh Scattering*. Berkeley, University of California Press. 548 pp.

Kourganoff, V., 1952: *Basic Methods in Transfer Problems*. Oxford, England, Clarendon Press, 281 pp.

van de Hulst, H. C., 1948: Scattering in a planetary atmosphere. *Astro. J.*, **107**, 220–246.

——, 1963: A new look at multiple scattering. New York, NASA Goddard Space Flight Center, 81 pp.

——, 1964: Diffuse reflection and transmission by a very thick plane-parallel atmosphere with isotropic scattering. *Icarus*, **3**, 336–341.

The Delta-Eddington Approximation for Radiative Flux Transfer

J. H. Joseph[1] and W. J. Wiscombe

National Center for Atmospheric Research,[2] Boulder, Colo. 80303

J. A. Weinman

Department of Meteorology, University of Wisconsin, Madison, Wis. 53706

(Manuscript received 30 March 1976, in revised form 16 August 1976)

ABSTRACT

This paper presents a rapid yet accurate method, the "delta-Eddington" approximation, for calculating monochromatic radiative fluxes in an absorbing-scattering atmosphere. By combining a Dirac delta function and a two-term approximation, it overcomes the poor accuracy of the Eddington approximation for highly asymmetric phase functions. The fraction of scattering into the truncated forward peak is taken proportional to the square of the phase function asymmetry factor, which distinguishes the delta-Eddington approximation from others of similar nature. Comparisons of delta-Eddington albedos, transmissivities and absorptivities with more exact calculations reveal typical differences of 0–0.02 and maximum differences of 0.15 over wide ranges of optical depth, sun angle, surface albedo, single-scattering albedo and phase function asymmetry factor. Delta-Eddington fluxes are in error, on the average, by no more than 0.5%, and at the maximum by no more than 2% of the incident flux. This computationally fast and accurate approximation is potentially of utility in applications such as general circulation and climate modeling.

1. Introduction

General circulation models and climate models, as well as certain agricultural and engineering applications, require the frequent evaluation of solar fluxes and heating rates at many locations. The most serious challenge facing radiative transfer theory in such applications is to devise simple and computationally fast analytic approximations having adequate fidelity to the more exact numerical solutions (which in turn incorporate approximations such as horizontal homogeneity). This problem is complicated by the wide ranges of values that radiative parameters may take on in realistic atmospheres. Many approximate methods are valid only for severely restricted ranges of one or more of these parameters.

The most simple and popular approximations to the complicated integro-differential equation of radiative transfer are the Eddington and the two-stream, and variants thereof. They both represent the angular dependence of the radiation intensity in terms of just two functions of optical depth, for which a pair of ordinary differential equations is derived. For homogeneous layers these differential equations have constant coefficients and, therefore, simple exponential solutions for fluxes. These solutions are framed in terms of the optical depth, the albedo for single scattering, and one or two moments of the scattering phase function.

Simple approximations, like the Eddington, are often incapable of coping with the highly asymmetric phase functions typical of particulate scattering. Fritz (1954, 1958), Irvine (1965) and Joseph (1968, 1970, 1971) noted that it is reasonable to replace the large forward peak in such phase functions by a Dirac delta function. They assumed that half of the scattered radiation goes into the forward peak. Fritz and Irvine then took the remaining scattering to be completely isotropic, while Joseph assumed isotropy separately in each hemisphere. Weinman (1968), Hansen (1969) and Potter (1970) truncated the peak by extrapolating the phase function from angles outside the peak; the remainder of the phase function was left intact. Excellent accuracy for albedo was achieved by this approximation except for nearly grazing angles. Wang (1972) approximated the untruncated part of the phase function by an Eddington-type phase function, but he chose the fraction scattered into the forward peak in such a way as to conserve the integral of the original phase function from 90° to 180°. He also uses the exponential kernel rather than the Eddington approximation and centers his interest on the albedo of semi-infinite atmospheres, for astrophysical applications.

2. The delta-Eddington approximation

We approximate the phase function by a Dirac delta function forward scatter peak and a two-term

---

[1] On leave from the Department of Geophysics and Planetary Sciences, Tel-Aviv University, Ramat-Aviv, Israel.

[2] The National Center for Atmospheric Research is sponsored by the National Science Foundation.

expansion of the phase function

$$P(\cos\theta) \approx P_{\delta-\text{Edd}}(\cos\theta) \equiv 2f\delta(1-\cos\theta)$$
$$+ (1-f)(1+3g'\cos\theta), \quad (1)$$

where f is the fractional scattering into the forward peak and g' is the asymmetry factor of the truncated phase function. It is easily shown that the delta-Eddington phase function is correctly normalized, i.e.,

$$\int_{4\pi} P_{\delta-\text{Edd}}(\cos\theta)\frac{d\Omega}{4\pi} = \frac{1}{2}\int_{-1}^{1} P_{\delta-\text{Edd}}(\mu)d\mu = f + (1-f) = 1.$$

We further require $P_{\delta-\text{Edd}}$ to have the same asymmetry factor (g) as the original phase function:

$$g = \int_{4\pi} \cos\theta P_{\delta-\text{Edd}}(\cos\theta)\frac{d\Omega}{4\pi} = f + (1-f)g', \quad (2a)$$

from which g' is determined as

$$g' = \frac{g-f}{1-f}. \quad (2b)$$

Finally, to determine f we require that the second moment of $P_{\delta-\text{Edd}}$,

$$\int_{4\pi} P_2(\cos\theta)P_{\delta-\text{Edd}}(\cos\theta)\frac{d\Omega}{4\pi} = f, \quad (3)$$

be identical to the second moment of the original phase function P which is g^2 if P is approximated by the Henyey–Greenstein phase function

$$P_{\text{H-G}}(\cos\theta) = \sum_{l=0}^{\infty} (2l+1)g^l P_l(\cos\theta). \quad (4)$$

(van de Hulst, 1968). Both van de Hulst (1968) and Hansen (1969) show that, for flux computations, Henyey–Greenstein phase functions may be used in place of the more realistic ones from Mie theory. Thus we require

$$f = g^2 \quad (5a)$$

and therefore

$$g' = \frac{g}{1+g} \quad (5b)$$

from Eq. (2b).[3] Note that Eq. (5b) implies $0 \leqslant g' \leqslant 0.5$

---

[3] Dual Henyey-Greenstein functions have sometimes been introduced to approximate phase functions peaked in the forward and backward directions. If $P_{\text{D-H-G}}(\cos\theta) = b\,P_{\text{H-G}}(g_1, \cos\theta) + (1-b)P_{\text{H-G}}(g_2, \cos\theta)$, it follows that $g' = [bg_1(1-g_1)+(1-b)g_2 \times (1-g_2)]/[1-bg_1^2-(1-b)g_2^2]$ and $f = bg_1^2+(1-b)g_2^2$. Although results will not be presented in this paper, agreement between fluxes computed rigorously and by the delta-Eddington approximation for $bg_1+(1-b)g_2>0$ is comparable to that shown in Figs. 1–3 of this study.

when $0 \leqslant g \leqslant 1$, so the two-term part of $P_{\delta-\text{Edd}}$ [Eq. (1)] applies to a range of g' for which the Eddington approximation is demonstrably accurate (Wiscombe et al., 1976).

To estimate the difference between the delta-Eddington and Henyey–Greenstein phase functions, we expand the delta function in Eq. (1) in a Legendre polynomial series (Morse and Feshbach, 1953):

$$P_{\delta-\text{Edd}}(\cos\theta) = f\sum_{l=0}^{\infty} (2l+1)P_l(\cos\theta)$$
$$+ (1-f)[1+3g'P_1(\cos\theta)]$$
$$= 1 + 3g\cos\theta + \sum_{l=2}^{\infty} (2l+1)g^2 P_l(\cos\theta), \quad (6)$$

where Eqs. (5a) and (5b) have been used to replace f and g'. The delta-Eddington phase function therefore agrees with that of Henyey–Greenstein [Eq. (4)] out to three terms, and their difference is

$$P_{\delta-\text{Edd}}(\cos\theta) - P_{\text{H-G}}(\cos\theta)$$
$$= \sum_{l=3}^{\infty} (2l+1)(g^2-g^l)P_l(\cos\theta). \quad (7)$$

Because the terms on the right-hand side of this equation approach zero as $g \to 1$, the delta-Eddington error tends to grow smaller rather than larger as $g \to 1$.

We now investigate the consequences of using the delta-Eddington approximate phase function in the radiative transfer equation. Since our interest is in fluxes only, it is sufficient to deal with the azimuthally averaged form of that equation, viz.,

$$\mu\frac{\partial I}{\partial\tau} + I = \frac{\omega}{2}\int_{-1}^{1} \bar{P}(\mu,\mu')I(\tau,\mu')d\mu', \quad (8)$$

where I is the azimuthally-averaged intensity and

$$\bar{P}(\mu,\mu') = \frac{1}{2\pi}\int_{0}^{2\pi} P_{\delta-\text{Edd}}(\cos\theta)d\phi$$
$$= 2f\delta(\mu-\mu') + (1-f)(1+3g'\mu\mu'). \quad (9)$$

Although thermal emission sources may be included in Eq. (8) and in our approximation, we omit them for reasons of simplicity. In Eq. (9) we have used the fact that when θ is the angle between the incident and scattering directions $\boldsymbol{\Omega} = (\mu,\phi)$ and $\boldsymbol{\Omega}' = (\mu',\phi')$ (in the usual notation), then

$$\cos\theta = \mu\mu' + (1-\mu^2)^{\frac{1}{2}}(1-\mu'^2)^{\frac{1}{2}}\cos(\phi-\phi'). \quad (10)$$

Also we have written the delta-function in $P_{\delta-\text{Edd}}$ in the form

$$\delta(1-\cos\theta) = 2\pi\delta(\mu-\mu')\delta(\phi-\phi'). \quad (11)$$

Inserting Eq. (9) into the transfer equation leads to

$$\mu \frac{\partial I}{\partial \tau'} + I = \frac{\omega'}{2} \int_{-1}^{1} (1 + 3g'\mu\mu') I(\tau',\mu') d\mu', \qquad (12)$$

which is similar to the original transfer equation except that the variables have been transformed as follows:

$$\tau' = (1 - \omega f)\tau \qquad (13)$$

$$\omega' = \frac{(1-f)\omega}{1 - \omega f}. \qquad (14)$$

Making the usual diffuse-direct transformation (where πF_0 is the monodirectional flux at $\tau' = 0$ incident at zenith angle $\cos^{-1}\mu_0$),

$$I = \begin{cases} i + \frac{1}{2}F_0\delta(\mu-\mu_0)\exp(-\tau'/\mu_0), & 0 < \mu \leqslant 1 \\ i, & -1 \leqslant \mu < 0 \end{cases} \qquad (15)$$

brings the transfer equation into the form

$$\mu \frac{\partial i}{\partial \tau'} + i = \omega'(i_0 + g'\mu i_1)$$
$$+ \frac{1}{4}F_0\omega'(1 + 3g'\mu_0\mu)\exp(-\tau'/\mu_0), \quad (16a)$$

where

$$i_n(\tau') = \frac{1}{2}(2n+1)\int_{-1}^{1} \mu^n i(\tau',\mu) d\mu, \quad n = 0, 1. \quad (16b)$$

If we now take the zeroth and first moments of the transfer equation by applying the operators $\int_{-1}^{1} d\mu$ and $\int_{-1}^{1} \mu d\mu$, and make the Eddington assumption that $i = i_0 + \mu i_1$, we arrive at a pair of differential equations for i_0 and i_1. The solution to these equations is given by Eqs. (12) or (16) of Shettle and Weinman (1970), with g', τ', ω' replacing g, τ, ω.

The delta-Eddington approximation is thus equivalent to the Eddington approximation with transformed parameters g', τ', ω' [Eqs. (5b), (13), (14)]. van de Hulst (1974) has developed several such similarity transformations in radiative transfer.

These transformations are

$$\frac{\tau}{\tau'} = \frac{1-\omega'}{1-\omega} = \frac{\omega'(1-g')}{\omega(1-g)} \qquad (17a)$$

and render it feasible to transform a problem with a strongly anisotropic phase function to one with $g' < g$. Eq. (17a) is satisfied by Eqs. (2b), (13) and (14). By further making the approximation outlined by Eqs. (3) and (4), we are led deductively to $f = g^2$ and also satisfy van de Hulst's empirical approximate similarity relations

$$\frac{1-\omega'}{1-\omega'g'} = \frac{1-\omega}{1-\omega g}, \qquad (17b)$$

$$\tau'(1-\omega'g') = \tau(1-\omega g). \qquad (17c)$$

We have thus obtained van de Hulst's approximate similarity relations in a physically consistent manner and demonstrated the latters' relationship to a two-term truncation of the phase function for a non-conservative medium.

3. Errors in the delta-Eddington approximation

We define the reflectivity, absorptivity and global transmissivity of an homogeneous layer of optical depth τ_0' to be, respectively,

$$r \equiv F\uparrow(0)/\mu_0\pi F_0, \qquad (18a)$$

$$a \equiv [F(0) - F(\tau_0')]/\mu_0\pi F_0, \qquad (18b)$$

$$t \equiv F_t^\downarrow(\tau_0')/\mu_0\pi F_0, \qquad (18c)$$

where $F\uparrow$ is upward flux; $F_t^\downarrow$ is global downward flux, consisting of a diffuse $F\downarrow$ and a "direct" component, i.e.,

$$F_t^\downarrow(\tau') = F\downarrow(\tau') + \mu_0\pi F_0\exp(-\tau'/\mu_0), \qquad (19)$$

and $F = F_t^\downarrow - F\uparrow$ is net flux. Note that the "direct" flux in Eq. (19), employing the scaled optical depth $\tau' \leqslant \tau$ [Eq. (13)], is larger than the actual direct flux. On account of the phase function truncation, this "direct" flux also includes scattered radiation traveling in very nearly the same direction as the incident beam. For example, in a dusty atmosphere, a substantial fraction of the solar aureole would be included in the delta-Eddington direct flux. This might actually be an advantage in applying delta-Eddington to the analysis of atmospheric radiation measurements, since it would remove the burden of carefully discriminating against the aureole in measuring direct flux. [In any case, because of the finite angular width of the sun ($\frac{1}{2}°$), it is impossible to entirely distinguish almost-forward-scattered radiation from direct radiation.]

The quantities r, a and t, on account of their definitions (18a)–(18c), satisfy the relationship

$$r + (1-A)t + a = 1, \qquad (20)$$

where A is the Lambertian albedo of the surface. Since Eq. (20) must hold for exact *and* for approximate solutions, then

$$\Delta r + (1-A)\Delta t + \Delta a = 0, \qquad (21)$$

where Δ denotes the difference between exact and approximate values. Thus Δr, Δt and Δa may not all have the same sign, and the various errors must compensate one another. This may be observed in all the comparison plots in Figs. 1–3.

The exact values of r, a and t (plotted as solid curves in Figs. 1, 2 and 3) are obtained from the doubling method using 16 Gaussian angles, the diamond initialization and Grant's renormalization method (cf. Wiscombe, 1976); the doubling results have an accuracy of better than 0.1%. These doubling values are compared with conventional Eddington and delta-Eddington values of the same quantities. (The conventional

Eddington approximation is sometimes used for highly asymmetric scattering and an indication of its usefulness is thus called for.) The parameter ranges for which these comparisons were made are shown in Table 1; they include the significant values one will encounter in a planetary atmosphere.

Fig. 1 shows r, a and t as functions of μ_0 for an asymmetry factor $g=0.8$ and a surface albedo $A=0$. Fig. 2 shows similar information for $g=0.95$. (Clouds and aerosols have asymmetry factors between 0.8 and 0.95 for most of the solar spectrum.) Since 90% of the phase function is being truncated in the $g=0.95$ case, because $f=g^2$ we felt it important to also show results for the less drastic $g=0.8$ case. Only $A=0$ is considered because the change in the delta-Eddington error with

g	0 – 0.95
τ_0	0.01–100
ω	0.1 – 0.99
μ_0	0.1 – 1.0
A	0 – 0.8

A is slight; increasing the albedo all the way to $A=0.8$ worsens that error for t and a by no more than 0.01–0.02 and actually decreases the error in r for small and intermediate optical depths (where the surface albedo strongly affects the system albedo). The optical depth τ_0 increases from 0.1 to 1 to 10 across the rows of Figs.

HENYEY–GREENSTEIN (g=0.8); SURFACE ALBEDO=0

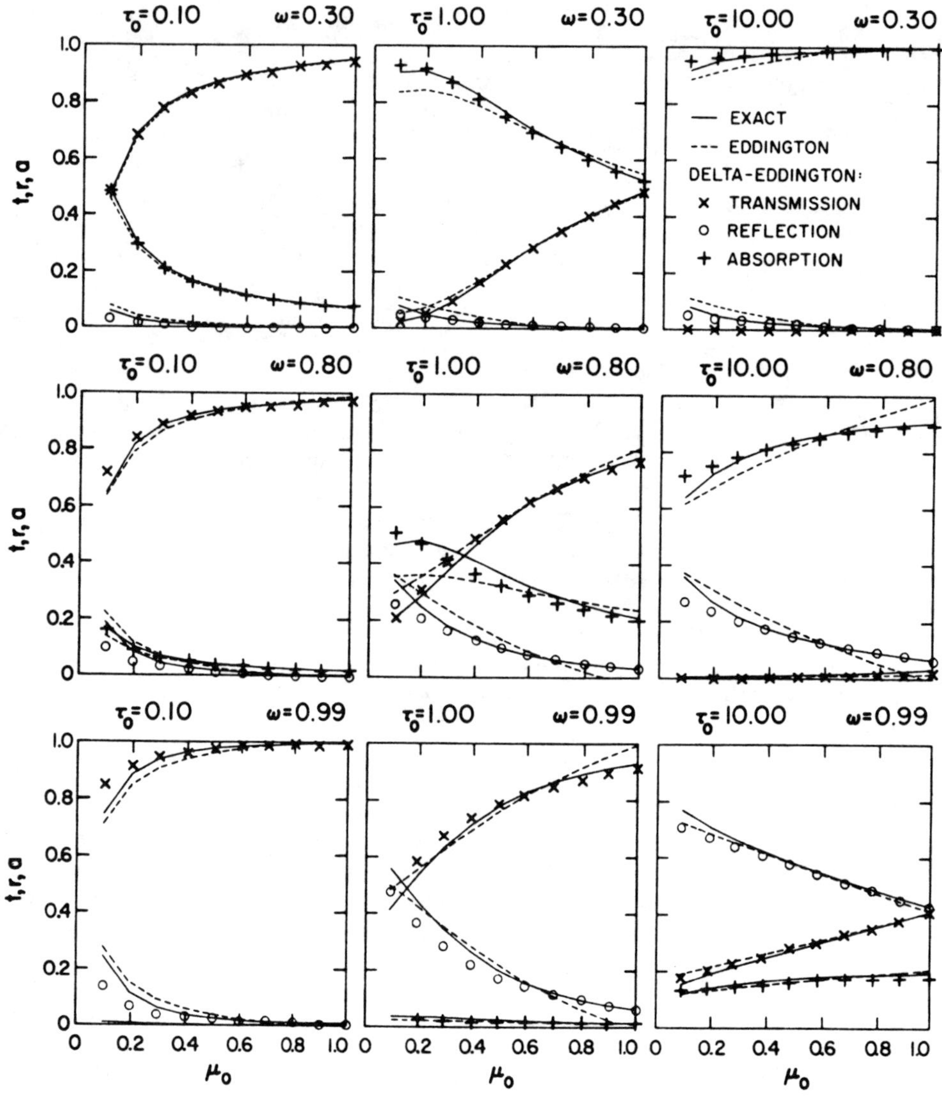

FIG. 1. Reflectivity (r), transmissivity (t) and absorptivity (a) as a function of sun angle μ_0 for various single-scattering albedos (ω) and layer optical depths (τ_0), comparing exact, Eddington and delta-Eddington methods for asymmetry factor $g=0.8$ and surface albedo $A=0$.

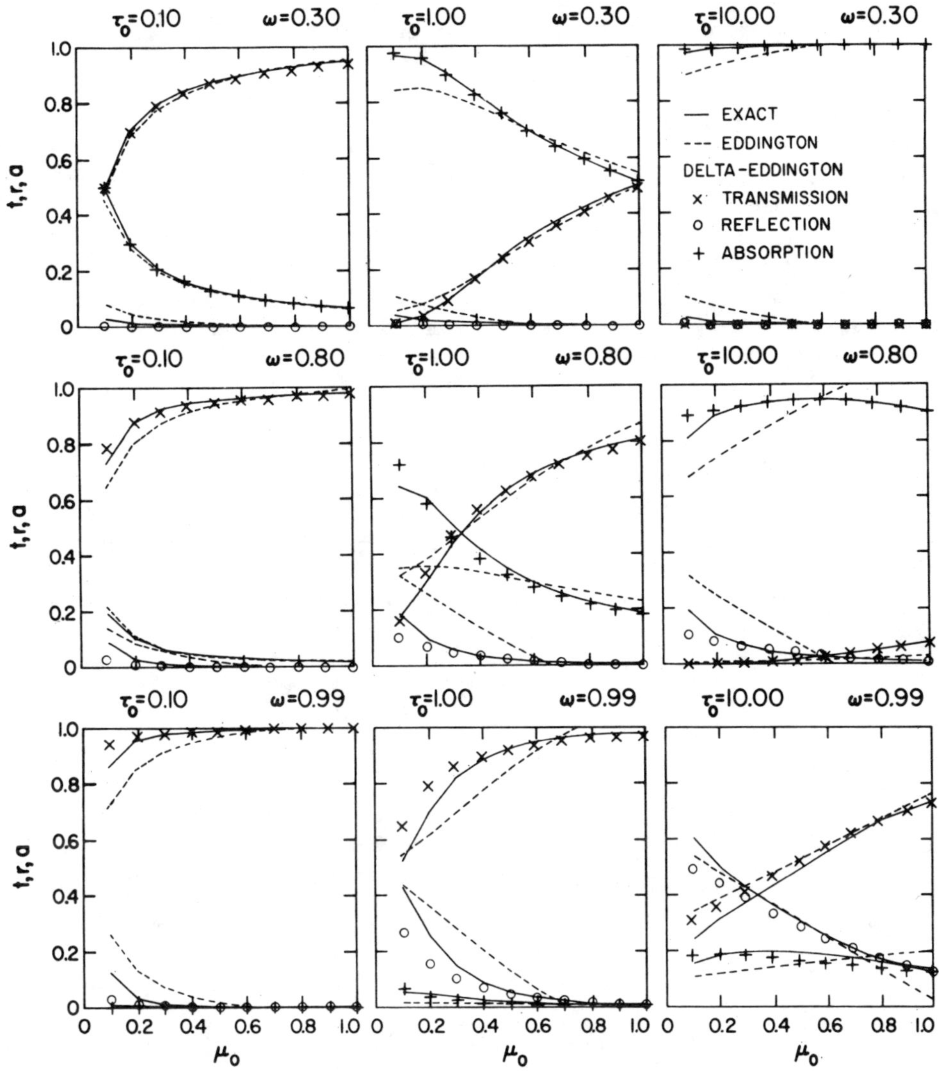

Fig. 2. As in Fig. 1 except for $g=0.95$.

1 and 2, and the single-scattering albedo ω increases from 0.3 to 0.8 to 0.99 down the columns. These values provide a representative sampling of the ω and τ_0 parameter ranges. Curves for $\omega = 0.1$ are omitted because in that case the Eddington, delta-Eddington and exact results are virtually indistinguishable except for small μ_0, where the delta-Eddington approximation is substantially better than the Eddington.

Figs. 1 and 2 show that the conventional Eddington approximation is unacceptable for optical depths <10. Eddington predictions of r in a majority of the cases become negative as $\mu_0 \to 1$, and in some cases values of t or a are predicted to be greater than 1. Comparison of corresponding plots in Figs. 1 and 2 furthermore shows the rapid deterioration of the Eddington approximation with increasing g. By contrast, the delta-Eddington approximation follows the exact curves very well except in the grazing-incidence limit $\mu_0 \to 0.1$. It should be remembered, however, that the actual radiation fluxes are the product of $\mu_0 \pi F_0$ with r, a and t [cf. Eqs. (18)] so that the delta-Eddington relative flux errors are scaled down by a factor μ_0 compared to the errors shown in Figs. 1 and 2. In our extensive intercomparisons, only a small sample of which are shown here, we have never observed a delta-Eddington flux error exceeding 2.5% of the incident flux πF_0.

Increasing error as $\mu_0 \to 0$ is typical of both the Eddington approximation (see Wiscombe *et al.*, 1976) and of truncated peak approximations (see, Hansen, 1969; Potter, 1970). Since these two approximations are employed in the delta-Eddington, it shares their deficiency. The reflected radiation in the case of near-grazing incident radiation normally contains a large component of single scattering into the forward peak, which is missed entirely when the forward peak is truncated. Furthermore, this singly-scattered reflected radiation corresponds least to the near isotropy assumed in the Eddington approximation. Because of these facts, the delta-Eddington prediction of r tends to be considerably more in error as $\mu_0 \to 0$ than its predictions of t and a for moderate surface albedos and for single-scattering albedos differing from unity.

The pattern of delta-Eddington error in Figs. 1 and 2 is different for t and r as compared to a. As ω increases (down the columns) t and r retain an accuracy of 0.02 or better for $\mu_0 \geqslant 0.5$, but for $\mu_0 < 0.5$ they deviate increasingly from the exact results. On the other hand, the absorptivity a retains an accuracy of 0.01 or better over almost the full range of μ_0; only for $\mu_0 < 0.2$ or for the mid-ranges of τ_0 and ω do we see errors in a occasion-

ally becoming as large as 0.08 (for $\mu_0 = 0.1$) or 0.04 (all other cases). This renders the computation of heating rates relatively reliable.

When $\mu_0 < 0.5$, the delta-Eddington reflectivity is systematically too low and the delta-Eddington transmissivity almost always systematically too high. The delta-Eddington absorption is systematically too high for $\mu_0 < 0.2$ and systematically too low for $\mu_0 > 0.3$.

In order to examine the problem of small μ_0 from a different viewpoint, we have plotted the doubling and delta-Eddington r, a and t versus optical depth τ_0 in Fig. 3, for an asymmetry factor $g = 0.85$. A surface albedo $A = 0.8$ is used in order to show an opposite extreme from Figs. 1 and 2, where $A = 0$. The sun zenith angle cosine μ_0 increases from 0.1 to 0.4 down the columns of Fig. 3, while the single-scattering albedo ω

HENYEY–GREENSTEIN (g=0.85); SURFACE ALBEDO=0.8

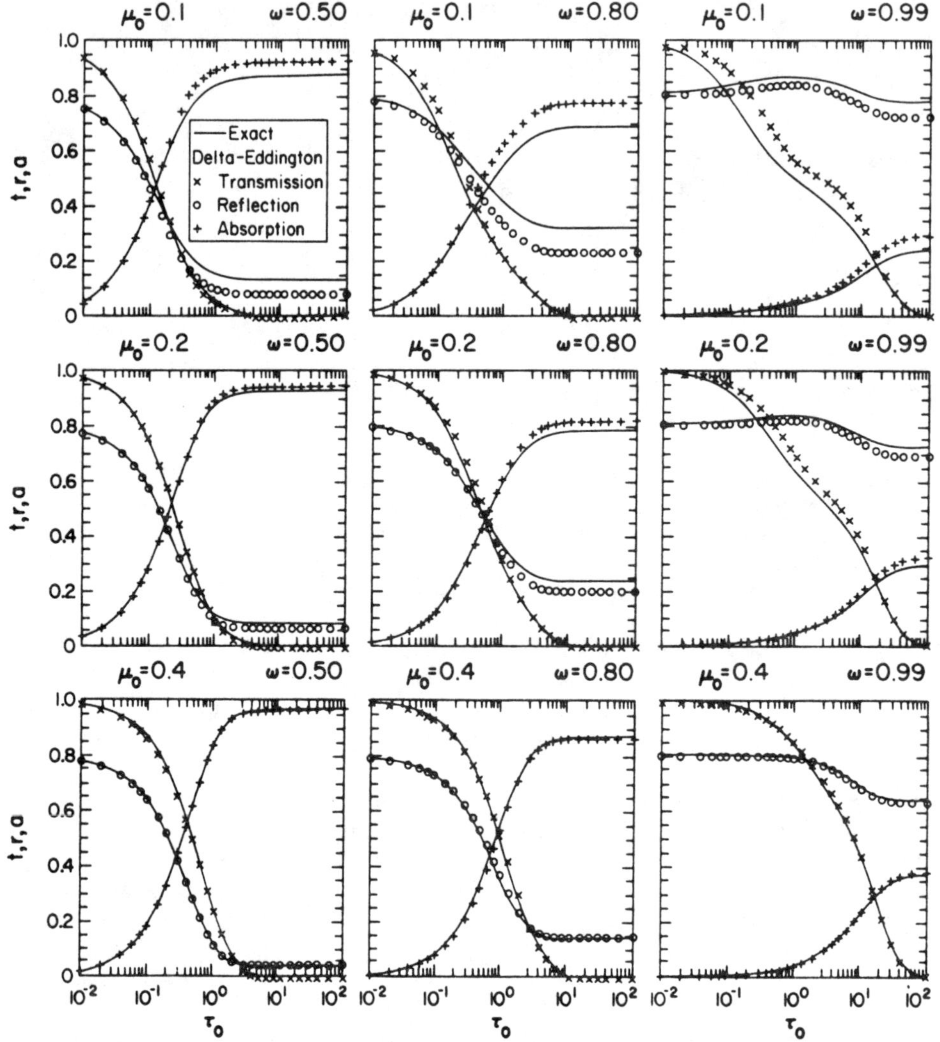

FIG. 3. Reflectivity (r), transmissivity (t) and absorptivity (a) versus layer optical depth τ_0 for a range of large solar zenith angles (μ_0) and larger single-scattering albedos (ω), comparing exact and delta-Eddington methods for asymmetry factor $g = 0.85$ and surface albedo $A = 0.8$.

increases from 0.5 to 0.8 to 0.99 across the rows. As we look down the columns of Fig. 3, we note the dramatic reduction in the delta-Eddington errors as μ_0 increases, for all values of optical depth. By the bottom row ($\mu_0=0.4$) the errors are at the 0–0.02 level, and they remain at this level from $\mu_0=0.4$ all the way to $\mu_0=1$. As ω decreases across the rows, the delta-Eddington errors decrease also; this improvement in accuracy is the more marked the smaller μ_0 becomes, while for $\mu_0 \geqslant 0.4$ it is barely perceptible to plotting accuracy. These trends continue all the way to $\omega=0$.

While the maximum errors for the large-surface albedo case of Fig. 3 are no greater than for the zero-surface albedo cases of Figs. 1 and 2, there has been a shift in the distribution of error. For $A=0$ the reflection r exhibited considerably larger errors than t or a, but for $A=0.8$ this is no longer true—the transmission, and especially the absorption, tend to be every bit as much in error as the reflection.

4. Average error

If the delta-Eddington approximation were applied to a planetary atmosphere, it would be exercised for wide ranges of optical depth and single-scattering albedo. Upon integration over wavelength, over a diurnal course, over a geographically extended area, or any combination of these, individual delta-Eddington errors will average out to some net error. In this section, we apply an averaging procedure which crudely mimics this effect.

Let us define the average relative reflected-flux error as

$$\frac{\langle F^{\uparrow}_{\text{doubling}}(0) - F^{\uparrow}_{\delta\text{-Edd}}(0)\rangle}{\pi F_0}$$

$$\equiv \frac{1}{N_{\tau_0} N_{\omega} N_{\mu_0}} \sum_{\tau_0} \sum_{\omega} \sum_{\mu_0} \mu_0 \left| r_{\text{doubling}} - r_{\delta\text{-Edd}} \right|.$$

Similar definitions obtain for the average relative transmitted- and absorbed-flux error. For the set of μ_0's, we chose successively increasing ranges $\{0.1\}$, $\{0.1 \rightarrow 0.3\}$, $\{0.1 \rightarrow 0.5\}$, $\{0.1 \rightarrow 0.7\}$ and $\{0.1 \rightarrow 1\}$ to simulate the increasing range of sun angle one encounters in going from pole to equator. We averaged over $N_{\omega}=11$ values of ω uniformly distributed between 0 and 1, and over $N_{\tau_0}=32$ values of τ_0. Nine values of τ_0 were taken in each of the decades $[0.1,1]$, $[1,10]$ and $[10,100]$, while only five values were taken in the decade $[0.01,0.1]$ because of its relative unimportance.

Selected but representative average errors are shown in Table 2 (some of the variations in this table seem erratic because all entries have been rounded to two digits). These average flux errors range between 0.1% and 0.5% of the incident flux. They do not exhibit the serious loss of accuracy for small solar elevations displayed by the flux ratios r, a and t, which bears out

TABLE 2. Average percent error in reflected (ref), transmitted (trn) and absorbed (abs) delta-Eddington fluxes for surface albedos $A=0$ and $A=0.8$, asymmetry factors $g=0.8$ and $g=0.95$, and various ranges of solar zenith angle cosine μ_0.

μ_0	$g=0.8$			$g=0.95$		
	ref	trn	abs	ref	trn	abs
			$A=0$			
0.1	0.42	0.13	0.30	0.40	0.15	0.28
0.1–0.3	0.31	0.20	0.23	0.25	0.19	0.20
0.1–0.5	0.27	0.22	0.28	0.21	0.18	0.22
0.1–0.7	0.29	0.21	0.33	0.20	0.16	0.23
0.1–1.0	0.29	0.17	0.32	0.18	0.13	0.21
			$A=0.8$			
0.1	0.33	0.12	0.31	0.31	0.14	0.29
0.1–0.3	0.22	0.16	0.21	0.18	0.16	0.18
0.1–0.5	0.22	0.18	0.23	0.18	0.18	0.19
0.1–0.7	0.28	0.25	0.28	0.22	0.26	0.24
0.1–1.0	0.33	0.41	0.35	0.28	0.42	0.33

our earlier comment about flux errors being scaled down by a factor μ_0; and they generally decrease as g increases, which bears out our discussion in Section 2 [see Eq. (7)]. The average reflected and absorbed flux errors are very nearly equal and usually exceed the average transmitted flux error. In general, for $A=0$ all the errors are fairly insensitive to the increasing range of sun angle (barring the $\mu_0=0.1$ case), but when $A=0.8$ all errors increase significantly with increasing range of μ_0. As A increases, errors for $\mu_0 \rightarrow 0$ are generally diminished, while those for $\mu_0 \rightarrow 1$ are enhanced.

5. Summary and conclusions

We have presented an extension of the Eddington approximation in which the forward peak of the phase function is approximated as a δ-function. The approximate phase function is constructed to have the same asymmetry factor and second moment as the actual phase function. For Henyey–Greenstein phase functions this leads to a scaling of the asymmetry factor, optical depth and single-scattering albedo which can be related to some recent "similarity relations" of van de Hulst. In particular, the transformed asymmetry factor assumes a value between 0 and 0.5 for which we demonstrated that the conventional Eddington approximation is good (see Wiscombe et al., 1976).

The transmission, reflection and absorption predicted by our approximation were compared with doubling method calculations for realistic ranges of optical depth, single-scattering albedo, surface albedo, sun angle and asymmetry factor. In all cases, all relevant fluxes are predicted to an accuracy of better than 2.5% of the incident flux with the "average" flux error (see Section 4) being no larger than 0.5%. The flux ratios— global transmissivity, reflectivity and absorptivity— are accurate to at least 0.02 when the solar zenith angle cosine $\mu_0 \gtrsim 0.4$, but these quantities experience errors increasing to a maximum of 0.15 (in the reflectivity) as $\mu_0 \rightarrow 0.1$. This loss of accuracy practically disappears, however, for fluxes rather than flux ratios. Also, our approximation never exhibits negative reflectivities, or transmissivities and absorptivities exceeding 100%, which may occur in the conventional Eddington approximation for large asymmetry factors.

The delta-Eddington approximation therefore furnishes a physically sound, accurate and analytically simple parameterization of radiation to replace the empiricism currently employed in many general circulation and climate models. While attention has been confined in this paper to homogeneous layers, vertical inhomogeneity may be simply treated by concatenating homogeneous layers and imposing flux continuity at the layer boundaries, in the manner of Shettle and Weinman (1970). Well-documented computer codes are available from the authors for monochromatic n-layer cases.

Acknowledgments. This study has in part been supported by NASA Contract 1057. We would like to thank Prof. van de Hulst for his helpful comments on the presentation of these results.

REFERENCES

Fritz, S., 1954: Scattering of solar energy by clouds of large drops. *J. Meteor.*, **11**, 291–300.

——, 1958: Absorption and scattering of solar energy in clouds of large water drops. *J. Meteor.*, **15**, 51–58.

Hansen, J. E., 1969: Exact and approximate solutions for multiple scattering by cloudy and hazy planetary atmospheres. *J. Atmos. Sci.*, **26**, 478–487.

Irvine, W. M., 1965: Multiple scattering by large particles. *Astrophys. J.*, **142**, 1563–1676.

——, 1968: Multiple scattering by large particles. II. Optically thick layers. *Astrophys. J.*, **152**, 823–834.

Jospeh, J. H., 1968: The effect of tenuous clouds on the flux of infrared radiation in the atmosphere. *Radiation Including Satellite Techniques*, WMO Tech. Note No. 104, 503–508. [Available from Unipub, Box 433, Murray Hill Station, New York, N. Y. 10016.]

——, 1970: Thermal radiation fluxes through optically thin clouds. *Israel J. Earth Sci.*, **19**, 51–67.

——, 1971: Thermal radiation fluxes near the sea surface in the presence of marine haze. *Israel J. Earth Sci.*, **20**, 7–12.

Morse, P. M., and H. Feshbach, 1953: *Methods of Theoretical Physics*. McGraw-Hill, 728–729.

Potter, J., 1970: The delta function approximation in radiative transfer theory. *J. Atmos. Sci.*, **27**, 943–949.

Shettle, E. P., and J. A. Weinman, 1970: The transfer of solar irradiance through inhomogeneous turbid atmospheres evaluated by Eddington's approximation. *J. Atmos. Sci.*, **27**, 1048–1055.

van de Hulst, H. C., 1968: Asymptotic fitting, a method for solving anisotropic transfer problems in thick layers. *J. Comput. Phys.*, **3**, 291–306.

——, 1974: The spherical albedo of a planet covered with a homogeneous cloud layer. *Astron. Astrophys.*, **35**, 209–214.

Wang, L., 1972: Anisotropic nonconservative scattering in a semi-infinite medium. *Astrophys. J.*, **174**, 671–678.

Weinman, J. A., 1968: Axially symmetric transfer of light through a cloud of anisotropically scattering particles. *Icarus*, **9**, 67–73.

Wiscombe, W. J., 1976: On initialization, error, and flux conservation in the doubling method. *J. Quant. Spectros. Radiat. Transfer* (in press).

——, J. H. Joseph and J. A. Weinman, 1976: The range of validity of the Eddington approximation. Submitted for publication.

Multiple scattering of light and some of its observable consequences

Craig F. Bohren
Department of Meteorology, Pennsylvania State University, University Park, Pennsylvania 16802

(Received 11 February 1986; accepted for publication 4 June 1986)

Many common observations are inexplicable by single-scattering arguments: the variation of brightness and color of the clear sky; the brightness of clouds; the whiteness of a glass of milk; the appearance of distant objects; the blueness of light transmitted in snow and other natural ice bodies; the darkening of sand upon wetting. Yet multiple scattering is seldom mentioned in optics textbooks. It is possible to understand many observable phenomena without invoking the complete theory of multiple (incoherent) scattering. A simple two-stream theory, in which photons are constrained to be scattered in only two directions, forward and backward, is adequate for interpreting many observations, even quantitatively, and it paves the way for advanced study.

I. INTRODUCTION

Multiple scattering of light gives rise to observable phenomena that cannot be explained by single-scattering arguments. For example, if single scattering prevailed in the atmosphere the sky would be uniform in color, which is contrary to what is observed. Clouds are white and bright mostly because of multiple scattering. Attenuation of visible light by ice grains is not spectrally selective, and yet crevasses and ice caves and even holes in ordinary snow may display hues more vivid than those of the bluest sky. Milk is a suspension of small particles that scatter blue light more than red, and yet a glass of milk is white. A white sandy beach—or salt, or sugar—has properties not shared by its grains. And who has not noticed sand darken after being washed by waves, or soil darken when wet by rain?

Single-scattering arguments are insufficient to explain these common observations. Yet multiple scattering is seldom mentioned in optics textbooks. A student or teacher who would learn something about multiple scattering must consult monographs such as those by Chandrasekhar[1] or by van de Hulst.[2] Although these are invaluable for specialists, neophytes are likely to find them formidable: They emphasize techniques for solving equations rather than developing physical intuition by explaining observations using simple models.

Because of this lack of suitable introductory treatments of multiple light scattering, I offer the following. My purpose is to convey as much physical understanding as possible from the least amount of mathematics. Approximate equations, which can be solved exactly, are derived, rather than exact equations which can be solved only approximately. This provides a simple theoretical framework for interpreting many observations. As an intended side effect, the terms and concepts found in advanced treatises are introduced, thereby smoothing the way for further study.

II. TWO-STREAM EQUATIONS OF RADIATIVE TRANSFER

Any scattering medium is composed of discrete scatterers, be they molecular or particulate. Since it is inconvenient to consider this discreteness explicitly, we usually replace discrete media with hypothetical continuous media, the scattering and absorption properties of which are determined by those of the former. The resulting continuum theories are applicable to discrete media provided they contain a great many scatterers in any volume of interest.

Throughout this paper *incoherent* scattering is assumed, that is, we do not take into account phase differences between scattered waves. *Coherent* scattering is treated in Refs. 3–7. There is no sharp boundary between coherently and incoherently scattering media. There are merely different *approximate* theories, in some of which phases are taken into account and in others they are not, which are applied with varying degrees of success to the prediction and interpretation of observations. An ordinary cloud is a typical medium in which incoherent scattering dominates what is observed, whereas a glass of pure water is a medium in which coherent scattering dominates (see, e.g., Ref. 8 for a good discussion of the scattering interpretation of Fresnel's equations).

We also assume that radiation can be scattered in only *two* directions, forward and backward. Finally, polarization is ignored.[9] These simplifying assumptions keep the mathematics manageable without greatly distorting our picture of reality. They underlie the elementary two-stream theory first set down by Schuster.[10] Subsequently, other

ber density and $1/\kappa$ is the absorption mean free path.

One further assumption will be made before proceeding: There are no time delays. That is, the dimensions of the medium are such that any change in the external illumination is felt instantaneously throughout it.

Now we apply conservation of radiant energy to Δz, first for the downward intensity:

$$I_{\downarrow}(z) + \beta\Delta z p_{\uparrow\downarrow} I_{\uparrow}(z + \Delta z)$$
$$= \kappa\Delta z I_{\downarrow}(z) + \beta\Delta z p_{\downarrow\uparrow} I_{\downarrow}(z) + I_{\downarrow}(z + \Delta z) , \qquad (1)$$

where $p_{\downarrow\uparrow}$ is the probability that a photon directed downward is scattered upward (and similarly for $p_{\uparrow\downarrow}$, $p_{\uparrow\uparrow}$, and $p_{\downarrow\downarrow}$). The terms on the left side of Eq. (1) are gains, while those on the right side are losses. If we divide both sides of Eq. (1) by Δz and take the limit as $\Delta z \to 0$, we obtain the following differential equation:

$$\frac{dI_{\downarrow}}{dz} = -\kappa I_{\downarrow} - \beta p_{\downarrow\uparrow} I_{\downarrow} + \beta p_{\uparrow\downarrow} I_{\uparrow} , \qquad (2)$$

and similarly for the upward intensity:

$$\frac{dI_{\uparrow}}{dz} = \kappa I_{\uparrow} + \beta p_{\uparrow\downarrow} I_{\uparrow} - \beta p_{\downarrow\uparrow} I_{\downarrow} . \qquad (3)$$

The sign reversal between Eqs. (2) and (3) occurs because the downward intensity is attenuated in the direction of increasing z, whereas the upward intensity is attenuated in the direction of decreasing z.

We take the medium to be isotropic, that is, $p_{\uparrow\downarrow} = p_{\downarrow\uparrow}$ and $p_{\downarrow\downarrow} = p_{\uparrow\uparrow}$. This is to be distinguished from *isotropic scattering* ($p_{\uparrow\downarrow} = p_{\uparrow\uparrow}$ and $p_{\downarrow\uparrow} = p_{\downarrow\downarrow}$) or an *isotropic radiation field* ($I_{\downarrow} = I_{\uparrow}$). As we shall see, isotropic scattering does not necessarily give rise to an isotropic radiation field, nor does anisotropic scattering necessarily give rise to an anisotropic radiation field. Examples of isotropic media are collections of spherical scatterers or randomly oriented nonspherical scatterers. A medium composed of nonspherical *oriented* scatterers is anisotropic.

In our model, photons must be scattered either forward or backward, which requires that

$$p_{\uparrow\downarrow} + p_{\uparrow\uparrow} = p_{\downarrow\uparrow} + p_{\downarrow\downarrow} = 1 . \qquad (4)$$

The *asymmetry parameter g*, defined as the mean cosine of the scattering angle (which for us has only two values, 1 and -1)

$$g = (1)p_{\downarrow\downarrow} + (-1)p_{\downarrow\uparrow} \qquad (5)$$

is a single number that specifies the degree of anisotropy of the scattering. It lies between 1 (strict forward scattering) and -1 (strict backward scattering); it is 0 for isotropic scattering. From Eqs. (4) and (5) it follows that the four probabilities in Eqs. (2) and (3) can be expressed in terms of g only:

$$p_{\uparrow\downarrow} = p_{\downarrow\uparrow} = (1-g)/2 , \quad p_{\uparrow\uparrow} = p_{\downarrow\downarrow} = (1+g)/2 . \qquad (6)$$

If we divide Eqs. (2) and (3) by $\kappa + \beta$, transform the variable from *physical depth z* to *optical depth τ* defined by

$$\tau = \int_0^z (\kappa + \beta)\, dz , \qquad (7)$$

and use Eq. (6), Eqs. (2) and (3) become

$$\frac{dI_{\downarrow}}{d\tau} = -I_{\downarrow} + \overline{\omega}_0 \frac{1+g}{2} I_{\downarrow} + \overline{\omega}_0 \frac{1-g}{2} I_{\uparrow} , \qquad (8)$$

$$\frac{dI_{\uparrow}}{d\tau} = I_{\uparrow} - \overline{\omega}_0 \frac{1+g}{2} I_{\uparrow} - \overline{\omega}_0 \frac{1-g}{2} I_{\downarrow} , \qquad (9)$$

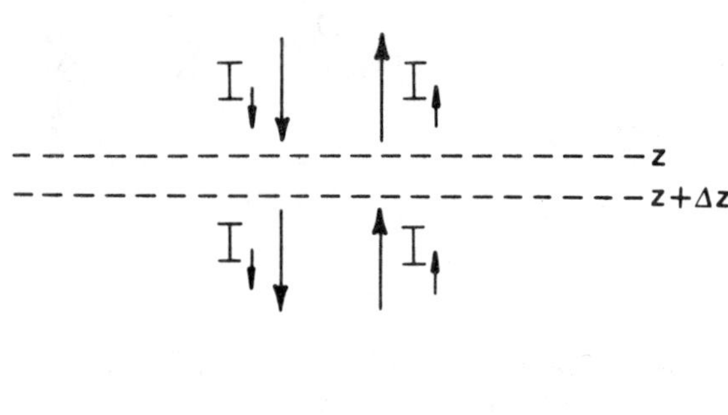

Fig. 1. Conservation of radiant energy applied to the region Δz in a scattering-absorbing medium of infinite lateral extent yields the two-stream equations of transfer for the downward and upward intensities $I_{\downarrow}$ and $I_{\uparrow}$.

two-stream equations have been derived (e.g., Refs. 11–13; see Refs. 14 and 15 for comparisons of various two-stream theories), usually by beginning with the integro-differential equation of radiative transfer and making various approximations. Our approach is to sidestep the exact equation and give a physical derivation similar to that of Schuster's.[10]

After a photon is emitted it suffers only one of two fates when it interacts with matter: (1) it is *absorbed*, that is, it ceases to exist, although its energy is taken up by whatever it interacts with; or (2) it is *scattered*, in which instance it survives the interaction intact but possibly changes direction. Consider a continuous scattering–absorbing medium, infinite in lateral extent, bounded by parallel planes (i.e., a *plane-parallel* medium). We assume that photons are emitted only by sources outside this medium. To derive the equations of radiative transfer for it we apply a radiant energy balance to a small region Δz (Fig. 1). Photons in a given direction incident on this region are lost by absorption and by scattering within it. But there is also a gain of photons because those in one direction are scattered into the opposite direction. By $I_{\downarrow}$ is mean the amount of radiant energy in a narrow frequency interval that crosses unit area per unit time in the downward direction. I shall call $I_{\downarrow}$ the monochromatic intensity in the downward direction, or simply the downward intensity.[16] $I_{\uparrow}$ is defined similarly for the upward direction.

$I_{\downarrow}$ and $I_{\uparrow}$ change, in general, with the depth z into the medium because of absorption and scattering, which are specified by the (volumetric) absorption coefficient κ and the (volumetric) scattering coefficient β. These cannot be obtained within the framework of radiative transfer theory (a *macroscopic* theory); recourse must be had to *microscopic* theories. For a collection of independent identical scatterers the scattering coefficient β is simply the individual scattering cross section times the number density (number per unit volume). Thus $1/\beta$ has the dimensions of length and may be interpreted as the scattering mean free path (i.e., the average distance between scattering events). Similarly, κ is the absorption cross section times the num-

where the *single-scattering albedo* $\overline{\omega}_0$ is defined as $\beta / (\kappa + \beta)$. If the medium is uniform, the optical depth, which is dimensionless, is $z(\kappa + \beta)$, and may be interpreted as the depth in units of total mean free path $1/(\kappa + \beta)$. Because the absorption and scattering coefficients depend on frequency, in general, so does the optical depth. The single-scattering albedo varies between 0 (no scattering) and 1 (no absorption). Both limits are idealizations never realized in practice.

Equations (8) and (9) are the two-stream equations of radiative transfer. Although we shall not do so, these equations can be extended to N streams, in which instance we obtain N coupled differential equations of the same form as Eqs. (8) or (9) but with $N + 1$ terms on the right side; one term represents attenuation and the N remaining terms represent all the possible ways in which light is scattered into one direction from all other directions. In the limit as N goes to infinity, sums become integrals and the set of equations collapses into a single integro-differential equation.[1,2]

A more compact form of Eqs. (8) and (9) is obtained by first adding and then subtracting them:

$$\frac{d}{d\tau}(I_\downarrow - I_\uparrow) = -(1 - \overline{\omega}_0)(I_\downarrow + I_\uparrow) , \qquad (10)$$

$$\frac{d}{d\tau}(I_\downarrow + I_\uparrow) = -(1 - \overline{\omega}_0 g)(I_\downarrow - I_\uparrow) . \qquad (11)$$

Equations (8) and (9) [or, equivalently, Eqs. (10) and (11)] are consequences merely of conservation of energy applied to streams of photons constrained to only two directions. Equations like these were first obtained by Schuster,[10] although he included emission and restricted himself to isotropic scattering ($g = 0$).

Now we shall try to obtain as much physical insight as possible from these simple equations by solving them subject to various boundary conditions, by interpreting the solutions, and by making a connection between them and observations.

III. CONSERVATIVE SCATTERING: NO ABSORPTION

No medium is strictly nonabsorbing. Nevertheless, we can sometimes ignore absorption and set the single-scattering albedo to unity without making great errors in calculated observable quantities (e.g., reflection of visible light by clouds). With this assumption Eqs. (10) and (11) have the simple solutions

$$I_\downarrow = D + C(1 - \tau^*) , \quad I_\uparrow = D - C(1 + \tau^*) , \qquad (12)$$

provided that g is independent of τ, where $\tau^* = (1 - g)\tau$ is the *scaled* optical depth. The constants C and D are determined by conditions at the upper ($\tau = 0$) and lower ($\tau = \overline{\tau}$) boundaries, where the (total) *optical thickness* of a medium with physical thickness h is

$$\overline{\tau} = \int_0^h (\kappa + \beta) \, dz . \qquad (13)$$

A. Equilibrium solution

Suppose that the medium is illuminated from above $[I_\downarrow(0) = I_0]$ and that a perfect reflector underlies it $[I_\uparrow(\overline{\tau}) = I_\downarrow(\overline{\tau})]$. For this case $C = 0$ and $D = I_0$, so that the radiation field is uniform and isotropic (i.e., $I_\downarrow = I_\uparrow$ for

all $\overline{\tau}_0$ and g). This is the *equilibrium solution*. To understand why this is so called, consider the medium to be enclosed by *two* perfect reflectors, above and below. The solution to Eq. (12) in this instance is $C = 0$ and D is arbitrary. That is, if photons are introduced into the medium, they rattle around (none are absorbed or leak from the medium) until the intensity is everywhere uniform and isotropic.

B. Reflection and transmission

Suppose that photons that leak out of the lower boundary of the medium are not returned to it, either because there is nothing to scatter them back or they are absorbed [i.e., $I_\uparrow(\overline{\tau}) = 0$]. In this case, solution of Eq. (12) yields the following expressions for the *albedo*[17] (or *reflection coefficient*) R and *transmission coefficient* T:

$$R = I_\uparrow(0)/I_0 = \overline{\tau}^*/(2 + \overline{\tau}^*) , \qquad (14)$$
$$T = I_\downarrow(\overline{\tau})/I_0 = 2/(2 + \overline{\tau}^*) , \qquad (15)$$

which satisfy $R + T = 1$.

Note that only the *scaled* optical thickness $\overline{\tau}^*$ determines these two observable properties of a (nonabsorbing) multiple-scattering medium. Thus two such media are optically similar if the product of $\overline{\tau}$ and $(1 - g)$ is the same for both rather than $\overline{\tau}$ and g separately. Therefore, it is not possible to determine both $\overline{\tau}$ and g uniquely by measuring transmission and reflection.

As the optical thickness increases, R and T approach the limits 1 and 0, respectively. R will be within 1% of its asymptotic value if $\overline{\tau} > 200/(1 - g)$, which provides a criterion for when a medium can be taken to be *optically thick* (i.e., effectively infinitely thick).

If $g = 1$ (forward scattering only), then $R = 0$ regardless of the optical thickness. But it does not then follow that R is necessarily 1 if $g = -1$ (backward scattering only).

We get more insight into Eq. (15) if we rewrite it using Eq. (6): $T = 1/(1 + p_{\downarrow\uparrow}\overline{\tau})$. The physical interpretation of this is that photons are lost (strictly) to the downward stream only if they are scattered in the opposite direction.

For small (scaled) optical thickness we can expand Eq. (15):

$$T \simeq 1 - \overline{\tau}^*/2 \simeq \exp(-\overline{\tau}^*/2) , \quad (\overline{\tau}^* \ll 1) .$$

Thus the incident intensity is attenuated exponentially only if multiple scattering is negligible. Suppose that once a photon has been scattered in the backward direction, it is removed from the medium. In this instance, there is no multiple scattering, and the corresponding transmission coefficient $\exp(-\overline{\tau}^*/2)$ is always less than that given by Eq. (15). All else being equal, therefore, attenuation is *less* in a multiple-scattering medium than in a single-scattering medium, which is perhaps contrary to what one expects because of the word "multiple." This is so because photons scattered out of a particular direction can find their way back into that direction by being scattered again one or more times.

C. Reflection and transmission by clouds

Multiple-scattering media such as clouds have high albedos (at visible wavelengths) because incident photons reemerge after having been scattered many times by particles that are only weakly absorbing. One cannot explain observations of clouds on the basis of single-scattering arguments.

Nevertheless, one encounters frequently the statement that clouds are white because they are composed of nonselective scatterers. It is true that cloud droplets are so large compared with the wavelengths of visible light that their scattering cross sections are nearly independent of wavelength. This is a *sufficient* condition for the whiteness of clouds, but it is not *necessary*. The converse is not true: A collection of selective scatterers is not necessarily colored. For example, milk is a suspension of small particles that scatter blue light more than red. You can demonstrate this by adding a few drops of milk to water and illuminating the resulting suspension with a collimated beam of white light. The scattered light will be bluish and that transmitted will be reddish. Yet a glass of milk is white. We can understand why by differentiating Eq. (14) with respect to the wavelength λ:

$$\frac{dR}{d\lambda} = \frac{2}{(2+\overline{\tau}^*)^2} \frac{d\overline{\tau}^*}{d\lambda} . \qquad (16)$$

If the optical thickness $\overline{\tau}^*$ (which is proportional to the scattering coefficient β) is independent of wavelength then so is R. But the converse is not true. For sufficiently large optical thickness, R is independent of wavelength regardless of the wavelength dependence of $\overline{\tau}^*$.

To proceed further we need to estimate the optical thickness of clouds. For simplicity, let us assume that they are uniform and are composed of droplets all of which have the same radius a. The optical thickness of a suspension of N identical particles per unit volume is

$$\overline{\tau} = NC_{\text{ext}}\, h = f(C_{\text{ext}}/v)\, h , \qquad (17)$$

where the extinction cross section C_{ext} is the sum of absorption and scattering cross sections (i.e., the particle's effective cross-sectional areas for removal of photons from a beam by absorption and scattering), v is the volume of a particle, h is the thickness of the suspension, and $f = Nv$ is the fraction of the total suspension volume occupied by the particles. For spheres much larger than the wavelength of the light illuminating them, C_{ext} is approximately twice the geometrical cross section πa^2 (see, e.g., Ref. 18, p. 107). Clouds are quite tenuous: f is typically around 3×10^{-7}, corresponding to a liquid water content of 0.3 g/m^{-3} (Ref. 19, p. 15). Cloud droplets are distributed in size (Ref. 19, pp. 13 and 14), but a diameter of $10\ \mu$m is representative. With these assumptions, the *approximate* optical thickness of a cloud is $\overline{\tau} = 100h$, where h is in km. But it is the scaled optical thickness that appears in Eqs. (14) and (15), so we need to estimate the asymmetry parameter g. From the calculations for water droplets tabulated by Irvine and Pollack,[20] it is evident that $g = 0.85$ is a representative value for cloud droplets at visible wavelengths, that is, scattering of light by such droplets is highly peaked in the forward direction. Thus we obtain $\overline{\tau}^* \simeq 14h$ as a rough estimate for the scaled optical thickness of a cloud of physical thickness h. It follows from this result and Eq. (16) that the albedo of clouds thicker than aboout 1 km would be nearly independent of wavelength regardless of the wavelength dependence of scattering by the individual droplets.

The preceding statements may seem, at first glance, to be incompatible with the colors seen often in thin clouds or at the edges of thick clouds when looking toward the sun. Such colors, called iridescence, are consequences of single scattering. Although *total* scattering by a cloud droplet is nearly independent of wavelength, the angular distribution of the scattered light (i.e., the *differential* scattering cross

section) is not. For more on iridescent clouds see Refs. 21–24.

It is rare for clouds overhead to be so thick that day becomes night. From Eq. (15) and our estimate for the optical thickness of clouds it follows that they would have to be more than 15-km thick to transmit less than 1% of the light incident on them. This would require the atmosphere to be filled uniformly with clouds from the surface to the tropopause.

D. Reflection by snow

Snow on the ground is one common example of a multiple-scattering medium that is often sufficiently deep for its albedo to be close to the asymptotic value. Ice grains in such snow are not necessarily spherical, although they are rarely the spatial dendrites favored by painters of Christmas cards (see Ref. 25 for a good discussion of how the shapes of snowflakes change with time after they settle). Like cloud droplets, these ice grains are nonselective scatterers of visible light. The volume fraction f of snow on the ground varies, but 0.3 is typical. The extinction cross section of any large particle is twice its geometrical cross section projected onto the beam illuminating it (Ref. 18, p. 107). Thus the extinction cross section of an ice grain is proportional to the square of a characteristic linear dimension d; its volume v is proportional to the cube of d. For ice grains in snow a representative value for d is about 1 mm. Because they are much larger than cloud droplets, scattering by the grains is more sharply peaked in the forward direction (i.e., g is closer to 1). Let us therefore take g to be 0.93, a reasonable estimate. Thus we estimate the (scaled) optical thickness $\overline{\tau}^*$ of a snowpack h meters deep to be $200h$ (for finer grained, denser snow the coefficient of h will be even greater). According to this estimate, snow about 1-m deep is optically thick (i.e., its albedo is within 1% of the asymptotic value). Absorption has been neglected; to include it would reduce the depth snow must be for it to be considered optically thick. In Sec. IV the observable consequences of not neglecting absorption by ice grains in snow will be discussed.

It may be inconvenient to tunnel deep into snow to verify my assertion about the optical thickness of snow. Observations are made more comfortably above snow. Suppose that a perfect reflector (i.e., a white surface) underlies the snow, in which case $R = 1$. Equation (14) applies to snow over a perfectly black surface. The difference between these two extreme albedos, which I shall denote as R_w and R_b, is

$$R_w - R_b = 2/(2+\overline{\tau}^*) .$$

Snow therefore does not have to be very deep, perhaps a few tens of centimeters, before it is not possible to tell what underlies it, which is often observable. Another observation you can make is of the comparative brightness of snow and clouds under similar illumination. Because deep snow is optically thicker than clouds, snow is usually brigher.

E. Diffuse radiation: Cloud light

Few of us are in the habit of staring directly at the sun, even when it is partially obscured by clouds. More often, we see *diffuse* radiation, that is, radiation that has been scattered (e.g., cloud light or skylight). If there is no illumination from below, the upward intensity $I_\uparrow$ is necessarily diffuse. To emphasize this, I shall henceforth denote this quantity as $D_\uparrow$. The downward intensity $I_\downarrow$ is the sum of two components, an unscattered component $I_\downarrow^u$ and a diffuse component $D_\downarrow$. We can determine these quantities by rewriting Eq. (2) as follows:

$$\frac{dI_\downarrow}{d\tau} + I_\downarrow = p_{\downarrow\uparrow}I_\downarrow + p_{\uparrow\downarrow}I_\uparrow \,.$$

The terms on the right side of this equation involve scattering. Thus the *unscattered* intensity satisfies the homogeneous equation

$$\frac{dI_\downarrow^u}{d\tau} + I_\downarrow^u = 0 \,,$$

which has the solution $I_\downarrow^u = I_0 \exp(-\tau)$ for the boundary condition $I_\downarrow^u(0) = I_\downarrow(0) = I_0$. Let us take what underlies the medium to be perfectly black [i.e., $I_\uparrow(\overline{\tau}) = 0$]. At any arbitrary optical depth τ into the medium, the two diffuse intensities are

$$D_\uparrow/I_0 = [\tfrac{1}{2}(1-g)(\overline{\tau}-\tau)]/[1+\tfrac{1}{2}(1-g)\overline{\tau}] \,, \quad (18)$$

$$D_\downarrow/I_0 = \{[1+\tfrac{1}{2}(1-g)(\overline{\tau}-\tau)]/$$

$$[1+\tfrac{1}{2}(1-g)\overline{\tau}]\} - \exp(-\tau) \,. \quad (19)$$

The upward intensity vanishes at the lower boundary, and has its greatest value at the upper boundary. The downward diffuse intensity vanishes at the upper boundary, increases to a maximum, and then decreases toward the lower boundary. It follows from inspection of Eqs. (18) and (19) that the two diffuse intensities are *approximately* equal for optical depths satisfying

$$2/(1-g) < \tau < \overline{\tau} - [2/(1-g)] \,. \quad (20)$$

This result has a simple physical interpretation, which follows from Eq. (6) and the definition of optical depth. The diffuse radiation field is highly anisotropic near boundaries, but because of scattering it tends toward isotropy as we move away from them. The quantity $\tau(1-g)/2$, which is (approximately) the fraction of photons that have been reversed in direction in traversing an optical depth τ, is 1 for $\tau = 2/(1-g)$. Thus at optical depths greater than $2/(1-g)$ from either boundary the diffuse radiation field is nearly isotropic. This isotropy of diffuse radiation deep within a multiple-scattering medium can be observed from an airplane descending through clouds. After entering the cloud, you soon will have no visual clues to tell up from down, but eventually you will see the upward intensity begin to decrease (unless you are flying over snow); this is the signal that your airplane is about to leave the cloud.

Now suppose that we are on the ground looking at clouds overhead.[26] The diffuse downward intensity $D_\downarrow(\overline{\tau})$, which from Eq. (19) is

$$D_\downarrow(\overline{\tau})/I_0 = \{1/[1+\tfrac{1}{2}(1-g)\overline{\tau}]\} - \exp(-\overline{\tau}) \,, \quad (21)$$

rises steeply with increasing optical thickness $\overline{\tau}$ beginning at $\overline{\tau} = 0$ to a maximum at $\overline{\tau} \simeq \ln[2/(1-g)]$, and then decreases more gradually with a further increase in $\overline{\tau}$ (Fig. 2). This has observable consequences. A thin cloud layer, say from a few tens to a few hundreds of meters thick, is brighter than the clear sky. You can sometimes see this when the sky is covered with thin clouds. If there are occasional breaks, you can compare the brightness of the clear and cloudy skies. Very thick clouds result in gloomy days, when the sky is less bright than on clear days.

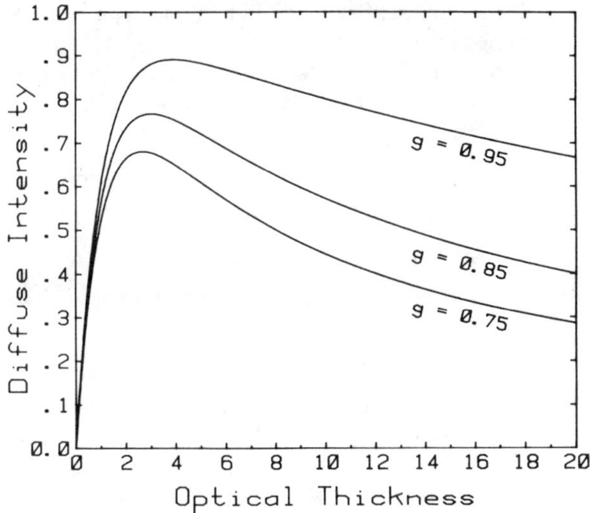

Fig. 2. Diffuse downward intensity of radiation below a nonabsorbing medium of finite optical thickness illuminated from above. g is the mean cosine of the scattering angle.

F. Diffuse radiation: Skylight

Let us now consider what we see when looking upward[26] on a very clear day. Even if the atmosphere were completely free of particles, we would still see a blue sky because of molecular scattering. Indeed, particles decrease the spectral purity of skylight.[27] Because the number density of molecules decreases with height, the molecular scattering coefficient is not uniform. But since this decrease is (approximately) exponential with a *scale height* $H = 8.4$ km, the optical thickness along a radial path from the surface to infinity is the same as that for an atmosphere extending from the surface to H with a constant density equal to the sea level value:

$$\beta_0 H = \int_0^\infty \beta_0 \exp(-z/H)\, dz \,,$$

where β_0 is the sea level molecular scattering coefficient. The optical thickness shown in Fig. 3 was obtained from values of β_0 tabulated by Penndorf.[28]

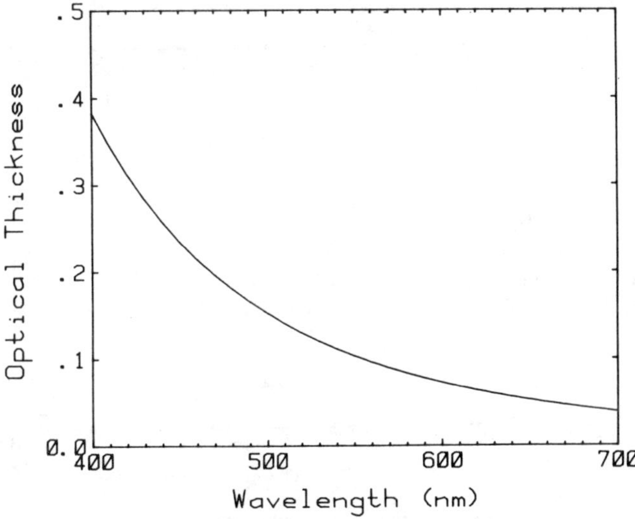

Fig. 3. Molecular optical thickness of Earth's atmosphere along a vertical path from the surface to infinity.

Molecular scattering of light is not isotropic, but it is symmetric about a scattering angle of 90°. Thus we may take $g = 0$ in Eq. (21) to obtain an approximate expression for the intensity of skylight overhead:

$$D_\downarrow/I_0 = [1/(1 + \tfrac{1}{2}\overline{\tau})] - \exp(-\overline{\tau}) \qquad (22)$$

for an atmosphere overlying black ground. If the ground is white (i.e., perfectly reflecting), the skylight intensity is

$$D_\downarrow/I_0 = 1 - \exp(-\overline{\tau}) . \qquad (23)$$

Since $\overline{\tau}$ is small (Fig. 3), we can expand Eqs. (22) and (23) to obtain

$$D_\downarrow^{b}/I_0 \simeq \tfrac{1}{2}\overline{\tau}, \quad D_\downarrow^{w}/I_0 \simeq \overline{\tau}, \qquad (24)$$

where the superscripts indicate the nature of the ground. In both instances we obtain a blue sky (provided we take into account the response of the human eye; for more about this see Ref. 27), but the sky brightness over black ground is about half that over white ground. This makes sense: Both direct sunlight and reflected groundlight illuminate the scatterers.

Only because $\overline{\tau}$ is small do we have a blue sky. If our atmosphere were much thicker, the sky would appear quite different. This is shown in Figs. 4 and 5 (see also Refs. 27 and 29). Skylight intensity is shown as a function of wavelength for both white and black ground. The optical thicknesses are for Earth's actual atmosphere, and for ones with ten and 30 times as many molecules. Note that increasing the number of molecules, which scatter very nearly according to the inverse fourth power of the wavelength, does not make the sky bluer. On the contrary, the intensity increases at the expense of spectral purity. Here is yet another place where multiple scattering must be invoked. In the absence of multiple scattering, an increase in optical thickness would make the sky brigher with *no* change in its spectrum.

Now that we have sharpened our physical intuition, let us be bold and briefly step outside of the one-dimensional domain imposed by the two-stream model. Suppose that the sun is overhead on an extraordinarily clear day, and that we have an unimpeded view of the horizon. What does the sky look like in different directions? It is not, as predicted by single scattering, uniform in color. The sky is bluest near the zenith; it is brightest, but of low spectral purity, near the horizon. The optical thickness (called the *normal*

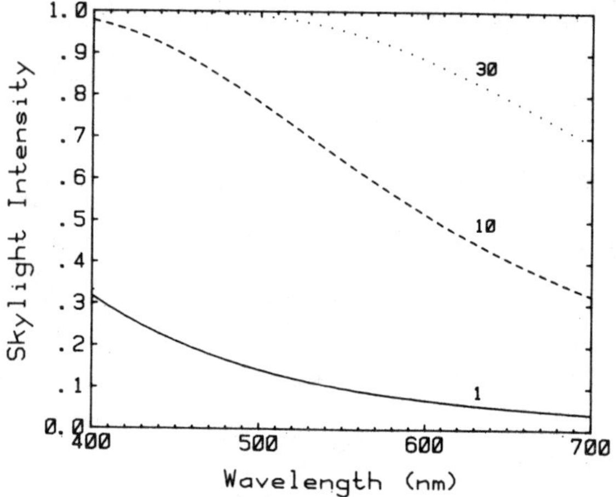

Fig. 4. Skylight intensity (overhead sky) over white ground. The numbers denote the optical thicknesses of hypothetical pure molecular atmospheres relative to the actual value for Earth's atmosphere.

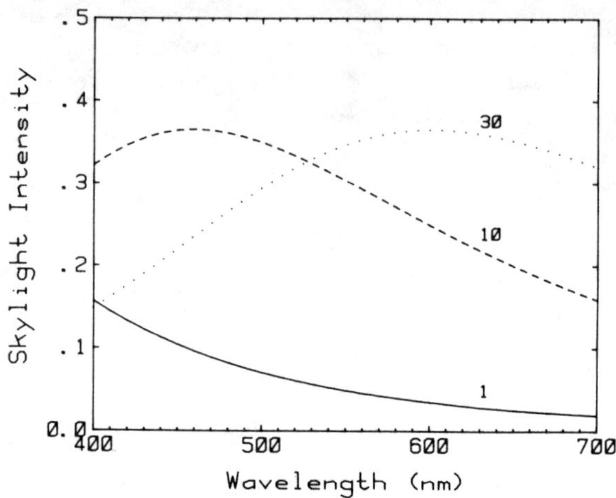

Fig. 5. Skylight intensity (overhead sky) over black ground. The numbers denote the optical thicknesses of hypothetical pure molecular atmospheres relative to the actual value for Earth's atmosphere. Note that the vertical scale is one-half that in Fig. 4.

optical thickness and denoted by $\overline{\tau}_n$) in Eq. (23) is that of a vertical path through the atmosphere, so it applies strictly to the zenith sky. But let us assume, on physical grounds, that this equation is valid even if $\overline{\tau}$ is that for a slant path through the atmosphere. Except for directions very close to the horizon, the optical thickness of a slant path is $\overline{\tau}_n/\cos\theta$, where θ is the angle measured from the zenith. With this assumption, the diffuse intensity in the direction θ is

$$D_\downarrow/I_0 = 1 - \exp(-\overline{\tau}_n/\cos\theta) . \qquad (25)$$

Equation (25), approximate to be sure, is in accord with observations of the color and brightness variation of clear skies. To further test its validity we can compare it with the exact calculations of Coulson *et al.*[30] for a pure molecular atmosphere (over a flat Earth). This is done in Fig. 6, from which it is evident that the simple theory [Eq. (25)] is adequate to explain the variation in brightness and color of the sky. The best place to observe a sky predicted by Eq. (25), which applies to a sky over a highly reflecting surface, is from an airplane flying over clouds. At the altitudes where commercial aircraft fly, there are not many particles: The scale height for atmospheric particles is a few kilometers. So we can be reasonably sure that what we see—dark blues near the zenith and a whitish horizon—is a property of an atmosphere in which scattering by molecules is predominant.

Before we consider absorption, there is another observation to which the preceding analysis may be applied: the appearance of distant objects. One receives not only light from such objects, but *airlight*—light scattered by everything along the line of sight—as well. If the object is black, then *all* the light one receives is airlight. Its intensity is proportional to $1 - \exp(-\tau)$, where τ is the optical thickness of the path from observer to object (see Ref. 31 for more details). If the scatterers are mostly molecules and very small particles, the optical thickness will be inversely proportional to the inverse fourth power of the wavelength. For small optical thicknesses ($\ll 1$) the airlight intensity is proportional to τ and hence is bluish. But as τ is increased, the airlight intensity increases in magnitude while at the same time becomes less dependent on wavelength. This can be observed in a series of parallel mountain ridges, one be-

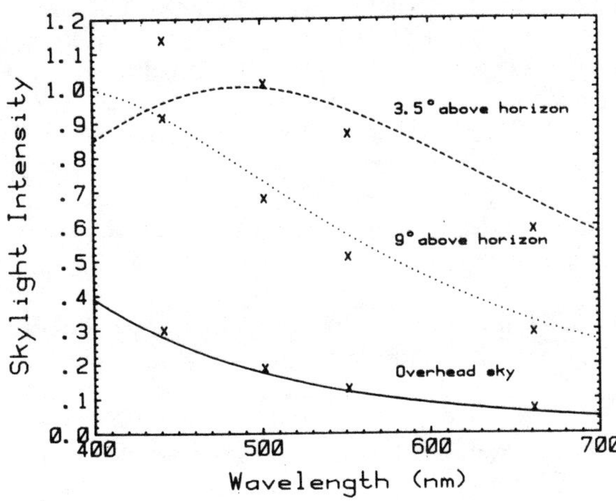

Fig. 6. Skylight intensity for a pure molecular atmosphere. The curves were calculated using the simple two-stream theory of this paper. The ground albedo 0.8 is sufficiently close to 1 that Eq. (25) is a good approximation, although the calculations were done using an expression for the diffuse downward intensity over ground with arbitrary albedo [this expression is not given, but it follows readily from Eq. (12)]. Here (——) is for the sky directly overhead; (- - -) is for that about 3.5° above the horizon; (···) is for that about 9 degrees; Xs show the detailed calculations of Coulson *et al.*[30] for the principal plane (determined by the direction of the incident sunlight and the normal to the ground) with the sun 23 degrees from the zenith. All intensities have been normalized by that at 500 nm for the near-horizon sky.

hind the other, covered with dark vegetation. The closest ridges will have their natural color; those somewhat farther will be bluish; the farthest ridges fade into the whitish horizon sky. To explain this common observation, one must invoke multiple scattering.

IV. MULTIPLE SCATTERING IN ABSORBING MEDIA

Let us now consider the more general case of multiple scattering in an absorbing medium ($\kappa > 0$ hence $\overline{\omega}_0 < 1$). By differentiating Eq. (10) [Eq. (11)] with respect to τ, substituting Eq. (11) [Eq. (10)] in the result, then adding and subtracting the two equations obtained, it follows that

$$\frac{d^2 F}{d\tau^2} = K^2 F,$$

$$K = \sqrt{(1 - \overline{\omega}_0)(1 - \overline{\omega}_0 g)}, \tag{26}$$

where F is either the sum or difference of the two intensities. Solutions to Eq. (26) are linear combinations of the two exponential functions $\exp(\pm K\tau)$. Since the intended applications are to optically thick media, we may take the medium to be infinitely thick. This will simplify the mathematics because the coefficients of $\exp(K\tau)$ must vanish in order to ensure finite intensities. By using Eqs. (10) or (11) and the boundary condition $I_\downarrow(0) = I_0$, a bit of algebra gives the downward and upward intensities:

$$I_\downarrow = I_0 \exp(-K\tau), \quad I_\uparrow = I_0 R \exp(-K\tau), \tag{27}$$

where the albedo $R = I_\uparrow(0)/I_0$ is

$$R = (\sqrt{1 - \overline{\omega}_0 g} - \sqrt{1 - \overline{\omega}_0})/(\sqrt{1 - \overline{\omega}_0 g} + \sqrt{1 - \overline{\omega}_0}). \tag{28}$$

As a check on the correctness of Eq. (28), we note that $R = 1$ when $\overline{\omega}_0 = 1$. Also, for $\overline{\omega}_0 \neq 1$, $R = 0$ for $g = 1$: With

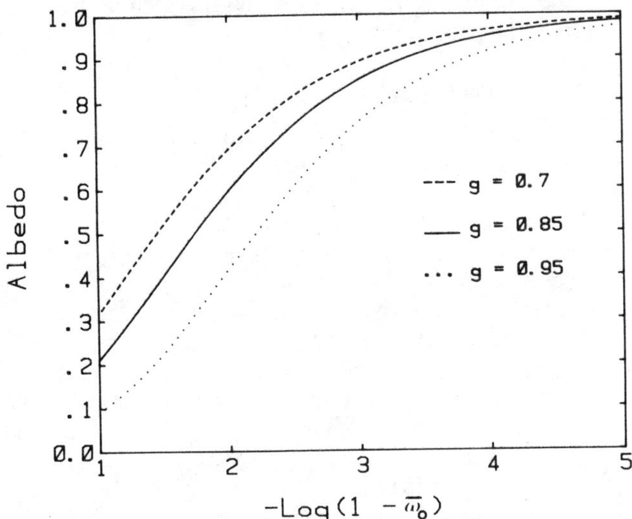

Fig. 7. Albedo of an infinitely thick medium as a function of single-scattering allbedo $\overline{\omega}_0$ for several asymmetry parameters: (- - -) $g = 0.7$; (——) $g = 0.85$; (···) $g = 0.95$.

no mechanism for redirecting incident photons, they can never reemerge from the medium. Note that the derivative of R with respect to $\overline{\omega}_0$ at $\overline{\omega}_0 = 1$ is infinite. This is a quantitative statement of what can be stated as an aphorism: In a multiple-scattering medium, a little bit of absorption goes a long way. The curves in Fig. 7 show the albedo R as a function of single-scattering albedo for a few values of g. We interpret $1 - \overline{\omega}_0$ as the probability that a photon is absorbed in a *single* interaction with a scatterer; $1 - R$ is the probability that it will be absorbed in *many* such interactions. If $1 - \overline{\omega}_0 = 0.001$, for example, then for $g = 0.9$, $1 - R = 0.18$. Thus there is an enormous amplification of absorption—nearly a factor of 200 in this example—in going from single to multiple scattering. For an incident downward photon to find its way into the upward intensity outside the medium, it must, on average, be scattered many times; the greater g is, the greater the number of scatterings (in the extreme case $g = 1$, incident photons never escape). In each scattering, a photon has a small chance of being absorbed, but in many scatterings its cumulative chance of being absorbed increases markedly.

In the context of multiple scattering, a single-scattering albedo of 0.9 (or even 0.99), which at first glance seems high, is in fact quite low, especially if the scatterers are larger than the wavelength and hence have large values of g (see, e.g., Ref. 32, pp. 385–388; Ref. 33, p. 183). This has consequences for the commonly made statement that temperatures are higher on cloudy nights than on clear nights because infrared radiation emitted by the Earth is "reflected" by the clouds. Although the observation is correct, this explanation is not. The single-scattering albedo of cloud droplets is quite high at visible and near-visible wavelengths, greater than 0.99999. But in the infrared it plunges because of the infrared absorption bands of water. The Earth emits radiation of all wavelengths, but most of it is between about 8 and 12 μm. At these wavelengths, the single-scattering albedo of typical cloud droplets is at most about 0.8 (see Ref. 20, Table V), which seems high, but is in fact quite low (see Fig. 7). As a consequence, clouds are nearly black to radiation emitted by the Earth. If clouds elevate nighttime temperatures, it is not because they reflect infrared radiation, but rather that they emit more of it than the clear night sky does.

A. Darkening of sand upon wetting

With Eq. (28) and Fig. 7 we can explain another common observation: the darkening of sand upon wetting. Sand is usually an optically thick multiple-scattering medium (i.e., it is so thick that the addition of another layer does not sensibly change its appearance). When dry sand is wetted, it becomes noticeably darker. At visible wavelengths, water is very weakly absorbing, so increased absorption by water is certainly not the reason for this. Grains in dry sand are surrounded by air. But when wet (strictly speaking, when the sand is saturated), they are surrounded by water. In going from air to water (or other liquids), the grains scatter more toward the forward direction. Indeed, when the refractive index of the grain is exactly that of the surrounding liquid, scattering is entirely in the forward direction. Thus when a large grain is surrounded by a liquid instead of by air, its asymmetry parameter g increases. As the average degree of forwardness of scattering increases, incident photons have to be scattered more times before reemerging from the sand and are therefore exposed to a greater probability of being absorbed. For more details about this explanation of the darkening of sand and experimental verification of it see Ref. 34.

While we are on the subject of sand, we can use Eq. (28) to explain another observation. The asymmetry parameter of very large particles does not depend strongly on their size or on the wavelength. If the particles are weakly absorbing, their absorption cross section is proportional to the product of the absorption coefficient α of the bulk parent material and the particle volume (i.e., the cube of a characteristic linear dimension d). But the extinction cross section (sum of absorption and scattering cross sections) is proportional to the square of d; hence $1 - \overline{\omega}_0$ is proportional to αd. If we expand Eq. (28) in powers of $\sqrt{\alpha d}$ and retain only the first term, we obtain

$$R \simeq 1 - C\sqrt{\alpha d}, \qquad (29)$$

where C is a wavelength-independent factor that depends on, among other things, the asymmetry parameter. Thus the smaller the grains, the closer the albedo is to 1. This can be observed on a beach with sands that have been size segregated by wind or water: The coarser sand is darker (see Ref. 35 for a photograph). If there is not a beach nearby you can verify the qualitative correctness of Eq. (29) by smashing colored beer bottles into small bits. According to Eq. (29), the smaller the grains, the higher the albedo. So if you smash the bottles into very small bits, the resulting heap of powdered glass will have a high albedo which will not be noticeably wavelength dependent. But by selecting sufficiently large glass particles (by sieving, for example), the heap of powdered glass will be less bright and may even display the color (although of lower spectral purity) of the parent bottles.

B. Colors in snow

I mentioned in Sec. III D that snow is another common multiple-scattering medium (for a good review of the optical properties of snow, see Ref. 36). In a recent popular article,[37] I found the following statement: "Snow is not white; it simply reflects the light, which is—usually—white." Rather than comment on this explanation, I shall try to give a more satisfactory one. Ice grains in snow scatter predominantly near the forward direction (i.e., g is close to 1). Incident photons that reemerge from a snowpack must therefore have been scattered many times. If they are to survive many scatterings, the probability of absorption $(1 - \overline{\omega}_0)$ must be small, which it is at visible wavelengths. Visible light must be transmitted through many meters of pure homogeneous ice to be appreciably absorbed.[38,39] Ice is intrinsically blue: It absorbs red light much more than blue light. But $1/\alpha$ for ice is of order meters over the visible spectrum, whereas grain sizes are of order millimeters. Thus the product αd for ice grains in snow is very small, and as a consequence [see Eq. (29)] the albedo of snow is high and does not vary appreciably over the visible spectrum.

But this does not mean that snow does not exhibit colors. Deep blues can be seen in crevasses, in ice caves, and even in holes in ordinary snow (see Ref. 40 and the cover of Ref. 41 for photographs of blue holes in snow). Raman[42] attributed this blueness to scattering by molecules, an incorrect explanation that is still extant,[43] although an essentially correct explanation was given in 1886 by Waltheré Spring.[44] Raman's[42] arguments apply only to ice of a transparency much greater than that usually found in nature. Molecular scattering in snow is overwhelmed by scattering by ice grains. Moreover, if Raman's explanation were correct, light transmitted in snow would be red, not blue. If K [see Eqs. (27) and (28)] were independent of wavelength, and τ inversely proportional to the fourth power of the wavelength, then the transmitted intensity would get redder with increasing depth into the snow. This is contrary to what is observed.

The optical thickness of snow is to good approximation independent of wavelength, from visible to infrared, because the extinction cross section of ice grains is nearly constant in this range. This statement seems to contradict the observation that sunlight transmitted into snow is blue. Here is a good example of the great difference between single and multiple scattering. For small optical thicknesses, multiple scattering is negligible, and all scattered photons are lost from an incident beam. Of course, photons can also be absorbed. It is the *sum* of scattering and absorption that is independent of wavelength. Multiple scattering in snow is not negligible. Attenuation in snow will therefore be less than it would otherwise be (i.e., with no multiple scattering) because photons scattered out of the direction of a beam can reappear in this direction. But it is only those photons that are *not* absorbed that do so. Visible light at the short wavelength end of the spectrum is less likely to be absorbed by an ice grain than light at the long wavelength end. Hence, it is only because of nonselective multiple scattering in a selectively absorbing medium that we see blue light in snow holes. Moreover, the depth to which white light must be transmitted in snow to become blue is considerably less, a factor of about 50, than it would have to be transmitted in homogeneous ice: Multiple scattering in snow increases the total path length a photon travels through ice before reaching a given depth. Although absorption by liquid water over the visible spectrum is quite similar to that by ice, we do not see vivid blue light beneath clouds because their optical thicknesses are smaller than those of deep snow and because cloud droplets are much smaller than ice grains.

As a final example, to inspire confidence in the simple-

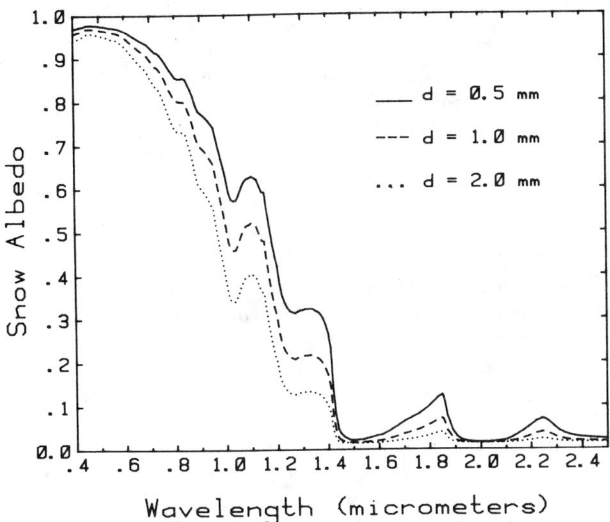

Fig. 8. Albedo of optically thick snow, calculated using the simple two-stream theory of radiative transfer, for various grain diameters: (——) $d = 0.5$ mm; (- - -) $d = 1.0$ mm; ($\cdots$) $d = 2.0$ mm.

two stream theory, I have used Eq. (28) to compute the albedo of snow over most of the solar spectrum. The grains are taken to be spheres, all with the same diameter d, and with an asymmetry parameter of 0.94. For the single-scattering albedo, I have used the approximation $\overline{\omega}_0 = 1 - Q_{abs}/2$, where the absorption efficiency Q_{abs} (absorption cross section normalized by geometrical cross section) is obtained from Ref. 45. The refractive index of ice is taken to be 1.3 over the solar spectrum; its absorption coefficient is obtained from the compilation by Warren.[39] The results are shown in Fig. 8. Despite all these simplifications, the calculated albedo agrees well with that measured by O'Brien and Munis[46] (see their Fig. 2). All the wiggles and bumps, corresponding to the absorption bands of bulk ice, are in their proper positions. Even the magnitude of the albedo is approximately correct, although theory tends to overestimate the visible albedo. The reason for this is probably contamination.[47] Absorption of visible light by carbonaceous materials (lumped under the heading of "soot") is more than a million times greater than absorption by ice. As a consequence, a few parts per million of soot in snow can markedly reduce its visible albedo. In the infrared, however, ice is so strongly absorbing that an absorbing contaminant in snow does not change its albedo.

Figure 8 illustrates a statement I make to get my students scratching their heads: Snow is the whitest natural substance on our planet; it is also the blackest. At visible wavelengths, the albedo of clean, fine-grained Antarctic snow[48] is as high as 0.97. But in the middle infrared, the albedo of snow is so low that it is nearly a perfect blackbody.

V. CONCLUDING REMARKS

Just as it is not necessary to begin with the Maxwell equations to understand coherent scattering (in the guise of reflection and refraction by homogeneous, optically smooth media), it is not necessary to begin with the exact equation of radiative transfer to understand incoherent scattering. I have approached multiple (incoherent) scattering by way of a simple two-stream theory. Its major defect is that it suppresses the full directionality of the radi-

ation field: 2π Sr are collapsed into a single direction. Despite this simplification, the results of the two-stream theory are applicable to a great many common observations, ones which cannot be understood by single-scattering arguments. My approach has been to first derive mathematical expressions, then apply them to observations, and finally to give a physical interpretation. Much of the terminology introduced in this simple context—single-scattering albedo, asymmetry parameter, optical depth, optical thickness, etc.—appears in more advanced treatises, so the way has been paved for further study by those whose appetites have been whetted. Some of the ideas in this paper, and simple ways to demonstrate them, are discussed at an elementary level (i.e., without mathematics) elsewhere.[35,49]

ACKNOWLEDGMENTS

I am grateful to William Doyle, Alistair Fraser, William Mach, John Olivero, and Warren Wiscombe for their critical comments on the first draft of this paper. This work was supported, in part, by the National Science Foundation under grant no. ATM-8303629.

[1] S. Chandrasekhar, *Radiative Transfer* (Dover, New York, 1960).

[2] H. C. van de Hulst, *Multiple Light Scattering* (Academic, New York, 1980), 2 vols.

[3] L. Foldy, Phys. Rev. **67**, 107 (1945).

[4] M. Lax, Rev. Mod. Phys. **23**, 287 (1951).

[5] M. Lax, Phys. Rev. **85**, 621 (1952).

[6] V. Twersky, J. Opt. Soc. Am. **52**, 145 (1962).

[7] A. Ishimaru, *Wave Propagation and Scattering in Random Media* (Academic, New York, 1978), Vol. 1.

[8] W. T. Doyle, Am. J. Phys. **53**, 463 (1985).

[9] If light incident on a collection of randomly oriented scatterers is unpolarized, then so is the forward and backward scattered light. For this reason, as much as for the additional mathematical complexity, it would be pointless to take polarization into account in a two-stream model.

[10] A. Schuster, Astrophys. J. **21**, 1 (1905); reprinted in *Selected Papers on the Transfer of Radiation*, edited by D. H. Menzel (Dover, New York, 1966).

[11] C. Sagan and J. B. Pollack, J. Geophys. Res. **72**, 469 (1967).

[12] W. E. Meador and W. R. Weaver, J. Atmos. Sci. **37**, 630 (1979).

[13] C. Acquista, F. House, and J. Jafolla, J. Atmos. Sci. **38**, 1446 (1981).

[14] W. G. Zdunkowski, R. M. Welch, and G. Kolb, Contribs. Atmos. Phys. **53**, 147 (1980).

[15] M. D. King and Harshvardhan, J. Atoms. Sci. **43**, 784 (1986).

[16] What I have chosen to call intensity because many physicists will be comfortable with this term is also called (radiant) flux or, in more modern terminology, *irradiance*. For the somewhat artificial model chosen here, there is no distinction between radiance, irradiance, and intensity [see E. J. McCartney, *Optics of the Atmosphere* (Wiley, New York, 1976), pp. 6–13 for definitions]. Although these terms now refer to somewhat different physical quantities, intensity was used in a loose way formerly to denote all three quantities, with its precise meaning determined by context. A great amount of energy has been expended in defining various radiometric quantities, inventing symbols for them, and then defending the merits of one set over another. It almost seems at times that radiometry is the science of terminology. See, for example, the *Selected Reprints on Radiometry* (AAPT, College Park, MD), most of which are concerned with terminology rather than with observations. For a wittier expression of my sentiments, I recommend the book review by R. C. Hilborn [Am. J. Phys. **52**, 668 (1984)]. He begins with his conviction that radiometry is an acronym for *r*evulsive, *a*rchaic, *d*iabolical, *i*nvidious, *o*dious, *m*ystifying, *e*xotic *t*erminology *r*egenerating *y*awns.

[17] In general, albedo and reflection coefficient (also called reflectance and reflection function) refer to different quantities: Albedo is the fraction of the light incident on the medium that is reflected in the entire backward hemisphere, whereas reflection coefficient is the fraction reflected

in a given direction in this hemisphere. In the two-stream model, however, these two quantities are the same.

[18]H. C. van de Hulst, *Light Scattering by Small Particles* (Wiley, New York, 1957).

[19]H. R. Pruppacher and J. D. Klett, *Microphysics of Clouds and Precipitation* (Reidel, Dordrecht, Holland, 1980).

[20]W. M. Irvine and J. B. Pollack, Icarus **8**, 324 (1968).

[21]M. Minnaert, *The Nature of Light and Color in the Open Air* (Dover, New York, 1954).

[22]R. A. R. Tricker, *Introduction to Meteorological Optics* (Elsevier, New York, 1970).

[23]R. Greenler, *Rainbows, Halos, and Glories* (Cambridge U.P., Cambridge, 1980).

[24]C. F. Bohren, Weatherwise **38**, 268 (1985).

[25]S. C. Colbeck, Weatherwise **38**, 312 (1985).

[26]Because of the limitations of the two-stream model, the diffuse downward intensity is in the same direction as the unscattered intensity and hence is indistinguishable from it. To interpret observations using our expressions for the diffuse downward intensity therefore requires that we take them to be valid (approximately) at directions away from that of the incident illumination.

[27]C. F. Bohren and A. B. Fraser, Phys. Teach. **23**, 267 (1985).

[28]R. Penndorf, J. Opt. Soc. Am. **47**, 176 (1957).

[29]N. A. Voishvillo and Yu. A. Anokhin, Iz. Atmos. Ocean. Phys. **20**, 760 (1984).

[30]K. L. Coulson, J. V. Dave, and Z. Sekera, *Tables Relating to Radiation Emerging from a Planetary Atmosphere with Rayleigh Scattering* (Univ. of Calif. Press, Berkeley, 1960).

[31]C. F. Bohren and A. B. Fraser, Am. J. Phys. **54**, 222 (1986).

[32]C. F. Bohren and D. R. Huffman, *Absorption and Scattering of Light by Small Particles* (Wiley-Interscience, New York, 1983).

[33]M. Kerker, *The Scattering of Light and Other Electromagnetic Radiation* (Academic, New York, 1969).

[34]S. A. Twomey, C. F. Bohren, and J. L. Mergenthaler, Appl. Opt. **25**, 431 (1986).

[35]C. F. Bohren, Weatherwise **36**, 197 (1983).

[36]S. G. Warren, Rev. Geophys. Space Phys. **20**, 67 (1982).

[37]The Economist, 21 December 1985, p. 103.

[38]T. C. Grenfell and D. K. Perovich, J. Geophys. Res. **86**, 7447 (1981).

[39]S. G. Warren, Appl. Opt. **23**, 1206 (1984).

[40]C. F. Bohren, J. Opt. Soc. Am. **73**, 1646 (1983).

[41]Science News **123** (1983); see, also, the article by I. Peterson on p. 364.

[42]C. V. Raman, Nature **111**, 13 (1923).

[43]P. V. Hobbs, *Ice Physics* (Oxford U.P., Oxford, 1972), p. 226.

[44]A translation of Spring's paper (and many others on the colors of water) was given by W. D. Bancroft, J. Franklin Inst. **187**, 249, 459 (1923).

[45]C. F. Bohren and T. J. Nevitt, Appl. Opt. **22**, 774 (1983).

[46]H. W. O'Brien and R. H. Munis, CRREL Res. rep. 332 (US Army Cold Reg. Res. and Eng. Lab., Hanover, NH, 1975).

[47]S. G. Warren and W. J. Wiscombe, J. Atmos. Sci. **37**, 2734 (1980).

[48]G. H. Liljequist, *Norwegian–British–Swedish–Antarctic Expedition, 1949–52, Scientific Results* (Norsk Polarinstitutt, Oslo, Norway, 1956), Vol. 2, part 1A.

[49]C. F. Bohren, Weatherwise **36**, 143 (1983).

Numerically stable algorithm for discrete-ordinate-method radiative transfer in multiple scattering and emitting layered media

Knut Stamnes, S-Chee Tsay, Warren Wiscombe, and Kolf Jayaweera

We summarize an advanced, thoroughly documented, and quite general purpose discrete ordinate algorithm for time-independent transfer calculations in vertically inhomogeneous, nonisothermal, plane-parallel media. Atmospheric applications ranging from the UV to the radar region of the electromagnetic spectrum are possible. The physical processes included are thermal emission, scattering, absorption, and bidirectional reflection and emission at the lower boundary. The medium may be forced at the top boundary by parallel or diffuse radiation and by internal and boundary thermal sources as well. We provide a brief account of the theoretical basis as well as a discussion of the numerical implementation of the theory. The recent advances made by ourselves and our collaborators—advances in both formulation and numerical solution—are all incorporated in the algorithm. Prominent among these advances are the complete conquest of two ill-conditioning problems which afflicted all previous discrete ordinate implementations: (1) the computation of eigenvalues and eigenvectors and (2) the inversion of the matrix determining the constants of integration. Copies of the FORTRAN program on microcomputer diskettes are available for interested users.

I. Introduction

The discrete ordinate method for radiative transfer is commonly ascribed to Chandrasekhar.[1] Computer implementations of that method were, however, plagued by numerical difficulties[2] to such an extent that researchers made little use of it. The purpose of this paper is to alert the community to a new numerical implementation of the discrete ordinate method for vertically inhomogeneous layered media which is free of these difficulties and to give a summary of its equations and its various advanced features. The resulting computer code represents the culmination of years of effort[3–9] on the part of ourselves and our collaborators to make it the finest algorithm available. Our intent is that the code be so well documented, so versatile, and so error-free that other researchers can easily and safely use it both in data analysis and as a component of large models.

Knut Stamnes is with University of Tromsø, Institute of Mathematical & Physical Sciences, Auroral Observatory, N-9001 Tromsø, Norway; W. Wiscombe is with NASA Goddard Space Flight Center, Greenbelt, Maryland 20771; the other authors are with University of Alaska—Fairbanks, Physics Department & Geophysical Institute, Fairbanks, Alaska 99775-0800.

Received 7 August 1987.

The problem to be solved is the transfer of monochromatic radiation in a scattering, absorbing, and emitting plane-parallel medium with a specified bidirectional reflectivity at the lower boundary. Section II summarizes the equations and boundary conditions. Section III discusses the numerical implementation of the theory, in particular, (a) how to compute eigenvalues and eigenvectors reliably and efficiently and (b) how to avoid fatal overflows and ill-conditioning in the matrix inversion needed to determine the constants of integration. Reference 9 provides a more detailed account as well as documentation, test problems, and a listing of the FORTRAN code.

II. Theory

A. Basic Equations

The purpose of this section is to present the basic radiative transfer formulas with a minimum of definition and explanation to establish notation and conventions. (For more comprehensive discussions of the subject, the reader is referred to textbooks[1,10] and review articles.[11–15])

The equation describing the transfer of monochromatic radiation at frequency ν through a plane-parallel medium is given by[1]

$$\mu \frac{du_\nu(\tau_\nu,\mu,\phi)}{d\tau} = u_\nu(\tau_\nu,\mu,\phi) - S_\nu(\tau_\nu,\mu,\phi), \tag{1}$$

where $u_\nu(\tau_\nu,\mu,\phi)$ is the specific intensity along direction μ,ϕ at optical depth τ_ν measured perpendicular to the surface of the medium (ϕ is the azimuthal angle, and μ is the cosine of the polar angle). S_ν is the source function given by

$$S_\nu(\tau_\nu,\mu,\phi) = \frac{\omega_\nu(\tau_\nu)}{4\pi} \int_0^{2\pi} d\phi' \int_{-1}^{1} d\mu' P_\nu(\tau_\nu,\mu,\phi;\mu',\phi')$$

$$\times\, u_\nu(\tau_\nu,\mu',\phi') + Q_\nu(\tau_\nu,\mu,\phi), \tag{2}$$

where $\omega_\nu(\tau_\nu)$ is the single-scattering albedo, and $P_\nu(\tau_\nu,\mu,\phi;\mu',\phi')$ is the phase function. For thermal emission in local thermodynamic equilibrium (LTE), the source term Q_ν is

$$Q_\nu^{(\text{thermal})}(\tau_\nu) = [1 - \omega_\nu(\tau_\nu)]B_\nu[T(\tau_\nu)], \tag{3}$$

where $B_\nu(T)$ is the Planck function at frequency ν and temperature T. If the usual diffuse–direct distinction is made (Ref. 1, p. 22), so that u_ν in Eqs. (1) and (2) describes the diffuse radiation only, then for a parallel beam incident in direction μ_0,ϕ_0 on a nonemitting medium,

$$Q_\nu^{(\text{beam})}(\tau_\nu,\mu,\phi) = \frac{\omega_\nu(\tau_\nu)I_0}{4\pi} P_\nu(\tau_\nu,\mu,\phi;-\mu_0,\phi_0)\exp(-\tau_\nu/\mu_0), \tag{4}$$

where $\mu_0 I_0$ is the incident flux. We take

$$Q_\nu(\tau_\nu,\mu,\phi) = Q_\nu^{(\text{thermal})}(\tau_\nu) + Q_\nu^{(\text{beam})}(\tau_\nu,\mu,\phi).$$

(Many radiative transfer models consider one or the other source term but not both.)

To simplify the notation we omit the ν subscript in the following. By expanding the phase function $P(\tau, \cos\theta)$ in a series of $2N$ Legendre polynomials and the intensity in a Fourier cosine series,[1,9]

$$u(\tau,\mu,\phi) = \sum_{m=0}^{2N-1} u^m(\tau,\mu)\cos m(\phi_0 - \phi), \tag{5}$$

we find that Eqs. (1) and (2) are replaced by $2N$ independent equations (one for each Fourier component)

$$\mu\frac{du^m(\tau,\mu)}{d\tau} = u^m(\tau,\mu) - \int_{-1}^{1} D^m(\tau,\mu,\mu')u^m(\tau,\mu')d\mu' - Q^m(\tau,\mu) \tag{6a}$$

$$(m = 0,1,2,\ldots,2N-1),$$

where

$$D^m(\tau,\mu,\mu') = \frac{\omega(\tau)}{2} \sum_{l=m}^{2N-1} (2l+1)g_l^m(\tau)P_l^m(\mu)P_l^m(\mu') \tag{6b}$$

$$Q^m(\tau,\mu) = X_0^m(\tau,\mu)\exp(-\tau/\mu_0) + \delta_{m0}Q^{(\text{thermal})}(\tau), \tag{6c}$$

$$X_0^m(\tau,\mu) = \frac{\omega(\tau)I_0}{4\pi}(2 - \delta_{m0}) \sum_{l=0}^{2N-1} (-1)^{l+m}(2l+1)$$

$$\times\, g_l^m(\tau)P_l^m(\mu)P_l^m(\mu_0), \tag{6d}$$

$$\delta_{m0} = 1 \text{ if } m = 0 \text{ (0 otherwise)}, \ g_l^m(\tau) = g_l(\tau)\frac{(l-m)!}{(l+m)!},$$

$$g_l(\tau) = \frac{1}{2}\int_{-1}^{1} P_l(\cos\theta)P(\tau,\cos\theta)d\cos\theta.$$

$P_l(\cos\theta)$ is the Legendre polynomial, $P_l^m(\mu)$ is the associated Legendre polynomial, and θ is the angle between

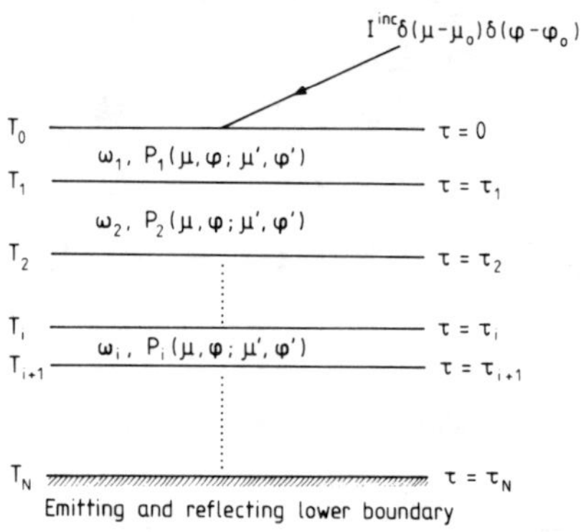

Fig. 1. Schematic illustration of a multilayered medium.

the direction vectors before and after scattering. Solutions to Eq. (6) give the azimuthal components, and then Eq. (5) gives the complete azimuthal dependence of the intensity. [When there is no beam source, the sum in Eq. (5) reduces to the $m = 0$ term, and μ_0, ϕ_0 are irrelevant.]

B. Discrete Ordinate Approximation—Matrix Formulation

The discrete ordinate approximation to (6) can be written[3,4]

$$\mu_i\frac{du^m(\tau,\mu_i)}{d\tau} = u^m(\tau,\mu_i) - \sum_{\substack{j=-N\\j\neq 0}}^{N} w_j D^m(\tau,\mu_i,\mu_j)$$

$$\times\, u^m(\tau,\mu_j) - Q^m(\tau,\mu_i) \qquad (i = \pm1,\ldots \pm N), \tag{7a}$$

where μ_i and w_i are quadrature points and weights.

Since the single-scattering albedo $\omega(\tau)$ and the phase function $P(\tau,\mu,\phi;\mu',\phi')$ are functions of τ in a vertically inhomogeneous medium, Eqs. (7) constitute (for each m) a system of $2N$ coupled differential equations with nonconstant coefficients for which analytic solutions do not exist. To obtain analytic solutions, the medium is assumed to consist of L adjacent homogeneous layers in which the single-scattering albedo and the phase function are taken to be constant within each layer (but allowed to vary from layer to layer, see Fig. 1), and the thermal source term is approximated by a polynomial in τ. For the time being, we shall consider a single homogeneous layer for which $\tau_{p-1} \leq \tau \leq \tau_p$. The τ arguments of D and X will be omitted since D and X are by assumption independent of τ in any one layer. We will also omit the p subscripts, which are implicit on all quantities, until later.

For any layer in Fig. 1, then, we write Eq. (7) in matrix form as

$$\begin{vmatrix} \dfrac{d\mathbf{u}^+}{d\tau} \\[2ex] \dfrac{d\mathbf{u}^-}{d\tau} \end{vmatrix} = \begin{vmatrix} -\alpha & -\beta \\ \beta & \alpha \end{vmatrix} \begin{vmatrix} \mathbf{u}^+ \\ \mathbf{u}^- \end{vmatrix} + \begin{vmatrix} \tilde{\mathbf{Q}}^+ \\ \tilde{\mathbf{Q}}^- \end{vmatrix}, \tag{7b}$$

where[3,16]

$$\mathbf{u}^\pm = [u^m(\tau,\pm\mu_i)] \qquad i = 1,\ldots,N,$$

$$\tilde{\mathbf{Q}}^\pm = \mathbf{M}^{-1}\mathbf{Q}^\pm,$$

$$\mathbf{Q}^\pm = Q^m(\tau, \pm\mu_i) \qquad i = 1, \ldots, N,$$

$$\mathbf{M} = \{\mu_i \delta_{ij}\} \qquad i,j = 1, \ldots, N,$$

$$\alpha = \mathbf{M}^{-1}(\mathbf{D}^+\mathbf{W} - \mathbf{I}),$$

$$\beta = \mathbf{M}^{-1}\mathbf{D}^-\mathbf{W},$$

$$\mathbf{W} = (w_i \delta_{ij}) \qquad i,j = 1, \ldots, N,$$

$$\mathbf{D}^+ = [D^m(\mu_i,\mu_j)] = [D^m(-\mu_i,-\mu_j)] \quad i,j = 1, \ldots, N,$$

$$\mathbf{D}^- = [D^m(-\mu_i,\mu_j)] = [D^m(\mu_i,-\mu_j)] \quad i,j = 1, \ldots, N.$$

C. Quadrature Rule

The use of Gaussian quadrature is essential because it makes phase function renormalization unnecessary, i.e.,

$$\sum_{\substack{j=-N \\ j\neq 0}}^{N} w_j D^0(\tau,\mu_i,\mu_j) = \sum_{\substack{i=-N \\ i\neq 0}}^{N} w_i D^0(\tau,\mu_i,\mu_j) = \omega(\tau), \qquad (7c)$$

implying that energy is conserved in the computation.[17]

The quadrature points and weights of the double-Gauss scheme adopted here satisfy $\mu_{-j} = -\mu_j$ and $w_{-j} = w_j$. Double-Gauss simply refers to a quadrature rule suggested by Sykes[18] in which the Gaussian formula is applied separately to the half-ranges $-1 < \mu < 0$ and $0 < \mu < 1$. The main advantage of this double-Gauss scheme is that the quadrature points (in even orders) are distributed symmetrically around $|\mu| = 0.5$ and clustered both toward $|\mu| = 1$ and $\mu = 0$, whereas in the Gaussian scheme for the complete range, $-1 < \mu < 1$, they are clustered toward $\mu = \pm 1$. The clustering toward $\mu = 0$ will give superior results near the boundaries where the intensity varies rapidly around $\mu = 0$. A half-range scheme is also preferred since the intensity is discontinuous at the boundaries. Another advantage is that upward and downward fluxes and average intensities are obtained immediately without further approximations.

D. Basic Solution

Equation (7b) is a system of $2N$-coupled ordinary differential equations with constant coefficients. These coupled equations are linear, and our goal is to uncouple them by using well-known methods of linear algebra.

Seeking solutions to the homogeneous version ($\tilde{\mathbf{Q}} = \mathbf{O}$) of Eq. (7b) of the form

$$\mathbf{u}^\pm = \mathbf{G}^\pm \exp(-k\tau), \qquad (8a)$$

we find

$$\begin{vmatrix} \alpha & \beta \\ -\beta & -\alpha \end{vmatrix} \begin{vmatrix} \mathbf{G}^+ \\ \mathbf{G}^- \end{vmatrix} = k \begin{vmatrix} \mathbf{G}^+ \\ \mathbf{G}^- \end{vmatrix}, \qquad (8b)$$

which is a standard algebraic eigenvalue problem of order $2N \times 2N$ determining the eigenvalues k and the eigenvectors $\mathbf{G}^\pm$.

Because of the special structure of the matrix in Eq. (8b), the eigenvalues occur in pairs ($\pm k$), and the order of the algebraic eigenvalue problem [Eq. (8b)] may be reduced by a factor of 2 as follows.[3] Breaking Eq. (8b) into

$$\alpha \mathbf{G}^+ + \beta\,\mathbf{G}^- = k\mathbf{G}^+,$$

$$\beta \mathbf{G}^+ + \alpha \mathbf{G}^- = -k\mathbf{G}^-,$$

and then adding and subtracing these two equations, we find

$$(\alpha - \beta)(\mathbf{G}^+ - \mathbf{G}^-) = k(\mathbf{G}^+ + \mathbf{G}^-), \qquad (8c)$$

$$(\alpha + \beta)(\mathbf{G}^+ + \mathbf{G}^-) = k(\mathbf{G}^+ - \mathbf{G}^-). \qquad (8d)$$

Combining Eqs. (8c) and (8d), we obtain

$$(\alpha - \beta)(\alpha + \beta)(\mathbf{G}^+ + \mathbf{G}^-) = k^2(\mathbf{G}^+ + \mathbf{G}^-), \qquad (8e).$$

which completes the reduction of the order. Following Stamnes and Swanson[3] we solve Eq. (8e) to obtain eigenvalues and eigenvectors $(\mathbf{G}^+ + \mathbf{G}^-)$. We then use Eq. (8d) to determine $(\mathbf{G}^+ - \mathbf{G}^-)$.

For beam sources $Q^{(\text{beam})}(\tau,\mu) = X_0(\mu)\exp(-\tau/\mu_0)$ [Eq. (6d)], it is easily verified that a particular solution of Eq. (7) is (omitting m superscript)

$$u(\tau,\mu_i) = Z_0(\mu_i)\exp(-\tau/\mu_0). \qquad (9a)$$

For thermal sources the emitted radiation is isotropic, so that $Q^m = 0$, $m > 0$, $Q(\tau) \equiv Q^0(\tau) = (1 - \omega)B(\tau)$. By approximating the Planck function for each layer by a polynomial in τ,[3,19] i.e.,

$$B[T(\tau)] = \sum_{l=0}^{K} b_l \tau^l, \qquad (9b)$$

we find that the particular solution may also be expressed as a polynomial in τ:

$$u(\tau,\mu_i) = \sum_{l=0}^{K} Y_l(\mu_i)\tau^l. \qquad (9c)$$

The coefficients $Z_0(\mu_i)$ and $Y_l(\mu_i)$ are readily determined by solving systems of linear algebraic equations (see Refs. 3 and 9 for details).

The general solution to Eq. (7a) consists of a linear combination, with coefficients C_j, of all the homogeneous solutions plus the particular solutions for beam and thermal emission sources (omitting the m superscript):

$$u(\tau,\mu_i) = \sum_{j=-N}^{N} C_j G_j(\mu_i)\exp(-k_j\tau)$$

$$+ \delta_{m0} \sum_{l=0}^{K} Y_l(\mu_i)\tau^l. \qquad (10a)$$

The k_j and the $G_j(\mu_i)$ are the eigenvalues and eigenvectors obtained as described in Sec. III.A, the μ_i the quadrature angles, and the C_j the constants of integration. We have for convenience included the beam solution in the sum by defining for $j = 0$

$$C_0 G_0(\mu_i) \equiv Z_0(\mu_i), \qquad (10b)$$

$$k_0 \equiv 1/\mu_0. \qquad (10c)$$

E. Boundary Conditions

Below we shall limit ourselves to a linear-in-optical-depth dependence of the Planck function within each layer, since this will give sufficient accuracy for most cases of practical interest. We shall discuss the inclusion of boundary conditions in some detail, since a proper formulation of this aspect of the problem (in terms of the bidirectional reflectivity) is rarely discussed in the open literature.

In general, Eq. (1) must be solved subject to the boundary conditions

$$u(\tau = 0, -\mu, \phi) = u_\infty(\mu, \phi), \tag{11a}$$

$$u(\tau = \tau_L, +\mu, \phi) = u_g(\mu, \phi), \tag{11b}$$

where u_∞ and u_g are the intensities incident at the top and bottom boundaries, respectively, and τ_L is the total optical depth.

Quite general boundary conditions can be accommodated in the discrete ordinate method. To be specific, we shall assume that the medium is illuminated from above by known diffuse radiation (incident parallel beams are treated as pseudosources) and that the bottom boundary has a known bidirectional reflectivity. Equation (11b) may now be rewritten as

$$u(\tau_L, +\mu, \phi) = \epsilon(\mu) B(T_g)$$

$$+ \frac{1}{\pi} \int_0^{2\pi} d\phi' \int_0^1 \rho_d(\mu, \phi; -\mu', \phi') u(\tau_L, -\mu', \phi') \mu' d\mu'$$

$$+ \frac{\mu_0}{\pi} I_0 \exp(-\tau_L/\mu_0) \rho_d(\mu, \phi; -\mu_0, \phi_0), \tag{12}$$

where $\epsilon(\mu)$ is the directional emissivity and T_g is the temperature of the bottom boundary; $\rho_d(\mu, \phi; -\mu', \phi')$ is the bidirectional reflectivity, and I_0 is the incident beam intensity at the upper boundary. Kirchhoff's law

$$\epsilon(\mu) + \frac{1}{\pi} \int_0^{2\pi} d\phi' \int_0^1 \rho_d(\mu, \phi; -\mu', \phi') \mu' d\mu' = 1$$

is applied to calculate the emissivity from the reflectivity. Note that some authors include the $(1/\pi)$ factor in ρ_d; we find our definition more useful, however, since when ρ_d is independent of angle it reduces to an albedo.

Below we will assume that the bidirectional reflectivity is a function only of the angle θ between the incident and the reflected radiation (i.e., there are no preferred directions at the lower boundary). Then it can be expanded in a series of $2N$ Legendre polynomials:

$$\rho_d(\mu, \phi; -\mu', \phi') = \sum_{l=0}^{2N-1} (2l+1) h_l P_l(\cos\theta), \tag{13a}$$

where

$$\cos\theta = -\mu\mu' + (1-\mu^2)^{1/2} (1-\mu'^2)^{1/2} \cos(\phi' - \phi). \tag{13b}$$

Using the addition theorem for spherical harmonics, we find

$$\rho_d(\mu, \phi; -\mu', \phi') = \sum_{m=0}^{2N-1} \rho_d^m(\mu, -\mu') \cos m(\phi' - \phi), \tag{13c}$$

$$\rho_d^m(\mu, -\mu') = (2 - \delta_{0m}) \sum_{l=m}^{2N-1} (2l+1) h_l^m P_l^m(-\mu') P_l^m(\mu), \tag{14}$$

where

$$h_l^m = h_l \frac{(l-m)!}{(l+m)!},$$

$$h_l = \frac{1}{2} \int_{-1}^1 P_l(\cos\theta) \rho_d(\cos\theta) d\cos\theta. \tag{15}$$

Substituting Eq. (13c) into Eq. (12) and using Eq. (5), we find that each Fourier component must satisfy the bottom boundary condition

$$u^m(\tau_L, +\mu) = u_g^m(\mu), \tag{16a}$$

where

$$u_g^m(u) = \delta_{0m} \epsilon(\mu) B(T_g)$$

$$+ (1 + \delta_{0m}) \int_0^1 \mu' \rho_d^m(\mu, -\mu') u^m(\tau_L, -\mu') d\mu'$$

$$+ \frac{\mu_0}{\pi} I_0 \exp(-\tau_L/\mu_0) \rho_d^m(\mu, -\mu_0). \tag{16b}$$

In a multilayered medium we must also require the intensity to be continuous across layer interfaces.[7] Thus Eq. (7) must satisfy boundary and continuity conditions as follows:

$$u_1^m(0, -\mu_i) = u_\infty^m(-\mu_i), \qquad i = 1, \ldots, N, \tag{17a}$$

$$u_p^m(\tau_p, \mu_i) = u_{p+1}^m(\tau_p, \mu_i), \qquad i = \pm 1, \ldots, \pm N;$$

$$p = 1, \ldots, L - 1; \tag{17b}$$

$$u_L^m(\tau_L, +\mu_i) = u_g^m(\mu_i), \qquad i = 1, \ldots, N, \tag{17c}$$

where

$$u_g^m(\mu_i) = \delta_{0m} \epsilon(\mu_i) B(T_g)$$

$$+ (1 + \delta_{0m}) \sum_{j=1}^N w_j \mu_j \rho_d^m(\mu_i, -\mu_j) u^m(\tau_L, -\mu_j)$$

$$+ \frac{\mu_0}{\pi} I_0 \exp(-\tau_L/\mu_0) \rho_d^m(\mu_i, -\mu_0). \tag{17d}$$

Evaluation of the integral in Eq. (16b) by numerical quadrature as in Eq. (17d) requires a quadrature rule which integrates separately over the downward and upward directions. As noted previously, the double-Gauss rule adopted here satisfies this requirement.

When discussing boundary conditions, it is convenient to write the discrete ordinate solution in the following form ($k_{jp} > 0$ and $k_{-jp} = -k_{jp}$):

$$u_p(\tau, \mu_i) = \sum_{j=1}^N [C_{jp} G_{jp}(\mu_i) \exp(-k_{jp}\tau)$$

$$+ C_{-jp} G_{-jp}(\mu_i) \exp(+k_{jp}\tau)] + R_p(\tau, \mu_i), \tag{18}$$

where the sum contains the homogeneous solution involving the unknown coefficients (the C_{jp}) to be determined, and $R_p(\tau, \mu_i)$ is the particular solution given by [Eq. (10a)]

$$R_p(\tau, \mu_i) = Z_0(\mu_i) \exp(-\tau/\mu_0) + \delta_{m0}[Y_0(\mu_i) + Y_1\tau]. \tag{19}$$

Insertion of Eq. (18) into Eqs. (17a)–(17c) yields (omitting the m superscript)

$$\sum_{j=1}^{N} [C_{j1}G_{j1}(-\mu_i) + C_{-j1}G_{-j1}(-\mu_i)]$$

$$= u_\infty(-\mu_i) - R_p(0,-\mu_i), \qquad i = 1,\ldots,N \qquad (20a)$$

$$\sum_{j=1}^{N} \{C_{jp}G_{jp}(\mu_i) \exp(-k_{jp}\tau_p)$$

$$+ C_{-jp}G_{-jp}(\mu_i) \exp(k_{jp}\tau_p)$$
$$- [C_{j,p+1}G_{j,p+1}(\mu_i) \exp(-k_{j,p+1}\tau_p)$$
$$+ C_{-j,p+1}G_{-j,p+1}(\mu_i) \exp(k_{j,p+1}\tau_p)]\}$$
$$= [R_{p+1}(\tau_p,\mu_i) - R_p(\tau_p,\mu_i)], \qquad i = \pm1,\ldots \pm N; p = 1,\ldots,L-1$$
$$(20b)$$

$$\sum_{j=1}^{N} [C_{jL}r_{jL}(\mu_i) \exp(-k_{jL}\tau_L)$$

$$+ C_{-jL}r_{-jL}(\mu_i) \exp(k_{jL}\tau_L)]$$

$$= \gamma(\tau_L,\mu_i), \qquad i = 1,\ldots,N, \qquad (20c)$$

where

$$\gamma(\tau_L,\mu_i) = \delta_{0m}\epsilon(\mu_i)B(T_g) - R_L(\tau_L,+\mu_i)$$

$$+ \frac{\mu_0}{\pi} I_0 \exp(-\tau_L/\mu_0)\, \rho_d(\mu_i,-\mu_0)$$

$$+ (1 + \delta_{0m}) \sum_{j=1}^{N} \rho_d(\mu_i,-\mu_j)w_j\mu_j R_L(\tau_L,-\mu_j), \qquad (21a)$$

$$r_{jL}(\mu_i) = G_{jL}(\mu_i) - (1 + \delta_{0m})$$

$$\times \sum_{n=1}^{N} \rho_d(\mu_i,-\mu_n)w_n\mu_n G_{jL}(-\mu_n). \qquad (21b)$$

Equations (20a)–(20c) constitute a $(2N \times L) \times (2N \times L)$ system of linear algebraic equations from which the $2N \times L$ unknown coefficients, the C_{jp} ($j = \pm1,\ldots,\pm N$; $p = 1,\ldots,L$) are determined. To solve this system of equations, we take advantage of the fact that the coefficient matrix is a $(6N - 1)$ diagonal band matrix.[7]

III. Numerical Implementation

A. Computation of Eigenvalues and Eigenvectors

Stamnes and Swanson[3] obtained the eigenvalues and eigenvectors of Eq. (8e) using subroutines in EISPACK[20] for a general real asymmetric matrix. Since such matrices may have complex eigenvalues/eigenvectors, the EISPACK routines use complex arithmetic. Yet invariably the eigenvalues returned were purely real as they should be.

As discussed by Stamnes et al.,[8] the advantage of solving the eigenvalue problem involving the asymmetric matrix $(\alpha - \beta)(\alpha + \beta)$ in Eq. (8e) is that only one matrix multiplication is necessary. Nakajima and Tanaka[16] showed that the eigenvalue problem can be reduced to finding eigenvalues and eigenvectors of the product of two symmetric matrices, one of which is positive definite. Stamnes et al.[8] used the Cholesky decomposition of this positive definite matrix to con-

vert the eigenvalue problem into one involving only one symmetric matrix. But the transformation to symmetric matrices (whose eigensolutions can be found faster and by using real arithmetic) and subsequent solution procedures[8,16] introduce matrix operations in which the effect of rounding error can seriously degrade the results. Stamnes et al.[8] compared the original Stamnes/Swanson method with the new procedure based on the Cholesky decomposition and with Nakajima and Tanaka's symmetrizing procedure. They summarize the outcome of these comparisons as follows:

(1) In single precision the Stamnes/Swanson procedure is more accurate than the Cholesky procedure, which in turn is slightly more accurate than the Nakajima/Tanaka procedure.

(2) The Cholesky procedure is slightly faster than Nakajima/Tanaka's, which is slightly faster than Stamnes/Swanson's.

Based on the findings, they decided to speed up the Stamnes/Swanson procedure since it is the most accurate. By eliminating complex arithmetic in the EISPACK subroutines, the speed became comparable with that of the Cholesky approach. We have adopted this faster algorithm in our code.

B. Scaling Transformation

As discussed by Stamnes and Conklin,[7] to avoid numerical ill-conditioning, it is necessary to remove the positive exponentials in Eq. (20) (remember $k_{jp} > 0$ by convention). This is achieved by the scaling transformation[7]

$$C_{+jp} = \tilde{C}_{+jp} \exp(k_{jp}\tau_{p-1}), \qquad (22a)$$

$$C_{-jp} = \tilde{C}_{-jp} \exp(-k_{jp}\tau_p). \qquad (22b)$$

Inserting Eqs. (22) in Eqs. (20) and solving for the $\tilde{C}_{jp}$ instead of the C_{jp}, we find that all the exponential terms in the coefficient matrix have the form $\exp[-k_{jp}(\tau_p - \tau_{p-1})]$; i.e., all the exponentials in the coefficient matrix have negative arguments ($k_{jp} > 0$, $\tau_p > \tau_{p-1}$). Consequently, numerical ill-conditioning is avoided implying that our numerical scheme for finding the $\tilde{C}_{jp}$ is unconditionally stable for arbitrary total optical depths and arbitrary individual layer thicknesses. [To solve Eqs. (20) for the $\tilde{C}_{jp}$ we use the LINPACK[21] routine SGBSL designed to solve a banded system of linear algebraic equations.]

C. Scaled Solutions—Angular Distributions

Since only the homogeneous solution is affected, we start by introducing the scaling transformation (22) into Eq. (18), i.e.,

$$u_p(\tau,\mu_i) = \sum_{j=1}^{N} \{\tilde{C}_{jp}G_{jp}(\mu_i) \exp[-k_{jp}(\tau - \tau_{p-1})]$$

$$+ \tilde{C}_{-jp}G_{-jp}(\mu_i) \exp[-k_{jp}(\tau_p - \tau)]\}. \qquad (23)$$

Since $k_{jp} > 0$ and $\tau_{p-1} \leq \tau \leq \tau_p$, all exponentials in Eq. (23) have negative arguments as they should to avoid fatal overflows.

For a slab of thickness τ_L, we may solve Eq. (1) formally to obtain ($\mu > 0$)

$$u(\tau,+\mu) = u(\tau_L,+\mu) \exp[-(\tau_L - \tau)\mu]$$

$$+ \int_\tau^{\tau_L} S(t,+\mu) \exp[-(t - \tau)/\mu] \, \frac{dt}{\mu}, \quad (24a)$$

$$u(\tau,-\mu) = u(0,-\mu) \exp(-\tau/\mu)$$

$$+ \int_0^\tau S(t,-\mu) \exp[-(\tau - t)/\mu] \, \frac{dt}{\mu}, \quad (24b)$$

where we have again omitted the m superscript. These two equations show that if we know the source function $S(t,\pm\mu)$, we can find the intensity at arbitrary angles by integrating the source function. The discrete ordinate solutions are used to derive explicit expressions for the source function, which can be integrated analytically.[3-5,9] This procedure is sometimes referred to as the iteration of the source-function technique, although as pointed out by Kourganoff[22] and discussed in some detail by Stamnes,[5] this procedure essentially amounts to an interpolation.

In a multilayered medium we may now integrate Eq. (24a) layer by layer[5] to obtain the following expression for the upward intensity at arbitrary τ and μ:

$$u_p(\tau,+\mu) = u(\tau_L,+\mu) \exp[-(\tau_L - \tau)/\mu]$$

$$+ \sum_{n=p}^{L} \left\{ \sum_{j=1}^{N} \left[\tilde{C}_{-jn} \frac{G_{-jn}(+\mu)}{1 - k_{jn}\mu} E_{-jn}(\tau,+\mu) \right. \right.$$

$$\left. + \sum \tilde{C}_{+jn} \frac{G_{+jn}(+\mu)}{1 + k_{jn}\mu} E_{+jn}(\tau,+\mu) \right]$$

$$+ \frac{G_{0n}}{1 + \mu/\mu_0} E_{0n}(\tau,+\mu)$$

$$+ \delta_{m0} \left[V_{0n}(\mu)F_{0n}(\tau,\mu) + b_{1n}F_{1n}(\tau,\mu) \right] \right\}, \quad (25a)$$

where the scaling given by Eqs. (22) has been incorporated and

$$E_{-jn}(\tau,+\mu) = \exp[-(k_{jn}\Delta\tau_n - \delta\tau/\mu)] - \exp[-(\tau_n - \tau)/\mu], \quad (25b)$$

$$F_{0n}(\tau,+\mu) = \exp(-\delta\tau/\mu) - \exp[-(\tau_n-\tau)/\mu], \quad (25c)$$

$$F_{1n}(\tau,+\mu) = (\tau_{n-1}+\mu) \exp(-\delta\tau/\mu)$$

$$- (\tau_n + \mu) \exp[-(\tau_n - \tau)/\mu], \quad (25d)$$

with

$$\delta\tau = \tau_{n-1} - \tau \text{ for } n > p \text{ and } \delta\tau = 0 \text{ for } n = p;$$

$$\Delta\tau_n = \tau_n - \tau_{n-1} \text{ for } n > p \text{ and } \Delta\tau_p = \tau_p - \tau \text{ for } n = p,$$

$$E_{+jn}(\tau,+\mu) = \exp[-(\tau_{n-1} - \tau)/\mu]$$

$$- \exp\{-[k_{jn}(\tau_n - \tau_{n-1}) - (\tau - \tau_n)/\mu]\}, \quad (25e)$$

for $n > p$ and

$$E_{+jp}(\tau,+\mu) = \exp[-k_{jp}(\tau - \tau_{p-1})]$$

$$- \exp\{-[k_{jp}(\tau_p - \tau_{p-1}) + (\tau_p - \tau)/\mu]\}. \quad (25f)$$

Similarly, we find that the downward intensity becomes

$$u_p(\tau,-\mu) = u(0,-\mu) \exp(-\tau/\mu)$$

$$+ \sum_{n=1}^{p} \left\{ \sum_{j=1}^{N} \left[\tilde{C}_{-jn} \frac{G_{-jn}(-\mu)}{1 + k_{jn}\mu} E_{-jn}(\tau,-\mu) \right. \right.$$

$$\left. + \tilde{C}_{+jn} \frac{G_{+jn}(-\mu)}{1 - k_{jn}\mu} E_{+jn}(\tau,-\mu) \right]$$

$$+ \frac{G_{0n}(-\mu)}{1 - \mu/\mu_0} E_{0n}(\tau,-\mu)$$

$$+ \delta_{m0}[V_{0n}(\mu)F_{0n}(\tau,-\mu) + b_{1n}F_{1n}(\tau,-\mu)] \right\}, \quad (26a)$$

where

$$E_{+jn}(\tau,-\mu) = \exp[-(k_{jn}\Delta\tau_n + \delta\tau/\mu)] - \exp[-(\tau - \tau_{n-1})/\mu]$$

$$(26b)$$

$$F_{0n}(\tau,-\mu) = \exp(-\delta\tau/\mu) - \exp[-(\tau - \tau_{n-1})/\mu] \quad (26c)$$

$$F_{1n}(\tau,-\mu) = (\tau_n - \mu) \exp(-\delta\tau/\mu)$$

$$- (\tau_{n-1} - \mu) \exp[-(\tau - \tau_{n-1})/\mu], \quad (26d)$$

with

$$\delta\tau = \tau - \tau_n \text{ for } n < p \text{ and } \delta\tau = 0 \text{ for } n = p,$$

$$\Delta\tau_n = \tau_n - \tau_{n-1} \text{ for } n < p$$

$$\text{and } \Delta\tau_p = \tau - \tau_{p-1} \text{ for } n = p,$$

$$E_{-jn}(\tau,-\mu) = \exp[-(\tau - \tau_n)/\mu]$$

$$- \exp\{-[k_{jn}(\tau_n - \tau_{n-1}) + (\tau - \tau_{n-1})/\mu]\}, \quad (26e)$$

for $n < p$ and

$$E_{-jp}(\tau,-\mu) = \exp[-k_{jp}(\tau_p - \tau)]$$

$$- \exp\{-[k_{jp}(\tau_p - \tau_{p-1}) + (\tau - \tau_{p-1})/\mu]\}. \quad (26f)$$

We see that all exponentials in Eqs. (25b)–(25f) and (26b)–(26f) have negative arguments as they should. This ensures that fatal overflow errors are avoided in the computations.

The G_{jn} terms and $\bar{V}_{0n}(\mu)$ in Eqs. (25a) and (26a) are given by the following explicit expressions (omitting the n subscript):

$$G_j(\mu) = \sum_{\substack{i=-N \\ i \neq 0}}^{N} w_i D(\mu,\mu_i) G_j(\mu_i) \text{ for } j \neq 0; \quad (27a)$$

$$G_0(\mu) \equiv Z_0(\mu) = \sum_{\substack{i=-N \\ i \neq 0}}^{N} w_i D(\mu,\mu_i) Z_0(\mu_i) + X_0(\mu); \quad (27b)$$

$$V_0(\mu) = \sum_{\substack{i=-N \\ i \neq 0}}^{N} w_i D^0(\mu,\mu_i) Y_0(\mu_i) + (1 - \omega)b_0; \quad (27c)$$

where b_0 is defined in Eq. (9b). Formulas (27a)–(27c) clearly reveal the interpolatory nature of Eqs. (25) and (26). Although we have generally omitted the m superscript above, we have explicitly written D^0 in Eq. (27c) to indicate that thermal radiation contributes only to the azimuth-independent part of the intensity.

D. Computational Shortcut for Highly Absorbing Media

For highly absorbing media without thermal emission, very little radiation will reach levels for which the absorbing optical depth is greater than, say, 10. In such cases we ignore the medium below this depth and set the surface reflection to zero. This may result in enormous computation savings. For example, in an atmosphere with clouds it prevents unnecessary scattering computations at wavelengths where sunlight never reaches the clouds.

E. Simplified Albedo and Transmissivity Computations

By appealing to the reciprocity principle we may derive simple expressions for the plane albedo $a(\mu)$ and the transmissivity $t(\mu)$ of a vertically inhomogeneous plane-parallel medium lacking thermal sources. Thus

(1) the plane albedo for a given angle of parallel incidence equals the azimuthally averaged reflected intensity $u^0(0,\mu)$ for isotropic illumination of unit intensity incident at the top boundary;

(2) the transmissivity equals the azimuthally averaged transmitted intensity $u^{0*}(0,\mu)$ for isotropic illumination of unit intensity incident at the bottom boundary.

Consequently, for an inhomogeneous medium the following duality relations apply[6]:

$$a(\mu) = u^0(0,\mu); \quad a^*(\mu) = u^{0*}(\tau_L,-\mu); \tag{28}$$

$$t(\mu) = u^{0*}(0,\mu); \quad t^*(\mu) = u^0(\tau_L,-\mu); \tag{29}$$

where τ_L is to the optical thickness of the inhomogeneous medium, and the asterisk refers to illumination from below.

Equations (28) and (29) offer substantial computational advantages when only albedo and transmissivity are required. The conventional procedure requires computation of the angular distribution of the diffuse intensity, which must then be integrated over angle to obtain the desired result. This necessitates the computation of a particular solution, which can be quite costly in an inhomogeneous (multilayered) medium, since such a solution is required for each layer and every direction of incidence considered. In contrast, by applying an isotropic boundary condition and using Eqs. (28) and (29), one may solve for all desired angles of parallel beam incidence simultaneously and avoid entirely the computation of a particular solution. (Details about the numerical implementation of this procedure are provided in Refs. 6 and 9.)

IV. Summary

We have described an advanced discrete ordinate algorithm for radiative transfer calculations, including multiple scattering, in vertically inhomogeneous nonisothermal plane-parallel media. This algorithm is the product of over seven years of development by ourselves and our collaborators. It has been designed to be as versatile and general as possible and should find applications from the UV through to the radar region of the electromagnetic spectrum.

The algorithm has the following important features:

(1) It is unconditionally stable for an arbitrarily large number of quadrature angles (streams) and arbitrarily large optical depths.

(2) It can be forced by any combination of parallel beam or diffuse incidence and thermal sources in the medium and at the boundaries.

(3) It allows for an arbitrary bidirectional reflectivity at the lower boundary; directional emissivity is computed from this reflectivity.

(4) It offers rapid computation of bulk albedo and transmissivity when thermal sources are absent.

(5) Unlike the popular doubling method, computing time for individual layers is independent of optical thickness.

(6) Because the solution is analytic, and by using the iteration-of-the-source-function method, intensities can be returned at arbitrary angles and optical depths, unrelated to the computational meshes for these quantities.

(7) It has been thoroughly tested against a wide variety of published solutions.

(8) It is thoroughly documented both in Ref. 9 and in the code itself (with extensive references to published equation numbers).

Copies of the FORTRAN program are available from the third author on microcomputer diskettes (IBM or MacIntosh).

This work was supported by the National Science Foundation through grant DPP 84-06093 to the University of Alaska. One of us (K.S.) would like to acknowledge partial support of this work from the Royal Norwegian Council on Science and Technology.

References

1. S. Chandrasekhar, *Radiative Transfer* (Dover, New York, 1960).
2. K.-N. Liou, "A Numerical Experiment on Chandrasekhar's Discrete-Ordinate Method for Radiative Transfer: Applications to Cloudy and Hazy Atmosphere," J. Atmos. Sci. **30**, 1303 (1973).
3. K. Stamnes and R. A. Swanson, "A New Look at the Discrete Ordinate Method for Radiative Transfer Calculations in Anisotropically Scattering Atmospheres," J. Atmos. Sci. **38**, 387 (1981).
4. K. Stamnes and H. Dale, "A New Look at the Discrete Ordinate Method for Radiative Transfer Calculation in Anisotropically Scattering Atmospheres, II: Intensity Computations," J. Atmos. Sci. **38**, 2696 (1981).
5. K. Stamnes, "On the Computation of Angular Distributions of Radiation in Planetary Atmospheres," J. Quant. Spectrosc. Radiat. Transfer **28**, 47 (1982a).
6. K. Stamnes, "Reflection and Transmission by a Vertically Inhomogeneous Planetary Atmosphere," Planet. Space Sci. **30**, 727 (1982b).
7. K. Stamnes and P. Conklin, "A New Multi-Layer Discrete Ordinate Approach to Radiative Transfer in Vertically Inhomogeneous Atmospheres," J. Quant. Spectrosc. Radiat. Transfer **31**, 273 (1984).
8. K. Stamnes, S.-C. Tsay, and T. Nakajima, "Computation of Eigenvalues and Eigenvectors for Discrete Ordinate and Matrix Operator Method Radiative Transfer," J. Quant. Spectrosc. Radiat. Transfer in press (1988).

9. K. Stamnes, S.-C. Tsay, W. J. Wiscombe, and K. Jayaweera, "An Improved, Numerically Stable Computer Code for Discrete-Ordinate-Method Radiative Transfer in Scattering and Emitting Layered Meda," NASA Report, in press (1988).

10. K. N. Liou, *An Introduction to Atmospheric Radiation* (Academic, Orlando, FL, 1980).

11. J. E. Hansen and L. D. Travis, "Light Scattering in Planetary Atmospheres," Space Sci. Rev. **16,** 527 (1974).

12. W. M. Irvine, "Multiple Scattering in Planetary Atmospheres," Icarus **25,** 175 (1975).

13. J. Lenoble, Ed., *Radiative Transfer in Scattering and Absorbing Atmospheres: Standard Computational Procedures* (A. Deepak, Hampton, VA, 1985).

14. W. J. Wiscombe, "Atmospheric Radiation: 1975–1983," Rev. Geophys. **21,** 957 (1983).

15. K. Stamnes, "The Theory of Multiple Scattering of Radiation in Plane Parallel Atmospheres," Rev. Geophys. **24,** 299 (1986).

16. T Nakajima and M. Tanaka, "Matrix Formulations for the Transfer of Solar Radiation in a Plane-Parallel Atmosphere," J. Quant. Spectrosc. Radiat. Transfer **35,** 13 (1986).

17. W. J. Wiscombe, "The Delta-M Method: Rapid Yet Accurate Radiative Flux Calculations for Strongly Asymmetric Phase Functions," J. Atmos. Sci. **34,** 1408 (1977).

18. J. B. Sykes, "Approximate Integration of the Equation of Transfer," Mon. Not. Roy. Astron. Soc. **11,** 377 (1951).

19. W. J. Wiscombe, "Extension of the Doubling Method to Inhomogeneous Sources," J. Quant. Spectrosc. Radiat. Transfer **16,** 477 (1976).

20. W. R. Cowell, Ed., *Sources and Developments of Mathematical Software* (Prentice Hall, Englewood Cliffs, NJ, 1980).

21. J. J. Dongarra, C. B. Moler, J. R. Bunch, and G. W. Stewart, LINPACK *User's Guide* (SIAM, Philadelphia, 1979).

22. V. Kourganoff, *Basic Methods in Transfer Problems* (Dover, New York, 1963).

Section Four
Clouds

Noctilucent Clouds

By F. H. LUDLAM, Imperial College, London

(Manuscript received December 12, 1956, revised March 11, 1957)

Abstract

Knowledge of noctilucent clouds is reviewed, and fresh observations made in Sweden are
described. The conditions under which the clouds may be seen are calculated and compared
with observations; it appears that the clouds are formed only in the summer in high latitudes.
The sizes and concentrations of the cloud particles are inferred, and the probsem of their composition and origin is discussed. The clouds are not likely to be formed by condensation of
water vapour, but are more probably dust clouds derived from great eruptions, very large
meteorites, and from interplanetary space. Evidence of dust in the stratosphere is assembled,
and the processes by which it appears in visible clouds are discussed.

1. Introduction; the properties of noctilucent clouds

In high latitudes during summer months
tenuous high clouds occasionally become visible about an hour after sunset. Only clouds
high in the stratosphere can remain in sunshine
so long after the sun has set on the ground.

Such clouds, called 'luminous night-clouds'
or *noctilucent clouds*, were first studied systematically by Jesse, in Germany, and by Tseraskii, in Russia, in 1885 and later years. It is
probable that similar clouds had been observed
previously, but the published accounts are not
sufficiently definite for this to be certain (see
ARAGO, 1854, and SCULTETUS, 1949, who discusses an observation by Lavoisier). There are
numerous reports of bright noctilucent clouds
in 1885, 1886 and 1887, but in succeeding years
there were fewer observations and the displays
became weaker, and since 1894 the clouds have
been observed only intermittently. A valuable
review of published reports was made by
VESTINE (1934), and more recently STØRMER
(1933, 1935) and PATON (1954) have on several
occasions observed and measured the height of
the clouds.

The problem of the nature of the clouds is
still unsolved. In this paper some new observations are described, previous reports are
summarized, and theories of the composition
and formation of the clouds are discussed,
with the aim of stimulating more widespread
and thorough observations during the approaching International Geophysical Year.

a) *The form and colour of noctilucent clouds*

In general appearance noctilucent clouds resemble a thin, decayed cirrostratus, without
any of the fallstreak forms characteristic of
many ice clouds (Figs. 1—3). They tend to occur
in extensive sheets, although the part of the
sky in which they can be seen is often restricted to a narrow segment above the horizon. The most careful classification of the
forms of noctilucent clouds is that of GRISHIN
(1955); he distinguishes the following:

1. The *nebula*, or flare. Between the more
prominent details of other forms there is frequently a nebulous, very tenuous and struc-
tureless veil, shining faintly with a delicately
white or bluish light. According to Grishin the
flare often can be detected by an experienced
observer as a slight luminescence of the twilight
sky, and may appear a half to one hour before
clouds with a definite structure.

2. *Bands.* Diffuse-edged, long streaks, known
as bands, are more or less straight or gently

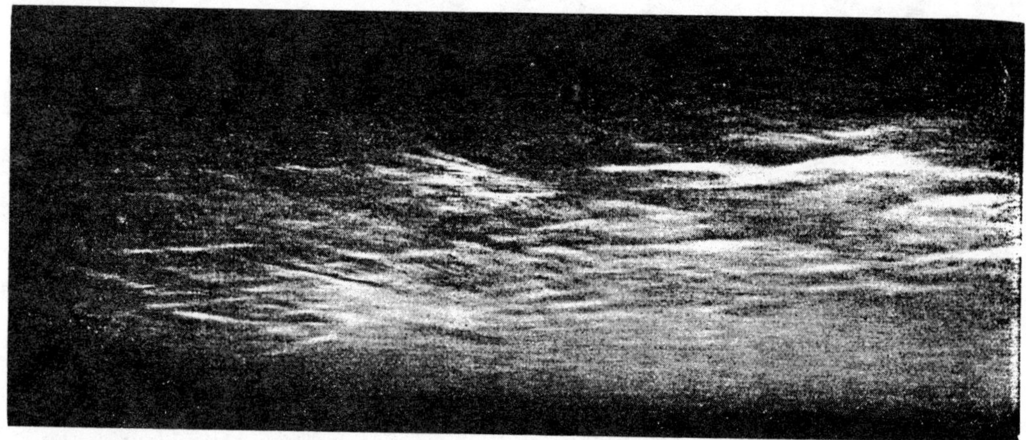

Fig. 1. Noctilucent clouds in the northern sky, Torsta, 13 August 1955, 0025 hours. A dark "shadow" can be seen above a bright detail in the upper right of the picture.

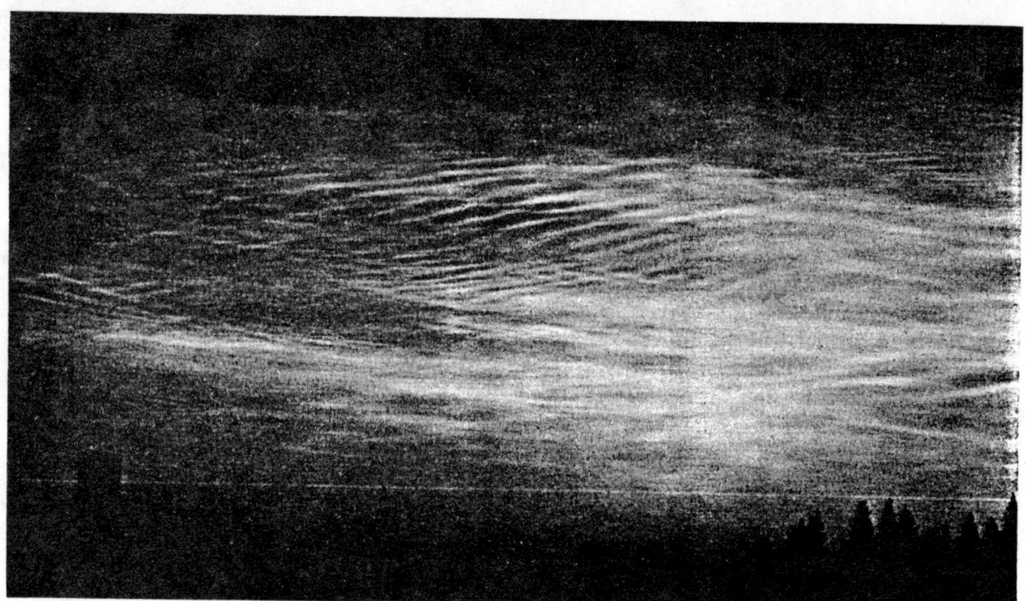

Fig. 2. Noctilucent cloud billows, Torsta, 13 August 1955, 0155 hours (photograph by F. Singleton).

curved, and sometimes hundreds of km long. They occur in extensive groups, the individuals of which are roughly parallel. They are persistent, with inappreciable change of detail. Smaller *streaks*, with twists or bends, may lie across the bands or branch out from them.

3. *Billows.* When the noctilucent clouds extend above low elevations, groups of rather sharply-defined billows may be seen. The distance separating pairs of billows is about 10 km. The billows tend to lie across the direction of the bands, but their alignment may differ noticeably in neighbouring parts of the sky. When the billows lie across a band part of the material of the band becomes gathered into the billow crests; at other times billows form within the flare. In contrast to the bands, the small streaks and billows may change their form and arrangement, or appear and disappear, within several minutes.

Fig. 3. Noctilucent clouds in NNE, 9 August 1955, 2340 hours (a), and 10 August 1955, 0010 hours (b).

The colour of the clouds is usually white; at high elevations they are often silvery, having a slight bluish tinge, while near the horizon they become yellowish, or even orange. Sometimes more closely-spaced *serrations* are seen, which to the unaided eye present an almost continuous cloud. Grishin also distinguishes individual wave-like curves or distortions in parts of noctilucent clouds, and

4. *Eddies,* usually of slight curvature, which may be seen in bands, serrations, and sometimes in flares. The eddies extend over very variable arcs, even to complete rings or vortices with dark centres.

Grishin's classification was developed during observations made in the years 1948 to 1953; in the last year a time-lapse film of noctilucent clouds was made, which clearly showed the wave-like form and evolution of many of the structural details. No satisfactory impression of the nature of the internal motion and transformations shown by these clouds can be gained except by the help of such films.

According to Helmholtz, Størmer, and Astapovich (see KHVOSTIKOV, 1952), their spectrum is similar to that of daylight; the Fraunhofer lines are clearly present, and there are no emission lines. The light from the clouds is evidently simply scattered sunlight, but with some excess of blue compared with red. PATON (1954) once observed a slow change of colour in portions of a noctilucent cloud from a vivid blue to white.

Occasionally the clouds are strikingly bright. In 1885 and 1886 they were bright enough at times to cast shadows and to allow the reading of ordinary print outdoors. A report in 1908 (WHIPPLE, 1934) speaks of 'an extraordinarily strong light, magical and very imposing in the fair summer night'. PATON (1951) mentions a display during which notes could be written without artificial light. Nevertheless the clouds are usually very tenuous, causing hardly any dimming of stars seen through them. During moderate displays in Sweden the clouds were photographed on fine-grain panchromatic film (speed 28° Sch.) with exposures varying from about 1 sec, with a solar depression of 6°, to as much as 100—500 sec, with a solar depression of 11—12.5°, at an aperture of f 5.6. The clouds are seen clearly through blue filters, but fade in red filters, and the latter, so useful in the photography of cirrus, should not be used when photographs are taken of noctilucent clouds.

The light of the noctilucent clouds is quite strongly polarized, in the same sense as the light of the sky. This property is sometimes useful in distinguishing faint noctilucent clouds from cirrus feebly illuminated by scattering from the twilight sky.

b) *The occurrence of noctilucent clouds*

According to previous accounts noctilucent clouds are observed only at places lying between latitudes of about 45° and 62° N (see, e.g. PATON, 1954). However, they were observed on ten nights during the months July and August, 1954 and 1955, at the field-station of the International Institute of Meteorology (Stockholm), which is at Torsta, near Östersund in central Sweden, and in latitude 63°15′ N. On some occasions the clouds were seen in the northern sky down to elevations of 1 to 2° above the horizon, corresponding to distances of 800 to 900 km. These clouds therefore extended northwards at least into the vicinity of North Cape, in latitude 71° N (only in the clean air so often enjoyed in this part of the world could it be expected that clouds would be visible at such a distance). The previous lack of reports from north of 62° N is therefore probably due to the scarcity there of keen observers.

It may equally be wondered if the complete lack of reports during some years reliably indicates the absence of the clouds. The decrease in the number of accounts after 1887 may have been due partly to waning interest. PATON (1954) remarks that the records from Ben Nevis Observatory, where hourly observations were made from 1883 to 1904, show that noctilucent clouds were frequently seen and noted as 'pearly cirrus', 'shining with a silverly blue light'. Paton himself, observing during the period 1939 to 1955, saw noctilucent clouds at least once each year, except in 1942 and 1944 (private communication). If these observations are added to those of Størmer, Malzew (quoted by STØRMER, 1935), and SPANGENBERG (1949), we find that only in these two years are reports missing in the years since 1932. It seems likely, therefore, that the clouds occur at least somewhere in high latitudes every summer, that often they are not seen for want of suitable weather conditions or interested observers, and that occasionally they are abnormally wide-spread and intense, compelling attention. The exceptional periods seem to be 1885—1892, 1908—1911, 1932—1935, and, possibly, 1953—1955.

The clouds have been seen in the northern hemisphere from late May to the middle of August, and, according to Vestine (who does not, however, quote references) in the southern hemisphere in the period of the year about six months later.

Fig. 4, from Vestine's review, shows that the clouds have been observed more frequently after than before the summer solstice.

During strong displays the clouds have appeared as early as 15 minutes after sunset over a large part of the sky, extending to near the SE horizon. As the sun sinks lower the illuminated area of the clouds recedes to a zone above the twilight arch, on either side of the vertical through the sun. This zone moves with the sun's vertical into the northern sky, where at midnight the upper edge of the illuminated area has reached a minimum elevation, and towards sunrise into the northeast before extending again across the zenith and fading away about half an hour before sunrise. In less striking displays the clouds appear in the northwest about an hour after sunset, and fade in the northeast about the same time before sunrise. Only rarely have the clouds been seen at elevations higher than 10°.

Except in very intense displays, the clouds have been seen more frequently after than before midnight, even though the number of observers is likely to have been less at the later hours. In the period 1889—1894 the clouds were observed on 6 occasions before, and on 33 occasions after midnight.

c) *The height of noctilucent clouds*

Tseraskii in 1895, and Pokrovskii in 1897 (see KHVOSTIKOV, 1952), JESSE (1896), STØRMER (1935) and PATON (1949) have measured the height of noctilucent clouds by photography from the ends of long base-lines. Tseraskii obtained an approximate height of 79 km, and Pokrovskii a value of 82 km. Jesse used a base-line 35 km long; in the summer of 1889 he determined the height of 108 cloud details identified on simultaneously exposed pairs of plates, and obtained an average of

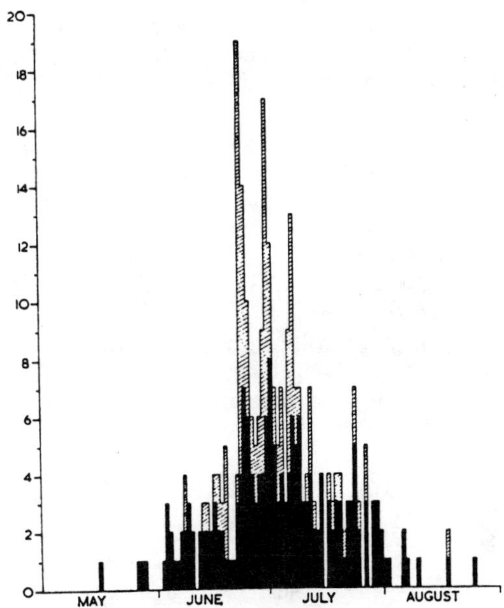

Fig. 4. Total number of times noctilucent clouds have occurred on a given date, 1885—1933 (solid histogram); total number of reports of noctilucent clouds on a given date (hatched histogram). (After Vestine, 1934.)

82.8 km. When these were combined with another 179 measurements made in the following summer, the mean became 82.1 km; the individual values varied from 79 to 90 km.

Størmer used his network of auroral cameras, which provides base-lines of 47, 65 and 105 km. On two occasions in 1932 he measured mean heights of 81.8 km (18 measurements) and 81.1 km (19 measurements). On another occasion in 1934 he obtained a mean height of 82.2 km from 41 measurements which varied from 78 to 85 km.

Paton also used auroral cameras on an occasion in 1949, with a base-line 27.6 km long, which, however, was unfavourably orientated with respect to the clouds, lying almost N—S; he measured a height of about 89 km.

The measurements are thus very consistent. Estimates based on the angular elevation of the nearest discernible details, and the assumption that they are lit by rays which graze the earth, give smaller height varying between about 30 and 70 km.

d) *The velocity of noctilucent clouds*

Jesse made a number of measurements of the movement of noctilucent clouds and found it to be almost always from some point between N and E, and to be rapid, with speeds of up to 100—200 m/sec (ARCHENHOLD, 1928). In 1932 Størmer observed the clouds to be moving from NNE at 44—55 m/sec, and in 1935 estimated a movement from E at about 80 m/sec. STØRMER (1935) quotes a measurement made in Russia in 1925 of a motion from 013° at 229 m/sec. KHVOSTIKOV (1952) mentions measurements in Russia of speeds varying from 50 to 135 m/sec.

According to Jesse's data the motion before midnight is usually from NE, while after midnight it tends to be from ENE. STØRMER (1933) suggests that the changes in velocity may occur near the earth's shadow.

2. Summary of observations at Torsta, 1954 and 1955

In the following summary of observations made at Torsta, near Östersund, a note is included of occasions when the clouds were not seen although sought under apparently favourable viewing conditions. The times quoted are in Swedish clock time (G.M.T. plus 1 hour), which within a few minutes is also the local time. The sun's declination is given in brackets following each date of observation. Cloud velocities are derived from theodolite observations, on the assumption that the cloud height is 80 km. Figures following bearings refer to elevations above the horizon.

18—19 July 1954 (21.0°)

The clouds were first noticed in the NE, 20° at 2322, and later appeared in various places between NNW and E, up to 30°. Billows were aligned E—W. About 0040 four successive measurements gave the following velocities: 003° 55 m/sec, 021° 70 m/sec, 009° 66 m/sec, 011° 65 m/sec.

The clouds were also observed in Stockholm from 2400 to 0100, between N and NNE (fig. 5).

20—21 July 1954 (20.8°)

A good display occurred after 2300, covering practically the whole sky north of the zenith, mainly as very long irregular streaks aligned N—S. Some billows in the east had the same alignment. At 0012 the clouds were visible in the NNW down to 6°, where they gleamed

Fig. 5. Noctilucent clouds in NNE, from Lidingö, Stockholm, 19 July 1954, about 0030 hours (photograph by Hans Höfer).

silver behind a rayed purple light extending up to about 10°; one measurement gave a velocity of 345° 86 m/sec. At 0035 the clouds extended to 18° above the SSW horizon; the display faded at 0200.

24 July 1954 (20.0°)

The clouds did not occur before 2330.

29—30 July 1954 (18.9°)

Some clouds were barely visible beyond stratocumulus, just E of N, 5°, at 2345. Later long streaks and billows appeared in the NE, about 10°.

16—17 August 1954 (13.9°)

Noctilucent clouds appeared at 2300 in an arch extending from WNW to NNE, with minimum and maximum elevations of 2.2 and 7° (where they faded sharply into a deep blue-black sky) in the NNW at 2335, and were still visible at 0010. One measurement at 2342 gave a velocity of 053° 55 m/sec.

17, 19 and 22 August 1954 (13.6, 13.0, 12.0°)

The clouds were not present before 2400.

18 August, 1954 (13.3°)

The clouds were not present before 2300.

29 August 1954 (9.6°)

The clouds were not present before 2200.

18—19 July 1955 (21.0°)

Between 2345 and 0035 faint streaks were visible near the zenith, down to 40° in the S and down to 10° between W and NNE. No appreciable movement of the streaks could be seen over the open sights of the theodolite during a ten-minute period.

23—24 July 1955 (20.0°)

At 2340 faint streaks lying N—S appeared in NE and at about 30° in the SE a faint streak was aligned E—W. At 2400 faint patches with billows appeared in SE, 60°; motion 355° 70 m/sec. At 0015 further streaks appeared in NW, 25° and ENE, 15°. Another measurement of the motion, more reliable than the first, gave 022° 80 m/sec. At 0110 the display had faded, but two long faint streaks were still visible in the E, 25°.

25—26 and 26—27 July 1955 (19.7 and 19.5°)

No noctilucent clouds before 0030.

27—28 July 1955 (19.2°)

At 2250 a thin streak appeared in N-NNE, 3°; soon another appeared at 6°, and a patch of billows formed just below 3°. At 2325 the clouds had extended up to 7.5°; they gave a strong impression of forming and dissolving with a life of about 20 min. From 2345 to 0040 there were only traces, from NNW to NNE, 3.5 to 4.6°.

29—30 July 1955 (18.8°)

No clouds present before 0030.

31 July—1 August 1955 (18.3°)

At 2245 pale streaks became visible in NNE, 7°; gradually they became brighter and more extensive; the longest streaks were aligned towards ENE, gradually bending towards NNW at low elevations, with many shorter streaks and some billows lying across their length. By 2330 the clouds extended from NNE to ENE, 1.7 to 10°. Immediately above bright streaks the sky was darker, as though they cast a shadow. At 2350 the motion was measured as 344° 36 m/sec. At 0019 and 0023 two further measurements in the same part of the sky, apparently no less reliable, gave 238° 45 m/sec and 243° 38 m/sec. The alignment of a billow was measured to be along 293°. After midnight the display became weaker, and only traces remained at 0040.

2 August 1955 (17.8°)

Clouds not present before 2315.

9—10 August 1955 (15.8°)

Noctilucent clouds appeared in the N at 2330, and by 2350 extended from NNE—ENE, 2.0 to 5.0°. Two measurements of the motion gave 325° 13 m/sec (at 2400) and 341° 16 m/sec (at 0025). By 0040 the clouds extended from 3.2 to 13.4°, but were fainter; throughout this period they were very sharply limited in the NNE. At 0145 traces remained in ENE, 11 and 6°. The clouds are illustrated in Fig. 3.

12—13 August 1955 (15.0°)

Clouds first observed at 2330; at midnight they extended from NNE to ENE, up to 4.3°; half an hour later they had spread into the N, with traces also in the NW, and the upper margin had increased to 6.2°. The area of clouds continued to shift westwards and the upper margin to rise; at 0135 the clouds extended between NNW and NNE and up to 12.8°. There were many groups of beautiful billows (Figs. 1, 2), and again marked 'shadows' occurred immediately above some of the brighter streaks (Fig. 1). After 0200 the clouds faded and shrank in extent, only traces in the NW remaining at 0225. Measurements of the motion gave the following results:
0100: 031° 36 m/sec; 0130: 052° 26 m/sec and 049° 60 m/sec; 0150: 054° 90 m/sec.

The acceleration seemed real, for in the later stages the motion in the field of the theodolite was obviously more rapid than previously.

21—22 August 1955 (12.1°)

Noctilucent clouds absent.

23, 24, 25, 26 August 1955 (11.4 to 10.4°)

Noctilucent clouds absent before 2330.

General remarks

Altogether during these two summers there were observed 2 faint, 3 moderate and 5 bright displays. It is remarkable that between successive displays the clouds are sometimes not to be seen under apparently perfect viewing conditions. Possibly on some of these occasions a watch, although maintained until about midnight, was not kept long enough, for in accordance with previous experience there is a clear tendency for the displays to be more marked after than before midnight. The clouds also are usually visible for a longer period after than before midnight.

It is noticeable that the clouds occur predominantly in the part of the sky which is east of N, and even for them to first appear there before midnight, when of course the sun lies west of N. On some occasions the clouds are clearly limited in extent, their boundaries not being determined simply by the border of the earth's shadow.

It is not easy to make a reliable measurement of the cloud motion by theodolite, because of the difficulty of finding a suitable detail which retains its identity and sharpness for more than a minute or two. Some of the prominent details, for example the bright places where two streaks cross or fuse, may not move with the wind velocity.

3. Conditions under which noctilucent clouds may be observed

It has been supposed that observations of the clouds are restricted to the summer months in middle and high latitudes because only then do the geometrical conditions allow the clouds to remain in sunshine for substantial periods after sunset on the ground. In low latitudes throughout the year the sun sinks rapidly below the horizon, greatly restricting these periods in which the clouds are suitably illuminated. The variation of viewing conditions throughout the seasons and over the earth can be calculated under some assumptions, as described in the following paragraphs.

An observer at o (Fig. 6) is considered able to see any clouds lying at the 80 km level over an arc extending in the vertical through the sun between angular elevations a, b, when the sun's apparent depression below the horizon is α. The sunlight grazes the earth at the point G,

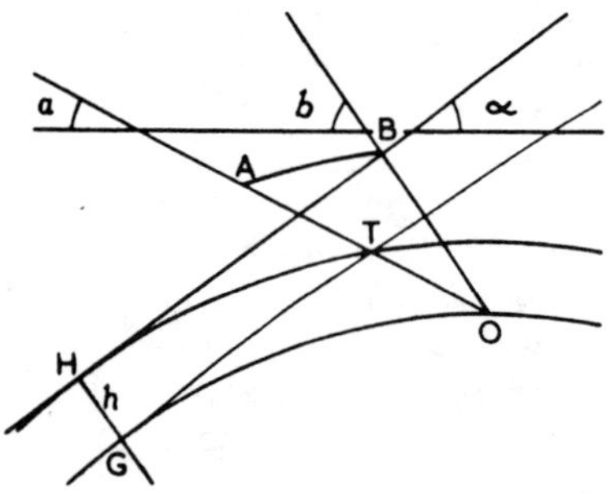

Fig. 6. Geometrical conditions under which noctilucent clouds may be seen.

but it is supposed that the light is much attenuated in a layer extending above the ground to a height h, so that only rays above that which grazes this layer at H illuminate the clouds sufficiently for them to become visible. On the other hand, it is assumed that the clouds cannot be seen beyond the point A because in the line of sight a part of the main scattering layer is still in direct sunshine, and the cloud details are swamped in the flood of scattered light which produces the twilight arch.

The angles a and b are functions of the sun's depression α, and of the heights assumed for the clouds and the top of the main scattering layer. They may be determined by calculating the paths through the atmosphere of the rays along OA and OB, and of the rays GT and HB which graze the earth's surface and the main scattering layer. Because these rays have long paths through the atmosphere it is necessary to consider the earth's curvature, and at the same time normal atmospheric refraction can be taken into account (although it is relatively unimportant).

It is desirable to determine a value of h from observations of the upper and lower boundaries of noctilucent clouds. Størmer (1933, 1935) has determined from his observations the height of the earth's shadow at the position B of the upper edge of the clouds. Although he neglected refraction he in this way obtains a close approximation to h, with results varying from 29 to about 50 km. Only the minimum values are likely to be significant, because on some occasions there may be no clouds extending into the position B; nevertheless in six measurements made on each of two occasions Størmer obtains values which rather consistently lie between 30 and 45 km. This astonishing result implies either that there is very substantial attenuation of sunlight even well in the stratosphere, or that the clouds do not quite extend to the position of the earth's shadow.

I have calculated values of h for the rays which illuminated the nearest and furthermost details of the clouds observed at Torsta, assuming the cloud height to be 80 km and taking into account refraction, with the results shown in Table 1. These confirm Størmer's values, and suggest that only rarely are the nearest details in a position corresponding to a value of h smaller than 30 km. The values of h

Table 1. Least height h above the ground of rays illuminating nearest and farthest details of noctilucent clouds

Date	Time	Sun's depression, deg.	Azimuth and Elevation of clouds, deg.	h, km
Nearest details:				
18.7.54	2400	5.7	360,30	54
21.7.54	0035	6.0	90	45
16.8.54	2335	12.8	330,07	12
19.7.55	0005	5.7	180,40	43
23.7.55	2400	6.8	90	34
27.7.55	2325	7.5	360,07.5	69
31.7.55	2330	8.5	360,10	55
9.8.55	2350	10.8	040,05	37
10.8.55	0025	11.0	031,06.3	34
	0040	10.7[1]	035,13.4	10
13.8.55	0040	11.4	033,06.3	32
	0115	10.6	015,09.9	30
	0140	9.7	007,12.8	29
	0215	8.0	353,10.0	45
Farthest details:				
19.7.54	0025	5.7	360,04	12
21.7.54	0015	6.0	340,06	16
16.8.54	2335	12.8	320,02.2	55
19.7.55	0005	5.7	360,10	16
27.7.55	2330	7.5	360,03.0	18
31.7.55	2330	8.5	020,01.7	23
9.8.55	2350	10.8	023,03.3	38
10.8.55	0025	11.0	023,02.9	37
	0040	10.7	024,03.2	36
13.8.55	0040	11.4	006,01.2	32
	0115	10.6	360,01.7	31
	0140	9.7	007,01.8	26
	0215	8.0	344,01.6	25
			008,04.7	25

[1] (very faint)

corresponding to the farthermost visible details seem significantly less, and bearing in mind that the minimum values are likely to be more significant, I have used a value of 16 km in the following calculations. It must be supposed, however, that the appropriate value may often be twice as much, with the effect of restricting the visible area of the clouds. It should be noted that the scattering of indirect light which limits the visibility of the clouds is that which occurs in the lower stratosphere, and that in the troposphere only attenuation by concentrations of haze or obscuration by clouds are likely to interfere with observations.

Fig. 7 shows the relations between the sun's apparent depression α and the angles a (for a value of $h = 16$ km) and b ($h = 30$ km), and Fig. 8 shows the relation between the distance of noctilucent clouds at 80 km and their angular elevation, used in the derivation of Fig. 7. This diagram shows that the clouds are likely to be visible only when the sun lies between about 5° and 16° below the horizon.

4. Variation of conditions of viewing according to latitude, season, and time of day

The true depression $-\gamma$ of the sun is a function of latitude, ϕ, sun's declination δ, and apparent solar time, τ, according to the equation:

$$\sin \gamma = \sin \phi \sin \delta + \cos \phi \cos \delta \cos \tau \qquad (1)$$

For a given latitude a diagram can be constructed with apparent solar time as ordinate

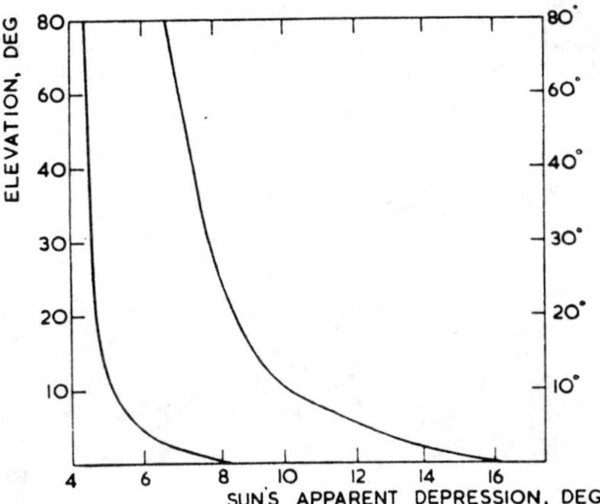

Fig. 7. Angular elevation of upper and lower borders of region, above sun's azimuth, in which noctilucent clouds may be seen, as a function of the apparent depression of the sun.

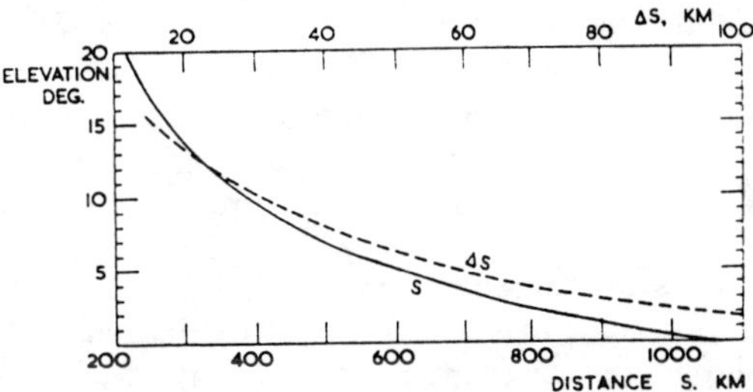

Fig. 8. Distance S of noctilucent clouds 80 km above the level of the observer, as a function of elevation above the horizon, and the change ΔS in S due to a change of 1° in the elevation.

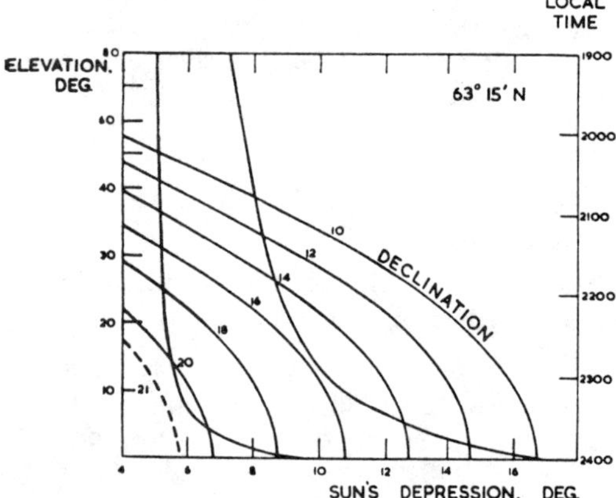

Fig. 9. Diagram for finding upper and lower border of noctilucent clouds on sun's azimuth for an observer in latitude 63° 15′ N, as a function of the sun's true depression (assuming 0.6° of refraction at the horizon), or as a function of local time and sun's declination.

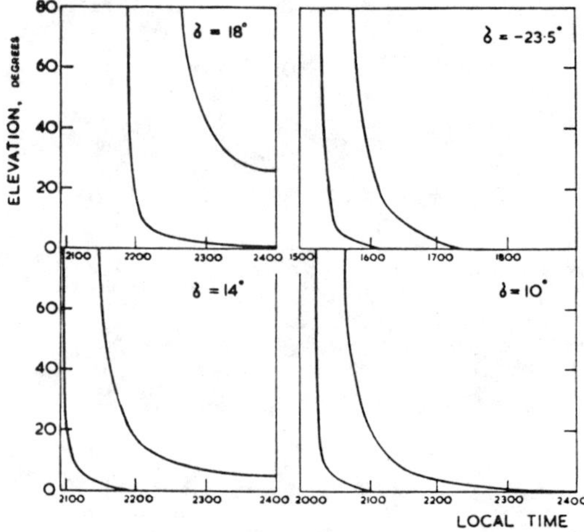

Fig. 10. Position of upper and lower borders of noctilucent clouds on the sun's azimuth for an observer in latitude 63° 15′ N, as a function of local time, for four times of the year.

and sun's true depression as abscissa, containing isopleths of declination. Such a diagram may have the curves of Fig. 7 superimposed, as for example in Fig. 9, since we can regard the true sun's depression to exceed the apparent by 0.6°, and it can then be used to find the elevations in the vertical through the sun where noctilucent clouds should be sought, as a function of season and apparent solar time. (The apparent solar time is found from the following relation: apparent solar time =

= G. M. T. + (4 min. for each degree of longitude east of Greenwich)—(the equation of time). Values of δ and of the equation of time are tabulated for each day of the year in the Nautical Almanac.) The diagram is used by determining the point corresponding to a given declination and time; the vertical through this point intersects the a, b- curves in two points defining the lower and upper borders of the part of the sky above the sun in which noctilucent clouds may be seen.

The form of the a—b curves in Fig. 7 implies that in the vertical above the sun the upper and lower borders of the visible area of clouds pass rapidly through the zenith, but descend rather slowly as they approach the horizon. This is illustrated in Fig. 10, which shows the position of these borders with time of day for several seasons in latitude 63° 15′ N. These diagrams also show the most favourable season for viewing to be near mid-summer.

As a measure of an observer's opportunity of seeing noctilucent clouds it is reasonable to consider the total period during which a cloud at an elevation of 10° on the sun's azimuth is suitably illuminated. From Fig. 9 it is seen that during this period the true depression of the sun changes from 5.75 to 10.7°; the period can therefore be calculated from eq. (1) by inserting these values of $-\gamma$, for particular latitudes and seasons. In this way we derive Fig. 11, which shows the length of the viewing period (before midnight) throughout the year in latitude 63° 15′ N. Over most of the year the period is nearly an hour, rising sharply to a peak of nearly three hours at a declination of 15°, and abruptly falling to zero at declinations exceeding 21°, corresponding to one month before and after the summer solstice, when the sun even at midnight is less than 5.75° below the horizon. At rather lower latitudes mid-summer is the most favourable viewing season, as shown in Fig. 12. In this diagram the curves show the variation of the viewing period (before midnight) for two months after the summer solstice, in latitudes 63° 15′ N, 56° N, 50° N and 30° N. In the lower latitudes the period changes little throughout the year, and amounts to between a half and three quarters of an hour.

Such a period must be considered ample for an attentive observer to detect the clouds if they are present. Nevertheless according to

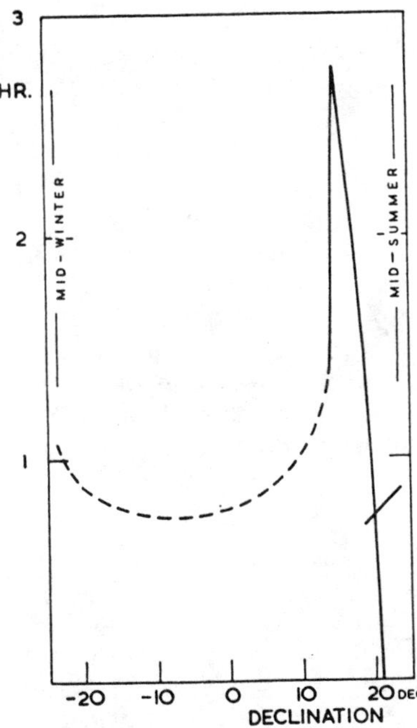

Fig. 11. Period before midnight during which noctilucent clouds on the sun's azimuth and at an elevation of 10° above the horizon are continuously visible, as a function of the sun's declination. The long curve is for an observer in latitude 63° 15′ N, and is drawn continuously over the season in which the clouds have been seen from Torsta. The short intersecting curve is the corresponding one for the latitude 50° N, and extends over the season in which the clouds have been reported from places between the latitudes 49° and 52° N.

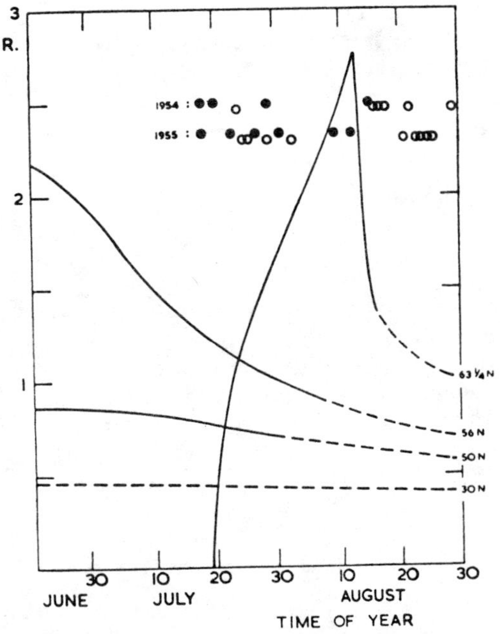

Fig. 12. Period before midnight during which noctilucent clouds on the sun's azimuth and at an elevation of 10° above the horizon are continuously visible, as a function of time of year. The continuous parts of the curves show the season during which the clouds have been observed at Torsta (63¼° N), Edinburgh (56° N), and at places between 49 and 52° N (50° N). The circles in the upper part of the diagram show the occasions when the clouds have been seen at Torsta (blocked-in circles), or not seen under apparently favourable viewing conditions (open circles).

Vestine the clouds have never been reported from latitudes lower than 45°; apparently the reports from as far south as this were made in Russia on an exceptional occasion, that of the fall of the Great Siberian Meteorite, in 1908, which is further discussed below. If this occasion is excluded it seems that there are very few reports from south of 50° N, and if we consider also that the clouds seen in places near this latitude have rarely extended more than a few degrees above the northern horizon, then the presumption is strong that noctilucent clouds do not exist in latitudes lower than about 55°.

In Figs. 11 and 12 the curves are drawn as continuous lines over the season within which reports of the clouds have been made. The Torsta observations were used for the curve corresponding to latitude 63° 15′ N, Mr. Paton's (Edinburgh) observations for 56° N,

and all those listed by Vestine from places between 49 and 52 1/2° N for the curve appropriate to 50° N. Evidently the observations of the clouds at Torsta and at Edinburgh cease when the viewing conditions, judged on this criterion, are still more favourable than ever occur in rather lower latitudes. At Torsta on clear days in mid-winter clouds 10° above the horizon would be visible for a whole hour in the morning and again in the afternoon, at times when they could hardly escape attention. It may also be significant that the clouds have been seen at Torsta rather later in the year than has ever been found at Edinburgh. On the basis of this evidence I conclude tentatively that the clouds, in addition to occurring mainly after rather than before mid-summer, are restricted in southern extent to about latitude 55° in July and to about 70° N in the latter part of August, and do not occur in these regions in seasons other than the summer. These circumstances suggest some meteorological control of their appearance.

5. The size and concentration of the particles in noctilucent clouds

The noticeable degree of polarization in the light from noctilucent clouds indicates that

$$\pi r < 0.5\,\mu,$$

where r is the predominant radius of the particles in the clouds. On the other hand, the scattering of light by the particles cannot be that which occurs in the Rayleigh regime, for if so the clouds would appear blue. They are observed to be white, although sometimes tinged with blue. This shows that

$$\pi r > 0.025\,\mu,$$

and combining these indications we have

$$10^{-6} < r < 10^{-5}\ \text{cm}.$$

In Sweden it was found that when the sun was about 6° below the horizon, the exposure needed to photograph the clouds on fine-grain panchromatic film (speed 28° Sch.) was about 5 sec at $f\,5.6$; if, however, the sun's depression was about 12°, the exposure needed was about 200 sec. It can be assumed, then, that an exposure of about 100 sec at this aperture is a measure of the brightness of typical clouds. This information allows the likely concentration of the cloud particles to be inferred, in the following manner.

It is assumed that the scattering particles are disposed in a layer of depth h, illuminated by the sun at an angle of incidence Θ. If a fraction $f(h)$, $\ll 1$, is scattered from a vertically incident solar beam, then a fraction $f(h)\sec\Theta$ is scattered from the beam incident at an angle Θ. Then the scattering fom a unit area of the cloud surface is independent of Θ and equal to $s \cdot f(h)$, where s is the flux/cm² in the solar beam. If the particles are small, half of the scattering is forward, and half backward.

If the cloud surface is assumed to scatter according to Lambert's law, its brightness B_c is independent of the direction of view and

$$B_c \sim s \cdot f(h)/2.$$

Over a surface of snow, or some other nearly perfect scatterer, which is illuminated vertically, the fraction of the incident beam scattered backwards is unity; thus if B_s is its brightness, $B_c \sim s$, and

$$B_c/B_s = f(h)/2.$$

We now suppose that the ratio of the brightness of these surfaces is given by the inverse ratio of the exposures needed to photograph them, which from our experience is about 10^{-6}, corresponding to exposures of 100 sec at $f\,5.6$ and $1/200$ sec at $f\,32$, respectively.

The scattering cross-section of particles in the size range inferred above is approximately equal to their geometrical cross-section πr^2, so that if n is their concentration we have

$$f(h) = hn\pi r^2 = 2 \times 10^{-6}.$$

The depth h of the layer occupied by noctilucent clouds is likely to be several km; for example, if the billows represent a convective process similar to that which produces the billows of tropospheric clouds, the depth of the layer involved is approximately $d/2.7$ (SCORER, 1951), where d is the separation of the billows, observed to be about 10 km. The billows, therefore, are likely to occur in a layer about 4 km deep, while the denser and much broader streaks and bands are probably even thicker. If we assume a value of about 6 km for h, then

$$nh\pi r^2 = 2 \times 10^{-6}, \text{ and}$$
$$n \approx 10^{-12}/r^2, \text{ or}$$
$$n \approx 10^{-2}/\text{cm}^3 \text{ if } r = 10^{-5}\ \text{cm},$$
$$n \approx 1/\text{cm}^3 \quad \text{if } r = 10^{-6}\ \text{cm}.$$

The corresponding space-densities of the material, assuming a particle density of 2.5, would be 10^{-16} and 10^{-17} g/cm³, respectively. These estimates can be regarded as representative of average conditions, and as unlikely to be in error by more than one power of ten. It would be very desirable to make better estimates by accurate quantitative study of particular clouds.

6. The substance of noctilucent clouds

Noctilucent clouds may be true ice clouds, or their particles may be composed of some other material.

In the ozone layer, and at higher levels up to about 55 km, the saturated vapour pressure of water at the rather high temperatures encountered exceeds the air pressure, and a condensation of vapour is therefore impossible. At still higher levels, however, the temperature falls with increasing height, to reach a minimum at a kind of second tropopause at about 80 km, approximately the level of noctilucent clouds, and a condensation is possible there if the air is moist enough and the temperature sufficiently low. In the lower stratosphere of middle latitudes the air is practically always very dry, and water clouds are never observed apart from the exceptional wave clouds sometimes formed over the Norwegian and some other mountains in winter. The soundings of the British Meteorological Research Flight have shown that 3 to 5 km above the tropopause the frost-point tends to become constant at about $-115°$ F ($193°$ A), corresponding to a relative humidity of less than 1 % (MURGATROYD AND OTHERS, 1955). If such air from these levels were raised or stirred up to the 80 km level, the mixing ratio, of order 10^{-3} g/kg, would be preserved, and the vapour density would become about 3×10^{-14} g/cm³ (at 80 km the air density is about 3×10^{-8} g/cm³ (WHIPPLE, 1952)). If this value is to exceed the saturated vapour density over ice, so that a condensation could occur, then the air temperature would need to be lower than about $145°$ A (accepting a value of the saturated vapour density extrapolated to this temperature from a formula regarded as thoroughly reliable in the range $173°$ to $273°$ A: O.M.I., 1951). Temperatures at the 80 km level have been deduced from instruments carried on rockets and from observations of meteors, and although they cannot be regarded as particularly accurate and were obtained mainly in latitudes lower than those in which the clouds are seen, they suggest a mean temperature of $220°$ to $230°$ A. WEXLER (1950) considers $170°$ A as a likely minimum value, and at this temperature the saturated vapour density is about 10^{-9} g/cm³. It must therefore be said that a condensation in the 80 km level could be expected only if the temperature there became abnormally low, or if the air were abnormally humid, containing a large proportion of water vapour; nevertheless if the clouds there are regarded as very rare phenomena, the possibility of their representing a condensation cannot quite be excluded on these grounds.

A further circumstance makes it seem improbable. The rapid changes of the internal details of the clouds and the presence of billows suggest that the layer in which the clouds occur is well stirred and has a strong lapse-rate; since the layer is a few km deep, the temperature in its upper parts must be twenty degrees or so below that in the lower part. As the saturated vapour pressure decreases very rapidly with temperature, the space-density of the condensate in a cloud extending through such a layer must be nearly the saturated vapour density at the lower surface, that is, at the condensation temperature. Even if this temperature were as low as $145°$ A, the corresponding mean density of the ice cloud would be of the order of 10^{-14} g/cm³, two or three orders of magnitude greater than the values deduced from the optical properties of the clouds.

Finally, the deduced concentrations of the cloud particles suggest a nucleated rather than a homogeneous condensation, so that if the clouds are regarded as ice clouds it is still necessary to consider upon what nuclei the particles formed. It is therefore natural to consider seriously other theories of the constitution of the clouds, which attribute them to concentrations of dust derived from the earth's surface or from outer space.

7. Noctilucent clouds and volcanic explosions

It has been suggested that the intensity and prevalence of noctilucent clouds in the years 1885 to 1894 are to be related to the eruption of Krakatoa in August 1883. There is no doubt that vast quantities of dust are introduced into the lower stratosphere by eruptive clouds. The final paroxysmal eruptions of Krakatoa occurred on the night of August 26, but already in the early afternoon the captain of the British ship *Medea*, over 75 miles away, could see the enormous cloud over the volcano, and from angular measurements estimated the height of its top as 17 mi, and later as 21 mi (34 km) (Report of the Krakatoa eruption, 1888). Since the wind at these levels carried the cloud away from the ship (towards the west), there is no reason to suppose that these are overestimates. Dust from the eruptive cloud spread across the earth and within three months was visible in western Europe as a tenuous layer even in full daylight, which produced spectacular twilight phenomena. Estimates of the height of this dust cloud, based on the sun's depression when the red glow faded from the cloud, varied between about 15 and 30 km, the higher values being obtained mainly in the early history of the cloud, and the lower values after some months; amongst the observers was Jesse, who found mean heights of 12—15 km, later becoming 10 km.

Similar, though less dense clouds of dust were recently observed at such heights by aircraft over western Europe in July and August 1953 (JACOBS 1953), following an eruption in Alaska, and again in April and May of 1955, following an eruption in Kamchatka (BULL and JAMES, 1956). The appearance of the clouds and estimates of their height were also recorded by ground observers (LUDLAM, unpublished). It therefore seems likely that dust from volcanoes is injected mainly into the lower stratosphere, but it must be supposed that a sufficient proportion to account for the appearance of noctilucent clouds could become diffused up to much higher levels.

It is remarkable that the first observations of noctilucent clouds in the years 1885—1894 were not made until 2 years after the arrival of the volcanic cloud in Europe, and Vestine points out that the period 1880—1887 was one of strikingly frequent and brilliant displays of comets and meteors, and that the great eruption of Katmai, in 1912, was not followed by any increase in the frequency of reports of noctilucent clouds. On the other hand, SPANGEN-BERG (1949) has tried to associate more frequent reports in the years 1932—1935 with an eruption in the Cordilleras in 1932, which threw up a cloud to 25 km. However, major volcanic eruptions occur sufficiently often for some to lie within any chosen period, and the reports of noctilucent clouds are made so unsystematically and depend so much on favourable weather in a particular short season, that it is hardly surprising no close relation can be demonstrated between volcanic activity and the occurrence of the clouds.

8. Noctilucent clouds and meteor showers

On the morning of June 30, 1908, a great meteorite fell in Siberia; its mass has been estimated at some tens of thousands of tons, and the explosion it produced was recorded by barographs in England. During the same night an unusually brilliant display of noctilucent clouds was reported by many observers in Russia, Sweden, Denmark, Germany and England. G. A. Clarke, at Aberdeen, observed the sudden arrival of the tenuous clouds, accompanied by a marked brightening of the twilight, about two hours after sunset. From this observation WHIPPLE (1930) deduced the eastward speed of the cloud movement to have been about 85 m/sec, supposing the cloud to have originated in the trail of the meteorite. Apparently it was on this exceptional occasion that noctilucent clouds were reported (over Russia) from as far south as latitude 45° N (VESTINE, 1934); no references to the clouds or to unusual twilight brightness could be found in the meteorological logs in the U.S.A., and no such reports were made in southern Europe.

A witness of the Great Siberian Meteorite said that it left in the atmosphere 'a light bluish trail', and the smaller streaks left by meteors large enough to be seen in daylight also have the pale silvery colour which is characteristic of noctilucent clouds; evidently, therefore, the particles in meteor trails have about the size of those composing the clouds. This conclusion is supported by the work of FESSENKOV (1949), who has discussed the properties of the trail left by another large meteorite, of estimated mass a few hundred tons, which also fell in Siberia, on 12 February 1947. Broken material was found in small craters, and with the help of mine detectors many small pieces, including a multitude of spheres of radii 15—$20\,\mu$ produced by thin sprays of solidifying iron, were found in the soil. At first the trail left in the atmosphere was about 1.5 km in diameter, and was so dense as to obscure the sun, but soon afterwards, as it diffused, the sun shone through as a red disc. The particles were therefore noticeably more transparent to red than to blue light. From these observations Fessenkov deduces that the diameter of the particles must have been less than 10^{-5} cm, and that their concentration in the original trail was about $7 \times 10^5/\mathrm{cm}^3$; since the total volume of the trail was originally about 70 km^3, it contained nearly 200 tons of iron. If the particles had diameters considerably smaller than 10^{-5} cm, these figures for the space density and total mass would be substantially increased. Since if noctilucent clouds are supposed to contain particles of diameter 10^{-5} cm their concentration would be about $10^{-2}/cm^3$, such a trail could diffuse horizontally to a diameter of 7,000 km before the mean particle concentration fell below that inferred for noctilucent clouds. Although the denser parts of a large meteorite trail lie below the 80 km level, clearly such trails could form a significant source of the substance of noctilucent clouds, divided into particles of the appropriate size. The particles are presumably produced by condensation of material boiling away from the incandescent meteorite, and perhaps also by the crumbling of its substance into the component cosmic particles. The much larger spherules sometimes found are drips from the molten but unvaporized parts of meteorites which reach the ground, and characteristically have radii of some tens to hundreds of microns.

Meteorites of the great size of those discussed above are very infrequent. Recently BOWEN (1953) has claimed that there is a very close relation between the occurrence of noctilucent clouds and recurrent meteor showers. Bowen compared the dates of the prominent annual meteor showers of June and July, according to LOVELL and CLEGG (1952), with the dates of maximum frequency of reports of noctilucent clouds, according to the diagram of VESTINE (1934) (Fig. 4). This diagram contains two curves, one of which shows, for reports collected by Vestine from the period 1885—1933, the total number of years in which reports have been made on particular dates, and the other of which shows the total number of reports on particular dates. The latter curve was alone considered by Bowen; it has several peaks which in seven cases out of eight correspond to or precede by 1 or 2 days the dates quoted for the prominent meteor showers. Some authors have since supposed that in this way Bowen has established a relation between the two phenomena. However, WHIPPLE and HAWKINS (1956) state that the dates of the principal meteor streams used by Bowen are inaccurate, and that when more modern data are used the only notable coincidence with the dates of the maxima on Vestine's curve occurs on June 30, corresponding to a weak meteor stream. On this date Vestine's curve indicates 17 reports, which an examination of his lists shows to contain probably 12 from the occasion of the fall of the Great Siberian Meteorite on June 30,

1908. Thus there is good reason to discount even this coincidence. Nor is the correspondence improved by considering Vestine's curve which shows the number of years, rather than the number of observers, when the clouds have been recorded, which might be regarded as the more appropriate curve to discuss. On this curve 5 occasions, and one the other curve 7 reports, would be sufficient to produce a peak, so that the significance of those shown in the diagram must be regarded as very doubtful. Moreover, it must be realized that the observations which it summarizes were made over a large range of latitudes, within which the viewing conditions vary considerably from place to place on the same date. For example, even excluding the peak of June 30, Vestine's diagram shows that the observations are notably most frequent about the end of June; however, if a sufficient number of observations from high latitudes were included, this maximum would disappear, for near midsummer the clouds there are invisible, even if present. Evidently only an extended series of observations in one locality or latitude, including reports of occasions when no clouds could be seen in apparently favourable viewing conditions, can provide a satisfactory basis for relating the occurrence of the clouds to any other phenomena.

The observations made in Sweden (Fig. 12) suggest that within the best viewing season there are clear nights when the clouds are definitely not observed, separating the occasions when they are seen. The occurrences are not obviously related to any meteor streams; the principal stream at this season is that of the Perseids, which begin to arrive about August 6 and reach a maximum between about August 10 and 14. This meteor shower, in common with the others, is spread over a few days, and cannot be associated with a particular date (LOVELL, 1954). On the basis of the available evidence, therefore, it cannot be concluded that there is any clear relation between the occurrence of noctilucent clouds and the arrival of meteor showers.

9. The accretion of interplanetary particles by the earth

The rate of arrival of visible meteors during the major showers rises by only a factor of about 2 above that of the sporadic meteors (LOVELL, 1954), so that the showers can be neglected in comparison with the sporadic meteors as a possible source of the material of noctilucent clouds.

Visible meteors are produced by particles of mainly stony or iron-nickel composition, and of radius exceeding about 100 μ, which enter the atmosphere at speeds of some tens of km/sec and become incandescent. Particles smaller than this are also fused, but emit too little light to be visible to the unaided eye, while particles of radius less than a few microns are decelerated in the atmosphere without reaching melting-temperatures ($1200-1700°A$) (WHIPPLE, 1950, 1951). The larger meteorites leave trails which often are visible for some minutes or even longer as a dark smoke, or, under favourable lighting, as pale streaks closely resembling the noctilucent clouds. It seems, also, that many meteors, particularly the fainter ones, have a low density (0.2 to 1.0 g/cm^3), and disintegrate under aerodynamic pressure and heating by a progressive crumbling (JACCHIA, 1955), so that they are regarded as porous and fragile structures, probably consisting of mineral fragments embedded in ices of such substances as H_2O, CO_2 and NH_3. However, particles of the size (10^{-5} to 10^{-6} cm) inferred to predominate in noctilucent clouds must be supposed to enter the atmosphere either directly with this size, or to be produced in the atmosphere by condensation of vapourized material in the wakes of meteorites having dimensions which exceed several microns. The visible paths of most meteors lie between heights of about 100 and 50 km above the ground, so that their condensation products are distributed over a layer extending somewhat above and below the level of noctilucent clouds.

The fall-speeds of small particles at these levels have been estimated by LINK (1950), using Stokes Law with Millikan's correction; he obtained values proportional to the particle radius r and density δ, and inversely proportional to the air density ϱ. At the 80 km level he found a fall speed of a few cm/sec for particles of radius 10^{-5} cm. On the other hand, the mean free path of the air molecules at this height is about 4 mm, and therefore so much greater than the particle sizes that it seems preferable to use Knudsen's (1934, p. 33) empirical data

for the resistance K of rarified air to the motion of a sphere. He measured K in air at pressures extending down to 0.14 dyne/cm^2, and expressed his results by a formula

$$K = 6\pi\eta r v (1 + 0.683\varkappa + 0.354\varkappa/e^{1.845\varkappa})^{-1},$$

where η is the coefficient of viscosity and $\varkappa = \lambda/r$, λ being the mean free path obtained from the expression

$$\lambda = \sqrt{(\pi/8)\eta/0.30967p}\sqrt{\varrho_1},$$

where p is the pressure and ϱ_1 is the density at 1 dyne/cm^2.
If $\varkappa \ll 1$ the formula reduces approximately to
$K = 6\pi\eta r v\varkappa$, and hence
$v = \varkappa v_s$,

where v_s is the Stokes Law fall-speed.

At the 80 km level we may assume $T = 200°A$, $\eta = 1.3 \times 10^{-4}$ g/cm sec, and $p = 17$ dynes/cm^2; with $\varrho_1 = 1.74 \times 10^{-9}$ g/cm^3 we then have $\lambda = 0.37$, and for particles of density 2.5 g/cm^3

$$v = 4.2 \times 10^6 r^2\varkappa.$$
If $r = 10^{-5}$ cm, $x = 3.7 \times 10^4$ and $v = 16$ cm/sec.
If $r = 10^{-6}$ cm, $x = 3.7 \times 10^5$ and $v = 1.6$ cm/sec.

These estimates differ by less than 50% from values deduced from ordinary kinetic theory of gases (Witt, private communication).

These values incidentally lend some support to the upper size limit deduced for the particles of noctilucent clouds on optical grounds, for it seems inconceivable that the clouds could possess persistent identifiable details if the fall speeds of their particles reached values in m/sec.

Given these fall speeds and the inferred space-density of noctilucent clouds, we can compute the density in space of meteoritic material sufficient to provide the source of the clouds and compare this with the density believed to exist, since meteoritic material arrives from interplanetary space with speeds within the range of about 11 km/sec (the speed of fall due to the earth's gravitation) and about 70 m/sec (compounded of the earth's orbital speed about the sun and the speed of escape from the sun at the distance of the earth).

From the observed properties of visible meteors, the size distribution of the particles over the range of radii from about 50 μ to 0.5 cm can be inferred. On the basis of this size distribution law, extrapolated to include both smaller and larger meteorites, LOVELL (1954) estimates the annual accretion of meteoritic material by the earth to amount to about 500 tons, corresponding to an average density in interplanetary space of between 10^{-25} and 10^{-24} g/cm³.

If material in the higher concentration were to enter the earth's atmosphere at 32 km/sec, and were wholly converted into particles of radius 10^{-6} or 10^{-5} cm, settling at their terminal speeds, its density at the 80 km level would become 2.10^{-18} or 2.10^{-19} g/cm³, respectively. If these values are compared with those deduced for noctilucent clouds composed of such particles (about 10^{-17} and 10^{-16} g/cm³, respectively), we see that they are smaller by nearly one to three orders of magnitude. Although the magnitudes of the compared quantities are each liable to be in error by as much as one power of ten, the values obtained for the density of the meteoritic material are likely to be too high rather than too low. The available evidence therefore cannot be considered with any confidence to present ordinary meteorites as a sufficient source of the material of noctilucent clouds. Other possible sources must also be considered.

10. The zodical light cloud

There is some evidence that in the meteor showers the numbers of very small particles are even less than indicated by extrapolation of the size distribution law amongst the visible meteors. One probable reason is that the small particles in meteorite swarms become separated from the larger. On the other hand there is also reason to suppose that the accretion of sporadic micrometeorites, too small to produce visible meteors, is very much greater than this law would suggest. This evidence comes mainly from the study of the sun's outer corona and the zodiacal light.

The zodiacal light is a wedge of diffuse, feeble light pointing upwards from the horizon along the position of the ecliptic during the dark hours, when the depression of the sun exceeds about $17°$; in intensity it is comparable with the light of the Milky Way (see, e.g., MINNAERT, 1940). It can be seen easily on clear, moonless nights in low latitudes, where it is always steeply inclined to the horizon, but it is more difficult to observe in middle and high latitudes.

The intensity of the light increases towards the sun, and VAN DE HULST (1947) suggested that this increase extends continuously into the outer corona of the sun. The light of the corona is composed of two parts, a K-component which is highly polarized and contains no spectral lines, and an F-component which has little or no polarization and contains Fraunhofer lines. The K-component decreases very rapidly in intensity away from the sun, but the F-component intensity decreases more slowly, and dominates the outer corona. The zodiacal light also contains the Fraunhofer lines, and both it and the F-component of the corona are attributed to scattering and diffraction of sunlight by a cloud of interplanetary particles disposed as a rather shallow disc lying in the plane of the ecliptic. At the time of van de Hulst's suggestion the only measurements available were those of the intensity of the zodiacal light in the range of solar elongations from 40 to $180°$, and those of the intensity of the F-component of the corona, out to about $1° 20'$ from the sun. Recently, however, RENSE and others (1953) have measured intensities in the range of solar elongations from 5 to $13°$ (from an aircraft flying at 10 km during a total solar eclipse); their results confirm van de Hulst's suggestion, forming a smooth link between the measurements of the corona and the zodiacal light.

ALLEN (1946) showed that the intensity of the outer corona and the zodiacal light, and also the distribution of intensity with solar distance and wavelength, could be explained by the diffraction and scattering of sunlight from particles of radius about 10^{-3} cm, whose concentration, assumed inversely proportional to the distance from the sun, is about 3×10^{-15}/cm³ at the distance of the earth, representing a space density of about 6×10^{-23} g/cm³. His measurements in the outer corona showed no indication of Rayleigh scattering such as would be produced by much smaller particles; on the contrary, there was some evidence of a reddening of the F-component near the sun. Recently BLACKWELL (1952) found a considerably greater increase towards the sun in the infra-red than in the violet, indicating the presence of particles of radius larger than about 2×10^{-3} cm. Van de Hulst assumed a size distribution of the scattering particles similar to that of meteoritic material, and a uniform concentration of the particles, considering that the increase of intensity towards the sun is due to the form of the scattering function, rather than to changes in the particle concentration. Since part of his scattering and diffracting material is distributed in larger, less efficient particles, he deduces a space-density, of 5×10^{-21} g/cm³, which is considerably greater than the value of about 6×10^{-23} g/cm³ corresponding to Allen's result. If van de Hulst's figure is accepted as probably more realistic and the thickness of the dust cloud perpendicular to the plane of the ecliptic is about 0.1 AU, the total mass of the particles within the earth's orbit becomes 5×10^{12} tons, or about 10^{-9} times the mass of the earth.

It is interesting to consider the origin and fate of this material. WHIPPLE (1955) considers four possible sources, of which only asteroidal collisions and cometary disintegrations appear significant. Of the particles which are produced in these ways those of radius less than about 2×10^{-5} cm are expelled from the solar system by the radiation pressure of sunlight. Larger particles spiral towards the sun under the Poynting-Robertson effect (see, for example, WYATT and WHIPPLE, 1950), at a cosmically rapid rate. Whipple deduces that to balance this loss and to maintain the outer corona and zodiacal light at their present brightness a source of about 1 ton/sec of meteoritic material is required, and quotes Fessenkov as having previously arrived at a similar conclusion by a less general theory. This figure may be compared with PIOTROWSKI's (1953) estimate of between 20 and 600 tons/sec of pulverized material which is produced by collisions between the bodies in the asteroidal belt. Probably, however, the bulk of this material is in the form of rather large fragments. According to WHIPPLE (1954) most of the small particles producing photographic meteors are derived from cometary debris (according to Whipple's model, a comet nucleus is composed of solid particles embedded in "ices" of substances such as H_2O, NH_3 and CO_2, which are partially vapourized on approach to the sun, so that the comet progressively disintegrates during successive orbits). Whipple estimates that particles in the debris of size greater than 10^{-2} cm are exposed to a high probability of destructive collisions (particles of size larger than this are also unlikely to escape accretion

with Jupiter during their passage of its orbit, according to ÖPIK, 1951), but that the cometary source of material for the zodiacal light cloud could readily amount to about 3 tons/sec, which is sufficient to sustain the cloud.

The available astronomical evidence, though rather tenuous, is thus consistent with the presence of an interplanetary dust cloud, in which the particles have radii mainly in the range 10^{-3} to 10^{-2} cm, and whose concentration corresponds to a space density of about 5×10^{-21} g/cm³. Such particles entering the atmosphere would be expected to produce very faint telescopic meteors in numbers about 10^4 greater than would be deduced from extrapolation of the size distribution law for meteors visible to the unaided eye. So far this great increase has not been demonstrated, but indirect evidence does suggest that the previously accepted accretion rates of meteoritic material are too small. For example, the impacts of small particles (of size a few microns) upon high-altitude rockets are greater than was anticipated (see WHIPPLE, 1952, p. 22), and so also is the abundance of iron and nickel in deep-sea sediments (PETTERSON and ROTSCHI, 1950; the significance of these deposits has been questioned, however: see, for example, ÖPIK, 1955), and the rate at which magnetic material of presumably meteoritic origin falls to the ground (THOMSEN, 1953; here also, however, the origin of the material is not certain: SCHAEFER, 1955, attributes most of it to industry).

If the estimate of the composition of the zodiacal light cloud is accepted, and the particles are supposed to evaporate in the high atmosphere and subsequently condense into particles of radius 10^{-6} or 10^{-5} cm, the space densities of these particles at the 80 km level estimated on the previous reasoning would be as follows:

for $r = 10^{-6}$ cm, space density = 10^{-14} g/cm³;
for $r = 10^{-5}$ cm, space density = 10^{-15} g/cm³.

Both these values exceed the space densities inferred for noctilucent clouds (10^{-17} and 10^{-16} g/cm³, respectively). Since we have used van de Hulst's rather than Allen's estimate of the density of the zodiacal light cloud, and since a proportion of its particles would be small enough to escape boiling in the atmosphere, the values are likely to be too high rather than too small. Nevertheless it may be concluded that interplanetary dust is a possible source of the material of noctilucent clouds. It also seems possible that if this dust contained a substantial proportion of iron, or if it were charged by solar radiation, it might enter the earth's atmosphere preferentially in high latitudes, thereby locally increasing in concentration.

11. Dust in the high atmosphere

Other evidence is available to suggest the presence of considerable quantities of dust in the stratosphere. For example, observations of the earth's shadow during lunar eclipses suggest that the atmosphere contains obscuring matter to levels well within the stratosphere, which LINK (1950) has proposed to be meteoritic dust extending from a height of 50 to 100 km down to the ground. Photographs of Mars and of Venus give ultra-violet images greater, by 3—6% and 2% respectively, than the infra-red images. According to LINK (1950), Fessenkov and Menzel have proved that an explanation in terms of molecular diffusion is not possible, and the discrepancies are tentatively explained by the presence of similar meteoritic dust layers. The smaller gravitational constants on these planets lead to a level of meteorite fusing which is farther above the ground than on earth, favouring a deeper dust layer. On the other hand on Jupiter, for example, the layer could be only some tens of km deep and could not produce noticeable effects. BOUSKA and SVESTKA (1950) have collected measurements of the size of the earth's shadow on the moon during lunar eclipses; the mean from 33 observations shows the shadow to be larger than the earth by about 2%, individual values varying from 1.7 to 2.5%. Bouska and Svestka were unable to find any significant relation between the enlargement and the declination of the sun or moon, the parallax of the sun, or the solar activity, but do claim a correlation with the occurrence of meteor showers.

From an analysis of observations of solar intensity at Mount Wilson in the period 1908—1920, ZACHAROV (1952) deduces an annual increase in atmospheric absorption by a few per cent over about four weeks including the latter half of August, which he associates with the arrival of the Perseids in

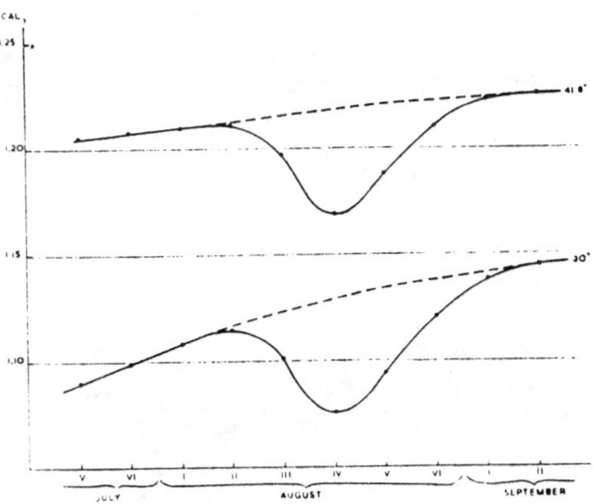

Fig. 13. Average solar intensity for each 5 days for a sun's elevation of 30° (lower curve) and 41.8° (upper curve), from 3282 and 2239 observations made at Yakutsk (1931—42), Pavlovsk (1913—37), Kursk (1928—35), Rostov (1930—37), Tbilissi (1928—37), Tashkent (1926—37) and Samarkand (1931—37). After Kalitin (1944).

mid-August. KALITIN (1944) showed that a similar decrease in the solar intensity at the same time of year is evident in the actinometric measurements of a number of widely-separated stations in Russia (Fig. 13); the coincidence is extraordinary. Presuming the effect is produced by the entry of a dust cloud into the atmosphere LINK (1955) infers the particle radius to be about 10^{-5} cm, and that the dust arrives at about the rate deduced above for accretion with the zodiacal light cloud. However, since the optical density of such a cloud would be several orders of magnitude greater than that of noctilucent clouds, it seems remarkable that its annual arrival is not a visible and celebrated event. Dust clouds from the catastrophic eruptions of Krakatoa (1883), M. Pelée (1902) and Katmai (1912), produced more prolonged decreases in solar intensity of about the same amount, but equally produced twilight phenomena of abnormal intensity (GRUNER and KLEINERT, |1927). Among these phenomena the purple light, which is attributed to particles, of dimensions a few microns, in the lower stratosphere, is especially enhanced in such disturbed periods. It may be significant that Fig. 14, taken from GRUNER and KLEINERT (1927), shows that abnormally intense purple lights are notably more frequent in August (when, however, convective lifting of dust from the ground is intense) than at other times of the year. Although, therefore, the evidence of Zacharov and Kalitin may need some other interpretation, their suggestions deserve further investigation. It may be noted that the recovery of the actinometer measurements after four weeks would suggest, if the disturbance is due to the arrival of a dust cloud, that the dust is substantially precipitated from the atmosphere within this period.

Direct confirmation of the presence of dust concentrations in the lower stratosphere comes from aircraft observations (e.g., PACKER and LOCK, 1950), and the photometric twilight observations of BIGG (1956) provide a strong indication of the persistent presence of concentrations of dust below the 80 km inversion. PENNDORF (1954) has assembled observations of particle concentration at various levels in the troposphere which show that the strong decrease with height ceases above about 4 km, and that thereafter the ratio between the concentration and the air density (a ratio we might expect to be preserved during the stirring of air to different levels) tends to increase with height, suggesting that at least a proportion of the particles in the upper troposphere have a source at a higher level. This result receives some support from the particle concentrations observed to occur in water clouds at the cirrus levels and in the spectacular mother-of-pearl clouds which occasionally form some 25 km above the Norwegian mountains. From coronae seen on freshly-

forming cirrus I deduced that the concentration of ice crystals in the clouds is about $7/cm^3$ (LUDLAM, 1956). Since these clouds were formed in moderately strong updraughts this probably also represents practically the total concentration of condensation nuclei at these levels, in accord with the values inferred by Penndorf from flight measurements of solar radiation. Thus at the cirrus levels (about 10 km) the ratio $C = n/\varrho$, where n is the concentration of small particles, has a value of about $7/5 \times 10^{-4}/g$, or $1.4 \times 10^4/g$. An observation by Størmer (1948) of the angular radius (15°) of the first red ring of a lunar corona in mother-of-pearl clouds, indicates spherical cloud particles of diameter about $2.5\,\mu$. At the temperature of these clouds (about $-80°$ C) the concentration of water condensed during a rise of a few hundred metres above the condensation level amounts to about 10^{-10} g/cm³; if this is divided uniformly (as justified by the purity of the diffraction colours) amongst droplets of the appropriate radius, their concentration becomes about $1/cm^3$. Since these wave clouds are also produced in updraughts of at least moderate strength, we can equally assume this value to represent the total concentration of condensation nuclei. At 25 km the air density is approximately 6×10^{-5} g/cm³, so that here the ratio C becomes 1.6×10^4. At the level of the noctilucent clouds the air density is about 3×10^{-8} g/cm³, and the inferred concentrations of particles vary from $10^{-2}/cm^3$ (for a radius 10^{-5} cm) to $1/cm^3$ (radius 10^{-6} cm), corresponding to values of C of 3×10^5 to 3×10^7. These values are rather higher than those deduced for the 10 km and 25 km levels, but the latter are so nearly the same and sufficiently close to the lower estimate for the 80 km level to encourage the view that the distribution of dust throughout the stratosphere is consistent with the existence of efficient mixing movements, and not inconsistent with a meteoritic source.

The occurrence of stirring motions in the lower stratosphere, where the air is statically very stable, and even in the upper stratosphere, where the air is stable for dry adiabatic ascent and condensation is impossible, is consistent with the uniform composition of the atmosphere but is difficult to reconcile with the usual thermodynamical arguments. RAKIPOVA (1947), however, has pointed out that dust

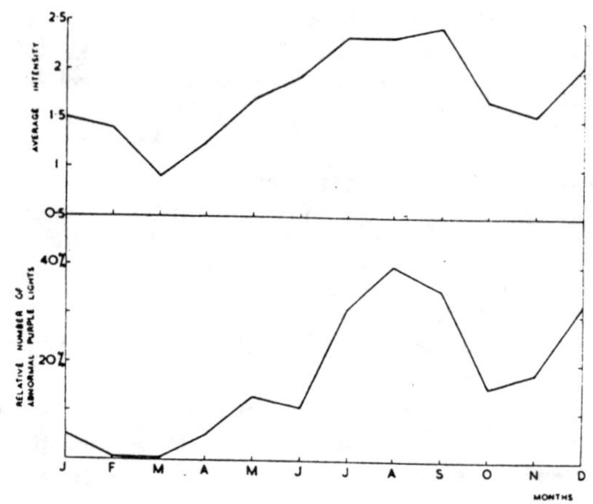

Fig. 14. Average intensity of purple light (on scale from 0 to 5), and relative number of abnormally strong purple lights, Berne, 1905—1912 (after Gruner and Kleinert, 1927).

clouds may be agencies for producing upward movement of air, by intercepting sufficient sunshine to become appreciably warmed. Air containing a dust cloud could thus commence to rise through a clearer environment, and, since it could carry its dust with it, may remain warmer and more buoyant than the surroundings. In these circumstances vertical movements are possible in statically stable, even isothermal atmospheres. Rakipova calculates the concentrations of dust particles required to produce significant effects and finds, for a radius 10^{-5} cm, a concentration of $1/cm^3$ at 80 km and about $10/cm^3$ at 30 km. These estimates may be obtained in the following way. At the 80 km level, for example, where the air density is 3×10^{-8} g/cm^3, solar energy is intercepted by particles (assumed to have a low albedo) at the rate of about $N\pi r^2$ cal/cm^3 min, where N is the concentration of particles of radius r, or about 200 Nr^2 cal/cm^3 hr. If this energy is communicated to the surrounding air it is sufficient to warm it at the rate of about 10^3 Nr^2/ϱ °C/hr, or about 2×10^{10} Nr^2 °C/hr at the 80 km level, so that for $r = 10^{-5}$ cm and $N = 1/cm^3$ the rate of heating becomes appreciable and equivalent to a temperature rise of 2° C/hr. In a lapse-rate of 6° C/km this could correspond to a vertical movement of a km in 3 hours, at a speed of about $1/3$ m/sec. Evidently for this process to be important the concentration of particles would need to be at least two orders of magnitude greater than that deduced to occur in noctilucent clouds ($10^{-2}/cm^3$ for $r = 10^{-5}$ cm). Since meteoritic material is unlikely to enter the atmosphere in a uniform stream, such high concentrations may sometimes occur in clouds which, if sunlit for a sufficient period, could by this means rise several km to the 80 km level. Above the base of the inversion there, the ascending speeds would be substantially reduced, and insufficient to lift the particles of the cloud, whose fall speeds amount to at least several cm/sec, so that the ascent would cease.

Particles which settle from the stratosphere into the troposphere aggregate with the condensation nuclei derived from the earth's surface and become involved in precipitation processes. The observations quoted above suggest that in the upper troposphere, where the total concentration of nuclei is about $10/cm^3$, solid meteoritic particles may often outnumber the nuclei of terrestrial origin. In the lower troposphere, where their concentration could not be significantly greater, they could comprise only an insignificant fraction of the condensation nuclei, although, in view of their silicate components, they might conceivably be an important source of freezing nuclei in clean air masses (on the other hand SCHAEFER, 1955, found that particles of magnetic dust collected in the open air, and particles from the smoke of a fused *metallic* meteorite, were not particularly effective ice nuclei).

12. The formation of noctilucent clouds

Some of the observations made in Sweden, particularly when noctilucent clouds developed after midnight, strongly suggested that the clouds formed within periods of several minutes in regions which previously, although apparently favourably illuminated, contained no discernible details. GRISHIN (1955) has remarked that the featureless "flare" can often be detected before the appearance of definite clouds, and thus also implies that clouds may form in the field of view, not necessarily simply arriving from places in which they already existed, or being revealed by improving viewing conditions.

The readiest interpretation of the formation in a restricted layer of clouds having definite details is in terms of a condensation process in air which is ascending or otherwise being chilled. However, we have already discussed difficulties in the way of this explanation. To reiterate, a condensation can occur only if the air were much moister or much colder than we have reason to expect, and even then the brightness of the clouds would suggest the condensation to be only partial, while the problem of accounting for the condensation nuclei would remain. If, moreover, the clouds were to appear by the growth of their particles, their first details might be expected to have a distinct blue colour which becomes steadily more white. Such a change of colour has only once been reported (PATON, 1954) and cannot be regarded as characteristic. Equally, then, it seems unlikely that the clouds become visible by a condensation (at vapour pressures below the saturation value) upon hygroscopic components of the nuclei.

If condensation of water vapour plays no part in the formation of the clouds, it must be considered how dust, probably present throughout the stratosphere, can become so disposed to produce distinct clouds which are always measured to be at a height of about 80 km. It seems likely that the formation of a *haze top* at the level of the inversion there is essential.

Pronounced haze tops are common in the troposphere, and are usually associated with inversions; in a layer of steep lapse-rates beneath the inversion particles of dust or other nuclei are stirred up from their sources near the ground, but above the base of the inversion stirring movements are suppressed by the strong static stability, and the concentrations of the particles are suddenly reduced by a factor of 10 to 100. Moreover, beneath the inversion the relative humidity increases with height, so that the size of the particles, which are all more or less hygroscopic, reaches a maximum at the haze top; immediately above, the air is usually very dry and the particles there are inappreciably swollen by condensation, so that the presence and distribution of water vapour are important in sharply defining the top of the haze layer. When wave motions disturb the layer, concentrations of haze become visible from below to an observer who looks along low angular elevations, for then adjacent lines of sight pass through layers of appreciably different optical thickness (Fig. 15). Patterns of dark haze billows are seen against a twilight background, or, if the haze top is still illuminated, bright billows are seen separated by darker spaces. Such billows are a feature of the dust cloud in the lower stratosphere, which cannot be seen overhead, but only at rather low elevations.

It is probable that the details of noctilucent clouds are produced in a similar way by disturbances in a haze top at the 80 km inversion.

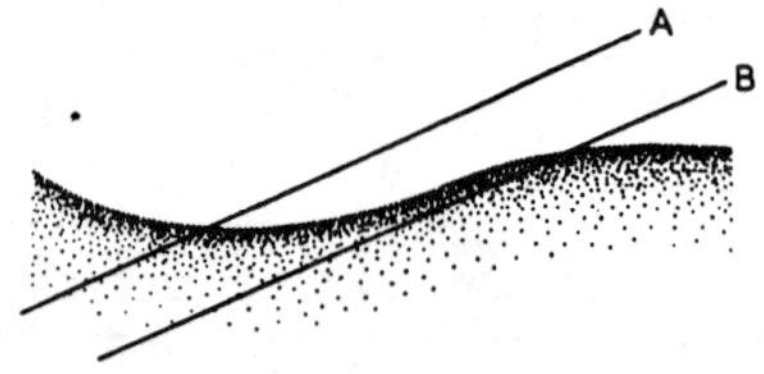

Fig. 15. A haze layer with a sharply-defined and undulating top presents visible features at low elevations, for there are marked differences in optical thickness along adjacent lines of sight, such as A, B.

It would not be surprising if all the sharply-defined clouds, which are naturally those selected for measurement, were necessarily associated with the haze top, and that other clouds are present which are considerably lower, but have more diffuse edges, and whose height is therefore never determined. The distinct "shadows" which sometimes seem to be cast by the denser cloud features (Fig. 1) almost certainly represent irregularities in a haze top, for the optical thickness of the clouds is so small that they could not possibly cast real shadows.

The production of a haze top at 80 km seems to require the ascent of dust clouds from below. WEXLER (1950) has discussed observational evidence of very steep, even superadiabatic lapse rates, in the layer between about 55 and 80 km. These arise during the daytime by solar heating of the ozone layer; although the ozone is concentrated at lower levels, the diurnal temperature change has a maximum at about 50 km (JOHNSON, 1953). It must be considered that the noctilucent clouds are observed only at a season and in latitudes such that a very prolonged heating of the ozone layer has occurred: at mid-summer in high latitudes the sun does not set on the top of the ozone layer. At this season, therefore, convective movements are particularly likely, and we may suppose that the ascending motions will occur preferentially in dust clouds. Consequently a concentration of dust may arise beneath the 80 km inversion, where the dusty air loses its buoyancy and spreads horizontally. If the dust has entered the stratosphere from below, as in eruptive clouds, a pronounced haze top is readily produced. If it has arrived from interplanetary space as the condensation product of fused meteors, it is not possible to assess the change of concentration which occurs across the 80 km inversion without information on the accumulation which occurs in the convective layer and on the levels at which the condensation takes place. Bright meteors produced by particles of mass several grams arriving at about 40 km/sec are first visible at a height of about 100 km and disappear at about 70 km; larger particles are able to penetrate to lower levels, and so considerable proportions of their condensation products are introduced below the 80 km level. This circumstance would help in the production of a well-defined haze top; on the other hand, estimates of the evaporation of particles of the small size inferred to compose the zodiacal light cloud, based upon the considerations summarized by WHIPPLE (1943), indicate that they must be evaporated entirely above the 80 km level. It therefore seems likely that only meteorological processes could account for the formation of a haze top or a marked concentration of small particles near the 80 km level.

It is interesting that dust clouds, which in sunlight may cause an appreciable warming of air, may equally be responsible for a significant cooling of air when they lie in the earth's shadow. Since the long-wave radiation emitted by the earth and its atmosphere has an equivalent black-body temperature of about 245° A, a particle in the high atmosphere (which radiates in all directions but receives radiation from the hemisphere below) is in radiative equilibrium at a temperature of about $245/2^{1/4}$, or 206° A. The air temperature in the upper stratosphere, at least to within a few km of the 80 km inversion, exceeds this value, so that dust clouds may behave as a cooling agent and promote sinking motions. Conceivably, in the high latitude summer, when the earth's shadow moves mainly latitudinally, there may be a tendency for a circulation in the vicinity of the shadow, in the layer from 50 to 80 km, as a result of a predominance of ascending motions in a belt a few hundred km broad in the sunlit region, and of descending motions in a similar belt inside the earth's shadow. Such a circulation would involve a component of the motion at the 80 km level directed away from the sun, as seems to be observed in the motions of noctilucent clouds.

13. Conclusion

It appears very unlikely that the condensation of water vapour plays any part in the formation of noctilucent clouds. They are probably composed of small solid particles, which may enter the stratosphere from below during major volcanic eruptions, or which may be produced in the stratosphere by the condensation of gases in the wakes of meteors. During the high latitude summer steep lapse rates develop in the upper stratosphere because of the prolonged heating of the ozone layer, and

convective motions, possibly aided by solar warming of particle clouds, distribute the particles throughout a layer extending up to the 80 km inversion. Irregularities and wave disturbances in the top of the dusty layer produce visible clouds. The clouds may appear predominantly in regions near the earth's shadow at midnight, when adjacent belts of the atmosphere remain for some hours on either side of the shadow, so that a circulation is established in the upper stratosphere in which the sunlit air rises and the shadowed air sinks. This circulation and the haze top at 80 km which favour the appearance of the clouds may not develop in other latitudes and seasons because of the absence of steep lapse rates and because of the rapidity with which the earth's shadow moves through the atmosphere. The tendency for clouds to appear after rather than before midnight may be related to the diurnal variation in the accretion of interplanetary particles, but this and some other cloud properties need more thorough investigation. Noctilucent clouds may interest astronomers as evidence of the accretion of interplanetary material, and are important to meteorologists as indicators of atmospheric processes affecting the stratosphere, and as reminders that a considerable proportion of the condensation and freezing nuclei of the upper troposphere may arrive from outer space.

In future observations of noctilucent clouds it will be important to record occasions when they are absent in apparently favourable viewing conditions. It would be most desirable to establish a network of observing stations, to examine the true areal extent of the clouds and the simultaneity or otherwise of their appearance in different parts of the world. The ethereal beauty of the clouds on the rare occasions of good displays is a recompense for many hours of attentive but fruitless observing.

Acknowledgments

The observations of noctilucent clouds were made as part of a research programme in cloud physics under development at the International Institute of Meteorology in Stockholm. This programme was supported by Statens Tekniska Forskningsråd, Sweden, and the Munitalp Foundation, U.S.A. Grateful acknowledgment is made of the assistance of staff of the

Institute, and thanks are due to Mr. J. Paton, for making available a summary of his observations, and to Dr. R. M. Goody for illuminating comments and discussions. G. Witt gave helpful advice on the calculation of fall speeds of particles in rarified gases.

REFERENCES

ALLEN, C. W., 1946: The spectrum of the corona at the eclipse of 1940. *Mon. Not. R.A.S.*, **106**, p. 137.

ARAGO, F., 1854: *Complete works.* **4**, p. 73.

ARCHENHOLD, F. S., 1928: *Die leuchtenden Nachtwolken...*, Das Weltall. 27. Jahrgang.

BLACKWELL, D. E., 1952: A comparison of the intensities of infra-red and violet radiation from the solar corona... *Mon. Not. R.A.S.*, **112**, p. 652.

BIGG, E. K., 1956: The detection of atmospheric dust and temperature inversions by twilight scattering. *J. Met.*, **13**, p. 262.

BOUSKA, J., and SVESTKA, Z., 1950: On the variation of the enlargement of the earth's shadow during lunar eclipses, *Bull. Astr. Inst. Czecho.*, **2**, p. 6.

BOWEN, E. G., 1953: The influence of meteoritic dust on rainfall. *Austral. J. Phys.*, **6**, p. 490.

BULL, G. A., and JAMES, D. G., 1956: Dust in the stratosphere over western Britain on April 3 and 4, 1956. *Met. Mag.*, **85**, p. 293.

FESSENKOV, V. G., 1949: The mass of the atmospheric residue of the Sikhote-Alin meteorite. *Dok. Akad. Nauk SSSR*, **66** (translation T 133 R, Defence Sci. Inf. Service, Canada, 1954).

GRISHIN, N. I., 1955: On the structure of noctilucent clouds. *Met. i Gidr.*, No. 1, p. 23 (in Russian).

GRUNER, P., and Kleinert, H., 1927: *Die Dämmerungserscheinungen*, Probleme der Kosmischen Physik, Henri Grand, Hamburg.

JACCHIA, L. G., 1955: Physical theory of meteors. *A. J.*, **121**, p. 521.

JACOBS, L., 1954: Dust cloud in the stratosphere. *Met. Mag.*, **83**, p. 115.

JESSE, O., 1896: Die Höhe der leuchtenden Nachtwolken. *Astr. Nachr.*, **140**, p. 161.

JOHNSON, F. S., 1953: High-altitude diurnal temperature changes due to ozone absorption. *Bull. Amer. Met. Soc.*, **34**, p. 106.

KALITIN, N. N., 1944: Cosmic dust according to actinometric measurements. *Comptes Rendus (Doklady) Acad. Sci. URSS*, **45**, p. 375.

KHVOSTIKOV, I. A., 1952: Silvery clouds. *Priroda*, Moscow, **5**, p. 49 (in Russian).

KNUDSEN, M., 1934: The Kinetic Theory of Gases. Methuen, London.

KRAKATOA COMMITTEE, 1888: *The eruption of Krakatoa and subsequent phenomena.* Trübner and Co., London.

LINK, F., 1950: Couche de poussières météoriques dans une atmosphère planétaire. *Bull. Astr. Inst. Czecho*,. **2**, p. 1.

— 1955: Contribution of meteoritic material to atmospheric absorption, "Meteors", Special Suppt. *J. Atm. Terr. Phys.*, **2**, p. 36.

LOVELL, A. C. B., 1954: *Meteor Astronomy*, Clarendon Press, Oxford.

— and Clegg, J. A., 1952: *Radio Astronomy*, Chapman and Hall, London.

LUDLAM, F. H., 1956: The forms of ice clouds, II. *Quart. J. R. Met. Soc.*, **82**, p. 257.

MINNAERT, M., 1940: *Light and colour in the open air*, G. Bell, London.

MURGATROYD, R. J., GOLDSMITH, P., and HOLLINGS, W. E. H., 1955: Some recent measurements of humidity from aircraft up to heights of about 50,000 ft over southern England. *Quart. J. R. Met. Soc.*, **81**, p. 533.

O.M.I., 1955: *Definitions and specifications of water vapour in the atmosphere*, Publ. 79, Lausanne.

ÖPIK, E. J., 1951: Collision probabilities with the planets and the distribution of interplanetary matter. *Proc. Roy. Irish Acad.*, A, **54**, p. 165.

— 1955: Cosmic sources of deep-sea deposits. *Nature*, **176**, p. 926.

PACKER, D. M., and LOCK, C., 1951: The brightness and polarisation of the daylight sky at altitudes of 18,000 to 38,000 ft. *J. Opt. Soc. Amer.*, **41**, p. 473.

PATON, J., 1949: Luminous night clouds. *Met. Mag.*, **78**, p. 354.

— 1951: Simultaneous occurrence of aurora and noctilucent clouds. *Ibid.*, **80**, p. 145.

— 1954: Direct evidence of vertical motion at about 80 km provided by photographs of noctilucent clouds, p. 31, Proc. Toronto Met. Conf., Roy. Met. Soc., London.

PENNDORF, R., 1954: The vertical distribution of Mie particles in the troposphere. *Geophys. Res. Paper* 25, G.R.D., Cambridge, U.S.A.

PETTERSSON, H., and ROTSCHI, H., 1950: Nickel content of deep-sea deposits. *Nature*, **166**, p. 308.

PIOTROWSKI, S. L., 1952: The collisions of asteroids. *A. J.*, **57**, p. 23.

RAKIPOVA, L. P., 1947: Possible effect of dust on vertical air movements and on isothermy in the stratosphere. *Izv. Akad. Nauk SSSR*, **11**, p. 15 (translation T 199 R, Defence Sci. Inf. Service, Canada, 1956).

RENSE, W. A., JACKSON, J. M., and TODD, B., 1953: Measurements of the inner zodiacal light during the total solar eclipse... *J. Geophys. Res.*, **58**, p. 369.

SCHAEFER, V. J., 1955: The question of meteoric dust in the atmosphere, unpublished.

SCORER, R. S., 1951: Billow clouds. *Quart. J. R. Met. Soc.*, **77**, p. 235.

SCULTETUS, H. R., 1949: Ältere Beobachtungen von Leuchtstreifen. *Z. f. Met.*, **3**, p. 272.

SPANGENBERG, W. W., 1949: Über die leuchtenden Nachtwolken, 1932—41. *Wetter und Klima*, **2**, p. 15.

STØRMER, C., 1933: Height and velocity of luminous night clouds observed in Norway, 1932, Publ. 6, Univ. Obs., Oslo.

— 1935: Measurements of luminous night clouds in Norway, 1933 and 1934. *Astrophys. Norvegica*, **1**, No. 3.

— 1948: Mother-of-pearl clouds. *Weather*, **3**, p. 13.

THOMSEN, J. W., 1953: The annual deposit of meteoritic dust. *Sky and Telescope*, **12**, p. 147.

VAN DE HULST, H. C., 1947: Zodiacal light in the solar corona. *Ap. J.*, **105**, p. 471.

VESTINE, E. H., 1934: Noctilucent clouds. *J. Roy. Astr. Soc. Canada*, p. 249.

WEXLER, H., 1950: Annual and diurnal temperature variations in the upper atmosphere. *Tellus*, **2**, p. 262.

WHIPPLE, F. J. W., 1930: The great Siberian meteorite and the waves, seismic and aerial, which it produced. *Quart. J. R. Met. Soc.*, **56**, p. 287.

— 1934: On phenomena related to the great Siberian meteorite. *Ibid.*, **60**, p. 505.

WHIPPLE, F. L., 1943: Meteors and the earth's upper atmosphere. *Rev. Mod. Phys.*, **15**, p. 246.

— 1950: The theory of micro-meteorites, I, *Proc. Nat. Acad. Sci.*, *Wash.*, **36**, p. 687.

— 1951: The theory of micro-meteorites, II. *Ibid.*, **37**, p. 19.

WHIPPLE, F. L., 1952: Results of rocket and meteor research, *Bull. Amer. Met. Soc.*, **33**, p. 13.

— 1954: Photographic meteor orbits and their distribution in space, *A. J.*, **59**, p. 201.

— 1955: A comet model. III. The Zodiacal light. *Ap. J.*, **121**, p. 750.

— and HAWKINS, G. S., 1956: On meteors and rainfall. *J. Met.*, **13**, p. 236.

WYATT, S. P., and WHIPPLE, F. L., 1950: The Poynting-Robertson effect on meteor orbits. *Ap. J.*, **111**, p. 134.

ZACHAROV, I., 1952: Influence des Perséides sur la transparence atmosphérique. *Bull. Astr. Inst. Czecho.*, **3**, p. 82.

Scattering and Polarization Properties of Water Clouds and Hazes in the Visible and Infrared

D. Deirmendjian

The extinction coefficient, albedo of single scattering, and differential scattering and polarization properties of water clouds and hazes in the visible and infrared have been computed using the complete Mie
series.   The results with three types of size distributions are presented and compared with observations.
These show a strong dependence of angular intensity and polarization patterns on the size distribution,
the size range, and the dielectric and absorbing properties of water droplets at each wavelength.   A
peculiarity of scattering at angles near 45°, observed experimentally and independently by two authors, is
corroborated by the numerical results.   Prominent observational features characteristic of natural fog,
such as an extremely bright and narrow aureole, cloudbows, and glories, are reproduced in a model cloud of
spherical water droplets, with a wide distribution in droplet radius and a maximum concentration at a 4-μ
radius.

Introduction

In studying the theoretical scattering properties of polydispersed suspensions, such as natural hazes and clouds in the atmosphere, two important facts should be considered.   (a) Because of the finite size of the particles, even if these are assumed spherical, the angular scattering characteristics cannot be approximated with sufficient accuracy by asymptotic expressions based on geometric optics or Green's function approximations for the internal field.   The complete Mie series must be used for each particle.   (b) The size distribution function used must reproduce as faithfully as possible the size spectrum of the suspension in question.

In this paper we shall present briefly examples simulating atmospheric water clouds and hazes with size distributions that are sufficiently realistic to allow comparisons with observational data.   The scattering properties of single Mie particles, both with real and complex index of refraction, have been adequately described previously.[1,2]   Therefore, we shall describe only the size-distribution functions used and justify their choice.

The author is with the RAND Corporation, Santa Monica, California.

Received 10 September 1963.

The research reported here was supported by a U.S. Air Force contract RAND.

Parts of this paper were originally presented at the International Symposium on Radiation of the Radiation Commission, IAMAP, Vienna, 1961.

Size Distribution Function

The function chosen is a generalization of that first proposed by Khrgian and Mazin[3,4] for clouds.   A survey of various proposed functions showed that the latter type seems to fit fairly well with measurements both on natural water clouds and aerosols, and it has the great advantage that its parameters have readily interpretable physical meaning.   In its most general form this function may be written

$$n(r) = ar^{\alpha}e^{-br^{\gamma}} \tag{1}$$

where $n(r)$ is the volume concentration at the radius r and $a,\ \alpha,\ b,$ and γ are positive constants.   For a particular choice of α and γ, the remaining constants can be uniquely determined by the total number of particles N per unit volume, and by the critical or mode radius r_c, where the concentration is at a maximum.   Thus the constant a can be determined from the integral

$$N = \int_0^{\infty} n(r)dr = \frac{a}{\gamma} b^{-\frac{\alpha+1}{\gamma}} \Gamma\left(\frac{\alpha+1}{\gamma}\right), \tag{2}$$

and b from the derivative, which can be written

$$\frac{d}{dr} n(r) = ar^{\alpha-1}e^{-br^{\gamma}}(\alpha - \gamma br^{\gamma}), \tag{3}$$

which vanishes when

$$b = \frac{\alpha}{\gamma r^{\gamma}}, \quad r = r_c, \tag{4}$$

as well as for $r = 0$ and ∞.

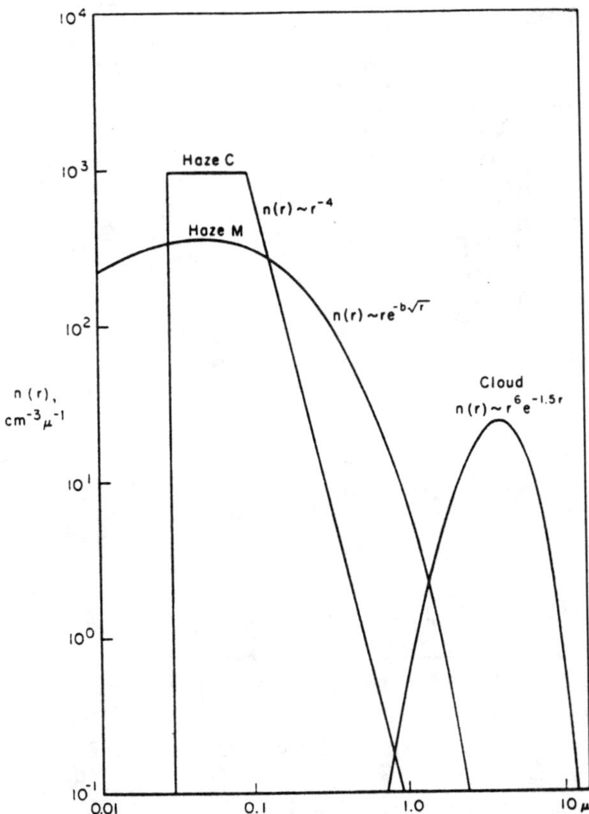

Fig. 1. Three size distribution functions used in the integration of the Mie functions. Total concentration 100 cm^{-3}.

In particular, with the choice $\alpha = 6$, $\gamma = 1$, $r_c = 4\ \mu$, and $N = 100$ cm^{-3} as a nominal concentration, one gets the distribution function

$$n(r) = 2.373\ r^6 e^{-1.5r}\ \text{cm}^{-3}\mu^{-1}, \tag{5}$$

when the radius is expressed in microns and the concentration per cubic centimeter. This model, marked "cloud" in Fig. 1, reproduces fairly well some of the distributions found by Durbin[5] for cumulus clouds having a depth between 230 m and 2100 m, as averaged and reproduced as "cumulus type I" in a diagram by Singleton and Smith.[6]

The discontinuous distribution called "haze C" in Fig. 1 roughly approximates the continental hazes measured by various workers and used by the author previously.[7-9]

The other continuous distribution in Fig. 1, marked "haze M" is obtained by putting $\alpha = 1$, $\gamma = \frac{1}{2}$ in (1) with $r_c = 0.05\ \mu$, resulting in the specific function

$$n(r) = 5.33 \times 10^4\ r e^{-8.944\ \sqrt{r}}. \tag{6}$$

This was found to agree with unpublished counts of aerosol particles made in Los Angeles by Gilbert,[10] and may be taken to represent coastal conditions. The differences from the model C haze are considerable, in that there is a much larger concentration of particles in the 0.15-μ to 3.5-μ range. The distribution (6) seems also to correspond fairly well to that of stratospheric aerosols found by Junge et al.[11] at least in the

range $0.2 < r < 1\ \mu$. Note that a linear representation of the continuous distributions, shown in Fig. 1 would result in bell-shaped curves which are skewed toward the larger sizes.

Integration of the Mie Functions with Size Distribution

To get the total scattering and absorption cross sections per unit volume of the cloud or haze, one has to integrate the Mie single-particle cross sections with respect to the appropriate distribution. Making the convenient change of variable $r = k^{-1} x$, where x is the Mie size parameter and $k = 2\pi/\lambda$ is the free-space propagation constant at the wavelength λ, one can rewrite the size distribution function (1) in the form

$$f(x,k) = ak^{-3-\alpha}x^\alpha \exp\left\{-b(x/k)^\gamma\right\}, \tag{7}$$

where the factor k^{-3} is introduced for convenience. It is easy to show then that the volume cross sections β may be obtained by the integration

$$\beta(m,\lambda;\ x_1,x_2) = \pi \int_{x_1}^{x_2} x^2 f(x,k)\ K(m,x)dx, \tag{8}$$

where m is the complex or the real index of refraction of the particles, K is the Mie extinction, scattering, or absorption cross section, as the case may be, divided by the geometrical cross section,[1] and x_1 and x_2 correspond to the lower and upper limits in the particle size.

The elements of the normalized scattering matrix or phase matrix as defined by Chandrasekhar[12] are given by the integrals

$$\frac{P_j}{4\pi}(m,\theta,\lambda;\ x_1,x_2) = \frac{1}{\beta_{sc}} \int_{x_1}^{x_2} f(x,k)\ i_j(m,x,\theta)dx, \quad j = 1,2,3,4, \tag{9}$$

where θ is the scattering angle, and $i_j(m,x,\theta)$ are the dimensionless Mie intensity functions computed previously,[1] related to the intensity or phase matrix operating on the Stokes vector of the incident radiation in the manner indicated by various authors.[13,14] β_{sc} is the integral (8) over the *scattering* cross section.

It is in the behavior of the integrals (8) and (9) as a function of x_2 that the advantage of the distribution functions of the type (1) becomes evident. Numerical tests using our IBM 7090 program show, in fact, that in this case, these integrals always converge to a constant value for all $x_2 > X$, where X is a finite number depending on the logarithmic derivative of the function $n(r)$ with respect to r. In particular, it appears that a sufficient condition for convergence to begin, for all integrals (8), (9) *including the forward scattering* case with $\theta = 0$, is that r_2 (or correspondingly x_2) be such that

$$-\frac{d}{d\log r}\log n(r) = \alpha\left[\left(\frac{r}{r_c}\right)^\gamma - 1\right] > 4. \tag{10}$$

Solving the inequality to the right, we get the condition for convergence to begin

$$r_2 > r_c\,(1 + 4/\alpha)^{1/\gamma}. \tag{11}$$

Table I. Volume Extinction Coefficient and Albedo

λ	Re$\{m\}$	Im$\{m\}$	Haze C (2300 cm^{-3}, $r_2 = 5\ \mu$)		Haze M (100 cm^{-3}, $r_c = 0.05\ \mu$)		Cloud (100 cm^{-3}, $r_c = 4.0\ \mu$)	
			β_{ext} km^{-1}	Albedo A (β_{sc}/β_{ext})	β_{ext} km^{-1}	Albedo A	β_{ext} km^{-1}	Albedo A
0.45 μ	1.34	0	0.1206	1.0	0.1056	1.0	16.33	1.0
0.70 μ	1.33	0	0.0759	1.0	0.1055	1.0	16.72	1.0
1.61 μ	1.315	0	0.0312	1.0	0.0691	1.0	17.58	1.0
2.25 μ	1.29	0	0.0194	1.0	0.0424	1.0	18.21	1.0
3.07 μ	1.525	0.0682	0.0289	0.620	0.0602	0.721	18.58	0.529
3.90 μ	1.353	0.0059	0.0128	0.930	0.0236	0.948	20.65	0.914
5.30 μ	1.315	0.0143	0.0075	0.800	0.0112	0.826	24.01	0.884
6.05 μ	1.315	0.1370	0.0129	0.260	0.0189	0.297	19.86	0.543
8.15 μ	1.29	0.0472	0.0050	0.385	0.0062	0.410	18.75	0.746
10.0 μ	1.212	0.0601	0.0032	0.202	0.0045	0.178	11.18	0.601
11.5 μ	1.111	0.1831	0.0064	0.051	0.0097	0.044	10.10	0.289
16.6 μ	1.44	0.4000	0.0082	0.104	0.0134	0.075	16.97	0.395

This is a very convenient property in regard to the contribution of the larger particles in a suspension, where the corresponding Mie series are very long and machine computation time becomes costly. It is also a physically justifiable feature of long-lived atmospheric aerosol and cloud particles, which are naturally limited in the maximum size. Distributions such as model C with $n(r) \sim r^{-4}$ are not convenient because the integrated volume extinction, and scattering in the forward area depend on the choice of the upper limit r_2.

The numerical results further showed that, for x in the region defined by (11), partial integrals of the type (8) and (9), when plotted as a function of $x < x_2$ are smooth and monotonic curves, with little change in value when the integration interval Δx is varied. This behavior is a result mainly of the properties of the *type of continuous size distribution* chosen here, since it is known[1,15] that the intensity for single particles at a constant θ, especially in the backward hemisphere, shows large and numerous fluctuations whose amplitude and frequency increase with x. This explains the absence of numerous coronas, cloudbows, and glories in sunlit natural clouds and fogs, which would certainly appear if the droplets were uniform in size or followed an extremely narrow distribution around a characteristic size.

It should be noted from the form of the integral (9) that the phase functions are independent of the total concentration N of particles per unit volume, as should be evident from physical considerations. This statement is, of course, valid when the concentration is small enough that only incoherent or independent scattering need be considered.

Volume Extinction and Albedo

Examples of the extinction per unit path for the three models and at selected wavelengths are shown in Table I. The particles are assumed to be made of liquid water, with the real or the complex index of refraction indicated in Table I, adapted from Centeno.[16] The table also lists the computed albedo A per unit volume of space, that is, the fraction of energy scattered in all directions weighted by the distribution function, compared to that scattered and absorbed. The numbers in Table I illustrate clearly the effects of changes in distribution function, in the relative size, and in the index of refraction of the particles. Noted particularly are the similarities and differences between the two haze models. The lower albedo for model C at $\lambda 3.07\ \mu$ and $\lambda 6.05\ \mu$ is explained by the larger proportion of small particles in it as compared to model M (see Fig. 1), such particles being relatively more efficient real absorbers than the larger ones for any given complex index. These differences are even more pronounced when comparing the haze and cloud distributions, especially at $\lambda 10\ \mu$.

The extinction values for the cloud model show the interesting fact that these should increase with wavelength in the near-infrared region, even though the relative size decreases, partly because of the increase in the absorption coefficient of liquid water.

A comparison of the corresponding extinction coefficients under Table I for the model C haze, with those approximated previously by Deirmendjian[9] by means of an empirical extension of van de Hulst's analytical approximation,[14] shows very good agreement, indicating the sound basis of van de Hulst's original analysis. Those corresponding to a cloud also agree in order of magnitude and wavelength dependence, but not in detail, because a different model, with $r_c = 3.5\ \mu$ and a narrower symmetrical distribution, was used in the earlier approximate calculation by Deirmendjian.[9]

Normalized Intensity Functions P_1 and P_2

Figures 2–9 represent examples of the integrated intensity functions or phase matrix elements $P_1/4\pi$ (solid line) and $P_2/4\pi$ (dashed lines) which are normalized so that they satisfy the condition:

$$\frac{1}{8\pi} \int_\Omega [P_1(\theta) + P_2(\theta)]\, d\omega = 1. \qquad (12)$$

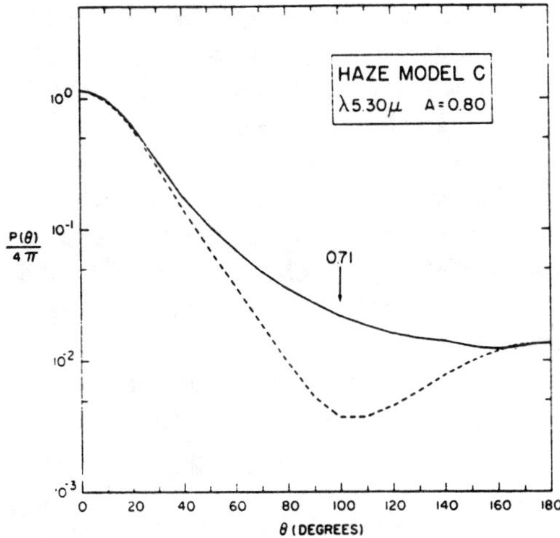

Fig. 2. Integrated and normalized intensity functions $P_1/4\pi$ (solid line) and $P_2/4\pi$ (dashed line) for haze particles with complex index (see Table I) and infrared illumination. Computed values at θ: $0(5)20(10)150(5)180°$ joined by straight lines.

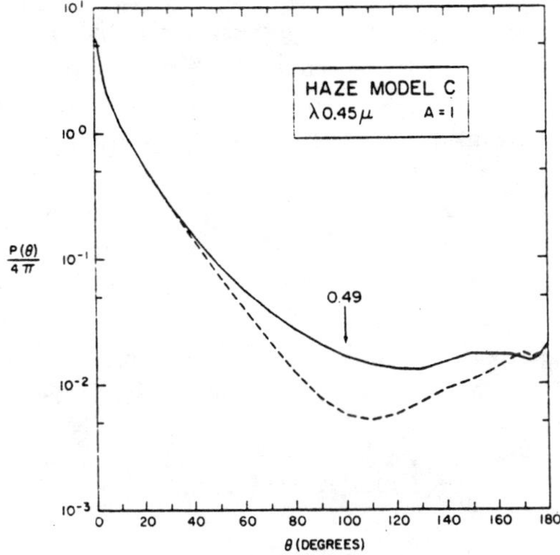

Fig. 3. Same as Fig. 2 but for particles with real index and visible light. Computed values at θ: $0(1)5$, $10(10)150(5)170(2)180°$.

The ordinates in all these figures are on a logarithmic scale, while the abscissa is linear in the scattering angle θ measured from the forward direction. For unpolarized and parallel monochromatic incident energy of unit flux, the degree of partial linear polarization is given by the ratio

$$\frac{P_1(\theta) - P_2(\theta)}{P_1(\theta) + P_2(\theta)},$$

with positive values corresponding to maximum electric-vector amplitude perpendicular to the scattering plane. The maximum numerical value and position of this parameter is indicated in Figs. 2–9.

Figure 2 corresponds to a model C water haze (maximum particle radius of $5\ \mu$) illuminated by infrared radiation at $\lambda 5.3\ \mu$. The scattering pattern is smooth

with strong anisotropy equivalent to a ratio of 10^2 between forward and backward scattering. A maximum positive polarization of 0.71 occurs at $\theta = 100°$, explained by the small relative size of the particles and their absorption properties.

Figure 3 shows the scattering pattern for the same haze when illuminated by visible light at $\lambda 0.45\ \mu$, where the refractive index of water is assumed real and equal to 1.34. If Fig. 3 is compared to Fig. 2, it is seen that the anisotropy has increased by one order of magnitude and the maximum polarization, though still at $\theta = 100°$, has decreased in value to 0.49. Both changes are due to the larger relative size of the particles. The pattern for $\lambda 0.70\ \mu$ radiation (not shown) is similar except that the maximum at $\theta = 0$ is reduced by about 27%.

Figures 4 and 5 correspond to the situation for a water haze with a distribution according to model M, at the wavelengths of $0.70\ \mu$ and $0.45\ \mu$, respectively. The intense aureole in the region $0° < \theta < 10°$ is more pronounced at the shorter wavelength here than in model C, with a reduction of 50% at $\lambda 0.70\ \mu$. Another difference is in the polarization pattern, which has migrated toward larger angles with a maximum of about 0.40 at $\theta = 152°$. In the region $20° < \theta < 100°$, the polarization is weak and negative. All these changes in going from a model C to a model M distribution are connected with the relatively larger amount of particles in the range $0.15 < r < 3.5\ \mu$ in the second model as compared to the first (see Fig. 1).

In general, the unit-volume scattering characteristics

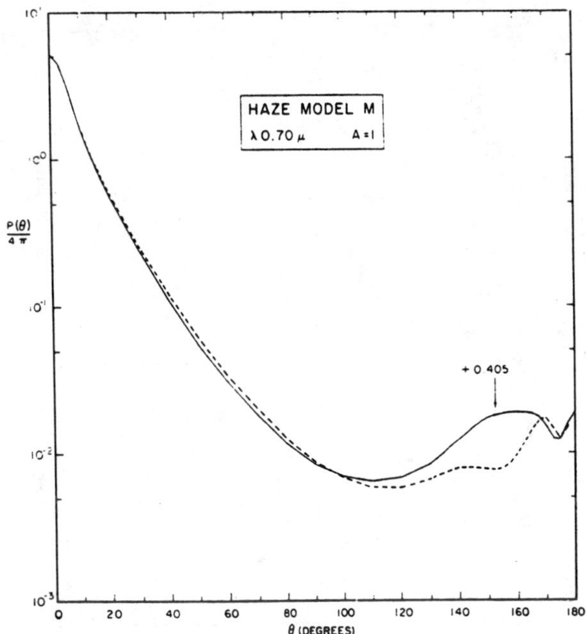

Fig. 4. Intensity functions for haze particles at $\lambda 0.70\ \mu$ and with real index 1.33 but a different distribution than in Fig. 3. Computed values at θ: $0(2.5)20(10)130(2.5)180°$.

of the haze particles must be closely related to the deviations of the observed degree of polarization and the position of the neutral points in the sunlit sky from those predicted by Rayleigh multiple scattering, as discussed by Sekera[13] and others. More particularly, it can be shown[8] that the most prominent deviations should appear in the brightness of the region around the sun called the aureole, which can be easily explained by the effect of primary scattering of sunlight on haze particles. In fact, the normalized intensity function $P_A(\theta)$ for a mixture of air and particles may be written in the form

$$P_A(\theta) = \frac{\beta_R P_R(\theta) + \beta_M P_M(\theta)}{\beta_R + \beta_M} \tag{13}$$

where the subscripts R and M stand for Rayleigh and Mie particles, respectively. Usually, even on clear days, the volume scattering coefficients β_R and β_M *in the visible range*, are of the same order of magnitude, whereas for small scattering angles say $\theta < 5°$, $P_M \simeq 100 P_R$, where $P_M = \frac{1}{2}(P_1 + P_2)$ and $P_R = \frac{3}{4}(1 + \cos^2\theta)$. Thus, at such angles, the large particle effect will dominate in the brightness (and hence also in the polarization).

In this connection, it is important to introduce the correct gradient and intensity of the aureole in estimates of the volume scattering coefficient based on an integration of the angular intensity over all solid angles. Extrapolations of the angular intensity in the region $0° \leq \theta < 10°$, when the corresponding data are unavailable, may lead to an incorrect estimate of both the

volume scattering coefficient and the anisotropy between forward and back scattering, and hence also of the wavelength dependence of these parameters.[17]

The exact integrations of the Mie functions can now be used to check a previous approximation by Diermendjian[7,8] based on the W.K.B. method. In that paper, the Mie phase functions were underestimated by a factor 4π [ref. 8, Eqs. (15), (16), and Table 3] because of an oversight in the definition of the amplitudes of the scattered field. Taking this into account, the following table gives a comparison of the approximate and exact calculations for a model C haze at $\lambda 0.45\ \mu$:

Phase Function $P_1(\theta)$, Model C Haze

	0.25°	1°	3°	6°	10°
W.K.B. method[8]	562	157	59.3	31.5	18.3
Exact Mie functions (this paper)	74.7	67.1	38.1	20.0	14.7

The large deviations appearing above are explained by the incorrect integration with respect to a model C distribution rather than by the W.K.B. approximation itself, which is very good for individual particles at small scattering angles.[7] In fact, if the original integration of the diffracted intensity were carried to a realistic finite upper limit of the particle size, instead of an infinite limit as in the Struve integral (ref. 8, p. 231), the use of the W.K.B. approximation should result in very good estimates of the aureole for this particular model. Furthermore, if a distribution of the type (1) is used instead, the convergence property, mentioned in the second section of this paper, suggests that, in the diffraction area, definite integrals of the form

$$\frac{1}{\sin^2\theta} \int_0^\infty x^{2+\alpha} e^{-bx} [J_1(x\sin\theta)]^2 dx$$

must exist, with a finite value even for $\theta = 0$.

Figures 6–9 present the normalized intensity obtained

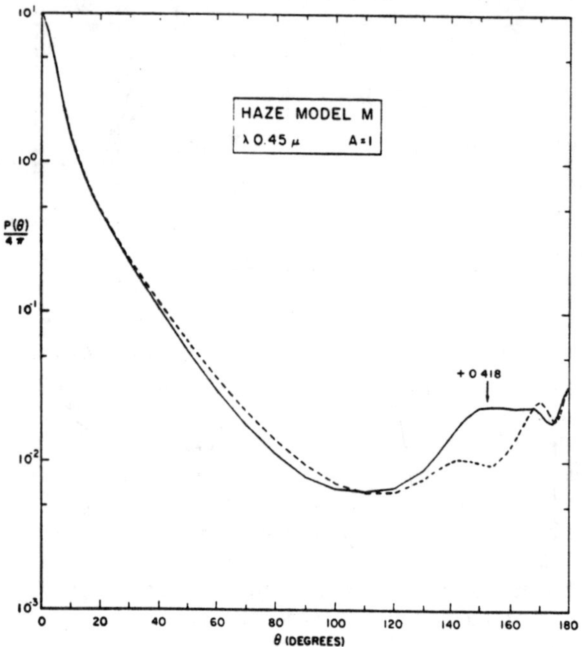

Fig. 5. Same as Fig. 4 but with illumination at $\lambda 0.45\ \mu$, with real index 1.34.

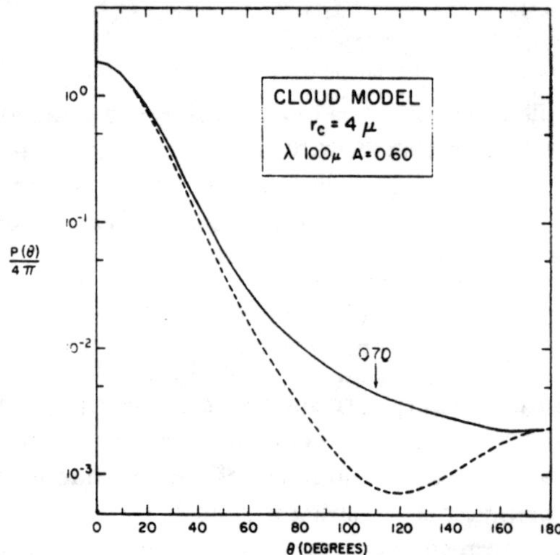

Fig. 6. Intensity functions for a water cloud illuminated by $\lambda 10.0$-μ infrared radiation. Computed values at θ: 0(5)180°.

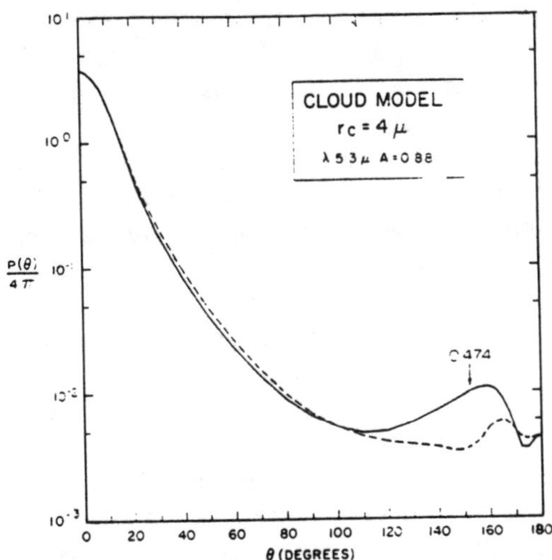

Fig. 7. Same as Fig. 6 but at $\lambda 5.3\ \mu$. Computed values at θ: 0(2.5)30(10)140(2.5)180°.

from a cloud-drop size distribution described by (5). Figure 6 corresponds to infrared illumination at $\lambda 10\ \mu$. A considerable anisotropy and strong maximum polarization at $\theta = 110°$ are evident, even for this long wavelength; otherwise the angular scattering pattern is quite smooth. The next case, shown in Fig. 7, can be compared with Fig. 2 for haze. It is seen that in the cloud model a maximum positive polarization occurs at $\theta = 152°$, in a broad, secondary maximum in the intensity, corresponding to an infrared rainbow. The anisotropy is high with a ratio of about 10^3 between forward and back scatter.

Going toward the visible, Figs. 8 and 9 show the very interesting patterns for a cloud illuminated by $\lambda 0.70$- and $\lambda 0.45$-μ radiation, respectively. Note first the very pronounced aureole or diffraction peak in the region $0° \leqq \theta < 5°$, which is twice as bright in the blue as in the red near 0°. This undoubtedly accounts for the very bright *"silver lining"* at the borders of certain clouds when near the sun. Next, in the region $20° < \theta < 0°$, the intensities fall exponentially with the scattering angle, with little or no polarization change. The region $80° < \theta < 110°$ corresponds to a minimum in intensity and change in the sign of the polarization.

The region $110° < \theta \leqq 180°$ contains distinct features which deserve discussion. The most salient of these is, of course, the pronounced and relatively narrow maximum around $\theta = 143°$. This is the fogbow or white rainbow mentioned by Minnaert.[18] Comparing Figs. 8 and 9, we see that there is no true dispersion, the blue bow maximum corresponding to the red in position. However, the blue bow is narrower and more intense, and its polarization *exceeds* that of the red bow. This must be explained in terms of the *larger* relative size of the cloud drops with respect to the wavelength. This trend indicates that the polarization in the true rainbow should be even greater. Furthermore, in the

true rainbow, which occurs at about $\theta = 138°$, the condition $x \rightarrow \infty$ applies to all wavelengths and the dispersion must be mostly due to the refrangibility of water. The interesting fact that emerges from the present study is that a wide size distribution in cloud droplets does give rise to *a single principal bow* (contributed mainly by the larger droplets in the distribution) so that all the secondaries and supernumeraries, which appear by scattering on single cloud droplets, are virtually suppressed by superposition. There is an indication of a weak secondary bow around $\theta = 122°$ at $\lambda 0.45\ \mu$ only.

Similarly, the appearance of a counter-corona or glory (cf. ref. 14, p. 249*ff*) of radius of about 2° at $\lambda 0.45\ \mu$ and 4° at $\lambda 0.70\ \mu$ is interesting for the same reason. The author has actually observed this visually on the vertical wall of a towering cumulus from a jet airliner at about 9 km, with the sun near the horizon. Note that the polarization in the counter-corona has the opposite sign from that of the rainbow in agreement with van de Hulst's (ref. 14, p. 256) analysis. The single glory is a result of superposition and is contributed by the larger droplets. Note also the existence of a counter-aureole or bright spot around $\theta = 180°$. No direct coronas were found in this model, however, because of the wide size distribution assumed.

Comparisons with Observations and Other Work

A most remarkable corroboration of the above cloud model comes from the careful angular scattering and polarization measurements made at night by Pritchard

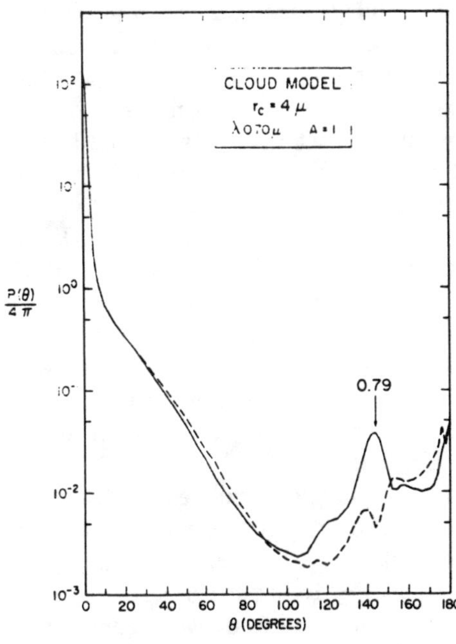

Fig. 8. Same as Fig. 6 but at $\lambda 0.70\ \mu$ and real index 1.33. Computed values at θ: 0(1)10(5)130(2)180°.

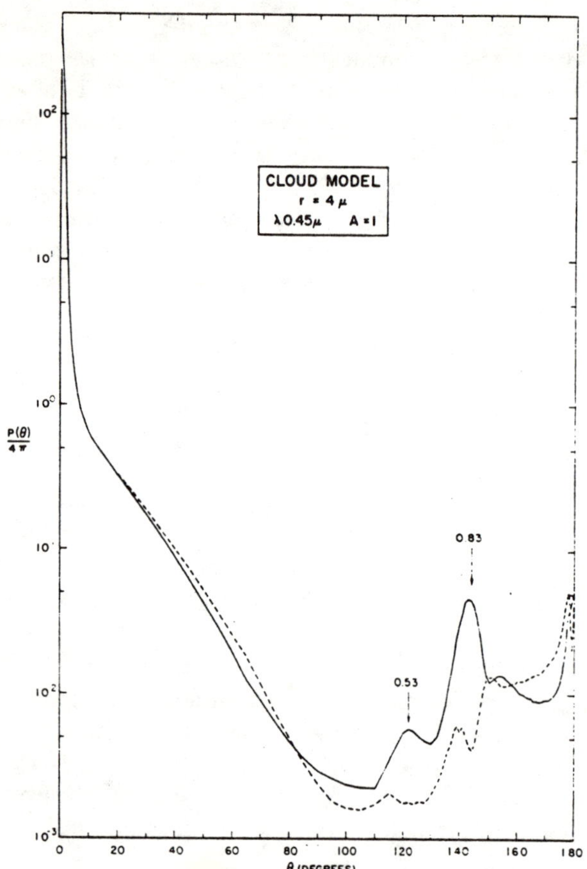

Fig. 9. Same as Fig. 6 but at λ0.45 μ and real index 1.34. Computed values at θ: 0(1)15(5)120(2)132(1)180°.

and Elliott[19] with their recording polar nephelometer on small volumes of natural fog (their Fig. 13). These measurements, made with a green filter, reproduce quite faithfully most of the salient features appearing in Figs. 8 and 9, including the intense aureole, the exponential decrease and other gradients, position of the minimum, position and intensity of the rainbow and its polarization (the components VV and HH of Pritchard and Elliott[19] are proportional to the elements P_1 and P_2, respectively). Apparently it was not possible to cover the range $170° < θ \leq 180°$; hence the glory and backward peak were not detected. Considering that the observations correspond to ground fog, in which the size distribution is not necessarily that of a cumulus cloud as in our model, the agreement is indeed remarkable. Also the measurements include the effects of molecular scattering which must have some influence on the polarization around $θ = 100°$ [see Eq. (13)].

The measurements on clear nights by Pritchard and Elliott[19] also show good agreement with our haze models except in the polarization, whose maximum does appear at $θ = 100°$ but has a larger value than shown in Figs. 3 or 5, for example, because of the strong influence of molecular scattering at this angle.

The theoretical curves should also be compared

with measurements of the angular intensity only, in the range $16° \leq θ \leq 164°$ reported by Barteneva,[20] whose work, however, is not as detailed as that of Pritchard and Elliott.

Integrations somewhat similar to ours, but not normalized with respect to the integrated scattering cross sections, and carried out to the constant upper limit of $x_2 = 40$, with an interval of $\Delta x = 1$, have been reported by Giese.[2] In contrast, for example, our integration intervals and range in x for Fig. 9 is 0.25-(0.25)60(0.50)160. Furthermore, Giese has used distribution functions of the type $n(r) \sim r^{-\alpha}$ with $\alpha = 2$, 2.5, 3 appropriate to zodiacal particles. It is interesting to note that samples of his results (ref. 2, Fig. 9a) show some of the features appearing in our cloud model, including a broad rainbow around $θ = 140°$, even though his size range corresponds to a haze model. This is explained by the size distribution function used by Giese, which is considerably less steep than in our models C and M.

One very interesting observational peculiarity of scattering on polydisperse Mie particles must be mentioned, whose theoretical corroboration could be provided only by computing the *normalized* intensities as in the present case. This is the apparent constancy of the quantity $(P_1 + P_2)/8\pi$ around $θ = 40°$, with a value of about 0.1 regardless of the size distribution

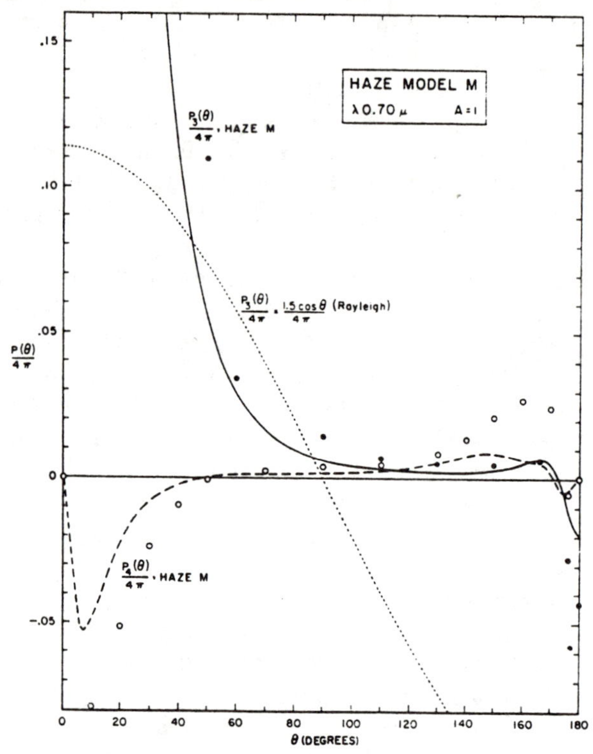

Fig. 10. Integrated and normalized functions $P_3/4\pi$ (solid line) and $P_4/4\pi$ (dashed line) for haze particles at λ0.70 μ. Comparable values, shown by solid and open dots, respectively, for a different size distribution and refractive index, adapted from Fraser (ref. 24, p. 123, "model D").

Table II. Cloud Phase Functions

$\lambda 0.45\ \mu$			$m = 1.34$		
θ	$P_1/4\pi$	$P_2/4\pi$	θ	$P_1/4\pi$	$P_2/4\pi$
0°	313.6	313.6	110°	0.002262	0.001740
1	140.8	140.9	115	0.003598	0.002083
2	21.94	21.92	120	0.005559	0.001784
3	5.369	5.315	122	0.005790	0.001796
4	2.590	2.537	124	0.005598	0.001778
5	1.616	1.575	126	0.005166	0.001819
6	1.171	1.136	128	0.004789	0.001809
7	0.9335	0.9018	130	0.004612	0.001987
8	0.7933	0.7678	132	0.005103	0.002396
9	0.7020	0.6778	134	0.006797	0.002971
10	0.6313	0.6163	136	0.01126	0.004090
11	0.5811	0.5636	138	0.01887	0.005499
12	0.5432	0.5262	140	0.03058	0.005605
13	0.5022	0.4986	142	0.04338	0.005211
14	0.4735	0.4650	144	0.04523	0.004129
15	0.4454	0.4420	146	0.03390	0.006071
20	0.3298	0.3337	148	0.01822	0.01060
25	0.2429	0.2522	150	0.01220	0.01326
30	0.1756	0.1875	152	0.01300	0.01284
35	0.1252	0.1384	154	0.01373	0.01172
40	0.08808	0.1008	156	0.01291	0.01177
45	0.06152	0.07262	158	0.01178	0.01217
50	0.04213	0.05193	160	0.01085	0.01261
55	0.02924	0.03650	162	0.01022	0.01262
60	0.01974	0.02551	164	0.009510	0.01309
65	0.01259	0.01835	166	0.009168	0.01350
70	0.009010	0.01186	168	0.008951	0.01396
75	0.006492	0.007668	170	0.009386	0.01550
80	0.004743	0.005057	172	0.009436	0.01749
85	0.003611	0.003358	174	0.01089	0.02082
90	0.002973	0.002407	176	0.01482	0.03004
95	0.002614	0.001817	178	0.03577	0.05156
100	0.002393	0.001639	180	0.05022	0.05022
105	0.002240	0.001604			

function, as evidenced by Figs. 4, 5, 8, and 9. A check into our tabulated values shows that this is true even if the integration is stopped at smaller sizes, provided these exceed the first resonance peak in the scattering cross section. This behavior suggests that, for small samples of a polydisperse suspension, *the concentration of the particles is directly proportional to the measured intensity scattered around $\theta = 40°$*, regardless of the size distribution and maximum size of the particles, provided the distribution is continuous and of the type (1), and the index of refraction is real and remains unchanged. The experimental facts are provided by the analyses of two independent sets of observations made on entirely different media, namely, those reported by Barteneva and Bashilov[21] for atmospheric aerosols, and those of Tyler[22] for hydrosols. Both papers reveal a very good linear correlation between the scattering function at $\theta = 45°$ and the volume scattering coefficient or degree of turbidity. This property is also mentioned in a very interesting review paper by Rozenberg.[23]

The Functions P_3 and P_4

The functions P_3 and P_4 integrated and normalized according to (9) have to do with a rotation of the plane of polarization and introduction of elliptical polarization by polydisperse suspensions, when illuminated by polarized light. These will not be discussed in detail here, except to show one example in Fig. 10 corresponding to model M at $\lambda 0.70\ \mu$. The functions $P_3 (\theta)/4\pi$ (solid line) and $P_4(\theta)/4\pi$ (dashed line) are plotted linearly against the scattering angle. For comparison, the corresponding Rayleigh value for $P_3(\theta)/4\pi$ is shown by the dotted line (the Rayleigh value for P_4 is zero everywhere). There are no experimental values that could be used for comparison here. Comparable values computed by Fraser[24] are shown by closed and open dots in Fig. 10. Although the latter were computed

Table III. Cloud Phase Functions

$\lambda 0.70\,\mu$			$m = 1.33$		
θ	$P_1/4\pi$	$P_2/4\pi$	θ	$P_1/4\pi$	$P_2/4\pi$
0°	133.7	133.7	110°	0.002518	0.001847
1	94.25	94.30	115	0.003849	0.002183
2	35.94	36.01	120	0.005132	0.001951
3	10.54	10.56	125	0.005546	0.002331
4	3.900	3.881	130	0.007422	0.003256
5	2.154	2.119	132	0.008564	0.004167
6	1.483	1.448	134	0.01181	0.004922
7	1.140	1.107	136	0.01648	0.006092
8	0.9355	0.9062	138	0.02281	0.006539
9	0.8009	0.7765	140	0.03086	0.006641
10	0.7092	0.6887	142	0.03618	0.005933
15	0.4642	0.4563	144	0.03772	0.004448
20	0.3368	0.3394	146	0.03217	0.005078
25	0.2524	0.2535	148	0.02251	0.007801
30	0.1758	0.1885	150	0.01449	0.01131
35	0.1246	0.1361	152	0.01060	0.01340
40	0.08793	0.09953	154	0.01032	0.01359
45	0.06190	0.07189	156	0.01129	0.01310
50	0.04315	0.05229	158	0.01170	0.01248
55	0.02941	0.03647	160	0.01104	0.01275
60	0.02077	0.02582	162	0.01071	0.01310
65	0.01372	0.01960	164	0.01034	0.01372
70	0.009539	0.01263	166	0.009988	0.01483
75	0.007007	0.008873	168	0.01018	0.01624
80	0.005156	0.006161	170	0.01077	0.01833
85	0.003944	0.004420	172	0.01142	0.02126
90	0.003322	0.003112	174	0.01494	0.02677
95	0.002798	0.002549	176	0.02506	0.04300
100	0.002587	0.002166	178	0.03623	0.02927
105	0.002320	0.002041	180	0.04972	0.04972

for $m = 1.40$ and a different size distribution and range, there is general agreement in the order of magnitude, the gradient and the position of the maxima. Fraser's results are based on a Legendre series expansion due to Sekera,[13] using the Mie scattering coefficients a_n and b_n, which are independent of θ. The general agreement mentioned above is proof of the correctness of the method. However, the possibility of accurately generating the complex scattered amplitudes by means of high-speed computers[1] allows a much more reliable direct integration of these parameters.

Some Conclusions

On the basis of this limited sample of our results and comparison with available observations, the following conclusions may be reached:

(a) The classical electromagnetic theory of scattering on homogeneous finite spherical particles, applied to a polydispersed suspension with a continuous size dis-tribution, can be used to reproduce rather accurately the observed angular intensity and linear polarization of samples of natural water clouds and hazes illuminated by visible light.

(b) No single particle in such a suspension exists, whose angular scattering and polarization properties correspond to the integrated effects from all the particles, which tend to suppress by superposition the extreme angular fluctuations characteristic of single particles. The introduction of such parameters as the "mean radius", the "mode radius", the "mean volume radius", etc., is not helpful in determining these properties for aggregates of particles, except perhaps in the case of very small particles relative to the wavelength. Our results further indicate that the angular volume scattering properties are uniquely determined by the type of size distribution, the range of sizes, and the dielectric and conducting properties of the particles in the suspension, independently of the total concentra-

Table IV. Cloud Phase Function $(P_1 + P_2)/8\pi$ for $\lambda 1.61\,\mu$ and $m = 1.315$

θ	Function	θ	Function
0°	26.4	100	0.00432
2	17.6	110	0.00413
5	5.94	120	0.00522
7	1.86	130	0.00897
10	0.936	140	0.0168
20	0.353	150	0.0169
40	0.0922	160	0.0152
60	0.0276	170	0.0309
80	0.00929	180	0.0604

Table V. Cloud Phase Function $(P_1 + P_2)/8\pi$ for $\lambda 3.07\,\mu$ and $m = 1.525 - 0.0682i$

θ	Function	θ	Function
0°	14.2	60	0.0170
2	12.6	80	0.00923
5	8.74	100	0.00584
7	4.92	120	0.00442
10	2.38	140	0.00410
15	0.555	150	0.00464
20	0.208	160	0.00605
30	0.0732	170	0.00527
40	0.0385	180	0.00984

tion. Hence, in the case of *optically thin* samples of such suspensions (where multiple scattering is negligible), these parameters may be deduced from careful measurements of the scattering properties in as many directions as possible and at various wavelengths.

(c) The existence of a solution to the analogous general problem in the case of optically thick media, i.e., that of the unique determination of the type, size, size distribution, and size range of the particles in suspension, from a knowledge of the angular intensity and polarization of the diffusely transmitted and reflected radiation, remains to be demonstrated. However, in the case of certain suspensions such as water-droplet clouds in the atmosphere, the existence of prominent features such as rainbows and counter coronas suggests that some information on the size and size distribution of the particles may be obtained from these features, which should be observable on diffuse reflection even against the background of the multiply scattered field.

(d) The calculation of the *normalized* scattered intensities corroborates the experimental result that, at least in the case of dielectric particles, the intensity at a fixed scattering angle around 45° is directly proportional to the total number concentration per unit volume independently of the size distribution and size range. Thus, under certain conditions and for optically thin media, the concentration or degree of turbidity may be obtained by this method as well as by the usual one based on the extinction of a collimated beam.

R. J. Clasen programmed the generation and integration with respect to size of the Mie functions on the IBM 7090 computer.

Appendix

To supplement the data presented graphically in Figs. 6–9, we have tabulated some of the computed cloud phase functions for $\lambda 0.45$, 0.70, 1.61, and 3.07 μ. The values are given at a sufficient number of scattering angles to allow for graphical interpolations when plotted as above. Tables II and III list the individual functions $P_1/4\pi$ and $P_2/4\pi$, rounded off to four significant figures, while Tables IV and V list their semi-sum, rounded off to three significant figures.

References

1. D. Deirmendjian, R. Clasen, and W. Viezee, J. Opt. Soc. Am. **51**, 620 (1961). See also D. Deirmendjian and R. J. Clasen, *RAND Reports* R-393-PR (1962) and R-407-PR (1963).
2. R. H. Giese, Z. Astrophys. **51**, 119 (1961).
3. A. Kh. Khrgian and I. P. Mazin, Tr. Tsentr. Aerolog. Observ. **7**, 56 (1952).
4. A. Kh. Khrgian and I. P. Mazin, Tr. Tsentr. Aerolog. Observ. **17**, 36 (1956).
5. W. G. Durbin, Tellus **11**, 202 (1959).
6. F. Singleton and D. J. Smith, Quart. J. Roy. Meteorol. Soc. **86**, 454 (1960).
7. D. Deirmendjian, Ann. Geophys. **13**, 286 (1957).
8. D. Deirmendjian, Ann. Geophys. **15**, 218 (1959).
9. D. Deirmendjian, Quart. J. Roy. Meteorol. Soc. **85**, 404 (1960).
10. J. Y. Gilbert, "Condensation Nuclei of the Los Angeles Region," Univ. of Calif., Department of Meteorology, Los Angeles (1954).
11. C. Junge, C. W. Chagnon, and J. E. Manson, J. Meteorol. **18**, 81 (1961).
12. S. Chandrasekhar, *Radiative Transfer* (Clarendon Press, Oxford, 1950).
13. Z. Sekera, Handbuch Phys. **48**, 288 (1957).
14. H. C. van de Hulst, *Light Scattering by Small Particles* (Wiley, New York, 1957).
15. R. Penndorf, J. Opt. Soc. Am. **52**, 402 (1962).
16. M. Centeno, J. Opt. Soc. Am. **31**, 244 (1941).
17. M. G. Gibbons, J. R. Nichols, F. I. Laughridge, and R. L. Rudkin, J. Opt. Soc. Am. **51**, 633 (1961).
18. M. Minnaert, *The Nature of Light and Color in the Open Air* (Dover, New York, 1954).
19. B. S. Pritchard and W. G. Elliott, J. Opt. Soc. Am. **50**, 191 (1960).
20. O. D. Barteneva, Bull. Acad. Sci. USSR, Geophys. Ser. No. 12, 1237 (1960), in AGU translation.
21. O. D. Barteneva and G. Ya. Bashilov, Bull. Acad. Sci. USSR, Geophys. Ser. No. 4, 395 (1961), in AGU translation.
22. J. E. Tyler, J. Opt. Soc. Am. **51**, 1289 (1961).
23. G. V. Rozenberg, Soviet Phys.—Uspekhi **3**, 346 (1960).
24. R. S. Fraser, Dissertation, Univ. of Calif., Los Angeles (1959). See also Sci. Rept. No. 2, Contract No. AF 19(604)-2429, AFCRC-TN-60-256 (1959).

The Influence of Pollution on the Shortwave Albedo of Clouds

S. Twomey

Institute of Atmospheric Physics, University of Arizona, Tucson 85721

26 October 1976 and 5 April 1977

ABSTRACT

By increasing droplet concentration and thereby the optical thickness of a cloud, pollution acts to increase the reflectance (albedo) of clouds; by increasing the absorption coefficient it acts to decrease the reflectance. Calculations suggest that the former effect (brightening of the clouds in reflection, hence climatically a cooling effect) dominates for thin to moderately thick clouds, whereas for sufficiently thick clouds the latter effect (climatically a warming effect) can become dominant.

1. Introduction

In the past few years interest has grown in the possibility that pollution can influence the planetary albedo. The SCEP Report (1970) referred briefly to the possible influence of pollution on cloud albedo and inferred that "dirty" clouds would be darker than "clean" clouds formed in uncontaminated air. The writer has pointed out (Twomey, 1974) that increasing pollution generally means increasing cloud nucleus concentrations, hence increasing numbers of cloud drops; this leads to increasing cloud optical thickness and hence for finite cloud thicknesses increasing cloud albedo. The present note will discuss the relative magnitude of these opposing influences on cloud albedo.

2. Optical thickness

Most lower and middle clouds are optically thick. When, for example, the sun's disc cannot be seen from the ground the optical thickness τ of the layer concerned must approach or exceed 10. A cloud of depth h containing $n(r,z)\Delta r$ drops of radius $r \rightarrow r+\Delta r$ per cubic centimeter at height z above cloud base possesses optical thickness τ at wavelength λ, with

$$\tau = \int_0^h k_E dz = \pi \int_0^h \int_0^\infty r^2 Q_E(r/\lambda) n(r,z) dr dz. \quad (1)$$

Here $Q_E(r/\lambda)$, the extinction efficiency, is given by Mie theory and k_E is the extinction coefficient. For solar

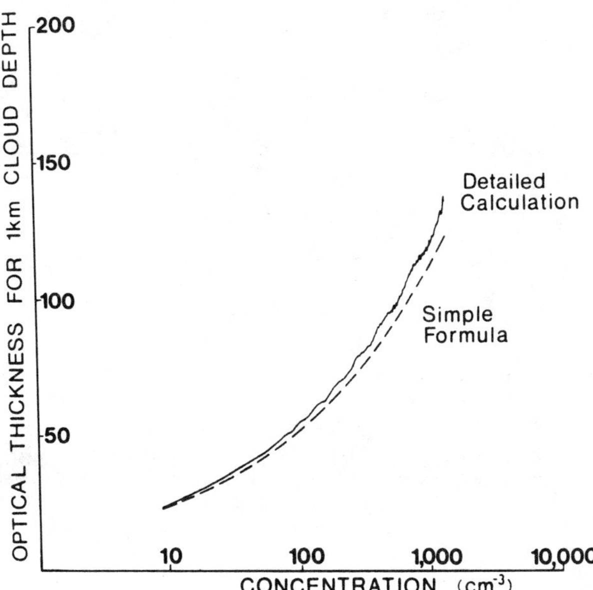

Fig. 1. Comparison of optical thickness calculated by the simple Trabert-type formula (2) and by exact Mie calculations integrated over a distribution of droplet sizes according to (1).

wavelengths the approximation $Q_E \equiv 2$ is quite adequate and for realistic (polydisperse) drop distributions we can eliminate the integration and adopt the simple formula

$$\tau = 2\pi N \bar{r}^2 h, \qquad (2)$$

where N is the drop concentration per cubic centimeter and $\bar{r}$ can be any representative mean or model radius. It is convenient and sufficiently accurate to use the volume-mean radius so that the liquid water content W is given by $(4\pi/3)\bar{r}^3 N$. Fig. 1 shows the extent of the approximations involved when (2) is used in place of the detailed calculations of (1). The simple formula is seen to be quite adequate.

Using either formula, one can calculate optical thickness for any depth of cloud as a function of drop concentration. A water content of $\frac{1}{3}$ g m⁻³ [which is typical at least of lower level clouds (see Warner and Squires, 1958)] was adopted to calculate values of extinction coefficient k_E (cm⁻¹) and optical thickness per kilometer of cloud depth t_{km}; the results are given in Table 1.

Clean air over the tropical oceans typically contains about 100 cloud nuclei per cubic centimeter, while over the cooler oceans concentrations can be as little as 10 cm⁻³ or even less; continental concentrations

typically run from 500–1000 cm⁻³ and in moderate to heavy pollution values of 5000 cm⁻³ and higher are often measured. The table therefore contains approximately the range of cloud nucleus and cloud drop concentrations in the present atmosphere. The criterion $\tau \gg 1$ which defines "optically thick" is satisfied for a 1 km depth of cloud for all concentrations within that range; a 100 m layer, on the other hand, would be optically thick only in continental to polluted air and would have $\tau \approx 1$ for clean conditions. However, even in extremely clean conditions ($N = 3$–10 cm⁻³) only a few hundred meters of geometric depth are needed to give a cloud which is optically thick.

3. Scattering properties of optically thick layers

When optical thickness becomes large, scattering of incoming radiation by a layer becomes relatively uniform, varying smoothly and slowly with angle. It is found that only a few parameters are needed to give the reflection, transmission and absorption quite precisely. These parameters are the optical thickness τ, the single scattering albedo $\bar{\omega}_0$ and the asymmetry factor g (which is the weighted average of the cosine of scattering angle; $g = 0$ for any symmetric scattering diagram and $g = 1$ for exactly forward scattering). Van de Hulst (1971) has shown that to high accuracy τ, $\bar{\omega}_0$ and g can be collected into two variables so as to further reduce the relevant dependences; in situations of moderate absorption van de Hulst's scaling principle reduces to the use of $\tau(1-g)$ to characterize optical thickness. An increase in drop concentration at constant water content results in smaller drops and hence a decrease in g as well as an increase in τ, so that $\tau(1-g)$ increases somewhat *faster* than τ. In the micron size range with which one is concerned, however, the change in g with size is rather slow; for most purposes its effect is second-order and it is sufficient to consider only the influence of τ. Nevertheless, it is relevant to note that if the effect of g is included, the dependence of reflectance, etc., on drop concentration is somewhat enhanced, rather than diminished.

For optically thick clouds one can therefore consider reflectance S and other optical properties of the layer to be functions of τ and $\bar{\omega}_0$ only. For fixed $\bar{\omega}_0$, S increases monotonically with τ but the rate of increase decreases monotonically with increase in τ; for fixed τ, S evidently decreases with decrease in $\bar{\omega}_0$ (i.e., with increasing absorption). Thus

$$\frac{\partial S}{\partial \tau} \geqslant 0, \quad \frac{\partial S}{\partial \bar{\omega}_0} \geqslant 0,$$

$$\frac{\partial S}{\partial \tau} \xrightarrow{\tau \to \infty} 0.$$

If the extent of pollution above a reference "clean" state (at which $\bar{\omega}_0 = 1$, $\tau = \tau_0$, $S = S_0$) is gaged by some

TABLE 1. Extinction coefficient k_E and optical thickness t_{km} for a 1 km depth.

	$\bar{r}(\mu m)$				
	2	5	10	20	30
N (cm⁻³)	9950	640	80	10	3
k_E (cm⁻¹)	0.0025	0.001	0.0005	0.00025	0.00017
t_{km}	250	100	50	25	17

variable x (the precise physical nature of which is irrelevant), then increasing pollution will increase the albedo if dS/dx is positive, i.e., if

$$\frac{\partial S}{\partial \tau}\frac{d\tau}{dx}+\frac{\partial S}{\partial \bar{\omega}_0}\frac{d\bar{\omega}_0}{dx}>0,$$

and vice versa.

The dependence of reflectance and albedo on τ and $\bar{\omega}_0$ may be represented by a set of graphs of S vs τ for a set of constant values of $\bar{\omega}_0$, as illustrated schematically in Fig. 2. With increase in x, S follows a trajectory represented by the broken line in the figure. The increase in τ with increase in the pollution index x will give a component along the curve $S(\tau)$, while a change in $\bar{\omega}_0$ will give a transverse component, which will be downward if $\bar{\omega}_0$ decreases with x.

To put Fig. 2 on a quantitative basis, one needs to prescribe $d\bar{\omega}_0/dx$ and $d\tau/dx$. Since $(1-\bar{\omega}_0)\tau$ equals the absorption optical thickness τ_a it follows that

$$\frac{1}{1-\bar{\omega}_0}\frac{d\bar{\omega}_0}{dx}=\frac{1}{\tau}\frac{d\tau}{dx}-\frac{1}{\tau_a}\frac{d\tau_a}{dx}.$$

Necessarily $\tau_a \leqslant \tau$ (and in any real cloud context $\tau_a \ll \tau$), so if absorption optical thickness τ_a increases with increasing pollution, $d\bar{\omega}_0/dx$ tends eventually to become negative. The simplest assumption which can be made (other than the assumption that scattering is conservative) is to take N and τ_a each to be directly proportional to x. This assumption amounts to assuming that all components in the aerosol increase together and in the same proportion, so that the increase in cloud nuclei and in aerosol absorption are proportional.[1] If W is to be maintained constant, independent of x, the total optical thickness τ, which is dominated by scattering, must then increase as $N^{\frac{1}{3}}$ or $x^{\frac{1}{3}}$. Thus $\tau^{-1}d\tau/dx = \frac{1}{3}\tau_a^{-1}d\tau_a/dx$.

Measurements of cloud nucleus concentration (Twomey and Wojchiechowski, 1969) suggest that $N \sim 10^3$ cm^{-3} is typical of the continental aerosol; measurements by de Luisi *et al.* (1976) suggest a value of around 0.05 for the absorption optical thickness for the entire atmospheric column, the aerosol scale height being of the order of 2 km. The latter measurements were made in geographical areas (Blythe, Calif., and Big Spring, Tex.) in which cloud nucleus concentrations around 1000 cm^{-3} could be expected with some confidence. The values 0.01 and 0.1 were therefore selected to bracket the measurements of de Luisi *et al.* and would

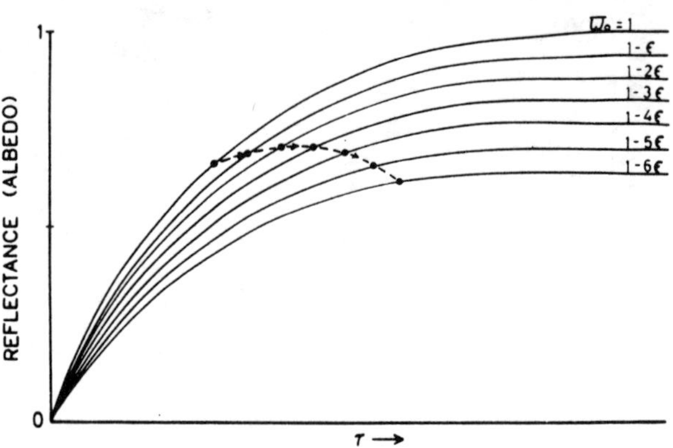

FIG. 2. Schematic diagram showing the variation of reflectance with optical thickness τ for different values of $\bar{\omega}_0$ and the trajectory (dashed curve) of the representative point when increasing pollution increases *both* τ and $1-\bar{\omega}_0$ (or τ_a); ϵ represents the increase in $1-\bar{\omega}_0$ from curve to curve.

roughly represent a reasonable value and an extreme value, respectively, for the absorption t_a per kilometer depth in continental conditions when N is ~ 1000 cm^{-3}. The arbitrary measure of pollution x can conveniently be defined to be unity when N is 1000 cm^{-3}; if a perfectly clean atmosphere is taken to have $N=25$ cm^{-3}, the following formulas ensue for the variation with x of N and t_a:

$$\left.\begin{array}{l} N=25+975x \ [\text{cm}^{-3}] \\[4pt] \text{Moderate absorption } t_a=0.01x \ [\text{km}^{-1}] \\[4pt] \text{Heavy absorption } t_a=0.1x \ [\text{km}^{-1}] \end{array}\right\}.$$

The clean value $N=25$ cm^{-3} was selected to give a representative starting point, but which starting point was used is of little consequence beyond its illustrative value. $N=25$ cm^{-3} is typical of very clean polar maritime air and in the present context all increases above this level are assumed to be occasioned by contamination which adds absorbing particles in the same proportion.

With the above assumptions, computations can be made for the variation with pollution index x of reflectance, albedo, etc. In Fig. 3 we have plotted trajectories (dashed curves) for moderate and heavy absorption for a thin (250 m) cloud layer (initial point A), a cloud layer 1 km thick (initial point B) and a thick (4 km) layer (initial point C). The quantity plotted was the spherical albedo A_s, which eliminates geometric variables and gives a reasonably representative global value of albedo. [It represents the fraction of incident radiation reflected by a sphere covered by a layer of the prescribed properties; for a reflection function $S(\mu,\mu_0)$ which relates intensity in the emergent μ direction to intensity in the incident μ_0 direction, the spherical albedo is given by

$$2\int_0^1\int_0^1 \mu S(\mu,\mu_0)d\mu d\mu_0.]$$

---

[1] One is also assuming tacitly that absorption is not greatly modified when a cloud forms; that will be the case if, for example, absorption is by particles which remain unchanged after cloud formation. The results of Prishivalko and Astafyeva (1974), and unpublished computations in this laboratory, suggest that even if drops form and grow around insoluble absorbing particles the overall absorption is not profoundly changed (the change being typically no more than 50%, and not necessarily an increase).

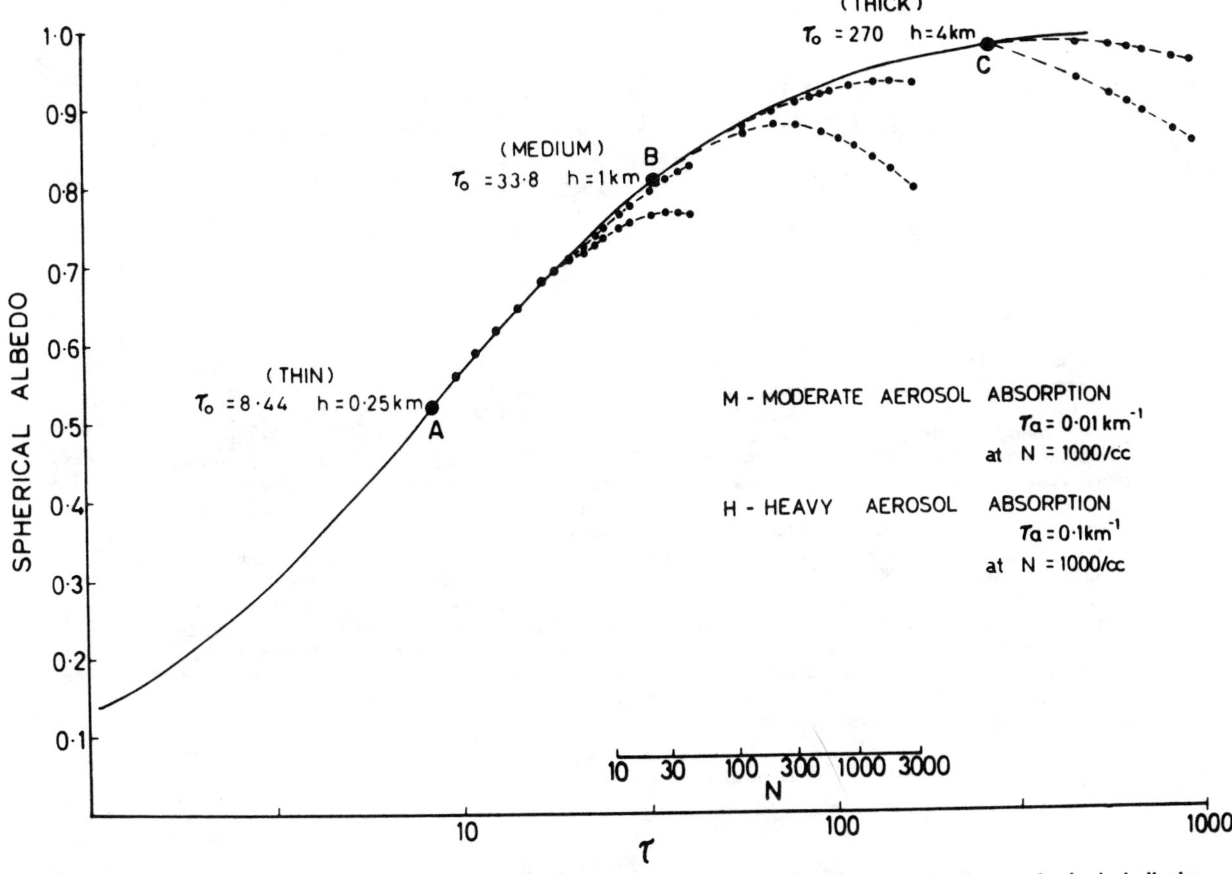

FIG. 3. Numerically computed trajectories corresponding to the schematic curves in Fig. 2, for the change of spherical albedo with increasing pollution for thin, moderate and thick clouds.

The solid curve in Fig. 3 represents the conservative ($\bar{\omega}_0 = 1$) value of A_s; if pollution did not increase absorption at all, trajectories of A_s vs x would lie on this curve, since then only τ would change with increasing pollution. The starting point for each pair of curves corresponds to $x = 0$, i.e., conservative scattering and $N = 25$ cm^{-3}. Fig. 3 shows that for all but the thickest clouds, the brightening influence of increasing τ outweighs the darkening influence of increasing τ_a, even though the latter increases in proportion to N while τ increases as $N^{\frac{1}{3}}$. One clearly would have had to postulate a much more rapid increase of τ_a relative to N to obtain a different result. It would be routine to carry out similar calculations for other assumed relationships between N and τ_a but no useful purpose would seem to be served by such computations. Measurements to determine how the relevant quantities N and τ_a vary in pollution are clearly what are required.

4. Conclusions

Pollution may increase or decrease the brightness of clouds depending on the optical thickness of the clouds and the way in which cloud nucleus concentration varies with absorption optical thickness. Plausible assumptions concerning the latter lead to results which suggest that in all but the thickest clouds the pollution increases the albedo. Since most of the earth's cloud cover is in the form of clouds which are not very thick this result suggests that the planetary albedo also will increase with increase of pollution.

REFERENCES

De Luisi, J. J., P. M. Furukawa, D. A. Gillette, B. G. Schuster, R. J. Charlson, W. M. Porch, R. W. Fegley, B. M. Herman, R. A. Rabinoff, J. T. Twitty and J. A. Weinman, 1976: Results of a comprehensive atmospheric aerosol-radiation experiment in the southwestern United States. Part I: Size distribution, extinction optical depth and vertical profiles of aerosols suspended in the atmosphere. *J. Appl. Meteor.*, **15**, 441–454.

Prishivalko, A. P., and L. G. Astafyeva, 1974: Absorption, scattering and extinction of light by water-coated atmospheric particles. *Izv. Atmos. Oceanic Phys.*, **10**, 815–818.

SCEP (Study of Critical Environmental Problems) 1970: *Mass Impact on the Global Environment*. The MIT Press, 319 pp.

Twomey, S., 1974: Pollution and the planetary albedo. *Atmos. Environ.*, **8**, 1251–1256.

——, and T. A. Wojchiechowski, 1969: The geographic variation of cloud nuclei. *J. Atmos. Sci.*, **26**, 684–688.

Van de Hulst, H. C., 1971: Some problems of anisotropic scattering in planetary atmospheres. *Planetary Atmospheres* (IAU Symposium No. 40) C. Sagan *et al.*, Eds., Reidel, 177–185.

Warner, J., and P. Squires, 1958: Liquid water content and the adiabatic model of cumulus development. *Tellus*, **10**, 390–394.

Solar Radiative Transfer in Cirrus Clouds. Part I: Single-Scattering and Optical Properties of Hexagonal Ice Crystals

YOSHIHIDE TAKANO AND KUO-NAN LIOU

Department of Meteorology, University of Utah, Salt Lake City, Utah

(Manuscript received 1 February 1988, in final form 13 June 1988)

ABSTRACT

We have developed an efficient light scattering and polarization program, based on a ray-tracing technique, for hexagonal ice crystals randomly and horizontally oriented in space. Improvements have been made on the ray-tracing computations through a proper treatment of the δ-forward transmission by geometric rays and incorporation of the effects of birefringence of ice. Using this program, computations of the scattering phase matrix are made from the observed ice crystal size distributions for four typical cirrus clouds. The results for single-scattering parameters, including the phase function, single-scattering albedo, extinction cross section, and asymmetry factor for five solar wavelengths, are presented and discussed. Moreover, we show that the assumption of equivalent spheres with the same surface areas as hexagonal ice crystals leads to larger asymmetry factors for all wavelengths and smaller single-scattering albedos for near IR wavelengths. The computed phase matrix elements compare reasonably well with experimental scattering results for laboratory ice crystal clouds. In particular, we illustrate that the neutral point (angle of zero linear polarization) is sensitive to the ice crystal shape. Thus, an observation of this position from space could provide a means for determining the aspect ratio of cloud particles. Light scattering computations are also made for horizontally oriented columns and plates of various sizes. Results are used to interpret the optical phenomena produced by cirrus clouds. The present light scattering program for hexagonal columns and plates identifies the positions of halos and arcs, as well as providing relative intensities for these optical features.

---

1. Introduction

Cirrus clouds have been identified as one of the major unsolved elements in weather and climate research (Liou 1986). There are significant problems in the development of satellite remote sensing techniques for the mapping of cirrus clouds due to their nonblackness, high altitude, and nonspherical ice crystals. Recently, intensive field observations of cirrus clouds have been conducted as a major component of the First ISCCP Regional Experiment (Starr 1987) to assist in the development of remote sensing methodologies for the determination of the temperature and optical properties of cirrus clouds. Moreover, the manner in which the formation of cirrus clouds and coupled radiative properties may be properly and effectively incorporated in large-scale numerical models is a subject unlikely to be resolved without considerable research effort.

To be successful in the development of retrieval methodologies for the detection of cirrus clouds, and parameterizations for their radiative properties for incorporation in dynamic models, we must have fundamental scattering and absorption data for nonspher-

ical ice crystals that are appropriate for cirrus clouds. From in situ aircraft observations, it has been determined that cirrus clouds are largely composed of nonspherical bullets, columns, and plates (see, e.g., Heymsfield and Platt 1984). For radiative transfer calculations, nonspherical ice crystals have been approximated by spherical particles (see, e.g., Plass and Kattawar 1968). Liou (1972) made the first attempt to model the single-scattering properties of nonspherical ice crystals using the scattering solution for long circular cylinders. Stephens (1980) adopted this cylindrical scattering solution to perform light scattering and radiative transfer calculations for ice clouds. Welch et al. (1980) derived the radiative properties of ice clouds using a semiempirical approach through a modification of the Mie solution for spherical particles. Although the aforementioned models can represent some aspects of the scattering and radiative characteristics of cirrus clouds, none of these approaches account for the hexagonal structure of ice crystals. Optical phenomena, such as halos and numerous arcs associated with cirrus, cannot be reproduced from scattering solutions for circular cylinders and spheres. Rainbows produced by spherical droplets are absent in cirrus. Moreover, the specific scattering and absorption features associated with the hexagonal structure, aspect ratio (length/diameter), and orientation of ice crystals could be important in the determination of bidirectional reflec-

Corresponding author address: Dr. Kuo-Nan Liou, Dept. of Meteorology, 819 Wm. C. Browning Building, University of Utah, Salt Lake City, UT 84112.

tance, cloud albedo and transmittance, sky polarization configurations, and atmospheric heating rates.

In recent years, light scattering and absorption programs for hexagonal columns and plates have been developed by Cai and Liou (1982) and Takano and Jayaweera (1985) based on a ray-tracing technique. In this paper, we have extended these programs by taking into account the ice crystal size distribution, as well as the possibility of horizontal orientation. In addition, improvements have been made on the light scattering program by proper summation of contributions due to geometric reflection and refraction and Fraunhofer diffraction, and by incorporation of the birefringence properties of ice. We shall demonstrate that the present scattering program for hexagonal particles is appropriate for modeling the scattering characteristics of ice crystal clouds. This is done by comparing the computed and measured scattering and polarization patterns, as well as by interpreting the observed optical features produced by cirrus clouds.

In section 2, we present single-scattering properties for randomly oriented hexagonal ice crystals. The computed scattering and polarization results are then compared with available measurement data in section 3. In section 4 we report single-scattering properties for horizontally oriented columns and plates and provide explanations for numerous halos and arcs that have been observed in the atmosphere. Finally, conclusions are given in section 5.

2. Single-scattering properties of randomly oriented hexagonal ice crystals

The basic formulation for the computation of reflection, refraction, and diffraction of light rays in the limit of geometric optics has been presented in our previous papers (Cai and Liou 1982; Takano and Jayaweera 1985). Thus, we shall briefly describe improvements and refinements that have been made in the current light scattering program, which are pertinent to the presentation of scattering and absorption results for oriented hexagonal ice crystals.

Ice is a uniaxial, doubly refracting, and optically positive crystal. The refracted rays split into ordinary and extraordinary components. The vibrational planes of the electric field vectors of these two rays are orthogonal to each other. This phenomenon is referred to as birefringence. Its effect on polarization and intensity distributions of halos has been investigated by Können (1983). We shall briefly discuss the manner in which this effect is incorporated in the present ray-tracing program.

According to Born and Wolf (1975), the effective index of refraction for extraordinary rays, n_{eff}, is given by

$$\frac{1}{n_{\text{eff}}^2} = \frac{\cos^2\kappa}{n_o^2} + \frac{\sin^2\kappa}{n_e^2}, \tag{1}$$

where n_o and n_e are, respectively, the indices of refraction for ordinary and extraordinary rays, and κ is the angle between the optical axis (c-axis) of the crystal and the light ray in the crystal, as shown in Fig. 1. The optical path lengths for ordinary rays, δ_o and extraordinary rays, δ_e may be expressed, respectively, by

$$\left.\begin{aligned}\delta_o{}^k &= -\frac{2\pi}{\lambda} n_o d_{k+1,k} \\[2ex] \delta_e{}^k &= -\frac{2\pi}{\lambda} n_{\text{eff}} d_{k+1,k}\end{aligned}\right\}, \quad k \geqslant 1 \tag{2}$$

where λ is the wavelength of the incident light and $d_{k+1,k}$ is the path length between points N_k and N_{k+1}, defined in Fig. 1.

To incorporate the effects of birefringence, the amplitude matrix, which is related to the electric field, requires modifications. After each refraction or reflection, the coordinate axis is rotated by an appropriate angle so that the reference plane for polarization coincides with the plane including the light ray and c-axis. The amplitudes of the extraordinary and ordinary rays are then multiplied, respectively, by the factors $e^{i\delta e}$ and $e^{i\delta o}$. The coordinate axis is subsequently rotated back to the original plane.

In the limit of geometric optics, half of the incident energy is associated with the diffracted rays. However, in the process of normalizing the phase function by numerical means, the 0° forward (hereafter referred to as the δ-function) transmission by two direct 0° refractions through parallel planes (e.g., a basal plane and opposite basal plane) cannot be properly accounted for. This δ-function transmission component, which differs from diffracted rays, is important for both absorption and nonabsorption cases. In the Appendix we describe a procedure by which the addition of geometric and diffracted rays and the associated normalization for the phase function may be performed correctly.

For randomly oriented ice crystals, integration of the phase matrix over all possible orientations must be carried out so that

$$\hat{P}_{kl}(\Theta) = \frac{6}{\pi} \int_0^{\pi/6} \int_0^{\pi/2} G_{kl}(\alpha, \beta, \Theta) \cos\alpha \, d\alpha \, d\beta,$$

$$k, l = 1\text{–}4 \tag{3}$$

where Θ denotes the scattering angle, α is the complementary angle between the incident ray and c axis, β the rotational angle about the c axis (see Fig. 1), and G_{kl} represents four-by-four phase matrix elements for a single crystal. Because of the symmetry of the hexagonal crystal, integration of the angle β is from 0 to $\pi/6$. The scattering cross section for randomly oriented ice crystals is given by

$$\hat{C}_s = \frac{6}{\pi} \int_0^{\pi/6} \int_0^{\pi/2} \hat{C}_s(\alpha, \beta) \cos\alpha \, d\alpha \, d\beta, \tag{4}$$

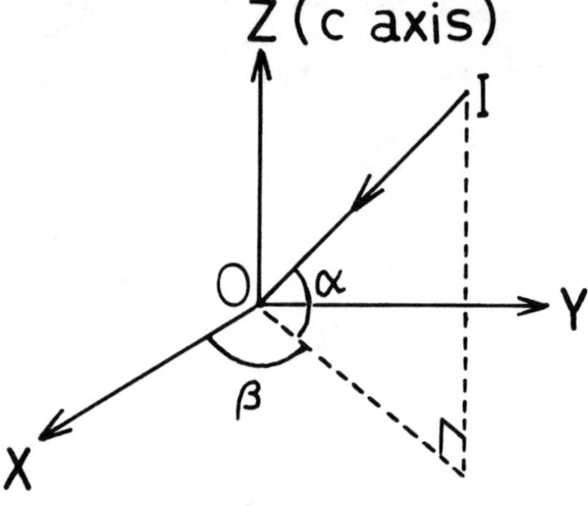

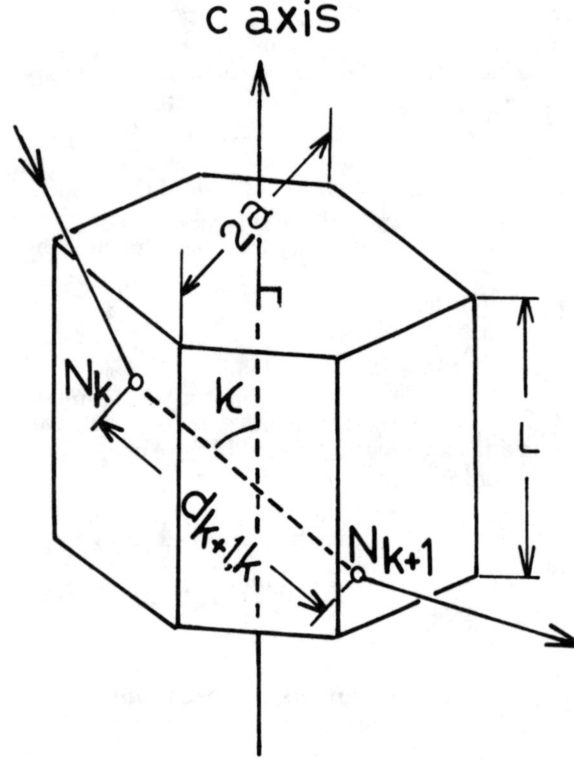

FIG. 1. Geometry of a light ray in an ice crystal.
All symbols are explained in the text.

where the scattering cross section for a single crystal $\hat{C}_s(\alpha, \beta)$ is computed by integration of the phase matrix element G_{11} over the 4π solid angle (van de Hulst 1957). According to the optical theorem, the extinction cross section of a single particle is twice the geometric cross section area in the limit of geometric optics. For an arbitrarily oriented hexagon, the geometric cross section area σ is

$$\sigma(\alpha, \beta) = \frac{3\sqrt{3}}{2} a^2 \sin\alpha + 2aL \cos\alpha \cos\left(\frac{\pi}{6} - \beta\right), \quad (5)$$

where L is the length of the hexagon and a the radius. Thus, the extinction cross section for a sample of randomly oriented ice crystals with the same sizes may be expressed in the form

$$\hat{C}_e = \frac{6}{\pi} \int_0^{\pi/6} \int_0^{\pi/2} 2\sigma(\alpha, \beta) \cos\alpha d\alpha d\beta$$

$$= \frac{3a^2}{2} [\sqrt{3} + 4(L/2a)] = A/2, \quad (6)$$

where A is the surface area of a hexagonal cylinder. The last expression was pointed out by Vouk (1948) for randomly oriented convex particles.

Next, we shall discuss the manner in which the ice crystal size distribution is incorporated in the computation of the single-scattering parameters. Based on observations by Ono (1969) and Auer and Veal (1970), the aspect ratio, $L/2a$, of ice crystals may be related to the crystal length L. If the ice crystal size distribution is denoted by $n(L)$, then the phase matrix for a sample of ice crystals of different sizes may be obtained from

$$P_{kl}(\Theta) = \frac{\int_{L_1}^{L_2} \hat{P}_{kl}(\Theta, L)\hat{C}_s(L)n(L)dL}{\int_{L_1}^{L_2} \hat{C}_s(L)n(L)dL}, \quad (7)$$

where L_1 and L_2 are the lower and upper limits of crystal lengths. The scattering and extinction cross sections for a sample of ice crystals of different sizes are then given by

$$C_{s,e} = \frac{1}{N_o} \int_{L_1}^{L_2} \hat{C}_{s,e}(L)n(L)dL, \quad (8)$$

where

$$N_o = \int_{L_1}^{L_2} n(L)dL,$$

the total number of ice crystals. The single-scattering albedo is defined by

$$\tilde{\omega} = C_s/C_e. \quad (9)$$

In the present computations, four ice crystal size distributions were used. The first two are for the cirrostratus (Cs) and cirrus uncinus (Ci) presented by Heymsfield (1975), while the other two are modified distributions given by Heymsfield and Platt (1984) corresponding to warm and cold cirrus clouds. These size distributions are displayed in Fig. 2. For scattering calculations, we have discretized these size distributions in five regions, as shown in Fig. 2. The aspect ratios, $L/2a$, used are 20/20, 50/40, 120/60, 300/100, and 750/160 in units of micrometers/micrometers, roughly

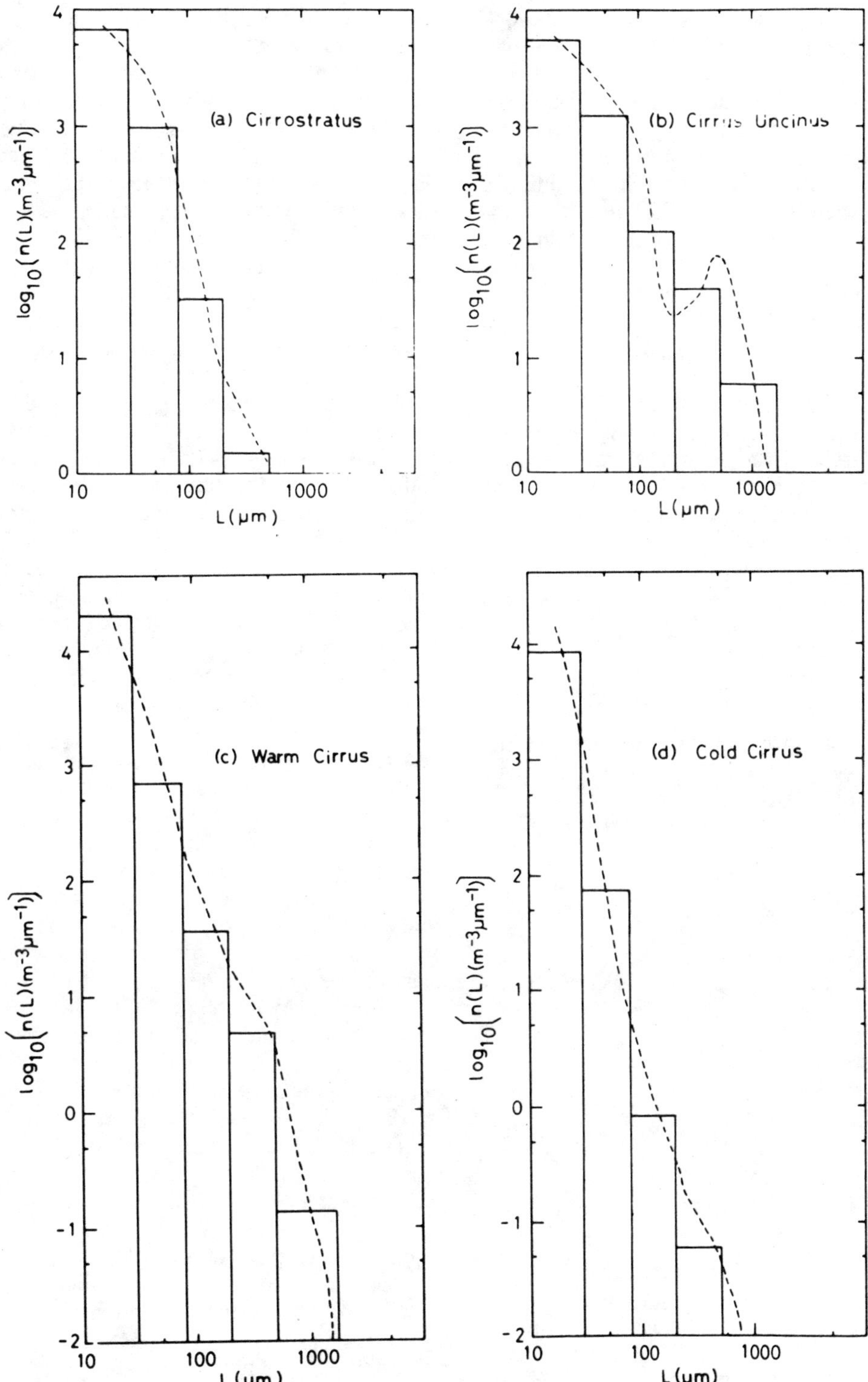

FIG. 2. Measured cirrus cloud particle size distributions (dashed lines) and the discretized size distributions (vertical bars) for (a) cirrostratus, (b) cirrus uncinus, (c) warm cirrus, and (d) cold cirrus.

corresponding to the observations reported by Ono (1969) and Auer and Veal (1970). The relationships between the crystal length L and diameter $2a$ derived by these researchers may not be universal, but these are the best experimental results that are available at the present time. The refractive indices for ice compiled by Warren (1984) were used in the scattering calculations.

Figure 3 shows the phase function as a function of the scattering angle for Cs and Ci at the 0.55 μm wavelength. In the numerical computations, we find that the intervals for orientation angles $\Delta\beta = 0.5°$ and $\Delta\alpha = 1°$ are sufficient to produce a smooth phase function curve. For the warm and cold Ci presented in Fig. 2, the phase functions are close to those for Cs and are not presented in the diagram. The 22° and 46° halos produced by two refracted rays are well illustrated in the diagram, in addition to the forward diffraction peak. For scattering angles between about 150° and 160°, there is also a maximum produced by rays undergoing two internal reflections. Table 1 lists the visible phase function values as a function of the scattering angle for Cs. From 0° to 2°, the values are given for every 0.1°, otherwise they are listed for every 1°. This table should be of use to researchers who are interested in investigating the effect of nonsphericity on the transfer of solar radiation, as well as developing an independent light scattering program for hexagonal ice crystals.

In Table 2 are shown the extinction coefficient $\beta_e = N_o C_e$, single-scattering albedo $\tilde{\omega}$, and asymmetry factor defined by

$$g = \frac{1}{2}\int_{-1}^{1} P_{11}(\Theta)\cos\Theta d\cos\Theta, \qquad (10)$$

for the four ice crystal size distributions shown in Fig. 2 and for five wavelengths in the solar spectrum. Cirrus (uncinus) contains a significant amount of large ice crystals in the second peak of the size distribution, as

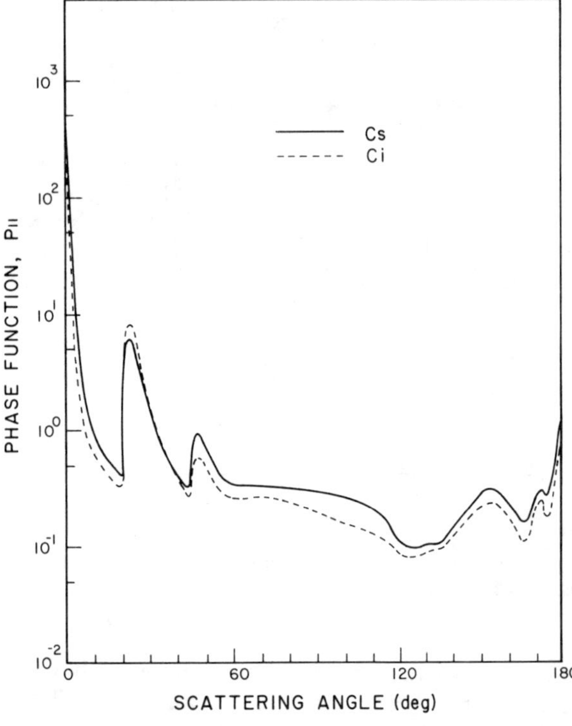

FIG. 3. Phase functions as a function of the scattering angle for cirrostratus and cirrus uncinus at the 0.55 μm wavelength.

shown in Fig. 2. For this reason, the extinction coefficient β_e for this cloud is much larger than those for the three other cloud types. In connection with the discussion in the Appendix, when the asymmetry factor is calculated from the phase function values listed in Table 1, the contribution from the δ-function transmission f_δ ($\Theta = 0°$) cannot be accounted for numerically. Using the similarity principle, the correct asymmetry factor may be expressed by $g = (1 - f_\delta)g^* + f_\delta$, where g^* represents the asymmetry factor without incorporation of the δ-function transmission. In the present case, $g^* = 0.751$ and $f_\delta = 0.126$. The correct g, including the contribution of f_δ denoted in Fig. A1 in the Appendix, is 0.782. The complex indices of refraction at each wavelength are averaged values over the wavelength band listed in the table, weighted by the solar irradiance (Thekaekara 1973). The extinction coefficients β_e in the limit of geometric optics for a given size distribution are the same regardless of the wavelength. The optical depth $\tau = \beta_e \Delta z$, corresponding to each size distribution, can be obtained if the cloud thickness Δz is given. There is significant absorption at the 3.0 μm wavelength with single-scattering albedos of about 0.53 that are almost independent of the size distribution. Considerable absorption is noted for the 1.6 and 2.2 μm wavelengths for cirrus uncinus consisting of large ice crystal sizes. At a given wavelength, the asymmetry factors for Cs, Ci (warm), and Ci (cold) are about the same since the phase functions for these clouds are very similar. For a given size distribution, the asymmetry factor increases toward the longer wavelength, where absorption is increased, leading to sharper diffraction patterns. The values listed in the table should be of use for applications to the calculation of the radiative transfer of cirrus clouds covering the solar spectrum.

As shown in Warren (1984), the imaginary part m_i of the refractive index for ice varies about six orders of magnitude in the near infrared region. Moreover, m_i is uncertain to about a factor of 2 in the near infrared. We wish to develop a parameterization equation to relate the single-scattering albedo $\tilde{\omega}$ to the absorption coefficient $k_i = 4\pi m_i/\lambda$ and aspect ratio $L/2a$. We note that $1 - \tilde{\omega} = C_a/C_e$, where C_a and C_e denote the absorption and extinction cross sections, respectively. The absorption cross section, C_a, is proportional to the product of the absorption coefficient and volume, i.e., $C_a \sim k_i V$ when the absorption is small, and C_e is given in Eq. (6). For a hexagonal cylinder, $V = 3\sqrt{3}a^2(L/2a)$. Thus, C_a/C_e must be proportional to the physical parameter defined by

$$z = k_i a \frac{3\sqrt{3}(L/2a)}{\sqrt{3} + 4(L/2a)}. \qquad (11)$$

Using 60 values of $k_i a$ and $L/2a$, computations of the single-scattering albedo were performed for a number of ice crystal sizes. We then performed a least squares fit to obtain the following parameterized equation:

TABLE 1. Scattering phase function for the cirrostratus cloud model at $\lambda = 0.55$ μm.

Θ	P_{11}	Θ	P_{11}	Θ	P_{11}	Θ	P_{11}
.00	1.083 +05*	32	1.13 +00	82	3.18 −01	132	1.04 −01
.10	6.037 +04	33	9.85 −01	83	3.14 −01	133	1.02 −01
.20	3.231 +04	34	8.54 −01	84	3.11 −01	134	1.02 −01
.30	1.809 +04	35	7.24 −01	85	3.11 −01	135	9.96 −02
.40	9.985 +03	36	6.26 −01	86	3.08 −01	136	1.02 −01
.50	5.477 +03	37	5.55 −01	87	3.03 −01	137	1.12 −01
.60	3.210 +03	38	5.55 −01	88	2.98 −01	138	1.21 −01
.70	2.106 +03	39	4.45 −01	89	2.95 −01	139	1.27 −01
.80	1.502 +03	40	4.12 −01	90	2.91 −01	140	1.35 −01
.90	1.095 +03	41	4.03 −01	91	2.89 −01	141	1.44 −01
1.00	7.875 +02	42	3.74 −01	92	2.87 −01	142	1.54 −01
1.10	5.550 +02	43	3.57 −01	93	2.85 −01	143	1.66 −01
1.20	3.885 +02	44	4.09 −01	94	2.83 −01	144	1.78 −01
1.30	2.758 +02	45	5.44 −01	95	2.82 −01	145	1.91 −01
1.40	2.027 +02	46	7.44 −01	96	2.81 −01	146	2.04 −01
1.50	1.563 +02	47	8.88 −01	97	2.78 −01	147	2.13 −01
1.60	1.267 +02	48	8.78 −01	98	2.75 −01	148	2.25 −01
1.70	1.069 +02	49	8.10 −01	99	2.72 −01	149	2.43 −01
1.80	9.266 +01	50	7.15 −01	100	2.68 −01	150	2.63 −01
1.90	8.151 +01	51	6.56 −01	101	2.62 −01	151	2.87 −01
2.00	7.210 +01	52	5.91 −01	102	2.56 −01	152	3.04 −01
3.00	2.223 +01	53	5.09 −01	103	2.53 −01	153	3.08 −01
4.00	9.666 +00	54	4.52 −01	104	2.48 −01	154	3.09 −01
5.00	5.198 +00	55	4.16 −01	105	2.42 −01	155	3.07 −01
6.00	3.208 +00	56	3.89 −01	106	2.35 −01	156	2.99 −01
7.00	2.182 +00	57	3.72 −01	107	2.27 −01	157	2.80 −01
8.00	1.598 +00	58	3.58 −01	108	2.21 −01	158	2.63 −01
9.00	1.236 +00	59	3.47 −01	109	2.16 −01	159	2.52 −01
10.00	1.013 +00	60	3.46 −01	110	2.11 −01	160	2.36 −01
11.00	8.296 −01	61	3.44 −01	111	2.05 −01	161	2.13 −01
12.00	7.494 −01	62	3.43 −01	112	1.99 −01	162	1.95 −01
13.00	6.753 −01	63	3.44 −01	113	1.96 −01	163	1.77 −01
14.00	6.220 −01	64	3.43 −01	114	1.89 −01	164	1.66 −01
15.00	5.681 −01	65	3.40 −01	115	1.79 −01	165	1.57 −01
16.00	5.248 −01	66	3.38 −01	116	1.64 −01	166	1.58 −01
17.00	4.883 −01	67	3.37 −01	117	1.42 −01	167	1.69 −01
18.00	4.598 −01	68	3.36 −01	118	1.27 −01	168	1.90 −01
19.00	4.409 −01	69	3.36 −01	119	1.18 −01	169	2.35 −01
20.00	4.227 −01	70	3.36 −01	120	1.11 −01	170	2.71 −01
21.00	1.935 +00	71	3.35 −01	121	1.04 −01	171	2.84 −01
22.00	5.333 +00	72	3.34 −01	122	9.93 −02	172	2.88 −01
23.00	6.137 +00	73	3.33 −01	123	9.86 −02	173	2.63 −01
24.00	6.043 +00	74	3.32 −01	124	9.79 −02	174	2.67 −01
25.00	4.660 +00	75	3.31 −01	125	9.70 −02	175	3.04 −01
26.00	3.665 +00	76	3.29 −01	126	9.67 −02	176	3.81 −01
27.00	2.955 +00	77	3.27 −01	127	9.68 −02	177	5.76 −01
28.00	2.404 +00	78	3.25 −01	128	9.76 −02	178	7.98 −01
29.00	1.982 +00	79	3.24 −01	129	9.97 −02	179	1.01 +00
30.00	1.638 +00	80	3.22 −01	130	1.01 −01	180	1.18 +00
31.00	1.342 +00	81	3.21 −01	131	1.03 −01		

* 1.083 +05 means 1.083×10^5.

$$\tilde{\omega} = 1 - b_1 z + b_2 z^2 - b_3 z^3 + b_4 z^4, \qquad (12)$$

where the empirical coefficients $b_1 = 1.1128$, $b_2 = 2.5576$, $b_3 = 5.6257$, and $b_4 = 5.9498$. Equation (12) is applicable for $\tilde{\omega}$ in the range from 0.75 to 1. It follows that if m_i is given, $\tilde{\omega}$ can be determined for a given size and wavelength. Equation (11) is valid for $0.4 \leqslant \lambda \leqslant 2.5$ μm.

Next, we investigate the effects of nonsphericity on the single-scattering parameters using two wavelengths of 0.55 (negligible absorption) and 2.2 μm (moderate absorption), and the five hexagonal particle sizes listed in Table 3. The single-scattering properties for equivalent ice spheres with the same surface areas as those for hexagons were computed by the Mie theory. The asymmetry factors for hexagonal ice crystals are uniformly smaller than those for equivalent ice spheres. This is especially evident for small sizes illuminated by visible light. The single-scattering co-albedos $1 - \tilde{\omega}$ for hexagonal ice crystals at 2.2 μm are smaller than those for equivalent ice spheres. The differences are especially evident for large particles. This is partly due to the fact that equivalent ice spheres have larger volumes and hence, absorb more incident radiation. For the extinc-

TABLE 2. Extinction coefficient β_e (km^{-1}),* single-scattering albedo $\tilde{\omega}$, and asymmetry factor g for four cirrus models and five solar wavelengths. The refractive index $m = m_r - im_i$, where m_r and m_i are averaged real and imaginary parts for the spectral band with limits λ_1 and λ_2.

λ (μm) (λ_1, λ_2)	m_r	m_i		Cs	Ci uncinus	Ci (warm)	Ci (cold)
0.55	1.311	3.110 -9**	β_e	0.3865	2.6058	0.6525	0.1662
			$\tilde{\omega}$	1.0000[†]	1.0000[†]	1.0000[†]	1.0000[†]
			g	0.7824	0.8404	0.7889	0.7724
1.0 (0.7, 1.3)	1.302	1.931 -6	β_e	0.3865	2.6058	0.6525	0.1662
			$\tilde{\omega}$	0.9995	0.9981	0.9994	0.9997
			g	0.7905	0.8448	0.7945	0.7780
1.6 (1.3, 1.9)	1.290	2.128 -4	β_e	0.3865	2.6058	0.6525	0.1662
			$\tilde{\omega}$	0.9658	0.9004	0.9628	0.9810
			g	0.8100	0.8778	0.8129	0.7925
2.2 (1.9, 2.5)	1.263	7.997 -4	β_e	0.3865	2.6058	0.6525	0.1662
			$\tilde{\omega}$	0.9185	0.7996	0.9154	0.9528
			g	0.8436	0.9116	0.8440	0.8220
3.0 (2.5, 3.5)	1.242	1.424 -1	β_e	0.3865	2.6058	0.6525	0.1662
			$\tilde{\omega}$	0.5321	0.5309	0.5327	0.5338
			g	0.9653	0.9728	0.9634	0.9583

* The number densities N_0 for Cs, Ci, Ci (warm), and Ci (cold) are 0.187, 0.213, 0.442, and 0.176 cm^{-3}, respectively.
** 3.110 -9 means 3.110 $\times$ 10^{-9}.
[†] $\tilde{\omega} = 0.99999$.

tion cross section at 0.55 μm, the results for randomly oriented hexagons computed from geometric optics are very close to those for equivalent spheres with the same surface areas. Note that in the limit of geometric optics, the extinction is independent of the wavelength. We also performed light scattering calculations using equivalent ice spheres with the same volumes as hexagons. The results show that the single-scattering albedos for equivalent volume spheres are closer to those for hexagons, but the extinction cross sections are smaller. The latter will lead to a smaller optical depth if spheres and hexagons have the same number density in a cloud. On the basis of the preceding comparisons, we conclude that neither the area-equivalent nor the volume-equivalent ice sphere can adequately approx-imate nonspherical ice crystals in terms of their scattering and absorption properties. This conclusion agrees with the criticism of using an "equivalent sphere" for nonspherical particles presented by Bohren (1986).

3. Comparisons with measurements

In this section, the computed phase matrix elements for hexagonal ice crystals are compared with experimental results. Dugin and Mirumyants (1976) measured the phase matrix elements for laboratory ice clouds containing 20–100 μm plates at the 0.57 μm wavelength. The ice crystals generated in the cold chamber were smaller than those occurring in cirrus

TABLE 3. Extinction cross section, single-scattering albedo, and asymmetry factor for hexagonal ice crystals and area equivalent ice spheres at $\lambda = 0.55$ and 2.2 μm.

$L/2a$ (μm/μm)	λ (μm)	C_e (cm^2) Hexagon	Sphere	$1 - \tilde{\omega}$ Hexagon	Sphere	g Hexagon	Sphere
20/20	0.55	8.598 -6*	8.918 -6	0	0	0.7704	0.8759
	2.2	8.598 -6	9.441 -6	4.085 -2	4.944 -2	0.8185	0.8729
50/40	0.55	4.039 -5	4.129 -5	0	0	0.7780	0.8837
	2.2	4.039 -5	4.270 -5	8.179 -2	9.550 -2	0.8416	0.8996
120/60	0.55	1.314 -4	1.334 -4	0	0	0.8155	0.8871
	2.2	1.314 -4	1.364 -4	1.246 -1	1.539 -1	0.8829	0.9174
300/100	0.55	5.150 -4	5.198 -4	0	0	0.8429	0.8895
	2.2	5.150 -4	5.273 -4	1.871 -1	2.504 -1	0.9165	0.9373
750/160	0.55	1.966 -3	1.978 -3	0	0	0.8592	0.8908
	2.2	1.966 -3	1.996 -3	2.503 -1	3.583 -1	0.9380	0.9566

* 8.598 -6 means 8.598 $\times$ 10^{-6}.

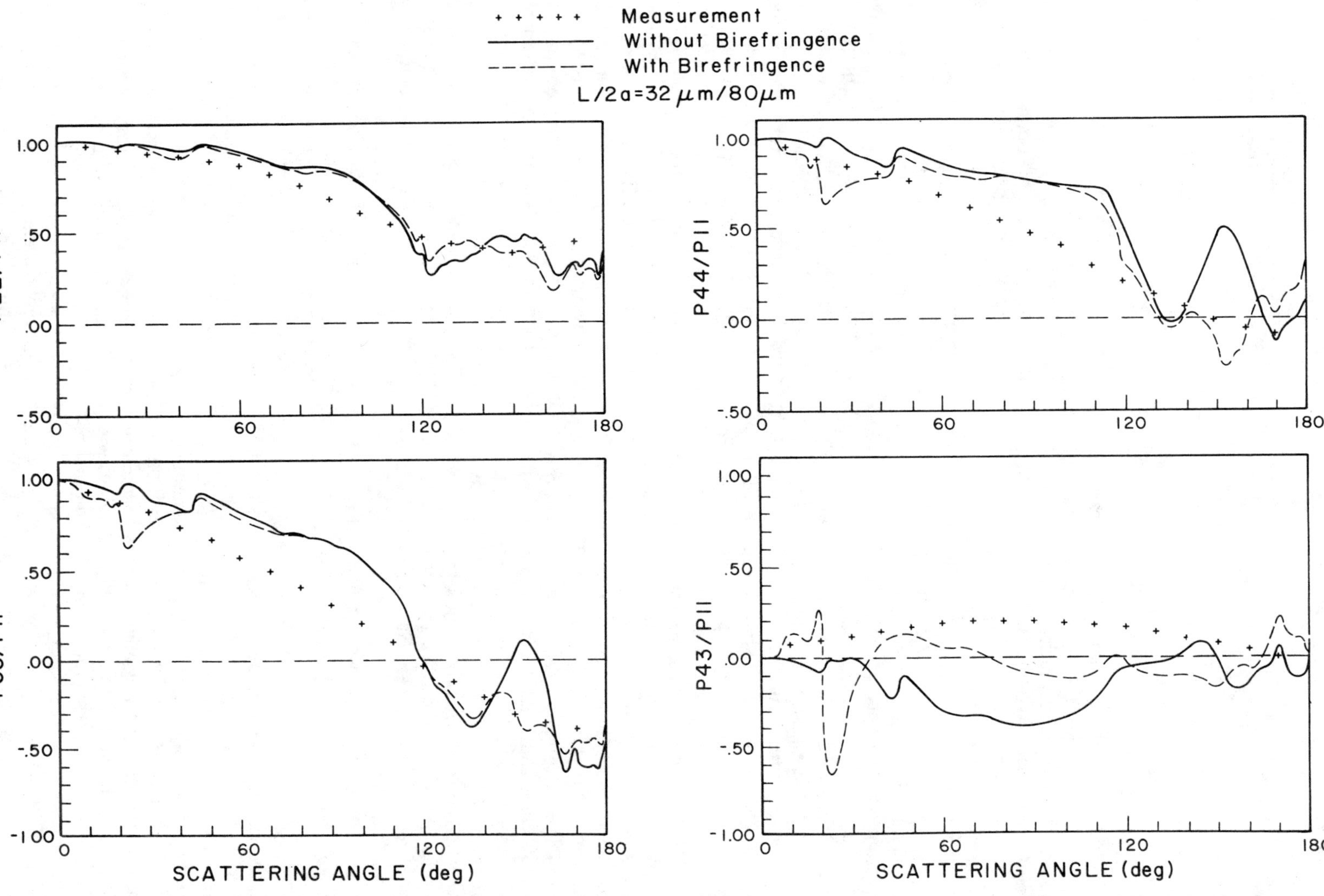

FIG. 4. Scattering phase matrix elements for randomly oriented hexagonal plate crystals with an aspect ratio $L/2a$ of 32 μm/80 μm at the 0.55 μm wavelength. The birefringence, $n_e - n_o$, is 0.0014. The plus symbols are experimental values reported by Dugin and Mirumyants (1976).

clouds. Small ice crystals tend to be randomly oriented (Fraser 1979). For randomly oriented nonspherical particles, the phase matrix consists of only six nonzero and independent elements in the form

$$
\mathbf{P} = \begin{bmatrix} P_{11} & P_{12} & 0 & 0 \\ P_{12} & P_{22} & 0 & 0 \\ 0 & 0 & P_{33} & -P_{43} \\ 0 & 0 & P_{43} & P_{44} \end{bmatrix} \tag{13}
$$

The laboratory data were normalized by the phase function P_{11}.

We first compare the computed and measured elements P_{22}/P_{11}, P_{33}/P_{11}, P_{43}/P_{11}, and P_{44}/P_{11}. A number of mean aspect ratios were used to interpret the observed data. We find that $L/2a$ of 32 μm/80 μm, as well as 8 μm/80 μm, closely match the measurements. Note that incorporation of the birefringence effect requires size information. The theoretical results with and without the effect of birefringence, along with the measured data are shown in Fig. 4. The effect of birefringence on the phase function and degree of linear polarization is insignificant, although this effect somewhat modifies the other phase matrix elements. The computed phase matrix element P_{22}/P_{11} agrees well with the measured value. Except for the element P_{43}/P_{11}, the incorporation of the birefringence effect brings the computed values closer to the measured values, especially in the scattering angle range between 140° and 160°. An agreement can be seen between the computed and measured values at the scattering angle Θ_{33} where the element P_{33} becomes zero. Table 4 lists Θ_{33} for several aspect ratios $L/2a$. Based on the laboratory experiment for plate crystals, $\Theta_{33} = 117°$, as shown in Fig. 4. The computed Θ_{33} for the aspect ratio of 32 μm/80 μm is about 120°, which agrees well with the experimental value. The scattering angle Θ_{33} varies with the aspect ratio and shows a maximum value when the length and width are about the same ($L/2a = 0.8$). Similar behaviors for needles and circular plates have been reported by Asano and Sato (1980) in light scattering calculations using spheroidal shapes. They showed that the Θ_{33} value for spheroids with an aspect ratio of 2 is larger than that with a larger aspect ratio of 5.

The computed values of P_{43}/P_{11} are also in general agreement with the measured values if the sign of P_{43} from the laboratory experiments is reversed. It appears that the sign of P_{43} is defined differently by Dugin and Mirumyants. This is likely a case where different conventions for right- and left-circular polarization are used. At the 22° scattering angle, there is a significant difference between the computed P_{43}/P_{11} values with and without the incorporation of the birefringence effect. This behavior can be explained as follows. The inner halo rays that undergo two refractions produce a maximum intensity at the 22° scattering angle. When the effect of birefringence is not considered, the element P_{43} basically arises from total reflection and has positive

TABLE 4. Scattering angle Θ_{33} where P_{33} becomes zero.

$L/2a$ (μm/μm)	Θ_{33} (°) Without birefringence	Θ_{33} (°) With birefringence
8/80	114.0 (99.7)*	115.0
16/80	116.1	116.4
32/80	119.5	119.6
64/80	127.9	126.9
200/80	119.2	114.1
400/80	103.5	100.5

* There are two Θ_{33} for this case.

values. When the birefringence effect is accounted for in the computation, the optical path length due to extraordinary rays δ_e is smaller than that due to ordinary rays δ_o. Since the element P_{43} for two refracted rays is proportional to $\sin(\delta_e - \delta_o)$, its value at the inner halo angle becomes negative.

The computed and measured patterns for the degree of linear polarization $-P_{12}/P_{11}$ are shown in Fig. 5. The upper diagram shows the computed values for a number of aspect ratios and the measured data presented by Dugin and Mirumyants (1976). The computed linear polarization for $L/2a = 0.1$ and 0.2 agrees well with the measured values. In the lower diagram, we compare the computed linear polarization with the measured data presented by Stahl et al. (1983) and Tomasko (personal communication). The experiments used a He–Ne (6328 Å) laser and 15 detectors spaced at 10° increments. The circles shown in Fig. 5b represent the linear polarization values for laboratory columns with a mean length of about 11 μm. Aspect ratios of 2.5 and 5 were used in the scattering calculations to interpret the observations. The computed values for the aspect ratio of 5 closely match the observed linear polarization pattern for columns.

The linear polarization pattern presented in Fig. 5a and b shows that polarization becomes zero (neutral point) at phase angles (180°-scattering angles) between 10° and 30°. For ice columns ($L/2a = 2.5$, 5) or ice crystals with $L \approx 2a$, the neutral point is 17°–19°, but it is 23°–24° for ice plates ($L/2a = 0.1$, 0.2, 0.4). As demonstrated in Part II (Takano and Liou 1988), the location of the neutral point is not affected by multiple scattering. Thus, an observation of the neutral point for clouds may provide a means for determining the shape (column or plate) of cloud particles.

4. Light scattering by horizontally oriented plate and columnar crystals: Interpretation of observed optical phenomena from cirrus

In this section, we present light scattering results for horizontally oriented ice crystals with the specific purpose of interpreting numerous optical features that have been observed in the presence of cirrus. The theory and computation of solar radiative transfer in cirrus

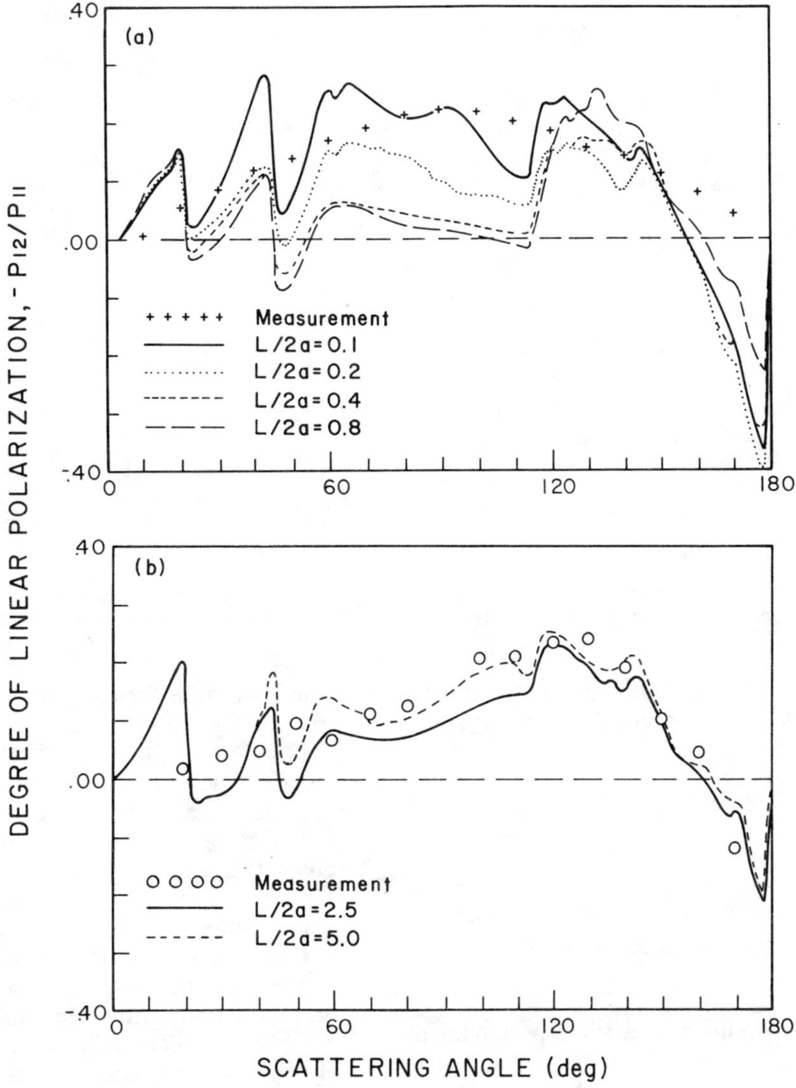

FIG. 5. Degree of linear polarization for randomly oriented ice crystals at $\lambda = 0.55\ \mu$m (a) for plate crystals, and (b) for columnar crystals. The plus and zero symbols are the experimental values reported by Dugin and Mirumyants (1976) and Stahl et al. (1983), respectively.

clouds composed of horizontally oriented ice crystals will be presented in Part II.

We shall begin our discussion with the manner in which the phase functions for horizontally oriented plates and columns were computed. When plate crystals are randomly oriented with their c axes vertical (referred to as 2-D plates), the phase matrix elements P_{kl} are obtained by integrating G_{kl} over the angle β as follows:

$$P_{kl}(\theta_0, \theta, \phi - \phi_0) = \frac{3}{\pi} \int_0^{\pi/3} G_{kl}(\theta_0, \theta, \phi - \phi_0; \beta)d\beta,$$

$$k, l = 1\text{--}4. \quad (14)$$

As shown in Fig. 6a, β is the angle denoting the rotation of the plate crystal around the c axis and θ_0 is the solar zenith angle, which is the complimentary angle of the

solar elevation angle, ϵ. In the case of 2-D plates, ϵ equals α. The scattering angle Θ and the azimuth angle Φ, whose reference plane is XOZ, are calculated by the ray-tracing method (see Cai and Liou 1982). Then, from the scattering geometry in Fig. 6a, θ and $\phi - \phi_0$ in Eq. (14) can be easily computed (see, e.g., Asano 1983). An angular interval $\Delta\beta = 0.5°$ is sufficient to converge the computations for the phase function.

When column crystals have both their c axes and a pair of prism faces horizontal (referred to as Parry columns), the phase matrix elements P_{kl} are expressed by

$$P_{kl}(\theta_0, \theta, \phi - \phi_0) = \frac{3}{\pi^2} \int_{-\pi/2}^{\pi/2} d\gamma \int_0^{\pi/3} G_{kl}(\theta_0, \theta, \phi$$

$$- \phi_0; \gamma, \beta)\delta(\beta - \beta^*)d\beta, \quad k, l = 1\text{--}4 \quad (15)$$

where $\delta = 1$ when $\beta = \beta^*$ and $\delta = 0$ otherwise. As

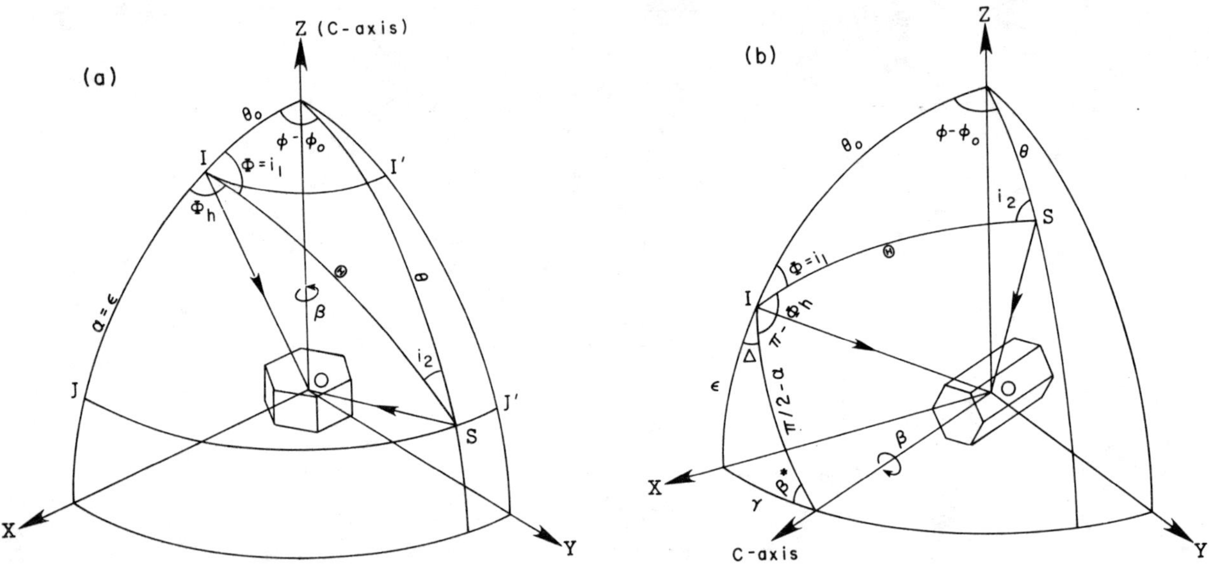

FIG. 6. Scattering geometries for horizontally oriented (a) plate crystals, and (b) columnar crystals. The terms IO and SO denote the incident and scattered directions, respectively. Other symbols are explained in the text.

shown in Fig. 6b, γ is the angle denoting the orientation of the c axis in the horizontal plane. The angles α and β^* are expressed by

$$\sin\alpha = \cos\epsilon \cos\gamma, \qquad (16)$$

$$\sin\beta^* = \sin\epsilon / \sin\alpha. \qquad (17)$$

There are two β^* satisfying Eq. (17) in the interval $(0, \pi/3)$. As shown in Fig. 6b, the azimuth angle Φ with respect to the plane XOZ is given by

$$\Phi = \Phi_h - \Delta, \qquad (18)$$

where Φ_h is the azimuth angle with respect to the plane containing the c axis of the column and the incident direction IO. The azimuth angle Φ_h is also calculated by the ray-tracing method. The angle Δ is given by

$$\cos\Delta = \frac{\cos\gamma \, \sin\epsilon}{\cos\alpha}. \qquad (19)$$

The set of angles $(\epsilon, \gamma, \Theta, \Phi)$ may be transformed to another set of angles $(\theta_0, \theta, \phi - \phi_0)$, as in the case of 2-D plates. In the numerical computations, we use angular intervals $\Delta\beta = 0.5°$ and $\Delta\gamma = 1°$, which are found to be adequate to obtain stable solutions.

In the case when column crystals orient with their c axes horizontal, as well as with random rotational orientation about the c axes (referred to as 2-D columns), the phase matrix elements P_{kl} may be computed from

$$P_{kl}(\theta_0, \theta, \phi - \phi_0)$$

$$= \frac{3}{\pi^2} \int_{-\pi/2}^{\pi/2} d\gamma \int_0^{\pi/3} G_{kl}(\theta_0, \theta, \phi - \phi_0; \gamma, \beta)d\beta,$$

$$k, l = 1\text{–}4. \qquad (20)$$

The procedure for computing Eq. (20) is similar to that for Eq. (15).

For interpretation of the optical features produced by cirrus clouds, it is sufficient to use the results from ray tracing involving reflections and refractions. In the following figures, relative scattered intensity rather than absolute intensity due to single scattering will be used for the identification of various halos and arcs. The abbreviations are defined in Table 5.

Figure 7a–c shows the relative intensity as a function of the azimuthal angle $\phi - \phi_0$ for plate crystals with an aspect ratio of 0.1 randomly oriented in the horizontal plane XOY. The solar zenith angles θ_0 ($\pi/2$-elevation angle ϵ) used in the computation are 75°, 50° and 25°. Due to geometry, the scattered light for horizontally oriented plate crystals is confined to four latitude belts denoted by II' and JJ' and their mirror images with respect to the horizontal plane (XOY).

Based on the ray-tracing geometry for plates, the emergent zenith angle θ may be computed from the incoming solar zenith angle θ_0 in the forms

$$\theta^* = \begin{cases} \pi/2 - \sin^{-1}\sqrt{m_r^2 - \sin^2\theta_0} \\ \quad \text{for} \quad \theta_0 > \sin^{-1}\sqrt{m_r^2 - 1} \approx 58° \\ \sin^{-1}\sqrt{m_r^2 - \cos^2\theta_0} \\ \quad \text{for} \quad \theta_0 < \cos^{-1}\sqrt{m_r^2 - 1} \approx 32° \end{cases} \qquad (21)$$

The two latitude belts are associated with $\theta_0(II')$ and its mirror image, $-\theta_0$, while the other two are associated with $\theta^*(JJ')$ and its mirror image, $-\theta^*$. The former two belts define the parhelic and subparhelic circles and the latter two belts are associated with the circumzenithal and subcircumzenithal arcs identified in Fig. 7a.

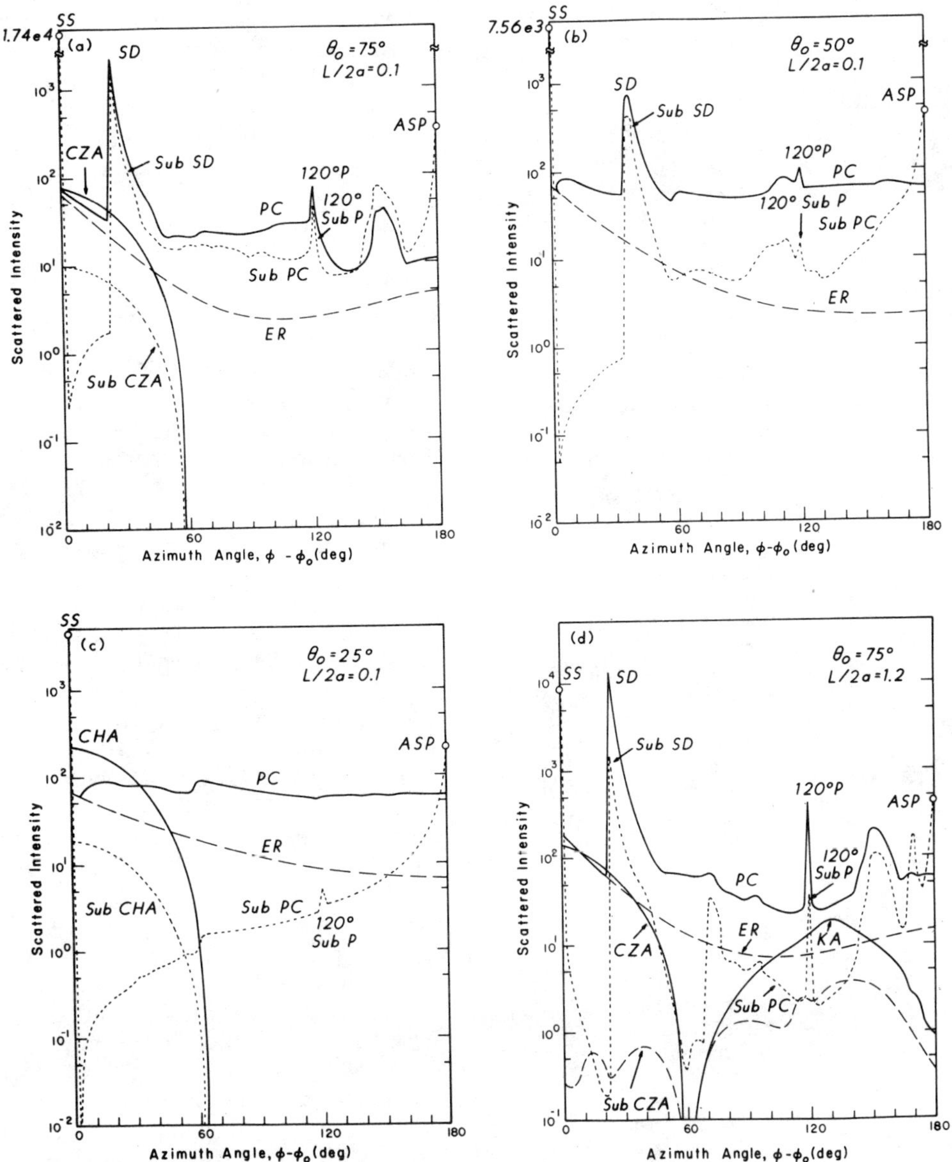

FIG. 7. Scattered intensities for 2-D plates with $L/2a = 0.1$ for (a) $\theta_0 = 75°$, (b) $\theta_0 = 50°$, (c) $\theta_0 = 25°$, and (d) with $L/2a = 1.2$ for $\theta_0 = 75°$ at $\lambda = 0.55$ μm. The abbreviations for the optical features shown in the figure are given in Table 5.

In Fig. 7, the scattered intensity due to external reflection (ER) is plotted for comparison with that produced by various arcs and halos. It is clear that the parhelic circle is a result of light rays undergoing refractions and internal reflections. The intensity pattern of the parhelic circle shows a discontinuity at the azimuthal angle $(\phi - \phi_0)$ of about 122°. This reduction in intensity results from the absence of total reflection for the internally reflected rays. From ray-tracing geometry, the critical angle $(\phi - \phi_0)$ for the total reflection is given by

$$(\phi - \phi_0)_c = 2 \sin^{-1}(\sqrt{m_r^2 - 1}/\sin\theta_0). \quad (22)$$

This formula is also applicable to the parhelic circles produced by horizontally oriented columns. For $\theta_0 = 75°$, we find $(\phi - \phi_0)_c = 122.3°$. The parhelic circle generated in Fig. 7a closely resembles the photograph taken by Evans and Tricker (1972, Fig. 2) for a solar zenith angle θ_0 of about 76°. Several specific features are also identified in this figure and listed in Table 5.

In reference to Eq. (21), when $32° < \theta_0 < 58°$, the scattered light is confined only to the two latitude belts corresponding to $\theta_0(II')$ and its mirror image, $-\theta_0$. If the sun's position is located in this angular range, only the parhelic and subparhelic circles will be produced, as shown in Fig. 7b, which also identifies a number of optical features listed in Table 5.

For a higher sun with a solar zenith angle of 25°,

TABLE 5. Optical phenomena caused by 2-D plates,
Parry columns, and 2-D columns.

Abbreviation	Definition
	2-D Plates
ASP	antisolar peak
CHA	circumhorizontal arc
CZA	circumzenithal arc
ER	external reflection
KA	Kern's arc
120°P	120° parhelion
120° Sub-P	120° subparhelion
PC	parhelic circle
SD	sundog (22° parhelion)
SS	subsun
Sub-CHA	subcircumhorizontal arc
Sub-CZA	subcircumzenithal arc
Sub-SD	subsundog (22° subparhelion)
	Parry Columns
ASA	antisolar arc
ASP	antisolar peak
CZA	circumzenithal arc
HA	heliac arc
HAA	Hastings' anthelic arc
LSCP	lower suncave Parry arc
LSVP	lower sunvex Parry arc
SHA	subhelic arc
SID	solar incident direction
SS	subsun
Sub-CZA	subcircumzenithal arc
Sub-PC	subparhelic circle
USCP	upper suncave Parry arc
USVP	upper sunvex Parry arc
	2-D Columns
ASA	antisolar arc
ASP	antisolar peak
DA	diffuse arc
ILA	infralateral arc
LTA	lower tangent arc
PC	parhelic circle
SHA	subhelic arc
SID	solar incident direction
SLA	supralateral arc
TAA	Tricker's anthelic arc
UTA	upper tangent arc
WAA	Wegener's anthelic arc

oriented ice plates produce circumhorizontal and subcircumhorizontal arcs instead of circumzenithal and subcircumzenithal arcs. This is shown in Fig. 7c. Using Eq. (21), the zenith angle θ from which these arcs may be observed is 71.1°. Generally, the intensity of the circumhorizontal arc is stronger than that of the circumzenithal arc. We have also used an aspect ratio of 0.4 in the calculation. All the features identified previously are reproduced with stronger scattered intensities.

In addition, an aspect ratio of 1.2 was used in the light scattering calculations, assuming that these crystals behave the same as plates. The results are presented in Fig. 7d for a solar zenith angle of 75°. The halos and arcs produced in this case are very similar to those presented in Fig. 7a with an addition of the Kern's arc identified in the diagram.

Figure 8 shows the scattered intensity for Parry columns with an aspect ratio of 2.5 and a solar zenith angle of 75°. The scattered intensity in the solar principal plane ($\phi - \phi_0 = 0°$ and 180°) is presented as a function of the zenith angle θ. Due to the specific orientation properties of Parry columns, the scattered intensity is confined to certain areas in the sky. Various halos and arcs are identified in Figs. 8a–d. Additional optical features produced by Parry columns are listed in Table 5. All the optical phenomena have been illustrated by Tricker (1979), Greenler (1980), and Tränkle and Greenler (1987). However, it is noted that the present light scattering program produces the intense antisolar peak (subanthelion) shown in Fig. 8c, which has not been quantitatively presented previously. Also, an arc (referred to as CA), which has not yet been observed, is produced below the parhelic circle shown in Fig. 8d.

Finally, Fig. 9 displays the scattered intensity for 2-D columns in the solar principal plane as a function of the zenith angle using the same aspect ratio for columns and solar zenith angle as in Fig. 8. The scattered light from 2-D columns covers almost the entire sky in contrast to the case of Parry columns. Several halos and arcs produced by 2-D columns are identified in Figs. 9a–d. Additional optical features that can be observed in regions other than the solar principal plane are listed in Table 5. It is noted that the subsun at θ = 75° cannot be identified definitely due to the overlapping intense lower tangent arc (LTA), as shown in Fig. 9a. The single-scattering patterns shown in Figs. 7–9 will be used for interpretation of the multiple scattering features presented in Part II.

5. Conclusions

A light scattering and polarization program has been developed for computations of the single-scattering properties of cirrus clouds containing hexagonal ice crystals. We have improved our previous ray-tracing program through the incorporation of the birefringence of ice and a proper treatment of the δ-transmission by geometric rays in the 0° scattering angle.

For the first time, the ice crystal size distribution consisting of nonspherical particles is accounted for in the light scattering computations. Four observed distributions for cirrus clouds are employed in the computations. We present the phase function, single-scattering albedo, extinction cross section, and asymmetry factor for wavelengths in the solar spectrum. We also derive a generalized equation for the single-scattering albedo as a function of the absorption parameter $k_i a$, where $k_i = 4\pi(m_i/\lambda)$, m_i is the imaginary part of the refractive index, a the radius of the ice crystal, and λ the incident wavelength. Since the phase function does not vary significantly with wavelength, the parameterized single-scattering albedo equation should be extremely useful for flux and heating rate calculations.

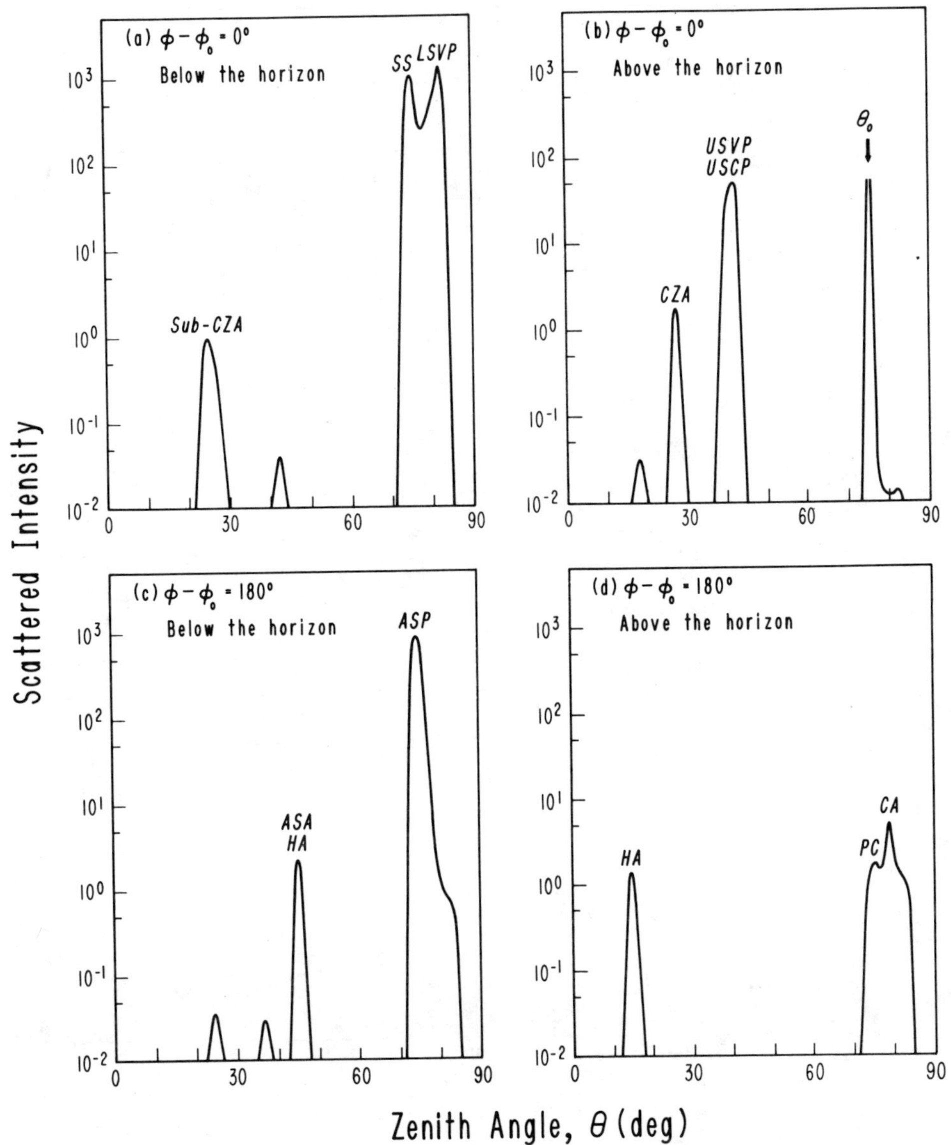

FIG. 8. Scattered intensities in the solar principal plane as functions of the zenith angle θ for Parry columns with $L/2a = 2.5$ for $\theta_0 = 75°$. (a) $\phi - \phi_0 = 0°$ below the horizon, (b) $\phi - \phi_0 = 0°$ above the horizon, (c) $\phi - \phi_0 = 180°$ below the horizon, and (d) $\phi - \phi_0 = 180°$ above the horizon. The abbreviations for the optical features depicted in the figure are given in Table 5.

We also examine the effect of nonsphericity on the single-scattering properties of ice crystals. By using two wavelengths of 0.55 (negligible absorption) and 2.2 μm (moderate absorption) and a number of aspect ratios for hexagonal ice particles, we show that the equivalent sphere with the same surface area or volume as the hexagonal ice crystal is inadequate to reproduce single-scattering properties of hexagonal ice crystals. In general, equivalent ice spheres generate larger asymmetry factors and smaller single-scattering albedos when compared to hexagonal ice particles. This is especially evident for wavelengths at which moderate absorption occurs.

Comparisons with available measurements for light scattering by ice crystals were made using a number of aspect ratios in the ray-tracing calculations. We show that the computed phase matrix elements P_{22}, P_{33}, P_{44}, and P_{43} for plates compare closely with those observed by Dugin and Mirumyants (1976). The effect of birefringence appears to be small, but the incorporation of this effect brings the computed values closer to the observed data. The computed degree of linear polarization patterns agrees well with the observed data presented by Dugin and Mirumyants (1976) and Stahl et al. (1983). We demonstrate that the scattering angle corresponding to zero polarization (neutral point) is

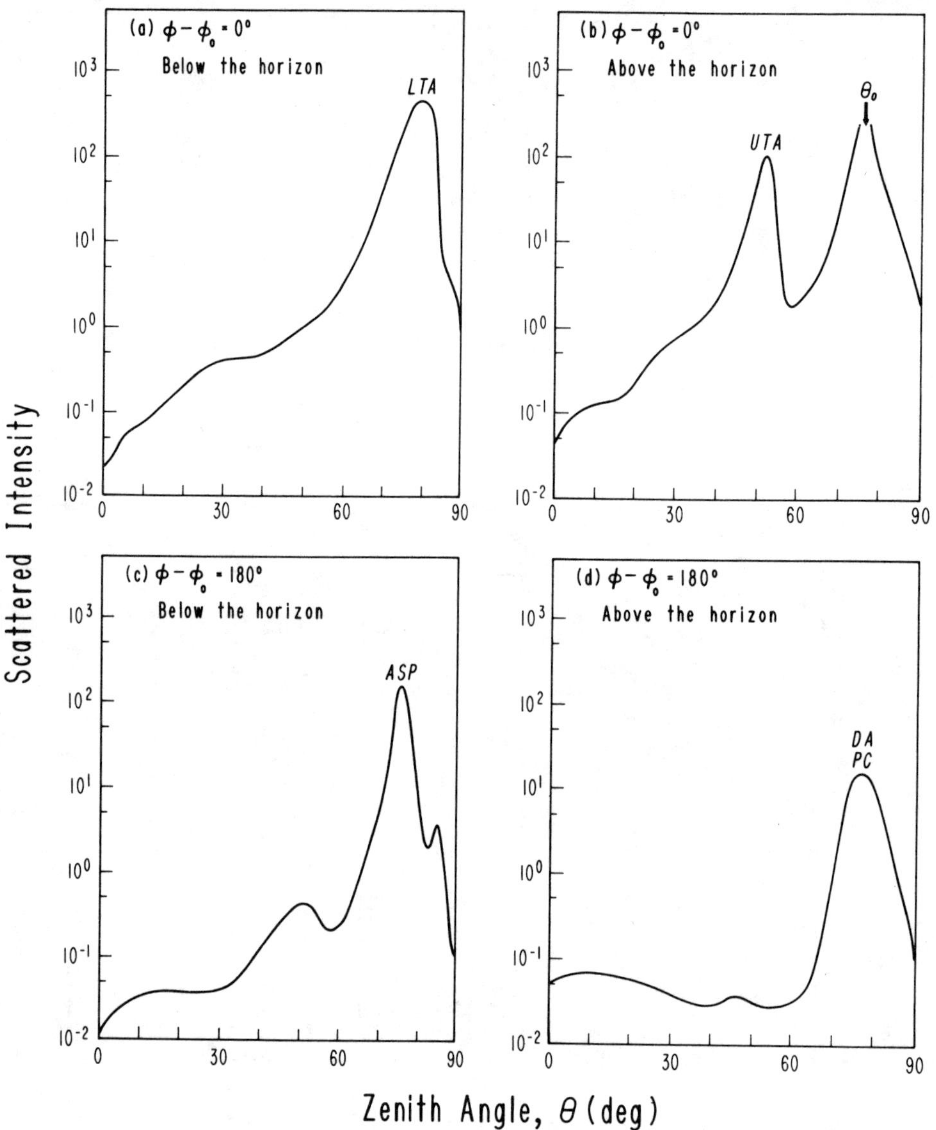

FIG. 9. As in Fig. 8, except for 2-D columns.

extremely sensitive to the aspect ratio of the ice crystal. Finally, to further verify the present light scattering program for hexagonal particles, the computed scattered intensities are used to identify various observed halos and arcs that are produced by horizontally oriented plates and columns. Our calculations are consistent with all the interpretations previously given by the other researchers cited in section 4.

The light scattering program developed in this paper is based on the laws of geometric optics. The computational method is an approximation based on the assumption that light may be thought of as consisting of separate localized rays that travel along straight line paths. It is an asymptotic approach that becomes increasingly accurate as the size parameter (size/wavelength ratio) approaches infinity. In view of the ob-

served ice crystal sizes in cirrus clouds ($\sim$20–2000 μm), the ray-tracing method should be valid for solar wavelengths (0.2–3.5 μm). However, for thermal infrared wavelengths (e.g., 10 μm), this method may not be appropriate for small ice crystals. In this case, a separate light scattering program should be developed. Moreover, the present program only deals with hexagonal columns and plates. A modification would be required to account for the scattering by bullet rosettes, which are also frequently observed in cirrus clouds.

Acknowledgments. All the computations contained in this research were carried out on the San Diego supercomputer, CRAY X-MP/48. The research was supported in part by NASA Grant NAG5-732 and NSF Grant ATM85-13975. We thank Graeme Stephens and the other anonymous reviewer for constructive comments on the paper. Sharon Bennett typed and edited the manuscript.

APPENDIX

Addition of Diffracted and Geometric Optics Rays

When light is scattered by polyhedral particles, such as hexagonal ice crystals, there is δ-function transmission through parallel planes (e.g., a basal plane and opposite basal plane) at $\theta = 0°$. Figure A1 shows a schematic representation of the phase function for this component, along with diffraction and geometric reflection and refraction. The delta forward peak does not contribute to the normalization of the scattering phase function. If the diffracted light rays and geometric optics rays are superimposed conventionally, the normalization integral for the phase function becomes 1 $- f_\delta$ where f_δ is the ratio of the δ function at $\theta = 0°$ to the entire scattered light. As shown in Table A1, f_δ becomes larger with increasing $L/2a$ due to a higher probability of plane parallel transmission. It becomes smaller as the wavelength increases due to stronger absorption.

TABLE A1. Fraction of the δ-forward transmission ($\theta = 0°$), f_δ.

$L/2a$	λ (μm)				
(μm/μm)	0.55	1.0	1.6	2.2	3.0
20/20	.12092	.11940	.11563	.10801	5.5296 −6
50/40	.12211	.12053	.11469	.10362	1.1296 −4
120/60	.14062	.13800	.12889	.11181	1.1024 −4
300/100	.15496	.15335	.13783	.11066	1.0831 −4
750/160	.16550	.16364	.13894	.09896	1.0670 −4

Let f_D be the ratio of the diffracted light to the entire scattered light; then the scattering phase matrix may be expressed by

$$G_{kl} = (1 - f_D) \sum_n G_{kl}^{(n)} + \delta_{kl} f_D G_D,$$

$$k, l = 1\text{-}4. \tag{A1}$$

Here, $\sum G_{kl}^{(n)}$ and G_D are scattering contributions from rays due to geometric optics and the Fraunhofer diffraction, respectively. Both $\sum G_{kl}^{(n)}$ and G_D are normalized to 1; δ_{kl} is 1 when $k = l$, and 0 otherwise.

In the case of polyhedral particles, f_D may be obtained by the following procedure. Since the extinction cross section C_e is twice the geometric cross section area, the extinction (scattering) cross section for the diffracted light is $C_e/2$. In general, light scattered at $\theta = 0°$ can be regarded as not being scattered at all (van de Hulst 1980). Thus, the cross section of light scattered substantially other than at $\theta = 0°$ is expressed by $C_s(1 - f_\delta)$. It follows that

$$f_D = \frac{1}{2\tilde{\omega}(1 - f_\delta)} \geq \frac{1}{2}, \tag{A2}$$

where the single-scattering albedo $\tilde{\omega} = C_s/C_e$. In the case of spherical particles, $f_\delta = 0$. If there is no absorption, $\tilde{\omega} = 1$, and if $f_\delta = 0$, $f_D = \frac{1}{2}$.

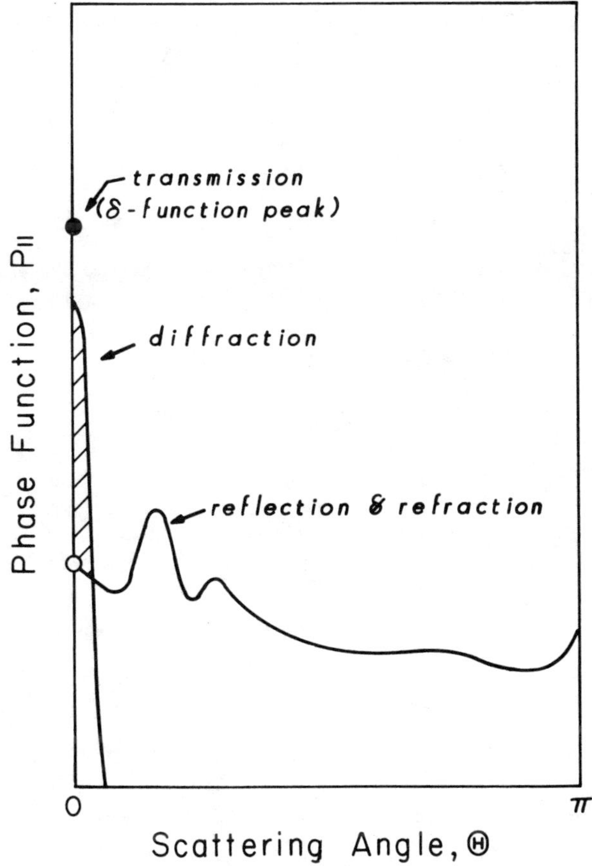

FIG. A1. A schematic representation of the components of the phase function, P_{11} for randomly oriented ice crystals.

REFERENCES

Asano, S., 1983: Transfer of solar radiation in optically anisotropic ice clouds. *J. Meteor. Soc. Japan*, **61**, 402–413.

——, and M. Sato, 1980: Light scattering by randomly oriented spheroidal particles. *Appl. Opt.*, **19**, 962–974.

Auer, A. H., Jr., and D. L. Veal, 1970: The dimension of ice crystals in natural clouds. *J. Atmos. Sci.*, **27**, 919–926.

Bohren, C. F., 1986: Absorption and scattering of light by nonspherical particles. *Proc., Sixth Conf. on Atmospheric Radiation*, Amer. Meteor. Soc., 1–7.

Born, M., and E. Wolf, 1975: *Principles of Optics.* Pergamon Press, 808 pp.

Cai, Q., and K. N. Liou, 1982: Polarized light scattering by hexagonal ice crystals: Theory. *Appl. Opt.*, **21**, 3569–3580.

Dugin, V. P., and S. O. Mirumyants, 1976: The light scattering matrices of artificial crystalline clouds. *Izv. Acad. Sci. USSR Atmos. Ocean Phys.*, **12**, 988–991.

Evans, W. F. J., and R. A. R. Tricker, 1972: Unusual arcs in the Saskatoon halo display. *Weather*, **27**, 234–238.

Fraser, A. B., 1979: What size of ice crystals causes the halos? *J. Opt. Soc. Am.*, **69**, 1112–1118.

Greenler, R. G., 1980: *Rainbows, Halos, and Glories.* Cambridge University Press, 195 pp.

Heymsfield, A. J., 1975: Cirrus uncinus generating cells and the evolution of cirriform clouds. *J. Atmos. Sci., 32,* 799–808.

——, and C. M. R. Platt, 1984: A parameterization of the particle size spectrum of ice clouds in terms of the ambient temperature and the ice water content. *J. Atmos. Sci., 41,* 846–855.

Können, G. P., 1983: Polarization and intensity distributions of refraction halos. *J. Opt. Soc. Am., 73,* 1629–1640.

Liou, K. N., 1972: Light scattering by ice clouds in the visible and infrared: A theoretical study. *J. Atmos. Sci., 29,* 524–536.

——, 1986: Influence of cirrus clouds on weather and climate processes: A global perspective. *Mon. Wea. Rev., 114,* 1167–1199.

Ono, A., 1969: The shape and riming properties of ice crystals in natural clouds. *J. Atmos. Sci., 26,* 138–147.

Plass, G. N., and G. W. Kattawar, 1968: Radiative transfer in water and ice clouds in the visible and infrared region. *Appl. Opt., 10,* 738–749.

Stahl, H. P., M. G. Tomasko, W. L. Wolfe, N. D. Castillo and K. A. Stahl, 1983: Measurements of the light scattering properties of water ice crystals. Topical Meetings, Meteorological Optics, Optical Society of America, Lake Tahoe.

Starr, D. O., 1987: A cirrus cloud experiment: Intensive field observations planned for FIRE. *Bull. Amer. Meteor. Soc., 68,* 119–124.

Stephens, G. L., 1980: Radiative properties of cirrus clouds in the infrared region. *J. Atmos. Sci., 37,* 435–446.

Takano, Y., and K. Jayaweera, 1985: Scattering phase matrix for hexagonal ice crystals computed from ray optics. *Appl. Opt., 24,* 3254–3263.

——, and K. N. Liou, 1988: Solar radiative transfer in cirrus clouds, Part II: Theory and computation of multiple scattering in an anisotropic medium. *J. Atmos. Sci. 45,* 20–36.

Thekaekara, M. P., 1973: Solar energy outside the earth's atmosphere. *Solar Energy, 14,* 109–127.

Tränkle, E., and R. G. Greenler, 1987: Multiple-scattering effects in halo phenomena. *J. Opt. Soc. Am. A4,* 591–599.

Tricker, R. A. R., 1979: *Ice Crystal Halos.* Optical Society of America, 53 pp.

van de Hulst, H. C., 1957: *Light Scattering by Small Particles.* Wiley, 470 pp.

——, 1980: *Multiple Light Scattering.* Academic Press, 739 pp.

Vouk, V., 1948: Projected area of convex bodies. *Nature, 162,* 330–331.

Warren, S. G., 1984: Optical constants of ice from ultraviolet to the microwave. *Appl. Opt., 23,* 1206–1225.

Welch, R. M., S. K. Cox and W. G. Zdundkowski, 1980: Calculations of the variability of ice cloud radiative properties at selected solar wavelengths. *Appl. Opt., 19,* 3057–3067.

Solar Radiative Transfer in Cirrus Clouds. Part II: Theory and Computation of Multiple Scattering in an Anisotropic Medium

YOSHIHIDE TAKANO AND KUO-NAN LIOU

Department of Meteorology, University of Utah, Salt Lake City, Utah

(Manuscript received 1 February 1988, in final form 13 June 1988)

ABSTRACT

We have developed a theoretical framework for the computation of the transfer of solar radiation in an anisotropic medium with particular application to oriented ice crystals in cirrus clouds. In the theoretical development, the adding principle for radiative transfer has been used with modifications to account for the anisotropy of the phase matrix. The single-scattering properties, including the phase function, single-scattering albedo, and extinction cross section, for randomly and horizontally oriented ice crystals are then used in the computation of reflected and transmitted intensities, planetary albedo, and polarization in multiple scattering. There are significant differences in the reflected and transmitted intensities between hexagonal ice crystals and equivalent ice spheres. In addition, it is found that ice spheres are inadequate to model the general pattern of reflected intensity. The orientation properties of ice crystals are also significant in the determination of the reflected and transmitted intensities. Various optical features can be produced only by horizontally oriented plates and columns. For the polarization of sunlight reflected by ice crystals, the neutral point is independent of the solar zenith angle as well as the optical depth. We have also closely matched the polarization patterns observed for Martian white clouds, as well as cirrus clouds, with the results from the present multiple-scattering computations for ice crystals. Finally, it is noted that the polarization configuration is extremely sensitive to the shape of the particles. Thus, its full information content should be explored for applications to the remote sounding of clouds.

1. Introduction

The interpretation of bidirectional reflectance and polarization patterns for planetary clouds observed from aircraft, satellites, or spacecraft requires multiple-scattering calculations for cloud particles. Multiple scattering in clouds is also important for the determination of cloud albedo, which is relevant to climate problems. There have been numerous methods developed for solving multiple scattering in planetary atmospheres. The adding method has been demonstrated to be a powerful tool for multiple-scattering calculations. The principle for the method was stated by Stokes (1862) in a problem dealing with reflection and transmission by glass plates. Peebles and Plesset (1951) developed the adding method theory for application to gamma-ray transfer. The adding equations for multiple scattering now commonly used are based on the formulation presented by van de Hulst in 1963 (van de Hulst 1980). We shall use the adding method for the formulation of multiple scattering in an anisotropic medium.

As pointed out in Part I (Takano and Liou 1988, hereafter referred to as Part I), cirrus clouds are com-

posed of hexagonal ice crystals, whose single-scattering properties differ significantly from those computed for spherical particles. Moreover, in the case of horizontally oriented ice crystals, the single-scattering parameters depend on the direction of the incident light beam. Thus, the conventional formulation for the multiple-scattering problem is no longer applicable. Liou (1980) formulated the basic equation for the transfer of solar radiation in an optically anisotropic medium in which the single-scattering properties vary with the incident angle of the light beam. Stephens (1980) and Asano (1983) discussed the transfer of radiation through optically anisotropic ice clouds. The latter author used a hypothetical cloud model in which the scattering phase function was expressed in terms of the incident angle. However, none of these authors included realistic scattering parameters for oriented ice crystals in the discussion and analysis, nor is the Stokes vector properly accounted for in the formulation.

This paper presents the theory and computations for multiple scattering in cirrus clouds containing oriented ice crystals. In section 2, radiative transfer in clouds composed of horizontally oriented ice crystals is formulated with the aid of the adding method. Using the single-scattering properties obtained in Part I, the reflected and transmitted intensities, planetary albedo, and polarization in multiple scattering by ice crystals are illustrated and discussed in section 3. In section 4,

Corresponding author address: Dr. Kuo-Nan Liou, Dept. of Meteorology, 819 Wm. C. Browning Building, University of Utah, Salt Lake City, UT 84112.

we interpret a number of polarization measurements based on computational results from randomly oriented ice crystals. Finally, conclusions are given in section 5.

2. Theory of radiative transfer in an anisotropic medium

In an anisotropic medium, the single-scattering properties depend on the direction of the incoming light beam. Let the directions of incoming and outgoing light beams be denoted by (μ', ϕ') and (μ, ϕ), respectively, where μ is the cosine of the zenith angle and ϕ the corresponding azimuthal angle. The scattering phase matrix $\mathbf{P}$ is a function of $(\mu, \phi; \mu', \phi')$ and cannot be defined by the scattering angle Θ alone as in conventional radiative transfer. Moreover, the extinction (σ_e or C_e in Part I) and scattering (σ_s or C_s in Part I) cross sections vary with the direction of the incoming light beam (μ', ϕ').

Consider an anisotropic medium consisting of ice crystals randomly oriented in a horizontal plane. Because of the symmetry with respect to the azimuthal angle for the incoming light beam, the phase matrix, and the extinction and scattering cross sections may be expressed by $\mathbf{P}(\mu, \phi; \mu', \phi') [=\mathbf{P}(\Theta, \Phi, \mu')]$, $\sigma_e(\mu')$, and $\sigma_s(\mu')$, respectively, where Φ is the azimuthal angle associated with the scattering angle Θ. In this case, we may define the differential normal optical depth in the form

$$\frac{d\tilde{\tau}}{dz} = -\tilde{\sigma}_e N_0, \qquad (1)$$

where the normal extinction cross section $\tilde{\sigma}_e = \sigma_e(\mu' = 1)$, N_0 is the number density of the particles, and z the distance. Let the Stokes vector intensity $\mathbf{I} = (I, Q, U, V)$. Following Liou (1980), the general equation governing the transfer of diffuse solar intensity may be written in the form

$$\mu \frac{d\mathbf{I}(\tilde{\tau}; \mu, \phi)}{d\tau} = \mathbf{I}(\tilde{\tau}; \mu, \phi)k(\mu) - \vec{\mathbf{J}}(\tilde{\tau}; \mu, \phi), \qquad (2)$$

where

$$k(\mu) = \sigma_e(\mu)/\tilde{\sigma}_e, \qquad (3)$$

and the source function,

$$\vec{\mathbf{J}}(\tilde{\tau}; \mu, \phi) = \frac{1}{4\pi} \int_0^{2\pi} \int_{-1}^1 \tilde{\omega}(\mu')\mathbf{P}(\mu, \phi; \mu', \phi')$$

$$\times \mathbf{I}(\tilde{\tau}; \mu', \phi')d\mu'd\phi' + \frac{1}{4\pi} \tilde{\omega}(-\mu_0)\mathbf{P}(\mu, \phi; -\mu_0, \phi_0)$$

$$\times \pi\vec{\mathbf{F}}_0 \exp[-k(-\mu_0)\tilde{\tau}/\mu_0]. \qquad (4)$$

In Eq. (4), μ_0 is the cosine of the solar zenith angle, ϕ_0 the corresponding azimuthal angle, and $-\mu_0$ denotes the downward solar incident direction. The first and second terms on the right-hand side represent contributions from multiple scattering and single scattering

of the direct solar intensity, respectively. The equivalent single-scattering albedo is defined by

$$\tilde{\omega}(\mu) = \sigma_s(\mu)/\tilde{\sigma}_e. \qquad (5)$$

The general phase matrix, with respect to the local meridian plane, is given by (Chandrasekhar 1960; Liou 1980)

$$\mathbf{P}(\mu, \mu'; \phi - \phi') = \mathbf{L}(\pi - i_2)\mathbf{P}(\Theta, \Phi, \mu')\mathbf{L}(-i_1), \qquad (6)$$

where i_1 and i_2 denote the angles between the meridian planes for the incoming and outgoing light beams, respectively, and the plane of scattering. The transformation matrix for the Stokes vector is given by

$$\mathbf{L}(\chi) = \begin{bmatrix} 1 & 0 & 0 & 0 \\ 0 & \cos 2\chi & \sin 2\chi & 0 \\ 0 & -\sin 2\chi & \cos 2\chi & 0 \\ 0 & 0 & 0 & 1 \end{bmatrix}, \qquad (7)$$

where $\chi = -i_1$ or $\pi - i_2$. From spherical geometry, these angles are given by

$$\cos i_1 = \frac{-\mu + \mu' \cos\Theta}{\pm(1 - \cos^2\Theta)^{1/2}(1 - \mu'^2)^{1/2}}, \qquad (8)$$

$$\cos i_2 = \frac{-\mu' + \mu \cos\Theta}{\pm(1 - \cos^2\Theta)^{1/2}(1 - \mu^2)^{1/2}}. \qquad (9)$$

If $P(\Theta, \Phi, \mu)$, $\sigma_e(\mu)$, and $\sigma_s(\mu)$ are known, then, in principle, Eq. (2) may be solved numerically.

We shall approach the multiple scattering problem by means of the adding method for radiative transfer. We define the reflection matrix $\mathbf{R}(\mu, \mu_0, \phi - \phi_0)$ and transmission matrix $\mathbf{T}(\mu, \mu_0; \phi - \phi_0)$ for radiation from above in the forms

$$\mathbf{I}_{\text{out, top}}(\mu, \phi) = \frac{1}{\pi} \int_0^1 \mu_0 d\mu_0 \int_0^{2\pi} d\phi_0$$

$$\times \mathbf{R}(\mu, \mu_0; \phi - \phi_0)\mathbf{I}_{\text{in, top}}(\mu_0, \phi_0), \qquad (10)$$

$$\mathbf{I}_{\text{out, bottom}}(\mu, \phi) = \frac{1}{\pi} \int_0^1 \mu_0 d\mu_0 \int_0^{2\pi} d\phi_0$$

$$\times \mathbf{T}(\mu, \mu_0; \phi - \phi_0)\mathbf{I}_{\text{in, top}}(\mu_0, \phi_0). \qquad (11)$$

Likewise, for radiation from below, the reflection and transmission matrices are defined by

$$\mathbf{I}_{\text{out, bottom}}(\mu, \phi) = \frac{1}{\pi} \int_0^1 \mu_0 d\mu_0 \int_0^{2\pi} d\phi_0$$

$$\times \mathbf{R}^*(\mu, \mu_0; \phi - \phi_0)\mathbf{I}_{\text{in, bottom}}(\mu_0, \phi_0), \qquad (12)$$

$$\mathbf{I}_{\text{out, top}}(\mu, \phi) = \frac{1}{\pi} \int_0^1 \mu_0 d\mu_0 \int_0^{2\pi} d\phi_0$$

$$\times \mathbf{T}^*(\mu, \mu_0; \phi - \phi_0)\mathbf{I}_{\text{in, bottom}}(\mu_0, \phi_0). \qquad (13)$$

To proceed with the adding principle for radiative transfer in an anisotropic medium, we shall utilize the reflection and transmission matrices defined in Eqs. (10)–(13) and consider an infinitesimal layer with a very small optical depth $\Delta\tilde{\tau}$, say 10^{-8}. Since the optical depth is so small, only single scattering takes place within the layer. From Eqs. (2) and (4), the analytic solutions for reflected and transmitted intensities undergoing single scattering may be derived. Subject to the condition that $\Delta\tilde{\tau} \to 0$, we find

$$\mathbf{T}(\mu, \mu_0; \phi - \phi_0) \approx \frac{\Delta\tilde{\tau}}{4\mu\mu_0} \tilde{\omega}(\mu_0)\mathbf{P}(\mu, \mu_0; \phi - \phi_0),$$
(14)

$$\mathbf{R}(\mu, \mu_0; \phi - \phi_0) \approx \frac{\Delta\tilde{\tau}}{4\mu\mu_0} \tilde{\omega}(\mu_0)\mathbf{P}(-\mu, \mu_0; \phi - \phi_0),$$
(15)

$$\mathbf{T}^*(\mu, \mu_0; \phi - \phi_0)$$
$$\approx \frac{\Delta\tilde{\tau}}{4\mu\mu_0} \tilde{\omega}(\mu_0)\mathbf{P}(-\mu, -\mu_0; \phi - \phi_0), \quad (16)$$

$$\mathbf{R}^*(\mu, \mu_0; \phi - \phi_0)$$
$$\approx \frac{\Delta\tilde{\tau}}{4\mu\mu_0} \tilde{\omega}(\mu_0)\mathbf{P}(\mu, -\mu_0; \phi - \phi_0), \quad (17)$$

where the phase matrix $\mathbf{P}$ is defined in Eq. (6).

Consider now two layers, one on top of the other. Let the subscripts a and b denote these two layers and let their optical depths be $\tilde{\tau}_a$ and $\tilde{\tau}_b$. Following the conventional adding principle for radiative transfer in an isotropic medium (see, e.g., Twomey et al. 1966; Lacis and Hansen 1974; Liou 1980), but with modifications to account for the dependence of the optical properties on the incoming direction, the procedure for computing the reflection and transmission matrices for the composite layer may be described by the following equations:

$$\mathbf{Q}_1 = \mathbf{R}_a^*\mathbf{R}_b,$$
(18)

$$\mathbf{Q}_n = \mathbf{Q}_1\mathbf{Q}_{n-1},$$
(19)

$$\mathbf{S} = \sum_{n=1}^{M} \mathbf{Q}_n,$$
(20)

$$\mathbf{D} = \mathbf{T}_a + \mathbf{S} \exp[-k(\mu_0)\tilde{\tau}_a/\mu_0] + \mathbf{S}\mathbf{T}_a, \quad (21)$$

$$\mathbf{U} = \mathbf{R}_b \exp[-k(\mu_0)\tilde{\tau}_a/\mu_0] + \mathbf{R}_b\mathbf{D}, \quad (22)$$

$$\mathbf{R}_{a,b} = \mathbf{R}_a + \exp[-k(\mu)\tilde{\tau}_a/\mu]\mathbf{U} + \mathbf{T}_a^*\mathbf{U}, \quad (23)$$

$$\mathbf{T}_{a,b} = \exp[-k(\mu)\tilde{\tau}_b/\mu]\mathbf{D}$$
$$+ \mathbf{T}_b \exp[-k(\mu_0)\tilde{\tau}_a/\mu_0] + \mathbf{T}_b\mathbf{D}. \quad (24)$$

In these equations, the product of two functions implies an integration over the appropriate solid angle so that all possible multiple scattering contributions are accounted for. For example,

$$\mathbf{R}_a^*\mathbf{R}_b = \frac{1}{\pi} \int_0^{2\pi} \int_0^1 \mathbf{R}_a^*(\mu, \mu'; \phi - \phi')$$
$$\times \mathbf{R}_b(\mu', \mu_0; \phi' - \phi_0)\mu'd\mu'd\phi'. \quad (25)$$

The term M in Eq. (20) is selected according to the convergence of the series, and varies from 5 to 12 in the present calculations. The exponential terms in the adding equations are the direct transmission through layer a or b without scattering, where the anisotropic factor $k(\mu)$ is given in Eq. (3). The total transmission for the combined layer is the sum of the diffuse transmission $\mathbf{T}_{a,b}$ and the direct transmission $\exp[-k(\mu_0)(\tilde{\tau}_a + \tilde{\tau}_b)/\mu_0]$ in the direction of the solar zenith angle θ_0.

In the numerical computations, it is economical to set $\tilde{\tau}_a = \tilde{\tau}_b$. This is referred to as the doubling method. We start with an optical depth $\tilde{\tau} \approx 10^{-8}$ and use Eqs. (14)–(17) to compute the reflection and transmission matrices. Equations (18)–(24) are subsequently employed to compute the reflection and transmission matrices for an optical depth of $2\tilde{\tau}$. The computations using these equations are repeated until the desired optical depth is obtained.

In order to compute the reflection and transmission matrices for the initial layer with a very small optical depth, via Eqs. (14)–(15), we need the phase matrix and single-scattering albedo, and directions for incoming and outgoing beams. The phase matrix elements must be expanded in terms of the incoming and outgoing directions denoted by μ, μ' and $\phi - \phi'$. For spherical particles, the phase matrix consists of four nonzero independent elements. These elements can be decoupled analytically in terms of functions associated with μ, μ' and $\phi - \phi'$ (Dave 1970; Kattawar et al. 1973). However, for nonspherical particles, the decomposition of the phase matrix elements has yet to be worked out. For randomly oriented nonspherical particles that have a plane of symmetry, the phase matrix contains six independent elements. In this case, there are seven symmetrical relationships for these elements based on the reciprocity principle (Hovenier 1969). From these relationships, the phase matrix elements can be expanded in terms of either cosine or sine Fourier components (Hansen 1971). For randomly oriented ice plates or columns in a horizontal plane, there are 16 elements in the phase matrix. It can be proven that these elements obey a number of symmetrical relationships, upon which they can be expanded in terms of either cosine or sine Fourier components in the manner described by Hansen (1971).

In view of the above discussion, the general phase matrix elements may be numerically expanded in the forms

$$\tilde{\omega}(\mu')P_{ij}(\mu, \mu'; \phi - \phi')$$
$$= P_{ij}^0(\mu, \mu') + 2 \sum_{m=1}^{N} p_{ij}^m(\mu, \mu') \begin{cases} \cos m(\phi - \phi'), c \\ \sin m(\phi - \phi'), s \end{cases}$$
(26)

where

$$c, ij = 11, 12, 21, 22, 33, 34, 43, 44$$
$$s, ij = 13, 14, 23, 24, 31, 32, 41, 42$$

and P_{ij}^m ($m = 0, 1, \cdots, N$) denote the Fourier expansion coefficients. With this expansion, each term in the Fourier series may be treated independently in numerical computations.

With respect to the normalization of the phase function P_{11}, the following procedures are followed. The phase function is normalized such that

$$\frac{1}{4\pi} \int_0^{2\pi} \int_{-1}^{1} P_{11}(\mu, \mu', \phi - \phi') d\mu d(\phi - \phi') = 1, \quad (27a)$$

where $d\mu d(\phi - \phi')$ denotes the differential solid angle. Using Eq. (26), we find

$$\frac{1}{2} \int_{-1}^{1} P_{11}^0(\mu, \mu') d\mu = \tilde{\omega}(\mu'), \quad (27b)$$

where $\tilde{\omega}(\mu')$ is defined in Eq. (5). In the case of randomly oriented nonspherical particles (or spherical particles), the single-scattering albedo $\tilde{\omega}$ is independent of μ' and is a constant. If there is no absorption, $\tilde{\omega} = 1$. In this case, Eq. (27b) can be derived from the expansion of the phase function in terms of the Legendre polynomial using the addition theorem for spherical harmonics (see, e.g., Liou 1980). However, for randomly oriented ice crystals in a horizontal plane, $\tilde{\omega}$ is a function of the incident angle. Normalization of the phase function must be performed for each μ'.

As shown in Part I, the phase function P_{11} for ice crystals has a common sharp diffraction peak. In order to properly account for this peak in numerical integrations, thousands of Fourier components are needed in the phase function expansion. To optimize the computational effort, we shall follow the procedure proposed by Potter (1970). In this procedure, the forward peak is truncated by extrapolating the phase function linearly from the scattering angles 10° to 0° in the logarithmic scale. Let the truncated phase function be P_{11}^t; then, the truncated fraction of scattered intensity is given by

$$f = \int_{4\pi} (P_{11} - P_{11}^t) d\Omega / 4\pi. \quad (28)$$

In our numerical computations, we find that $f \approx f_D$, the ratio of the diffracted light to the entire scattered light, described in the appendix of Part I. In the limits of geometric ray optics, it is apparent that the scattered energy contained in the truncated forward peak, which is the shaded area in Fig. A1 of Part I, is approximately equal to the scattered energy associated with Fraunhoffer diffraction.

To use the truncated phase function in multiple scattering computations, but at the same time to achieve the "equivalent" result as in the case when the sharp diffraction peak was included in the computations, an adjustment must be made for the optical depth and single-scattering albedo. Since the forward peak is associated only with scattering, the adjusted scattering

and absorption optical depth must be $\tilde{\tau}'_s = (1 - f)\tilde{\tau}_s$ and $\tilde{\tau}'_a = \tilde{\tau}_a$. Thus, the adjusted optical depth should be

$$\tilde{\tau}' = \tilde{\tau}'_s + \tilde{\tau}'_a = (1 - f\tilde{\omega})\tilde{\tau}, \quad (29)$$

where the single-scattering albedo $\tilde{\omega} = \tilde{\tau}_s / \tilde{\tau}$. The adjusted single-scattering albedo is

$$\tilde{\omega}' = \frac{\tilde{\tau}'_s}{\tilde{\tau}'} = \frac{(1 - f)\tilde{\omega}}{1 - f\tilde{\omega}}. \quad (30)$$

In addition to the preceding adjustment, we must also account for the contribution of the δ-transmission part f_δ, associated with the forward scattering at $\Theta = 0°$ described in the appendix of Part I, which cannot be included in numerical computations. Following the above procedure, the optical thickness $\tilde{\tau}''$ and single-scattering albedo $\tilde{\omega}''$, which are corrected for this δ-function peak, are given by

$$\tilde{\tau}'' = (1 - f_\delta \tilde{\omega}')\tilde{\tau}', \quad (31)$$

$$\tilde{\omega}'' = \frac{(1 - f_\delta)\tilde{\omega}'}{1 - f_\delta \tilde{\omega}'}. \quad (32)$$

For conservative scattering, the adjustment is required only for the optical depth. In the case of 2-D crystals, f and f_δ depend on the incident direction θ_0, as well as the extinction and scattering cross sections, σ_e and σ_s. Equations (29)–(32) constitute the generalized similarity principle for radiative transfer in horizontally oriented polyhedral particles. The similarity principle has been discussed by Sobolev (1975) for isotropic scattering and van de Hulst (1980), who included the asymmetry factor in the discussion.

3. Computational results

In this section we present computational results for reflected and transmitted intensities, planetary albedo, and reflected polarization for randomly (3-D) and horizontally (2-D) oriented ice crystals using a visible wavelength of 0.55 μm. In the computations, the number of emergent angles μ used for 2-D columns is 20. For 2-D plates, Parry columns, and 3-D columns and plates, it is 40. Azimuthal angular intervals of 1° were used. These angular intervals are adequate to produce smooth curves for the reflected and transmitted intensities.

Figure 1 shows the reflected and transmitted (diffuse) intensities for 3-D ice columns ($L/2a = 125 \ \mu$m$/50$ μm) and area-equivalent ice spheres as a function of the zenith angle θ for an overhead sun ($\theta_0 = 0°$). The reflected intensity increases with increasing optical depth. Significant differences between the reflected intensities for ice columns and spheres are seen. Ice spheres produce a peak intensity at $\theta = 45°$, associated with a combination of primary and secondary rainbow features due to single-scattering. But ice columns have larger reflected intensities in other zenith angle regions. In the transmitted intensity pattern, the 22° and 46°

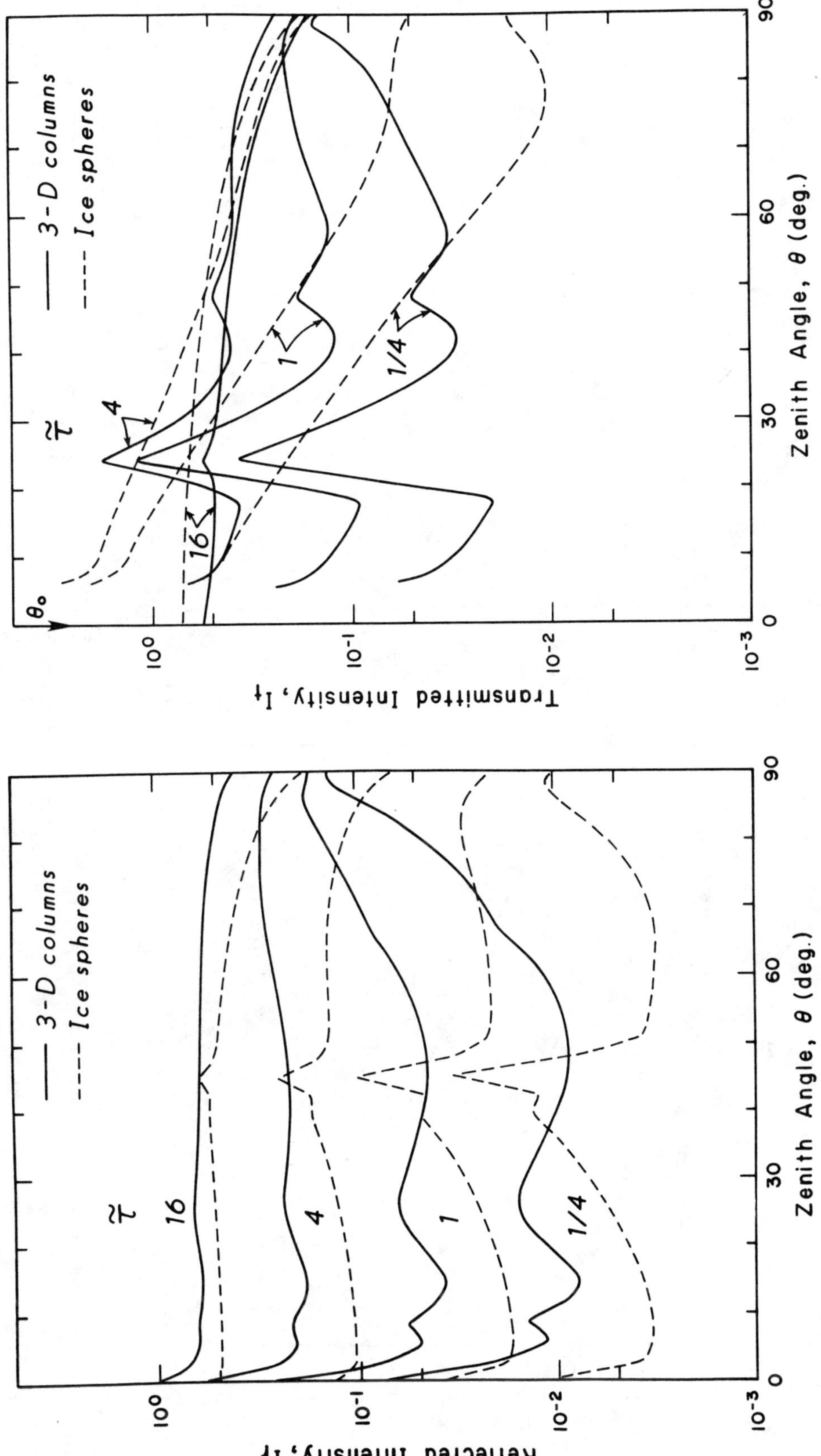

FIG. 1. Intensity reflected and transmitted by 3-D columns ($L/2a = 125\ \mu m/50\ \mu m$) and area-equivalent spheres with an overhead sun ($\theta_0 = 0°$) at $\lambda = 0.55\ \mu m$. The values of $\tilde{\tau}$ in 3-D and sphere cases denote the conventional optical depth.

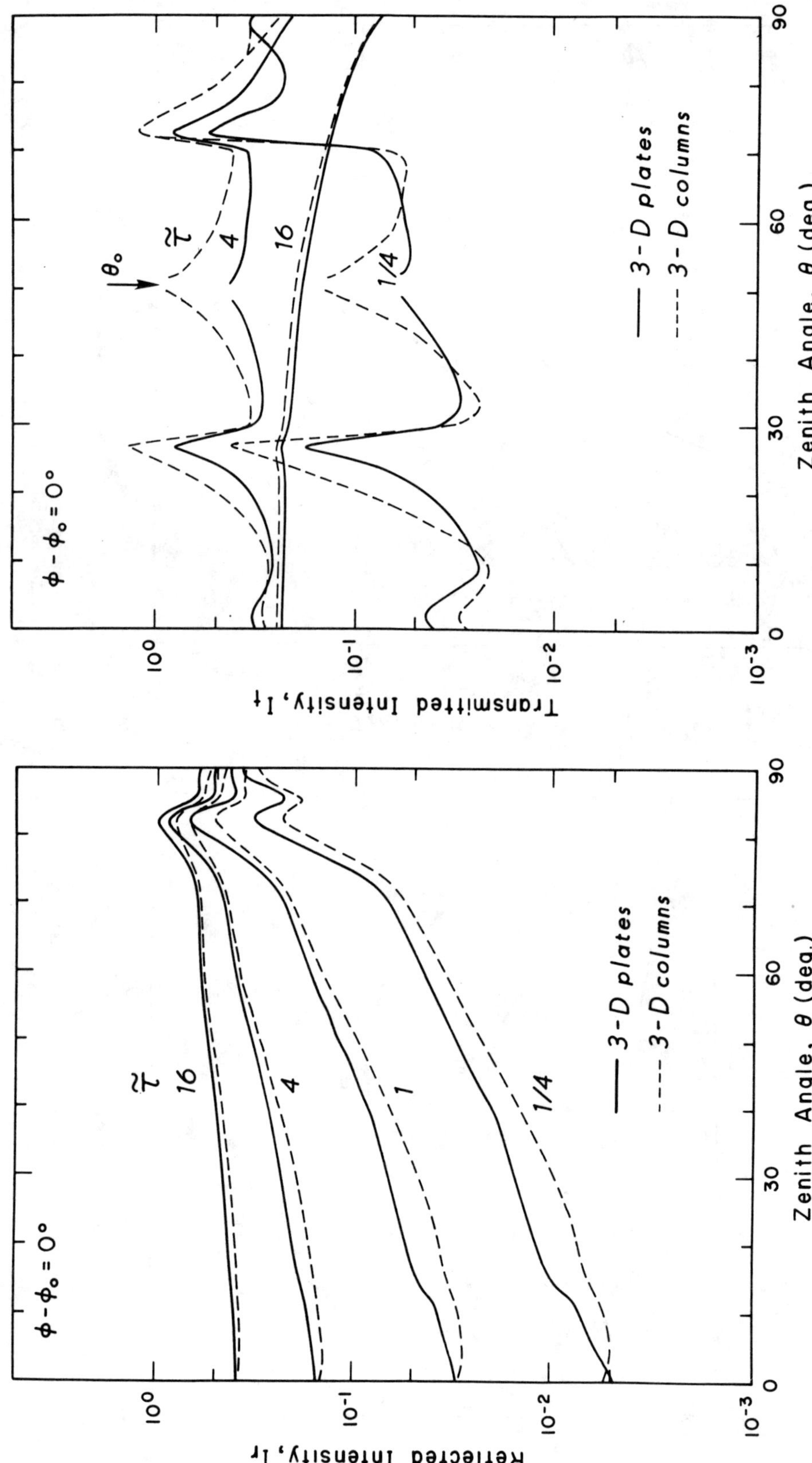

FIG. 2. Intensity reflected and transmitted by 3-D plates ($L/2a = 0.4$) and 3-D columns ($L/2a = 2.5$) in the solar principal plane ($\phi - \phi_0 = 0°$) at $\lambda = 0.55\ \mu m$. The solar zenith angle is 50°.

halo features produced by ice columns are very distinct for small optical depths. However, they disappear when the optical depth is greater than about 16. The transmitted intensities for ice spheres are generally larger than those for ice columns for zenith angles between 0° and ~40°, but are smaller between ~60° and 90°. We have also carried out computations for 3-D ice plates using an aspect ratio of 32 μm/80 μm. Results for the reflected and transmitted intensities are extremely similar to those for 3-D ice columns, except that the 22° halo feature for optical depths less than 4 is less pronounced.

In the preceding presentation, we have used the same crystal sizes, randomly oriented in space, in multiple-scattering computations. The scattered intensities from these computations, however, should not deviate significantly from those using the observed ice crystal size distributions shown in Fig. 2 of Part I. The reasons are as follows. First, at the 0.55 μm wavelength, absorption of ice is practically negligible so that the single-scattering albedo $\tilde{\omega} \approx 1$. Second, in multiple-scattering calculations, the optical depth is fixed. It follows that the scattered intensities depend only on the phase function employed in the calculations. The phase functions for 3-D columns (125 μm/50 μm) and 3-D plates (32 μm/80 μm), which approximately represent mean sizes of the observed ice crystal size distribution, are about the same as those presented in Fig. 3 of Part I. Thus the multiple-scattering results illustrated in this section using a single size ice crystal should be substantially similar to those using an observed ice crystal size distribution. Following the preceding reasoning, the use of area- or volume-equivalent ice spheres would produce similar multiple-scattering results. In Table 3 of Part I, we listed the asymmetry factors for area-equivalent ice spheres. For the 0.55 μm wavelength, the asymmetry factors for volume-equivalent ice spheres do not deviate substantially from those values.

In Fig. 2 we compare the reflected and transmitted intensities for 3-D plates and columns for a solar zenith angle θ_0 of 50°. These intensities are plotted as a function of the zenith angle θ on the plane $\phi - \phi_0 = 0°$. For the reflected intensity, 3-D plates reflect slightly more than 3-D columns. This is because the asymmetry factor for 3-D columns is larger than that for 3-D plates. Both columns and plates show a peak at $\theta \approx 83°$, which is associated with greater intensity at the outer halo. The patterns of the transmitted intensity are similar for the two cases. However, the transmitted intensity for plates (32 μm/80 μm) is generally smaller than that for columns (125 μm/50 μm).

The planetary albedo (referred to as reflection in radiative transfer) is defined by

$$R(\mu_0) = \frac{1}{\pi} \int_0^{2\pi} \int_0^1 I_r(\mu, \mu_0; \phi - \phi_0)\mu d\mu d\phi/(\mu_0 \pi F_0),$$

$$(33)$$

where $\mu_0 \pi F_0$ denotes the solar flux perpendicular to the plane-parallel atmosphere. Figure 3 shows reflection of sunlight as a function of the cosine of the solar zenith angle μ_0 for randomly oriented ice columns and plates. Comparisons are also made with area-equivalent ice spheres. Reflection by 3-D crystals is generally larger than that of spheres, and the differences increase with increasing optical depth. This is because ice spheres have larger forward scattering than ice crystals.

We next present the reflected and transmitted intensity patterns for horizontally oriented ice crystals. Figure 4 illustrates these patterns for 2-D plates when the solar zenith angle θ_0 is 75°. Note that the optical depth presented in this graph is a mean value averaged over the zenith angle θ. In the case of 2-D plates, scattered sunlight is confined to four latitude belts, due to specific geometry. For $\theta_0 = 75°$, these latitude belts correspond to zenith angles of ±75° and ±27°, with negative values representing mirror images. This is described in Eq. (21) and displayed in Fig. 7a of Part I for single-scattering analyses. If the incident angle is 27°, the four latitude belts are ±27° and ±75°. Due to the symmetrical property of 2-D plates with respect to incoming light beams, all multiple-scattered light is also confined to the four latitude belts. The reflected and transmitted intensities for optical depths of ¼, 1, 4, and 16 are displayed as a function of the azimuthal angle $\phi - \phi_0$. The reflected intensity increases with increasing optical depth. For the 75° emergent zenith angle, the subsun, subsundog, 120° subparhelion, and antisolar peak optical features are shown distinctly for an optical depth of ¼, where single-scattering dominates. For an optical depth of 16, only the subsun and subparhelion are observed. For the transmitted intensity, the sundog is visible even for large cloud optical depths. In addition to these optical phenomena, the anthelion (AN) located at the 180° azimuthal angle is seen for the small optical depth of ¼, due to double scattering, viz., the coupling of the subsun and antisolar peak. For an optical depth of 16, the 44° parhelion produced by double scattering (denoted as 44°P) is also observed. At the 27° zenith angle, the Kern's arc (KA) appears for $\tau \gtrsim 1$ due to the effects of multiple scattering. This arc has been observed by Ripley and Saugier (1971) and simulated by Tränkle and Greenler (1987) using the Monte Carlo method for multiple scattering.

Figures 5 and 6 display the reflected and transmitted intensity patterns as a function of the zenith angle on the plane $\phi - \phi_0 = 0°$ for Parry and 2-D columns, respectively. The solar zenith angle used for the computation is 75°. These patterns for an optical depth of ¼ are basically similar to those from single-scattering computations displayed in Figs. 8 and 9 of Part I, except for the subpeak at $\theta = 82°$ in the transmitted intensity in Fig. 5. This peak is caused by the lower sunvex Parry arc of the subsun. For Parry columns, several optical

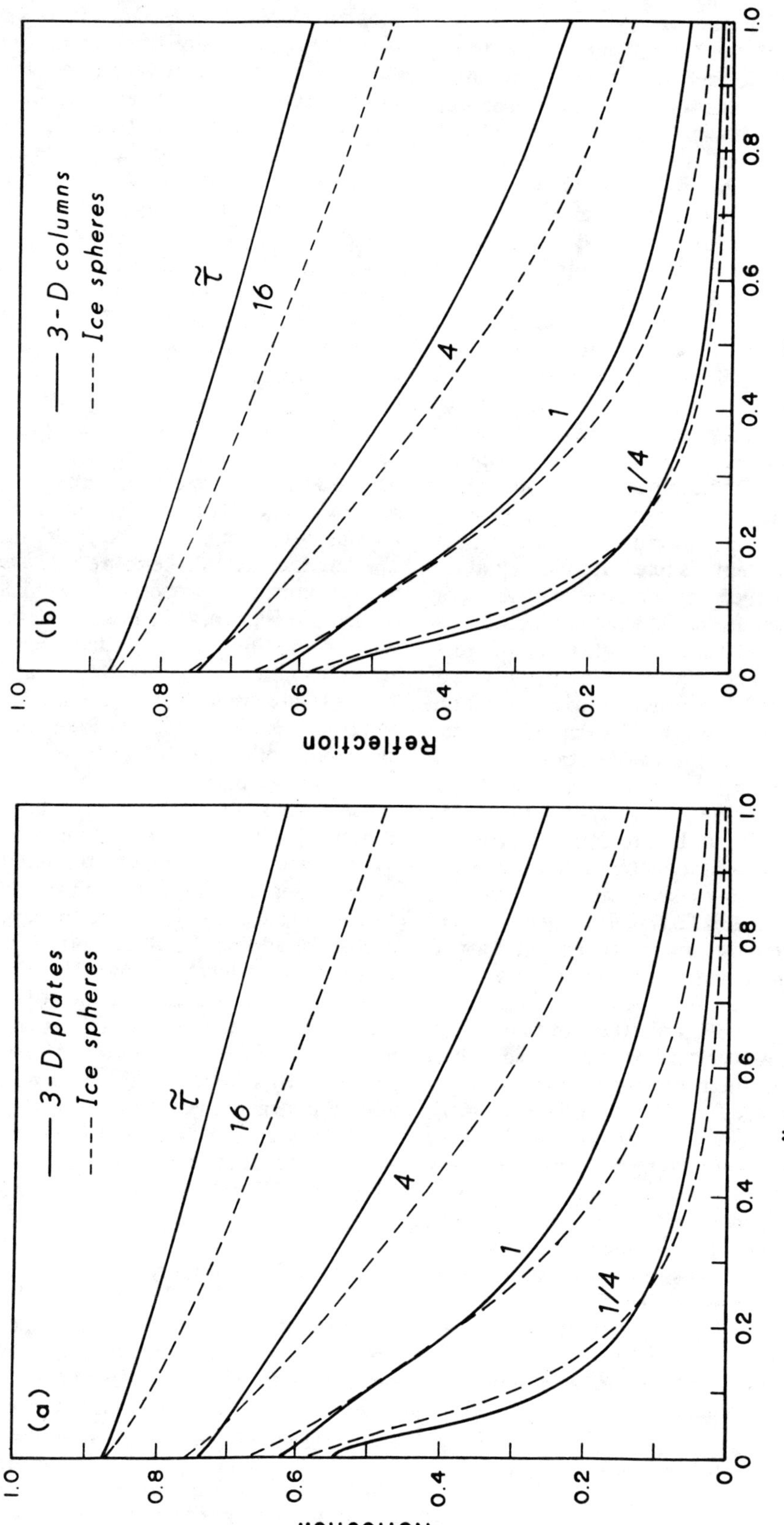

Fig. 3. Reflection for (a) 3-D plates ($L/2a = 32\ \mu m/80\ \mu m$) and area-equivalent spheres as a function of the cosine of the solar zenith angle at $\lambda = 0.55\ \mu m$, and (b) 3-D columns ($L/2a = 125\ \mu m/50\ \mu m$).

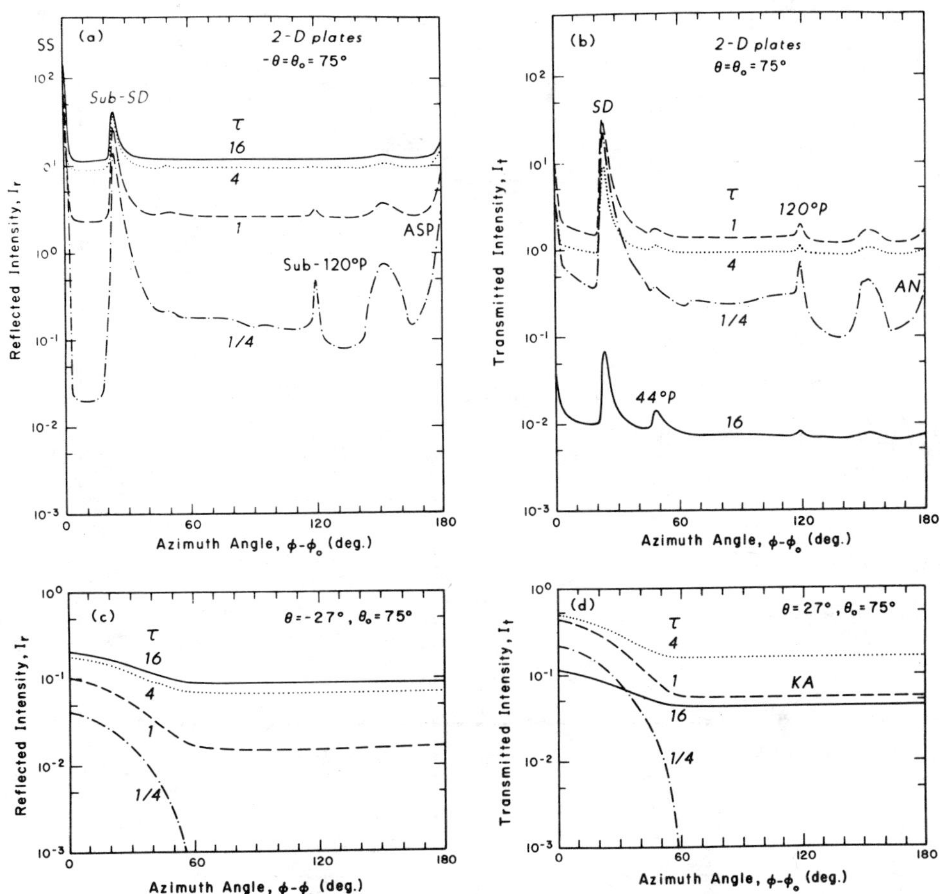

FIG. 4. Intensity of sunlight reflected and transmitted by 2-D plates ($L/2a = 0.1$) as a function of the azimuth angle, $\phi - \phi_0$ for (a) $-\theta = \theta_0 = 75°$, (b) $\theta = \theta_0 = 75°$, (c) $(\theta, \theta_0) = (-27°, 75°)$, and (d) $(\theta, \theta_0) = (27°, 75°)$. The optical depth τ in 2-D cases is a value averaged over all directions ($0 \leq \mu \leq 1$).

features are observed for optical depths less than 4. These include the subsun (SS) and lower sunvex Parry arc (LSVP) in the reflected intensity (these are also visible for an optical depth of 16), and the circumzenithal arc (CZA) and upper suncave (USCP) and sunvex (USVP) Parry arcs in the transmitted intensity. For 2-D columns, the lower tangent arc (LTA) is noticeable in the reflected intensity for all optical depths. The upper tangent arc (UTA) is observable in the transmitted intensity for optical depths less than 4. When the optical depth is large, the transmitted intensities have the lowest values in the zenith and nadir directions, as shown in Figs. 5 and 6. Except for these features, the reflected and transmitted intensities of 2-D columns are similar to those of randomly oriented columns. Also, it is noted that the reflected and transmitted intensity distributions of Parry columns are similar to those of 2-D columns for large optical depths ($\tau = 16$) because sufficient multiple scattering is present.

The effects of orientation on reflection are shown in Fig. 7. As shown in Fig. 7a, plates reflect more solar flux when they are horizontally oriented, as compared

with random orientation in space. An exception is when the sun is nearly overhead. In this case, the forward transmission by horizontally oriented plates is most significant, thereby reducing the reflection values. The reflection patterns for Parry and 3-D columns also show large differences. For large μ_0, Parry columns reflect less solar flux because of large forward transmission. However, for small μ_0, due to a longer effective optical pathlength, the reverse is true. The reflection patterns for 2-D and 3-D columns are extremely similar, however.

Finally, we discuss the polarization pattern for ice crystal clouds computed from the present program. Figure 8 shows the linear polarization ($-Q/I$) of sunlight reflected by 3-D plates ($L/2a = 0.1$) and 3-D columns ($L/2a = 2.5$) with an overhead sun ($\theta_0 = 0°$) for optical depths of ¼, 1, 4, and 16. In this case, polarization is azimuth independent. When the optical depth is ¼, polarization deviates only slightly from that for single scattering shown in Fig. 5 of Part I. With increasing optical depth, polarization approaches zero. However, the neutral point (angle of zero polarization)

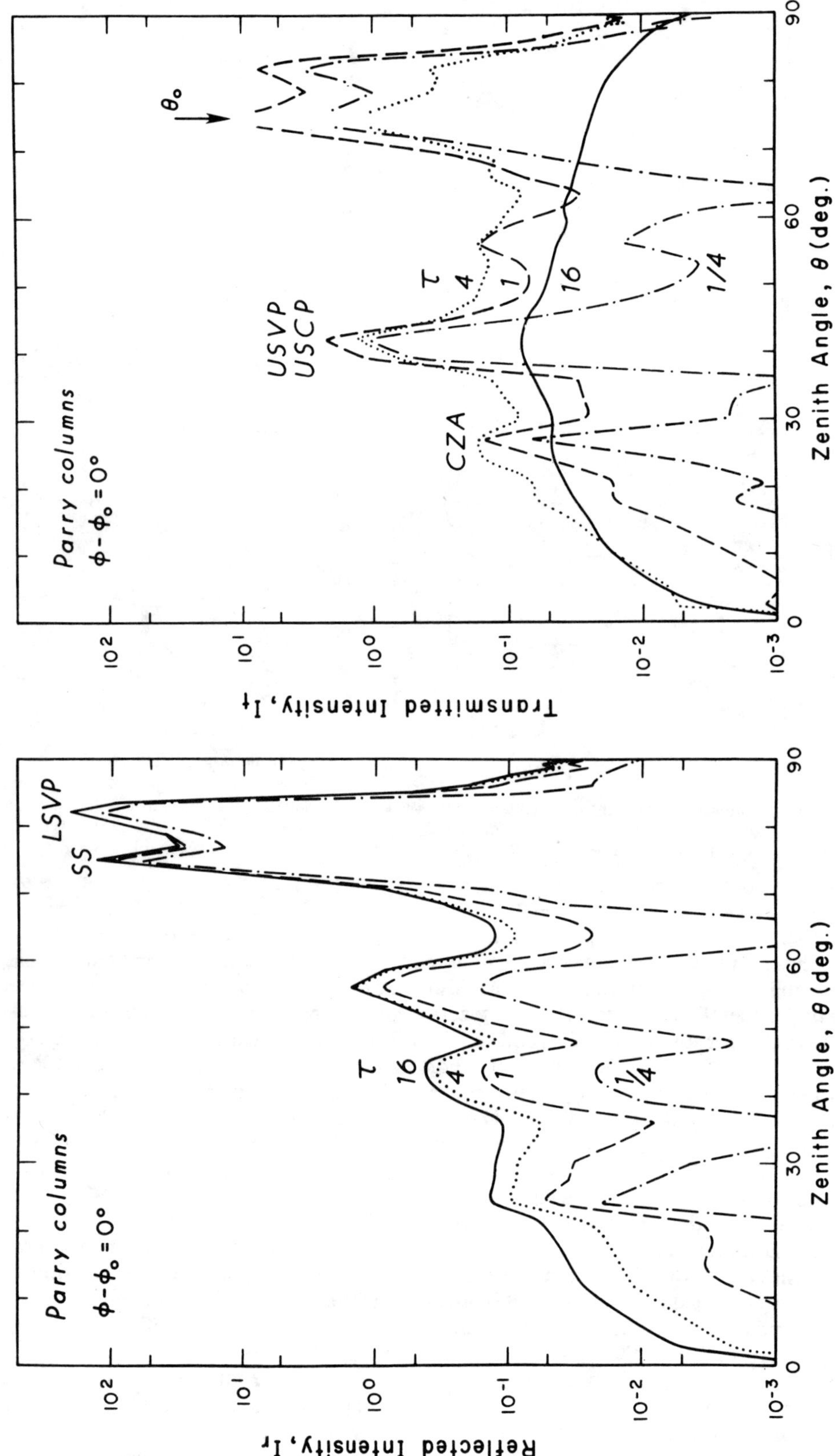

FIG. 5. Intensity of sunlight reflected and transmitted by Parry columns ($L/2a = 2.5$) in the solar principal plane ($\phi - \phi_0 = 0°$) at $\lambda = 0.55$ μm. The solar zenith angle is 75°.

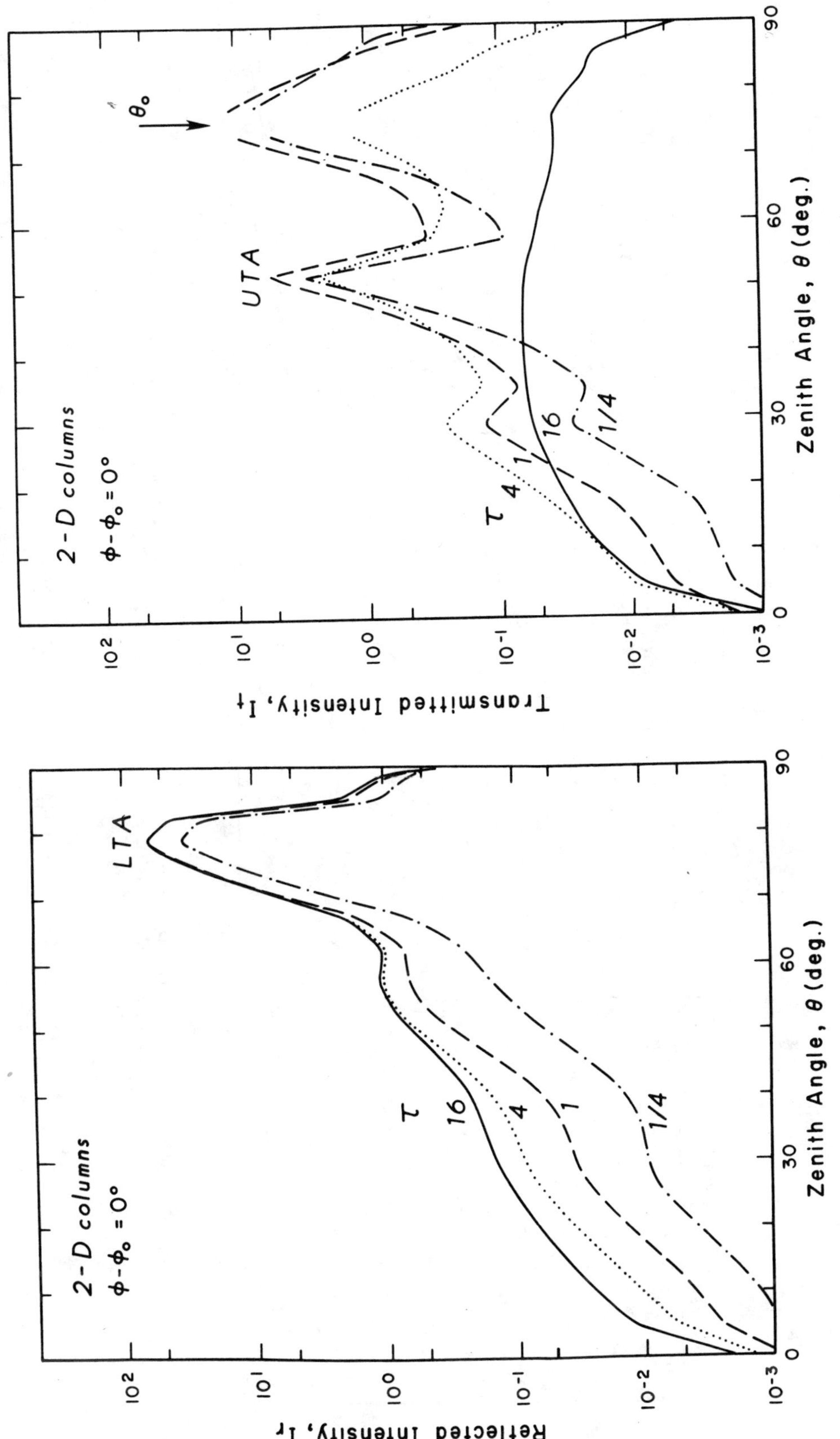

FIG. 6. As in Fig. 5, except for 2-D columns ($L/2a = 2.5$).

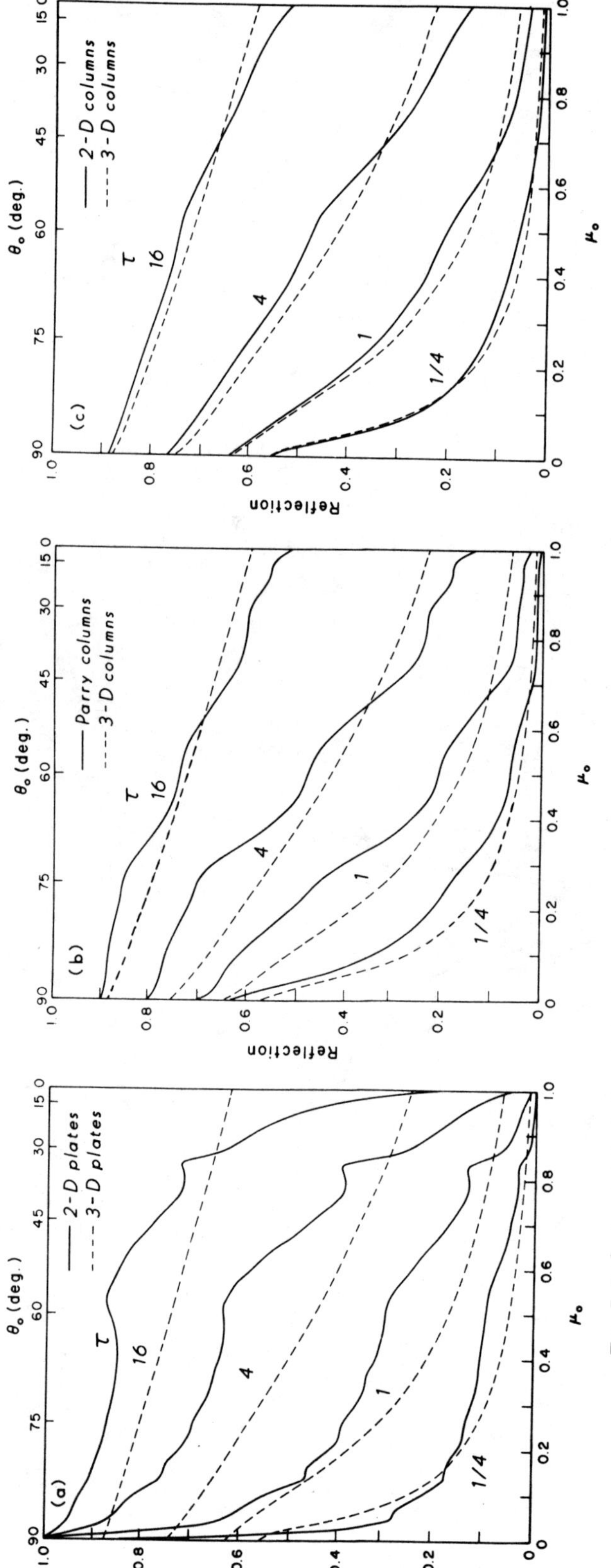

FIG. 7. Reflection (a) for 2-D and 3-D plates ($L/2a = 0.4$) as a function of the cosine of the solar zenith angle at $\lambda = 0.55\ \mu m$, (b) for Parry and 3-D columns ($L/2a = 2.5$), and (c) for 2-D and 3-D columns ($L/2a = 2.5$).

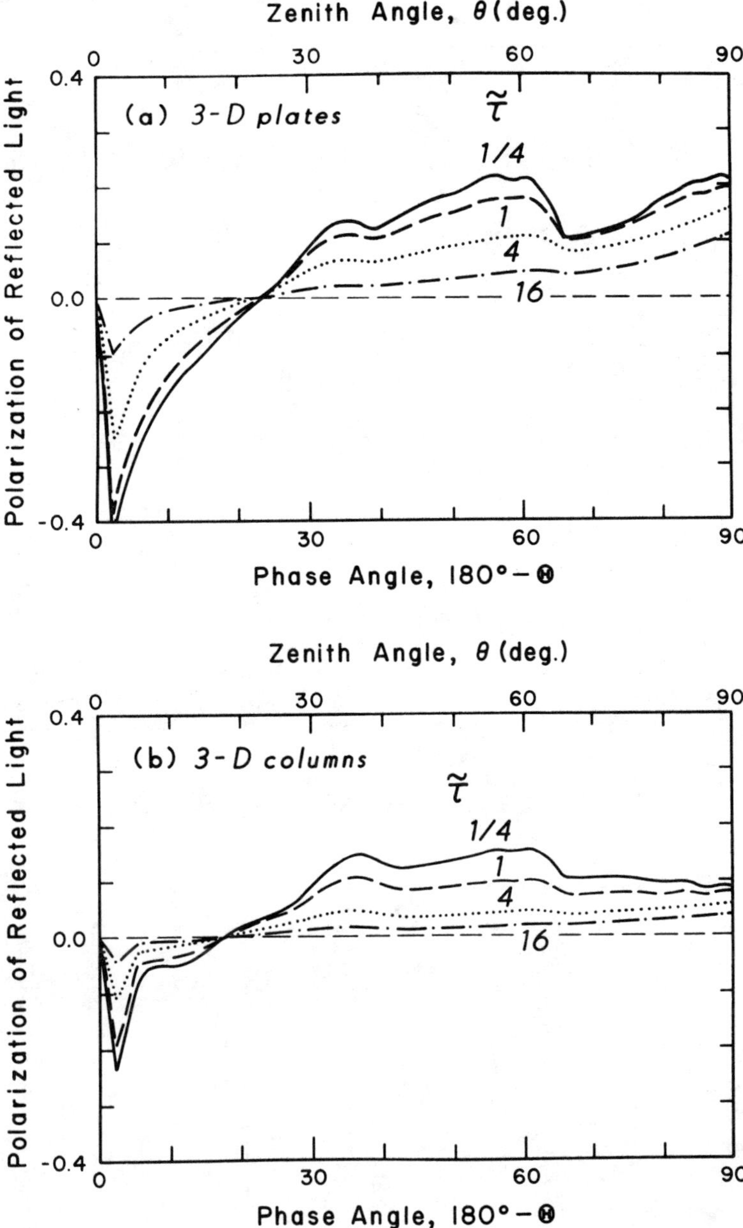

FIG. 8. Polarization, $-Q/I$ of sunlight reflected by (a) 3-D plates ($L/2a = 0.1$) and (b) 3-D columns ($L/2a = 2.5$) as a function of the phase angle, $180° - \Theta$ for an overhead sun ($\theta_0 = 0°$) at $\lambda = 0.55$ μm.

is not affected by multiple-scattering processes. This has been discussed by Hansen (1971) in the case of spherical particles.

Figure 9 shows the polarization of sunlight reflected by 3-D crystals and area-equivalent spheres in the solar principal plane ($\phi - \phi_0 = 0°/180°$) when the position of the sun is at $\theta_0 = 50°$. For ice spheres, a maximum polarization of about 80% is shown at the ~45° phase angle for an optical depth of 1 (Fig. 9a). This maximum polarization is associated with the rainbow features produced by spherical particles and is absent for ice crystal clouds. Ice plates and columns both show negative polarization maxima at phase angles close to 0°. In general, plates produce larger polarization than columns. Using an optical depth of 16, polarization decreases significantly, as shown in Fig. 9b (note that a different scale is used here). However, the polarization pattern does not vary with increasing optical depth. In particular, it is noted that the neutral points at phase angles of ~18° for columns and ~23° for plates re-

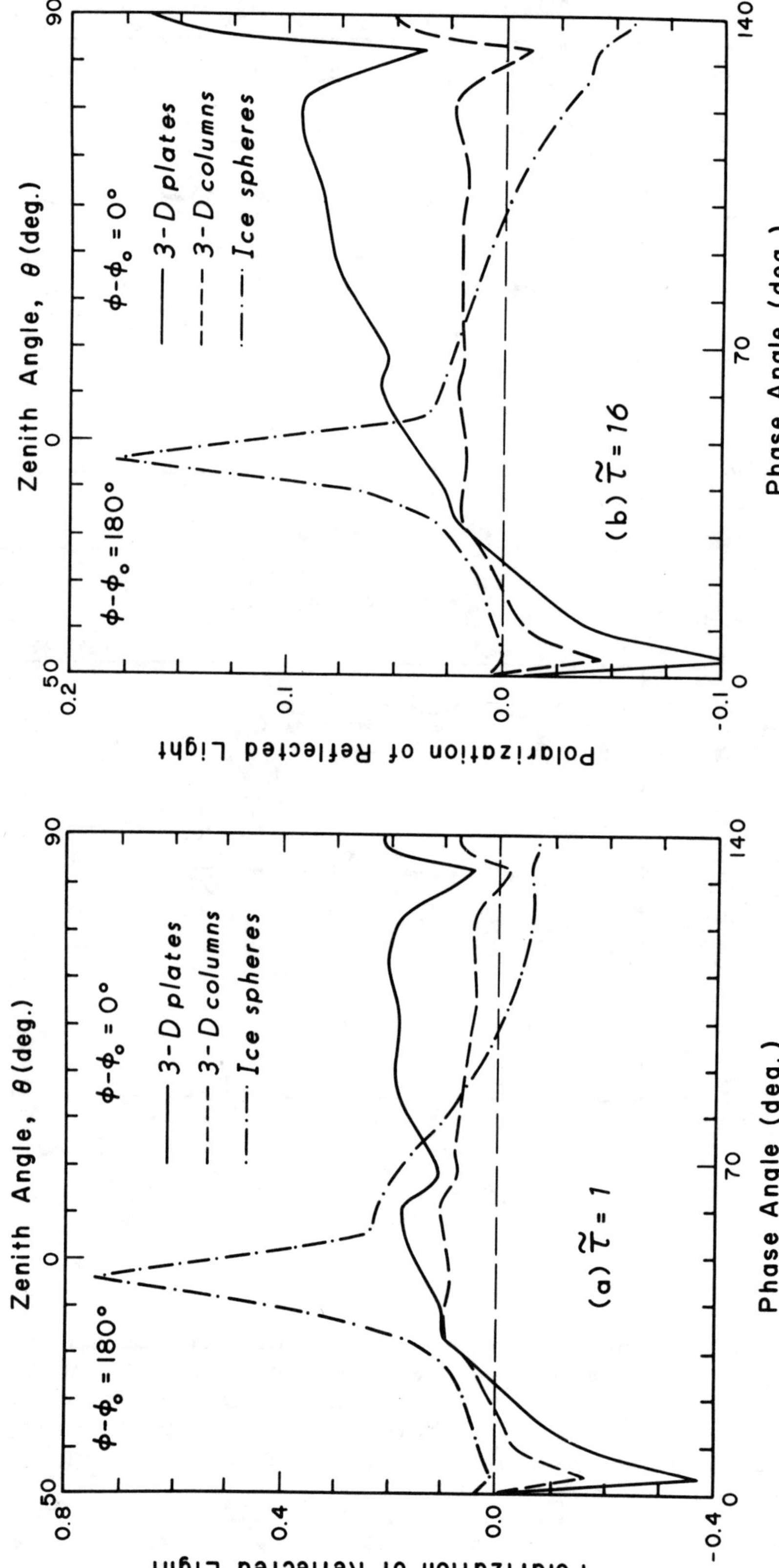

FIG. 9. Polarization of sunlight reflected by 3-D plates ($L/2a = 0.1$), 3-D columns ($L/2a = 2.5$), and area-equivalent spheres as a function of the phase angle in the solar principal plane ($\phi - \phi_0 = 0°/180°$) at $\lambda = 0.55\ \mu m$. The solar zenith angle is 50°. The optical depths considered are (a) 1, and (b) 16.

main nearly constant, regardless of the optical depth. These neutral points are also independent of the incident solar angle. The preceding discussions, however, do not apply to sky polarization due to Rayleigh scattering. In the case of Rayleigh scattering, negative polarization, and hence the neutral point, are produced by double scattering (van de Hulst 1980). Thus multiple scattering, which depends on the optical depth and incident solar angle, would significantly affect the position of neutral points.

4. Interpretation of polarization measurements for ice crystal clouds

Santer et al. (1985) measured the polarization of sunlight reflected by Martian white clouds using the photopolarimeter equipment on the spacecraft MARS-5. They also presented polarization measurements using a ground-based telescope reported by Dollfus. These results are shown in Fig. 10. The two dash-dotted curves denote the fittings of polarization measured by MARS-5 at $\lambda = 0.592$ μm, assuming optical depths of ∞ and 0.3, reported by these authors. Since there is no positive peak around the 45° phase angle, the presence of water droplets is excluded. If the white clouds on Mars consist of dry ice (cubes), the neutral point should have been at $\sim$40°, according to the computations given by Liou et al. (1983). Since this is not present in the measurements, dry ice is not a likely candidate for the particles in Martian white clouds. For interpretation purposes,

randomly oriented plates and columns with an optical depth of 64 are used. As shown in Fig. 10, the neutral point of 24° for 3-D plates closely matches the observed value. It is noted that the neutral point does not vary significantly with the crystal aspect ratio (see Fig. 5a of Part I), nor with the optical depth, as illustrated in Figs. 8 and 9. The differences around the phase angle of $\sim$5° between the observed and computed values could be explained by deviations from the exact plate- or column-like shape in the ice crystal types occurring in Martian white clouds. These differences have been noted between laboratory and model ice crystals observed by Cai and Liou (1982). On the basis of our theoretical interpretation of the neutral point, we speculate that Martian white clouds are composed of randomly oriented plate-like ice crystals.

Coffeen (1979) measured the polarization of sunlight reflected by cirrus using an infrared polarimeter aboard the NASA Convair 990. The measurement was performed on the solar principal plane, $\phi - \phi_0 = 0°/180°$ at $\lambda = 2.22$ μm. Polarization results are shown in Fig. 11. Since the observed cirrus thickness is about 5 km, we used an optical depth of 64 in the theoretical calculations. To match the observed phase angles between 0° and 160°, we used a solar zenith angle of 70°. The negative values around the 140° and 160° phase angles result from the outer and inner halos, respectively. Around the $\sim$0° phase angles, the magnitude of negative polarization is larger than at other phase angles. This is because these directions are close to the horizon and, when the optical depth is large, multiple scattering

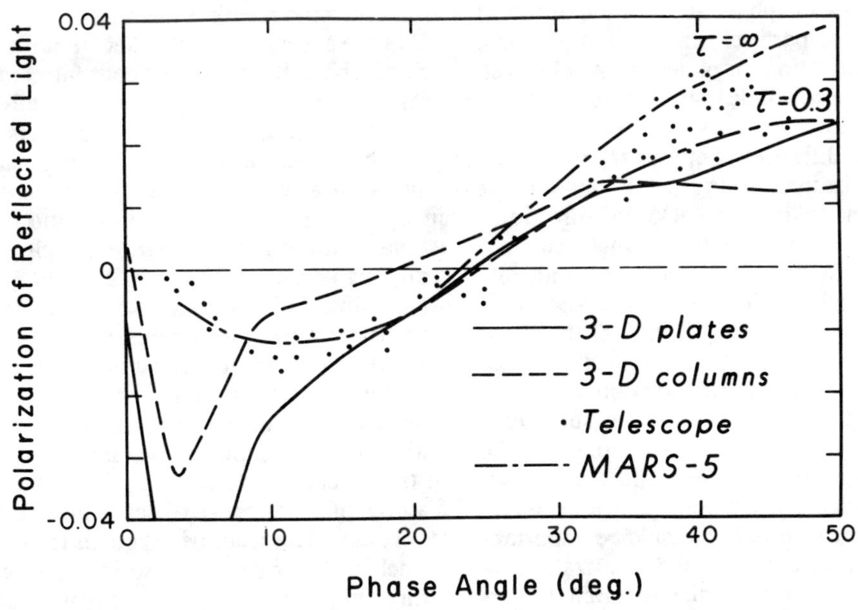

FIG. 10. Comparison of polarization of sunlight reflected by Martian white clouds between the measurements reported by Santer et al. (1985) and the present computation. The ice crystal models used for interpretation are 3-D plates ($L/2a = 0.1$) and 3-D columns ($L/2a = 2.5$) at $\lambda = 0.55$ μm with an optical depth of 64.

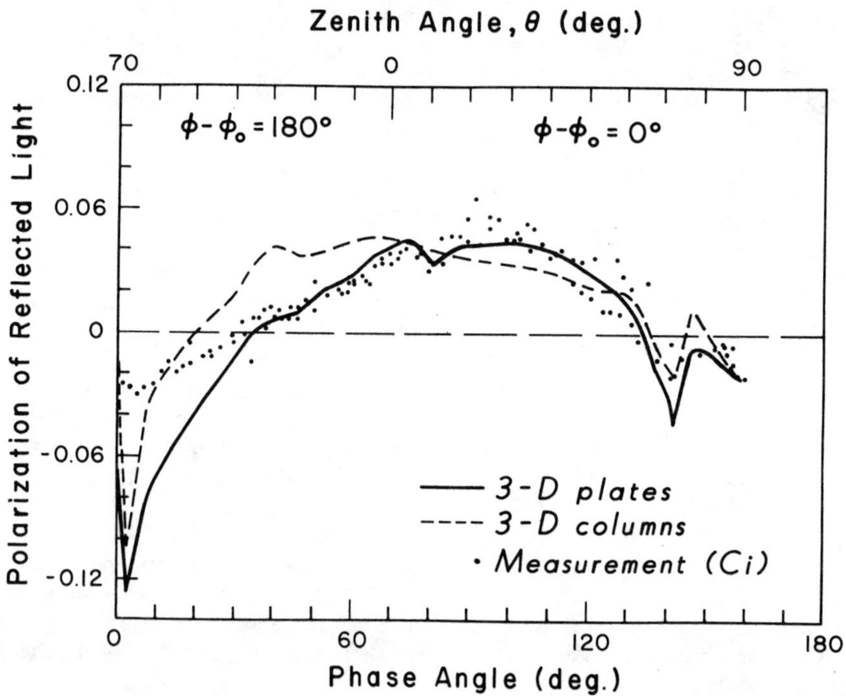

FIG. 11. Comparison of polarization of sunlight reflected by cirrus at $\lambda = 2.22$ μm between measurements reported by Coffeen (1979) and the present computation. The ice crystal models used are 3-D plates ($L/2a = 4$ μm$/40$ μm) and 3-D columns ($L/2a = 100$ μm$/40$ μm) with an optical depth of 64.

is relatively less important. Thus, polarization of the reflected sunlight is produced mostly by single scattering. Since there is no sharp peak due to the rainbow feature around the ~45° phase angle in the observed polarization, the cloud particles must be nonspherical. The computed polarization for plates fits the observed values quite well. We conclude that the cloud particles near the cloud top must be randomly oriented plate-like ice crystals. The differences between the computed and observed polarization around the 0° phase angle may be explained as follows: in backward directions, i.e., the 0° phase angle, it is possible that singly scattered light is blocked by the body of the airplane and only multiple-scattered light with small polarization can reach the polarimeter. Also, it is likely that the cloud contains irregular ice crystals, which specifically reduce polarization in the backscattering directions.

The interpretation of polarization measurements presented here, however, is meant to demonstrate the applicability of the multiple-scattering program developed for hexagonal ice crystals. For absorption wavelengths, ice crystal size distributions could be important in the determination of scattering and polarization results. This is an area requiring further research. In particular, concurrent cloud physics measurements are needed in order to develop remote sounding techniques for the detection of ice crystal clouds using the principle of multiple scattering and polarization.

5. Conclusions

In this paper, radiative transfer of polarized light in an anisotropic medium has been formulated with the aid of the adding principle. The iterative equations developed have been used to compute bidirectional intensities for randomly and horizontally oriented ice crystals. The single-scattering parameters, including the phase function, single-scattering albedo, and extinction cross section derived for ice crystals, were incorporated in the multiple-scattering computations. In addition, we have utilized the similarity principle for radiative transfer to account for the diffraction peak and δ-forward transmission due to two refractions in the multiple scattering computations.

For the reflected and transmitted intensities, we show that there are significant differences in these intensities between hexagonal ice crystals and equivalent spheres with either the same surface area or volume. A peak in the reflected intensity associated with the rainbow features of spherical particles is absent in the case of ice crystals. This leads us to conclude that the spherical model is inadequate for use in the interpretation of bidirectional reflectance from cirrus clouds. The transmitted intensity patterns for ice crystals differ distinctly from those for ice spheres, principally due to the 22° and 46° halo maxima produced by the hexagonal structure. In regard to the crystal orientation, horizontally oriented plate and column crystals produce numerous optical features in the reflected and transmitted intensity patterns, which are absent if crystals are randomly oriented in space. The planetary albedo for horizontally oriented crystals is lower than that for randomly oriented crystals when the sun is near the zenith, whereas the reverse is true when the sun is close to the horizon.

With respect to the polarization of sunlight reflected by ice crystal clouds, we show that the neutral point is independent of the solar zenith angle, as well as the optical depth. The computed polarization patterns for hexagonal ice crystals reveal notable differences from those for ice spheres and other particle shapes, such as cubes. In particular, we find that the neutral point of the reflected sunlight is very sensitive to the particle shape. This suggests that a detection of the neutral point in the polarization pattern could provide a means for the identification of the particle shape.

Finally, we interpret the polarization patterns for Martian white clouds observed from the MARS-5 spacecraft in the visible wavelength and for a cirrus cloud observed from a near infrared polarimeter aboard the NASA Convair 990. We show that the computed polarization for randomly oriented ice plates can match the observed data for a large range of phase angles. This leads us to speculate that the top of these clouds is composed of randomly oriented plate-like ice crystals. In summary, there is significant information content in the polarization pattern for sunlight reflected by clouds. It appears that polarization measurements could be effectively utilized for the identification of the particle characteristics of clouds.

Acknowledgments. All the computations contained in this research were carried out on the San Diego supercomputer, CRAY X-MP/48. The research was supported in part by NASA Grant NAG5-732 and NSF Grant ATM85-13975. We thank Graeme Stephens and the other anonymous reviewer for constructive comments on the paper. Sharon Bennett typed and edited the manuscript.

REFERENCES

Asano, S., 1983: Transfer of solar radiation in optically anisotropic ice clouds. *J. Meteor. Soc. Japan,* **61,** 402–413.

Cai, Q., and K. N. Liou, 1982: Polarized light scattering by hexagonal ice crystals: Theory. *Appl. Opt.,* **21,** 3569–3580.

Chandrasekhar, S., 1960: *Radiative Transfer.* Dover, 393 pp.

Coffeen, D. L., 1979: Polarization and scattering characteristics in the atmosphere of Earth, Venus, and Jupiter. *J. Opt. Soc. Am.,* **69,** 1051–1064.

Dave, J. V., 1970: Coefficients of the Legendre and Fourier series for the scattering functions of spherical particles. *Appl. Opt.,* **9,** 1888–1896.

Hansen, J. E., 1971: Multiple scattering of polarized light in planetary atmospheres. Parts I and II. *J. Atmos. Sci.,* **28,** 120–125, 1400–1426.

Hovenier, J. W., 1969: Symmetry relationships for scattering of polarized light in a slab of randomly oriented particles. *J. Atmos. Sci.,* **26,** 488–499.

Kattawar, G. W., S. J. Hitzfelder and J. Binstock, 1973: An explicit form of the Mie phase matrix for multiple scattering calculations in the I, Q, U, and V representation. *J. Atmos. Sci.,* **30,** 289–295.

Lacis, A. A., and J. E. Hansen, 1974: A parameterization for the absorption of solar radiation in the earth's atmosphere. *J. Atmos. Sci.,* **31,** 118–133.

Liou, K. N., 1980: *An Introduction to Atmospheric Radiation.* Academic Press, 392 pp.

——, Q. Cai, J. B. Pollack and J. N. Cuzzi, 1983: Light scattering by randomly oriented cubes and parallelepipeds. *Appl. Opt.,* **22,** 3001–3008.

Peebles, G. H., and M. S. Pleeset, 1951: Transmission of gamma rays through large thicknesses of heavy materials. *Phys. Rev.,* **81,** 430–439.

Potter, J. E., 1970: The delta function approximation in radiative transfer theory. *J. Atmos. Sci.,* **27,** 945–951.

Ripley, E. A., and B. Saugier, 1971: Photometeors at Saskatoon on 3 December 1970. *Weather,* **26,** 150–157.

Santer, R., M. Deschamps, L. V. Ksanfomaliti and A. Dollfus, 1985: Photopolarimetric analysis of the Martian atmosphere by the Soviet MARS-5 orbiter. *Astron. Astrophys.,* **150,** 217–228.

Sobolev, V. V., 1975: *Light Scattering in Planetary Atmospheres.* Pergamon, 256 pp.

Stephens, G. L., 1980: Radiative transfer in a linear lattice: Application to anisotropic ice crystal clouds. *J. Atmos. Sci.,* **37,** 2095–2104.

Stokes, G. G., 1862: On the intensity of the light reflected from or transmitted through a pile of plates. *Proc. Roy. Soc. London,* **11,** 545–556.

Takano, Y., and K. N. Liou, 1988: Solar radiative transfer in cirrus clouds. Part I: Single-scattering and optical properties of hexagonal ice crystals. *J. Atmos. Sci.* **46,** 3–19.

Tränkle, E., and R. G. Greenler, 1987: Multiple-scattering effects in halo phenomena. *J. Opt. Soc. Am., A,* **4,** 591–599.

Twomey, S., H. Jacobowitz and H. B. Howell, 1966: Matrix method for multiple scattering problems. *J. Atmos. Sci.,* **23,** 289–296.

van de Hulst, H. C., 1980: *Multiple Light Scattering.* Academic Press, 739 pp.

Section Five
Polarization

THE ILLUMINATION AND POLARIZATION OF THE SUNLIT SKY ON RAYLEIGH SCATTERING

S. Chandrasekhar and Donna D. Elbert

CONTENTS

		PAGE
1.	Introduction	643
2.	The solution of the fundamental problem	643
3.	The effect of reflection by the ground	647
4.	Description of the tables	647
5.	The polarization of the sunlit sky: the theory of neutral points and lines	649
6.	The tables	655

1. INTRODUCTION

Since 1871 when Lord Rayleigh first accounted for the principal features of the brightness and polarization of the sunlit sky in terms of the laws of scattering now associated with his name, it has been generally recognized that a problem of fundamental importance both for meteorological optics and for theories of planetary illumination is the following:

A parallel beam of radiation in a given state of polarization is incident on a plane-parallel atmosphere of optical thickness τ_1 in some specified direction. Each element of the atmosphere scatters radiation in accordance with Rayleigh's laws. It is required to find the distribution of intensity and polarization of the light diffusely transmitted by the atmosphere below $\tau = \tau_1$ and of the light diffusely reflected by the atmosphere above $\tau = 0$.

In the theory of planetary illumination one is principally interested in the reflected light while in the theory of sky illumination one is similarly interested in the transmitted light. In this paper we shall be concerned only with the latter.

It is clear that an exact treatment of the foregoing problem in the theory of diffuse reflection and transmission will require the formulation and solution of the appropriate equations of radiative transfer. This was accomplished six years ago[1] and the theory is described and briefly illustrated in the book *Radiative Transfer* (Oxford, 1950) by one of us. A general account of the theory, together with a comparison of its predictions with observations particularly those relating to the polarization of the sunlit sky, was published in 1951 in a brief article.[2]

In this paper we shall present the calculations which we have made (at intervals) during the past five years with the object of giving the theory a concrete form. At one time it was our hope to present our calculations based on the exact mathematical solution of the prob-

lem with detailed comparisons not only with the available observational data but also with the calculations of the earlier investigators based on approximations of various kinds. But pressure of time and circumstance have forced us to abandon this plan: this paper will be restricted to giving the results of our calculations with only such comparisons with observations as seemed to us of particular interest.

2. THE SOLUTION OF THE FUNDAMENTAL PROBLEM

As we have stated, the problem in the theory of diffuse reflection and transmission formulated in § 1 has been exactly solved; the solution is given in *Radiative Transfer* (§§ 69–73). We shall not describe in any detail how the solution was obtained. But a few explanatory remarks on the parameters in terms of which the solution was obtained and on the structure of the solution itself may be useful in the present connection.

Since on Rayleigh's laws light gets partially plane-polarized whenever it is scattered, it is clear that in formulating the equations of radiative transfer we must allow for the partial plane-polarization of the radiation field. Now to describe a radiation field which is partially plane-polarized we need three parameters to specify the intensity, the degree of polarization, and the plane of polarization. It would scarcely be expected that one could include such diverse quantities as an intensity, a ratio, and an angle in any satisfactory way in formulating the basic equations of the problem. It appears that for these latter purposes the most convenient representation of polarized light is a set of parameters first introduced by Stokes[3] in 1852. The meaning of these parameters for a partially plane-polarized beam is simple: Let l and r refer to two arbitrarily chosen directions at right-angles to one another in the plane transverse to the direction of propagation of the beam. The intensity $I(\psi)$ in a direction making an angle ψ (measured clock-wise) to the direction of l can be expressed in the form

$$I(\psi) = I_l \cos^2 \psi + I_r \sin^2 \psi + \tfrac{1}{2} U \sin 2\psi. \quad (1)$$

The coefficients I_l, I_r and U in this representation are the Stokes parameters. In terms of these parameters the angle χ which the plane of polarization makes with the direction l and the degree of polarization, δ, are given by

$$\tan 2\chi = U/(I_l - I_r) \quad (2)$$

---

[1] S. Chandrasekhar, On the radiative equilibrium of a stellar atmosphere. XXII. (V. Rayleigh scattering), *Astrophys. Jour.* **107**: 199, 1947.

[2] S. Chandrasekhar and Donna Elbert, Polarization of the sunlit sky, *Nature* **167**: 51, 1951.

[3] G. G. Stokes, On the composition and resolution of streams of polarized light from different sources, *Trans. Camb. Phil. Soc.* **9**: 399, 1852.

and

$$\delta = (I_l - I_r) \sec 2\chi / (I_l + I_r). \tag{3}$$

The additive property of the Stokes parameters which makes them so convenient for treating problems of radiative transfer is evident from the representation (1): If two independent streams of polarized light are mixed, then the Stokes parameter characterizing the mixture is the sum of the Stokes parameters of the individual streams.

In terms of the Stokes parameters a law of scattering is specified by a matrix, since an elementary act of scattering results in a linear transformation of the parameters. Consequently, by considering the intensity as a vector I with the components I_l, I_r and U (where l and r from now on refer to directions parallel and perpendicular, respectively, to the meridian through the point under consideration and in the plane containing the directions of the beam and of the normal to the plane of stratification of the atmosphere) and by replacing the "phase function" commonly introduced to describe the angular distribution of the scattered radiation by a *phase matrix*, P, we can formulate the basic equation of transfer without any difficulty of principle. In this manner we find that the equation we have to solve is[4]

$$\mu \frac{dI(\tau, \mu, \varphi)}{d\tau} = I(\tau, \mu, \varphi)$$

$$- \frac{1}{4\pi} \int_{-1}^{+1} \int_0^{2\pi} P(\mu, \varphi; \mu', \varphi') I(\tau, \mu', \varphi') \, d\mu' \, d\varphi'$$

$$- \tfrac{1}{4} e^{-\tau/\mu_0} P(\mu, \varphi; -\mu_0, \varphi_0) F. \tag{4}$$

where

$$F = (F_l, F_r, F_U) \tag{5}$$

is the (Stokes) vector which represents the parallel beam of radiation incident on the atmosphere in the direction $(-\mu_0, \varphi_0)$: πF_l, πF_r and πF_U denote the net fluxes per unit area normal to the beam in the three Stokes parameters. Further, in equation (4) μ denotes the cosine of the angle to the outward normal and the azimuthal angle. And, finally, for the case of Rayleigh scattering the phase matrix $P(\mu, \varphi; \mu', \varphi')$ has the explicit form:

$$P(\mu, \varphi; \mu', \varphi') = Q[P^{(0)}(\mu, \mu') + (1 - \mu^2)^{\frac{1}{2}}(1 - \mu'^2)^{\frac{1}{2}} P^{(1)}(\mu, \varphi; \mu', \varphi') + P^{(2)}(\mu, \varphi; \mu', \varphi')], \tag{6}$$

where

$$Q = \begin{bmatrix} 1 & 0 & 0 \\ 0 & 1 & 0 \\ 0 & 0 & 2 \end{bmatrix}, \tag{7}$$

$$P^{(0)}(\mu, \mu') = \frac{3}{4} \begin{bmatrix} 2(1 - \mu^2)(1 - \mu'^2) + \mu^2\mu'^2 & \mu^2 & 0 \\ \mu'^2 & 1 & 0 \\ 0 & 0 & 0 \end{bmatrix}, \tag{8}$$

$$P^{(1)}(\mu, \varphi; \mu', \varphi') = \frac{3}{4} \begin{bmatrix} 4\mu\mu' \cos(\varphi' - \varphi) & 0 & 2\mu \sin(\varphi' - \varphi) \\ 0 & 0 & 0 \\ -2\mu' \sin(\varphi' - \varphi) & 0 & \cos(\varphi' - \varphi) \end{bmatrix} \tag{9}$$

and

$$P^{(2)}(\mu, \varphi; \mu', \varphi') = \frac{3}{4} \begin{bmatrix} \mu^2\mu'^2 \cos 2(\varphi' - \varphi) & -\mu^2 \cos 2(\varphi' - \varphi) & \mu^2\mu' \sin 2(\varphi' - \varphi) \\ -\mu'^2 \cos 2(\varphi' - \varphi) & \cos 2(\varphi' - \varphi) & -\mu' \sin 2(\varphi' - \varphi) \\ -\mu\mu'^2 \sin 2(\varphi' - \varphi) & \mu \sin 2(\varphi' - \varphi) & \mu\mu' \cos 2(\varphi' - \varphi) \end{bmatrix}. \tag{10}$$

The solution of equation (4) appropriate to the problem on hand must satisfy the boundary conditions

$$I(0, -\mu, \varphi) \equiv 0 \quad (0 < \mu \leqslant 1, \ 0 \leqslant \varphi \leqslant 2\pi)$$

and

$$I(\tau_1, +\mu, \varphi) \equiv 0 \quad (0 < \mu \leqslant 1, \ 0 \leqslant \varphi \leqslant 2\pi), \tag{11}$$

since there is no diffuse radiation in any inward direction at $\tau = 0$ and in any outward direction at $\tau = \tau_1$. And the solution to the problem of diffuse reflection and transmission will be completed when we specify the angular distribution and the state of polarization of the diffuse light which emerges from $\tau = 0$ and $\tau = \tau_1$.

The laws of diffuse reflection and transmission by a plane-parallel atmosphere are generally expressed (cf. *Radiative Transfer*, 44) in terms of a *scattering matrix*, $S(\mu, \varphi; \mu_0, \varphi_0)$ and a *transmission matrix*, $T(\mu, \varphi; \mu_0, \varphi_0)$ such that the reflected and the transmitted intensities are given by

$$I(0; \mu, \varphi; \mu_0, \varphi_0) = \frac{1}{4\mu} S(\mu, \varphi; \mu_0, \varphi_0) F$$

and

$$I(\tau_1; -\mu, \varphi; \mu_0, \varphi_0) = \frac{1}{4\mu} T(\mu, \varphi; \mu_0, \varphi_0) F. \tag{12}$$

Our problem, then, is to specify S and T for an atmosphere scattering radiation in accordance with Rayleigh's laws.

---

[4] The derivation of the equations which follow will be found in *Radiative Transfer* § 16: 35–45.

The elements of S and T are clearly functions of the four variables μ, φ, μ_0 and φ_0 in addition, of course, to the optical thickness τ_1 which may, however, be treated as a parameter. If the variables μ, φ, μ_0 and φ_0 were not separable the problem of tabulating S and T may indeed be considered as impracticable. But the essential feature of the solution for S and T (which we shall presently write down) which makes the problem a practicable one is that S and T involve only four pairs of functions $X_l(\mu)$, $Y_l(\mu)$; $X_r(\mu)$, $Y_r(\mu)$; $X^{(1)}(\mu)$, $Y^{(1)}(\mu)$; and $X^{(2)}(\mu)$, $Y^{(2)}(\mu)$ all of the single variable μ. Further, these four pairs of functions belong to a general class (the X- and Y-functions) which satisfy a simultaneous pair of integral equations of the form[5]

$$X(\mu) = 1 + \mu \int_0^1 \frac{\Psi(\mu')}{\mu + \mu'}$$
$$\times \left[X(\mu)X(\mu') - Y(\mu)Y(\mu') \right] d\mu' \quad (13)$$

and

$$Y(\mu) = e^{-\tau_1/\mu} + \mu \int_0^1 \frac{\Psi(\mu')}{\mu - \mu'}$$
$$\times \left[Y(\mu)X(\mu') - X(\mu)Y(\mu') \right] d\mu', \quad (14)$$

where the *characteristic function* $\Psi(\mu)$ is in problems of radiative transfer, an even polynomial in μ satisfying the condition

$$\int_0^1 \Psi(\mu)\, d\mu \leqslant \tfrac{1}{2}. \quad (15)$$

The case when equality occurs in (15) (the so-called *conservative case*) is special: The solutions of equations (13) and (14) are, then, no longer unique; they form instead a one-parameter family. In conservative cases one therefore defines what are called *standard solutions* which have the property:

$$\int_0^1 X(\mu)\Psi(\mu)\, d\mu = 1$$

and

$$\int_0^1 Y(\mu)\Psi(\mu)\, d\mu = 0. \quad (16)$$

As we have already stated, the solutions for S and T (for an atmosphere scattering radiation in accordance with Rayleigh's laws) involve only four pairs of X- and Y-functions: X_l, Y_l; X_r, Y_r; $X^{(1)}$, $Y^{(1)}$; $X^{(2)}$, $Y^{(2)}$; and the characteristic functions in terms of which

these are defined are:

$$\Psi_l(\mu) = \tfrac{3}{4}(1 - \mu^2);$$
$$\Psi_r(\mu) = \tfrac{3}{8}(1 - \mu^2),$$
$$\Psi^{(1)}(\mu) = \tfrac{3}{8}(1 - \mu^2)(1 + 2\mu^2)$$

and

$$\Psi^{(2)}(\mu) = \tfrac{3}{16}(1 + \mu^2)^2, \quad (17)$$

respectively. The function $\Psi_l(\mu)$ belongs to the conservative class; accordingly, in this case we define $X_l(\mu)$, $Y_l(\mu)$ as the standard solutions having the property

$$\tfrac{3}{4} \int_0^1 X_l(\mu)(1 - \mu^2)\, d\mu = 1$$

and

$$\int_0^1 Y_l(\mu)(1 - \mu^2)\, d\mu = 0. \quad (18)$$

After these explanatory remarks we shall now write down the solutions for S and T given in *Radiative Transfer*:

The scattering and the transmission matrices allow a decomposition into azimuth independent and azimuth dependent terms in the same manner as the phase matrix [equation (6)] and have the forms:

$$S(\mu, \varphi; \mu_0, \varphi_0) = Q\left[\tfrac{3}{4}S^{(0)}(\mu; \mu_0)\right.$$
$$+ (1 - \mu^2)^{\frac{1}{2}}(1 - \mu_0^2)^{\frac{1}{2}}S^{(1)}(\mu, \varphi; \mu_0, \varphi_0)$$
$$\left. + S^{(2)}(\mu, \varphi; \mu_0, \varphi_0)\right] \quad (19)$$

and

$$T(\mu, \varphi; \mu_0, \varphi_0) = Q\left[\tfrac{3}{4}T^{(0)}(\mu; \mu_0)\right.$$
$$+ (1 - \mu^2)^{\frac{1}{2}}(1 - \mu_0^2)^{\frac{1}{2}}T^{(1)}(\mu, \varphi; \mu_0, \varphi_0)$$
$$\left. + T^{(2)}(\mu, \varphi; \mu_0, \varphi_0)\right]. \quad (20)$$

The dependence of the azimuth dependent terms $(S^{(1)}, T^{(1)})$ and $(S^{(2)}, T^{(2)})$ on $\varphi_0 - \varphi$ are essentially the same as $P^{(1)}$ and $P^{(2)}$; indeed, we have

$$\left(\frac{1}{\mu_0} + \frac{1}{\mu}\right) S^{(i)} = \left[X^{(i)}(\mu)X^{(i)}(\mu_0)\right.$$
$$\left. - Y^{(i)}(\mu)Y^{(i)}(\mu_0)\right]P^{(i)}(\mu, \varphi; -\mu_0, \varphi_0)$$

and

$$\left(\frac{1}{\mu_0} - \frac{1}{\mu}\right) T^{(i)} = \left[Y^{(i)}(\mu)X^{(i)}(\mu_0)\right.$$
$$\left. - X^{(i)}(\mu)Y^{(i)}(\mu_0)\right]P^{(i)}(-\mu, \varphi; -\mu_0, \varphi_0)$$
$$(i = 1, 2). \quad (21)$$

In contrast, the solutions for the azimuth independent terms $S^{(0)}$ and $T^{(0)}$ are very complicated. They are given by

$$\left(\frac{1}{\mu_0} + \frac{1}{\mu}\right) S^{(0)}(\mu; \mu_0) = \begin{pmatrix} \psi(\mu) & 2^{\frac{1}{2}}\phi(\mu) & 0 \\ \chi(\mu) & 2^{\frac{1}{2}}\zeta(\mu) & 0 \\ 0 & 0 & 0 \end{pmatrix} \begin{pmatrix} \psi(\mu_0) & \chi(\mu_0) & 0 \\ 2^{\frac{1}{2}}\phi(\mu_0) & 2^{\frac{1}{2}}\zeta(\mu_0) & 0 \\ 0 & 0 & 0 \end{pmatrix}$$

$$- \begin{pmatrix} \xi(\mu) & 2^{\frac{1}{2}}\eta(\mu) & 0 \\ \sigma(\mu) & 2^{\frac{1}{2}}\theta(\mu) & 0 \\ 0 & 0 & 0 \end{pmatrix} \begin{pmatrix} \xi(\mu_0) & \sigma(\mu_0) & 0 \\ 2^{\frac{1}{2}}\eta(\mu_0) & 2^{\frac{1}{2}}\theta(\mu_0) & 0 \\ 0 & 0 & 0 \end{pmatrix} \quad (22)$$

---

[5] For the theory of the X- and Y-functions see *Radiative Transfer*, chap. VIII.

and

$$\left(\frac{1}{\mu_0} - \frac{1}{\mu}\right) T^{(0)}(\mu;\mu_0) = \begin{pmatrix} \xi(\mu) & 2^{\frac{1}{2}}\eta(\mu) & 0 \\ \sigma(\mu) & 2^{\frac{1}{2}}\theta(\mu) & 0 \\ 0 & 0 & 0 \end{pmatrix} \begin{pmatrix} \psi(\mu_0) & \chi(\mu_0) & 0 \\ 2^{\frac{1}{2}}\phi(\mu_0) & 2^{\frac{1}{2}}\zeta(\mu_0) & 0 \\ 0 & 0 & 0 \end{pmatrix}$$

$$- \begin{pmatrix} \psi(\mu) & 2^{\frac{1}{2}}\phi(\mu) & 0 \\ \chi(\mu)\cdot & 2^{\frac{1}{2}}\zeta(\mu) & 0 \\ 0 & 0 & 0 \end{pmatrix} \begin{pmatrix} \xi(\mu_0) & \sigma(\mu_0) & 0 \\ 2^{\frac{1}{2}}\eta(\mu_0) & 2^{\frac{1}{2}}\theta(\mu_0) & 0 \\ 0 & 0 & 0 \end{pmatrix}, \quad (23)$$

where ψ, ϕ, χ, etc., are eight functions expressible in terms of the two pairs of X- and Y-functions, X_l, Y_l and X_r, Y_r in the forms

$$\psi(\mu) = \mu[\nu_1 Y_l(\mu) - \nu_2 X_l(\mu)],$$

$$\xi(\mu) = \mu[\nu_2 Y_l(\mu) - \nu_1 X_l(\mu)],$$

$$\phi(\mu) = (1 + \nu_4\mu)X_l(\mu) - \nu_3\mu Y_l(\mu),$$

$$\eta(\mu) = (1 - \nu_4\mu)Y_l(\mu) + \nu_3\mu X_l(\mu),$$

$$\chi(\mu) = (1 - u_4\mu)X_r(\mu) + u_3\mu Y_r(\mu) + Q(u_4 - u_3)\mu^2[X_r(\mu) - Y_r(\mu)],$$

$$\sigma(\mu) = (1 + u_4\mu)Y_r(\mu) - u_3\mu X_r(\mu) - Q(u_4 - u_3)\mu^2[X_r(\mu) - Y_r(\mu)],$$

$$\zeta(\mu) = \tfrac{1}{2}\mu[\nu_1 Y_r(\mu) - \nu_2 X_r(\mu)] + \tfrac{1}{2}Q(\nu_2 - \nu_1)\mu^2[X_r(\mu) - Y_r(\mu)],$$

$$\theta(\mu) = \tfrac{1}{2}\mu[\nu_2 Y_r(\mu) - \nu_1 X_r(\mu)] - \tfrac{1}{2}Q(\nu_2 - \nu_1)\mu^2[X_r(\mu) - Y_r(\mu)], \quad (24)$$

where the constants ν_1, ν_2, ν_3, ν_4, u_3, u_4 and Q are to be determined by the following formulae:

$$\nu_2 + \nu_1 = 2\Delta_1(\kappa_1\delta_1 - \kappa_2\delta_2); \quad \nu_2 - \nu_1 = 2\Delta_2(\kappa_1\delta_1 - \kappa_2\delta_2),$$

$$\nu_4 + \nu_3 = \Delta_1(d_1\kappa_1 - d_0\kappa_2); \quad \nu_4 - \nu_3 = \Delta_2[c_1\delta_1 - c_0\delta_2 - 2Q(d_0\delta_1 - d_1\delta_2)],$$

$$u_4 + u_3 = \Delta_1(c_1\delta_1 - c_0\delta_2); \quad u_4 - u_3 = \Delta_2(d_1\kappa_1 - d_0\kappa_2),$$

$$\Delta_1 = (d_0\delta_1 - d_1\delta_2)^{-1}; \quad \Delta_2 = [c_0\kappa_1 - c_1\kappa_2 - 2Q(d_1\kappa_1 - d_0\kappa_2)]^{-1},$$

$$Q = (c_0 - c_2)[(d_0 - d_2)\tau_1 + 2(d_1 - d_3)]^{-1},$$

$$c_0 = A_0 + B_0 - \frac{8}{3}; \quad d_0 = A_0 - B_0 - \frac{8}{3},$$

$$c_n = A_n + B_n; \quad d_n = A_n - B_n; \quad \kappa_n = \alpha_n + \beta_n; \quad \delta_n = \alpha_n - \beta_n \quad (n = 1, 2, 3, \cdots), \quad (25)$$

α_n, β_n, A_n and B_n are the moments of order n of X_l, Y_l, X_r and Y_r, respectively. (It may be recalled here that X_l and Y_l are the standard solutions for the case.)

In the theory of the illumination of the sky we are interested in the transmitted light in the case of incident natural light. In this latter case $F_l = F_r = \frac{1}{2}F$ (where πF denotes the net flux of the incident natural light) and $F_U = 0$. The equations governing the intensity and polarization of the sky as witnessed by an observer at $\tau = \tau_1$ in these circumstances readily follow from the solutions already given; thus by setting $F = \frac{1}{2}(F_l, F_r, 0)$ in equation (12) and combining equations (20), (21), and (23) appropriately, we find:

$$I_l(\tau_1; -\mu, \varphi; \mu_0, \varphi_0) = \tfrac{3}{32}[\{\psi(\mu_0) + \chi(\mu_0)\}\xi(\mu) + 2\{\phi(\mu_0) + \zeta(\mu_0)\}\eta(\mu) - \{\xi(\mu_0) + \sigma(\mu_0)\}\psi(\mu)$$

$$- 2\{\theta(\mu_0) + \eta(\mu_0)\}\phi(\mu) + 4\mu\mu_0(1 - \mu^2)^{\frac{1}{2}}(1 - \mu_0^2)^{\frac{1}{2}}\{X^{(1)}(\mu_0)Y^{(1)}(\mu) - Y^{(1)}(\mu_0)X^{(1)}(\mu)\}\cos(\varphi_0 - \varphi)$$

$$- \mu^2(1 - \mu_0^2)\{X^{(2)}(\mu_0)Y^{(2)}(\mu) - Y^{(2)}(\mu_0)X^{(2)}(\mu)\}\cos 2(\varphi_0 - \varphi)]\frac{F\mu_0}{\mu - \mu_0},$$

$$I_r(\tau_1; -\mu, \varphi; \mu_0, \varphi_0) = \tfrac{3}{32}[\{\psi(\mu_0) + \chi(\mu_0)\}\sigma(\mu) + 2\{\phi(\mu_0) + \zeta(\mu_0)\}\theta(\mu) - \{\xi(\mu_0) + \sigma(\mu_0)\}\chi(\mu)$$

$$- 2\{\theta(\mu_0) + \eta(\mu_0)\}\zeta(\mu) + (1 - \mu_0^2)\{X^{(2)}(\mu_0)Y^{(2)}(\mu) - Y^{(2)}(\mu_0)X^{(2)}(\mu)\}\cos 2(\varphi_0 - \varphi)]\frac{F\mu_0}{\mu - \mu_0}$$

and

$$U(\tau_1; -\mu, \varphi; \mu_0, \varphi_0) = \tfrac{3}{16}[2(1 - \mu^2)^{\frac{1}{2}}(1 - \mu_0^2)^{\frac{1}{2}}\mu_0\{X^{(1)}(\mu_0)Y^{(1)}(\mu) - Y^{(1)}(\mu_0)X^{(1)}(\mu)\}\sin(\varphi_0 - \varphi)$$

$$- \mu(1 - \mu_0^2)\{X^{(2)}(\mu_0)Y^{(2)}(\mu) - Y^{(2)}(\mu_0)X^{(2)}(\mu)\}\sin 2(\varphi_0 - \varphi)]\frac{F\mu_0}{\mu - \mu_0}. \quad (26)$$

3. THE EFFECT OF REFLECTION BY THE GROUND

Before we can apply the solution for S and T given in § 2 to the problem of the illumination of the sky, we must consider the effect of the ground at $\tau = \tau_1$. The solution for S and T given in § 2 was derived on the assumption that at $\tau = \tau_1$ there is no diffuse radiation in the outward direction [*cf.* equation (11)]. The presence of the ground will alter this. However, if the law of reflection by the ground is specified then it is not a difficult matter to relate the solution of the problem when there is a ground to the solution of the problem when there is no ground. This reduction is particularly simple if the ground reflects according to Lambert's law with a certain albedo λ_0; that is, if the light reflected by the ground is unpolarized and uniform in the outward hemisphere independently of the state of polarization and the angular distribution of the incident light, and if, further, the outward flux of the reflected light is always a certain fixed fraction, λ_0, of the inward flux of the radiation incident on the surface. Under these latter circumstances it can be shown (*Radiative Transfer*, § 73) that the effect of the ground is to increase the diffuse intensities emergent at $\tau = 0$ and directed inward at $\tau = \tau_1$ by amounts $I^*(0; \mu, \varphi)$ and $I^*(\tau_1; -\mu, \varphi)$ given by

$$I^*(0; \mu, \varphi) = \frac{\lambda_0 \mu_0}{2(1 - \lambda_0 \bar{s})} \, \mathbf{\Gamma}(\mu; \mu_0) F \qquad (27)$$

and

$$I^*(\tau_1; -\mu, \varphi) = \frac{\lambda_0 \mu_0}{2(1 - \lambda_0 \bar{s})} \, \mathbf{\Lambda}(\mu; \mu_0) F, \qquad (28)$$

where $\bar{s}$ is a constant (to be defined presently) and

$$\mathbf{\Gamma}(\mu; \mu_0) = \begin{bmatrix} \gamma_l(\mu)\gamma_l(\mu_0) & \gamma_l(\mu)\gamma_r(\mu_0) & 0 \\ \gamma_r(\mu)\gamma_l(\mu_0) & \gamma_r(\mu)\gamma_r(\mu_0) & 0 \\ 0 & 0 & 0 \end{bmatrix} \qquad (29)$$

and

$$\mathbf{\Lambda}(\mu; \mu_0) = \begin{bmatrix} \{1 - \gamma_l(\mu)\}\gamma_l(\mu_0) & \{1 - \gamma_l(\mu)\}\gamma_r(\mu_0) & 0 \\ \{1 - \gamma_r(\mu)\}\gamma_l(\mu_0) & \{1 - \gamma_r(\mu)\}\gamma_r(\mu_0) & 0 \\ 0 & 0 & 0 \end{bmatrix} . \qquad (30)$$

In equations (29) and (30) $\gamma_l(\mu)$ and $\gamma_r(\mu)$ are two functions which are related to X_l, Y_l, X_r and Y_r by

$$\gamma_l(\mu) = \tfrac{3}{8} Q(\nu_2 - \nu_1)(d_0 - d_2)[X_l(\mu) + Y_l(\mu)], \qquad (31)$$

and

$$\gamma_r(\mu) = \tfrac{3}{8} Q(d_0 - d_2)$$
$$\times \big[(u_4 - u_3)\{X_r(\mu) + Y_r(\mu)\}$$
$$- u_5 \mu \{X_r(\mu) + Y_r(\mu)\} \big], \qquad (32)$$

where

$$u_5 = \Delta_2(c_0 \kappa_1 - c_1 \kappa_2), \qquad (33)$$

and the remaining constants have the same meanings as in equations (25). Finally, the constant $\bar{s}$ in equations (27) and (28) is given by

$$\bar{s} = 1 - \tfrac{3}{8} Q(d_0 - d_2)$$
$$\times \big[(\nu_2 - \nu_1)\kappa_1 + (u_4 - u_3)c_1 - u_5 d_2 \big]. \qquad (34)$$

Again, when the incident light is natural the corrections which have to be made to the intensities given by equations (26) to allow for a ground surface at $\tau = \tau_1$ which reflects according to Lambert's law with an albedo λ_0 are given by

$$I_l^*(\tau_1; -\mu, \varphi; \mu_0, \varphi_0)$$
$$= \frac{\lambda_0}{4(1 - \lambda_0 \bar{s})} \{\gamma_l(\mu_0) + \gamma_r(\mu_0)\}$$
$$\times \{1 - \gamma_l(\mu)\}\mu_0 F,$$

$$I_r^*(\tau_1; -\mu, \varphi; \mu_0, \varphi_0)$$
$$= \frac{\lambda_0}{4(1 - \lambda_0 \bar{s})} \{\gamma_l(\mu_0) + \gamma_r(\mu_0)\}$$
$$\times \{1 - \gamma_r(\mu)\}\mu_0 F,$$

$$U^*(\tau_1; -\mu, \varphi; \mu_0, \varphi_0) = 0. \qquad (35)$$

4. DESCRIPTION OF THE TABLES

The solution of the fundamental problem in the theory of the illumination of the sky given in the two preceding sections was obtained some six years ago. A detailed examination of its predictions had to await the tabulation of the basic eight functions X_l, Y_l, X_r, Y_r, $X^{(1)}$, $Y^{(1)}$, $X^{(2)}$, and $Y^{(2)}$ of the variable μ for various values of τ_1. This tabulation has now been completed for $\tau_1 = 0.05, 0.10, 0.15, 0.20, 0.25, 0.50$, and 1.00 with the cooperation of the Watson Scientific Computing Laboratory (New York).[6] The solutions were obtained by a direct process of iteration applied to the governing integral equations. The iterations were started with the solutions in the corrected second approximation described in *Radiative Transfer*, Chapter VIII [§ 60, see particularly equations (117), (118), and (120)]. The corrected second

[6] While these calculations were in progress, Dr. Z. Sekera initiated a similar program at the Department of Meteorology of the University of California at Los Angeles in cooperation with the Institute for Numerical Analysis of the National Bureau of Standards at Los Angeles. Their Report No. 3 (prepared by the Air Material Command, Air Force Cambridge Research Center) provides some calculations for $\tau_1 = 0.15, 0.25$, and 1.0. Their calculations, while they are much less extensive than ours, do provide a valuable check.

approximations were computed by one of us (D. D. E.) at the Yerkes Observatory. The iterations were carried out at the Watson Scientific Laboratory with IBM pluggable sequence relay calculators by Miss Ann Franklin to whom and to Dr. Wallace Eckert we are very greatly indebted. The functions obtained after the iterations showed, however, a certain "raggedness" between $\mu = 0.9$ and 1.0. The solutions have therefore been "smoothed" by plotting the deviations of the iterated solutions from the corrected second approximations. Table 1 presents these smoothed solutions. While the solutions have been tabulated to five decimals the last place is definitely not reliable. But it is expected that if the tabulated solutions are rounded to one less place the solution may be trusted to two or three units in the surviving place. The solutions for $\tau_1 \leqslant 0.20$ are very probably more accurate than this while for $\tau_1 = 0.5$ and 1.0 they may be less accurate. Nevertheless, the solutions are given to five places since the functions as tabulated do have smooth differences and as such they can be used for further iterations to improve their accuracy if the need for it should arise.

The moments α_n, β_n, A_n and B_n of order n of X_l, Y_l, X_r and Y_r, respectively, which are needed in the evaluation of the various terms of S and T are given in table 2. The theory of the X- and Y-functions leads to a number of identical relations which must exist between their moments; these relations among the moments listed in table 2 have been verified within the accuracy of the tabulated values. Table 2 also includes the values of the constants v_1, v_2, v_3, v_4, u_3, u_4 and Q which occur in the definitions of the functions ψ, ϕ, etc. [equations (24) and (25)]; the value of the constant δ [equation (34)] which occurs in the expressions for the ground corrections [equations (27) and (28)] is also listed in this table.

According to equations (22) and (23) the calculation of S and T in a given case can be most easily carried out in terms of the auxiliary functions ψ, ϕ, χ, ζ, ξ, η, σ and θ. These functions computed with the aid of tables 1 and 2 are given in table 3. Similarly, table 4 gives the functions $\gamma_l(\mu)$ and $\gamma_r(\mu)$ which are needed to allow for reflection by the ground [cf. equations (31) and (32)].

With the basic functions tabulated we can calculate the theoretical illumination and polarization of the sky on Rayleigh scattering for plane-parallel atmospheres. To illustrate the use of tables 1–4 we have made some model calculations to which the remaining tables are devoted.

The most extensive calculations were made for $\tau_1 = 0.15$; this is approximately the value of the optical thickness at $\lambda 4500$A. For $\tau_1 = 0.15$, angles of incidence (or, zenith distances) $\theta_0 = 90°$, $85.4°$, $76.1°$, $58.7°$, $50.1°$, $43.9°$, $36.9°$, $19.95°$, and $0°$ (corresponding to $\mu_0 = 0$, 0.08, 0.24, 0.52, 0.64, 0.72, 0.80, 0.94, and 1.00) and azimuthal differences $\varphi_0 - \varphi$ (denoted,

simply, by φ in the tables) $= 0°(10°)90°$ were considered. The results of the calculations are summarized in table 5; it gives the total intensity, $I_l + I_r$, in units of F, the inclination, χ, of the plane of polarization with the meridian through the direction of observation and the degree of polarization, δ. Less extensive calculations were made for $\tau_1 = 0.10$ and 0.20. For these two values of the optical thicknesses the total intensity $(I_l + I_r)$ and the degree of polarization, δ, were found only in the principal meridian $(\varphi_0 - \varphi = 0°)$ containing the sun and in the plane at right-angles $(\varphi_0 - \varphi = 90°)$ for zenith distances $\theta_0 = 90°$, $80.8°$, $60°$, $30.7°$, and $0°$ (corresponding to $\mu_0 = 0$, 0.16, 0.50, 0.86, and 1.00). The results of the calculations are given in tables 6 and 7.

In the calculations presented in tables 5–7 no allowance has been made for the reflection by the ground surface. This is taken into account in tables 8–10 in accordance with equations (35). Again, the most extensive calculations were made for $\tau_1 = 0.15$. For angles of incidence corresponding to $\mu_0 = 0$, 0.24, 0.52, 0.64, 0.72, 0.80, and 1.00 the effect of a ground reflecting according to Lambert's law for two values of the albedo, λ_0, on the intensity and the degree of polarization in the principal meridian $(\varphi_0 - \varphi = 0°)$ was determined. For $\mu_0 = 0.64$ the effect on the intensity for $\varphi_0 - \varphi = 0°(10°)90°$ and $\lambda_0 = 0.10$ was determined; and for the same angle of incidence the effect on the intensity for $\varphi_0 - \varphi = 90°$ was also found for $\lambda_0 = 0.25$. The results of all these calculations are given in table 8. For $\tau_1 = 0.10$ and 0.20 the effect of ground reflection on the intensity and polarization in the principal meridian is illustrated for $\lambda_0 = 0.10$ and 0.20 and for angles of incidence corresponding to $\mu_0 = 0$, 0.16, 0.50, 0.86, and 1.00. The results of these calculations are given in tables 9 and 10.

Table 11 gives the positions of the neutral points. The results given in this table will be described and discussed in the following section.

Finally the supplementary table 12 gives the functions ψ, ϕ, χ, ζ, ξ, η, σ, θ, $X^{(1)}$, $Y^{(1)}$, $X^{(2)}$, $Y^{(2)}$, γ_l, and γ_r for $\tau_1 = 0.01(0.01)0.20(0.05)0.50(0.10)1.0$ and for representative values of μ. At the head of the table for each value τ_1, the value of δ (needed for the evaluation of the ground correction) is also given. The values of the functions listed in this table are *not* based on the solutions of the basic X- and Y-functions derived from the integral equations they satisfy; they are based, instead, on the corrected second approximations (cf. *Radiative Transfer*, § 60) for the relevant X- and Y-functions. However, for $\tau_1 \leqslant 0.25$ the table should suffice to calculate S and T to well within a fraction of a per cent; for the larger values of τ_1 accuracy within a few per cent may be expected. But one could, if one wished, obtain from these tables values of considerably higher precision by differencing the values of the functions for $\tau_1 = 0.05$, 0.10, 0.15,

0.20, 0.25, 0.50, and 1.0 given in tables 3 and 12 and interpolating among these differences to estimate the corrections (to the values given by the corrected second approximation) for any other intermediate value of τ_1. With these supplementary tables, then, the theory developed and described in *Radiative Transfer* has been, finally, brought to a point where it is capable of giving numerical values for any of the desired quantities under most conditions in which they are likely to be of interest.

5. THE POLARIZATION OF THE SUNLIT SKY: THE THEORY OF THE NEUTRAL POINTS AND LINES

As we have already stated in the introductory section we shall not attempt in this paper any detailed comparison between the calculations presented here

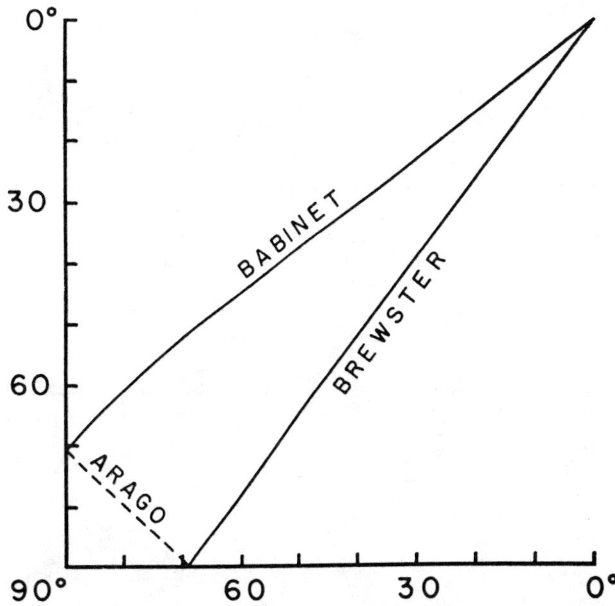

FIG. 2. Calculated positions of the neutral points for various angles of incidence for an atmosphere of optical thickness $\tau_1 = 0.15$. The abscissa and the ordinate have otherwise the same meanings as in fig. 1.

such neutral points. For angles of incidence not exceeding 70° these neutral points occur between 0° and 20° above and below the sun; these are the neutral points of Babinet and Brewster, respectively. But when the sun is low, the neutral point occurs about 20° above the anti-solar point in the opposite sky: this is the Arago point. These facts concerning the

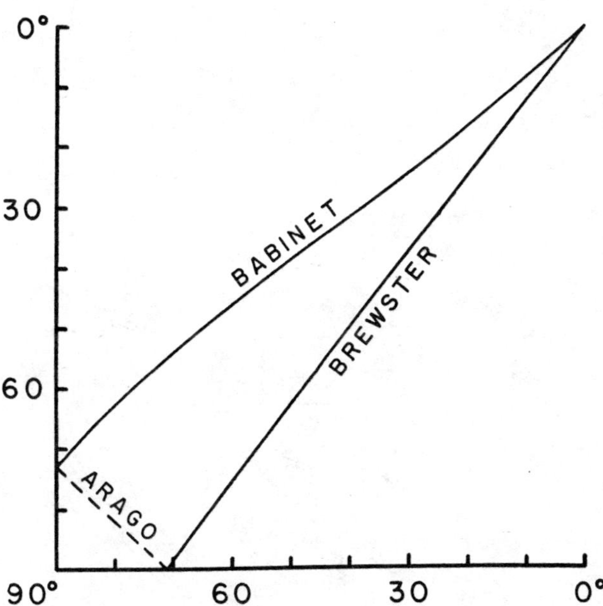

FIG. 1. Calculated positions of the neutral points for various angles of incidence for an atmosphere of optical thickness $\tau_1 = 0.10$. The abscissa gives the zenith distance of the sun and the ordinate gives the corresponding positions of the neutral points. The Arago point occurs on the side of the horizon opposite the sun; to emphasize this its position on the sky is indicated by the dashed curve.

and those of earlier investigators based on approximations of various kinds. But an exception might be made with regard to the quantitative explanation which the present theory affords for the phenomena associated with the *neutral points* of Arago, Babinet, and Brewster. The phenomena in question are these:

The neutral points are the points of zero polarization; from symmetry we should, of course, expect them to occur in the principal meridian. And for a long time it has been known that there are in general two

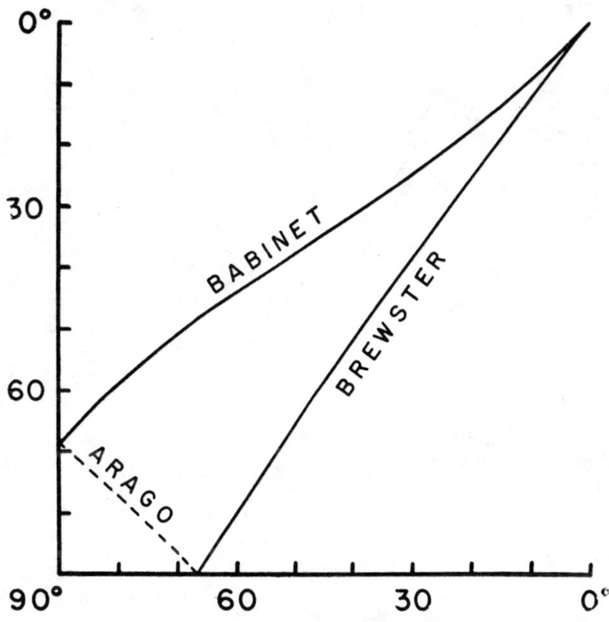

FIG. 3. Calculated positions of the neutral points for various angles of incidence for an atmosphere of optical thickness $\tau_1 = 0.20$. The abscissa and the ordinate have otherwise the same meanings as in fig. 1.

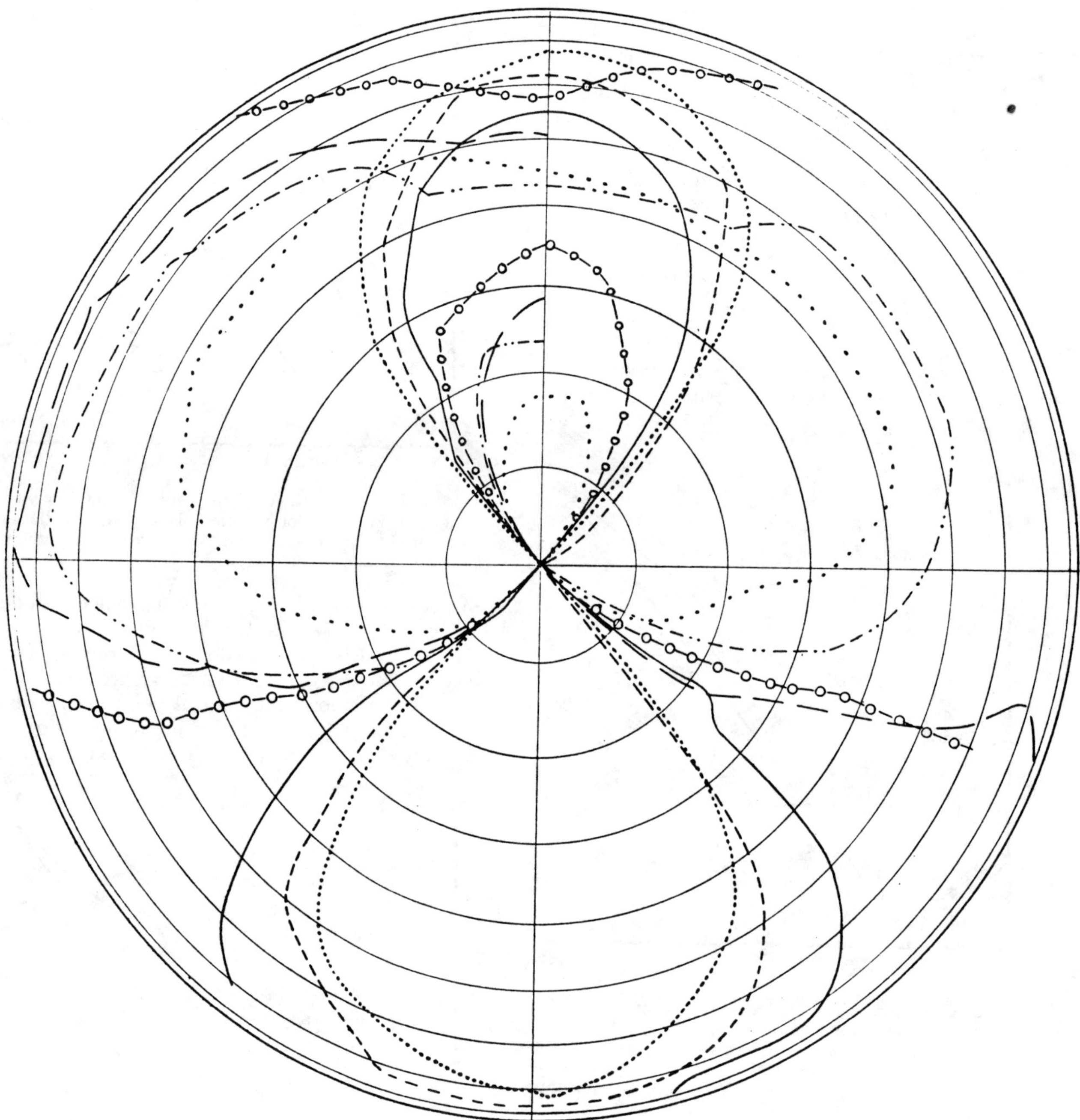

Fig. 4. Dorno's observations of the neutral lines on May 17, 1917, at Davos. The various curves were determined during the following times when the zenith distance of the sun varied by the amounts given:

....	7ᵖ 19– 7ᵖ 28	$\theta_0 = 87°\ 10'–\ 0°$
- - - - - -	6ᵖ 13– 6ᵖ 33	$\theta_0 = 78°\ 14'–81°\ 30'$
————	5ᵖ 14– 5ᵖ 34	$\theta_0 = 68°\ 20'–71°\ 44'$
–o–o–o–o–	3ᵖ 15– 3ᵖ 42	$\theta_0 = 48°\ 12'–52°\ 40'$
– – – –	9ᵃ 9– 9ᵃ 34	$\theta_0 = 44°\ 21'–40°\ 33'$
– · – · – ··	9ᵃ 44–10ᵃ 11	$\theta_0 = 39°\ 5'–35°\ 26'$
· · · · · · · ·	12ᵖ 56– 1ᵖ 14	$\theta_0 = 29°\ 50'–31°\ 26'$

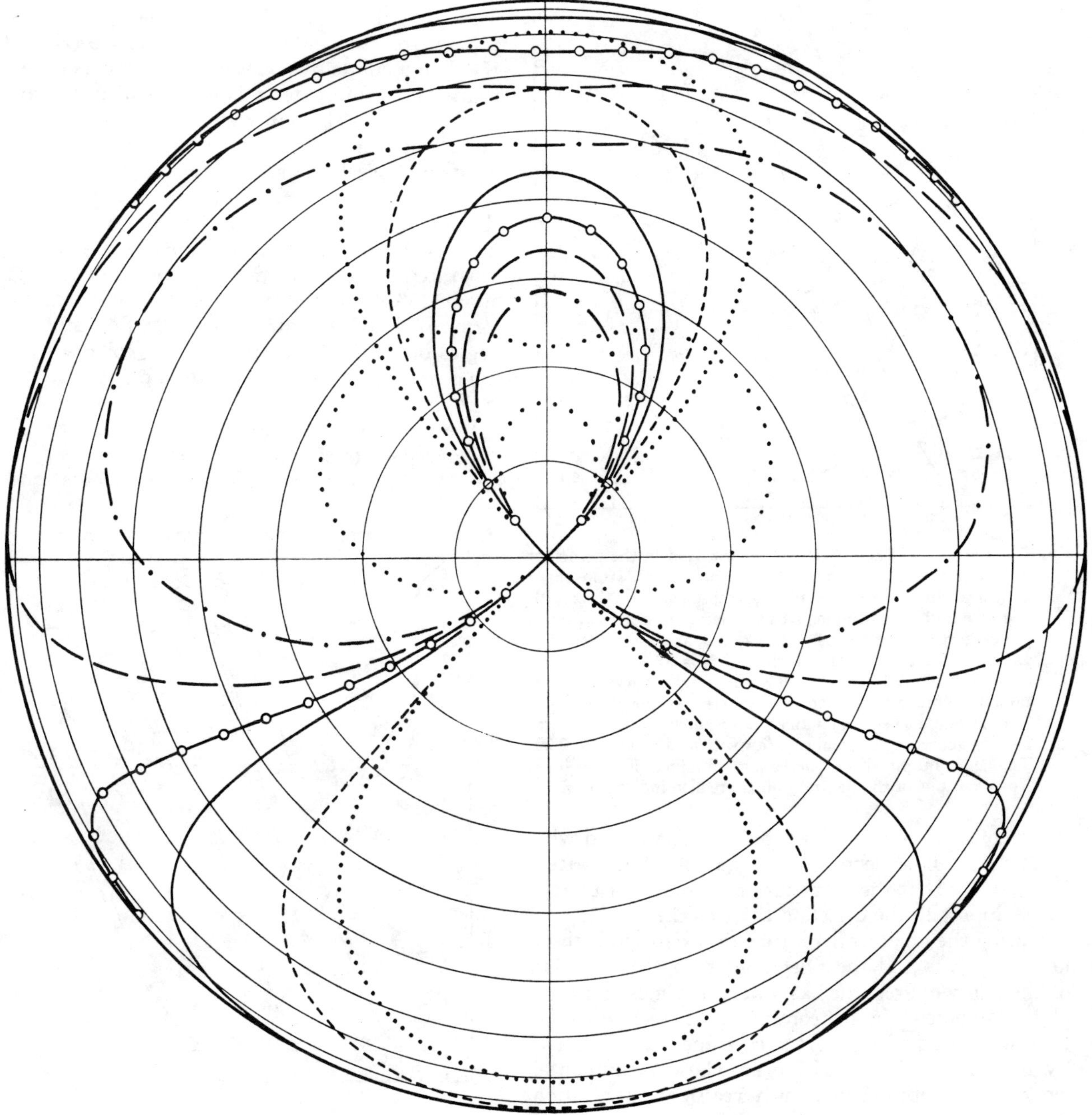

Fig. 5.   The neutral lines as predicted by the theory.   The various curves refer to the following zenith distances of the sun:

The curves in this figure which roughly correspond to Dorno's observations are marked similarly.

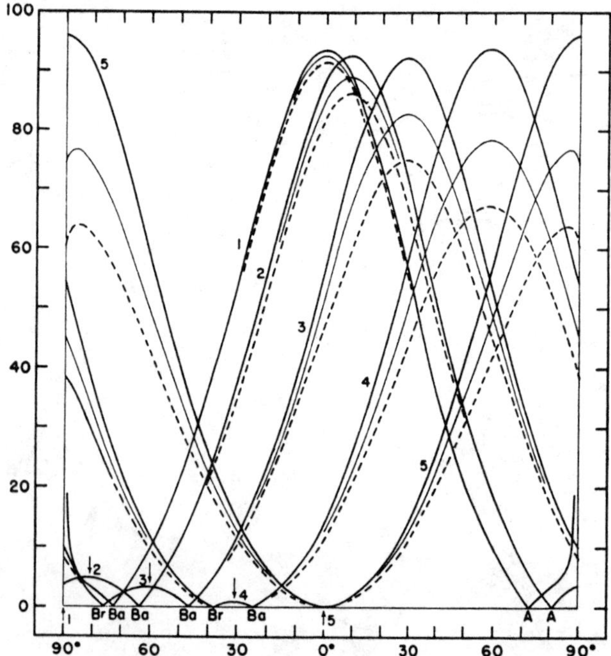

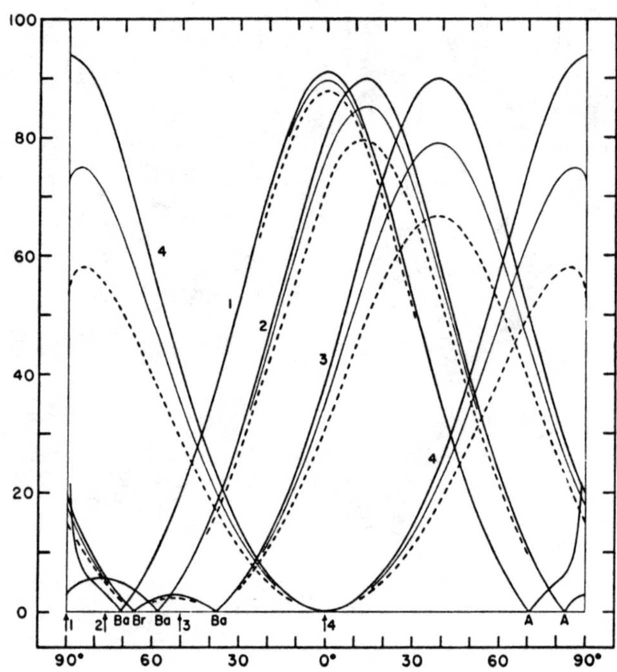

FIG. 6. Variation of the degree of polarization in the principal meridian for various angles of incidence for an atmosphere of optical thickness $\tau_1 = 0.10$. The abscissa gives the zenith distance and the ordinate gives the degree of polarization in per cent. The curves marked 1, 2, 3, 4, and 5 represent the variation for the angles of incidence $\theta_0 = 90°$, $80.8°$, $60.0°$, $30.7°$ and $0°$, respectively. The thick solid curves are obtained before any ground corrections have been applied. The dashed curves are obtained if we allow for a ground reflecting according to Lambert's law with an albedo $\lambda_0 = 0.20$. The thin intermediate curves are obtained if $\lambda_0 = 0.10$. The positions of the neutral points are also indicated.

points of zero polarization should be contrasted with what should be expected on the laws of single scattering, namely that the polarization should tend to zero as we approach the direction towards the sun.

During the nineteenth century the existence of these neutral points and their behavior with the direction of the sun were regarded as among the most remarkable phenomena in meteorological optics. As such they were studied with great care and attention and by none more than Carl Dorno whose monumental work on the subject[7] contains a wealth of information painstakingly gathered. Dorno not only observed the neutral points on the principal meridian, but he also investigated in detail the continuation of these neutral points over the entire hemisphere along what he called the neutral lines. These lines separate the regions of positive from the regions of negative polarization;[8] they show a remarkable dependence on the

---

[7] C. Dorno, Himmelshelligkeit, Himmelspolarisation und Sonnenintensität in Davos 1911 bis 1918, *Veröffentl. Preuss. Met. Inst.*, No. 303, Berlin, 1919.

[8] The polarization is assumed positive if I_l is less than I_r and negative if the reverse is true. On this convention the polarization is negative on the principal meridian between the neutral points of Babinet and Brewster; these points, therefore, separate

direction of the sun. A diagram representing Dorno's principal results is reproduced in figure 4. It will be seen from this figure that when the sun is nearly on the horizon the neutral line connects the Babinet and the Arago points by a closed symmetrical curve of the shape of a lemniscate: this is the so-called lemniscate of Busch. As the sun rises the lemniscate becomes more and more asymmetrical and when the angle of incidence exceeds about 70° the lemniscate opens out and a part of the locus appears on the horizon below the sun and passes through the Brewster point which has now risen. The neutral line consists of two such separated curves until the angle of incidence becomes about 45° when the opposite ends join together to form a closed re-entrant curve. For still smaller angles of incidence the neutral line collapses towards the center and finally reduces to a point when the sun is at the zenith.

We shall now see how this entire range of phenomena associated with the neutral points and lines are faithfully reproduced by our calculations. In

FIG. 7. Variation of the degree of polarization in the principal meridian for various angles of incidence for an atmosphere of optical thickness $\tau_1 = 0.15$. The abscissa gives the zenith distance and the ordinate gives the degree of polarization in per cent. The curves marked 1, 2, 3 and 4 represent the variation for the angles of incidence $\theta_0 = 90°$, $76.1°$, $50.2°$ and $0°$, respectively. The thick solid curves are obtained before any ground corrections have been applied. The dashed curves are obtained if we allow for a ground reflecting according to Lambert's law with an albedo $\lambda_0 = 20.5$. The thin intermediate curves are obtained if $\lambda_0 = 0.10$. The positions of the neutral points are also indicated.

the regions of positive from the regions of negative polarization on this meridian. The neutral lines of Dorno do the same for the entire hemisphere.

table 11 we have collected all the information contained in tables 5–10 regarding the points at which the polarization changes sign for various zenith distances of the sun; they are further illustrated in figures 1, 2, 3, and 5. Considering first figures 1–3 (which refer to the principal meridian), we observe that the calculations predict the occurrence of the neutral points as observed. In particular it will be noticed that the Brewster point sets when the angle of incidence is about 70°; its dependence on the values of the optical thickness in the range of interest is not pronounced. Also as the Brewster point sets the Arago point rises in the opposite sky. And as the sun sinks lower, the Arago point continues to rise until, when the sun sets, the Babinet and the Arago points are both at an equal

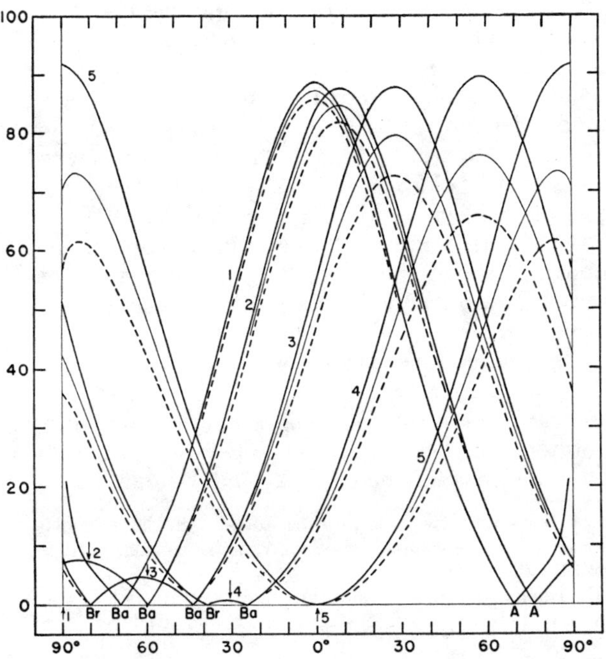

Fig. 8. Variation of the degree of polarization in the principal meridian for various angles of incidence for an atmosphere of optical thickness $\tau_1 = 0.20$. The abscissa gives the zenith distance and the ordinate gives the degree of polarization in per cent. The curves marked 1, 2, 3, 4, and 5 represent the variation for the angles of incidence $\theta_0 = 90°$, $80.8°$, $60.0°$, $30.7°$, and $0°$, respectively. The thick solid curves are obtained before any ground corrections have been applied. The dashed curves are obtained if we allow for a ground reflecting according to Lambert's law with an albedo $\lambda_0 = 0.20$. The thin intermediate curves are obtained if $\lambda_0 = 0.10$. The positions of the neutral points are also indicated.

elevation of about 20° (it varies from 17° to 21° for τ_1 in the range $0.10 \leqslant \tau_1 \leqslant 0.20$) from the horizon.

With the calculations for the different values of $\varphi_0 - \varphi$ for $\tau_1 = 0.15$ given in table 5, we can draw an entire system of calculated neutral lines. This has been done in figure 5. Comparing it with the results of Dorno's observations (fig. 4) we observe how well the two sets of curves match.

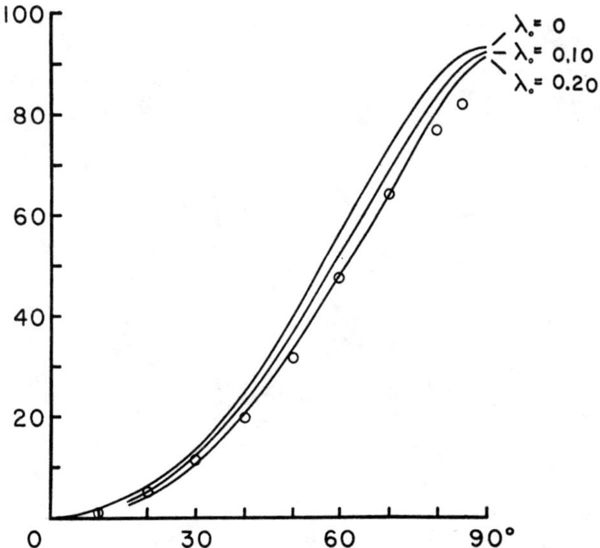

Fig. 9. Variation of the degree of polarization at the zenith for various angles of incidence and for an atmosphere of optical thickness $\tau_1 = 0.10$. The curve $\lambda_0 = 0$ allows for no ground reflection. The other two curves allow for reflection by a ground with the albedos indicated. The circles represent the observations of Tousey and Hulburt for values of τ_1 and λ_0 estimated at 0.10 and 0.20, respectively.

Turning next to the calculated degrees of polarization, we have illustrated its variation on the principal meridian for various angles of incidence and for the three values of the optical thickness for which calculations have been made in figures 6, 7, and 8. It will be noticed from these figures that the effect of a ground surface with an albedo even as high as 0.25 does not

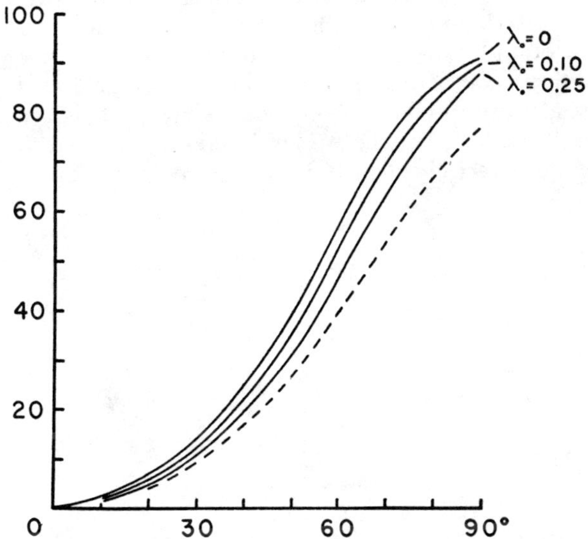

Fig. 10. Variation of the degree of polarization at the zenith for various angles of incidence and for an atmosphere of optical thickness $\tau_1 = 0.15$. The curve $\lambda_0 = 0$ allows for no ground reflection. The other two curves allow for reflection by a ground with the albedos indicated. The dashed curve represents the observations of Tichanowsky.

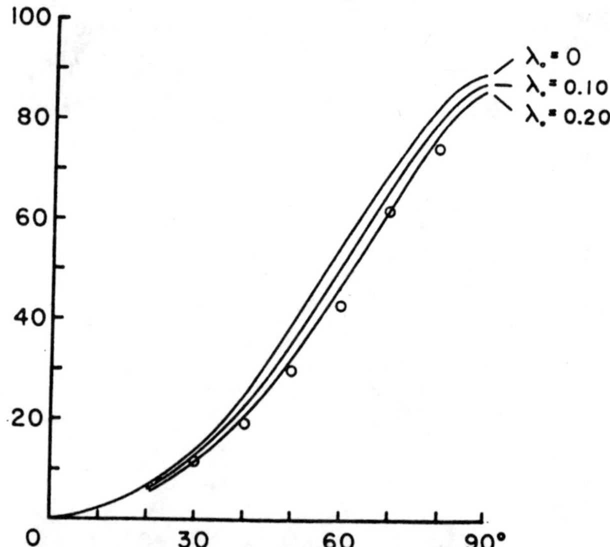

FIG. 11. Variation of the degree of polarization at the zenith for various angles of incidence and for an atmosphere of optical thickness $\tau_1 = 0.20$. The curve $\lambda_0 = 0$ allows for no ground reflection. The other two curves allow for reflection by a ground with the albedos indicated. The circles represent the observations of Richardson and Hulburt for a value of τ_1 estimated at 0.3.

make any essential difference to the predicted positions of the neutral points. This independence of the neutral points (and lines) on ground reflection is not difficult to understand. As is well known, the laws of Rayleigh scattering give the maximum polarization for the scattered light; all other laws give much less polarization. A ground reflecting according to Lambert's· law can, therefore, hardly compete with Rayleigh scattering for producing polarization. The effect of reflection by the ground is therefore essentially one of adding a component of natural light to the polarized light already present. This last statement is, of course, not strictly true. The difference between γ_l and γ_r is precisely a measure of the polarization of the ground contribution to the sky brightness; but as is apparent from the tables the difference between γ_l and γ_r is generally very small. For these same reasons we should not expect that the direction of maximum polarization will be influenced by ground reflection. This is, indeed, the case: the direction of maximum polarization always occurs in a direction which is very nearly at right-angles to the direction of the sun. In contrast, ground reflection has a very pronounced effect on the *degree* of maximum polarization: In the absence of ground corrections, the maximum polarization varies between 94 and 89 per cent depending on τ_1 but is very nearly independent of the altitude of the sun (though there is a slight increase for higher altitudes). However, when the effect of ground reflection is taken into account the maximum polarization shows a very marked decrease with altitude. Observations do show such a behavior and we conclude that this is probably due to the effect of ground reflection.

Finally, in figures 9, 10, and 11 we have compared the degree of polarization at the zenith for various angles of incidence with the observations of Tichanowsky,[9] Tousey and Hulburt,[10] and Richardson and Hulburt.[11] It will be seen that the agreement between the theory and the observations is as good as one might expect.

In concluding this paper we should again like to record our thanks to Dr. Wallace Eckert and Miss Ann Franklin of the Watson Scientific Computing Laboratory for their generous co-operation in obtaining the solutions for the basic X- and Y-functions.

[9] J. J. Tichanowsky, Resultate der Messungen der Himmelspolarisation in verschiedenen Spektrumabschnitten, *Meteor. Ztschr.* **43**: 288, 1926.

[10] R. Tousey and E. O. Hulburt, Brightness and polarization of the daylight sky at various altitudes above sea level, *Jour. Opt. Soc. Amer.* **37**: 78, 1947.

[11] R. A. Richardson and E. O. Hulburt, Sky-brightness measurements near Bocaiuva, Brazil, *Jour. Geophys. Research* **54**: 215, 1949.

Brightness and Polarization of the Daylight Sky at Various Altitudes above Sea Level

R. TOUSEY AND E. O. HULBURT

Naval Research Laboratory, Washington, D.C.

INTRODUCTION

IN the following pages is described an experimental and theoretical investigation of the light of the daylight sky. The brightness and polarization of the sky at various points around the compass from the horizon to the zenith for various altitudes of the sun were measured during eleven flights in an airplane from sea level to 10,000 feet and one flight to 20,000 feet. The weather was selected, not always with success, to be clear and free from clouds. The earth reflectivity and the solar illumination were also measured.

A quantitative theory of the sky brightness and polarization was developed on the assumption that the atmosphere was composed of pure air molecules with scattering according to Rayleigh. For such an atmosphere the attenuation β of sunlight viewed with the light adapted eye was 0.0126 km^{-1}. The "U. S. standard atmosphere," the reflectivity of the earth, the polarization defect of air, the absorption of atmospheric ozone, and the solar illumination entered into the theory.

It was found that the sky data at 10,000 feet for the five flights when the air was the clearest agreed well with each other and disagreed consistently with the theoretical calculations for pure air for which $\beta = 0.0126$ km^{-1}. They agreed with the theory within experimental error when β was increased to 0.017 km^{-1}. This indicated that in these five clearest cases the air above the observer was not pure air, but was air which contained impurities in the form of scattering particles small with respect to the wave-length of visible light sufficient to increase the theoretical scattering of pure air by about 35 percent. This was again found to be the case in the one flight to 20,000 feet. The data of the other seven flights indicated less clear air than in the above five flights and required β's from about 0.018 to 0.021 km^{-1} in the theory in order to obtain agreement.

OBSERVATIONS OF SKY BRIGHTNESS AND POLARIZATION

The sky brightness and polarization were measured with a Macbeth illuminometer modified slightly for the purpose. The field of view was a cone of about 4° angular diameter. To obtain a color match with the sky a blue filter was placed over the comparison lamp of the illuminometer; the match was good, but rarely perfect. The sky illuminometer was in a mount clamped to the airplane with motions in altitude and azimuth. Over the front of the instrument was placed a polarizing plate and a tube with diaphragms to reduce stray light. Measurements could be made to within about 8° of the sun, but at smaller angles direct sunshine in the instrument caused errors.

The sky illuminometer was calibrated by standard procedure to read the brightness of a source of the spectral distribution of sky light in candles per square foot. The calibration was obtained by a comparison of the instrument without the blue filter against a lamp standardized by the National Bureau of Standards, together with the photometric transmission of the blue filter determined from the I.C.I. spectral luminosity curve, and the measured spectral curves of the transmission of the blue filter and the intensity of the illuminometer lamp.

In order to measure the solar illumination the direct sunlight was allowed to pass through an open port in the plane, thereby screening off nearly all of the sky light; it illuminated a diffuse white plaque held perpendicular to the rays of the sun. The brightness of the white plaque was measured with another calibrated illuminometer equipped with a yellow filter over the external field to give a color match.

All observations were made over a region within about 200 miles of Washington, D. C., during twelve airplane flights usually to altitudes of about 10,000 feet; in Flight 9 an altitude of 20,000 feet was reached. All observations were made through an open port in the plane. The

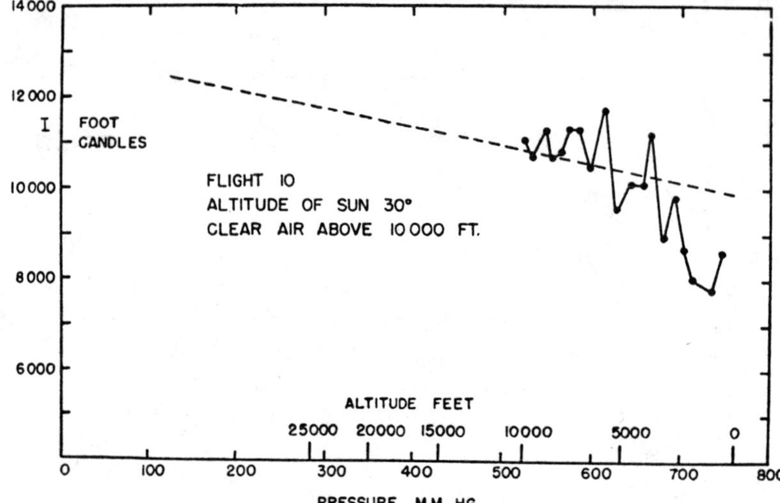

Fig. 1. Solar illumination for "clear" air above 10,000 feet.

usual program of sky observations included measurements of the zenith sky as the plane climbed from the surface to 10,000 feet, and measurements at various points in the sky at 10,000 feet. On several flights solar illumination measurements were made and special programs were carried out. The programs were modified, and often partially interrupted, by weather, clouds, and conditions of flight.

Viewed from the surface in cloudless weather with a normally clear atmosphere the region of the sky about 90° from the sun and not too near the horizon is the darkest, bluest, and most polarized region of the sky. As the angular distance from the sun decreases, the sky increases gradually in brightness and becomes whiter, and within 10 or 15 degrees of the sun the brightness and whiteness are usually very noticeable. The region near the sun is described as an "aureole" and is caused by haze of foreign particles of relatively large size. The angular extent and the brilliancy of the aureole vary with the amount and type of haze. A familiar manifestation of aureole is the glow surrounding lights on a foggy night.

At the surface around Washington, and probably in most localities, a condition of no aureole rarely occurs, indicating that the atmosphere at sea level is hardly ever free of haze. The whiteness of the sky near the horizon is usually caused by large particle haze. The smaller the size of the haze particle the greater the angular extent of the aureole, the less its whiteness and the less

conspicuous it becomes. For particles small with respect to the wave-length of visible light an aureole does not appear since approximately the same type of scattering occurs as for air molecules. Such a haze can be revealed by a brightness measurement; it causes an increase in brightness but no great change in polarization, angular distribution, and blueness. Evidence of this type of haze was found at the higher altitudes of the present experiments. In brief, although the presence of an aureole is conclusive proof of haze, the fact that no aureole can be readily seen does not mean that the atmosphere consists only of air molecules.

As one ascends through a normally clear atmosphere the aureole becomes less and at a sufficient altitude may become imperceptible. It was noticed that often, but not always, one passed through most of the haze and could see the top of the haze appearing as a dark line around the horizon with relatively blue sky above. The dark line was termed the "haze level"; near this level the aureole usually, but not always, disappeared. In most of the present sky measurements the haze level and the aureole were recorded in order to give some indication of the amount of haze prevailing above the observer.

Representative observational data of four flights are plotted in Figs. 1 to 10; the data of all the flights were too numerous to publish. Since each series of measurements extended over a time interval during which the altitude of the

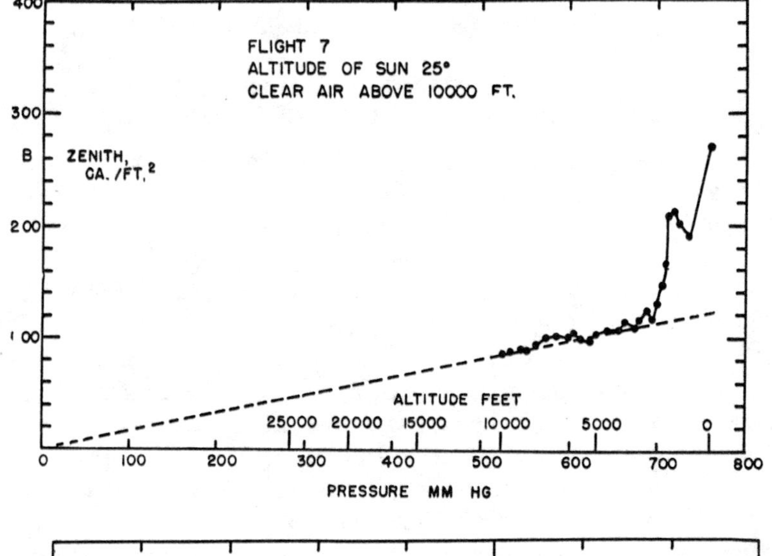

FIG. 2. Zenith sky brightness for "clear" air above 10,000 feet.

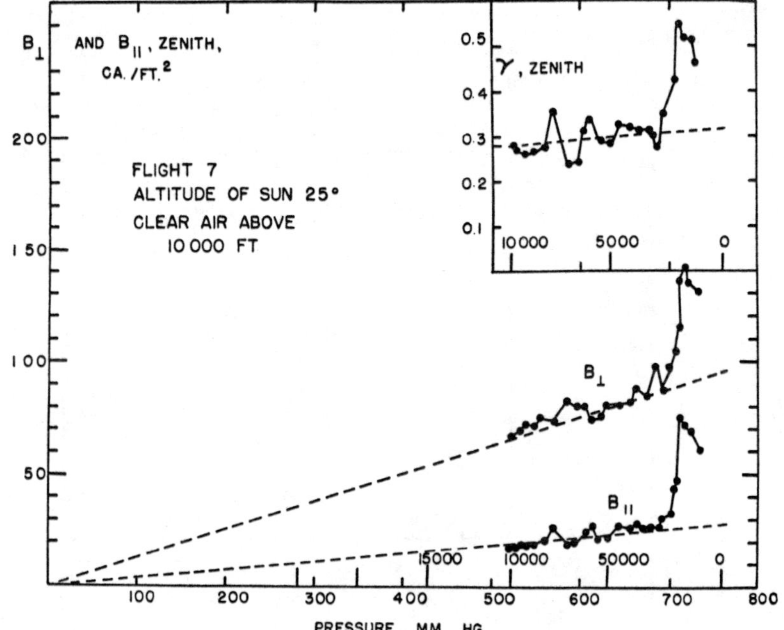

FIG. 3. Polarized components and polarization of zenith sky brightness for "clear" air above 10,000 feet.

sun changed slightly, they were corrected to refer to a single altitude by means of the slopes of the experimental or theoretical curves. The corrections were small, being usually less than 2 percent and always less than 5 percent. With these corrections the data were plotted exactly as they were taken and show evidence of experimental vicissitude due mainly to motion of the plane and lack of opportunity to repeat readings. In the figures the dots and heavy lines are the observations and the dotted curves are theoretical from Table I.

The following notation is used:

ζ zenith angle of the sun, the altitude of the sun is $90° - \zeta$;

ξ zenith angle of the point in the sky under observation, the altitude of the point above the horizon is $90° - \xi$;

Z the bearing of the point in the sky relative to the sun;

B candles per square foot, brightness of the point in the sky, $B = B_{\|} + B_{\perp}$;

$B_{\|}$ candles per square foot, the polarized component of B with electric vector in the plane passing through the sun, observer, and point in the sky;

$B_{\perp}$ candles per square foot, the polarized component of B with electric vector perpendicular to the plane;

$\gamma = B_{\|}/B_{\perp}$; the polarization γ ranged from 0 for complete polarization to 1 for no polarization, and hence determined how much the sky may be darkened by a polarizing plate;

I footcandles, the solar illumination at the observer.

The data of Figs. 1 to 10 were selected to illustrate the cases of "clear" and "less clear" air above 10,000 feet; the terms are defined in the last section. "Clear" refers to skies of brightness nearest to Rayleigh theory, while "less clear" applies to skies with obvious haze. It is seen from Figs. 1 to 5, which refer to "clear" air, and from Figs. 6 to 10, which refer to "less clear" air, that I is greater and B is less for clear air than for less clear air, and γ is about the same for the two cases. The conditions of the four flights of Figs. 1 to 10 were: Flight 6, November 4, 1315 to 1430, scattered cirrus clouds, surface horizontal visibility greater than 10 miles, haze level indistinct 4000 to 7000 feet, aureole at 10,000 feet: Flight 7, November 18, 1330 to 1445, cloudless, surface horizontal visibility 6 miles, haze level 3000 feet, no aureole at 10,000 feet; Flight 10, January 16, 1140 to 1300, cloudless, surface horizontal visibility greater than 10 miles, haze level 6500 feet, faint aureole at 10,000 feet; Flight 12, February 14, 1130 to 1300, cloudless, northwest wind after four days of dry northwest winds, surface horizontal visibility 8 miles, haze level about 6000 feet but not sharply defined, wide aureole at 10,000 feet.

THEORY OF SKY BRIGHTNESS AND POLARIZATION

Several theoretical treatments of sky brightness and polarization have been developed, as, for example, those of Tichanowski,[1] Pokrowski,[1] and King.[2] The developments were not adequate for application to the present measurements for they did not consider completely the effects of haze, ozone, the reflectivity of the earth, and the imperfection of polarization of the air molecules, and in no case were the theories carried through to the point of calculating the brightness and polarization of various points in the sky at various altitudes above sea level. Therefore, in the following paragraphs are presented theoretical formulas which, in some respects, are more complete than those hitherto available.

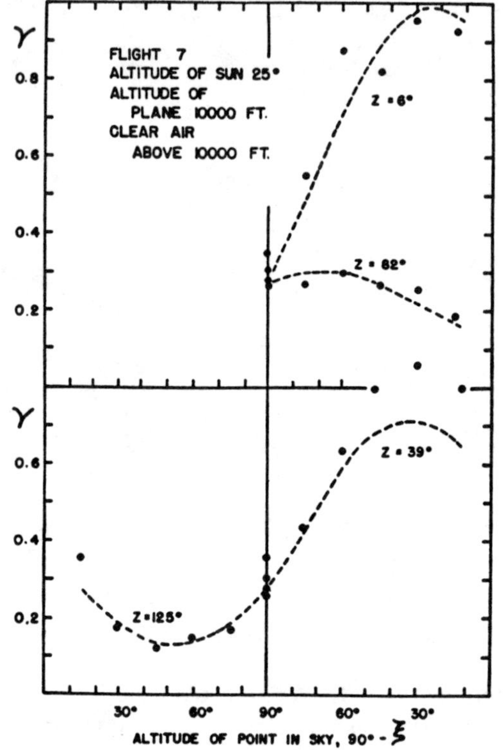

FIG. 5. Sky polarization at 10,000 feet for "clear" air.

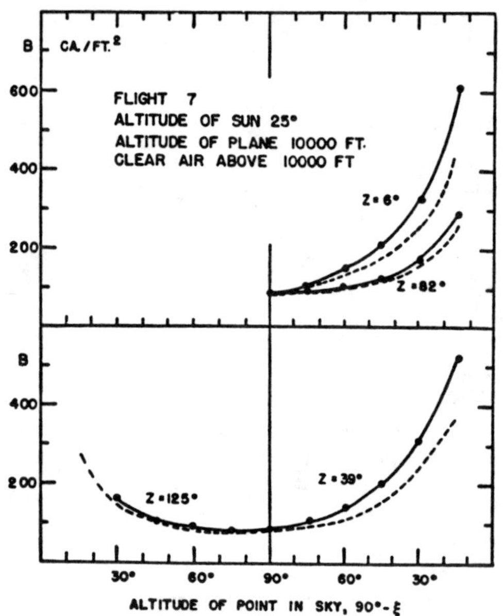

FIG. 4. Sky brightness at 10,000 feet for "clear" air.

[1] See H. Geiger and K. Scheel, *Handbuch der Physik* (1928), Vol. 19, Chapter 4.

[2] L. V. King, Phil. Trans. 212, 375–433 (1913).

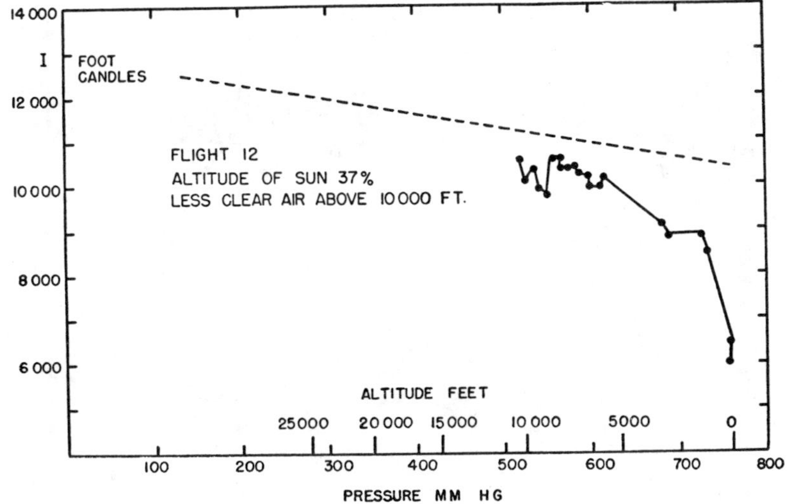

FIG. 6. Solar illumination for "less clear" air above 10,000 feet.

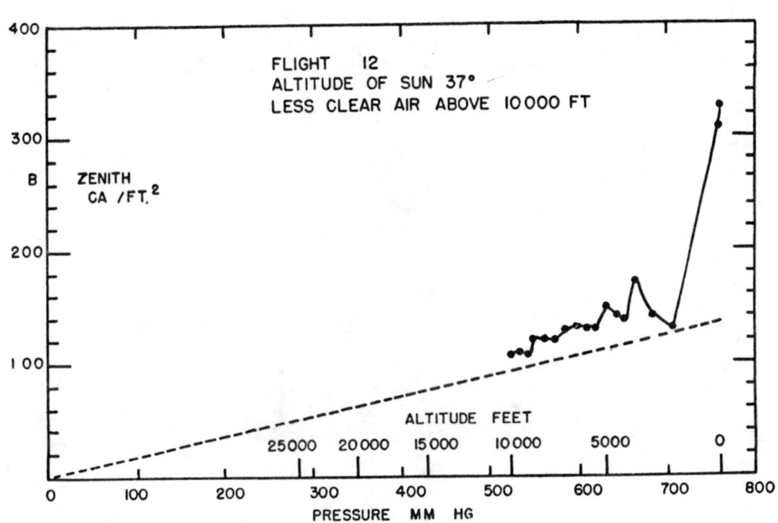

FIG. 7. Zenith sky brightness for "less clear" air above 10,000 feet.

Let B be the brightness of a point in the sky as it appears to an observer. It is assumed that

$$B = B_p + B_m. \tag{1}$$

B_p is the brightness arising from the direct rays of the sun scattered to the observer by all the air particles along his line of sight; B_p is *primary* scattered sunlight. B_m is the brightness arising from all other rays scattered to the observer by the air particles along his line of sight. Since these rays come from the surrounding atmosphere and from the surface of the earth which owe their illumination to sunlight, B_m is *multiply* scattered sunlight. We proceed to evaluate B_p and B_m.

Scattering, Polarization, and Attenuation Formulas

The optical scattering coefficient $\sigma_{\phi\lambda}$ is defined by

$$i_{\phi\lambda} = \sigma_{\phi\lambda} i_\lambda, \tag{2}$$

where i_λ is the flux per unit area of a beam of unpolarized light included in the wave-length interval λ to $\lambda + d\lambda$ which traverses unit volume of the scattering material, and $i_{\phi\lambda}$ is the flux per unit solid angle per unit volume scattered at angle ϕ to the beam. For a gas $\sigma_{\phi\lambda}$ may be derived from the Rayleigh[3] theory of the scat-

[3] Lord Rayleigh, Phil. Mag. **41**, 107 (1871); *ibid.* **47**, 375–384 (1899); Proc. Roy. Soc. **84A**, 25–46 (1911); *ibid.* **90A**, 219–225 (1914).

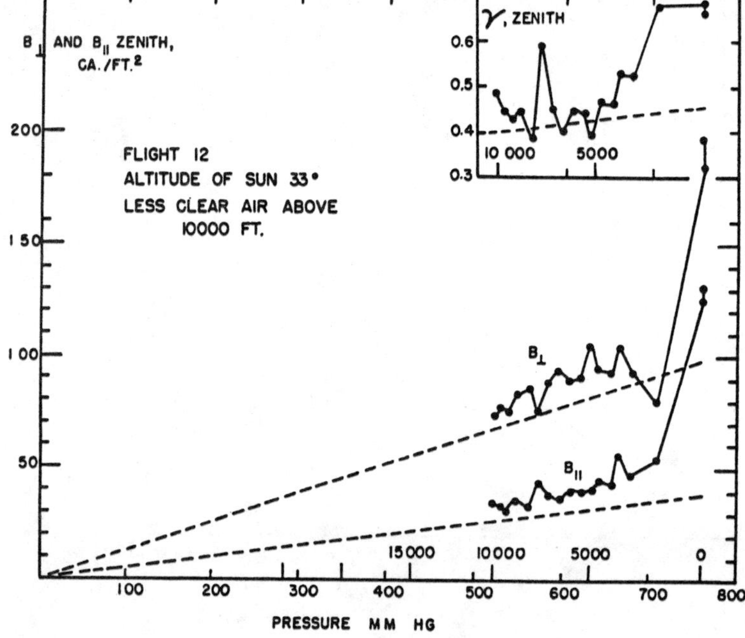

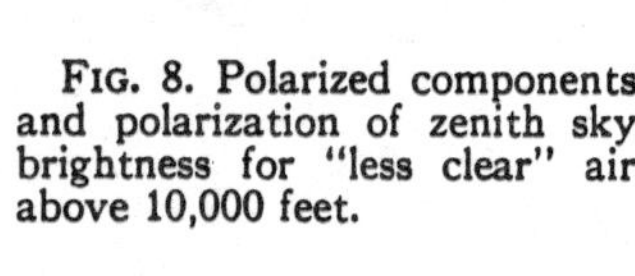

Fig. 8. Polarized components and polarization of zenith sky brightness for "less clear" air above 10,000 feet.

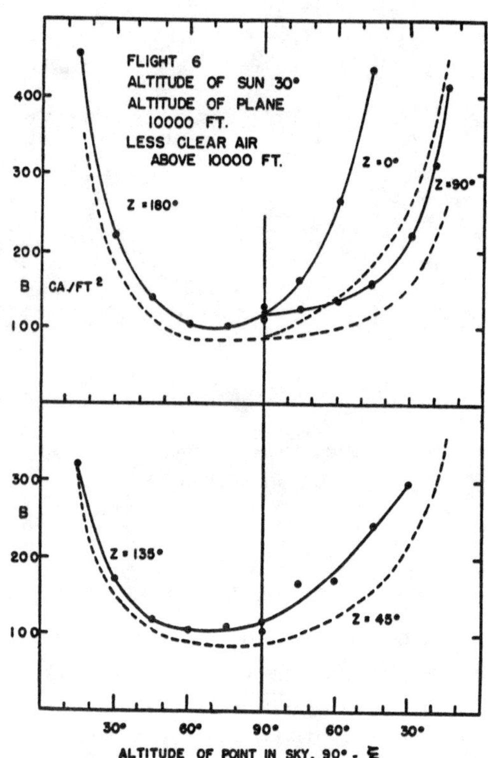

Fig. 9. Sky brightness at 10,000 feet for "less clear" air.

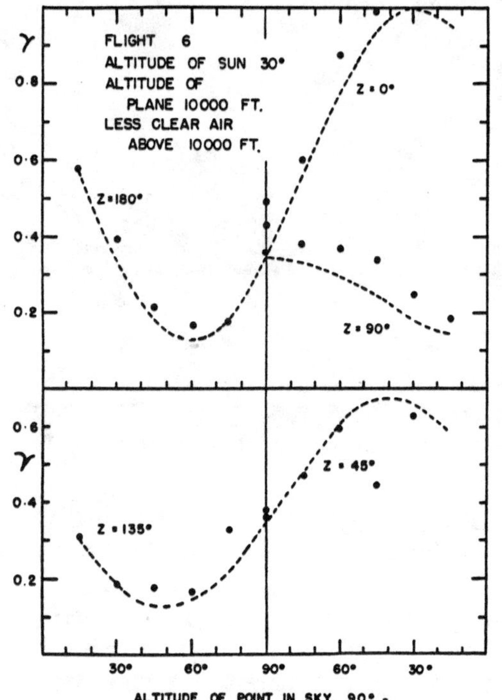

Fig. 10. Sky polarization at 10,000 feet for "less clear" air.

tering of light by particles. For a gas one finds

$$\sigma_{\phi\lambda} = \frac{2\pi^2}{n\lambda^4}(\mu_\lambda - 1)^2(1 + \cos^2\phi),\qquad (3)$$

where μ_λ and n are, respectively, the refractive index and the molecular density of the gas.

From the scattering theory it follows that light scattered at $\phi = \pi/2$ is completely polarized.

However, observations[4] have shown that for pure air the light scattered at $\phi=\pi/2$ was not completely polarized, the ratio δ of the weak to the strong polarized component being about 0.04 and approximately constant with λ. δ was termed the "polarization defect." When this term was introduced into the scattering theory (3) became[5]

$$\sigma_{\phi\lambda}=\frac{2\pi^2}{n\lambda^4}(\mu_\lambda-1)^2\frac{6(1+\delta)}{6-7\delta}\left(1+\frac{1-\delta}{1+\delta}\cos^2\phi\right). \quad (4)$$

In the present experiments the polarization γ has been measured. Referring to (1) B_m is assumed to be unpolarized and B_p to be polarized according to the linear oscillator theory underlying (3) and (4). Then it may be shown[5] that

$$\gamma=\left[D+\frac{B_m}{2B_P}(1+D)\right]\Big/\left[1+\frac{B_m}{2B_p}(1+D)\right], \quad (5)$$

where

$$D=\delta+(1-\delta)\cos^2\phi. \quad (6)$$

Let $B_\perp$ and $B_{||}$ be the two polarized components of B with electric vectors perpendicular and parallel, respectively, to the plane containing the observer, the sun, and the point in the sky under observation. Then

$$B=B_\perp+B_{||}, \quad (7)$$

and

$$B_\perp=\frac{1}{1+\gamma}B, \quad (8)$$

$$B_{||}=\frac{\gamma}{1+\gamma}B. \quad (9)$$

Ozone exists in the atmosphere mainly in levels above 20 km, and, because of the Chappuis band, it absorbs sunlight in the visible portion of the solar spectrum. The atmosphere above the ozone amounts to less than one percent of the total atmosphere, and, therefore, the simplifying assumption is made, valid to a close approximation, that all the ozone is above the "top" of the atmosphere. As a result of the assumption we calculate the degradation of the sunlight in its passage through the ozone region and then determine the effects of the sunlight in the atmosphere.

The attenuation of visible light in the natural atmosphere below the ozone region, whether clear or rendered unclear by haze, fog, rain, and snow, is mainly caused by scattering and absorption. The only two permanent constituents of the lower atmosphere which have bands or lines of absorption in the visible spectrum are oxygen and water vapor, but estimates from available and none too complete data indicate that their attenuations due to true absorption are considerably less than those due to scattering. This is not true for certain types of smokes, but smoke is omitted from the present considerations. Therefore, it is assumed that the atmospheric attenuation for visible light arises mainly from scattering.[6]

The attenuation coefficient β_λ due to scattering is the integral of $\sigma_{\phi\lambda}$ over a sphere, or

$$\beta_\lambda=2\pi\int_0^\pi\sigma_{\phi\lambda}\sin\phi d\phi. \quad (10)$$

In the case of pure air molecules $\sigma_{\phi\lambda}$ is given by (4), and (10) becomes

$$\beta_\lambda=\frac{8\pi^3}{3}\frac{(\mu_\lambda-1)^2}{n\lambda^4}\frac{6(1+\delta)}{6-7\delta}\left(3+\frac{1-\delta}{1+\delta}\right). \quad (11)$$

Calculation of B_p

Referring to Fig. 11 let O be the position of the observer and X his zenith. Let OS be the direction of the rays of the sun and OP the line of sight of the observer. Let the zenith angles, i.e., the angles to the vertical of OS and OP, be ζ and ξ, respectively. Let the angle between OS and OP be ϕ and the projection of this in the horizontal plane be Z; Z is the bearing of the line of sight from the sun. Then

$$\cos\phi=\cos\zeta\,\cos\xi+\sin\zeta\,\sin\xi\,\cos Z. \quad (12)$$

It is assumed that the earth is flat and that the characteristics of the atmosphere vary only in the vertical direction. The results which we shall obtain will differ from those derived for a curved earth[7] by less than 2 percent for $\zeta=\xi$

[4] See M. Born, *Optik* (Verlagsbuchhandlung, Julius Springer, Berlin, 1933), p. 388.
[5] L. H. Dawson and E. O. Hulburt, J. Opt. Soc. Am. **31**, 554–558 (1941).
[6] E. O. Hulburt, J. Opt. Soc. Am. **25**, 125–130 (1935).
[7] E. O. Hulburt, Phys. Rev. **55**, 639–645 (1939); see page 641.

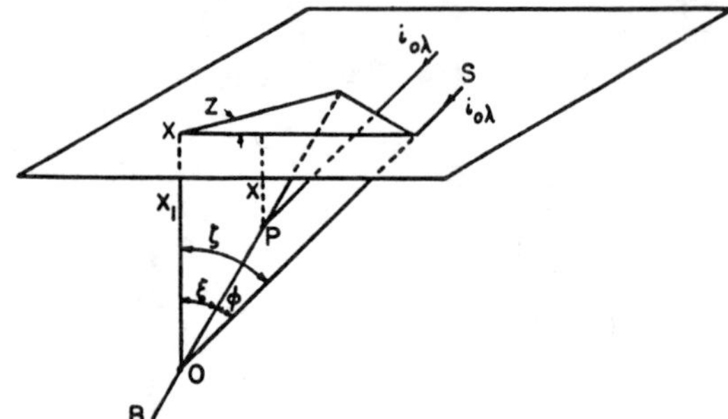

FIG. 11. Geometry of primary scattering.

$<80°$. Let x_1 be the vertical thickness of the atmosphere above the observer and x the thickness above a point P on his line of sight. The thicknesses are reduced to standard conditions at 1 atmosphere and 0°C; this introduces no error. Refraction of rays of light in the atmosphere is small, less than $\frac{1}{2}$ degree, and is neglected.

Let β_λ' be the absorption coefficient of the ozone region for light of wave-length λ passing through the entire ozone region in a vertical direction. The amount of ozone in a vertical column through the region is known to vary with the year, latitude, season, and weather. But since there are no data on the amount of ozone for the times and places of this investigation, possible variances in the present calculations due to changes in ozone cannot be determined. As mentioned later the variances, although not inappreciable, are not large, probably less than 3 percent.

Let $i_{0\lambda}'$ be the intensity of sunlight outside the atmosphere for wave-length λ, and $i_{0\lambda}$ the intensity after passing through the ozone region; $i_{0\lambda}$ is the intensity which falls on the "top" of the atmosphere. Then

$$i_{0\lambda} = i_{0\lambda}' \exp(-\beta_\lambda' \sec\zeta). \quad (13)$$

The intensity reaching P, Fig. 11, is

$$i_{0\lambda} e^{-\beta_\lambda x \sec\zeta}.$$

The amount scattered toward O is this quantity multiplied by $\sigma_{\phi\lambda}$, and is attenuated by a factor

$$e^{-\beta_\lambda(x_1-x)\sec\xi}$$

in reaching the observer. The total intensity $i_{p\lambda}$ which reaches the observer from all elements along OP is

$$i_{p\lambda} = \int_0^{x_1} i_{0\lambda}\sigma_{\phi\lambda} e^{-\beta_\lambda x \sec\zeta} e^{-\beta_\lambda(x_1-x)\sec\xi} dx \sec\zeta. \quad (14)$$

Integrating

$$i_{p\lambda} = i_{0\lambda}\frac{\sigma_{\phi\lambda}}{\beta_\lambda} \frac{e^{-\beta_\lambda x_1 \sec\zeta} - e^{-\beta_\lambda x_1 \sec\xi}}{1 - \sec\zeta \cos\xi}. \quad (15)$$

Introducing (4), (11), and (13) into (15) yields

$$i_{p\lambda} = \frac{3}{4\pi} i_{0\lambda}' \exp(-\beta_\lambda' \sec\zeta) \frac{1 + \dfrac{1-\delta}{1+\delta}\cos^2\phi}{3 + \dfrac{1-\delta}{1+\delta}}$$

$$\times \frac{e^{-\beta_\lambda x_1 \sec\zeta} - e^{-\beta_\lambda x_1 \sec\xi}}{1 - \sec\zeta \cos\xi}. \quad (16)$$

The brightness is $\psi_\lambda i_{p\lambda}$, where ψ_λ is the luminosity function of the light adapted eye. Then the total brightness B_p is

$$B_p = \int_0^\infty \psi_\lambda i_{p\lambda} d\lambda. \quad (17)$$

To calculate B_p by means of (17) was tedious and therefore an approximation was made based on the use of β' the visual attenuation coefficient of the ozone layer, of β the visual attenuation coefficient of the atmosphere, and of I_0' and I_0 the illuminations of sunlight on the top of the ozone layer and the top of the atmosphere, respectively. One has

$$\beta' = \int_0^\infty i_{0\lambda}'\psi_\lambda\beta_\lambda' d\lambda \Big/ \int_0^\infty i_{0\lambda}\psi_\lambda d\lambda, \quad (18)$$

$$\beta = \int_0^\infty i_{0\lambda}\psi_\lambda d\lambda \Big/ \int_0^\infty i_{0\lambda}\psi_\lambda d\lambda, \quad (19)$$

$$I_0' = \int_0^\infty i_{0\lambda}'\psi_\lambda d\lambda, \quad (20)$$

$$I_0 = \int_0^\infty i_{0\lambda}\psi_\lambda d\lambda. \quad (21)$$

From (13), (18), (20), and (21), one finds to a close approximation

$$I_0 = I_0' \exp(-\beta' \sec\zeta). \quad (22)$$

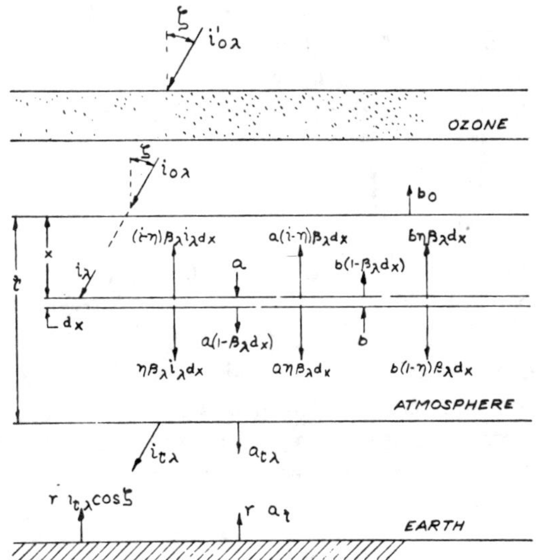

FIG. 12. Geometry of multiple scattering.

When (18)–(22) were introduced into (16) one obtained approximately

$$B_p = \frac{3}{4\pi} I_0' \exp(-\beta' \sec\zeta) \frac{1 + \dfrac{1-\delta}{1+\delta}\cos^2\phi}{3 + \dfrac{1-\delta}{1+\delta}}$$

$$\times \frac{e^{-\beta x_1 \sec\zeta} - e^{-\beta x_1 \sec\xi}}{1 - \sec\zeta\,\cos\xi}, \quad (23)$$

which is the expression used to calculate B_p.

Analysis showed that (23) departed from the exact expression (17) by an amount that increased with ξ and ζ to about 4 percent for $\xi = \zeta = 75°$. It was concluded that (23) was sufficiently accurate for the present purpose.

Calculation of B_m

In order to calculate B_m one must evaluate the flux of light, other than the direct rays of the sun, which impinges upon the atmosphere along the line of sight of the observer. The flux comes from the surrounding atmosphere and from the surface of the earth, which in turn owe their illumination to the sunlight. As the rays of the sun pass into the atmosphere a fraction of their light is scattered downward and upward, thereby giving rise to two streams of diffuse radiation, one stream of intensity a moving downward and stream of intensity b moving upward. The rays of the sun finally pass through the atmosphere and are assumed to be diffusely reflected upward by the surface of the earth. The reflected light passes through the atmosphere, a fraction of it is scattered and contributes to the a and b streams. The calculation of the values of a and b is given below. Actually, the calculation has already been made in an earlier paper[8] by dealing first with the case of a scattering and absorbing medium and then reducing the results to the simpler case of a scattering and non-absorbing medium. For the sake of clearness the direct calculation of this simpler case is given here.

In Fig. 12 is shown a vertical section of the atmosphere with, to avoid confusion, the ozone region separated from the "top" of the atmosphere, and the "bottom" of the atmosphere separated from the surface of the earth. Let t be the total thickness of the atmosphere and x the distance from the top of the atmosphere to a horizontal layer of thickness dx. The intensity of sunlight on the top of the atmosphere is $i_{0\lambda}\cos\zeta$ and on the layer is $i_\lambda \cos\zeta$. Assuming no true absorption on passage through the layer dx an amount $\beta_\lambda i_\lambda \cos\zeta dx/\cos\zeta = \beta_\lambda i_\lambda dx$ is scattered. A fraction η of this is scattered downward and contributes to the a stream and a fraction $(1-\eta)$ is scattered upward and contributes to the b stream. Similarly for the a and b streams in their passage through the layer. The details of the various attributes are shown in Fig. 12. In general a and b are functions of λ. For a Rayleigh atmosphere $\eta = \frac{1}{2}$ for both diffuse and collimated radiation.

Then

$$da = a(1-\beta_\lambda dx) + a\eta\beta_\lambda dx + b(1-\eta)\beta_\lambda dx + \eta\beta_\lambda i_\lambda dx - a, \quad (24)$$

$$db = b - b(1-\beta_\lambda dx) - b\eta\beta_\lambda dx - a(1-\eta)\beta_\lambda dx - (1-\eta)\beta_\lambda i_\lambda dx, \quad (25)$$

which reduce to

$$da/dx = (-a+b)(1-\eta)\beta_\lambda + \eta\beta_\lambda i_\lambda, \quad (26)$$

$$db/dx = (-a+b)(1-\eta)B_\lambda - (1-\eta)\beta_\lambda i_\lambda. \quad (27)$$

Let subscripts 0 and t denote the values of the quantities at the upper and lower surfaces of the

---

[8] E. O. Hulburt, J. Opt. Soc. Am. **33**, 42–45 (1943).

atmosphere, respectively. Let r be the diffuse reflectivity of the surface of the earth and assume that r is independent of λ. We note that

$$i_\lambda = i_{0\lambda}e^{-\beta_\lambda x \sec\zeta}, \quad i_{t\lambda} = i_{0\lambda}e^{-\beta_\lambda t \sec\zeta}, \quad (28)$$

and

$$a_0 = 0, \quad b_t = r(a_t + i_{t\lambda}\cos\zeta). \quad (29)$$

Introducing (28) and (29) into (26) and (27) and solving eads to

$$a = \frac{i_{0\lambda}\cos\zeta}{1+gt}\{(1+gt)(C-CX) - gx(C-CT+T)\}, \quad (30)$$

$$b = \frac{i_{0\lambda}\cos\zeta}{1+gt}\{(1+gt)(C-CX+X) - (gx+1-r)(C-CT+T)\}, \quad (31)$$

where

$$g = (1-r)(1-\eta)\beta_\lambda, \quad X = e^{-\beta_\lambda x \sec\zeta},$$
$$C = \eta + (1-\eta)\cos\zeta, \quad T = e^{-\beta_\lambda t \sec\zeta}. \quad (32)$$

Equations (30) to (32) are the same as Eqs. (22) to (27) of the former paper.[8]

Referring to Fig. 11 an element $dx/\cos\xi$ of the atmosphere at a point P on the line of sight of the observer is subjected to downward and upward streams of radiation of intensities a and b, respectively. Assuming that these streams of radiation are unpolarized and are uniformly distributed over the upper and lower hemispheres, the amount scattered by the element in any direction per unit solid angle is

$$\frac{1}{4\pi}\beta_\lambda(a+b)\sec\xi\,dx. \quad (33)$$

The intensity which reaches the observer is (33) multiplied by

$$e^{-\beta_\lambda(x_1-x)\sec\xi},$$

and the total intensity $i_{m\lambda}$ along the line of sight is the integral from 0 to x_1. Then

$$i_{m\lambda} = \frac{1}{4\pi}\int_0^{x_1}\beta_\lambda(a+b)\sec\xi e^{-\beta_\lambda(x_1-x)\sec\xi}dx, \quad (34)$$

and the brightness B_m is

$$B_m = \int_0^\infty \psi_\lambda i_{m\lambda}d\lambda. \quad (35)$$

It was tedious to calculate B_m from (35) and a sufficiently close approximation was obtained by replacing β_λ and $i_{0\lambda}$ in (30)–(32) by β of (19) and I_0 of (21), and calculating B_m from

$$B_m = \frac{1}{4\pi}\int_0^{x_1}\beta(a+b)\sec\xi e^{-\beta(x_1-x)\sec\xi}dx. \quad (36)$$

The direct integration of (36) resulted in a cumbersome expression. A further approximation was made which led to little departure from exactness and to considerable simplification. Since in the present investigation β was small, average values of a and b were used and taken from under the integral sign. Let a_1, a_0, and b_1,

TABLE I. Theoretical sky brightness and polarization at 10,000 feet for clear air. $\beta = 0.017$.

Zenith angle of sun $\zeta = 0°$

ξ	B_p	B_m	γ
0°	133	24	1.00
10°	133	24	0.98
20°	133	25	0.90
30°	134	27	0.79
40°	137	31	0.66
50°	145	36	0.52
60°	163	46	0.39
70°	204	64	0.28
75°	246	81	0.24
80°	330	111	0.21

Zenith angle of sun $\zeta = 30°$

ξ	$Z=0°$ B_p	B_m	γ	$Z=45°$ B_p	B_m	γ	$Z=90°$ B_p	B_m	γ	$Z=135°$ B_p	B_m	γ	$Z=180°$ B_p	B_m	γ
0°	116	21	0.79	116	21	0.79	116	21	0.79	116	21	0.79	116	21	0.79
10°	126	21	0.91	123	21	0.86	116	21	0.77	110	21	0.69	107	21	0.66
20°	138	22	0.98	132	22	0.90	117	22	0.72	105	22	0.57	101	22	0.51
30°	151	24	1.00	141	24	0.89	119	24	0.64	102	24	0.44	97	24	0.38
40°	166	27	0.98	152	27	0.83	123	27	0.54	103	27	0.33	97	27	0.27
50°	188	32	0.91	169	32	0.74	133	32	0.43	111	32	0.24	106	32	0.20
60°	221	40	0.79	196	40	0.62	154	40	0.33	132	40	0.18	130	40	0.17
70°	283	56	0.66	251	56	0.50	199	56	0.24	184	56	0.17	189	56	0.20
75°	338	71	0.59	300	71	0.44	242	71	0.22	233	71	0.18	246	71	0.23
80°	440	98	0.52	401	98	0.41	325	98	0.19	329	98	0.20	352	98	0.27

Zenith angle of sun $\zeta = 60°$

ξ	$Z=0°$ B_p	B_m	γ	$Z=45°$ B_p	B_m	γ	$Z=90°$ B_p	B_m	γ	$Z=135°$ B_p	B_m	γ	$Z=180°$ B_p	B_m	γ
0°	79	14	0.35	79	14	0.35	79	14	0.35	79	14	0.35	79	14	0.35
10°	90	14	0.50	86	14	0.43	80	14	0.34	74	14	0.26	72	14	0.24
20°	105	15	0.64	97	15	0.54	82	15	0.33	72	15	0.19	70	15	0.16
30°	125	16	0.78	111	16	0.61	86	16	0.30	75	16	0.15	74	16	0.14
40°	150	18	0.90	129	18	0.66	94	18	0.26	83	18	0.14	85	18	0.16
50°	185	21	0.98	154	21	0.68	107	21	0.22	100	21	0.15	108	21	0.24
60°	233	26	1.00	189	26	0.65	128	26	0.19	130	26	0.20	149	26	0.35
70°	325	37	0.98	260	37	0.62	176	37	0.16	197	37	0.28	237	37	0.49
75°	404	46	0.94	321	46	0.58	220	46	0.15	260	46	0.32	317	46	0.57
80°	541	64	0.90	429	64	0.55	300	64	0.14	371	64	0.37	460	64	0.64

Zenith angle of sun $\zeta = 75°$

ξ	$Z=0°$ B_p	B_m	γ	$Z=45°$ B_p	B_m	γ	$Z=90°$ B_p	B_m	γ	$Z=135°$ B_p	B_m	γ	$Z=180°$ B_p	B_m	γ
0°	60	8	0.16	60	8	0.16	60	8	0.16	60	8	0.16	60	8	0.16
10°	67	9	0.26	65	9	0.22	61	9	0.15	58	9	0.11	58	9	0.10
20°	78	9	0.40	72	9	0.30	63	9	0.15	60	9	0.10	60	9	0.10
30°	94	10	0.55	84	10	0.38	68	10	0.14	65	10	0.11	69	10	0.16
40°	117	11	0.71	100	11	0.47	75	11	0.13	77	11	0.16	84	11	0.25
50°	150	13	0.84	123	13	0.53	87	13	0.12	95	13	0.21	111	13	0.39
60°	200	16	0.94	159	16	0.60	109	16	0.11	129	16	0.29	157	16	0.54
70°	284	22	0.99	221	22	0.59	149	22	0.10	190	22	0.37	240	22	0.70
75°	366	28	1.00	283	28	0.58	198	28	0.10	252	28	0.42	322	28	0.77
80°	500	39	0.99	386	39	0.57	262	39	0.10	357	39	0.46	459	39	0.84

TABLE II. Thickness of atmosphere x_1.

Altitude above surface, feet	x_1 km	Altitude above surface, feet	x_1 km
0	8.00	30000	2.37
2000	7.45	32000	2.16
4000	6.91	34000	1.97
6000	6.40	36000	1.79
8000	5.94	38000	1.63
10000	5.50	40000	1.48
12000	5.08	42000	1.35
14000	4.70	44000	1.23
16000	4.34	46000	1.11
18000	4.00	48000	1.00
20000	3.68	50000	0.91
22000	3.38	52000	0.83
24000	3.10	54000	0.75
26000	2.84	56000	0.68
28000	2.60	60000	0.62

b_0 be the values of a and b at x_1 and 0, respectively, from (29) $a_0 = 0$. The average values over the interval of integration are $a_1/2$ and $(b_1+b_0)/2$. Then (36) reduces to

$$B_m = \frac{1}{8\pi}(a_1+b_1+b_0)(1-e^{-\beta x_1 \sec \xi}), \qquad (37)$$

which is the expression used to calculate B_m.

The assumption made in obtaining (33), namely, that the radiation streams a and b were unpolarized and uniformly distributed over their respective hemispheres deserves comment. Although the streams are in general not completely unpolarized and not uniformly distributed, consideration indicates that the assumption introduces an error in B_m of probably less than 30 percent. Since B_m is usually less than $\frac{1}{4}$ of B (see Table I) it will produce errors in B probably less than 10 percent. However, the assumption has the effect of obliterating the points of "neutral," i.e., zero, polarization which are observed in the sky. In other words, the present theory does not lead to the neutral points, and the cause of this inadequacy appears to be in the above assumption. This is of negligible importance in the present connection, since the neutral points are in the region within 30° of the sun, a region where the sky is but slightly polarized anyway.

NUMERICAL VALUES

The following numerical values were used in the theoretical equations.

1. Atmospheric thickness x_1.—The equivalent vertical thickness x_1 of the atmosphere above each altitude above the surface, reduced to km of air at N.T.P., was calculated from the pressure altitude values[9] of the "U.S. Standard Atmosphere." The values of x_1 for various altitudes above the surface are given in Table II.

2. Earth reflectivity r.—The diffuse reflectivity, or albedo, r of the surface of the earth was measured in flight at low altitude by comparing the brightness of the surface with that of a horizontal white plaque exposed to the sun and sky. The following values of r were obtained:

forest	4 to 10 percent
green fields	10 to 15
dry grass fields	15 to 25
dry ploughed fields	20 to 25
bay and rivers	6 to 10
sea	3 to 7

Snow, clouds, and low altitude haze and fog below the observer increase r; gloss of vegetation covered fields and shine of water modify r. The flights of the present program occurred in moderately clear winter weather, no snow, and mostly over land. r was observed to be between 0.10 and 0.25 during all the flights. We use

$$r = 0.20.$$

3. Polarization defect δ of air.—For pure air the observed values of δ were*: Rayleigh, 0.042, 0.050; Cabannes, 0.041; Raman, 0.0437; Rao, 0.0415. We use

$$\delta = 0.04.$$

4. Absorption coefficient β' of atmospheric ozone.—The values of β_λ' for atmospheric ozone in the Chappuis band of the visible spectrum have been given by Fowle[10] for the atmosphere above Mount Wilson, the total amount of ozone in a vertical column being 0.32 cm at N.T.P. From the values β_λ', together with standard values[11] of ψ_λ and Abbot's[12] values of $i_{0\lambda}'$, the coefficient β' was calculated by means of (18). It was

[9] W. G. Brombacher, National Advisory Committee for Aeronautics, Report No. 538 (1935).
* Reference 4, page 384.
[10] F. E. Fowle, Smithsonian Miscellaneous Collections **81**, 1 to 27 (1929).
[11] *Smithsonian Physical Tables* (1932), eighth edition.
[12] C. G. Abbot, *Measurement of Radiant Energy* (McGraw-Hill Book Company, Inc., New York, 1937); see also A. C. Hardy, *Handbook of Colorimetry* (Massachusetts Institute of Technology Press, Cambridge, 1936), p. 18.

$\beta' = 0.023$. Fowle noted variations amounting to about 0.01 in β' caused by changes in the amount of ozone. Changes of this amount will cause changes of less than 3 percent in the present calculations. The amount of atmospheric ozone during the present experiments was not known. We take for a vertical column through the atmosphere

$$\beta' = 0.023.$$

5. Attenuation coefficient β of air.—β_λ was calculated throughout the visible spectrum from (11) with the values of $i_{0\lambda}$ determined from Abbot's values[12] of $i_{0\lambda}'$ by means of (13) and with standard values[11] of μ_λ. β was then calculated from (19) and was 0.101 per atmosphere or 0.0126 per km of air at N.T.P. The correctness of this theoretical value of β for pure air has never been tested rigorously. However, observations of the atmosphere above Mount Wilson corrected for haze and dust were shown[13] to agree with the Rayleigh formula within a few percent for wave-lengths from about 4500 to 3600A outside of the ozone bands.

We take for *pure* air the value

$$\beta = 0.0126 \text{ per km.}$$

It may be stated in advance that we have encountered an atmosphere that is not *pure* air, and have been concerned with a "clear" atmosphere, defined in an arbitrary way, for which

$$\beta = 0.017 \text{ per km.}$$

6. Illumination of sun I_0'.—The solar illumination I_0' outside of the atmosphere was determined in several ways.

(a) Calculation from (20) with Kimball's[14] values of the solar illumination at ground level for an average clear atmosphere at Washington, D. C., and with Abbot's[12] average spectral transmission coefficients of the Washington atmosphere, reduced to a zenith sun, gave $I_0' = 13600$ footcandles. The value is open to some uncertainty since it is not known whether the "average clear atmosphere" of Kimball was the same

as that of Abbot. However, the value appears to be the most accurate available at present.

(b) Calculation from Abbot's[12] solar spectral energy curve and solar constant gave $I_0' = 11300$ footcandles. It was apparent, however, that the value was open to considerable uncertainty because of the lack of spectral energy data in the ultraviolet and infra-red regions of the spectrum.

(c) During the stratosphere flight[15] of Explorer II a value $I_0 = 12000$ footcandles was obtained when the zenith angle of the sun was about 60°. The thickness of the ozone was observed to be 1.9 mm from measurements made during the flight.[16] These values in (22) gave $I_0' = 12300$. The value seems low.

It is possible that I_0' may not be constant since long and short period fluctuations of a few percent in various regions of the visible solar spectrum have been mentioned.[17] We take

$$I_0' = 13,600 \text{ footcandles.}$$

The value refers to the mean solar distance.

COMPARISON OF EXPERIMENT AND THEORY

Since haze was always present at the surface there was little point in comparing the theory of a pure air atmosphere with surface observations. Therefore, the theoretical calculations were made for an altitude of 10,000 feet on the idea that at this altitude the atmosphere above the observer might approach pure air, as indeed it was found to do although not completely.

The numerical values for pure air, $\delta = 0.04$, $\beta = 0.0126$, $\beta' = 0.023$, $r = 0.20$, and $I_0' = 13,600$, were introduced into (5), (23), and (37), and B_p, B_m, and γ were calculated for an altitude of 10,000 feet, the values being listed in a table which is not printed here. The values were plotted in several families of curves, and from the curves, which are not included here, the pure air theoretical values of B, $B_\perp$, $B_{||}$, and γ could

[13] F. E. Fowle, Astrophys. J. **38**, 392–406 (1913); **40**, 435–442 (1914).
[14] H. H. Kimball, Trans. Ill. Eng. Soc. 16, 255–283 (1921).

[15] R. P. Teele, Nat. Geog. Soc. U. S. Army Air Corps Stratosphere Flight of 1935, pages 133–138.
[16] B. O'Brien, F. L. Mohler, and H. S. Stewart, Nat. Geog. Soc. U. S. Army Air Corps Stratosphere Flight of 1935, pages 70–93.
[17] C. G. Abbot, F. E. Fowle, and L. B. Aldrich, Annals Astrophys. Obs. Smithsonian Institution 3, 135 (1913); 4, 17 and 207 (1922); C. G. Abbot, Smithsonian Miscellaneous Collections 101, No. 5, preface (1941).

TABLE III. Zenith brightness B at 10,000 feet.

Flight	Observed B	Theoretical pure air B	Percent difference	β
1	153 c ft.$^{-2}$	96 c ft.$^{-2}$	59	0.0201
2	130	89	46	.0184
3	103	73	41	.0178
4	124	87	42	.0179
5	(101)	87	(16)	(.0146)
6	114	69	55	.0195
7	87	63	38	.0174
8	91	68	34	.0169
9	83	64	30	.0164
11	92	69	33	.0168
12	110	73	51	.0190

be read off for any point in the sky for solar altitudes between 15° and 90°. The pure air values of B obtained in this way were found in all cases to be less than the observed values at 10,000 feet by amounts ranging from 30 to 60 percent. This indicated that the atmosphere above 10,000 feet was not pure air but contained scattering particles in addition to the air molecules.

Thereupon, for reasons given in the following paragraphs, the attenuation β was increased to 0.017, an increase of about 35 percent over the pure air value 0.0126, and B_m, B_p, B, and γ were recalculated using the numerical values $\delta = 0.04$, $\beta = 0.017$, $\beta' = 0.023$, $r = 0.20$ and $I_0' = 13,600$. The new values of B_m, B_p, and γ are given in Table I; B_m and B_p exceeded the pure air values by 30 to 37 percent, the exact amount depending on ζ and ξ. The values of γ of Table I were approximately the same as the pure air values. With $\beta = 0.017$ new theoretical curves were calculated for comparison with the observations and are plotted in the dotted curves of Figs. 1 to 10. An atmosphere above 10,000 feet for which $\beta = 0.017$ is defined as "clear."

Referring to the sky data at 10,000 feet of Figs. 4, 5, 9, and 10 it is seen that the "clear" air theoretical values of B are not far from, and are usually less than, the observed values, and that the theoretical values of γ agree with the observed values, the deviations appearing to be accidental, due to conditions of flight.

Turning to the zenith values of B, $B_\perp$, $B_\parallel$, and γ at various altitudes of Figs. 2, 3, 7, and 8, the observed values were considerably above the "clear" air theoretical curves at low altitudes due to prevalent haze near the surface. It is seen that the effect of the haze was to add to the polarized zenith light a considerable amount of unpolarized light. This increased $B_\parallel$ and $B_\perp$ by approximately equal increments and caused corresponding increases in γ. With increasing altitude, particularly after the haze level was passed, the observed quantities approached the theoretical curves showing that the observer was rising above the haze into purer air. In five of the flights, of which Figs. 2 and 3 are an example, they were approximately coincident with the theoretical curves at, or below, 10,000 feet showing that the observer had reached "clear" air for which β was 0.017. In seven of the flights, of which Figs. 7 and 8 are an example, an aureole was visible around the sun and the observed values were greater than the theoretical values showing that the air above 10,000 feet on those occasions contained large particle haze.

In none of the flights did the observer reach theoretically "pure" air, i.e., air for which $\beta = 0.0126$. This is brought out in Table III which gives the observed and pure air theoretical values of B at the zenith and the percent difference for an altitude of 10,000 feet. The five lowest values of the difference averaged 0.30, corresponding to an increase of β to 0.017. The low value of B, and hence the small percent difference of Flight 5, was probably due to observational error, for the other data of this flight did not indicate an unusually low sky brightness. In the last column of Table III are given the values of β required to obtain agreement between theory and observation.

The only flight much above 10,000 feet was Flight 9 of the present experiments. In this case clouds interfered with an extended program of measurement, but a few values of B for the zenith sky were obtained at 18,800 feet. These were about 30 percent above the theoretical values for pure air. This meant that β for this case was about 0.017, and hence that the optical purity, or rather impurity, of the air above 18,800 feet was the same as that of the air above 10,000 feet for the condition described as "clear." It would be important to obtain more observations at altitudes above 10,000 feet.

Measurement of the solar illumination I are

plotted in Figs. 1 and 6. In the figures the dotted theoretical curves were calculated from the relation

$$I = I_0' \exp \left(-\beta' \sec \zeta\right) e^{-\beta z \sec \zeta}, \qquad (32)$$

where $I_0' = 13,600$, $\beta' = 0.023$, and $\beta = 0.017$, which are the values used for calculating the sky brightness and polarization for clear air. It is seen that the theoretical curve approaches agreement with the observations at the higher altitudes for the "clear" air case of Fig. 1, and deviates from the observations for the "less clear" air case of Fig. 7 as expected. At the lower altitudes the observed values of I were below the theoretical curves due to increasing haze near the surface.

It is of interest to compare the observations of Abbot, Fowle, and Aldrich[17] with the present results. They measured the transmission for various wave-lengths in the visible spectrum through a clear atmosphere above Mount Wilson (altiude 5680 feet). Calculation from their average data of 1905 and 1906 gave $\beta = 0.016$, in good agreement with the present result 0.017. The agreement is considered to indicate that the atmosphere above Mount Wilson described as "clear" was of about the same clarity as the atmosphere above 10,000 feet of the present investigation also described as "clear."

The conclusion that the attenuation and scattering of the "clear" air above 10,000 feet was about 35 percent above that of theoretically pure air, is not appreciably disturbed by uncertainties in the value of the earth reflectivity $r = 0.20$ used in the theoretical calculations. This may be seen from Table IV which gives the theoretical clear air ($\beta = 0.017$) values of B_p, B_m, B, and γ for r from 0 to 1 for two points in the sky. r was observed to be between 0.10 and 0.25 during the present flights, and Table IV shows that for this range of r the variances in B and γ were less than about 10 percent, respectively, from their values for $r = 0.20$. Furthermore, the value of γ of Flight 11 at the point of maximum polarization, taken purposely for this comparison, was 0.13, which agreed well with the value 0.14 of Table IV for $r = 0.20$. Table IV suggests, however, that B and γ might be noticeably increased by large areas of white clouds or fresh snow beneath the

TABLE IV. Sky values at 10,000 feet as a function of r for "clear" air, $\beta = 0.017$.

	$\zeta = 60°$		$\xi = 30°$		$Z = 180°$	
r	0	0.2	0.4	0.6	0.8	1.0
B_p	74	74	74	74	74	74
B_m	5	16	25	36	47	56
B	79	90	99	110	121	130
γ	0.08	0.14	0.20	0.24	0.29	0.34

	$\zeta = 60°$		$\xi = 0°$			
r	0	0.2	0.4	0.6	0.8	1.0
B_p	79	79	79	79	79	79
B_m	4	14	22	32	42	50
B	83	93	101	111	121	129
γ	0.31	0.35	0.39	0.42	0.45	0.47

observer, for in these cases r might be greater than 0.4. Over the sea with a clear atmosphere r is usually less than 0.10. This would result in slightly lowered values of B and γ.

The result is that we assume that the "clear" atmosphere above 10,000 feet is an atmosphere composed of pure air molecules with a number of foreign particles of size small with respect to λ sufficient to increase the light scattering and attenuation of pure air by 35 percent. This amounts to a definition of a *"clear atmosphere above 10,000 feet."* True, the definition is arbitrary but, being based on experiment, is practical and permits one to say that if the additional light scattering is greater or less than 35 percent of that of pure air, the atmosphere is less clear, or clearer, than *"clear,"* respectively.

Now an atmosphere whose attenuation and scattering is only about 35 percent greater than that calculated for pure air molecules is very pure indeed in comparison with the amount of optical contamination of the usual clear atmosphere at sea level. It is possible that the theory may be incorrect to a certain extent and that a portion of the 35 percent difference may be unreal. However, for the present we regard the entire 35 percent difference as real and due to scattering particles. For want of a better term we may refer to these as "haze" particles, in spite of the fact that because of their extreme tenuousness the appellation "haze" may be misleading, since haze ordinarily refers to an easily perceptible obscurity in the lower atmosphere.

The data, of which Figs. 4 and 9 are a sample, showed that the values of B at 10,000 feet, especially toward the sun, were in many cases greater than the theoretical curves for $\beta = 0.017$.

This was taken to mean that in these cases there was additional haze which varied from flight to flight. A detailed analysis of the differences between the observed and theoretical values indicated that the haze particles were not all of the same size and that the size distribution varied from flight to flight. This seemed reasonable from a meteorological standpoint, for one would expect the composition of the haze, as water droplets, dust, or whatever it was, to vary with the previous history and condition of the atmosphere. Small haze particles will produce no visible aureole, and large particles may produce a visible aureole. Apparently both types of particles were encountered in the flights of the present experiments.

Interpretation of the Polarization of Venus

James E. Hansen

Goddard Institute for Space Studies, New York, N. Y. 10025

J. W. Hovenier'

Dept. of Physics and Astronomy, Free University, Amsterdam, Netherlands

(Manuscript received 20 November 1973, in revised form 15 January 1974)

ABSTRACT

The linear polarization of sunlight reflected by Venus is analyzed by comparing observations with extensive multiple scattering computations. The analysis establishes that Venus is veiled by a cloud or haze layer of spherical particles. The refractive index of the particles is 1.44 ± 0.015 at $\lambda = 0.55$ μm with a normal dispersion, the refractive index decreasing from 1.46 ± 0.015 at $\lambda = 0.365$ μm to 1.43 ± 0.015 at $\lambda = 0.99$ μm. The cloud particles have a narrow size distribution with a mean radius of ~ 1 μm; specifically, the effective radius of the size distribution is 1.05 ± 0.10 μm and the effective variance is 0.07 ± 0.02. The particles exist at a high level in the atmosphere, with the optical thickness unity occurring where the pressure is about 50 mb.

The particle properties deduced from the polarization eliminate all but one of the cloud compositions which have been proposed for Venus. A concentrated solution of sulfuric acid (H_2SO_4-H_2O) provides good agreement with the polarization data.

---

1. Introduction

Venus is our nearest planetary neighbor, yet one of the most mysterious. To a large extent this is due to the veil of clouds surrounding the planet. These clouds not only mask Venus, but their own composition is unknown. Many possible compositions have been suggested in the literature, including water, H_2O ice, solid CO_2, carbon suboxide (C_3O_2; Sinton, 1953; Kuiper, 1957; Harteck *et al.*, 1963), hydrated ferrous chloride ($FeCl_2 \cdot 2H_2O$; Kuiper, 1969), NaCl (Hunten, 1968), formaldehyde (CH_2O; Wildt, 1940), hydrocarbons (Velikovsky, 1950; Hoyle, 1955; Kaplan, 1963), hydrocarbon-amide polymers (Robbins, 1964), polywater (Donahue, 1970), ammonium nitride ($NH_4 \cdot NO_2$; Dauvillier, 1956), calcium and magnesium carbonates (Öpik, 1961), NH_4Cl (Lewis, 1968; Hunten and Goody, 1969), mercury and mercury compounds (Lewis, 1969; Rasool, 1970) and aqueous solutions of hydrochloric acid ($HCl \cdot nH_2O$; Lewis, 1972; Hapke, 1972) and sulfuric acid ($H_2SO_4 \cdot nH_2O$; Sill, 1972; Young and Young, 1973; Young, 1973). Although a large amount of theoretical and observational effort has been expended on this problem, no consensus on the cloud composition has evolved.

Our best means for investigating the clouds of Venus still is through measurements of reflected sunlight. The spectral reflectivity of Venus exhibits strong absorption features in the near-infrared ($\lambda \approx 3$ μm) and in the ultraviolet ($\lambda \lesssim 0.35$ μm) which a proposed cloud material should be consistent with. However, these features have proved insufficient for either a specific identification of the cloud particles or eliminating most proposed compositions, in part because it is always possible to hypothesize a gaseous absorber, an admixture of another cloud material, or a lower cloud layer to provide the observed absorption. There have been concerted attempts to link several weaker features in the spectral reflectivity to specific cloud compositions, particularly H_2O ice (Sagan and Pollack, 1967) and $FeCl_2 \cdot 2H_2O$ (Kuiper, 1969); however, the features which were associated with ice (other than the absorption at $\lambda \approx 3$ μm) have also been credited to gaseous CO_2 absorption (cf. Rea and O'Leary, 1968), and a number of the features associated with ferrous chloride are of doubtful reality (cf. Cruikshank and Thomson, 1971).

A different sort of evidence on the cloud composition is provided by the angular distribution of reflected light, both the distribution of brightness over the planetary disk and the disk-integrated brightness as a function of planetary phase angle. Arking and Potter (1968) analyzed such observations for Venus by comparing them with multiple scattering computations for spherical cloud particles. They found agreement with the observations for micron-sized or larger transparent spheres with a real refractive index $1.33 \lesssim n_r \lesssim 1.7$ for visible wavelengths. This conclusion depends on the *assumption* that the particles are spherical, and furthermore the derived range for the refractive index includes practically all of the compositions which have been suggested for the Venus cloud particles.

Polarization observations are more sensitive to cloud particle characteristics than are brightness observations (cf. Hansen, 1971b) and thus the polarization offers a potentially powerful tool for investigating the nature of the Venus clouds. High quality polarization observations of Venus were obtained by Lyot (1929) for visual light and extended throughout the range $0.35 \lesssim \lambda \lesssim 1 \ \mu m$ by Dollfus (1966) and Coffeen and Gehrels (1969). Semi-quantitative analyses of the polarization were made by Lyot (1929) on the basis of laboratory comparisons and by Coffeen (1968, 1969), Sobolev (1968) and Loskutov (1971) on the basis of single scattering computations for spheres. The most complete of these analyses is that of Coffeen, in which he found that the available observations were compatible with spheres of radius $1.25 \pm 0.25 \ \mu m$ and refractive index $1.43 \lesssim n_r \lesssim 1.55$. Nonspherical particles were not excluded by the analysis of Coffeen and no attempt was made to investigate the influence of the particle size distribution on the conclusions.

The major task required in a quantitative analysis of the polarization is a proper accounting for multiple scattering. Horak (1950) made accurate multiple scattering computations for Rayleigh scattering, based on the theory of Chandrasekhar (1950), but he found that Rayleigh scattering did not resemble Lyot's polarization observations of Venus. Hansen and Hovenier (1971) made accurate computations with the doubling method (Hansen, 1971a; Hovenier, 1971) for scattering by cloud particles; their results demonstrated the large effect of multiple scattering on the polarization. Kattawar *et al.* (1971) made multiple scattering computations with the approximate Monte Carlo method for a spherical atmosphere and concluded that the cloud particles in the highest cloud layer on Venus have a refractive index $1.45 \lesssim n_r \lesssim 1.60$. Hansen and Arking (1971) made computations with the doubling method for comparison with observations of Lyot (1929) and Coffeen and Gehrels (1969) at three wavelengths; the results showed that the cloud particles on Venus are spherical with refractive index 1.45 ± 0.02 and mean radius $\sim 1 \ \mu m$ and that the cloud-top pressure level is ~ 50 mb.

In this paper we report the results of a series of computations much more extensive than those of Hansen and Arking (1971). Many observations not included by Hansen and Arking (Kuiper, 1957; Marin, 1965; Dollfus, 1966; Dollfus and Coffeen, 1970; Forbes, 1971; Veverka, 1971) are incorporated in this study, and observations at all available wavelengths are analyzed. In addition, a detailed documentation of the method of analysis is included, and the refractive index dispersion derived from the polarimetric data is compared with dispersions of proposed cloud compositions.

2. Single scattering

Theoretical calculations for scattering by a planetary atmosphere are conveniently divided into two parts: single scattering by small volume elements in the atmosphere with optical thickness much less than unity and multiple scattering by the complete model atmosphere with any optical thickness. For comparison to disk-integrated observations, the multiple scattering computations must also be integrated over the visible part of the disk, as described in Section 3.

Exact solutions for single scattering are feasible for spheres and particles of a few other shapes (cf. van de Hulst, 1957; Kerker, 1969; Kratohvil, 1964; Coffeen and Hansen, 1973). For practical applications computations are almost always made using spherical particles, and it is essential to realize the limitations that this places on conclusions which may be obtained. However, in some applications the observations are sufficient to prove that the particles being examined are in fact spherical. This turns out to be the case for the Venus clouds.

The solution for scattering of a plane wave by an isotropic homogeneous sphere was obtained by Mie (1908). The results for a single sphere depend on n_c and x, where $n_c = n_r - i n_i$ is the complex refractive index of the sphere relative to the surrounding medium, with $i = (-1)^{\frac{1}{2}}$, and $x = 2\pi r/\lambda$ is the size parameter, with r the particle radius and λ the wavelength of the incident radiation. The theory for single scattering is described in detail by van de Hulst (1957); here we give only information required in the application to follow.

The single scattering quantities required for the multiple scattering computations are the single scatterings albedo $\tilde{\omega}_0$ and the phase matrix $\mathbf{P}(\alpha)$, where α is the scattering angle ($\alpha = 0$ for light scattered exactly in the forward direction). In the clouds of Venus $\tilde{\omega}_0$ must be very close to unity in the optical window to yield the observed high spherical (Bond) albedo, and an approximate value for $\tilde{\omega}_0(\lambda)$ can be estimated from the requirement of matching the observed spherical albedo. Thus, the major requirement for interpreting the polarization of Venus is a thorough knowledge of the dependence of the phase matrix on the particle refractive index and size distribution.

The phase matrix for a single sphere at a particular wavelength may be described by four functions of the radius and scattering angle, $M_1(r,\alpha)$, $M_2(r,\alpha)$, $S_{21}(r,\alpha)$ and $D_{21}(r,\alpha)$, where we employ the notation of van de Hulst (1957). In the case of a size distribution of particles which scatter independently, the corresponding functions are obtained by integrating over all particles in the size distribution, e.g.,

$$M_1(\alpha) = \int_{r_1}^{r_2} M_1(r,\alpha) n(r) dr, \qquad (1)$$

where $n(r) dr$ is the number of particles per unit volume

with radius between r and $r+dr$, and r_1 and r_2 are the smallest and largest particles in the size distribution.

The normalized phase matrix for spheres, defined with respect to the Stokes parameters $\{I,Q,U,V\}$, has the form

$$\mathbf{P}(\alpha)=\begin{bmatrix} P^{11} & P^{21} & 0 & 0 \\ P^{21} & P^{22} & 0 & 0 \\ 0 & 0 & P^{33} & -P^{43} \\ 0 & 0 & P^{43} & P^{33} \end{bmatrix}, \qquad (2)$$

where

$$\left.\begin{aligned} P^{11}(\alpha)&=c[M_2(\alpha)+M_1(\alpha)]/2 \\ P^{21}(\alpha)&=c[M_2(\alpha)-M_1(\alpha)]/2 \\ P^{33}(\alpha)&=cS_{21}(\alpha) \\ P^{43}(\alpha)&=cD_{21}(\alpha) \end{aligned}\right\}, \qquad (3)$$

and c is a constant defined such that the phase function $P^{11}(\alpha)$ is normalized as

$$\frac{1}{4\pi}\int_{4\pi} P^{11}(\alpha)d\omega=1, \qquad (4)$$

where $d\omega$ is an element of solid angle.

The phase matrix $\mathbf{P}(\alpha)$ may be computed from Mie theory for any refractive index, wavelength and size distribution of spheres. But in the case of remote measurements of scattered light, such as Earth-based observations of Venus, the size distribution of particles is unknown. Thus to "invert" such measurements and obtain cloud particle properties it is essential to make a systematic analysis of the effect of the size distribution on the scattered light, as we describe here.

It is common practice to describe a distribution function by its moments, or parameters simply related to the moments, e.g., the mean, the variance, the skewness, etc. To facilitate the inversion of radiation measurements these parameters should be chosen such that the number required to describe an arbitrary particle size distribution is as small as possible. Clearly the first parameter should be some measure of the mean particle size. The simple arithmetic mean is

$$\langle r\rangle=\frac{\displaystyle\int_{r_1}^{r_2} rn(r)dr}{\displaystyle\int_{r_1}^{r_2} n(r)dr}=\frac{1}{N}\int_{r_1}^{r_2} rn(r)dr, \qquad (5)$$

where N is the total number of particles per unit volume. However, a sphere of size $r\gtrsim\lambda$ scatters an amount of light approximately proportional to its area, πr^2, and for smaller particles the amount of scattered light is proportional to an even higher power of the particle radius, reaching r^6 for Rayleigh scatterers.

Thus, as the first parameter describing the size distribution, we use the *effective radius*, defined as

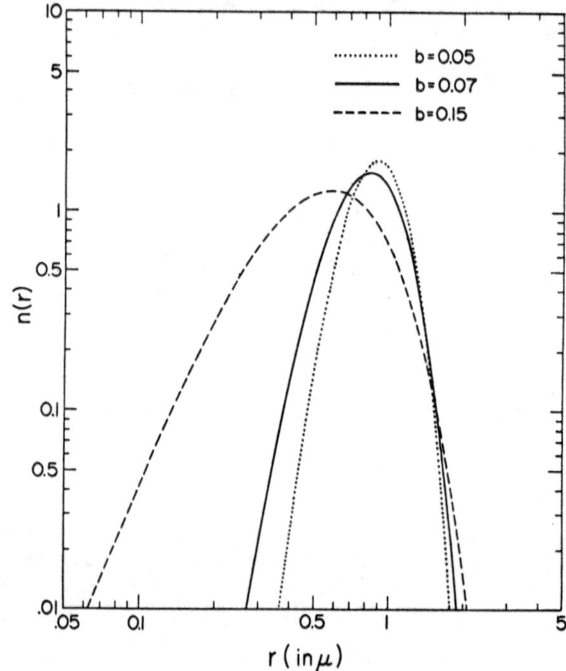

FIG. 1. Size distribution (8) for three values of b, where b is the effective variance. All three curves have $a=1.05\ \mu$m, where a is the effective radius. The mode radius, where $n(r)$ is a maximum, is given by $a(1-3b)$. The distributions are normalized such that the integral over all sizes is unity.

$$r_{\text{eff}}=\frac{\displaystyle\int_{r_1}^{r_2} r\pi r^2 n(r)dr}{\displaystyle\int_{r_1}^{r_2} \pi r^2 n(r)dr}. \qquad (6)$$

Similarly, as a measure of the width of the size distribution, we use the *effective variance*, defined as

$$v_{\text{eff}}=\frac{\displaystyle\int_{r_1}^{r_2} (r-r_{\text{eff}})^2\pi r^2 n(r)dr}{r_{\text{eff}}^2\displaystyle\int_{r_1}^{r_2} \pi r^2 n(r)dr}, \qquad (7)$$

where r_{eff}^2 in the denominator makes v_{eff} dimensionless.

This procedure can be continued to higher moments. However, for many purposes r_{eff} and v_{eff} are adequate for describing the size distribution. This has been demonstrated, for example, in computations by Hansen (1971b) for measured size distributions of terrestrial water clouds and for the analytic distribution

$$n(r)=\text{constant}\times r^{(1-3b)/b}e^{-r/ab} \qquad (8)$$

with the same values of r_{eff} and v_{eff}. The close similarity of the results for the measured and analytic distributiony shows that these two parameters define the major characteristics in the intensity and polarization as a function of scattering angle.

In the computations for this paper we also use the distribution (8) because it has the simple property that

$$\left.\begin{array}{l} a = r_{\text{eff}} \\ b = v_{\text{eff}} \end{array}\right\},$$

provided the integrations in (6) and (7) extend over all particles ($r = 0, \infty$). Distribution (8) is a form of the gamma distribution (cf. Kendall and Stuart, 1963); other variations have been extensively used for cloud particle size distributions, e.g., by Khrgian (1961) and Deirmendjian (1964). The constant in (8) is related to

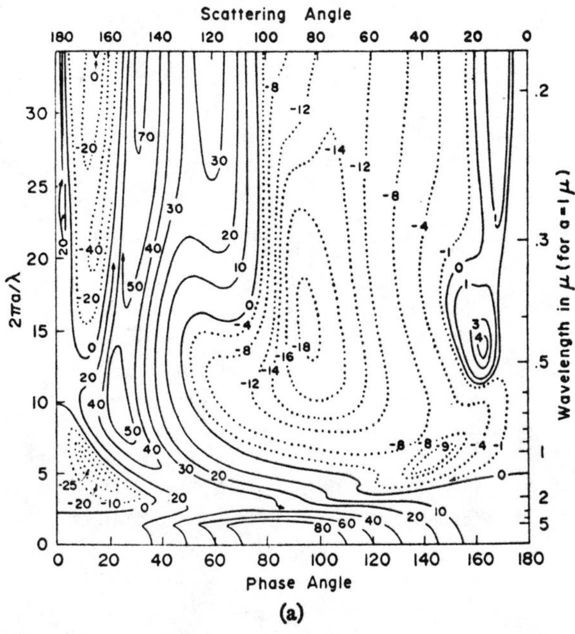

(a)

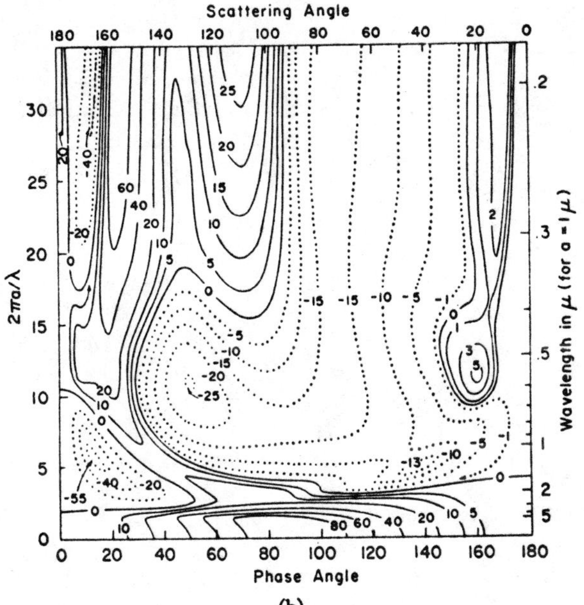

(b)

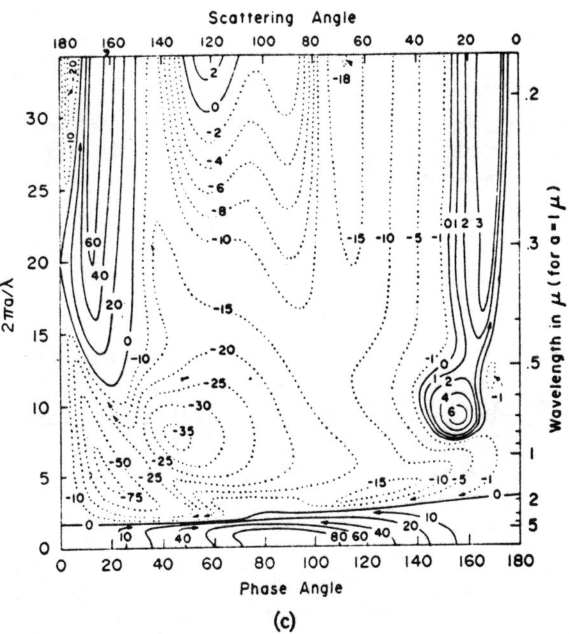

(c)

FIG. 2. Percent polarization, $-100P^{21}/P^{11}$, for single scattering of unpolarized incident light by a size distribution of spheres. Solid lines indicate positive polarization and dotted lines negative polarization. The three parts of the figure show results for three real refractive indices, $n_r = 1.33$(a), 1.40(b) and 1.50(c). The size distribution is that given by (8) with $b = 0.05$ in all three cases. The wavelength scale applies for the choice $a = 1$ μm.

the total number density of particles per unit volume, N, by

$$\text{constant} = N(ab)^{(2b-1)/b}/\Gamma[(1-2b)/b], \qquad (9)$$

where Γ is the gamma function.

Fig. 1 shows three examples of distribution (8). All three have $a = 1.05\,\mu$m, but $b = 0.05$, 0.07 and 0.15, respectively. The maximum of $n(r)$ occurs at $r_m = a(1-3b)$. The standard deviation for the distribution (8) is $\sigma = a[b(1-2b)]^{\frac{1}{2}}$. For $a = 1.05\,\mu$m and $b = 0.07$, for example, we find $r_m = 0.83\,\mu$m and $\sigma = 0.26\,\mu$m.

The contour diagrams in Figs. 2 and 3 demonstrate the dependence of the single scattering polarization on the refractive index and size distribution. These diagrams show the percent polarization for single scattering of incident unpolarized light, $-100P^{21}/P^{11}$, as a function of phase angle (the supplement of the scattering angle) on the horizontal axis and as a function of the effective size parameter $2\pi a/\lambda$ on the vertical axis. By choosing any fixed value for a the vertical scale can be converted to a scale for λ. This is done on the right side of each figure for the choice $a = 1\,\mu$m; the wavelength scale for a different choice of a can be obtained by multiplying the given scale by a (in μm). For any fixed wavelength λ the effective size parameter scale on the left can be converted to a scale for a by means of the multiplication factor $\lambda/(2\pi)$.

Fig. 2 illustrates the dependence of the polarization on the refractive index for three refractive indices in the

291

range previous studies have shown to be relevant to the clouds of Venus. Each of the three parts in Fig. 2 is for size distribution (8) with $b=0.05$. This distribution is broad enough to smooth out most of the interference maxima and minima which occur for a single particle, but not broader than naturally occurring distributions.[1]

The integration over size parameters extends from $x_1=0$ to $x_2=45.7$. For the top part of Fig. 2 this is not equivalent to the range $(0,\infty)$ which is required to make exactly $r_{\mathrm{eff}}=a$ and $v_{\mathrm{eff}}=b$. For example, at $2\pi a/\lambda=30$, $2\pi r_{\mathrm{eff}}/\lambda$ is ~ 29.3 and v_{eff} is ~ 0.044; however, the deviations of r_{eff} from a and v_{eff} from b are not sufficient to alter the conclusions we draw from the figure. The computations were made for 130 phase angles, 0.5(1)39.5(2)139.5(1)179.5, and 61 equally spaced effective size parameters in the interval (0,34.3). In the proximity of isolated maxima and minima of the polarization some additional computations were made. The accuracy with which the contours could be drawn is estimated to be close to the width of the lines in the figures.

For the smallest size parameters there is the strong positive polarization of Rayleigh scattering, with the maximum polarization at phase angle 90°. The Rayleigh scattering region is similar for the different refractive indices, but it is more compressed for the larger values of n_r; this is understandable since the conditions for Rayleigh scattering are $x \ll 1$ and $|n_c x| \ll 1$.

For the largest size parameters in Fig. 2 the polarization approaches that for geometrical optics (cf. van de Hulst, 1957; Liou and Hansen, 1971). At small scattering angles the polarization is small because of the predominance of unpolarized diffracted light. Other than diffraction, most of the light scattered in the forward hemisphere is due to rays passing through the particle with two refractions. This light is negatively polarized, as follows from Fresnel's equations. Reflection from the outside of the particles contributes a positive polarization at all phase angles; although the intensity of these rays is small, it is sufficient to cause the long peninsula of positive polarization at scattering angles $\alpha \approx 15°$. This feature becomes stronger as n_r increases, because the Fresnel reflection coefficients increase with n_r.

The steep ridge and positive polarization maximum at scattering angles $\sim 150°$ (for $n_r=1.33$) is the primary rainbow. This arises from rays internally reflected one time in spheres. These rays tend to be concentrated at a given scattering angle, as can be shown from Snell's law and the Fresnel reflection coefficients. Similarly, the weaker feature at $\alpha \approx 120°$ (for $n_r=1.33$) is the second rainbow, due to rays undergoing two internal reflections. Still higher rainbows contain a negligible fraction of the scattered light, and they do not contribute any noticeable feature in Fig. 2. The location of the rain-

bows in scattering angle varies with n_r in accordance with Snell's law.

The sharp maximum in the polarization in the back-scattering direction ($\alpha \approx 180°$) is the so-called glory. This is due in large part to incident edge rays (i.e., grazing rays) which set up surface waves on the scattering particle (van de Hulst, 1957; Bryant and Cox, 1966; Fahlen and Bryant, 1968). These surface waves spew electromagnetic energy in all directions, but it is focused in the forward direction, where it is lost in the stronger diffracted light, and in the backward direction, where it gives rise to the glory. For refractive indices in the range $2^{\frac{1}{2}} \lesssim n_r \lesssim 2$ there is also a large contribution to the glory from rays internally reflected one time.

For scattering angles $\sim 20°$ and size parameters ~ 15 (for $n_r=1.33$) there is a hill of positive polarization, which nearly forms an island but is connected to the peninsula of positive polarization for larger particles. This feature is a manifestation of what van de Hulst (1957) calls "anomalous diffraction." It is due to optical interference between diffracted light and light reflected and transmitted by the particle in the near-forward direction. The phase shift of a ray traveling through the center of the sphere is $\rho = 2x(n_r-1)$, which accounts for the location of this feature in size parameter varying approximately as $1/(n_r-1)$.

In the transition region between large-particle scattering and Rayleigh scattering the polarization is a complicated function of size parameter. As the size parameter decreases the degree to which the paths of separate light rays can be localized decreases. Thus the second rainbow, with a more detailed ray path, is lost from the polarization before the primary rainbow is. With decreasing size parameter the primary rainbow becomes blurred and its peak first moves toward larger scattering angles due to the asymmetric shape of the rainbow. For $n_r=1.33$ the peak of the primary rainbow shifts to smaller scattering angles for size parameters $\lesssim 10$ and merges with Rayleigh scattering. This effect is less pronounced for the larger values of n_r because the Rayleigh region is more depressed and the rainbow is at a scattering angle further from the Rayleigh maximum. The negative polarization features for size parameters ~ 5 are due to edge rays and resulting surface waves, i.e., they are "glory" phenomena. The angular size of the region into which the glory is focused increases inversely with the size parameter.

Fig. 3 shows the polarization for $n_c=n_r=1.44$ and the size distribution (8) with $b=0.05$, 0.07 and 0.15. The integration limits on x for these three cases were (0,34.3), (0,34.3) and (0,68.6), respectively. For the top part of the figure these integration limits are not equivalent to $(0,\infty)$. Nevertheless, the differences which would exist in the polarization if we used the limits $(0,\infty)$ are small enough that none of our conclusions are affected by the finite integration limits. The calculations were for the same phase angles and size parameters as for Fig. 2.

---

[1] Typical values of v_{eff} for terrestrial atmospheric particles are (Hansen and Travis, 1974): 0.05–0.4 for water clouds, 0.5–20. for tropospheric hazes, and 0.05–0.1 for the particles in the aerosol layer near 20 km (Junge layer).

Fig. 3 thus illustrates the effect of the width of the size distribution on the polarization. The qualitative effect of broadening the distribution is easy to understand: it roughly corresponds to taking averages along vertical lines. Thus, with increasing b hills tend to be smoothed out, holes are filled in, and corners are rounded off; straight vertical lines, however, remain essentially unchanged.

The feature of anomalous diffraction (at $2\pi a/\lambda \approx 10$, $\alpha \approx 20°$) is one of the most sensitive to the particle size distribution. Its maximum polarization is almost cut in half as b increases from 0.05 to 0.07, and the feature is washed away for $b=0.15$. This is understandable since only a narrow distribution of sizes has the phase shift required for the interference feature.

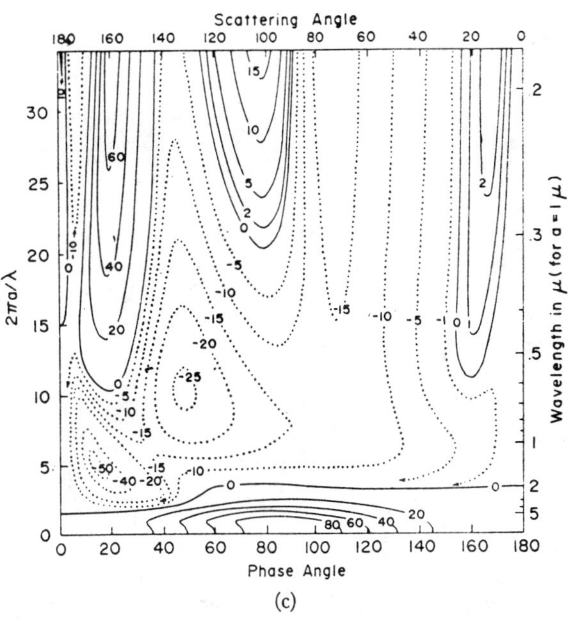

(c)

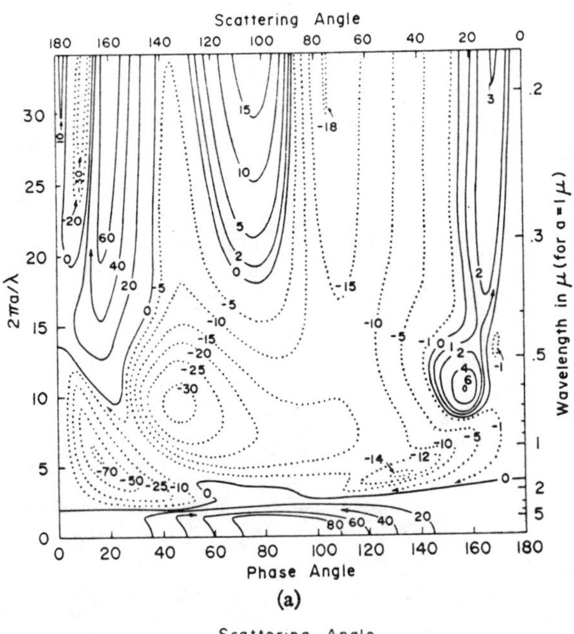

(a)

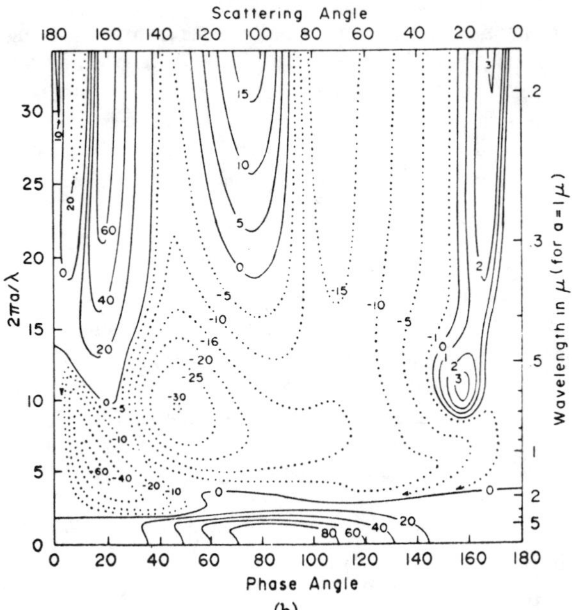

(b)

FIG. 3. Percent polarization for single scattering of unpolarized incident light by a size distribution of spheres, $-100P^{21}/P^{11}$. As in Fig. 2 except that the three parts of the figure show results for three effective variances, $b=0.05$(a), 0.07(b) and 0.15(c) with $n_r=1.44$ in all three cases.

The bridge of positive polarization formed by the merging of the primary rainbow and Rayleigh scattering (e.g., for $n_r=1.33$ and 1.40 in Fig. 2) is also strongly affected by the width of the size distribution. For $n_r=1.33$ and $b=0.05$ the polarization is positive for all size parameters at scattering angles $\sim 140°$; thus, for any broader size distribution the polarization for single scattering must be positive at $\alpha \approx 140°$. On the other hand, it is apparent that for $n_r=1.40$ the bridge will be eroded away for a broad distribution. We have verified this with computations for $b=0.16$ which show a breach of negative polarization with a vertical extent of ~ 1.3 in size parameter. The extent of this breach increases for larger b.

The existence or non-existence of the bridge of positive polarization is important for the application to Venus. Coffeen (1969) has noted that at $\lambda=1\,\mu\mathrm{m}$ the polarization of Venus is negative for all phase angles. By making the assumption that multiple scattering does not change the sign of the polarization from that which exists for single scattering, Coffeen concluded that for the Venus cloud particles $n_r \gtrsim 1.43$, since his computations showed the existence of the bridge for smaller n_r. However, his computations were all for a very narrow size distribution, with $v_{\mathrm{eff}}=1/48 \approx 0.02$. If v_{eff} is permitted to be as large as for terrestrial clouds and the other assumptions of Coffeen are kept, it can only be concluded that the lower limit on the refractive index is $n_r \gtrsim 1.37$.

Lyot (1929) and Coffeen (1969) noted that the qualitative effect of multiple scattering on the polarization is to reduce the degree of polarization without changing its sign. Thus, the features in the single scattering described above are useful for interpreting the polarization of scattered light, even for a thick atmosphere. Hansen and Hovenier (1971), however, found from accurate computations that the polarization of multiple-scattered photons is not negligible, particularly for photons which are scattered only two or three times. In the case of cloud particles which are comparable to or larger than the wavelength, the polarization of photons emerging from the atmosphere after just a few scatterings is qualitatively similar to the polarization for single scattering. This is a result of the fact that many photons are scattered in the forward direction on the first one or two scatterings at which time they are still nearly unpolarized, and then they are scattered out of the atmosphere on their next scattering. Such photons have a polarization similar to that for single scattering, but with the features somewhat smeared out as a function of phase angle [cf. Figs. 24 and 25 of Hansen (1971b) and the accompanying explanation]. Thus, multiple scattering, in addition to reducing the degree of polarization, causes a smoothing of the polarization along horizontal lines in Figs. 2 and 3. This smoothing is less pronounced than that along vertical lines due to the distribution of particle sizes; furthermore, it is fully accounted for by an exact multiple scattering theory.

All the computations we have illustrated pertain to spheres. For small nonspherical particles there would also be a region of Rayleigh scattering. For large size parameters the same division of rays into diffracted, reflected and refracted components can be made as in the case of spheres. However, the polarization for most of these components will, in general, be quite different than for spheres. Rainbows, for example, depend on the particle having a circular cross section; thus, long circular cylinders cause a rainbow, and for such cylinders oriented perpendicular to the incident light the polarization of the rainbow is very similar to that for spheres (cf. Liou, 1972). The glory requires a spherical particle shape; this is clear from the physical origin of the glory and it is illustrated, for example, by computations for circular cylinders (Liou, 1972) and by the absence of a glory for terrestrial ice clouds. The feature of anomalous diffraction would be smoothed away by an irregular particle shape or by a random orientation of any nonspherical particles in the same way that it is smoothed away by a broad size distribution; this is because the path length through the particle must have a fixed relation to the path length outside the particle.

Nonspherical particles with a regular shape may give rise to specific polarization features of their own. For example, large hexagonal ice crystals ($n_r = 1.31$) yield a concentration of negatively polarized light at $\alpha \approx 22°$, the so-called 22° halo (cf. Minnaert, 1954; O'Leary, 1966). Similarly a 90° crystal interface causes a 46° halo for ice crystals. Usually, however, there is a distribution of particle orientations and particle shapes; this tends to average out sharp angular features in the polarization. Laboratory observations (e.g., Lyot, 1929; Huffman, 1970) and airborne observations of ice clouds (Coffeen and Hansen, 1973; Coffeen et al., 1974) tend to verify that the polarization for nonspherical particles is a smoother function of scattering angle than it is for spheres.

The computations we have illustrated are for a real refractive index, i.e., for $n_i = 0$. This is sufficient for application to the visible clouds of Venus, as is shown in the next section.

3. Multiple scattering

The computations which we present in this paper are for the simplest possible relevant model: a homogeneous locally-plane-parallel atmosphere. It is preferable to fully investigate this model before adding complications such as vertical and horizontal inhomogeneities. Indeed, with a thorough understanding of this model in hand the general effects of such complications can be anticipated, as is discussed below and in Section 5.

Our multiple scattering computations were made with the doubling method. van de Hulst (1963) developed this method in essentially the form which we use, but without polarization; it was extended to include polarization by Hansen (1971a) and Hovenier (1971). The method provides a prescription for obtaining the reflection and transmission matrices for an atmosphere composed of two layers from the reflection and transmission matrices for each of the component layers. Thus, by choosing the two layers to be identical a thick atmosphere can be built up geometrically. The dependence on azimuth was handled by means of Fourier series expansions. Complete formulas are given in the above references.

In most of our computations we began the doubling at an optical thickness $\tau_0 = 2^{-15}$, with the reflection and transmission matrices for the layer of this thickness obtained from equations for first-order and second-order scattering (Hovenier, 1971). In some computations τ_0 was chosen as 2^{-20} which is sufficiently small for only single scattering to be employed for this layer. These initial optical thicknesses and the number of Gauss points used in the integrations over angle were adequate to allow a final accuracy comparable to the thickness of the curves in the figures for polarization versus phase angle. Some of the checks which we have made on the accuracy are listed by Hansen and Hovenier (1971).

In principle, the computations should employ phase matrices of four rows and four columns [cf. (2)]. However, for the case of spherical particles and incident unpolarized light, Hansen (1971b) has shown that the error in the degree of polarization for multiple scattered light is $\lesssim 0.00002$ for the approximation obtained by setting $P^{43} = 0 = P^{34}$. This error, of course, is negligible for our purposes, and thus in our computations we included only the first three rows and columns in the phase matrices and reflection and transmission matrices.

Most of the polarization observations of Venus refer to the total light from the visible part of the planetary disk. Thus, we integrated the results of our multiple scattering computations over the planetary disk, assuming a spherical but locally-plane-parallel atmosphere. Only at large phase angles is the neglect of atmospheric curvature of possible significance, and computations with the Monte Carlo method (Kattawar and Adams, 1971; Collins *et al.*, 1972) indicate that even at these phase angles the difference in polarization is small between the locally-plane-parallel model and the model accounting for curvature. Cloud bumpiness or waviness may have an effect at large phase angles at least comparable to that of atmospheric curvature (cf. Öpik, 1962; van Blerkom, 1971); if such effects exist it is not clear whether a model accounting for curvature or a locally-plane-parallel model is more accurate. However, single scattering strongly dominates in the polarization for large phase angles and thus the choice of model is probably not very important. The disk integration was performed at about 50 phase angles in the range 0–180°, the exact values of the phase angles being more concentrated near sharp features in the polarization.

We basically followed the disk-integration method of Horak (1950), which involves a double-quadrature, or "cubature," over the visible disk. The Stokes parameters of the reflected radiation were evaluated at each of the cubature points by linear interpolation from the Stokes parameters computed at the Gauss divisions with the doubling method. At small phase angles ($\alpha \lesssim 20°$) the interpolation did not yield sufficiently accurate results at the cubature points; this difficulty was overcome by employing certain integral equations satisfied by the reflection matrix [cf. Chandrasekhar (1950, p. 169) and correction indicated by Hovenier (1969, p. 493)] to improve the accuracy of the Stokes parameters at the cubature points. The procedure is similar to that described by Horak and Little (1965) for unpolarized light. The accuracy of the disk integration was tested by comparison with Horak's results, by comparison with the analytic solution for a Lambert surface, by computations for cases involving only single scattering, and by varying the number of cubature and Gauss points. Most of the final computations were made with 50-point cubature over half of the visible disk. In the figures which we present, the accuracy should be comparable to the thickness of the lines.

In most of our computations we took the optical thickness of the atmosphere to be $\tau = \infty$ and chose the single scattering albedo $\tilde{\omega}_0$ to yield the observed spherical albedo of Venus.[2] The value of $\tilde{\omega}_0$ required to

yield the spherical albedo A can be found for any phase matrix without iteration as follows. Table A1 of Chamberlain and Smith (1970) gives the value $\tilde{\omega}_0^{\mathrm{iso}}$ which will yield the spherical albedo A for isotropic scattering with $\tau = \infty$. The similarity relation (Hansen, 1969; van de Hulst and Grossman, 1968)

$$\tilde{\omega}_0 = 1 - (1 - \tilde{\omega}_0^{\mathrm{iso}})(1 - \langle \cos\alpha \rangle) \tag{10}$$

then yields the required single scattering albedo; $\langle \cos\alpha \rangle$ is the asymmetry parameter of the phase matrix for which $\tilde{\omega}_0$ is required. The simple form of the similarity relation (10) is sufficient if $(1 - \tilde{\omega}_0) \ll 1$ as is the case for Venus.

We numerically verified that the polarization in the above case is practically indistinguishable from the case $\tilde{\omega}_0 = 1$ with the ground albedo zero and τ chosen to yield the observed spherical albedo. With a homogeneous atmosphere, and the phase matrix and spherical albedo of the planet fixed, the only way a significantly different polarization could be obtained is with a thin atmosphere (and a high ground albedo). But τ must be large for the Venus atmosphere; measurements by the Soviet spacecraft Venera 8 of sunlight transmitted by the atmosphere of Venus (Avduevsky *et al.*, 1973) indicate that the cloud optical thickness is $\tau_c \gtrsim 10$ (Lacis and Hansen, 1974). The model-insensitivity of the polarization for a fixed phase matrix and planetary albedo is easy to understand. The degree of polarization is the ratio of the intensity of polarized light to the total intensity, I_{pol}/I. The total intensity I is essentially determined by the spherical albedo and the phase function in the upper part of the atmosphere; I_{pol} is determined by photons scattered not more than a few times, i.e., by the phase matrix in the upper part of the atmosphere.

Most of the phase matrices we employed were computed for real refractive indices, i.e., for $n_i = 0$. This special case is sufficient for the following reasons. The high albedo of Venus throughout the region $0.4 \lesssim \lambda \lesssim 2.5$ μm requires that the cloud particles have either a very small value of n_i or such a large refractive index that they are essentially "metallic." In the latter case, however, the polarization would be entirely different from that which is observed for Venus, as has been noted by Coffeen (1969). To test the possible effect of small values of n_i on the polarization we found by iterative calculations the value of n_i required to yield the $\tilde{\omega}_0$ corresponding to the assumed spherical albedo of Venus. This exercise was performed once for $\lambda = 0.55$ μm and once for $\lambda = 0.365 \,\mu$m. At all phase angles the polarization, graphed as in Figs. 4–12, was indistinguishable from the case in which the phase matrix was computed for $n_i = 0$.

Thus, the phase matrix for the cloud particles on Venus depends on the real refractive index n_r and the particle size distribution, as discussed in Section 2. For the size distribution (8), which is sufficient to represent the major characteristics of most naturally occurring

---

[2] Computations with the doubling method were actually stopped at $\tau = 256$, which for all practical purposes was equivalent to $\tau = \infty$.

distributions, the two parameters are the effective radius a and the effective variance b. In addition, the phase matrix for a unit volume of the atmosphere depends on the ratio of the Rayleigh scattering coefficient (per unit length) to the cloud particle scattering coefficient, i.e.,

$$f = \frac{k_{\text{sca},R}}{k_{\text{sca},c}}, \qquad (11)$$

and the complete phase matrix is

$$\mathbf{P} = \frac{1}{1+f}\,\mathbf{P}_c + \frac{f}{1+f}\,\mathbf{P}_R, \qquad (12)$$

where $\mathbf{P}_c$ and $\mathbf{P}_R$ are the phase matrices for the cloud particles and for isotropic Rayleigh scattering.[3] In order to represent the Rayleigh contribution to the phase matrix in terms of a single wavelength-independent parameter we let $f_R \equiv f(\lambda = 0.365\ \mu m)$. At wavelengths other than $0.365\ \mu m$ we use[4]

$$f = \left[\frac{0.365}{\lambda(\text{in }\mu m)}\right]^4 f_R. \qquad (13)$$

Thus, for our homogeneous model atmosphere the polarization at a particular wavelength is a function of n_r, a, b and f_R. Although the assumption of a homogeneous mixture of cloud particles and Rayleigh scatterers may be very inaccurate, this significantly affects the polarization only in the ultraviolet. The values derived for n_r, a and b are independent of this assumption; only the number density of particles is uncertain.

We have made several hundred separate multiple scattering computations for comparison to available observations of Venus. In obtaining $\tilde{\omega}_0$ from (10) the assumed spherical albedo of Venus was based mainly on the intermediate-bandwidth photometric observations reported by Irvine (1968), and for $\lambda = 0.55\ \mu m$ also on the observations of Knuckles *et al.* (1961). The uncertainty in the spherical albedo is $\sim 10\%$. This translates into a comparable uncertainty in the degree of polarization, I_{pol}/I, since its only significant effect is on I. Thus, in most cases the variability in the theoretical polarization (for given n_r, a, b, f_R) due to the uncertainty in the spherical albedo is not more than

---

[3] With this phase matrix we are still able to account for the effects of anisotropic Rayleigh scattering, as discussed below and in Appendix A.

[4] Eq. (13) follows from the assumption that σ_R is proportional to λ^{-4} and σ_c is independent of wavelength for the region in which polarization observations are available and f is significant ($0.34 \lesssim \lambda \lesssim 0.6\ \mu m$). The first assumption is sufficiently accurate for most gases including CO_2. The accuracy of the second assumption depends on the particle size distribution; however, we made several computations in which the variation of σ_c with λ was accounted for (using the Mie theory) and we found that this affected the final polarization by at most a few tenths percent polarization.

TABLE 1. Examples of the single scattering albedo which approximately yield a given spherical albedo for the case $a = 1.05$ μm, $b = 0.07$, $f_R = 0$. $\langle\cos\alpha\rangle$ was computed from Mie theory and $\tilde{\omega}_0$ from (10).

λ (μm)	n_r	$\langle\cos\alpha\rangle$	A	$\tilde{\omega}_0$
0.99	1.43	0.715	0.90	0.99941
0.55	1.44	0.718	0.87	0.99897
0.365	1.46	0.761	0.55	0.98427

a few tenths percent polarization. Table 1 shows the values of $\tilde{\omega}_0$ employed at three wavelengths for the case $a = 1.05\ \mu m$, $b = 0.07$ along with the computed spherical albedos for the case $f_R = 0$. Since $\tilde{\omega}_0$ was not changed for other values of f_R, the computed spherical albedo in the ultraviolet increased slightly with increasing f_R; for example, for $f_R = 0.045$ the computed spherical albedo at $\lambda = 0.365\ \mu m$ is $\sim 58\%$. The wavelengths in Table 1 are those employed in most of the graphs we present and they span the region containing most of the polarization observations. In the remainder of this section we present a number of the results selected to illustrate the dependence of the polarization on wavelength and on the parameters n_r, a, b and f_R.

a. Wavelength $\lambda = 0.55\ \mu m$

We first present results for $\lambda = 0.55\ \mu m$ because this wavelength region is the most sensitive to the particle size and it is also sensitive to the refractive index. In Figs. 4–6 we include a Rayleigh contribution $f_R = 0.045$ as derived from the ultraviolet observations. We have not varied f_R in these figures because the effect of Rayleigh scattering is sufficiently small at $\lambda = 0.55\ \mu m$ (it increases the polarization by $\sim 1\%$ at phase angle $90°$, compared with $f_R = 0$) that variations due to the uncertainty in $f_R(\sim 0.01)$ are negligible.

The observations in Figs. 4–6 include those made by Lyot (1929) with a visual polarimeter in the 1920's; these have been reproduced in many publications during the last 50 years. Although they refer to a rather broad wavelength region (~ 800 Å full-width at half-maximum for a completely dark-adapted eye), they are in good agreement with the other observations which were made with intermediate-bandwidth filters [~ 600 Å at $\lambda = 0.55\ \mu m$ (cf. Coffeen and Gehrels, 1969)]. All of the calculations are for a single wavelength. The finite bandwidth of the filters is essentially equivalent to an integration over size parameter, but the width of the intermediate bandwidth filter is sufficiently small that it should not have a major impact on the interpretations. Furthermore, a quantitative analysis shows that accounting for the finite bandwidth of the observations modifies the derived value of b by $\lesssim 0.01$.

Figs. 4 and 5 illustrate the effect of the size distribution on the polarization. In the computations of the phase matrix for these and all following figures the integration over particle radii was for the interval

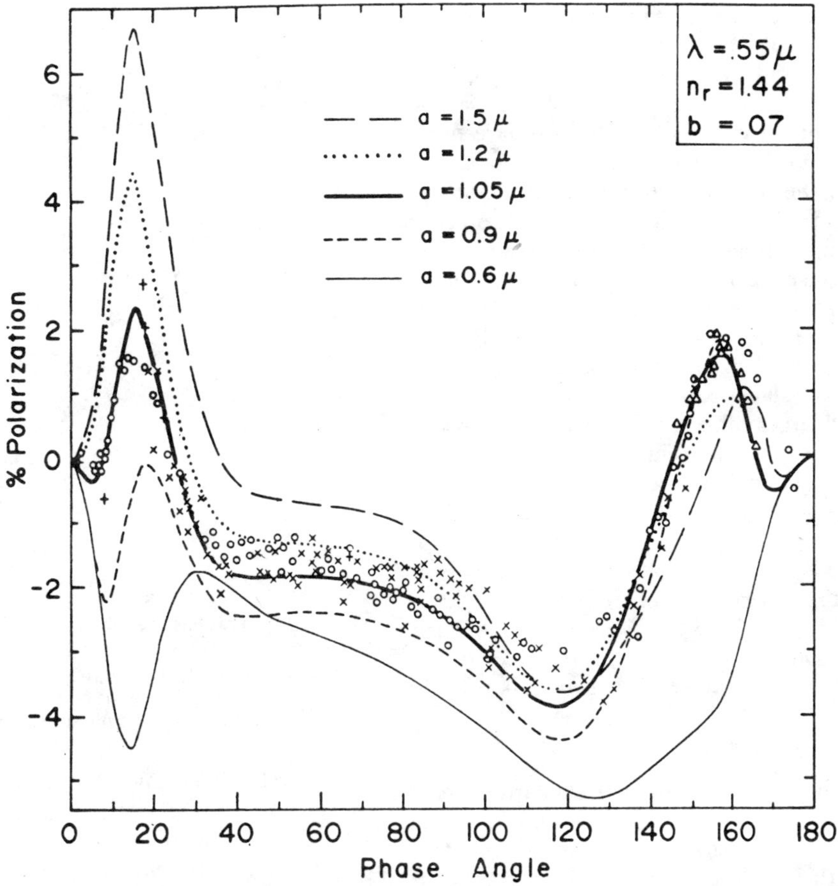

FIG. 4. Observations of the polarization of sunlight reflected by Venus in the visual wavelength region and theoretical computations for $\lambda = 0.55$ μm. The $\circ$'s are wide-band visual observations by Lyot (1929) while the other observations are for an intermediate bandwidth filter centered at $\lambda = 0.55$ μm; the $\times$'s were obtained by Coffeen and Gehrels (1969), the $+$'s by Coffeen (cf. Dollfus and Coffeen, 1970), and the Δ's (which refer to the central part of the crescent) by Veverka (1971). The theoretical curves are all for a refractive index 1.44, the size distribution (8) with $b = 0.07$, and a Rayleigh contribution $f_R = 0.045$. The different curves show the influence of the effective radius on the polarization.

(0, 5 μm), which was sufficient that $a = r_{\text{eff}}$ and $b = v_{\text{eff}}$. In Figs. 4 and 5 the refractive index is $n_r = 1.44$, a value which provides good agreement with the observations. In Fig. 4 b is held constant at 0.07 and a is allowed to vary from 0.6 to 1.5 μm. For values of a outside this range the discrepancies with the observations increase. In Fig. 5 a is held constant at 1.05 μm and b is allowed to vary over a wide range. For b smaller than 0.02 or larger than 0.25 the discrepancies with the observations are still larger than the extremes illustrated.

In the theoretical curves the maximum in the polarization at phase angles $\sim 20°$ is the primary rainbow. The maximum at $\sim 155°$ is the feature of anomalous diffraction. If observations were only available in this one wavelength region, there would be no assurance that the features in the observations were actually due to a rainbow and anomalous diffraction. However, as is demonstrated below, large variations of these features occur with changing wavelength in precise agreement

with the theory for spheres, including a changeover toward Rayleigh scattering at wavelengths in the infrared. These wavelength variations are sufficient to demonstrate conclusively that our interpretation of the polarization features is valid.

Figs. 4 and 5 indicate that, if the refractive index is ~ 1.44, the polarization is consistent with the effective radius, ~ 1.05 μm± 0.1 μm, and the effective variance, $\sim 0.07 \pm 0.02$. Results at other wavelengths confirm these values, though the polarization at most other wavelengths is not as sensitive to the size distribution as it is at $\lambda = 0.55$ μm. Note that the observations of Veverka (1971), obtained in search of a possible halo effect, are very useful for defining the anomalous diffraction feature and thus the width of the size distribution; Lyot (1929) also observed that feature but his observations were for a broader bandwidth.

Since the refractive index may, in principle, vary significantly with wavelength, it was necessary to actually do a huge number of computations to try to investigate all possible values of a, b and $n_r(\lambda)$. Fig. 6 shows theoretical results for the refractive indices, 1.33, 1.4 and 1.5 at $\lambda = 0.55\,\mu m$. For each value of n_r the value of a was chosen which yields the best agreement with the observations, the main criterion for agreement being the fit to the positive polarization maximum at phase angle $\sim 15°$. It is possible to get the theoretical rainbow at the appropriate phase angle for a fairly wide range of n_r, because the angular location of the rainbow varies somewhat with a (cf. Fig. 2); thus, the larger values of n_r require a larger value of a to fit the observations. We used $b = 0.05$ in Fig. 6 so that the results can be related to the single scattering contour diagrams of Fig. 2. With a larger value of b the results for $n_r = 1.40$ can be brought into fair agreement with the observations. However, by making the assumption that n_r does not vary by more than ~ 0.01 over several hundred angstroms in the visible region (cf. Section 4), we are able to conclude from the observations shown in Figs. 4–6 and the observations of Coffeen and Gehrels (1969) at $\lambda = 0.52$ and $0.655\,\mu m$, Veverka at $0.655\,\mu m$ and Dollfus (cf. Dollfus and Coffeen, 1970) at $\lambda = 0.527$, 0.593 and $0.617\,\mu m$ that $n_r(\lambda = 0.55\,\mu m) = 1.44 \pm 0.015$.

b. Wavelength $\lambda = 0.99\,\mu m$

At wavelengths $\sim 1\,\mu m$ the observed polarization of Venus is negative at all phase angles, in contrast to the results for shorter wavelengths. The variation of the polarization with wavelength has a smooth transition as indicated by observations of Dollfus (1966), Coffeen

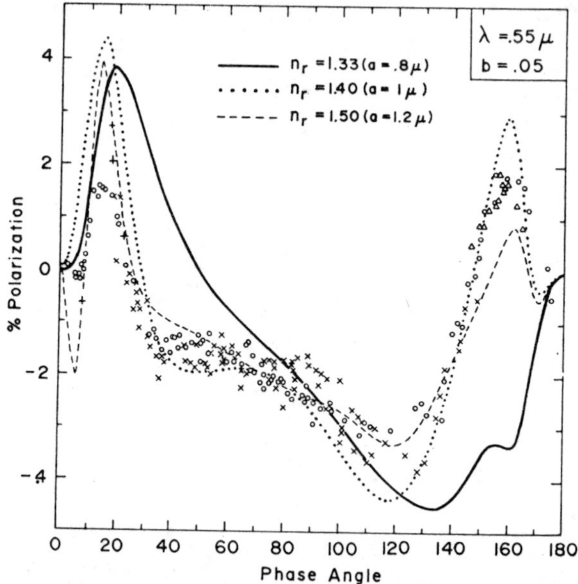

FIG. 6. Observations are the same as in Fig. 4. The theoretical curves are for three refractive indices with the effective particle radius chosen in each case to yield the best agreement with the observations. The size distribution is (8) with $b = 0.05$. The Rayleigh contribution to the phase matrix is given by $f_R = 0.045$

and Gehrels (1969) and Dollfus and Coffeen (1970) at several wavelengths in the region $0.34 \lesssim \lambda \lesssim 1\,\mu m$. The large difference in the polarization between visible and near-infrared wavelengths is a qualitative indication that the cloud particles must be on the order of the wavelength in size for this spectral region. If the particles were much larger (or smaller) than the wavelength the general shape of the polarization curve as a function of phase angle would not change so drastically with wavelength; this is illustrated in Figs. 2 and 3.

Single scattering computations for spheres qualitatively agree with the observations at $\lambda = 0.99\,\mu m$ for refractive indices $1.37 \lesssim n_r \lesssim 2$ [cf. Section 2 and Coffeen (1969)]. As illustrated by Figs. 2 and 3 the polarization for the relevant range of the effective size parameter is very sensitive to both the refractive index and the particle size. Thus, with accurate multiple scattering computations it is possible to find fairly narrow limits for these parameters.

Figs. 7 and 8 show observations for $\lambda = 0.99\,\mu m$, which is the wavelength in the near infrared with the greatest number of observations. The theoretical curves for these figures were computed with values of $\tilde{\omega}_0$ obtained from (10) with the spherical albedo of Venus taken as 0.90 (cf. Irvine, 1968). A Rayleigh contribution specified by $f_R = 0.045$ was included in the computations, but it had a negligible effect on the polarization. In Fig. 7 the theoretical curves are for different refractive indices, the particle size being chosen for each refractive index to yield the best agreement *at all wavelengths*. In choosing a special emphasis was placed on fitting the rainbow, which is present in the observed

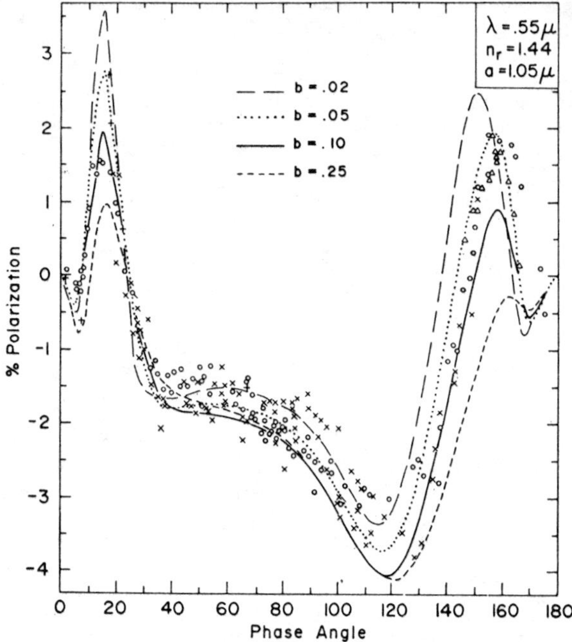

FIG. 5. As in Fig. 4 except that all of the theoretical curves are for $a = 1.05\,\mu m$, while the effective variance is allowed to range over the values 0.02 to 0.25.

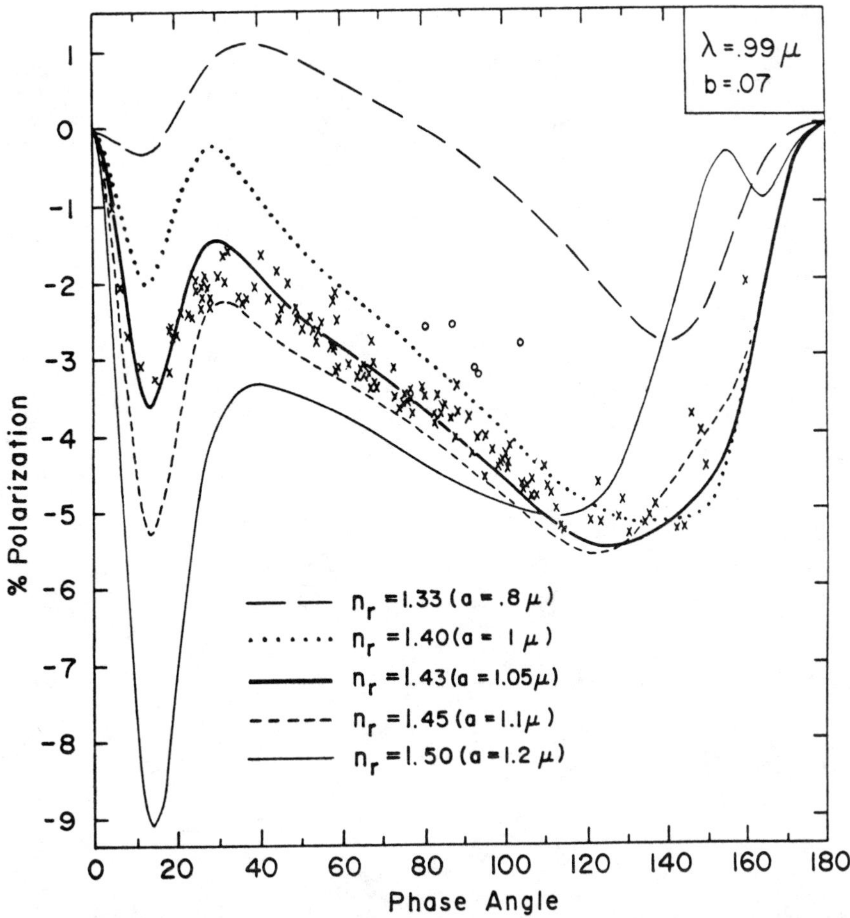

FIG. 7. Observations and theoretical computations of the polarization of sunlight reflected by Venus at $\lambda = 0.99$ μm. The observations were made with an intermediate bandwidth filter, the ×'s being obtained by Coffeen and Gehrels (1969) in 1959–67 and by Coffeen (cf. Dollfus and Coffeen, 1970) from 1967 to March 1969, and the ○'s being obtained by Coffeen (cf. Dollfus and Coffeen, 1970) in May–July, 1969. The theoretical curves are for spherical particles having the size distribution (8) with $b = 0.07$. The different theoretical curves are for various refractive indices, the effective particle radius being selected in each case to yield closest agreement with the observations for all wavelengths.

polarization for wavelengths $\lesssim 1$ μm, and it was assumed that the dispersion of refractive indices between the visible region and $\lambda = 0.99$ μm is $\lesssim 0.04$ (cf. Section 4).

Fig. 7 illustrates the sensitivity of the polarization to the refractive index and indicates that $n_r \approx 1.43$ provides the best fit. All of the curves in this figure are for $b = 0.07$; however, as is shown in Fig. 3, the polarization at $\lambda \approx 1$ μm (for $a \approx 1$ μm) is less sensitive to b than it is in the visible region. Furthermore, since the size distribution must be narrow enough for the anomalous diffraction feature to exist (at $\lambda \approx 0.5$–0.6 μm), there is little room for varying b.

Fig. 8 shows the sensitivity of the polarization to the effective particle radius a. A rather narrow range of acceptable sizes is indicated, with $a \approx 1$ μm. For larger particles the rainbow becomes much too pronounced to agree with the observations. For smaller particles the theoretical polarization becomes less negative than that

observed, and it finally shifts to positive polarization at all phase angles as the Rayleigh region is approached.

The most recent observations of Coffeen (cf. Dollfus and Coffeen, 1970), shown as circles in Fig. 7 and 8, indicate a less negative polarization than the more numerous earlier observations of Coffeen and Gehrels (1969). Observations of Dollfus at $\lambda = 0.95$ and 1.05 μm (cf. Dollfus and Coffeen, 1970) taken during some of the same years as the observations of Coffeen and Gehrels also tend to fall above the observations of Coffeen and Gehrels, although on the average they are not as far above as the circles in Figs. 7 and 8. The spread of the observations is thus somewhat greater than would be suggested by examination of only the early observations of Coffeen and Gehrels. However, the derived refractive index, $n_r(\lambda = 0.99$ μm$) = 1.43 \pm 0.015$, was obtained with cognizance of all the observations.

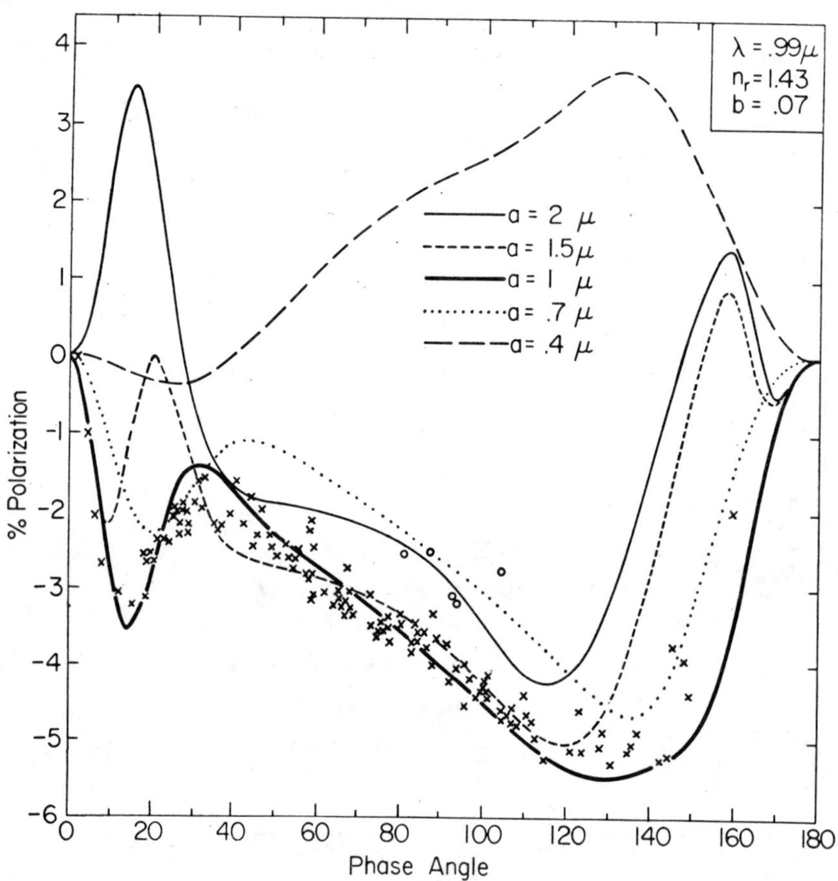

FIG. 8. The observations are the same as in Fig. 7. The theoretical curves are for spheres with refractive index $n_r = 1.43$. The particle size distribution is (8) with $b = 0.07$ and with the different curves for five values of the effective radius a.

c. Wavelength $\lambda = 0.365\ \mu m$

The polarization in the ultraviolet is less sensitive to the particle size distribution than it is in the visible or infrared regions. This can be understood from the single scattering contour diagrams (Figs. 2 and 3) along with knowledge of the fact that the observations at longer wavelengths require a to be $\sim 1\ \mu m$. However, the UV observations are very useful for establishing the optical thickness of Rayleigh scatterers in or above the clouds. In addition the refractive index can be rather accurately obtained in this wavelength region since the rainbow has a sharp maximum.

The most extensive UV observations are for an intermediate bandpass centered at $\lambda \approx 0.365\ \mu m$. In Figs. 9 and 10 the crosses are observations of Coffeen and Gehrels (1969) obtained from 1959 through 1967. The maximum polarization obtained in the rainbow during those years was $\sim 7\%$ with the peak at phase angle $\sim 17°$; however, all of the points near the rainbow peak were measured during one apparition in 1965 except one point at phase angle 21.1° for which a polarization of 5.6% was obtained in 1967. The circles are observations obtained in 1967–69 by Coffeen (cf.

Dollfus and Coffeen, 1970). These show a maximum polarization $\sim 10\%$ in the rainbow; two of the three points with polarization greater than 9% were obtained in January 1967 and the third during a different apparition in May 1968. The triangles are observations of Dollfus obtained in 1969 (cf. Dollfus and Coffeen, 1970); these measurements refer to the central portion of the crescent. We include these local observations in our figure because there are no published disk-integrated observations at large phase angles for $\lambda \approx 0.365\ \mu m$. Below we illustrate that at large phase angles the theoretical polarization for this spot on the planet differs little from that for the light integrated over the planet.

The calculations in Fig. 9 are for the size distribution (8), with $a = 1.05\ \mu m$ and $b = 0.07$ being selected to obtain good agreement with the polarization observations at all wavelengths. The refractive index, $n_r = 1.45$, was chosen to give a good fit to the location of the rainbow for $\lambda = 0.365\ \mu m$. Fig. 9 thus illustrates the effect of Rayleigh scattering on the polarization in the ultraviolet. As described above, f_R is the ratio of the Rayleigh scattering coefficient to the cloud particle scattering

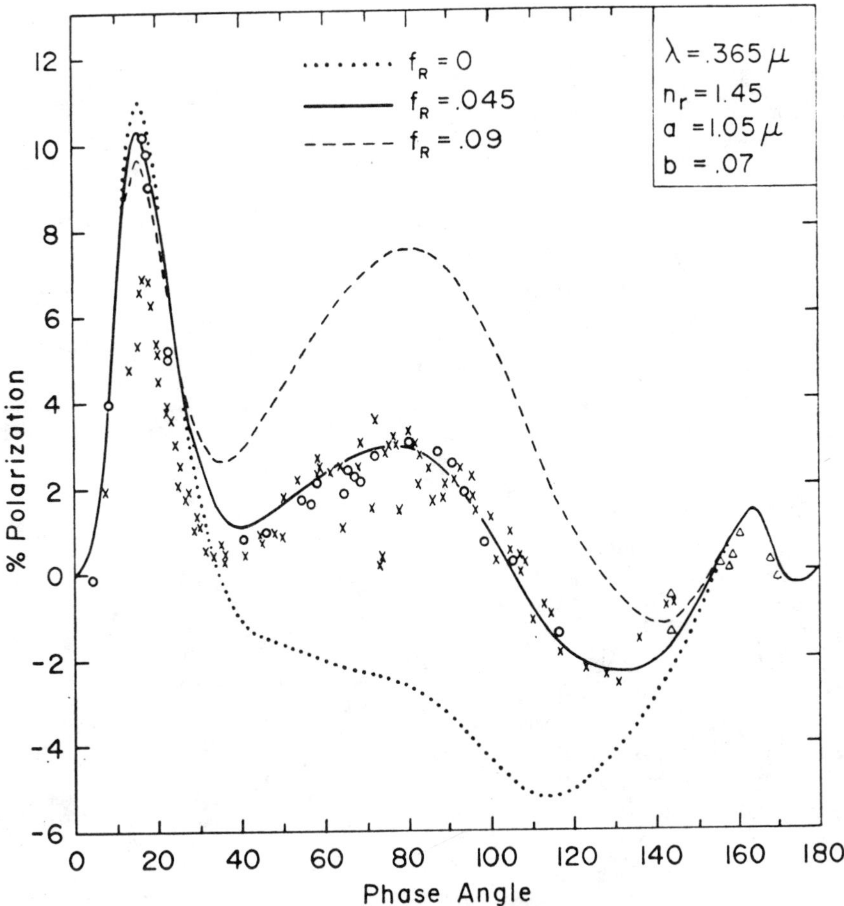

FIG. 9. Observations and theoretical calculations of the polarization of sunlight reflected by Venus at $\lambda = 0.365$ μm. The observations were made with intermediate bandwidth filters centered at $\lambda = 0.365$ μm, the ×'s being obtained by Coffeen and Gehrels (1969) in 1959–67, the ○'s by Coffeen in 1967–69 (cf. Dollfus and Coffeen, 1970), and the △'s by Dollfus in 1969 (cf. Dollfus and Coffeen, 1970). The observations of Coffeen and Gehrels refer to the entire visible planetary disk, while those of Dollfus refer to an area on the disk midway between limb and terminator. The theoretical curves are the results after integration over the visible disk. They are for a homogeneous atmosphere containing spherical particles of the size distribution (8) with $a = 1.05$ μm and $b = 0.07$. The refractive index of the particles is 1.45. The three theoretical curves are for different Rayleigh contributions to the phase matrix.

coefficient. Fig. 9 indicates that the best value of f_R for a homogeneous atmosphere is $f_R \approx 0.045$. Using the relation

$$p_1 \approx 1.16 f_R \quad [p_1 \text{ in bars}], \tag{14}$$

where p_1 is the pressure level at cloud optical depth unity (see the Appendix),[5] we find that the atmospheric pressure at cloud optical depth unity is ~ 50 mb.

This derived pressure level depends significantly on our assumption of a homogeneous atmosphere. However, even if the atmosphere of Venus is vertically inhomogeneous (as is probable) the cloud optical depth unity must nevertheless be at a pressure on the order of 50 mb. This is a result of the fact that, roughly stated, the polarization arises from optical depths

between zero and unity. It is unlikely that the error in the pressure at cloud optical depth unity is more than ~ 25 mb. However, this quantity and the vertical distribution of particles should be more thoroughly investigated by means of computations for an inhomogeneous atmosphere and comparisons to observations of high spatial resolution.

Fig. 10 illustrates the sensitivity of the polarization in the ultraviolet to the refractive index. For each of the three refractive indices, 1.33, 1.4 and 1.5, that value of a is shown which yields best agreement with the observations for all wavelengths; in choosing a it was assumed that the dispersion of the refractive index between the UV and visible regions is $\lesssim 0.04$. Parameters $b = 0.07$ and $f_R = 0.045$ are the same for all three curves. For $n_r = 1.40$ the agreement with observations at phase angles $\sim 80°$ can be improved by choosing $f_R = 0.035$,

---

[5] We also show in the Appendix that the effect of molecular anisotropy on this relation is negligible.

while for $n_r = 1.50$ the agreement can be improved at phase angles $\sim 160°$ by choosing $b = 0.10$. However, such variations of f_R and b do not significantly affect the rainbow, which is the primary determinant of the refractive index at this wavelength.

Figs. 9 and 10 show that the refractive index for $\lambda = 0.365\ \mu$m is ~ 1.45. The value $n_r = 1.46$ fits at least as well as 1.45, the observations on the sides of the rainbow being approximately equidistant from the calculated curve for $n_r = 1.46$. The movement of the rainbow for small changes of n_r can be estimated by interpolating between the results in Figs. 9 and 10; also, Hansen and Arking (1971) and Coffeen and Hansen (1973) have published computations for $n_r = 1.46$. As is indicated below, the best agreement at $\lambda = 0.34\ \mu$m appears to occur with n_r between 1.46 and 1.47. Taking this into account we conclude that $n_r(\lambda = 0.365\ \mu\text{m}) = 1.46 \pm 0.015$.

The UV observations in the rainbow region during 1965 show a polarization smaller than that measured in two later apparitions and smaller than that obtained theoretically. Dollfus and Coffeen (1970) discussed the possibility of a "red leak" in the observing systems, which would have decreased the polarization, but they concluded that the time variation of the polarization was probably real. The UV polarization was also lower at most other phase angles during the 1965 apparition (cf. Fig. 3 of Dollfus and Coffeen, 1970), and in general the polarization is more variable in the ultraviolet than

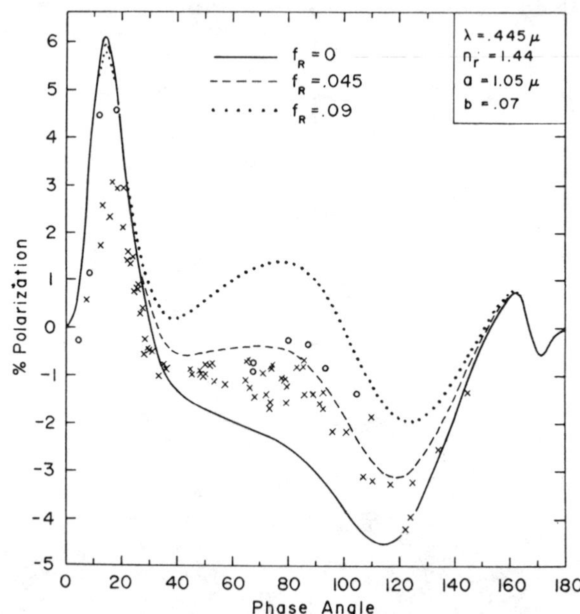

Fig. 11. Observations and theoretical calculations of the polarization of sunlight reflected by Venus at $\lambda = 0.445\ \mu$m. The observations were made with intermediate bandwidth filters, the ×'s being obtained by Coffeen and Gehrels (1969) in 1959–67 and the ○'s by Coffeen in 1967–69 (cf. Dollfus and Coffeen, 1970). The theoretical curves are for a homogeneous atmosphere containing spherical particles of the size distribution (8) with $a = 1.05\ \mu$m and $b = 0.07$. The refractive index of the particles is 1.45. The three theoretical curves are for different Rayleigh contributions to the phase matrix.

at longer wavelengths (cf. Fig. 17 of Coffeen and Hansen, 1973). It seems likely that the variability of the UV polarization is related to the variable UV markings on the planet (Boyer and Camichel, 1961; Dollfus, 1968; Boyer and Guerin, 1969; Scott and Reese, 1972).

Variations in the polarization could arise, for example, from changes in the cloud height, the number density of cloud particles, the UV cloud albedo, or the fraction of the planet covered by high clouds. It is probable that the effects of both vertical and horizontal atmospheric inhomogeneities are more significant in the ultraviolet than at longer wavelengths. Indeed, UV polarization and intensity measurements with high spatial and temporal resolution, e.g., from an orbiting spacecraft, would be ideally suited for investigating the atmospheric structure.

d. Other wavelengths

The three wavelengths for which results of calculations are presented above are those with the greatest number of observations, and they practically span the region $(0.34–1\ \mu\text{m})$ in which measurements have been concentrated. However, we have examined all of the observations mentioned in Section 1 and compared them with theoretical computations. Below we give a

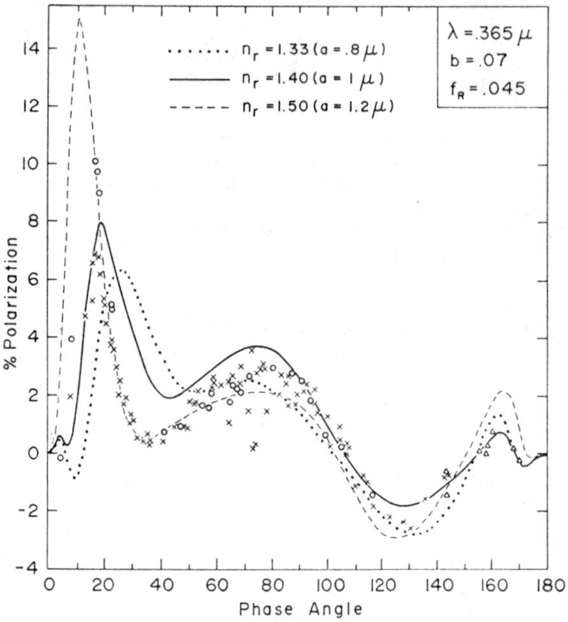

Fig. 10. The observations are the same as in Fig. 9. The theoretical curves are for a homogeneous atmosphere containing spherical particles of the size distribution (8) with $b = 0.07$. The Rayleigh contribution to the phase matrix is specified by $f_R = 0.045$. The different theoretical curves are for different refractive indices, in each case with a chosen to yield the best agreement at all wavelengths.

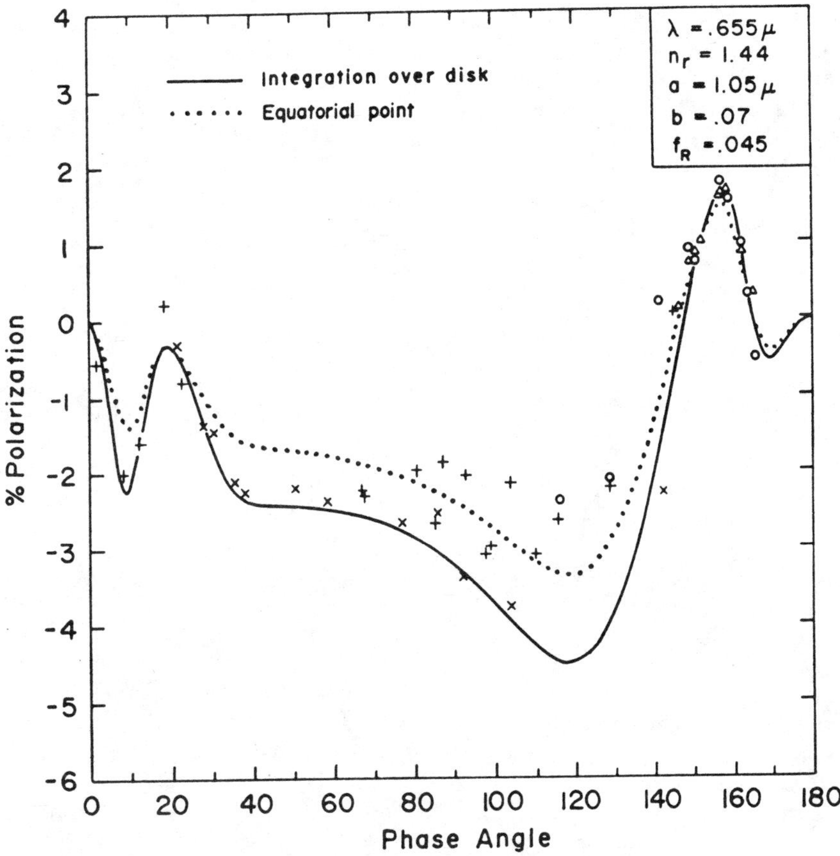

FIG. 12. Observations and theoretical calculations of the polarization of sunlight reflected by Venus at $\lambda = 0.655\ \mu$m. The observations were made with intermediate bandwidth filters. The ×'s (Coffeen and Gehrels, 1969) and +'s (Dollfus and Coffeen, 1970) refer to the entire visible planetary disk. The ○'s (Dollfus and Coffeen, 1970) and △'s (Veverka, 1971) refer to the central portion of the crescent. The theoretical computations are for a homogeneous atmosphere containing spherical particles of the size distribution (8) with $a = 1.05\ \mu$m and $b = 0.07$. The refractive index of the particles is $n_r = 1.44$ and the Rayleigh contribution is $f_R = 0.045$. The solid curve refers to light integrated over the planetary disk and the dotted curve refers to the point on the equator with $\theta_0 = \theta = \rho/2$.

summary of results at other wavelengths. Except where indicated otherwise, the wavelengths mentioned below refer to intermediate passband filters of Coffeen and Gehrels. The observations of Dollfus, which are in essentially the same region as the observations of Coffeen and Gehrels, are in rather good agreement with those of Coffeen and Gehrels (cf. Dollfus and Coffeen, 1970).

At $\lambda = 0.34\ \mu$m the observations are similar to those at $\lambda = 0.365\ \mu$m. Under the assumption that $a \approx 1.05\ \mu$m, $b \approx 0.07$, the best fit to the rainbow is for $n_r = 1.46$ or 1.47. The observations at intermediate phase angles are fit best with a Rayleigh scattering contribution $f_R \approx 0.04$.

At $\lambda = 0.4\ \mu$m the observations are entirely those of Dollfus (cf. Dollfus and Coffeen, 1970). Calculations with $a = 1.05\ \mu$m, $b = 0.07$ and $n_r = 1.45$ are in good agreement with the observations. All of the observations for phase angles 35°–110° were obtained during one apparition (1967). For those observations the best fit for the contribution of Rayleigh scattering is $f_R \approx 0.05$.

Observations for $\lambda = 0.445\ \mu$m are shown in Fig. 11. The crosses were obtained by Coffeen and Gehrels (1969) in 1959–67; the circles were obtained by Coffeen in 1967–69 (cf. Dollfus and Coffeen, 1970). Most of the points in the rainbow region were measured during one apparition in 1965; of the two points with polarization exceeding 4%, one was obtained in 1967 and the other during a different apparition in 1968. The theoretical curves in Fig. 11 are for $n_r = 1.44$, $a = 1.05\ \mu$m and $b = 0.07$. These were computed with $\bar{\omega}_0$ obtained from (10) for an assumed spherical albedo of Venus of 0.75 (cf. Irvine, 1968). The theoretical curves illustrate that a Rayleigh contribution $f_R \approx 0.045$ fits the 1967–69 observations, while $f_R \approx 0.035$ is a better fit to the older observations. Dollfus' observations at $\lambda = 0.44\ \mu$m (cf. Dollfus and Coffeen, 1970) are fit best by $f_R \approx 0.045$. The rainbow at $\lambda = 0.445\ \mu$m can be fit about equally well for any refractive index in the range 1.43–1.46.

At $\lambda=0.52\,\mu$m the observations are similar to those at $\lambda=0.55\,\mu$m. Several measurements are available in the region of anomalous diffraction. These are useful for establishing the width of the size distribution and they are in agreement with $b\approx0.07$.

Observations for $\lambda=0.655\,\mu$m are shown in Fig. 12. Those obtained by Coffeen and Gehrels (1969) in 1966–67 and those by Coffeen in 1967–1969 (cf. Dollfus and Coffeen, 1970) refer to the entire visible planetary disk; those obtained by Coffeen in 1969 (cf. Dollfus and Coffeen, 1970) and Veverka (1971) in 1969 refer to the central portion of the crescent. The theoretical curves in Fig. 12 are for $n_r=1.44$, $a=1.05\,\mu$m, $b=0.07$ and $f_R=0.045$. These were computed with $\tilde{\omega}_0$ obtained from (10) with the spherical albedo of Venus taken as 0.95 (cf. Irvine, 1968). The solid curve is the theoretical polarization for the entire visible planetary disk, while the dotted curve is for the point on the equator with $\theta_0=\theta=\rho/2$, where θ_0 and θ are the zenith angles for the incident and emergent light, respectively, and ρ is the phase angle. The absolute value of the polarization is generally higher for the disk-integrated light than for the light from the equatorial midpoint. This is easy to understand since a significant fraction of the disk-integrated light is from the limb or terminator, where single scattering dominates and the polarization is high. Thus, the polarization for the area observed about the equator should also be somewhat less than that for the entire visible planetary disk. The available observations at $\lambda=0.655\,\mu$m are primarily useful for determining the width of the size distribution; they indicate that $b\approx0.07$.

At $\lambda=0.685\,\mu$m and $\lambda=0.74\,\mu$m the observations fit calculations for $n_r=1.43$ or 1.44, $a=1.05\,\mu$m, $b=0.07$ and $f_R=0.045$. There are no observed points in the region of anomalous diffraction and only a few at phase angles less than 20°.

At $\lambda=0.875\,\mu$m the observations are in good agreement with calculations for $n_r=1.43$, $a=1.05\,\mu$m, $b=0.07$ and $f_R=0.045$.

A few observations have been made in the infrared between wavelengths 1.25 and $3.6\,\mu$m. Forbes (1971) has made several observations at $\lambda=1.25$, 1.65, 2.25 and $3.6\,\mu$m. These observations contain large fluctuations, sometimes varying by 2 or 3% polarization in a few days. Kuiper (1957) made measurements at wavelength $2\,\mu$m, reporting results for the phase angle 80° and for several phase angles in the range 141°–162°. The infrared observations are thus too sparse and uncertain to allow a precise refractive index or other detailed information to be derived. However, it is significant that the observations yield negative polarization at all observed phase angles for $\lambda=1.25$, 1.65 and $2\,\mu$m, but primarily positive polarization for $\lambda=2.2\,\mu$m and only positive polarization for $\lambda=3.6\,\mu$m. The contour diagrams in Figs. 2 and 3 show that this is just the behavior expected of cloud particles with $a\approx1\,\mu$m, and thus our basic interpretation of the polarization is

reconfirmed in a region of the spectrum where the polarization has an entirely different nature than that in the visual region.

4. Refractive indices

We have shown that the real refractive index of the Venus cloud particles is 1.44 ± 0.015 at $\lambda=0.55\,\mu$m, and that the refractive index varies from ~1.46 at $\lambda=0.365\,\mu$m to ~1.43 at $\lambda=0.99\,\mu$m. The main value of this information is the precise criterion it provides for the cloud particle composition. Different materials which have been proposed for the composition of the Venus clouds can be examined to see if they have the appropriate $n_r(\lambda)$; to be in agreement with the polarization of Venus the material must also have a negligible absorbtivity in the visual region and it must form spherical particles at the conditions in the Venus cloudtops.

a. Interpolation formulas

The refractive indices of most materials vary significantly with temperature and wavelength, but in many cases laboratory measurements of n_r exist for at most a few values of T and λ. Thus, it is important to have a method to interpolate and/or extrapolate from a few measured points. For this purpose we use two formulas of classical physics, given, for example, by Born and Wolf (1965). These formulas are applicable to isotropic substances in a spectral region with negligible absorbtivity, provided also that the frequency is high enough ($\gtrsim 10^{11}$ Hz) to ensure that the molecules behave as nonpolar molecules.

The formula we use for the temperature dependence of the refractive index is the Lorentz-Lorenz equation written in the form

$$\frac{n_r{}^2-1}{n_r{}^2+2}=\frac{M}{W}\rho(T), \qquad (15)$$

where M is the molar refractivity of the material, W the molecular weight, and $\rho(T)$ the density at temperature T. If the molecules keep their identity, M is nearly independent of temperature and density for most materials. For example, a typical variation of M is an increase of 0.01% per degree Kelvin (cf. Batsanov, 1961); it has also been experimentally verified at 20C for sodium light that dn_r/dT for water and CS_2 is almost entirely accounted for by the change in density with temperature (cf. Jaffé, 1928). Thus (15) gives the temperature dependence of n_r at any specific wavelength, provided reliable data are available for $\rho(T)$.

We also need a formula for the dispersion, i.e., for the wavelength dependence of the refractive index. The refractive index of a material at a particular density satisfies

$$\frac{n_r{}^2-1}{n_r{}^2+2}=\sum_i \frac{s_i}{\nu^2-\nu_i{}^2}, \qquad (16)$$

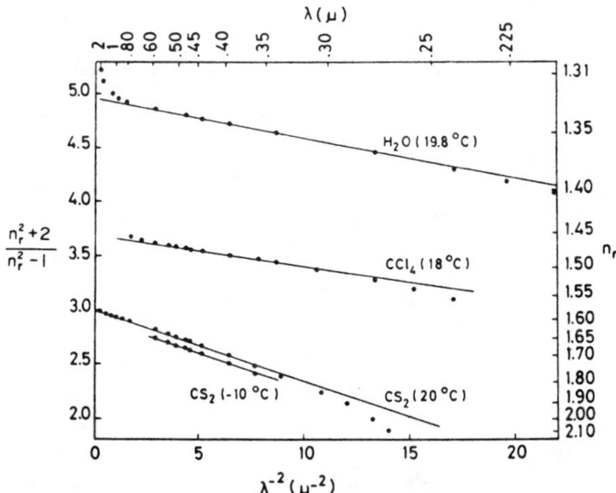

Fig. 13. Refractive indices of several liquids as a function of wavelength. Experimental values are represented by dots. The data for water were taken from Jaffé (1928) and the data for CCl$_4$ and CS$_2$ were derived from tables of Hellwege (1962). The straight lines are eyeball fits to the near-linear domains.

where ν is the frequency and each ν_i is a resonance frequency corresponding to a "strength" s_i. Damping is neglected in (16); this is equivalent to the assumption that absorption is negligible, as is true at frequencies far from any ν_i. Thus, the formula is particularly useful in the visible part of the spectrum for substances transparent to the eye. The resonance frequencies for such materials fall in the ultraviolet and in the infrared, and the refractive index for visible light is primarily determined by the UV resonances.

We would like an interpolation formula which has only a few parameters, but which is still accurate enough for our purposes. Thus, we replace (16) with the simple formula

$$\frac{n_r^2-1}{n_r^2+2}=\frac{a}{\nu^2-\nu_a^2},\qquad(17)$$

where the two parameters a and ν_a can be determined from values of n_r measured for a given temperature at two or more wavelengths. Eq. (17) can be thought of as lumping all resonance frequencies into an average resonance frequency ν_a. Although many empirical dispersion formulas have been published, we have not found the above formula used in the literature as a simple interpolation tool. If the denominator on the left-hand side of (17) is omitted, we obtain a one-term Sellmeier dispersion formula which has often been used (cf. Born and Wolf, 1965, Sec. 2.3). However, one advantage of (17) is that, according to (15), a is simply proportional to the density.

We have used laboratory measurements of n_r to check the accuracy of (17) and to demonstrate the significance of deviations from that equation. We rewrite (17) in the form

$$\frac{n_r^2+2}{n_r^2-1}=\frac{c^2}{a}\left(\frac{1}{\lambda^2}-\frac{1}{\lambda_a^2}\right),\qquad(18)$$

where c is the speed of light, and plot the measured values as $(n_r^2+2)/(n_r^2-1)$ vs $1/\lambda^2$. We have made such plots for a number of colorless liquids at various temperatures. Fig. 13 shows results for water, CCl$_4$ and CS$_2$. In each case we find that the linear dependence implied by (18) is indeed a good approximation over a wide range of wavelength. To bring this out more clearly we have included in Fig. 13 straight lines which are eyeball fits to the near-linear domains. We conclude that for colorless liquids $(n_r^2+2)/(n_r^2-1)$ is approximately linearly dependent on $1/\lambda^2$ for visible wavelengths.

The deviations from straight lines in Fig. 13 arise from strong absorption in the ultraviolet and infrared. For water the increase in $(n_r^2+2)/(n_r^2-1)$ for $\lambda \gtrsim 1\,\mu$m is a result of the fundamental vibration-rotation band in the $\lambda \approx 3\,\mu$m region. As shown by the data of Irvine and Pollack (1968), the absorbtivity of water has a strong maximum at $\lambda \approx 2.95\,\mu$m with a corresponding minimum in n_r at $\lambda \approx 2.75\,\mu$m. At the other end of the spectrum, $(n_r^2+2)/(n_r^2-1)$ for CS$_2$ bends below the straight line, indicating appreciable absorption in the near-ultraviolet.

Fig. 13 also illustrates a check on the temperature dependence of n_r. According to Eq. (15), $(n_r^2+2)/(n_r^2-1)$ should shift by some constant factor when the temperature is changed. This is demonstrated by data for CS$_2$. The ratio of the experimental values for $(n_r^2+2)/(n_r^2-1)$ at -10C and $+20$C varies only from 0.97149 at $\lambda=0.3612\,\mu$m to 0.97087 at $\lambda=0.5893\,\mu$m.

b. Application to Venus

The open circles in Fig. 14 are the refractive indices deduced from the polarization of Venus, with the "error bars" representing the maximum uncertainty in n_r. The three open circles happen to fall almost exactly on a straight line, as indicated by a heavy line. It is gratifying that the dispersion obtained for the Venus cloud particles is "normal," i.e., of the same sign and small magnitude as for colorless liquids. This result is consistent with the observed high albedo of Venus from $\lambda \approx 0.35\,\mu$m to $\lambda \approx 2\,\mu$m.

Hansen and Arking (1971) found that none of the materials proposed for the Venus clouds prior to 1970 were in good agreement with the polarization. However, carbon suboxide (C$_3$O$_2$) and an aqueous solution of hydrochloric acid (HCl·nH$_2$O) are sufficiently near the required n_r to warrant examination. In Fig. 14 the dots for C$_3$O$_2$ at 273K represent measurements of Diels and Blumberg (1908) for the Fraunhofer C, D and G lines ($\lambda=0.656$, 0.589 and 0.434 μm, respectively). The point for $T=261$K, $\lambda=0.589\,\mu$m was also measured by Diels and Blumberg; we obtained the other points for that temperature from the assumption that the ratio of $(n_r^2+2)/(n_r^2-1)$ for the two temperatures is the same at all wavelengths. Observations of several types indicate that the temperature of the Venus cloud tops is $\sim$220–250K: absorption bands in reflected sunlight suggest $T \approx 250$K (Young, 1972); thermal radiation in the 8–13μ m window indicates $T \approx 220$–250K (cf. Hanel *et al.*, 1968); the radiometric albedo of 77% (Irvine, 1968) corresponds to an effective temperature $\sim$237K; atmospheric models based on Mariner 5 measurements yield a temperature $\sim$225K at the 50-mb level (Fjeldbo *et al.*, 1971). Thus, we conclude that the clouds of Venus are not C_3O_2.

Lewis (1971, 1972) has suggested that the clouds of Venus may be an aqueous solution of hydrochloric acid with 25–30% HCl by weight at a temperature near 200K. In Fig. 14 we have plotted the refractive index for a 27.6% HCl solution at 20C; the data were obtained by interpolating between measurements of Howell (cited by Timmermans, 1960) for proximate concentrations. In addition, we have used density determinations of Garret and Woodruff (1951) and the Lorentz-Lorenz formula (15) to reduce the values of $(n_r^2+2)/(n_r^2-1)$ to 200K; a similar procedure has been used by Lewis. Fig. 14 illustrates that this solution of HCl is not compatible with the refractive index of the Venus cloud particles, especially since the temperature in the Venus clouds is probably greater than 200K. A stronger HCl concentration might have the appropriate refractive index, but such strong concentrations are apparently not consistent with observed abundances of gaseous HCl and H_2O (Young, 1973).

Recently, Sill (1972) and Young (1973) have independently suggested that the cloud particles on Venus may be a strong aqueous solution of sulfuric acid [cf. also Young and Young (1973) and Young (1974)]. Young proposes a freezing solution which he specifies as 75.9% H_2SO_4 by weight at 250K; Sill suggests an 86% concentration at $T=235$K. We have derived values of n_r at several wavelengths for a 75.9% solution at $T=15$C by interpolation between measurements of Veley and Mauley (cf. Timmermans, 1960) for proximate concentrations. The resulting values for $(n_r^2+2)/(n_r^2-1)$ are shown as dots in Fig. 14. Young (1973) has given the value $n_r(\lambda=0.589\,\mu m)=1.44193$ for $T=250$K which he obtained by extrapolating density measurements for higher temperatures and by using the Lorentz-Lorenz relation. From this value and the assumption of a constant scale factor for $(n_r^2+2)/(n_r^2-1)$, we obtained $n_r(\lambda)$ at that temperature. The result is in excellent agreement with the refractive index of the Venus cloud particles. Somewhat different concentrations and temperatures would also be compatible with the polarization, e.g., concentrations $\gtrsim 75\%$ may have the required refractive index if they are supercooled.

None of the other materials which have been suggested in the literature as composing the Venus clouds and clearly specified are compatible with the polarization. We mention here two substances for which there have been claims of conclusive identification with the visible clouds of Venus: H_2O and hydrated $FeCl_2$.

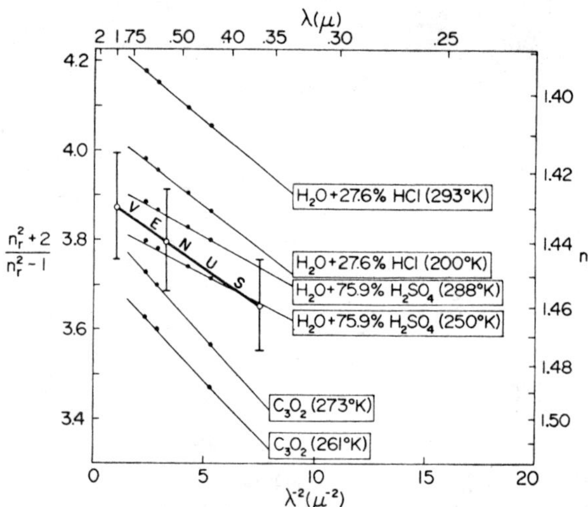

FIG. 14. Refractive indices of the Venus cloud particles (open circles) deduced from the polarization. The error bars represent the maximum uncertainty, not a probable error. The experimental values (dots) for the indicated liquids are based on laboratory measurements and, in some cases, interpolation formulas, as explained in the text.

Water has a refractive index much smaller than that of the particles in the visible Venus clouds, as illustrated in Fig. 13. Freezing the water only increases the disparity with the polarization of Venus, because ice tends to form nonspherical crystals. Iron chloride has a refractive index much larger than that of the particles in the visible Venus clouds (Kuiper, 1969); in addition, iron chloride would probably be in the form of platy crystals at the conditions in the Venus clouds.

5. Conclusions

Analysis of the polarization of sunlight reflected by Venus leads to several specific conclusions on the nature of the visible clouds of Venus. This information is remarkable because of the precision and certainty with which it defines physical properties of the cloud particles. These properties are:

REFRACTIVE INDEX AND DISPERSION. The index of refraction of the particles in the visible clouds is $n_r=1.44\pm0.015$ at $\lambda=0.55\,\mu$m. The indicated uncertainty refers to the limits for acceptable values, not to a probable error. The refractive index has a normal dispersion, decreasing from 1.46 ± 0.015 at $\lambda=0.365\,\mu$m to 1.43 ± 0.015 at $\lambda=0.99\,\mu$m.

PARTICLE SHAPE. The particles in the visible clouds of Venus are spherical. There is a clear signature of the particle shape in the polarization as a function of scattering angle, and its interpretation is confirmed by the variation of the polarization with wavelength.

SIZE DISTRIBUTION (MEAN SIZE AND VARIANCE). The effective radius [cf. (6)] for the size distribution of the cloud particles is $1.05\pm0.10\,\mu$m and the effective variance [cf. (7)] is 0.07 ± 0.02. By terrestrial standards this is a narrow size distribution. Under the assumption that the shape of the Venus size distribution resembles the distribution (8) [cf. Fig. 1], the above values correspond to a mode radius $r_m = 0.83 \pm 0.08$ μm and a standard deviation $\sigma = 0.26$ μm.

PRESSURE AT THE CLOUD TOPS. The atmospheric pressure at cloud optical depth unity is ~ 50 mb. This value, unlike the properties above, depends on the model which we use for the vertical distribution of particles and gas. We estimate the uncertainty in this number to be ± 25 mb.

There is, in addition, an important corollary to the derived physical properties of the cloud particles: *the particle size, shape and refractive index are very uniform over most of the illuminated part of the planet.* This refers to the visible clouds down to an optical depth at least ~ 1. It has been demonstrated (e.g., Hansen and Hovenier, 1971) that the polarization is primarily due to single-scattered photons, with qualitatively similar but smaller contributions from photons scattered two or three times. Thus, if there is more than one type of particle present, due to a mixture or layering of particles at a given location or to variations across the planet, each type of particle contributes to the polarization essentially according to its share of the "single" scattered light. If there were two types of spherical particles, differing say in refractive index, the polarization of each would appear in the observations; for example, in the case of transparent spheres larger than the wavelength, the rainbows for both types of particles would be present in the polarization. However, all of the features observed in the polarization of Venus are due to the particles described above. It is possible that a small fraction of the planet is covered by particles of another type, especially if those particles are of an irregular shape such that their polarization is rather featureless. The polar regions of Venus, for example, contribute only a small fraction of the light from the total disk, so the cloud properties could be quite different there. Also, the deviations which sometimes exist between the observations in the ultraviolet and the theoretical polarization could be due to a variation of cloud particle properties over part of the planet.

It is surprising that the particle *size* is so uniform over the planet. The value derived for the effective variance of the size distribution refers to an average size distribution over the illuminated part of the planetary disk down to optical depth ~ 1. Such a uniform particle size is uncharacteristic of terrestrial clouds in which the effective radius varies from a few micrometers to about 100 μm, and the effective variance for individual clouds varies from ~ 0.05 to ~ 0.40. However, the size distribution of particles in the stratospheric aerosol (Junge) layer on Earth has been measured by Mossop (1965) and Friend (1966) who both found $a_{\mathrm{eff}} \approx 0.06$–$0.08$. Since the Junge layer exists at the pressure level ~ 50 mb, this suggests a close analogy with the atmospheric particles on Venus, as we have previously pointed out (cf. Hunten, 1971).

The above properties can be used for additional inferences about the nature of the Venus atmosphere. The most important of these concerns the *cloud composition.* Of all the materials proposed in the literature for the Venus clouds and clearly specified, the only material in agreement with the polarization is a concentrated solution of sulfuric acid. The similarities mentioned above between the Venus particles and the Junge layer on Earth, together with the fact that the Junge layer contains a large fraction of sulfuric acid (Rosen, 1971; Lazrus *et al.*, 1971; Toon and Pollack, 1973) support the conclusion that the composition of the cloud particles is sulfuric acid. Since the particles in the Junge layer are smaller [$r_{\mathrm{eff}} = 0.3$–0.4 μm; Friend (1966)] and have a much smaller optical thickness [$\sim 2 \times 10^{-2}$; Elterman *et al.* (1973)], the mass of particles above unit area on Venus must be at least a factor of 100 greater than for the Junge layer. However, Prinn (1973) has argued that a significant photochemical production rate for H_2SO_4 is possible in the atmosphere of Venus. In addition, Samuelson (private communication) and Young (1974) have found that the thermal infrared spectrum of Venus is in good agreement with sulfuric acid cloud particles. We thus conclude, primarily on the basis of the refractive index, that the clouds of Venus are probably composed of a strong sulfuric acid solution.

It is no doubt possible to also obtain the appropriate refractive index in the case of some other substances by judiciously adding certain impurities. Hapke (1972), for example, has suggested the possibility of "dirty" hydrochloric acid. But in any event we have obtained a stiff criterion from the polarization which can be used to test any proposed substance which is chemically specified, since $n_r(\lambda)$ can always be obtained from experiments and theory. To be consistent with the polarization the substance must also be in the form of spheres which are essentially homogeneous. This latter requirement is particularly evident from the rainbow, which becomes increasingly sharp as the wavelength decreases. Thus, undissolved mixtures such as dust and water are excluded, as are particles with a liquid coating on a nucleus which has a different refractive index.

The clouds examined by means of the polarization of reflected solar radiation are the visible clouds of Venus. This cloud layer (or thick haze) occurs high in the atmosphere by terrestrial standards, at a pressure corresponding to the altitude ~ 20 km on Earth. It is of course possible that there are other cloud layers deeper in the atmosphere with quite a different composition. Theoretical calculations of the polarization for a multi-layered atmosphere will still agree with the observations of Venus if the top layer of particles has an optical thickness $\tau_c \gtrsim 1$ and particles with the physical properties specified above. This lower limit on τ_c can be converted to a lower limit on the total number N_c of particles in the "polarization clouds" above unit area. Since the size distribution is narrow we may write $N_c Q_{ext}\pi a^2 \approx \tau_c \gtrsim 1$, where $Q_{ext}\approx 2$ (van de Hulst, 1957) and $a \approx 1\,\mu$m; thus, $N_c \gtrsim 2\times10^7$ cm^{-2}. With a density $\rho \approx 1.7$ gm cm^{-3}, appropriate for an approximate 75% H_2SO_4 solution, this corresponds to a cloud particle mass $\gtrsim 10^{-4}$ gm cm^{-2}. We do not yet have a reliable measure of the linear thickness of this cloud or haze region. However, it is worth emphasizing that its optical thickness is substantial; the "polarization clouds" are the visible clouds of Venus, not a tenuous upper haze.

Finally, we would like to point out two types of polarization observations which could considerably refine our knowledge of the composition and structure of the Venus clouds. The extension of accurate polarization observations into the infrared ($\lambda = 1$–4 μm) and ultraviolet ($\lambda < 0.34\,\mu$m) is needed in order to obtain the refractive index in the spectral regions where significant variations are probable. This would allow a definite identification of the cloud particle composition. In addition, observations in the UV and visible regions with a high spatial resolution could be used to obtain the vertical and horizontal distributions of cloud particles. To obtain the complete potential information would require observations of a given point on the planet from several different zenith angles, as could be obtained from an orbiting spacecraft.

Acknowledgments. We would like to thank D. Coffeen, H. Loman, H. C. van de Hulst and A. Young for helpful discussions and R. Jastrow for his hospitality at the Institute for Space Studies.

One of us (J.E.H.) was supported during the course of part of this work by NASA Grant 33-008-012 through Columbia University; the other (J.W.H.) was supported by the Netherlands Organization for the Advancement of Pure Research during a one-year stay at the Institute for Space Studies in 1970–71 when this work was originated.

APPENDIX

Anisotropic Rayleigh Scattering

The scattering by molecules was approximated by isotropic Rayleigh scattering in the calculations for the graphs presented in this paper. Although many molecules, particularly CO_2, are not isotropic, we show here that anisotropy has only a slight effect on the cloud-top pressure which we deduce from the polarization and also a negligible effect on our other conclusions. Anisotropy reduces the degree of polarization for single scattering, but the effect of this is practically canceled by a comparable increase in the molecular scattering cross section.

a. Phase matrix

The phase matrix for anisotropic Rayleigh particles in random orientation is

$$\mathbf{P}_R(\alpha) = \Delta \mathbf{P}_{IR}(\alpha) + (1-\Delta)\mathbf{P}_I, \tag{A1}$$

where

$$\mathbf{P}_{IR}(\alpha) = \begin{Bmatrix} \frac{3}{4}(1+\cos^2\alpha) & -\frac{3}{4}\sin^2\alpha & 0 \\ -\frac{3}{4}\sin^2\alpha & \frac{3}{4}(1+\cos^2\alpha) & 0 \\ 0 & 0 & \frac{3}{2}\cos\alpha \end{Bmatrix} \tag{A2}$$

is the phase matrix for isotropic Rayleigh scattering,

$$\mathbf{P}_I = \begin{Bmatrix} 1 & 0 & 0 \\ 0 & 0 & 0 \\ 0 & 0 & 0 \end{Bmatrix} \tag{A3}$$

is the phase matrix for isotropic scattering,

$$\Delta = \frac{1-\delta}{1+\delta/2}, \tag{A4}$$

and δ, the so-called depolarization factor, is the ratio of intensities parallel and perpendicular to the plane of scattering (I_l/I_r) for light single scattered at $\alpha = 90°$ with the incident light unpolarized. The above phase matrix is referred to the Stokes parameters $\{I,Q,U\}$. A derivation of the phase matrix is given, for example, by Chandrasekhar (1950), though for a different set of Stokes parameters. Measured values of δ are given by Penndorf (1957) for a number of gases; some values are $H_2 \sim 0.02$, $N_2 \sim 0.03$, air ~ 0.03, $O_2 \sim 0.06$ and $CO_2 \sim 0.09$.

From the above equations we find that for anisotropic Rayleigh scattering the degree of linear polarization is

$$\frac{\sin^2\alpha}{1+\cos^2\alpha+\left(\dfrac{2\delta}{1-\delta}\right)} \tag{A5}$$

for single scattering of unpolarized incident light. Thus, it is clear that anisotropy reduces the degree of polarization at all scattering angles, while the general shape as a function of scattering angle remains about the same.

b. Scattering coefficient

The scattering coefficient per unit length for anisotropic molecules in random orientation is

$$k_{sca} = \frac{8\pi^3}{3}\frac{(n_g^2-1)^2}{\lambda^4 N}\frac{6+3\delta}{6-7\delta}, \tag{A6}$$

where N is the number of molecules per unit volume, n_g the refractive index of the gas, and the last factor arises from the anisotropy. A derivation of (A6) is also given by Chandrasekhar (1950). For a mixture of gases

$$k_{sca} = \frac{8\pi^3}{3\lambda^4 N}\sum_i v_i(n_{g,i}^2-1)^2\frac{6+3\delta_i}{6-7\delta_i}, \tag{A7}$$

where $n_{g,i}$, δ_i and v_i are respectively the refractive index, depolarization factor, and fraction by volume of gas i.

c. p-τ relation

A relation between atmospheric pressure and the optical thickness due to Rayleigh scattering can be derived as follows. Outside of absorption bands the Rayleigh optical thickness due to the gaseous atmosphere above height h is

$$\tau_R(h) = \int_h^\infty \kappa_{sca}\rho \, dh', \qquad (A8)$$

where ρ is the density of gas and κ_{sca} the scattering coefficient of the molecules per unit mass. Assuming hydrostatic equilibrium, the pressure at height h is

$$p(h) = \int_h^\infty g\rho \, dh', \qquad (A9)$$

where g is the acceleration of gravity. Throughout the bulk of the atmosphere the dependence of κ_{sca} and g on height can be neglected; thus,

$$p = \frac{g\tau_R}{\kappa_{sca}}. \qquad (A10)$$

Since $k_{sca} = \kappa_{sca}\rho$ and $\rho = \bar{\mu}N$, where $\bar{\mu}$ is the mean molecular mass,

$$p = g\bar{\mu}\tau_R \left[\frac{8\pi^3}{3\lambda^4 N^2}\sum_i v_i(n_{g,i}^2-1)^2\frac{6+3\delta_i}{6-7\delta_i}\right]^{-1}. \qquad (A11)$$

Values for the temperature and pressure dependent quantities in the brackets can be taken for any set of conditions (e.g., for STP, in which case N is equal to Loschmidt's number).

d. Application to Venus

If we neglect molecular anisotropy (i.e., if we take $\delta = 0$) and assume a pure CO_2 atmosphere, (A11) leads to

$$p(\text{in bars}) \approx 1.16\tau_R(0.365\ \mu\text{m}), \qquad (A12)$$

where $\tau_R(0.365\ \mu\text{m})$ is the Rayleigh optical thickness at $\lambda = 0.365\ \mu$m. In obtaining (A12) we took $N = 2.687 \times 10^{19}$ cm^{-3}, $g = 870$ cm sec^{-2}, $\bar{\mu} = 44 \times 1.66 \times 10^{-24}$ gm and $n_g \approx 1.00046$ (cf. Allen, 1963, p. 87).

Eq. (A12) corresponds to (14) which gives the pressure p_1 at the level where the cloud optical depth is unity. Since the total polarization arises primarily from the region where the total $\tau \lesssim 1$, p_1 is also approximately the pressure at the $\tau = 1$ level even if the atmosphere of Venus does not approximate a homogeneous mixture of particles and gas. Thus, p_1 can be termed the cloud-top pressure.

Now consider the case of anisotropic Rayleigh scattering. For a given number of molecules, and thus for a given pressure, anisotropy increases the molecular optical thickness by the factor

$$\frac{1+\frac{1}{2}\delta}{1-\frac{7}{6}\delta}. \qquad (A13)$$

This corresponds to ~ 1.17 for the case of pure CO_2, assuming $\delta(CO_2) = 0.09$. However, (A5) shows that anisotropy also decreases the degree of polarization for single scattering by the factor

$$\frac{1+\cos^2\alpha}{1+\cos^2\alpha+\dfrac{2\delta}{1-\delta}}. \qquad (A14)$$

This corresponds to ~ 0.83 for $\delta = 0.09$ and $\alpha = 90°$, and it depends little on α. The product of the above two factors is within a few percent of unity for all relevant phase angles, i.e., $\sim 40°$–$140°$.

Thus, molecular anisotropy should have little effect on our determination of the atmospheric pressure level at the cloud top. We have verified this by making sample computations with the complete anisotropic Rayleigh phase matrix (A1). The results obtained with anisotropic Rayleigh scattering with $\delta = 0.09$ do not differ from those for isotropic Rayleigh scattering with the same number of molecules by more than the thickness of the lines in Figs. 9 and 11.

REFERENCES

Allen, C. W., 1963: *Astrophysical Quantities*. London, Athlone Press, 291 pp.

Arking, A., and J. Potter, 1968: The phase curve of Venus and the nature of its clouds. *J. Atmos. Sci.*, **25**, 617–628.

Avduevsky, V. S., M. Ya. Marov, B. E. Moshkin and A. P. Ekonomov, 1973: Venera 8: Measurements of solar illumination through the atmosphere of Venus. *J. Atmos. Sci.*, **30**, 1215–1218.

Batsanov, S. S., 1961: *Refractometry and Chemical Structure*. New York, Consultants Bureau, 250 pp.

Born, M., and E. Wolf, 1965: *Principles of Optics*. London, Pergamon Press, 808 pp.

Boyer, C., and H. Camichel, 1961: Observations photographiques de la planète Vénus. *Ann. Astrophys.*, **24**, 531–535.

——, and P. Guerin, 1969: Etude de la rotation rétrograde, en 4 jours, de la couche extérieur nuageuse de Vénus. *Icarus*, **11**, 338–355.

Bryant, H. C., and A. J. Cox, 1966: Mie theory and the glory. *J. Opt. Soc. Amer.*, **56**, 1529–1532.

Chamberlain, J. W., and G. R. Smith, 1970: Interpretation of the Venus CO_2 absorption bands. *Astrophys. J.*, **160**, 755–765.

Chandrasekhar, S., 1950: *Radiative Transfer*. London, Oxford University Press, 393 pp.

Coffeen, D. L., 1968: A polarimetric study of the atmosphere of Venus. Ph.D. thesis, University of Arizona, 116 pp.

——, 1969: Wavelength dependence of polarization. XVI. Atmosphere of Venus. *Astron. J.*, **74**, 446–460.

——, and T. Gehrels, 1969: Wavelength dependence of polarization. XV. Observations of Venus. *Astron. J.*, **74**, 433–445.

——, and J. E. Hansen, 1973: Polarization studies of planetary atmospheres. *Planets, Stars, and Nebulae Studied with Photopolarimetry*. Tucson, University of Arizona Press, 1133 pp.

——, —— and L. P. Whitehill, 1974: Polarization of sunlight reflected by terrestrial clouds: Observations and interpretation. To be submitted to *J. Atmos. Sci.*

Collins, D. G., W. G. Blättner, M. B. Wells and H. G. Horak, 1972: Backward Monte Carlo calculations of the polarization characteristics of the radiation emerging from spherical-shell atmospheres. *Appl. Opt.*, **11**, 2684–2696.

Cruikshank, D. P., and A. B. Thomson, 1971: On the occurrence of ferrous chloride in the clouds of Venus. *Icarus*, **15**, 497–503.

Dauvillier, A., 1956: Sur la nature des nuages de Vénus. *Comp. Rend.*, **243**, 1257–1258.

Deirmendjian, D., 1964: Scattering and polarization properties of water clouds and hazes in the visible and infrared. *Appl. Opt.*, **3**, 187–196.

Diels, O., and P. Blumberg, 1908: Über das Kohlensuboxyd. *Chem. Ber.*, **41**, 82–86.

Dollfus, A., 1966: Contribution au Colloque Caltech-JPL sur la lune et les planètes: Venus. Jet Propulsion Laboratory Tech. Memo. No. 33–266, 187–202..

——, 1968: Synthesis on the ultraviolet survey of clouds in Venus' atmosphere. *The Atmospheres of Venus and Mars*, J. C. Brandt and M. E. McElroy, Eds., New York, Gordon and Breach, 133–146.

——, and D. L. Coffeen, 1970: Polarization of Venus. I. Disk observations. *Astron. Astrophys.*, **8**, 251–266.

Donahoe, F. J., 1970: Is Venus a polywater planet? *Icarus*, **12**, 424–430.

Elterman, L., R. B. Toolin and J. D. Essex, 1973: Stratospheric aerosol measurements with implications for global climate. *Appl. Opt.*, **12**, 330–337.

Fahlen, T. S., and H. C. Bryant, 1968: Optical back scattering from single water droplets. *J. Opt. Soc. Amer.*, **58**, 304–310.

Fjeldbo, G., A. J. Kliore and V. R. Eshleman, 1971: The neutral atmosphere of Venus as studied with the Mariner V radio occultation experiments. *Astron. J.*, **76**, 123–140.

Forbes, F. F., 1971: Infrared polarization of Venus. *Astrophys. J.*, **165**, L21–L25.

Friend, J. P., 1966: Properties of the stratospheric aerosol. *Tellus*, **18**, 465–473.

Garret, A. B., and S. A. Woodruff, 1951: A study of several physical properties of electrolytes over the temperature range of 25°C to −73°C. *J. Phys. Colloid Chem.*, **55**, 477–490.

Hanel, R., M. Forman, G. Stambach and T. Meilleur, 1968: Preliminary results of Venus observations between 8 and 13 microns. *J. Atmos. Sci.*, **25**, 586–593.

Hansen, J. E., 1969: Absorption-line formation in a scattering planetary atmosphere: A test of van de Hulst's similarity relations. *Astrophys. J.*, **158**, 337–349.

——, 1971a: Multiple scattering of polarized light in planetary atmospheres. Part I. The doubling method. *J. Atmos. Sci.*, **28**, 120–125.

——, 1971b: Multiple scattering of polarized light in planetary atmospheres. Part II. Sunlight reflected by terrestrial water clouds. *J. Atmos. Sci.*, **28**, 1400–1426.

——, and A. Arking, 1971: Clouds of Venus: Evidence for their nature. *Science*, **171**, 669–672.

——, and J. W. Hovenier, 1971: The doubling method applied to multiple scattering of polarized light. *J. Quant. Spectros. Radiat. Transfer*, **11**, 809–812.

——, and L. D. Travis, 1974: Light scattering in planetary atmospheres. Submitted to *Space Sci. Rev.*

Hapke, B., 1972: Venus clouds: A dirty hydrochloric acid model. *Science*, **175**, 748–751.

Harteck, P., R. R. Reeves and B. A. Thompson, 1963: Photochemical problems of the Venus atmosphere. NASA TN D-1984, 39 pp.

Hellwege, A. M., 1962: Flüssigkeiten. *Landolt-Börnstein Zahlenwerte und Funktionen*, Band II, Teil 8. Berlin, Springer-Verlag, 901 pp.

Horak, H. G., 1950: Diffuse reflection by planetary atmospheres. *Astrophys. J.*, **112**, 445–463.

——, and S. J. Little, 1965: Calculations of planetary reflection. *Astrophys. J.* Suppl., **11**, 373–428.

Hovenier, J. W., 1969: Symmetry relationships for scattering of polarized light in a slab of randomly oriented particles. *J. Atmos. Sci.*, **26**, 488–499.

——, 1971: Multiple scattering of polarized light in planetary atmospheres. *Astron. Astrophys.*, **13**, 7–29.

Hoyle, F., 1955: *Frontiers of Astronomy*. New York, Harper, 360 pp.

Huffman, P., 1970: Polarization of light scattered by ice crystals. *J. Atmos. Sci.*, **27**, 1207–1208.

Hunten, D. M., 1968: The structure of the lower atmosphere of Venus. *J. Geophys. Res.*, **73**, 1093–1095.

——, 1971: Composition and structure of planetary atmospheres. *Space Sci. Rev.*, **12**, 539–599.

——, and R. M. Goody, 1969: Venus: The next phase of planetary exploration. *Science*, **165**, 1317–1323.

Irvine, W. M., 1968: Monochromatic phase curves and albedos for Venus. *J. Atmos. Sci.*, **25**, 610–616.

——, and J. B. Pollack, 1968: Infrared optical properties of water and ice spheres. *Icarus*, **8**, 324–360.

Jaffé, G., 1928: Dispersion und Absorption. *Handbuch der Experimentalphysik*, XIX, W. Wien and F. Harms, Eds., Leipzig, Akad. Verlag. M.B.H., 430 pp.

Kaplan, L. D., 1963: Spectroscopic investigation of Venus. *J. Quant. Spectrosc. Radiat. Transfer*, **3**, 537–539.

Kattawar, G. W., and C. N. Adams, 1971: Flux and polarization reflected from a Rayleigh-scattering planetary atmosphere. *Astrophys. J.*, **167**, 183–192.

——, G. N. Plass and C. N. Adams, 1971: Flux and polarization calculations of the radiation reflected from the clouds of Venus. *Astrophys. J.*, **170**, 371–386.

Kendall, M. G., and A. Stuart, 1963: *The Advanced Theory of Statistics. I. Distribution Theory*. New York, Hafner, 433 pp.

Kerker, M., 1969: *The Scattering of Light and Other Electromagnetic Radiation*. New York, Academic Press, 666 pp.

Khrgian, A., Kh., 1961: *Cloud Physics*. Israel Program Scientific Translation, Jerusalem, 392 pp.

Knuckles, C. F., M. K. Sinton and W. M. Sinton, 1961: UBV photometry of Venus. *Lowell Obs. Bull.*, **5**, 153–156.

Kratohvil, J. P., 1964: Light scattering. *Anal. Chem.*, **36**, 458R–472R.

Kuiper, G. P., 1957: The atmosphere and cloud layer of Venus. *Threshold of Space*. M. Zelikoff, Ed. New York, Pergamon, 342 pp.

——, 1969: Identification of the Venus cloud layers. *Comm. Lunar Planet. Lab.*, No. 101, 1–21.

Lacis, A. A., and J. E. Hansen, 1974. Atmosphere of Venus: Implications of Venera 8 sunlight measurements. Submitted to *Science*.

Lazrus, A. L., B. Gandrud and R. D. Cadle, 1971: Chemical composition of air filtration samples of the stratospheric sulfate layer. *J. Geophys. Res.*, **76**, 8083–8088.

Lewis, J. S., 1968: Composition and structure of the clouds of Venus. *Astrophys. J.*, **152**, L79–L83.

——, 1969: Geochemistry of the volatile elements on Venus. *Icarus*, **11**, 367–385.

——, 1971: Refractive index of aqueous HCl solutions and the composition of the Venus clouds. *Nature*, **230**, 295–296.

——, 1972: Composition of the Venus cloud tops in light of recent spectroscopic data. *Astrophys. J.*, **171**, L75–L79.

Liou, K. N., 1972: Light scattering by ice clouds in the visible and infrared: A theoretical study. *J. Atmos. Sci.*, **29**, 524–536.

——, and J. E. Hansen, 1971: Intensity and polarization for single scattering by polydisperse spheres: A comparison of ray optics and Mie theory. *J. Atmos. Sci.*, **28**, 995–1004.

Loskutov, V. M., 1971: Interpretation of polarimetric observations of the planets. *Sov. Astron.—AJ*, **15**, 129–133.

Lyot, B., 1929: Recherches sur la polarisation de la lumière des planètes et de quelques substances terrestres. *Ann. Observ. Paris (Meudon)*, **8**, 161 pp. [available in English as NASA TT F-187, 1964].

Marin, M., 1965: Mesures photoélectriques de polarization à l'aide d'un télescope coudé de 1 m. *Rev. Optique*, **44**, 115–144.

Mie, G., 1908: Beiträge zur Optik Trüber Medien, speziell kolloidaler Metallösungen. *Ann. Phys.*, **25**, 377–445.

Minnaert, M., 1954: *Light and Colour*. New York, Dover, 362 pp.

Mossop, S. C., 1965: Stratospheric particles at 20 km altitude. *Geochim. Cosmochim. Acta*, **29**, 201–207.

O'Leary, B. T., 1966: The presence of ice in the Venus atmosphere as inferred from a halo effect. *Astrophys. J.*, **146**, 754–766.

Öpik, E. J., 1961: The aelosphere and atmosphere of Venus. *J. Geophys. Res.*, **66**, 2807–2819.

——, 1962: Atmosphere and surface properties of Mars and Venus. *Progress in the Astronautical Sciences*, S. F. Singer, Ed., Amsterdam, North-Holland, 416 pp.

Penndorf, R., 1957: Tables of the refractive index for standard air and the Rayleigh scattering coefficient for the spectral region between 0.2 and 20.0 μ and their application to atmospheric optics. *J. Opt. Soc. Amer.*, **47**, 176–182.

Prinn, R. G., 1973: A photochemical haze model for the clouds of Venus. *Bull. Amer. Astron. Soc.*, **5**, 300.

Rasool, S. I., 1970: The structure of Venus clouds—Summary. *Radio Sci.*, **5**, 367–368.

Rea, D. G., and B. T. O'Leary, 1968: On the composition of the Venus clouds. *J. Geophys. Res.*, **73**, 665–675.

Robbins, R. C., 1964: The reaction products of solar hydrogen and components of the high atmosphere of Venus—A possible source of the Venusian clouds. *Planet. Space Sci.*, **12**, 1143–1146.

Rosen, J. M., 1971: The boiling point of stratospheric aerosols. *J. Appl. Meteor.*, **10**, 1044–1046.

Sagan, C., and J. B. Pollack, 1967: Anisotropic nonconservative scattering and the clouds of Venus. *J. Geophys. Res.*, **72**, 469–477.

Scott, A. H., and E. J. Reese, 1972: Venus: Atmospheric rotation. *Icarus*, **17**, 589–601.

Sill, G. T. 1972: Sulfuric acid in the Venus clouds. *Comm. Lunar Planet. Lab.*, No. 171, 191–198.

Sinton, W. M., 1953: Distribution of temperatures and spectra of Venus and other planets. Ph.D. thesis, The Johns Hopkins University, 123 pp.

Sobolev, V. V., 1968: An investigation of the atmosphere of Venus. II. *Sov. Astron.-AJ*, **12**, 135–140.

Timmermans, J., 1960: *The Physico-Chemical Constants of Binary Systems in Concentrated Solutions*, Vol. 4. New York, Interscience, 1332 pp.

Toon, O. B., and J. B. Pollack, 1973: Physical properties of the stratospheric aerosols. *J. Geophys. Res.*, **78**, 7051–7056.

van Blerkom, D. J., 1971: Diffuse reflection from clouds with horizontal inhomogeneities. *Astrophys. J.*, **166**, 235–242.

van de Hulst, H. C., 1957: *Light Scattering by Small Particles*. New York, Wiley, 470 pp.

——, 1963: A new look at multiple scattering. Tech. Rept., Goddard Institute for Space Studies, NASA, New York, 81 pp.

——, and K. Grossman, 1968: *The Atmospheres of Venus and Mars*, J. C. Brandt and M. B. McElroy, Eds., New York, Gordon and Breach, 288 pp.

Velikovsky, I., 1950: *Worlds in Collision*. New York, Macmillan, 401 pp.

Veverka, J., 1971: A polarimetric search for a Venus halo during the 1969 inferior conjunction. *Icarus*, **14**, 282–283.

Wildt, R., 1940: On the possible existence of formaldehyde in the atmosphere of Venus. *Astrophys. J.*, **92**, 247–255.

Young, A. T., 1973: Are the clouds of Venus sulfuric acid? *Icarus*, **18**, 564–582.

——, 1974: Venus clouds: Structure and composition. Submitted to *Science*.

Young, L. D. Gray, 1972: High-resolution spectra of Venus—A review. *Icarus*, **17**, 632–658.

——, and A. T. Young, 1973: Comment on "The composition of the Venus cloud tops in light of recent spectroscopic data." *Astrophys. J.*, **179**, L39–L43.

Light Scattering in the Atmosphere and the Polarization of Sky Light*

Z. Sekera

Department of Meteorology, University of California, Los Angeles, California

(Received July 2, 1956)

The characteristic properties of scattered light in the atmosphere can be derived from the equations of radiative transfer, conveniently formulated in terms of four Stokes polarization parameters. For the case of Rayleigh scattering, by a method outlined by Chandrasekhar, it is possible to obtain an exact solution of this equation in the sense that the effect of all orders of scattering is included. With the use of this method, the intensity and the polarization of the diffuse sky radiation in a pure, molecular atmosphere were computed. These theoretical values are compared with the results of measurements of sky-light polarization, performed by means of a new photoelectric polarimeter specially constructed for such measurements. The deviations of the observed sky-light polarization from that of a molecular atmosphere are discussed with respect to the degree of atmospheric turbidity. These deviations show a systematic character which can be deduced already from the form of the equation of radiative transfer for a turbid atmosphere. The method of solving the problem of radiative transfer for a turbid atmosphere is outlined, with emphasis on the possible determination of the type of scattering from the measurements of sky-light polarization.

THERE is no doubt that the problem of light scattering in the atmosphere represents an important part of atmospheric optics. Its importance has increased recently because of many direct or indirect applications in several other fields of science and technology.

There are several ways to study light scattering in the atmosphere. The most logical one is the study of diffuse sky radiation, which results from the scattering of the sun radiation illuminating the atmosphere. Since the scattered radiation is in general polarized, it is convenient to represent quantitatively the diffuse sky radiation by a set of four Stokes parameters[1]: the components of the specific intensity I_l, I_r parallel and normal, respectively, to the vertical plane containing the direction of the radiation, and two additional parameters, U, V, defining the position of the plane of polarization and the ellipticity of the polarized light.† These four parameters can be considered as the components of a four-dimensional vector or as the elements of a one-column matrix $\mathbf{I}$.

The earth atmosphere can be considered in a very good approximation to be plane parallel and illuminated at its top by the parallel sun radiation of the flux πF_0 normal to its direction, specified by the zenith distance θ_0. In such a case the vertical coordinate can be replaced by the optical thickness τ, a nondimensional quantity,

$$\tau = \int_z^\infty k_s \rho dz,$$

where k_s denotes the mass scattering coefficient and ρ the density of the scattering material. For the study of scattering it is preferable to choose that wavelength of the studied radiation for which the absorption coefficient is negligible compared to the scattering coefficient.

If the direction of the radiation is specified by the parameter μ and by the azimuth φ ($\mu = \cos\theta$, θ being the zenith distance), the intensity $\mathbf{I}$ satisfies the integro-differential equation of the radiative transfer[1,2]

$$\mu \frac{d\mathbf{I}(\tau;\mu,\varphi)}{d\tau} = \mathbf{I}(\tau;\mu,\varphi) - (1/4)e^{-\tau/\mu_0}\mathbf{P}(\mu,\varphi;-\mu_0,\varphi_0)\cdot\mathbf{F}$$

$$- (1/4\pi)\int_{-1}^1\int_0^{2\pi}\mathbf{P}(\mu,\varphi;\mu',\varphi')\cdot\mathbf{I}(\tau;\mu',\varphi')d\mu'd\varphi', \quad (1)$$

where $\mathbf{F}$ denotes the corresponding matrix for the net flux of the extraterrestrial sun radiation. For a neutral light it contains the elements $\mathbf{F}\equiv(\frac{1}{2}F_0,\frac{1}{2}F_0,0,0)$. The first term on the right-hand side expresses the effect of the attenuation due to the scattering; the negative sign is incorporated in the definition of the optical thickness and of the parameter μ. The other two terms contain the matrix $\mathbf{P}$, which expresses mathematically the law of scattering. The intensity of the scattered radiation in any particular direction (defined by μ,φ) is obtained from the intensity of the incident radiation from the direction (μ',φ') by multiplication by this four-by-four matrix $\mathbf{P}(\mu,\varphi;\mu',\varphi')$. Its elements are functions of the scattering angle between the directions (μ,φ) and (μ',φ') and of other parameters defining the intrinsic nature of the scattering. The matrix $\mathbf{P}$ in Eq. (1) is normalized to unity, i.e., the total intensity of the scattered radiation from a neutral source of unit intensity, integrated over all solid angles yields unity.

---

* Presented at the Fourth Congress of the International Commission of Optics, held in Cambridge-Boston, Massachusetts, March 28–April 3, 1956. Published with financial assistance from UNESCO and the International Union of Pure and Applied Physics.

[1] S. Chandrasekhar, *Radiative Transfer* (Clarendon Press, Oxford, 1950); D. L. Falkoff and J. E. MacDonald, J. Opt. Soc. Am. 41, 861 (1951).

† If ψ denotes the deviation of the plane of polarization from the direction of I_l, and a the major and b the minor axis of the ellipse described by the electric vector, then

$$U = (I_l - I_r)\tan2\psi, \quad V = (I_l + I_r)\sin2\omega, \quad \tan\omega = b/a.$$

[2] Z. Sekera, "Polarization of skylight," *Encyclopedia of Physics* (Springer Publishing Company, New York, 1957), Vol. 48, p. 288; *Advances in Geophysics* (Academic Press, Inc., New York, 1957), Vol. III, p. 43.

The second term on the right-hand side of Eq. (1) represents the effect of primary scattering, i.e., the contribution due to the scattering of the direct sun radiation, reduced by the proper attenuation factor. The last term in this equation expresses the scattering of the scattered radiation and, when included, serves to incorporate higher orders of scattering.

For Rayleigh type of scattering, the matrix **P** assumes a simple form and the equation of radiative transfer can be solved by an ingenious method developed by Chandrasekhar.[1] The solution of this equation for the emerging intensities at the bottom and at the top of the atmosphere can be related to the incident flux by means of two matrices. Chandrasekhar called these the scattering and transmission matrices. From the definition of these matrices, Chandrasekhar derived a series of principles of invariance, which together with the reciprocity principle reduce the integrodifferential equation (1) to four systems of simultaneous integral equations. From the solutions of these integral equations, obtained by a successive iteration, it is possible to compute the intensity of the diffuse sky radiation in different directions as a function of a single independent parameter, the optical thickness τ.

In general, the polarization of the scattered light is a very good indicator of scattering processes. In the case of the diffuse sky radiation, the study of its polarization has an additional advantage in that it is not necessary to know the extraterrestrial sun radiation. As a part of an extensive study of sky-light polarization‡ the polarization parameters of the diffuse sky radiation were computed by Chandrasekhar's method. The results of the computation of the degree of polarization along the sun vertical§ for different values of the optical thickness[3] are given in Fig. 1. Over a larger part of the sun vertical the degree of polarization is positive; the plane of polarization is normal to the sun vertical. In this part a maximum degree polarization appears about 90° from the sun. The sign of the degree of polarization, and thus the orientation of the plane of polarization, changes at two points, so-called neutral points, where the polarization disappears. Between the Babinet neutral point above the sun and the Brewster neutral point below the sun, the degree of polarization is negative and the plane of polarization is in the sun vertical. If the sun is closer to the horizon,

---

‡ At the Department of Meteorology, University of California at Los Angeles, sponsored by the Air Force Cambridge Research Center, Contracts AF 19(122)-239 and AF 19(604)-1303. Results of the computations were published in the Scientific Report No. 3, November, 1952.

§ Since along the sun vertical $U = V = 0$, the degree of polarization P is simply given as

$$P = (I_r - I_l)/(I_r + I_l).$$

The computation for $\tau = 0.50$ is based on values published by Chandrasekhar and Elbert [Trans. Am. Phil. Soc. **44** (6), 643 (1954)].

[3] K. L. Coulson, Sci. Rept. No. 4, Contract AF 19(122)-239 Department of Meteorology, University of California, Los Angeles (1952).

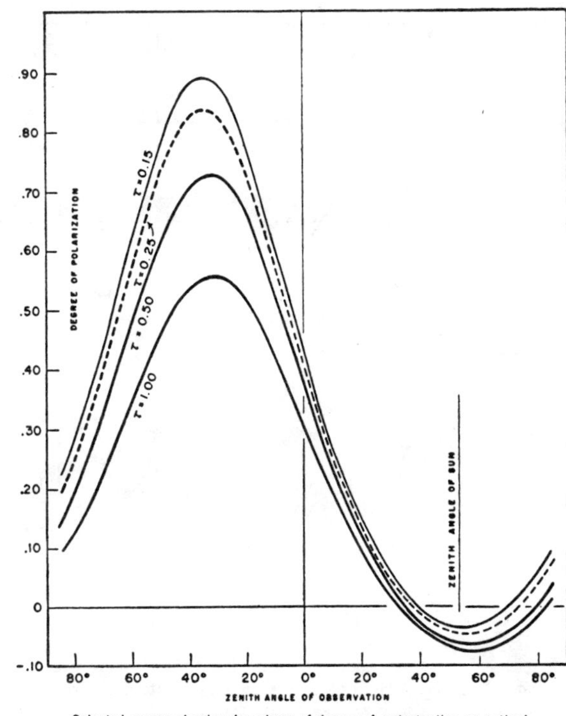

FIG. 1. Distribution of the degree of polarization along the sun vertical for different optical thicknesses of a Rayleigh atmosphere for sun zenith distance $\theta_0 = 53.1°$.

another neutral point appears on the antisolar side of the sun vertical, called Arago point. The agreement of this theoretical distribution of the degree of polarization with the observation is very good and supersedes the results of all previous theoretical computations. It gives the measured positions of the neutral points, as well as the maximum polarization, and it reproduces correctly their variations with optical thickness. In the theoretical computation it is also possible to include the effect of a ground reflection, especially of the Lambert type, in which it is assumed that the reflected radiation is isotropic and neutral. Such type of ground reflection decreases the maximum polarization with increasing albedo; the positions of the neutral points are practically uneffected.

In order to apply these theoretical results to the earth's atmosphere, it is necessary to compute first the optical thickness. This can be easily done for a pure molecular atmosphere. The results of the computation based on the most recent data of the composition and density distribution in the upper atmosphere[4] are reproduced in Fig. 2. This figure gives the optical thickness of the molecular atmosphere at any elevation above the sea level and for any wavelength. Once the optical thickness is known, the theoretical computations yield the distribution and the magnitude of the sky-light polarization for a pure molecular atmosphere for any

---

[4] D. Deirmendjian, Arch. Meteorol. Geophys. u. Bioklimatol., Ser. **B6**, 452 (1955).

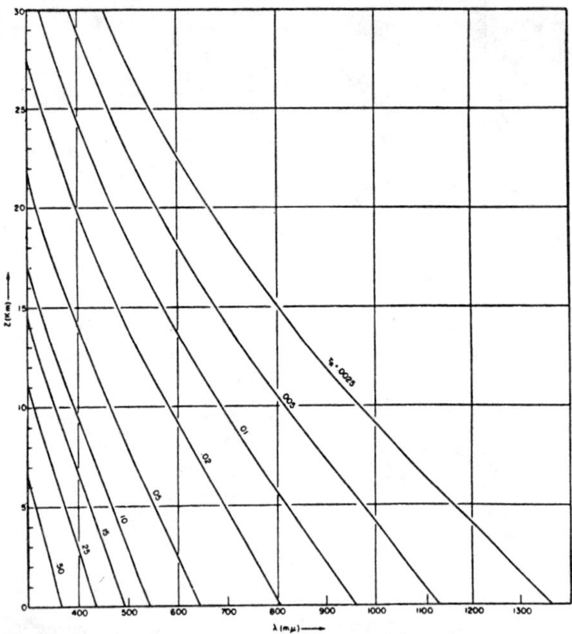

FIG. 2. Optical thickness of a molecular atmosphere for different elevations above sea level and for different wavelengths.

the deviation of the theory from the observations, will give an indication of the scattering of light on all other particles in the atmosphere besides the air molecules. These particles constitute the so-called atmospheric aerosol.

In order to secure proper measurements for such comparison, a photoelectric polarimeter[5] was constructed (Fig. 3). The basic element of the instrument is a retardation plate which rotates with a uniform speed (10 rps) in front of a fixed analyzer. By a simple analysis it can be shown that the luminous flux leaving the analyzer and measured by the photomultiplier has the character of a composite alternating current. The dc component is proportional to the total intensity of the measured light. The amplitude of the first (20 rps) harmonics of the speed of revolution of the plate is proportional to the Stokes parameter V, and thus measures the ellipticity. The amplitude of the third (40 rps) harmonics is proportional to the product of the total intensity and the degree of linear polarization, while the phase of this component equals double the angle between the plane of polarization of the measured light and the plane of transmission of the analyzer. With properly tuned amplification, these components can be separated and their amplitudes and phases can be measured. In order to prevent the shift of the working point from the linear characteristic of the photomultiplier due to the large change of the intensity during the

location and for any particular wavelength. These values are compared with the direct measurements; and the difference between these two values, which is

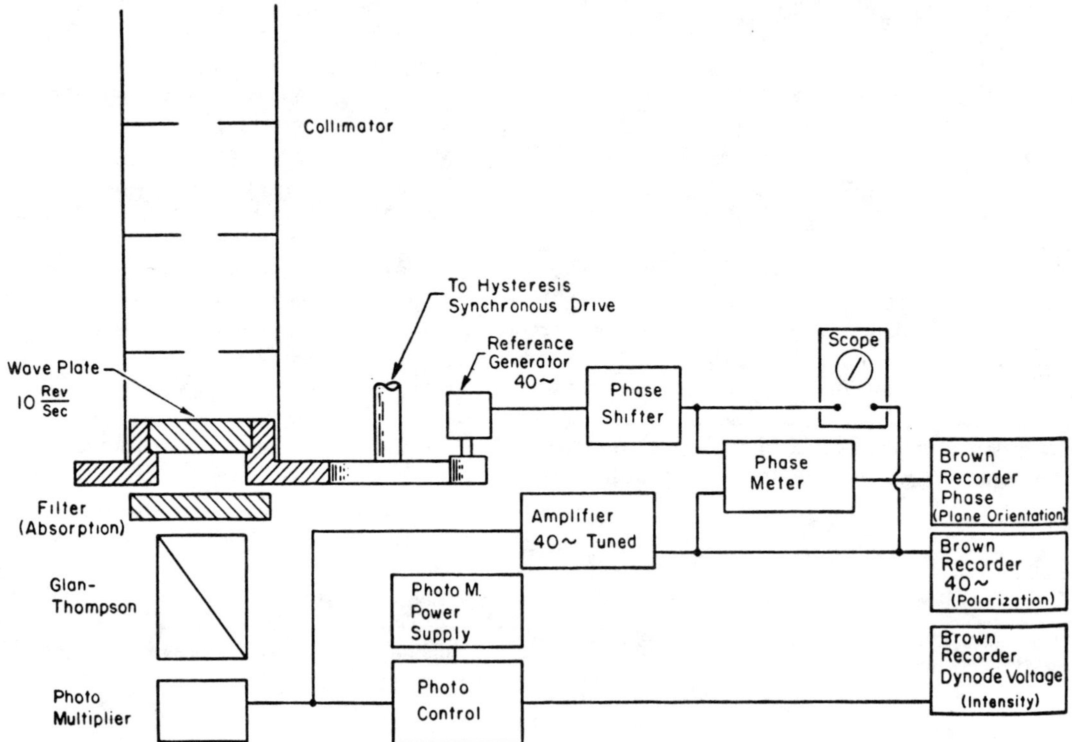

FIG. 3. Diagram of the photoelectric polarimeter for the measurements of sky-light polarization.

[5] C. H. Seaman and Z. Sekera, Appendix A, Final Report, Contract AF 19(122)-239, Department of Meteorology, University of California, Los Angeles (1955).

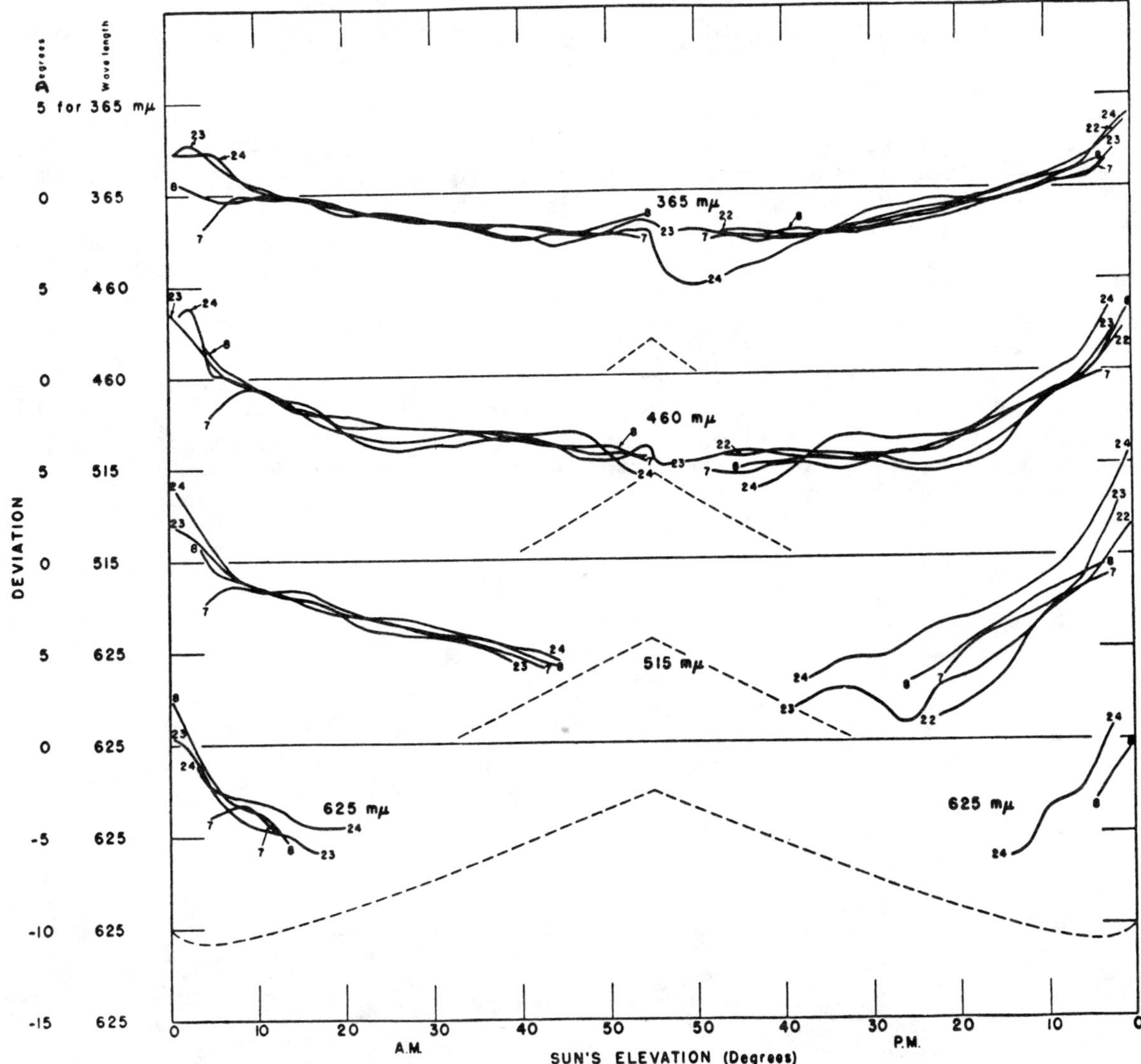

Fig. 4. Deviations of the positions of the Babinet neutral point from its theoretical value for several days with similar turbidity conditions as measured during September, 1953, at Los Angeles (UCLA campus).

scan along the sun vertical, a special arrangement was necessary. The dynode voltage is regulated to give a constant low-level output and the varying voltage is used as the measure of the total intensity. This arrangement offers two advantages: a logarithmic scale for the intensity and the amplitude of the third harmonics becomes proportional only to the degree of polarization. In order to facilitate the phase measurement, a generator of the frequency of the fourth harmonics is attached to the gear rotating the plate. The auxiliary signal is used for the calibration of the amplifier and after passing through an adjustable phase shifter, it is used as the reference signal for the oscilloscope or phasemeter. The optical components are built in two identical units, each provided with a set of three monochromatic filters, exchangeable by remote control. These units are scanned in a vertical plane, held in the sun vertical by means of a photoelectric sun follower. Its collimator carries a shield with neutral filters of high density to protect the photosensitive surface of the photomultipliers against direct sun radiation. The angular distance of the axes of the two units from the sun is measured and recorded by means of electrical contacts attached to the protractor on the scanning axis.

After a proper calibration, the degree of polarization can be obtained directly from the recording of the amplitude of the fourth harmonics. The position of the neutral points is given by a minimum in the recorded curve. This minimum can be made by proper amplification as sharp as possible and the position of the neutral point can be determined with an accuracy of 3/10 of a degree. Moreover the sudden change of the plane of polarization by 90° at the neutral point during the scan can be easily recognized in the recording of the phase and thus can serve as another indication of the position of the neutral point.

As an example, the measured deviations in the positions of neutral points are shown in Fig. 4, obtained from several measurements of the Babinet neutral point made at the University of California (Los Angeles) during the month of September, 1953 on days with similar turbidity conditions. Each line represents the results of measurements from one day, the date of which is indicated in attached numbers. Measurements were made at four different wavelengths. The lines of zero deviations for each wavelength are staggered along the ordinate axis. The dotted line outlines the region in which the neutral points cannot be observed, being obscured by the shield. Except for low sun elevations, for which the theory may not give correct results because of the assumption of a plane-parallel atmosphere, the deviations for all wavelengths are negative, i.e., the neutral points were measured to be closer to the sun than the theory predicts. The magni-tude of these deviations is definitely beyond the limit of observational errors. The deviations show quite a systematic character. Daily variations, the asymmetry of the curves with respect to the noon, the magnitude of the deviations, all these quantities increase with increasing wavelength. In Fig. 5 the corresponding deviations are shown for the Brewster and for the Arago points. The deviations of the Arago point are all positive, i.e., the measured distances of the Arago point from the antisolar point are larger than those given by the theory. The deviations of the Brewster point, mostly negative, show much greater asymmetry with respect to the noon than the deviations of the Babinet point. Also the variations from day to day are much larger. However, the reason for larger daily variations in the shorter wavelength is most likely the varying reflection over the sea surface in the vicinity of the observational site. The measurements made on other

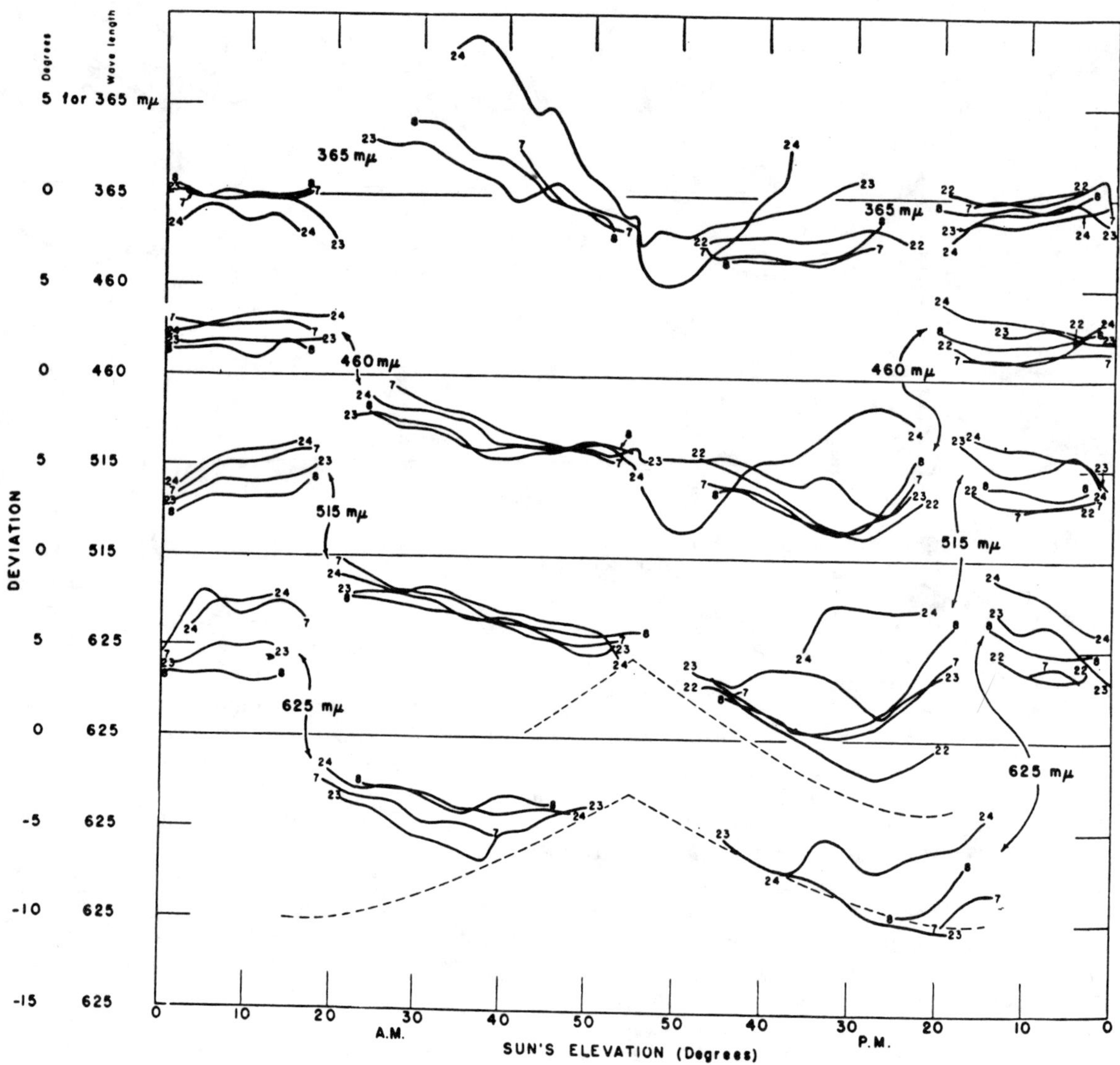

FIG. 5. Deviations of the positions of the Brewster and Arago neutral points for the same time and location as in Fig. 4.

316

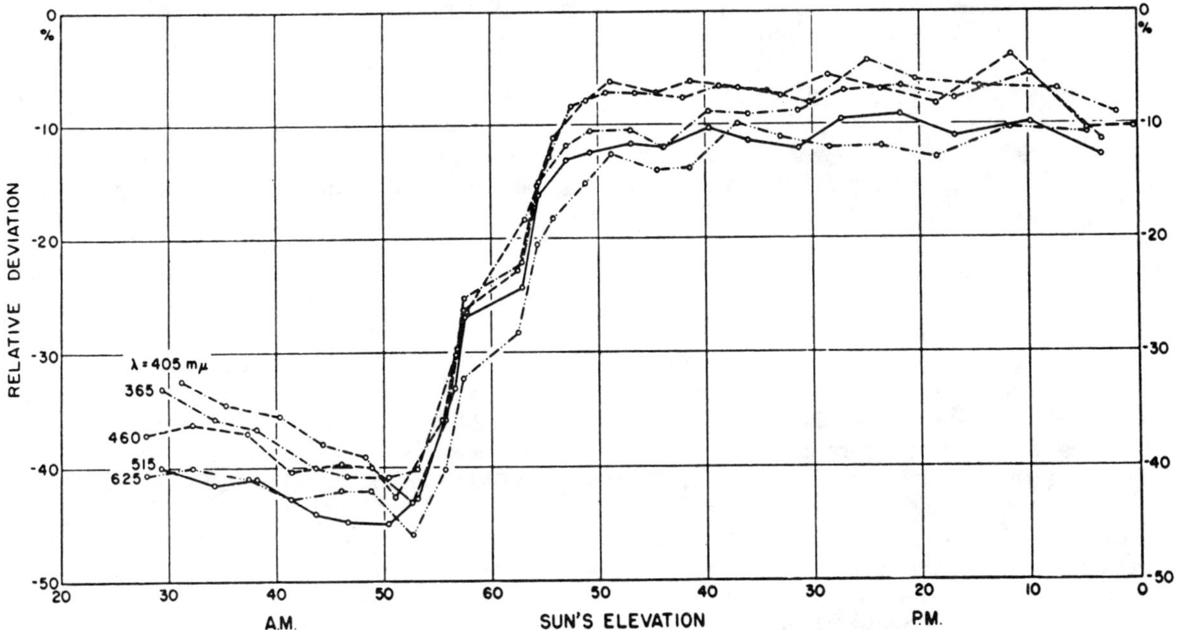

FIG. 6. Relative deviations of the maximum degree of polarization as measured at Los Angeles on September 17, 1954.

locations confirm completely these characteristic features of the deviations in the positions of the neutral points.

The measurements of the maximum degree of polarization gave a similar result. The agreement of the theory with the results of measurements can be conveniently studied from the relative deviations, defined as the ratio $(P_0 - P_t)/P_t$, where P_0, P_t denote the observed and the theoretical degree of polarization. The relative deviations computed from the measurements performed on a location with a very low turbidity (Cactus Peak, 5500 ft above sea level, far from any inhabited areas) are definitely beyond the limits of observational errors, varying within the limits -10% to -20%. An interesting variation of the degree with the turbidity is demonstrated in Fig. 6, where the relative deviations are shown for measurements on a particular day at Los Angeles. In the morning hours a strong haze lowered the horizontal visibility to 2 miles. During the noon hours, the haze was removed from the area by a sudden increase of wind. The rapid change in the turbidity was recognized by a sudden increase of visibility to 12 miles, as well as by a sudden increase of the relative deviations from -30% to -40% to -10%. The deviations in the haze increased with the wavelength. In the clear air, however, the deviations were of the same magnitude for all wavelengths. As another example, the effect of smog in the Los Angeles area on the skylight polarization is shown in Fig. 7, which gives the results of measurements made at the California Institute of Technology in Pasadena, on November 4, 1954. During this day the horizontal visibility remained constant, about 8 miles, but the increased wind in the afternoon brought a moderate smog into the Pasadena area. The sudden drop of the deviations that followed, up to 50–60%, was the only indication of the presence of pollutants in the air. In the morning hours, the magnitude of the deviations shows quite a distinct increase with increasing wavelength, in agreement with the similar effect in the positions of the neutral points.

These examples illustrate quite adequately the character and magnitude of the deviations of the observed values of sky-light polarization from the

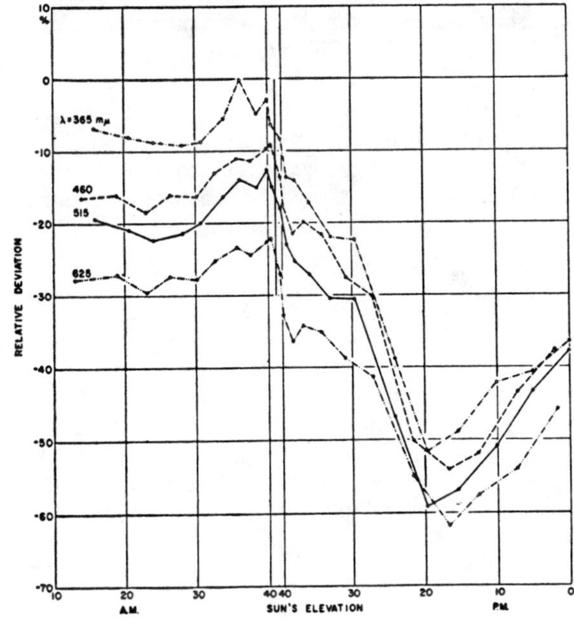

FIG. 7. Relative deviations of the maximum degree of polarization as measured at California Institute of Technology in Pasadena, California, on November 4, 1954.

theoretical values computed for a pure molecular atmosphere. The asymmetry with respect to the noon, the increase of the deviations and their daily variations with longer wavelengths and with increasing turbidity, confirm fully the previous statement. They also confirm the expectation that these deviations are due to the presence of a highly varying content of aerosol particles. A large part of the observed characteristics of these deviations can be explained, so far at least, qualitatively.

The presence of aerosol particles, the dimensions of which are comparable or larger than the wavelength, is to be incorporated in Eq. (1) in two places: in the optical thickness τ and in the matrix $\mathbf{P}$.[2] If τ_R denotes the optical thickness of the molecular atmosphere, then $\tau = \tau_r + \tau'$, where τ' is the increase of the optical thickness due to the presence of the aerosol particles. This quantity can be computed without difficulty[6] provided that the size distribution of the particles and their refractive index is known‖ and that their optical behavior can be approximated by that of a dielectric sphere. Similarly, the matrix $\mathbf{P}$ can be split into two parts, one expressing the scattering by air molecules, the other the scattering by aerosol particles. The matrix corresponding to the scattering by aerosol particles can be again split in two parts, namely, the matrix of Rayleigh scattering $\mathbf{P}_R$ and the matrix $\mathbf{P}'$, containing terms responsible for the deviations of the large particle scattering from the scattering of Rayleigh type. The proper normalization of the matrices leads to the following expression for the matrix $\mathbf{P}$ in Eq. (1).

$$\mathbf{P} = \mathbf{P}_R + T(\tau)\mathbf{P}', \quad T(\tau) = \beta_M/(\beta_R + \beta_M), \qquad (2)$$

where β_R denotes the volume scattering coefficient of the air molecules. The factor $T(\tau)$ has a character of a turbidity factor, vanishing for a pure molecular atmosphere and approaching unity for large turbidity, when $\beta_R \ll \beta_M$.

The form of the matrix $\mathbf{P}$ in Eq. (2) suggests two different effects of the aerosol particles. The first term leads to the change of the Rayleigh scattering and of the polarization of skylight in a molecular atmosphere due to the increase of the optical thickness. This effect can explain the observed shift of the Arago neutral point from the antisolar point, as well as the decrease of the maximum polarization (negative relative deviations). The second term will include the effect of non-Rayleigh scattering. It is generally accepted that aerosol particles are, under normal conditions, most likely coated with a sufficient layer of water so that they can be considered equivalent to water droplets. In such case the matrix $\mathbf{P}'$ will include primarily the effect of forward scattering, and will be thus more effective close to the sun than on the antisolar side. This latter effect will then explain the shift of the Babinet and Brewster points towards the sun. Since the volume scattering for Rayleigh scattering β_R varies with the wavelengths as λ^{-4}, while for the aerosol particles β_M varies as λ^{-1} or is independent of the wavelength,[7] it is evident that the increase of the optical thickness due to the presence of aerosol particles is much larger for the longer than for the shorter wavelengths. For the same reason the factor $T(\tau)$ will be larger for longer wavelengths and thus the second effect will be also more pronounced for longer wavelengths, in agreement with the observed evidence. However, if $T(\tau)$ is large also for shorter wavelengths, as may happen in the case of high turbidity, the effect of the presence of aerosol particles will result in the same magnitude of the deviatons for all wavelengths.

The definite proof of these observations can, of course, be given after the complete solution of the equation of radiative transfer in a turbid atmosphere is available. Recent measurements of the size distribution of aerosol particles have given a convincing evidence at least of the type of the size distribution of particles that can be expected in the atmosphere. With the use of Mie theory it is thus possible to compute the elements of the matrix $\mathbf{P}'$. Such computation is, however, very complicated and quite tedious. The current research [conducted at the Department of Meteorology, University of California at Los Angeles, under the Contract AF 19(604)-1303 with the Air Force Cambridge Research Center] is directed toward the possible simplifications which will make such computation a little more practical.

The theoretical studies of the light scattering in a turbid atmosphere and the complete solution of the equation of radiative transfer [Eq. (1)] will lead to the correct quantitative values for the observed deviation in the sky-light polarization. Then, the light scattering properties of the atmosphere will be known to an accuracy sufficient for other applications.

---

[6] For details see Final Report, Contract AF 19(122)-239, Department of Meteorology, University of California, Los Angeles (1955).

‖ $\tau' = \int_z^\infty \beta_M(z)dz$, where the volume scattering coefficient is given by

$$\beta_M(z) = \sum_m \int_{a0}^{\infty} K(m,ka)N(m,a;z)\pi a^2 da,$$

$K(m,ka)$ denotes the total scattering cross section, $N(m,a;z)$ the number of particles of the refractive index m and of the radius a in the interval $\langle a, a+da \rangle$ at a given elevation z, $k = 2\pi/\lambda$, λ being the wavelength.

[7] D. Deirmendjian and Z. Sekera, J. Opt. Soc. Am. **46**, 565 (1956).

Skylight polarization during a total solar eclipse: a quantitative model

G. P. Können

Royal Netherlands Meteorological Institute, P.O. Box 201, 3730 AE De Bilt, The Netherlands

Received April 23, 1986; accepted October 9, 1986

The polarization distribution in the sky during a total solar eclipse is calculated with a simple secondary light-scattering model. This model uses the light-intensity measurements near the horizon during the eclipse and the pretotality and posttotality skylight polarization observations as input. It is found that the model can explain various observations during totality, including the quantitative measurements of Shaw [Appl. Opt. 14, 388 (1975)] of the polarization distribution of the sky in the solar vertical during the 1973 total eclipse.

1. INTRODUCTION

When the totality phase of a solar eclipse starts, the appearance of the sky changes dramatically. In a rapid transition the circumstances change from full daylight to a situation comparable to twilight, which is accompanied by a sudden drop in the sky intensity of about 3 orders of magnitude.[1–3] The remaining lighting of the sky is caused by multiple scattering of sunlight, which starts in the region outside the lunar umbra,[3] instead of by singly scattered light from a well-defined point source during nontotality conditions. (The light of the solar corona is 6 orders of magnitude weaker than that of the uneclipsed Sun[3] and therefore makes only a negligible contribution to the illumination.) Therefore the illumination of the sky during totality must be described basically in a two-step process: (1) at least one scattering in the region outside the umbra, followed by some absorption; and (2) at least one scattering inside the umbra.[4] Light resulting from the first step is visible during totality as a reddish band about 10° in width above the horizon all around us; this band acts as the light source for the illumination of the sky above us. Above 10–20°, light from the two-step process dominates the scenery.[5]

Since the polarization for singly scattered light and multiply scattered light differs completely, the polarization pattern of the sky also changes abruptly at totality. It has been known since at least 1905 (Ref. 6) that the polarization of the sky decreases drastically, at least at 90° from the Sun. Starting in 1961, a few instrumental records have been published of the polarization change of the sky during eclipses.[7–11] However, the information obtained by these early observers about this state of polarization of the sky is rather limited, since they restricted their measurements to one single point in the sky, located at 90° from the Sun in the solar vertical. This situation lasted until 1973, when Shaw recorded the polarization of the sky as a function of the zenith angle during the June 30th eclipse.[1] He found a symmetry in the polarization with respect to the zenith, with a minimum value in the zenith and maximum polarization rather close to the horizon. His data are complete enough to justify the development of a simple model as a first attempt to come to a quantitative understanding of the polarization of the sky during solar eclipses.

In this perspective, we present in this paper a simple two-step scattering model for the polarization of the sky during totality. It is based on secondary scattering by a degraded Rayleigh scatterer in which the depolarization factor is deduced from the pretotality and posttotality measurements, and it uses the observed intensity distribution near the horizon during totality as input. Despite the simplicity of this approach, its numerical results compare satisfactorily with observations made during various eclipses.

2. FORMULATION OF THE MODEL

The propositions of the model are the following:

1. The polarization is described as result of a two-step scattering process.
2. Step 1 is the scattering of sunlight in a region outside the umbra, followed by some absorption and depolarization and possibly by additional scattering by aerosols.
3. In the numerical evaluations, the polarization of light resulting from step 1 must be neglected, since the available data do not include measurements of it.
4. Step 2 is single scattering to the observer of light produced in step 1 by a degraded Rayleigh scatterer.
5. The paths of light from step 1 to the secondary scattering centers are parallel to the ground.
6. The secondary scattering centers are close to the observer.
7. The scattering matrix $\mathcal{M}$ of step 2 is identical to the scattering matrix during pretotality and posttotality and is given by a linear combination of a pure Rayleigh scatterer and an unpolarized isotropic scatterer.[12]

Some comments must be made on the above-mentioned propositions and the handling of them.

a. Light resulting from step 1 is concentrated near the horizon. Its intensity is taken from the observations.
b. The depolarization factor of the degraded Rayleigh scatterer is taken from the pretotality and posttotality skylight polarization observations.

c. Neglect of polarization resulting from step 1 is justified if the optical thickness along the line between the observer and the edge of the umbra is large enough. In that case additional scatterings by aerosols are important, and they destroy the polarization. This depolarization can be expected to be more effective at small wavelengths. As we will see below, there is some indirect evidence for this depolarization in Shaw's 400-nm measurements. In Appendix A, a quantitative estimate is made from the effect of relaxing proposition 3 and hence introducing polarization in step 1.

d. It should be noted that the propositions of our model are close to the ones used by Soret[13] and by Ahlgrimm[14] in their models to describe the polarization of sunlit sky, taking into account secondary scattering.

e. The validity of the model is restricted to the regions of the sky where single scattering can be neglected, i.e., above a height of about 20° over the horizon.[5]

3. CALCULATION OF THE POLARIZATION DISTRIBUTION

Since circularly polarized light does not show up in the two-step process, we can describe the polarization of light by a three-dimensional Stokes vector $\mathbf{S}$:

$$\mathbf{S} = \begin{bmatrix} I \\ Q \\ U \end{bmatrix} \equiv \begin{bmatrix} I \\ IP \cos 2\phi \\ IP \sin 2\phi \end{bmatrix}. \tag{1}$$

Here I denotes the intensity, P denotes the degree of polarization, and ϕ denotes the angle of polarization with respect to a plane of reference.[15,16] We take the vertical as the plane of reference for Stokes vectors $\mathbf{S}$. However, for scattering matrices $\mathcal{M}$, the scattering plane is taken to be the plane of reference. Let $\mathbf{S}_1$ be the Stokes vector of light after step 1, thus entering the secondary scattering center from the horizon, and let $\mathcal{T}(\phi)$ be the rotation matrix defined by

$$\mathcal{T}(\phi) = \begin{bmatrix} 1 & 0 & 0 \\ 0 & \cos 2\phi & \sin 2\phi \\ 0 & -\sin 2\phi & \cos 2\phi \end{bmatrix}. \tag{2}$$

The Stokes vector $\mathbf{S}_2$ after step 2 is then found by the matrix multiplication[16]

$$\mathbf{S}_2 = \mathcal{T}(-\phi_3)\mathcal{M}\mathcal{T}(\phi_2)\mathbf{S}_1. \tag{3}$$

Here ϕ_2 denotes the angle of the scattering plane with the vertical in step 2 as seen from the secondary scattering center in the direction of the light ray incoming from the horizon, and ϕ_3 denotes the angle of this scattering plane with the vertical as seen by the observer looking to the secondary scattering center.

From proposition 7, the scattering matrix is given by[12,16]

$$\mathcal{M} = \begin{bmatrix} 1 + \cos^2\theta + I_D & -\sin^2\theta & 0 \\ -\sin^2\theta & 1 + \cos^2\theta & 0 \\ 0 & 0 & 2\cos\theta \end{bmatrix}, \tag{4}$$

where the factor I_D results from the unpolarized isotropic

scattering matrix and θ denotes the scattering angle. A substitution of this matrix $\mathcal{M}$ into Eq. (3) yields the Stokes vector $\mathbf{S}_2$ in arbitrary units, as normalization constants have been omitted in Eq. (4).

In pretotality and posttotality, single scattering is dominating, and $\mathcal{M}$ is acting on $\mathbf{S}_0 = (1, 0, 0)$. When this matrix multiplication is carried out for single scattering, the degree of polarization becomes

$$P \equiv \frac{\sqrt{Q^2 + U^2}}{I} = \frac{\sin^2\theta}{1 + \cos^2\theta + I_D}, \tag{5}$$

which gives, for $\theta = 90°$,

$$P = \frac{1}{1 + I_D}. \tag{6}$$

So, with the aid of Eq. (5) or (6), the factor I_D in Eq. (4) that determines the degradation with respect to pure Rayleigh scattering can be fixed from measurements outside totality.

The evaluation of the two-step process with Eq. (3) requires the introduction of some angles. We define z as the solar elevation, h as the height in the sky where the observer is looking (height of the secondary scattering center), ψ as the azimuth of the secondary scattering center minus the azimuth of the Sun, α as the azimuth of a light ray coming in to the secondary center and measured relative to the line connecting the observer with secondary scattering center, and ψ_1 as the azimuth of a light ray coming in to the secondary center minus the azimuth of the Sun.

All angles ψ_1, ψ, α, ϕ_2, ϕ_3 are taken to be positive in the anticlockwise direction. Under proposition 6, the relation

$$\psi_1 = \psi + \alpha \tag{7}$$

holds. Figure 1 displays the geometry of the problem.

The Stokes vector $\mathbf{S}_1$ is a function of ψ_1 and depends on several factors, among them the distance to the edge of the umbra, the cloud decks, and the reflectivity of the Earth. If we assume the latter two factors to be constant around the observer, $\mathbf{S}_1$ will be a symmetrical function of ψ_1 during midtotality. Relaxing proposition 3 for a moment, $\mathbf{S}_1$ can be expressed as

$$\mathbf{S}_1 = \mathcal{T}(-\phi_1)\mathcal{M}'\mathbf{S}_0 f(\psi_1), \tag{8}$$

in which $\mathcal{M}'$ is given by Eq. (4) with an unknown factor I_D' in it, ϕ_1 is the angle of the primary scattering plane relative to the vertical, $f(\psi_1)$ is the azimuthal dependence of $\mathbf{S}_1$ after integration of all primary scattering centers in the direction ψ_1 outside the umbra, and $\mathbf{S}_0 = (1, 0, 0)$ is the Stokes vector of the Sun. Note that the factorization of $\mathbf{S}_1$ in Eq. (8) in a matrix multiplication and an azimuth-dependent intensity function $f(\psi_1)$ acting equally on all Stokes parameters is an approximation, one that is completely true if step 1 contains just a single scattering.

From spherical geometry one finds the relations

$$\cos \theta_1 = \cos z \cos \psi_1$$

and

$$\tan \phi_1 = -\sin \psi_1/\tan z, \tag{9}$$

where θ_1 denotes the scattering angle in step 1.

A straightforward calculation of the intensity function

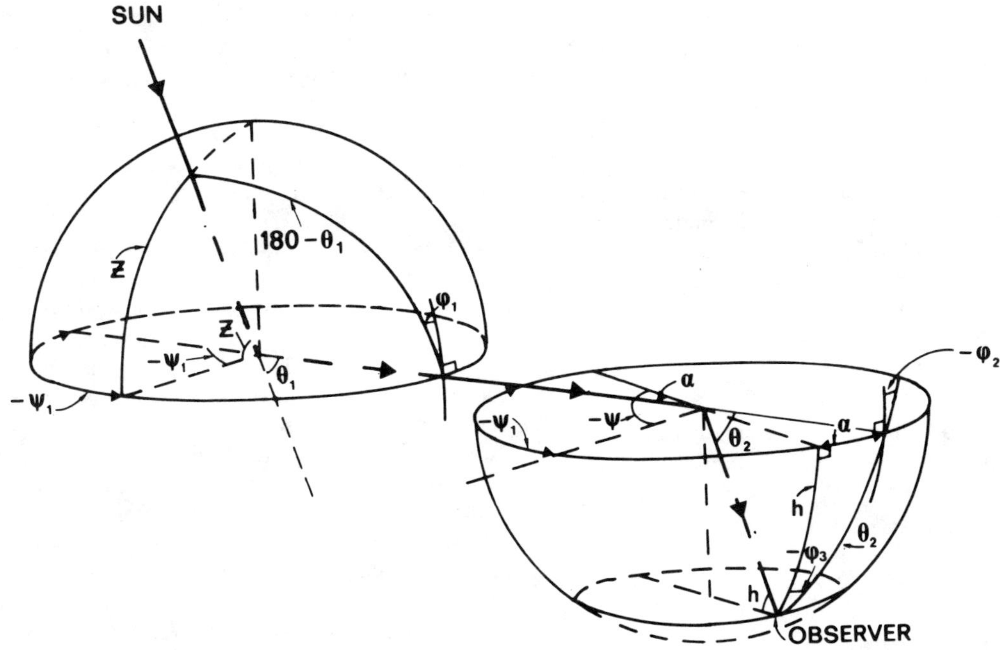

Fig. 1.   Geometry of the two-step scattering model.   The definitions of the angles are given in the text.

$f(\psi_1)$ from the geometry of the umbra is permitted only for an optically thin atmosphere and for homogeneous meteorological conditions and surface reflectivity around the observation site.   If we neglect the extinction and the curvature of the Earth,[5] we get simply

$$f(\psi_1) \propto r^{-1}(\psi_1), \tag{10}$$

where r denotes the distance of the observer from the edge of the umbra (r is not too small).   Of course, the shape of the umbra is an ellipse.   Hence, if the observer is the central point of this ellipse, one finds from Eq. (10) that

$$f(\psi_1) = (1 - \cos^2 z \cos^2 \psi_1)^{1/2} \simeq 1 - \tfrac{1}{2} \cos^2 z \cos^2 \psi_1. \tag{11}$$

The latter approximation holds if $\cos^2 z$ is not too large. When the observer is in one of the foci of the ellipse, one finds that

$$f(\psi_1) = 1 + \cos z \cos \psi_1. \tag{12}$$

When the optical thickness along r is sufficiently large, $f(\psi_1)$ may deviate from expression (10).   This will be more likely if the wavelength is shorter.   However, a decrease of $f(\psi_1)$ with increasing r can be expected anyhow.   For this case, we use instead of expression (11) the approximate expression

$$f(\psi_1) = 1 - a^2 \cos^2 \psi_1, \tag{13}$$

where the empirical factor $a^2 < 1$ is determined from the intensity measurements near the horizon during eclipse. Because of the uncertainty in the processes in step 1 (among them the depolarization that is due to processes such as small-angle scattering), such an empirical approach should always be preferred for the determination of $\mathbf{S}_1$ above a direct calculation from Eq. (8).

Under proposition 3, one has, with Eq. (7), for $\mathbf{S}_1$

$$\mathbf{S}_1 = \begin{bmatrix} f(\psi + \alpha) \\ 0 \\ 0 \end{bmatrix}. \tag{14}$$

Furthermore, one finds from spherical geometry for step 2 the following expressions from Fig. 1:

$$\cos \theta_2 = \cos \alpha \cos h,$$

$$\tan \phi_2 = -\sin \alpha / \tan h,$$

$$\tan \phi_3 = -\tan \alpha / \sin h. \tag{15}$$

Here, θ_2 is the scattering angle in step 2.

Carrying out Eq. (3) leads to the following expressions for the Stokes parameters in $\mathbf{S}_2$:

$$I_2(h, \psi, \alpha) = (1 + I_D + \cos^2 h \cos^2 \alpha) f(\psi + \alpha),$$

$$Q_2(h, \psi, \alpha) = (1 - [1 + \sin^2 h]\cos^2 \alpha) f(\psi + \alpha),$$

$$U_2(h, \psi, \alpha) = \sin 2\alpha \sin h \, f(\psi + \alpha). \tag{16}$$

Integration of Eqs. (16) over α from $0 \to 2\pi$ yields the desired Stokes vector at (h, ψ) in the sky.   We denote this end result by a barred symbol $\bar{\mathbf{S}}_2 = (\bar{I}_2, \bar{Q}_2, \bar{U}_2)$.   Equations (16) imply that the relation

$$\bar{\mathbf{S}}_2(h, \psi) = \bar{\mathbf{S}}_2(h, \psi + 180°) \tag{17}$$

holds.   Thus, in a given vertical plane, the polarization at either side of the zenith is equal.

To calculate $\bar{\mathbf{S}}_2$, we first take the intensity function $f(\psi_1)$ of the form of Eq. (12) (the observer is at one of the foci of the ellipse-shaped umbra).   Integration of Eqs. (16) then yields

$$\bar{I}_2(h, \psi) = \pi[2(1 + I_D) + \cos^2 h],$$

$$\bar{Q}_2(h, \psi) = \pi \cos^2 h,$$

$$\bar{U}_2(h, \psi) = 0, \tag{18}$$

which shows that the direction of the plane of polarization is always vertical ($\bar{Q}_2 > 0$, $\bar{U}_2 = 0$) and the degree of polarization P is independent of the azimuth and independent of the ellipticity of the umbra and hence of z. Therefore the degree of polarization becomes essentially the same as for a circularly shaped umbra and is given by

$$P(h, \psi) = \frac{\bar{Q}_2}{\bar{I}_2} = \frac{\cos^2 h}{2(1 + I_D) + \cos^2 h}, \tag{19}$$

which ranges from zero in zenith to maximally 33% near horizon. If one applies Eq. (13) for $f(\psi_1)$ (midtotality and the observer at the central line of eclipse), one has

$$\bar{I}_2(h, \psi) = \pi[(2 - a^2)(1 + I_D)$$
$$+ (1 - \tfrac{1}{4}a^2 \cos 2\psi - \tfrac{1}{2}a^2)\cos^2 h]$$

$$\bar{Q}_2(h, \psi) = \pi[\tfrac{1}{2}a^2 \cos 2\psi + (1 - \tfrac{1}{4}a^2 \cos 2\psi - \tfrac{1}{2}a^2)\cos^2 h]$$

$$\bar{U}_2(h, \psi) = \tfrac{1}{2}\pi a^2 \sin 2\psi \sin h, \tag{20}$$

which remains azimuth dependent. In the solar vertical ($\psi = 0, 180°$), $\bar{U}_2 = 0$ and $\bar{Q}_2 > 0$, so that the polarization is vertical again. The degree of polarization is given by

$$P(h, 0) = P(h, 180°) = \frac{\bar{Q}_2(h, 0)}{\bar{I}_2(h, 0)}$$

$$= \frac{\tfrac{1}{2}a^2 + (1 - \tfrac{3}{4}a^2)\cos^2 h}{(2 - a^2)(1 + I_D) + (1 - \tfrac{3}{4}a^2)\cos^2 h}. \tag{21}$$

In the plane perpendicular to the solar vertical containing the zenith ($\psi = 90°, 270°$), $\bar{U}_2$ is zero again, but $\bar{Q}_2$ changes sign at

$$\cos^2 h_n = \frac{2a^2}{4 - a^2}, \tag{22}$$

indicating the existence of neutral points at either side of the solar vertical during mideclipse. Since $a^2 < 1$, these neutral points will always be higher in the sky than 35°.

If the observer is not in the center of the umbra anymore, or if $f(\psi_1)$ is otherwise irregularly distributed around the observer, then $\bar{U}_2(h, 0)$ may be nonzero. This indicates some tilt in the direction of polarization. However, Eqs. (16) indicate that $\bar{U}_2(h, 0)$ is usually much closer to zero than is $\bar{Q}_2(h, 0)$ if the properties of $f(\psi_1)$ are not too extreme. Therefore the model does not yield much change in the direction of polarization in the solar vertical when the eclipse proceeds. Of course, if proposition 3 is relaxed, larger tilts of the polarization plane become possible.

4. COMPARISON WITH OBSERVATIONS

A. Polarization Distribution in the Solar Vertical

The only measurements of the polarization distribution in the sky that have come to our attention are those taken by Shaw during the 1973 eclipse.[1] He scanned during totality the degree of polarization in the solar vertical as a function of the zenith angle for a wavelength of 400 nm (presented in his Fig. 10). At the same wavelength he measured the degree of polarization as a function of time for a fixed point in the sky, chosen in the solar vertical and at 90° from the Sun (his Fig. 9). Moreover, he performed intensity scans at 400 and 600 nm in the solar vertical and in the plane perpendicular to it (his Figs. 2–5).

Although it can be inferred from his Figs. 9 and 10 that his polarization scan in the solar vertical did not take place at midtotality, it is not possible to reconstruct its exact timing. For this reason, and because the intensity scans in Figs. 2 and 4 of Ref. 1 provide only a few points of the radiance during the eclipse near the horizon (necessary input for our model), it is also not possible to give an exact experimental value for the intensity function $f(\psi_1)$ in Eq. (14) after step 1, although it is obvious from Shaw's measurements that $f(\psi_1)$ changed during the course of the eclipse.

Fortunately, however, Shaw observed for 400 nm only a weak dependence of the intensity as a function of ψ_1. Therefore, it is possible to choose for $f(\psi_1)$ the simple form of formula (13). If we take for the intensity after step 1 the measured intensity at $h = 10°$ as the standard, we find that $a^2 = 0.1$. This is considerably less than the value of $a^2 = 0.3$, expected from geometry alone [formulas (10) and (11)]. This low value of a^2 can be considered an indication that proposition 3 is largely fulfilled at 400 nm. From Shaw's pretotality and posttotality measurements at 90° from the Sun, one finds that the degree of polarization without eclipse would have been 42%, and hence $I_D = 1.22$ [Eq. (6)].

Figure 2 compares Shaw's observations with the theory, Eq. (21). The agreement at $h > 25°$ is satisfactory, although our model generates a slightly lower polarization. However, from his time series, Shaw reported at midtotality a polarization of only 4% at 90° from the Sun, as compared with 9% during his solar vertical scan. So, rather than giving an underestimate, our model slightly overestimates the polarization, but such small differences are well within the uncertainties of the model results.

The experimental curve of Shaw shows some asymmetry with respect to the zenith, which Shaw attributes to differences in surface albedo around the observing site. However, our model predicts for every intensity function $f(\psi_1)$ a symmetric behavior of the polarization with respect to the zenith [see Eq. (17)], while during Shaw's eclipse proposition 3 seems to be largely fulfilled. Therefore we attribute the observed asymmetry chiefly to the change of the eclipse geometry and hence of $f(\psi_1)$ during the scan, taking into account that this scan would probably have taken at least 60 sec, i.e., 20% or more of the time of totality, and that the eclipse geometry is rapidly changing.

B. Direction of Polarization

In the solar vertical our model predicts at midtotality a vertical polarization. Unfortunately, Shaw did not present measurements of the angle of polarization. Therefore, at my request, Jannink[17] observed visually the direction of skylight polarization with the aid of a simple Minnaert polariscope[18] during the 1981 Siberian eclipse ($z = 25°$) near the Sun (so $h = 25°$). The observed direction of the easily visible polarization was within 10° of vertical. This visual observation was completed by a set of two slides that he

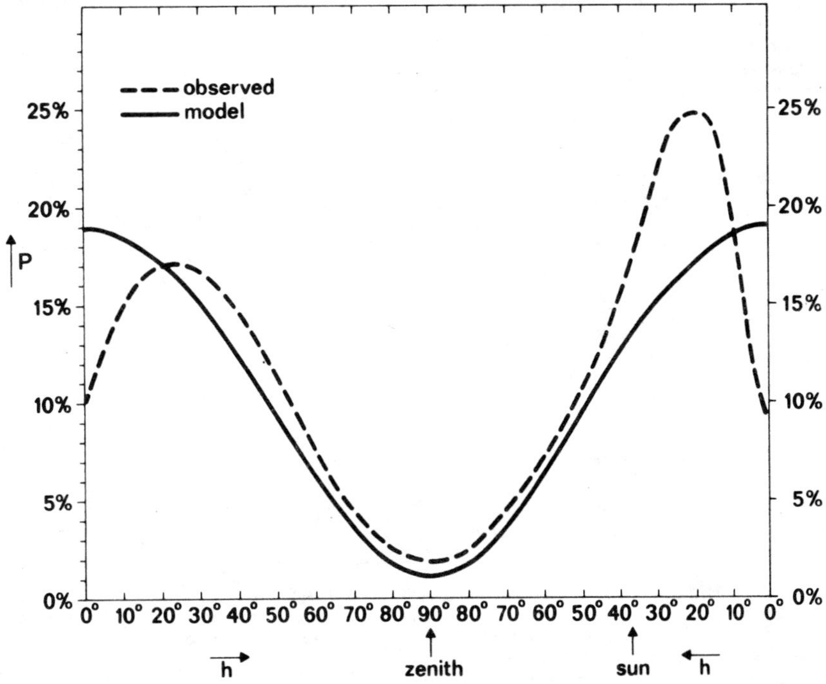

Fig. 2. Polarization distribution in the solar vertical during totality. The dashed line is the observed polarization reported by Shaw.[1] The solid line is the calculated polarization of the present two-step scattering model with parameters $I_D = 1.22$ and $a^2 = 0.1$, taken from Shaw's pretotality, posttotality, and intensity observations. Note that the model is essentially unable to describe the behavior for low h, where singly scattered light becomes dominant.

made with a polarizer before the camera (horizontal and vertical axes, respectively), showing both the polarization of the solar corona and that of the sky around it. The combination of these two data (together with the known shape of the corona during that particular eclipse) independently confirms his visual observation. From this report we conclude that in general the dominating aspect of the direction of polarization is vertical during eclipse, in agreement with our model.

By combining Jannink's observations with the measurements of Shaw, a further conclusion can be drawn on the applicability of the present theory. Since singly scattered light arriving in the solar vertical is horizontally polarized, a switch in the direction of polarization is expected at some height in the sky where multiple scattering starts to dominate. This switch should be accompanied by a local minimum of the degree of polarization as a function of height. However, in Shaw's scan no trace of such a minimum is apparent. This can be considered a second indication that for Shaw's observations the optical thickness is so large at 400 nm that light coming from outside the umbra is largely depolarized by additional (forward) scatterings in its path to the observer (comment c in Section 2). Again, this means that the application of our model with proposition 3 (neglecting the polarization of $\mathbf{S}_1$) is justified for Shaw's observational conditions.

If one assumes that only light from step 2 is responsible for the polarization observed by Shaw at $h = 0$ (so $U_1 = Q_1 = 0$), and if one accepts our model result that the Stokes parameters after step 2 are weakly dependent on h for low h [see Eqs. (16), (18), and (20)], then it is possible to infer from the drop in the polarization near the horizon the relative intensities of light from step 1 and light from step 2. The decrease

in the polarization by a factor of 2 between $h = 20°$ and $h = 0°$ indicates that near the horizon these intensities at 400 nm are of comparable strength, which is consistent with the outcomes of the various intensity scans of Shaw and which does not conflict with the theoretical findings of Gedzelman.[5]

C.   Other Observations

Apart from Shaw's measurements, there are a few instrumental observations of the polarization of the sky during eclipses.[7–11] All the measurements have in common that they are made of a single fixed place in the sky, located in the solar vertical and at 90° from the Sun. Scans as a function of time have been reported, from which the value at maximum eclipse can be deduced. Since intensity scans near the horizon are not reported by these authors and the eclipse and observing conditions are different for each eclipse, a detailed comparison with theory cannot be carried out. However, to get an impression of it we have developed the following procedure. First, as a standard polarization distribution, we adopt the azimuth-independent polarization distribution given by Eq. (19) with $I_D = 1$. Second, each observed degree of polarization at mideclipse P_{tot} is transformed to P_{red}, which is the degree of polarization that would have been observed if $I_D = 1$ (which means that the observed degree of polarization at noneclipse conditions P_{pret} is replaced by a standard value of 0.5). This is done with aid of Eq. (19), so that

$$P_{\text{red}} = \frac{2(1 + I_D) + \cos^2 h}{4 + \cos^2 h} P_{\text{tot}} = \frac{2 P_{\text{pret}}^{-1} + \cos^2 h}{4 + \cos^2 h} P_{\text{tot}}. \quad (23)$$

The values of P_{red} obtained in this way are compared with

Table 1. Summary of Instrumental Observations of Skylight Polarization at Mideclipse[a]

Observer(s)	Observation Number	Year	Wavelength (nm)	h (°)	P_{pret} (%)	P_{tot} (%)	P_{red} (%)
de Bary et al.[7]	1	1961	green	78	78	0	0
Moore and Rao[8]	2	1965	475	49	46	0.5	0.5
	3	1965	601	49	33	4.5	6.6
Miller and Fastie[9]	4	1965	558	25	62	31	26.0
	5	1965	578	25	66	35	28.0
	6	1965	610	25	49	28	28.5
	7	1965	630	25	47	26	27.4
Rao et al.[10]	8	1966	475	20	60.5	19	16.3
	9	1966	601	20	62.5	21	17.6
Dandekar and Turtle[11]	10	1970	475	44	42	4	4.7
	11	1970	600	44	42	<0.5	<0.6
Shaw[1]	12	1973	400	53	45	4	4.4

[a] P_{pret} is the polarization at noneclipse conditions (obtained from interpolation of the pretotality and posttotality measurements), P_{tot} is the polarization at mideclipse, and h is the height above the horizon of the point of observation. Since all measurements are performed in the solar vertical at 90° from the Sun, the solar elevation z can be found from $z = 90° - h$. P_{red} is the degree of polarization during eclipse after transformation to standard circumstances with $P_{\mathrm{pret}} = 50\%$ ($I_D = 1$; see the text). The numbering corresponds to that in Fig. 3.

the standard polarization curve, which is Eq. (19) with $I_D = 1$.

Table 1 summarizes the observations (including that of Shaw), and Fig. 3 compares P_{red} with the standard curve. Although this method is quite rough, and the observations are done at very different conditions (e.g., observations 2–9 are from a high-flying aircraft), there is clearly a correlation between the curve and the observations, although the ground-based observations systematically yield a lower degree of polarization than does the theory (the same holds in

principle for Shaw's polarization distribution, as outlined above). As shown in Appendix A, such behavior is to be expected if the polarization of light resulting from step 1 is completely neglected.

A few further remarks must be made. In the solar vertical, our model predicts essentially a vertical polarization during mideclipse. Moore and Rao,[8] however, reported for red a reversed polarization (their observation for blue is in agreement with our model). We attribute this effect to the larger contribution of polarization in $\mathbf{S}_1$ for red light, which

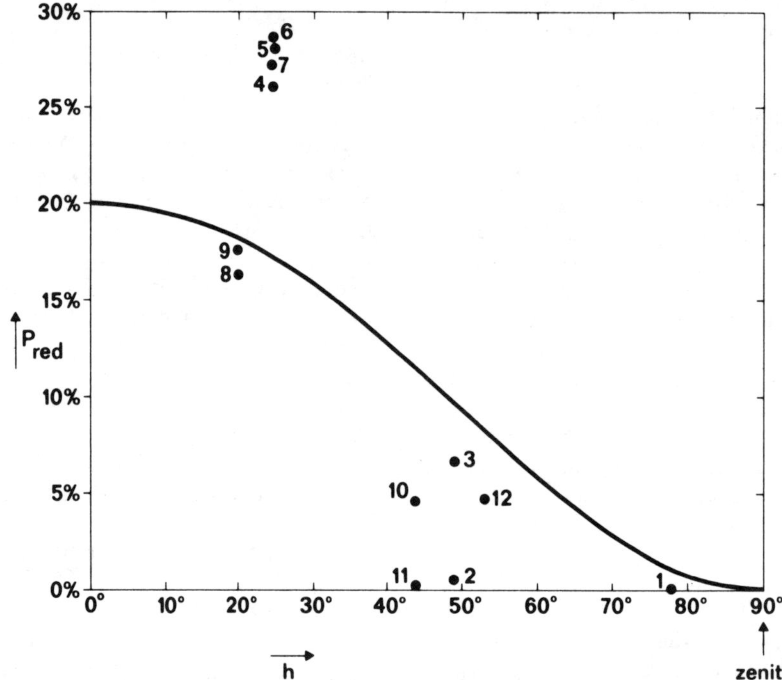

Fig. 3. Comparison of the observed degree of polarization during several eclipses with the theory. The observations are transformed into a degree of polarization P_{red} for standard circumstances. The theoretical curve is Eq. (19) with $I_D = 1$. The numbers at the points correspond to those in Table 1.

should result in a reduction of the vertical polarization of the sky (see Appendix A). The fact that Shaw's intensity measurements near the horizon at 600 nm largely satisfy expression (10) is consistent with this explanation. Furthermore, the polarization phase scan of Dandekar and Turtle[11] also displays a wavelength dependence. The fact that their measurements do not show a complete reversal of the polarization plane may be attributed to the eclipse geometry [the shape of $f(\psi_1)$], combined with a relatively large contribution of polarization in $\mathbf{S}_1$.

5. CONCLUSION

The previous section indicates that, despite its simplicity, our model seems to be able to describe the observations of the polarization of the sky during eclipses at least qualitatively and, in the case of Shaw's observations, even more than that. The conclusion might be drawn that, at least for the shorter wavelengths, it is justified to postpone propositions 1–7. For longer wavelengths, the analysis indicates that proposition 3 is the weakest of the propositions. The analysis in Appendix A indicates that our model with proposition 3 would overestimate the polarization for many solar heights; this finding is consistent with Subsection 4.C.

What remains to be discussed is the good agreement between Shaw's observations and the theory (Subsection 4.A). In the light of the findings in Appendix A, one may even wonder how far this close agreement is a lucky coincidence, caused by the short wavelength and by the fact that at $z = 37°$ the model is rather insensitive to the neglect of polarization of $\mathbf{S}_1$. Anyhow, one would expect at best only qualitative agreement if one calculates the polarization in this way from the limited data of $\mathbf{S}_1$. The close agreement was therefore also a surprise to us.

Within the limited set of existing observations there is no possibility to test the model further at present. This has to wait until more detailed observations are available. Such observations should include the polarization distribution of the eclipsed sky, preferably in the solar vertical plane and in the plane perpendicular to the solar vertical containing the zenith, together with simultaneous alumcanter scans of intensity and polarization near the horizon, all of them preferably at various wavelengths. Only if such a complete set of measurements is available will a rigorous test of models like the present one be possible.

APPENDIX A: ANALYSIS OF THE EFFECT OF RELAXING PROPOSITION 3

To get an impression of the effect of proposition 3 (neglecting the polarization of $\mathbf{S}_1$), we calculate in the solar vertical for $h = 0$ and $h = 90°$ the polarization of the sky with and without proposition 3. The calculation is carried out explicitly for the special case in which the observer is located in the central point of a circularly shaped umbra, $f(\psi_1)= 1$. We note, however, that for an ellipse-shaped umbra the results are identical if the observer is at one of the foci of the ellipse.

In this appendix we maintain our notation, but for the model without proposition 3 a prime is added to the relevant symbols. Thus $P'(h)$ is the degree of polarization without proposition 3, $P(h)$ is the degree of polarization with proposition 3, and so on. In the present analysis, we take $I_D = 1$ in

the scattering matrix $\mathcal{M}$ for step 2, as we did also in the standard curve used in Subsection 4.C.

If the Stokes vector after step 1 is given by $\mathbf{S}_1' = (I_1', Q_1', U_1')$, then the matrix multiplication of Eq. (3) results in

$$
\mathbf{S}_2'(h = 90°) = \begin{bmatrix} 2I_1' - Q_1' \\ (-I_1' + Q_1')\cos 2\alpha \\ (I_1' - Q_1')\sin 2\alpha \end{bmatrix}, \tag{A1}
$$

$$
\mathbf{S}_2'(h = 0°) = \begin{bmatrix} I_1'(2 + \cos^2\alpha) + Q_1' \sin^2\alpha \\ Q_1'(1 + \cos^2\alpha) + I_1' \sin^2\alpha \\ 2U_1' \cos\alpha \end{bmatrix}, \tag{A2}
$$

in which the solar vertical plane is the reference. By using Eqs. (8) and (9), $\mathbf{S}_1'$ can be specified. To make the results comparable to those obtained with Eq. (19), we use for step 1 a simplified scattering matrix $\mathcal{M}'$ in which the matrix element $m_{11}' = 1 + I_D'$ instead of $m_{11}' = 1 + I_D' + \cos^2\theta$. I_D' (the measure of depolarization in step 1) remains unspecified for the moment. For $f(\psi_1) = 1$, we have, from Eqs. (8) and (9),

$$
\mathbf{S}_1' = \begin{bmatrix} I_1' \\ Q_1' \\ U_1' \end{bmatrix} = \begin{bmatrix} 1 + I_D' \\ \cos 2z - \cos^2\alpha \cos^2 z \\ \sin\alpha \sin 2z \end{bmatrix}. \tag{A3}
$$

By substituting Eq. (A3) into Eqs. (A1) and (A2), one finds that after integration over α the third Stokes parameter in $\bar{\mathbf{S}}_2'$ is zero again for $h = 0°$ and $h = 90°$. The degree of polarization can be calculated from the integration of I_2' and Q_2' over α and results in

$$
P'(h = 90°) = \frac{-\tfrac{1}{2}\cos^2 z}{4(1 + I_D') - 3\cos^2 z + 2}, \tag{A4}
$$

$$
P'(h = 0°) = \frac{(1 + I_D') + \tfrac{17}{4}\cos^2 z - 3}{5(1 + I_D') + \tfrac{7}{4}\cos^2 z - 1}. \tag{A5}
$$

Here the sign of P' corresponds to that of $\bar{Q}_2'$. Thus $P' > 0$ means that the direction of polarization is parallel to the solar vertical plane. Under proposition 3, $Q_1 = U_1 = 0$ and we have $P = 0$ and $\tfrac{1}{5}$ for $h = 90°$ and $0°$, respectively. By taking now a lower limit of 1 for I_D', one has for $h = 90°$

$$
0 < P(90°) - P'(90°) < 0.07 \cos^2 z, \tag{A6}
$$

indicating that proposition 3 tends to overestimate the polarization in the zenith by at the most a few percent [for Shaw's eclipse, $P(90°) - P'(90°) < 0.04$; for de Bary's[7] eclipse, $P(90°) - P'(90°) < 0.07$; and for Moore's[8] and Dandekar's[11] eclipses, $P(90°) - P'(90°) < 0.03$].

For $h = 0$, one has, from Eq. (A5) and $P(0°) = \tfrac{1}{5}$,

$$
P(0°) - P'(0°) = \frac{\tfrac{14}{5} - \tfrac{39}{10}\cos^2 z}{5(1 + I_D') + \tfrac{7}{4}\cos^2 z - 1}, \tag{A7}
$$

which indicates that for solar elevations below 32°, the introduction of proposition 3 tends to cause underestimation of the polarization. In the set of available observations (Table 1) none represents this case with low z but observation near the horizon. For $z > 32°$, Eq. (A7) indicates that assumption 3 causes overestimation of the polarization. If $I_D' > 1$, the upper limit of this overestimation ranges from 0.03 and

0.05 for the eclipses observed by Shaw[1] and Moore and Rao,[8] respectively, to 0.25 for those observed by Rao *et al.*[10] and by Miller and Fastie.[9]

REFERENCES

1. G. E. Shaw, "Sky brightness and polarization during the 1973 African eclipse," Appl. Opt. **14,** 388–394 (1975).
2. W. E. Sharp, S. M. Silverman, and J. W. F. Lloyd, "Summary of sky brightness measurements during eclipses of the sun," Appl. Opt. **10,** 1207–1210 (1971).
3. S. M. Silverman and E. G. Mullen, "Sky brightness during eclipses: a review," Appl. Opt. **14,** 2838–2843 (1975).
4. G. E. Shaw, "Sky radiance during a total solar eclipse: a theoretical model," Appl. Opt. **17,** 272–276 (1978).
5. S. D. Gedzelman, "Sky color near the horizon during a total solar eclipse," Appl. Opt. **14,** 2831–2837 (1975).
6. N. Piltschikoff, "Sur la polarisation du ciel pendant les éclipses du soleil," C. R. Acad. Sci. Paris **142,** 1449 (1906).
7. E. de Bary, K. Bullrich, and D. Lorenz, "Messungen der Himmelsstrahlung und deren Polarisationgrad während der Sonnenfinsternis am 15.2.1961 in Viareggio (Italien)," Geofis. Pura Appl. **48,** 193–198 (1961).
8. J. G. Moore and C. R. N. Rao, "Polarization of the daytime sky during the total solar eclipse of 30 May 1965," Ann. Geophys. **22,** 147–150 (1966).

8a. R. Gerharz, "Appearance of the atmospheric scatter field during a solar eclipse," J. Geophys. **42,** 163–167 (1976).

9. R. E. Miller and W. G. Fastie, "Skylight intensity, polarization and airglow measurements during the total solar eclipse of 30 May 1965," J. Atmos. Terr. Phys. **34,** 1541–1546 (1972).
10. C. R. N. Rao, T. Takashima, and J. G. Moore, "Polarimetry of the daytime sky during solar-eclipses," J. Atmos. Terr. Phys. **34,** 573–576 (1972).
11. B. S. Dandekar and J. P. Turtle, "Day sky brightness and polarization during the total eclipse of 7 March 1970," Appl. Opt. **10,** 1220–1224 (1971).
12. H. C. van de Hulst, *Multiple Light Scattering, Tables, Formulas and Applications,* Vol. 2 (Academic, New York, 1980).
13. J. L. Soret, "Sur la polarisation atmosphérique," Ann. Chim. Phys. **14,** 503–541 (1888).
14. F. Ahlgrimm, "Zur Theorie der atmosphärischen Polarisation," Jahrb. Hamburger Wiss. Anst. **32,** 1–66 (1914).
15. K. Serkowski, "Polarimeters for optical astronomy," in T. Gehrels, *Planets, Stars and Nebula, Studied with Photopolarimetry* (University of Arizona, Tucson, Ariz., 1974), pp. 135–174.
16. E. Collet, "The description of polarization in classical physics," Am. J. Phys. **36,** 713–725 (1968).
17. D. W. Jannink, Royal Netherlands Meteorological Institute, De Bilt, The Netherlands (personal communication, 1981).
18. G. P. Können, *Polarized Light in Nature* (Cambridge U. Press, Cambridge, 1985).

Naval Research Laboratory, Washington, D.C.

Section Six
Rainbows, Coronas, and Glories

Reprinted from *Transactions of the Cambridge Philosophical Society,* Vol. VI, 397–403 (1838); Addendum, Vol. VIII, 595–600 (1849).

XVII. *On the Intensity of Light in the neighbourhood of a Caustic. By* George Biddell Airy, *Esq. A.M., Astronomer Royal: Late Fellow of Trinity College, and Plumian Professor of Astronomy and Experimental Philosophy in the University of Cambridge.*

[Read *May* 2, 1836, and *March* 26, 1838.]

When a great physical theory has been established originally on considerations and experiments of a simple kind, which by degrees have been exchanged for comparisons of more distant results of the theory with more complicated cases of experiment, it has always been considered a matter of great interest, to trace out accurately by mathematical process the consequences, according to that theory, of different modifications of circumstances: which can then be compared with measures that have been made, or that may easily be made in future. It is with this view that I solicit the indulgence of the Society, for the following investigation of the Intensity of Light in the neighbourhood of a Caustic, as mathematically estimated from the Undulatory Theory.

The investigation which I present here belongs, ostensibly, only to the case of reflection. The introductory part of it will, however, (with the proper modifications) apply equally well to all cases of refraction and all combinations of reflection and refraction. There seems also to be no reason why the latter part (the estimation of the intensity of light, by considering the wave of light when it leaves the last surface to be divided into a great number of small parts, whose separate effects are then to be compounded,) should not apply to those cases. For though, strictly speaking, we ought to consider the wave to be thus broken up where it leaves the first surface, in order to find the intensity of vibration at every point of the second; yet it seems clear, that those reasonings which establish the definite reflection or refraction of a wave, (and which are founded upon the consideration above alluded to,) point out that there will be as to sense a mutual destruction of all vibrations at the second surface, (supposed to be not distant from the first,) excepting those which would be fully taken into account on the ordinary laws of Geometrical Optics. Where the light meets the second surface in the state of convergence, this conclusion perhaps is not so clear: but even there I believe that it may easily be shewn to be correct. I have mentioned these points because one of the most interesting cases of natural caustics (the rainbow) is affected by them; the exterior bow involving the first-mentioned condition, and the interior bow involving both the first and the second.

---

1. The notion of a caustic, and its mathematical definition, are essentially founded upon the laws of Geometrical Optics; and to these, therefore, we must refer in order to discover a representation of the conditions adapted to the investigations of Physical Optics. For simplicity we shall confine our diagrams to the plane of reflection, and shall consider the reflecting surface as symmetrical (to a sensible extent) with respect to that plane, so that the portion of the caustic formed by that part of the surface will be in the same plane.

2. In fig. 1., let the origin of light S be the origin of co-ordinates; x, y, the co-ordinates of a point X of the reflecting surface; p, q the co-ordinates of a point P in the reflected ray; V the length of the path of light from S to any point of the reflecting surface and thence to the point P. The ordinary law of reflection informs us that the angles of incidence and reflection are equal; and therefore, that, if we take a point X' on the reflecting surface very near to X, and join it with the origin and the point P, the lengthening ZX' of one

of these lines will be equal (ultimately) to the shortening XZ' of the other, and their sum (ultimately) will not be altered; or that, putting $\dfrac{d(V)}{dx}$ for the differential coefficient of V with regard to x, considering y also as a function of x, $\left(\text{which is otherwise written } \dfrac{dV}{dx} + \dfrac{dV}{dy}\cdot\dfrac{dy}{dx}\right)$, $\dfrac{d(V)}{dx} = 0$. This is the condition which holds at the point of reflection.

3. Now if p, q, be the co-ordinates of a focus, since in that case it is a point in the paths of rays reflected from every point of the surface, $\dfrac{d(V)}{dx} = 0$ at every point, and therefore $V = $ constant, and $\dfrac{d^2(V)}{dx^2}$, $\dfrac{d^3(V)}{dx^3}$, &c. are $= 0$ at every point. This is the condition for the reflection of rays to a focus.

4. But though the condition $V = C$ and all its consequences are necessary for the convergence of reflected rays to a focus, yet this condition is not necessary for the convergence of a very small pencil of rays incident on the reflecting surface. It is only necessary for this, that the equations $\dfrac{d(V)}{dx} = 0$, and $\dfrac{d(V')}{dx'} = 0$ should hold at the same time, when $x' = x + \delta x$ and V' has the corresponding value; that is, that the following equations should be true at the same time,

$$\frac{d(V)}{dx} \;\dotfill\; = 0,$$

$$\frac{d(V)}{dx} + \frac{d^2(V)}{dx^2}\cdot\frac{\delta x}{1} + \frac{d^3(V)}{dx^3}\cdot\frac{(\delta x)^2}{1.2} + \&\text{c.} = 0.$$

From this we obtain $\dfrac{d^2(V)}{dx^2} + \dfrac{d^3(V)}{dx^3}\cdot\dfrac{\delta x}{1.2} + \&\text{c.} = 0$, as the equation expressing that the rays incident at the points x and $x + \delta x$ intersect: and making δx indefinitely small, this reduces itself as nearly as we please to the equation $\dfrac{d^2(V)}{dx^2} = 0$. This then is the equation which must hold for the ultimate convergence of rays.

5. Now the definition of a caustic in Geometrical Optics, is "the locus of the ultimate intersections of reflected rays:" and therefore, for every point of a caustic, $\dfrac{d(V)}{dx} = 0$ and $\dfrac{d^2(V)}{dx^2} = 0$ when that value of x is used which corresponds to the point of the reflecting surface, from which the light is reflected to each particular point of the caustic. But $\dfrac{d^3(V)}{dx^3}$ is not necessarily $= 0$: and in general its value is finite. For if, in fig. 2, we take a point P' of the caustic nearer to the reflecting surface than P, and if X' is the corresponding point of the reflecting surface; then we know from the geometrical theory of caustics, that

$$SX' + X'P' + P'P = SX + XP.$$

Now if we join X' with P, it will be evident that

$$X'P < X'P' + P'P.$$

Therefore, $\qquad\qquad SX' + X'P < SX + XP,$

$$\text{or}\quad V' < V.$$

Similarly, if we take a point P'' on the caustic further from the reflecting surface than P, and X'' for the corresponding point on the reflecting surface,

$$SX'' + X''P'' = SX + XP + PP''.$$

$$\text{But}\quad X''P + PP'' > X''P''.$$

$$\text{Therefore}\quad SX'' + X''P + PP'' > SX + XP + PP''.$$

$$\text{Or}\quad SX'' + X''P > SX + XP,$$

$$\text{or}\quad V'' > V;$$

consequently the first differential coefficient of V which has a finite value is of an odd order: and as, in the general case, we must, from the very meaning of the word *general*, take those conditions which require the smallest number of peculiar equations, we must fix on the first coefficient of an odd order which has not yet been fettered by any equation; and therefore in the general case, for a point in a caustic,

$$\frac{d^3(V)}{dx^3} \text{ has a finite value.}$$

It may possibly happen at singular points that $\frac{d^3(V)}{dx^3}$ and $\frac{d^4(V)}{dx^4}$ vanish, and that $\frac{d^5(V)}{dx^5}$ has a finite value: but of these peculiar cases I intend to take no further notice.

6. By pursuing this train of investigation we should find that at a cusp of a caustic $\frac{d(V)}{dx} = 0$, $\frac{d^2(V)}{dx^2} = 0$, $\frac{d^3(V)}{dx^3} = 0$, and $\frac{d^4(V)}{dx^4}$ has a finite value. I shall not however pursue this subject further.

7. The conditions then which hold, with reference to any point of a caustic in general, are these: If V be measured from the origin of light to any point of the reflecting surface and then to the given point of the caustic: in the case of the point of the reflecting surface coinciding with the corresponding point of reflexion,

$$\frac{d(V)}{dx} = 0,$$

$$\frac{d^2(V)}{dx^2} = 0,$$

$$\frac{d^3(V)}{dx^3} = C,$$

C being a finite function of x, y, p, and q. The sign of C may be thus found. In the case assumed in (5) and represented in fig. 2, V' was $< V$ and $V'' > V$; if then V' implies that x is diminished, or if x is measured from the convexity of the caustic, C is positive. If x is measured towards the convexity of the caustic, in that case C is negative.

8. The value of C may thus be found. Draw $P'Q'$ perpendicular to PX': then $Q'X'$ may be considered equal to $P'X'$ (it will differ from it only by quantities depending on the fourth power of PP' or XX');

and therefore $V - V'$, which in (5) was found $= X'P' + P'P - X'P$, is $= P'P - Q'P$. Let x be measured nearly perpendicular to the caustic at P; put ρ for the radius of curvature of the caustic at P, and ϕ for the small angle made by PX and $P'X'$. Then $\delta x = PX \cdot \phi$, therefore $\phi = \dfrac{\delta x}{PX}$. And $P'P - Q'P = \rho \cdot \dfrac{\phi^3}{1.2.3} = \dfrac{\rho}{(PX)^3} \cdot \dfrac{\delta x^3}{1.2.3}$. Making this equal to the corresponding term in Taylor's series for V', we find

$$C = \frac{d^3(V)}{dx^3} = \frac{\rho}{(PX)^3}.$$

9. Now take a point near the caustic, whose co-ordinates are $p + \delta p$ and q (δp being measured from the convexity of the caustic parallel to x, or nearly perpendicular to the caustic at P). Let $V_{\prime}$ be the length of the path of light from the origin to any point of the reflector and thence to the point $p + \delta p$, q. Then we have

$$V = \sqrt{x^2 + y^2} + \sqrt{(x-p)^2 + (y-q)^2},$$

$$V_{\prime} = \sqrt{x^2 + y^2} + \sqrt{(x-p-\delta p)^2 + (y-q)^2};$$

$$\text{or} \quad V_{\prime} = V + \sqrt{(x-p-\delta p)^2 + (y-q)^2} - \sqrt{(x-p)^2 + (y-q)^2},$$

which, if we expand to the first power of δp, becomes

$$V_{\prime} = V - \frac{x-p}{\sqrt{(x-p)^2 + (y-q)^2}}\, \delta p,$$

and therefore, in the general case of measuring V through any point of the reflecting surface,

$$\frac{d(V_{\prime})}{dx} = \frac{d(V)}{dx} + A \cdot \delta p,$$

$$\frac{d^2(V_{\prime})}{dx^2} = \frac{d^2(V)}{dx^2} + B \cdot \delta p;$$

$$\frac{d^3(V_{\prime})}{dx^3} = \frac{d^3(V)}{dx^3} + D \cdot \delta p,$$

A, B, and D, being finite functions of x, y, p, and q.

In the particular case of measuring V through the point which reflects rays (according to the ordinary rules) to p, q,

$$\frac{d(V_{\prime})}{dx} = A \cdot \delta p,$$

$$\frac{d^2(V_{\prime})}{dx^2} = B \cdot \delta p;$$

$$\frac{d^3(V_{\prime})}{dx^3} = C + D \cdot \delta p,$$

or, as we shall always suppose δp small,

$$\frac{d^3(V_{\prime})}{dx^3} = C.$$

10.   Consequently, if $V'_{\prime}$ be put for the length of the path through $x + \delta x$, $y + \delta y$, to $p + \delta p$, q, (δp being entirely independent of δx),

$$V'_{\prime} = V_{\prime} + A\delta p\,\frac{\delta x}{1} + B\delta p \cdot \frac{(\delta x)^2}{1.2} + C \cdot \frac{(\delta x)^3}{1.2.3},$$

or, putting z for δx,

$$V'_{\prime} = V_{\prime} + A\delta p \cdot \frac{z}{1} + B\delta p \cdot \frac{z^2}{1.2} + C \cdot \frac{z^3}{1.2.3},$$

omitting the following terms.

11.   The value of C has been found: that of A (the only other quantity which interests us) will be obtained by actual differentiation of the expression for $V_{\prime}$:   Thus we find

$$A\delta p = \delta p \cdot \frac{(x-p)(y-q) \cdot \dfrac{dy}{dx} - (y-q)^2}{\{(x-p)^2 + (y-q)^2\}^{\frac{3}{2}}}.$$

Or, as $x - p$ is supposed to be very small, $A = -\dfrac{1}{y-q} = -\dfrac{1}{PX}.$

12.   There is one case so peculiar, and which seems so likely to cause a failure of these expressions, that it merits a particular investigation: the more so as it occurs in the rainbow.   It is the case in

which, on ascertaining the deviation (from a fixed direction) of the rays reflected or refracted, we find, on proceeding in the same direction along the reflecting or refracting surface, that the deviation increases to a certain amount and then diminishes, or *vice versá*. In this case the caustic consists of two unconnected infinite branches in opposite directions, with a common asymptote parallel to the position of maximum or minimum deviation of the rays. To investigate this case, we shall examine the form of the front of the wave immediately after leaving the reflecting or refracting surface, and shall measure the lengths of paths of light from that front. In fig. 3, let A be the point at which the asymptote intersects the front of the wave (which will be the same as the point of the front where the deviation is maximum or minimum) whose co-ordinates are 0 and b: let X be any other point in the front, whose co-ordinates are x and y, (x being measured from the asymptote and y parallel to it:) and p and q the co-ordinates, similarly measured, of any point P near the asymptote. If the length AX be called s, and the angle made by the tangent at X with the tangent at A be called θ, then the condition that the deviation of the direction of the rays from a fixed direction (or the deviation of the tangent to the front of the wave from another fixed direction,) is maximum or minimum gives $\theta = \dfrac{s^2}{a^2}$, a being some constant. Observing that $\dfrac{dx}{ds} = \cos\theta$, and $\dfrac{dy}{ds} = \sin\theta$, we get with sufficient approximation $x = s$, $y = b + \dfrac{s^3}{3a^2} = b + \dfrac{x^3}{3a^2}$, and the front of the wave is therefore a cubical parabola. The distance of the point P from X

$$= \sqrt{(x-p)^2 + (y-q)^2} = \sqrt{p^2 - 2px + x^2 + (b-q)^2 + \dfrac{2(b-q)}{3a^2}x^3},$$

and, expanding this to the third power of x, and putting c^2 for $p^2 + (b-q)^2$,

$$PX = c\left\{1 - \dfrac{p}{c^2}x + \tfrac{1}{2}\cdot\dfrac{c^2-p^2}{c^4}x^2 + \left(\dfrac{b-q}{3a^2c^2} + \tfrac{1}{2}\cdot\dfrac{p(c^2-p^2)}{c^6}\right)x^3\right\}.$$

If we take only the principal part of each coefficient (which for any practical case will be abundantly sufficient, observing that $\dfrac{p}{b-q}$ will probably never, in observation, amount to $\tan 2^\circ$), this becomes

$$PX = (b-q)\left\{1 - \frac{p}{(b-q)^2}\,x + \tfrac{1}{2}\cdot\frac{1}{(b-q)^2}\,x^2 + \frac{1}{3\,a^2(b-q)}\,x^3\right\},$$

$$= b-q-\frac{p}{b-q}\,x+\tfrac{1}{2}\cdot\frac{1}{b-q}\,x^2+\frac{1}{3\,a^2}\,x^3.$$

In the applications of this, it will be important to notice that the coefficient of x^3 is independent of p and q, (depending only on the dimensions of the rain-drop or other refracting or reflecting body,) and that the coefficient of x depends only on the angle made by PX with the asymptote.

13.   It appears, therefore, that in both the cases considered, the form of the expression for the length of the path of the wave to the point under consideration near the caustic, passing through the general point of the reflecting surface or of the front of the wave, is that of a general formula of the third order; in which the coefficient of the first power of the ordinate of the point on the front of the wave is proportional to the distance of the illuminated point from the caustic or the asymptote, and in which the coefficient of the third power is independent of that distance.   If in the first instance we make $z + \dfrac{PX^3.B.\delta p}{\rho} = z'$, the first expression becomes (putting E for a term independent of z, and in the coefficient of z omitting the term involving δp^2 in comparison with δp,)

$$V' = E + \frac{\rho}{6\,PX^3}\left\{z'^3 - \frac{6.PX^2}{\rho}\,\delta p\,.\,z'\right\}.$$

And if in the second instance we make $x + \tfrac{1}{2}\cdot\dfrac{a^2}{b-q} = x'$, and observe that for the rainbow a is a very small fraction of an inch while p may be many feet, and that a may therefore be omitted in comparison with p,

$$PX = F + \frac{1}{3\,a^2}\left\{x'^3 - \frac{3\,a^2}{b-q}\,p\,.\,x'\right\}.$$

3 D

14. To proceed now with the intensity of light at the illuminated point. I shall omit entirely the integration for the ordinate perpendicular to the plane of x, y, because it would only introduce a factor common to every part, and therefore would not modify the proportion of intensity at different points. The great wave of light being supposed to be divided into indefinitely small parts, each of which is the origin of a small wave spreading in all directions: the disturbance of ether at the illuminated point produced by this small wave, on the Undulatory Theory, will be estimated by

$$\text{portion of surface of small wave} \times \sin \frac{2\pi}{\lambda}\,(vt - \text{whole path})$$

which in the first case becomes

$$\delta z' \times \sin \frac{2\pi}{\lambda}\left\{vt - E - \frac{\rho}{6PX^3}\left(z'^3 - \frac{6.PX^2}{\rho}\,\delta p\,.\,z'\right)\right\},$$

and in the second case

$$\delta x' \times \sin \frac{2\pi}{\lambda}\left\{vt - F - \frac{1}{3a^2}\left(x'^3 - \frac{3a^2}{b-q}\,p\,.\,x'\right)\right\}.$$

15. In the first case therefore, the expression for the whole disturbance is

$$\int_{z'} \sin \frac{2\pi}{\lambda}\left\{vt - E - \frac{\rho}{6PX^3}\left(z'^3 - \frac{6.PX^2}{\rho}\,\delta p\,.\,z'\right)\right\}.$$

The limits through which the integration is to be performed are from z' a sensible quantity negative to z' a sensible quantity positive, and on account of the minuteness of the divisor λ, and the inefficiency of the rays whose paths differ from E by many multiples of λ, this will be the same as taking it between the limits $-$ infinity, $+$ infinity. Now the integral is the same as

$$\sin \frac{2\pi}{\lambda}\,(vt - E) \int_{z'} \cos \frac{2\pi}{\lambda}\,.\,\frac{\rho}{6PX^3}\left(z'^3 - \frac{6.PX^2}{\rho}\,.\,\delta p\,.\,z'\right)$$

$$-\cos \frac{2\pi}{\lambda}\,(vt - E) \int_{z'} \sin \frac{2\pi}{\lambda}\,.\,\frac{\rho}{6PX^3}\left(z'^3 - \frac{6.PX^2}{\rho}\,.\,\delta p\,.\,z'\right).$$

But between $-$ infinity and $+$ infinity it is evident that

$$\int_{z'} \sin \frac{2\pi}{\lambda}\,.\,\frac{\rho}{6PX^3}\left(z'^3 - \frac{6.PX^2}{\rho}\,.\,\delta p\,.\,z'\right) = 0,$$

because every positive value is balanced by an equal negative value; and therefore the expression for the disturbance of ether at the illuminated point is

$$\sin \frac{2\pi}{\lambda}(vt-E)\int_{z'} \cos \frac{2\pi}{\lambda}\cdot\frac{\rho}{6PX^3}\left(z'^3 - \frac{6\cdot PX^2}{\rho}\cdot\delta p\cdot z'\right),$$

the integral being taken between $-$ infinity, $+$ infinity;

$$\text{or, } 2\sin \frac{2\pi}{\lambda}(vt-E)\int_{z'} \cos \frac{2\pi}{\lambda}\cdot\frac{\rho}{6PX^3}\left(z'^3 - \frac{6PX^2}{\rho}\cdot\delta p\cdot z'\right),$$

the integral being taken from 0 to infinity.

Making $\dfrac{2\pi}{\lambda}\cdot\dfrac{\rho}{6PX^3}\cdot z'^3 = \dfrac{\pi}{2}w^3$, or $z' = PX\left(\dfrac{3\lambda}{2\rho}\right)^{\frac{1}{3}}\cdot w$, and putting m for $\delta p \times \left(\dfrac{96}{\lambda^2\rho}\right)^{\frac{1}{3}}$, and omitting the constant factor, we find as the expression for the disturbance of ether at the illuminated point

$$\sin \frac{2\pi}{\lambda}(vt-E)\int_w \cos \frac{\pi}{2}(w^3 - m\cdot w),$$

and therefore the expression for the intensity of light is

$$\left[\int_w \cos \frac{\pi}{2}(w^3 - m\cdot w)\right]^2,$$

the integral being taken from $w = 0$ to $w =$ infinity.

It will be observed that m is proportional to δp, and therefore the intensity of light at the Geometrical Caustic, or where $\delta p = 0$, is found by making $m = 0$ in this formula.

16. In the second case, the expression for the whole disturbance of ether is

$$\int_{x'} \sin \frac{2\pi}{\lambda}\left\{vt - F - \frac{1}{3a^2}\left(x'^3 - \frac{3a^2}{b-q}p\cdot x'\right)\right\},$$

which, as above, is shewn to be equal to

$$2\sin \frac{2\pi}{\lambda}(vt-F)\cdot\int_{x'} \cos \frac{2\pi}{\lambda}\cdot\frac{1}{3a^2}\left(x'^3 - \frac{3a^2}{b-q}p\cdot x'\right),$$

from $x' = 0$ to $x' =$ infinity.

Making $\dfrac{2\pi}{\lambda} \cdot \dfrac{1}{3a^2} x'^3 = \dfrac{\pi}{2} w^3$, or $x' = \left(\dfrac{3a^2\lambda}{4}\right)^{\frac{1}{3}} . w$, and putting m for $\dfrac{p}{b-q} \times \left(\dfrac{48a^2}{\lambda^2}\right)^{\frac{1}{3}}$, and omitting the constant factor, the intensity of light, as above, is shewn to be

$$\left[\int_w \cos \dfrac{\pi}{2} (w^3 - m.w)\right]^2,$$

the integral being taken from $w = 0$ to $w =$ infinity.

It will be remarked that m, in this case, is proportional to $\dfrac{p}{b-q}$; it is therefore 0 for points in the direction of the asymptote, and for other points it is proportional to the angle made by the line from the center of the wave with the asymptote.

17. The values of $\int_w \cos \dfrac{\pi}{2} (w^3 - m.w)$, from $w = 0$ to $w =$ infinity, and the squares of these numbers, for every 0·2 from $m = -4·0$ to $m = +4·0$, are contained in the following table: for the calculation of which I refer to the Appendix.

Values of m	Corresponding values of $\int_w \cos \dfrac{\pi}{2}(w^3 - m.w)$ from 0 to $\dfrac{1}{0}$	Squares of the last Numbers	Values of m	Corresponding values of $\int_w \cos \dfrac{\pi}{2}(w^3 - m.w)$ from 0 to $\dfrac{1}{0}$	Squares of the last Numbers
− 4·0	+ 0·00298	0·0000089	+ 0·6	+ 0·91431	0·83597
− 3·8	+ ·00431	·0000186	+ 0·8	+ 0·97012	0·94114
− 3·6	+ ·00618	·0000382	+ 1·0	+ 1·00041	1·00082
− 3·4	+ ·00879	·0000773	+ 1·2	+ 0·99786	0·99572
− 3·2	+ ·01239	·0001536	+ 1·4	+ ·95606	·91406
− 3·0	+ ·01730	·000299	+ 1·6	+ ·87048	·75773
− 2·8	+ ·02393	·000573	+ 1·8	+ ·73939	·54670
− 2·6	+ ·03277	·001074	+ 2·0	+ ·56490	·31912
− 2·4	+ ·04442	·00197	+ 2·2	+ ·35366	·12508
− 2·2	+ ·05959	·00355	+ 2·4	+ ·11722	·013741
− 2·0	+ ·07908	·00625	+ 2·6	− ·12815	·016422
− 1·8	+ ·10377	·01077	+ 2·8	− ·36237	·13131
− 1·6	+ ·13461	·01812	+ 3·0	− ·56322	·31721
− 1·4	+ ·17254	·02977	+ 3·2	− ·70874	·50231
− 1·2	+ ·21839	·04769	+ 3·4	− ·78018	·60868
− 1·0	+ ·27283	·07444	+ 3·6	− ·76516	·58547
− 0·8	+ ·33621	·11304	+ 3·8	− ·66054	·43631
− 0·6	+ ·40839	·16678	+ 4·0	− 0·47446	0·22511
− 0·4	+ ·48856	·23869			
− 0·2	+ ·57507	·33071			
0·0	+ ·66527	·44259			
+ 0·2	+ ·75537	·57059			
+ 0·4	+ 0·84040	0·70628			

The extent of this table for the positive values of m is not so great as I could wish; but it goes far enough to enable us to point out the most remarkable circumstances of the distribution of illumination.

18. From $m = -4\cdot0$ to $m = -1\cdot6$, the illumination is almost insensible. (In fact it appears to diminish, as the negative value of m increases, in a nearly geometrical or perhaps hyper-geometrical progression). It then increases rapidly, and acquires its maximum value when $m = +1\cdot08$ nearly; its value is then nearly $1\cdot001$. It then diminishes rapidly till $m = +2\cdot48$ nearly, when the illumination is zero. It then increases till $m = +3\cdot47$, when the illumination is nearly $0\cdot615$, or about three-fifths of its former maximum. It then diminishes rapidly to the end of the table: and appears likely to become zero for a value of m differing little from $+4\cdot4$.

19. One of the most important points to be remarked is, that the maximum illumination does not take place at the Geometrical Caustic, or where $m = 0$, but where $m = +1\cdot08$, that is, on the external side of the convexity of the caustic, or on the luminous side of the geometrical position of the rainbow, that is, (for the primary bow,) within it. The following rule derived from the numbers above, will suffice, in practice, to determine the geometrical position. When the first spurious bow is visible, measure the distance of its maximum intensity from that of the brilliant bow; then the geometrical bow is exterior to the brilliant bow by $\dfrac{11}{24}$ of this distance.

20. It is a matter of curiosity to ascertain the relation of the intensities, or at least of the places of maximum and minimum intensity, as determined thus by a complete investigation on the theory of undulations, with those which would be found on the imperfect theory that light proceeds in straight rays according to the laws of Geometrical Optics, and that rays of light are capable of interfering according to the simple rules of interference. We have first to discover the position of the two rays which interfere at any point. Now the

length of the path of any ray to the illuminated point is $E + \frac{\lambda}{4}(w^3 - m.w)$: and, by (2), the first differential coefficient of this quantity with regard to w will be zero for those rays which pass according to the ordinary rules of reflection and refraction. Performing this differentiation, $3w^2 - m = 0 : w = \pm \sqrt{\frac{m}{3}}$: the lengths of path therefore of the two rays are $E - \frac{\lambda}{4}\sqrt{\frac{4m^3}{27}}$ and $E + \frac{\lambda}{4}\sqrt{\frac{4m^3}{27}}$: and the difference of these is $\frac{\lambda}{2}\sqrt{\frac{4m^3}{27}}$. The destruction of light would therefore take place, on this imperfect theory, when $\sqrt{\frac{4m^3}{27}} = 1$, or $= 3$, or $= 5$, &c.; that is, when $m = \sqrt[3]{\frac{27}{4}}$, or $= \sqrt[3]{\frac{27.9}{4}}$, &c.; or when $m = 1\cdot89$, $3\cdot93$, &c.: and there would be no light whatever for negative values of m. We have found above, on the complete theory, that there is sensible light for negative values of m, and that the destruction of light takes place when $m = 2\cdot48$, $4\cdot4$ (nearly). According to the imperfect theory, the intensity would be infinite when $m = 0$, and the next maximum would be nearer to $1\cdot89$ than to $3\cdot93$: perhaps when $m = 2\cdot7$: we have found above that the intensity is nowhere infinite, that the first maximum takes place when $m = 1\cdot08$, and the second when $m = 3\cdot47$.

21. In figure 4, I have represented the intensity of the light by the ordinates of a curve, of which the abscissa represents different values of m. The strong line corresponds to the determination of the complete theory: the dotted line, to that of the old theory of emission (supposing the intensity inversely as the square root of the distance from the caustic): and the faint line, to that of the imperfect theory of interference mentioned above, giving to the maxima values in some degree proportionate to the ordinates of the dotted line. The absolute values of the ordinates in the faint and the dotted line are not to be understood as necessarily referred to the same unit as those in the strong line: but the abscissæ correspond exactly in all.

APPENDIX.

ON the numerical computation of the definite integral $\int_w \cos \frac{\pi}{2}(w^3 - m.w)$,

between the limits 0 and $\frac{1}{0}$.

The simplicity of the form of this differential coefficient induces me to suppose that the integral may possibly be expressible by some of the integrals whose values have been tabulated. After many attempts however, I have not succeeded in reducing it to any known integral: and I have therefore computed its value by actual summation to a considerable extent and by series for the remainder.

The summation was continued {for each of the values of m in the table given in (17)}, as far as $w = 2$. This extent was divided into eight sections, corresponding nearly to quadrants of the circular function when $m = 0$, as follows:

$$
\begin{aligned}
&\text{1st section, from } w = 0 \quad \text{ to } w = 1\cdot00, \\
&\text{2nd} \qquad\qquad \text{from } w = 1\cdot00 \text{ to } w = 1\cdot26, \\
&\text{3rd} \qquad\qquad \text{from } w = 1\cdot26 \text{ to } w = 1\cdot44, \\
&\text{4th} \qquad\qquad \text{from } w = 1\cdot44 \text{ to } w = 1\cdot58, \\
&\text{5th} \qquad\qquad \text{from } w = 1\cdot58 \text{ to } w = 1\cdot70, \\
&\text{6th} \qquad\qquad \text{from } w = 1\cdot70 \text{ to } w = 1\cdot82, \\
&\text{7th} \qquad\qquad \text{from } w = 1\cdot82 \text{ to } w = 1\cdot92, \\
&\text{8th} \qquad\qquad \text{from } w = 1\cdot92 \text{ to } w = 2\cdot00.
\end{aligned}
$$

These sections were divided into small intervals corresponding to uniform increments in the value of w, as follows:

$$
\begin{aligned}
&\text{In the 1st section the increment of } w \text{ was } 0\cdot04, \\
&\text{2nd} \dotfill \cdot02, \\
&\text{3rd} \dotfill \cdot018, \\
&\text{4th} \dotfill \cdot014, \\
&\text{5th} \dotfill \cdot012, \\
&\text{6th} \dotfill \cdot010, \\
&\text{7th} \dotfill \cdot010, \\
&\text{8th} \dotfill \cdot008.
\end{aligned}
$$

As the increments in the 6th and 7th sections were the same, the calculations for these two sections were conducted without the interruption which was necessary at the separation of the other sections.

The values of w and $\dfrac{\pi}{2}(w^3 - m.w)$ corresponding to the middle of each interval were then computed, and the appropriate value of

$$\cos \frac{\pi}{2}(w^3 - m.w)$$

was formed. Thus, in the first section the computations were made for $w = {\cdot}02,\ {\cdot}06,\ {\cdot}10$, &c: in the 4th section, for $w = 1{\cdot}447,\ 1{\cdot}461,\ 1{\cdot}475$, &c. For a reason that will be shortly mentioned, the values of $\cos \dfrac{\pi}{2}(w^3 - m.w)$ were also computed for two values of w preceding the first, and two values following the last, of each section. These values were then differenced as far as the 4th order: which operation, besides giving a means of checking most severely the accuracy of the computations, supplied the numbers necessary (in the next process) for converting the *sum* into an *integral*.

Now, suppose, that u_x is a function of x, and that the quantity h is so small that the functions u_{x+h}, u_{x+2h}, admit of being expressed with sufficient accuracy by the formula $u_x + b.h + c.h^2 + d.h^3 + e.h^4$, $u_x + b.(2h) + c.(2h)^2 + d.(2h^3) + e.(2h)^4$. This assumption is abundantly accurate for the numbers of which we are treating here. Set down two values preceding u_x and two following it, and take their differences as far as the 4th order, thus

	1st Differences.	2d Differences.	3d Differences.	4th Diff.
$u_x - 2bh + 4ch^2 - 8dh^3 + 16eh^4$				
	$bh - 3ch^2 + 7dh^3 - 15eh^4$			
$u_x - bh + ch^2 - dh^3 + eh^4$		$2ch^2 - 6dh^3 + 14eh^4$		
	$bh - ch^2 + dh^3 - eh^4$		$6dh^3 - 12eh^4$	
u_x		$2ch^2 + 2eh^4$		$24eh^4$
	$bh + ch^2 + dh^3 + eh^4$		$6dh^3 + 12eh^4$	
$u_x + bh + ch^2 + dh^3 + eh^4$		$2ch^2 + 6dh^3 + 14eh^4$		
	$bh + 3ch^2 + 7dh^3 + 15eh^4$			
$u_x + 2bh + 4ch^2 + 8dh^3 + 16eh^4$				

These expressions correspond to the following quantities:

$$
\begin{array}{c|c|c|c|c}
u_{x-2h} & & & & \\
& \Delta'_{-\frac{3h}{2}} & & & \\
u_{x-h} & & \Delta''_{-h} & & \\
& \Delta'_{-\frac{h}{2}} & & \Delta'''_{-\frac{h}{2}} & \\
u_x & & \Delta'' & & \Delta'''' \\
& \Delta'_{\frac{h}{2}} & & \Delta'''_{\frac{h}{2}} & \\
u_{x+h} & & \Delta''_{h} & & \\
& \Delta'_{\frac{3h}{2}} & & & \\
u_{x+2h} & & & &
\end{array}
$$

Equating the quantities on the middle line, or the sums of the quantities next above and next below,

$$
eh^4 = \frac{\Delta''''}{24},
$$

$$
dh^3 = \frac{\Delta'''_{-\frac{h}{2}} + \Delta'''_{\frac{h}{2}}}{12},
$$

$$
ch^2 = \frac{\Delta''}{2} - eh^4 = \frac{\Delta''}{2} - \frac{\Delta''''}{24},
$$

$$
bh = \frac{\Delta'_{-\frac{h}{2}} + \Delta'_{\frac{h}{2}}}{2} - dh^3 = \frac{\Delta'_{-\frac{h}{2}} + \Delta'_{\frac{h}{2}}}{2} - \frac{\Delta'''_{-\frac{h}{2}} + \Delta'''_{\frac{h}{2}}}{12}.
$$

Now the integral with respect to z of the function u_{x+z} or

$$
u_x + bz + cz^2 + dz^3 + ez^4,
$$

is

$$
u_x \cdot z + \frac{b}{2} \cdot z^2 + \frac{c}{3} \cdot z^3 + \frac{d}{4} \cdot z^4 + \frac{e}{5} \cdot z^5 :
$$

and this between the limits $z = -\dfrac{h}{2}$ and $z = +\dfrac{h}{2}$ is

$$
u_x \cdot h + \frac{c}{12} h^3 + \frac{e}{80} h^5
$$

$$
= u_x \cdot h + \frac{\Delta''}{24} h - \frac{\Delta''''}{288} h + \frac{\Delta''''}{1920} h
$$

$$
= h \left\{ u_x + \frac{\Delta''}{24} - \frac{17 \cdot \Delta''''}{5760} \right\}.
$$

This formula applies to the order of the calculations described above, because we have computed the value of u_x for the middle of each of the intervals into which w is divided. And as the same formula applies to every one of the intervals, it applies to the sum of all. And the sum of all the partial integrals through each section is the whole integral through each section. Thus we find,

$$\text{Integral through each section} = \text{interval} \times \left\{ \begin{array}{l} \text{sum of computed values of } \cos\frac{\pi}{2}(w^3 - m \cdot w) \\[2ex] +\dfrac{1}{24}\text{ sum of corresponding 2d differences} \\[2ex] -\dfrac{17}{5760}\text{ sum of corresponding 4th differences} \end{array} \right\}$$

$$= \text{interval} \times \left\{ \begin{array}{l} \text{sum of computed values of } \cos\frac{\pi}{2}(w^3 - m \cdot w) \\[2ex] +\dfrac{1}{24}\left\{ \begin{array}{l}\text{1st difference following the last term} - \text{1st difference} \\ \quad\text{preceding the first term}\end{array}\right\} \\[2ex] -\dfrac{17}{5760}\left\{ \begin{array}{l}\text{3d difference following the last term} - \text{3d difference} \\ \quad\text{preceding the first term}\end{array}\right\} \end{array} \right\}.$$

This process was used throughout. For the purpose of forming the 3rd difference following the last term and that preceding the first term, it was necessary to compute two values of $\cos\frac{\pi}{2}(w^3 - m \cdot w)$ following the end of each section and two preceding its beginning.

The values of $\frac{\pi}{2}(w^3 - m \cdot w)$ were computed by means of Delambre's *Tables Trigonométriques Décimales*. The centesimal division of the circle is, in every instance in which I have used it, far more convenient than the sexagesimal: but in an instance like the present, where there is continual addition or subtraction of arcs, and where the whole arc amounts to several circumferences, the labour and liability to error would be so great with the sexagesimal division, as to make the operation almost impracticable. The numbers were thus formed in each section. The first four values of $\frac{\pi}{2}w^3$ were computed independently and differenced, and with these differences the rest of the series of $\frac{\pi}{2}w^3$ was formed: the last was also computed independently as a check. The first term of each series of $\frac{\pi}{2}(w^3 - m.w)$ was formed by applying $-\frac{\pi}{2}m.w$ to the first term of the series of $\frac{\pi}{2}w^3$: and the first difference was formed, all through the series, by applying $-\frac{\pi}{2}m.\delta w$ to the corresponding 1st difference of $\frac{\pi}{2}w^3$. The last terms of all the series were compared together, as a check. Then for every term the arc less than $\frac{\pi}{4}$ was taken whose sine or cosine (with proper sign) represented $\cos\frac{\pi}{2}(w^3 - m.w)$. The natural numbers were taken to the 7th decimal place, and differenced, as has been mentioned. The number of arguments thus computed is 5166; and the number of natural terms for the summation is the same; and the whole of these have been differenced on paper to the third order, and mentally to the fourth order.

The integration as far as $w = 2\cdot00$ being thus completed, and with the utmost accuracy, the next step was to compute the integral from $w = 2\cdot00$ to $w = $ infinity. Let $u = \frac{\pi}{2}(w^3 - m.w)$: the problem is now to find $\int_w \cos u$. If we make $\dfrac{1}{\dfrac{du}{dw}} = v$, this integral

$$= \int_w v . \cos u \, \frac{du}{dw} = v \sin u - \int_w \sin u \, \frac{dv}{dw} = v \sin u - \int_w v \, \frac{dv}{dw} . \sin u \, \frac{du}{dw},$$

where the last term may be integrated by parts as before. Proceeding with this operation, and putting

$$v_0 = v,$$
$$v_1 = v \, \frac{dv}{dw},$$
$$v_2 = v \, \frac{d}{dw}\left(v \, \frac{dv}{dw}\right),$$
$$v_3 = v \, \frac{d}{dw}\left[v \, \frac{d}{dw}\left(v \, \frac{dv}{dw}\right)\right],$$
$$v_4 = v \, \frac{d}{dw}\left\{v \, \frac{d}{dw}\left[v \, \frac{d}{dw}\left(v \, \frac{dv}{dw}\right)\right]\right\},$$

and so on, we find for the integral generally

$$(v_0 - v_2 + v_4 - v_6 + \&\text{c.}) \sin u + (v_1 - v_3 + v_5 - v_7 + \&\text{c.}) \cos u.$$

The first limit of the integration being w, and the last being infinity, and the quantities v_0, v_1, &c. vanishing for $w = $ infinity, the value of the integral between these limits is,

$$(- v_0 + v_2 - v_4 + v_6 - \&\text{c.}) \sin u + (- v_1 + v_3 - v_5 + v_7 - \&\text{c.}) \cos u.$$

It will be observed here that $v = \dfrac{2}{\pi} \cdot \dfrac{1}{3w^2 - m}$.

The following are the expressions for v_0, v_1, &c.

$$v_0 = \frac{2}{\pi} \cdot \frac{1}{3w^2 - m},$$

$$v_1 = -\frac{24}{\pi^2} \cdot \frac{w}{(3w^2 - m)^3},$$

$$v_2 = \frac{240}{\pi^3} \cdot \frac{1}{(3w^2 - m)^4} + \frac{288m}{\pi^3} \cdot \frac{1}{(3w^2 - m)^5},$$

$$v_3 = -\frac{11520}{\pi^4} \cdot \frac{w}{(3w^2 - m)^6} - \frac{17280m}{\pi^4} \cdot \frac{w}{(3w^2 - m)^7},$$

$$v_4 = \frac{253440}{\pi^5} \cdot \frac{1}{(3w^2 - m)^7} + \frac{725760m}{\pi^5} \frac{1}{(3w^2 - m)^8} + \frac{483840m^2}{\pi^5} \frac{1}{(3w^2 - m)^9},$$

$$v_5 = -\frac{21288960}{\pi^6} \cdot \frac{w}{(3w^2 - m)^9} - \frac{69672960m}{\pi^6} \cdot \frac{w}{(3w^2 - m)^{10}} - \frac{52254720m^2}{\pi^6} \cdot \frac{w}{(3w^2 - m)^{11}},$$

$$v_6 = \frac{723824640}{\pi^7} \cdot \frac{1}{(3w^2 - m)^{10}} + \frac{3413975040m}{\pi^7} \frac{1}{(3w^2 - m)^{11}} + \frac{4981616640m^2}{\pi^7} \frac{1}{(3w^2 - m)^{12}} + \frac{2299207680m^3}{\pi^7} \frac{1}{(3w^2 - m)^{13}},$$

$$v_7 = -\frac{86858956800}{\pi^8} \cdot \frac{w}{(3w^2 - m)^{12}} - \frac{450644705280m}{\pi^8} \cdot \frac{w}{(3w^2 - m)^{13}} - \frac{717352796160m^2}{\pi^8} \frac{w}{(3w^2 - m)^{14}} - \frac{3586763980800m^3}{\pi^8} \cdot \frac{w}{(3w^2 - m)^{15}}.$$

Making $w = 2{\cdot}00$, computing these expressions for every value of m, and substituting them in the expression for the integral from $w = 2{\cdot}00$ to $w = $ infinity, the numerical value of the integral for each value of m was found.

This process was exceedingly accurate for all the negative values of m, and for the positive values about as far as $m = + 3{\cdot}0$, when a difficulty presented itself. It must be remarked that when $3w^2 - m$ (or in the present instance $12 - m$,) amounts to several integers, the values of the successive terms decrease at first with great rapidity: yet, in all cases, they increase at some part or other in hyper-geometrical proportion, and finally become greater than any assignable quantity. This will be seen most readily on observing the law of the terms when $m = 0$.

$$\text{We have then } v_0 = \frac{2}{3\pi} \cdot \frac{1}{w^2}: \quad v_1 = \left(\frac{2}{3\pi}\right)^2 \cdot \frac{-2}{w^5}: \quad v_2 = \left(\frac{2}{3\pi}\right)^3 \cdot \frac{2.5}{w^8}:$$

$$v_3 = \left(\frac{2}{3\pi}\right)^4 \cdot \frac{-2.5.8}{w^{11}}: \quad v_4 = \left(\frac{2}{3\pi}\right)^5 \cdot \frac{2.5.8.11}{w^{14}}: \quad \&c.$$

It is evident that these terms, however small may be the quantity $\dfrac{2}{3\pi.w^3}$, will at some stage receive in succession new multipliers, greater than $\dfrac{3\pi.w^3}{2}$; that after this they will increase, and that the rapidity of their proportionate increase will go on continually increasing. From this point then the magnitude of the terms will increase hyper-geometrically.

The value of the integral however will be finite; and a limit for the value remaining after the computation of any number of terms in the series v_0, v_1, &c. may be found. For, wherever we stop, the residual term will be of the form $\int_w \cos u . \dfrac{dv_n}{dw}$ or $\int_w \sin u . \dfrac{dv_n}{dw}$: where v_n is the term last found in the series. Now it is evident that either of these quantities is less than $\int_w \dfrac{dv_n}{dw}$; for the magnitudes of the quantities to be integrated are always smaller, except in the particular cases when $\cos u$ or $\sin u = \pm 1$; and their signs are constantly varying as the value of w varies: whereas the sign of $\dfrac{dv_n}{dw}$ is always the same. The residual integral, therefore, is certainly less than v_n, the last term found in the series, and is probably much less: and therefore, if the last term computed consist only of integers in the last place of decimals which we wish to retain, even though the divergence of the series be just beginning, the use of these terms will give the integral required with the utmost practical accuracy.

Now when $m = +3\cdot0$ nearly, it is found that the divergence commences at v_7, or before it; and that the term v_7 is not so small that a quantity which is likely to be a sensible portion of it can be safely neglected. To approximate here, I have used the following consideration. It is known that the slowly converging series

$$A - B + C - D + E - F + \text{\&c.}$$

may be converted into a series of the same kind with much smaller terms by putting it in this form,

$$\tfrac{1}{2}A$$
$$+ \tfrac{1}{2}(A - B) - \tfrac{1}{2}(B - C) + \tfrac{1}{2}(C - D) - \tfrac{1}{2}(D - E) + \tfrac{1}{2}(E - F) - \text{\&c.}$$

In the instance before us, such a series would be produced by commencing the integration by parts with $\tfrac{1}{2}(v_0 - v_2)$ instead of v_0; and the residual term will be of the form of

$$\tfrac{1}{2}\int_w \cos u \cdot \frac{d}{dw}(v_n - v_{n+2}) \quad \text{or} \quad \tfrac{1}{2}\int_w \sin u \frac{d}{dw}(v_n - v_{n+2}).$$

Now if the progression is stopped at such a point that $\dfrac{dv_n}{dw}$ is, for all the following values of w, greater than $\dfrac{dv_{n+2}}{dw}$, then the quantity $\dfrac{d}{dw}(v_n - v_{n+2})$ has always the same sign, and the reasoning above shews that the residual integral will be less than $\tfrac{1}{2}(v_n - v_{n+2})$. A close approximation therefore will be obtained by summing the series as if we had begun with $\tfrac{1}{2}(v_0 - v_2)$ instead of v_0: and this, it is easily seen, will be effected by taking half of the last term in each of the series multiplying $\sin u$ and $\cos u$. The multipliers which I have used are, in fact,

$$- v_0 + v_2 - v_4 + \tfrac{1}{2}v_6 \ldots\ldots\ldots\text{for } \sin u,$$
$$- v_1 + v_3 - v_5 + \tfrac{1}{2}v_7 \ldots\ldots\ldots\text{for } \cos u.$$

The doubt which remains extends, I apprehend, to digits in the fifth place of decimals, but no higher.

The number of terms computed by logarithmic process for this part of the integral is about 900.

I could have wished to extend the computation as far as $m = +6\text{·}0$, so as to include perhaps one or two more maxima of the values of intensity. Indeed, if it had been possible to foresee the approximate places &c. of maxima, I should probably have commenced the computations to that extent. The trouble however would be great, as the summation must be extended as far as $w = 2\text{·}5$. Should any person be disposed to go to that extent, I would recommend that the summation of the values computed in this Paper should be also extended, for the values of m beginning with $+3\text{·}0$.

I subjoin a Table of the different sections of the summation and remaining integration for all the values used in this Paper.

TABLE of the different parts of $\int_w \cos \frac{\pi}{2}(w^3 - m.w)$, from $x = 0$ to $x = \infty$.

| Value of m. | Values of the integral found by actual summation. | | | | | | | Values of the remaining integral. | | Sum. |
	From $x=0.00$ to $x=1.00$.	From $x=1.00$ to $x=1.26$.	From $x=1.26$ to $x=1.44$.	From $x=1.44$ to $x=1.58$.	From $x=1.58$ to $x=1.70$.	From $x=1.70$ to $x=1.92$.	From $x=1.92$ to $x=2.00$.	Term depending on sin u.	Term depending on cos u.	
−4·0	+·0929154	−·1621545	+·1280207	−·0757713	−·0036556	−·0150365	+·0374784	·0000000	+·0011788	+·0029754
−3·8	+·0899076	−·1537134	+·1071428	−·0318602	−·0507057	+·0210294	−·0021156	+·0236322	+·0009898	+·0043069
−3·6	+·0783010	−·1250013	+·0663779	+·0200012	−·0852129	+·0538698	−·0412717	+·0387234	+·0003926	+·0061800
−3·4	+·0591033	−·0787271	+·0125367	+·0683706	−·0979422	+·0721698	−·0655350	+·0392214	−·0004079	+·0087896
−3·2	+·0342832	−·0200615	−·0450110	+·1023750	−·0852309	+·0684171	−·0658255	+·0245557	−·0011101	+·0123920
−3·0	+·0066555	+·0439416	−·0959701	+·1141579	−·0500345	+·0418171	−·0418367	·0000000	−·0014269	+·0173039
−2·8	−·0202801	+·1051981	−·1309732	+·1006947	−·0012464	−·0008297	−·0022201	−·0252118	−·0012011	+·0239304
−2·6	−·0426027	+·1556300	−·1433068	+·0645600	+·0485494	−·0467686	+·0385325	−·0413456	−·0004776	+·0327706
−2·4	−·0562773	+·1882352	−·1301966	+·0134390	+·0863295	−·0810887	+·0653953	−·0419123	+·0004974	+·0444215
−2·2	−·0574983	+·1980667	−·0934064	−·0415027	+·1020423	−·0915539	+·0683497	−·0262631	+·0013569	+·0595912
−2·0	−·0430375	+·1829652	−·0390242	−·0880664	+·0912823	−·0729091	+·0461214	·0000000	+·0017488	+·0790805
−1·8	−·0105661	+·1439504	+·0235453	−·1157328	+·0565228	−·0291543	+·0067189	+·0270131	+·0014761	+·1037734
−1·6	+·0410858	+·0852018	+·0832149	−·1180453	+·0065563	+·0271234	−·0354520	+·0443405	+·0005884	+·1346138
−1·4	+·1117252	+·0136169	+·1291863	−·0941134	−·0457342	+·0783376	−·0648587	+·0449910	−·0006146	+·1725361
−1·2	+·1996990	−·0620069	+·1529169	−·0488959	−·0866990	+·1074010	−·0705636	+·0282196	−·0016814	+·2183897
−1·0	+·3019144	−·1320681	+·1497066	+·0078255	−·1054914	+·1034152	−·0502948	·0000000	−·0021732	+·2728342
−0·8	+·4139711	−·1873967	+·1196276	+·0635668	−·0969417	+·0656568	−·0113461	−·0290842	−·0018395	+·3362141
−0·6	+·5303921	−·2204699	+·0675952	+·1058949	−·0629782	+·0044341	+·0320494	−·0477897	−·0007356	+·4083923
−0·4	+·6449407	−·2264423	+·0025723	+·1252092	−·0122017	−·0616677	+·0639213	−·0485430	+·0007707	+·4885595
−0·2	+·7510400	−·2038501	−·0639887	+·1169208	+·0422963	−·1114242	+·0724424	−·0304813	+·0021152	+·5750704
0·0	+·8422086	−·1548861	−·1201662	+·0825417	+·0863120	−·1278013	+·0543205	·0000000	+·0027427	+·6652719
+0·2	+·9125150	−·0852037	−·1557205	+·0294203	+·1082374	−·1037569	+·0160657	+·0314867	+·0023295	+·7553735
+0·4	+·9570127	−·0032555	−·1639662	−·0308393	+·1021275	−·0450631	−·0283460	+·0517992	+·0009347	+·8404040
+0·6	+·9721117	+·0807460	−·1430312	−·0848876	+·0693085	+·0309509	−·0625827	+·0526798	−·0009826	+·9143128
+0·8	+·9558650	+·1561216	−·0962649	−·1206499	+·0181032	+·1005011	−·0739646	+·0331201	−·0027067	+·9701249
+1·0	+·9081446	+·2131038	−·0317049	−·1299775	−·0382828	+·1408098	−·0581618	·0000000	−·0035225	+1·0004087
+1·2	+·8306993	+·2441213	+·0393109	−·1105332	−·0851711	+·1375753	−·0208408	−·0343016	−·0030028	+·9978573
+1·4	+·7270641	+·2448168	+·1041461	−·0663663	−·1102398	+·0899919	+·0243670	−·0565062	−·0012095	+·9560641
+1·6	+·6023587	+·2146627	+·1511250	−·0070683	−·1067638	+·0115856	+·0608471	−·0575460	+·0012765	+·8704775
+1·8	+·4629601	+·1570802	+·1716481	+·0543138	−·0754212	−·0736006	+·0751107	−·0362310	+·0035295	+·7393896
+2·0	+·3160989	+·0790543	+·1617709	+·1041549	−·0241758	−·1383955	+·0617838	·0000000	+·0046115	+·5649030
+2·2	+·1693705	−·0097331	+·1229644	+·1312871	+·0337502	−·1611914	+·0256330	+·0376336	+·0039467	+·3536610
+2·4	+·0302520	−·0981192	+·0619006	+·1295037	+·0832918	−·1331501	−·0201404	+·0620890	+·0015960	+·1172234
+2·6	−·0944017	−·1748910	−·0107004	+·0989853	+·1114670	−·0615225	−·0587219	+·0633290	−·0016907	−·1281469
+2·8	−·1987892	−·2302290	−·0819605	+·0463043	+·1107817	+·0321479	−·0758654	+·0399339	−·0046921	−·3623684
+3·0	−·2785332	−·2569831	−·1391417	−·0169919	+·0812275	+·1185093	−·0651540	·0000000	−·0061508	−·5632179
+3·2	−·3309399	−·2516145	−·1719364	−·0769201	+·0303310	+·1696336	−·0304033	−·0416103	−·0052779	−·7087378
+3·4	−·3551338	−·2146670	−·1743344	−·1201569	−·0287648	+·1680727	+·0156965	−·0687550	−·0021380	−·7801807
+3·6	−·3520567	−·1507154	−·1457300	−·1370101	−·0807012	+·1128004	+·0562188	−·0702322	+·0022653	−·7651611
+3·8	−·3243480	−·0677906	−·0910614	−·1235960	−·1119003	+·0200155	+·0762169	−·0443170	+·0062719	−·6605390
+4·0	−·2761063	+·0236462	−·0199552	−·0827357	−·1141216	−·0815972	+·0682405	·0000000	+·0081698	−·4744595

XLI.[*] *Supplement to a Paper " On the Intensity of Light in the neighbourhood of a Caustic." By* GEORGE BIDDELL AIRY, ESQ., *Astronomer Royal.*

[Read *May* 8, 1848.]

IN a Paper "On the Intensity of Light in the neighbourhood of a Caustic" communicated to the Cambridge Philosophical Society about ten years ago, and printed in the 6th Volume of their *Transactions*, I shewed that the expression for the intensity of light near a caustic would depend on the infinite integral

$$\int_w \cos \frac{\pi}{2} (w^3 - m.w)^* \left\{ \text{ from } w = 0 \text{ to } w = \frac{1}{0} \right\},$$

where m is a quantity proportional to the distance of a point from the geometrical caustic, measured in a direction perpendicular to the caustic, and estimated positive towards the bright side of the caustic: and I gave a detailed account of the method of quadratures by which I had computed the numerical value of this infinite integral for the values of $m - 4{\cdot}0$, $- 3{\cdot}8$, &c. as far as $+ 4{\cdot}0$; and I exhibited in a table the computed values of the integral.

The computation by quadratures was exceedingly laborious, and I did not resort to it without trying other methods of a more refined nature. But in every attempt at expansion of the formula I was met by the integral of a sine or cosine with infinite limits. The reasonings upon which several mathematicians have attempted to establish the value of such an integral appeared to me so little conclusive, that I preferred at once to abandon the expansions which introduced them, and to rely only on the infallible but laborious method of quadratures.

On my stating to Professor De Morgan, after terminating the calculations, the scruples which had led me to reject the expansions, he expressed himself so strongly confident of the correctness of the conclusions upon the point which I had considered doubtful, that I was induced to undertake the numerical computation of the series given by expansion of the formula. I proceeded at once as far as it was possible to go with 7-figure logarithms, when I was interrupted, and the computations were laid aside for some years. I have lately taken them up again, and have completed them as far as they can be carried with 10-figure logarithms. It is the result of this calculation, and the comparison of this result with that formerly obtained from quadratures, that I now beg leave to present to this Society.

Before entering upon the numerical investigations, I will transcribe a letter which Professor De Morgan at my request has written to me, and which he has permitted me to publish. It contains an explanation of his views upon the evidence for the numerical certainty of the results obtained by such integrals as those to which I have alluded.

---

* I retain this notation in preference to that which is commonly employed, partly because it is familiar to me, and because I have used it in the paper to which I refer, partly because I think that any notation which requires the expression of a differential at the end is for that reason objectionable.

" In reply to your request that I would send you a sketch of the method which I communicated to you some years ago, for finding the numerical value of $\int_0^\infty \cos (w^3 - mw)\, dw$, I send you the following. I am not aware that there is anything about it peculiarly my own, or other than what would suggest itself as a matter of course to any one familiar with the current methods in definite integrals.

" The series which I furnished depend ultimately upon the following formulæ:—

$$\int_0^\infty \epsilon^{-r\cos\theta . w} . \cos (r \sin \theta . w) . w^{n-1} dw = \Gamma_n \frac{\cos n\theta}{r^n},$$

$$\int_0^\infty \epsilon^{-r\cos\theta . w} . \sin (r \sin \theta . w) . w^{n-1} dw = \Gamma_n \frac{\sin n\theta}{r^n},$$

in which r and θ are independent of w, $r \cos \theta$ is positive, n is positive, and Γ_n stands for $\int_0^\infty \epsilon^{-x} x^{n-1} dx$, as usual. Under these conditions the theorems do not or need not rely upon any notion of algebraical as distinguished from numerical equality. Calling either of them $\int \phi w . dw$, common arithmetical calculation would establish any degree of approximation between the convergent series $\phi 0 . a + \phi a . a + \phi 2a . a + \ldots$ and the asserted value of the definite integral, if a were taken small enough. And this for any value of θ, from $\theta = 0$ to $\theta = \frac{\pi}{2} - \beta$, β being of any degree of smallness. But when $\theta = \frac{\pi}{2}$, the numerical character of the equivalence is lost, and the equations assume the same character as $1 - 1 + 1 - 1 + \ldots = \frac{1}{2}$, and are subject to the same discussion.

" The above equations were first obtained by substituting $a + b \sqrt{-1}$ for a in

$$\int_0^\infty \epsilon^{-aw} w^{n-1} dw = \frac{\Gamma_n}{a^n},$$

which is an equivalence of numerical character even after the substitution, if a be positive, and b (be it positive or negative) numerically less than a. For the use of the expansions of $\epsilon^{-bw\sqrt{-1}}$ and $(a + b \sqrt{-1})^{-n}$ in powers of b would produce an equivalence such as

$$A_0 + A_1 k + A_2 k^2 + \ldots = B_0 + B_1 k + B_2 k^2 + \ldots$$

where $k = \sqrt{-1}$, $A_n = B_n$ is a numerical equivalence, and ΣA_n is a convergent series. But, when b is numerically greater than a, a convergent series would be rendered divergent *in integration*: and, when this happens, I do not see any way to place the divergent series so obtained upon the same footing as those of ordinary algebra.

" It is not however necessary to depend upon this introduction of divergency. If we call the two integrals C_n and S_n, and differentiate both with respect to θ, we have

$$\frac{dC_n}{d\theta} = r \sin \theta . C_{n+1} - r \cos \theta . S_{n+1},$$

$$\frac{dS_n}{d\theta} = r \sin \theta . S_{n+1} + r \cos \theta . C_{n+1};$$

whence –
$$rC_{n+1} = \frac{dC_n}{d\theta}\sin\theta + \frac{dS_n}{d\theta}\cos\theta,$$

$$rS_{n+1} = \frac{dS_n}{d\theta}\sin\theta - \frac{dC_n}{d\theta}\cos\theta,$$

from which it easily appears that for what value soever of n the equations first given are true, they remain true when that value is increased by a unit. And that they are true when $n = 1$ is proved by common integration by parts.

" If instead of w we write w^k, k being positive, and then for kn write n, we have

$$\int_0^\infty \epsilon^{-r\cos\theta\,.\,w^k}\,.\,\cos\left(r\sin\theta\,.\,w^k\right).\,w^{n-1}\,dw = \frac{1}{k}\Gamma_{\left(\frac{n}{k}\right)}\,.\,\cos\frac{n\theta}{k}\,.\,r^{-\frac{n}{k}},$$

$$\int_0^\infty \epsilon^{-r\cos\theta\,.\,w^k}\,.\,\sin\left(r\sin\theta\,.\,w^k\right).\,w^{n-1}\,dw = \frac{1}{k}\Gamma_{\left(\frac{n}{k}\right)}\,.\,\sin\frac{n\theta}{k}\,.\,r^{-\frac{n}{k}}.$$

" If $r = 1$, and we call these integrals C_n and S_n, let us take

$$\cos\left(\sin\theta\,.\,w^3 - mw\right) = \cos\left(\sin\theta\,.\,w^3\right).\left\{1 - \frac{m^2 w^2}{2} + \ldots\ldots\right\}$$

$$+ \sin\left(\sin\theta\,.\,w^3\right).\left\{mw - \frac{m^3 w^3}{2.3} + \ldots\ldots\right\}.$$

Multiplying by $\epsilon^{-\cos\theta\,.\,w^3}\,.\,dw$, and integrating, we have

$$\int_0^\infty \epsilon^{-\cos\theta\,.\,w^3}\,.\,\cos\left(\sin\theta\,.\,w^3 - mw\right)dw = C_1 + S_2\,.\,m - C_3\frac{m^2}{2} - S_4\frac{m^3}{2.3} + \ldots\ldots.$$

" If we now make $\theta = \dfrac{\pi}{2}$, and observe that in this case C_n vanishes whenever n is an odd multiple of 3, and S_n whenever n is an even multiple, we obtain

$$\int_0^\infty \cos\left(w^3 - mw\right)dw = C_1 - S_4\frac{m^3}{2.3} - C_7\frac{m^6}{2.3\ldots\ldots 6} + S_{10}\frac{m^9}{2.3\ldots\ldots 9} - \ldots\ldots$$

$$+ S_2 m + C_5\frac{m^4}{2.3.4} - S_8\frac{m^7}{2.3\ldots\ldots 7} - C_{11}\frac{m^{10}}{2.3\ldots\ldots 10} + \ldots\ldots$$

$$= \frac{1}{3}\Gamma_{\frac{1}{3}}\cos\left(\frac{1}{3}\,.\,\frac{\pi}{2}\right) - \frac{1}{3}\Gamma_{\frac{4}{3}}\sin\left(\frac{4}{3}\,.\,\frac{\pi}{2}\right).\frac{m^3}{2.3} - \frac{1}{3}\Gamma_{\frac{7}{3}}\cos\left(\frac{7}{3}\,.\,\frac{\pi}{2}\right)\frac{m^6}{2.3\ldots 6} + \ldots\ldots$$

$$+ \frac{1}{3}\Gamma_{\frac{2}{3}}\sin\left(\frac{2}{3}\,.\,\frac{\pi}{2}\right)m + \frac{1}{3}\Gamma_{\frac{5}{3}}\cos\left(\frac{5}{3}\,.\,\frac{\pi}{2}\right).\frac{m^4}{2.3.4} - \frac{1}{3}\Gamma_{\frac{8}{3}}\sin\left(\frac{8}{3}\,.\,\frac{\pi}{2}\right)\frac{m^7}{2.3\ldots 7} - \ldots\ldots$$

$$= \frac{1}{3}\Gamma_{\frac{1}{3}}\,.\,\cos\frac{\pi}{6}\,.\,\left\{1 - \frac{1}{3}\,.\,\frac{m^3}{2.3} + \frac{4}{3}\,.\,\frac{1}{3}\,.\,\frac{m^6}{2.3\ldots 6} - \frac{7}{3}\,.\,\frac{4}{3}\,.\,\frac{1}{3}\,.\,\frac{m^9}{2.3\,,\ldots 9} + \ldots\right\}$$

$$+ \frac{1}{3}\Gamma_{\frac{2}{3}}\,.\,\cos\frac{\pi}{6}\,.\,\left\{m - \frac{2}{3}\,.\,\frac{m^4}{2.3.4} + \frac{5}{3}\,.\,\frac{2}{3}\,.\,\frac{m^7}{2.3\ldots 7} - \frac{8}{3}\,.\,\frac{5}{3}\,.\,\frac{2}{3}\,.\,\frac{m^{10}}{2.3\ldots 10} + \ldots\right\}.$$

" I may observe that the precautions which I have taken, to shew that the *algebraical* cases are limits of arithmetical ones, are not absolutely necessary in this instance. For if we resolve

$\int_0^\infty \cos w^3 dw$ into its successive positive and negative portions, we have $A_0 - A_1 + A_3 - A_5 + \ldots$ in which by A_{2n+1} is meant the portion of the integral (taken positively) which occurs from $w^3 = (2n+1)\dfrac{\pi}{2}$ to $w^3 = (2n+3)\dfrac{\pi}{2}$. The greater n is made, the smaller is the interval of this partial integration; and these successive portions diminish, and diminish without limit, so that the series is convergent, and the error always less than the first term rejected. And $\int_0^\infty \sin w^3 dw$ may be treated in the same way.

" A. DE MORGAN."

The following numerical values occurring in the application of Professor De Morgan's final series may be conveniently placed here:—

$$\text{Log } \Gamma_{\frac{1}{3}} = 0 \cdot 4279627493.$$

$$\text{Log } \Gamma_{\frac{2}{3}} = 0 \cdot 1316564916.$$

With these series I have computed the values of $\int_w \cos \dfrac{\pi}{2} (w^3 - m \cdot w) \left(m = 0 \text{ to } m = \dfrac{1}{0}\right)$; for $w = -5.6, -5.4$, &c. as far as $+5.6$: and I now exhibit a table of the results, compared with those deduced from quadratures as far as the latter were carried. Each term of the series was computed to 6 decimals, and one figure was struck off in the sum.

Values of m.	Values of Integral by Quadratures.	Values of Integral by Series.	Values of m.	Values of Integral by Quadratures.	Values of Integral by Series.	Values of m.	Values of Integral by Quadratures.	Values of Integral by Series.
− 5·6		+ 0·00011	− 1·8	+ 0·10377	+ 0·10377	+ 2·0	+ 0·56490	+ 0·56490
− 5·4		+ 0·00018	− 1·6	+ 0·13461	+ 0·13462	+ 2·2	+ 0·35366	+ 0·35366
− 5·2		+ 0·00028	− 1·4	+ 0·17254	+ 0·17254	+ 2·4	+ 0·11722	+ 0·11722
− 5·0		+ 0·00041	− 1·2	+ 0·21839	+ 0·21839	+ 2·6	− 0·12815	− 0·12815
− 4·8		+ 0·00063	− 1·0	+ 0·27283	+ 0·27283	+ 2·8	− 0·36237	− 0·36237
− 4·6		+ 0·00093	− 0·8	+ 0·33621	+ 0·33622	+ 3·0	− 0·56322	− 0·56323
− 4·4		+ 0·00138	− 0·6	+ 0·40839	+ 0·40839	+ 3·2	− 0·70874	− 0·70876
− 4·2		+ 0·00204	− 0·4	+ 0·48856	+ 0·48856	+ 3·4	− 0·78018	− 0·78021
− 4·0	+ 0·00298	+ 0·00297	− 0·2	+ 0·57507	+ 0·57507	+ 3·6	− 0·76516	− 0·76516
− 3·8	+ 0 00431	+ 0·00429	0·0	+ 0·66527	+ 0·66527	+ 3·8	− 0·66054	− 0·66044
− 3·6	+ 0·00618	+ 0·00621	+ 0·2	+ 0·75537	+ 0·75537	+ 4·0	− 0·47446	− 0·47419
− 3·4	+ 0·00879	+ 0·00878	+ 0·4	+ 0·84040	+ 0·84040	+ 4·2		− 0·22645
− 3·2	+ 0·01239	+ 0·01239	+ 0·6	+ 0·91431	+ 0·91431	+ 4·4		+ 0·05193
− 3·0	+ 0·01730	+ 0·01730	+ 0·8	+ 0·97012	+ 0·97012	+ 4·6		+ 0·32258
− 2·8	+ 0·02393	+ 0·02393	+ 1·0	+ 1·00041	+ 1·00041	+ 4·8		+ 0·54475
− 2·6	+ 0·03277	+ 0·03277	+ 1·2	+ 0·99786	+ 0·99786	+ 5·0		+ 0·68182
− 2·4	+ 0·04442	+ 0·04442	+ 1·4	+ 0·95606	+ 0·95607	+ 5·2		+ 0·70818
− 2·2	+ 0·05959	+ 0·05959	+ 1·6	+ 0·87048	+ 0·87048	+ 5·4		+ 0·61515
− 2·0	+ 0·07908	+ 0·07908	+ 1·8	+ 0·73939	+ 0·73939	+ 5·6		+ 0·41460

It is impossible to make the calculation for larger values of m, positive or negative, even with 10-figure logarithms, on account of the divergence of the first terms of the series. For the values

± 5.6, the largest term in the series is $169\cdot044826$: and it is necessary to proceed as far as the 45th power of m. The result $+ 0\cdot000114$ for $m = -5\cdot6$ is obtained by combining the sum of positive terms $+ 614\cdot149962$ with the sum of negative terms $- 614\cdot149848$: and the result $+ 0\cdot414595$ for $m = + 5\cdot6$ is obtained by combining the sum of positive terms $+ 614\cdot357203$ with the sum of negative terms $- 613\cdot942608$. For values of m greater than $\pm 5\cdot6$, the calculation must be made in natural numbers.

The agreement of the values of the integral, computed by methods so totally different, is not a little remarkable. On the one hand, it may be received by some persons as a proof of the correctness of that part of the theory of the series which asserts the evanescence of the integral of a cosine when the limits are 0 and $\dfrac{1}{0}$: on the other hand it may be considered to afford evidence of the great care with which the quadrature computations had been made.

For the last two or three sets of numbers compared, there is a trifling discordance. It will be remarked that in my account of the computation by quadratures I have shewn that difficulties begin to arise in the accurate computation for the values of m approaching to $4\cdot0$, (unless the actual summation were carried to higher values of w than I carried it in those computations). That the source of the discordances is in these difficulties and the consequent inaccuracy of the quadratures, and not in the inaccuracy of the series, is evident from the following consideration. The numbers computed by the two methods agree well for the values of $m - 4\cdot0$, $- 3\cdot8$, $- 3\cdot6$: and as the quadratures there present no difficulty, it is reasonable to suppose that both sets of numbers are accurate (within such limits as are possible for the sums of numerous figures). Now the terms of the series combined to form the value of the integral for $m = + 4\cdot0$, $+ 3\cdot8$, $+ 3\cdot6$, are exactly the same as those by which the value of the integral for $m = - 4\cdot0$, $- 3\cdot8$, $- 3\cdot6$, is formed: the only difference being that they are combined in a different manner, and therefore, from the evident accuracy of the series for $m = - 4\cdot0$, $- 3\cdot8$, $- 3\cdot6$, we are entitled to infer the accuracy of the series for $m = + 4\cdot0$, $+ 3\cdot8$, $+ 3\cdot6$.

Reprinted from *Quarterly Journal of the Royal Meteorological Society,* Vol. 38, 291–301 (1912).

CORONÆ AND IRIDESCENT CLOUDS.

By GEORGE C. SIMPSON, D.Sc., F.R.Met.Soc.,
Meteorologist to the British Antarctic Expedition, 1910.

PRELIMINARY.—During September 1911 I was one of a party led by Captain Scott to survey the western coast of M'Murdo Sound, and the following is a note from the meteorological log kept on that journey :—

"*Sunday, September 24.*—The fog which enveloped us at 11 a.m. continued into the evening. During the afternoon march a fine fog-bow appeared. It was opposite the sun, and a measurement of the radius with a theodolite gave 38°. The bow was practically white, but a reddish tinge could be seen on the outer side. Within the arch the sky appeared whiter than outside, but just within the sky was nearly as blue as outside, so that there appeared to be a second arch within the main one.

"When the arch was at its brightest the reddish colour on the outside was clearly visible, and one could imagine that other colours were present, but one could not be sure of this. During this period the sun shone faintly through the fog, brightening up the fog in its neighbourhood; but no colours or rings were visible in this half of the sky.

"As the fog dissipated the upper sky became clearer, and the sun shone over the top of a heavy bank of fog. For some minutes the sun had a brilliant corona with bright colours, and the diameter of this corona seemed unusually large, but there was no opportunity to make a measurement. As the fog still further cleared away glimpses of the corona appeared again, and the fog under the sun became fairly brilliantly illuminated with iridescent colours, which did not appear to be part of the corona, but in places blended into it. During the whole period the temperature was between $-15°$ and $-21°$ F. The fur of the sleeping bags and the wool of sweaters became covered with hoar frost."

The above description was written in the field; on my return to winter quarters I looked up the literature to identify the bow. According to Pernter (*Meteorologische Optik.* vol. **3.**) there are two bows which may be seen opposite the sun, having diameters of approximately 38°.

The first of these is the Bouguer Bow, which is caused by ice-crystals, but theory shows that this bow is pure white and can show no colours, nor is it accompanied by a secondary bow within the main one. Again, it is practically impossible for this bow to appear without being part of an extensive halo system, for it necessitates regular crystals which could not fail to give rise to the more common halos. There can be no doubt that the phenomenon observed was the second bow described by Pernter, viz. a white rainbow commonly called by English writers "fog-bow." The following is a description of such a bow given by Pernter (*l.c.* page 247) :—

"The colour mixture which results from the combination of the different colours caused by drops of radius 25 μ (0·025 mm.) appears white from 40° 40' to 38° 40', and this white middle band of 2° has on the outside a yellow which appears weak orange) border, and on the inside a violet border. The whole white rainbow, together with its two borders, extends from 41° 20' to 37° 20', *i.e.* 4° broad. Inside this bow is a colourless space extending to 35° 40', where a secondary bow commences which is blue outside and red inside."

This observation proves that the fog was composed of water-drops having a radius smaller than ·025 mm., and this with a temperature of $-21°$ F. ($-29°$ C.). Support is lent to this conclusion by the observation that the hair of sweaters and fur bags became covered with hoar frost, which is a sure sign of supercooled water.

A similar bow had previously been observed in Spitzbergen by Carlheim-Gyllenskiold with a temperature of $-14°$ C.; these two observations show that water can exist in a liquid state in the atmosphere at much lower temperature than has generally been supposed by meteorologists.

It is now generally admitted that while halos are caused by the refraction and reflection of ice-crystals, coronæ are due to diffraction effects of either small drops of water or thin ice needles. From certain observations made in the Antarctic I was led to doubt the possibility of ice-crystals ever forming diffraction effects. This is an important question for meteorology, for if it is true we have a powerful instrument for determining the constitution of a cloud: if there is a corona the cloud must be composed of water, while if there is a halo it must be composed of ice.

Before examining the theory of the formation of diffraction effects by ice-needles, it will be as well to consider why meteorologists up to the present have been so certain that ice-crystals do give rise to coronæ and other diffraction effects. In order not to have to refer to a large number of writers I shall take Pernter's book as being the most complete exposition of modern views on meteorological optics, and shall quote him throughout, although he himself may not have been the originator of the theories considered.

Pernter's reasons for believing that coronæ are produced by ice-crystals may be summed up in the three following statements :—

(a) Coronæ are seen on clouds having temperatures much below the freezing point.

(b) The most beautiful coronæ appear on light white cirro-cumulus or fine cirro-stratus clouds ; and these clouds are always composed of ice-crystals.

(c) Halos and coronæ have been observed at the same time ; and as halos are a sure sign of ice-crystals the coronæ must, therefore, be formed in ice-clouds.

With regard to the first of these, we have seen in the preliminary part of this paper that water-drops have been actually observed to exist at a temperature of $-29°$ C. No reason is apparent why this should be the lower limit of temperature at which water can exist; and until physics throws more light on this question temperature cannot be taken as any criterion as to the nature of a cloud.

In statement (b) it is assumed that cirrus clouds are always composed of ice-crystals. But what evidence have we as to the icy nature of cirrus clouds ? It appears to me that the evidence reduces itself to the following :—Cirrus clouds are always at temperatures far below the freezing-point, and as halos prove that these clouds are *often* ice-clouds, the presumption is that they are *always* ice-clouds. This, however, is not sound logic ; for exactly the same can be said of winter Antarctic fogs which occur at temperatures similar to those of the cirrus clouds and often exhibit halos, yet white fog bows show that they are sometimes composed of water-drops.

Statement (c) is by far the most important, for if it is true Pernter's position is proved. But is it true ? Pernter quotes from the Ben Nevis and other meteorological logs several cases in which coronæ and halos are entered together. But this is not sufficient ; it must first be proved that one cloud produces at one time both halo and corona. To do this necessitates a trained observer who is directing his attention to this special point. During the whole of my stay in the Antarctic I watched the clouds to decide this question. On one occasion I saw some clouds near the sun tinged with colours which were parts of a corona and at the same time part of a halo on other clouds ; but in this case there was no doubt whatever that entirely different clouds were concerned ; those causing the corona were light A. Cu. or Ci. Cu. clouds high in the sky, while the halo was caused by dark clouds much lower and nearer the horizon. On several occasions it happened that a thin cloud spread over the sky showing a corona. Later this cloud became thicker and denser, the corona disappeared and a halo was seen. The two, however, did not appear at the same time. On no occasion was a corona and halo seen at the same time on the same cloud.

If this reasoning is accepted, then we are no longer compelled to admit that ice-crystals produce coronæ, for there is nothing in the physics of the atmosphere to prohibit the possibility of water-drops being the cause. This leaves us in a position to discuss the optical question from a different standpoint from that taken up by Pernter. We will now consider whether crystals or drops are best able to produce the phenomena observed, and an attempt will be made to prove that diffraction effects are probably never produced by ice-crystals.

The problem we are faced with is as follows :—

Coronæ are produced by low clouds which are known to consist of water-drops, and for these the optical theory is satisfactory. Coronæ are also produced on high clouds of a cirrus character, and in Pernter's words "the latter are the most fully developed and the most beautiful" (*Met. Optik.* vol. **3**, p. 424). There must therefore be some reason why the high clouds produce finer coronæ than low clouds. Pernter ascribes it to the ice-crystals, and develops a theory which has generally been accepted. Before proceeding further this theory must be explained and criticised.

Fig. 1 shows the diffraction pattern produced in monochromatic light $\lambda = \cdot 000571$ by a circular obstacle (sphere) having a diameter of $\cdot 02$ mm., and Fig. 2 the similar pattern produced by a long narrow obstacle (needle) having a breadth of $\cdot 02$ mm. The curves are drawn to the same scale, the intensity of the main beam being taken as 1. It will be noticed that the successive maxima of intensity produced by the needles are larger and fall off more slowly than those produced by the spheres ; hence, if ice-needles produce a diffraction pattern it will be more intense and show more bands of colours than a similar one produced by water-drops. The only question is, Can a cloud of ice-needles produce a diffrac-

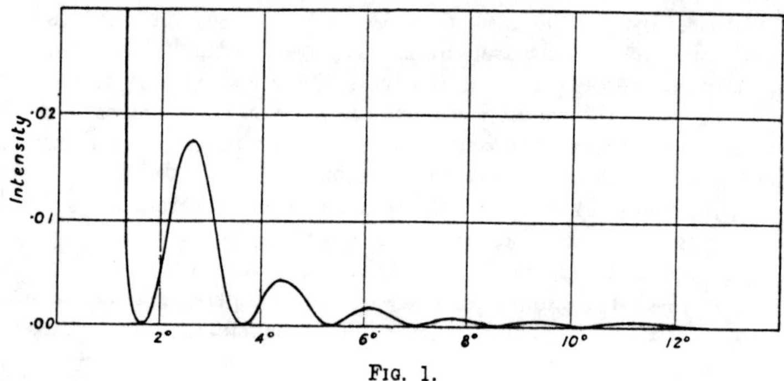

Fig. 1.

tion pattern ? Pernter reasons in this way. Consider a cloud consisting of long thin ice-needles. If these needles were all parallel, say with their axes vertical, a horizontal band of colour would extend on each side of

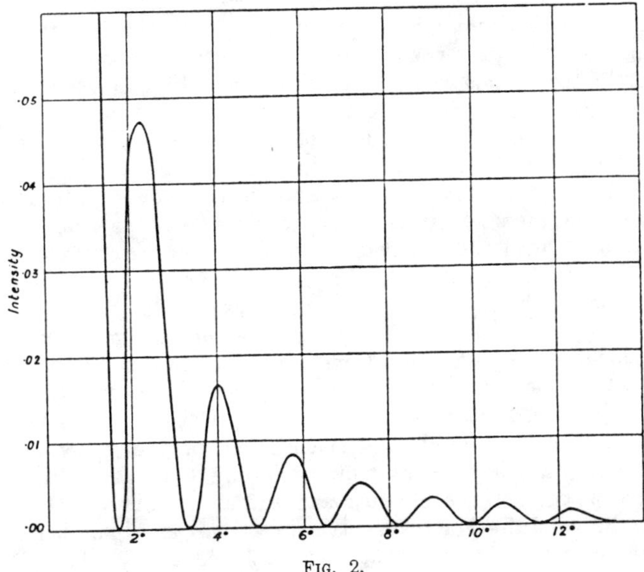

Fig. 2.

the sun [1] showing the diffraction colours. Similarly, if the needles were parallel and horizontal they would show a vertical beam of colour. In the same way a beam of colour could be produced in any direction by placing the ice-crystals at right angles to that direction. Now if beams were produced in all directions simultaneously, the result would be coloured circles surrounding the sun, *i.e.* a corona would be seen. Pernter considers a cirrus cloud to consist of "an infinite number of ice-needles orientated in all directions." Hence, if any direction be taken from the sun there will be needles with their axes perpendicular to this direction, and these will produce their diffraction effect. As this is true for all directions, a cloud of ice-needles of uniform size will produce a corona, and for the reason given above this will be a finer corona than one produced by water-drops.

The fallacy of this argument lies in neglecting all the ice-crystals but those which lie tangential to circles concentric with the sun. The crystals which are only a few degrees from tangential will produce a diffraction effect similar to the one produced by the tangential crystals, but the colours of the two systems will not fall together. The further the crystals depart from the tangential direction the greater will be the difference between the colour patterns. As there are as many crystals in any one direction as in any other, and as each direction produces a different diffraction effect, it is obvious that the colour mixture cannot possibly be pure. Without a difficult mathematical investigation it cannot be said whether the different colour systems will exactly compensate one another and produce only white light or not; but it cannot be doubted that the mixture of light will be far too impure for anything but the faintest tinge of colour to be seen.

Now the coronæ produced on high clouds are "the most fully developed and the most beautiful"; it must therefore be admitted that these clouds are not composed of ice-crystals.

There is another fact which has not been sufficiently recognised. No other ice-crystals but long thin prisms are able to produce a diffraction effect, and Pernter postulates a cloud in which there are "an infinite number of ice-needles orientated in all directions" (*Met. Optik.* p. 449). Now, no matter how disturbed the air in which a cloud of such crystals is formed, there will always be a large excess of crystals having their axes in a horizontal direction, for this is the position which any long thin body tends to assume when floating in the atmosphere (see *Met. Zeit.* vol. **25**, 1908, p. 557). According to Pernter's theory these would cause a vertical band of colour, which would be much brighter than the

---

[1] In this paper the source of light is always spoken of as the sun; but in nearly all cases the same effect would be produced by the moon. For purposes of calculation the sun is also assumed to be a point source of light.

corresponding horizontal band produced by the few crystals which at any moment are in the unstable position of vertical axes. Thus a corona formed by ice-crystals would *always* be much brighter above and below the sun than at the sides. This, however, does not agree with observation, for the beautiful coronæ seen on so-called ice-clouds are remarkable for the uniformity of the illumination of their rings. The finest corona we saw during the winter in the Antarctic was wonderfully uniform and did not show the slightest excess of colour above or below the moon. It may be pointed out that it is the tendency for ice-crystals to arrange themselves in horizontal directions, which gives rise to the crosses and mock suns which practically always accompany halos.

We will now consider what sort of a corona we might expect if the high clouds consist of water-drops. We have already seen that low temperatures need not be considered a bar to the existence of water-drops even though it falls as low as that of the cirrus clouds. These water-drops are the result of condensation in the upper atmosphere, where condensation proceeds more uniformly than near the ground on account of the absence of dust and eddies. Whatever the nuclei on which the condensation takes place in the upper atmosphere, they are sure to be very uniform, so that the drops produced will be uniform over a large area. Then again the low temperature and consequent scarcity of water vapour will keep the drops small even when the condensation is far advanced and a large cloud has been produced. At higher temperatures such small drops will only exist during the unsettled periods at the commencement and end of the condensation process, and, therefore, practically never occur with a cloud sufficiently extended to exhibit the whole of a large corona.[1]

Thus in a high cloud we have the conditions for the formation of large brilliant coronæ, viz., small drops, equality of size and distribution over a large area. In fact, it appears that, granted water-drops can exist at all temperatures, the higher the cloud and the lower its temperature the more perfect will be its corona.

This reasoning has led to the conclusion that ice-crystals cannot produce a corona, or at least one which could be seen on the brilliantly illuminated mass of ice-crystals, while clouds of water-drops existing at high altitudes and low temperatures would exhibit more brilliant and larger coronæ than those actually observed on low clouds. Considering further that a cloud of water-drops exhibiting a brilliant corona has been observed at a temperature as low as − 29° C., it does not seem unreasonable to conclude that any high cloud exhibiting a corona, in whole or part, must be composed of water-drops and not of ice-crystals.

IRIDESCENT CLOUDS

When thin white clouds of the Ci. S. or Ci. Cu. types exist within a region extending to approximately 25° from the sun, a more or less brilliant tinge of colour can often be seen upon them. The arrangement of the colours differs from that of coronæ in that while the latter are concentric to the sun the iridescent colours show no signs of concentricity, but follow rather the outlines of the clouds on which they appear.

The colours of iridescent clouds are generally supposed to be diffraction effects caused by ice-needles, but an entirely satisfactory explanation has never been given. The reader will find in Pernter's book on meteorological optics a good discussion of the problem and a clear statement of the difficulties; there will, however, be no need to enter into these here, as the explanation given below is based on different principles from those used by Pernter.

We have seen above good reasons for believing that ice-crystals can never produce diffraction effects, and, therefore, wherever we see diffraction colours we must assume the presence of water-drops. Now there can be little doubt that iridescent colours are diffraction effects, for examples were often seen in the Antarctic in which iridescent clouds blended into a

---

[1] Fragments of large coronæ may sometimes be seen on the edges of low clouds which are either forming or dissolving, for then the drops may be very small.

corona, so that it was not possible to say where the iridescent colours
ended and the corona commenced.

Fig. 1 shows the diffraction pattern produced by a water-drop of ·02 mm.
diameter in monochromatic light of wave-length $\lambda = ·000571$.   This wavelength was chosen because experience has shown that where this light has
its first minimum the first red ring of the corona appears, and where it has
its second minimum the red band of the second order spectrum occurs, so
that between these two minima all the colours of the first order appear.
Hence if a cloud with drops of ·02 mm. diameter covers the sun or moon
the first series of colours will extend from 1° 38′ to 3° 16′.   The position
of these two minima is a function of the radius of the drops, and in Fig.
3 the position of each is plotted against the radius of the drops.

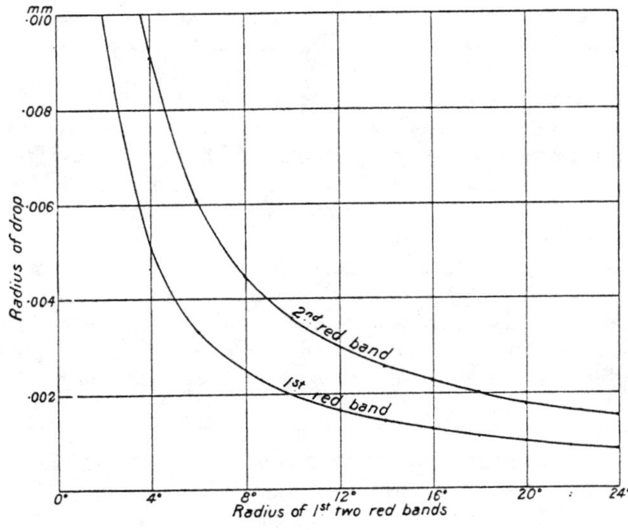

Fig. 3.

This figure is very instructive, for (1) it shows the extent of the first
order colours for any size drops; for example, if a cloud consists of drops
having a radius of ·005 mm. the first red band will be 4° away from the
sun or moon, while the second red band will appear at 7·2°, that is, the
first series of colours extends over 3·2°; (2) it shows what size drops are
necessary to produce first order colours at any distance from the sun,
e.g. at 12° from the sun any drops of radius between ·0017 and ·0030 mm.
will be able to produce first order colours while no other size drops will.

Let us now fix our attention on an element of a cloud at any given
distance from the sun, for example, 12°.   Now if this element consists of
drops of ·0017 mm. radius it will appear brightly coloured red; near to
this element there may be another having drops slightly larger, then this
element will be coloured green; if there is another element of ·0030 mm.
it will be pink or red; larger drops, however, would only produce second
order colours which would probably be too faint to be seen, for their
intensity is less than a quarter of the first order colours.

We will now turn our attention to the structure of the clouds which
show iridescent colours.   These clouds are generally Ci. S. or Ci. Cu.
The latter are wave-clouds and they will be discussed in detail for
illustration.

Ci. Cu. clouds are formed at the crests of air waves which form between
two currents.   The air ascending from a wave hollow cools as it rises to
the wave crest, and condensation takes place; as the air again descends
it is heated and the products of the condensation disappear.   The air
which has risen to the crest of the wave has been cooled the most so
that the drops produced by the condensation will be larger there than in
any other part of the cloud.   On each side of the crest smaller and
smaller drops will be met with until they entirely disappear in the
hollow between two crests.   In short, the drops are arranged in the
cloud in an ordered sequence from the edge to the centre, and bands
could be described following the contour of the cloud in which only drops

of a given size will be found.   If now the size of the drops varies within
the limits between the two lines plotted in Fig. 3 at the proper distance
of the cloud from the sun, then the cloud will be coloured with bands
approximately following the outline of the cloud; in other words the
cloud will show iridescent colours.   To take a concrete example: let a
Ci. Cu. cloud appear at 12° from the sun, and suppose that in the centre
of the cloud the drops are ·004 mm. in radius.   In this case the centre
would show little or no colour because the drops are too large, but
encircling it would be a region in which all the drops were approximately
·0030 mm. radius; these drops would give rise to a red band; outside
this region there would be another of somewhat smaller radius, and it
would be coloured with a slightly different colour, and so on through
all the diffraction colours until the region is reached in which the drops
have a radius of ·0017 mm.; this region would encircle the edge of the
cloud and be coloured red; outside this region the drops would be too
small to produce a diffraction colour or even to be seen at all.

An assemblage of Ci. Cu. clouds, each coloured in this way, exhibits
that mother-of-pearl-like effect which has been so often described, but the
explanation of which has given so much difficulty (see Pernter's suggestion which necessitates the presence in the cloud of mathematically
regular ice-crystals consisting of many branched stars, each of which acts
as an optical grating).

According to the theory sketched above, every position in the half of
the sky containing the sun is able to show diffraction colours, provided
that the appropriate size drop is present, and those clouds are iridescent
in which an assemblage of drops of the size required to produce first
order diffraction colours exists over an area sufficiently large to be
visible.

Iridescent colours are best seen at about 10° from the sun, for there
the eye is not blinded by the direct rays of the sun, and the region over
which a given colour can extend is wider there than at positions nearer
the sun.   But this necessitates very small drops of approximately ·004
mm. radius.   Such drops would be very unstable at high temperatures,
and may possibly only be able to exist at low temperatures.   This is
no doubt the reason why iridescent colours are almost entirely limited to
cirrus and other high clouds in temperate regions, while they may be seen
on low-lying mists in cold climates.

Conclusion

Just before leaving England with Captain Scott's Antarctic Expedition
Dr. W. N. Shaw very kindly discussed with me some of the meteorological
problems which our stay in the Antarctic might help to elucidate.
During our conversation he strongly impressed on me the importance of
determining the lowest temperature at which liquid water can exist in
the atmosphere.   While it cannot be said that that question has received
a definite answer, yet, if the reasoning set out in this paper is accepted,
an advance has been made in that direction; for we now know that
water drops exist at much lower temperatures than was ever suspected
by meteorologists, and a means has been found for carrying on the
investigation without proceeding to polar regions.   It only remains to
determine the temperature of clouds exhibiting either corona or iridescent
clouds, which are seen in all parts of the world, to collect a mass of data
from which a conclusion can be drawn.

DISCUSSION

Captain D. WILSON-BARKER said he was very glad to be able to welcome Dr. Simpson back from the Antarctic Regions, and that he had listened with great pleasure to Dr. Simpson's paper, but could not agree with some of his conclusions. He would not attack the paper on the theoretical ground, but he had never thought that halos were produced by any other than cirrus clouds, not by cirro-cumulus. They got certain coronal effects from cirro-cumulus clouds, but the best and most brilliant effects were produced by fleecy cumulus clouds, sometimes seen in a cloud break in the equatorial West winds accompanying a gale. With regard to halos, these were very faintly coloured unless under certain conditions of atmosphere, when full of very fine ice dust, which gave the sky a milky-looking appearance. Coronæ were formed by high floccules of cirro-cumulus clouds and seemed less brilliant than those produced by cumulus clouds. He considered iridescent clouds to be produced by refraction of light striking the ice crystals, the surfaces of which were melting and covered with a thin film of moisture which set up refraction and reflection effects. There was another form of corona: when the sky was perfectly clear they sometimes got a green corona round the moon; this indicated a prevalence of polar West wind in either hemisphere. As a sailor he had had plenty of opportunities of noticing these phenomena, and had always been greatly interested in them.

Dr. H. R. MILL joined in the expression of pleasure at seeing Dr. Simpson back again with his power of exposition and his scientific imagination in no way impaired by his sojourn in polar regions. He doubted, however, whether it was sufficient to imagine a possible explanation of any phenomenon without proceeding to demonstration. He would like to enquire as to the physical state of liquid water in the atmosphere at a temperature many degrees below zero Fahrenheit. Did it affect the hygrometer? Did it wet a surface with which it came in contact, or did it freeze immediately on touching anything? Could not these optical phenomena be produced by transparent spherules of ice? It seemed so unlikely that a cloud of water drops could exist at a temperature of 14° or 15° below zero that he felt the assumption of a cloud of globules of clear vitreous ice was the less violent improbability; would this produce the same effect as globules of water?

Dr. W. N. SHAW said that with regard to super-cooled water, there was no doubt that water could exist below the freezing point. In the introduction to a certain book, he had urged the desirability of getting away from 0° C. as being a cardinal point in meteorological work, because there were very many cases in which water was not ice when it got below 0° C. Anybody could make experiments on these matters. There was an instrument known as Bunsen's Calorimeter, and in one process this was filled with water freed from air, and then the tube inside was surrounded by ice by putting a freezing mixture through it. You could pass this through the tube as long as you liked and get ice everywhere except inside the instrument where it was wanted. There were a considerable number of other experiments with water in a mixture of essential oils about the same density, and, the temperature of the water reduced below freezing point, the water still remained liquid globules of large size. In order to get these changes of state, it was necessary to have something in the water which promoted the first formation of ice. One of the first things you learned in the experimental science of heat was that there were two definite points—boiling and freezing points of water; and the two things that remained most indefinite were the boiling and freezing points of water. He was perfectly certain that super-cooled water could be obtained. There was the phenomenon called glatteis; when this occurred it seemed as though the touch of any body, such as a branch or leaf, were sufficient to alter the condition. The ice probably fell as water and was transformed into ice later, and this caused the breaking of trees upon which it fell. The juices of plants remained liquid below the ordinary freezing point, and in the capillary tubes the freezing point was also changed. The question was how far was it possible to get the temperature down. He had always regarded a halo as conclusive evidence of the presence of ice crystals. It was hardly credible that the transformation from a water globule to ice would solidify in the form of a sphere; it would probably take the form of an ice crystal. It was a matter of gratification to him that Dr. Simpson had pursued the subject, and that he had sent out a definite challenge on the subject. As regards halos, artificial halos could be formed in the laboratory by means of alum. A corona could also be seen by travelling in a cab on a cold night with the windows shut. The condensation of the globules on the glass would produce a corona round every gas lamp seen. He had always wanted to pursue the experiment further and see if the corona could be seen through a frosted pane, but he had never succeeded in getting the condensation with ice. He thought the paper most interesting. There were two main subjects—the existence of water drops in the atmosphere and the transformation from a single experiment to the effect produced by an aggregate of a large number of globules in the atmosphere. He was interested in the account of iridescent clouds. If a cloud were formed artificially it was possible to get coronal effects, but not in the first cloud formed, which was always a grey mist. If this mist were allowed to settle and then condensed again, more brilliant colours would form, and the more often this process were repeated, the more brilliant the colours would become. Possibly the brilliance was due to the completeness of the sorting process. He thought the iridescent clouds were old ones, with fine drops at the top and larger ones at the bottom, and the iridescent effect was produced by the mixing, probably while the old cloud was disappearing.

Mr. W. W. BRYANT referred to a parhelic circle which he had observed on May 28. This, he said, was a complete circle of about 60° diameter passing horizontally through the sun. He enquired whether Dr. Simpson could suggest any explanation of it more satisfactory than that given in the new edition of the *Encyclopædia Britannica* attributing it to ice needles floating vertically, while Dr. Simpson implied that such crystals would always be horizontal.

The PRESIDENT remarked that he was rather surprised that Dr. Simpson had not said anything about the special iridescent clouds associated with volcanic dust. Possibly the gradual sifting of the atmosphere after an eruption had occurred, accounted for the brilliant colouring noticed ; and the finer particles left behind would make the iridescent effect more marked.

Dr. G. C. SIMPSON, in reply to the various speakers, said that he agreed with Dr. Shaw that the sorting of the drops in iridescent clouds might be due to gravity ; but this might be only one way amongst others by which the drops were arranged in definite sizes. With regard to the horizontal beam of light seen by Mr. Bryant, this was formed by reflection from ice crystals with vertical faces. With regard to the question of mixed drops, every sized drop, had its own diffraction pattern, and when a cloud contained drops of various sizes the colours from different patterns would combine to form white light. In reply to Captain Wilson-Barker and Dr. Mill, he summed up his paper by saying he had conclusively proved that water could exist at very low temperatures and ice crystals could not produce diffraction colours ; hence the inference was that any cloud which showed diffraction colours (coronæ and iridescent colours) must be composed of water drops.

Colonel H. E. RAWSON wrote that he much regretted that he had been called away by a telegram and had been unable to remain for the discussion of the paper. He had been recently making an investigation into the physical cause of colour in natural and artificial bodies and the relation between structure and colour, and he thought that the iridescence which the author had observed in these phenomena gave a clue upon which to found a sound theory in explanation of them. All iridescent objects such as birds' feathers, opal, mother-of-pearl, etc., change from red towards violet in the order of the colours of the spectrum as the angle of incidence of the illuminating light increases, that is, as the direction of the light becomes more oblique. The author had observed iridescent clouds with red changing to yellow and then to green, no doubt in consequence of the angles of incidence and reflection becoming greater. But when such colour phenomena are seen they naturally suggest the questions, what is the thickness of the plate, or the diameter of the drop reflecting or transmitting the light ? or is the surface of the structure composed of a series of fine lines ? The author was led to examine critically the diameter of the drop of liquid which could give an explanation of what he saw, but he does not seem to have considered the thickness of any frozen structure which could have produced the same effects. " Pigmentary " or absorption colours may be left out of this question. The colours seen were iridescent, and from Dr. Hodgkinson's investigations it appears that almost without exception the colours of natural iridescent objects are due to interference produced by thin plates. A thin plate of mica appears red when viewed with light falling on it perpendicularly ; by inclining the plate it will appear orange, then yellow, then yellowish-green, green and bluish-green. The same thing may be seen in crystals of chlorate of potash. Ice occurs like many solids in a non-crystalline, amorphous form, but water in a solid state also forms crystals of the hexagonal system seen in snow. Colonel Rawson has also seen snow in thin discs with a series of lines so regularly disposed that each disc was of the nature of a small diffraction grating. The subject of this paper well deserves special discussion on another occasion.

A Theory of the Anti-Coronae

H. C. van de Hulst
*Astronomical Observatory, Utrecht, Holland**
(Received October 14, 1946)

The anti-coronae, or luminous rings around the anti-solar point, are explained as a peculiar diffraction phenomenon in small water drops. The optical and the electromagnetic theories give results in agreement with each other and with the observations. The most important features are those which relate to the polarization.

THE diffraction coronae, or colored rings, which surround the sun or moon, when covered by a thin veil of cloud, are well known. A similar phenomenon can occasionally be observed in the opposite direction. A person standing on a high point observes his shadow projected on low clouds or on a layer of mist. He observes a gradual increase of the intensity of reflected light towards the shadow's head and, if conditions are favorable, some colored rings appear around the head. We shall call this phenomenon the anti-corona.[1] It is often seen in the mist on mountain tops and occasionally in lower country, when the sun is near the horizon. An air-pilot related that he saw the anti-corona around the shadow of his plane on the clouds nearly every day. According to the position of the observer, the center of the rings shifts from the shadow of the head of the plane to the shadow of its tail. Once or twice the anti-corona has been studied in laboratory experiments.

A satisfactory explanation has not so far been available. The old explanation by which the anti-corona is ascribed to common diffraction of light somehow reflected around the foremost drops of the cloud, does not appear to be sound. The more recent view is that these peculiar fluctuations in the intensity of nearly backward scattered light are already present in the scattering diagram of one single drop. Thus far, however, a more specific theory explaining the differences between the common coronae and the anti-coronae, has not been given.

For a general investigation of the scattering of light by spherical particles, we refer to a former publication,[2] which gives an extensive treatment of dielectrical and absorbing particles on the basis of Mie's electromagnetical theory. Data from other authors are compiled, various limiting cases are discussed and numerical results are given. The present problem deals with the limiting case of very large particles, where the scattering approaches the classical theories of geometrical optics and of optical interference. We shall first derive the results from the optical theory and then from the rigorous electromagnetical formulae.

OPTICAL THEORY

The intensity of light scattered by a sphere which is large in comparison with the wavelength can be computed by means of the laws of geometrical optics. The additional scattering by diffraction around the sphere is irrelevant for the present subject. The resulting formula is (T. 5, 14)

$$2\pi I_{1,2}(\theta) = \sum \epsilon_{1,2}{}^2 |a|^2. \qquad (1)$$

Here I is the emergent flux per unit solid angle, divided by the incident flux on the whole sphere; θ is the angle between the scattered and proceeding rays; finally the subscripts 1, 2 refer to the components the electric vectors of which are perpendicular and parallel, respectively, to the plane through the incident and scattered rays.

The intensities given by Eq. (1) consist of sums that have to be extended over all different light paths along which rays can be reflected, or refracted, into the direction θ. One light path can be specified by the parameters τ and p

---

* Now at the Yerkees Observatory, University of Chicago, Williams Bay, Wisconsin.

[1] The frequent use of the common term "glory" (cf. Webster) was considered unsuitable in an article of this kind.

[2] H. C. van de Hulst, "Optics of spherical particles," Thesis Utrecht, 1946; also published in Rech. Astr. de l'Observatoire d'Utrecht, 11, Part I. Formulae from this work will be quoted as (T. 5, 14) etc.

which are, respectively, the angle between the incident ray and the surface of the sphere, and the number of internal reflections added to 1. The case $p=0$ means external reflection. For a given light path, $\epsilon_{1,2}$ (T. 5, 13) denotes the decrease of the amplitude caused by the partial reflection and refraction on account of Fresnel's theory. Furthermore $|a|^2$ accounts for the decrease in intensity caused by the geometrical divergence of the rays; the value of $|a|$ is (T. 5, 15):

$$|a| = \left| \frac{\sin \tau \cos \tau}{\sin \theta} \frac{d\tau}{d\theta} \right|^{\frac{1}{2}}. \qquad (2)$$

It has been observed long ago that the intensity given by (1) and (2) would become ∞ for a light path for which

$$d\theta/d\tau = 0. \qquad (3)$$

This condition defines the *rainbows*. The two main rainbows lie near:

$$p=2, \quad \tau=31°, \quad \theta=138°, \quad \pi-\theta=42°;$$
$$p=3, \quad \tau=18°, \quad \theta=129°, \quad \pi-\theta=51°.$$

A strict infinity does not occur if interference is taken into account. The rays defined by (3) and the adjacent rays with equal p but with slightly different τ form a cubic wave front and a classical calculation[3] yields the correct intensity curve for the rainbows and supernumerary bows.

A second case in which the expression (2) diverges occurs when

$$\sin \theta = 0, \text{ while } \sin 2\tau \neq 0. \qquad (4)$$

No attention has been paid to this case, though it is quite similar to the case of the rainbow. Again, the divergence can be removed by taking into account the interference with the adjacent rays; a determinate intensity distribution will then result.

Condition (4) implies the cases $\theta=0$ and $\theta=\pi$. All peculiarities in the intensity distribution near $\theta=0$ are blended by the much stronger coronae caused by the diffraction of light around the sphere. Near $\theta=\pi$, however, a peculiar phenomenon may be expected. Indeed, as we shall see, the observed anti-coronae are explained in this way.

---

[3] Airy's theory. See any textbook on meteorological optics.

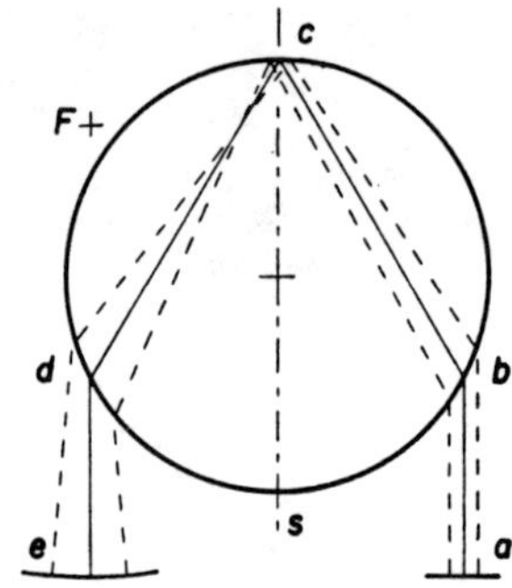

FIG. 1. Origin of toroidal wave front.

Figure 1 shows the simplest case of a light ray, *abcde*, satisfying the conditions required. Two adjacent rays, which emerge under slightly different angles, are also drawn. The linear front of the incident plane wave transforms into the circular front of the emergent wave, which has a virtual focus at F. However, still other rays must be taken into account. The whole figure must be rotated around the axis cs, and all the outgoing rays, thus defined, will interfere with each other. They define a *toroidal wave front* which seems to emerge from the focal circle described by F. We now have to calculate by means of Huygens' principle the interference pattern corresponding to this particular wave front.

The most interesting feature of the present problem is that the two directions of polarization cannot be treated separately. Let the incident wave be plane polarized with its electrical vector vibrating in the plane of Fig. 1. Then, for the rays drawn, it possesses a parallel vibration and emerges with an amplitude containing the factor ϵ_2. In the perpendicular plane, however, the same wave appears as a perpendicularly vibrating wave so that it is transmitted with an amplitude proportional to ϵ_1. Both transmitted waves are *still* vibrating in equal directions and are capable of interference. It is thus seen that the interference of rays in different azimuthal planes *simulates* an interference of rays with different directions of polarization.

Figure 2 explains the symbols we shall use in our analysis. The circle represents the focal circle from which the toroidal wave seems to emerge; let its radius be r'. We will compute the intensity radiated in a direction which makes a small angle γ with the axis and the projection of which on the plane of Fig. 2 is directed toward

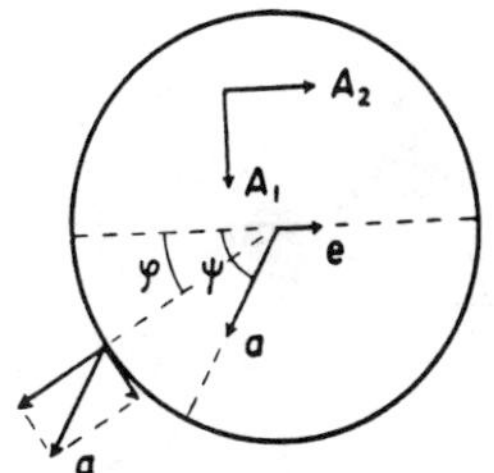

FIG. 2. Decomposition of light vectors.

the right, as shown by the arrow e. The consistent notations for the components of the light vector A of the emergent wave which vibrate parallel to and perpendicularly to the plane through the axis and e, are A_2 and A_1, respectively.

Let the incident light first be linearly polarized in a direction a making an angle ψ with the

fixed direction mentioned. The light emerging from an arbitrary point of the focal circle, as specified by the angle φ, consists of two components which in the incident beam have the amplitudes $\cos(\psi-\varphi)$ in the outward radial direction and $\sin(\psi-\varphi)$ in the counterclockwise tangential direction. The amplitudes with which these components emerge are:

$$\text{radial}: C_2 \cos(\psi-\varphi) \qquad (5)$$
$$\text{tangential}: C_1 \sin(\psi-\varphi)$$

where the constants C_1 and C_2 are proportional to ϵ_1 and ϵ_2, respectively. To derive the amplitude vectors of the total emergent light wave, we must decompose the vector (5) into components with directions parallel to A_1 and A_2. These are:

$$a_1(\varphi) = (C_1 \cos^2\varphi + C_2 \sin^2\varphi)\sin\psi - (C_1-C_2)\sin\varphi\cos\varphi\cos\psi,$$
$$a_2(\varphi) = (C_1-C_2)\sin\varphi\cos\varphi\sin\psi - (C_1\sin^2\varphi + C_2\cos^2\varphi)\cos\psi. \qquad (6)$$

Writing $r'\sin\gamma = r'\gamma = u$, we now find the total amplitudes of the emergent light to be

$$A_{1,2} = \frac{1}{2\pi}\int_0^{2\pi} e^{-iu\cos\varphi} a_{1,2}(\varphi)d\varphi. \qquad (7)$$

The definite integrals needed to reduce this expression,

$$\frac{1}{2\pi}\int_0^{2\pi} e^{-iu\cos\varphi}\cos^2\varphi\,d\varphi = \tfrac{1}{2}\{J_0(u)-J_2(u)\},$$

$$\frac{1}{2\pi}\int_0^{2\pi} e^{-iu\cos\varphi}\sin^2\varphi\,d\varphi = \tfrac{1}{2}\{J_0(u)+J_2(u)\},$$

$$\frac{1}{2\pi}\int_0^{2\pi} e^{-iu\cos\varphi}\sin\varphi\cos\varphi\,d\varphi = 0,$$

are related to Sommerfeld's integral defining the Bessel functions.

Incident natural light can be considered as a superposition of non-coherent linearly polarized waves with randomly distributed directions of vibration. Averaging, therefore, the intensities in respect to ψ, we find a factor $\langle\sin^2\psi\rangle_{Av} = \langle\cos^2\psi\rangle_{Av} = \tfrac{1}{2}$ both in A_1^2 and in A_2^2. With omission of the factor $\tfrac{1}{8}$, the final intensities are:

$$I_1 = [C_1\{J_0(u)-J_2(u)\} + C_2\{J_0(u)+J_2(u)\}]^2$$
$$I_2 = [C_2\{J_0(u)-J_2(u)\} + C_1\{J_0(u)+J_2(u)\}]^2 \qquad (8)$$

and in both directions of polarization together:

$$I = 2(C_1+C_2)^2 J_0^2(u) + 2(C_1-C_2)^2 J_2^2(u).$$

These formulae show the full implications of the interference effect described. If no interference between rays in different azimuthal planes took place, the intensities $I_{1,2}$ would contain only the coefficients $C_{1,2}$ with the *same* index. This is the case for the rainbow; it holds also for large angles in the present theory: with increasing distance from the center of the interference pattern the factor

$J_0(u)+J_2(u)$ decreases rapidly, like $u^{-\frac{3}{2}}$, leaving

$$J_0(u)-J_2(u)=(8/\pi u)^{\frac{1}{2}}\cos\,(u-\pi/4)$$

with the same index coefficient as the predominant term.[4] For small u, however, the terms of (8) are of the same order of magnitude; this region will be discussed in the last section.

ELECTROMAGNETICAL THEORY

The general expression for the intensity distribution of light scattered by a homogeneous sphere was first derived by Mie in 1908.[5] The resulting formulae in a somewhat changed notation (T. 2, 12) are:

$$\pi I_{1,2}(\theta)=\frac{1}{2x^2}|\textstyle\sum_{1,2}|^2,\tag{9}$$

where

$$\sum_1=\sum_{n=1}^{\infty}\frac{2n+1}{n(n+1)}\{a_n\pi_n(v)+b_n\tau_n(v)\},$$

$$\sum_2=\sum_{n=1}^{\infty}\frac{2n+1}{n(n+1)}\{b_n\pi_n(v)+a_n\tau_n(v)\}.\tag{10}$$

By x we denote the ratio $2\pi r/\lambda$, the circumference of the sphere in terms of the wave-length. The coefficients a_n and b_n refer to the electric and magnetic 2^n-poles, respectively; they are complex functions (T. 2, 7) of x and of the refractive index, m. Finally, the formula contains the spherical harmonics (T. 2, 10):

$$\pi_n(v)=\frac{1}{\sin\theta}P_n{}^1(v)\quad\text{and}\quad\tau_n(v)=\frac{d}{d\theta}P_n{}^1(v),$$

where $v=\cos\theta$.

Anti-coronae in visible light are observed on clouds consisting of drops with diameters of the order of 25μ, which corresponds to a value for x of about 150. For such a large value of x the evaluation of expressions (9) is impracticable so that we have to use their asymptotic forms. We denote the small angle $\pi-\theta$ by γ. The spherical harmonics have then the asymptotic forms (T. 5, 5):

$$\pi_n(v)=(-1)^{n-1}\frac{n(n+1)}{2}\{J_0(z)+J_2(z)\},$$

$$\tau_n(v)=(-1)^{n}\frac{n(n+1)}{2}\{J_0(z)-J_2(z)\},\tag{11}$$

where we have written $z=n\gamma$.

As was stated by Debye, the terms of (10) are of comparable size up to $n=x$ and decrease rapidly as soon as n exceeds x. Most of these complex terms will nearly cancel each other. Only when the terms of the order

$$N-3,\ N-2,\ N-1,\ N,\ N+1,\ N+2,\ N+3,\ \cdots$$

have nearly equal phases, does an appreciable total amplitude result. In particular, the amplitudes of the nearly backward scattered light will be determined by the terms near the order N for which the sums

$$c_1=\sum_n(2n+1)(-1)^nb_n,\quad c_2=\sum_n(2n+1)(-1)^{n-1}a_n,\tag{12}$$

consist of terms with stationary phases. Neglecting the effect of the further terms, we find from

---

[4] By comparing this asymptotic behavior of (8) with Eq. (1) the relative coefficients may be expressed in absolute units.
[5] G. Mie, Ann. d. Physik **25**, 377 (1908).

(10) and (11) the total amplitudes

$$\sum_1 = \tfrac{1}{2}c_2\{J_0(u)+J_2(u)\}+\tfrac{1}{2}c_1\{J_0(u)-J_2(u)\},$$
$$-\sum_2 = \tfrac{1}{2}c_1\{J_0(u)+J_2(u)\}+\tfrac{1}{2}c_2\{J_0(u)-J_2(u)\},\qquad(13)$$

where u is written for $N\gamma$. These formulae, when substituted into (9), agree with the optically derived formula (8). Only the values of c_1, c_2 and N have still to be calculated.

Before we proceed, attention may be drawn to the following point. In Debye's asymptotic formulae (T. 5, 8) the coefficients a_n and b_n refer to *one* direction of polarization. Yet they appear simultaneously in the amplitudes for *each* direction of polarization. This apparent paradox can now be completely explained. In the asymptotic form the term with the different index drops out for most directions because $\pi_n(v)$ has a smaller order of magnitude than $\tau_n(v)$. Only near the forward or backward directions the mixture of both terms remains and can be fully explained on the basis of Huygens' principle, as was shown in the preceding section.

The final step consists of the determination of N, c_1, and c_2 from (12). We may evaluate them either analytically or numerically.

a. By the analytical procedure we can formally demonstrate the complete equivalence of the optically and electromagnetically derived formulae for drops which are very large in comparison with the wave-length. Similar proofs have been given for the special case of the rainbow[6] and for the general case[7] in which expression (2) has no discontinuity. The modification for the present special case would present no difficulties. In each of the three cases we start by replacing the coefficients a_n and b_n by their asymptotic forms for very large order, according to Debye. Different expressions are needed for $n<x$ and for $n>x$, as can be visualized by means of the "localization principle," which assigns the terms of the order n to rays passing the center at a distance $n\lambda/2\pi$. Terms with $n>x$ correspond to rays passing along the sphere and give a vanishing contribution. Terms with $n<x$ correspond to rays hitting the sur-face under an angle $\tau = \arccos(n/x)$. Further, the asymptotic form for each term can be decomposed into subterms corresponding to different numbers of internal reflections, each of them with the proper coefficients ϵ_1, or ϵ_2. It appears that b_n refers to the perpendicularly vibrating wave (index 1) and a_n refers to the parallel wave (index 2). Finally, the subterms have a stationary phase for the order N which corresponds exactly with the location of the classical light paths. The sum of the terms of orders near N can be approximated by a Fresnel integral or, in the case of the rainbow, by an Airy integral. The resulting intensities, polarizations and phases agree with the optically derived results. For the anti-corona we would conclude that $N\lambda/2\pi$ is equal to the radius r' of the focal circle and that c_1 and c_2 are, like C_1 and C_2, proportional to ϵ_1 and ϵ_2.

b. The analytical derivation of N, c_1, and c_2 of which we gave an outline holds for the general case defined by Eqs. (4). However, our assumption that the actual anti-coronae are caused by light rays satisfying these conditions is not correct: a light path as illustrated in Fig. 1 does not exist for a waterdrop. When we pass from central rays to edge rays the deviation γ of the ray with a single internal reflection increases from $0°$ to $42°$ (rainbow) and decreases again to the final value of $14°$. The value $\gamma=0°$ is not again reached. This would seem to invalidate our explanation. On the other hand, it must be pointed out that centrally incident rays, though they give $\gamma=0$, give a perfectly smooth intensity distribution since the discontinuity of expression (2) is removed by the fact that also $\cos\tau=0$.[8] Further, a rough computation shows that the more complicated light paths give rise to anti-coronae that are too weak to be observed, just as is the case for rainbows caused by more than two internal reflections.

The following simple solution is suggested as the most probable: though x can be as large as

<hr>

[6] Balth. van der Pol and H. Bremmer, Phil. Mag. **24**, 141 and 825 (1937). Further articles on an analogous problem, *ibid.* **25**, 817 and **27**, 261.

[7] The outline given here follows the treatment by H. C. van de Hulst, reference 2, Chapter V.

[8] The explanation by B. Ray, Proc. Ind. Ass. for the Cultivation of Science **8** (1923), is not correct, for this reason.

200, the actual drops are still small enough to give strong deviations from the results for the optical case, $x \to \infty$. The following considerations tend to support this explanation. A light path as shown by Fig. 1 exists for refractive indices between 2 and $\sqrt{2}$. The angles of incidence required are given in Table I. It is obvious that $m = 1.33$ just fails to satisfy the conditions, if the optical theory is to be rigorously valid. For a finite value of x, however, the change at $m = 1.41$ will not be so abrupt. From the values of $\cos \tau$, which is equal to N/x, we see that the anti-corona is caused by rays near the edge of the droplet. We therefore infer that also for $m = 1.33$ the *radius of the focal ring is nearly equal to the radius of the droplet*.

Furthermore, we know that Debye's semi-convergent formulae for the cylindric functions, on which the proof of the equivalence between the optical and the electromagnetical results was based, are *not* valid for n near x. E.g., for $x = 150$ they are useless for the terms ranging from about $n = 144$ to 156.[9] This makes the optical theory for the anti-coronae completely unreliable for spheres with $x = 150$ and $m = 1.6$ to 1.41. Consequently, it is probable that drops of this size and of refractive index 1.33 still give a strong anti-corona.

The values of c_1 and c_2 cannot now be derived from the analytical theory. Instead, we should compute the values of a_n and b_n from their rigorous definitions and add the sums (12) *numerically*. The optical considerations only predict a stationary phase of the terms near $n = x$. We have not made such a computation and, accordingly, we cannot give a theoretical prediction about the ratio of c_1 to c_2.

TABLE I. Rays that would give rise to anti-coronae if the refractive index were m.

m	τ	$\cos \tau$
2.0	90°	0
1.9	54°	0.59
1.8	38°	0.78
1.7	26°	0.90
1.6	16°	0.96
1.5	7°	0.99
1.41	0°	1.00
1.33	—	—

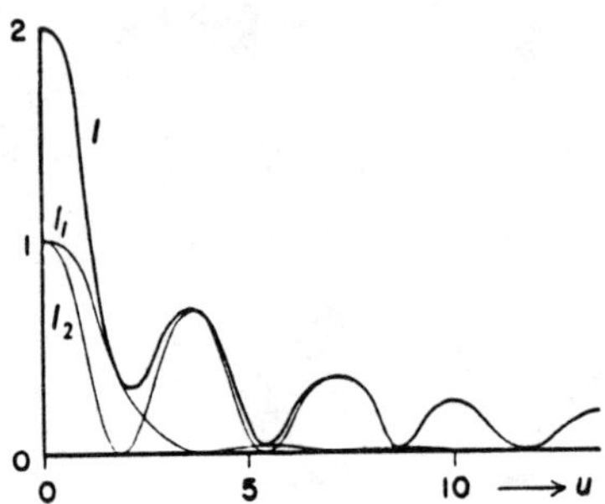

FIG. 3. Intensity distribution of the anti-coronae in both directions of polarization, in case $C_1 = 0$.

COMPARISON WITH THE OBSERVATIONS

For any given ratio of C_1 to C_2 the relative intensities in both directions of polarization can be readily evaluated from (8). The results for a few cases are:

1. $C_1 = C_2$. The anti-corona is wholly unpolarized. The central field is very luminous and dark rings appear at $u = 2.5,\ 5.6,\ 8.7,\ 11.8,\ \cdots$.

2. $C_1 = -C_2$. The anti-corona is again unpolarized. The anti-solar point is dark and is surrounded by a luminous ring at $u = 3.1$. Dark rings are situated at $u = 5.2,\ 8.5,\ 11.6,\ \cdots$.

3. $C_1 = 0$. The intensities for this case are shown by Fig. 3. In the central field the "alien" polarization, index 1, is slightly preponderant except at the anti-solar point itself. At $u = 2.3$ we find a fairly dark ring in which the polarization changes its sign. The bright ring at $u = 3.5$ and all further rings are nearly completely polarized in the "proper" direction, index 2. They are separated by dark rings at $u = 5.4,\ 8.6,\ 11.7,\ \cdots$.

We shall now compare these predictions with the available *observations*. Data concerning anti-coronae on natural clouds have been collected by Pernter-Exner.[10] Measurements on artificial mist, together with qualitative observations of the polarization, have been published by Mierdel.[11] These data, though not obtained with modern observational technique, suffice to give some checks on the theory and a preliminary determination of C_1/C_2.

The anti-coronae show distinct *differences from common coronae*, all of which can be explained on the basis of the present theory. (a) One feature is

[9] See Jahnke-Emde, *Tables of Functions* (B. G. Teubner, Leipzig, 1933 edition), especially Fig. 105.

[10] Pernter-Exner, *Meteorologische Optik* (W. Braumüller, Wien, 1910), p. 413 ff.
[11] F. Mierdel, Beiträge z. Physik d. freien Atmosphäre 8, 95 (1919).

TABLE II. Ratios of the radii of the dark rings.

Observer	Light	Cloud	γ_1/γ_2	γ_3/γ_2
Mierdel	white	artificial	0.34 ± 0.05	1.66 ± 0.03
	red	artificial	—	1.68 ± 0.04
	photographic	artificial	—	1.59
Average of several ob-servations	white	natural	0.46 ± 0.05	1.67 ± 0.06
Wegener	photographic	natural	0.41	(1.74)
Theory	$C_1 = 0$		0.43	1.60
	$C_1/C_2 = -0.25$		0.35	1.61

their variability. It is obvious that the interference of rays which are refracted by opposite sides of a waterdrop will be much more sensitive to slight deformations of the droplets than are the rainbows in which only adjacent rays interfere, or the common coronae in which non-refracted rays interfere. (b) The outer rings of anti-coronae are much more pronounced than are the outer rings of common coronae. On one occasion as much as 5 minima could be observed. The explanation is that the intensity in the anti-coronae decreases proportionally to γ^{-1}, whereas common coronae follow a θ^{-2}-law. (c) A final striking feature is the haziness of the first dark ring (which for normal sunlight is observed as the first red ring). Mierdel even calls it a slight depression separating the inner and outer parts of the central field. This haziness can be understood if the intensity distribution of Fig. 3 is approximately correct. A further increase of the first bright ring is obtained if we approach case 2, where C_1 and C_2 have opposite signs. Estimating from Mierdel's description that

$$0.30 < \frac{\text{brightness of first ring}}{\text{brightness of central field}} < 0.80,$$

we find the value of either C_2/C_1 or C_1/C_2 to lie between 0 and -0.25.

Further information is obtained from Mierdel's observations of the *polarization*. From the fact that the picture rotates with the analyzing nicol we infer circular symmetry to exist so that the glass plates in Mierdel's experiment cannot have had a serious effect. The complementary colors of the central field and the rings indicate different planes of polarization, in agreement with Fig. 3. The preponderant polarization in all bright rings is the one in which the electric vector vibrates

in the radial direction. This shows that $C_2 > C_1$, disregarding signs. Still, the tangential component is also visible. Estimating its intensity to exceed 4 percent of the total intensity of the rings, we find $C_1^2/C_2^2 > 0.04$. Together with the former data this indicates that C_1/C_2 must be about $-\frac{1}{4}$ or $-\frac{1}{5}$. This estimate is based on laboratory data; no observations of the polarization of natural anti-coronae seem to have been published.[12]

Finally we use the *radii* of the rings. As is true for common coronae, the radii of the rings are inversely proportional to the sizes of the droplets. However, it was noticed long ago that the rings of anti-coronae do not give consistent sizes when interpreted with the older theory. Let us denote the radii of the dark rings by γ_1, γ_2, γ_3, etc. For $C_1 = 0$, the present theory gives

$$\gamma_1 : \gamma_2 : \gamma_3 : \gamma_4 = 2.3 : 5.4 : 8.6 : 11.7$$
$$= 0.43 : 1.00 : 1.60 : 2.17.$$

For other values of C_1/C_2 the ratios are not very different. Table II shows the observed and computed ratios.

The agreement between theory and observations is as good as can be expected. No correction has been applied for the size of the sun and it was assumed that the edges of the red rings, which were measured by the observers, indicate the positions of the *dark* rings in the monochromatic picture for yellow light.[13] This introduces a considerable uncertainty but the ratio $\theta_2/\theta_1 = 1.80$, valid for the common diffraction coronae, differs strongly enough from the tabulated values to indicate again that this theory does not apply to the anti-corona.

On the basis of the present theory, measurements of the various dark rings give consistent sizes for the droplets. From the average diameters, which are about $2\gamma_1 = 1°\ 30'$, $2\gamma_2 = 3°\ 50'$, $2\gamma_3 = 6°\ 30'$, we find 0.027 mm for the average diameter of the water drops in the clouds on which anti-coronae have been observed.

---

[12] Dr. O. Struve has related to me that he observed an anti-corona through a nicol prism on a transatlantic flight, Sept. 15, 1946. It was strongly polarized and the dark and bright features rotated when the nicol was rotated. No radii or intensities were however estimated.

[13] Early experiments by Fraunhofer indicated that this rule was correct for the first three rings of common coronae. The "wave-length of white light" is 0.57μ. See Pernter-Exner, p. 460.

Mie Theory and the Glory*

H. C. Bryant and A. J. Cox†

Department of Physics and Astronomy, The University of New Mexico, Albuquerque, New Mexico 87106

(Received 21 April 1966)

Results are presented of the predictions of Mie's series solution to the scattering of an electromagnetic plane wave from a dielectric sphere of index 1.333. Cases of size parameters near 200 and 500 are examined in detail in an attempt to understand the mechanism responsible for the backscattering of light. These calculations demonstrate that the backscattering is due mainly to the last few significant terms in the Mie expansion, which can be associated with geometrical rays grazing the droplet surface, or "surface waves." Sharp periodic spikes in the dependence of the backscattered intensity are shown to be associated with very slightly damped surface waves of both polarizations involving hundreds of circumvolutions. The geometrically computed axial contribution to the backscattered light is shown to be present in the Mie results.

Index Headings: Scattering; Atmospheric optics.

I. INTRODUCTION

THE glory, or pronounced backscattering of light by water droplets, is not amenable to explanation in terms of ray optics, as is, for instance, the rainbow. For refractive indices less than $\sqrt{2}$, the case of water in the visible region, no off-axis rays with less than four internal reflections can contribute, and the contributions from multiply reflected rays are negligible.[1] From the ray-optics point of view, in fact, only the axial rays appear to contribute appreciably to the glory. This contribution is due to the ray reflected externally and the ray which undergoes one internal reflection, both along the axis of symmetry. See Fig. 1. The scattered intensity arising from these rays may be easily calculated

$$I_{\text{axial}} = [Rx^2/2(2-n)^2]$$
$$\times [(n^2-2n+2)+n(2-n)\cos 4nx], \quad (1)$$

where x is the size parameter, defined as the ratio of the circumference of the sphere to the wavelength of the incident light, R is the reflection coefficient at normal incidence, and n is the index of refraction. This axial contribution is inadequate to explain the glory phenomenon, as we see below, and, as van de Hulst points out,[2] some explanations in terms of surface waves must be advanced.

The series-expansion solution of G. Mie gives exact predictions for the ideal case of a plane electromagnetic wave interacting with a homogeneous sphere. In this paper we examine these predictions relating to the glory phenomenon and draw some conclusions from our calculations about the properties of surface waves.

II. MIE CALCULATION MECHANICS

The Mie calculations[3] we report in this paper are based on the exposition of van de Hulst,[4] using certain recursion relations of Deirmendjian, Clasen, and Viezee.[5] Checks were made with several published results,[6] and in every case agreement was good. The calculations were performed on the CDC 1604 and CDC 3600 computers. Using double-precision arithmetic, 22 significant figures were carried for all the computed quantities, in order to minimize errors from loss of significant figures due to subtractions in the recursion relations. As a check on this possibility, random bits were introduced into the last two places in the computation, at various stages in its progress, with no effect whatsoever on the final figures.[7]

III. RESULTS FROM MIE CALCULATIONS

Figure 2 displays several intensities calculated in the region of size parameter 200. The interval between cal-

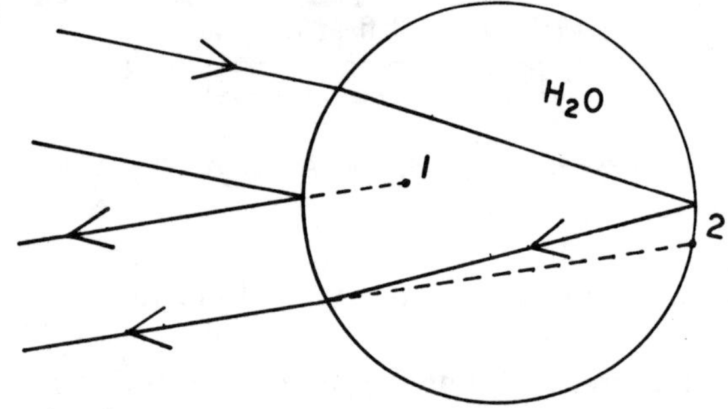

Fig. 1. Paths of axial rays in a water droplet. Points 1 and 2 are the loci of virtual images of the source.

* This work is supported by the Sandia Corporation, a Prime Contractor to the Atomic Energy Commission.

† Present address: Physics Department, University of Arizona, Tucson, Arizona 87521.

[1] H. C. van de Hulst, *Light Scattering from Small Particles* (John Wiley & Sons, Inc., New York, 1957), p. 231.

[2] Reference 1, p. 375.

[3] For a full description of the formulae used and procedure followed see A. J. Cox, Master's thesis, University of New Mexico (1965).

[4] Reference 1, Chap. 9.

[5] D. Deirmendjian, R. J. Clasen, and W. Viezee, J. Opt. Soc. Am. **51**, 620 (1961).

[6] See Ref. 3.

[7] The authors wish to thank Dennis Hayes for suggesting and carrying out this computational check.

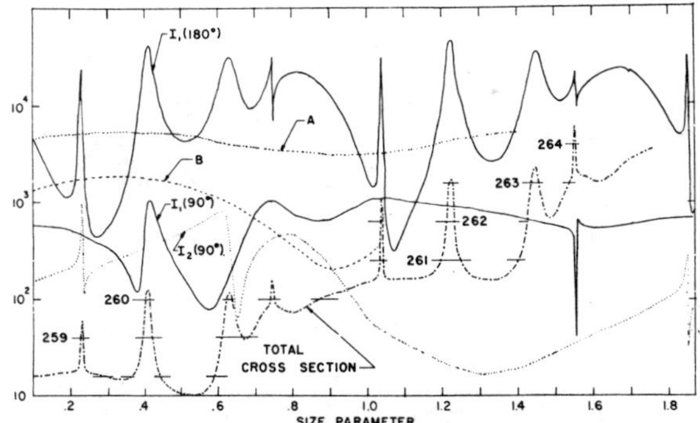

FIG. 2. Plots of $I_1(180°)$, $I_1(90°)$, $I_2(90°)$, A, B, and the total cross section vs the size parameter. All but the total cross section are plotted on a log ordinate scale. Curves A and B are, respectively, the upper limit on the contribution from axial rays predicted by Mie theory, and the result of the geometrical calculation, Eq. (1). Note that 200.0 must be added to the abscissae. The index is 1.333.

culated points is 0.005. All of the curves, except the total cross section, are on a logarithmic ordinate scale. The total cross section represents the area of the incident wave front undergoing scattering, and is a little larger than twice the geometrical cross section of the sphere. Table I displays the periods in x of the four sets of peaks in the total cross section in the region illustrated. The presence of these very small peaks is an important clue in the interpretation of the large fluctuations at 180° as we shall see below. The curves marked $I_1(90°)$, $I_2(90°)$, $I_1(180°)$ are, respectively, the scattered intensity at 90°, with the incident light polarized normal to the scattering plane; at 90°, with the incident light polarized in the scattering plane; and at 180° where the polarization does not matter. The small peaks in the total cross section are enormously amplified spikes in the 180° intensity; a sinusoidally varying component of period near 0.8 appears also. The intensities at 90° display the same sort of fluctuations as seen at 180°, but much reduced in magnitude, with each type of spike showing up in one of the polarizations, but not in the other. Perhaps the most interesting feature to point out about these 90° curves is that the periodicity in x of the spikes is twice that of the total cross section and the 180° intensity. Curve B is a plot of Eq. (1) for $n=1.333$, the expected axial contribution to $I_1(180°)$, and curve A is an estimated upper limit on the axial contribution to $I_1(180°)$ as given by the "growth" curves we shall describe below. These have a period of 1.18 in x.

TABLE I. Periods in x of the four prominent types of peaks seen in the cross section in the region $x=200$.

x	Period	Polarization
200.2280	0.810	radial
200.4075	0.815	azimuthal
200.6300	0.818	radial
200.7460	0.809	azimuthal

Figure 3 displays three growth curves at the value $x=200.400$, which is a smooth region, and three growth curves at a spike at $x=500.235$. Each of these is a plot of the calculated intensity as a function of the number of terms included in the Mie expansion. The series converges shortly after the number of terms exceeds the size parameter. The growth of $I_1(180°)$ is especially noteworthy; the features of C and F are quite typical of all such curves at this angle. In these, the intensity rapidly rises to a maximum value which remains quite constant until the last few significant terms, although the oscillations are pronounced. We call the maximum value in this region before the last significant terms the

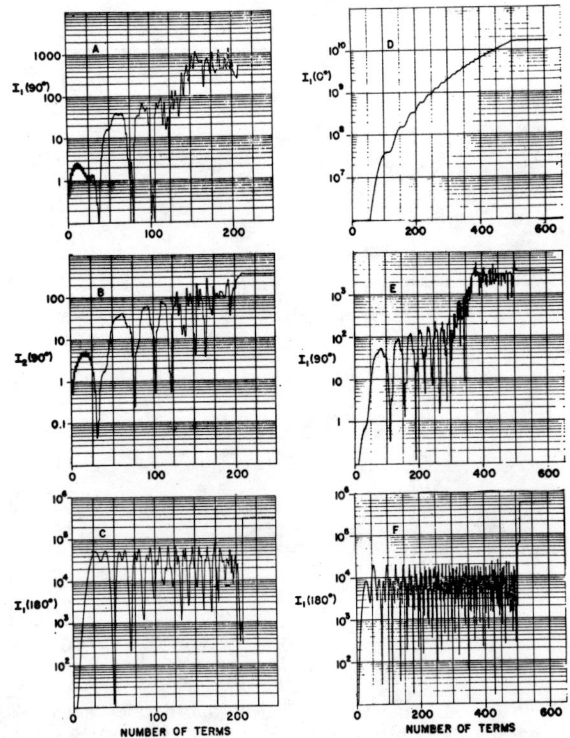

FIG. 3. "Growth curves": plots of the calculated intensity vs the number of terms included in the Mie expansion. Curves A, B, C correspond to droplets of $x=200.4$ and D, E, F to droplets of $x=500.235$. The index is 1.333.

"axial contribution" to the 180° intensity, and this is plotted as curve B in Fig. 2. The abrupt change, usually an increase, just before the series has converged to its final value is clearly associated with the spikes. The axial contribution contains only a smooth sinusoidal variation.

Figure 4 shows a detailed view of $I_1(180°)$ and the extinction coefficient Q (defined as the total cross section divided by the geometrical cross section) in the region of the spike at $x=200.746$.

Figure 5 is a plot of $I_1(180°)$ in the region $x=500$. Note the similarity to $I_1(180°)$ in Fig. 2. The sinusoidal component in the $I_1(180°)$ at $x=200$ appears here to be resolved into two. Included for comparison in this figure is the intensity that would be expected if light incident on the sphere were scattered isotropically, e.g., a perfectly reflecting sphere, with no diffraction effects.

IV. DISCUSSION

The growth curves presented in Fig. 3 can be interpreted in terms of the localization principle[8] which says that a term of order n in the Mie expansion corresponds to a ray passing the origin at a distance $n\lambda/2\pi$. The basis for this principle is the same as in partial-wave analysis in quantum theory: the nth term is associated with an orbital angular momentum $nh/2\pi$, which corresponds classically to a particle of momentum h/λ at an impact

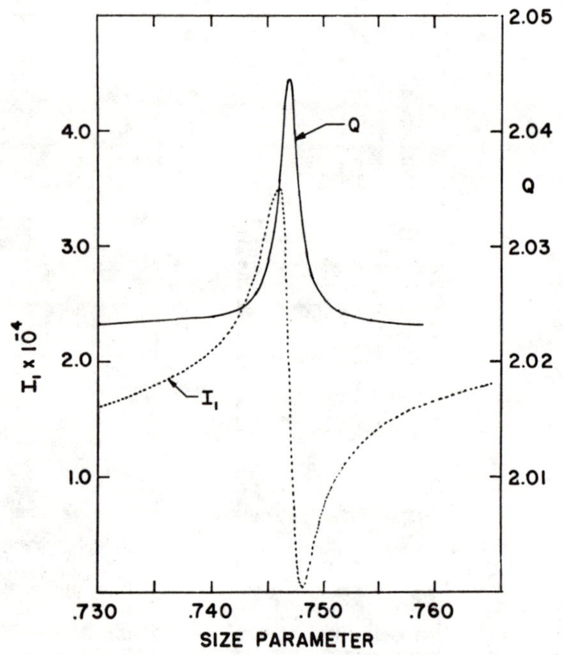

FIG. 4. Extinction coefficient (linear scale) and $I_1(180°)$ (log scale) vs size parameter. 200.0 must be added to the abscissae. The index is 1.333.

parameter of $n\lambda/2\pi$. Thus the convergence of the series shortly after n exceeds x results from the fact that the rays are no longer passing through the sphere. In the $I_1(180°)$ we can see that the results are determined almost entirely by rays nearly tangent to the surface, with a contribution from rays passing nearly through the center of the sphere: surface rays and axial rays, respectively.

We note, parenthetically, that the growth curves at 90° in Fig. 3 provide another interesting example of the applicability of the localization principle. From geometrical optics we would expect that the main contribution here would be a single external reflection. However, the reflection coefficients for water at 45° are $R_1=0.053$

---

[8] H. C. van de Hulst, Ref. 1, p. 208.

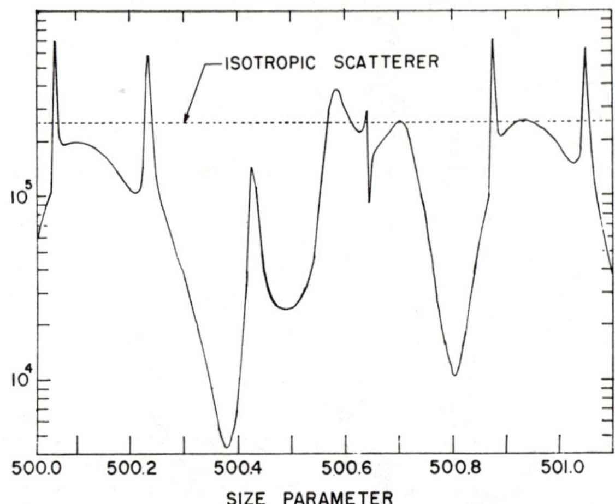

FIG. 5. $I_1(180°)$ for H_2O droplet, $n=1.333$, and for an isotropic scatterer, vs size parameter near 500.

and $R_2=0.003$, so that the polarization 1 should benefit from this mechanism much more than polarization 2. The contributions from this mechanism are easily calculated and are given by

$$I_1(90°)=0.053x^2/4$$
$$I_2(90°)=0.003x^2/4. \tag{2}$$

For $x=200$, the rays in question should correspond to terms of order near 141, and for $x=500$, terms of order near 353. The graphs in Fig. 3 confirm these predictions.

Further evidence regarding the axial ray is gained by comparing the two curves A and B in Fig. 2. Although they differ in amplitude, they have the same period and are in phase. Their differences are perhaps due to the rather arbitrary way in which B was defined.

By comparing the 90° curves with the 180° curve in Fig. 2 we see that the spikes can be associated with definite polarizations: radial or azimuthal, as indicated in Table I. The doubling of the periods at 90° can be explained in terms of the simple phenomenological model described below.

V. PHENOMENOLOGICAL MODEL OF THE SURFACE WAVES

Assume that an incident ray bundle tangent to the sphere is trapped on the surface of the sphere. This "surface wave" propagates around the sphere with a constant attenuation due to reradiation. Consider its amplitude to be 1 at the point of entry at 0° (see Fig. 6), then, if at 90° its amplitude is A, its amplitude will be A^2 at 180°, A^3 at 270°, A^4 at 360°, etc., where $|A|<1$. Since at zero degrees we have a focal point at infinity, we may write for the scattering amplitude at zero degrees

$$S(0°)=\alpha(1+A^4+A^8+A^{12}+\cdots)+B, \tag{3}$$

where α is a real constant, and B is the contribution to zero-degree scattering from other mechanisms.

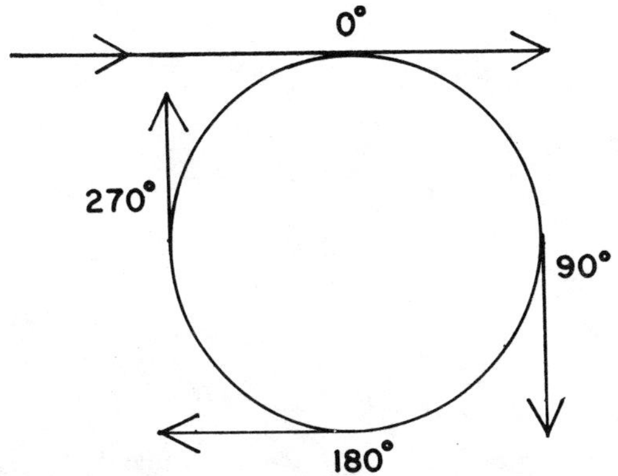

FIG. 6. Path of surface wave produced by ray bundle from the left.

Equation (3) may be written

$$S(0°)=\alpha/(1-A^4)+B. \tag{4}$$

Putting $A^4=ae^{i\phi}$, and applying the optical theorem,[9] namely $Q=4\,\mathrm{Re}[S(0°)]/x^2$, we get

$$Q=Q_0+(4\alpha/x^2)\left(\frac{1-a\cos\phi}{1+a^2-2a\cos\phi}\right), \tag{5}$$

where Q_0 is the part of the extinction coefficient due to the term B.

Consider the region of a resonance near $\phi=2m\pi$, where m is an integer. Put $a=e^{-g}$, and replace ϕ by ϕ modulo 2π, assuming ϕ and g small:

$$Q(\phi)=\frac{C_1}{1+(\phi/g)^2}+C_2, \tag{6}$$

where C_1 and C_2 are not sensitive to small changes in ϕ. Since the period of this peak is 0.809, $\phi=2\pi x/0.809$. Equation (6) can then be fit well to the plot of Q in Fig. 4

____

[9] H. C. van de Hulst, Ref. 1, p. 30.

to give $g=0.008$. This small value for the attenuation around one circumference of the sphere means that this peak is an interference maximum involving hundreds of circumvolutions of the surface wave.

The same considerations may be applied at $\theta=180°$:

$$S(180°)=\alpha A^2/(1-A^4)+D, \tag{7}$$

where D is assumed effectively constant. The scattered intensity, given by the absolute square of $S(180°)$, becomes, near resonance:

$$I(180°)=\frac{(\alpha/g)^2}{1+(\phi/g)^2}$$

$$\times[1+(2/\alpha)(\pm g\,\mathrm{Re}D+\phi\,\mathrm{Im}D)]+D^2. \tag{8}$$

The sign of g alternates from one resonance to the next. Considering the case $\theta=90°$ we find that

$$S(90°)=\delta(A+A^3+A^5+\cdots)+E. \tag{9}$$

We can see that the resonances at 90° have twice the period that Q and $I(180°)$ exhibit, since neighboring amplitudes composing the former differ in phase by one-half the phase difference for neighboring amplitudes in the latter two quantities.

CONCLUSION

The calculations presented in this paper demonstrate that the backscattered light from spherical water droplets arises chiefly from the last few significant terms in the Mie expansion, which by means of the localization principle, can be associated with rays grazing the surface of the droplet. These "surface waves" are responsible for several types of periodic fluctuations of period about 0.815 in the size parameter. The sharper peaks in the scattered-light intensity can be interpreted as being due to a very slightly damped surface wave involving hundreds of circumvolutions. The geometrically computed axial contribution to the backscattered light is shown to occur in the Mie result.

Infrared Rainbow

Abstract. *Radiation in the near-infrared spectral region should produce a rainbow that is not visible to the human eye. An infrared photograph is shown which displays the primary bow, the secondary bow, and two supernumerary bows inside the primary bow.*

The rainbow has long been a cause for wonder and delight to observers. The explanations of its origin and of a variety of its subtle features have paralleled the development of the principles of geometrical and physical optics. To a person who takes pleasure in contemplating the rainbow, the question "Does there exist an infrared rainbow?" would seem to be one of natural interest. I have been unable to find the question raised in any published literature and, although the subject may not be profound, it raises an interesting speculation.

For the infrared rainbow to exist, the radiation from the sun, after having traveled through a long atmospheric path, must still have an infrared compo-nent. Measurements show (1) that, even for long slant paths through the atmosphere, there is appreciable energy in the near-infrared region out to wavelengths of at least 2.5 μm (2500 nm). Another requirement is that raindrops must be transparent to the radiation forming the bow. For path lengths as great as a centimeter, water transmits appreciably for wavelengths out to about 1.3 μm (2).

The conclusion from these considerations is that there should indeed exist a rainbow of near-infrared radiation. To record this bow, I used Eastman Kodak infrared film IR 135, in conjunction with a Kodak 87C filter. This combination of film and filter made it possible to isolate a band pass, centered at 865 nm, which is separated from the visible part of the spectrum. Figure 1 shows the film sensitivity and filter transmittance along with the spectral sensitivity of the human eye, which serves to define the visible region of the spectrum.

The infrared photograph (Fig. 2) shows that the primary infrared rainbow (resulting from one internal reflection in the raindrops) is quite bright. Outside this primary arc, the secondary rainbow (resulting from the two internal reflections) is visible. Just inside the primary bow are two supernumerary bows which can be explained as an interference phenomenon arising from rays which emerge from the raindrop at the same angle but which have taken different paths through the drop [see Humphreys (3) for a more complete discussion of this effect].

These effects are, of course, known and understood from their occurrence in visible light (and one could predict their occurrence in the near-ultraviolet spectral region). There is, however, a

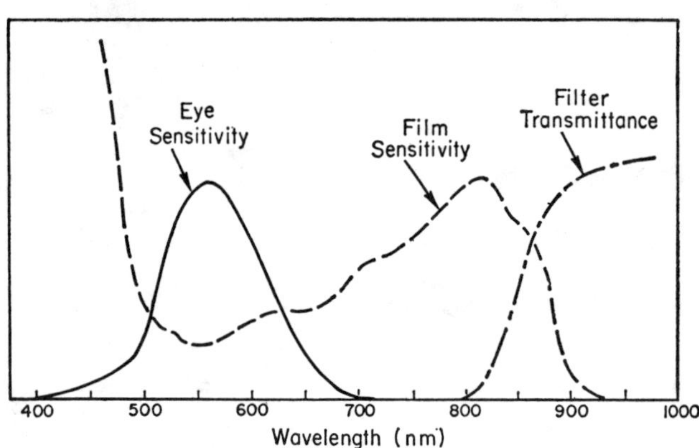

Fig. 1 (left). Relative spectral sensitivity of the human eye (——) [data for foveal cones taken from Wald (4)]. Relative spectral sensitivity of Eastman Kodak infrared film IR 135 (———) (data from Eastman Kodak Co.). Transmittance of the Eastman Kodak 87C infrared transmitting filter (- —-) (data from Eastman Kodak Co.).   Fig. 2 (right). Photograph of the infrared rainbow.

fascination in "seeing" for the first time an infrared rainbow which has hung in the sky undetected since before the presence of man on this planet.

ROBERT G. GREENLER

Department of Physics, University of Wisconsin–Milwaukee, Milwaukee

References

1. D. M. Gates, *Science* **151**, 523 (1966).
2. J. Strong. *Procedures in Experimental Physics* (Prentice-Hall, Englewood Cliffs, N.J., 1938), p. 369.
3. W. J. Humphreys, *Physics of the Air* (Dover, New York, 1964), chap. 3.
4. G. Wald, *Science* **101**, 653 (1945).

3 May 1971

Complex angular momentum theory of the rainbow and the glory

H. M. Nussenzveig

Instituto de Física, Universidade de São Paulo, São Paulo, Brazil
(Received 31 January 1979)

A survey is given of the applications of complex angular momentum theory to Mie scattering, with special emphasis on the recent treatments of the rainbow and the glory. The theory yields uniform asymptotic expansions of the scattering amplitudes for rainbows of arbitrary order, for size parameters $\gtrsim 50$, in close agreement with the exact results. The Airy theory fails for parallel polarization in the primary bow and for both polarizations in higher-order rainbows. The theory provides for the first time a complete physical explanation of the glory. It leads to the identification of the dominant contributions to the glory and to asymptotic expressions for them. They include a surface-wave contribution, whose relevance was first conjectured by van de Hulst, and the effect of complex rays in the shadow of the tenth-order rainbow. Good agreement with the exact results is obtained. Physical effects that play an important role include axial focusing, cross polarization, orbiting, the interplay of various damping effects, and geometrical resonances associated with closed or almost closed orbits. All significant features of the glory pattern found in recent numerical studies are reproduced.

INTRODUCTION

Analyzing exact solutions for the diffraction of light by an object, Sommerfeld points out in his lectures on optics[1]:

"The simplest example of such an object is the sphere. The field outside a sphere can be represented by series of spherical harmonics and Bessel functions of half-integer indices. These series have been discussed by G. Mie for colloidal particles of arbitrary compositions. But even there a mathematical difficulty develops which quite generally is a drawback of this method of series development: for fairly large particles ($ka > 1$, a = radius, $k = 2\pi/\lambda$) the series converge so slowly that they become practically useless. Except for this difficulty we could in this way obtain a complete solution of the problem of the rainbow, the difficulty of which was pointed out ..."

The purpose of this review is to outline how this problem has finally been solved, through the application of complex angular momentum techniques originally introduced by Poincaré and Watson essentially for this purpose, and to which Sommerfeld himself gave important contributions. These techniques enable us to solve not only the problem of the rainbow, "that most impressive of celestial phenomena" (in Sommerfeld's expression), but also a much more difficult one, posed by an equally beautiful, though more elusive, phenomenon: the glory. While the physical origin of the rainbow was well understood, this was not so for the glory,

although a conjectured explanation by van de Hulst[2,3] included some of the relevant effects.

The essential role played by the extension to complex variables in the solution deserves some discussion. An early example of such an extension is Sommerfeld's own solution of the half-plane diffraction problem. It would be misleading to view this extension as some contrived mathematical device: it has a much deeper significance.

We are accustomed to picturing optics in terms of real rays. It will be seen, though, that "complex rays," which represent the analytic continuation of real rays to complex values of some associated parameters, play an important role in both the rainbow and the glory.

Complex rays are already well known in optics in total reflection, where they describe the exponentially damped penetration into the rarer medium associated with surface waves traveling along the boundary. This typical wave effect is responsible, in quantum mechanics, for the tunneling through a potential barrier. It is very natural, from this viewpoint, to describe diffraction in terms of complex rays, since diffraction always represents the penetration of light into regions that are forbidden to the real rays of geometrical optics.

We begin with a brief survey of the complex angular momentum theory, as applied to the present problem (Sec. I).

The applications to the rainbow and the glory are described in Secs. II and III, respectively. The conclusions and some additional applications to meteorological optics are summed up in Sec. IV.

I. COMPLEX ANGULAR MOMENTUM THEORY OF MIE SCATTERING

The Mie solution

The exact Mie solution[4] for the scattering amplitudes when a monochromatic plane wave is incident on a homogeneous sphere of radius a and refractive index N may be written as follows:

$$S_j(\beta,\theta) = \frac{1}{2} \sum_{l=1}^{\infty} \{[1 - S_l^{(j)}(\beta)]t_l(\cos\theta)$$
$$+ [1 - S_l^{(i)}(\beta)]p_l(\cos\theta)\}, \quad (i,j = 1,2; i \neq j), \quad (1.1)$$

where $S_1(\beta,\theta)$ and $S_2(\beta,\theta)$ are the scattering amplitudes associated with perpendicular and with parallel polarization, respectively, θ is the scattering angle and $\beta = ka$ is the size parameter.

The angular functions are defined by

$$p_\nu(\cos\theta) = [P_{\nu-1}(\cos\theta) - P_{\nu+1}(\cos\theta)]/\sin^2\theta, \quad (1.2)$$

$$t_\nu(\cos\theta) = -\cos\theta p_\nu(\cos\theta) + (2\nu + 1)P_\nu(\cos\theta), \quad (1.3)$$

where $P_\nu(\cos\theta)$ is the Legendre function of the first kind (Legendre polynomial, when $\nu = l$ is an integer). The functions $S_l^{(j)}(\beta)$ are S-matrix elements associated with magnetic ($j = 1$) and electric ($j = 2$) multipoles of order l, respectively. They are given by

$$S_l^{(j)}(\beta) = -\frac{\zeta_l^{(2)}(\beta)}{\zeta_l^{(1)}(\beta)} \left(\frac{\ln' \zeta_l^{(2)}(\beta) - N\eta_j \ln'\psi_l(\alpha)}{\ln' \zeta_l^{(1)}(\beta) - N\eta_j \ln'\psi_l(\alpha)} \right), \quad (1.4)$$

where

$$\eta_1 = 1, \quad \eta_2 = N^{-2}, \quad \alpha = N\beta \quad (1.5)$$

and ψ_l and $\zeta_l^{(1,2)}$ are the Ricatti-Bessel and Ricatti-Hankel functions, respectively.

The scattered intensities for the two polarizations are given by

$$i_j(\beta,\theta) = |S_j(\beta,\theta)|^2, \quad (1.6)$$

and $\delta(\beta,\theta) = \arg S_1 - \arg S_2$ determines the state of polarization of the scattered light.

Except near $\theta = 0$ and $\theta = \pi$, the contribution from t_l is dominant over that from p_l in (1.1), so that $S_1(S_2)$ is dominated by magnetic (electric) multipole contributions.

In the domain of interest ($\beta \gg 1$), the lth "partial wave" in (1.1) is associated with incident rays having an impact parameter b_l given by the "localization principle"

$$b_l = [l + (1/2)]/k. \quad (1.7)$$

We expect that only rays hitting the sphere ($b_l \lesssim a$) are significantly scattered, so that the number of terms that must be retained in the Mie series to get an accurate result should be of the order of β. For visible light scattered by water droplets in the atmosphere, β ranges up to values of several thousand. This justifies Sommerfeld's statement quoted in the Introduction.

Numerical calculations of backscattering[5,6] based on (1.1) have shown an extremely rapid variation of the intensity with β. To resolve the details, the computations are made at intervals $\delta\beta = 10^{-2}$, with double precision arithmetic, using 11 significant decimal digits. The intensities are also rapidly varying functions of N and θ. Thus, even the availability of large computers has not solved the practical problem of extracting the information contained in the Mie solution.

The Watson transformation

Complex angular momentum techniques were introduced to deal with this problem in the early years of this century.[7] The Watson transformation starts by rewriting a partial-wave series such as (1.1) as a contour integral around the positive real half-axis in the λ plane, where

$$\lambda = l + 1/2 \quad (1.8)$$

is now regarded as a complex variable ("complex angular momentum"), and a suitable factor is introduced to generate poles at the physical values of λ (integral l), so that the corresponding residues reproduce the partial-wave series.

The advantage of rewriting the series as a path integral is that one gains the freedom to deform the path in the λ-plane. One looks for a deformation (depending on the problem) such that the dominant high-frequency contributions to the resulting expression arise from the neighborhood of a small number of "critical points" (e.g., saddle points on the deformed path or singularities swept through in this process), rather than being distributed among a large number of partial waves.

Watson's interest was in radio wave propagation, and he confined his attention to the deep shadow of a highly absorbing sphere (the Earth), where the above mentioned critical points are complex poles, now known as Regge poles. The corresponding residues are "creeping waves"[8] generated by incident rays tangent to the sphere, and traveling around it as surface waves, rapidly damped by radiation in tangential directions. The imaginary part of the poles, which determines the damping, increases rapidly, leading to a rapidly convergent residue series in the deep shadow region. The results can also be interpreted in terms of "diffracted rays," as shown by Keller[9] in his geometrical theory of diffraction.

In an illuminated region, Regge poles, as a rule, are not dominant critical points any longer. Such a region is accessible to geometrical-optic rays, which correspond to stationary optical paths by Fermat's principle. Since the phase of the path integral along the real λ axis is directly related with the optical path, these real rays are associated with stationary-phase points on the real axis. Such points are at the same time real saddle points, and one looks for a different deformation of the original contour so that the resulting path integral, often called a "background integral," will include portions of the associated steepest-descent paths going through these saddle points, which are new dominant critical points. Fock[10] was one of the main proponents[7] of this idea. He also studied the transition from light to shadow ("penumbra") on the surface of the sphere, and showed that it is described by a new mathematical function: the Fock function.

Applications of Watson's transformation to diffraction up

to this point, including Fock's reformulation, were limited in scope to only a few disconnected regions of space. It was applied to the rainbow problem by van der Pol and Bremmer,[11] who employed it to recover Airy's approximation, but they did not go any further.

The modified Watson transformation

In 1965, the author[12] developed a modified form of the Watson transformation that can be applied in any region of space. For an impenetrable sphere, one starts by applying to the partial-wave series the Poisson sum formula

$$\sum_{l=0}^{\infty} \phi\left(l + \frac{1}{2}, \mathbf{r}\right)$$
$$= \sum_{m=-\infty}^{\infty} (-)^m \int_0^{\infty} \phi(\lambda, \mathbf{r}) \exp(2im\pi\lambda) \, d\lambda, \quad (1.9)$$

where the "interpolating function" $\phi(\lambda,\mathbf{r})$ reduces to $\phi(l + 1/2,\mathbf{r})$ at the physical points. The term $m = 0$ in (1.9) corresponds to approximating the sum by an integral.[13] Terms with $m \neq 0$ are associated with paths that wind $|m|$ times around the center of the sphere.

Making use of reflection properties of the integrand, one can rewrite (1.9) in terms of integrals over the whole real axis, which are then judiciously deformed into the complex λ plane. Different deformations are employed in different regions of space. They usually lead to background integrals and to residues at Regge poles, with interpretations similar to those given above. Finally, one must consider the transitional domains between different spatial regions, where diffraction effects appear. We refer to a previous survey[14] for an account of this work.

The method succeeds for an impenetrable sphere because of the rapid damping of the surface waves associated with the Regge-pole contributions. This damping is determined by a purely geometrical property, the curvature of the surface, which gives rise to radiation as the waves travel around it.

The Debye expansion

If the sphere is penetrable, the waves get inside, leading to resonance effects with much weaker damping. Correspondingly, many Regge poles are located close to the real axis, spoiling the rapid convergence. (In fact, the residue series at these poles converges about as slowly as the original Mie series.)

In order to recover it, the solution must be rewritten in terms of surface interactions. This can be done by a procedure similar to the multiple-reflection treatment of the Fabry-Perot interferometer[1]: each spherical multipole wave undergoes successive internal reflections from the surface and from the center of the sphere (which simulates a perfect reflector, converting incoming waves into outgoing ones). The resulting expansion was employed by Debye[15] for a cylinder, so that we call it the Debye expansion. For the S-matrix elements in (1.1), it is of the form

$$S_l^{(j)}(\beta) = \frac{\zeta_l^{(2)}(\beta)}{\zeta_l^{(1)}(\beta)} R_{22}^{(j)} + \frac{\zeta_l^{(2)}(\beta)}{\zeta_l^{(1)}(\beta)} \frac{\zeta_l^{(1)}(\alpha)}{\zeta_l^{(2)}(\alpha)}$$
$$\times T_{21}^{(j)} T_{12}^{(j)} \left(\sum_{p=1}^{P} \rho_j^{p-1} + \frac{\rho_j^P}{1 - \rho_j} \right), \quad (1.10)$$

where

$$\rho_j = [\zeta_l^{(1)}(\alpha)/\zeta_l^{(2)}(\alpha)]R_{11}^{(j)} \quad (1.11)$$

and $T_{rs}^{(j)}$, $R_{rs}^{(j)}$, $(r,s = 1,2)$ are spherical transmission and reflection coefficients between region 1 (inside the sphere) and region 2 (outside the sphere). They are given by expressions similar to (1.4), except that they involve $\zeta_l^{(1,2)}(\alpha)$ instead of $\psi_l(\alpha)$ (i.e., traveling waves instead of standing waves inside the sphere).

The corresponding Debye expansion of (1.1) is

$$S_j(\beta,\theta) = S_{j,o}(\beta,\theta) + \sum_{p=1}^{P} S_{j,p}(\beta,\theta) + \text{remainder}, \quad (1.12)$$

where $S_{j,0}$ is associated with direct reflection from the surface and $S_{j,p}$, the pth term of the Debye expansion, is associated with transmission after $(p - 1)$ internal reflections at the surface. Since $|\rho_j| < 1$, we can even let $P \to \infty$ in (1.10) and (1.12), but the above form yields the remainder after P terms.

We can now apply the modified Watson transformation to each term in (1.12). Since only surface interactions are involved, the corresponding poles in the λ plane, which we call Regge-Debye poles, are all associated with rapidly damped surface waves, so that we get rapidly convergent asymptotic expansions for each Debye term.[16]

On the other hand, $|\rho_j(l,\beta)|$, as defined by (1.11), is approximately equal to the Fresnel reflection coefficient at the corresponding impact parameter (1.7), which is fairly small below the "edge domain"

$$\beta - c\beta^{1/3} \lesssim l \lesssim \beta + c\beta^{1/3}, \quad c = \mathcal{O}(1). \quad (1.13)$$

Thus, when the dominant contributions to (1.12) do not come from this domain, the Debye expansion converges rapidly. For water droplets, at most values of θ, we can stop at $P = 2$ in (1.12). A conspicuous exception, as will be seen below, is the glory (neighborhood of $\theta = \pi$).

Results for penetrable sphere

For each Debye term, the critical points in the λ plane are Regge-Debye poles and saddle points (real or complex) associated with background integrals.

A real saddle point $\bar{\lambda}$ in the range $0 \leq \bar{\lambda} \leq \beta$ for the pth Debye term corresponds to a geometrical-optic ray incident at the impact parameter $\bar{\lambda}/k$ [cf. (1.7)], that undergoes $(p - 1)$ internal reflections before emerging in the direction θ, so that the Debye expansion also parallels the geometrical-optic ray-tracing method. For a given p, as the impact parameter ranges from 0 to a, each domain of scattering angles may be covered one or more times by geometrically scattered rays. Thus, for each p, the domain $0 \leq \theta \leq \pi$ is subdivided into angular sectors, at the level of geometrical optics, corresponding to 0-ray (shadow) regions, 1-ray regions, 2-ray regions, etc. Each region usually requires a different path deformation to go through the associated saddle points.

The residues at Regge-Debye poles again correspond to surface waves excited by tangentially incident rays. In contrast with the impenetrable sphere problem, however, these waves not only travel around the sphere, shedding "diffracted rays" tangentially at each point, but also, at each point, they

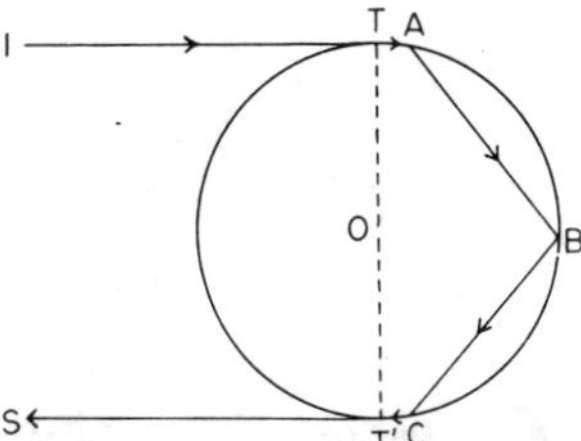

FIG. 1. Tangential incident ray *IT* gives rise to surface wave along *TA*, critically refracted to the inside at *A*, "totally" reflected at *B*, and critically refracted to the outside at *C*, traveling as surface wave along *CT'* and reemerging tangentially as the scattered ray *T'S*. * This path with two shortcuts contributes to the $p = 2$ Debye term.

undergo critical refraction into the sphere, taking a "shortcut" through it. At the opposite side, they are internally reflected as well as critically refracted to the outside, to reemerge tangentially as surface waves. These various processes are governed by associated diffraction, reflection and transmission coefficients. For the pth Debye term, there are p shortcuts. An illustrative example for $p = 2$ is shown in Fig. 1.

The boundary between two different angular sectors for a given Debye term is a shadow boundary, in the sense that one or more geometrical-optic rays disappear as we cross it, so that the intensity would be discontinuous according to geometrical optics. This is prevented, as usual, by diffraction, which broadens each such boundary into a transition region. The most interesting effects are found in such regions. The forward and backward directions may be regarded as special boundaries in this sense, because they are associated with focal lines, along which geometrical optics would predict an infinite intensity.

In the neighborhood of the forward direction, $0 \leqslant \theta \lesssim \beta^{-1}$, one finds the forward diffraction peak, where the intensity is dominated by the well-known Airy pattern of Fraunhofer diffraction by a circular disk.

A common transition region is associated with the disappearance of one real ray. In the λ plane, one type of critical point, a real saddle point, is replaced by a set of Regge-Debye poles. We call the corresponding transition a Fock-type transition, because the amplitudes in the transition region are described by Fock-type functions. These functions are defined by integrals whose asymptotic behavior is determined by a saddle point on one side of the transition and by a residue series on the other side; in between, they interpolate smoothly between the two types of behavior. The angular width of Fock-type transition regions is, typically, of the order of $\beta^{-1/3}$.

The other types of transition regions found in this problem are rainbow regions and the glory region. Their treatment, the main topic of the present review, is discussed below.

II. THE RAINBOW

Previous theories

A rainbow occurs when the scattering angle, as a function of impact parameter, goes through an extremum. The "folding back" of the corresponding scattered ray takes place at the extremal scattering angle, the rainbow angle. Thus, two rays scattered in the same direction with different impact parameters on the illuminated side of the rainbow fuse together at the rainbow angle and then disappear as we go over

to the dark side. This is one of the simplest examples of a "fold catastrophe" in the sense of Thom.[17]

Rainbows of different orders are associated with Debye terms of different orders; multiple reflection renders them fainter as their order increases. The most common rainbows are the primary bow, associated with $p = 2$ in (1.12), and the secondary bow, associated with $p = 3$.

Descartes' geometrical theory of the rainbow explained the primary and secondary bows in terms of the intensity enhancements at the extremal scattering angles for $p = 2$ and $p = 3$, respectively; it also explained the dark band in the sky between the two bows (Plate 107) as a shadow side common to both bows. Since a rainbow angle corresponds to a caustic for the scattered rays, geometrical optics predicts infinite intensity at a rainbow angle, as well as zero intensity on the dark side.

The supernumerary arcs (Plate 107) were explained by Young as one of the earliest examples of interference phenomena: they arise from the interference between the two scattered rays on the lighted side that merge together at the rainbow angle.

Classical diffraction theory was applied to the rainbow by Airy.[18] He applied the Huygens-Fresnel principle to a virtual wave front associated with geometrically scattered rays near the rainbow angle, making an assumption similar to Kirchhoff's approximation: constant amplitude along this wave front. The result was his celebrated "rainbow integral," now known as the Airy function, which plays in rainbow phenomena a role similar to that of Fresnel's integral in Fresnel diffraction. It shows oscillatory behavior on the lighted side, corresponding to the supernumerary arcs, and it is damped out faster than exponentially on the dark side.

Light from the rainbow is strongly polarized perpendicular to·the observation plane, as was discovered by Biot and Brewster. The suppression of the parallel component is due to a coincidence: the angle of incidence, for the primary bow, is close to Brewster's angle. A more detailed survey of the historical development of the theory of the rainbow can be found elsewhere.[18,19]

The domain of validity of Airy's theory was investigated by van de Hulst,[3] who concluded that it is limited, by the constant-amplitude assumption, to $\beta \gtrsim 5000$ and to $|\epsilon| \lesssim 0.5^0$, where

$$\epsilon = \theta - \theta_R \tag{2.1}$$

is the deviation from the rainbow angle θ_R. He pointed out at the time that no quantitative rainbow theory was available outside of this domain (apart from the numerical summation of the Mie series).

Complex angular momentum theory of the rainbow

The complex angular momentum theory allows us to derive the asymptotic behavior of the exact Mie solution in a rainbow region,[20,21] bridging the gap towards lower values of β and larger $|\epsilon|$.

For a Debye term of given order p, a rainbow is characterized, in the λ plane, by the occurrence of two real saddle points $\bar{\lambda}'$ and $\bar{\lambda}''$ between 0 and β in some domain of scattering angles θ, corresponding to the two scattered rays on the lighted side.

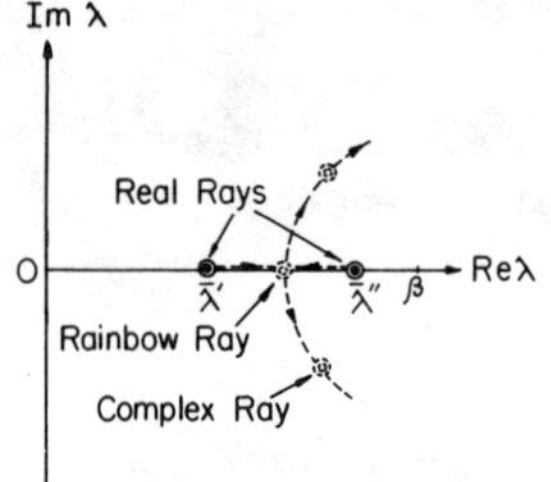

FIG. 2. As θ approaches θ_R from the lighted side, the two real saddle points λ' and λ'' move towards confluence; for θ in the dark side, they separate along complex conjugate directions.

When θ approaches θ_R from this side, the two saddle points move toward each other along the real axis, merging together at $\theta = \theta_R$. As θ goes over to the dark side, the two saddle points become complex, moving away from the real axis along complex conjugate directions. Thus, from a mathematical point of view, a rainbow can be defined as a collision between two saddle points in the complex angular momentum plane (Fig. 2).

It is well known in the saddle-point method that each saddle point has a characteristic range, the region of the complex plane around it that yields the main contribution to the integral. Within the rainbow region, the ranges of the saddle points $\bar{\lambda}'$ and $\bar{\lambda}''$ overlap, so that the usual method cannot be applied.

Uniform asymptotic expansions of integrals with overlapping saddle points were first derived by Chester, Friedman and Ursell.[22]

According to (1.7) and (1.8) and to the relation $b = a \sin\theta_1$ between the impact parameter b and the angle of incidence θ_1, the saddle points $\bar{\lambda}'$ and $\bar{\lambda}''$ may be written as

$$\bar{\lambda}' = \beta \sin\theta_1' = N\beta \sin\theta_2'; \quad \bar{\lambda}'' = \beta \sin\theta_1'' = N\beta \sin\theta_2'', \quad (2.2)$$

where, on the lighted side, θ_1' and θ_1'' are the angles of incidence of the two scattered rays that merge at the rainbow angle (θ_2' and θ_2'' are the corresponding angles of refraction); on the shadow side, θ_1' and θ_1'' become complex.

The angle of incidence $\theta_{1R,p}$ associated with the rainbow of order $p - 1$ (in the pth Debye term) is given by

$$\sin\theta_{1R,p} = (p^2 - N^2)^{1/2}/(p^2 - 1)^{1/2} \quad (p = 2,3,\ldots), \quad (2.3)$$

and θ_1' and θ_1'' may be regarded as functions of ϵ, where ϵ is given by (2.1), with θ_R now referring to the pth term.

The Chester-Friedman-Ursell method leads to expressions of the following form for $S_{j,p}^{(R)}$, the rainbow contribution to the pth term in (1.12):

$$S_{j,p}^{(R)}(\beta,\epsilon) = \beta^{7/6} \exp[2\beta A_p(\epsilon)]$$
$$\times \{c_{j,p}(\beta,\epsilon)Ai[(2\beta)^{2/3}\zeta_p(\epsilon)]$$
$$+ d_{j,p}(\beta,\epsilon)\beta^{-1/3}Ai'[(2\beta)^{2/3}\zeta_p(\epsilon)]\}, \quad (2.4)$$

where $Ai(z)$ is the Airy function,

$$\left\{\begin{array}{c} A_p(\epsilon) \\ \frac{2}{3}[\zeta_p(\epsilon)]^{3/2} \end{array}\right\} = \frac{1}{2}[pN(\cos\theta_2' \pm \cos\theta_2'')$$

$$- (\cos\theta_1' \pm \cos\theta_1'')], \quad (2.5)$$

and

$$c_{j,p}(\beta,\epsilon) = c_{j,p}^{(o)}(\epsilon) + \beta^{-1}c_{j,p}^{(1)}(\epsilon) + \cdots,$$
$$d_{j,p}(\beta,\epsilon) = d_{j,p}^{(o)}(\epsilon) + \beta^{-1}d_{j,p}^{(1)}(\epsilon) + \cdots, \quad (2.6)$$

are asymptotic expansions in inverse powers of β.

The right-hand side of (2.5) is proportional to the sum (difference) of the optical paths through the sphere associated with the rays corresponding to θ_1' and θ_1''. The difference vanishes at the rainbow angle: $\zeta_p(0) = 0$.

The coefficients $c_{j,p}^{(k)}$ and $d_{j,p}^{(k)}$ in (2.6) are given by expressions involving the Fresnel transmission and reflection coefficients (including multiple reflections) and their derivatives of various orders with respect to the angle of incidence, evaluated at θ_1' and θ_1''. For small enough $|\epsilon|$, these coefficients, as well as the functions $A_p(\epsilon)$ and $\zeta_p(\epsilon)$, can be expanded as power series in ϵ; in particular, $\zeta_p(\epsilon) = \mathcal{O}(\epsilon)$ for small $|\epsilon|$.

The Airy theory corresponds to the lowest-order approximation to (2.4) in several senses: (i) $A_p(\epsilon)$ and $\zeta_p(\epsilon)$ are replaced by terms up to $\mathcal{O}(\epsilon)$ of their power series expansion in ϵ; (ii) all coefficients $d_{j,p}^{(k)}$ are set equal to zero, so that the Ai' correction in (2.4) is neglected; (iii) all coefficients $c_{j,p}^{(k)}$ for $k \geq 1$ are set equal to zero; and, (iv) $c_{j,p}^{(o)}(\epsilon)$ is approximated by $c_{j,p}^{(o)}(0)$.

Predictions of complex angular momentum theory

Let us now discuss the main features predicted by the complex angular momentum result (2.4) and compare them with previous theories:

(i) Geometrical-optic contributions to the scattering amplitudes are, typically, of order β, whereas (2.4) reaches values of order $\beta^{7/6}$ near the rainbow angle. Thus, the maximum rainbow enhancement for the amplitudes is, typically,[23] of order $\beta^{1/6}$.

(ii) The main rainbow peak is characterized by $(2\beta)^{2/3}|\zeta_p(\epsilon)| \lesssim 1$, where $\zeta_p(\epsilon) = \mathcal{O}(\epsilon)$; its width, therefore, is $\mathcal{O}(\beta^{-2/3})$. As a function of p, the width increases,[24] growing linearly with p for $p \gg 1$, so that the peak becomes flatter as the order of the rainbow increases.

(iii) Within the main rainbow peak, the Ai and Ai' functons in (2.4) are of the same order, so that the dominant contribution would be expected to come from $c_{jp}^{(o)}$ in (2.6). However, $c_{jp}^{(o)}$ is proportional to the Fresnel reflection coefficient for polarization j at angles of incidence near $\theta_{1R,p}$. For the primary bow ($p = 2$), $\theta_{1R,2}$ is close to Brewster's angle, so that $c_{22}^{(o)}$ is very small, rendering the Ai' correction to Airy's theory much more important for polarization 2 (parallel). The smallness of $c_{22}^{(o)}$ as compared with $c_{12}^{(o)}$ also accounts for the strong perpendicular polarization of the primary bow. Since the Fresnel reflection coefficient is always larger for perpendicular polarization, we may expect this polarization to be dominant also for higher-order rainbows.

(iv) Since $Ai'(z)/Ai(z) = \mathcal{O}(z^{1/2})$ for $|z| \gg 1$, the Ai' terms become of the same order as the Ai terms in (2.4) outside of the main rainbow peak. Thus, the validity of the Airy theory is limited, at best, to small values of $|\epsilon|$, within the main peak, in agreement with van de Hulst's criticism. For the primary bow, the worst disagreement should be found for parallel polarization: due to the dominance of the Ai' term, (2.4) predicts supernumerary peaks where the Airy theory predicts zeros, and minima at the peaks of the Airy theory. Thus, maxima and minima are interchanged for the two polarizations. This effect, which has been observed by Bricard[25] at large angles, is related to the change of sign in the reflection coefficient as it goes through the zero at Brewster's angle. This affects one of the two scattered rays that interfere at large angles, leading to destructive rather than constructive interference.

(v) The result (2.4), in contrast with Airy's theory, is a uniform asymptotic expansion. Thus, for large $|\epsilon|$ on the illuminated side, where the two real saddle points are outside each other's range, it goes over smoothly into the sum of their contributions, yielding the interference between real scattered rays that gives rise to the supernumeraries (oscillatory behavior of the Airy function).

On the shadow side, for large $|\epsilon|$, we may employ the asymptotic expansion of $Ai(z)$ for large positive z,

$$Ai(z) \approx z^{-1/4} \exp\left(-\frac{2}{3} z^{3/2}\right) / 2\sqrt{\pi}, \quad (z \gg 1). \quad (2.7)$$

Although both complex saddle points contribute to (2.5) and (2.6), only the contribution from the saddle point in the lower half of the λ plane survives for large $|\epsilon|$, in agreement with the prediction of the steepest-descents method. Thus, (2.4) also merges smoothly with the wide-angle result on the dark side of the rainbow. This is a typical complex-ray contribution on the shadow side of a caustic,[9] similar to the quantum barrier-penetration effect.

(vi) Rainbows of order higher than the second are usually masked by background glare in the sky, although they can be observed in the laboratory.[26] While their intensity becomes progressively damped by multiple reflections, the damping weakens as the order increases, because by (2.3) they are formed by rays closer and closer to glancing incidence. On the other hand, they become progressively broader.

The fact that high-order rainbows are formed by rays incident close to the edge has another consequence: since the Fresnel reflection coefficients are not only close to unity but also vary rapidly with the angle in this region, their derivatives with respect to angle become larger, thus enhancing the contribution of higher-order coefficients in (2.6). Therefore, the Airy approximation becomes progressively worse (for both polarizations) for higher-order rainbows.

For a rainbow formed near the backward direction, there is an additional enhancement factor arising from axial focusing, as will be seen in Sec. III. We will see that light diffracted into the shadow of the tenth-order rainbow contributes significantly to the glory, exemplifying a (rather indirect) natural manifestation of a higher-order rainbow.

Comparison with the exact solution

Numerical comparisons between (2.4) and the exact Mie solution (1.1), as well as with the Airy approximation, have been carried out[21,27,28] for $N = 1.33$, $50 \leqslant \beta \leqslant 1500$, for the primary rainbow $p = 2$ ($136° \leqslant \theta \leqslant 142°$), and also, in connection with the glory,[27–29] for the tenth-order rainbow ($p = 11$). The results agree with the above predictions in all cases.

In a rainbow region, (2.4) should be the dominant contribution to the scattering amplitudes. However, to carry out

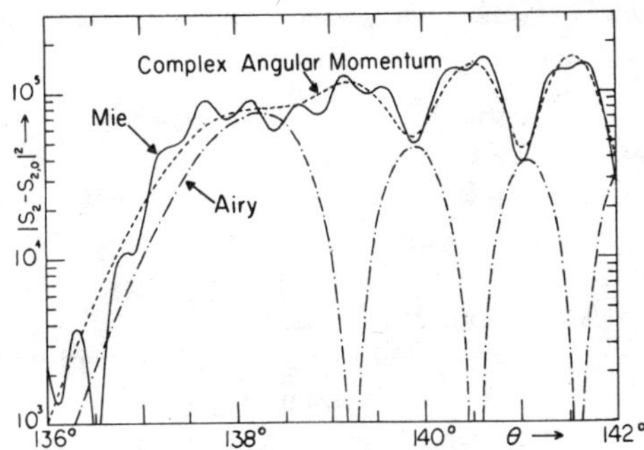

FIG. 3. Comparison between the exact Mie solution for $|S_2 - S_{2,0}|^2$ (——), the intensity $|S_{2,2}^{(R)}|^2$ of the rainbow term according to complex angular momentum theory (- - -), and the Airy approximation (– ·· –), for $N = 1.33$ and $\beta = 1500$.

comparisons with (1.1), one must take into account the effect of other Debye terms that may still yield significant (though smaller) contributions. For the primary bow, besides the rainbow term (2.4) with $p = 2$, one should also include the direct reflection term $S_{j,o}$ in (1.12). Its interference with the rainbow term gives rise to fast, small-amplitude oscillations superimposed on the rainbow oscillations.

As an illustration of the results, we show in Fig. 3 a comparison of the Mie, Airy, and complex angular momentum results for $\beta = 1500$ and parallel polarization. To avoid the rapid oscillations, we have subtracted out the direct reflection term from the Mie solution, comparing $|S_2 - S_{2,0}|^2$, where S_2 is given by (1.1), with $|S_{2,2}^{(R)}|^2$ given by (2.4) and with the Airy approximation. In agreement with the above discussion, the Airy approximation fails badly for this polarization, except in a small neighborhood of the main rainbow peak; its out-of-phase character in the supernumeraries is apparent. The complex angular momentum result for the rainbow term agrees closely with the exact result. The deviations, which have an oscillatory character, are due to interference with higher-order Debye terms, the effects of which will be discussed in connection with the glory.

The complex angular momentum theory, in contrast with the Airy approximation, is also in excellent agreement[21] with the exact Mie results for the phase difference $\delta = \arg S_1 - \arg S_2$. Its domain of applicability extends down to values of β of the order of 50.

III. THE GLORY

van de Hulst conjecture

Although it offers a display hardly less impressive than the rainbow, the glory has remained a considerably more recondite phenomenon, ever since its first reported observation[30] in 1735. Its sighting (Plate 108) requires locating the antisolar point, usually through the shadow of some object on the clouds (most often an airplane). Typical sightings occur over thin clouds or mist, with size parameters ranging from the order of 100 to about 1000 (the average β in reported observations[3] is ~160). The intensity decrease is slower than in coronae, so that as many as five sets of colored rings have been seen. There is considerable variability in the observations; some observers have reported finding parallel polarization in the rings.

The first significant theoretical contribution towards the explanation of the glory is due to van de Hulst.[2,3] Central or near-central rays, backscattered by direct reflection or after undergoing one or more internal reflections, cannot provide an explanation: they do not lead to an intensity enhancement and their contribution is much weaker than the observed effects. Suppose, however, that a nonparaxial incident ray were to emerge in the backward direction, e.g., after one internal reflection ($p = 2$). Such a ray is usually called a "glory ray." It was pointed out by van de Hulst that there would indeed be an intensity enhancement associated with a glory ray.

This enhancement arises from the axial symmetry. A narrow pencil of scattered rays emerging in a nonaxial direction is usually associated with a virtual focus and with a portion of a spherical wave front. For a pencil of nonparaxial rays scattered in an axial direction, however, the axial symmetry spreads out the virtual focal point into a virtual focal circle, corresponding to toroidal wave fronts. Constructive interference along the axis of the rays emanating from this focal circle produces the enhancement. We will call this the axial focusing effect.

Another peculiarity of the axial direction is that the associated scattering plane is undefined. In other directions, as was noted following (1.6), the scattering amplitude for magnetic (electric) polarization is dominated by magnetic (electric) contributions, but both become comparable in the axial direction, simulating an interference between parallel and perpendicular polarization (contributions from orthogonally situated points of the virtual focal circle). This effect was also pointed out by van de Hulst, who called it the cross-polarization effect.

For $p = 2$, however, there are no glory rays, except for refractive indices in the range $\sqrt{2} < N < 2$. For the refractive index of water, the closest approach to such a ray is provided by a tangentially incident ray, which, after one internal reflection, emerges some 15° away from the backward direction. It was conjectured by van de Hulst that this angular gap may be bridged by surface waves, taking paths similar to that shown in Fig. 1. We will call the effective contribution of all such paths the van de Hulst contribution.

The asymptotic evaluation of this contribution was not at hand at the time when van de Hulst formulated his conjecture, so that, although it seemed to account for some qualitative features of the glory then known, a quantitative test was missing.

Glory features revealed by numerical studies

More recently, several new numerical studies of the Mie series in the glory region have been made, revealing a remarkable degree of complexity and bringing out many new features to be explained.

Bryant and co-workers[5,31] computed the back-scattered intensity for $200.0 \leqslant \beta \leqslant 201.8$ and $500 \leqslant \beta \leqslant 501$ with $N = 1.333$, and for $3000.0 \leqslant \beta \leqslant 3001.4$, with $N = 1.333$ and $N = 1.333 + 2 \times 10^{-6} i$. Dave[32] computed the scattered intensity and the degree of polarization for natural incident light as a function of θ near 180°, with $N = 1.342$, for

$\beta = 98.2$, 196.3, 392.7, and 785.4. Shipley and Weinman[6] computed the normalized phase function

$$P(N,\beta,\theta) = 2\beta^{-2}[i_1(N,\beta,\theta) + i_2(N,\beta,\theta)]/Q_{sca}(N,\beta),$$

$$(3.1)$$

where Q_{sca} is the total scattering efficiency,[3] with $N = 1.333$ and $\theta = \pi$, for several size parameter ranges in the interval $200 \leqslant \beta \leqslant 4520$. They also evaluated average values of (3.1) over certain ranges of size parameters. Such averages are needed in lidar (laser radar) studies of rainfall, where the droplet populations are polydisperse.

The following features of glory scattering have emerged from these numerical studies:

(i) The scattered intensities in the glory region are rapidly varying functions of β,θ, and N. Within the usual range of observation of natural glories, they are, typically, about one order of magnitude larger than the geometrical-optic axial-ray contributions.

(ii) As a function of β, the backscattered intensity shows extremely rapid quasiperiodic oscillations. For refractive indices in the range associated with water, the quasiperiod $\Delta\beta$ of these oscillations ranges from about 0.81 to about 0.83.

These oscillations have been experimentally observed by Fahlen and Bryant[31] in laser light backscattered by a single water droplet, with β in the range 10^3 to 10^4: continuous evaporation of the droplet produced the variation in β. Similar observations, with β ranging from about 50 to about 300, have been made by Saunders.[33]

(iii) Within a single quasiperiod, there are rapid intensity fluctuations superimposed on a relatively slowly varying "background." The latter goes through two to three humps per quasiperiod.

(iv) The rapid fluctuations include spikes where the intensity varies by one to two orders of magnitude for size parameter variations $\delta\beta \sim 0.01$. (This corresponds to a change in average droplet radius of a few thousandths of a wavelength, i.e., of atomic dimensions.)

(v) The inclusion of a small absorptive term in the refractive index tends to smooth down the sharp spikes, but not the humps.

(vi) Fourier analysis of the backscattered intensity as a function of β reveals, besides the basic quasiperiod, strong periodic components with the periods $\Delta_1\beta \approx 0.41$, $\Delta_2\beta \approx 1.1$, and $\Delta_3\beta \approx 14$ (for $N = 1.333$).

(vii) A plot of the normalized backscattering phase function, with $N = 1.333$, averaged over the largest period $\Delta_3\beta \approx 14$, as a function of β, in the range $200 \leqslant \beta \leqslant 4520$, shows a broad peak, with the maximum average backscattering located around $\beta \sim 10^3$.

(viii) The angular distribution and polarization of the scattered intensity for $10^2 \lesssim \beta \lesssim 10^3$ vary considerably with β. For $\beta \sim 10^2$, the first dark ring is rather hazy and the outer rings are predominantly parallel-polarized. For $\beta \sim 10^3$, they tend to become perpendicular-polarized.

(ix) Studies[5] of numerical convergence of the Mie series show that the dominant contributions to the intensity in the glory region arise from the "edge domain" (1.13).

General considerations

The complex angular momentum theory of the glory[20,27–29,34] explains all of the above features. Let us first consider some general features of near-backward scattering.

As was mentioned following (1.6), the terms in t_l in (1.1) are dominant in nonparaxial directions; on the other hand, for $\theta = \pi$, since $t_l(-1) = -p_l(-1)$, we get

$$S_1(\beta,\pi) = S^M(\beta) + S^E(\beta) = -S_2(\beta,\pi), \qquad (3.2)$$

where S^M and S^E, the magnetic and electric contributions, are comparable, so that their interference has to be taken into account. This is van de Hulst's cross-polarization effect.

The axial focusing effect may be discussed in terms of the behavior of the angular functions (1.2) and (1.3), where $\nu = \lambda - 1/2$. For nonparaxial ray contributions, we have $\lambda = \mathcal{O}(\beta)$, so that we may employ the asymptotic expansion of these functions for large index. For scattering angles not very close to axial directions, this introduces an extra factor $\mathcal{O}(\lambda^{-1/2}) = \mathcal{O}(\beta^{-1/2})$, as compared with the neighborhood of axial directions. Thus, the amplitude enhancement associated with axial focusing is $\mathcal{O}(\beta^{1/2})$.

Let us concentrate our attention on the exact backscattering direction, $\theta = \pi$. The axial ray contributions arise from saddle points at $\lambda = 0$ in the even-order terms of the Debye expansion. They are readily evaluated, yielding the geometrical-optics result (plus higher-order WKB corrections). The main contributions are those from $p = 0$ and $p = 2$,

$$S_j^{\text{axial}}(\beta,\pi) \approx S_{j,0}^{(g)}(\beta,\pi) + S_{j,2}^{(g)}(\beta,\pi), \qquad (3.3)$$

where the superscript (g) stands for "geometrical." Higher-order axial Debye contributions are strongly damped by multiple internal reflections. The contributions (3.3) are of the form $a_j\beta$, where $|a_j|$ is of the order of the Fresnel reflection coefficient at perpendicular incidence. As was mentioned above, they are completely unable to account for the glory. They provide a smoothly varying background, representing only a small fraction of the glory intensity.

Analyzing the Debye expansion, we find that only the edge domain (1.13) can yield contributions capable of explaining the glory. This agrees with feature (ix) above. In this domain, the magnitude of the spherical reflection amplitude (1.11) is

$$|\rho_j| = |R_{11}^{(j)}| = 1 - b_j\beta^{-1/3}, \qquad (3.4)$$

where b_j is of the order unity; total reflection is not reached, due to the surface curvature.

The reflection damping factor associated with the $(p + 1)$th Debye term in the edge domain, for large p, is therefore

$$|\rho_j|^p \approx \exp(-pb_j\beta^{-1/3}), \qquad (3.5)$$

so that Debye terms of order up to $\mathcal{O}(\beta^{1/3})$ can still give appreciable contributions. Thus, for incident rays in the edge domain, an effect similar to "orbiting"[13] (the existence of paths winding indefinitely around a scattering center) takes place. Higher-order contributions from outside of the edge domain are damped out by multiple reflections and may be neglected.

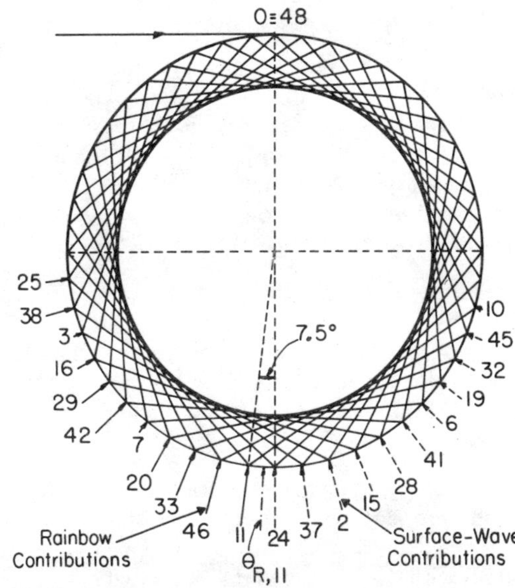

FIG. 4. The closed orbit associated with tangential incidence for $N \approx 1.33007$. The numbers are the values of p at the vertices. The directions of the adjoining arrows are those of the angles ζ_p defined by (3.7) and (3.9). Their lengths give a qualitative indication of the ordering of surface-wave contributions (- - -) by increasing ζ_p and of rainbow contributions (——) by increasing $\epsilon_{R,p}/p$ [cf. (3.8)]. The rainbow angles $\theta_{R,p}$ are shifted from the corresponding ζ_p (this is indicated for $\theta_{R,11}$). This ordering does not take into account reflection damping, which suppresses high-p contributions.

Classification of leading Debye contributions

The basic problem, when so many Debye terms may contribute, is to find out which are the leading ones and to classify them in order of decreasing importance.

In view of the axial focusing effect, it should be expected that one significant contribution would be that of a glory ray. Although such a ray does not exist for $p = 2$ in water, it may exist (at least to a close approximation) for higher p, although very high values of p, for a given β, are excluded by the reflection damping factor (3.5). If

$$N = [\cos(11\pi/48)]^{-1} \approx 1.33007, \qquad (3.6)$$

which is within the refractive index range for water in the visible domain, there exists a glory ray for $p = 24$ [cf. (3.7) below for $p = 24$]. It is associated with a tangentially incident ray, which produces the closed orbit shown in Fig. 4, a regular stellated polygon of 48 sides. At each vertex, of course, there is transmission to the outside, and the orbit is repeatedly described, the values of p recurring with periodicity $\Delta p = 48$.

The discussion could equally well be carried out for neighboring values of N, associated with nearly closed orbits around this or other values of p. However, it is somewhat simpler to illustrate the various effects for a value of N such that a closed orbit exists, so that we now fix N at the value (3.6). Closed or nearly closed orbits lead to "geometrical resonance effects,"[20] implying quasiperiodic features which, as we have seen, are indeed manifested in the glory [feature (ii) above].

Apart from the glory ray contribution, what other contributions from the edge domain do we expect to be important at $\theta = \pi$? We know that the edge rays give rise to surface waves with p shortcuts for the pth Debye term and that, for $p = 2$ (van de Hulst term), the "missing angle" ζ_2 that must be described along the surface before emerging in the backward direction after two shortcuts is small [Fig. 1; $\zeta_2 = \pi/24$ for N given by (3.6)]. The angular damping exponent of the surface waves, given by the imaginary part of the Regge-Debye poles, is $\mathcal{O}(\beta^{1/3})$ [cf. (3.10) below], so that, for the range of β relevant to the glory, the damping along the missing angle is not very large. Furthermore, this damping is counteracted by the enhancement due to axial focusing.

We expect, therefore, that the van de Hulst term is indeed significant. This was first quantitatively confirmed[20] for scalar scattering; however, it was also pointed out that the van de Hulst term alone cannot account for the glory; there must be other significant contributions from higher-order Debye terms.

If the radiation damping associated with surface-wave propagation were the only damping effect, the magnitude of surface-wave contributions of order p would decrease roughly exponentially with the increase of the associated missing angle ζ_p. Restricting ourselves to surface waves that emerge less than one shortcut away from $\theta = \pi$, the angle ζ_p is given by

$$\zeta_p \equiv \pi - p\theta_t \pmod{2\pi}, \quad 0 \leqslant \zeta_p < \theta_t = 2\cos^{-1}(1/N). \quad (3.7)$$

These considerations lead to the "naive" ordering of surface-wave contributions illustrated in Fig. 4 by the decreasing length of the arrows that point at increasing values of ζ_p. Besides the values of p shown in Fig. 4, we would get additional solutions of (3.7) by adding to each p multiples of the basic period $\Delta p = 48$.

The true ordering will be affected by the reflection damping, which according to (3.5) increases exponentially with p for high p. This cuts off higher values of p, and the cutoff point decreases when β decreases, so that this effect is of special importance within the range of β relevant to the glory. For instance, while the contributions from $p = 24$ and $p = 37$ should be dominant over the van de Hulst $p = 2$ term for large β, the reverse should be true for lower β.

Besides these surface-wave contributions, the edge domain also generates higher-order rainbows [cf. Sec. II, (vi)], some of which lie close to the backward direction, turning their dark side towards it. Let $\theta_{R,p}$ denote the rainbow scattering angle for the pth Debye term. Then, the angular distance from the rainbow angle to the backward direction, for large p, is given by

$$\epsilon_{R,p} = \pi - \theta_{R,p} \approx |\zeta_p| - (N^2 - 1)^{1/2}/p, \quad (p \gg 1), \quad (3.8)$$

where ζ_p is defined by (3.7), now extended to negative angles.

The penetration of light within the dark side of the rainbow is described by a complex ray (Sec. II), with a damping exponent which, for large p, according to (2.4) and (2.7), is proportional to $\beta(\epsilon_{R,p}/p)^{3/2}$ at $\theta = \pi$. Although this is faster than the surface-wave damping, these contributions are enhanced not only by the axial focusing factor $\mathcal{O}(\beta^{1/2})$, but also by the rainbow enhancement $\mathcal{O}(\beta^{1/6})$.

Thus, we also expect significant contributions to the glory from complex rays on the shadow side of higher-order rain-

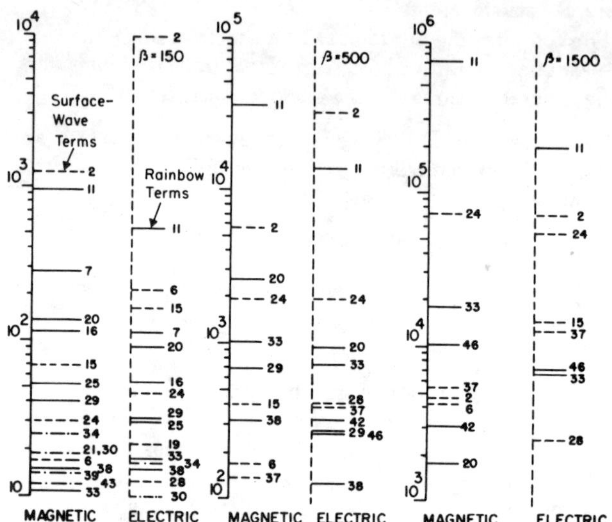

FIG. 5. Contributions to $|S^M(\beta)|^2$ and to $|S^E(\beta)|^2$ from Debye terms of various orders p (the values of p are indicated) for $N \approx 1.330\,07$ and $\beta = 150, 500,$ and 1500: —— rainbow terms; - - - surface-wave terms. For $\beta = 150$, there are contributions (-·-) from values of p that do not appear in Fig. 4.

bows formed at angular distances from $\theta = \pi$ given by (3.8), for values of p such that

$$\zeta_p \equiv \pi - p\theta_t \pmod{2\pi}, \quad -\theta_t \leqslant \zeta_p < 0. \quad (3.9)$$

The "naive" ordering of these contributions classifies them in order of decreasing importance according to increasing values of $\epsilon_{R,p}/p$. The arrows in Fig. 4 point at the directions ζ_p, and their lengths decrease according to this "naive" ordering. The leading contribution arises from the 10th-order rainbow, for which $\epsilon_{R,11} \approx 0.05$ rad $\approx 3°$.

Again, the true ordering will be affected by reflection damping, the more strongly the lower the value of β. Thus, for low β, $p = 33$ should dominate over $p = 46$, while the reverse should be true for large β.

An additional consideration that affects the ordering is that the width of the main rainbow peak is $\mathcal{O}(p\beta^{-2/3})$ for large p [Sec. II, (ii)], so that, for a given p, the backward direction, while lying deep within the rainbow shadow for high β, may lie within the main rainbow peak for low β, thus yielding a larger contribution, because the damping given by (3.7) has not yet set in.

Verification

To check the validity of the above discussion, the contributions from different Debye terms to $|S^M(\beta)|^2$ and $|S^E(\beta)|^2$ in (3.2) were numerically evaluated[29,34] by summing the corresponding partial-wave series for $\beta = 150, 500,$ and 1500. The results, including all terms that contribute up to $\sim 0.1\%$ to the total, are shown in Fig. 5. They are in complete agreement with our expectations.

For $\beta = 150$, which is close to the average size parameter in reported observations of the glory,[3] the van de Hulst term is the leading one, followed by the 10th-order rainbow term. The two contributions become comparable at $\beta = 500$, and the 10th-order rainbow predominates at $\beta = 1500$, where the glory-ray contribution ($p = 24$) is already larger than the van de Hulst term for magnetic polarization.

In order to carry out a quantitative comparison with the exact Mie solution, as well as to justify the physical interpretation ascribed to the various terms in the above discussion, we must be able to asymptotically evaluate the leading Debye contributions.

For the $p = 11$ rainbow term, the complex angular momentum theory leads to an expression of the type (2.4), except that the $\mathcal{O}(\beta^{1/2})$ axial focusing enhancement replaces the factor $\beta^{7/6}$ by $\beta^{5/3}$. This should be compared with the geometrical-optic axial-ray contributions (3.3), which are $\mathcal{O}(\beta)$. The evaluation of the coefficients (2.6) shows, in agreement with our expectations [Sec. II, (vi)], that the Airy theory fails for both polarizations at this higher value of p. A numerical comparison of the complex angular momentum prediction with the exact Mie result for the 11th Debye term[27–29,34] shows good agreement in both amplitude and phase, justifying the interpretation of this term as a complex-ray contribution on the dark side of the 10th-order rainbow. This is also a good test of the complex-angular momentum theory of the rainbow for a high-order rainbow.

For the $p = 2$ van de Hulst term, we must, according to the complex angular momentum theory, evaluate a series of residues at the Regge-Debye poles, which are of the third order for this term.[20,27] The evaluation leads to an expression of the form

$$S_{j,2}^{(\mathrm{res})}(\beta,\pi) = \beta^{4/3} \sum_n r_{nj} \exp(i\lambda_{nj}\,\zeta_2)$$

$$\times [1 + c_{nj}\beta^{-2/3} + \mathcal{O}(\beta^{-1})], \quad (3.10)$$

where ζ_2 is the missing angle (3.7) for $p = 2$, and λ_{nj} is the nth Regge-Debye pole for polarization j, in order of increasing imaginary part; $\mathrm{Im}\,\lambda_{nj} = \mathcal{O}(\beta^{1/3})$ yields the radiation damping of the surface waves.

Since $\mathrm{Im}\,\lambda_{nj}$ increases rapidly with n, the residue series is rapidly convergent, and it usually suffices, in practice, to keep only one or two poles. The coefficients r_{nj} are proportional to ζ_2, and therefore small, so that the $\mathcal{O}(\beta^{-2/3})$ correction inside the square brackets in (3.10) is important, especially for low β. Again, a numerical comparison with the exact Mie result[28] shows good agreement in amplitude and phase.

The asymptotic evaluation of the $p = 24$ glory-ray contribution is somewhat more complicated,[27,34] because $\theta = \pi$ lies within a Fock transition region for this term. Numerically, the results again compare favorably with the exact Mie solution.[29,34]

Complex angular momentum theory of the glory

Having at hand both the classification of the dominant Debye terms and asymptotic expressions for the leading contributions, we can finally discuss the explanation of the characteristic features of the glory listed before.

The glory arises from the interference among Debye terms of various different orders, whose relative importance varies with β. The phase factors relevant to the interference pattern originate mainly from the shortcuts through the sphere. Each shortcut contributes $2(N^2-1)^{1/2}\beta$ to the phase. With N given by (3.6), a change in β by an amount

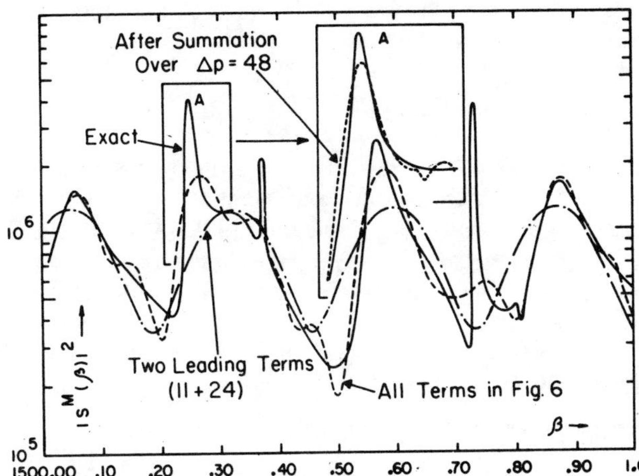

FIG. 6.   Behavior of $|S^M(\beta)|^2$ for $N \approx 1.330\,07$ near $\beta = 1500$: —— exact; –·– approximation by the two leading Debye terms in Fig. 5 ($p = 11 + p = 24$); - - - - approximation by the sum of all Debye terms shown in Fig. 4, without summing over $\Delta p = 48$. The inset shows an amplification (——) of the spike at position A, together with the result obtained by summing over $\Delta p = 48$ (- - -).

$$\Delta\beta = 5\pi/[22(N^2-1)^{1/2}] \approx 0.814 \quad (3.11)$$

changes the phase of the dominant Debye terms in Fig. 5 either by π or by an amount close to π (mod 2π), thus leaving the intensity almost (but not quite) invariant. This explains the quasiperiodicity of the backscattered intensity, as well as the observed value of the quasiperiod [feature (ii) above].

Within a quasiperiod, one can consider, as a first approximation to the amplitudes, the sum of the two leading Debye terms in Fig. 5, one of which is a surface-wave term and the other a rainbow term. Comparing the resulting approximation to $|S^M(\beta)|^2$ with the exact results near $\beta = 150, 500,$ and 1500, one finds[29,34] that the approximation yields the relatively slowly varying "background," with two to three humps per quasiperiod, which accounts for most of the average intensity in the glory [feature (iii)]. Figure 6 illustrates this near $\beta = 1500$.

If one includes the contributions from all the values of p shown in Fig. 4, without summing over the period $\Delta p = 48$, the result is very close to the exact curve, failing to reproduce only the sharp spikes (Fig. 6).

The spikes are obtained when we carry out the summation over $\Delta p = 48$. This is illustrated for one spike by the inset in Fig. 6. Thus, the spikes represent geometrical resonance effects associated with the existence of quasiperiodic orbits. Their sharpness arises from their association with very long optical paths with low damping [feature (iv)]. If one includes a small amount of absorption, they must therefore be the first to disappear [feature (v)].

The spikes, as well as other peaks, occur at different values of β for S^M and S^E. The cross-polarization effect [interference between S^M and S^E in the backscattered intensity (3.2)] therefore gives rise to additional interference features in the total intensity.

Within the range $10^2 \lesssim \beta \lesssim 10^3$ that is most relevant to observations of the glory, the dominant contributions arise from the van de Hulst term (3.10), from the 10th-order rainbow term [given by (2.4) for $p = 11$ with an extra axial focusing factor $\beta^{1/2}$], and from the geometrical-optic axial-ray contributions (3.3):

$$S_j \approx S_{j,2}^{(\text{res})} + S_{j,11}^{(R)} + S_{j,0}^{(g)} + S_{j,2}^{(g)}. \qquad (3.12)$$

In the intensity $|S_j|^2$, the interference among the terms in (3.12) gives rise to oscillations [whose amplitudes are slowly varying on the scale of the basic quasiperiod (3.11)], with quasiperiods given by: $\Delta_1\beta \approx 0.41$ (interference between $S_{j,2}^{(\text{res})}$ and $S_{j,11}^{(R)}$), $\Delta_2\beta \approx 1.1$ (interference between $S_{j,2}^{(\text{res})}$ and $S_{j,0}^{(g)}$) and $\Delta_3\beta \approx 14$ (interference between $S_{j,2}^{(\text{res})}$ and $S_{j,2}^{(g)}$); the quasiperiods $\Delta_2\beta$ and $\Delta_3\beta$ had already been predicted earlier.[20] These results are in complete agreement with feature (vi).

If we average $|S_j|^2$, as given by the approximation (4.12), over the largest period $\Delta_3\beta$, we find the sum of the squared moduli of the four terms in (3.12). The last two (geometrical-optic) terms contribute only a small constant background to the average normalized phase function. The van de Hulst term (3.10) yields a contribution of order $\beta^{2/3}\exp(-c\beta^{1/3})$ (where c is of order unity). This is the leading term for $\beta \sim 10^2$, but it decreases as β increases, and is surpassed by the 10th-order rainbow contribution at values of β of the order of a few hundred. This contribution to the phase function contains a factor $\beta^{4/3}$ and terms proportional to Ai^2 and Ai'^2 [cf. (3.4)] with rather small coefficients. Due to the effect of increased width for higher-order rainbows discussed above, the backward direction falls within the main peak of the tenth-order rainbow up to $\beta \gtrsim 10^3$; thereafter, it starts to get deep within the rainbow shadow, where the strong complex-ray attenuation appears. This explains[28] why the average backscattering phase function goes through a peak around $\beta \sim 10^3$ [feature (vii)]. The four-term approximation (3.12) accounts for 80% to 90% of the computed exact results.[6]

Finally, let us consider the angular distribution and polarization of the scattered intensity for natural incident light. We restrict our consideration to the first few glory rings, where

$$u = \beta(\pi - \theta) \qquad (3.13)$$

is not $\gg 1$. In this region, as a first approximation, we find[27,28]

$$S_1(\beta,\theta) \approx 2S^M(\beta)J_1'(u) + 2S^E(\beta)J_1(u)/u, \qquad (3.14)$$

where J_1 is Bessel's function of the first order; we obtain $-S_2$ by interchanging S^M and S^E. At $\theta = \pi$, this reduces to (3.2). Van de Hulst[2,3] had proposed an expression of this form with unknown coefficients c_1 and c_2, corresponding respectively to $2S^M$ and $2S^E$.

The polarized intensities associated with S_1 and S_2 follow from (1.6). For natural incident light, the angular distribution and polarization are proportional, respectively, to $i_1 + i_2$ and to $(i_1 - i_2)/(i_1 + i_2)$, so that they depend on the relative magnitude and phase of S^M and S^E in (3.14). For large u, S_1 is dominated by S^M and S_2 by S^E, as expected. Inspection of the graphs of $J_1^2(u)$ and $J_1^2(u)/u^2$ shows[34] that the first dark ring will be hazy when the orders of magnitude of S^M and S^E are comparable. Also, the polarization of the outer rings tends to be mainly perpendicular when S^M predominates and mainly parallel when S^E predominates.

According to (3.4), magnetic polarization is dominant in the 10th-order rainbow contribution, in agreement with Sec. II,

(iii) and with Fig. 5. It follows from (3.10) that electric polarization is dominant in the van de Hulst term; this may also be seen in Fig. 5.

The nature of the leading contribution to the glory changes with β in the range 10^2–10^3; the angular distribution and polarization undergo corresponding changes. In particular, near $\beta \sim 10^2$, where the van de Hulst term predominates, the outer rings tend to be parallel-polarized and the first dark ring is rather hazy (this is the situation in most reported observations of natural glories[3]); near $\beta \sim 10^3$, the rainbow contribution is dominant and the outer rings tend to be perpendicular-polarized [feature (viii)].

IV. CONCLUSION

The complex angular momentum theory of Mie scattering allows us to account accurately for both rainbow and glory effects. In both cases, contributions from diffracted light appear as a kind of analytic continuation from ray optics, representing the penetration into regions forbidden to real rays. Thus, use of the complex angular momentum theory leads to new insights and allows us to treat a new domain in optics.

The characteristic feature of rainbow scattering is the coalescence of real rays that become complex; this allows us to separate it from other interfering contributions that emerge in the same direction. We obtain a uniform asymptotic expansion for a rainbow contribution of arbitrary order. The Airy theory represents a crude approximation to this expression, which already fails badly for parallel polarization in the primary bow, and for both polarizations in higher-order bows. The domain of applicability of the complex angular momentum theory also extends to much lower values of β.

The glory arises from incident rays in the edge domain (1.13). After penetrating inside the sphere close to the critical angle, they are almost totally reflected (not quite, because of surface curvature), so that orbits involving many shortcuts through the sphere still contribute appreciably ("orbiting").

While geometrical-optic paraxial-ray contributions are of only marginal importance, nonparaxial contributions to near-backwards scattering, in particular those from the edge domain, are enhanced by axial focusing. Cross-polarization effects are also significant.

The leading contributions to the glory come from the complex domain. Those of surface-wave type, such as the van de Hulst term, are associated with Regge-Debye poles. As a limiting case, one may have higher-order glory rays [such as $p = 24$ for N given by (3.6)], whose contributions are of Fock type. Complex rays within the shadow of higher-order rainbows formed near the backward direction, especially the tenth-order rainbow, represent the other type of leading contribution.

The quasiperiodic features of the glory as a function of size parameter correspond to the existence of closed or nearly closed orbits, which give rise to narrow spikes through geometrical resonance effects.

Several types of damping with different β dependence play a role in the glory: the radiation damping of surface waves, the damping of complex-ray rainbow contributions by "tunneling," which is also affected by the broadening of the main rainbow peak as the order of the rainbow increases, and the damping due to multiple internal reflections, which affects both types of contributions.

The nature and the ordering of the leading Debye contributions change with β due to the competition among these various damping effects, producing changes in the angular distribution and polarization of the glory. The same effects are responsible for the peak in the average backscattering phase function around $\beta \sim 10^3$.

Quasiperiodic intensity fluctuations with size parameter, similar to those in the glory, also occur at other angles, but their amplitude decreases rapidly as one moves away from the backward direction[20] due to the absence of axial focusing. Their treatment is analogous to that of the glory.

In particular, they give rise to the "ripple" in the extinction efficiency[35] and in the radiation pressure; the latter has been observed[36] by optical-levitation techniques.

The average efficiency factors over a range of size parameters (where the "ripple" component is averaged out), including both extinction and absorption, play an important role in meteorological optics. The complex angular momentum theory of Mie scattering remains valid for complex refractive indices, and it leads to asymptotic expressions for the average efficiency factors[37] that are accurate over a wide range of size parameters and absorptive parts of the complex refractive index.[38]

ACKNOWLEDGMENTS

The author wishes to thank Professor Alistair B. Fraser for making available the color photographs of the rainbow and the glory reproduced in Plates 107 and 108.

[1] A. Sommerfeld, *Optics* (Academic, New York, 1954), p. 247.

[2] H. C. van de Hulst, "A Theory of the Anti-coronae," J. Opt. Soc. Am. **37**, 16–22 (1947).

[3] H. C. van de Hulst, *Light Scattering by Small Particles* (Wiley, New York, 1957).

[4] G. Mie, "Beiträge zur Optik trüber Medien, speziell kolloidaler Metallösungen," Ann. Phys. (Leipzig) **25**, 377–445 (1908).

[5] H. C. Bryant and A. J. Cox, "Mie theory and the glory," J. Opt. Soc. Am. **56**, 1529–1532 (1966).

[6] S. T. Shipley and J. A. Weinman, "A numerical study of scattering by large dielectric spheres," J. Opt. Soc. Am. **68**, 130–134 (1978).

[7] N. A. Logan, "Survey of Some Early Studies of the Scattering of Plane Waves by a Sphere," Proc. IEEE **53**, 773–785 (1965).

[8] W. Franz, *Theorie der Beugung Elektromagnetischer Wellen* (Springer-Verlag, Berlin, 1957).

[9] J. B. Keller, "A Geometrical Theory of Diffraction," in *Calculus of Variations and its Applications, Proceedings of Symposia in Applied Mathematics*, edited by L. M. Graves (McGraw-Hill, New York, 1958), Vol 8.

[10] V. A. Fock, *Diffraction of Radio Waves Around the Earth's Surface* (Publishers of the USSR Academy of Sciences, Moscow, 1946).

[11] B. van der Pol and H. Bremmer, "The Diffraction of Electromagnetic Waves from an Electrical Point Source round a Finitely Conducting Sphere, with Applications to Radio-Telegraphy and the Theory of the Rainbow," Phil. Mag. **24**, 141–176, 825–864 (1937); **25**, 817–837 (1938).

[12] H. M. Nussenzveig, "High-Frequency Scattering by an Impenetrable Sphere," Ann. Phys. (N.Y.) **34**, 23–95 (1965).

[13] K. W. Ford and J. A. Wheeler, "Semiclassical Description of Scattering," Ann. Phys. (N.Y.) **7**, 259–286 (1959).

[14] H. M. Nussenzveig, "Applications of Regge Poles to Short-Wavelength Scattering," in *Methods and Problems of Theoretical Physics*, edited by J. E. Bowcock (North-Holland, Amsterdam, 1970), p. 203–232.

[15] P. J. Debye, "Das Elektromagnetische Feld um Einen Zylinder und die Theorie des Regenbogens," Physik. Z. **9**, 775–778 (1908).

[16] H. M. Nussenzveig, "High-Frequency Scattering by a Transparent Sphere. I. Direct Reflection and Transmission," J. Math. Phys. **10**, 82–124 (1969).

[17] M. V. Berry, "Waves and Thom's Theorem," Adv. Phys. **25**, 1–26 (1976).

[18] C. B. Boyer, *The Rainbow: From Myth to Mathematics* (Thomas Yoseloff, New York, 1959).

[19] H. M. Nussenzveig, "The Theory of the Rainbow," Sci. Am. **236**, 116–127 (1977).

[20] H. M. Nussenzveig, "High-frequency scattering by a transparent sphere, II. theory of the rainbow and the glory," J. Math. Phys. **10**, 125–176 (1969).

[21] V. Khare and H. M. Nussenzveig, "Theory of the rainbow," Phys. Rev. Lett. **33**, 976–980 (1974).

[22] C. Chester, B. Friedman and F. Ursell, "An Extension of the Method of Steepest Descents," Proc. Camb. Phil. Soc. **53**, 599–611 (1957).

[23] For rainbows formed very close to the backward direction there is an additional enhancement factor of $\mathcal{O}(\beta^{1/2})$ due to axial focusing (cf. Sec. III).

[24] The contrary statement in Ref. 19 is an uncalled-for editorial insertion.

[25] J. Bricard, "Contribution à l'Étude des Brouillards Naturels," Ann. Phys. **14**, 148–236 (1940).

[26] J. D. Walker, "Multiple rainbows from single drops of water and other liquids," Am. J. Phys. **44**, 421–433 (1976).

[27] V. Khare, "Short-Wavelength Scattering of Electromagnetic Waves by a Homogeneous Dielectric Sphere," Ph.D. thesis, University of Rochester (1975) (unpublished).

[28] V. Khare and H. M. Nussenzveig, (unpublished).

[29] V. Khare and H. M. Nussenzveig, "Theory of the glory," Phys. Rev. Lett. **38**, 1279–1282 (1977).

[30] For the early history of the subject, cf. J. M. Pernter and F. M. Exner, *Meteorologische Optik* (Braumüller, Vienna, 1910).

[31] T. S. Fahlen and H . C. Bryant, "Optical back scattering from single water droplets," J. Opt. Soc. Am. **58**, 304–310 (1968).

[32] J. V. Dave, "Scattering of visible light by large water spheres," Appl. Opt. **8**, 155–164 (1969).

[33] M. J. Saunders, "Near-field backscattering measurements from a microscopic water droplet," J. Opt. Soc. Am. **60**, 1359–1365 (1970).

[34] V. Khare and H. M. Nussenzveig, "The Theory of the Glory," in *Statistical Mechanics and Statistical Methods in Theory and Application*, edited by U. Landman (Plenum, New York, 1977), 723–764.

[35] P. Walstra, "Light Scattering by Dielectric Spheres: Data on the Ripple in the Extinction Curve," Proc. Koninkl. Nederl. Akad. Wetensch. B **67**, 491–499 (1964).

[36] A. Ashkin and J. M. Dziedzic, "Observation of resonances in the radiation pressure on dielectric spheres," Phys. Rev. Lett. **38**, 1351–1354 (1977).

[37] H. M. Nussenzveig (unpublished).

[38] H. M. Nussenzveig and W. J. Wiscombe, (unpublished).

PLATE 107. (H. M. Nussenzveig, p. 1068). Both the primary and secondary bows are visible in this photograph taken in Sweden on 16 July 1978. That the sky is darkest between the two bows (Alexander's dark band) is easily visible as is a single pronounced supernumerary bow lying to the inside of the primary bow. Photograph courtesy of Alistair B. Fraser.

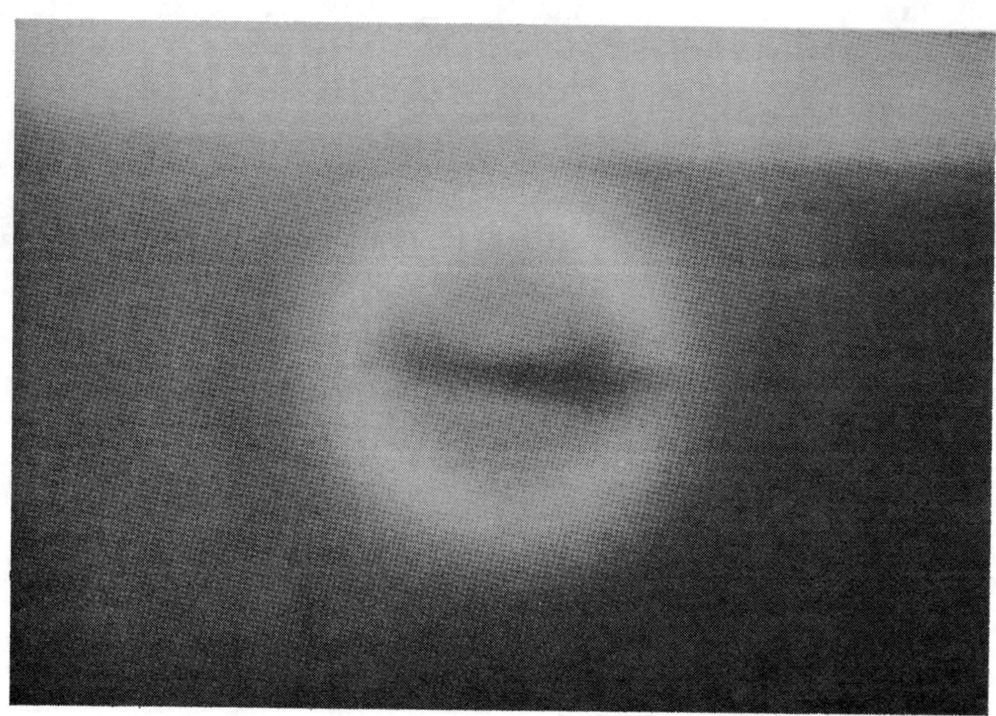

PLATE 108. (H. M. Nussenzveig, p. 1068). The glory is seen here around the shadow of an airplane. Three orders can be seen (on the original). Photograph courtesy of Alistair B. Fraser.

Why can the supernumerary bows be seen in a rain shower?

Alistair B. Fraser

Department of Meteorology, Pennsylvania State University, University Park, Pennsylvania 16802

Received June 3, 1983

Although the spectra of drop radii in rain showers are broad, the supernumerary bows are caused by only those drops with radii of about 0.25 mm. The angle of minimum deviation, the rainbow angle, is a function of drop size, being large for big drops, owing to drop distortion, and large for small drops, owing to interference. Between these extremes, there is a minimum rainbow angle. The drops that cause it give rise to the supernumerary bows.

INTRODUCTION

The supernumerary bows are an interference phenomenon. Indeed, Young, who in 1803 first showed that light was capable of interference, provided this explanation for the supernumerary bows.[1] In 1838 Airy showed that the spacing between adjacent bows, and the width of the primary bow itself, increased with decreasing drop radius.[2] Although it was not apparent at the time, this nice result presented as many problems as it raised. Any normal rain shower has a spectrum of drop radii that spans more than an order of magnitude. Because the position of the supernumerary bows for each drop size is different, no consistent pattern should emerge, and thus no supernumeraries should be seen. And yet they regularly appear! It is the purpose of the paper to explain why and to predict the spacing of the supernumerary bows that should be seen in a rain shower. The validity of the prediction is then tested using data extracted from photographs of rainbows in showers. Because, in rain showers, good supernumerary bows form only near the top of the bow, this work is continued to a study of the bows that are caused when light passes through the vertical cross section of a drop.

DEVIATION ANGLE AND PHYSICAL OPTICS

It has been known, ever since it was demonstrated by Descartes in 1637, that the rainbow is a phenomenon of the minimum angle of deviation. This can be seen in Fig. 1, which shows the cross section through a spherical drop. The light rays falling upon the upper half of the drop are deviated through an angle that exceeds 138 deg after they have undergone two refractions and one internal reflection. This diagram is usually used to argue that the two sets of rays traveling in the same direction, in the vicinity of the minimum angle ray, will give rise to interference. Although this is correct, a more compelling visualization is obtained if the rays are replaced by waves, as has been done in Fig. 2. Here the interference, and thus a supernumerary bow, is simulated with a moiré pattern in a manner similar to Young's original conjecture. It is clear that, although the ray diagram would remain unchanged as the drop size changes, the wave pattern does not. Big drops [Fig. 2(a)] give tightly spaced supernumeraries with narrow maxima, whereas little drops [Fig. 2(b)] do the reverse. In Fig. 3 the positions of the primary bow and the first two supernumerary bows are plotted on the basis of

Airy's theory using a wavelength of 0.55 μm.[3] Much has been written about the inadequacies of Airy's theory to represent the intensities for such small drops. Airy's theory does, however, adequately specify the positions of the maxima for radii greater than 0.05 mm, which is all that concerns us here.

DEVIATION ANGLE AND DROP DISTORTION

It has long been known that large raindrops are not spherical but are flattened to assume a shape like that of a hamburger bun.[4,5] Moebius showed that the nonspherical shape would increase the deviation angle of the Descartes ray.[6] This is shown in Fig. 4, where a drop with an exaggerated ellipticity is used better to illustrate the effect. If an actual drop had an aspect ratio as large as that shown in Fig. 4, the drop would no longer be an oblate spheroid but would be markedly flattened on the bottom. However, for the smaller sizes of drops, which will be of interest here, an ellipsoidal drop is an excellent approximation both to a more detailed theory and to measurements.[7] In consequence, the minimum deviation angles for drops are shown in Fig. 5, but instead of the ellipticities' being plotted, the corresponding drop radius[8] (of an equivalent volume sphere) is presented by using an expression derived by Green. The deviation angle of this Descartes ray is a strong function of drop radius (drop ellipticity) but only a weak function of solar elevation. Consequently, in the calculations that follow, only one solar elevation, 0 deg, was used.

COMBINED EFFECTS

Airy's theory specifies the position of the bows as a departure from the position of the Descartes ray. For small ellipticities, there is no reason to expect that the use of the slightly larger value of the deviation angle of the Descartes ray would produce any egregious errors in the application of Airy's theory. The larger values were used, and the results are presented in Fig. 6. Each curve evinces a minimum, and thus a drop radius for which the bow appears closest to the sun. The bow that will be seen will be caused only by those drops that produce the minimum on the curve. Further away from the minimum, the bows caused by one drop size will be displaced from the bows caused by another, and no consistent pattern will appear. Thus, near the top of the bow where drop dis-

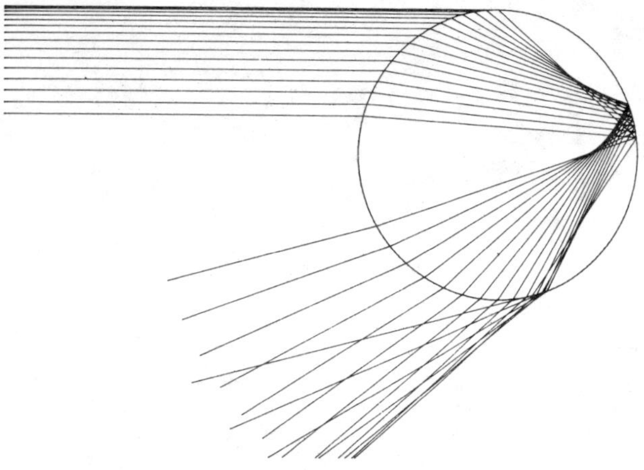

Fig. 1. Light rays entering a spherical raindrop are deviated by 138 deg or more, so that rays striking near the edge of the drop will emerge in the same direction as rays that strike closer to the center of the drop.

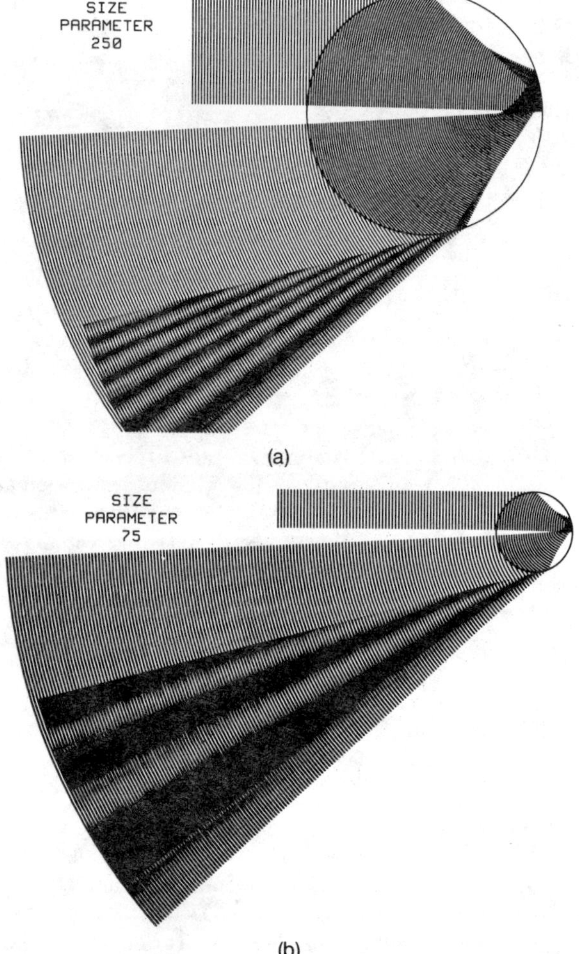

Fig. 2. The interference pattern, produced because the emerging wave front folds over on itself, is simulated with a moiré pattern. Large drops (a) produce closely spaced supernumerary bows, whereas small drops (b) produce widely spaced ones.

tortion is important, the primary bow will be caused only by drops that have a radius between about 0.1 and 0.2-mm. When there is a broad drop distribution, the top of the bow will be seen only if these 0.1–0.2-mm drops are also present.

Similarly, the first two supernumerary bows require drops of about 0.2–0.3-mm radius. Other drops make no contribution. The angular separation of the minima in the curves of the supernumerary bows in Fig. 6 is easy to extract; it is 0.75 deg. A prediction can then be made: If supernumerary bows are seen in a rain shower, their separation will be ~0.75 deg, independent of the drop-size distribution of the rain. Indeed, this was a prediction; it was made before the examination of any data.

THE DATA

In my file of rainbow pictures I found shots from five different showers that showed the first two supernumeraries: four of

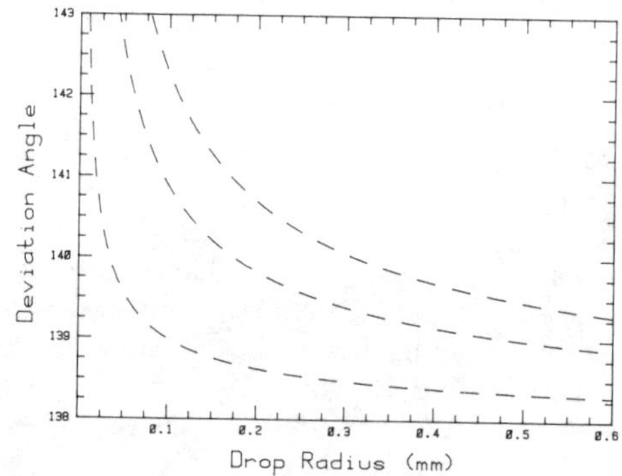

Fig. 3. The angular deviation of the maximum brightness for the primary rainbow (lowest curve) and the first two supernumerary bows as given by Airy's theory.

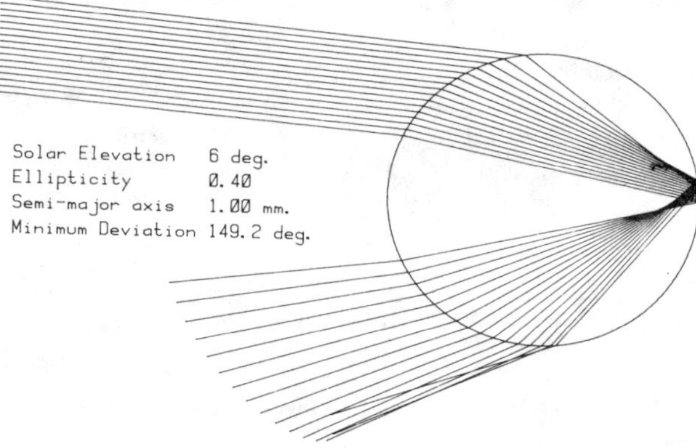

Fig. 4. Ray tracing though an ellipsoidal drop shows that the minimum deviation is now greater that 138 deg.

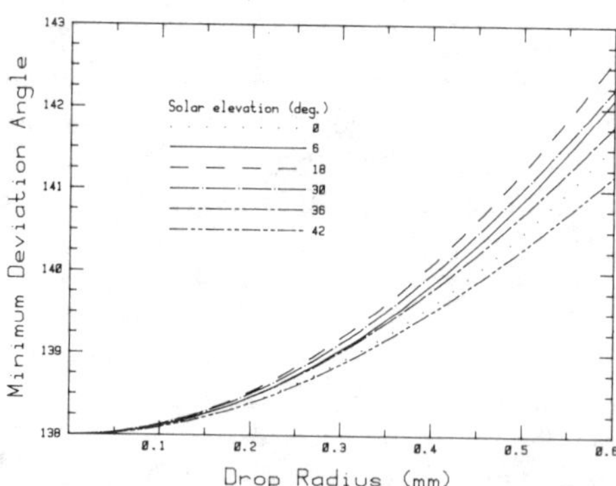

Fig. 5. The minimum deviation angle for ellipsoidal drops as a function of drop radius (of an equivalent volume sphere) and solar elevation.

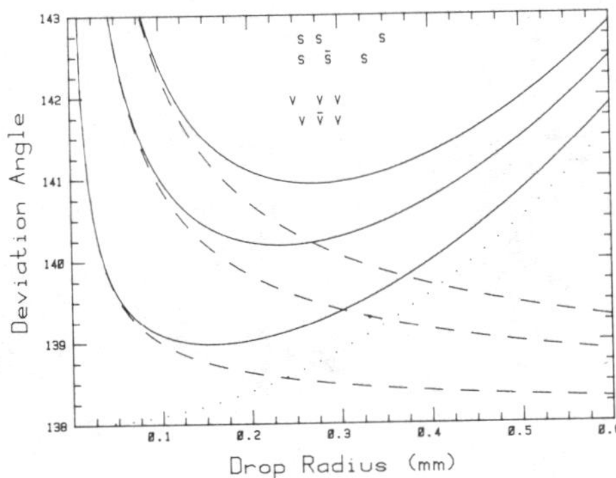

Fig. 6. The dashed lines are the three curves from Fig. 3. The dotted line is the 0-deg solar-elevation curve from Fig. 5. The solid curves show the position of the primary bow (lowest curve) and the first two supernumerary bows when both wave interference and drop ellipticity are taken into account. The V's mark the drop radii that would cause the measurements extracted by Vaucher; the S's are for those of Sardie. The overbar indicates the mean of the five measurements. Each datum point is plotted with an arbitrary ordinate (an angular separation was measured, not an angular deviation). Only the position along the abscissa is of interest.

my own pictures and one taken by my father, R. T. Fraser. One of mine is shown in Plate I. To eliminate my possible bias, the pictures were analyzed independently by two graduate students, Joseph Sardie and Christopher Vaucher. Neither of the students was told of the prediction nor of the purpose of the analysis. They both measured the angular spacing between the first two supernumerary bows on each picture. Sardie's data (0.73, 0.63, 0.73, 0.70, 0.60 deg) had a mean of 0.68 deg, while Vaucher's (0.73, 0.66, 0.74, 0.70, 0.66) had a mean of 0.70 deg. Considering the difficulty in specifying the center of a supernumerary bow, it is interesting that the two analyses agree as closely as they do. Both are within 10% of the predicted value (the prediction probably should not have been trusted to 10% anyway). The drop radii that would have caused these angular measurements were calculated using Airy's theory, and then they were plotted (with arbitrary ordinates) on Fig. 6. It is clear that they all lie near the position of the minimum.

CONCLUSIONS

In spite of the broad drop-size spectrum found in rain showers, supernumerary bows can still be seen. They are seen near the top of the bow and are caused only by drops with a radius of about 0.25 mm. For smaller sizes, no consistent pattern emerges owing to the displacement of the bows by the interference between two portions of a wave front. For larger sizes, no consistent pattern emerges owing to the displacement of the Descartes ray by the ellipticity of the drop. It can be noted that even though drops with radii between 0.1 and 0.3 mm are difficult to detect in a shower by conventional methods, they are nevertheless present if the top of the bow and its supernumeraries are seen.

ACKNOWLEDGMENTS

I am indebted to R. T. Fraser for providing some of the data and to J. Sardie and C. Vaucher for analyzing the data.

REFERENCES

1. T. Young, "Experiments and calculations relative to physical optics," Philos. Trans. R. Soc. London 1–16 (1804). This is the publication of the Bakerian Lecture of 1803.
2. G. B. Airy, "On the intensity of light in the neighbourhood of a caustic," Trans. Cambridge Philos. Soc. **VI**, 397–403 (1838); **VIII**, 595–600 (1849).
3. H. C. van de Hulst, *Light Scattering by Small Particles* (Wiley, New York, 1957), p. 245.
4. P. Lenard, "Ueber Regen," Meteorol. Zeit. **21**, 248–262 (1904).
5. J. E. McDonald. "The shape and aerodynamics of large raindrops," J. Meteorol. **11**, 478–494 (1954).
6. W. Moebius, "Zur Theorie des Regenbogens un ihrer experimentellen Prüfung," Ph.D. Dissertation, Leipzig, 1907; Abh. Wiss. Kgl. Ges. Sachsen Math. Phys. Kl. (1907); also see F. E. Volz, "Some aspects of the optics of the rainbow and the physics of rain," in *The Physics of Precipitation*, H. Weickmann, ed., Vol. 5 of American Geophysical Union Monograph Series (American Geophysical Union, Washington, D.C., 1960), pp. 280–286.
7. H. R. Pruppacher and J. D. Klett, *Microphysics of Clouds and Precipitation* (Reidel, The Netherlands, 1978), p. 315.
8. A. W. Green, "An approximation for the shapes of large raindrops," J. Appl. Meteorol. **14**, 1578–1583 (1975), Eq. (6).

Plate I. (Alistair B. Fraser, p. 1626). The top of a rainbow, which shows two supernumerary bows. © Alistair B. Fraser.

Section Seven
Ice Crystal Phenomena

A GENERAL THEORY OF HALOS.

By Charles S. Hastings.

[Yale University, May, 1920.]

SYNOPSIS.

The general theory of halos developed in this paper rests on the assumptions that two kinds of simple ice-crystals—elongated hexagonal rods and hexagonal plates—are occasionally present in a tolerably transparent atmosphere; moreover, that these crystals subsiding in quiescent air would necessarily fall into four groups.

The first portion of the paper establishes the validity of the assumptions by reference to well-recorded observations.

The second portion is devoted to a development of the consequences from the presence of each of these groups for various altitudes of the sun. It is there shown that all the authenticated features of complex halos are naturally explained (excepting certain rare multiple concentric circles) as inevitable consequences of the hypotheses. In addition, this portion gives a new means of classifying the various phenomena, showing unsuspected relationships as well as essential diversity in certain other cases where common origin was formerly assured.

I.

During the 72 years which have elapsed since Bravais published his celebrated and comprehensive work on halos many observations have been accumulated — some even by means of photography—and much has been written in the effort to improve questionable points in the theory presented by that admirable writer. As regards the efforts of the theorists it does not seem unfair to say that they have been quite futile; at least, no solution of a difficulty left by Bravais, as far as known to me, has ever commanded general acceptance. The elaborate mathematical discussions by Pernter of the tangent arcs to the 46° circle and of the, so called, arcs of Lowitz, perfectly illustrate the rather sweeping statement: Each of these is a logical conclusion from premises which no instructed meteorologist can possibly accept.

Before advancing any new views regarding the highly complex phenomena involved it will be well to summarize what was known when Bravais finished his work. The number of features which he considered and attempted to account for was about twenty. Of them we may ignore one or two as not being sufficiently authenticated, but we must add two which are of unquestionable authenticity; thus the total number remains nearly the same. Unfortunately, a small minority only of these were satisfactorily explained. We may catalogue these here and escape an undue lengthening of this paper by unnecessary repetition.

(1) The ordinary circle about the sun of 22° radius, attributed by Mariotte to the action of ice crystals suspended in the air and having faces inclined at 60°, the directions of their crystallographic axes being entirely fortuitous. This explanation of the commonest of all halos is thoroughly satisfactory and universally accepted.

(2) The 22°-parhelia, often called sun-dogs, are prismatic images of the sun right and left of it and at the same altitude. With a low sun they are at the angular distance named, but at a higher altitude the angular separation increases. They are not seen higher than 50°. At high latitudes they are more frequently noted than any other feature and the explanation—also first advanced by Mariotte—as due to hexagonal ice crystals with persistently vertical axes leaves nothing to be desired.

(3) The parhelic circle—a faint, colorless circle everywhere equally distant from the zenith and passing through the sun. This was attributed by Thomas Young to reflection from the faces of hexagonal prisms falling vertically. Bravais improved this theory by the remark that crystals with their principal axes persistently horizontal would also contribute to this feature. I shall show that probably only such reflection as is total, hence from the interior of the crystals, is generally effective.

(4) Upper and lower tangent arcs to the 22°-circle. These are due to the presence of crystals whose principal axes are horizontal, the lateral faces having any direction in space. As the sun rises to an altitude of about 45° these two arcs unite and form a ring inclosing the 22°-halo and touching it at its highest and lowest points. At very high sun this ring, called the circumscribed halo by Bravais, approaches more and more a true circle. This ring may exist alone. Admirable photographs taken at New Haven, Conn., and at Chester, Pa., of the halo of March 20, 1915, have been published in the Monthly Weather Review.[1] Bravais gave a very complete analysis of these features with tables which may be used to find the position of any desired point

---

[1] May, 1915, 43: 213–216 and October, 1915, 43: 498–499.

with any assigned altitude of the sun. In the calculations which are at the basis of the solutions given below I have saved myself unnecessary labor by recourse to these tables.

(5) Upper and lower tangent arcs to the 46°-halo. These are exactly tangent only in the cases of a solar altitude of 22.1° for the former and 67.9° for the latter, but they may be conspicuous when the sun departs several degrees from these stations. Their forms will appear in certain of the solutions given in this paper. The theory and analysis of this feature by Bravais appears to be faultless.

All theorists since the time of Bravais would doubtless include as a sixth feature of complex halos finally and completely accounted for by him the 46°-halo.

This feature was explained by Cavendish as similar in origin to that of the 22°-halo except that it is produced by prisms having faces at angles of 90° instead of 60°, such as would be formed by a base and any side of a simple hexagonal prism. These refracting edges are assumed to have purely fortuitous directions in space and, as in the case of the 60° prisms, only those which happen to be near the position of minimum deviation are effective. This explanation has been universally accepted and it certainly is recommended by its simplicity; it presents, however, difficulties which are not easily to be disposed of. If it were true, one would expect to see the 46°-circle almost as frequently as the 22°, at least when the latter is brilliant; but this is far from being the case. Again, this accepted explanation would make the appearance as likely with the sun high in the heavens as when the sun is low, which is also very far from true. There is no record of the 46°-circle having attained the zenith or even come very near it—in short, one might assert with considerable confidence that this feature does not appear with the sun more than 32° above the horizon. Again, were this explanation correct, the circle in its ordinary exhibitions ought to appear uniformly bright, as does the 22°-circle. Such a condition may indeed occur, but not according to my rather limited experience. In every case which I have observed the brightness has been far from uniform, but has—a very significant fact—been distributed in arcs symmetrically placed as regards the vertical through the sun. We shall see in Parry's famous record an exactly similar condition. In view of these facts I prefer an explanation offered below which, however, does not exclude the possibility that fortuitously directed crystals play a part in some cases unlike those which have come under my observation.

It will not be profitable to dilate upon the reasons why the explanations of other features of complex halos are held by me as untenable. The only reason of moment is, of course, the belief that the explanations offered here are better, and any reader can easily resolve any doubt as to that by recourse to easily accessible works; I shall, however, permit myself to call attention to those cases where an unforced explanation of a feature as derived in this paper stands without any alternative.

The method adopted by all investigators has been to establish the existence of a feature by reference to the records and then to invent a form of ice crystal which was though adequate to produce it. These inventions are very numerous and some of them of admirable ingenuity. There are two serious objections to such procedure. One is perfectly obvious, namely, the objection to assuming at any particular time the presence of rare, or even unrecognized crystal forms, in such over-

whelming numbers as to be effective. The other is a little less evident although even more formidable; the presence of any crystal not immediately effective in the production of a given feature obscures it. The conditions of development of rainbows and of halos are, in a sense, antithetic. In the former the more opaque the background, provided that the opacity is due to raindrops alone, the more brilliant the phenomena; in the latter a considerable transparency of the sky is a primary requisite. Thus economy in the number of types of crystals assumed in any theory is of the highest importance.

In the theory presented here only two form of crystals are postulated both of which are the simplest types of ice crystals and both of which are familiar to observers. The first is that of a hexagonal prism with a short principal axis—in short, a hexagonal plate of which the thickness is a small fraction of its diameter. Such crystals I shall style the A type. The other form is that of a hexagonal prism of which the length is much greater than the diameter; it might be described as a hexagonal rod. This second form I shall designate as the B type. Both types have dihedral edges of 90° and of 120°, the former being the angle at which the basal and side faces meet and the latter that of the lateral faces; but optically (as regards transmitted light) they yield only prisms of 90° and of 60°, the prism of smaller angle being truncated by the plane which forms the intermediate face of the hexagonal crystal.

Such small bodies falling through a quiescent resisting medium would have a decided tendency to assume positions such as to offer a maximum resistance to the relative motion, hence the A crystals would tend to remain horizontal—that is, with their short principal axes vertical—while the B crystals would maintain their principal axes horizontal. This assumption is exactly opposite to that of Bravais and his followers who supposed that such bodies would set themselves so as to meet with the minimum resistance. These assumptions constitute the fundamental difference between the old theories and the present one and should therefore be carefully noted by the critical reader. It is difficult to understand how the earlier view could be taken, notwithstanding the lack of a knowledge of the mechanical principles involved which we now possess, for every one knew of the difficulties met in keeping an elongated projectile in end-on flight.

This tendency to stability of direction of the principal crystalline axis of the two types is far from being a strong one; its effect is as remote as possible from producing a pendulum-like oscillation such as Pernter has assumed as the basis of certain of his explanations. In the latter kind of motion the moment of restitution increases proportionally with the displacement from the position of equilibrium, while in this case of fluid constraint the moment of restitution decreases with departure from equilibrium and wholly vanishes at some indeterminate angle. Thus, although we may have recurrent motion, simulating harmonic motion, in cases restricted to very small amplitudes, the usual result would be a continuous rotation about some major axis. Examples of such motion are familiar to every observer, e. g., as shown by the fall of petals of fruit blossoms, by that of small bits of paper.

With this enlarged view as to the phenomena presented by small, regularly formed bodies in falling through quiescent air, we find that we have, with two types of ice crystals only, four different groups. These may be designated and defined as follows:

A group: those hexagonal plates which fall with their principal crystaline axis continuously vertical.

A′ group: similar plates which, in falling, rotate continuously about a major diagonal.

B group: those elongated hexagonal crystals which fall not only with their principal axes continuously horizontal but also their maximum cross-section horizontal.

B′ group: crystals like the last but rotating continuously about their principal crystaline axis. Some, or even many, of these may be assumed to have a motion about their centers of mass so that these axes describe cones in space, always, however, of small angular opening.

With these simple postulates the problem of halos divides into two parts, first, to demonstrate from the records the occasional existence of these four groups of crystals in our atmosphere; and, second, to deduce all of the optical consequences from the presence of such crystals and compare these consequences with recorded observations.

There exists one record (only one, unfortunately) which admirably meets our requirements for the first step; it is the halo figured and described in Parry's First Voyage,[2] p. 164–165. This record is peculiar in several respects; first, it was observed in common by two skilled observers, Parry and Sabine, and had a complexity involving three of the groups of crystals named above; and, second, because of the remarkable duration of slowly changing phases due to the high latitude, which was 74° north.

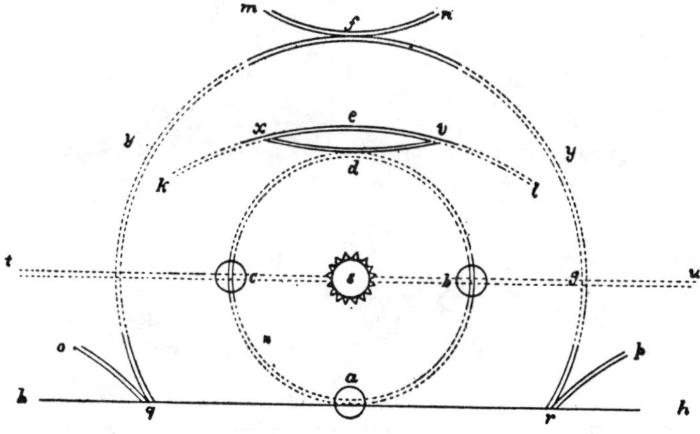

FIG. 1.—The halo of Parry and Sabine.

The diagram and description as given by Parry and Sabine are as follows:

From half-past six till eight A. M., on the 9th, a halo, with parhelia, was observed about the sun, similar in every respect to those described on the 5th. At one P. M. these phenomena re-appeared, together with several others of the same nature, which, with Captain Sabine's assistance, I have endeavoured to delineate in the annexed figure.

s, the sun, its altitude being about 23°, h, h, the horizon.

t, u, a complete horizontal circle of white light passing through the sun.

a, a very bright and dazzling parhelion, not prismatic.

b, c, prismatic parhelia at the intersection of a circle a, b, d, c, whose radius was 22½° with the horizontal circle t, u.

x, d, v, an arch of an inverted circle, having its centre apparently about the zenith. This arch was very strongly tinted with the prismatic colours.

k, e, l, an arch apparently elliptical rather than circular, e being distant from the sun 26°; the part included between x and v was prismatic, the rest white. The space included between the two prismatic arches, x e v d was made extremely brilliant by the reflection of the sun's rays, from innumerable minute spiculae of snow floating in the atmosphere.

q f r, a circle having a radius from the sun, of 45°, strongly prismatic about the points f q r, and faintly so all round.

m n, a small arch of an inverted circle, strongly prismatic, and having its centre apparently in the zenith.

r p, q o, arches of large circles, very strongly prismatic, which could only be traced to p and o; but on that part of the horizontal circle t u, which was directly opposite to the sun, there appeared a confused white light, which had occasionally the appearance of being caused by the intersection of large arches coinciding with a prolongation of r p, and q o.

The above phenomenon continued during the greater part of the afternoon; but at six P. M., the distance between d and e increased considerably, and what before appeared an arch, x, d, v, now assumed the appearance given in fig. 12, plate 287, of Brewster's Encyclopaedia, resembling horns, and so described in the article "Halo," of that work. At 90° from the sun, on each side of it, and at an altitude of 30° to 50°, there now appeared also a very faint arch of white light, which sometimes seemed to form a part of the circles q o, r p; and sometimes we thought they turned the opposite way. In the outer large circle, we now observed two opposite and corresponding spots y, y, more strongly prismatic than the rest, and the inverted arch m, f, n, was now much longer than before, and resembled a beautiful rainbow.

This sketch is not drawn according to any geometric projection and does not admit of quantitative analysis; but if we accept the position angles as referred to the sun and the angular distances as proportional to the linear distances, we shall be able to redraw it according to any system of projection preferred, since the scale is given by the known angular dimensions of the inner circle. The projection which I shall choose for all of the diagrams in this paper is that known as the spherical projection. Moreover, with a single exception, I shall choose the plane of the paper as that of the horizon, the center of the circle representing the visible horizon being the projection of the zenith. The advantages of this particular system are many; every circle in the heavens is represented by a circle on the plane (which becomes a straight line when a great circle passing through the zenith) and the angle of all intersections is preserved unaltered. Moreover, the coordinates of every point as represented by azimuth and zenith distances are readily found, the first by direct reading from horizon circle and the second by means of a simple trigonometric formula or by a scale constructed for that purpose. The only serious fault is that of a considerable distortion in the neighborhood of the horizon.

Figure 2 represents the observations of Parry and Sabine thus reduced to a spherical projection. I shall proceed to construct a halo, according to the same laws of projection, which would result as a consequence of the fundamental assumptions above. A comparison of the two may be expected to validate these assumptions or the contrary. To do this it is necessary to define certain constants and symbols.

The acepted mean index of refraction of ice is 1.31, which is the value adopted by Bravais and his followers. It is that of yellow-green light, the most brilliant portion of the spectrum. This constant yields 49.76° for the critical angle of interior reflection, 45.74° for minimum deviation for 90° prisms and 21.84° for that of 60° prisms.

In calculating the optical effect of a prism, the prism will be regarded as placed at the center of the celestial hemisphere of which the area within the circle of the horizon is the projection; the aspect of the prism with respect to the celestial sphere will be defined by the positions of the poles of its eight faces. A convenient

---

[2] John Murray, London, 1821.

notation which will be adopted is the following: o and o mark the poles of one base and its opposite; p, p', p'' give the places of the poles of three successive lateral

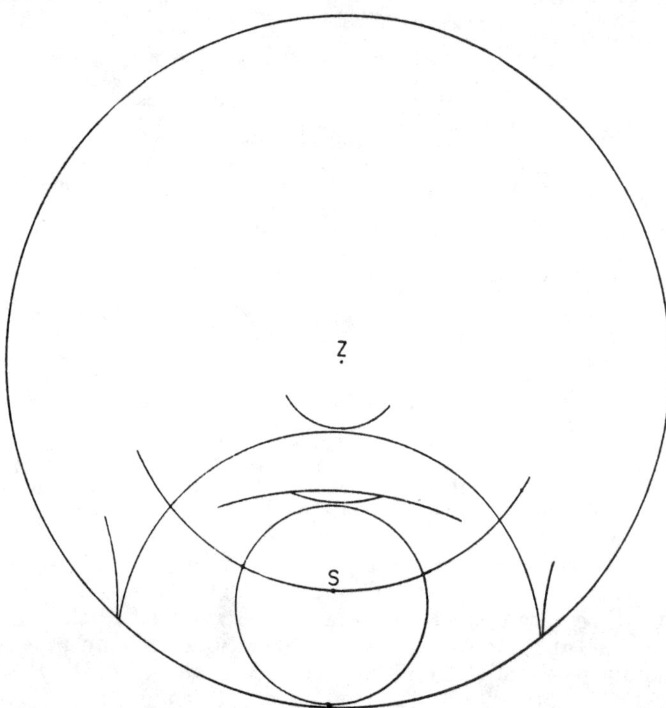

FIG. 2.—Halo of Parry and Sabine drawn in spherical projection with zenith at pole of plane of projection.

faces and p, p', p'' those of their opposite faces, respectively. The spherical coordinates of points on the sphere are the zenith distance, symbol z, and the dif-

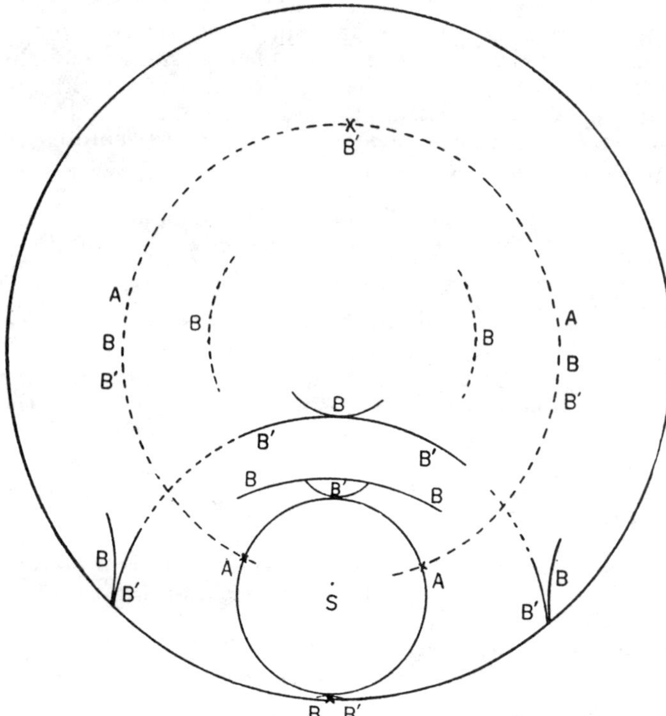

FIG. 3.—Halo of Parry and Sabine according to theory here presented. The letters indicate particular group of crystals involved.

ference in azimuth between that of the point to be defined and that of the sun. The latter angle will be called the amplitude and designated by the symbol u.

The accompanying figure 3 shows the results of such calculations; it includes two features which are outside the limits of the drawing by Parry and Sabine but are described in the text. These are the faint luminous spot opposite the sun in the parhelic circle called the anthelion, and the pair of faint arcs between the zenith and the parhelic circle having a mean amplitude of approximately 90°. The origin of each of the details is indicated by the lettering which shows the group or groups of crystals concerned in its production. The close resemblance between figures 2 and 3, the former being merely a record of eye observations and the latter a mathematical deduction from the fundamental assumptions for a sun altitude of 23°, is a quite sufficient proof that of the groups of crystals assumed all but the A' group are occasionally effective in causing halos. The proof as regards this last group will appear later.

A somewhat complete analysis of this solution embodied in figure 3 will save many words in descriptions of corresponding features characteristic of higher altitudes.

The effects of the A group are rather insignificant in this particular manifestation for, besides the familiar 22°-parhelia, they only add a small portion of the total light in the parhelic circle and to the inverted arc above the 46°-circle.

The B group, particularly prominent in this halo, although far from rare in other cases, is the cause of the arc whose highest point is about 39.7° from the zenith, equivalent to 5.5° above the vertex of the 22°-circle; of the brilliant spot of light at the horizon which is merely the complement of the preceding feature; of all of the light in the oblique arcs springing from the horizon at the lower ends of the interrupted 46°-circle; of most of the light in the upper tangent arc to the 46°-circle; of the faint arcs between the zenith and parhelic circle which we shall later find cause for denominating the higher oblique arcs through the anthelion; and, finally, to much of the light which comes from the parhelic circle.

The B' group produces the upper and lower tangent arcs to the 22°-circle, the latter being mostly below the horizon; of the three brilliant portions of the 46°-circle which are tangent to three arcs mentioned above as due to B crystals and, possibly, to the totality of the 46°-circle; of much of the parhelic circle; and, finally, of the faint spot of light opposite the sun in the parhelic circle. These various details of the halo in question will be taken up in the order named, ignoring, however, the 22°-circle, concerning which everyone is agreed.

That the A group is relatively inconspicuous in the Parry halo is proved by the moderate intensity and extension of the 22°-parhelia, very different in these respects from the famous halo of Hevelius. In the latter, which will be discussed later, the presence of the A' group, here wanting, will be demonstrated.

The crystals of the B group have the upper and lower lateral faces persistently horizontal; in other words, they subside in the atmosphere with a maximum cross-section constantly horizontal. The two basal planes are constantly vertical. Light from the sun which falls upon the upper face, p, will emerge after refraction through the surface p', provided that the amplitude of the pole o of the base is not too far removed from ±90°. A sufficient number of places of images of the sun produced thus with different values of the amplitude of o were calculated so that the long arc depicted between the 22° and the 46° circles could be accurately constructed. Since none of my predecessors has considered this highly important feature I venture to call it Parry's upper arc, and hereafter I shall refer to it under that name.

Light which enters the same B crystals at a p'' face and emerges also at a p' face forms an arc below the sun convex upward and mostly below the horizon. Its vertex is 23° from the sun and very brilliant. At first thought it appears contradictory that the observers noted this as white, but a recognition of the facts that the less refrangible portion of its spectrum is combined with a more refrangible portion of the ordinary lower tangent arc of the 22°-circle, and that the distinctive colors of short wave lengths are invisible on account of falling below the horizon, readily disposes of the contradiction. This arc, which will recur in other cases with a higher sun, will be called Parry's lower arc.

In this particular halo the arc tangent at the vertex of the 46°-circle is chiefly due to the B group, although in many cases only the A group is concerned, as in the Hevelian halo, which follows; indeed, Bravais emphasizes the relation of this arc to the 22°-parhelia inasmuch that they occasionally exist together as the whole of the manifestation. Of course that writer, recognizing neither the A group nor the B group as defined here, but only the elongated prisms assumed to fall with vertically directed axes as providing horizontal rectangular edges, was obliged to regard them as always associated, whereas in this particular halo the association is only partial. However, given such persistently horizontal refracting edges of 90°, the theory of Bravais is complete and the topic might be left without further discussion were it not for a significant remark in the description which is worth consideration. The observers remark extraordinary purity of the colors in this arc when the sun was much lower, although during the earlier period the incidence of the light was almost exactly that corresponding to minimum deviation and maximum brightness. The reason is not far to seek. The visible spectrum produced by a right angle prism of ice is a short one—we may rate it at about 1.5° in neglecting the fainter terminal colors; but the diameter of the sun is much too considerable a portion of this angle to yield a spectrum approaching purity of colors. This effect due to angular magnitude of the source diminished with increasing angle of incidence and the spectrum becomes an absolutely pure one at the limit of 90° incidence. As this effect is reversed when the angle of incidence is less than that proper to minimum deviation it is likely that the arc has been more frequently recorded with excessive incident angles than when the emergence angles were equally in excess.

The curious arcs springing from the horizon and tangent to the 46°-circle come from light which, incident on a base of a B crystal, emerges from a lateral face. It is easy to see that the conditions necessary for their production are very unusual, and they would also be very evanescent, except in polar regions; they have not been considered, as far as known to me, by any previous writer except Bravais, who classed them with certain tangent arcs due to B' group. In this he was certainly in error, as will appear when I discuss the features attributable to the latter group. They may be called Parry's lateral tangent arcs to the 46°-circle.

A portion of the light which enters the upper face of a B crystal would fall on a basal plane, and after reflection from that plane would emerge from the p' surface. In those cases in which the amplitude of the crystal is such that this interior reflection is total this light is significant. Such is the origin of the arcs near the zenith and drawn as broken lines in figure 3. With a higher sun this feature is sometimes conspicuous and it will be convenient to defer a theoretical consideration until such cases come

under review. They may be styled Upper oblique arcs passing through the anthelion.

As a final feature, due in part only to the B group, should be named the parhelic circle. All light reflected from the bases of both groups B and B' would appear to come from the parhelic circle as also that from the sides of the A group; but especially important would be that portion which has been totally reflected from the interior. Such total reflection ceases at amplitudes not far from 130°. Beyond these limits the circle would be fainter, a character which is not noted in the description unless the extreme faintness of the anthelion clearly implies it.

The effects of the B' group in the immediate vicinity of the 22°-circle are so perfectly understood from the discussion of Bravais that it is not necessary to consider them further here, especially as all the solutions of these details in this paper are deduced from the tables given by him.

The contributions of the B' group to the features near the 46°-circle can not be so easily dismissed. They consist of a number of arcs, more or less perfectly tangent to the 46°-circle, sometimes concave toward the sun, but occasionally having the opposite curvature. The conditions of their appearance are easily defined. Suppose a crystal of this group at the center of the celestial sphere; change the amplitude of its principal axis, at the same time rotating it about this axis, until a principal plane of a rectangular edge passes through the sun, then, if the angle of incidence of the sunlight is that, or nearly that, corresponding to minimum deviation there will result an image of the sun at a point in the 46° circle or just outside of it. In this case all crystals having nearly the orientation defined would contribute to the formation of an arc passing through this image no point of which could be nearer the sun, hence the arc would appear to be tangent to the 46°-circle.

There is another highly instructive method of attaining the result as applied to the Parry and Sabine halo. Having shown that crystals of the B group, which have a single degree of freedom as regards orientation, produce three tangent arcs to the 46°-circle and knowing that the crystals which are supposed to produce the 46°-circle have complete freedom in this respect, that is, three degrees, it follows at once that the B' group, possessing two degrees of freedom, must form arcs more closely adjusted to the circle than the former arcs. The observations are in complete accord.[3]

It will be noted that these three arcs, although departing but little from the 46°-circle, are limited in extent and leave portions of this circle vacant. If, however, we attribute to some of this group a moderate rocking motion about the center of mass of the character described above, this vacancy would disappear and the circle would appear unbroken although not uniformly bright. Calculation shows that crystals departing 14° from the horizontal would perfectly replace the hypothetical random crystals to which the accepted theory attributes the 46°-circle, while half this angular deviation would be quite sufficient to give rise to a ring which could not be distinguished from a circle except by careful measurement. These considerations lead me to prefer this explanation of the 46°-circle to the current one which has long been accepted. It is adequate—even to ex-

---

[3] The previous theories of these arcs are somewhat tangled. Bravais gives his theory in three lines, a theory which is not clear to me. Pernter rejects this theory and replaces it by another which so acute a critic as Besson finds untenable: nor does the latter theory appear to me to accord with the mechanical laws to which falling crystals are subject or the records.

plaining the unique observation of Besson, who found a visible separation between the circle and the closely-agreeing tangent arc—and also answers the puzzling question as to why the circle has never been seen complete; that is, wholly above the horizon.

There remain the short arcs through the anthelion and the anthelion itself. These I attribute to the action of the B′ group and explain as follows: Imagine a crystal of this kind at the center of the celestial sphere with its p face vertical, the p'' and p′ being respectively above and below it, and the sun near the horizon. Light from the sun entering this face near the end of the prism would, after successive reflection from the vertical base and opposite side in either order, emerge at the surface of entry as coming from a point in the parhelic circle exactly opposite the sun, in short, from an anthelion. This would be true for all angles of incidence, but in those cases where both interior reflections are partial the returning light would be entirely insignificant. On the other hand, when the reflection from the basal surface is total the quantity of light returned would be vastly greater. If, however, the angle of incidence is small the dimensions of the reflected beam of light would be small on account of the foreshortening of the totally-reflecting surface; as this angle increases, the quantity of light would continuously increase until it reached its maximum at the critical angle of interior incidence. With a higher sun, approximating to 30° for example, light entering at the p'' face and emerging, after having experienced a similar double reflection, at the p′ face would also appear to come from the anthelion. But the assumed position of the crystal is not a stationary one according to the mechanical principles governing the falling of light bodies through a resisting medium, hence the effects produced would be less simple than this. In fact, the rotating B′ crystals would yield two arcs passing through the anthelion; the outer edges would be tolerably well defined and much the brightest portions, so that they would appear as two short arcs crossing at the anthelion under a determinate angle depending upon the altitude of the sun. With the sun at the horizon calculation shows that this angle would be about 20°, the angle increasing rapidly with increasing altitude. With the altitude of the sun at 30° the short arcs due to light entering and emerging at different faces the angle of inclination may be rated at 70°. With a certain range of intermediate altitudes both pairs of arcs may coexist, although such occurrences must be infrequent.

A review of the last paragraph shows that the present theory does not allow for a true anthelion, that is, there is no stationary image produced by a host of crystals of widely varying orientations; on the other hand, there is often a brighter portion of the parhelic circle exactly opposite the sun which is the locus of the intersection of three or of five arcs, as the case may be, and which still may bear the name. This node is accentuated by the fact that the short arcs are brightest just at their middle points, which fall on the parhelic circle. Necessary deductions are that these phenomena are only associated with a low sun. The explanation is advanced with some confidence not only because it seems to fit admirably with the records but also because there is no alternative other than one which attributes the arcs to diffraction from highly fantastic crystals which have never been observed and which are supposed to fall edgewise.

The famous halo of Hevelius, observed and recorded by that astronomer in 1661, presents a new set of phenomena. Unfortunately the record is very imperfect, since the

drawing and the description are very discordant just where we demand precision. But in the admirable collection of recent observations described by Dr. Louis Besson [4] there is one, a drawing by Orin Parker, of a halo seen by him at Bentonville, Ark., November 1, 1913, which is almost a replica of that recorded by Hevelius except that the paranthelia and the oblique arcs through them are placed in their proper positions, namely, at amplitudes of 120°, plus and minus, respectively. At any rate, the following explanation will rest upon the assumption of the essential identity of the two.

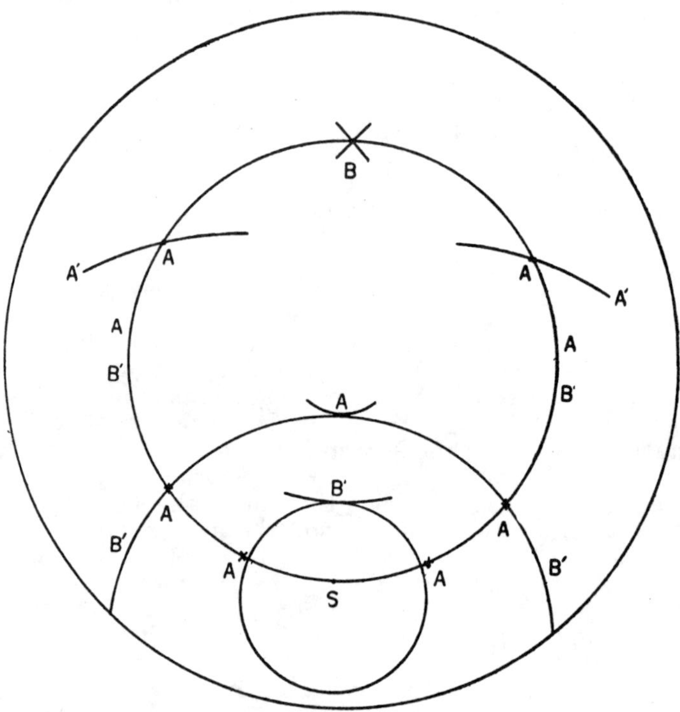

FIG. 4.—Halo as deduced from present theory, similar to that of Hevelius. The B group supposed to be absent. Zenith distance of sun 65°.

Both the halos are characterized by the brilliancy of the effects due to the A type of crystals, the other type being represented only by the B′ group. The evidence in favor of this statement lies in the intense brightness of the 22° parhelia and the absence of all traces of the Parry arcs and of their attendant consequences. As the B′ group produces nothing not already considered in the Parry halo, there is no reason for discussing their effects—the geometric drawing of figure 4 will enable one not only to compare the solution with Parker's drawing, but to find coordinates for all points desired with sufficient precision.

Let an A crystal be placed at the center of the celestial sphere and consider the course of sunlight falling upon it. Most of the light entering the upper surface will emerge from the under surface in an unchanged direction, but a portion will fall upon a vertical face and be reflected—totally reflected if the zenith distance of the sun is not too great—thence, emerging from the lower base, the light would come from some point in the parhelic circle. In many cases, however, a portion of the light reflected from the vertical face would, before being transmitted through the lower base, fall on an adjacent face; such light would also appear to come from the parhelic circle, but from one of two points only, each at 120° from the sun. This assertion does not require proof here because it is contained in the familiar theory of the kaleidoscope, but it is equiva-

<hr>

[4] Monthly Weather Review July, 1914, 42: 436–446.

lent to a statement that these common and puzzling features are an immediate consequence of our theory. Nor is it merely a few of the crystals which contribute light to the paranthelia of 120°, for theoretically just half of them are thus involved, although those of importance at any one instant are as maller portion. It is clear that these paranthelia can not appear when the sun is very near the horizon or the zenith, whence we may fairly conclude that they appear more frequently at mid altitudes, from 25° to 50°, for example, a conclusion wholly accordant with the records. The only other theoretical explanations of the paranthelia known to me involve columnar crystals with their principal axes continuously vertical, in one case demanding stellate cross sections, and in the other two interior partial reflections. They appear quite untenable.

If a sufficient number of the crystals belong to the A' group, there results a curious addition to the 120°-paranthelia, namely, a short oblique arc passing through each of them. These are due to that portion of the A' group which rotates about that axis, which is equally inclined to those faces which, when the hexagonal plate is horizontal, would contribute to the paranthelion itself. I have calculated the form of these arcs for a sun altitude of 30° and represented them in the figure 4. The sun appears as sensibly at the center of these arcs, since the 120°-paranthelia being at 97.2° from the sun, the arcs from 21.0° below the parhelic circle to 12° above are more distant by the inappreciable amount of two and a half degrees.

Without doubt these arcs constitute the famous 90° halo of Hevelius which has baffled all explanation. The obvious objections to this declaration are, first, that they are fragmentary arcs—but that is true of the drawing by Hevelius; second, that the recorder described them as 90° from the sun, but his sketch places them at 78°, which is a notably greater divergence.[5]

Figure 4 shows the general halo for the zenith distance of 60° for the sun in the absence of the B group, but the effective presence of the other three; it may be compared with the figures of Hevelius and of Parker.

II.

The first section of this paper may be accepted as demonstrating the occasional existence in the atmosphere of some or all of four groups of crystals, these groups being necessary consequences of the two types of familiar crystal forms which are known to exist. The present section will concern itself with an investigation of the extension of the theory to halos accompanying the sun at higher altitudes. This does not involve a great deal of description, since the illustrative figures are all geometrical, so the amplitude of any point on the diagram can be determined directly, by means of a protractor and the zenith distance, from the formula

$$Z = R \; tg \; 1/2 \; z$$

where R is the radius of the circle representing the horizon, Z the linear distance from the center of the circle to the point in question, and z the zenith distance. It will be noted that, as in the preceding projections, the lines show the loci of that particular color corresponding to a refraction index of 1.31 and neglect the angular dimensions of the sun. The modifications necessary to involve other colors and the dimensions of the sun are easily sup-

<hr>

[5] One might add to these objections the fact that Bravais has cited three more recent observations of a 90° halo, but I have, by referring to the original sources, persuaded myself that the citations were founded upon misapprehensions. Since one of the references was to the account of Perry and Sabine, printed in full above, the reader may judge for himself in respect to that one.

plied; moreover, I shall give later a list of the uncolored features of halos, the others being prismatic.

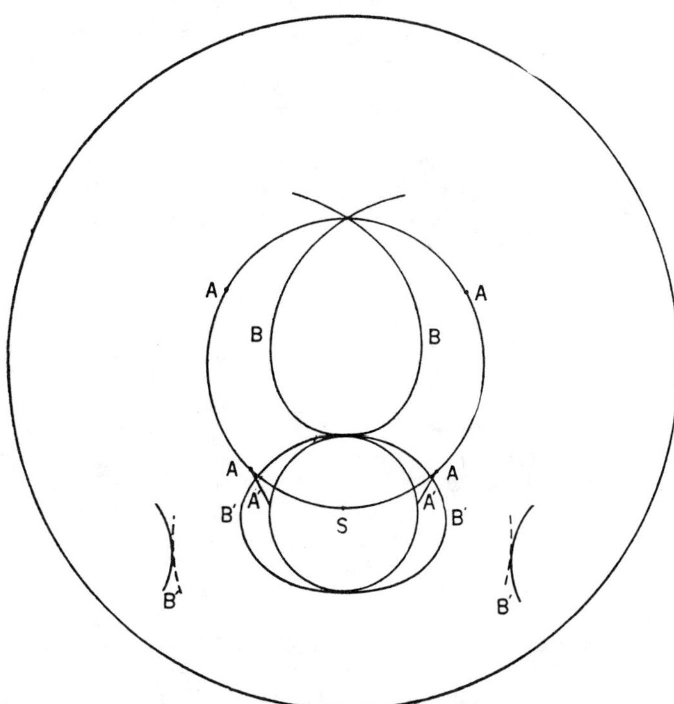

FIG. 5.—A highly developed halo with sun at a zenith distance of 45°, closely resembling the St. Petersburg halo recorded by Lowitz.

Figure 5 shows a possible halo with the sun at an altitude of 45° when all four groups are effective; the 22°-circle is added for comparison, although not necessarily present. The letters attached to each detail indicate the origin with sufficient definiteness with the exception of the pair of long arcs crossing at the anthelion and the pair connecting the ordinary parhelia with the 22°-circle. The former pair I shall style the Upper Oblique Arcs passing through the anthelion, in order to distinguish them from another pair of similar origin which occasionally attend the sun at very high altitudes; the second pair has already been named the arcs of Lowitz from the observer and recorder of the famous St. Petersburg halo of 1790. The long arcs have heretofore been confused with the short arcs confined to lower altitudes of the sun and produced by a different group of crystals, while the arcs of Lowitz have given rise to much theoretical discussion. Let us consider them in the order named.

The upper arcs passing through the anthelion are simply the development of those faint arcs which we found in the halo of Parry and Sabine between the zenith and the parhelic circle. They are produced by light which, falling near the ends of the B crystals and undergoing total reflection, emerges through the same surface as that effective in the upper Parry's arc. The ends toward the sun are at the middle of the latter arc, although it is obvious that it would be impossible to trace them very near that point; they would ordinarily be confounded with the Parry arc, as is so well exemplified in figure 8 in Besson's paper cited above. The theoretical limits in the opposite direction are set by the approach to the critical angle of incidence on the emergent surface, although it is easy to see that the visible limit must be reached before that.

The arcs of Lowitz are of special theoretical interest on account of their extreme rarity with unquestionable authenticity and the fact that theorists have given them so much attention. According to the theory here presented, they are caused by a certain portion of the A' group. Such crystals rotate about one of three major diagonals; for suppose a host of such crystals at the center of the celestial sphere and we confine our attention to those of them which when horizontal contribute to the light of a 22°-parhelion. Dominate the face of entry by p and that of exit by p'', then one-third of these crystals will rotate about an axis which causes p and p'' to alter their direction at an equal rate; a second third causes displacements of p and p'' in opposite directions, the latter being at the greater rate, and, finally, the remainder cause displacements of unequal rate but that of p being greater. These may be called the first, second, and third modes. The first mode does not produce a visible effect at this altitude of the sun, but, as shall be proved later, it is the occasional cause of a curious feature with the sun at the horizon. So, too, the third mode is ineffective, but the second mode gives rise to the short arcs connecting the parhelia with the 22°-circle and which are in closest agreement with the record of Lowitz. The fact that the first mode alone produces tangent arcs to the 22°-circle when the sun is at the horizon, while only the second is effective at the altitude of 45° is very suggestive of the reason why these arcs are so rare and are not seen at intermediate altitudes.

At first thought it appears improbable that so small a number of effective crystals could produce a visible effect, but calculation shows that an enormous change in the angle of rotation about the axis shifts the image of the sun along the arc by a very small amount. Thus the condition is an approximation to the "stationary" state of a parhelion and the brightness is correspondingly enhanced.

Aside from these points, the diagram may be regarded as explaining itself except, perhaps, the apparent fragments of the 46°-circle—not always present, it is true— which are attributable to that fraction of the B' crystals which have a restricted oscillation about their centers of mass.

With a zenith distance much less than that of figure 5 a new feature appears which has been rarely seen in the latitudes of northern Europe but less uncommonly in the United States and which has not yet been discussed. This feature consists of a pair of arcs lying chiefly outside of the parhelic circle but crossing at the anthelion point and from there curving toward the zenith. These arcs have been observed by Lea, Melville, and probably by Meriwether, all of whom are cited in the work by Bravais; but by far the best record known to me is that by H. W. Crawley, described and figured in the Report Brit. Assn. 1861 (2), p. 63. One suspects that only the B and B' groups were present in this interesting phenomenon and that the circle drawn about the sun was in reality the circumscribing oval, otherwise the darkness within it could not have so impressed the observer. This observation seems to have escaped the notice of meterologists. This feature under discussion is due to the B group of crystals and are therefore complimentary to the oblique arcs exhibited in figure 5. The accompanying figure 6 shows the form which these arcs have when the zenith distance of the sun is 30°. They may be styled the Lower Oblique Arcs through the anthelion since they are produced chiefly by light which emerges from the lower horizontal surface after refraction at a p' surface and total reflection at a basal surface. The figure shows only these curves due to the B crystals and the ordinary 22°-circle to afford a ready scale of dimen-

sions. It will be recalled that the upper oblique arcs, as shown in figure 5, are due to light which enters the upper horizontal face. In this particular case the inner extension from the anthelion are also due to light entering this face and are therefore properly a portion of the upper oblique arcs. It is easy to state in general terms what the added features would be if the other groups were present. The A group would yield its share of the light to the parhelic circle and possibly exhibit paranthelia, although these would be unquestionably very faint; if the arc below the 46°-circle were apparent, as it often is with the sun at a somewhat greater altitude, this too would be in part due to the A group. The B' group, if present, would also contribute to the parhelic circle and form a circumscribed oval about the 22°-circle lying very close to it in all its parts.

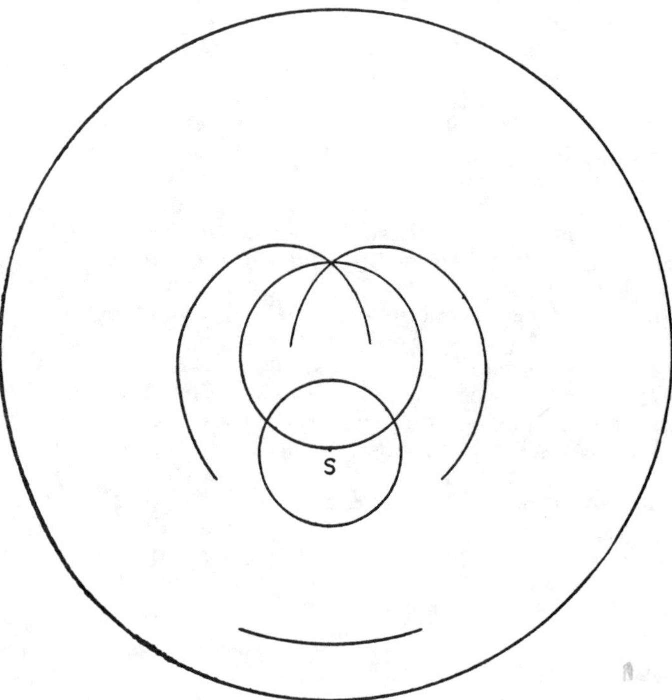

Fig. 6.—A halo due to B group of crystals for 30° zenith distance of sun. The 22° circle is added for scale.

A feature which has a single record is worthy of consideration here, not only because the record is photographic and therefore unimpeachable but also because it verifies the occasional existence of the A' group in which we have found the explanation of the rare phenomenon of the Lowitz arcs and the Hevelius 90°-arcs.

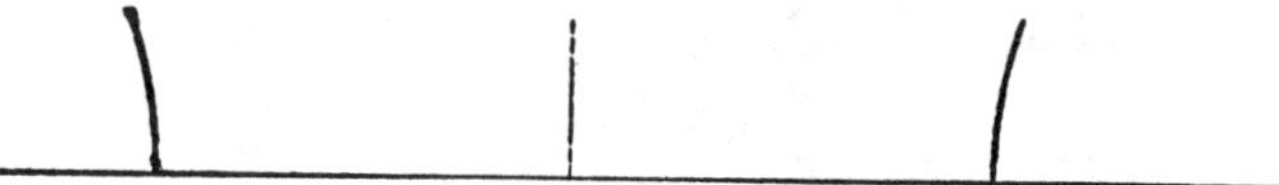

Fig. 8.—Halo of Schultz according to theory. The projection is that of a rectilinear camera so that it may be compared directly with the photograph.

The photograph is found in Archbishop Stuck's "Ten Thousand Miles with a Dog Sled," p. 388,[6] and is here reproduced in figure 7. Figure 8 gives the forms of the curves as calculated on the hypothesis that they are due to A' crystals rotating on that diameter of the hexagonal plates which is symmetrical to the incident and emergent faces of the crystal. The projection is that of a rectilinear camera, but the scale is altered to

<hr>

[6] Charles Scribner's Sons, New York, 1914.

agree with the photograph and its center is a point on
the true horizon supposed to be the place of the sun.
The slight eccentricity of the sun in the photograph is
due to a fault of direction in the camera. The fainter
vertical column of light directly above the sun is a
secondary phenomenon due to these same crystals;
in other words, it is the sum of images of the two tangent
arcs formed by A′ crystals and, of course, colorless.
This is the only certainly established secondary phenomenon excepting the parhelia at approximately 90°
right and left of the sun, which are represented in figure
5 above and which have frequently been recorded.

The question as to whether a feature is prismatic or
without color is easy to answer from the mode of its
production. In general, if produced by refraction we
may expect attendant color unless the refraction at
entry and emergence from the crystal is compensatory,
as in the parhelic circle, the paranthelia and the anthelia.
Of course, all very faint arcs would fail to betray colors
for quite the same reason that the lunar rainbow does so.

In the above theory explanations for all the well
authenticated features of halos are given with the single
exception of the rarely observed concentric circles about
the sun of radii differing from 22°. These may be due
to the presence of crystals of which the rhombohedral
faces are developed—and indeed, Bravais has attempted
to explain them in this way—but such crystals, although
next in crystalographic simplicity to the hexagonal
prisms assume as the bases of the theory here developed,
have never been observed and therefore lie outside the
proper scope of this paper.[7]

Photo by Paul Schultz.

THE SUN DOGS.

FIG. 7.—A reproduction of the remarkable photograph by Paul Schultz in Archdeacon Stuck's book "Ten Thousand Miles with a Dog Sled," page 388, Scribner's, 1914.

Circumscribed Halos

Circumscribed halos which appear around the sun are simulated by a computer treatment of a simple model.

Robert G. Greenler and A. James Mallmann

Ice crystals in the atmosphere are responsible for a variety of optical phenomena in the sky. One of the most familiar of these phenomena is a circle of light around the sun, with an angular radius of 22°, as shown in Fig. 1. This halo results from sunlight deviated by about 22° by refraction through randomly oriented hexagonal ice crystals. Occasionally this 22° halo is circumscribed by another halo which is tangential to the 22° halo at the top and bottom. Unlike the circular 22° halo, the shape of the circumscribed halo varies markedly with the altitude of the sun (as can be seen in Fig. 7). In this article we analyze the source and form of this circumscribed halo.

Dr. Greenler is professor of physics at the University of Wisconsin–Milwaukee, Milwaukee 53201. Until September 1972, he is on leave at the School of Chemical Sciences, University of East Anglia, University Plain, Norwich NOR 88C, England. Mr. Mallmann is an assistant professor of physics at the Milwaukee School of Engineering, Milwaukee, Wisconsin 53201.

The Model

Ice crystals in the air commonly exist in the form of hexagonal prisms. If the length of the crystal along the axis is short as compared with its width, the crystal will have the form of a flat hexagonal plate. If the axial dimension is long, the crystal will have the shape of a hexagonal column which we refer to as a pencil crystal, after its resemblance to the shape of the common wooden pencil. These crystals, falling in still air, tend to assume the orientation which provides maximum air resistance. This means that flat plate crystals will tend to fall with their axes vertical and the pencil crystals with their axes horizontal. The details of the orientations of a group of falling ice crystals will determine the pattern of light in the sky which results from sunlight refracted by these crystals. Our approach has been to calculate the path of light rays through a distribution of ice crystals in order to find that distribution which produces the observed forms of the circumscribed halo.

The Calculation

Instead of dealing with a multitude of crystals at different locations in the sky, we have used one crystal, at one location, and have assigned to it all the desired orientations. With the invaluable aid of the computer, we have, for each crystal orientation, calculated the direction of the light ray leaving the crystal. Figure 2 shows that light passing through alternate faces of a pencil crystal is refracted as if it had passed through a 60° prism. If a light ray from the sun reaches an observer after being deviated through an angle δ by an ice crystal, the observer will have to look an angle δ away from the sun to see that light ray (Fig. 3). The angle of minimum deviation for a 60° ice prism is 22°; rays with either greater or smaller angles of incidence on the crystal face will be deviated by more than 22° in passing through the prism. The nature of this minimum in the deviation angle determines two important features of a halo: (i) the angular radius of the inside of a halo is equal to the minimum deviation angle, and (ii) the halo is brightest around its inside edge. The case of refraction by a prism treated in most optics textbooks is for a light ray that is perpendicular to the axis of the prism, and for this case the angle of minimum deviation is easily calculated. For rays that are not perpendicular to the prism axis (skew rays) the angle of minimum deviation depends on the skewness of the rays. Earlier investigators (1) have calculated the pattern of light in the sky that would result only from the rays with minimum deviation. Such an approach necessarily omits intensity information and the contributions of all other refracted rays.

Our calculation includes intensity information resulting from the dependence of deviation of a light ray on the orientation of the prism. In addition, we have accounted for three other factors that influence the intensity distribution. First, the intensity of the exit ray relative to the incident ray is diminished because of reflection losses at each surface. The intensity loss due to reflection depends on the angle of incidence and on the index of refraction of the crystal. Using Fresnel reflection coefficients (2), we have accounted for this reflection factor. The other two intensity factors that we have included are geometrical factors. For the circumscribed halo we

Fig 1. Photograph of a 22° halo.

are concerned with light that is refracted through two surfaces that make an angle of .60° with each other. The cross-sectional area of the light beam incident on the first crystal face depends on the cosine of the angle of incidence; when the projected area of the crystal face in the direction of the incident light is small, the relative intensity of the exit beam is low. These first two intensity factors have been treated exactly in our formulation. A third factor allows for the fact that only for the case in which the light passes through the crystal at minimum deviation, and normal to the crystal axis, will all the light that enters the first face pass through the second face. For all other paths some of the light transmitted by the first face will strike other faces of the crystal. We

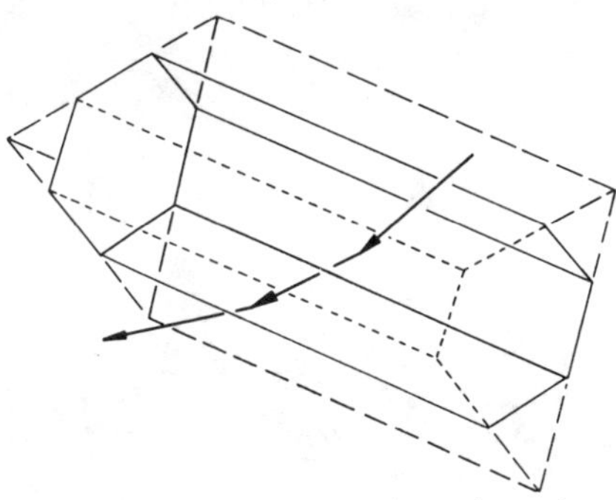

Fig. 2. Light ray passing through a hexagonal ice crystal.

have accounted for this last geometrical factor by applying the correction appropriate to light rays that are perpendicular to the crystal axis. By using this approximate correction we have also neglected any effects due to the ends of the pencil crystals. Using these intensity-influencing factors, we calculated relative intensities for the exit rays which ranged between zero and unity. Before plotting the data we compared the relative intensity of the exit ray with a random number between zero and unity. Only if the relative intensity of the exit ray was greater than the random number did that ray contribute to the plot. According to this procedure, the effect would be, for example, that only one-tenth of the spots associated with rays having a relative intensity of 0.1 would be plotted.

The orientation of a crystal is determined by the specification of three angles. We assign sets of random values to the three angles with appropriate weighting factors to produce the desired distribution. The following approach is used to make a plot of the

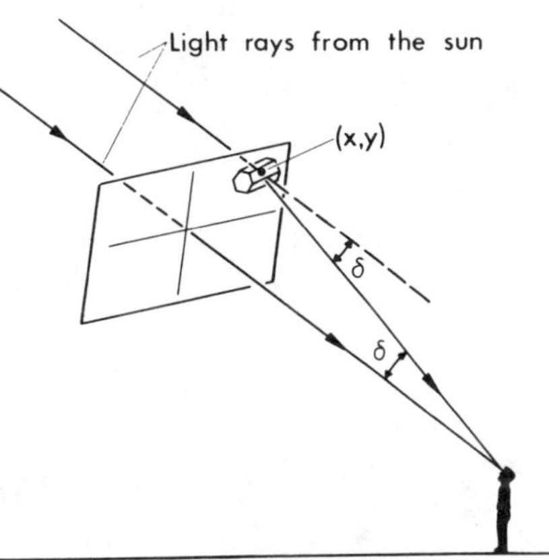

Fig. 3. An observer viewing a light ray refracted by an ice crystal.

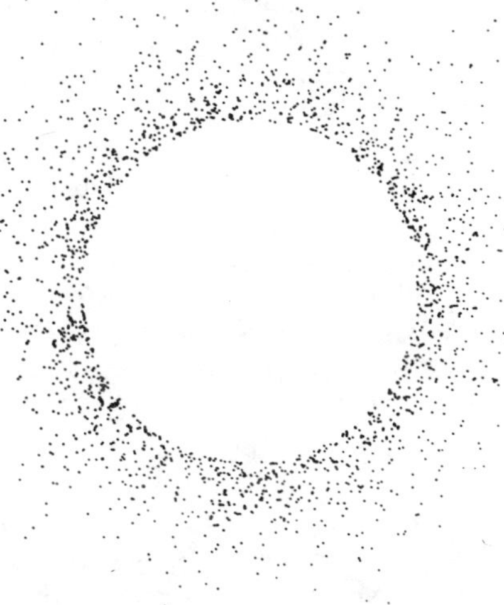

Fig. 4. Simulation of a 22° halo.

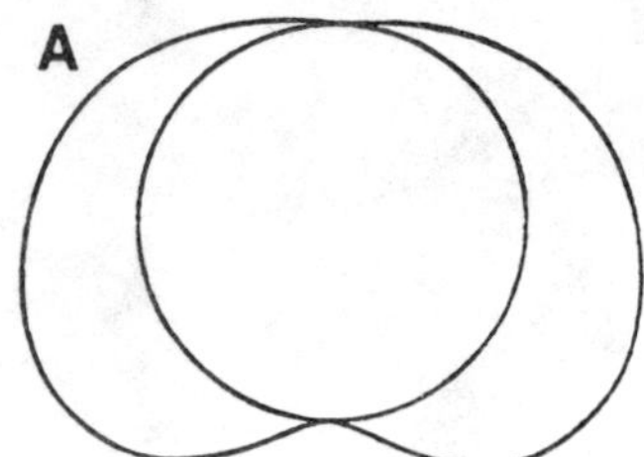

Fig. 5. (A) Description of a circumscribed halo for a sun altitude of 35°, taken from Pernter and Exner (*1*). (B) Simulation of a circumscribed halo for a sun altitude of 35°.

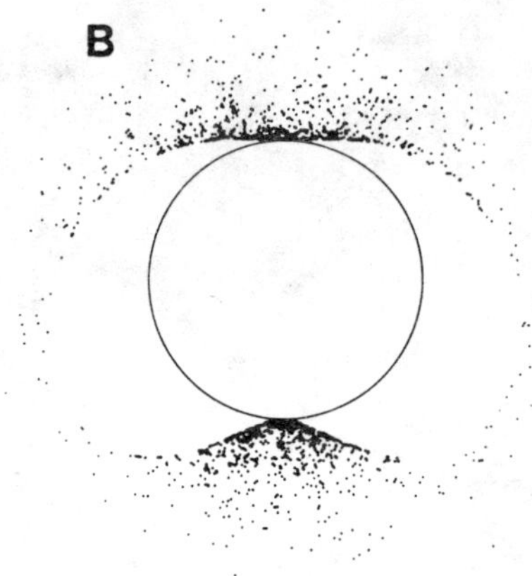

simulated halo: after the direction of the exit ray is determined, we calculate the coordinates of the point in a plane (normal to a line between the observer and the sun) at which the crystal can be located to give an exit ray which points toward the observer. The situation is depicted in Fig. 3. We calculate the coordinates of several thousand points for each simulation and plot them automatically. The number of points per unit area is proportional to the intensity of the light.

Results and Discussion

As a special case of our formulation, a random distribution of orientations of the ice crystals should produce a simulation of the 22° halo. The result is shown in Fig. 4, which may be compared with the photograph of an actual halo shown in Fig. 1. The simulated halo shows the intensity distribution for red light. In all of our simulations we have ignored the diameter of the sun; our results are those predicted for a point sun. To take the diameter of the sun (0.5°) into account, we could smear the pattern over a 0.5° circle, but this operation would not change any of the essential features we describe.

The simplest distribution of crystals that should produce a circumscribed halo is one in which all the pencil crystals have horizontal axes, including all orientations resulting from rotation about that axis and from rotation of that axis in the horizontal plane. Figure 5B shows the pattern resulting from such a distribution for the sun at an altitude of 35°. About 20,000 different crystal orientations were used to produce this figure. In this simulation we have drawn a circle representing the inner edge of the 22° halo. Figure 5A, taken from Pernter and Exner (*1*),

shows their locus of rays with minimum deviation for horizontal crystals for this same sun altitude. The locus of minimum deviation matches the inner boundary of our intensity distribution.

It is clear, however, that the intensity information contained in the dot diagrams adds a great deal of predictive value to the simulation.

Figure 6 shows the resulting intensity distribution for pencil crystals with horizontal axes for several different values of the sun altitude. A horizon has been added to each of the figures along with a circle of 22° radius. For sun altitudes less than about 30°, there is no light at the sides of the halo and the halo separates into two segments. These are generally referred to as the upper and lower tangential arcs. For sun altitudes between 30° and about 40°, the circumscribed halo extends completely around the sun, in principle, but our plots predict such a low intensity at the sides of the halo as to be unobservable. For higher sun altitudes, the

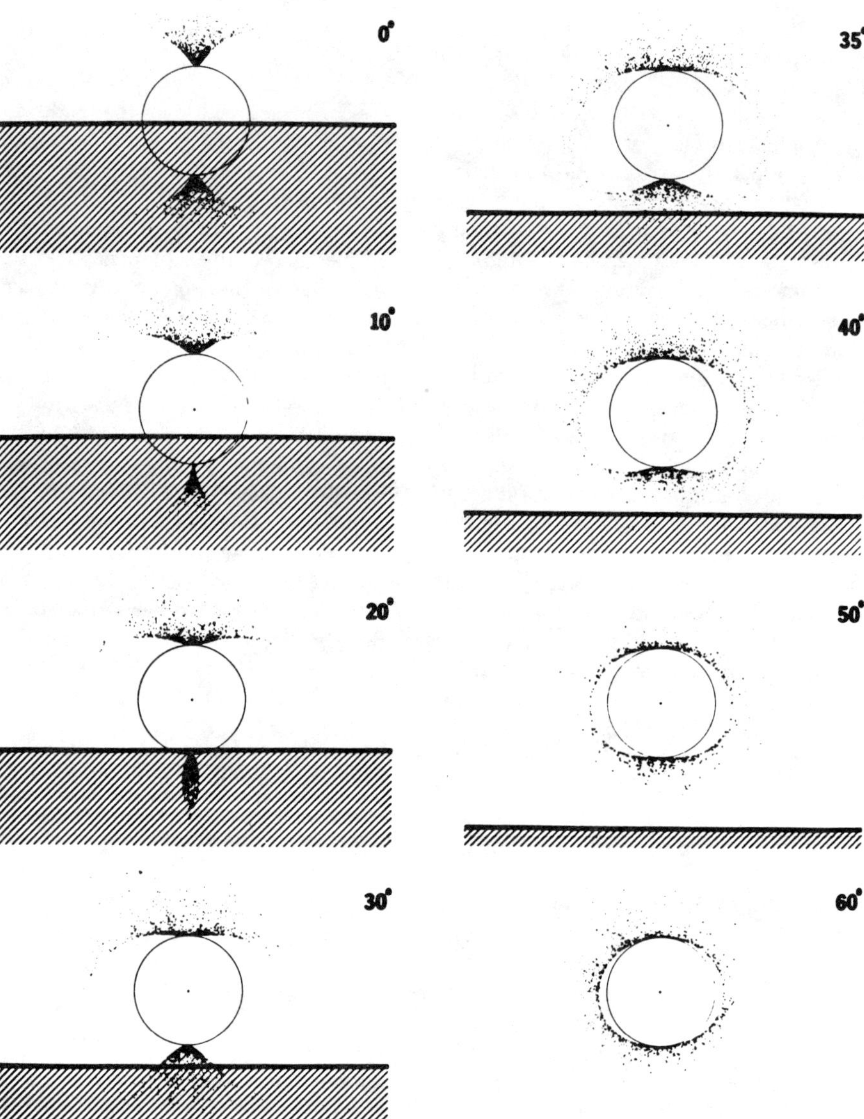

Fig. 6. Simulation of circumscribed halos for various altitudes of the sun, based on the assumption that the ice crystal axes are horizontal. The heavy line represents the horizon, and the circle marks the inner edge of the 22° halo.

circumscribed halo becomes more uniformly intense around its perimeter and shrinks to match the circular shape of the 22° halo.

We have taken photographs with which to compare these predictions, some of which are shown in Fig. 7. Figure 7, C and F (sun altitudes of 25° and 60°), show the 22° halo not as a circle but somewhat horizontally elongated. These two photographs were taken with the camera facing a reflecting glass sphere (3), and the resulting distortion is the price one pays for obtaining a photograph which originally covered the entire sky. For altitudes of 0° and 25° (Fig. 7, A and B) the upper tangential arc is essentially complete.

It is not expected, even in the most intense display, to continue around to the sides of the 22° halo. For sun altitudes less than 22° the lower tangential arc will be below the horizon, but an observer in an airplane may see it.

The 22° halo and the circumscribed halo can occur independently. Figure 1 illustrates the more common situation in which only the 22° halo is seen. Five of the photographs in Fig. 7 (A, B, C, E, and F) show both halos, but Fig. 7D (for a sun altitude of 40°) shows the upper arc of the circumscribed halo with no trace of the 22° halo. The simultaneous occurrence of both halos could result from groups of ice crystals with different orientation distributions located at different altitudes in the sky; or they might result from two different distributions in the same region of the sky resulting from crystals of different sizes or different length-to-width ratios. Flat plate crystals oriented with their axes nearly vertical will give rise to bright spots located at about 22° away from the sun on either side. These parahelions or "mock suns" are commonly observed. The techniques used in this investigation appear to produce satisfactory simulations of these effects also (4).

Conclusion

The detailed agreement between the form of the circumscribed halo and our simulations for a range of sun altitudes would seem to provide convincing evidence that hexagonal ice crystals with nearly horizontal axes are responsible for this phenomenon. In general, slight deviation from the horizontal will tend to smear the upper and lower arcs out along the 22° halo, the extreme case of deviation from the horizontal resulting in a turning of the circumscribed halo into the 22° halo. Our simulation shows that the features of the circumscribed halo are reproduced by crystals with exactly horizontal axes, no tilt from the horizontal being necessary.

References and Notes

1. J. M. Pernter and F. M. Exner, *Meterologische Optik* (Braumüller, Vienna and Leipzig, 1910). For a review, in English, of the work by Pernter and Exner, see W. J. Humphreys [*Physics of the Air* (Dover, New York, 1964), pp. 519–524] and M. Minnaert [*The Nature of Light and Colour in the Open Air* (Dover, New York, 1954), p. 197]. For a continuation of this treatment, see R. A. R. Tricker, *Introduction to Meteorological Optics* (American Elsevier, New York, 1970), pp. 108–117.
2. See, for example, M. Born and E. Wolf, *Principles of Optics* (Pergamon, Oxford, ed. 4, 1970), pp. 38–42.
3. The reflecting sphere used to obtain Fig. 7, C and F, is actually a Christmas tree ornament. The composition of Fig. 7C is not intended to imply anything about the character of the senior author who appears in it, but results from an attempt to reduce a lens flare by blocking out the sun.
4. See R. G. Greenler, M. Drinkwine, A. J. Mallmann, G. Blumenthal, *Amer. Sci.*, in press.

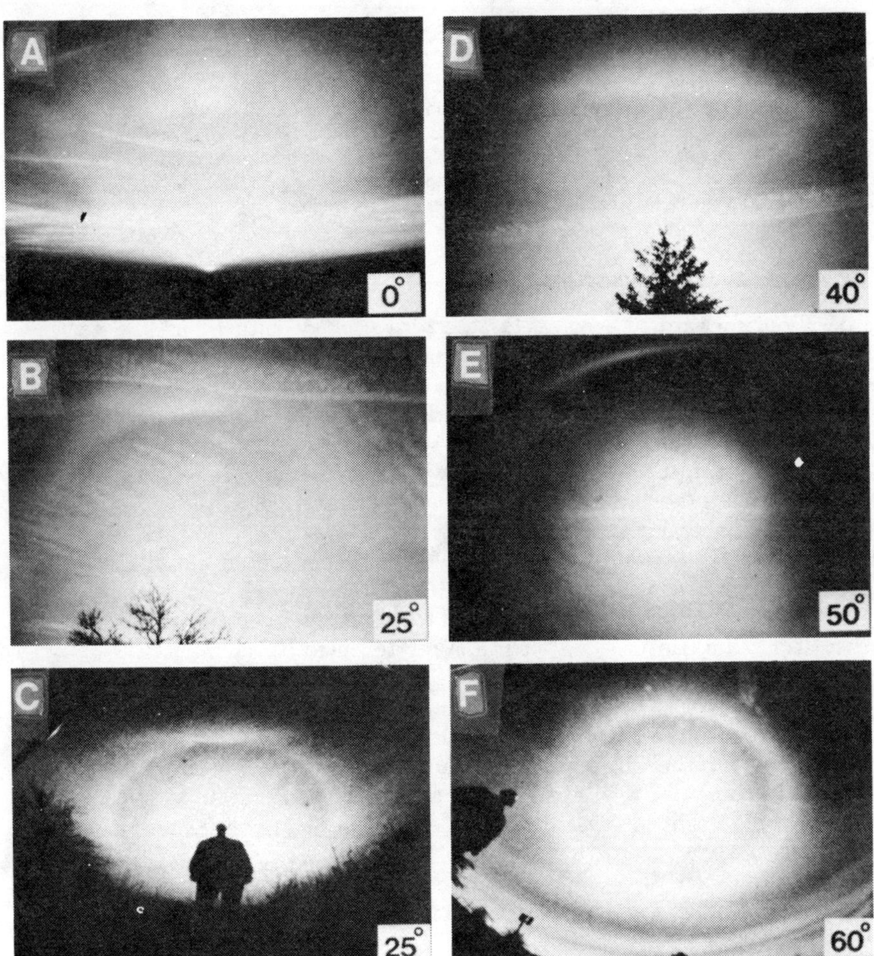

Fig. 7. Photographs of circumscribed halos for various altitudes of the sun.

A simple theory of certain heliacal and anthelic halo arcs. The long hexagonal ice prism as a kaleidoscope

By R. A. R. TRICKER
Freshwater Bay, Isle of Wight

(Manuscript received 18 December 1972; in revised form 8 May 1973)

SUMMARY

G. H. Liljequist and J. R. Blake have both recorded anthelic halo arcs on several occasions in Antarctica, with the sun at comparatively low elevations, which exclude on some occasions, and almost exclude on others, the mechanism put forward by Alfred Wegener to account for such arcs. A particularly complete display reported by J. R. Blake enables these arcs to be identified with those predicted by a very simple theory given in the following paper, to which are added some remarks about the formation of an anthelion.

1. DISCUSSION

Two theories have been put forward to account for the formation of anthelic arcs. Both employ long hexagonal ice crystals sinking through the air with their long axes horizontal, and consider rays of light which enter by a side face, are reflected off the vertical end face and emerge through another or the same side face. Probably the more satisfactory of these theories was the work of Wegener (1925) who showed that the rays which give rise to the upper arc of contact to the halo of 22°, would, if internally reflected off the end face, produce arcs through the anthelion. They would be tangential to the 22° halo on one side of the celestial sphere and intersect at the anthelion on the other. No such arcs could arise, however, if the solar elevation was less than 14° and they would be unlikely to be observable unless it was considerably higher. The other theory was put forward by Hastings (1920). He employed the same group of crystals as Wegener and pointed out that when they also had a pair of sides vertical, a ray reflected off such a side face and the end face would emerge from the face of entry, coming from the anthelion. If such a crystal rotated about its horizontal axis the emergent ray would trace out an arc through the anthelion. A crystal rotating about the horizontal axis would pass through this position and give rise to an arc through the anthelion, the combination of those arising from all positions of the horizontal axis giving rise to resultant arcs, though it would seem very diffuse ones. The question of the descent of crystals with side faces vertical will be discussed at the end of this article.

It is remarkable that a much simpler theory, based on nothing more difficult than the solution of the schoolboy problem of images in inclined mirrors, should not have been considered. The nature of the arcs which this theory would appear to explain can be understood from accounts of two comparatively recent observations in Antarctica, one by Liljequist (1956) and the other by Blake (1961), both at low solar elevations.

Fig. 1 shows Liljequist's anthelic cross observed on 30 September 1950, with the sun at an elevation of 16°. The arms of the cross were estimated to intersect at an angle of 60°. Fig. 2 is by Blake and shows the much more complicated display he recorded on 2–3 December 1958, with the sun at an elevation of 13°. The arcs in which we are at present interested are those labelled 1, 2, 3, and 4. They were white in colour and Blake records that 'although not bright, these curves were certainly not faint, and were quite distinct.'

The present theory assumes the presence of long hexagonal crystals floating with their long axes horizontal and possessing flat ends – as with the other two theories. Light entering by an end face and lying within a cone of semi-vertical angle 58° round the normal, will be totally reflected from the side faces inside the crystal (Fig. 3) and it will be

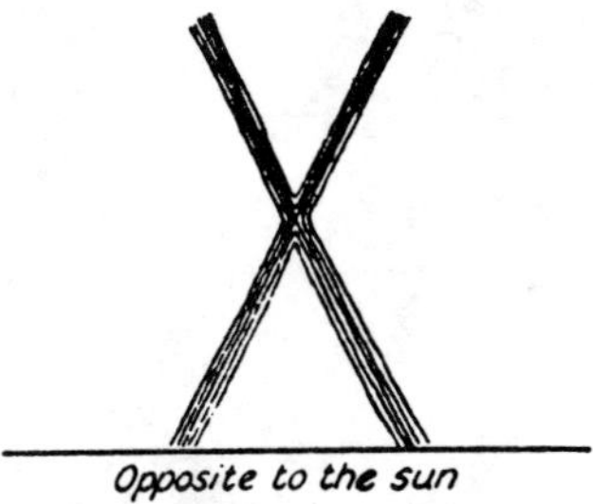

Figure 1.   Liljequist's Anthelic Cross (Liljequist 1956).

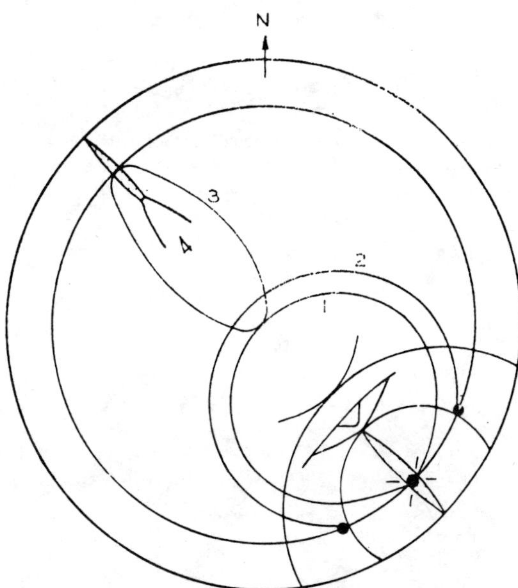

Figure 2.   Fig. from J. R. Blake's ' Solar halos in Antarctica ' (Blake 1961). The present article deals with the arcs labelled 1, 2, 3, and 4, and with the anthelic pillar.

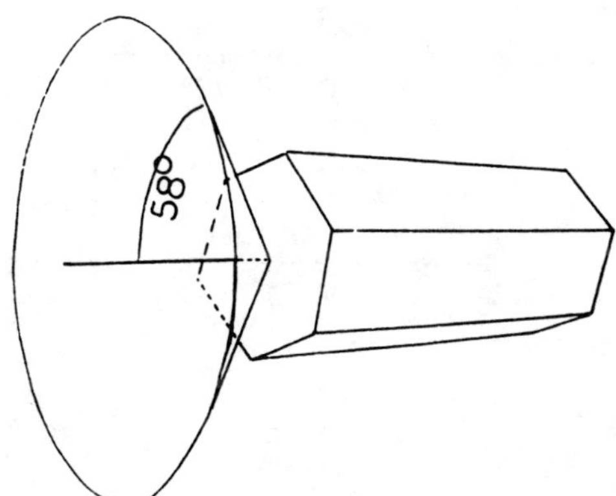

Figure 3.   A ray of light entering a long hexagonal ice crystal from within the cone of semi-vertical angle 58° will be totally reflected from the side faces inside.

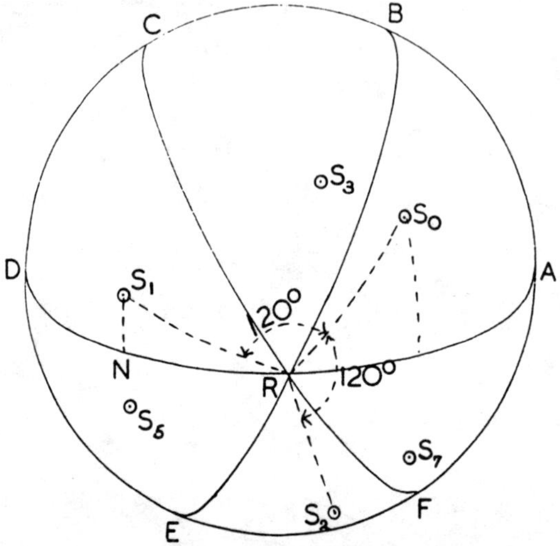

Figure 4. On emerging by the end face opposite to that of entry, the ray will proceed in one of the six directions SO, S_1O, S_2O, S_3O, S_5O, or S_7O (O being the centre of the celestial sphere – not marked in the Fig.).

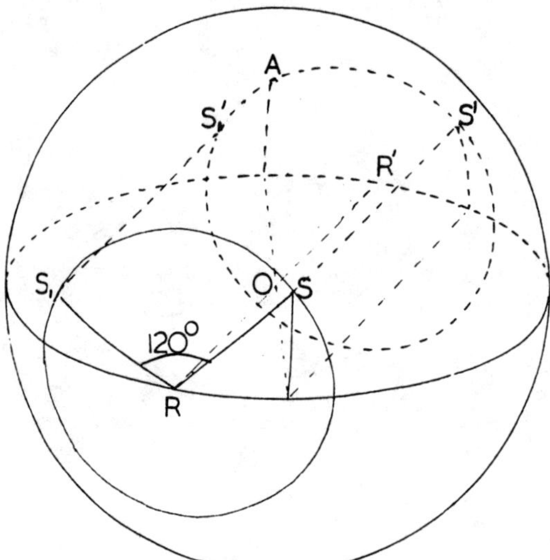

Figure 5. After reflection at the end face and emergence through the face of entry, the ray will lie in directions such as $S_1'O$, etc.

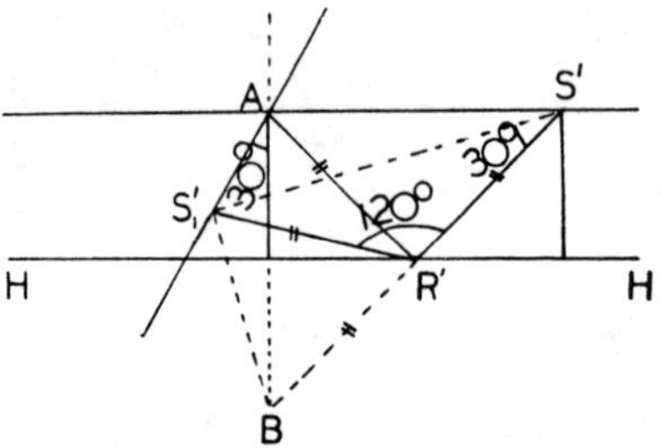

Figure 6. The figure on the back of the celestial sphere when solar elevation and distances from the anthelion are small.

' piped ' from one end of the crystal to the other, by a series of total reflections, with no loss of intensity on that account.

If the light emerges from the flat end face at the opposite end of the crystal from that by which it entered, the ray will lie in one of the six directions S_0O; S_1O; S_3O; S_2O; S_7O; S_5O – O being the centre of the celestial sphere – as in Fig. 4. The centre O has not been marked to avoid overcrowding the Figure. In this Figure R gives the direction of the axis of the crystal, S_0 the position of the sun on the celestial sphere, and RA, RB, RC . . . are the reflecting planes formed by the sides of the hexagonal crystal prism. In this Figure, S_3 is, of course, the image of S_0 in the plane RB, while S_1 is the image of S_3 in the plane RC and gives the direction from which the beam of light proceeds after two reflections. Similarly S_7 is the image of S_0 in the plane RA, and S_2 is the image of S_7 in the plane RF, and, like S_1, gives the direction from which the light proceeds after two reflections. The image S_5 is the image of the points S_1 and S_2 in the planes RD and RE, respectively, and is the origin of the beam after three reflections.

It is evident that the angles S_0RS_1 and S_0RS_2 are both equal to twice $S_0RB + 60°$ $- S_0RB = 60°$. That is

$$S_0RS_1 = S_0RS_2 = 120°.$$

An additional remark is required about the refraction which occurs when the light enters by one end face and leaves by the other. Reflection off the side faces of the prism does not alter the angle of incidence of the light on to the end face within the crystal, apart from rotating the plane of incidence. Thus the refraction which occurs as the light emerges from the crystal is equal in magnitude to that which occurred on entry, but, of course, away from the normal instead of towards it. Thus S_0O (Fig. 4) is the same as the direction of the incident light on to the crystal (i.e. the direction of the sun on the celestial sphere) while the other images lie on a small circle whose pole is at R, passing through the sun. The theory is identical to that of the simple kaleidoscope.

So long as the horizontal crystal axis is unaltered in position, the images S_1 and S_2 are fixed in position whatever the orientation of the crystal about its axis. As the axis R moves over the horizontal plane, S_1 and S_2 describe arcs in the sky which we shall seek to identify with Arc 2 in Blake's diagram. If the ray is reflected in the rear face and finally emerges from the front face by which it entered, an arc passing through the anthelion will be formed, which we shall seek to identify with Blake's Arc 4.

Fig. 5 shows the effect of the reflection in the end face (at right angles to OR). We obtain the origins, S′ and S_1′, of the rays SO and S_1O after reflection, by drawing SS′ and S_1S_1′ parallel to OR. When the elevation of the sun is small and the image S_1′ lies close to the anthelion, the figure on the back of the celestial sphere may be represented by straight lines, as in Fig. 6. If we drop a perpendicular from the anthelion, A, on to the horizontal plane and produce S′R′ to meet it at B, we have,

$$R'S' = R'A = R'S_1' = R'B.$$

The points S_1′, A, S′, and B, thus lie on a circle with centre R′, and the angle S_1′AB is half of the angle S_1′R′B, and therefore equal to 30°.

Thus S_1′ lies on a line through A making an angle of 30° with the vertical. When S′ falls on the other side of A we have a similar arc generated by the image S_2′. This, therefore, provides an explanation of Liljequist's anthelic cross, the two arcs intersecting at an angle of 60°.

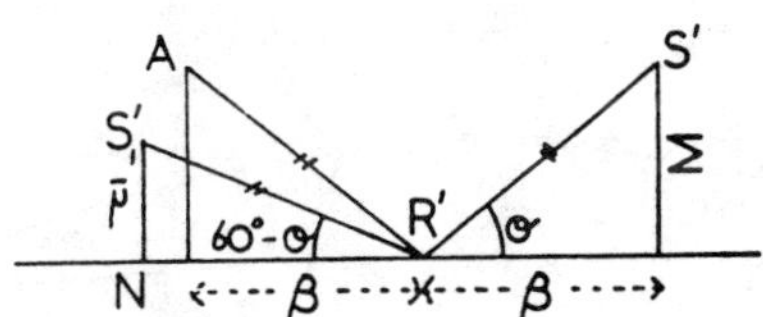

Figure 7.   Figure for calculating the arcs generated.

If β is the azimuth of the crystal axis R relative to the solar vertical, we can easily calculate the position of the image S_1' for any given elevation of the sun, Σ. There is no point in doing this by plane trigonometry, assuming the lines on the celestial sphere to be straight. Using spherical trigonometry instead we have from Fig. 7

$$\cos R'S' = \cos \Sigma \cos \beta$$

$$\sin \sigma = \frac{\sin \Sigma}{\sin R'S'}$$

$$\sin S_1'N = \sin \bar{\mu} = \sin (60°-\sigma) \sin R'S'$$

$$\cos NR' = \frac{\cos R'S'}{\cos \bar{\mu}}$$

Measured from the vertical through the anthelion the azimuth of S_1' is

$$\lambda = NR' - \beta.$$

By giving β a series of values between $-90°$ and $+90°$ the entire arc generated by S_1' can be plotted. That generated by S_2' (the reflection of S_2 of Fig. 4) is continuous with it and the two together produce the curves plotted for a series of solar elevations in Fig. 8. When the sun is on the horizon the anthelion is at A_0 and we have the pear shaped

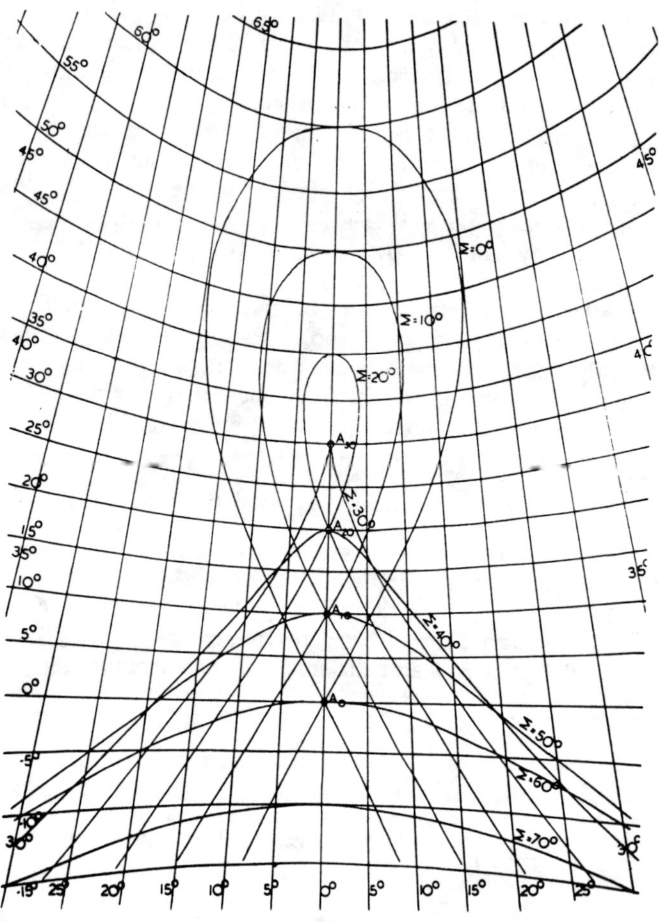

Figure 8.   Anthelic arcs formed by the images S_1' and S_2'.

curve labelled $\Sigma = 0$, which also has an identical loop lying inverted below the horizon. When the sun is at an altitude of $10°$ the anthelion is at A_{10} and the upper loop has shrunk in size while the lower one below the horizon has grown. With the sun at an elevation of $30°$ the arc just reaches the anthelion A_{30} at a cusp and for greater elevations no loop is formed at all and the arc does not reach up to the anthelion.

The existence of the Parry arc shows that long crystals sometimes fall not only with the long axis horizontal but also with a pair of faces horizontal as well. In such crystals the inclinations of the reflecting planes ARD, BRE, and CRF (Fig. 4) are fixed and the images S_3 and S_5 can also give rise to arcs. One of the planes (ARD) will be horizontal and S_3 and S_5 are the reflections of S_2 and S_1 in it. The resulting arcs will thus be simply the reflections of those plotted previously in Fig. 8, in the horizontal plane. The resulting series of anthelic arcs formed by the reflections of these images in the end faces of the crystals have been plotted in Fig. 9 for a similar range of solar elevations.

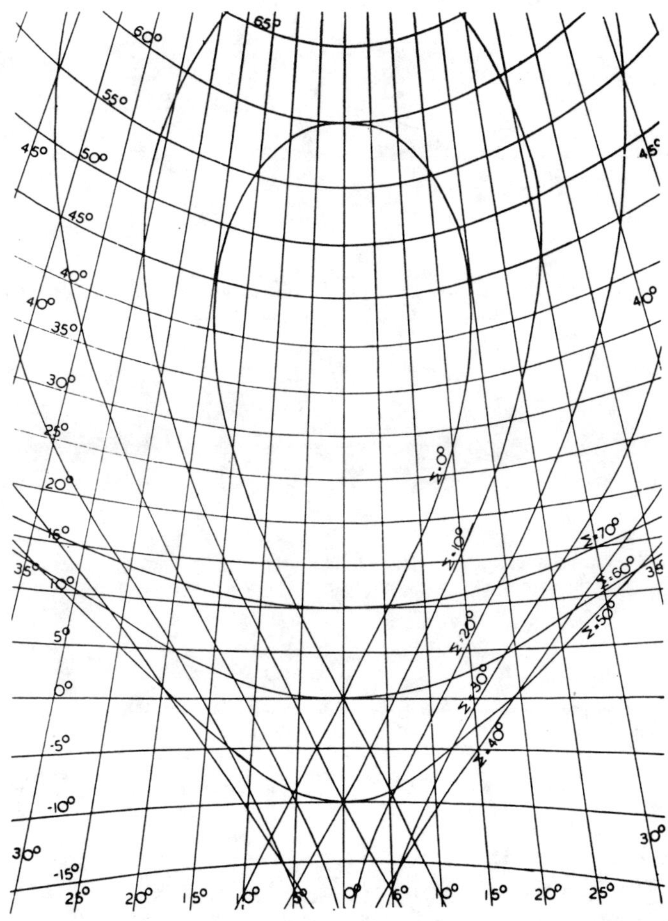

Figure 9.   Sub-anthelic arcs formed by the images S_3' and S_5'.

Those rays which emerge from the rear face instead of being reflected in it, give rise to heliacal arcs. The corresponding figure on the front of the celestial sphere is identical to Fig. 7 except that the anthelion does not appear and S' becomes S, the position of the sun itself. The azimuth of S_1 (in place of S_1') measured from the point on the solar vertical underneath the sun (and not under the anthelion as before) is $NR + \beta$, while its altitude remains the same. The effect of this is to produce a much larger loop than before. The

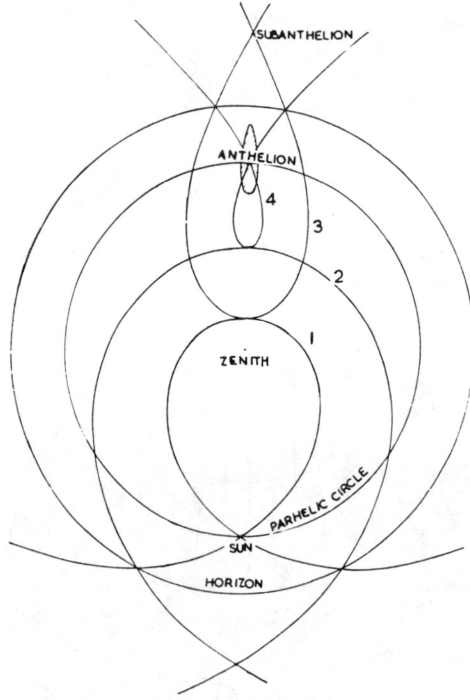

Figure 10. Stereographic plot of the four arcs generated by reflections in the faces of long hexagonal prisms floating with their long axes horizontal.

Figure 11. A possible crystal shape, with one pair of side faces smaller than the others, which would float through the air with this pair of side faces vertical.

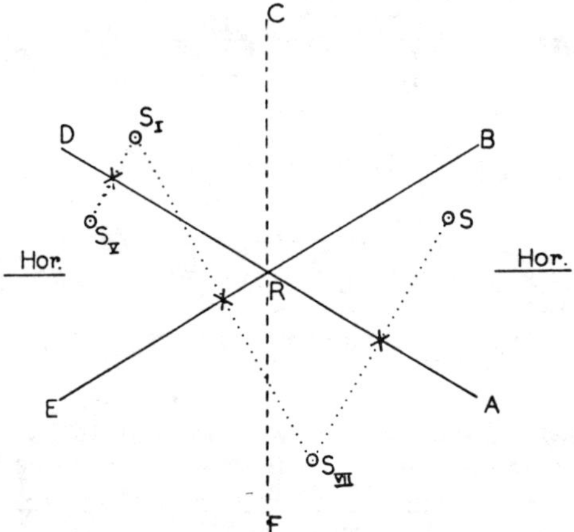

Figure 12. Indicating possible reflections in the inclined faces of such a crystal as that in Fig. 11, which would give rise to an anthelion and an anthelic pillar.

arc crosses the solar vertical above the sun when $\beta = 90°$, at an azimuth of $180°$ from the sun, or $0°$ measured from the anthelion. The anthelic arc also crosses the solar vertical when $\beta = 90°$, and since the altitude is the same for both arcs, they touch at this point.

On the same side of the celestial sphere as the sun the heliacal arc again crosses the solar vertical but at the sub-sun and not at the sun itself. It is, in fact, a sub-heliacal arc. Again Parry type crystals give rise to an arc arising from the images S_3 and S_5, which is the reflection of the S_1S_2 arc in the horizontal plane. Since the S_1S_2 arc passes through the sub-sun its reflection (the S_3S_5 arc) will pass through the sun itself. A similar argument to that given for the tangency of the heliacal and anthelic S_1S_2 arcs applies to the two S_3S_5 arcs, which are also tangential where they cross the solar vertical. For any given elevation of the sun below $30°$ there will thus be an heliacal and a sub-heliacal arc together with an anthelic and a sub-anthelic arc. The two heliacal arcs are too large to be plotted on the same projection as that used for the anthelic arcs in Figs. 8 and 9. All four arcs, however, have been plotted on a stereographic projection in Fig. 10 for a solar elevation of $15°$. The resemblance to Blake's diagram (Fig. 2) is striking, especially when it is remembered that the observed elevation of the sun was somewhat lower ($13°$) and this would cause the outer sub-heliacal and sub-anthelic loops to contract and the inner heliacal and anthelic loops to expand.

Blake's observations on this halo are very numerous and detailed and in spite of being made from a mobile station, appear highly accurate according to the theory. No photographic record was taken of this unique display and so no further check of such differences as exist between theory and observation is now possible.

It is satisfactory to note that such differences are comparatively few. They occur near to the anthelion through which arc 3 was recorded as passing. Perhaps this could be ascribed to the presence of the parhelic circle, into which the arc could have appeared to merge. Otherwise the differences are remarkably small.

There remains to be discussed the presence of an anthelic pillar, which both Blake and Liljequist have recorded on other occasions as well, and we will conclude this article with some remarks about the formation of an anthelion and an anthelic pillar.

Attempts to explain the anthelion have usually been based upon the postulation of crystals sinking with the end face and one pair of side faces vertical. Reflection of a ray in two such vertical faces at right angles would reverse its direction in the horizontal plane. However, a crystal is only likely to sink in such an orientation if it is broader than it is deep, as a result of less development of the vertical side faces (Fig. 11) and this would seriously diminish the reflecting mechanism. However, three total reflections in the inclined side faces (or an odd number of three such reflections) together with one off the back face, would give rise to an anthelion, the ray emerging from the front face as before. The position of the image S_V in question, and three possible reflections in the inclined faces, are indicated in Fig. 12, (where only construction lines, and not the paths of actual rays, have been drawn). Small variations in the alignment around the horizontal axis would cause S_V to move over a small arc of a circle centred on R. For low solar elevations the predominant displacement, as R moves over the horizontal plane, would be vertical, thus producing a vertical pillar through the anthelion. Such an anthelic pillar is plotted in Fig. 10, corresponding to a maximum departure from the preferred orientation of the vertical side faces, of $5°$.

References

Blake, J. R.	1961	' Solar halos in Antarctica,' *ANARE Reports.* Series A, **IV**, Melbourne, pp. 1–45.
Hastings, C. S.	1920	' A general theory of halos,' *Mon. Weath. Rev.,* **48**, pp. 322–330.
Liljequist, G. H.	1956	' Halo phenomena and ice crystals. Norwegian-British-Swedish Antarctic Exp 1945–52,' *Scientific Results,* **II**, (2) pp. 1–110.
Wegener, A.	1925	' Theorie der Haupthalos,' *Arch. der Deutschen Seewarte,* **XLIII** (2), pp. 1–32.

What size of ice crystals causes the halos?

Alistair B. Fraser

Meteorology Department, Penn State University, University Park, Pennsylvania 16802

(Received 16 January 1979)

It is shown that, contrary to classical theory, the circular halos need not be caused by randomly oriented crystals. Furthermore, if Brownian motion is the disorienting mechanism then the circular halos cannot be caused by the randomly oriented crystals, which are too small to produce a reasonably sharp diffraction pattern. However, the circular halos can be caused by crystals that are in the region where there is a transition between randomness and high orientation. These crystals have diameters between about 12 and 40 μm. Larger crystals produce the parhelia and tangent arcs. It is shown that the 46° halo is rare because it can be produced only by solid columns, and then for only a restricted range of sun heights.

INTRODUCTION

One of the most firmly held canons of meteorological optics is that the circular (22° and 46°) halos are caused by randomly oriented ice crystals. By contrast, most other halo phenomena are credited to highly oriented crystals. Although the only significant orienting mechanism experienced by falling crystals is an aerodynamic couple, three disorienting mechanisms come to mind: Brownian rotations, turbulence, and tumbling at large Reynolds' numbers. The tumbling becomes important only for ice crystals much larger ($R_e > 100$, $D > 1000\ \mu$m) than those normally encountered. The effect of turbulence in disorienting the crystals is unknown, but it is not expected to be significant except in high winds in the first few meters above the ground. Be that as it may, the conclusions drawn in this paper are based on the assumption that there is no significant turbulence.

In addition to these disorientation mechanisms there is a crystal form that would not require them, that is, clusters (rosettes) of bullets. If three or more bullets are joined at their points then they need not all lie in the same plane so that a particular member of the cluster could have any orientation regardless of the aerodynamic couple. The possible role of this interesting crystal form will be speculated on later in the paper. In the meantime, the study will be limited to simple columns or plates.

This paper will consider the transition between the disorientation of small crystals by Brownian rotations and the progressive orientation of large crystals by aerodynamic couples. The object of the study was to determine the size at which the transition occurs, so that halos could be used to determine remotely an approximate crystal size in a cloud.

While this object was accomplished, the results proved far stranger than expected.

Tricker considered the issue briefly.[1] He hints that the transition occurs somewhere between a diameter of 5 and 30 μm by noting that a 5.4 μm diameter sphere would experience a Brownian rotation of one radian per second while at 30 μm the dominant factor in determining the motion is viscosity. This study confirms the bounds set by Tricker and goes on to examine the behavior of the transition and its influence on halos.

I. AERODYNAMIC COUPLE AND BROWNIAN ROTATION

There is no known analytic expression that describes the aerodynamic couple on an arbitrary body moving through the air. We are thus constrained to replacing the hexagonal column crystal with a prolate ellipsoid and the hexagonal plate with an oblate ellipsoid. The particular ellipsoid chosen will be the one with the same mass, aspect ratio, and terminal velocity as the crystal it replaces. This compromise is not expected to affect significantly the validity of the results as they apply to crystals, because, as will be seen, the transition occurs so rapidly that slight shifts in the assumptions have little effect on the final answer.

Cox[2] has considered the couple on an ellipsoidal particle that is moving through a fluid at small Reynolds' numbers. The surface of the particle was described by

$$r = c \left\{ 1 + \epsilon \left[r^3 \frac{\partial^2}{\partial z^2} \left(1/r \right) \right] \right\},$$

where c is the amplitude of the surface harmonic of order one (radius of undistorted sphere) and ϵ is the amplitude of the second harmonic. The axis of symmetry is along z so that when $\epsilon > 0$, a prolate ellipsoid is obtained; while when $\epsilon < 0$, an oblate ellipsoid is obtained. Since $r^2 = x^2 + z^2$, the particle surface can be rewritten

$$r = c \left[1 + \epsilon \left(\frac{3z^2}{r^2} - 1 \right) \right], \tag{1}$$

so that the diameter of the ellipsoid is $2c\,(1 - \epsilon)$ while the length of the prolate or thickness of the oblate ellipsoid is $2c\,(1 + 2\epsilon)$.

For such an ellipsoid Cox showed that when there was an angle, ψ, between the translation velocity, v, and the axis of symmetry, the couple, G, on the ellipsoid was

$$G = q \sin 2\psi, \tag{2}$$

where $q = (-87/40)\,\pi\,\rho\,c^3\,v^2\,\epsilon$ and ρ is the air density. The expression can be applied only at small Reynolds' numbers and small ellipticities. The first restriction is easily met as Reynolds' numbers are generally less than 0.1 for the "crystals" considered. The value of ϵ chosen will be either 0.25 (prolate) or -0.25 (oblate). These will correspond to a column with a length to diameter ratio of 2.0 and a plate with diameter to thickness ratio of 2.5. Although the theory would be deficient for larger aspect ratios, it can be asserted with reasonable confidence that such crystals would, if anything, become oriented more rapidly. This is considered further in the summary.

An examination of Eq. (2) confirms that both oblate and prolate ellipsoids are in equilibrium when $\psi = 0$ and $\psi = \pi/2$ but that the equilibria are stable only for the oblate ellipsoids when $\psi = 0$ and for the prolate ellipsoids when $\psi = \pi/2$. The energy required to rotate an ellipsoid from its stable equilibrium up to a tip angle of θ is, when prolate,

$$E(\theta) = -q \int_{\pi/2}^{\pi/2+\theta} \sin 2\psi\, d\psi = Q \sin^2 \theta,$$

where $Q = q$, and when oblate,

$$E(\theta) = -q \int_{0}^{\theta} \sin 2\psi\, d\psi = Q \sin^2 \theta,$$

where now $Q = -q$. As q changes sign with ϵ, Q remains positive for both cases and is given by

$$Q = (87/40)\,\pi\,\rho\,c^3 v^2\,|\epsilon|.$$

We now must consider the probability that a ellipsoid will have its axis of symmetry pointing within some incremental solid angle, $d\omega$, of a particular azimuth angle and elevation angle. The probability that the ellipsoid will be tipped up to an angle θ will be a function of its size c. For Brownian rotations it is given by

$$P(c,\theta) = A(c) \exp[-E(\theta)/kT],$$

where k is Boltzmann's constant and the normalization constant $A(c)$, is given by

$$1 = A(c) \int_{0}^{\pi/2} \exp[-E(\theta)/kT] d\theta.$$

A good way to illustrate the behavior of the ellipsoids is to plot the cumulative probability, as is done in Fig. 1. The line labeled $\theta_{0.5}$, or median angle, is the angle below which 50% of the crystals will be tipped and so is defined by

$$0.5 = \int_{0}^{\theta_{.5}} P(c,\theta) d\theta.$$

It is seen that when $Q/kT \ll 1$, the ellipsoids are dominated by Brownian rotations and are randomly oriented, being ev-

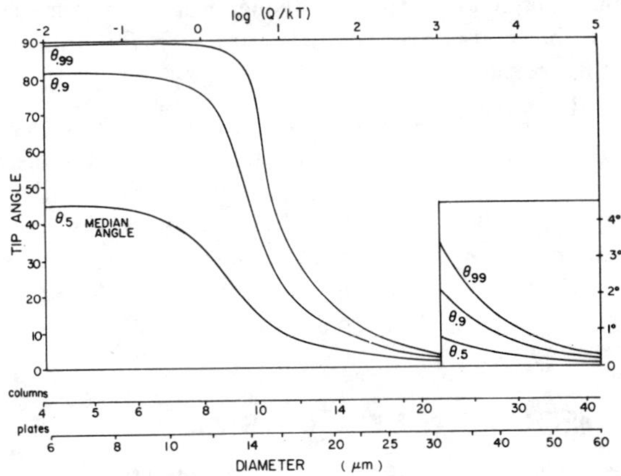

FIG. 1. Cumulative probability distribution for crystal tip angle as a function of plate or column diameter. To the left are crystals that are randomly oriented owing to Brownian rotations. Here the median angle $\theta_{0.5} = 45°$. To the right are aerodynamically oriented crystals, where for a 40-μm column, $\theta_{0.99} < 1/2°$ or 99% of the crystals are confined below a tip angle of 1/2°.

idenced by the median angle approaching 45°. When $1 \ll Q/kT$ the ellipsoids are dominated by the aerodynamic orienting forces and so the median angle approaches 0°. Also shown are the $\theta_{0.9}$ line (90% of the crystals are tipped to angles less than $\theta_{0.9}$) and $\theta_{0.99}$.

The lower abscissa of Fig. 1 is labeled in terms of an equivalent ice crystal diameter (a-axis). This equivalent diameter was obtained by taking $c^3 v^2$ equal for a ellipsoid and hexagonal ice crystal of equal eccentricity ($\epsilon = \pm 0.25$ as mentioned above). The fall velocities for the hexagonal crystals were calculated from the results of Jayaweera and Cottis.[3] The temperature chosen was $-10°C$ while the air density was calculated by assuming an air pressure of 100 kpa.

It might be assumed that a series of plots should be made for various levels in the atmosphere but this proved to be unnecessary. Recalling that in the Stokes region $v \propto r^2/\mu$, where μ is the viscosity, which varies as $T^{1/2}$, it is clear that Q/kT varies as $|\epsilon|\, p\, c^7/T^3$ where p is the air pressure obtained from $\rho = p/RT$. Thus a decrease in pressure to 30 kpa would be compensated by increasing c (or the diameter) by only 19%. Similarly, a decrease in temperature of 15% would decrease in c by less than 7%. In practice a decrease in the pressure level considered would also be accompanied by a decrease in temperature so that the effects would partially compensate. The values chosen are not unrepresentative as a combination, and so, in going to the upper troposphere, one would shift the abscissa by only $1.19 \times 0.93 = 1.11$ or 11%. This is hardly significant.

It is also interesting to note that if the ellipticity ϵ had been changed by a factor of 2, then c would only change by 10%.

Figure 1 can thus be considered to be fairly general. No other reasonable assumptions could change characteristic diameters in the transition region by as much as a factor of 2.

Figure 1 strikingly confirms Tricker's suggestions. For a population of columns to be oriented randomly they all must have a diameter smaller than about $6\,\mu$m, while a population of plates must have diameters smaller than about $9\,\mu$m. There follows a rapid transition (D^7) so that before a column diameter of $20\,\mu$m, or a plate diameter of $30\,\mu$m, is reached, 99% of the crystals are oriented within 5° of the stable equilibrium position.

Thus if we are to accept the oft-quoted contention that the circular halos are caused by randomly oriented crystals, and if the disorientation is caused by Brownian motion, then we must assume that when these halos are seen, the columns (plates) have diameters of less than $6\,\mu$m ($9\,\mu$m). This result will be accepted tentatively while we calculate a lower bound for the crystal sizes based on diffraction theory.

II. DIFFRACTION BY A CRYSTAL

Each ice crystal, as it acts as a prism, effectively produces a small slit through which the light is diffracted (as well as having been refracted by the prism). The effective slit width depends on the deviation angle, but if we choose the minimum deviation for a 22° halo, it is about 37% of the crystal diameter, D. The result does not depend on whether we are dealing

with plates or columns. The shape of the diffraction pattern along a halo radius, can be specified by the intensity, I, as a function of the angular departure from the minimum deviation, β. Thus,

$$I = [0.37\, DL(\sin x/x)]^2,$$

where

$$x = \pi(0.37D/\lambda)\sin\beta,$$

and L is the crystal length, or thickness, and λ is the wavelength of light.

The angular width $\beta_{1/2}$ from the maximum intensity (at the minimum deviation angle) to the half power point (one photographic stop down) is specified by $x = 0.44\,\pi$. For a $2\,\mu$m crystal diameter, $\beta_{1/2} = 19°$. Clearly such a small crystal could not produce a recognizable maximum for the 22° halo. When $D = 9.5\,\mu$m, $\beta_{1/2} = 4°$, giving a barely recognizable and very poor halo. When $D = 19\,\mu$m, $\beta_{1/2} = 2°$, and a passable halo is seen.

We thus have the situation where the crystals must have a diameter greater than about $10\,\mu$m to form a recognizable halo, but they have to be smaller than about $6\,\mu$m (columns) or $9\,\mu$m (plates) to be randomly oriented. One is forced to the untenable conclusion that the 22° halo cannot exist.

Possibly a more detailed diffraction model would help. Figure 2 shows the results of integrating over all deviation angles and including the Fresnel coefficients. The results are essentially as before. A 10-μm crystal would give a barely detectable pattern; a 20-μm crystal, an easily recognizable one. (The "supernumerary" halos that show up in these calculations would be substantially eliminated by the angular width of the sun.)

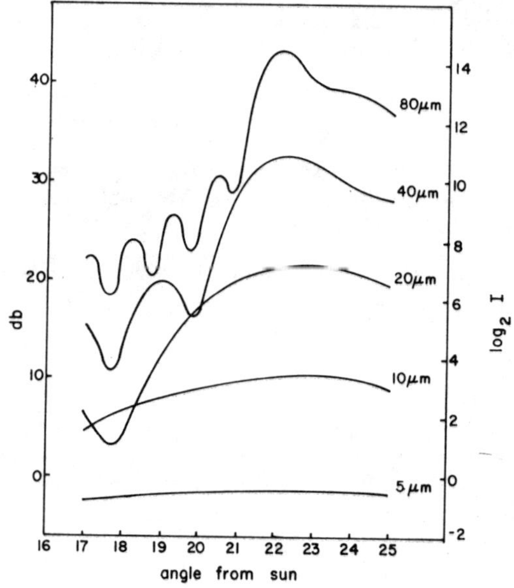

FIG. 2. Intensity distribution across the 22° halo based on a diffraction model that includes the effects of the Fresnel coefficients. A 5-μm diameter crystal shows no halo while an 80-μm crystal shows a good one. The calculations, which were done for a point source, show supernumerary halos, which would substantially vanish in nature owing to the 1/2° angular diameter of the sun.

The situation is, if anything, worse when the 46° halo is considered. The calculations are essentially identical. Curiously, the diffraction slit width is still approximately 37% of the diameter if we are considering minimum deviation refraction by columns. Thus columns will, as before, not produce the 46° halo. For plates, the slit width is about 37% of the thickness so that for a 10 μm thickness we are dealing with a diameter of, say, 20 to 30 μm such that a plate is highly oriented. Plates cannot even come close to satisfying the requirements of producing the 46° halo.

The theory presented to this point assures us that neither columns nor plates can produce either halo. If the crystals are small enough to be randomly oriented, they are too small to produce a sharp diffraction pattern. In short, it has been demonstrated that neither the 22° halo or the 46° halo can exist in nature! (I sympathize with the proverbial scientist who proved that bumble bees cannot fly.)

Clearly, one or more of the assumptions that went into the theory are in error. One likely candidate would seem to be that Brownian motion is not an important mechanism for disorienting crystals and that turbulence is required. It may well be that turbulence is important, but as noted earlier, it will not be treated in this paper. Rather it will be seen whether there exists a solution that is consistent with Brownian motion.

It may be that Cox's expression for the aerodynamic couple is in error or that it cannot be satisfactorily applied to hexagonal crystals. This part of the theory is unquestionably approximate but, as the previous error analysis illustrated, even a substantial change in the assumptions (or indeed the coefficients) would not markedly change the predictions.

Surprisingly, the most suspicious assumption made up to this point is the one that requires the circular halos to be caused by randomly oriented crystals. This will be tested by examining the contribution of nonrandom crystals to the halos. This, in turn, is broken into two separate questions. What is the contribution of a population of crystals of diameter D to a particular tip angle θ? Secondly, how does a crystal tipped to an angle θ contribute to a particular portion of a halo?

III. LIGHT INTENSITY VERSUS SIZE AND TIP

For a given sun height and position on a halo, the light is caused by refraction through a crystal with a uniquely specified tip angle. The relation between these variables will be considered in the next section. Here will be considered the intensity of the light produced by a population of crystals of a known tip angle.

If $n(D) = dN/dD$ is the number density of crystals of diameter D, then $P(D,\theta)n(D)$ is the number of these crystals that are tipped to some angle θ. (As the conversion between c and D has already been presented, P is now expressed in terms of D.) For the minimum deviation ray, the intensity of the light at the peak of the diffraction pattern is proportional to the area of the slit opening squared, which in turn is proportional to D^4. The intensity per unit number density of crystals is thus given by,

$$[I/n(D)] \propto D^4 P(D,\theta) \; .$$

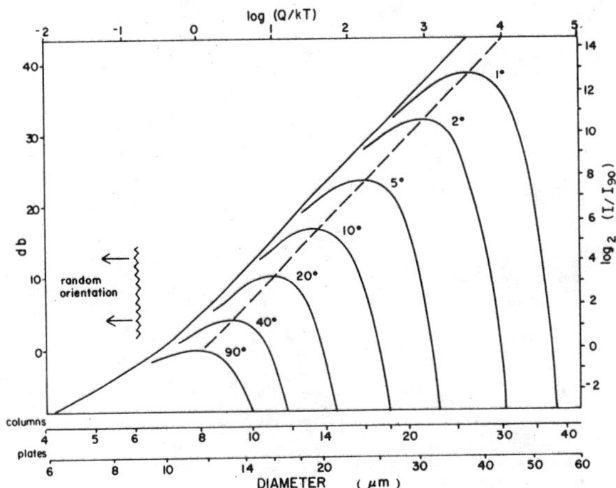

FIG. 3.   Intensity per unit number density versus crystal diameter for various tip angles.   Each tip angle has a local maximum.

which is presented in Fig. 3 as a series of lines of constant tip angle θ.

For very small diameters, all the θ isopleths converge so that all tip angles contribute an equal intensity as they should when the crystals are randomly oriented. For larger sizes, each tip angle has a local maximum as a function of crystal diameter. Smaller crystals contribute very little light while larger crystals have a low probability of being tipped as high as θ. For each tip angle, and thus for each position on the halo, there is a particular crystal size that makes the greatest contribution.

(Such strongly peaked functions suggest that it should be possible to measure the intensity variation around a halo and reconstruct the unknown size distribution, thus using the halo as a means of remote sensing.)

The dashed line on Fig. 3 is the locus of the maxima of the θ isopleths and reveals a simplified relation between tip angle and crystal size. Thus if we are looking at a portion of a halo that requires a plate to be tipped to an angle of, say, 10°, then that portion of the halo will be caused (mainly) by 20-μm diameter crystals.

Clearly, if there were a broad size distribution which included both the small, randomly oriented crystals, and the larger, becoming-oriented crystals, the latter class would make the only significant contribution. Thus, even if the randomly oriented population were not excluded by virtue of its broad diffraction pattern; it still would not be the crystal group that would produce the circular haloes. The larger crystals not only can produce (portions of) the halo, but are in fact preferred. The assumption that the circular haloes are produced by randomly oriented crystals is wrong!

It now remains to be seen what portion of a halo is produced by crystals tipped to θ.

IV. TIP ANGLES FOR HALOS

When it was assumed that the circular halos were due to randomly oriented crystals, there was available a crystal of the

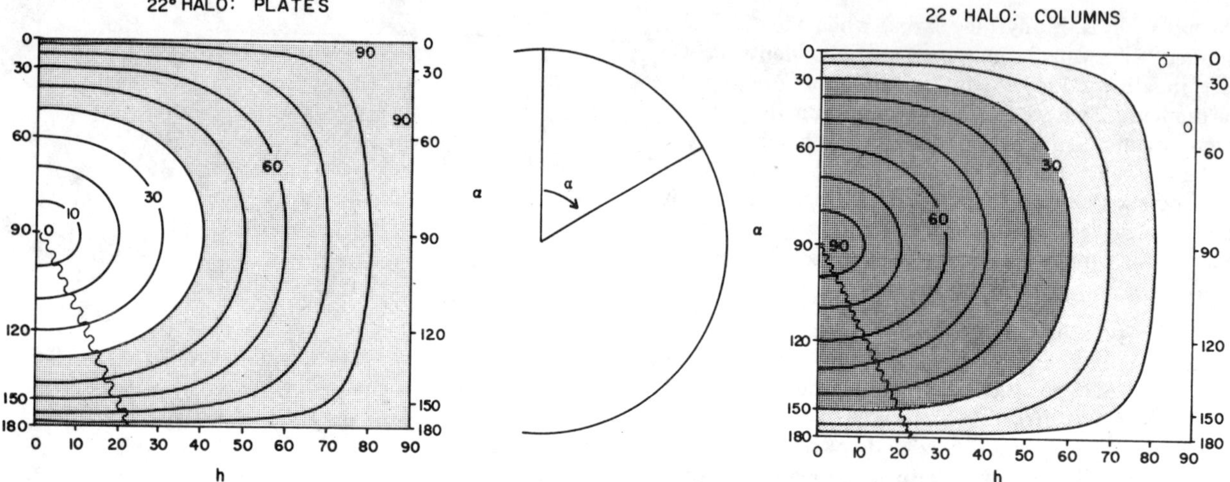

22° HALO: PLATES   22° HALO: COLUMNS

FIG. 4.   Isopleths of crystal tip angle that are required to produce a spot of light at the angular position, α, on a 22° halo for a particular sun height h.   The ordinate is linear in $\cos\alpha$ so that the results can easily be transferred to the halo in the center.   The dark gray region has crystals that are too small to form halos, the light gray region shows poor halos while the white region shows fair to good halos.   The wiggly diagonal line shows the position of the horizon.

requisite tip for every portion of the halo so there was scant need to inquire what that tip angle was.   We now have a relation between the crystal size and the tip angle and a knowledge that some crystals are too small to give a recognizably narrow diffraction pattern.   The larger crystals will give various portions of the halo that will depend on their size and the solar elevation.

If the angle α describes an azimuth position on the halo such that $\alpha = 0$ at the 12 o'clock position and $\alpha = 90°$ at the 3 or 9 o'clock position, and the angle h is the solar elevation, then for the 22° halo caused by plates,

$$\cos\theta = \sin\alpha\,\cos h$$

and for the 22° halo caused by columns,

$$\sin\theta = \sin\alpha\,\cos h.$$

These relations, which follow simply from the spherical trigonometry of the celestial sphere, are plotted in Fig. 4.   The tip angles that are caused by crystals with diameters of less than 10 μm are dark gray to illustrate that, for these combinations of h and α, a halo would be too diffuse and faint to see.

The region caused by crystals from 10 μm to about 14 μm is light gray indicating that here the halo would be poor but recognizable.   The white region would give passibly good halos of reasonable intensity.

Clearly, when the sun is low, plates give better halos than columns, while the reverse is true when the sun is high.   With the exception of high sun angles, the best 22° halos would be caused by a combination of plates and columns.   They might even be the same crystal with an aspect ratio sufficiently close to unity that it was bistable.   The diagrams clearly illustrate why for various combinations of sun height and crystal type and size, only portions of the halo are seen.

It should be noted that, now and then, (usually when the sun is high) a complete 22° halo is seen which appears uniformly bright around the circumference.   Although bistable crystals might account for it, nevertheless it would appear more promising to invoke either clusters of bullets or else turbulence.   At this point there does not seem to be any way to choose which of these two is the more promising, but after considering the 46° halo it will be seen that indirect evidence seems to favor the bullet clusters.

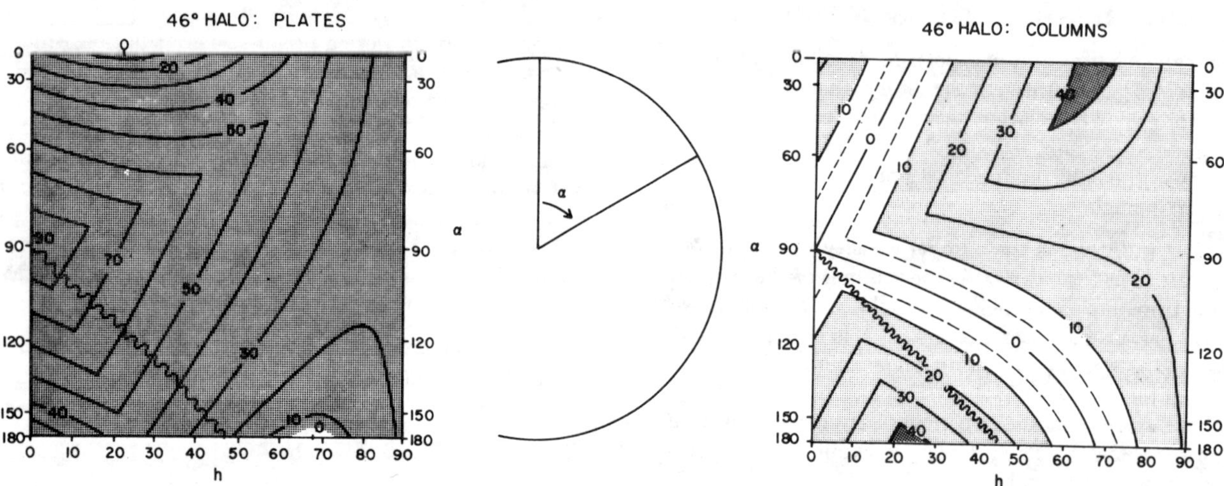

46° HALO: PLATES   46° HALO: COLUMNS

FIG. 5.   Same as Fig. 4 only for a 46° halo.

The comparable diagrams for the 46° halo are given in Fig. 5. The trigonometric relations, from which they are derived, are now more complex. If the sunlight enters a prism face and exits a basal face, the tip angle θ' of the c axis of the crystal (i.e., the tip angle for a column) is given by

$$\sin\theta' = \sin\beta\sin h - \cos\beta\cos h\cos\alpha,$$

where $\beta \simeq 68°$ is the incident angle for the entering ray (or exit ray) at minimum deviation. If the sunlight enters a basal face and exits a prism face, the tip angle θ of the c axis is given by

$$\sin\theta = \cos\beta\sin h + \sin\beta\cos h\cos\alpha.$$

As there will be generally many more crystals at the smaller tip angles, the smaller of θ or θ' is chosen and plotted. The sharp corners in the diagram reflect a transition between θ and θ'.

The tip of the crystal c axis is appropriate for the tip of columns but its complement is required for plates. As with the 22° halo, the above expressions apply to plates if $\sin\theta$ and $\sin\theta'$ are replaced by $\cos\theta$ and $\cos\theta'$, respectively. Again the smaller value is chosen.

The forbidden and poor halo regions are marked for the 46° halo as they were for the 22° halo. Because it is now the thickness that counts for a plate, a 10-μm thickness or greater would have a diameter of about 25 μm or greater and thus a most favorable tip angle of less than 4°. Consequently, essentially all of the diagram is forbidden. The 46° halo cannot be caused by plates! This is one reason why the 46° halo is rarer than the 22° halo which can be caused by both plates and columns.

Of the four halo diagrams presented in Figs. 4 and 5, the most interesting one is unquestionably the one for the 46° halo produced by columns. When the sun is on the horizon, the halo is sharply defined and bright only on the vertical portions, $60° < \alpha < 120°$ where generally the region below $\alpha = 90°$ would not be seen, being blocked by the ground. As the sun rises, the "good" portion splits and part rises up the halo while the other part descends. A solar elevation of about 12° would give the most complete "good" halo above the horizon (wiggly line). No solar elevation would show a "good" complete halo which explains Hastings'[4] comment. In 1920 he noted that "the circle has never been seen complete; that is wholly above the horizon."

Examination of many 46° halo photographs shows that the halo does indeed have an intensity variation around its circumference as predicted, for a given sun height, by Fig. 5.

The 46° halo (or portions thereof) is not rare only because it cannot be caused by plates. The type of columns that can cause it are themselves rare in that they must be solid. Most columns grow at near water saturation and as a result have hollow ends. These "hollow" columns are nevertheless solid near the center and can contribute to the 22° halo. Less common in the atmosphere are the near ice-saturation conditions that will grow the solid columns that are necessary to give the 90° prism ends required by the 46° halo.

It was noted earlier that the 22° halo does occasionally display a uniformly bright complete circle but the weight of the somewhat tenuous evidence is that the 46° halo does not.

If these "good" 22° halos had been caused by turbulence, then there seems no way of avoiding the conclusion that there should be "good" complete 46° halos also. However if the "good" 22° halos were caused by bullet clusters, then it is easy to understand why analogous "good" 46° halos do not occur. Bullet clusters grow under conditions of near water saturation and so like the hollow columns do not have the necessary 90° prism ends. Although admittedly tenuous, this argument suggests that bullet clusters are a more likely cause of good 22° halos than turbulence.

V. SUMMARY

It was stated in the Introduction that the motivation for this study was to be able to determine the size of ice crystals from observations of halos. Figure 6 provides a summary of what the work has revealed. The diameter of a crystal is plotted along the abscissa while the length (column) or thickness (plate) along the crystal c axis is given as an ordinate. Columns therefore lie above the diagonal while plates lie below it. All possible aspect ratios do not occur in nature so the approximate boundaries of natural crystals have been drawn, based on the work of Auer and Veal.[5]

Because the crystal tip angle can be identified with a dominant crystal diameter, these tip angles are marked on the diagram. The angle is labeled at the position of the aspect ratios D/L chosen for this study (2.5 for plates, 0.5 for columns). Short isopleths of constant tip angle are drawn through these points based on the approximation

$$Q/kT \propto c^7|\epsilon| \propto L^7(1 + 2D/L)^6|1 - D/L|$$

where the last proportionality used Eq. (1). L is the crystal length. The lines are not extended very far because for large

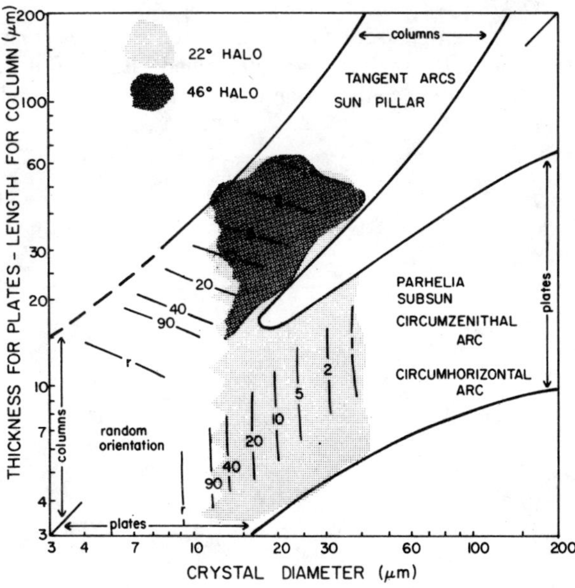

FIG 6.   With crystal diameter on the abscissa and length on the ordinate, columns lie above the diagonal while plates lie below it. Shown are the regions where natural crystals occur, based on Auer and Veal, and where various halo phenomena are predicted. Also shown are short isopleths of the tip angle that prefer a particular crystal size. The r marks the boundary of randomly oriented crystals.

$|\epsilon|$ the expression for the aerodynamic couple is invalid while for small $|\epsilon|$ the representation of the crystal as an ellipsoid is bad.

It is seen that, for plates, the tip angle isopleths roughly follow crystal diameter regardless of aspect ratio D/L. The abscissa for plates on Figs. 1 and 3 and the shading for plates on Figs. 4 and 5 are thus essentially independent of the aspect ratio chosen. The same is not true for columns which show a marked tendency to orient earlier for smaller aspect ratios, i.e., greater lengths. The foregoing diagrams should be read with this in mind.

The region occupied by the circular halos is marked, being bounded on the left by diffractive broadening and on the right, somewhat arbitrarily, by the 1° tip angle. Greater orientation than about 1° leads to such things as the parhelia and tangent arcs. The 46° halo is seen confined to the column branch of the diagram. What halos, or portions of them, that would be seen on any given occasion would of course depend upon the sun height, which is not represented in any way in Fig. 6.

It is seen that the region occupied by the crystals with random orientations lies well to the left of that which gives the circular halos.

VI. CONCLUSIONS

It has been shown that the classical assumption requiring the circular halos to be caused by randomly oriented simple crystals is wrong. Indeed, if the disorientation is due to Brownian rotations—the most likely case—then the randomly oriented crystals do not even contribute significantly to the halo as we know it. This was determined by modeling the transition between random and aerodynamically oriented crystals. It was found that to be randomly oriented the crystals would be too small to give a sufficiently sharp diffraction pattern.

As each position around a halo is caused by an ice crystal with a known tip angle, it was determined what crystal size made the greatest contribution to any particular tip angle.

These halo-causing crystals were found to be in the transition region between the random and highly oriented crystals and in general had diameters between about 12 and about 40 μm.

The relative rarity of the 46° halo was accounted for by the fact that it can be produced only by an uncommon crystal, the solid column. The 22° halo on the other hand can be formed by not only solid and hollow columns but also plates.

It can be noted in passing that the very poor colors that usually accompany the 22° halo are not due just to the overlapping of the colors resulting from it being a minimum angle of deviation phenomenon. If that were the only explanation, one would be hard pressed to explain the very much better (but still not good) colors that are found in the tangent arcs and parhelia. Obviously the diffractive broadening through the small crystals involved plays a major role in degrading the color.

This study dwelt on the role of Brownian motion in disorienting the ice crystals. The next step is clear; the role of turbulence needs study.

ACKNOWLEDGMENTS

The author gratefully acknowledges the input to the very early stages of this work made by Richard C. Easter, who studied the problem for an undergraduate term paper when the author was at the University of Washington in 1972. This work was sponsored in part by a National Science Foundation Grant No. ATM 77-00911.

[1] R. A. R. Tricker, *Introduction to Meteorological Optics* (Elsevier, New York, 1970), p. 81.

[2] R. G. Cox, "The steady motion of a particle of the arbitrary shape at small Reynolds numbers," J. Fluid Mech. **23**, 625–643 (1965).

[3] K. O. L. F. Jayaweera and R. E. Cottis, "Fall velocities of plate-like and columnar ice crystals," Q. J. R. Meteorol. Soc. **95**, 703–709 (1969).

[4] Charles S. Hastings, "A General Theory of Halos," Monthly Weather Review, 322–330 (June 1920).

[5] A. H. Auer and D. L. Veal, "The dimensions of ice crystals in natural clouds," J. Atmos. Sci. **27**, 919–926 (1970).

Some ice crystals that made halos

Walter Tape

Department of Mathematical Sciences, University of Alaska, Fairbanks, Alaska 99701

Received May 11, 1983

During low-level halo displays, ice crystals in the atmosphere at ground level were collected and studied. I discuss the crystals in connection with the halos present at the time of collection.

Atmospheric halos are supposed to be caused by ice crystals. It is natural, then, to examine ice crystals when halos occur. I describe several crystal samples collected in Fairbanks, Alaska, during halo displays in the winter of 1982–1983 and discuss the samples with regard to the halos present.

PROCEDURES

During low-level halo displays, ice crystals in the atmosphere at ground level were collected in Petri dishes containing hexane or silicone oil. The crystals were studied with a binocular microscope and photographed, usually an hour or two after collection. Crystals were also collected and replicated using acrylic spray.

The halos were photographed if they were strong enough. All halos mentioned were verified as occurring at ground level—that is, where the crystals were being collected. In the case of nearby streetlight halos the verification is automatic. Daytime halos can be so verified by getting the halo to show up between the observer and a building or other object, as shown in Plate III.

Temperatures were measured at the collecting sites, but, because of severe temperature inversions and common warm moisture sources, the temperature readings may be a poor guide to the actual temperatures where the crystals were formed, perhaps only a few tens of meters higher.

RESULTS

Figures 1–7 show small portions of crystal samples collected on seven different occasions. The crystals are real in all figures except 2 and 6, which show replicas. Because time was short, the collection dishes for the crystals were usually not waved through the air during collection, so the samples, both replicas and real, are probably biased toward the faster-falling crystals. For most samples I have given approximate percentages of the various types of crystals present, as measured from the corresponding replicas. The percentages are difficult to estimate and should be taken only as a crude guide. In particular, it is easy to overestimate the number of hexagonal plates, since short columnar crystals resting on their basal faces can masquerade as plates.

Figure 1(a) shows crystals collected March 4, 5 a.m., at −26°C. These crystals produced only pillars, shown in Fig. 1(b). Theoretically, pillars have been attributed to hexagonal columnar crystals with principal axes horizontal or to hex-

agonal plate crystals with principal axes vertical. (See Ref. 1 for a description of halos and crystal models on which theoretical halo explanations are based.) Such hexagonal plates are also supposed to cause parhelia. It is fairly easy to believe that the crystals in Fig. 1(a), which are somewhat ornate and perhaps lack good, sufficiently large prism faces, would not produce parhelia. Note that the light sources are 100 m or so from the collection site, where the photograph was taken, so there is a slight chance that the pillars were caused by crystals unlike those in the sample.

Figure 2 shows a replica of a crystal collected March 1, 8 p.m., at −18°C. This is a typical crystal from another sample of crystals that produced only pillars, shown in Plate IV. Figures 1 and 2, together with many similar observations in Alaska and Wisconsin, show that pillars and subsuns can be produced by many types of crystals that cause no other halos. For pillar formation it is enough to have a concentration of nearly horizontal crystal faces.

Figure 3 shows crystals collected February 15, 4 a.m., at −27°C (89% columns, most with large indentations in the ends; 10% irregular spatial crystals; 1% ornate plates). These crystals produced a 22° halo with very slight enhancement at the top and even less enhancement at the parhelion locations. Apparently there was a slight tendency for the principal axes of these columnar crystals to align horizontally; this would account for the enhancement at the top of the halo. By and large, however, the orientations of these relatively large (about 100-μm) columns must have been sufficiently random to produce the 22° halo. It is easy to believe, in view of the large hourglass indentations in the ends, that the columnar crystals produced no 46° halo.

Figure 4 shows crystals collected February 17, 11 a.m., at −26°C (29% short, simple, well-formed columns; 27% thin, simple plates; 29% ornate plates; 10% spatial crystals; 5% unclassified). These crystals produced fairly strong parhelia and side portions of the 22° halo (Plate V), a circumzenithal arc, a subsun, subparhelia, a weak pillar, and perhaps parts of a faint parhelic circle. The simple, nearly equidimensional hexagonal prisms are intriguing. (Figure 4 shows several; two appear together at the upper right, one resting on a basal face and one on a prism face.) One can hardly imagine better-shaped halo makers. Yet I saw neither the 46° halo nor the top and bottom of the 22° halo, even though the subsun and, at times, the subparhelia were visible in the few meters between my eye and the ground, partially in shadow. This suggests that these equidimensional crystals were falling with

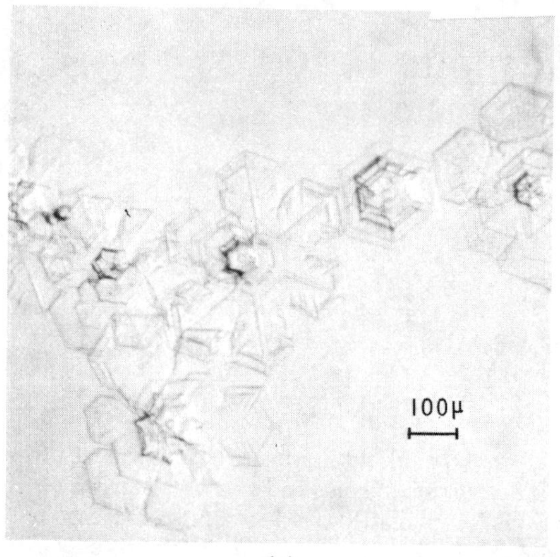

(a)

(b)

Fig. 1.   (a) Some ice crystals collected during the light pillar display of (b).

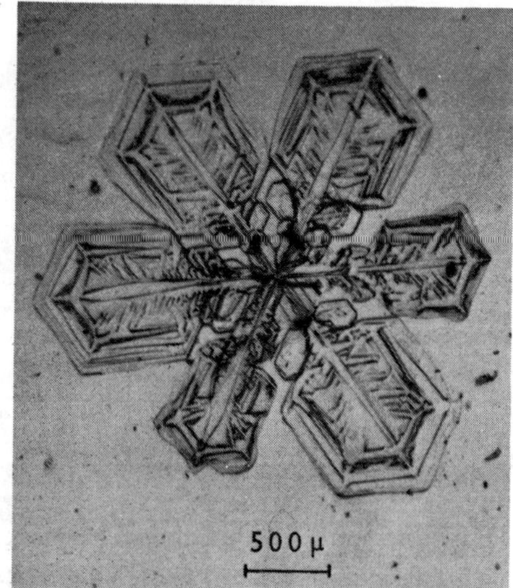

Fig. 2.   Replica of typical crystal collected during the light pillar display in Plate IV.   The other large crystals were similar to this one. Some much smaller plate crystals were also present.

principal axes nearly vertical and were producing the halos.

Several factors weaken the above unorthodox conclusion. First, the sample is not homogeneous, unfortunately, and is probably biased toward the faster-falling crystals.   Second, more crystals are required to produce a detectable 22° halo than a parhelion.   Third, looking nearly horizontally at the parhelia (Plate V), we are looking through a larger thickness of crystals than when looking up toward the top of the 22° halo region.

On the other hand, scattered through the halo literature[2-7] are indications that parhelia may sometimes be caused by

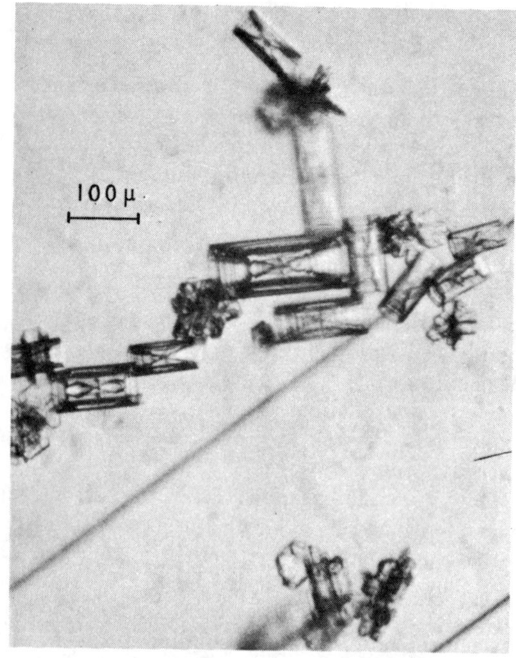

Fig. 3.   Some crystals that produced a 22° halo that was very slightly enhanced at the top and sides.

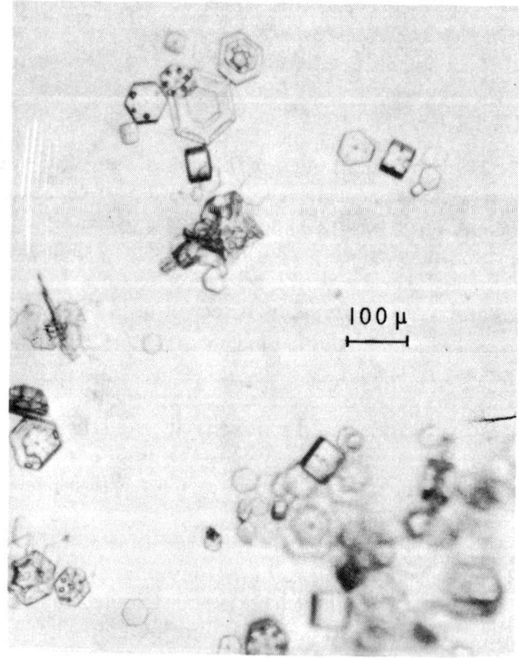

Fig. 4.   Some crystals that produced the halo display of Plate V.

428

crystals other than thin hexagonal plates. Liljequist (Ref. 5, p. 83), for example, once observed a 22° halo, strong parhelia (actually paraselenae), a 46° halo, and a pillar. The ice crystals at the time were only short columns. I used to think that perhaps Liljequist had not checked to be sure that the halos were occurring at low level, where his crystals were collected. But perhaps short columns do sometimes fall with vertical principal axes.

Figures 5(a) and 5(b) show crystals collected February 20, 6:15 a.m., at −31°C (50% polyhedral, nonprismatic crystals; 17% simple, well-formed short columns; 2% plates; 31% unclassified). These crystals produced a weak circular halo, which I assumed at the time to be a 22° halo, a weak circumzenithal arc or portion of the 46° halo, weak parhelia, and pillars.

The replicas (not shown) contain considerable numbers of exotic, somewhat planar crystals like the two in the lower

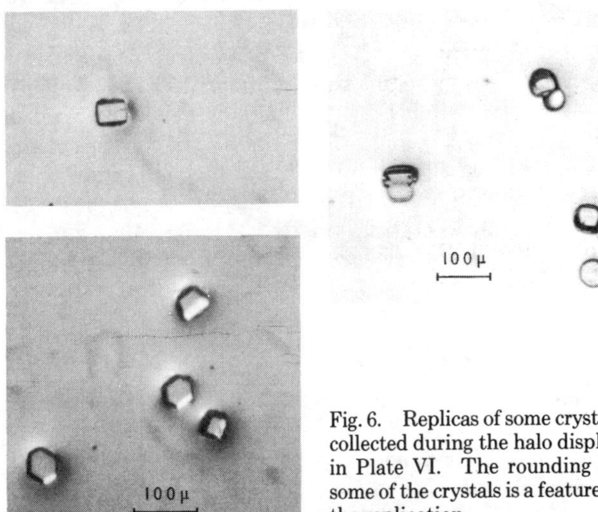

Fig. 6. Replicas of some crystals collected during the halo display in Plate VI. The rounding on some of the crystals is a feature of the replication.

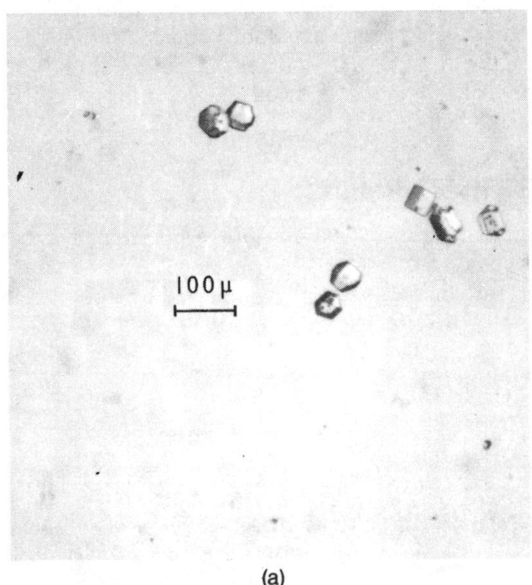

(a)

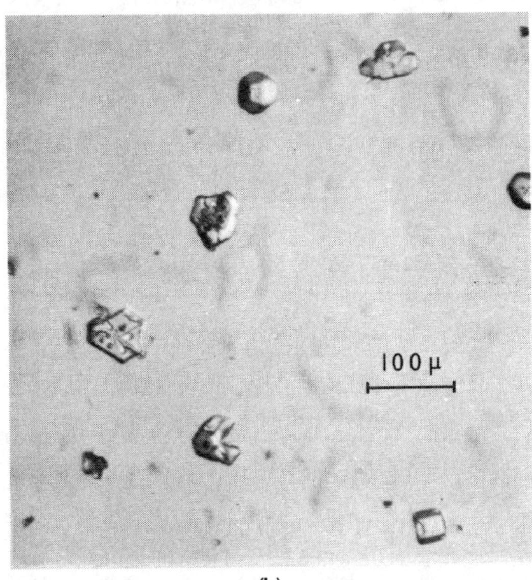

(b)

Fig. 5. (a) and (b) Some crystals that produced a weak 22° halo, weak parhelia, a weak circumzenithal arc or portion of the 46° halo, and pillars.

Fig. 7. Some crystals that produced no halos.

left-hand part of Fig. 5(b). Such crystals perhaps contributed to the pillars. The replicas also contain an occasional hexagonal plate crystal, but the plates are rare enough so that I wonder again whether the equidimensional simple prisms [one shown in each of Figs. 5(a) and 5(b)] might have tended to align with their principal axes vertical, so as to cause the parhelia and circumzenithal arc. Such crystals would probably also contribute to the apparent 22° halo if they oriented more randomly.

The most interesting crystals are the polyhedral, nonprismatic crystals. Four appear in Fig. 5(a) and two in Fig. 5(b). They apparently are not rare in Fairbanks; I found them in quantity on at least three occasions during the winter of 1982–1983, and they have been seen previously,[8] although at much lower temperatures. Circular halos of odd radius[1,9,10] have often been attributed to such crystals. I noticed no such halos, and halo photographs taken at the time show only pillars. But the halos mentioned above were weak even to the eye, and the crystal swarm was not dense, so failure to notice odd-radius halos is probably not significant, especially since the light source was nearby. Unfortunately, the sky was cloudy when the sun rose later in the morning.

Figure 6 shows replicas of crystals collected December 22, about 12:30 a.m., at −29°C (44% polyhedral, nonprismatic crystals; 30% plates; 16% short columns; 10% unclassified). These crystals produced a 22° halo, parhelia, and a pillar (shown in Plate VI); and a 46° halo and a circumzenithal arc. Again, I do not believe it is significant that odd-radius halos were not noticed in this streetlight display.

Figure 7 shows crystals collected February 19, 2 a.m., at −19°C. These crystals produced no halos, not even pillars or subsuns.

NEARBY-SOURCE HALOS

The halos associated with Figs. 1–3, 5, and 6 were produced at night by streetlights or other artificial light sources. At one time or another, and not restricted to Alaska, I have seen all the common halos in this way. Nearly all of them are strikingly three dimensional, since the light source is not at infinity. The 22° halo becomes a cigar-shaped[11] hollow in the crystal swarm, the 46° halo becomes a larger hollow,[12] the parhelia become bands—easily mistaken for a parhelic circle—that appear to line the 22° hollow, etc. Plate VI, showing the 22° halo and parhelia, in no way does justice to the beautiful spatial character of these halos. In the planar photograph the cigar is flattened to a disk and the bands are flattened into a horizontal line segment.

Incidentally, determination of the relative mix between 22° halo and upper and lower tangent arcs, as in the halos associated with Fig. 3, can be tricky unless the halos are strong. Usually the light source is higher than the observer, so, as the observer moves closer to, and hence under, the source to get a better look, the upper and lower tangent arcs evolve into a circumscribed halo that becomes more and more nearly circular. Another suggestion to those who try to observe nearby-source halos: one should choose a source that is not overly directional, as are most streetlights and other outdoor lighting.

FURTHER WORK

I have presented the limited results of Figs. 1–7 partly to encourage others who live in cold climates to look at crystals during low-level halo displays. Establishing an empirical correlation between halo types and crystal types is a long-term project that needs to be carried out in diverse regions. The conditions in Fairbanks tend to be unnatural, since the moisture is often coming from warm power-plant exhaust or from open water downstream from the plant.[13] Of course, in all areas it would be desirable to complement the low-level work with studies from aircraft.

Apart from intrinsic interest, there are several reasons for looking at ice crystals during halo displays:

1. The theory of halos[1,14] uses ray tracing to associate specific halos with specific idealized ice-crystal models. An essential assumption is the specification of the orientation of (usually) two crystal faces. Examining crystals during halo displays serves as a partial test of the theory and, perhaps more importantly, reveals what actual physical crystals can supply the oriented faces assumed by the theory.

2. By looking at the crystals, we may be able to establish an association of halo type with crystal type that is more re-fined than that provided by the ray-tracing theory. For example, the 22° halo can probably be caused by many types of crystals. Is it possible to infer the crystal type from the character of the resulting 22° halo?

If we understood *in detail* the relations between crystal type and halo type, and between atmospheric conditions and resulting crystal type, we might be able to infer from halos the atmospheric conditions in the cloud where the crystals grew.

3. Examination of ice crystals that produce halos leads to knowledge of the ice crystals themselves, particularly their natural falling attitudes.

Liljequist has made the most serious effort in the field to correlate halo types with crystal types. His fundamental work[5] lists halo and crystal observations gathered during 18 months in Antarctica in 1950–1951. Currie[15] and Sassen[16] looked at crystals causing night light pillars. Ohtake and Jayaweera[13] discuss some parhelia and subsun displays in Fairbanks. Hattinga-Verschure[17] and Mattsson[18] describe some halos from nearby sources. From an aircraft Weickmann[19] collected crystals from clouds, many of which were producing halos.

ACKNOWLEDGMENTS

I would like to thank Takeshi Ohtake and Kolf Jayaweera for helpful conversations.

This research was partially supported by a grant from the University of Wisconsin, Eau Claire.

REFERENCES

1. R. Greenler, *Rainbows, Halos, and Glories* (Cambridge U. Press, 1980).
2. A. Dobrowolski, "Les cristaux de glace aeriens et le phenomene des halos," Ciel Terre **28**, 183–342 (1907).
3. A. Dobrowolski, "Über neue Beobachtungen von Eiskristallen, welche die Haloerscheinungen bewirken," Meteorol. Z. **26**, 433–437 (1909).
4. F. Heim, "Diamantstaub und Schneekristalle in der Antarktis (Wedellsee)," Meteorol. Z. **31**, 232–235 (1914).
5. G. Liljequist, "Halo phenomena and ice crystals," in *Norwegian-British-Swedish Antarctic Expedition 1949–1952, Scientific Results* (Norsk Polarinstitutt, Oslo, Norway, 1956), Vol. II, Part 2A, pp. 1–85 and plates.
6. M. Kuhn, "Ice crystals and solar halo displays, Plateau Station, 1967," in *International Symposium on Antarctic Glaciological Exploration* (International Association of Scientific Hydrology, 1970).
7. V. Schaefer and J. Day, *A Field Guide to the Atmosphere* (Houghton Mifflin, Boston, Mass., 1981), Color Plate 15.
8. T. Ohtake, "Unusual crystal in ice fog," J. Atmos. Sci. **27**, 509–511 (1970).
9. R. Tricker, "Arcs associated with halos of unusual radii," J. Opt. Soc. Am. **69**, 1093–1100 (1979).
10. F. Turner and L. Radke, "A rare observation of the 8° halo," Weather **30**, 150–156 (1975).
11. M. Minnaert, *The Nature of Light and Color in the Open Air* (Dover, New York, 1954), p. 206.
12. Theoretically, the boundary surface of each hollow is a surface of revolution formed as follows: Let E be the eye and let S be the light source. Choose a point O so that angle EOS is $2D$, where D is the minimum deviation angle of the circular halo in question. In the plane of EOS draw the circular arc through E and S with center O. Revolve this arc about the axis ES to generate a surface. The resulting surface is the boundary of the hollow.
13. T. Ohtake and K. Jayaweera, "Ice crystal displays from power plants," Weather **27**, 271–277 (1972).

14. W. Tape, "Analytic foundations of halo theory," J. Opt. Soc. Am. **70,** 1175–1192 (1980).

15. B. Currie, "Ice crystals and halo phenomena," Mon. Weather Rev. **63,** 57–58 (1935).

16. K. Sassen, "Light pillar climatology," Weatherwise **33,** 259–262 (1980).

17. P. Hattinga-Verschure, "Streetlamp halos," Weather **38,** 48–52 (1983).

18. J. Mattsson, " 'Subsun' and light-pillars of street lamps," Weather **28,** 66–68 (1973).

19. H. Weickmann, "Formen und Bildung atmospharischer Eiskristalle," Beitr. Phys. Atmos. **28,** 12–52 (1945).

Plate III.   (Walter Tape, p. 1641).   Broad parhelion seen between the photographer and the building.   Such a situation establishes that the halo is being formed by crystals at low level.   The sun is out of the picture at the right.

Plate IV.   (Walter Tape, p. 1641).   Light pillar display associated with Fig. 2.

Plate V. (Walter Tape, p. 1641). Halo display associated with Fig. 4.

Plate VI. (Walter Tape, p. 1641). Nighttime 22° halo, parhelia, and pillar formed by nearby light source. The source is hidden by the dark oval cardboard in the center. A 15-mm lens was used in an unsuccessful attempt to record the 46° halo, which was visible to the naked eye. Replicas of the corresponding crystals are shown in Fig. 6.

Multiple-scattering effects in halo phenomena

E. Tränkle

Freie Universität Berlin, FB Physik, WE 4, Arnimallee 14, D-1000 Berlin 33, Federal Republic of Germany

Robert G. Greenler

Department of Physics, The University of Wisconsin-Milwaukee, Milwaukee, Wisconsin 53201

Received August 23, 1986; accepted November 19, 1986

Using a hit-and-miss Monte Carlo method, we followed many paths of Sun rays through a layer of hexagonal ice crystals. The angle of scattering by a single crystal is determined by tracing the path of the Sun rays through the crystal with its particular orientation. In this way we produced point plots of the intensity of the scattered sunlight for a specified population of crystals. By comparing the intensity patterns from single and multiple scattering, we looked for additional effects caused by multiple scattering. Several new secondary structures appear; however, their intensities are significant only when they result from spotlike primary structures. It appears that secondary structures from two combinations should be observable: from thick-plate crystals, the sun dogs of the sun dogs; and for long-column crystals in the Parry-arc orientation, the subsun of the lower tangent arc for Sun elevations around 15°. We show that multiple scattering is essential to explain the Saskatoon display.

INTRODUCTION

To our knowledge all previous calculations of halo phenomena are based on single scattering.[1–3] A single-scattering event may consist of several internal reflections in the same ice crystal. When sunlight is scattered by a layer of ice crystals, some rays are scattered by two or more ice crystals in the layer, and some pass through without any scattering.

Because of multiple scattering the background intensity increases, and the sharp structures of single scattering are smeared out. A more interesting effect is that additional structures should appear, e.g., the 22° parhelia of the 22° parhelia. The question, of course, is whether the intensities of the secondary and possible tertiary structures are large enough to be observable.

The Monte Carlo approach, as developed by Pattloch and Tränkle,[2] yields a point plot of the intensity of the reflected and refracted sunlight for a specified population of ice crystals. In particular, the relative intensities of all circles, arcs, and spots are predicted even for those effects that result from multiple reflections within one ice crystal. We extend this approach, taking into account scattering from more than one ice crystal, to look for additional phenomena.

In the region near the Sun, multiple-scattering effects of reflection and refraction may be superimposed upon diffraction effects. Takano and Asano showed that randomly oriented crystals produce a spotlike diffraction structure around the Sun and that oriented crystals produce a cross-like or diamondlike structure in the Sun region.[4]

As is well known, most of the halo phenomena are caused by three different kinds of oriented crystal: randomly oriented crystals, oriented plates, and oriented columns with the main *c* axis lying horizontal. Photos recently taken by Tape clearly show that the columns can be further oriented with respect to the rotation angle (Parry-arc orientation).[5]

In this paper we first give an overview on possible multiple-scattering effects by showing the intensity patterns of a typical example for each orientation. Then we present a simulation of the extraordinary Saskatoon halo display of 1970.[6,7]

THE MONTE CARLO METHOD

Earlier investigations of halo phenomena used ray-tracing methods with paper and pencil. Today the computer is a very powerful tool to investigate these effects. The hit-and-miss Monte Carlo procedure is a conceptually simple computer method, which has been successfully applied to many problems. We introduced this method to simulate single scattering of sunlight by ice crystals in the framework of geometrical optics.[2] In geometrical optics a ray is split at a surface into a reflected and transmitted ray according to the Fresnel equations. In the spirit of hit-and-miss Monte Carlo, we take the ray to be either reflected or transmitted at each encounter with a surface, with the probabilities R^2 and $1 - R^2$, respectively. One follows the path of a ray through the crystal until it leaves the crystal.

The hit-and-miss Monte Carlo procedure is also appropriate to simulate multiple scattering of sunlight by a layer of ice crystals (Fig. 1). We follow the path of a Sun ray through a horizontal, plane-parallel layer of optical depth D (in units of the mean free path). The incident angle is given by the Sun elevation. We draw a random free path l from the distribution $p(l) = \exp(-l)$ and a new direction from the single-scattering algorithm. We repeat the last steps until the ray leaves the layer, above or below, and plot a point at the intersection of the ray with the drawing plane.

By repeating this random process for many incoming rays we produce a point plot of the intensity of the scattered sunlight from a population of ice crystals in a layer of depth

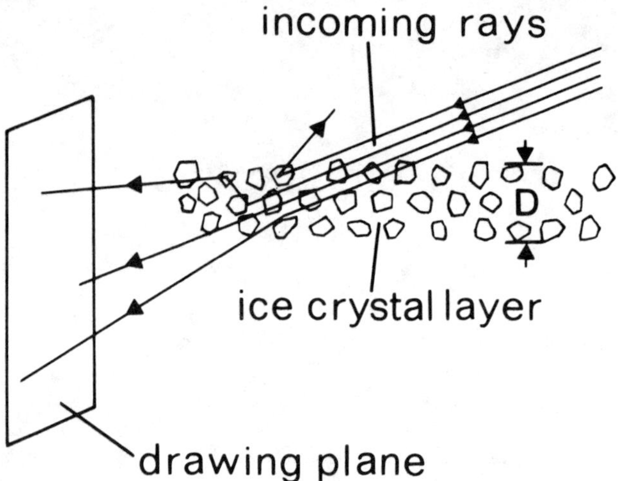

Fig. 1.   Model of ray tracing through the ice crystal layer.

D. In particular, we obtain the relative intensities of all circles, arcs, and spots. We also calculate the mean scattering multiplicity, the reflection and transmission coefficients of the layer, and the fraction of directly transmitted light.

The simplified geometry of the layer can be easily changed in our algorithm, and a second layer can be modeled. Different kinds of crystal can be put in the same layer or in different layers. It is well known that the hit-and-miss Monte Carlo method is well suited even for complex geometries of the medium.

The main shortcoming of the present version of the algorithm is that it is based on geometrical optics and does not include diffraction scattering. Therefore it does not describe the additional structures caused by diffraction in the Sun region and possible smearing out of sharp reflection and refraction structures due to diffraction.

The calculations were performed on the Cray 1 M computer of the Konrad-Fure Computer Center in West Berlin. A typical simulation with 500,000 incoming rays requires a memory of about 140,000 words and 22 min of CPU time. The point plot was stored in a matrix by using bit arithmetic.

ADDITIONAL STRUCTURES

In order to look for additional structures we did simulations for randomly oriented crystals, oriented plates, singly oriented columns, and columns with the Parry-arc orientation.[1]

The orientations are specified by distribution functions of three Euler angles [cf. Eqs. (4) and (5) of Ref. 2]. Instead of the box distribution for the maximum tilting angle of oscillations [Eq. (5b) of Ref. 2] we now use a Gaussian distribution. The largest maximum tilting angle θ_{lm} is now the width of the Gaussian distribution. For each orientation we present the results for a typical example that is supposed to have most intense primary halo structures. The depth of the layer is chosen to be 0.5, which corresponds to a mean scattering multiplicity of approximately 2. Therefore the secondary structures and even possible tertiary structures should show up. The parameter values and the calculated scattering coefficients are given in Table 1.

Figure 2 shows the intensity plots for random orientation. For this value of the aspect ratio (a = diameter/height), a strong 22° halo and a very faint 46° halo occur in the case of single scattering. The main difference in the plot including multiple scattering is an increased background intensity. No additional spot in the center and no additional circle at 44° (the 22° halos of the 22° halo) show up. The vanishing of the halo at the horizon is due to the infinite extension of the horizontal layer.

Sometimes small-angle scattering (giving the appearance of an enlarged Sun) is observed together with a halo (plate 2-1 of Ref. 1). This scattering cannot be explained by single or multiple refraction scattering. Most likely it is produced by diffraction scattering by ice crystals[5] or possibly by water droplets or aerosols in a low-lying layer of fog or haze.

Figure 3 shows the results for oriented thick plates. In this case additional structures that are due to multiple scattering clearly show up. The 22° parhelia produce two secondary effects: a diamondlike structure in the Sun region and additional parhelia at 44°. They should be observable. Their intensity is relatively small compared with that of their parent parhelia. Three other multiple-scattering effects can be seen in the simulation: subparhelia at 44°, a subcircumzenithal arc, and a continuation of the circumzenithal arc (Kern's arc). These three structures are probably too faint to be seen. The most intense tertiary structures would be expected to be parhelia at 66°. They do not appear for this choice of the parameters.

By varying the parameters we found that the intensity of the 44° parhelia rapidly decreases with increasing aspect ratio (thinner plates) and vanishes for $a = 5$. Therefore an observation of the 44° parhelia implies the presence of thick plates in the air.

Table 1.   Parameter Values for Four Orientations

Parameter	Value for the Following Orientation			
	Random	Plates	Columns	Parry Columns
Incoming rays	500 000	500 000	500 000	500 000
Aspect ratio a	0.25	1.5	0.25	0.25
Tilting angle θ_{lm} (°)	—	5	0.5	0.5
Rotation angle ϕ_{lm} (°)	—	—	—	2.0
Sun elevation (°)	—	11.5	11.5	15.0
Depth of layer D^a	0.5	0.5	0.5	0.5
Mean multiplicity	2.0	2.5	1.9	1.8
Direct transmittance (%)	8	8	8	14
Reflectance (%)	38	34	42	42

^{a} Expressed in units of the mean free path.

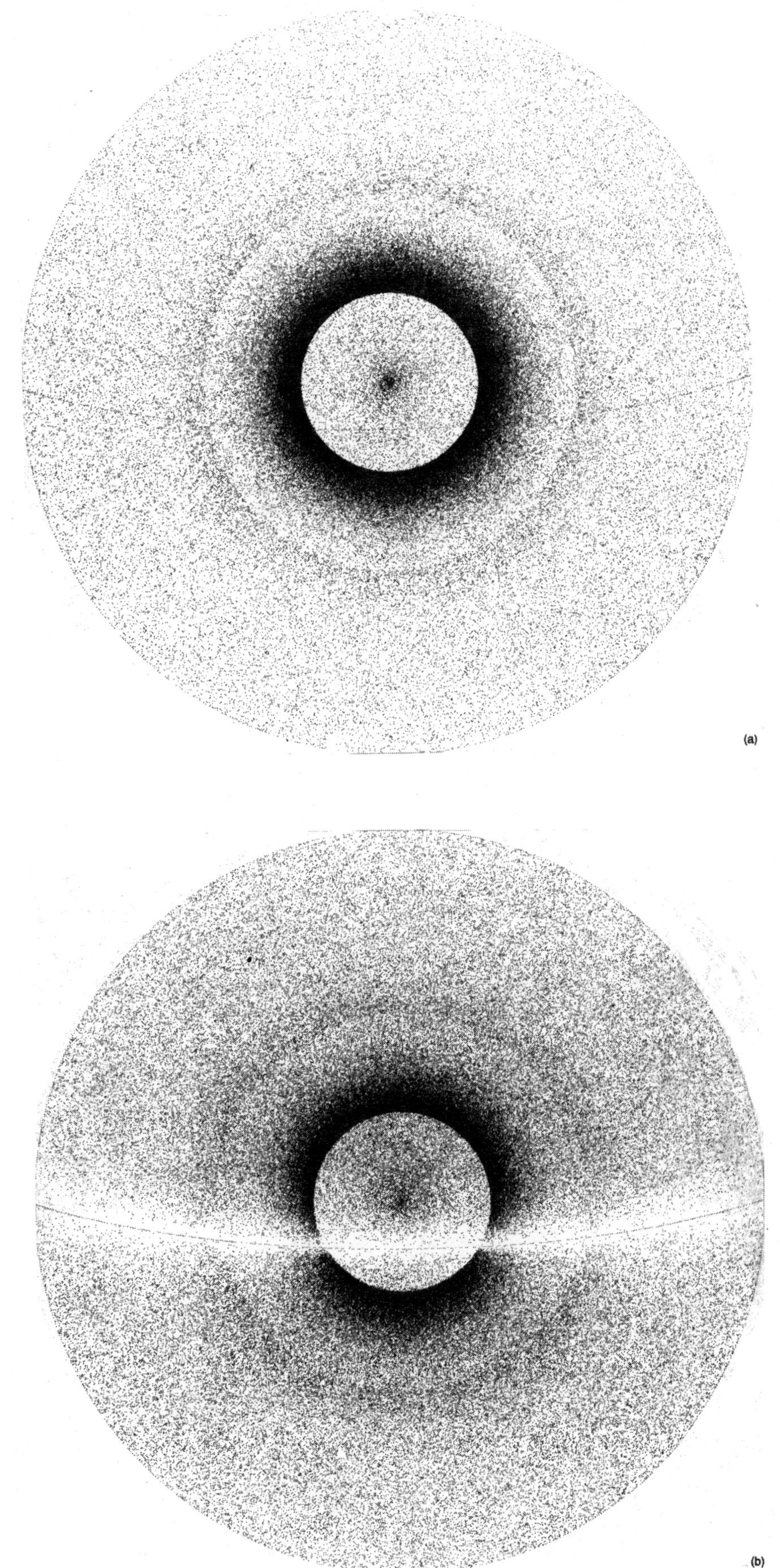

Fig. 2. Intensity plots for randomly oriented crystals: (a) single scattering, (b) multiple scattering.

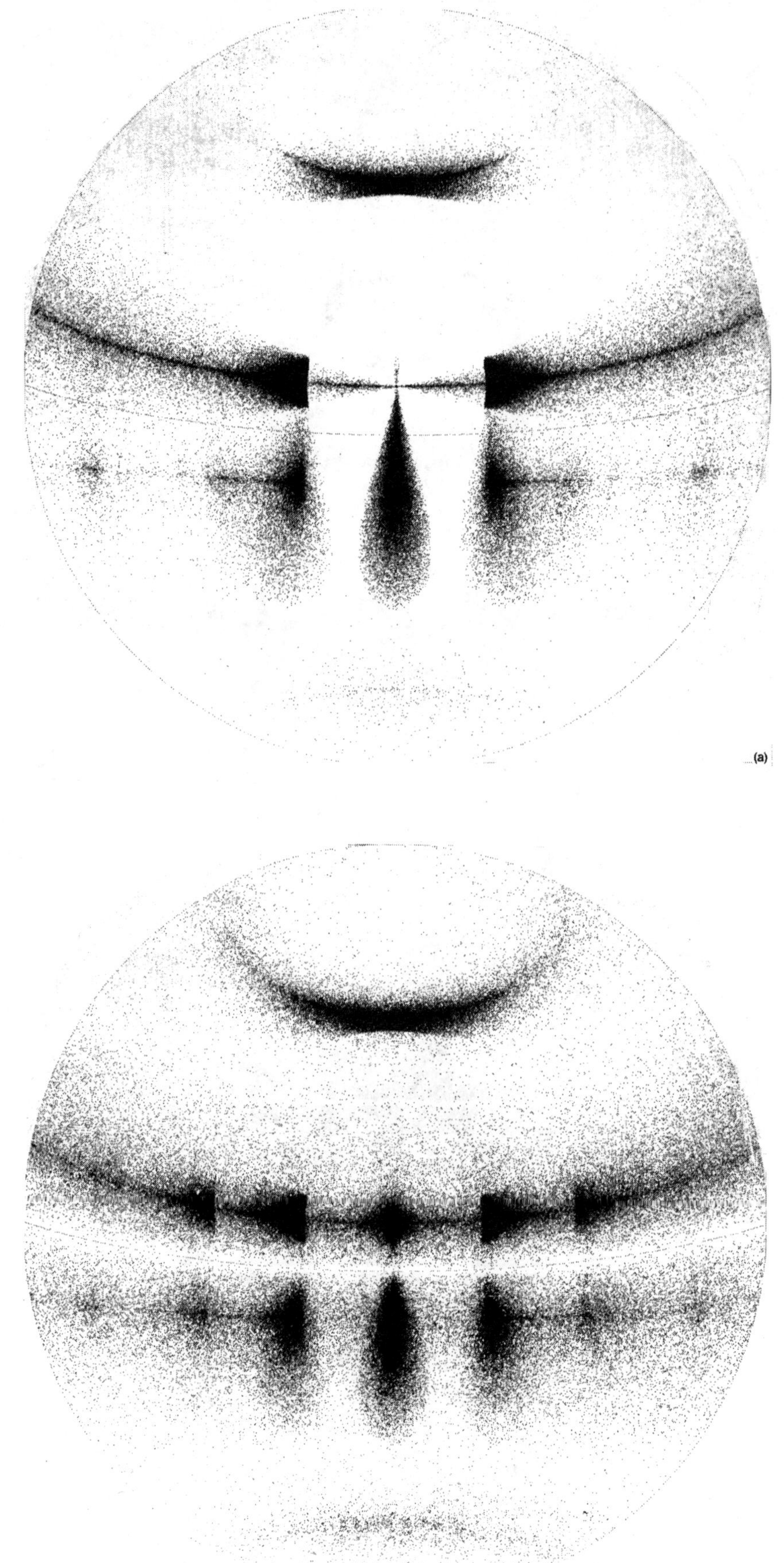

Fig. 3.   Intensity plots for thick, oriented plates: (a) single scattering, (b) multiple scattering.

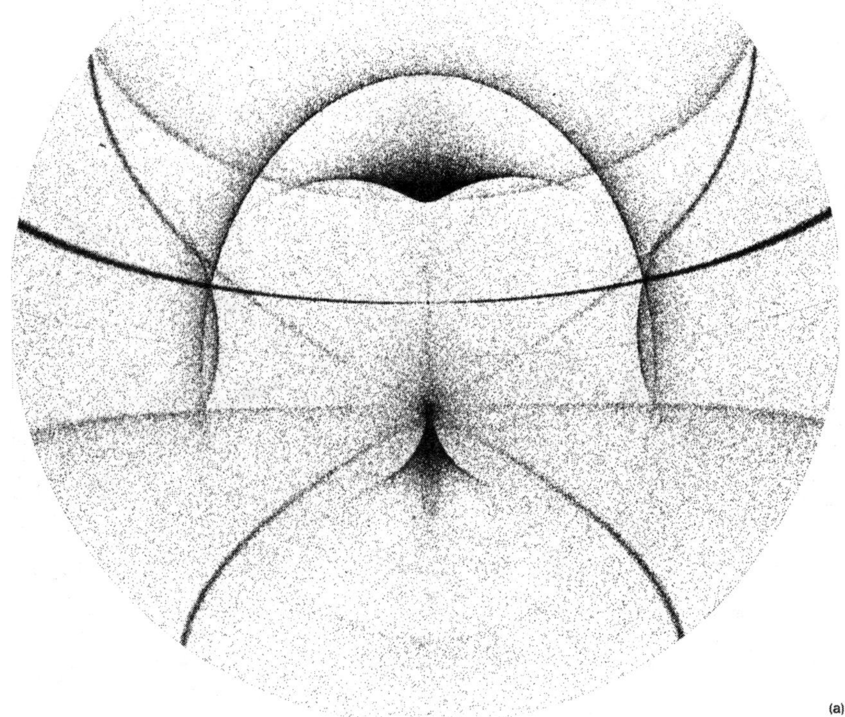

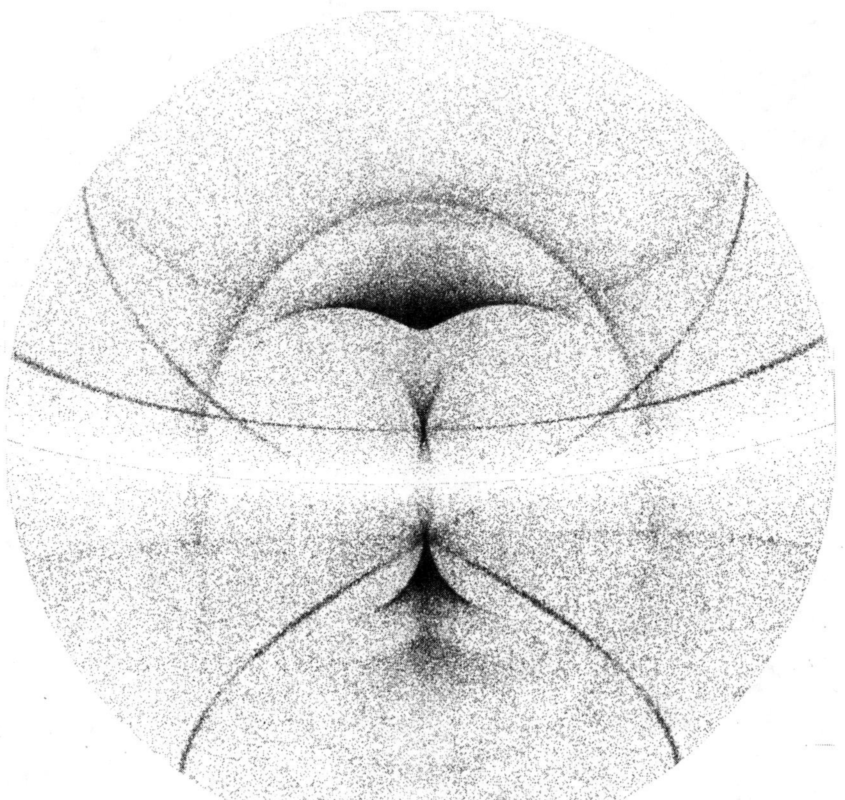

Fig. 4. Intensity plots for long, singly-oriented columns: (a) single scattering, (b) multiple scattering.

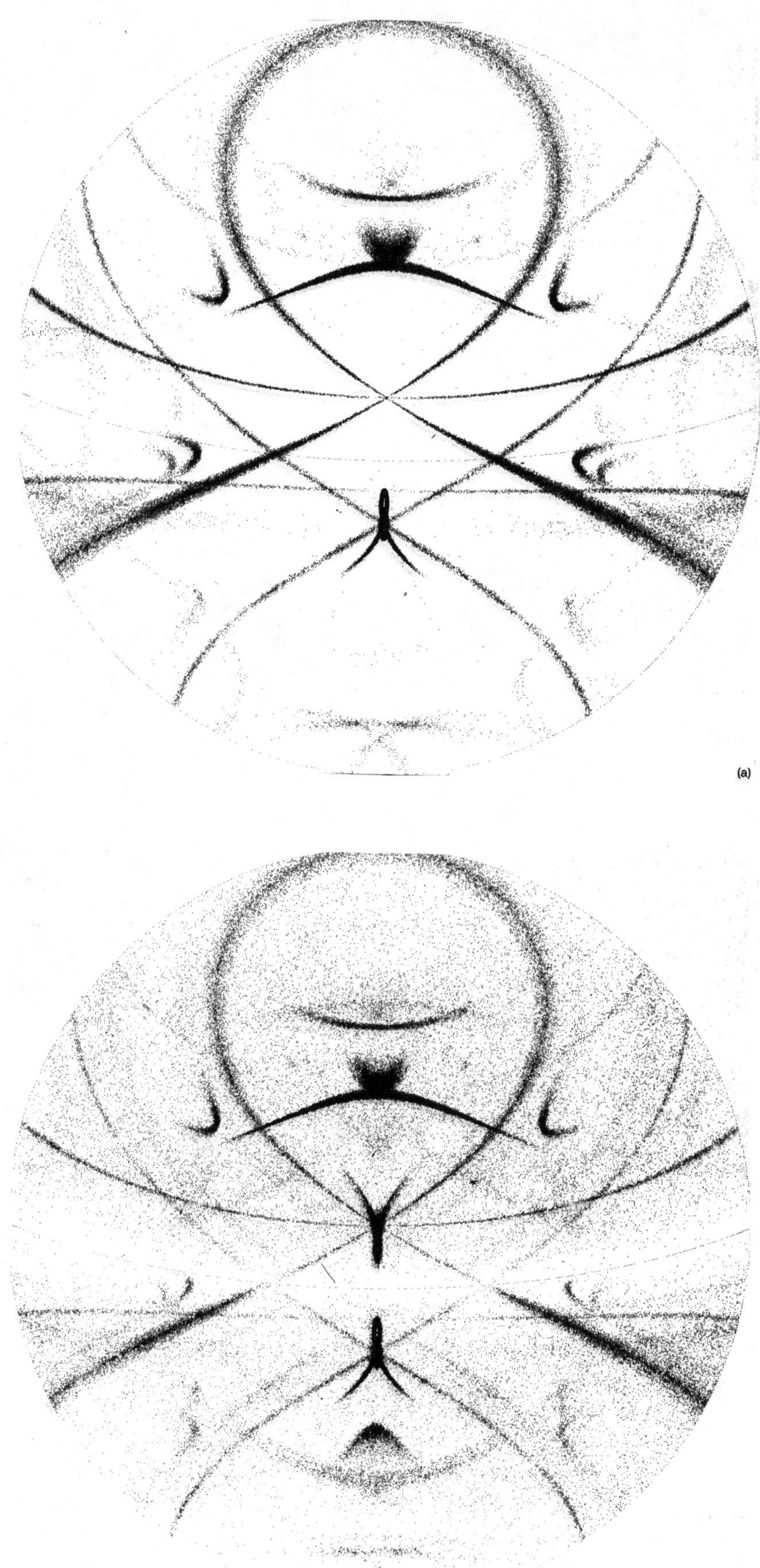

Fig. 5.   Intensity plots for long, Parry-arc-oriented columns: (a) single scattering, (b) multiple scattering.

Some of the best published photographs of parhelia at about 44° are from the Saskatoon display of 1970.[6,7] The authors call these spots parhelia of the 46° halo. However, the column crystals that produce the 46° halo would have to be oriented in a specific manner to produce 46° parhelia. The orientation would require a horizontal axis and a vertical pair of side faces. Such an orientation has been investigated (see Ref. 1, Chap. 2) as an alternative explanation of the Parry arcs. All the observations seem to indicate that this is not the correct distribution for Parry arcs, and there is no evidence that it is an aerodynamically stable orientation. Therefore we think that what have been called 46° parhelia in the past are 44° parhelia resulting from double scattering by oriented plates. This change of interpretation can be supported by our prediction that the 44° parhelia will occur only with the bright parent 22° parhelia and a fairly bright background. We did produce a simulation for the alternative Parry-arc orientations, and we saw an intensity pattern that is completely different from that in Fig. 3.

Sometimes crosslike or diamondlike structures are observed, superimposed upon the Sun, similar to that shown in Fig. 3. Plate XIV shows the effect. In the photograph the cross in the Sun region appears brighter than the 22° parhelion, which is not consistent with the simulation. The contribution from diffraction scattering[4] allows the diamond structure to be brighter than the 22° parhelia. If the diamond structure comes primarily from double scattering, then there should appear 44° parhelia, which are predicted to be of comparable intensity to the diamond-Sun structure.

Figure 4 shows the results for singly oriented columns. In the case of multiple scattering the background intensity is approximately constant apart from the horizon effect. An additional crosslike structure of fairly large intensity shows up in the Sun region. It is a superposition of the lower tangent arc of the upper tangent arc and the upper tangent arc of the lower tangent arc. Two other faint multiple-scattering effects also occur: the upper tangent arc of the upper tangent arc at 44° above the Sun and the lower tangent arc of the lower tangent arc at 44° below the Sun.

By varying the aspect ratio we found that the intensity of all three structures decreases with increasing a. For long columns, at least, the crosslike structure in the Sun region might be observable. For small crystals the observation can be hindered by the superposition of diffraction-scattering effects. Plate 2-21 of Ref. 1 shows the diamondlike diffraction shape when most of the oriented crystals present are column crystals.

Figure 5 shows the intensity plots for Parry-arc-oriented columns with a small rotation angle ($\phi_{lm} = 2°$). Because of the double orientation of the crystals, only a low background intensity shows up for single scattering. The "strange structure" in the subsun region is a superposition of the subsun and the lower sunvex Parry arc.[1]

In this case quite a few additional effects appear that are due to multiple scattering. The most intense structure is the looplike arc in the Sun region that is a superposition of the subsun and the lower sunvex Parry arc of the strange structure of single scattering. The upper Parry arcs of the strange structure show up at 44–66° below the Sun.

The looplike arc should be observable in the case of the

Table 2. Parameter Values for the Saskatoon Display

Parameter	Value for	
	Thick Plates	Thin Plates
Incoming rays	400 000	755 000
Aspect ratio	1.2	10
Tilting angle θ_{lm} (°)	10	15
Sun elevation (°)	14	14
Depth of layer D^a	0.5	0.5
Mean multiplicity	2.1	1.7
Direct transmittance (%)	13	19
Reflectance (%)	26	85

a Expressed in units of the mean free path.

Parry-arc-oriented medium and long columns. Because of the slight preference for singly oriented columns, Parry-arc orientation is suggested to occur only for rather large columns. In this case the diffraction pattern shrinks to a very small region in the direction of the Sun and should not disturb the observation of the looplike structure very much.

THE SASKATOON DISPLAY

We applied our multiple-scattering algorithm to analyze the Saskatoon display, which is an unusual display in several aspects. Parhelia at about 44° and a superposition of a Sun pillar and a diamondlike structure in the Sun region are sketched and photographed.[6,7] The intensities of these structures are small compared with those of the 22° parhelia, in agreement with our prediction. Two other unusual arcs are sketched by Ripley and Saugier: the continuation of the circumzenithal arc around the zenith (Kern's arc) and parhelia at about 66°. Two different unusual structures are reported by Evans: the anthelion and lateral tangent arcs to the 22° halo.

There were two layers of ice crystals in the atmosphere at the time of the most brilliant halo display. At 1130 hours the parhelia at 44° and 66° and the Kern arc vanished, although the 22° parhelia, the parhelic circle, the sun pillar, and the circumzenithal arc persisted until about 1400 hours.

To describe these observations we introduce two populations of oriented plates of small and large aspect ratio. The parameters of a good fit to the observations are given in Table 2. The thin plates ($a = 10$) do not produce 44° parhelia because the probability of two subsequent parhelia paths is small for this value of a (Fig. 6). In contrast to this, the probability of two subsequent reflections at the front faces (subsun of the subsun) is large and gives rise to the sun pillar in the Sun region.

In addition to the 44° parhelia and the diamondlike structure in the Sun region, the thick plates ($a = 1.2$) produce Kern's arc. It is consistent with the observation that this population disappeared at 1130 hours, whereas the thin plates stayed in the atmosphere until 1400 hours.

In the Saskatoon display faint 22° and 46° halos also appear. Assuming a single value of the aspect ratio for the whole population is, of course, an idealization. Some of the thick plates may have values $a \approx 1$ and tumble through the atmosphere with a nearly random orientation.

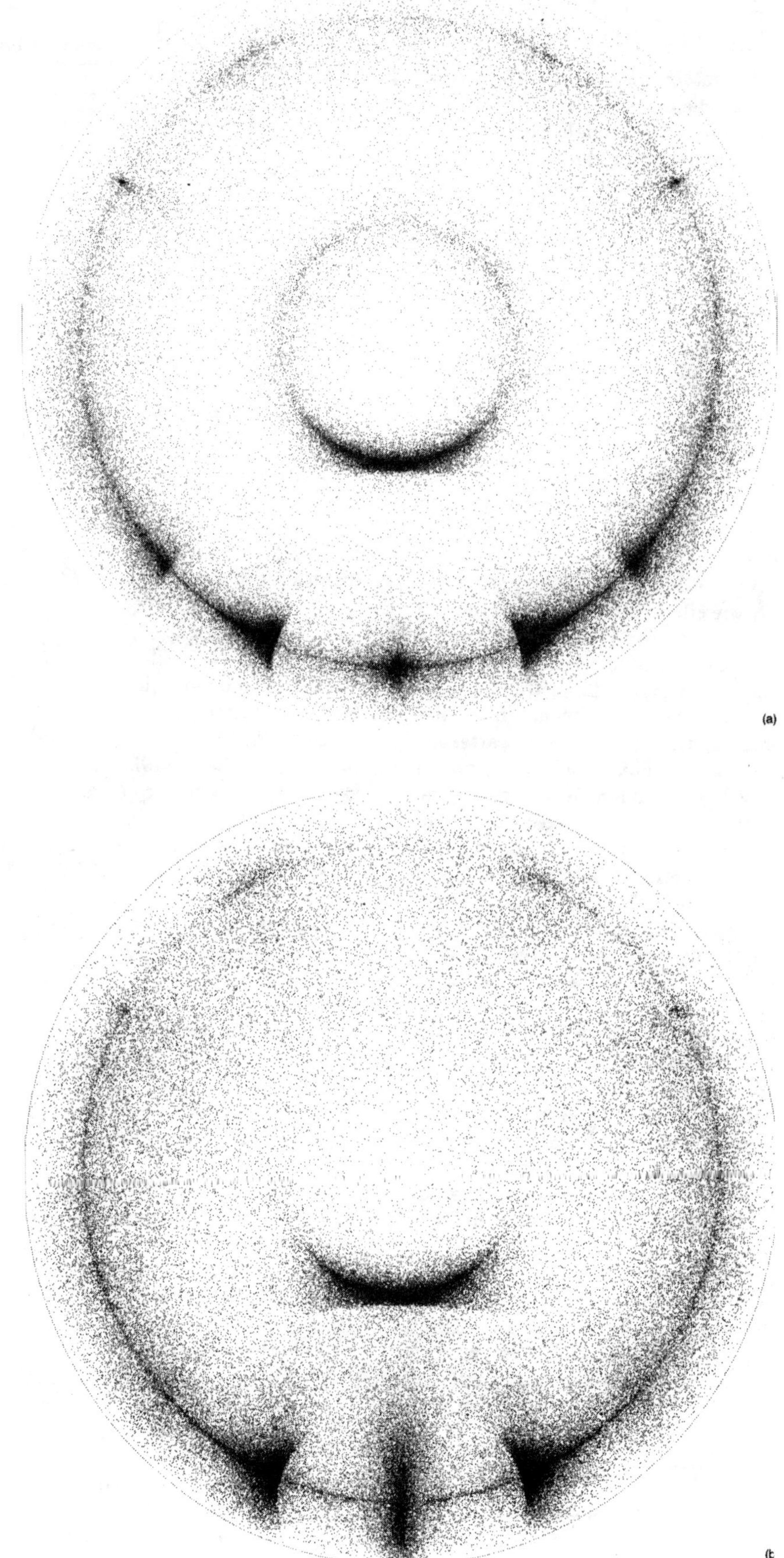

Fig. 6. (a) Contribution from thick plates ($a = 1.2$) to the Saskatoon display, (b) contribution from thin plates ($a = 10$) to the Saskatoon display.

ACKNOWLEDGMENTS

We thank W. Tape for helpful discussions. The kind assistance of the Computer Center of the Free University of Berlin is also gratefully acknowledged.

REFERENCES

1. R. G. Greenler, *Rainbows, Halos, and Glories* (Cambridge U. Press, New York, 1980).
2. F. Pattloch and E. Tränkle, "Monte Carlo simulation and analysis of halo phenomena," J. Opt. Soc. Am. A **1,** 520–526 (1984).
3. R. G. Greenler and E. Tränkle, "Anthelic arcs from airborne ice crystals," Nature (London) **311,** 339–343 (1984).
4. Y. Takano and S. Asano, "Fraunhofer diffraction by ice crystals suspended in the atmosphere," J. Met. Soc. Jpn. **61,** 289–300 (1983).
5. W. Tape, Department of Mathematics, University of Alaska, Fairbanks, Alaska 99701 (personal communication, 1986).
6. E. A. Ripley and B. Saugier, "Photometeors at Saskatoon on 3 December 1970," Weather **26,** 150–157 (1971).
7. W. F. J. Evans and R. A. R. Tricker, "Unusual arcs in the Saskatoon halo display," Weather **27,** 234–236 (1972).

Plate XIV. (E. Trankle and Robert G. Greenler, p. 591.) Sun with 22° parhelion. The diamond-shaped structure on the Sun in this photo probably results from diffraction rather than from double scattering. Photographed in Wisconsin by Robert G. Greenler.

Mirages and the Green Flash

3. *Elementare Theorie der atmosphärischen Spiegelungen;*
von Alfred Wegener.

Die häufigen Luftspiegelungen, sowohl nach oben wie nach unten, die ich während meines im ganzen vierjährigen Aufenthaltes in Grönland zu beobachten Gelegenheit hatte[1), und die ein ausgezeichnetes Material für eine Prüfung der Theorie abgeben werden, veranlaßten mich zu einer Beschäftigung mit den vorhandenen, wesentlich auf Tait und Biot basierenden theoretischen Arbeiten, und führten mich dadurch bald zu der Einsicht, daß diese mehr für experimentelle Untersuchungen im Laboratorium zugeschnitten sind als für die atmosphärischen Spiegelungen. Dies hat auch einer der neueren Theoretiker auf diesem Gebiete, W. Hillers, selber ausgesprochen: „Für die Untersuchung der wirklichen atmosphärischen Luftspiegelungen mit ihren labilen Verhältnissen, die kaum eindeutig aus gegebenen Bedingungen zu entwickeln sind wie jene im Experiment, wird die Behandlung wohl noch „phänomenologisch" bleiben müssen. Eine Spiegelung wird „erklärt" sein, wenn es geglückt ist, die Erscheinungen aus einer beobachteten Temperaturfunktion berechnet zu haben. Einer solchen Durchrechnung stellen sich aber meist außerordentliche mathematische Schwierigkeiten entgegen, auch wenn die zur Berechnung notwendige Temperaturfunktion genügend genau bekannt wäre. Schon die jetzt in den meisten Handbüchern wiedergegebene Darstellung nach Tait der Vinceschen Luftspiegelung läßt das

---

1) Einen Teil dieser Beobachtungen (mit zahlreichen Abbildungen) findet man im Wetterjournal der „Meteorologische Terminbeobachtungen am Danmarks-Havn", Danmark-Ekspeditionen til Grónlands Nordóstkyst 1906—1908. Bd. II. No. 4 (Meddelelser om Grónland, Bd. XLII), Kóbenhavn 1911. Das (meist photographische) Material von der Grönlanddurchquerung mit Koch 1912/13 ist noch nicht veröffentlicht.

erkennen."[1]) Der Grund liegt darin, daß man meint, der Temperaturverlauf mit der Höhe müsse unter allen Umständen ein stetiger sein, und es dürften also keine Temperatursprünge vorkommen. Man suchte also nach einer solchen stetigen Funktion, welche imstande war, Spiegelbilder zu liefern, und der Erfolg der theoretischen Behandlung hing dann davon ab, ob sich diese Funktion integrieren ließ, so daß man die Gleichung der Lichtkurve ableiten konnte. Dieser Ansatz erscheint für atmosphärische Spiegelungen, namentlich solche nach oben, die durch eine über dem Beobachter gelegene Temperaturinversion erzeugt werden, recht ungeeignet. Auf der Danmark-Expedition habe ich für 7 Fälle mit Spiegelung nach oben Höhenlage und Betrag der Inversion mit Drachen oder Fesselballon ermittelt und die vertikale Temperaturverteilung durch Zustandskurven nach den Registrierungen dargestellt.[2]) Wie nach unserer heutigen Kenntnis der Inversionen ohnehin zu erwarten war, zeigte sich, daß das Aussehen dieser Zustandskurven von Fall zu Fall und auch zeitlich (zwischen Aufstieg und Abstieg) sehr variierte. Es erscheint gänzlich ausgeschlossen, hier eine mittlere Form zu finden, die man etwa versuchen könnte, analytisch darzustellen. Aber etwas Gemeinsames ist in diesen Kurven doch zu erkennen: sie erscheinen sämtlich als Unstetigkeiten des Temperaturverlaufes, durch welche die Luft in zwei Schichten, eine untere kalte und eine obere warme, zerlegt wird. Freilich ist dieser Temperatursprung nicht scharf, sondern durch teilweise Mischung der Schichten verwaschen; aber diese Mischung ist so verschieden weit vorgeschritten und zeigt im einzelnen ein so verschiedenes Aussehen, daß es unzweifelhaft das Naturgemäße ist, sie in erster Näherung zu vernachlässigen und mit einem scharfen Temperatursprung zu rechnen, den die bisherige Theorie gerade prinzipiell ausschloß. Man ist ja auch auf anderen Gebieten längst an diese Auffassung gewöhnt, daß die Inversionen Diskontinuitätsflächen zwischen verschieden temperierten Luftschichten sind, ein

---

1) W. Hillers, Photographische Aufnahme einer mehrfachen Luftspiegelung. Physik. Zeitschr. 14. p. 718 u. 719. 1913.

2) A. Wegener, Drachen- und Fesselballonaufstiege. Danmark-Ekspeditionen til Grónlands Nordóstkyst 1906—1908. Bd. II. Nr. 1 (Meddelelser om Grónland, Bd. XLII), Kóbenhavn 1909.

Begriff, der zuerst von Helmholtz eingeführt wurde und sogleich zur Erklärung der Luftwogen und Wogenwolken führte. Auch auf dem Gebiete der atmosphärischen Spiegelungen bringt diese Auffassung der Inversionen als Diskontinuitätsflächen, wie gezeigt werden wird, eine ungeheure Vereinfachung des Problems und gestattet, mit elementaren Hilfsmitteln eine Theorie abzuleiten, welche eine weitgehende Vergleichung mit den Beobachtungen ermöglicht.

Noch in einem zweiten wichtigen Punkte weicht die vorliegende Theorie von der bisherigen ab: die Krümmung der Erdoberfläche, durch deren Mitnahme die alte Theorie hoffnungslos kompliziert geworden wäre, konnte jetzt ohne Schwierigkeit berücksichtigt werden. Es zeigte sich dabei, daß gerade durch sie manche Erscheinungen überhaupt erst ihre Erklärung finden, so daß schon wegen ihres Fehlens die alte Theorie ganz unzulänglich erscheint. Man kann sich dies leicht an einigen Zahlen klarmachen, nämlich wenn man berücksichtigt, wieviel stärker die Erdkrümmung ist als die der Lichtstrahlen. Der Krümmungsradius der (allein in Frage kommenden) nahe horizontalen Lichtstrahlen ist nämlich bei einem Temperaturgefälle von $3,4^0$ pro 100 m Höhe unendlich, bei Isothermie beträgt er immer noch ca. 4 Erdradien, und erst bei einem umgekehrten Gefälle von 11^0 pro 100 m Höhe wird er gleich dem Erdradius; sein gewöhnlichster Wert ist gleich 7 Erdradien. Unter diesen Umständen müssen also die Erscheinungen durch die Erdkrümmung ganz wesentlich beeinflußt werden, und es ist nicht zulässig, dieselbe zu vernachlässigen.[1])

Im folgenden wird der Versuch gemacht werden, die

---

1) In der Tat hat sich diese Vernachlässigung bereits gerächt; denn noch niemand, auch nicht Pernter in seiner „Meteorologische Optik" (Wien und Leipzig 1902) ist auf die richtige Erklärung der so häufig beschriebenen Verzerrung der Sonnenscheibe im Horizont gekommen, — eben weil sich die Theorie der Spiegelung nach oben nicht auf außerirdische Objekte anwenden läßt, solange man von der Erdkrümmung absieht. Die erste, qualitative Erklärung gab ich in meiner Abhandlung „Über die Ursache der Zerrbilder bei Sonnenuntergängen" in Beitr. z. Phys. d. freien Atmosphäre. Bd. 4. Heft 1. Die dort der Einfachheit halber gemachte Annahme, daß innerhalb jeder der beiden Schichten der Lichtstrahl geradlinig verläuft, wird in der vorliegenden Arbeit aufgegeben.

atmosphärischen Spiegelungen unter der Annahme zu berechnen, daß an der Schichtgrenze ein scharfer Temperatursprung herrscht, so daß der Lichtstrahl hier eine Brechung bzw. Reflexion erleiden kann. Dagegen wird vorausgesetzt, daß innerhalb der beiden Schichten normale Brechungsverhältnisse herrschen, d. h. der Lichtstrahl einen flachen Kreisbogen beschreibt, wie es in der Theorie der terrestrischen Refraktion gewöhnlich angenommen wird. Und endlich wird die Krümmung der Erdoberfläche und die konzentrische Krümmung der Schichtgrenze (selbstverständlich unter der vereinfachenden Annahme der Kugelgestalt) berücksichtigt werden.

Das Verhältnis des Erdradius zum Krümmungsradius des Lichtstrahles wird bekanntlich als Koeffizient der terrestrischen Refraktion (k) bezeichnet und kann entweder aus geodätischen Messungen oder aus dem Temperaturgefälle innerhalb der Schicht berechnet werden. Die Einführung von k in unsere Gleichungen gestattet also, auch das Temperaturgefälle innerhalb der Schichten zu berücksichtigen, wenn Beobachtungen darüber vorliegen. Ist dies nicht der Fall, so nimmt man für k einen mittleren Wert, worüber weiter unten nähere Angaben gemacht werden.

Wir werden in drei Abschnitten die Spiegelung nach oben, diejenige nach unten und die Zerrbilder der Sonne im Horizont behandelt.

1. Spiegelung nach oben.

Bei dieser Spiegelung liegt, wie erwähnt, eine Inversion oberhalb des Beobachters. Den Schlüssel zum Verständnis der Erscheinungen bietet der Satz, daß der *am Auge horizontal eintreffende Strahl an der Schichtgrenze mit dieser den kleinsten Winkel bildet*, d. h. den größten Einfallswinkel hat. Der Beweis hierfür läßt sich leicht aus Fig. 1 ableiten.

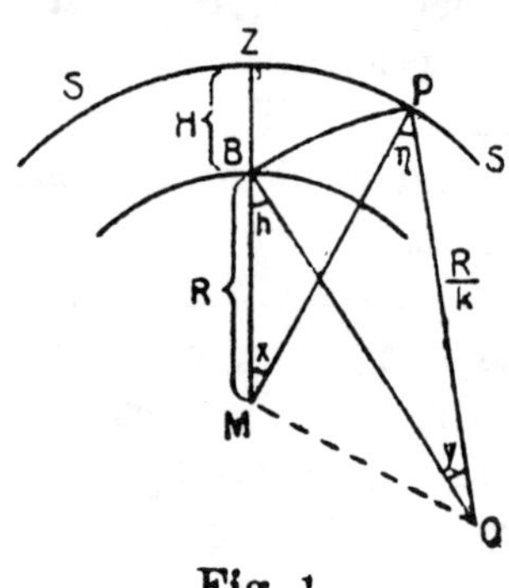

Fig. 1.

SS sei die Schichtgrenze. Der Beobachter B befinde sich an der Erdoberfläche. Vom Punkte P der Schichtgrenze erreicht ihn ein beliebiger Strahl PB, dessen Krümmungsmittelpunkt O

sei. M ist der Erdmittelpunkt. H sei die Höhe der Schichtgrenze, also $MP = R + H$, wo R der Erdradius ist. Nach der Definition von k ist $OB = OP = R/k$. $\measuredangle MBO = h$ ist gleich dem Höhenwinkel des Strahles beim Beobachter, $\measuredangle MPO = \eta = 90 - \alpha$ das Komplement des Einfallswinkels an der Schichtgrenze.

Das Quadrat der Seite MO ist nach dem Kosinussatz in den beiden Dreiecken MOB und MOP:

$$R^2 + \frac{R^2}{k^2} - 2\frac{R^2}{k}\cos h = (R + H)^2 + \frac{R^2}{k^2} - 2\frac{R}{k}(R + H)\cos \eta,$$

$$2\frac{R}{k}(R + H)\cos \eta = (R + H)^2 - R^2 + 2\frac{R^2}{k}\cos h,$$

$$\frac{R + H}{R}\cos \eta = \cos h + \frac{k}{2}\left[\left(\frac{R + H}{R}\right)^2 - 1\right]$$

Oder wenn wir $\dfrac{H}{R} = \varepsilon$ setzen:

$$(1) \qquad (1 + \varepsilon)\cos \eta = \cos h + \frac{k}{2}(2\varepsilon + \varepsilon^2),$$

worin wir $\cos \eta$ auch durch $\sin \alpha$ ersetzen können. Da das zweite Glied der rechten Seite ebenso wie der Faktor $(1 + \varepsilon)$ auf der linken nur von den vorgegebenen meteorologischen Bedingungen abhängt, ist ohne weiteres ersichtlich, daß der Einfallswinkel α seinen größten oder η seinen kleinsten Wert erreicht, wenn $\cos h = 1$ oder $h = 0$ ist. Damit ist die Behauptung erwiesen.

Für jeden Wert von $h > 0$ wird der Einfallswinkel α einen kleineren Wert annehmen, und da $\cos(-h) = \cos h$, so erhält man denselben Wert α auch für den gleichen Höhenwinkel mit entgegengesetztem Vorzeichen. Bezeichnen wir der Kürze halber den Strahl, der beim Beobachter horizontal eintrifft, als horizontalen Strahl, so können wir sagen: zu jedem Strahl oberhalb des horizontalen gehört ein symmetrischer unterhalb desselben, welcher an der Schichtgrenze den gleichen Einfallswinkel besitzt; der Einfallswinkel des horizontalen aber ist der größte.

Eine Spiegelung kann erst dann eintreten, wenn dieser Einfallswinkel den Grenzbetrag für Totalreflexion überschreitet. Vom Beobachter gesehen, muß sich dann die Spiegelung auf einen Streifen bestimmter Breite beschränken, *dessen Mitte genau im astronomischen Horizont liegt.*

In Fig. 2 stellt $P_1 Z P$ die Schichtgrenze dar, und der flache Bogen $P_1 P$ einen Strahl, welcher in P (und also auch in P_1) an der Schichtgrenze gerade den Grenzeinfallswinkel

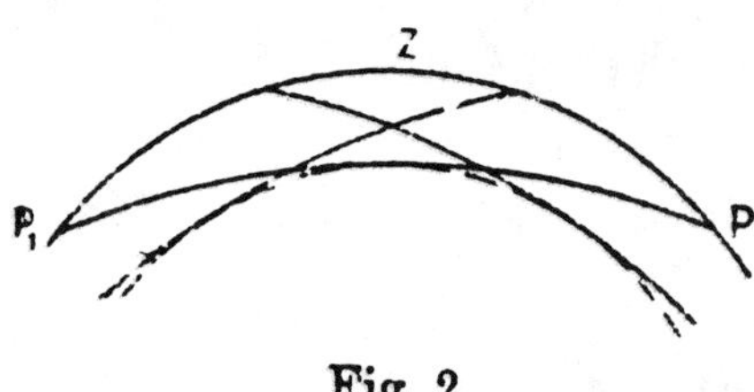

Fig. 2.

besitzt. Ziehen wir von allen Punkten der Schichtgrenze diese Grenzstrahlen, so erfüllen sie in ihrer Gesamtheit eine Kugelschale, in welcher sich der Beobachter und sein Objekt jedenfalls befinden müssen, wenn eine Spiegelung zustande kommen soll; denn alle Strahlen, die von einem Punkte unterhalb der punktierten Kreislinie gezogen werden können, treffen die Schichtgrenze unter kleinerem Einfallswinkel und passieren sie also, ohne zurückgeworfen zu werden. Die Mächtigkeit dieser Kugelschale wollen wir deshalb die *Wirkungstiefe* der Inversion nennen. Sie ist nur abhängig von den Temperaturverhältnissen derselben.

Über den *Einfluß der Entfernung auf die Art der gespiegelten Bilder* läßt sich einiges aus Fig. 3 entnehmen.

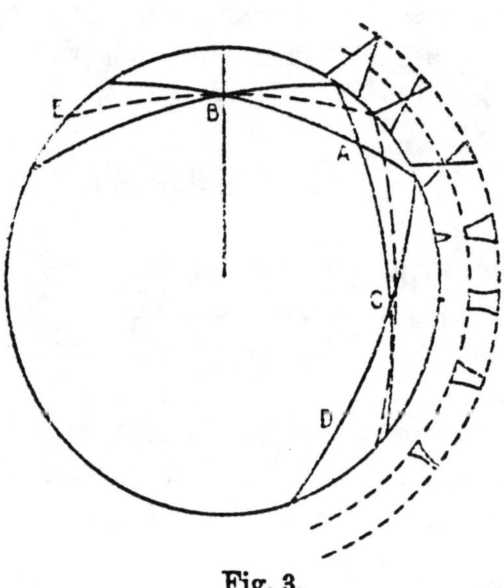

Fig. 3.

Vom Beobachter B sind darin die beiden Grenzstrahlen der Totalreflexion und der horizontale Strahl (dieser gestrichelt) gezeichnet. Rechnet man die rückwärtigen Verlängerungen bis E hinzu, so sind die beiden Grenzstrahlen offenbar gleich lang; der horizontale Strahl aber ist kürzer.

Infolgedessen müssen auch die beiden Reflexionspunkte der Grenzstrahlen bei E, A und D stets gleich weit voneinander entfernt sein, während der Reflexionspunkt des horizontalen Strahles sich bei jeder neuen Reflexion mehr gegen sie verschiebt. Der Strahlengang ist also nicht genau symmetrisch zur Reflexionsstelle A, und bei C schneiden sich nur die beiden Grenzstrahlen, während der horizontale hier etwas oberhalb dieses Schnittpunktes verläuft. Immerhin ist aber die Symmetrie noch so weit gewahrt, daß alle total reflektierten Strahlen, die bei A einen erheblichen Raum einnehmen, sich bei C knotenartig zusammenziehen. BC ist demnach diejenige Entfernung, in welcher die Objekte die größte Vertikalverzerrung erfahren; denn ein Objekt von geringer Höhe, das sich hier befindet, füllt für den Beobachter die ganze Breite des spiegelnden Streifens aus. Dicht jenseits von A, vom Beobachter aus gerechnet, ist dagegen die geringste Verzerrung des Spiegelbildes zu erwarten. Auch zwischen C und D nimmt die Verzerrung wieder ab; doch hält sie sich hier im oberen Teile des Spiegelbildes noch bis zur Reflexionsstelle. In der Figur sind direktes Bild und Spiegelbild an dem Beispiel einer steilen Pyramide schematisch erläutert: Während das direkte Bild mit zunehmender Entfernung vom Beobachter mehr und mehr unter den Horizont versinkt, sind die Spiegelbilder an den total reflektierenden Streifen gebunden, der stets in gleicher Höhe liegt. Die Spiegelung beginnt zunächst im oberen Teile dieses Streifens; bei zunehmender Entfernung dehnt sich das Spiegelbild nach unten aus, so daß die Pyramidenspitze schließlich unterhalb des spiegelnden Streifens liegen würde und also unsichtbar ist. Bei der in C stattfindenden Maximalphase der Vertikalverzerrung tritt nun eine interessante Erscheinung ein: jenseits derselben erscheint das Spiegelbild wieder aufrecht. *Die Vertikalverzerrung bildet also einen stetigen Übergang vom umgekehrten zum aufrechten Spiegelbilde.* Damit ist das Wesen dieser Vertikalverzerrung, die bei den Luftspiegelungen eine so überaus große Rolle spielt, erklärt.

Wir gehen nun dazu über, die Gleichungen für die quantitativen Berechnungen abzuleiten. Vorausgesetzt sei dabei die Berechnung des Brechungsquotienten der Schichtgrenze sowie des Refraktionskoeffizienten, da diese Größen fortwährend in den Gleichungen auftreten.

Um den *Brechungsquotienten der Schichtgrenze* aus dem dort herrschenden Drucke p und den absoluten Temperaturen T_1 und T_2 der unteren und oberen Schicht zu berechnen, berücksichtigen wir, daß sich nach einem empirischen Gesetz[1]

1) Vgl. z. B. Chwolson, Lehrbuch d. Physik. 2. Bd. p. 372. Braunschweig 1904.

der absolute Brechungsquotient ν der Luft (gegen den leeren
Raum) mit der Luftdichte ϱ so verändert, daß der Ausdruck
$\dfrac{\nu-1}{\varrho}$ konstant bleibt.

Es ist also

$$\nu = \frac{\varrho}{\varrho_0}(\nu_0 - 1) + 1\,,$$

wo sich ν_0 und ϱ_0 auf Luft von 0^0 C und 760 mm Druck beziehen. Ferner ist nach einem bekannten Gesetz der Brechungsquotient der Schichtgrenze gleich dem Verhältnis der absoluten Brechungsquotienten der unteren und der oberen Schicht:

$$n = \frac{\nu_1}{\nu_2} = \frac{\dfrac{\varrho_1}{\varrho_0}(\nu_0 - 1) + 1}{\dfrac{\varrho_2}{\varrho_0}(\nu_0 - 1) + 1}\,.$$

Da nun

$$\frac{\varrho_1}{\varrho_0} = \frac{p}{760} \cdot \frac{273}{T_1} \quad \text{und} \quad \frac{\varrho_2}{\varrho_0} = \frac{p}{760} \cdot \frac{273}{T_2}$$

ist, so wird

$$n = \frac{\dfrac{p}{760} \cdot \dfrac{273}{T_1}(\nu_0 - 1) + 1}{\dfrac{p}{760} \cdot \dfrac{273}{T_2}(\nu_0 - 1) + 1}\,.$$

ν_0 ist gleich 1,000294; wir können daher $(\nu_0 - 1)$ als kleine Größe betrachten, und folglich auch den ganzen Ausdruck

$$\frac{p}{760} \cdot \frac{273}{T_2}(\nu_0 - 1);$$

daher können wir genähert schreiben:

$$n = \left[\frac{p}{760} \cdot \frac{273}{T_1}(\nu_0 - 1) + 1)\right] \cdot \left[1 - \frac{p}{760} \cdot \frac{273}{T_2}(\nu_0 - 1)\right]\,.$$

Multiplizieren wir aus und vernachlässigen $(\nu_0 - 1)^2$, so wird

$$n = 1 + \frac{p}{760} \cdot 273\,(\nu_0 - 1)\left(\frac{1}{T_1} - \frac{1}{T_2}\right),$$

oder, wenn wir für ν_0 seinen Zahlenwert einführen und $T_2 - T_1 = \varDelta T$ setzen:

$$(2) \quad \begin{cases} n = 1 + 0{,}000106\,\dfrac{p}{T_1 T_2} \cdot \varDelta T, \\[2mm] \text{oder wenn man } n - 1 = \delta \text{ setzt:} \\[2mm] \delta = 0{,}000106\,\dfrac{p}{T_1 T_2} \cdot \varDelta T, \end{cases}$$

wobei p in Millimetern auszudrücken ist. Da der Faktor $p/T_1 T_2$ stets nahezu gleich $^1/_{100}$ ist, so genügt für Überschlagszwecke meist die vereinfachte Formel:

$$\delta = 10^{-6} \cdot \varDelta T \, . \qquad \text{(2a)}$$

Die genauere Formel (2) gibt z. B. für $p = 750$ und $T_1 = 273$ folgende Werte von δ für eine Inversion von

$\varDelta T =$	2^0	5^0	10^0	15^0
$\delta =$	0,0000021	0,0000052	0,0000103	0,0000152

Die Ermittelung des *Refraktionskoeffizienten* k ist eine bekannte Aufgabe der Geodäsie. Er wird dort aus gleichzeitigen gegenseitigen Zenitdistanzmessungen ermittelt. Auf diesem Wege, den wir empirisch nennen können, hat man folgende Werte gefunden[1]):

Landestriangulation, über Land	$k = 0{,}12$
Struve	$k = 0{,}12$
Baeyer, über Land	$k = 0{,}12$
Caraboeuf	$k = 0{,}13$
Landestriangulation nahe der Küste	$k = 0{,}13$
Gauss	$k = 0{,}13$
Dépôt de la Guerre	$k = 0{,}14$
Baeyer, nahe der Küste	$k = 0{,}14$
Ostpreußische Gradmessung	$k = 0{,}14$
Danmark-Expedition, über Land	$k = 0{,}16$
Delambre	$k = 0{,}17$
Baeyer, über Wasser	$k = 0{,}18$
Danmark-Expedition { über Wasser	$k = 0{,}52$
über Wasser zum Meereshorizont	$k = 0{,}58$

Wie stark k variieren kann, zeigt die folgende Zusammenstellung kleinster und größter Werte:

	Max.	Min.
Delambre	$+0{,}60$	$-0{,}01$
Ostpreußische Gradmessung	$+0{,}15$	$+0{,}13$
Nivellement Swinemünde-Berlin	$+0{,}27$	$+0{,}08$
Küstenvermessung	$+0{,}39$	$+0{,}10$
Landestriangulation	$+0{,}17$	$+0{,}06$
Gauss	$+1{,}21$	$-0{,}11$
Danmark-Expedition { über Land	$+0{,}36$	$+0{,}02$
„ Wasser	$+1{,}72$	$+0{,}25$
„ Wasser zum Meereshorizont	$+1{,}22$	$+0{,}17$

---

1) Nach I. P. Koch, Survey of Northeast Greenland, Danmark-Expeditionen til Grönlands Nordóstkyst 1906—1908. Bd. VI. Nr. 2 (Meddelelser om Grónland, Bd. XLVI), Kóbenhavn 1916. p. 164.

Die Werte schwanken also zwischen $-0{,}11$ (einem nach oben konkaven Strahl entsprechend) und $+1{,}7$, während etwa $1{,}15$ dem Mittelwerte aus den geodätischen Messungen entspricht. Es ist hieraus ersichtlich, daß man bei bloßer Schätzung von k große Fehler begehen kann, und daß man womöglich versuchen muß, diesen Faktor aus Temperaturbeobachtungen zu ermitteln. Ist man genötigt, ohne derartige Anhaltspunkte seine Wahl zu treffen, so wird man jedenfalls einen Wert benutzen müssen, der wesentlich höher ist, als der eben genannte Mittelwert aus den geodätischen Messungen, da bei den letzteren sorgfältig diejenigen Tage ausgeschlossen sind, bei denen die Refraktion infolge von Inversionen abnorm war. Man kann in unserer Tabelle sehr leicht diese Beobachtungsreihen von solchen unterscheiden, bei welchen der Refraktionskoeffizient nicht nur als Nebenresultat abgeleitet wurde, sondern welche gerade zur Bestimmung desselben unternommen wurden. Da bei diesen auch die abnormen Zustände berücksichtigt sind, so fällt der Wert für k hier erheblich höher aus. Ganz besonders ist dies der Fall bei den Beobachtungen der Danmark-Expedition, auf welcher gerade auch recht häufig Luftspiegelungen wahrgenommen wurden. Man wird hiernach Werte zwischen $k = 0{,}2$ und $k = 0{,}5$ annehmen dürfen, wenn keine Anhaltspunkte für eine andere Wahl vorliegen.

Ist das Temperaturgefälle innerhalb der unteren Schicht etwa durch Drachenaufstiege bekannt, so läßt sich der Refraktionskoeffizient durch eine Rechnung, die wir, im Gegensatz zu seiner Ableitung aus geodätischen Messungen, als eine theoretische bezeichnen können, in folgender Weise ermitteln. Wir benutzen eine aus der Geodäsie bekannte Formel, ohne hier auf ihre Ableitung einzugehen[1]:

$$k = 0{,}000293 \, \frac{p}{760} \cdot \frac{273}{T} \cdot \frac{g}{G} \cdot R \cdot F,$$

$$\text{wo} \quad F = \frac{\tau}{T} + \frac{1 - \dfrac{3e}{8p}}{7993} \cdot \frac{273}{T} \cdot \frac{g}{G}$$

$$\text{und} \quad \frac{g}{G} = 1 - 0{,}00265 \cos 2\varphi - \frac{2h}{R} \, .$$

---

1) Fr. H. Helmert, Die mathematischen und physikalischen Theorien der höheren Geodäsie. II. Teil. p. 577. Nach I. P. Koch.

Hierin bedeutet p den Druck in Millimetern, g/G das Verhältnis der örtlichen Schwere zu derjenigen in 45° Breite am Meeresspiegel, R den Krümmungsradius des Ellipsoids, τ das Temperaturgefälle pro Meter, e den Dampfdruck, φ die geographische Breite und h die Höhe über dem Meere.

Für die roheren Zwecke unserer Rechnungen läßt sich diese Gleichung ganz erheblich vereinfachen. Wir setzen $\frac{g}{G} = 1$, vernachlässigen $\frac{3\,e}{8\,p}$ und betrachten R als Erdradius gleich 6370 km. Damit wird die Gleichung

$$(3) \qquad k = 670{,}4\, \frac{p}{T^2}\, (\tau + 0{,}0342).$$

Wenn wir zur weiteren Vereinfachung noch ein für alle Mal $p = 750$ und $T = 273$ setzen, so wird

$$(3\,\mathrm{a}) \qquad k = 6{,}75\, (\tau + 0{,}034).$$

Hiermit erhält man z. B. folgende Zahlenwerte für k:

Temp.-Gefälle pro 100 m	$-3{,}4^{\circ}$	$-1{,}0^{\circ}$	$-0{,}5^{\circ}$	$0{,}0^{\circ}$	$+6{,}9^{\circ}$	$+11{,}4^{\circ}$
$k =$	0	0,16	0,20	0,23	0,50	1,00

Nach dieser Vorbereitung sollen nunmehr die Gleichungen für die Spiegelung nach oben abgeleitet werden.

Um zu entscheiden, *ob eine Schichtgrenze* von gegebener Höhe über dem Beobachter und von gegebenen Temperaturverhältnissen *eine Luftspiegelung erzeugen kann*, berechnen wir aus Gleichung (1) den Einfallswinkel α_0 des horizontalen Strahles, indem wir $h = 0$ setzen:

$$(1 + \varepsilon)\sin\alpha_0 = 1 + \frac{k}{2}\,(2\,\varepsilon + \varepsilon^2),$$

$$\sin\alpha_0 = \frac{1}{1+\varepsilon} + \frac{k}{2}\cdot\frac{2\,\varepsilon + \varepsilon^2}{1+\varepsilon}.$$

Soweit ist die Formel streng. Wir betrachten nun ε als kleine Größe und setzen also $\varepsilon^2 = 0$ und

$$\frac{1}{1+\varepsilon} = 1 - \varepsilon$$

und erhalten:

$$\sin\alpha_0 = 1 - \varepsilon\,(1 - k).$$

Nun lautet die Bedingung für Totalreflexion

$$\sin\alpha_0 \geqq \frac{1}{n},$$

worin n der Brechungsquotient der Schichtgrenze, also gleich $1 + \delta$ ist. Also

$$\sin \alpha_0 \gtreqqless \frac{1}{1 + \delta} \,.$$

Zusammen mit der vorigen Gleichung ergibt dies

$$1 - \varepsilon(1 - k) \gtreqqless \frac{1}{1 + \delta} \,,$$

oder unter Rücksicht darauf, daß auch δ eine kleine Größe ist:

$$1 - \varepsilon(1 - k) \gtreqqless 1 - \delta \,, \quad \text{und also}$$

$$(4) \qquad\qquad \varepsilon(1 - k) \lesseqqgtr \delta \,.$$

Dies ist die Bedingungsgleichung für Luftspiegelung nach oben. Um einige Zahlenwerte zu geben, seien in der folgenden Tabelle für einige Werte von k und H diejenigen Werte von δ zusammengestellt, welche mindestens erreicht sein müssen, wenn Luftspiegelung eintreten soll.

Minimalwerte für $\delta \times 10^6$

H	50 m	100 m	150 m	200 m
$k = 0,2$	6,3	12,6	18,8	25,1
$k = 0,5$	3,9	7,8	11,8	15,7

Mit Rücksicht auf Gleichung (2a) kann der Inhalt dieser Tabelle offenbar näherungsweise gleich dem Inversionsbetrage an der Schichtgrenze gesetzt werden.

Wenn beide Seiten der Ungleichung (4) einander gleich sind, so besagt dies offenbar, daß nur für den horizontalen Strahl gerade noch Totalreflexion eintritt. Dies ist der Fall, wenn die Höhe H der Inversion gerade gleich ihrer Wirkungstiefe ist. Da

$$\varepsilon = \frac{H}{R} \,,$$

ergibt sich so die *Wirkungstiefe* nach

$$(5) \qquad\qquad W = R \frac{\delta}{1 - k} \,.$$

Eine Inversion um $10^{\,0}$ $(\delta = 10^{-5})$ wird also bei $k = 0,2$ eine Wirkungstiefe von 80 m haben, bei $k = 0,5$ eine solche von 130 m.

Um die *Breite des spiegelnden Streifens* zu erhalten, berechnen wir den Höhenwinkel h des Grenzstrahles für Totalreflexion, indem wir in Gleichung (1) die Beziehung

$$\sin \alpha = \frac{1}{1 + \delta}$$

berücksichtigen:

$$\cos h_0 = \frac{1 + \varepsilon}{1 + \delta} - \frac{k}{2}\,(2\,\varepsilon + \varepsilon^2),$$

oder, wenn wir δ und ε als kleine Größen betrachten:

$$\cos h_0 = 1 - \delta + \varepsilon\,(1 - k)\,.$$

Da die Berechnung des kleinen Winkels h durch den Cosinus ungenau ausfällt, bilden wir hieraus den Sinus:

$$\sin h_0 = \sqrt{1 - \cos^2 h_0} = \sqrt{2\,\delta - 2\,\varepsilon(1 - k)},$$

oder in Minuten:

$$h_0{}' = \frac{\sqrt{2}}{\sin 1'}\,\sqrt{\delta - \varepsilon(1 - k)}\,.$$

Die Breite b des spiegelnden Streifens ist gleich dem doppelten Betrag, also

$$(6) \qquad\qquad b' = 9700\,\sqrt{\delta - \varepsilon(1 - k)};$$

z. B. erhält man für $H = 50$ m und $\delta = 10^{-5}$ die Breite 4,2', wenn $k = 0,2$ ist, oder 24,0', wenn $k = 0,5$ ist.

Die *Entfernung des Objekts* vom Beobachter kann, wie bereits Fig. 3 lehrt, stark variieren. Wir können aber einen Begriff davon erhalten, wenn wir uns die Entfernung berechnen, in welcher ein beliebiger, vom Beobachter ausgehender Strahl die Schichtgrenze trifft. Wählen wir dann den Strahl so, daß er der oberen Grenze des spiegelnden Streifens entspricht

$$\left(\sin \alpha = \frac{1}{n}\right),$$

so haben wir die Entfernung des nächstgelegenen Reflexionspunktes überhaupt; setzen wir statt dessen $h = 0$, so haben wir die Entfernung des Reflexionspunktes des horizontalen Strahles. Die Objekte müssen jedenfalls jenseits der so gefundenen Abstände liegen.

Wir greifen zurück auf Fig. 1 und führen die Bezeichnung ein: $x = \sphericalangle\,BMP$ und $y = \sphericalangle\,BOP$. Ferner denken wir uns die Sehne BP gezogen, welche mit den Tangenten in B und P den Winkel $y/2$ bildet, so daß die Zenitdistanz dieser Sehne in B gleich

$$90 - h + \frac{y}{2}$$

und der Winkel bei P zwischen der Sehne und dem Radius MP
gleich

$$\alpha - \frac{y}{2}$$

wird. Dann ist der Außenwinkel ZBP des Dreiecks MBP,

$$90 - h + \frac{y}{2} = x + \alpha - \frac{y}{2},$$

oder

$$(90 - a) - h = x - y\,.$$

Bezeichnet D die gesuchte Entfernung BP in Kilometern,
so können wir genähert sowohl

$$\sin x = \frac{D}{R} \quad \text{als} \quad \sin y = \frac{D}{R} \cdot k$$

und also $y = k\,x$ setzen, so daß wird: $(90 - a) - h = x\,(1 - k)$.
Da alle drei Winkel $(90 - a)$, h und x klein sind, kann man
schreiben:

$$\cos \alpha - \sin h = (1 - k) \sin x = (1 - k)\frac{D}{R},$$

so daß wird:

$$D = \frac{R}{1 - k}(\cos \alpha - \sin h).$$

Dies ist die allgemeine Gleichung, die für einen beliebigen
Strahl gilt. Spezialisieren wir sie für die obere Grenze des
spiegelnden Streifens, so haben wir hinzuzunehmen die Gleichung

$$\sin \alpha = \frac{1}{1 + \delta},$$

woraus man leicht die genäherte Formel findet: $\cos \alpha = \sqrt{2\delta}$.
Ferner wird nach der allgemeinen Gleichung (1):

$$\cos h = (1 + \varepsilon) \sin a - k\varepsilon$$

oder für $\sin \alpha = \dfrac{1}{1 + \delta}$:

$$\cos h = 1 - \delta + \varepsilon(1 - k).$$

Hieraus findet man

$$\sin h = \sqrt{2\,\delta - 2\,\varepsilon(1 - k)}.$$

Setzen wir beide Werte ein, so folgt

$$(7) \qquad D = \frac{R\sqrt{2}}{1 - k}\left[\sqrt{\delta} - \sqrt{\delta - \varepsilon(1 - k)}\right].$$

Diese Gleichung gestattet, die Entfernung des Reflexionspunktes des oberen Grenzstrahls des spiegelnden Streifens

zu finden. Z. B. wird sie für $H = 50$ m und $\delta = 10^{-5}$ gleich 14 km, wenn $k = 0.2$ ist, und gleich 12 km, wenn $k = 0.5$ ist.

Um andererseits den Reflexionspunkt des horizontalen Strahles zu finden, setzen wir in unserer allgemeinen Gleichung $k = 0$ und drücken α nach Gleichung (1) durch h aus:

$$\sin \alpha = \frac{\cos h + \varepsilon k}{1 + \varepsilon} = (1 + \varepsilon k)(1 - \varepsilon) = 1 - \varepsilon(1 - k).$$

Hieraus findet man

$$\cos \alpha = \sqrt{2\,\varepsilon(1 - k)},$$

so daß wird

$$(8) \qquad\qquad D = R\sqrt{\frac{2\,\varepsilon}{1 - k}}\,.$$

In diesem Falle ist D natürlich unabhängig vom Brechungsquotienten der Schichtgrenze. Für $H = 50$ m erhalten wir nach dieser Gleichung $D = 28$ km, wenn $k = 0.2$, oder 36 km, wenn $k = 0.5$ ist.

2. Spiegelung nach unten.

Die Spiegelung nach unten verlangt ein umgekehrtes Dichtegefälle in der Luft, so daß die leichte Luft unten, die schwere oben liegt. Da diese Schichtung nicht stabil ist, so kann sie längere Zeit nur unmittelbar über dem Boden bestehen, wo sie zwar auch unausgesetzt durch vertikalen Luftaustausch zerstört, aber gleichzeitig durch Wärmezufuhr vom Boden her immer wieder erneuert wird. Es herrscht dann als stationärer Zustand im untersten Höhenmeter oder Halbmeter eine Abnahme der Temperatur nach oben von ca. 1—5°. Die Bedingung hierzu ist offenbar, daß der Boden wärmer ist als die Luft, was besonders über sonnenbestrahlten Wüsten, im Polargebiete auch über dem herbstlichen Neueis und sehr häufig auch auf See zutrifft. Es ist also klar, daß hier nicht eine scharfe Schichtgrenze wirksam ist, sondern daß das Dichtegefälle hier in der untersten Luftschicht stetig verläuft. Bei der Spiegelung nach unten läßt sich also der Ansatz der alten Theorie eher rechtfertigen als bei der nach oben, indessen bleibt auch hier die Vernachlässigung der Erdkrümmung durchaus unzulässig. Es wäre also, streng genommen, nötig, die alte Theorie unter Berücksichtigung der Erdkrümmung zu verbessern. Es ist aber wenig wahrscheinlich,

daß sich bei dieser Komplikation der Aufgabe ein Resultat
erzielen läßt, welches eine bequeme Vergleichung von Theorie
und Beobachtung ermöglicht.

Da sich aber der intensive Luftaustausch vom Boden her
nur in einer sehr dünnen Schicht geltend macht, die nur
selten mehr als 1 m dick zu sein pflegt, und oft noch unter
dieser Größe bleibt, so werden wir jedenfalls eine für die
meisten Fälle sehr brauchbare Annäherung erhalten, wenn
wir auch hier eine scharfe Schichtgrenze annehmen, die mit
dem Erdboden zusammenfällt. Die Ungenauigkeit, die man
damit begeht, ist so gering, daß sie reichlich durch den Vorteil
aufgewogen wird, Gleichungen zu erhalten, die eine bequeme
Vergleichung mit den Beobachtungen ermöglichen.

Fig. 4 erläutert die wesentlichen Erscheinungen dieser
Spiegelung. B ist der Beobachter; der flache Kreisbogen BP

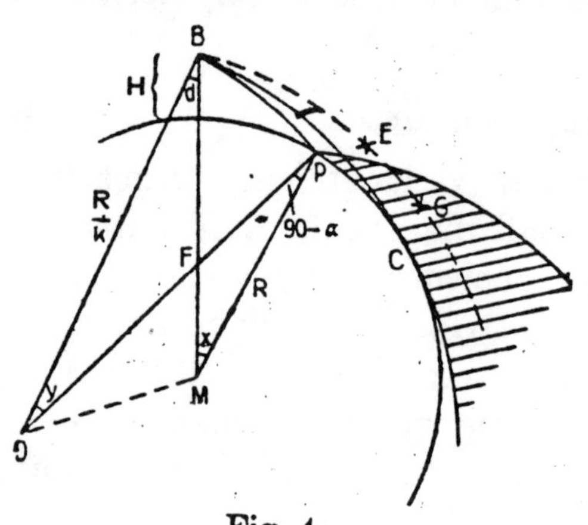

Fig. 4.

sei der untere Grenzstrahl
des spiegelnden Streifens.
Der mit demselben Krüm-
mungsradius gezeichnete
Kreisbogen BC, der bei C
die Erdoberfläche tangiert,
stellt die Sicht nach der
Kimm und also zugleich die
obere Begrenzung des spiegelnden Streifens dar. Die
Breite des spiegelnden Streifens ist also der Winkel
zwischen den Kreisbögen BP und BC bei B. Bei der
Spiegelung nach unten liegt, wie man sieht, der spiegelnde
Streifen nicht im astronomischen Horizont, sondern er-
streckt sich von der Kimm abwärts. Ferner ist klar, daß
hier gar keine andere Bedingung für das Auftreten der Spiegelung besteht, als daß überhaupt ein Temperatursprung vorhanden ist. Jeder noch so kleine Temperatursprung erzeugt
bereits eine Spiegelung, gleichgültig, in welcher Höhe sich
der Beobachter befindet. Hierin liegt die Erklärung für die
Tatsache, daß Spiegelungen nach unten so viel häufiger zur
Beobachtung gelangen als solche nach oben. Da die Schichtgrenze als konvexer Spiegel wirkt, ist auch einleuchtend,
daß das Spiegelbild in vertikaler Richtung etwas geschrumpft

erscheinen muß, im Gegensatze zur Spiegelung nach oben, bei welcher das gespiegelte Bild stets mehr oder weniger in der Vertikalen gezerrt ist.

Man sieht leicht, daß alle Gegenstände, die sich in dem schraffierten Raume der Fig. 4 befinden, gespiegelt werden. Zwei Berge, deren Gipfel z. B. in E und G liegen, also von B aus in derselben Höhe erscheinen, verhalten sich verschieden, indem von dem näher gelegenen nur der untere Teil, der entferntere aber ganz gespiegelt wird. Diese Verhältnisse werden durch Fig. 5 erläutert.

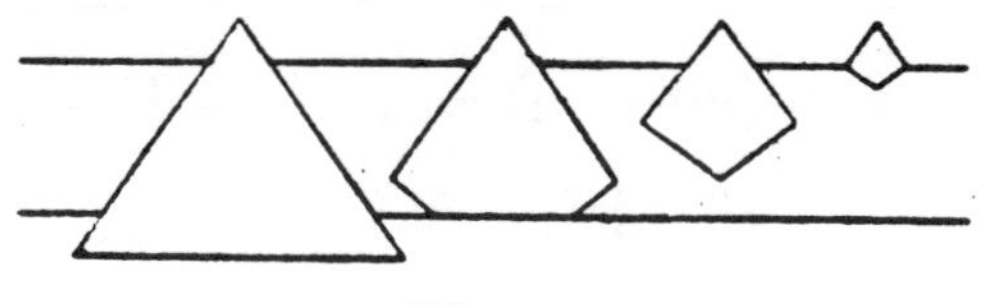

Fig. 5.

Um die Gleichungen der Luftspiegelung nach unten zu berechnen, fassen wir zunächst den Strahl BP in der Fig. 4 als einen beliebigen auf und leiten die Gleichungen für den Depressionswinkel d und die Länge D desselben ab. Indem wir diese Gleichungen sodann einerseits für die untere, andererseits für die obere Grenze des spiegelnden Streifens spezialisieren, erhalten wir alle Elemente der Spiegelung.

Wenn a der Einfallswinkel des Strahles an der Schichtgrenze ist, so ist offenbar $\angle OPM = 90 - a$. Ferner ist $\angle OBM = d$. Wir wenden auf die gemeinsame Seite OM der beiden Dreiecke OBM und OPM den Cosinussatz an und erhalten:

$$\left(\frac{R}{k}\right)^2 + (R + H)^2 - \frac{2R}{k}(R + H)\cos d = \left(\frac{R}{k}\right)^2 + R^2 - \frac{2R^2}{k}\sin a,$$

$$\text{oder} \qquad \cos d = \frac{k}{2}\left(\frac{R + H}{R} - \frac{R}{R + H}\right) + \frac{R}{R + H}\sin a;$$

setzen wir wieder

$$\frac{H}{R} = \varepsilon, \text{ so wird: } \cos d = \frac{k}{2}\left(1 + \varepsilon - \frac{1}{1 + \varepsilon}\right) + \frac{\sin a}{1 + \varepsilon}.$$

Soweit ist die Formel streng. Behandeln wir ε als kleine Größe, so können wir genähert schreiben:

$$(9) \qquad \cos d = k\varepsilon + (1 - \varepsilon)\sin a.$$

Damit haben wir die allgemeine Gleichung für d. Wir leiten gleich auch diejenige für D ab: $\measuredangle BFP$ ist offenbar einerseits gleich $x + (90 - a)$ und andererseits gleich $y + d$, also:

$$d - (90 - a) = x - y .$$

Nun können wir, wie schon früher, genähert schreiben:

$$\sin x = \frac{D}{R} \quad \text{und} \quad \sin y = \frac{D}{R} \cdot k$$

und also

$$y = k : x ,$$

da x und y kleine Winkel sind. Damit erhalten wir

$$d - (90 - a) = x \, (1 - k) .$$

Da auch d und $(90 - a)$ kleine Winkel sind, können wir statt dessen schreiben

$$\sin d - \cos \alpha = (1 - k) \sin x = (1 - k) \frac{D}{R} ,$$

oder

(10)
$$D = \frac{R}{1 - k} \, (\sin d - \cos \alpha).$$

Damit haben wir auch die allgemeine Gleichung für die Entfernung des Reflexionspunktes.

Wir spezialisieren nun (9) und (10) zunächst für die *untere Grenze* des spiegelnden Streifens, indem wir

$$\sin \alpha = \frac{1}{n} = \frac{1}{1 + \delta}$$

setzen, oder da δ klein ist,

$$\sin a = 1 - \delta .$$

Damit erhalten wir aus Gleichung (9):

$$\cos d_u = 1 - \delta - \varepsilon \, (1 - k)$$

oder, da d_u als kleiner Winkel besser aus dem Sinus berechnet wird:

$$\sin d_u = \sqrt{2 \, \delta + 2 \, \varepsilon (1 - k)},$$

oder in Minuten

(11)
$$d_u' = \frac{\sqrt{2}}{\sin 1'} \, \sqrt{\delta + \varepsilon (1 - k)} .$$

Führen wir die gleiche Spezialisierung in (10) ein, indem wir $\cos a = \sqrt{2 \, \delta}$ setzen, so wird

$$D_u = \frac{R}{1 - k} \, (\sin d_u - \sqrt{2 \, \delta}),$$

oder, wenn wir für $\sin d_u$ seinen Wert einführen:

$$(12) \qquad D_u = \frac{R\sqrt{2}}{1-k}\left(\sqrt{\partial + \varepsilon(1-k)} - \sqrt{\delta}\right).$$

Nunmehr spezialisieren wir (9) und (10) andererseits für die *obere Grenze* des spiegelnden Streifens, d. h. für den nach der Kimm gerichteten Strahl, indem wir $\sin a = 1$ bzw. $\cos a = 0$ setzen, und erhalten

$$\cos d_0 = 1 - \varepsilon(1-k)$$

oder

$$\sin d_0 = \sqrt{2\,\varepsilon(1-k)},$$

oder in Minuten

$$(13) \qquad d_0' = \frac{\sqrt{2}}{\sin 1'}\sqrt{\varepsilon(1-k)}.$$

Und für die Entfernung

$$D_0 = \frac{R}{1-k}\sin d_0,$$

oder, wenn wir für $\sin d_0$ seinen Wert einsetzen

$$(14) \qquad D_0 = R\sqrt{\frac{2\,\varepsilon}{1-k}}.$$

Die Gleichungen (11) bis (14) gestatten, alle wichtigen Elemente der Spiegelung aus dem Temperatursprung, der Höhe des Beobachters und dem Refraktionskoeffizienten zu berechnen. Insbesondere ist die Breite des spiegelnden Streifens gleich $d_u - d_0$ und die kleinste Entfernung spiegelnder Objekte D_u.

Wenn über den Refraktionskoeffizienten k keine Anhaltspunkte vorliegen, so wird man ihn bei Spiegelung nach unten etwa gleich dem Mittel aus den geodätischen Werten, d. h. ca. 0,15 wählen. Mit diesem Werte geben unsere Gleichungen für eine Höhe des Beobachters von 10 m (dem Verdeck eines Schiffes entsprechend) und für einen Temperatursprung von 5^0 ($\delta = 0{,}000005$) das Resultat $d_u = 12{,}2'$, $d_0 = 5{,}6'$, also eine Breite des spiegelnden Streifens von $6{,}6'$, ferner $D_u = 3{,}0$ km und $D_0 = 12{,}2$ km.

3. Zerrbilder der Sonne (und des Mondes) im Horizont.

Die Zerrbilder und Spiegelungen der Sonne bei ihrem Auf- oder Untergang bilden das siderische Gegenstück der terrestrischen Luftspiegelungen und werden durch die gleichen

Ursachen hervorgerufen. Es leuchtet ohne weiteres ein, daß die Spiegelung nach unten für die Sonne in derselben Weise in Erscheinung treten muß wie für terrestrische Objekte, so daß wir diese hier nicht näher zu behandeln brauchen. Die Spiegelung nach oben muß sich bei der Sonne in der Weise geltend machen, daß sie die Sonnenscheibe teilt, indem nur oberhalb und unterhalb des spiegelnden Streifens überhaupt Licht von außen zum Beobachter gelangen kann. Der spiegelnde Streifen der terrestrischen Luftspiegelung wird hier also zu einem *blinden Streifen* im Sonnenbilde.

Qualitativ läßt sich das hierbei wirksame Gesetz ohne weiteres aus Fig. 6 ablesen. Wenn BAA' und BCC' die den

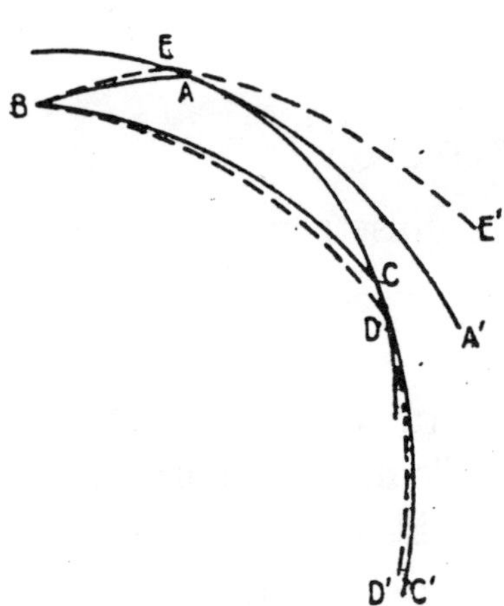

Fig. 6.

blinden Streifen begrenzenden Grenzstrahlen der Totalreflexion sind, bei denen also die äußeren Strahlenteile gerade die Schichtgrenze in A und C tangential verlassen, so sind bei den benachbarten Strahlen BE und BD sowohl Einfallswinkel wie Austrittswinkel an der Schichtgrenze bereits kleiner geworden. Bei dem oberen Strahle (BEE') bewirkt nun diese Verkleinerung des Austrittswinkels eine Hebung des Astes EE' sie wirkt daher in gleichem Sinne wie auch die Hebung des inneren Astes BE. Die beiden äußeren Äste EE' und AA' werden also stärker divergieren als die beiden inneren BE und BA. Mit anderen Worten: *Oberhalb des blinden Streifens herrscht Vertikalschrumpfung.* Anders bei den Strahlen unterhalb des spiegelnden Streifens. Denn hier wirkt die Hebung des äußeren Astes entgegen der Senkung des inneren Astes, so daß die äußeren Äste CC' und DD' jedenfalls weniger divergieren als die inneren BC und BD. *Es herrscht also unterhalb des blinden Streifens Vertikalzerrung.* Diese Vertikalzerrung lernten wir schon früher als Übergang zwischen aufrechtem und umgekehrtem Bilde kennen, und in der Tat ist dies auch hier der Fall, da, wie gezeigt werden wird, dicht unterhalb des blinden Streifens sich die benachbarten Strahlen infolge Überkompensierung der Senkung schneiden und also ein umgekehrtes Bild geben. Wir wollen diese noch wenig bekannte Erscheinung als *Nachspiegelung* bezeichnen, weil sie bei Sonnenuntergängen erst dann deutlich in Erscheinung tritt, wenn die Sonne bereits den blinden Streifen ganz passiert hat und sich von seinem Unterrande loslösen will, also zu einem Zeitpunkte, zu dem der Beobachter keine weitere Refraktionsstörung mehr erwartet. Wir werden auf diese interessante Erscheinung der Nachspiegelung weiter unten zurückkommen.

Unsere vorangehenden Ausführungen über die terrestrische Luftspiegelung nach oben geben uns auch die Mittel an die Hand, die quantitative Berechnung der Zerrbilder bei Sonnenuntergängen durchzuführen.

In Fig. 7, welche bis auf die Fortsetzung des Strahles außerhalb der Schichtgrenze identisch mit Fig. 1 ist, bedeutet wieder B den Beobachter und P einen Punkt der Schichtgrenze, M den Erdmittelpunkt und O den Krümmungsmittelpunkt des Strahles BP. Verfolgen wir den Lichtstrahl von B bis

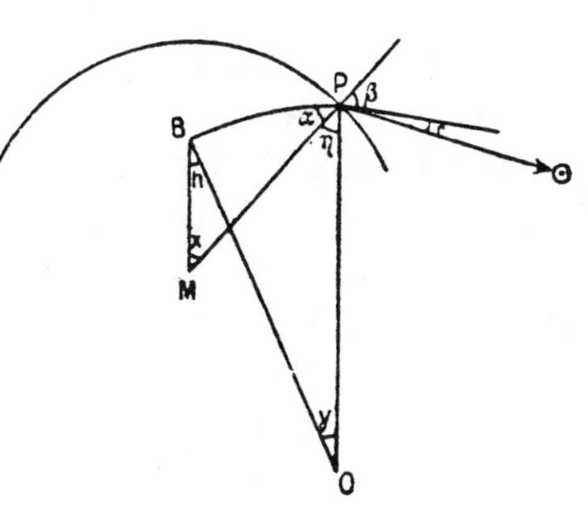

Fig. 7.

außerhalb der Atmosphäre, so wird er bis P um den Winkel y gesenkt, in P um den Winkel $\beta - \alpha$ und sodann noch um die astronomische Refraktion r weitergesenkt. Letztere kann aus Bessels Refraktionstafeln entnommen werden, wobei aber als scheinbare Höhe der Winkel $90 - \beta = \zeta$ zu benutzen ist. Wir wollen deshalb diese Refraktion mit r_ζ bezeichnen. Die Gesamtablenkung nach unten ist also

$$\Delta = y + \beta - \alpha + r_\zeta,$$

oder wenn wir statt α und β ihre Komplemente η und ζ einführen:

$$\Delta = y + \eta - \zeta + r_\zeta.$$

Bezeichnet h die scheinbare Höhe der Sonne beim Beobachter, $h_\odot$ die astronomische Sonnenhöhe, so ist

$$h_\odot = h - \Delta,$$

oder

$$h_\odot = h - y - \eta + \zeta - r_\zeta.$$

Aus der Figur läßt sich unmittelbar die Beziehung ablesen

$$y + \eta = x + h\,.$$

Benutzen wir die früher abgeleitete genäherte Beziehung

$$x = \frac{y}{k}\,,$$

so wird hiernach

$$y\left(1 - \frac{1}{k}\right) = h - \eta\,.$$

Setzen wir diesen Wert von y in die Gleichung für $h_\odot$ ein, so wird

$$(15) \qquad h_\odot = \frac{h - \eta}{1 - k} + \zeta - r_\zeta\,.$$

Offenbar ist das Problem gelöst, wenn wir für jede Höhe h beim Beobachter die zugehörige astronomische Höhe angeben können. Wir müssen also der Reihe nach die drei Größen η, ζ und r_ζ berechnen.

Zur Berechnung von η benutzen wir die allgemeine Gleichung (1), welche sich unter Berücksichtigung, daß $\varepsilon\,(=H/R)$ eine kleine Größe ist, schreibt:

$$\cos \eta = \varepsilon k + (1 - \varepsilon) \cos h\,.$$

Da $\quad \cos \eta = 1 - 2 \sin^2 \dfrac{\eta}{2} \quad$ und $\quad \cos h = 1 - 2 \sin^2 \dfrac{h}{2}$, so wird

$$1 - 2 \sin^2 \frac{\eta}{2} = \varepsilon k + 1 - \varepsilon - 2(1 - \varepsilon)\sin^2 \frac{h}{2}\,,$$

oder $\qquad \sin^2 \dfrac{\eta}{2} = (1 - \varepsilon)\sin^2 \dfrac{h}{2} + \dfrac{\varepsilon}{2}(1 - k)\,.$

Da η und h kleine Winkel sind, schreiben wir statt

$$\sin^2 \frac{\eta}{2} \quad \frac{\eta^2}{4}\sin^2 1'\,,$$

wobei η in Minuten ausgedrückt ist, und ebenso statt

$$\sin^2 \frac{h}{2} \quad \frac{h^2}{4}\sin^2 1'\,,$$

so daß wir erhalten

$$(16) \qquad \eta^2 = (1 - \varepsilon) h^2 + \frac{2\,\varepsilon\,(1 - k)}{\sin^2 1'}\,.$$

Wenn man nur auf Zehntelminute rechnet, kann man hierbei unbedenklich ε im ersten Gliede vernachlässigen, so daß wird

$$(16\,\mathrm{a}) \qquad \eta^2 = h^2 + \frac{2\,\varepsilon\,(1 - k)}{\sin^2 1'}\,.$$

Diese Gleichung gestattet eine bequeme Berechnung des Winkels η.

Zur *Berechnung von* ζ benutzen **wir** die Definitionsgleichung

$$\frac{\sin \beta}{\sin \alpha} = 1 + \delta,$$

welche wir auch schreiben können

$$\cos \zeta = (1 + \delta) \cos \eta.$$

Wenden wir wieder auf beide Kosinusse den Satz

$$\cos \zeta = 1 - 2 \sin^2 \frac{\zeta}{2}$$

an, so wird

$$1 - 2 \sin^2 \frac{\zeta}{2} = (1 + \delta) - 2(1 + \delta)\sin^2 \frac{\eta}{2},$$

oder

$$\sin^2 \frac{\zeta}{2} = (1 + \delta)\sin^2 \frac{\eta}{2} - \frac{\delta}{2},$$

wofür wir wieder schreiben können

$$(17) \qquad \zeta^2 = (1 + \delta)\eta^2 - \frac{2\,\delta}{\sin^2 1'}.$$

ζ und η sind hierin wieder in Minuten auszudrücken.

Rechnet man nur auf Zehntelminuten, so kann auch hier δ im ersten Gliede vernachlässigt werden, so daß wird

$$(17\,\text{a}) \qquad \zeta^2 = \eta^2 - \frac{2\,\delta}{\sin^2 1'}.$$

Die *Refraktion* r_ζ endlich entnehmen wir der Besselschen Tabelle, von der hier ein Auszug gegeben sei:

ζ	0′	10′	20′	30′	40′	50′	60′
r_ζ	34,90′	32,82′	30,87′	29,06′	27,38′	25,83′	24,41′

Wünscht man hierbei noch Druck und Temperatur zu berücksichtigen, so ist der Wert für r noch mit dem Faktor $(1-a-b)$ zu multiplizieren, wo a und b aus den folgenden beiden abgekürzten Tabellen entnommen werden können:

Luftdruck	720	730	740 mm	Temp.	-10^0	0^0	$+10^0$	$+20^0$
a	0,042	0,029	0,015	b	$-0,073$	$-0,034$	$+0,002$	$+0,036$

Als Druck und Temperatur müssen natürlich diejenigen dicht oberhalb der Schichtgrenze gewählt werden.

Man hat also die folgenden Gleichungen durchzurechnen, wenn man aus der Höhe der Inversion (oder ihrem Verhältnis zum Erdradius), dem Inversionsbetrag ($\delta \times 10^6$) und dem Refraktionskoeffizienten k der unteren Schicht den Verlauf eines Sonnenunterganges vollständig berechnen will:

$$(15)\text{—}(17)\qquad \begin{cases} \eta^2 = h^2 + \dfrac{2\,\delta\,(1-k)}{\sin^2 1'}\,, \\[2ex] \zeta^2 = \eta^2 - \dfrac{2\,\delta}{\sin^2 1'}\,, \\[2ex] h_\odot = \dfrac{h-\eta}{1-k} + \zeta - r_\zeta\,. \end{cases}$$

An der Hand unserer Gleichungen läßt sich auch die Notwendigkeit der Nachspiegelung leicht erweisen. Differentiieren wir die Gleichung für $h_\odot$ nach h, so erhalten wir

$$\frac{\partial h_\odot}{\partial h} = \frac{1}{1-k} - \frac{\partial \eta}{\partial h}\cdot\frac{1}{1-k} + \frac{\partial \zeta}{\partial \eta}\cdot\frac{\partial \eta}{\partial h} - \frac{\partial r_\zeta}{\partial \zeta}\cdot\frac{\partial \zeta}{\partial \eta}\cdot\frac{\partial \eta}{\partial h}\,,$$

oder

$$\frac{\partial h_\odot}{\partial h} = \frac{1}{1-k} - \frac{\partial \eta}{\partial h}\cdot\frac{1}{1-k} + \frac{\partial \zeta}{\partial \eta}\cdot\frac{\partial \eta}{\partial h}\left(1 - \frac{\partial r_\zeta}{\partial \zeta}\right).$$

$\partial r_\zeta/\partial \zeta$ können wir für kleine Winkel ζ als konstant betrachten und aus der Besselschen Tabelle zu ca. $-0,2$ entnehmen, so daß die Klammer gleich $+1,2$ wird. $\partial \eta/dh$ ist endlich und negativ. $\partial \zeta/\partial \eta$ aber ist für $\zeta = 0$ gleich $+\infty$. Dies sieht man leicht, wenn man die Definitionsgleichung

$$\cos \zeta = (1 + \delta)\cos \eta$$

differentiiert:

$$\frac{\partial \zeta}{\partial \eta} = (1 + \delta)\frac{\sin \eta}{\sin \zeta} = \frac{(1 + \delta)\sin \eta}{0}\,.$$

Infolgedessen wird auch

$$\frac{\partial h_\odot}{\partial h} = -\infty \quad \text{für } \zeta = 0\,,$$

d. h. für die untere Grenze des blinden Streifens, oder mit anderen Worten: In unmittelbarer Nachbarschaft des blinden Streifens wird trotz der Senkung des Strahles beim Beobachter der außerhalb der Atmosphäre liegende Ast gehoben und kommt also zum Schnitt nicht nur mit dem Grenzstrahl selbst, sondern auch mit dem späteren bei fortschreitender Senkung, bei denen dann auch außerhalb der Atmosphäre die Senkung wieder überwiegt. Man kann ohne Schwierigkeit die Differentialquotienten $\partial \eta / \partial h$ und $\partial \zeta / \partial \eta$ bilden und durch Nullsetzen von $\partial h_\odot / \partial h$ die Gleichung desjenigen Höhenwinkels h beim Beobachter berechnen, bis zu welchem die Umkehrung des Bildes herabreicht. Man erhält so die Gleichung für die untere Grenze der Nachspiegelung

$$(18) \qquad \frac{1}{h} = \frac{1}{\eta} - \frac{A}{\zeta},$$

wo die Konstante

$$A = \left(1 - \frac{\partial r_\zeta}{\partial \zeta}\right)(1 - k) \text{ ist.}$$

Wenn man hieraus mit Hilfe der ersten beiden Gleichungen des Systems (15) — (17) η und ζ eliminiert, erhält man den gesuchten Höhenwinkel h für die untere Grenze der Nachspiegelung. Es ist mir indessen bisher nicht gelungen, die Elimination in einer Weise auszuführen, welche eine für die praktische Rechnung brauchbare Formel liefert.

Die Berechnung eines Sonnenunterganges möge an einem Beispiel erläutert werden. Gegeben sei die Höhe der Schichtgrenze zu 50 m ($H = 0{,}05$ km) ferner der Inversionsbetrag zu $7^{\,0}$ ($\delta = 0{,}000007$) und der Refraktionskoeffizient der unteren Schicht $k = 0{,}2$. Der Erdradius wird zu 6370 km angenommen.

Zunächst berechnen wir zur Orientierung die Breite des blinden Streifens nach Gleichung (6). Sie ergibt sich zu 8,28'. Die Grenzstrahlen liegen also bei $h = + 4{,}14'$ und bei $h = - 4{,}14'$. Da in der Nähe des blinden Streifens die Verzerrung am größten ist, werden wir $h_\odot$ für folgende Wertreihe von h berechnen: $h = \pm 4{,}14'$, $\pm 4{,}2'$, $\pm 5{,}0'$, $\pm 6{,}0'$, $\pm 8{,}0'$, $\pm 10{,}0'$, $\pm 15{,}0'$. Zunächst ergeben sich die konstanten Glieder von η^2 und ζ^2 zu

$$\frac{2\,s\,(1 - k)}{\sin^2 1'} = 148{,}42$$

und

$$\frac{2\,\delta}{\sin^2 1'} = 165{,}45 \,.$$

Damit finden wir folgende Werte (r_ζ aus der oben stehenden Refraktionstabelle):

h	$\pm 4{,}14'$	$\pm 4{,}2'$	$\pm 5{,}0'$	$\pm 6{,}0'$	$\pm 8{,}0'$	$\pm 10{,}0'$	$\pm 15{,}0'$
η	12,9'	12,9'	13,2'	13,6'	14,6'	15,8'	19,3'
ζ	0,0'	0,8'	2,8'	4,4'	6,8'	9,1'	14,4'
r_ζ	34,9'	34,8'	34,3'	34,0'	33,4'	33,0'	31,9'

15*

und hieraus

h	$-15,0'$	$-10,0'$	$-8,0'$	$-6,0'$	$-5,0'$	$-4,2'$	$-4,14'$
$h_\odot$	$-60,4'$	$-56,1'$	$-54,7'$	$-54,1'$	$-54,2'$	$-55,3'$	$-56,1'$

h	$+4,14'$	$+4,2'$	$+5,0'$	$+6,0'$	$+8,0'$	$+10,0'$	$+15,0'$
$h_\odot$	$-45,9'$	$-44,8'$	$-41,7'$	$-39,1'$	$-34,7'$	$-31,1'$	$-22,9'$

Mit Hilfe dieser Tabelle läßt sich der Verlauf des Sonnenunterganges sehr einfach graphisch konstruieren, indem wir in zwei Figuren mit Minutenskalen — die eine für $h_\odot$ und die

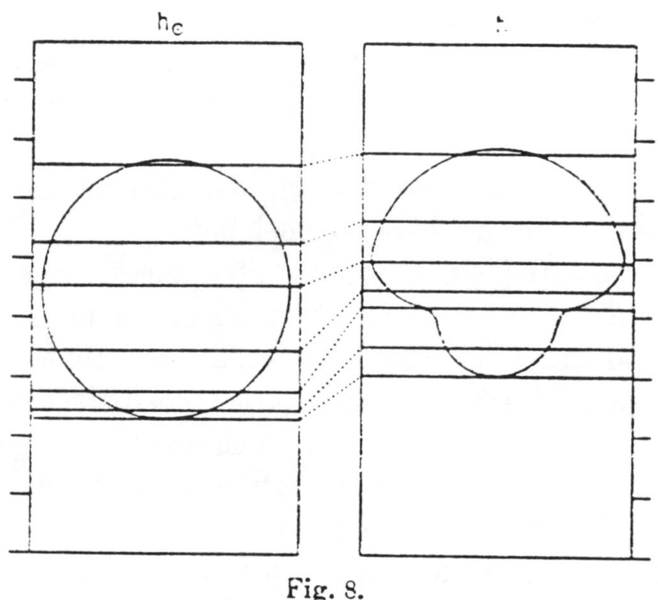

Fig. 8.

andere für h — diejenigen horizontalen Minutenlinien hervorheben, welche den berechneten zusammengehörigen Wertepaaren unserer Zahlentabelle entsprechen. Zeichnet man dann in der Figur für $h_\odot$ die Sonnenscheibe als Kreis vom Radius 16′ in beliebiger Höhe, so lassen sich, wie Fig. 8 schematisch zeigt, alle Schnitte mit den hervorgehobenen Linien unmittelbar mit dem Zirkel in die Figur für h übertragen und die vollständigen Konturen des Zerrbildes dann leicht ergänzen.

In dieser Weise sind die in Fig. 9 dargestellten Zerrbilder für die oben angegebenen Zahlenwerte erhalten. Die Bilder zeigen deutlich die Schrumpfung oberhalb und die Zerrung unterhalb des blinden Streifens. Auch ist die Nachspiegelung besonders auf den letzten beiden Figuren deutlich zu erkennen. Sie äußert sich aber auch schon in der zweiten Figur dadurch,

daß die untere Bildhälfte nicht am Unterrande des blinden
Streifens, sondern in derjenigen Höhe zu erscheinen beginnt,
welche der unteren Grenze der Nachspiegelung, also der

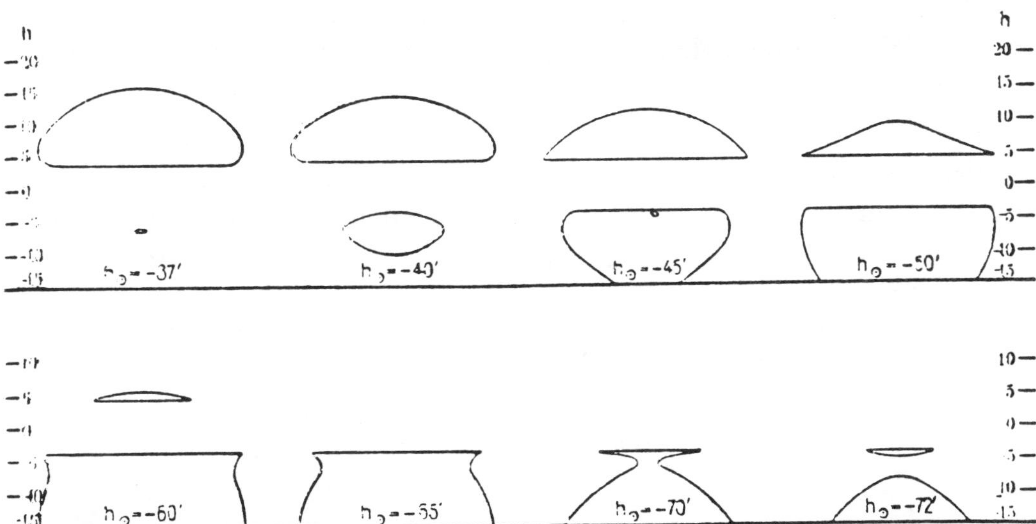

Verlauf des Sonnenunterganges bei einer Inversion um 7° in 50 m Höhe
über dem Beobachter, wenn $k = 0,2$ ist. $h_\odot$ gilt für den Sonnenmittelpunkt.

Fig. 9.

Grenze zwischen umgekehrtem und aufrechtem Bilde entspricht. Die Nachspiegelung reicht in unserem Falle etwa
von $h = -4',14$ bis $h = -6'$.

Es sei nur kurz erwähnt, daß
auch solche Inversionen, welche keine
terrestrischen Luftspiegelungen erzeugen können, doch noch eine Verzerrung der Sonnenscheibe ergeben.
Der blinde Streifen fehlt dann allerdings; aber es tritt am horizontalen
Strahle eine Unstetigkeit im Ver-
laufe der |Refraktion auf, und zwar
wieder in dem Sinne, daß oberhalb
Schrumpfung, unterhalb Zerrung
herrscht, beide mit dem Maximum
am horizontalen Strahle selbst. Fig. 10
gibt das Konstruktionsergebnis für
ein Zahlenbeispiel; sie ist mit den-
selben Zahlen wie das frühere Beispiel berechnet, nur mit
$\delta = 0.000006$ statt 0.000007. Nach unserer früheren Tabelle

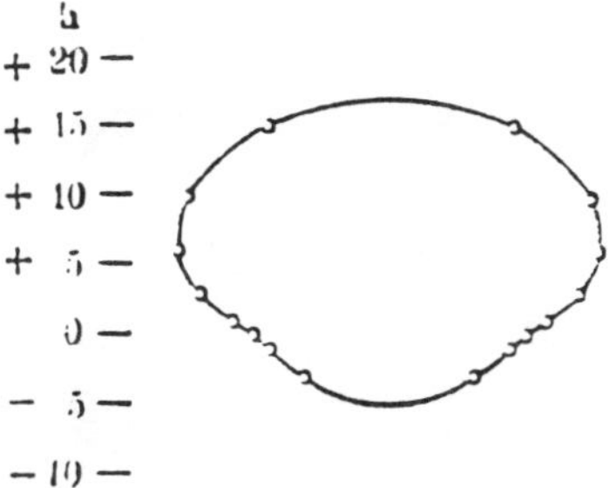

Verzerrung der Sonnen-
scheibe für eine Inversion
von 6,0° in 50 m Höhe über
dem Beobachter bei $k = 0,2$.
Wahre Höhe des Sonnen-
mittelpunktes $- 35'$.

Fig. 10.

wäre bei den gegebenen übrigen Bedingungen eine Inversion von $6,3^0$ ($\delta = 0,0000063$) nötig, um Luftspiegelung zu geben, daher muß hier der blinde Streifen fehlen. Auch die Nachspiegelung ist hier bereits verschwunden.

Wenn umgekehrt der blinde Streifen so breit wird, daß sein Unterrand unter der Kimm liegt und also nicht mehr gesehen werden kann, so beschränkt sich der ganze Verlauf der Erscheinung auf die oberen Teile unserer Fig. 8, d. h. die Sonne geht hier hinter einer unsichtbaren Wand, nämlich dem blinden Streifen, unter, in erheblicher Höhe über der Kimm.

Im Felde, im Juni 1918.

(Eingegangen 18. Juni 1918.)

Normal Atmospheric Dispersion as the Cause of the " Green Flash " at Sunset, with Illustrative Experiments.

By Lord RAYLEIGH, F.R.S.

(Received November 20, 1929.)

The phenomenon of the " green flash " or " green ray " at sunset has often been discussed. The reader may be reminded that this effect is sometimes seen just as the sun's upper limb vanishes below the horizon ; for the last 2 or 3 seconds, the vanishing upper limb appears of a more or less vivid green colour. The most obviously necessary condition is a clear sky at the horizon, a condition much more often fulfilled in the tropics than in this country. For a more detailed description reference may be made to a paper by A. A. Rambaut,* and some other authors will be referred to in the sequel. It is not necessary to go back to the earlier literature.

The theory which attributes the green sensation to retinal fatigue, though still sometimes met with, is conclusively disposed of by the simple observation that the green flash occurs at sunrise as well as at sunset. It is now generally recognised that the correct explanation depends on atmospheric refraction. Refraction keeps the sun's upper limb in view when its position, geometrically considered, is below the observer's horizon. Refraction is accompanied by dispersion, though the latter effect is comparatively small, and therefore the more refrangible part of the spectrum of the sun's upper limb should be the last to disappear. The green usually seen is in fact more refrangible than the general light of the disc at the horizon.

Any doubt that remains is rather as to whether the normal dispersion of the atmosphere at the horizon is large enough to explain the effect, or whether we must appeal to special conditions such as cause mirage. It is to this aspect of the question, and its experimental investigation, that the present discussion is directed.

The refraction of the atmosphere at the horizon is given in Bessel's tables of refractions as 35 minutes of arc. All observers from the time of Tycho Brahe onwards have given values very near this. According to the data of Meggers and Peters† we have for the refractivity ($\mu - 1$) of air : —

$$
\begin{aligned}
&\text{Fraunhofer line } \mathrm{F}\ (\text{green}) \ldots\ldots\ldots\ldots\ldots\ldots\ 2939 \cdot 3 \times 10^{-7} \\
&\qquad\quad ,, \qquad\quad ,, \ \ \mathrm{D}\ (\text{yellow}) \ldots\ldots\ldots\ldots\ldots\ 2917 \cdot 6 \times 10^{-7} \\
&\qquad\quad ,, \qquad\quad ,, \ \ \mathrm{C}\ (\text{red}) \ldots\ldots\ldots\ldots\ldots\ldots\ 2908 \cdot 9 \times 10^{-7}
\end{aligned}
$$

* Symons, ' Meteorological Magazine,' vol. 41, pp. 23 and 40 (1906).

† ' Bulletin of the Bureau of Standards,' vol. 14, p. 697 (1918).

This gives for the relative dispersion

$$\frac{\mu_F - \mu_C}{\mu_D - 1} = 1\cdot04 \times 10^{-2}.$$

The earlier work of Scheel* gives a result in substantial agreement with the above.

Thus the angular dispersion between the C and F lines, now more generally called H_α and H_β is $1\cdot04 \times 10^{-2}$ of 35′, or 22″.   Such a layer would only separate the red and green hydrogen lines by 22 seconds of arc.   A theoretically perfect eye at the extreme aperture of 5 mm. could only resolve two points of light if they were at least 25 seconds apart, and in practice Helmholtz found that at least 1 minute was necessary.†

These considerations show in accordance with direct experience that the spectroscopic resolving power of the atmospheric prism used with the aperture of the naked eye, should not give visible chromatic effects when a star or planet, or the sun's limb, is seen near the horizon.   With a magnifying telescope, the aperture, dispersion, and resolving power are increased, and chromatic effects may then be observed.

In the light of these considerations it was not altogether obvious, to me at least, how the atmospheric prism could be quantitatively adequate to show the green flash to the naked eye.   It is true that the green flash is not always seen, even when the atmospheric clearness seems good ; and I was tempted to take refuge in the assumption of exceptionally great dispersion. I shall return to this point later ; but it will appear, both theoretically and experimentally, that after all the normal amount of dispersion is adequate. It is the screening off by the distant horizon that makes it so.

To experiment upon the subject quantitatively, it is convenient to make use of a crown glass prism of such an angle as to give a dispersion of 22″ between C and F, equal therefore to the dispersion of the atmosphere near the horizon. A crown glass prism of 45′ angle was ordered.   Such a prism should give a deviation of about 23′, and on arrival the deviation was found to be in fact very near this value.   Taking the relative dispersion $\frac{\mu_F - \mu_C}{\mu_D - 1}$, as $1\cdot6 \times 10^{-2}$, which is the approximate value for this kind of glass, the angular dispersion between C and F is 23′ $\times$ $1\cdot6 \times 10^{-2}$ or 22″, the exact value required.

An illuminated aperture was used to represent the sun and the glass prism

* ' Verh. Deutsch. Phys. Ges.,' vol. 9, p. 24 (1907).
† See for instance Schuster's ' Optics,' 2nd ed., p. 149.

at a distance from it to represent the dispersing atmosphere. The natural conditions of sunset would be most exactly imitated if an obscuring edge to represent the horizon were placed between the prism and the eye of a distant observer. As will be explained in detail later on this method requires greater distances than are usually available in the laboratory, and to avoid them a modified experiment was first tried. A telescope was used, and the obscuring edge placed in the focal plane of the eyepiece. But as already emphasised, it is not legitimate to use a telescope which increases the angle of dispersion under which the spectrum is viewed. This objection is avoided if we use a " telescope " of magnifying power unity, *i.e.*, one with the objective and eyepiece of the same focal length. In that case the angular dispersion of the spectrum is the same as if it were viewed with the naked eye, but we have the advantage that the obscuring edge is in focus with it, allowing the point of cut-off to be precisely defined.

In carrying out the experiment, a slit 1 mm. wide was placed in the shutter, and could be backed either with the diffuse light of the sky, or with an incandescent lamp, with opal globe, as preferred. The illuminated aperture should be seen on as dark a background as possible. Four metres away was placed the prism giving the same dispersion as the atmosphere, and behind the prism a telescope made up of a cemented achromatic objective of 6 cm. focal length, and a Ramsden eyepiece, also of 6 cm. focal length. At the focal plane was placed an obscuring edge. It consisted of a piece of Gillette razor blade, the sides of which were smoked by burning camphor, and the edge alone wiped clean. The razor edge was vertical, and thus parallel to the edge of the source and of the prism. Exactly half of the field of view was covered by it.

In making the experiment, the eyepiece is carefully focussed on the obscuring edge, and then both are moved relative to the objective lens until the source is seen in sharp focus, without parallax relative to the knife edge. The special prism is then placed in position, and the telescope turned slightly if necessary so as to view the illuminated slit source. This slit is fairly wide* and represents the sun before it has completely set. No chromatic effects can be seen, the red fringe at one side of the white image, and the blue one at the other, being so narrow under this small dispersion that they cannot be seen while the broader and brighter white image is in view. If we turn the " telescope "

* 1 mm. is suitable. The sun is, of course, of much greater angular width, but as we are imitating what happens when nearly all of it is covered up, this is not important. The only objection to having a much wider source in the experiment is that there is more trouble with false light.

(preferably by a screw motion) so that the image is gradually obscured by the knife edge, we have a representation of sunset, and a flash of colour comes into view at the last moment. If the prism is so placed that its thin edge is on the same side of the observer as the knife edge (*e.g.*, on his left) the disappearance of the light is marked by a blue flash. Reversing the prism, so that its thick end, or base, is on the same side as the knife edge, the disappearing flash is red.

It is important for the validity of the experiment to make sure that this change really occurs when the prism is reversed. It is possible to be deceived by chromatic effects produced by the telescope lenses, if we observe away from the axis of the (extemporised) telescope. If this is the cause, the effects will persist when the prism is removed, and will not be reversed when the prism is reversed. In my own final observations of this kind, the reversal was repeatedly verified, the results being as definite and convincing as could be wished.

I have spoken of the disappearing flash, but in the laboratory experiment it is, of course, easy to arrest the movement at the point where the colour effect is at its best, and to observe it at leisure. When diffuse daylight is used behind the slit, the colour is blue. Using the incandescent lamp with opal bulb, which is probably a better approximation to the colour of the setting sun, the flash might be described as greenish blue. It must be admitted, however, that nothing was seen experimentally like the emerald green which (I think) has been recorded at sunset by some observers.

It has seemed worth while to describe this form of experiment, because it can be carried out without the use of long distances, with only the angular dispersion of the atmospheric prism, unmagnified. It is not, however, an altogether satisfactory imitation of the natural effect. In the latter the obscuring edge is not viewed close under a magnifying lens, nor on the other hand is the spectroscopic aperture that of the naked eye, as in this experiment.

I pass now to a better form of experiment, which closely imitates the natural phenomenon, and affords convincing evidence that the normal atmospheric dispersion is large enough to cause it. This experiment is easy to carry out, given sunshine and a long room.

The acute-angled prism, giving dispersion equal to the atmospheric, is placed in an aperture in the shutter with its refracting edge vertical, and is limited at the thick end, and preferably at the thin end also, by a vertical straight edge. Sunlight is reflected in by a flat silvered on its front surface, and the beam is received on a white screen 20 metres away. The luminous patch is then seen to be conspicuously fringed with colour, red on the less refrangible side, blue-green on the more refrangible. The red fringe is the most conspicuous, but the blue-green one is also very conspicuous, and could not escape even casual observation. If we wish to have the blue-green colour in the form of a "flash," a narrow strip of paper, say 1 mm. wide, may be substituted for the screen, its length being parallel to the coloured fringe. Then, as the diurnal motion carries the coloured edge of the luminous patch across the strip, the latter is momentarily seen illuminated with blue-green only. The duration of this flash should be about the same as that of the natural one, since both depend on the angular velocity of the sun. In accordance with this the experimental flash is found in fact to last about 3 seconds.

It may be asked, why not approximate still more closely to the natural conditions, and look directly at the obscuring edge, placing the eye in the coloured border of the illuminated area. The objection to doing this is that the coloured border is too narrow to fill the pupil of the eye satisfactorily. The whole length of the spectrum from C to F, with 22″ dispersion at 20 metres distance is only 2·1 mm.; the eye would, therefore, more than take in the whole of it. A minor objection is that the vertical height of the prism, 5 cm., subtends only 9 minutes of arc at a distance of 20 metres away, which is somewhat small. Moreover, the light of the sun when high up is undesirably bright for direct observation. By using the paper screen we avoid all these objections. The coloured fringe is seen under a good visual angle, in contrast to the neighbouring white light, and without undesirable brightness. The imitative experiment made in this way is, in fact, much more striking and easy to observe than the natural phenomenon, though the angular dispersion employed is the same.

Let us now pass, in imagination, to the latter. The obscuring edge, which has been used, is now replaced by the observer's horizon, at, say, 10 kilometres distance.* The dispersing prism is replaced by the atmospheric layer, and is situated, not at the obscuring edge, but behind it. This, however, makes no material difference. It is the distance of the obscuring edge which determines the length of the spectrum, which, from C to F, is now 107 cm., and therefore very large in comparison with the pupil of the eye.

The conditions are also more advantageous in another way. We have said

* This corresponds to an elevation of the observer of 8 metres, thus agreeing approximately with the ordinary conditions of observation from a ship's deck. A sea horizon does not lend itself conveniently to observations of the minimum distance required. Mr. C. T. Whitmell has observed the flash due to the sun going down behind a terrestrial obstacle 300 yards distant. In this case an opera glass was used. See ' J. British Astronomical Association,' vol. 12, p. 289 (1902).

nothing so far about the definition of the shadow cast by the obscuring edge. Experimentally it is observed that we get the coloured fringe at the edge of the shadow, and therefore the shadow must be sharp enough in the conditions of the short distance experiment. There is not, however, a great deal to spare in this respect.

The diffraction pattern of an obscuring edge, as investigated theoretically and experimentally by Fresnel, consists of alternate maxima and minima of decreasing amplitude outside the geometrical shadow, and a gradient of intensity within it which approaches asymptotically to zero. It is not necessary for the present purpose to refine to the utmost on this matter. Numerical calculation by Fresnel's methods shows that where the intensity gradient is steepest, that is, between the geometrical shadow and the first maximum the intensity falls to half in a distance d given by

$$d = 0 \cdot 52 \times \sqrt{\tfrac{1}{2} p \lambda},$$

where p is the distance from the obscuring edge to the screen, and λ the wavelength. We may call d the diffusion of the shadow.

Since the spectrum increases in length as the direct distance, whereas the diffraction pattern lengthens only as the square root of the distance, the sharpness of cut-off, measured in Ångströms, continually improves with distance.

It is scarcely necessary for our purpose to consider the variation of diffusion of the shadow at different wave-lengths. Taking a representative wave-length of 5×10^{-5} cm. in the blue-green we get

Distance.	Length of spectrum C–F.	Diffusion of shadow.
20 metres	0·21 cm.	0·12 cm.
10 kilometres	107 cm.	2·60 cm.

We see then that in the laboratory experiment the diffusion of the shadow occupies more than half the length of the spectrum. In the natural phenomenon with a horizon 10 kilometres distant, it occupies only 1/40th of the length of the spectrum.

It may seem at first sight that this is scarcely consistent with the ordinary theory of spectroscopic resolving power, for, as we have seen the aperture of the eye, applied to the atmospheric prism, is barely sufficient to separate C and F. But the relevant aperture in this case is not that of the observer's eye. It is a much larger aperture in the plane of the obstructing edge. Since the aperture is limited by a screen on one side only, its effective breadth is somewhat ill-defined. If, however, the observer's eye is placed in the position of most rapid fall of intensity, and the obscuring edge is imagined to be moved forward $2 \cdot 6$ cm. so as to cover a strip of $2 \cdot 6$ cm. wide adjoining its former position, the whole diffraction pattern at the observation plane will be moved forward $2 \cdot 6$ cm., and at the same time the intensity at the observer's eye will fall to half. We may, therefore, consider that most of the light comes from an area of this order of breadth, and in fact that the effective aperture is of this order. It is therefore intelligible that we get much more spectroscopic resolution than when the effective spectroscopic aperture is that of the eye alone.

A spectroscope in which the effective aperture is not covered with a lens is nowadays unfamiliar, but, after all, in Newton's first experiments with the spectrum, no lens was used, and it is obvious that when the spectrum is received on a screen, the aperture of the eye cannot have anything to do with the resolving power. My father, the late Lord Rayleigh,* pointed out long ago that, given a sufficiently long screen distance a " pinhole " image of the sun could be formed with full definition, however large the pinhole might be. With a " pinhole " larger than the pupil of the eye, the image on the screen at the necessary very long distance is better than that seen direct. Similarly, if the large " pinhole " is covered with a prism, the spectrum of a star will be better resolved than the spectrum of the same star seen by placing the prism over the eye, and looking at the spectrum directly.

This discussion seems to show as clearly as could be wished, that the normal dispersion and resolving power of the atmospheric spectroscope are ample to account for the " green flash."

The natural phenomenon is, however, notoriously capricious in its appearance. It has been suggested by R. W. Wood† that the occasions when it is well seen may be those of exceptionally large atmospheric dispersion near the horizon. The refraction, and therefore also the dispersion, near the horizon are known by practical navigators to be somewhat uncertain ; but if the variations are of much importance in this connection, we must suppose the refraction on some occasions to be at least double the ordinary. This would defer the time of sunset by 2 minutes or more. I have not been able to learn of any exact series of observations of the time of sunset, as compared with the predicted time, but there should apparently be no particular difficulty about making them, and the green flash might be looked for at the same time. I am

* ' Phil. Mag.,' vol. 11, p. 214 (1881) ; ' Scientific Papers,' vol. 1, p. 513.
† ' Nature,' vol. 111, p. 501 (1923).

hoping to get such observations made in tropical seas through the good offices of the Astronomer Royal.

Since the normal atmospheric dispersion is shown to be adequate, I rather incline at present to attribute the variability almost wholly to variable atmospheric clearness. It is generally agreed that a very clear horizon is essential. It is said, moreover, that in some places, *e.g.*, the island of Grenada, near Barbadoes, the green flash can nearly always be seen, and we cannot well suppose that the atmospheric refraction is always exceptionally high at sunset there.

It is a matter of experience that the green flash is best seen with a telescope. According to the views developed above, the effective spectroscopic power is not dependent on the aperture of the eye, and the advantage of, at any rate, a small telescope cannot be in an increase of spectroscopic resolving power. It is probably rather to be sought in an increase of the apparent size of the luminous source which makes its colour easier to distinguish. On the other hand, when we look at a star or planet near the horizon with a telescope, and it appears drawn out into a short spectrum* there *is* an increase in spectroscopic dispersion and resolving power, which explains the observation in that case.

With regard to the colour being usually green rather than blue, I have nothing new to offer. In the laboratory experiment at 20 metres the colour would be described as greenish blue, but the spectroscopic purity in that case is rather inadequate. In any case the flash has to assert itself in competition with scattered light, and it may be that the inferior visual luminosity of the pure blue of the spectrum, together with the large toll taken from it by atmospheric scattering, prevent its asserting itself, after the green is below the apparent horizon.

Summary.

The ordinary dispersion of the atmosphere near the horizon appears adequate to explain the " green flash." It can be imitated experimentally with a prism of the same dispersive power as the normal value at the horizon, and without making use of lenses.

* See C. J. P. Cave, ' Nature,' vol. 123, p. 607 (1929), also J. Evershed, ' Nature,' vol. 120, p. 876 (1927).

Observations and Theoretical Reconstruction
of the Green Flash

By GLENN E. SHAW[1])

Summary – The phenomenon of the green flash at sunset (or sunrise) is reviewed. Several possible mechanisms that have been put forth to explain the green flash are discussed. A quantitative model for the phenomenon is then described, which includes parameters that are representative for a polar atmosphere (low humidity and small aerosol optical depth). It is supposed that the primary mechanism responsible for causing the green flash are natural molecular dispersion and the filtering action imposed by the atmosphere for low elevation angles. Results from the model indicate that a green rim of vertical extent $\simeq 0.15$ milliradians would appear at the upper limb of the sun during sunset or sunrise. The theoretical results are compared with observations of a green flash made at wintertime in interior Alaska.

I. Introduction

The last remaining glint of light from the upper limb of the sun at sunset is occasionally observed to take on a greenish hue. The transitory phenomenon, which can also be observed at the first light of sunrise, has been variously referred to in English as the green flash or the green ray; in French it is known as *le rayon vert*. The very existence of the spectacle has, in times past, been the subject of controversy; however, photographic documentation obtained over the past decade or so has proven beyond any doubt that the phenomenon exists, at least on occasions when atmospheric conditions are favorable.

The relative rarity of the green flash may be appreciated from the small number of reported early observations. Strangely enough, there is evidently no indication that early astronomers of Babylonia, Egypt or the Yucatan, although they were known to be diligent observers of the sun, ever recorded seeing the green flash. However, as the number of trained solar observers increased in the 17th and 18th centuries, the recorded references to a low sun green hue or flash increased. A considerable amount of attention has been devoted to the subject over the past half century or so by astronomers and other scientists, and an excellent account of the phenomenon, in its various forms, along with a complete bibliography has been prepared by O'CONNELL [8][2]).

---

[1]) Geophysical Institute, University of Alaska, College, Alaska 99701, U.S.A.
[2]) Numbers in brackets refer to References, page 234.

It is clear that the green flash is intimately associated with conditions existing in the atmosphere of the earth, but it is not obvious exactly what set of atmospheric conditions favors its occurrence. Upon inspecting the observational reports, it seems that an unusually large number of observations have come from various locations in, or close to, the polar regions. This suggests that the unusual atmospheric conditions which prevail in the arctic and subarctic regions may in some way be conducive for the occurrence of the green flash.

It is known that the polar regions have persistent and strong temperature inversions with associated thermocline structure. This unusual temperature structure can, at times, lead to anomalous refraction and cause spectacular mirages or distortions of distant objects to occur. It is also probable that the increased atmospheric refractivity is related to the common association between green flashes and polar locations.

In addition to enhanced atmospheric refraction, the extreme clarity of the air in polar regions, especially during wintertime when the ground is frozen, apparently contributes to the unusually large number of polar observations and associated low-sun phenomena. Measurements made at McCall Glacier in northern Alaska (latitude 69°18′N, longitude 143°48′W), for example, show that the total optical depth from aerosols is several times smaller than that reported for temperate latitudes; this results in great clarity of the air and can have a pronounced effect on the transmission of the atmosphere through the long paths taken by solar rays at sunset.

An additional factor which may enhance the chances of observing a green flash is the oblique angle that the solar trajectory makes across the sky at polar locations. Because of this, the limb of the sun remains close to the horizon for considerable amounts of time with an attendant increase in the probability of observing a green flash.

According to photographic records taken through long focal length lenses, the sun's upper limb has a green or yellowish-green hue when the sun is very low on the horizon. It is clear that the green 'flash' is actually the last remaining remnant of this green 'rim' as the sun moves under the local horizon. The transitory nature of the phenomenon can be appreciated when it is realized that the observed width of the green rim, as determined by scaling the photographs, would cause the green light to remain above the horizon for only about one second of time, provided the sun's apparent path on the sky is nearly vertical as it would be in a low latitude. As alluded to previously, the low slant trajectory of the sun in the polar sky would result in this duration of time being considerably lengthened.

II. A review of factors which may contribute to the green flash

A number of 'explanations' for the cause of the greenish hue have been put forth at various times. Among these are theories involving various combinations of dispersion, refraction and absorption by the atmosphere. W. H. JULIUS [6] first suggested that the green flash was perhaps due to anomalous dispersion by the earth's atmosphere which could occur in select wavelength intervals located near telluric absorption features to cause the green rays to be refracted more than the longer or shorter ones. Although there was some attempt made to experimentally test this theory, it is now clear that this mechanism is not sufficient to explain the presence of the green flash.

It is likely that normal dispersion introduced by the atmosphere plays a predominant role in the formation of the green flash. The normal dispersion of atmospheric air results in a greater refractive bending for the shorter wavelengths, and hence the sun's upper limb consists of short-wave radiation that is lofted upwards with respect to the longer and redder wavelengths, Lord Rayleigh, in 1930, carried out a series of investigations and demonstrated that the dispersion of air is probably large enough in magnitude to at least partially explain the mechanism of the flash.

An additional parameter to consider in the theory is the possible selective absorption of light by specific gases, which, depending upon the wavelengths involved, results in a modification of the spectral characteristics of the sunlight. The effect is especially large at low sun elevation angles where long atmospheric path-lengths are involved. As an example, it has been suggested that the color index might be modified by the well known Brewster, or rain bands, due to absorption by water vapor. The broad Chappuis bands of ozone and various absorption features caused by oxygen and perhaps O_4 might also have some bearing on the color phenomena seen at low sun angles. A review of telluric absorption features in the visible region and their possible effects on color index has been prepared by DUFAY [2].

Undoubtedly, the molecular scattering of light (Rayleigh scattering) plays a very important role in determining the apparent color of luminous objects whose rays enter through the atmosphere at small elevation angles. The well known inverse fourth power wavelength dependency of Rayleigh scattering is responsible, qualitatively speaking, for the 'blue' color of the sky and the 'red' color of the setting sun.

Lastly, scattering and absorption by particulate material in the atmosphere might be of some importance in explaining the green flash. It has been found out experimentally that optical depths in the visible arising from aerosol extinction can be of a magnitude quite comparable with extinction caused by the total of other effects such as scattering by molecules and absorption by gases. Furthermore, the wavelength dependence of extinction by atmospheric aerosols depends upon the composition and size distribution of the particulates, and hence is highly variable.

In the light of present knowledge, it appears that of the many factors possible, some of which were reviewed above, the normal dispersion of air along with molecular scattering and aerosol extinction are the most likely causes for the green flash.

III. Description of a theoretical reconstruction of the green flash

The factors that might contribute to the formation of a green flash or other associated low-sun phenomena were enumerated in the last section. It is apparent that there are a variety of *possible* mechanisms; however, a quantitative treatment has apparently never been presented. In this section, a model atmosphere is used along with relevant radiative transfer theory to assess the spectral distribution of direct sunlight as seen at low angles. Resultant wavelength distribution of intensities for different portions of the solar disk are then used to calculate, with the aid of colorimetry theory, the dominant wavelengths and the degree of saturation.

Briefly, the mechanism proposed to account for the green flash involves a combination of atmospheric refraction and extinction of light. The atmospheric refractivity, which is a maximum at sunset, causes the blue and green rays from the upper limb to undergo a greater amount of bending than the longer red and orange rays. During their long path through the atmosphere the solar rays undergo varying amounts of transmission, depending upon their wavelength. The combination of the two effects can explain the major phenomena that are observed during a green flash of the sun.

Transmission of the atmosphere in the visible region can be expressed through the Lambert Beer law, which can be written as

$$T(\lambda, \delta) = e^{-m(\delta)\,\tau_t(\lambda)} \tag{1}$$

where the symbols have the following meaning:

$T(\lambda, \delta)$ = Transmission through the atmosphere for a light ray at wavelength λ and at apparent vertical elevation angle δ.

$m(\delta)$ = Atmospheric airmass relative to a ray entering from the zenith direction, evaluated at elevation angle δ.

$\tau_t(\lambda)$ = The total optical depth at wavelength λ and referred to the vertical direction.

Furthermore, the optical depth, $\tau_t(\lambda)$, may be written in terms of its additive constituents:

$$\tau_t(\lambda) = \tau_r(\lambda) + \tau_d(\lambda) + \tau_g(\lambda) \tag{2}$$

where in equation 2, the terms have the following significance:

$\tau_r(\lambda)$ = Rayleigh optical depth caused by scattering of light by the atmospheric molecules.

$\tau_d(\lambda)$ = Optical depth arising from a combination of absorption and scattering from particulate material in the air.

$\tau_g(\lambda)$ = Optical depth due to absorption by specific gaseous constituents in the atmosphere.

Each optical depth term, τ_i, in equation 2, $\tau_i(\lambda)$, can be expressed in terms of the volume extinction coefficients, β_i, as,

$$\tau_i(\lambda) = \int_0^{\infty} \beta_i(\lambda, h)\, dh, \tag{3}$$

where

$\beta_i(\lambda, h) =$ Volume extinction coefficient at wavelength λ and at height h for the ith species.

The values for the terms displayed in equation 2 can be estimated numerically for a model atmosphere. For instance, the Rayleigh optical depth term, $\tau_r(\lambda)$, is given by

$$\tau_r(\lambda) = \frac{0.00895}{\lambda^4} \cdot \frac{p(h)}{p(0)},\tag{4}$$

where

$\lambda =$ Wavelength of radiation in microns.

$p(h) =$ Atmospheric pressure at observer's location.

$p(0) =$ Atmospheric pressure at sea level for standard temperature conditions.

The optical depth arising from gaseous absorption $\tau_g(\lambda)$, occurs in isolated bands and lines throughout the spectrum. For the visible, the most serious absorption is contributed by ozone and water vapor. The ozone extinction within the broad Chappuis bands has been estimated for typical ozone conditions existing at a latitude of 65°N. Water vapor absorption in the visible region is small and will be ignored; this approximation is valid for regions where the precipitable water overhead is small, as at arctic locations during winter time.

Finally, the extinction arising from scattering and absorption from particulate material must be estimated. Since many observations of the green flash have been reported from the polar regions, and since our observations were made in interior Alaska during winter time, it is appropriate to use a model corresponding to the clear air found in the Arctic. According to pyroheliometer measurements that we have made in northern Alaska on McGall Glacier (SHAW and WENDLER [11]) and from measurements reported from other polar stations (HOLMGREN [5]), we feel that a value of $\tau_d = 0.05$ at $\lambda = 0.55$ μm is reasonable to represent the clear atmosphere in Alaska.

The variation of $\tau_d(\lambda)$ with wavelength is dependent upon the micro size distribution of the atmospheric particulates. Experimentally, one finds that the atmospheric particles span a size range extending from a low limit of approximately 0.1 microns radius to an upper limit of a few microns. The small size limit is determined primarily by the rate of coagulation of small atmospheric ion clusters due to Brownian motion; the upper limit is brought about by the balance between gravitational settling and eddy turbulence. Between the upper and lower cutoff limits, it is found that continental aerosols are often distributed (assuming them to be spheres, at least statistically) according to a power-law relationship of the form (JUNGE [7]),

$$\psi(r)\, dr = c\, r^{-\nu}\tag{5}$$

where

$\psi(r)\, dr =$ The number density of particulates contained between radius limits r and $r + dr$.

$c =$ A normalizing constant.

$\nu =$ Junge power-law exponent.

The exponent v ranges between values extending from about two to four, with values near three most commonly found. Assuming a Junge size distribution expressed by equation 5 with a value of $v = 3.0$, it can be shown from use of the Mie theory that the wavelength dependency of extinction by aerosols can be approximately expressed as,

$$\tau_D(\lambda) = k \, \lambda^{-v+2}.$$ (6)

Computerized modeling shows that the relationship in equation 6 holds to better than five percent when the aerosol particles are contained between radii limits of 0.01 and 3.0 microns.

As was mentioned before, the power law exponent, v, is typically found to have a numerical value of approximately three. Employing this value in equation 6 yields a λ^{-1} dependence on wavelength. Although too few measurements of $\tau(\lambda)$ have been made in arctic and subarctic locations, we have determined that a λ^{-1} wavelength dependence of aerosol optical depth holds at least on some occasions. This result is based on radiometric measurements made through narrow band optical filters of the direct solar radiation at College, Alaska.

The atmospheric model which will be employed for the reconstruction of the green flash and other low-sun phenomena is analogous to a large-scale optical filter whose transmission as a function of wavelength depends upon contributions from molecular scattering, absorption by ozone and scattering and absorption by particulate material. From the preceding discussion, it was indicated that all of these optical depth factors can be estimated and quantitatively evaluated for a model atmosphere.

The air mass term, $m(\delta)$, in equation (1), represents the amount of atmospheric material traversed by a hypothetical bundle of incoming rays having a cross sectional area equal to unity. Air mass is dimensionless and is expressed in terms of the amount of material contained in a vertical column of unit area. For solar elevation angles larger than about 20°, the curvature of the atmospheric layers can be neglected, and the atmospheric layer can be approximated as plane parallel. In this case, the air mass term is given by the cosecant of the elevation angle δ. When the sun is close to the horizon, however, the curvature of the atmosphere becomes important and a numerical estimation for the air mass term becomes dependent on the vertical distribution of absorbing and scattering particles and gases.

A numerical method of evaluating the airmass, $m(\delta)$, for a given height distribution of gases and aerosols was implemented for this study. This computerized routine was employed to calculate the air mass for aerosols distributed with height in accordance with Elterman's tables (ELTERMAN [4]). The variation in air mass with elevation angle was found to be in quite good agreement with an empirical formula originally developed by ROZENBURG [10]. The form of this expression is given by,

$$m(\delta) = \left[\sin(\delta) + 0.025 \, e^{-11.0 \, \sin(\delta)}\right]^{-1}$$ (7)

where

δ = apparent elevation angle,

$m(\delta)$ = atmospheric airmass.

Equation (7) agrees to within an accuracy of several percent with the values of $m(z)$ obtained by using the aerosol variation with heights given in Elterman's tables.

Finally, the amount of refraction to which an incoming light ray is subjected to must be estimated, since this is intimately associated with the mechanism of the green flash. The total atmospheric refraction depends upon knowledge of atmospheric density at each height which, in turn, may be obtained from the vertical temperature distribution. Although it is suspected that the high atmospheric refractivity often found over the polar regions can act to enhance the low sun phenomena, our model incorporates only standard values of atmospheric refraction which, for zero degree elevation, is listed as being about thirty-five minutes of arc or, equivalently, 9.8 milliradians.

The variation of refraction with wavelength is caused by the normal dispersion of air. This can be expressed quantitatively (EDLEN [3]) as,

$$(m - 1) \times 10^8 = \left\{ 6432.8 + \frac{2949810}{146 - (\lambda^{-2})} + \frac{25540}{41 - (\lambda^{-2})} \right\} \tag{8}$$

(λ in micron).

Differential refractive bending for different wavelengths is determined by using,

$$\frac{d\theta}{d\lambda} = \frac{\theta(\lambda)}{m(\lambda) - 1} \cdot \frac{dm(\lambda)}{d\lambda}, \tag{9}$$

where

θ = vertical angular displacement of incoming light ray at wavelength λ.

The upper limb of the sun is of particular interest, since it is here that the green rim forms which is basically responsible for the green flash. To obtain an indication of the possible angular extent of the colored rim at the top of the solar disk we can use equation (9) to calculate the vertical angular separation between a red and violet ray due to differential refraction. When this is done, a numerical value equal to about 0.25 milliradians or of the order of 1/30th of the apparent solar diameter is found. Because of the fact that the short violet rays are refracted most, these rays will form the top portion of the rim which will then consist of spectrally pure light. Similarly the extreme bottom limb of the sun will consist of spectrally pure red light.

For the purposes of determining the colors existing within the rim at the top of the solar disk at sunset, we illustrate in Figure 1 the solar disks as they would be seen by a ground-based observer in blue light and in red light; the two disks are slightly separated because of the dispersion introduced by the atmosphere. An image of the unrefracted solar disk is also drawn in Figure 1. As can be seen, the solar disk, when on the horizon, is displaced vertically by approximately one solar diameter due to refraction.

Because of the dispersion of air the differential angular displacement of the sun's upper limb is a function of wavelength, and this dependency is shown in Figure 2. Thus the radiant energy falling into the rim segment at a distance of $\Delta\theta$ above the limb formed by light at some reference wavelength consists only of light bluer than some critical wavelength λ_c. For example, at a vertical distance of $1.012\,\theta_{REFT}$ from the sun's limb in $\lambda = 0.7\,\mu$m light, the critical wavelength, λ_1, is found to be (from

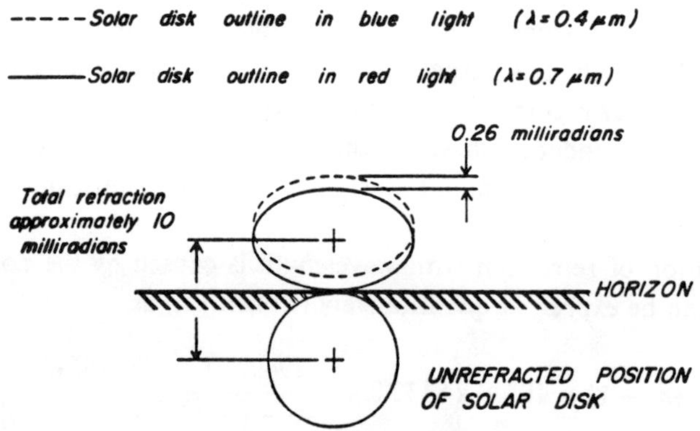

Figure 1

Illustration of the solar disk as it would appear at sunset to a ground-based observer in the monochromatic light. The two disks, corresponding to red and blue wavelengths, are slightly displaced to each other because of the dispersion of air. The lower disk corresponds to the unrefracted location of the sun. The oblate shape arises due to the difference in atmospheric refraction between the top and bottom limb of the sun

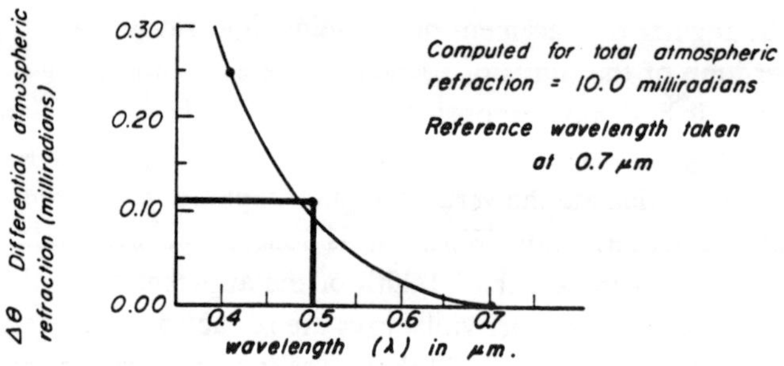

Figure 2

Atmospheric differential refraction $\Delta\theta$ for a standard atmosphere (in milliradians) plotted against wavelength. The values are referenced to an incoming light ray at $\lambda = 0.70\,\mu$m. The critical refraction angle, $\theta_1(\lambda)$, is shown corresponding to a wavelength, λ, of $0.50\,\mu$ meters is illustrated

Figure 2) equal to about 0.50 micron. Thus, radiant energy illuminating the vertical segment extending from $\Delta\theta_1$ to $\Delta\theta_1 + d\theta_1$ consists entirely of light energy bluer, or shorter, than λ critical.

Figure 3 shows the wavelength dependence of the solar radiation, $I(\lambda)$ (THEKAEKRA [12]) for air masses of zero, 2, 20 and 40. The curves in Figure 3 were calculated

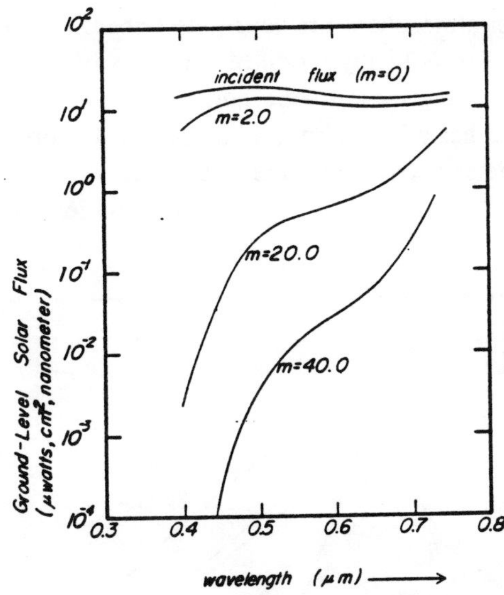

Figure 3

Ground-based solar flux as a function of wavelength computed for a model atmosphere for several values of solar elevation angle, δ

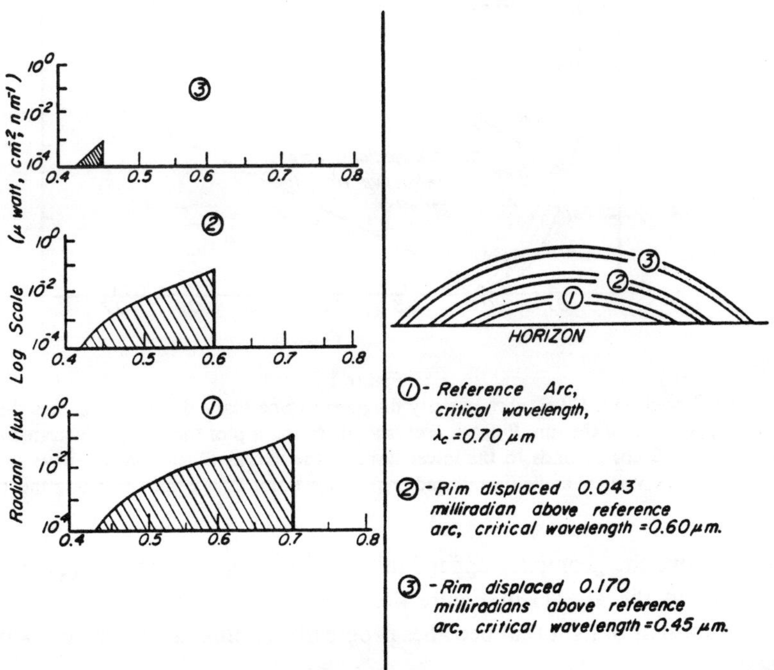

Figure 4

4a) Illuminating radiant flux as a function of wavelength for different values of vertical elevation angle in the vicinity of the rim at the top of the solar disk. 4b) indicates the relative position of the rims corresponding to the different spectral curves in Figure 4a by using the model atmosphere previously described. As can be seen, the spectral distribution of the sun is highly modified for cases of large air masses which correspond to the sun being close to the horizon. It should be pointed out that the ordinate in Figure 3 has a logarithmic scale, so that factors of differences of several orders of magnitude exist from one case to another.

Finally, Figure 4 illustrates the spectral distribution of intensity at 3 angular

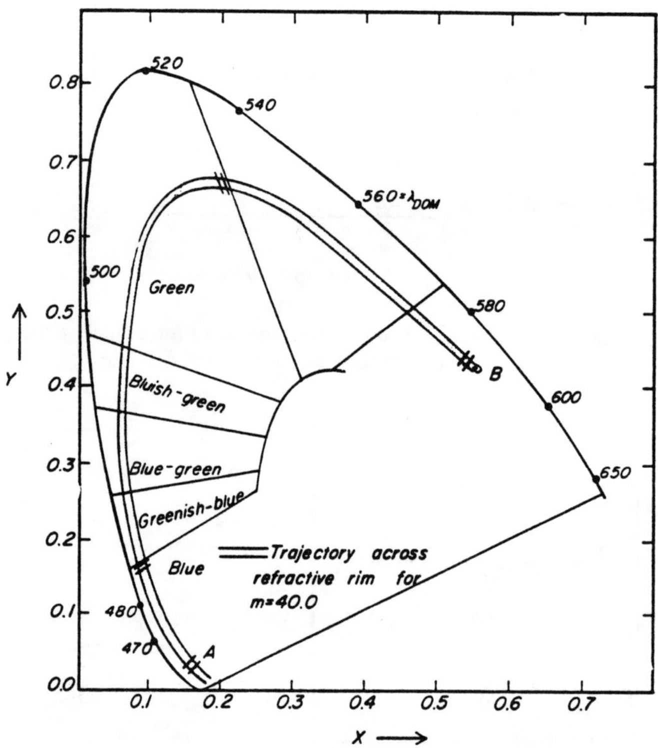

Figure 5

Trajectory followed on a C.I.E. chromaticity diagram as one scans downward across the refractive rim at the upper limb of the sun. Puint A corresponds to the top of the rim as illustrated in Curve 3, Figure 4b, Point B corresponds to the lower limit of the rim as illustrated in Curve 1, Figure 4b. Values of $\Delta\theta$, the vertical angle (in milliradians) above Point B, are marked along the trajectory

distances above the geometric upper limb. The distribution labeled (1) in Figure 4a corresponds to the arc (1) in Figure 4b. The light illuminating the upper regions of the rim consists of light that becomes progressively bluer as the critical wavelengths decrease.

In order to establish the colors at different locations in the rim of the sun, the illumination profile was calculated for twenty individual critical wavelengths in the visible region, each corresponding to a given vertical angle within the refractive rim at the solar upper limb. The C.I.E. chromaticity coordinates, x, y, z, were calculated for each spectral distribution corresponding to the separate regions within the refractive rim by the usual methods of colorimetry (BILLMEYER and SALTZMAN [1]). Figure 5 illustrates the trajectory that is followed on a chromaticity diagram as one traces downward across the refractive upper rim of the sun. Point A is the extreme top of the rim, whereas point B is at the region corresponding to the solar disk. The dominant wavelength, λ_D, which indicates the visual aspect of hue, has been calculated for several points along the trajectory in Figure 5 and is displayed in Table 1 along with the corresponding angle, $\Delta\theta$, and color purity or chroma.

The important conclusion which may be drawn from the colorimetry calculations is that the rim at the upper limb of the sun, at sunset, would appear predominately green in hue. The visual aspects of hue are also tabulated in Table 1, and it is seen that the blue-green to yellow-green region extends over a vertical angular distance of approximately 0.15 milliradians (about 0.5 minutes of arc). Strangely enough, the topmost blue-appearing region is spatially small with respect to the green hue and, in addition, is reduced in intensity by several orders of magnitude and hence would not be apparent – a comforting fact, since qualitative arguments would lead one to expect a blue flash.

Table 1

Values of dominant wavelength, λ_d, and color purity for different regions, $\Delta\theta$, within the refractive rim at the upper limb of the sun. The angle $\Delta\theta$ is taken above the top of a reference disk formed by light at a wavelength of 0.7 micron

$\Delta\theta$ (milliradians)	λ_d (μ meters)	Color Purity	Relative Intensity	Color
0.000	0.59	0.85	1.00	Orange
0.025	0.58	0.86	0.19	Orange
0.050	0.56	0.83	0.62×10^{-1}	Greenish
0.075	0.53	0.78	0.26×10^{-1}	Green
0.100	0.51	0.74	0.91×10^{-2}	Green
0.125	0.48^+	0.80	0.31×10^{-2}	Green-Blue
0.150	0.46	0.91	0.80×10^{-3}	Blue

IV. Discussion

The major reported features of the green flash in accordance with the results of the model. As is generally agreed, the green flash phenomenon is actually the visual remnants of the green rim atop the sun. The green rim, according to our model, should appear to have a vertical extent of approximately 0.15 milliradians, or about 1/60th the solar diameter.

The inclination of the plane containing the sun's path across the sky with the horizon is small at high latitude locations, and as a result the solar disk remains near the horizon for rather extended periods of time. To estimate the length of time during which the green rim remains above the horizon we use the parameters for our location at College, Alaska. Here the green flash was observed on several occasions during the 2 weeks preceding and following winter solstice. Thus, taking a latitude of 64°52′N (College, Alaska), a solar declination of $-22°$ and a green rim of 0.15 milliradians in width, computations yield a value of time equal to about 15 seconds during which the green rim of the stated angular width would remain above the horizon. This time period is considerably larger than the corresponding time increment that would occur at lower latitudes and probably explains, at least in part, why there seem to be so many reports of the green flash from high latitude locations.

During one occasion, on the 22 January 1972, the sun was observed skirting the horizon and progressively sinking lower. Several seconds before the final disappearance of all sunlight, the remaining rim of light took on a brilliant emerald green appearance that lasted for fully 4 or 5 seconds. Eventually, the green rim was occulted by Mt. McKinley (which is in a SW direction from College at a distance of 260 kilometers) only to reappear several seconds later on the other side of the mountain. Again, the remaining light was tinged a beautiful green color which lasted for another several seconds. Besides the observation just recounted the green flash has been observed from College on at least five other occasions during the months extending from November 1971 to January 1972.

Photographs of the solar disk have been taken with a one-meter focal-length Schmidt-Cassegrain telescope. Eyepiece projection was used to magnify the solar image to a diameter of five centimeters. The photos show the presence of a green tinted thin rim at the extreme top edge of the disk which is of a width corresponding to our modeled calculations. It is apparent, especially when observing the projected image on a screen, that atmospheric turbulence degrades the rim shape and causes it to scintillate rapidly.

During the winter months, strong polar temperature inversions normally exist over the regions of interior Alaska, and these are responsible for strong atmospheric refraction phenomena. Mirages and distortions of distant objects, due to the thermocline and strong temperature gradients, are commonly observed in interior Alaska. During times of unusual temperature structure, the shape of the solar disk is often observed to become highly distorted. Simultaneously, one often can see green regions develop in the vicinity of the distorted regions on the solar disk. The explanation seems to lie simply in the fact that the greenish regions are caused by atmospheric refraction, which itself is enhanced in the vicinity of the regions of distortion on the solar disk.

REFERENCES

[1] BILLMEYER and SALTZMAN, *Principle of Color Technology*, Interscience Press, 180 pp., 1966.
[2] J. DUFAY, *Notes sur l'absorption selective dans l'atmosphère terrestre*, Ann. d'Astrophys, Vol. 5, 93–1131, 1942.

[3] B. EDLEN, *The Dispersion of Standard Air*, Journal of the Optical Society of America, Vol. 43, May 1953.

[4] L. ELTERMAN, *U.V., Visible, and IR Attenuation for Altitudes to 50 km, 1968*, Air Force Cambridge Research Laboratories Environmental Research Papers, No. 285 (AFCRL-68-0153), 1968.

[5] B. HOLMGREN, *Climate and Energy Exchange on a Sub-Polar Ice Cap in Summer, Part E, Radiation Climate*, Uppsala Universitet, 1971.

[6] W. H. JULIUS, *Le Rayon Vert*, Arc. Nearl. des sc. (2), Vol. 6, 385–389, 1901.

[7] C. E. JUNGE, E. ROBINSON and F. L. LUDWIG, *A Study of Aerosols in Pacific Air Masses*, Journal of Applied Meteorology, Vol. 8, June 1969.

[8] D. J. K. O'CONNELL, *The Green Flash*, North Holland Publishing Co., Amsterdam, 192 pp., 1968.

[9] LORD RAYLEIGH, *Normal Atmospheric Dispersion as the Cause of the 'Green Flash' at Sunset, with Illustrative Experiments*, Proc. Roy. Soc., A, Vol. 126, 311–318, 1930.

[10] G. V. ROZENBERG, *Twilight, A Study In Atmospheric Optics*, Plenum Press, New York (Translated from Russian) 1966.

[11] G. E. SHAW and G. WENDLER, *Atmospheric Turbidity Measurements at McCall Glacier in Northern Alaska*, To be submitted at the Conference on Atmospheric Radiation to be held at Fort Collins, Colorado, August 7, 1972.

[12] M. D. THEKAEKARA, *Solar Irradiance Measurements From a Research Aircraft at 38,000 Feet* Report X-322-66-304 (Goddard Space Flight Center, Greenbelt, Maryland) August 1968.

(Received 9th May 1972)

The Green Flash and Clear Air Turbulence[1] [2]

Alistair B. Fraser

*Department of Meteorology,
The Pennsylvania State University*

[Manuscript received 16 December 1974; in revised form 24 April 1975]

ABSTRACT

There has long been speculation about why the green flash can be seen on one day, while on another apparently similar day it cannot. Although the green flash can be produced by a number of different refractive structures in the atmosphere, only one is of consequence when the viewing is done over the irregular terrain of most land surfaces. A combination of extensive observations and simple theory suggests that this refractive structure is formed by gravity waves which have wavelengths between about 0.2 and 2.0 kilometers. In the atmosphere such waves derive their energy from wind shear and are the same waves that are associated with clear air turbulence. This model not only explains the frequent observations of multiple green flashes, but also, by demonstrating a dependence on atmospheric dynamics, it accounts for the variability in the occurrence of the green flash on otherwise comparable days.

"What I want to know, do you ever get to see that flash of green? People say they've seen it, the minute the sun disappears."
"Now, it stands to reason, mister, any damn fool stares into the sun long enough, he'll end up seeing exactly what some other damn fool tells him he's going to see."

(from *A Flash of Green* by John D. MacDonald)

1 Introduction

Anyone who enjoys watching the sunset will appreciate that the sun can assume some very strange shapes just before it dips below the horizon. Sometimes the edges of the sun will actually become convex, such as the example in Fig. 1 taken over Lake Winnipeg. This type of distortion is produced by strong temperature inversions. Even stranger are the little spikes that occasionally seem to grow out of the sides of the sun (Fig. 2). As you watch, the spikes appear to drift up the sides until they detach and leave an island of light floating above the top of the sun. Suddenly, all the yellow color drains away from this island and for an instant it flashes green before summarily vanishing. This is the green flash! (Actually, this represents one of four different ways the flash is manifested.)

The green flash has attracted the occasional attention of physicists ever since Jules Verne published a novel called *Le Rayon Vert* in 1882. Early interest in the phenomenon centered on the issue of "why green?". The history of this is discussed in O'Connell's book (1958). The final resolution of that problem is remarkably simple. The atmosphere is dispersive (it acts like a prism) so that the setting sun should always be seen with a red rim on the bottom and a blue rim on the top. Atmospheric scattering will usually remove the blue so that a

[1] Paper presented at the 8th Annual Congress of the Canadian Meteorological Society, May 30, 1974, at York University, Toronto.
[2] Work supported by NSF Grant No. GA-38522.

Fig. 1   This view of the setting sun was made from Grand Beach on Lake Winnipeg on
August 26, 1973. The drawing was made from the projected image of a 35 mm
slide that was photographed by the author. A portion of the sun, exhibiting the
reverse curvature, is greatly compressed, while below that the image is greatly
magnified. Such an image is seen when the temperature profile over the lake has
an inflection point, that is a lifted inversion. This temperature profile is frequently
found above enclosed bodies of water late in the day. The grey area below the sun
is the opposite shore of the lake.

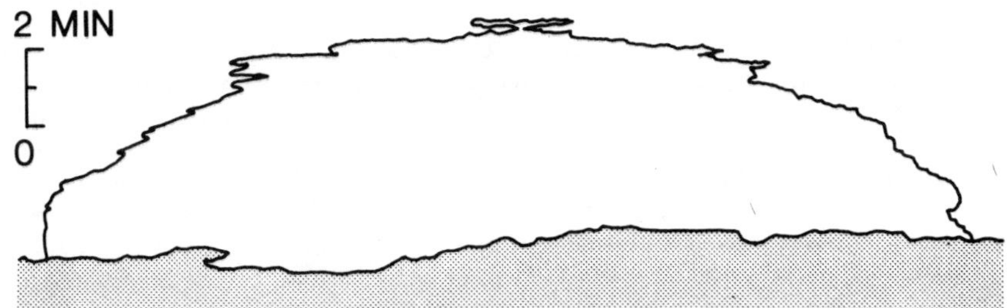

Fig. 2   This view of the setting sun was made on December 26, 1971, from Seattle,
Washington, while looking out over the Olympic Mountains. The drawing was
made from the projected image of a 35 mm slide photographed by the author. The
grey area at the bottom is a cloud. The vertical scale shows two minutes of arc.

green rim is seen on the top of the sun. Shaw (1973) discusses this aspect of
the problem.

This explanation seems perfectly satisfactory (and is indeed correct) but
when you calculate the angular size of the green rim under normal atmospheric
conditions you discover that it is below the angular resolution of the human
eye. And yet the green flash is regularly seen by naked-eye observers! There are
two ways to get around this difficulty. We might ask that the green rim be
magnified by the proper "mirage" conditions. One way that this can be done is
the topic of much of this paper. Alternately, we could require that the main part
of the sun be hidden from view, by the horizon, a mountain, or a cloud. Then
if the green rim remained above the obstacle, it could be seen although it could
not be resolved. (The effect is similar to viewing a star. You cannot resolve a
star, but you see it because it is isolated.)

If we leave aside this rather artificial device for seeing the green rim, we
are left to consider how the atmosphere can occasionally produce a short-lived
magnification of a portion of the rim (the green flash generally lasts for about
a second) and in particular how the phenomenon is related to the spikes that
seem to extend from the sides of the sun.

2 Multiple Images

Appearances to the contrary, the spikes did not "grow" out of the sides of the
sun, but are in fact multiple images of a small segment of the sun's disk. If you
were to draw a vertical line through the region of these spikes (as in Fig. 4b),
then each time that the line intersects the edge of a spike we are seeing another
image of the same point on the edge of the sun. The spikes do not actually move
up the sun, but rather, the sun as it sets moves down through the region that is
producing the multiple images. The island of light that can be left temporarily
"floating" above the rest of the sun is composed of two images of the upper
rim of the sun. The bottom of the island is the upper rim seen inverted. As the

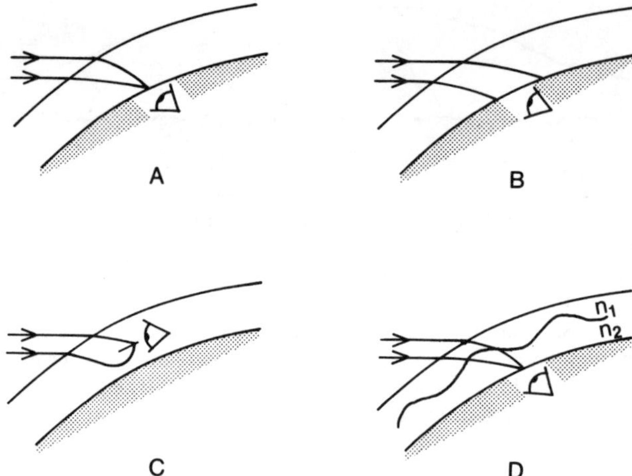

Fig. 3   Light rays that have originated from the same point on the sun pass down through the atmosphere to an observer on the earth. A. In a horizontally homogeneous atmosphere this situation is impossible. Two images of the same point cannot be seen at different elevation angles above the horizon. B. Parallel rays (originating from the same point on the sun) must always diverge as they pass down through a horizontally homogeneous atmosphere. C. Two images are possible in a horizontally homogeneous atmosphere if one image is seen below the astronomical horizon (short straight line). D. Two, or more, images can be seen above the astronomical horizon if we allow horizontal inhomogeneities such as the gravity waves on a sharp inversion surface illustrated here.

sun moves lower in the sky, the island gets smaller and smaller until it is composed only of the two images of the green rim seen back to back. Then this vanishes as the sun gets too low in the sky to produce images in that location.

This mechanism, which produces the type of green flash that is most frequently seen over land, simultaneously doubles the angular size of the green region by combining two images and also displaces the green away from the main portion of the sun so that it is easier to distinguish. The problem of what causes this type of green flash is now reduced to the problem of what atmospheric conditions give rise to multiple images of portions of the sun.

Theorem

In a horizontally (spherically) homogeneous atmosphere it is impossible for more than one image of an extraterrestrial object (sun) to be seen above the astronomical horizon.

The mathematics of the proof of this theorem is given in Appendix I. Physically, however, if two rays of light from the same point on the sun arrive at different locations at the top of the atmosphere, then Snell's Law applied to a spherically homogeneous refracting medium requires that the two rays continue to diverge from each other as long as both rays are still passing downward through the medium (as long as the sun appears above the astronomical horizon). Thus, the two rays will not intersect anywhere to allow an eye placed at this point to see that portion of the sun at two distinct angular elevations (Fig. 3a, b).

There are two ways to get around the restrictions imposed by this theorem. We could have one of the light rays (images) approach our eye from below the astronomical horizon. When this happens, we have an *inferior mirage* (Fig. 3c). The effect is identical to the second inverted image of distant automobiles and of the sky (which looks like water) that one sees when driving along a sunbaked road. When observing the sun, one generally sees the effect over oceans or large lakes when the water is warmer than the overlying air. Green flashes produced by the combination of these two images are often seen from

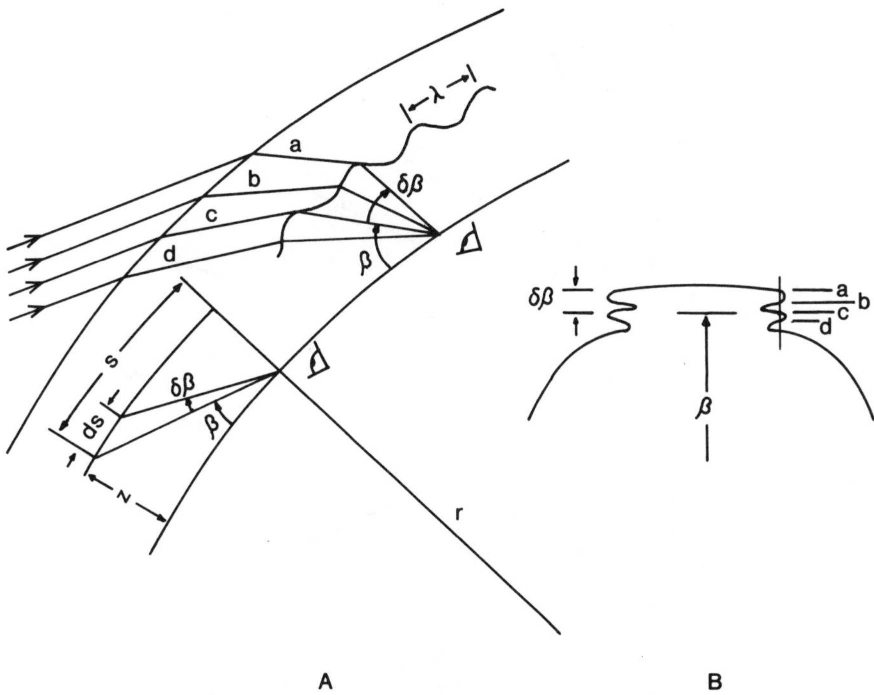

A   B

Fig. 4   A. Assume that an inversion at height z separates two layers of the atmosphere that have different refractive indices. If there is a gravity wave on this inversion surface with wavelength λ, then initially parallel rays of light a, b, c, d (coming from a single point on the sun) can all arrive at the eye. The view as seen by the eye is shown in B. The geometry of the situation is simplified on the left of A for calculative purposes (see Appendix II).

shipboard and probably constitute the most common form of the flash seen over oceans. Under these circumstances the atmosphere can remain horizontally homogeneous.

3 Gravity Waves

The other way around the theorem is to examine an atmosphere with a horizontal inhomogeneity. Consider a stable layer which has gravity waves propagating along it. Light rays from a point on the sun that fall on a portion of the wave that is tipped towards the observer will be bent very strongly while those that fall on the portion of the wave that is tipped away from the observer will be bent weakly. It is possible in this way for both of the light rays to converge on a single location. If your eye is placed there, then two separate images of that point on the sun can be seen, one above the other (Fig. 3d). A more detailed analysis reveals that one of the images is always inverted. The same thing is repeated as we pass the crest of the next wave in the train.

Thus, each wavelength of the gravity waves is able to produce two images, a correct and an inverted one. We can speak of the angular wavelength of the images; the angular distance between two correct, or alternately, two inverted images. One angular wavelength corresponds to (is caused by) one space wavelength of the gravity waves so we now wish to calculate the numerical relation between the two. The question can be phrased: Are the spikes on the sun and the green flash caused by mountain waves with their characteristic scale of from 5 km to 25 km, or are they caused by billows (Kelvin–Helmholtz waves) with their characteristic scale of from about 100 m to a few km?

We will assume that there is a stable layer (inversion) in the atmosphere at height z (Fig. 4). If we are looking at an angular elevation β, then we need only calculate how far our view would shift along the inversion, ds, if we were to shift our view to $\beta + \delta\beta$. We identify $\delta\beta$ with the angular wavelength of the gravity waves. The calculation can only be performed simply if we assume that there

497

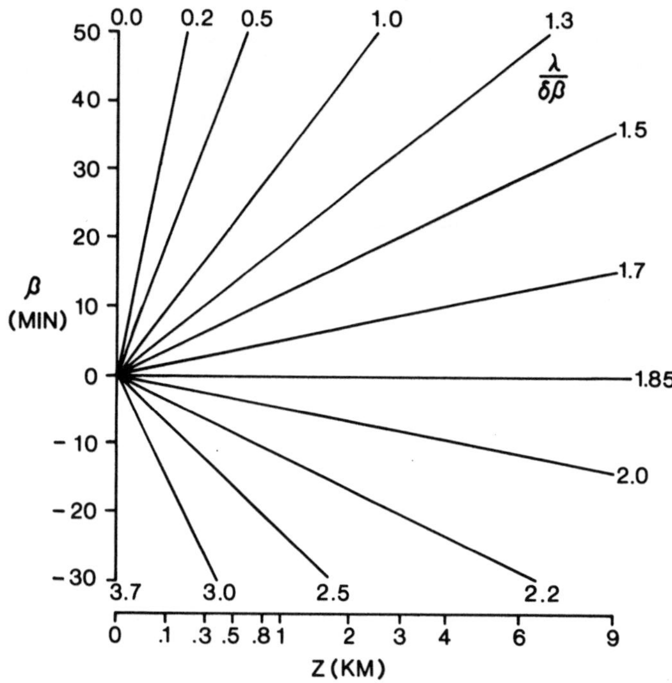

Fig. 5   The function $\lambda/\delta\beta$ is the ratio of the space wavelength λ in kilometers to the angular wavelength between similar images $\delta\beta$ in minutes of arc. The height of the inversion above the eye is z in kilometers and the angular elevation of the images above the astronomical horizon is β in minutes of arc.

is negligible refraction occurring on the ray between our eye and the inversion. As great accuracy is not required on the final answer, this is a perfectly satisfactory assumption. The mathematics of the calculation are confined to Appendix II while the results are presented in Fig. 5.

The function plotted in Fig. 5, $\lambda/\delta\beta$, is the ratio of the space wavelength, λ, in kilometers to the angular wavelength of the images, $\delta\beta$, in minutes of arc. The abscissa shows the height of the inversion in kilometers, while the ordinate shows the angular elevation of the image (above the astronomical horizon) in minutes of arc. If we consider cases where the inversion height is a kilometer or more above our eye, then it can be seen that the ratio $\lambda/\delta\beta$ has values that range between 1 and 2. This is the most significant thing that the graph shows; if the angular spacing between pairs of images is about one minute (the sun's diameter is about 32 minutes), then the multiple images are caused by gravity waves on an inversion with a wavelength of about one to two kilometers.

In both Figs. 2 and 6 the bottom of the sun was nearly touching the horizon and the multiple images were occurring near the top of the sun or at an angular elevation of about 30 minutes. In Fig. 2 the spacing between pairs of images is about a third of a minute, which would give a space wavelength of between 200 and 700 meters, depending on the assumed height of the inversion. In Fig. 6 a similar calculation yields about three times those values. Extensive observing of the setting sun and its distortions indicates that these values are typical.

Because billows generally occur as a train of waves, the solar distortions often take the form of a number of spikes that move up the sides of the setting sun. It is then not unusual to see a sequence of green flashes as successive island images appear above the sun. On another apparently similar day, an observer might not see even one green flash if there were no billows present. The largest number of discrete green flashes that the author has seen during a single sunset is twenty-one and while such displays are infrequent, three or four flashes are common.

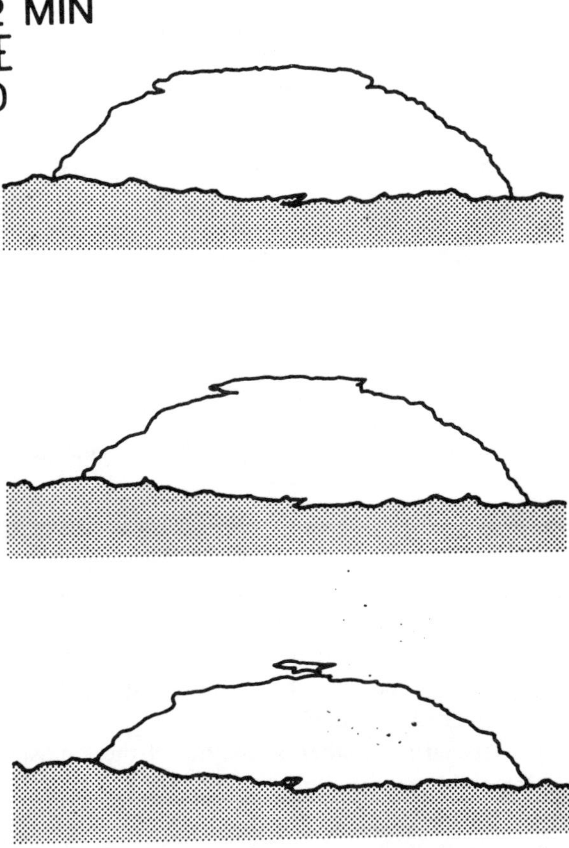

Fig. 6   This sequence (from top to bottom) of views of the setting sun was made at intervals of about three or four seconds. The drawings were made from the projected image of 35 mm slides that were taken by the author at State College, Pennsylvania, in October 1972. The scale shows two arc minutes. The sequence shows the development and appearance of the green flash when it originates with multiple images caused by gravity waves on an inversion surface. The island of light floating above the sun in the bottom picture is composed of a correct and an inverted image of the upper (green) rim of the sun. In the photograph this island is a deep green. The green lasted about a second before the island vanished. The grey area below the sun is a wave cloud formed by distant hills. The photographs discussed in this paper were taken through a catadioptric telescope (Questar).

4 Conclusions

It has been shown that the spikes that sometimes extend from the sides of the sun and the green flash that is often associated with them must be caused by horizontal inhomogeneities in the atmosphere. When these inhomogeneities take the form of gravity waves propagating along an inversion layer, then observations of the angular spacing of the spikes indicate that they are generally caused by the wavelengths of between 200 m and 2 km. This is the range for shear induced gravity waves that are known to give rise to clear air turbulence. Thus we come to the somewhat surprising conclusion that the green flash, when seen above the horizon and associated with spikes on the sides of the sun, is symptomatic of clear air turbulence.

Appendix I

Theorem

In a horizontally (spherically) homogeneous atmosphere it is impossible for more than one image of an extraterrestrial object (sun) to be seen above the astronomical horizon.

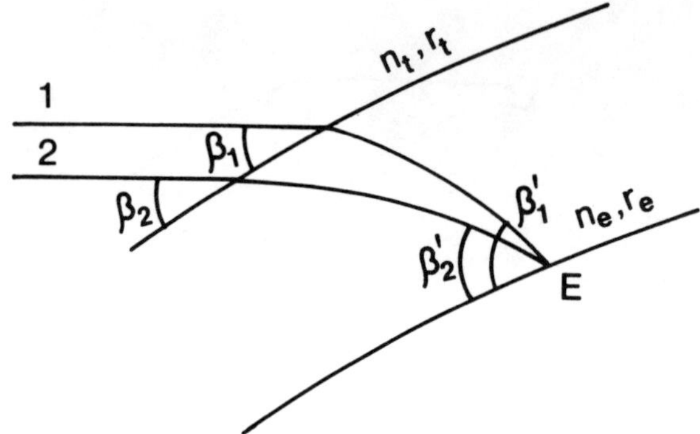

Fig. 7   These are the assumed ray paths through the atmosphere if the theorem is false. The index of refraction n, and the distances to the center of symmetry (the center of the earth) r are subscripted t (for the *top* of the atmosphere) and e (for the *e*arth or *e*ye).

Proof

Let us assume that the theorem is false and that it is possible for two parallel rays of light (coming from the same point at infinity) to impinge on the top of the atmosphere and to be refracted such that they both meet at a point (E in Figure 7). An eye at this location could thus see images at two different angles β_2' and β_1' which are both positive.

If the lower ray is labeled ray 2, then at the top of the atmosphere $\beta_2 > \beta_1$ and as we have horizontal homogeneity

$$n_t\, r_t \cos \beta_2 < n_t\, r_t \cos \beta_1$$

In a horizontally homogeneous atmosphere Snell's Law is written $nr \cos \beta =$ constant along a ray, so that it follows that

$$n_e\, r_e \cos \beta_2' < n_e\, r_e \cos \beta_1'$$

or

$$\cos \beta_2' < \cos \beta_1'$$

Now as $\beta_1' > 0$ and $\beta_2' > 0$ this requires that $\beta_1' < \beta_2'$. But as ray 2 is the lower ray, then clearly at point E, $\beta_2' < \beta_1'$ which contradicts the assumption that the theorem is false. (Strictly, we have considered only the first intersection of the rays and while on a subsequent intersection $\beta_1' < \beta_2'$ in apparent agreement with the assumption, it could not occur as the first intersection is shown to be impossible.)

Note: Although the theorem is true in general, the proof just given breaks down when an object is located at the zenith. In this case the rays can converge on the eye but this does not give two images but rather a single very slightly magnified image. This detail is of little interest in the present context, however.

Corollary

In a horizontally homogeneous atmosphere, two images of an extra-terrestrial object are not prohibited if one image is seen below the astronomical horizon.

Proof

We have shown that $\cos \beta_2' < \cos \beta_1'$. Now if $\beta_2 < 0$ and $\beta_1 > 0$ we have no contradiction so that the rays can meet at a point.

Appendix II

Let the length of the ray traveling from the eye to the inversion be d. (See Figure 4A.) Then consider the triangle formed by the eye, the center of the earth, and the intersection of the ray and the inversion. Then,

$$(r + z)^2 = r^2 + d^2 + 2rd \cos (\beta + \pi/2) \tag{1}$$

Now β is small so $\sin \beta \approx \beta$ and $d \approx s$ and $z \ll r$ so

$$(r + z)^2 \approx r^2 + 2rz. \text{ We get}$$

$$s^2 + 2rs\beta - 2rz = 0 \tag{2}$$

Differentiation gives

$$\frac{ds}{d\beta} = \frac{-rs}{s + r\beta} \tag{3}$$

where s is eliminated to give

$$\lambda/\delta\beta = r[1 - N (1 + 2z/r\beta^2)^{-1/2}] \tag{4}$$

by using the expression

$$s = -r\beta + Nr\beta (1 + 2z/r\beta^2)^{1/2}$$

where $N = +1 \qquad$ if $\beta > 0$
$\qquad\quad N = -1 \qquad$ if $\beta < 0$

and $\lambda \equiv ds$, $\delta\beta \equiv d\beta$.

Equation (4) is plotted in Fig. 5.

References

O'CONNELL, D.J.K., 1958: *The Green Flash.* North Holland Publishing Co., Amsterdam.

SHAW, GLEN E., 1973: Observations and Theoretical Reconstruction of the Green Flash. *Pure and Applied Geophysics,* **102**, 223–235.

Section Nine
Extinction

Reprinted from *Geografiska Annaler*, Vol. 11, 156–166 (1929).

ON THE ATMOSPHERIC TRANSMISSION OF SUN RADIATION AND ON DUST IN THE AIR.

By *ANDERS ÅNGSTRÖM.*

In recent years the elaborate investigations of Abbot and his collaborators have in a high degree attracted the interest of meteorologists to the determinations of the solar constant and its variations. This interest is well deserved and will probably in the future result in considerable progress especially as regards long range forecasts of temperature.

At the present time, however, attention must be drawn to the fact that the variations in the radiation reaching the surface of the earth are to a much larger extent due to fluctuations in the transmission power of the atmosphere than to the variations of the solar constant itself. In Fig. 1 we have only, in order to refer to an instant, given a graphical representation both of the variations of the solar constant according to Abbot and also of the variations of the total incoming radiation at Stockholm during the years 1923—1928. While the changes of the atmosphere — clouds, water vapour, scattering particles etc — cause variations in the incoming radiation amounting to 25 per cent from minimum to maximum, the corresponding variation of the solar constant is only slightly more than a fiftieth of that, viz. about 0,5 per cent. The prime importance of the conditions of the atmosphere for the transport of energy to the surface of the earth would appear to be clear from this brief consideration, and numerous other instances might be given.

From a meteorological point of view, therefore, it is of great interest to study more in detail the ways in which the atmosphere acts on the radiations penetrating it. In fact, the investigations of Abbot have also in this respect furnished a very rich material of observations, a fact which perhaps has been overlooked to some degree as a consequence of the overwhelming interest of the investigator himself in the variations of the sun. Through the spectrobolometric registrations of Langley, Abbot and Fowle and those of K. Ångström and Lindholm at Upsala, together with the spectrophotometric observations of Müller at Potsdam, the foundations have been laid for a study of atmospheric diffusion and absorption.

As is well known, we may divide the influences which cause a diminution in the incident energy passing the atmosphere into two principal groups, viz.

I. The selective absorption by the gases of the atmosphere. This absorption gives rise to the terrestrial lines or bands and is generally localized to special limited regions in the solar spectrum. It is practically negligible at intermediate intervals.

The largest absorption occurs in the infra-red: it is due especially to water vapour and carbon dioxide but also in an appreciable degree to oxygen and ozone. Abbot and Fowle express the total percentual absorption exerted by the water vapour through the equation:

$$(1) \ldots \ldots \ldots \ldots \ldots \ldots \quad F = 5{,}1 + 2{,}8 \ e_0 \cdot m$$

where e_0 is the pressure of the water vapour in mm at the place of observation and m is the air mass with the vertical air mass as unit. Westman expresses the water vapour absorption in gram-calories per minute in a column of one square centimeter cross-section through the equation:

$$(2) \ldots \frac{dQ}{de} = a + b \cdot m$$

The values of these constants are, according to

Westman:
$$a = 0{,}0114 \quad b = 0{,}0020$$

Lunelund:
$$a = 0{,}0123 \quad b = 0{,}0019$$

Consequently an increase of the water vapour pressure by 1 mm causes in general a diminution of the direct sun radiation of about 0,03 gram-calories per square centimeter and minute at vertical incidence.

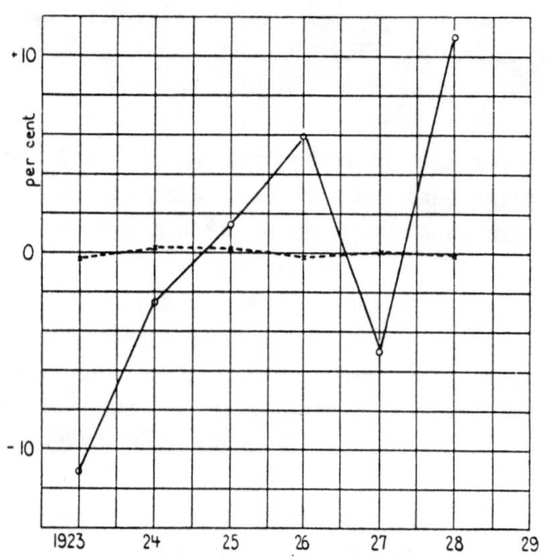

Fig. 1. Circles give deviation of total annual radiation at Stockholm from mean value; crosses give the same variations of solar constant according to Abbot.

In this connection it is of interest to refer to an expression deduced by Knut Ångström from his spectrobolometric researches into the solar spectrum in order to include the *total absorption* by the gases of the atmosphere, namely:

$$(3) \ldots \ldots \ldots \ldots \ldots \ldots \quad A = Q_0 \cdot 0{,}1 \ (e_0 \cdot m)^{0{,}275} \ (5 < e_0 \cdot m < 30)$$

where Q_0 is the solar constant.

II. The scattering or diffusing effect of the atmosphere causes also a diminution of the incident energy. This scattering effect is due especially to two causes. One is the scattering by the molecules, which has been theoretically treated by Lord Rayleigh, Schuster, Einstein and Keesom and recently by Oseen and Lundblad.

According to the current theory, the molecular scattering of radiations passing through a gas is a result of two combined phenomena, namely a »proper» scattering by the molecules themselves and a scattering due to the variations of density as a result of the variable speed of the molecules, according to the theory of Maxwell.

The combined effect of the molecular scatterings in a homogeneous gas may be expressed mathematically through the well-known law of Rayleigh, viz:

$$(4) \dots \dots \dots \dots \dots \dots Q = Q_{o\lambda}\, e^{-k \cdot a}$$

where

$$k = \frac{32\,\pi^3\,(\mu - 1)^2}{3\,N\,\lambda^4}$$

(μ = index of refraction, λ = wave length of incident radiation, N = number of molecules per cubic centimeter). In the case of the sun radiation passing through the atmosphere in a vertical direction equation (4) takes the form:

$$(5) \dots \dots \dots \dots \dots \dots Q_\lambda = Q_{o\lambda}\, e^{-\frac{a}{\lambda^4}}$$

where

$$a = \frac{32\,\pi^3\,(\mu - 1)^2}{3\,N_o} \cdot \frac{H \cdot p}{760}$$

Here H is the height of the homogeneous atmosphere, at 760 mm, $0°$ C, N_o the number of molecules in a cubic centimeter at the same standard pressure and temperature, and p is the pressure at the place of observation. From (5) we may compute for every special wave-length the fraction of the incident light which is lost through molecular scattering. It has been shown by Fowle that for high altitude stations, where almost no dust is present, the radiation computed according to formula (5) is in very close agreement with the values actually measured. As an instance we may refer to the transmission values for Washington (at 35 m height) and for Mount Wilson (at 1 730 m) computed from the molecular scatter-

Table I. **Molecular transmission.**

Wave length	0,37	0,40	0,43	0,46	0,50	0,60	0,75	1,00	1,50
Washington	0,648	0,726	0,785	0,830	0,875	0,936	0,973	0,982	0,997
Mount Wilson	0,680	0,755	0,808	0,850	0,890	0,946	0,977	0,987	0,998

ing alone (Table I). In this connection, however, it must be emphasized, that the diminution in the direct sun radiation caused by the scattering, does not imply that all this energy is lost from the point of view of the heat economy of the earth; a considerable part of it is namely radiated against the surface of the earth in the form of diffused sky radiation.

Another part of the incident energy is scattered by the dust in the atmosphere. Several writers have assumed this scattering to be constant for various wave-lengths.

In that case we should be able to express the combined effect of the scattering by molecules and dust through the equation:

$$(6) \ldots \ldots , \ldots \ldots \quad I = I_o \, e^{-\frac{a}{\lambda^4}} \cdot e^{-\gamma} = I_o \, e^{-\frac{a}{\lambda^4} - \gamma}$$

where γ would be independent of the wave-length. In fact the total atmospheric scattering has been expressed by Fowle through this equation.

A closer examination of the great material of transmission coefficients given by Abbot and his collaborators, however, seems to indicate with a great amount of certainty that γ also, like the coefficient of molecular scattering, is a function of the wave-length and that γ, in fact, is larger for short waves than for long ones.

This is to be expected also from other experiments on the absorption of light of powders and colloidal substances. Lundholm[1] studied the absorption of thin layers of soot, MgO, SbO and ZnO. His investigation is of special interest, because, as we shall see later, the size of the particles constituting his absorbing substances was in general of about the same magnitude which is found for the particles of the dust in the atmosphere. He found that the absorption coefficient γ could in general be expressed in the form:

$$(7) \ldots \ldots \ldots \ldots \ldots \ldots \ldots \quad \gamma = \frac{\beta}{\lambda^a}$$

where a assumes values between 0 and 4. The larger the particles, the smaller was the value found for a. For very small particles a must approach the value demanded by the theory of Rayleigh, viz. 4. In the case of Lundholm's experiments the particles had a size varying from $0{,}25 \, \mu$ diameter in the case of MgO to about $2 \, \mu$ for ZnO. In the former case a had a value of $3{,}6$, against a value of $0{,}7$ in the latter case.[2]

The dust absorption in the atmosphere is in fact of a similar character.

Let us pick out two wave-lengths λ_1, and λ_2 at regions of the spectrum where the selective absorbtion is nil or negligible. Let the vertical transmission computed from the molecular scattering alone be T_1 and T_2 respectively. Then the actually observed transmissions U_1 and U_2 are given by:

$$(8) \ldots \ldots \ldots \ldots \ldots \ldots \quad \begin{cases} U_1 = T_2 \, e^{-\frac{\beta}{\lambda_1{}^a}} \\[2mm] U_2 = T_2 \, e^{-\frac{\beta}{\lambda_2{}^a}} \end{cases}$$

---

[1] Spektrofotometriska undersökningar över diffusa ämnens absorption. Uppsala 1912.

[2] The theoretical and experimental investigations of Mie show (Annalen d. Physik (4) 25: 377 (1908)) that the formula (7) must be regarded as a special case of a more complicated law. In fact the formula (7) is only valid for a limited range of values on the diameter of the particles and for a limited intervall of wave lengths.

From which we get:

$$\frac{\beta}{\lambda_1{}^a}\,\log c = \log \frac{T_1}{U_1}$$

$$\frac{\beta}{\lambda_2{}^a}\,\log e = \log \frac{T_2}{U_2}$$

We obtain:

$$(9)\ \dots\dots\dots\dots\quad a = -\frac{\log \dfrac{\log T_1 - \log U_1}{\log T_2 - \log U_2}}{\log \lambda_2 - \log \lambda_1}$$

Inserting this value of a in one of the equations given above, we may compute β also. Evidently β is the absorption coefficient corresponding to the wave length $\lambda = 1$, independently of of the value of a.

From the transmission values given in the Annals of the Astrophysical Observatory II—IV have now computed a for a number of cases, according to the procedure given above, and in general starting from the transmission values corresponding to the wave-lengths $\eta = 0{,}45\ \lambda$ and $\lambda = 0{,}90\ \mu$.

It is evident that, if the dust absorption were independent of the wave-length, we should get a value for a which would be equal to, or differ only slightly from, zero. This is found not to be the case. On the contrary a has in general a value between $1{,}0$ and $1{,}5$. Only in extreme cases does it go below the former or above the latter value.

How well the transmission values in general are satisfied by an equation of the form adopted may be seen from Fig. 2, where the observed

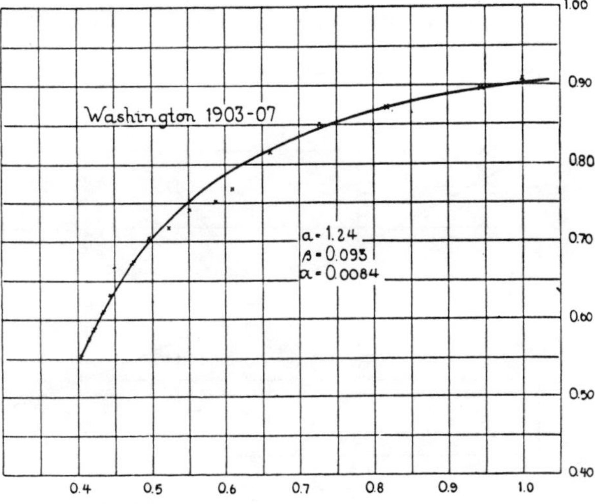

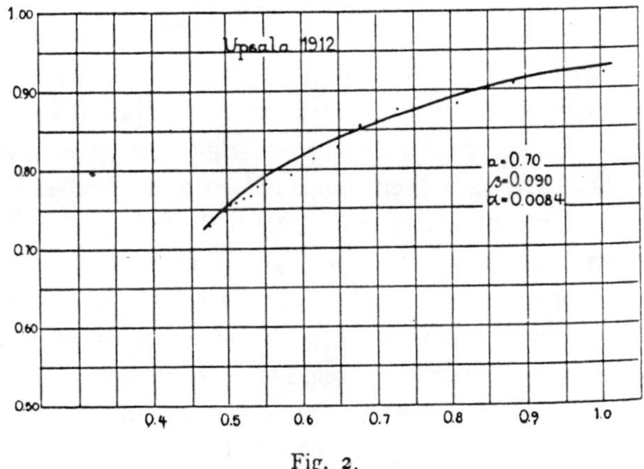

Fig. 2.

Table II.

	H	Average conditions		Haziest days Aug. 1912		Observer
		β	a	β	a	
Washington	35	0,098	1,24	—	—	Smithsonian Inst.
Upsala	35	(0,090)	(0,70)	0,362	0,515	Lindholm 1912.
Bassour	1 160	0,031	1,22	0,255	0,53	Smithsonian Inst.
Hump Mountain	1 500	0,031	1,33	—	—	»
Mount Wilson	1 780	0,018	1,26	0,205	0,70	»
Calama	2 250	0,023	1,33	—	—	»

transmission for Washington and Upsala is indicated by points, the computed transmission being given through the curve. A small but noticeable deviation occurs within the region of the so called *Chappuis* band at 0,5 to 0,65 μ.

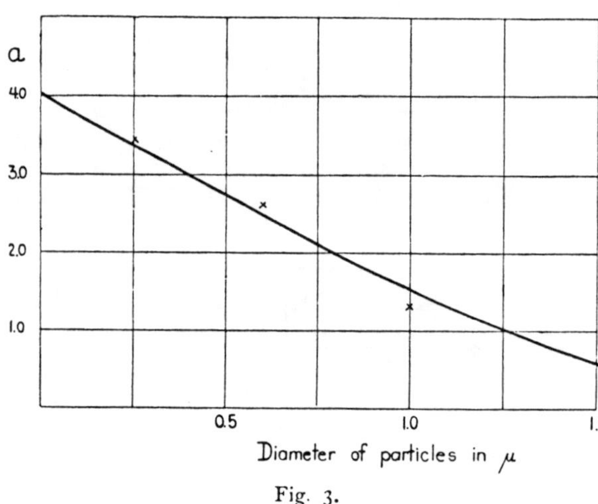

Fig. 3.

In Table II we have given the mean values for a and β computed for various stations. No definite variation of a with the time of the year is found to exist. Further a shows, as is seen 'from the table, a remarkable constancy from place to place. This is so much more worthy of attention as the places are situated at very different heights above sea-level, the altitude varying from 35 m for Washington up to 2 250 for Calama.

An exception from this rule of the constancy of a is apparent the hazy days of the summer of 1912, when the eruption of Mount Katmai caused an optical disturbance of a special nature. The values of a corresponding to disturbances of volcanic origin are given in the last columns of Table II. We find that a is then subjected to a marked decrease down to 0,50—0,70.

Confronting these results with the laboratory measurements referred to above, we may draw the conclusion that the size of the dust particles is comparatively independent of the height. At first sight this result may perhaps seem astonishing. we should perhaps expect the effect of gravity to cause a more rapid falling of the large particles relatively to the small ones, thus producing a concentration of large particles in the lowest layers of the atmosphere. If, however, we consider the extremely small falling velocity of these minute particles — for a par-

ticle with $1\,\mu$ diameter it would take about one month to fall a distance of
1 000 meters — we must conclude that the process which carries back the dust
to the surface of the earth is essentially different from gravity. In fact the condensation and precipitation must below 2 000 m altitude be the chief regulator of
the cleaning of the atmosphere from dust, which is also evident from considerations of the equilibrium of sedimentation of dust with due regard to the eddy
convectivity in the atmosphere.

As has been already mentioned, Lundholm found a very marked dependence
of a on the size of the scattering particles, but not, in the case of uncoloured
substances like MgO, ZnO, SbO, on the substance itself. Averaging his values,
we obtain a relation between a and diameter, which is represented by the curve
in Fig. 3. From the value for a obtained from the spectrobolometric registrations
we should thus expect an average magnitude of $1\,\mu$ diameter for the dust particles in the atmosphere.

It is interesting to compare this result with some conclusions reached by J. S. Owen through his elaborate studies by means of the dust counter invented by himself. Dr Owen states:

»During the fog of 5th November 1920 I obtained a sample of the solid particles by causing a small jet of air to impinge upon a glass mi-croscope slide; this resulted in a small dark patch of fog, particles being deposited
upon the slide. On examining these under the microscope, magnification about
1 000 diameter, I found a large mass of black particles scattered over the slide.
There was an indication that the larger of these were made up of smaller particles, which had stuck together. — — — — — — — — — — — — — — —
There where plenty of small individual particles visible and I measured the diameter of these and found that they varied between about 10^{-5} to $0,5\ 10^{-4}$ of
an inch».

Reducing from inches to microns, we find that the small particles measured by
Owen had a diameter of between 1,3 to 0,26 microns.

The same result is indicated by the elaborate photographs published by H. H.
Kimball (Monthly Weather Review, March 1924) and also by the measurements

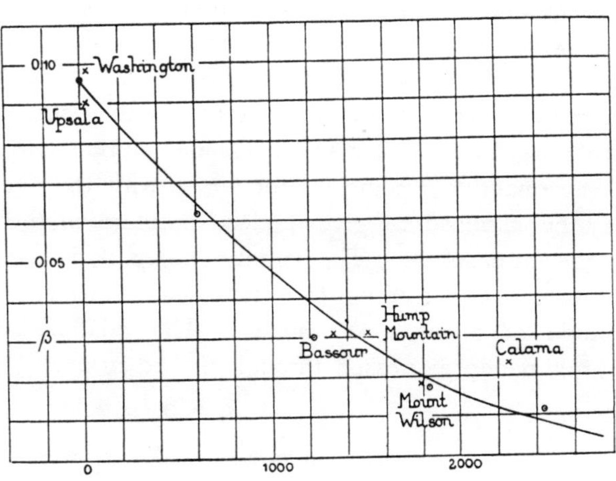

Fig. 4.

at Stockholm discussed by myself.[1]   Here also the average size of the particles
seems to vary between about 0,2 to 1,5 microns.

Kimball states *(loc. cit.)* that the size of the particles seems to decrease with
the altitude.   »It is roughly estimated that the average diameter of the particles
collected at the surface is about four times that of those collected at 3 000 meters.»
Such a variation is not, as we have stated above, indicated by the actinometric
measurements.   The reason for this may perhaps be that the size of the various
particles comes into play in the actinometric results in proportion to the optical
efficiency of the particles and not in proportion to their number.   The optically
measured diameter can therefore not be expected to show more than an approximate coincidence with the average diameter directly measured.

Another possible explanation of this discrepancy may perhaps be included in
the fact that the actinometric stations are all situated at the surface of the earth,
from which the dust is transported to the atmosphere through convection through
the air.   The properties of the dust in the air above such stations may be considerably different from the dust collected in the free atmosphere, as in the case
with the samples collected by Kimball during air-plane flights.

The quantity β coincides for $\lambda = 1$ with the coefficient of dust absorption.   It
may be assumed to give a close measure of the total quantity of dust present
in the atmosphere.   The variation of β with altitude is demonstrated in fig. 4.
It decreases, according to these measurements, to half its value with a rise in
altitude of about 1 000 meters.   The relation between β and altitude may be
approximately expressed through the equation:

$$(10) \quad \ldots \ldots \ldots \ldots \ldots \ldots \quad \beta_h = 0{,}094 \; e^{-\partial \cdot h}$$

$$\text{where } \partial = 0{,}69 \cdot 10^{-5}, \text{ if } h \text{ is given in cm.}$$

A similar expression may be obtained from Kimball's observations during airplane flights as to the number of dust particles at different heights.   After some
approximations and smoothing out occasional variations we may put:

$$(11) \quad \ldots \ldots \ldots \ldots \ldots \ldots \quad N_h = N_0 \; e^{-\varphi \cdot h}$$

$$\text{where } \varphi = 0{,}7 \cdot 10^{-5}$$

For the number of particles in the total vertical column of 1 square centimeter
cross-section we obtain:

$$\sum_h^\infty N = \bar{N} = N_0 \int_h^\infty e^{-\varphi \cdot h} \, dh = N_0 \; 1{,}4 \cdot 10^5 \cdot e^{-\varphi \cdot h}$$

At Washington we have as a general mean:
$N_0 =$ about 500, at Stockholm $N_0 = 340$.   If we assume $N_0 = 400$ corresponding

---

[1] Investigations of the dust content of the atmosphere in Sweden.   Procès-Verbaux des séances de
la section de Météorologie de l'Union Géodésique et Géophysique Internationale.   Madrid, 1924.

to the average conditions under which the spectrobolometric measurements are made, we find that a value of β equal to 0,094 corresponds to a total number of dust particles equal to $5,6 \cdot 10^7$, and that in general:

$$\beta = 1,7 \cdot 10^{-9} \, \Sigma N$$

If, according to the measurements referred to above, we assume the diameter of the particles equal to 1 micron (0,0001 cm), we may also express this result by the statement that every dust particle causes a decrease of about 20 per cent in the energy of wave-length $1 \, \mu$ passing through its own cross section, the particles assumed to be spherical. If compared with the results of measurements of the absorption of soot and powders, this result seems a probable one.

Table III.

	Washington	
	$- \log T\beta$	β
J	0,0727	0,062
F	0,0888	0,075
M	0,1296	0,110
A	0,1662	0,141
M	0,1864	0,157
J	0,1818	0,154
J	0,1519	0,129
A	0,1215	0,103
S	0,0915	0,077
O	0,0706	0,060
N	0,0605	0,051
D	0,0605	0,051

The yearly variation of β in Washington is apparent from Table III. There is a pronounced maximum at the end of May and a minimum in November or December. This variation does not agree with the variation of the number of dust particles per unit volume measured directly at the earth's surface at Washington. It is therefore evident that the value of the quantity φ in equation (11) must take on highly different values on different occasions, and that the constants introduced above into equations (10) and (11) are only valid for mean values during considerable intervals of time.

The total number of dust particles in the atmosphere above a certain place is probably to a large extent dependent upon the vertical convection currents or the »Austausch». As now convection probably has a maximum in the spring or early summer, when the vertical temperature gradient reaches its maximum on account of the rapid heating of the earth's surface, the variation of β seems already from this reason a natural one. The distribution of dust in the atmosphere, however, is also dependent upon the way in which the atmosphere is cleaned through condensation and precipitation. But a general treatment of this problem falls outside the limits of the present paper.

From the low value of a corresponding to the optical disturbance caused by the eruption of Mount Katmai in June 1912, we may conclude that the size of the volcanic particles must have been considerable larger than the ordinary size of dust in the atmosphere.

A comparison with Lundholm's values makes it probable that the average size of the former particles ought to have been about $1,5—2,0$ μ in diameter.

The falling velocity of such particles of spherical size may be computed according to the modified form of Stokes formula given by Schmidt.[1] It comes out to be about $0,035$ cm per second for particles 2 μ in diameter, if we assume the specific gravity equal to $2,6$, an assumption which is founded upon an analysis of the dust referred to by Abbot and Fowle (2). This velocity corresponds to a fall of about 1 000 meters in the month.

No data as regards the height of the volcanic dust layer are, so far as I have found, available from the Mount Katmai eruption, but Abbot and Fowle refer to a statement indicating a decrease of 16 km in 12 months of the altitude of the dust layer from the Krakatoa eruption (From 32 to 16 km height). This gives a falling velocity of 1 300 meters in the month or $0,045$ cm in the second, which values are in astonishingly good agreement with the values computed above.

Summarizing some of the results of the considerations brought forward in the present paper, we may finally say that it seems probable that by means of energy measurements at two different parts of the solar spectrum, where no selective absorption occurs, we may obtain (1) a value for the average size of the dust particles present in the atmosphere and (2) an idea of the total amount of scattering particles in the air.

On the other hand, information as regards the values of a and β in our formula (7) is sufficient to give us means to compute from the known distribution of energy in the solar spectrum outside the atmosphere the actual distribution at the bottom of the atmosphere, as far as non-selective regions of the spectrum are concerned. As the selective absorption, on the other hand, in general may be approximately computed from the water vapour pressure at the place of observation, we see that a very limited number of observations is commonly enough to give a general idea of the total energy distribution within the spectrum.

These considerations, however, do not apply to the region of the extreme ultra-violet, where the ozone absorption enters as a fundamental factor. This region constitutes only a small part of the total *energy*, but on account of its important

---

(1) Sitzungsber. der Akad. d. Wissenschaften in Wien. Bd. 118 II a und auch Meteor. Zeitschrift 1909, S. 183.

2 Volcanoes and climate. Smithsonian Misc. Collections. Vol. 60. N. 29.

12. *Geografiska Annaler 1929.*

effects of a biological and therapeutic character, it can not be neglected in actinometric studies of a climatological nature which have completeness in view. In
order to obtain an idea of the intensity between $0,29$ to $0,35$ μ direct measurements are necessary and an extrapolation of the formula (6) to this region is
not allowed.

The previous considerations may perhaps be sufficient to indicate, on the basis
of discussions of the spectrobolometric material, the chief outlines of a practical
program for actinometric investigations of climatological character.

The most important application of the foregoing discussion seems, however, to
be included in the fact, that the determination of the quantity a gives us an idea
of the size of the diffusing particles in the atmosphere and that we are probably
in many cases able to conclude from this size to the origin of the optical disturbance itself and to the origin of the air masses. The volcanic eruptions seem
to give larger particles than the ordinary size of dust brought by convection from
the earth to the atmosphere. As regards the disturbances of cosmic origin, which
according to several authors seem to be frequent at the times of sun spot maxima,
they are probably characterized by smaller particles than the ordinary size. Further
investigations along the lines here indicated are now in progress.

Reprinted from *Geografiska Annaler*, Vol. 12, 130–159 (1930).

ON THE ATMOSPHERIC TRANSMISSION OF SUN RADIATION. II.

By *ANDERS ÅNGSTRÖM*.

The spectrobolometric studies of sun radiation have given a very detailed view of the different ways in which radiation is influenced by the atmosphere. Radiation is scattered and absorbed by molecules and minute particles of solid or fluid impurities in the atmosphere and by the atmospheric gases, of which especially the water vapour and the carbon dioxide play an important part. From two spectrobolograms taken at different heights of the sun, it is possible to determine the way in which the intensity of every separate wavelength is diminished in passing through the atmosphere provided that no changes occur in the atmospheric conditions.

Already the earliest examinations of such bolograms showed that there are two principal ways in which the incident energy is diminished. One affects all wave-lengths, and in fact the short waves more than the long ones: this effect is generally ascribed to the *scattering and general absorbtion* in the atmosphere. The other effect is practically limited to special regions of the spectrum: it is the common way in which the band — or line — absorption by gases affects the radiation passing through the gas.

Theoretical considerations, as well as experimental work in the laboratory, show that the scattering and general absorbtion must be regarded as due to two separate causes, namely the scattering by the molecules on the one hand and the scattering and absorbtion by minute particles in the air on the other. We thus have three different causes of the diminution of light waves passing through the atmosphere, and in all researches which aim not only at a climatological survey but also at a knowledge of the physical state of the atmosphere, it is important that the effects of these separate causes should be kept apart.

Apart from the spectrobolometric researches, which alone are able to give a detailed idea of the scattering and absorption at different wave-lengths, some attempts have already been made to carry out, from actinometric measurements of a more or less simple and limited character, an analysis of the physical conditions of the atmosphere in respect of scattering power and selective absorption. Thus K. Ångström in 1906 in his paper »Méthode Nouvelle pour l'étude de la radiation solaire» (1) showed how to arrive from measurements of the total solar radiation at different air masses not only at a value of the solar constant but also at a value of a certain quantity ∂, which latter is a measure on the scattering

power of the atmosphere. The absorption by the water vapour is computed
from the vapour pressure at the place of observation. In this way Ångström
obtained a separation of the effect of turbidity from the effect of selective absorption — 98 per cent of the selective absorption being due to the water vapour
— and could on the basis of actinometric measurements indicate a value of the
degree of turbidity or »trübung» of the atmosphere. In this turbidity coefficient
however, were included as well the scattering by the molecules as also the scattering
and absorption by impurities suspended in the atmosphere (1).

Fowle computes from the spectrobolometric measurements of the Astrophysical
Observatory at Washington the effects of the two kinds of extinction[1] separately
(2) and expresses the extinction through the formula:

$$(1) \quad \ldots \ldots \ldots \ldots \ldots \quad I_\lambda = I_{o\lambda}\, e^{-(a_1 + a_2)\cdot m}$$

where I_λ is the energy of the wave-length λ at the surface of the earth, $I_{o\lambda}$ is
the energy of the same wave-length outside the atmosphere, a_1 is the coefficient
of molecular scattering; a_2 is the coefficient of scattering and general absorption
due to other causes; and m is the air mass. In the case of the upper comparatively dry atmosphere, where a_2 becomes zero or almost negligible, Fowle has
shown that a_1 converges towards a value which closely coincides with the coefficient of molecular scattering computed according the theory of Rayleigh. Hence
it seems allowable to compute *in general* a_1 from this theory and regard a_2 as
an index of the extinction due to other causes. If we write: $e^{-a_1} = q_1$, we may
compute a_2 from the equation:

$$(2) \quad \ldots \ldots \ldots \ldots \ldots \quad a_2 = -\frac{1}{m}\cdot {}^e\!\log \frac{I_\lambda}{I_{o\lambda}\cdot q_1{}^m}$$

As the radiation $I_{o\lambda}$ outside the atmosphere seem to be comparatively constant
and, through Abbot's measurements, relatively well known, and as q_1 may be
computed from the theory of Rayleigh, as we have just mentioned, we may
consequently compute a_2 from a single measurement of I_λ, the air mass m being
known from observation on the sun's altitude or from a computation from the
time equation.

Fowle has shown that the coefficient a_2 thus defined is correlated to the water
vapour pressure in such a way that much water vapour in general corresponds
to a higher value of a_2 and vice-versa. Fowle himself seems to infer from this
that a certain part of the general extinction *is due to* the water vapour and
several authors have since adopted that point of view. They have even gone so
far as to try to compute from the correlation with the water vapour pressure the

---

[1] In the following we use the word *extinction* to indicate the *general absorption and scattering* by
solid or fluid particles in the atmosphere and by the molecules. The selective absorbtion is not included in the extinction.

part of the extinction due to water and the part due to dust. To the mind of the present author the legitimacy of these attempts is not satisfactory, and they seem even liable to lead to a wrong interpretation of the physical facts. The correlation existing between general extinction and water vapour pressure does not, in fact, include the necessity of a physical connection between the two phenomena. It is scarcely probable that such a connection, if it exists, is as pronounced as the correlation has been considered to indicate. The high extinction and the relatively high water vapour pressure corresponding to it are probably to a large extent due to a common cause, namely the atmospheric circulation, which at times brings in warm and humid air from the south, where a vivid dust production often occurs at the surface of the earth. At other times the cold and dry air, purified through the contact with the snow-surfaces, is introduced from the north. These processes have led to the wrong idea that the extinction to a considerable extent *is due* to the water. It is a very instructive instance of the necessity of being careful with conclusions from correlation studies. I will later return to these questions.

Proceeding further in this short historical survey, we may finally mention that Linke has recently derived another mathematical formula for a separation of the extinction by dust from the scattering by the molecules. He writes the equation (1) of Fowle in the following way:

$$(3) \cdot \cdot \cdot \cdot \cdot \cdot \cdot \cdot \cdot \cdot \cdot \cdot \cdot \cdot \cdot \quad I_\lambda = I_{o\lambda}\, e^{-a_1 \cdot T \cdot m}$$

where a_1 is the coefficient of molecular scattering and T is a factor which can be regarded as a measure on the turbidity or »trübung» of the atmosphere. Apart from the fact that the original »trübungsfaktor» of Linke included not only the extinction by dust but also, which does not seem appropriate, the selective absorption by the gases, I cannot find Linke's mode of expressing the influence of dust to be a happy one. Linke's coefficient T expresses the total extinction with the molecular scattering as unit. His T becomes unity for an ideally pure atmosphere. A value of T of, for instance, 2,o tells us that the extinction is twice that of a perfectly clear atmosphere with no selective absorption. But the reason for attaching so much weight to the molecular scattering, which is far from constant from place to place, is not at all clear. In order to express the scattering of dust in terms which gives us a possibility of comparing the conditions at various places, it seems far more appropriate to go back to the original expression of Fowle and let the »trübung» be expressed by the coefficient a_2, which gives us the extinction due to the dust and *which is independent of the value of* a_1. The coefficient a_2 probably gives us, as I will discuss later, a very good measure of the total amount of dust present in the atmosphere above the place of observation, while the »Trübungsfaktor» can only through troublesome reductions and corrections be brought to give a measure of the amount of dust at a place with a certain air-pressure relative to a place with another pressure.

We now recur to the original expression of Fowle for the non-selective extinction. In my first paper on the atmospheric transmission (4), I have shown that a_2 also is a function of the wave-length, and that, in virtue also of some results of laboratory experiments on the absorption by dust and powders, it can be expressed by the equation:

$$(4) \quad \ldots \ldots \ldots \ldots \ldots \ldots \quad a_2 = \frac{\beta}{\lambda^a}$$

Here a is an exponent which, for very small particles like those demanded by the theory of Rayleigh, may be as large as 4,0. For large particles it may go down to a value between 0 and 1,0. As regards the atmospheric impurities a has in general a value between 1,0 and 1,5. Under abnormal conditions, as for instance after violent volcanic outbreaks, it decreases to 0,5 or even less. An interesting result is derived from Abbot's spectrobolometric measurements, namely that a on the average keeps an almost constant value for stations at different heights above sea-level, which seems to indicate that the average size of the scattering particles in the atmosphere is almost constant for different layers of the atmosphere. An approximate computation of the size of the scattering particles is attempted by means of a comparison with laboratory measurements on particles of various sizes. A value of about 1,0 μ for the diameter seems probable.

It must here be emphasized, however, that the size of the scattering particles is quite certainly *not constant within the same layer*. Even a superficial investigation under the microscope of the particles collected by, for instance, the OWEN dust-counter show that such a constancy does not exist. The frequencies of the different sizes do not, in general, continuously decrease on both sides of a single maximum frequency, but are often concentrated on two or perhaps several preferred sizes. As, however, the extinction effect referred to a given number of particles, decreases rapidly as the size of the particles diminishes, it seems as if the effect of the 1,0 μ particles is *so predominant at all levels* as to overshadow the effect of the much smaller but certainly very numerous particles which also occur.

In other words, if we regare a_2 as being the extinction coefficient made up of the effects of particles of a great number of different sizes, it would strictly be expressed by

$$(5) \quad \ldots \ldots \ldots \ldots \ldots \quad a_2 = \frac{\beta_1}{\lambda^{a_1}} + \frac{\beta_x}{\lambda^{a_x}} + \frac{\beta_\gamma}{\lambda^{a_\gamma}} +$$

Here it seems, however, from the spectrobolometric measurements, as if β_x, β_y etc. would *at all levels* be small as compared with β_1, the latter referring to the diameter $1,0 \mu$. The consequence of this is that optical measurements yield the result that the size is constant at all levels. This result, however, as is evident from the way of treatment, only refers to the *optically effective size*. It ought not to be generalized to mean anything else.

The extinction equation of Fowle may thus be written in the following way:

$$(6) \quad \ldots \ldots \ldots \ldots \ldots \quad I_\lambda = I_{o\lambda}\, q^m\, e^{-\frac{\beta}{\lambda^a} \cdot m}$$

Here β is a coefficient dependent upon the amount of dust present in the atmosphere. For a constant value of a we must expect β to be proportional to the amount of dust in the air. In the following we will for β use the notation the *turbidity coefficient;* for a it seems adequate to introduce the name: the *size exponent*, because it seems to vary chiefly with the size of the scattering particles.

If we turn *to averages*, it seems as if the turbidity coefficient could be expressed as an exponential function of the height above sea-level:

$$(7) \quad \ldots \ldots \ldots \ldots \ldots \ldots \quad \beta_h = \beta_o\, e^{-\partial h}$$

On different occasions ∂ assumes highly different values. If no cleansing of the atmosphere took place through precipitation, the value of ∂ would be determined chiefly by two factors: namely, the turbulent movements which bring dust from the surface to the air and transports it to higher levels, and on the other hand the falling velocity of the particles, which tends to bring the dust back to the surface of the earth.

From the spectrobolometric measurements at different high-level stations we derive a value for ∂ of about $0{,}7 \cdot 10^{-5}$, the height h being given in cm.

For constant size of the dust particles we now have for the number of dust particles N_h in unit volume at the altitude h cm:

$$(9) \quad \ldots \ldots \ldots \ldots \ldots \quad N_h = k\, \frac{d\,\beta_h}{d\,h} = k\,\beta_o \cdot \partial \cdot e^{-\partial h};$$

or

$$(10) \quad \ldots \ldots \ldots \ldots \ldots , \ldots \quad N_h = N_o\, e^{-\partial h}$$

It thus appears as if the dust in the atmosphere on the average is distributed in the atmosphere according to a law similar to that which determines the decrease of pressure or water-vapour content with an increase in altitude. Expressing with Hann the water-vapour pressure by a formula similar (7), namely:

$$(8) \quad \ldots \ldots \ldots \ldots \ldots \ldots \quad e_h = e_o \cdot e^{-\partial_1 h}$$

where $\partial_1 = 0{,}37 \cdot 10^{-5}$, we find that the coefficient ∂_1, determining the percentual decrease of water-vapour pressure at a certain altitude is only about half that of ∂ in the previous formula for the dust.

While the water-vapour content decreases to half its value with a rise in altitude of about 2 000 meters, the content of large dust particles has decreased by half already with a rise of about 1 000 meters. The atmospheric pressure decreases by half with a rise of about 5 000 m. As regards the dust here in question, we must refer to the discussion presented above. Fig. 1 gives graphically an idea of the decrease of air pressure, water vapour and large dust particles with height.

If we go back now to Equation (6) for the atmospheric transmission, we may, with due regard to the decrease of the dust with height, write it in the following way:

$$(11) \quad I_\lambda = I_{o\lambda}\, q^m\, e^{-\frac{\beta_0}{\lambda^a}} \cdot e^{-\partial h} \cdot m$$

For the total energy I of the whole spectrum we get

$$(12) \quad I = \int_0^\infty I_\lambda\, d\lambda = \int_0^\infty I_{o\lambda}\, q^m \cdot e^{-\frac{\beta_0}{\lambda^a}}\, e^{-\partial h} \cdot m_{d\lambda}$$

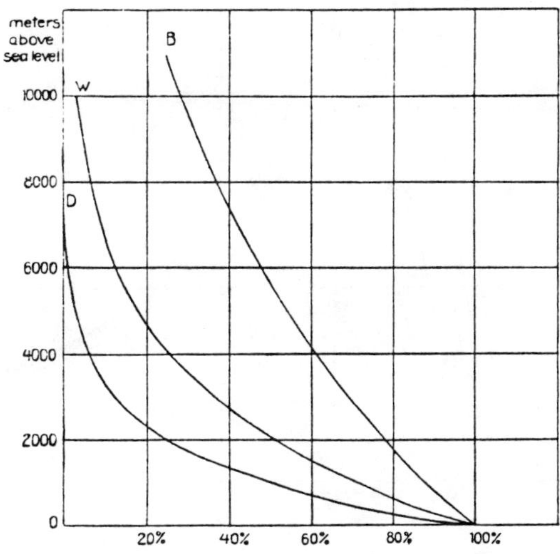

Fig. 1. Schematical design showing the decrease of air pressure (B), water vapour pressure (W) and content of large dust particles (D) with height above sea level.

The quantity I would be the total direct solar radiation reaching the surface of the earth for the case that we had no selective absorption but only scattering by molecules and dust.

As regards the water-vapour absorption it can according to Abbot and Fowle (5) be expressed by the equation:

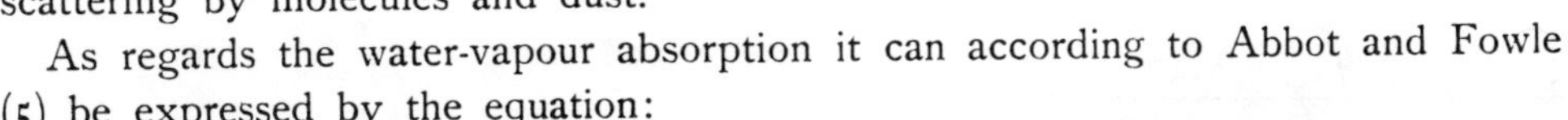

$$(13) \quad \ldots\ldots\ldots\ldots\ldots \quad F = 0{,}10 + 0{,}0054\, e_0 \cdot m \qquad \left\{ \begin{array}{l}\text{(gram calories per square centi-}\\ \text{meter and minute)}\end{array}\right.$$

the solar constant is hereby assumed equal to 1,95. e_0 is the water vapour pressure in mm at the place of observation.

For the total solar energy reaching the surface of the earth, with due regard to water vapour absorption and extinction by molecules and dust we thus obtain:

$$(14) \quad \ldots\ldots\ldots\ldots\ldots \quad Q = \int_0^\infty I_{o\lambda}\, q_1^m\, e^{-\frac{\beta_0}{\lambda^a} \cdot m}\, d\lambda - F$$

Here q_1 is naturally a function of h and λ determined by the decrease of air-pressure with height above sea-level.

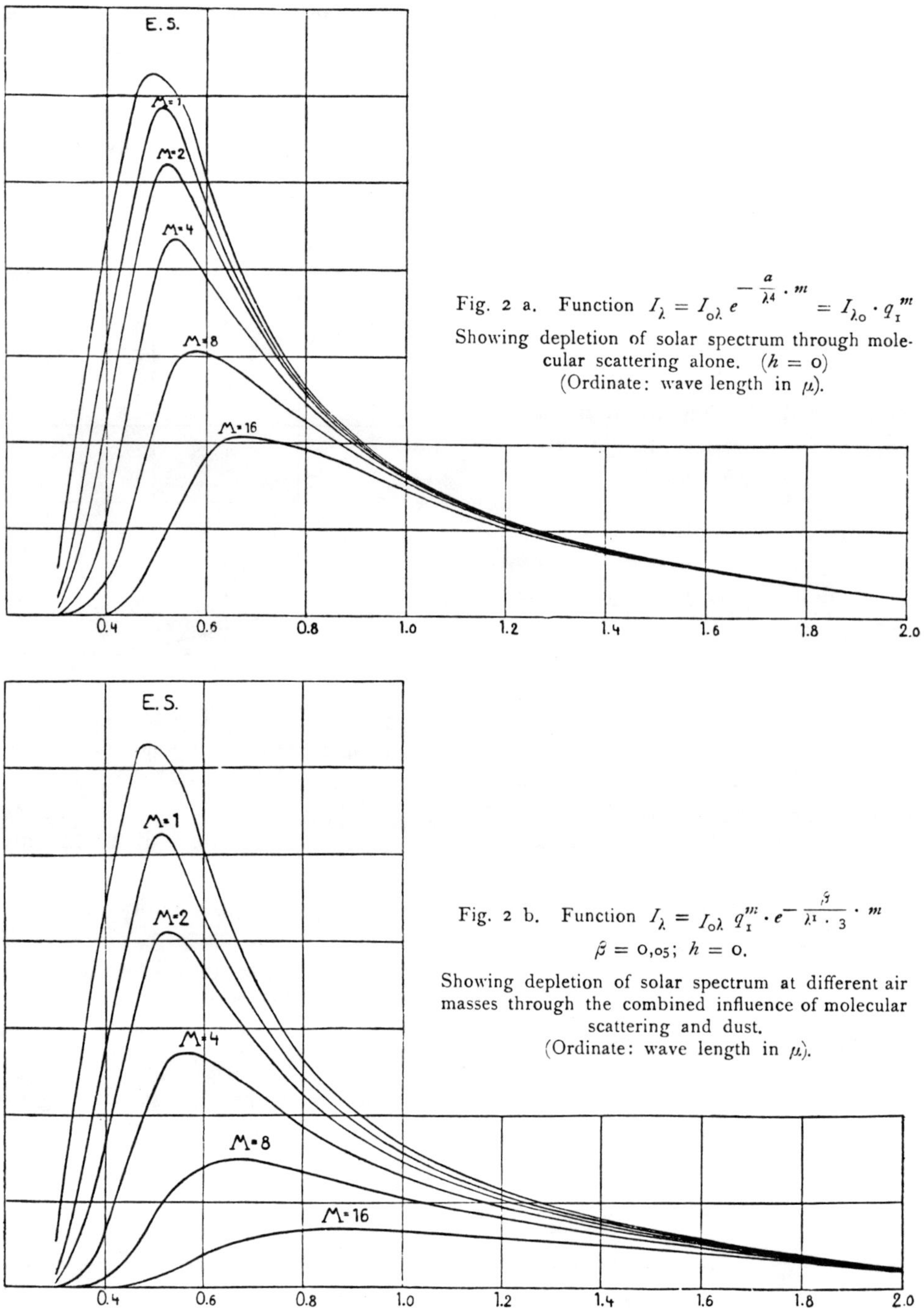

Fig. 2 a. Function $I_\lambda = I_{o\lambda}\, e^{-\frac{a}{\lambda^4}\cdot m} = I_{\lambda o}\cdot q_{\rm I}^{m}$

Showing depletion of solar spectrum through molecular scattering alone. $(h = 0)$
(Ordinate: wave length in μ).

Fig. 2 b. Function $I_\lambda = I_{o\lambda}\, q_{\rm I}^{m}\cdot e^{-\frac{\beta}{\lambda^{1\cdot 3}}\cdot m}$

$\beta = 0{,}05;\ h = 0.$

Showing depletion of solar spectrum at different air masses through the combined influence of molecular scattering and dust.
(Ordinate: wave length in μ).

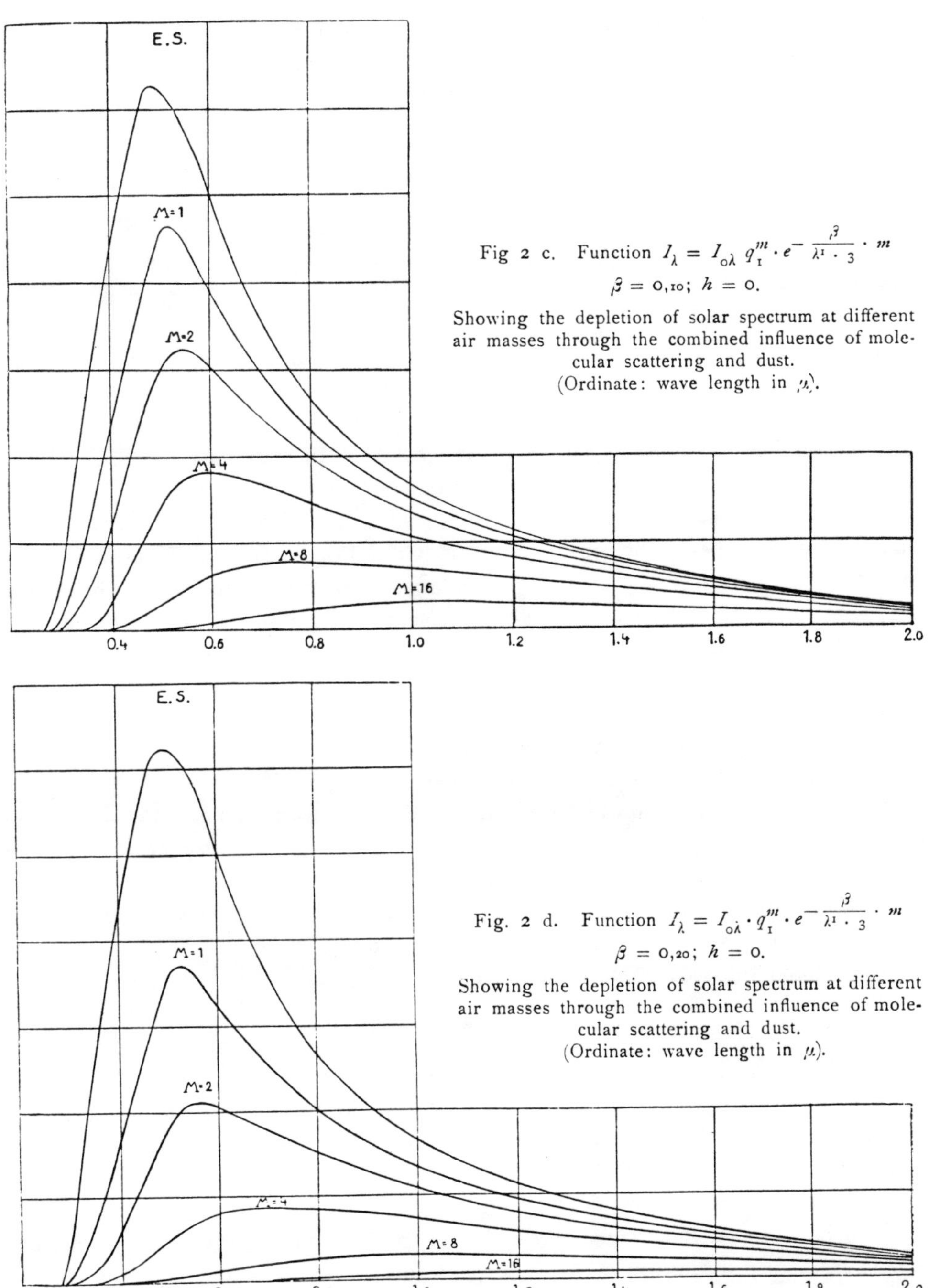

Fig. 2 c. Function $I_\lambda = I_{o\lambda}\, q_{\text{I}}^{m} \cdot e^{-\frac{\beta}{\lambda^{\text{I}} \cdot 3} \cdot m}$

$\beta = 0{,}10;\ h = 0.$

Showing the depletion of solar spectrum at different air masses through the combined influence of molecular scattering and dust.
(Ordinate: wave length in μ).

Fig. 2 d. Function $I_\lambda = I_{o\lambda} \cdot q_{\text{I}}^{m} \cdot e^{-\frac{\beta}{\lambda^{\text{I}} \cdot 3} \cdot m}$

$\beta = 0{,}20;\ h = 0.$

Showing the depletion of solar spectrum at different air masses through the combined influence of molecular scattering and dust.
(Ordinate: wave length in μ).

The formula (14) may now be taken as a basis for a discussion of the radiation measurements. It is evident that it is highly important that we should make clear what information can really be obtained from radiation measurements of a certain kind and a certain limitation and also to what extent radiation measurements are needed in order that we may obtain certain information which is demanded.

The formula (14) contains a certain number of variables, of which some may be determined directly, while others may be computed from a number of equations of the given kind.

One of the greatest and most important achievements of Abbot and his collaborators certainly is, that they have fixed the values of $I_{o\lambda}$ throughout the whole spectre, that is to say, determined with a comparatively high amount of accuracy the intensity distribution of the extra-terrestrial solar spectrum (5). Further they have shown, or it has been shown on the basis of their researches, that this intensity distribution is only subjected to relatively small variations. It is at present generally agreed that variations from the mean value of the solar constant larger than about $\pm$ 2 per cent occur very seldom. For studies of variations in the terrestrial atmosphere, therefore it is, in general permitted to regard the values of $I_{o\lambda}$ as constant and as those given by the investigations of the Astrophysical Observatory of the Smithsonian Institution.

The values of q_1^m, for various wave-lengths may further, as we have already shown, be computed according to the theory of Rayleigh. The air mass m may be computed from the height of the sun.

The remaining variables are β, a and F.

As I have mentioned above the quantity a seems to be subjected only to small variations under ordinary conditions. It keeps a value of very nearly 1,3 and the variations with the time of the year aud the height above sea-level generally keep within the limits $\pm$ 0,2, as long as we have not to do with impurities of volcanic origin in the atmosphere. It is of interest, therefore, to study a little more in detail the variations of the quantity

$$I_\lambda = I_{o\lambda}\, q_1^m\, e^{-\frac{\beta_0}{\lambda^a} \cdot m}$$

for the case that we assign a value for a of 1,3.

Now the curves of Fig. 2 a— c show the variation of I_λ for a number of different values of β and m, all referring to the conditions at sea-level (air-pressure 760 mm). For $\beta = 0$, and $m = 0$ we get $I_\lambda = I_{o\lambda}$ and we thus obtain the curve of ABBOT for the extra-terrestrial spectrum.

If we have a constant value of a and know β, we are evidently able to construct the energy distribution on all wave-lengths for the parts of the spectrum where no selective absorption occurs.

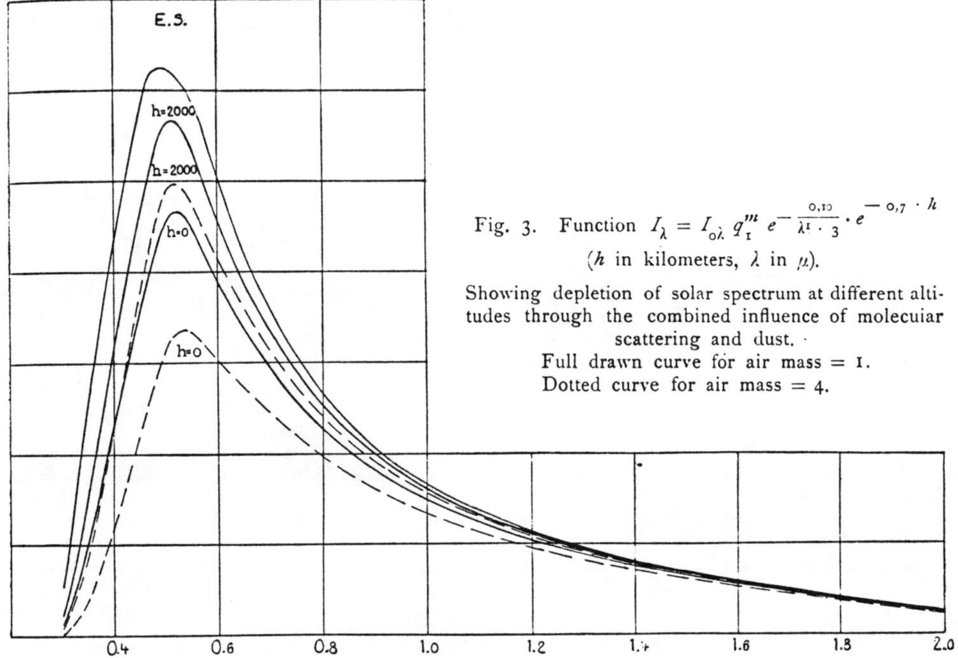

Fig. 3. Function $I_\lambda = I_{o\lambda}\, q_{\scriptscriptstyle 1}^{m}\, e^{-\frac{0,10}{\lambda^{1}\cdot 3}}\cdot e^{-0,7\,\cdot\,h}$

(h in kilometers, λ in μ).

Showing depletion of solar spectrum at different altitudes through the combined influence of molecuiar scattering and dust. ·
Full drawn curve for air mass = 1.
Dotted curve for air mass = 4.

A method of computing β from more or less limited actinometric measurements is therefore of great importance. In the first place, because in this way we shall be able to determine from measurements in a special region or over a special range of wave-lengths, the intensities also for other regions which have not been directly measured.

In the second place, however, β is also an important characteristic of the air mass itself and may furnish material for conclusions as to its origin.

We may now consider some cases of practical importance separately, namely:

I. The total radiation from the sun Q is measured together with the water-vapour pressure e_o; the air mass m is obtained through measurements, or computations from the time equation.

This represents the most common case of actinometric measurements.
From (14) we obtain:

$$Q_m + F = \int\limits_o^{\infty} I_{o\lambda}\, q_\lambda^m\, e^{-\frac{\beta}{\lambda^a}}\cdot m_{d\lambda} = \int\limits_o^{\infty} I_{o\lambda}\, \Psi\,(m,\,\beta,\,\lambda)\,d\lambda$$

The value of the integral is graphically represented in Fig. 4 for various air masses and various values of β. The integral has been graphically evaluated from curves of the kind represented in Fig. 2.

Table 1.

$$F = 0,10 + 0,0054\ e_0 \cdot m$$

e_0	1	2	3	4	5	6	7	8	10	12	14
$m = 1,0$	0,105	0,111	0,116	0,122	0,127	0.132	0,138	0,143	0,154	0,165	0,176
1,5	0,108	0,116	0,124	0,132	0,140	0,149	0,157	0,165	0,181	0,197	0,213
2,0	0,111	0,122	0,132	0,143	0,154	0,165	0,176	0,186	0,208	0,229	0,251
2,5	0,113	0,127	0,140	0,154	0,167	0,181	0,194	0,208	0,235	0,262	0,289
3,0	0,116	0,132	0,149	0,165	0,181	0,197	0,213	0,229	0,262	0,294	0,326
3,5	0,119	0,138	0,157	0,176	0,194	0,213	0,232	0,251	0,289	0,327	0,364
4,0	0,122	0,143	0 165	0,185	0,208	0,229	0,251	0,273	0,316	0,359	0,402
5,0	0,127	0,154	0,181	0,208	0,235	0,262	0,289	0,316	0,370	0,424	0,478
6,0	0,132	0,165	0,197	0,229	0,262	9,294	0,326	0,359	0,424	0,489	0,553
7,0	0,138	0,176	0,213	0,251	0,289	0,326	0,364	0.402	0,478	0,553	0,629
8,0	0,143	0,186	0,229	0,273	0,316	0,359	0,402	0,445	0,532	0,618	0,705
9,0	0,149	0,197	0,246	0,294	0,343	0,391	0,440	0,489	0,585	0,683	0 780
10,0	0,154	0,208	0,262	0,316	0,370	0,424	0,478	0,532	0,640	0,748	0,855

The evaluation of β from actinometric observations of the total radiation alone can thus be made in the following way with the aid of Fig. 4.

Add to the observed value of the total radiation the water-vapour correction, computed from the formula of Fowle, and tabulated in Table 1. Enter for the air mass corresponding to the time of observation into the diagram of Fig. 4 with this corrected value. Through interpolation between the β-curves, we then obtain the corresponding value for β.

If β is given from other kinds of observations, we may in a reversed way find the corresponding value of the total radiation.

The method suffers from the weakness that we must assume the water vapour absorption to be given by the water-vapour pressure at the earth's surface. Even if we assume the formula of Fowle to be valid at all localities for the average conditions, it must be realised that the vapour-pressure at the earth's surface is only a rough indicator of the total water content, and it is naturally this latter quantity that determines the absorption.

II. **The total radiation Q_m is measured and also the red and infra-red radiation by means of the well-known Schott filter F 4512.**

The transmission of the filter has been very accurately determined by Kaempfert (6) at the observatory of Linke. As is shown by the curve in Fig. 8, which is reproduced from Kaempfert's paper, the filter transmits almost 90 per cent of

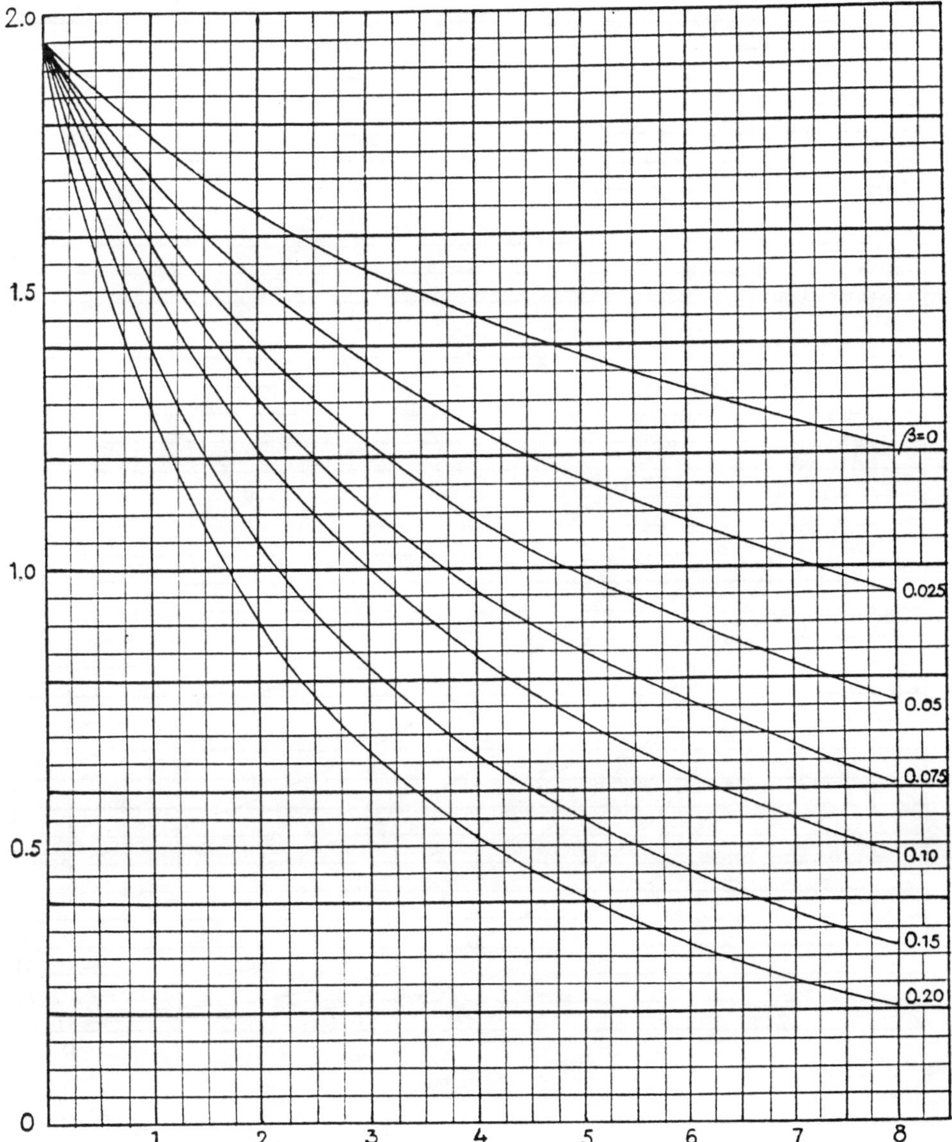

Fig. 4. Value of integral

$$\int_0^\infty I_{o\lambda} \cdot \Psi\,(m,\ \beta,\ \lambda)\ d\lambda$$

for different air masses m and different values on β. Total radiation in gram calories per square centimeter per minut.

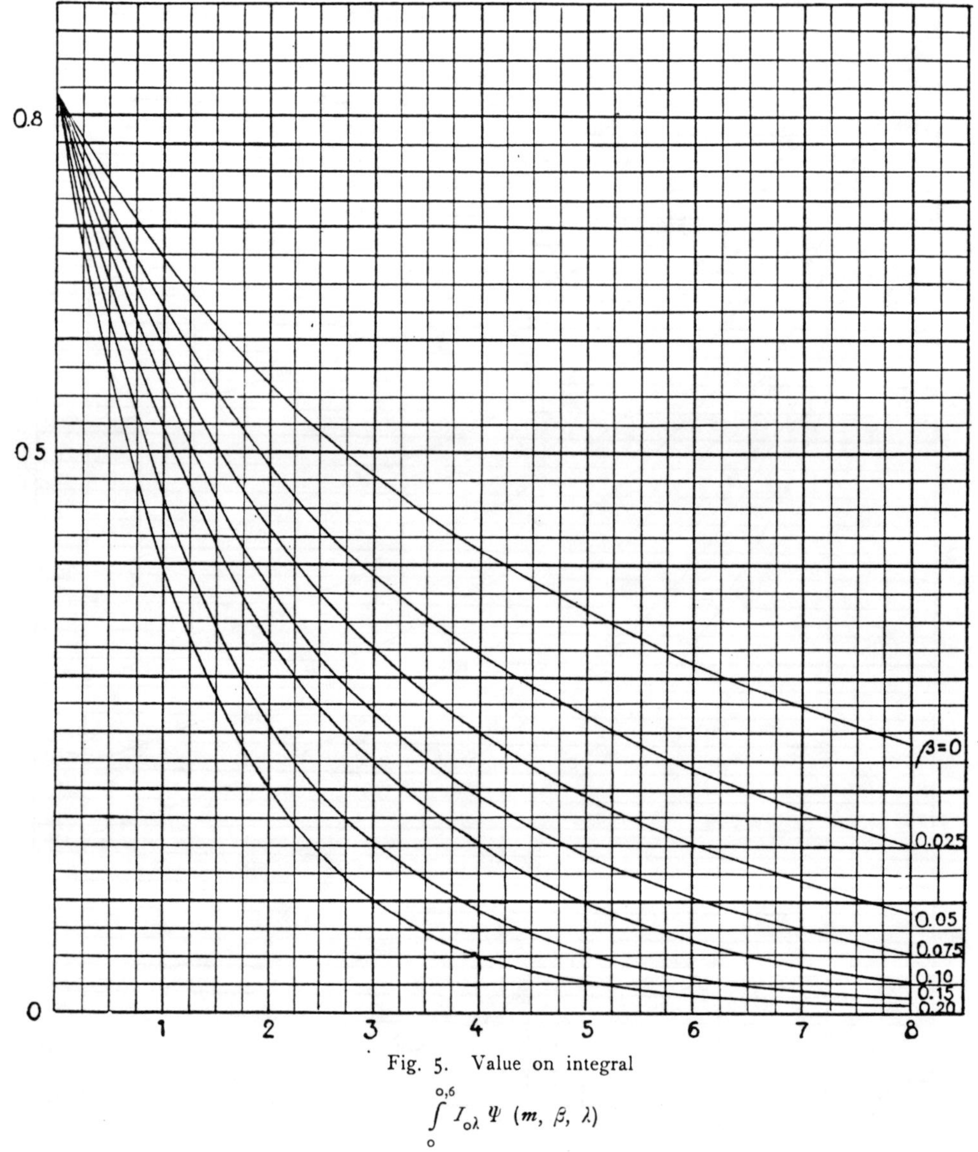

Fig. 5. Value on integral

$$\int_0^{0,6} I_{o\lambda}\,\Psi\,(m,\ \beta,\ \lambda)$$

for different m and β.
Radiation in gram calories per square centimeter per minut.

all wave-lengths above 0,6 μ. An important fact for our present purposes is that practically the total water-vapour absorption in the solar spectrum falls on the wave-lengths which are transmitted by the red filter. Putting the almost constant transmission of the filter equal to γ, we thus obtain for the filtered radiation Q_r:

$$(16) \ \ldots \ldots \ldots \ldots \quad Q_r = \gamma \int_{0,5}^{\infty} I_{o\lambda} \cdot \Psi\,(m,\ \beta,\ \lambda)\ d\lambda - \gamma \cdot F.$$

Combining this equation with (14), we obtain

$$Q_m - \cdot \frac{1}{\gamma}\,Q_r = \int_{o}^{\infty} I_{o\lambda}\ \Psi\,(m,\ \beta,\ \lambda)\ d\lambda - \int_{0,6}^{\infty} I_{o\lambda}\ \Psi\,(m,\ \beta,\ \lambda)\ d\lambda$$

which gives

$$(17) \ \ldots \ldots \ldots \ldots \quad Q_m - \frac{1}{\gamma} \cdot Q_r = \int_{\infty}^{0,6} I_{o\lambda}\ \Psi\,(m,\ \beta,\ \lambda)\ d\lambda$$

As we see, the water-vapour absorption F is here eliminated. Observing Q_m and Q_r, we consequently get a mean to determine the quantity β from the connection existing between β and the value of the integral of equation (17).

The variation of this integral for various values of m and β are graphically represented in Fig. 5. The method of determining β from simultaneous measurements of Q_m and Q_r is then clear from (17).

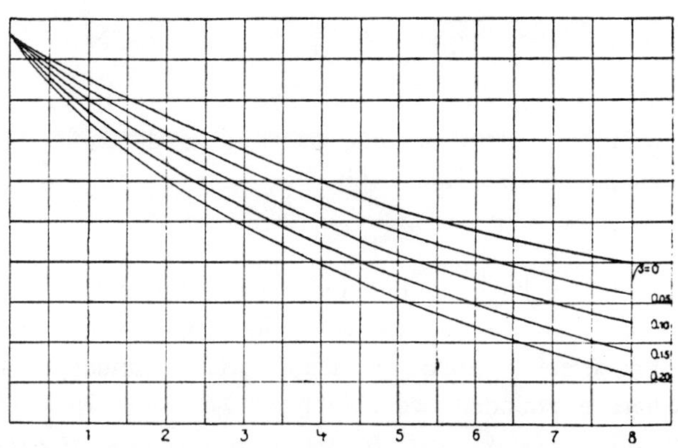

Fig. 6. Value of $\dfrac{I_{0,45}}{I_{0,55}}$ for different values on m and β.

The quantity β having been determined in this way, we may, on the basis of equation (14) and with the aid of the diagram Fig. 4, determine the water-vapour absorption F.

Remark. On account of the fact that various red glasses of different thickness vary somewhat as regards the transmission especially in the region in the neighborhood of $0,6\ \mu$, the upper limit of the integral in (17) may possibly not be exactly equal to $0,6$, but may vary to some extent. Practically this may be taken into account by substituting for the formula (17) the more empirical expression:

$$(18) \ \ldots \ldots \ldots \ldots \quad b\,Q_m - c\,Q_r = \int_{o}^{0,6} I_{o\lambda}\ q_1^{m} \cdot e^{-\frac{\beta}{\lambda^a} \cdot m}\ d\lambda$$

where b and c are filter constant. For the red filter F 4512 used in Stockholm we have $b = 0{,}95$; $c = 1{,}15$ (Thickness 3,5 mm).[1] The introduction of these constants means that we apply a certain correction to the difference between Q_m and Q_r in order to reduce to the exactly defined spectral range between 0 μ and 0,6 μ.

III. The ratio between the intensities at two different wave-lengths for which no selective absorption occurs, are given.

Practically such a case is given when observations or registrations with the potassium cell and also of the luminous radiation with the aid of potassium cell and filter are available. The effective wave-length of the radiation measured in the former way is about 0,45 μ; the effective wave-length of the latter is about 0,55. The diagram in Fig. 6 shows the value of the ratio $\dfrac{E_{0,45}}{E_{0,55}}$ for different values of m and β. The diagram may be used either for a determination of β from the ratio named, or if β is determined in other ways, to find the ratio $\dfrac{E_{0,45}}{E_{0,55}}$.

As we have pointed out before, the value of determining β lies chiefly in the fact that we in this way obtain, in a single quantity, a definition of the whole non-selective solar spectrum. As instances of the applicability of the methods here expounded, we may briefly consider the following questions:

A. *The contrasts in the solar spectrum at different heights of the sun and for different turbidity coefficients.* As a measure of the contrast we may take the ratio $\dfrac{E_{0,4}}{E_{0,7}}$, i. e. the ratio between the energy values at the two ends of the visible spectrum respectively. In diagram (7) the variation of this ratio with air-mass m and turbidity coefficient β is given.

B. *The wave-length corresponding to maximum intensity in the solar spectrum under various conditions.* The diagram (9) shows how the maximum is shifted to the longer wave-lengths with an increase in β or an increase in the air-mass, i. e. with decreasing height of the sun. As long as the sun's height is above about 30° (corresponding to $m = 2{,}0$), the variation of the positions of the maximum is not large. It varies between about 0,47 μ for very clear skies to about 0,56 for hazy ones.

---

[1] Investigations carried out by Dr Rebbestad and the author show that the corresponding constants for the red glass RG 2, F 20 702 (thickness 3 mm) now furnished by the Meteorological Observatory, Potsdam, are 0,95 aud 1.09 resp.

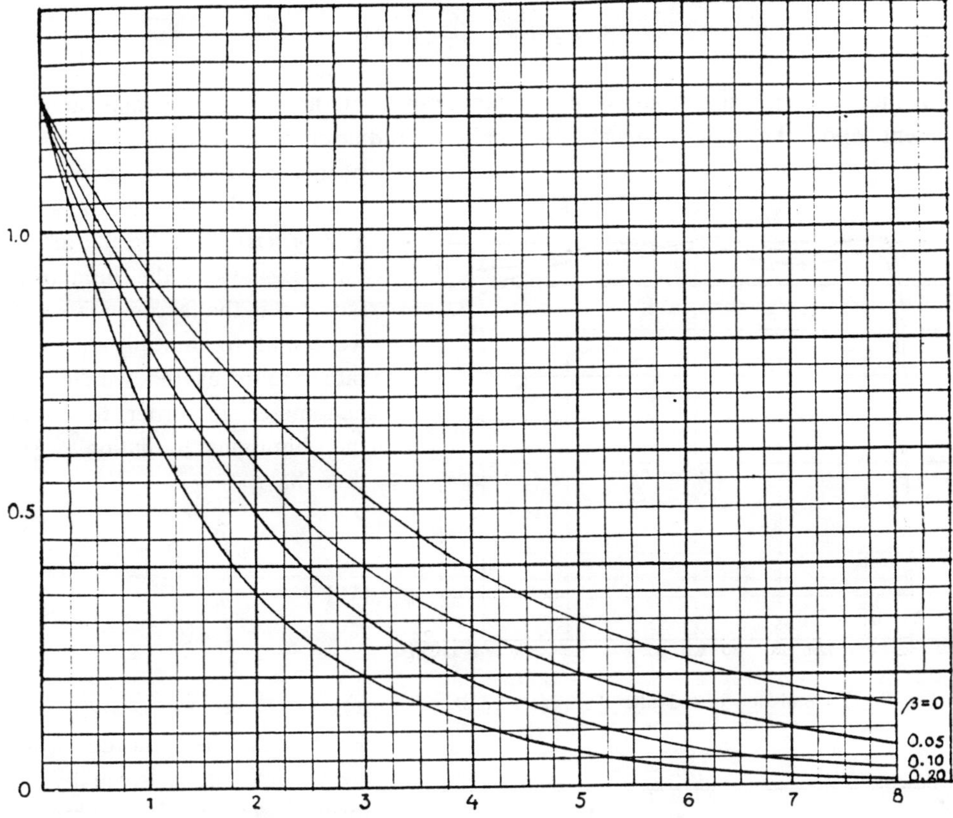

Fig. 7. Contrast ratio: $\dfrac{I_{0,4}}{I_{0,7}}$ for different values on m and β.

C. *The variation of the luminous radiation.* The eye being sensitive to the radiation between 0,4 and 0,7 μ, with a maximum sensitiveness at about 0,56 μ, it is of a certain interest to see how the radiation within this special range of wave-lengths varies under different conditions. The problem is simplified by the fact that no selective absorption of any appreciable amount occurs within the region named.

Let the intensity of the different wave-lengths in the solar spectrum be expressed by $f(\lambda)$. We then have:

$$f(\lambda) = I_{o\lambda}\, q_1^m\, e^{-\frac{\beta}{\lambda^a}\cdot m} \quad \text{(for non-selective regions).}$$

Further let the relative sensitiveness of the eye be expressed by the function $\varphi(\lambda)$. Then the function:

$$S = c \int_0^\infty f(\lambda) \cdot \varphi(\lambda) \, d\lambda$$

is a measure on the intensity of the luminous radiation, c being a constant dependent upon the unit in which the luminous radiation is measured. For many purposes it seems now convenient and practical to define a *luminosity factor* ρ in the following way: the luminosity factor ρ is the factor with which we ought to multiply the total radiation measured with a pyrheliometer in order to obtain the luminous radiation. We then have:

$$(19) \quad \rho = \frac{c \int_0^\infty f(\lambda) \cdot \varphi(\lambda) \, d\lambda}{\int_0^\infty f(\lambda) \, d\lambda - F},$$

where F is the selective absorption in the infra-red, which we assume to be given by the formula of Fowle. Through graphical evaluations of the integrals of equation (19) with the application of the values of the sensitiveness of the eye given, for instance, by Kaempfert (6) on p. 175 in his paper, we now obtain relative values for ρ under various conditions. For the sake of simplicity we put $\rho = 100$ for the extra terrestrial spectrum. The diagram (10) then shows the variation of ρ under various conditions as regards humidity and turbidity. The figure brings out the fact, already known from direct actinometric and photometric measurements combined, that the factor is not subjected to very great variations, at least not for the greater

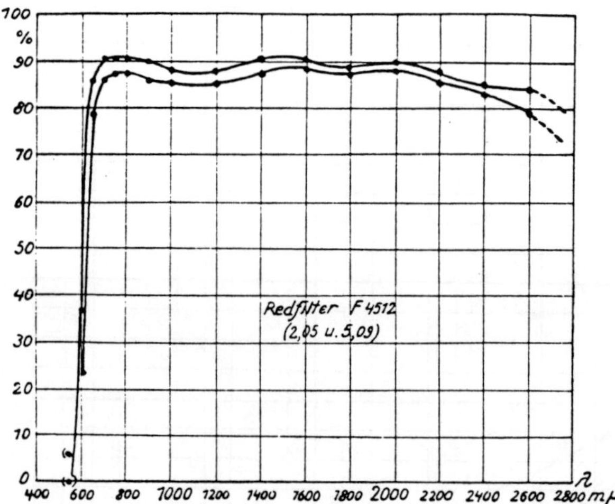

Pig. 8.   Transmission of red filter F. 4512 according to Kaempfert.

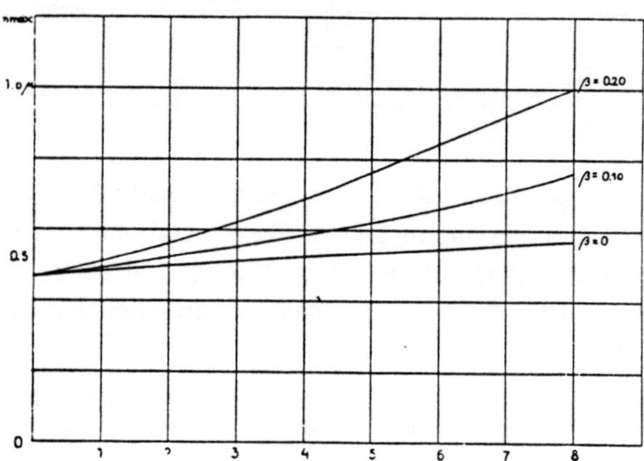

Fig. 9.   Position of λ_{max} corresponding to different values on m and β.

Table 2.

Jan.	Febr.	March	April	May	June	July	Aug.	Sept.	Oct.	Nov.	Dec.
96	87	91	92	99	102	101	95	91	87	95	$109 \cdot 10^3$

heights of. the sun. Other conditions being equal, ρ increases with an increase in water-vapour pressure, but decreases with an increase in β. With an increase in solar height it decreases for large water-vapour pressures ($e = 10$ mm), but increases when the atmosphere contains a small amount of water-vapour ($e = 5$ mm). ρ is almost constant for cases when $\beta = 0{,}20$ and $e = 10$ mm or $\beta = 0{,}05$ and $e = 5$ mm. As a high value for β in general occurs under conditions when e is also large, we see that the constancy of the luminosity factor is maintained through the combined effect of the variations of water-content and turbidity.

Table 2 gives the number of normal candles corresponding to an intensity of 1 gram calorie per square centimeter per minute according to computations from Lunelund's (7) measurements at Helsingfors. They are in general agreement with the considerations set forth above.

D. In this connection it is also of interest to consider briefly another problem, which at one time attracted great interest in connection with certain computations of the solar constant through extrapolation. It was assumed that the solar radiation could be practically treated as a homogeneous radiation, in which case we ought to be able to express the radiation at the earth's surface for various air-masses by the equation:

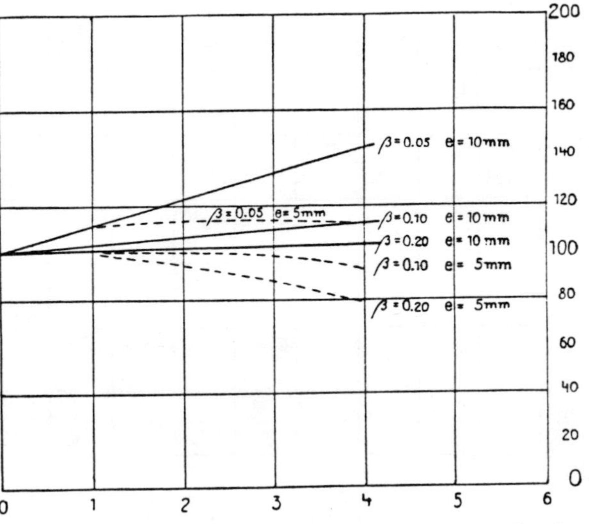

Fig. 10.   Value of luminosity factor ρ for different m, β and e.

$$(20) \quad \ldots \ldots \ldots \ldots \ldots \quad Q_m = Q_0\, e^{-a \cdot m}$$

or

$$(21) \quad \ldots \ldots \ldots \ldots \ldots \quad Q_m = Q_0 \cdot q^m$$

where q would be the transmission for the air mass m equal to unity. From (20) we should obtain:

$$(22) \quad \ldots \ldots \ldots \ldots \quad \log\, Q_m = \log\, Q_o + m \cdot \log\, q.$$

Assuming the equation (20) to be valid, we should thus obtain a linear relation between the logarithm of the radiation Q_m and the air-mass.

Now the observations of the total radiation at different air-masses have shown that the formula (20) is certainly not valid as long as no correction is applied for the selective absorption. But it is still sometimes claimed that, if such a correction is applied, i. e. if the selectively absorbed energy is added to the measured one, we should obtain results which would practically be satisfied by the equation (20). From the values of the transmitted radiation computed above on the basis of Abbot's transmission values, we may now get a general view upon this question. Thus we may from the diagram in Fig. 4 compute the logarithms of the radiation values and plot them against the corresponding air-mass. The diagram of Fig 11 shows the relations thus obtained. Instead of the straight lines of equation (22) we obtain slightly curved lines, showing that we cannot under conditions of constant atmospheric transmission expect to get a linear relation between logarithms of radiation and air-mass. *The deviation from the straigh line is larger, the larger the turbidity.*

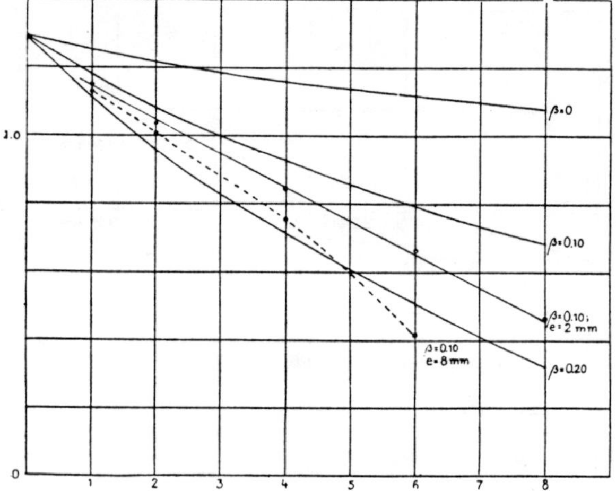

Fig 11. Shows relation between $\log\,Q + 1$ and m for different values on β and e.

Further complications arise if we consider also the influence of the water-vapour absorption. In Fig. (11) we have also introduced some values where the water-vapour absorption has been assumed to take place according to the formula of Fowle. We see that for the special case that the turbidity coefficient has a value: $\beta = 0,10$ and the water-vapour pressure is 2 mm, the straight line relation is almost exactly satisfied. But for instance for the case: $\beta = 0,10$; $e_o = 8$ mm, the deviation from a linear relation is very large.

We thus conclude that a linear relation of the kind just named is no criterion of the constancy of the atmospheric transmission. On the other hand, the atmospheric conditions may very vell remain unchanged and at the same time we may find a large deviation from the straight line equation. All here depends upon the simultaneous values of β and e_o on the occasion when the measurements are made.

It is of a certain interest to compute for these cases the »Trübungsfaktor», as it has been defined by Linke for the total radiation:

$$
\begin{array}{l l c c c c c}
 & m = & 1 & 2 & 4 & 6 & 8 \\
T & \beta = 0{,}20;\ e = 0 \text{ mm} & 4{,}5 & 4{,}5 & 4{,}4 & 4{,}6 & 4{,}6 \\
 & \beta = 0{,}10;\ e = 8 \text{ mm} & 4{,}0 & 3{,}7 & 4{,}2 & 5{,}1 & \text{---}
\end{array}
$$

We see that for the case of no water-vapour absorption the »Trübungsfaktor» under constant conditions keeps an almost constant value. Here only the case of $\beta = 0{,}20$ has been referred to as an instance, but the same can be shown to be the case independent of the value of β. But for the cases where the water-vapour absorption comes in as an additional factor, the conditions get more complicated, *and then the* »*Trübungsfaktor*» *is in general not constant with variation in the air-mass under constant atmospheric conditions.*

If, in other words, we compute the »Trübungsfaktor» from observations of the total radiation at different times of the day, we shall not, as a rule, find that factor to be constant, even if the atmosphere is perfectly unchanged. This shows that this entire method, at least in its application to the total radiation measurements, is not very well adapted for studies of atmospheric changes. For such investigations it is further highly important to get a separation between the effects of selective absorption and diffusion by dust. As we have shown above, such a separation may be effected by other methods.

Now the discussions presented above have almost entirely been based upon a treatment of the very comprehensive material of transmission coefficients determined by Abbot and his collaborators (5). The question then naturally arises whether it is allowed to apply these values to other conditions to the extent which here has been suggested.

A possibility of testing this applicability we now have in the measurements of the total and red radiation available at different localities. As we have shown above, these measurements give us two different ways of computing the atmospheric turbidity. The one is based upon the measurements of the total radiation alone. We then assume the water-vapour absorption to take place according to the formula of Fowle and the dust extinction to follow a certain law, which we have deduced from Abbot's measurements $\left(d I = - I \dfrac{\beta}{\lambda^{1{,}3}}\, d m\right)$.

On the other hand, the turbidity may also, as we have shown, be determined from measurements of the total radiation combined with measurements of the red radiation $(\lambda > 0{,}6\ \mu)$. In the latter case we are able to eliminate the water-vapour absorption, but the method is still founded upon the validity of our diffusion law.

A comparison between the β-values obtained through the two methods respectively gives us consequently a means to test to what degree our assumptions are valid in a more special case.

Table 3.

Year	Date	m	c_o	F	O_t	Q_r	$Q_t + F$	$cQ_t - bQ_r$	β_t	β_r
1921	3. 3.	2,90 a	3,2	0,15	1,005	0,591	1,155	0,265	0,070	0,080
»	3. 3.	3,19 p	3,1	0,15	1,115	0,645	1,265	0,307	0,038	0,049
»	7. 3.	4,72	3,0	0,18	0.910	0,568	1,090	0,211	0,038	0,048
»	11. 3.	2,97	4,3	0,17	0,930	0,538	1,100	0,263	0,080	0,081
»	12. 3.	3,05	4,3	0,17	0,834	0,513	1,004	0,201	0,097	0,110
»	13. 3.	2,77	4,2	0,16	0,962	0,567	1,122	0,263	0,083	0,089
»	14 3.	2,16	5,3	0,16	1,050	0,591	1,210	0,318	0,092	0,099
»	15. 3.	2,12	5,3	0,16	1,210	0,635	1,370	0,419	0,051	0,049
[» [1]	15. 3.	2,84	5,2	0,18	0,700	0,331	0,880	0,284	(0,142	0,075)]
[» [1]	21. 3.	2,55	4,8	0,17	1,010	0,575	1,180	0,300	(0,076	0,081)]
»	25. 3.	2,31	3,7	0,15	1,220	0,654	1,370	0,409	0,046	0,044
»	31. 3.	3,25	2,8	0,15	1,079	0,620	1,229	0,430	0,037	0,045
»	1. 4.	2,16	3,0	0,135	1,092	0,601	1,227	0,345	0,087	0,082
»	7. 4.	1,69	3,2	0,13	1,284	0,674	1,414	0,445	0,064	0,064
»	9. 4.	2,09	5,2	0,16	1,114	0,605	1,274	0,365	0,076	0,075
»	12. 4.	2,77	4,7	0,17	0,958	0,562	1,128	0,265	0,079	0,088
»	22. 4.	2,45	4,4	0,16	1,095	0,612	1,255	0,336	0,064	0,069
»	29. 4.	1,97	3,4	0,14	1,260	0,657	1,400	0,445	0,051	0,045
»	30. 4.	1,91	5,0	0,15	1,229	0,638	1,379	0,433	0,061	0,056
»	1. 5.	1,72	3,8	0,135	1,081	0,587	1,216	0,354	0,125	0,115
»	4. 5.	1,33	4,4	0,13	1,200	0,640	1,330	0,405	0,128	0,133
»	10. 5.	1,28	9,3	0,16	1,190	0,587	1,350	0,455	0.125	0,102
»	26. 5.	2,26	7,0	0,19	1,064	0,583	1,254	0,340	0,074	0,077
»	31. 5.	1,41	6,3	0,15	1,186	0,628	1,336	0,403	0,117	0,126
1923	29. 9.	2,16	5,6	0,16	1,230	0,631	1,390	0,445	0,047	0,036
1924	23. 1.	5,12	1,1	0,13	0,902	0,580	1,032	0,198	0,042	0,047
»	15. 2.	3,40	1,4	0,13	0,818	0,493	0,948	0,211	0,098	0,089
»	29. 2.	2,65	2,0	0,13	0,965	0,575	1,095	0,256	0,095	0,100
»	21. 3.	1,98	1,9	0,12	1,050	0,570	1,170	0,341	0,112	0,099
»	24. 4.	1,46	3,0	0,12	1,182	0,616	1,302	0,415	0,123	0,108
[» [1]	14. 6.	1,44	5,2	0,14	0,955	0,613	1,095	0,201	(0,020	0,26)]
1925	7. 4.	1,64	3,0	0,12	1,200	0,635	1,320	0,410	0,100	0,087
»	19. 5.	1,29	6,1	0,14	1,178	—	1,318	—	(0,138)	—
»	17. 11.	9,40	3,7	0,28	0,550	0,374	0,924	0,176	0,022	0,012
»	7. 12.	6,90	3,8	0,24	0,849	0,532	1,089	0,193	0,018	0,022
1926	24. 3.	1,93	3,1	0,13	1,418	0,750	1,548	0,486	0,022	0,028
»	3. 9.	2,32	7,8	0,20	1,226	—	1,426	—	(0,033)	—
»	14. 9.	2,12	5,2	0,16	1,292	0,669	1,452	0,458	0,033	0,029
»	11. 11.	5,81	6,4	0,30	0,760	—	1,060	—	(0,030)	—
1927	11. 1.	6,48	2,2	0,18	0,825	0,593	1,005	0,103	0,055	0,067
1928	31. 5.	1,26	4,3	0,13	1,493	0,741	1,623	0,567	0,037	0,037
»	26. 6.	1,24	8,2	0,155	1,390	—	1,545	—	(0,046)	—
»	3. 9.	1,62	7,2	0,16	1,371	0,672	1,531	0,567	0,038	0,012
»	5. 10.	3,10	5,1	0,185	1,125	0,606	1,310	0,392	0,033	0,022
1929	15. 6.	1,45	10,8	0,185	0,771	0,475	0,956	0,186	0,27	0,32
»	8. 10.	2,36	6,8	0,19	0,930	0,592	1,020	0,203	0,135	0,158

[1] Light Ci or Al—St.

In Table 3 we have now collected the data referring to a number of simultaneous observations of the total and red radiation during the years 1921—1929 at Stockholm. The columns of the table contain the following data: (1) the year; (2) the day; (3) the air-mass m, computed from the solar height according to the tables of Bemporad; (4) the absolute humidity e_0 in mm Hg; (5) the water-vapour absorption F, computed according to the formula of Fowle; (6) the total radiation Q_t from the sun against a surface perpendicular to the sun's rays, measured by means of the Ångström compensation pyrheliometer N° 158 or the Michelson actinometer 129; (7) the red radiation Q_r, measured by means of the Michelson actinometer 129 (The radiation intensities are all reduced to mean solar distance); (8) the quantity $Q_t + F$; (9) the expression $cQ_t - bQ_r$ of equation 18, c and b being filter constants (In our case we have put: $c = 0,95$; $b = 1,15$); (10) the turbidity coefficient β_t, computed from the total radiation alone; (11) the same coefficient β_r, computed from the measurements of the total and the red radiation.

As is seen from the table, a very close agreement exists between the values of β obtained through the two methods. This agreement is also apparent from Fig. 12 and from the fact that the correlation coefficient computed from the given values is as high as $0,993 \pm 0,003$. From the figure it is evident that the relation between β_t and β_r is very nearly a linear one. There are, however, indications of a very slight but noticeable curvature, in the direction that for high β-values, the method of computing β_r gives values slightly lower than the method for computing β_t. The reason for this is probably that the filter constants c and b, which are introduced into the equation for the purpose of computing β_r, have been assumed to be independent of the transmission, that is to say of the composition of the incident radiation. This is not strictly true. The observations, however, show that this assumption, as well as the other ones which have been made and referred to above, must be very closely fulfilled in actual practice. *We have*

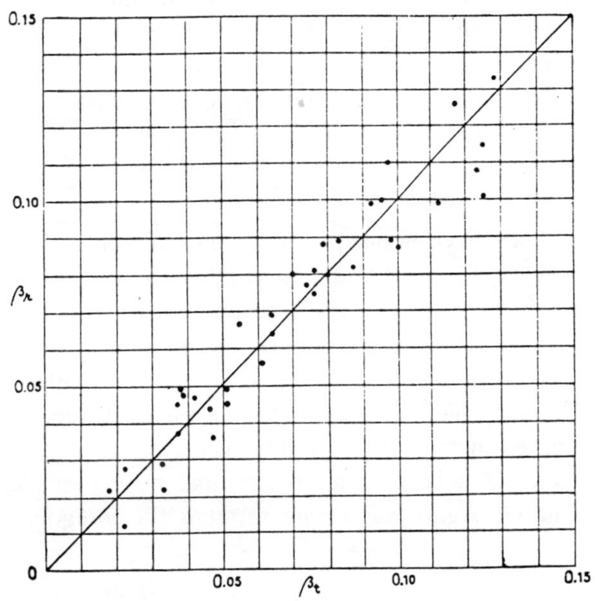

Fig. 12. Shows relation between simultaneous values on β_t and β_r.

consequently here a simple method to obtain both the turbidity and the water-vapour absorption from simultaneous measurements of (1) the total and (2) the red and infra-red radiation.

In individual cases, however, considerable differences have occured between the values of β computed according to the two methods. It is clear from the foregoing how such differences generally arise. Suppose the water-vapour absorption to be *larger* than the value given by the formula of Fowle. Then we obtain too small a value for $Q_t + F$ and consequently *too large* a value for β_t. The value of β_r, however, is independent of the assumption as regards the water-vapour absorption and is correct, if other assumptions hold.

Suppose, on the other hand, that the formula for the dependence of the extinction of the wave-length is not fulfilled. On the basis of Abbot's results we have expressed the extinction due to dust by the formula:

$$dI = -I\frac{\beta}{\lambda^a}dm$$

and we have in our computations counted with a value on a equal to 1,3. Suppose now a to be different from this value. A larger value for a evidently gives a larger contrast in the spectrum, arising through the fact that the short waves are then scattered comparatively more than is provided by the formula. The consequence of this is that for β_r, which is computed from the extinction of the waves shorter than 0,6, we obtain a larger value than for β_t, which is obtained from the extinction effect in the whole spectrum, including longer waves also. As a value of a larger than the normal means smaller particles than the ordinary size and vice-versa, we obtain through the foregoing considerations the following rules:

I) $\beta_r = \beta_t$

water-vapour absorption according to the formula of Fowle.   and $a = 1,3$ i. e. scattering particles having ordinary size.

II) $\beta_r > \beta_t$

water-vapour absorption below value given by the formula of Fowle.   or $a > 1,3$ i. e. scattering particles smaller than ordinary size

III) $\beta_r < \beta_t$

water-vapour absorption higher than value given by formula.   or $a < 1,3$ i. e. scattering particles larger than ordinary size.

An additional case of $\beta_r = \beta_t$ arises, when both the water-vapour absorption and the size of particles differ from their ordinary values, but the effects of the deviations exactly balance one another.

As we have seen before, the values of β_t and β_r are generally in close agreement with another and practically Case I is that which most frequently occurs. Either β_t and β_r may therefore be taken as a basis for further considerations as regards the turbidity of the atmosphere; in both cases the considerations will give practically the same results as far as the observations at Stockholm are concerned.

I. Variation of atmospheric turbidity with the time of the year.

The observations are not numerous enough to enable a very conclusive discussion. A comparison between results at different stations, however, gives so many indications of a certain regularity that it seems worth while to enter briefly upon the question.

Table 4.

Jan.	Febr.	March	April	May	June	July	Aug.	Sept.	Oct.	Nov.	Dec.
0,057	0,095	0,068	0,075	0,098	(0,158)	—	—	0,035	0,090	0,021	0,022

In Table 4 I have tabulated the monthly mean values of the turbidity β as computed from the values β_t of Table 3. The smoothed curve giving the annual variation is marked SS in Fig. 13. For comparison the curve WW for Washington based upon the values for β given in my first paper (table 3) is also added.

From the values given by Lindholm (8) for the dust absorption in the atmosphere above *Davos* one is able to compute approximate values of β for that place with aid of the graphical representation in Fig. 4 of the present paper giving the relation between intensity, air-mass and β values. To the values for the percentual absorption by dust at air-mass 2,5, computed by Lindholm, we must add the small percentual extinction, which by Lindholm, in accordance with the considerations of Fowle, is ascribed to the scattering by moist air (8), but which is here regarded as an extinction by minute fluid and solid particles and consequently in no way different from the extinction process in general. The reasons for this opinion have been set forth above.

From this procedure the values of β for Davos represented graphically in Fig. 13 are obtained. The values refer to the period 1914—1926. (Compare page 135 in Lindholm's paper.)

From the various observations on the depletion of sun radiation by the atmosphere, we may draw the following conclusions.

The dust content of the atmosphere shows at all stations here considered a very marked variation with the time of the year. It has a very pronounced maximum in May or June, i. e. in the spring or early summer, while it goes down to a minimum in the early winter, i. e. in November or December. At the minimum, the turbidity coefficient is only a third or a fourth of its value at the maximum, and the dust content consequently varies in about the same proportions.

The dust content is much smaller at the high-level station Davos, but shows a similar variation as at the low-level stations. This shows that the cause of the variation in the dust content ought to be sought at the surface of the earth and in the conditions of the lower layers of the atmosphere. It seems a fact of some importance for the explanation that the time of maximum of dust content corresponds very closely with the time of the largest temperature gradient in the lower atmospheric layers. At that time the lower air layers probably are very often characterized by an unstable equilibrium; air heated at the contact with the earth's surface rises especially in the middle of the day and maintains, through the effect of the »Austausch», a transport of the dust from low levels to high ones.

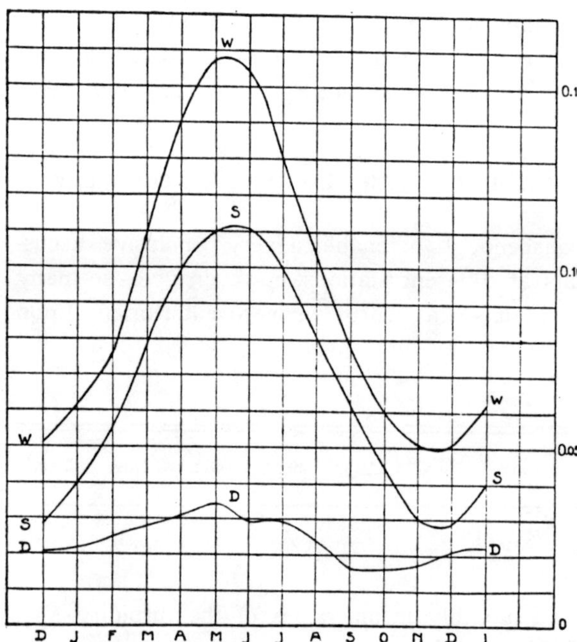

Fig. 13. Annual variation of turbidity coefficient (β), for Washington (W), Stockholm (S) and Davos (D).

II. Variation of atmospheric turbidity with latitude.

A more close examination of this important question on the basis of the large material of observations now available well deserves to be made. Here we only aim at a preliminary orientation on the basis of a more limited material.

Westman (9) in his paper on insolation has tabulated a number of total radiation values from Upsala, Spitzbergen and Modena not only for different air-masses but also as regards the water-vapour pressure. If β_m^p denotes the turbidity computed from radiation values at air-mass m and water-vapour pressure p, we find for Modena at 44° 54 N $\beta_{1,2}^{12} = 0{,}14$; $\beta_{1,2}^{6} = 0{,}12$ and $\beta_{2}^{3} = 0{,}088$, with a mean value of about 0,11; for Upsala at 59° 51 N $\beta_{2}^{3} = 0{,}06$; $\beta_{2}^{6} = 0{,}05$; $\beta_{2}^{9} = 0{,}06$, with a mean value of about 0,05, against yearly a mean value $\beta = $ about 0,065 computed from our radiation values in Stockholm (59° 21 N). Westman's values from Spitzbergen give (9): $\beta_{2}^{3} = 0{,}05$; $\beta_{4}^{3} = 0{,}04$; $\beta_{4}^{1} = 0{,}05$ with a mean value of about 0,045 for the 80th latitude. The mean turbidity value 0,11 at Modena corresponds very well to the mean value 0,10 computed from Abbot's transmission values at Washington at 38° 54 N.

On the other hand, the values of Gorczynski (10) from the city of Bangkok 13° 44 N seem to give a β-value of about 0,125 (computed from the total and red radiation according to the procedure given above), and his values from the oceans around the equator vary between about 0,075 and about 0,125. In general, it seems from the results of Gorczynski as if the equatorial oceans in general at clear sky are characterized by a considerably smaller turbidity coefficient than the land stations. This, if it should prove valid, is a very important result and may be applied to considerations as regards the radiation climate of maritime stations in comparison with continental ones. As a high dust content corresponding to a high value of β, necessarily causes an intense reduction of especially the short waves in the sun radiation, it would probably be allowed to draw the conclusion that winds from the ocean at maritime stations in low latitude ought to be characterized by a radiation climate comparatively rich in violet and ultra-violet rays, and *vice-versa*. Further investigation along these lines in different latitudes are necessary, however, before it is allowed to draw definite conclusions.

On the other hand, it is evident that on account of the decrease of the turbidity with latitude, the northerly winds in the northern hemisphere must in general be characterized by a lower β-value than the air coming from the southern regions. For the months March, April and May, for which we have a considerable number of measurements at Stockholm I have computed mean values of β for Polar and Tropical air respectively and find in this way: $\beta_P = 0{,}059$; $\beta_T = 0{,}088$. The consequence of this difference as regards the differences in the composition of sun radiation in corresponding cases is clear from the foregoing.

When the material of observations at different stations has been treated more in detail, we may on the basis of similar considerations put up a system of the following kind, where the numbers here introduced have only a preliminary character:

I. Polar air.

a. Continental   b. Maritime
$\beta = 0,02 - 0,06$ $\beta = 0,02 - 0,06$

II. Tropical air.

a. Continental   b. Maritime
$\beta = 0,10 - 0,20$ $\beta = 0,07 - 0,12$

The determination of β may serve, after a larger material of observations is awailable, to determine the origin of various air-masses.

The importance of such a determination for weather-forecasting has been emphasized especially by the Bergen school of meteorologists and for further details on the subject we may refer to an important and systematic treatise by T. Bergeron (11). Also to the numerous works on radiation by F. Linke and his collaborators we owe that the importance of determinations of the »trübung» is now generally acknowledged (3).

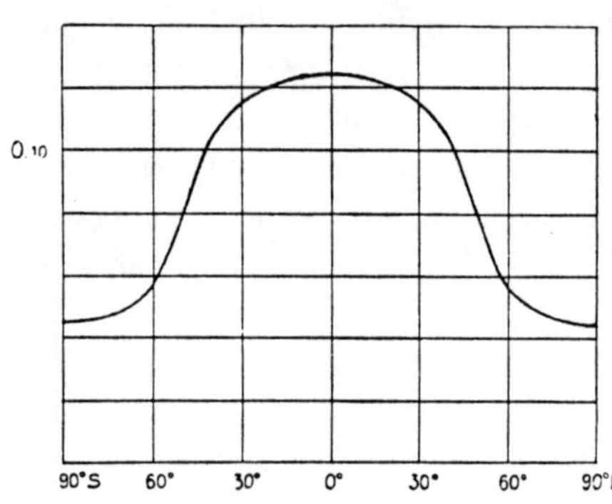

Fig. 14. Preliminary curve showing variation of β with latitude.

III. Variation of β with the height above sea-level.

The question has been touched on above in connection with our treatment of the observations of Abbot and his collaborators. From their transmission values it appears as if the dependence of β of altitude *on the average* could be expressed by an exponential formula of a kind similar to that which gives the decrease of water-vapour with altitude, is to say, we have:

$$\beta_h = \beta_0 \cdot e^{-0,7 \cdot H \cdot 10^{-3}}$$

H is here given in meters.

The value of the factor $e^{-0,7 \cdot H \cdot 10^{-3}}$ is given in Table 5 for different values of H.

Table 5.

$H =$	200	400	600	800	1 000	1 500	2 000	2 500
$e^{-0,7 \cdot H \cdot 10^{-3}} =$	0,86	0,76	0,66	0,58	0,50	0,35	0,25	0,17

The observations especially in high latitudes are not numerous enough to allow us to draw very definite conclusions as regards the combined influence of latitude and hight above sea-level. As a preliminary attempt we may try the validity of the formula:

$$(23) \ldots \ldots \ldots \ldots \quad \beta = (A + B \cos^2 \gamma) \cdot e^{-0,7 \cdot H \cdot 10^{-3}}$$

where we put $A = 0,040$; $B = 0,085$.

The observed mean values are given, together with the computed ones, in the following Table 6.

Table 6.

Station	Lat.	Altitude	β_{ber}	β_{obs}	Diff.	Authority
Equator	0—15° N a S	0	0,125	0,075—0,14	—	Gorczynski
Washington	38° 54 N	34 m	0,091	0,098	— 0,007	Abbot
Modena	44° 54 N	51 m	0,083	0,10	— 0,027	Chistoni
Stockholm	59° 21 N	20—57 m	0,062	0,065	— 0,008	Ångström
Upsala	59° 51 N	24 m	0,061	0,055	+ 0,006	Westman
Treurenberg	79° 55 N	8 m	0,043	0,045	— 0,002	»
Hump Mountain	36° 8 N	1 500	0,034	0,031	+ 0,003	Abbot
Davos	46° 48 N	1 600	0,026	0,024	+ 0,002	Lindholm
Mount Wilson	34° 13 N	1 730	0,030	0,018	+ 0,012	Abbot
Calama	22° 28 S	2 250	0,024	0,023	+ 0,001	»

The agreement between observed and computed values is generally fairly good. The deviations may partly be due to a real difference. It must be kept in mind, however, that large parts of the observations upon which the computation here is based are limited to a few years or even, as in the case of Gorczynski's observations during expeditions to the southern oceans, to a small part of the year.

It is interesting to note how regular the conditions are as regards the turbidity especially at the high level stations.

Hump Mountain, Davos and Calama, come all very near to the computed values. Mount Wilson shows a slightly lower value on the turbidity than is computed from the formula. The discrepancy may be accidental, but may also possibly be due to the neighburhood of the ocean, tending to purify the air from dust particles of mineral origin.

In a previous paper (12) we have given some reasons, which speak in favour of the view that the scattering particles in the *perfectly* clear atmosphere chiefly consist of solid particles and not of fluid or partly fluid ones. It seems natural if we adhere to that point of view, that the neighbourhood of large surfaces of water, where no real dust production occurs, shall have a consequence as regards the turbidity of the kind actually observed.

It is true that the researches of Köhler have made it probable that the nuclei of condensation to large extent consist of small salt particles, naturally having their origin in the waters of the oceans, and it could from these researches be suspected that the oceans to a certain extent are producers of diffusing particles in the atmosphere. Without doubt this is also the case. But on the other hand, it seems probable from the actinometric measurement that the optical efficiency of the salt particles, having a diameter of only about 10^{-4} mm, must be small compared with that of the ordinary dust particles, which seem to be in general of a much larger size.

It seems probable that further actinometric researches will lead us to distinguish between at least two kinds of dry dust, namely the ordinary dust, which generally seems to be of mineral origin, and the finer and, from actinometric point of view, less active dust consisting of salt particles from the oceans, the general existence of the latter being a consequence of the investigations of Köhler. It would mean an important advancement if this latter pollution could also be detected and followed synoptically through actinometric measurement.

In the present paper we have tried to simplify the treatment of actinometric observations through introduction of a turbidity coefficient β of the atmosphere, which is similar to the turbidity factor of Linke, but defined in a somewhat different way. The advantage of its introduction is in the first place, that it is independent of the wave length under ordinary conditions and also independent of the coefficient of molecular scattering and of the selective absorption.

The chief cause of the variations in the intensity of various wave-lengths, for which no selective absorption takes place, now undoubtedly must be sought in the variations of the turbidity of the atmosphere which may be quantitatively defined through the value of the coefficient β.

Parallel to investigations of the variations of for instance the ultraviolet light it is therefore to be recommended that studies are made of the variations of the turbidity. In this way we may hope to be able to eliminate the chief cause of the variations in the intensity of various wave-lengths. Thus we may find new causes for variations and give a quantitative measure on their influence.

In what extent the actually observed variations of the intensity of certain given wave-lengths with altitude and latitude can be explained chiefly through the normal dependence of β of these variables, is a question of high interest, whose closer examination, however, we must postpone to a following study.

Summary.

The present paper contains the following main results and conclusions.

(1) The atmospheric turbidity can be defined under average conditions through the value of a coefficient β, wich is independent of selective absorption, of molecular scattering, of wave-length and air mass.

(2) Analytical and graphical methods are developed to determine β from common actinometric measurements.

(3) Analytical and graphical methods are developed to determine *the total water content* of the atmosphere from common actinometric measurements.

(4) The variation of β with the latitude, with the time of the year and with height above sea level has been subjected to a preliminary study. The dependence of latitude and height has been expressed through an empirical formula.

a) In general β decreases with an increase in latitude and is at the 60th latitude only slightly more than half the value at the equatorial regions.

b) β decreases to half its value with a rise in altitude of about 1 000 meters.

c) β has at the stations, here studied, a pronounced maximum in the spring or early summer, a minimum in November or December.

(5) The dust transport from the earth to the atmosphere stays in close relation to the heating of the lowest air layers through sun radiation.

(6) It has been emphasized that the variations of the intensity especially of the sun radiation of short wave-length are determined chiefly by variations in the turbidity of the atmosphere.

Statens Meteorologisk-Hydrografiska Anstalt. March 1930.

Litterature.

(1) ÅNGSTRÖM, K.   Méthode nouvelle pour l'étude de la radiation solaire: Nova acta regiæ Soc. Scient. Upsaliensis. Ser. IV. Vol. 1. N. 7.

(2) FOWLE, F. E.   The non-selective transmissibility of radiation through dry and moist air. Astrophysical Journal, Vol. XXXVIII, No. 4, november 1913.

(3) LINKE, F.   Beiträge z. Physik der freien Atmosphäre. *10* 91.

MILCH, W.   Optik der Atmosphäre. Handbuch der Experimentalphysik, Bd. 25. 1 Teil, pag. 230, 1928.

(4) ÅNGSTRÖM, A.   On the atmospheric transmission of sun radiation and on dust in the air Geografiska Annaler 1929. H. 2.

(5) Annals of the Astrophysical Observatory of Smithsonian Institution. Vol. I—IV.

(6) KAEMPFERT, W.   Über die Durchlässigkeit von Strahlungsfiltern. Gerlands Beiträge zur Geophysik. Bd. 23. S. 167.

(7) LUNELUND, H.   Über die Wärme und Lichtstrahlung in Finland. Societ. Scient. Fennica Commentationes Physico-Mathem. II. No. 11. 1924.

(8) LINDHOLM, F.   Über die Staubtrübung der Atmosphäre 1909 bis 1926, aus Davoser Strahlungsmessungen. Gerlands Beiträge zur Geophysik. Bd XVIII. H. 1/2. 1927.

(9) WESTMAN, J.   Die Verteilung der Insolation in Schweden. Nova acta Reg. Soc. Scient. Upsal. Ser. IV. Vol. 2. N. 7. 1910.

(10) GORCZYNSKI, WL.   Rapport sur les résultats de mission scientifique polonaise au Royaume de Siam et dans les Indes en printemps et en été 1923.

(11) BERGERON, T.   Über die dreidimensional verknüpfende Wetteranalyse. I Teil. Geof. Publ. Vol. V. No 6.

(12) ÅNGSTRÖM, A.   The transportation of dust through the circulation of the atmosphere. Geogr. Annaler. 1930.

THE BLUE SUN OF 1950 SEPTEMBER

R. Wilson

(Communicated by the Astronomer Royal for Scotland)

(Received 1951 June 13)

Summary

This paper deals with the " blue " Sun observed at Edinburgh on 1950 September 26. Spectrophotometric observations of the " blue " and normal Suns give an extinction curve of the layer causing the phenomenon. This covers the wave-length range 3800–6300 A. and shows blueing on the long wave-length side, and reddening on the short wave-length side, of a minimum situated at $\lambda 4350$. This affords an experimental verification of an outstanding feature of scattering theory. It is concluded that the scattering particles producing the minimum are dielectric, and uniform in size and constitution. A relation between the size and refractive index of the dielectric particles is deduced. In addition to the selective effect, there is a heavy non-selective extinction which may arise from another system of absorbing particles in the layer. Other workers have established the source of the layer as the extensive forest fires burning in Alberta on 1950 September 23.

1. *Observational.*—In Edinburgh on the afternoon of 1950 September 26, the Sun was observed to be a deep indigo blue. This remarkable phenomenon was first perceived at about $15^h 00^m$ U.T., when a thinning out of cloud enabled the solar disk to become visible. From then it could be seen through the cloud for considerable periods up to sunset. The phenomenon continued into the night when a "Blue Moon" was also observed. Next morning, the Sun was normal.

During the blueing period three spectrograms of varying density (exposure times, 10, 30 and 180 seconds) were secured of the Sun's centre on a Kodak P.25 panchromatic plate. The 36-inch Cassegrain reflector of the Royal Observatory was used with the two-glass prism train of the attached Hilger universal spectrograph, giving the respective dispersions 132, 45 and 28 A./mm. at $H\alpha$, $H\beta$ and $H\gamma$. The plate was obtained between $15^h 20^m$ and $15^h 25^m$ U.T., at a solar altitude of 19°. The author was extremely fortunate to obtain this plate. It was $15^h 00^m$ U.T. when the phenomenon was observed, and by $15^h 15^m$ U.T. the instrument was ready for use and trained on the Sun. At $15^h 20^m$ U.T. there was a gap in the cloud, and the above plate was taken as an exposure test. At the end of the 15 minutes required for development and fixing, the cloud had again closed and no more exposures were possible. The spectrophotometric procedure normal to this observatory was not therefore carried out. The emulsion was not calibrated on the same plate as the spectrum to be investigated and the plate was not "brush" developed. Through force of habit, however, the author took the temperature of the developer as well as the duration of development, so that the development of separate calibration plates could be standardized as well as possible.

It was now necessary to obtain an equal altitude comparison with the normal Sun. The difficulty immediately arose as to how to bridge the large intensity

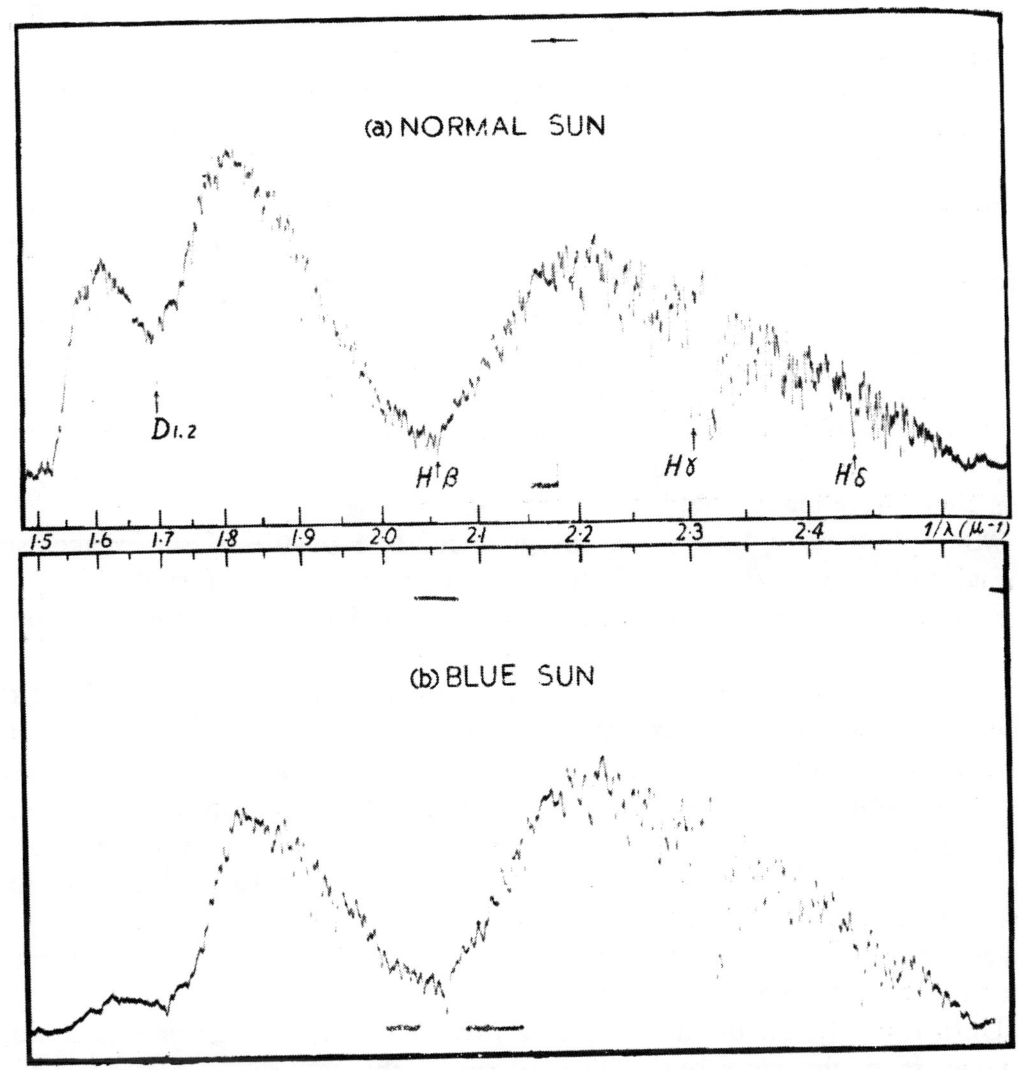

Microphotometer tracings of (a) normal and (b) " blue " solar spectrograms, obtained with 36-inch reflector and attached spectrograph.
(a) Exposure time 30 seconds: aperture $\frac{1}{24}$ square inch.
(b) Exposure time 10 seconds: full aperture.

range of the two sources being compared (a ratio of about 10,000/1). It was decided to stop down the telescope aperture and use a single pencil passing through a mean thickness of the prisms in the spectrograph. This was achieved by means of a cardboard disk into which a hole of the requisite area could be cut. The same three exposure times used for the "blue" Sun spectrograms were adopted, and two sets of normal solar spectrograms of the centre of the disk were obtained, each on a single plate, on the dates 1950 October 26 and 1950 November 15. The respective solar altitudes were $21°$ and $17°$ and the respective apertures used were $\frac{1}{24}$ and $\frac{1}{8}$ square inch. These apertures were adopted to give "good" spectrograms for the respective exposure times 30 seconds and 10 seconds. It is interesting to note that the greater difficulties were experienced in obtaining the normal solar spectrograms. The delay was caused by the scarcity of good cloudless skies and by the accuracy required in the cutting of such small apertures to give good densities on high contrast plates. Unlike the case of the "blue" Sun, there were many failures.

Microphotometer tracings of the 30 seconds exposure of the weaker set of normal Sun spectrograms, and the 10 seconds exposure of the "blue" Sun spectrograms, are reproduced in Plate 8. These indicate clearly the extinction of red light in the case of the "blue" Sun.

The emulsion was calibrated on three separate plates, for each of the three exposure times involved, by two exposures in a multiple slit spectrograph. It was considered desirable to measure the denser spectrograms in order to extend the results into the relatively insensitive region of the Hilger spectrograph beyond $1/\lambda = 2 \cdot 47 \mu^{-1}$ where the instrumental absorption is heavy (see Plate 8). This necessitated a brighter source in the calibration spectrograph, which is normally used for stellar spectroscopy work involving exposures of about one hour. This was accomplished by placing an extra projector lamp in the calibrator source and reducing the diffusing screen to a filter paper.

Unfortunately, all of the six plates involved in the investigation could not be taken from the same box. Two of these (the calibration plates for the 10 seconds and 180 seconds exposure times) had to be taken from a new box of a different batch. A photometric comparison of the two batches showed reasonable agreement, however, as can be seen from Table I. This gives the values of the contrast factor f, determined from two separate plates taken from the different batches, for various values of the function Δ and the wave-length λ, where

$$f = d\Delta/d(2 \cdot 5 \log_{10} I)$$
$$\text{and} \quad \Delta = \log_{10}\{(1 - T)/T\}.$$

I is the light intensity at λ and T the transparency of the negative, defined as the ratio of the galvanometer deflections observed with the microphotometer for the photographic deposit and the clear plate. The values of f for the two batches are paired together, the first figure referring to the earlier batch.

TABLE I

Values of contrast factor f for different values of Δ and λ, for the two batches of plates used in the present work

Δ	$-1 \cdot 0$	$-0 \cdot 5$	$0 \cdot 0$	$0 \cdot 5$	$1 \cdot 0$
$\lambda 6340$	$1 \cdot 57$; …	$1 \cdot 38$; $1 \cdot 35$	$1 \cdot 34$; $1 \cdot 34$	$1 \cdot 57$; $1 \cdot 52$	$2 \cdot 15$; $1 \cdot 86$
$\lambda 5250$	$1 \cdot 49$; …	$1 \cdot 64$; $1 \cdot 49$	$1 \cdot 71$; $1 \cdot 58$	$1 \cdot 76$; $1 \cdot 61$	$1 \cdot 78$; $1 \cdot 64$
$\lambda 4630$	$1 \cdot 54$; $1 \cdot 40$	$1 \cdot 43$; $1 \cdot 36$	$1 \cdot 43$; $1 \cdot 40$	$1 \cdot 51$; $1 \cdot 48$	$1 \cdot 63$; $1 \cdot 60$
$\lambda 4210$	$1 \cdot 42$; $1 \cdot 34$	$1 \cdot 34$; $1 \cdot 24$	$1 \cdot 30$; $1 \cdot 20$	$1 \cdot 31$; $1 \cdot 20$	$1 \cdot 37$; $1 \cdot 25$

It should be noted that the conditions of observation involving the photographing of the spectra of the "blue" and normal Suns on different plates, at different times and with different optical trains, do not favour spectrophotometry of the highest precision. It so happens, however, that the main conclusions of this paper represent very large effects (as will be evident later), and that they cannot be materially affected by the enhanced photometric errors which necessarily arise in an unpredicted "snap" observation, for which no preparation could be made.

The solar and calibration spectrograms were photometered at a number of wave-lengths chosen to avoid deep lines in the solar spectrum. A non-recording microphotometer of Moll type was used with an analysing slit of $25\,\mu$, and each measurement was the mean of ten readings $25\,\mu$ apart. This instrument was used in preference to a recording microphotometer, because of its greater flexibility in measuring high densities. Wave-lengths were determined from an identification of solar lines.

The normal and "blue" solar spectrograms with the same exposure times were paired together, and the measurements reduced to the quantity

$$y = 2 \cdot 5 \log_{10}(cI_N/I_B),$$

where I_N and I_B are the respective intensities at wave-length λ of the normal and "blue" Suns. The constant c is the ratio of the apertures used for the "blue" and normal Suns. The factor $2 \cdot 5$ gives a "magnitude scale" which possesses the convenient property that one unit in the second place of decimals is reasonably related to the observational error. The results are given in Table II. Column 1 gives the wave-length in angstroms. Column 2 gives the reciprocal of the wave-length in micron^{-1}. Columns 3, 4 and 5 give the respective values of $2 \cdot 5 \log_{10}(c_1 I_N/I_B)$ for the exposure times 10, 30 and 180 seconds, determined from the combination of the "blue" solar spectra with the denser set of normal solar spectra. Column 6 gives the values of $2 \cdot 5 \log_{10}(c_2 I_N/I_B)$ determined from the combination of the 30 second "blue" solar spectrum with the 30 second exposure of the weaker set of normal solar spectra. These represent all the measurements possible. There are six possible pairs of spectra, but only these four were measurable in each case at corresponding wave-lengths. The constants c_1 and c_2 were determined from the values of the telescope aperture used in each case and given above, and the results combined in the form $2 \cdot 5 \log_{10}(I_N/I_B)$ in column 8.

A comparison of columns 3 and 6 gives an idea of the accuracy of the results. These two sets of values are determined from different "blue" and normal solar spectra and also different calibration plates. Further, one of the calibration plates (used for the values in column 3) was taken from a separate batch, as was mentioned above, and we have a further check on the agreement of the two batches. Since there is a systematic variation in the differences of the values in these two columns, the random error derived from them will have little significance and the variation is expressed in column 7 in terms of α, where α represents the departure of the difference of columns 3–6 from the mean. The agreement shown by the figures is regarded as satisfactory.

The combined results of column 8, representing the extinction of the absorbing layer causing the "blue" Sun, are plotted against $1/\lambda$ (μ^{-1}) in Fig. 1. This shows blueing on the long wave-length side, and reddening on the short wave-length side, of a minimum situated at $1/\lambda = 2 \cdot 3\,\mu^{-1}$ $(\lambda\,4350)$.

It must be remembered that this curve represents a combination of what are actually different extinction curves, since they are obtained from separate "blue" and normal solar spectrograms, taken at different times. Hence there is the possibility of varying conditions, e. g. a variation of the thickness of the absorbing layer in the case of the "blue" Sun, or a variation of atmospheric conditions in either case. The essential nature of the curve will, however, remain unchanged.

TABLE II
Observed extinction values of scattering layer

λ (A.)	$\dfrac{1}{\lambda}$ (μ^{-1})	$\dfrac{5}{2}\log_{10}\left(c_1\dfrac{I_N}{I_B}\right)$			$\dfrac{5}{2}\log_{10}\left(c_2\dfrac{I_N}{I_B}\right)$	α	$\dfrac{5}{2}\log_{10}\dfrac{I_N}{I_B}$
(1)	(2)	(3)	(4)	(5)	(6)	(7)	(8)
6338	1·578				0·03		10·92
6202	1·612				$\bar{1}$·96		10·85
6042	1·655				$\bar{1}$·88		10·77
5967	1·676	1·07			$\bar{1}$·87	−0·01	10·76
5803	1·723	0·93			$\bar{1}$·77	−0·05	10·64
5642	1·772	0·82			$\bar{1}$·68	−0·07	10·54
5535	1·807	0·74			$\bar{1}$·67	−0·14	10·50
5501	1·818	0·73			$\bar{1}$·66	−0·14	10·48
5382	1·858	0·73			$\bar{1}$·52	0·00	10·42
5246	1·906	0·66			$\bar{1}$·36	+0·09	10·30
5070	1·972	0·53			$\bar{1}$·28	+0·04	10·19
4900	2·041	0·46			$\bar{1}$·20	+0·05	10·12
4835	2·068	0·43			$\bar{1}$·16	+0·06	10·09
4744	2·108	0·41			$\bar{1}$·15	+0·05	10·07
4627	2·161	0·38			$\bar{1}$·12	+0·05	10·04
4550	2·198	0·35			$\bar{1}$·11	+0·03	10·02
4428	2·258	0·33			$\bar{1}$·09	+0·03	10·00
4370	2·289	0·32			$\bar{1}$·09	+0·02	9·99
4332	2·308	0·32			$\bar{1}$·09	+0·02	9·99
4206	2·378	0·33					10·00
4110	2·433	0·40	0·48				10·07
4050	2·469	0·42	0·52				10·09
4028	2·483		0·54				10·12
3981	2·512		0·61				10·19
3955	2·529		0·67				10·25
3951	2·531		0·68				10·26
3916	2·554		0·73	0·83			10·31
3893	2·569			0·97			10·45
3865	2·587			1·09			10·57
3845	2·601			1·16			10·64
3811	2·624			1·29			10·77

The points obtained from the extinction values given in columns 3 and 6 are denoted by full circles. These values can be combined, since they have been shown to have a reasonable correspondence by the values of α in column 7. The points obtained from the extinction values given in columns 4 and 5 are denoted by open circles and crosses respectively. These must necessarily be given less weight than the full circles. They cover a spectral region that was never meant to be utilized in the instrument used. The instrumental absorption is becoming heavy, as can be seen from the microphotometer tracings in Plate 8, and consequently the gradient of photometric density is large; the spectrum is no longer in sharp focus and is in close proximity to the edge of the plate holder; the spectrum is over-exposed in the visible and hence the errors due to scattered light are magnified.   Further, there is no check on the correspondence of these points with the full circles, and atmospheric variations are known to be violent in this region.   The value of these two series of points lies, however, in the circumstance that each series indicates that the extinction is increasing with decreasing wave-length in the violet and near ultra-violet, thereby confirming the existence of the extinction minimum at $1/\lambda = 2 \cdot 3\,\mu^{-1}$.

The full curve in Fig. 1 represents a theoretically determined curve, the computation of which is explained later in the text.

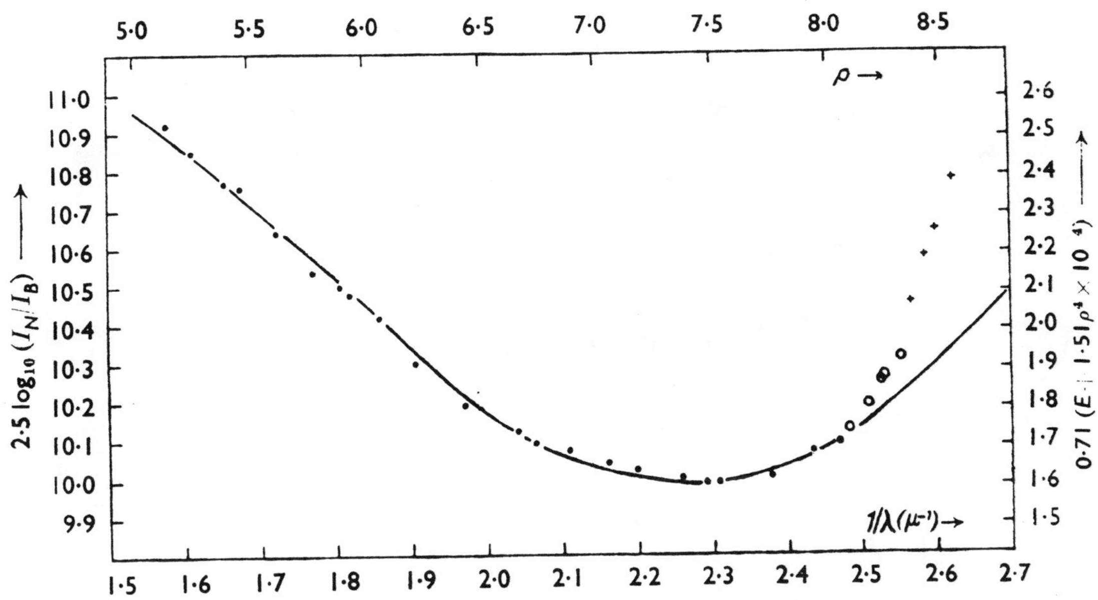

FIG. 1.—*Plot of observed extinction values against* $1/\lambda$.
Full curve represents theoretically determined extinction plotted against ρ.

2. *Resumé of theory.*—The problem of light extinction by a system of particles of known dimensions and electric and magnetic properties can in principle be solved on the basis of Maxwell's electromagnetic light theory.   Due to the great labour involved, however, the computation of extinction curves has mostly been confined to the following case—scattering of a plane wave by a single isotropic sphere, the complete analytical solution of which was given by Mie (**1**).   To apply these results to a cloud of isotropic spheres we have the added conditions that the source and observer must be at a large distance from the cloud, and the particles at a low enough density not to cause any interference with each other. If there is a dispersion in size and constitution of the individual isotropic spheres, the resulting extinction curve may be found by taking the mean of the various extinction curves for every type of particle and weighting with respect to population.

The results depend upon the two values r, the radius of the sphere and m, its complex refractive index relative to the surrounding medium.

$$m = \sqrt{(\epsilon - i\sigma/\omega)},$$

where ϵ is the dielectric constant and σ the conductivity of the material, i the imaginary unit and $\omega = 2\pi c/\lambda$, c being the velocity of, and λ the wave-length of the incident light in the surrounding medium.   The magnetic permeability is assumed

to be unity.   The extinction E is defined as the ratio of scattering to geometrical
cross-section i.e. for a single spherical particle, E is the ratio of the scattered
light to the light incident on the sphere. E is usually determined in terms of
m and of the parameter $x = 2\pi r/\lambda$.

The computations are divided into three groups (a) m real; (b) $m \to \infty$;
(c) m complex.   Group (a) has received most attention and three curves are
reproduced in Fig. 2. E is plotted against the parameter $\rho = 2x(m-1)$, which

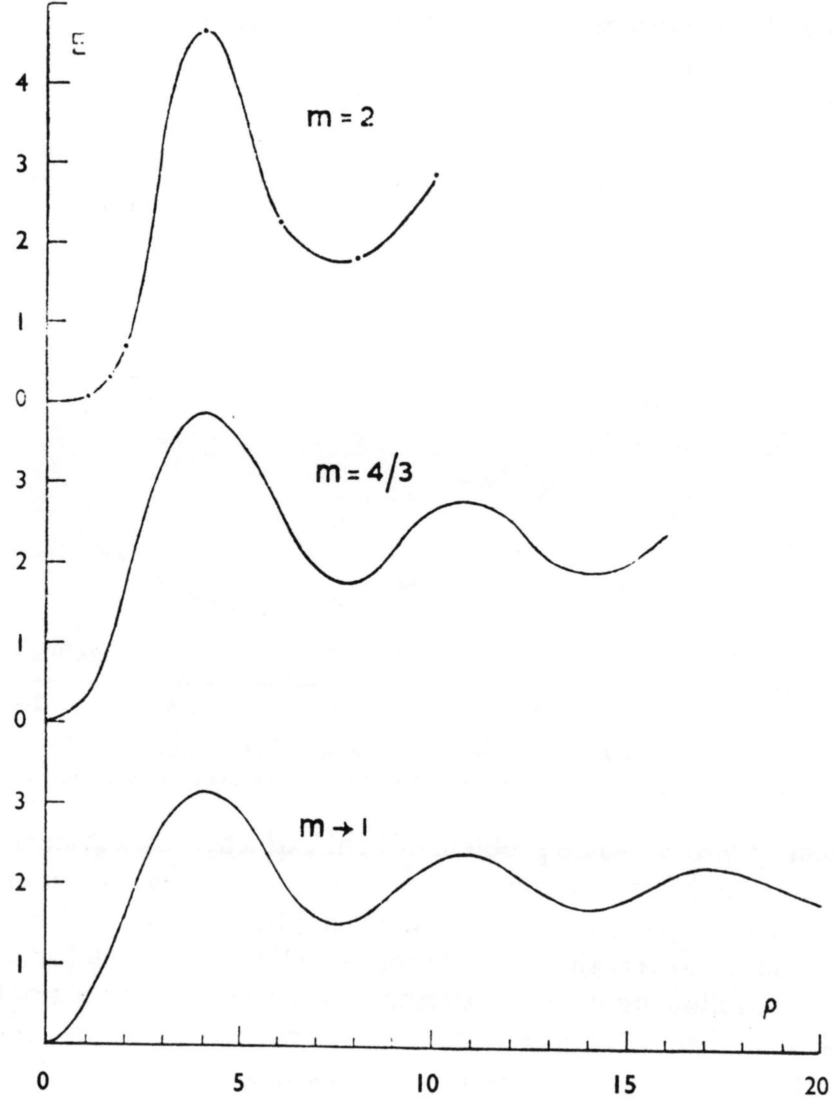

FIG. 2.—*Theoretical extinction curves for values of m real.*
E plotted against $\rho = 4\pi r(m-1)/\lambda$.
$m \to 1$: *Analytical formula by Van de Hulst.*
$m = \frac{4}{3}$: *Smoothed curve of computations by Houghton and Chalker.*
$m = 2$: *Computations by Greenstein.*

was shown by Van de Hulst (**2**) to be of more basic significance than **x**.   It
represents the phase shift suffered by a ray of light traversing the sphere along a
diameter.   Using this parameter as abscissa, the extinction curves for various **m**
in the neighbourhood of unity are very similar.   This is shown by Fig. 2 which is
based on Fig. 21 of Part I of Van de Hulst's papers.   The curve for $m \to 1$ is
obtained from the formula

$$E = 2 - 4 \sin \rho/\rho + 4(1 - \cos \rho)/\rho^2$$

derived by Van de Hulst. This formula is for the actual limiting case $m \to 1$. For values of m near unity the formula gives a good representation of E except near the origin for those values of ρ small compared to $(m-1)$. For small values of x, Van de Hulst's formula gives $E = 2(m-1)^2 x^2$, whereas the actual expression for E is of course the Rayleigh formula $E = 32(m-1)^2 x^4/27$.

The curve for $m = \frac{4}{3}$ is the smoothed curve obtained from the extensive computations of Houghton and Chalker (3). The unsmoothed curve shows numerous minor fluctuations in the vicinity of the major maxima and minima, and it would require only a small dispersion in r to smooth these out. The computed points for the third curve ($m = 2$) are due to Greenstein (4). These are only sufficient, however, to give a smoothed version of the curve which preserves the major features. The more extensive computations of La Mer show minor fluctuations. The actual calculations of La Mer were not available to the author, but Van de Hulst plots them graphically in Fig. 3 of Part 2 of his papers. The curves start at $E = 0$ for $\rho = 0$, rise to a first maximum at $\rho \simeq 4$, reach a first minimum at $\rho \simeq 7.5$ and then go through a series of diminishing undulations, which asymptotically approach the value $E = 2$.

The region for $m > 2$ is almost devoid of practical interest and hence few investigations have been carried out. The form of the extinction curve is therefore uncertain. As m increases above 2, the maxima depicted in Fig. 2 will be crowded towards the origin when x (not ρ) is used as abscissa. Moreover, Van de Hulst (2) has shown that an additional series of "resonance" maxima will develop for large m and small x. These approach the positions $x = k\pi/m$ (where k is an integer) and the values $E = \frac{1}{2}$, but the interrelation of these with the "interference" maxima, referred to above, is obscure. Van de Hulst, in his examination of the extinction curves obtained for a number of values of $m < 2$ from the numerous calculations of La Mer, has pointed out the small "bump" in the initial upward slope of the curves. In the case of $m = 2$, this is a definite subsidiary maximum. He shows that these "bumps" correspond to the position $x = \pi/m$ and suggests that they are related to the resonance maxima. It does indeed seem possible that they represent the development of the first resonance peak.

For the limiting case of group (b) $m \to \infty$, computations have been made by Götz (5) and Greenstein (4). This is the case of totally reflecting spheres. The curve based on the combined calculations of Götz and Greenstein is of a similar nature to those for m real and less than 2. It shows a rise to a first maximum and then undulates as it asymptotically approaches $E = 2$. The undulations are very much reduced, however, the first maximum having a value $E = 2.29$ at $x = 1.2$ and the first minimum having $E = 2.12$ at $x = 1.6$. Again, the relation of this region with that for m real and large (but finite) is obscure. The resonance maxima at $x = k\pi/m$ for large m and small x must still exist, but as m increases they all run together at $x = 0$ and disappear in the limiting case $m \to \infty$. This is a case of non-uniform convergence, the curves for m large converging non-uniformly to the limiting case $m \to \infty$.

For group (c) m complex, Van de Hulst has investigated analytically the case $m \to 1$ along the same lines as his investigations for dielectric spheres. He shows that the extinction is similar to the dielectric case on a ρ basis, but only a small conductivity is required to smooth out the undulations entirely. As the conductivity increases further, the first maximum also disappears and the curve rises asymptotically from the origin to the value $E = 2$.

3. *Comparison of observational material with theory.*—Returning to the present problem, it is reasonable to place the "blue" Sun extinction in the class *m* real. An extinction minimum can be produced by conducting particles, if the conductivity is small (as was indicated above), but the minimum will be much more shallow than for dielectric particles. It is therefore proposed that the observed extinction curve can be represented by curves of the nature depicted in Fig. 2. For the superposition of the extinction curves of the individual dielectric particles in the absorbing layer to produce a minimum, we must have small dispersion in the quantity ρ, i.e. small dispersion of $r(m-1)$, otherwise the various maxima and minima of the individual curves would overlap and the undulations be smoothed out. The only physical reason for $\delta\{r(m-1)\}$ being small is of course that δr and δm are small, i.e. the dielectric particles are uniform in size and constitution. An upper limit of the dispersion, assuming a uniform distribution, is given by the variation required to superimpose the first maximum and first minimum, smaller dispersions being required to superimpose any other consecutive maxima and minima. The respective values of ρ for the first maximum and minimum are 4·1 and 7·6 and hence the upper limit of dispersion is given by

$$\Delta\rho/\rho_{\text{mean}}=0\cdot6.$$

This gives the upper limit as 60 per cent for $r(m-1)$, or assuming m to be constant, 60 per cent for r. The actual dispersion is probably much less, the minimum becoming very shallow as the dispersion approaches 60 per cent.

The observed minimum must represent the first minimum of the theoretical curve, since any other minimum would require the preceding maximum to be within the observed range. But a close comparison of the observed and theoretical curves did not show agreement. The slope of the observed curve on the short wave-length side of the minimum is steeper than would be expected from theory. Such an effect would be caused if, in addition to the uniform dielectric particle extinction, there was an overall reddening. Such an effect is to be expected in the comparison of the atmosphere at different times, the variation of normal atmospheric reddening having been shown (**6**) to be considerable.

There is also the possibility of instrumental colour effects. The effect of using different optical trains was examined in the laboratory. Since by far the greatest thickness of glass that the light passes through in the spectrograph is due to the two prisms, the effect of these alone was investigated. Light was passed through varying thicknesses of one of these prisms and the distribution of intensity throughout the spectrum measured by means of an Evans blocking layer photo-cell and galvanometer. From these measurements the absorption coefficient of the glass was determined at a number of wave-lengths. The absorption of a single pencil passing through a mean thickness of prism (the normal Sun) was then determined, and the absorption of a beam using the full aperture ("blue" Sun) computed. A comparison of the two showed agreement up to $1/\lambda=2\cdot3\,\mu^{-1}$, beyond which the normal Sun is reddened with respect to the "blue" Sun. Hence the correction for this effect will increase the discrepancy in the observed curve. The correction a to be applied to the slope $dy/d(1/\lambda)$ of the observed curve, is given in the following table:

$1/\lambda(\mu^{-1})$	1·6–2·3	2·4	2·5	2·6
a	0·0	+0·3	+0·6	+0·9

The normal solar spectrograms were taken at later dates, at intervals of 30 and 50 days after the "blue" Sun. There is, therefore, the possibility of a change in colour of the instrument. This would be caused mainly by a deterioration of the silver coat on the primary mirror (the Cassegrain mirror of the Edinburgh reflector is aluminium coated), which has been shown by Greaves, Davidson and Martin (6) to redden with time. These workers, in an exhaustive survey of such effects, give a reddening of $+0.2$ in terms of $dy/d(1/\lambda)$ for 30 days, this figure applying to a wave-length range 4400 A. to 6500 A. and to an optical train involving four silvered surfaces. Again, this will aggravate the distortion of the observed curve, which would require an instrumental blueing for explanation.

The possibility of vignetting can be ruled out. All the rays over the wavelength range examined enter the camera lens.

The instrumental reddening will cause an apparent blueing of the uniform dielectric particle extinction, and hence the atmospheric reddening of the "blue" Sun relative to the normal Sun must have been large enough to overcome this. Assuming an effective overall reddening, and that this can be represented by a $(1/\lambda)^n$ law, we have, in general,

$$y = 2.5 \log_{10}(I_N/I_B) = 2.5 (\log_{10} e)\pi\Sigma(r^2 NE) + p(1/\lambda)^n,$$

where N is the total number of scattering particles, of radius r, in a beam of unit cross-section and E is the corresponding extinction.

With no dispersion in r and since $\rho = 4\pi r(m-1)/\lambda$, the above equation becomes

$$y = k(E + q\rho^n),$$

where $\rho = h/\lambda$ and k, q and h are constants. It is now required to choose values of k, q, h and n, so as to fit the observed curve. The values of E are taken from the smoothed curve drawn through the computations of Houghton and Chalker for $m = \frac{4}{3}$. This curve is given in Fig. 2. The actual value of m for the scattering particles is unknown, but the similarity of the extinction curves for $m \to 1$ to $m = 2$ has been pointed out by Van de Hulst and is illustrated in Fig. 2. Accordingly, the curve for $m = \frac{4}{3}$ can be taken as roughly representing the extinction curve for the unknown m, when plotted on a ρ basis. In particular, the value of ρ, corresponding to the first minimum, varies little for different values of m, and hence the value of the constant h, obtained from fitting the observations to theory for $m = \frac{4}{3}$, should be closely correct (see below).

Values of y were computed and, by trial and error, the best fit was obtained firstly, for all the observations, and secondly, neglecting the observations beyond $1/\lambda = 2.47\,\mu^{-1}$, which are of doubtful accuracy (see Section 1 above). The computations were confined to $n = 1$ and $n = 4$. No reasonable fit was possible in the first case (all observations), but in the second case a good fit was obtained (apart from a difference of zero on which comment will be made below) for $k = 0.71$, $q = 1.51 \times 10^{-4}$, $h = 3.26$ and $n = 4$. This theoretical curve

$$y = 0.71(E + 1.51\rho^4 \times 10^{-4})$$

is plotted in Fig. 1 as the full curve.

The real overall reddening is easily explainable by normal atmospheric variation. The departure of the doubtful points from the theoretical curve seems great, even allowing for the experimental difficulties, but extra atmospheric reddening (the points are obtained from spectra taken at different times) at this altitude ($19°$), and in this spectral region, can become quite violent.

The value of ρ for the first minimum of the smoothed version of the calculations by Houghton and Chalker is found to be 7·6. The minimum of the reddened theoretical curve, obtained to fit theory and observation, is shifted to $\rho = 7·5$, a shift of less than two per cent. Having regard to the general similarity of the curves for different m in the range $m \to 1$ to $m = 2$, it is reasonable to suppose that a similar shift (of the order of two per cent) would be found by using other values of m. Now the value $h = 3·26$, quoted above, was obtained from the consideration that the value of $\rho(7·5)$, for the first minimum of $E + q\rho^n$, corresponds to $1/\lambda = 2·3\,\mu^{-1}$ for the minimum of the observed curve. It would therefore appear that the value $h = 3·26$ is correct to a few per cent. But since $\rho = 4\pi r(m-1)/\lambda$, then

$$h = 2\pi d(m-1),$$

where d is the diameter of the scattering particles. Substituting for h gives the relation between the diameter and refractive index of the scattering particles to be

$$d(m-1) = 0·519 \text{ microns.}$$

Values of d for various m are given in the following table:

m	1·10	1·25	1·33	1·50	1·75	2·00	3·00
$d(\mu)$	5·19	2·08	1·57	1·04	0·69	0·52	0·26

The value of k obtained above gives

$$k = 2·5\,(\log_{10} e)\pi r^2 N_0 = 0·71.$$

It is pointed out later that m is probably in the vicinity of 1·5 and therefore r is probably of the order of $0·5\,\mu$. Using this, the order of value of the total number of scattering particles N_0 in a beam of unit cross-section, is determined as

$$N_0 \simeq 8 \times 10^7 \text{ cm.}^{-2}.$$

This value is for an altitude 19°. Hence, if n_0 is the average density of the scattering particles and S the thickness of the layer,

$$N_0 = n_0 S \operatorname{cosec} 19°.$$

The thickness of the layer was determined to be 12,000 feet by an aircraft sent from Leuchars. This gives the order of density n_0 of the scattering particles to be

$$n_0 \simeq 70 \text{ cm.}^{-3}.$$

It will be seen from Fig. 1 that there is a considerable difference of zero, amounting to 8·4 magnitudes (a light intensity ratio of 2300/1), between the observed y and the theoretical $y = 0·71\,(E + 1·51\rho^4 \times 10^{-4})$. This indicates a heavy non-selective obscuration in addition to the selective extinction. Such a neutral obscuration, over the observed wave-length range, could be caused by particles of large size (i.e. large ρ) or by heavily absorbing (i.e. conducting) particles of smaller size (see Section 2). Part or possibly all of the effect may have been caused by cloud (water droplets with large ρ) at the time of the "blue" Sun for, although the "blue" solar spectrograms were obtained through a gap in low-lying cloud, cloud at a higher level may have produced obscuration. On the other hand, it is possible that part of the non-selective obscuration may be localized in the layer itself. The experimental evidence in favour of this interesting possibility is weak. It rests solely on the large value of obscuration (8·4 magnitudes) to be explained and the visual observation of the author that, at the time of observation, the "blue" Sun seemed clear of cloud.

4. *General discussion.*—To summarize the conclusions briefly, it is found that the layer causing the "blue" Sun has an extinction minimum, thus affording an experimental verification of an outstanding feature predicted by scattering theory. The results are shown to agree with theory on the assumption of an overall reddening. The scattering particles producing the minimum are dielectric in nature and remarkably homogeneous in size and constitution. A relation between the diameter and refractive index of the particles is deduced, and an order of their density is given. The possibility is also indicated that, in conjunction with the uniform dielectric particles, there is another system of particles in the layer producing a neutral extinction.

The source of the layer has been determined by Bull (**7**) as the extensive forest fires burning in Alberta on 1950 September 23. He states that the smoke from these fires reached the large cities of eastern Canada on September 24, and was thick enough there to blot out the Sun. The upper winds over the Atlantic were then favourable for bringing the smoke to Europe. A trajectory starting from the Toronto area on September 24, calculated from synoptic charts and including the upper wind observations of the Ocean Weather ships, would bring the smoke to Scotland on September 26. The smoke then travelled over to the Continent where the effects were seen until September 30.

Over North America the effect of the smoke was quite different owing to its great density. The fires occurred in muskeg country and are characterized by tremendous volumes of smoke. Elsley (**8**) gives an account of the visual effects and shows from wind observations that the smoke came from the fires. He states that at times there was complete darkness and the Sun, when visible, was purple or blue. He describes the colours of the sky (i.e. the scattered light) as shades of orange, pink and yellow. He also states that aircraft flying through the smoke were found to be covered with an oily substance on landing. Unfortunately, however, no one seems to have taken some of the oil into the laboratory for analysis.

The smoke was also observed by Dr H. S. Hogg (**9**) from southern Ontario, 2000 miles from the fires, on September 24 and 25. She states that the Sun could be seen at periods on both days, and describes it as a "pale, bluish-mauve disc".

Over Edinburgh the smoke was observed on the afternoon of September 26, when the Sun could be seen quite clearly through the haze and was a deep indigo blue. The extent of the layer was determined by an aircraft sent from Leuchars, which localized the smoke between 31,000 and 43,000 feet.

It seems, in the light of Elsley's (**8**) information, that the scattering particles are globules of oil, caused by the distillation of wood. Their uniform character is yet to be explained, and may have been established by combustion in the fire itself or by an atmospheric filtering process favouring particles of one size. As regards size, most organic oils have a refractive index near $m = 1 \cdot 5$ which would correspond to a diameter of the order of one micron. If, as has been indicated in Section 3, there is another system of particles in the scattering layer, producing a non-selective obscuration, their most likely constitution is carbon. This material is highly absorbent in the visible spectral range, and hence the extinction of such particles can be represented by a curve which rises asymptotically from the origin to the value $E = 2$, the extinction becoming practically neutral in the region of $\rho \simeq 4$ (see Section 2). The interesting possibility therefore arises that they can be of the same size as the uniform dielectric particles.

The occurrence of this "blue" Sun is not unique. There have been other cases, usually connected with dust storms in a desert, or a volcanic eruption. The most famous case is the eruption of Krakatoa in 1883, the effects of which were observed all over the globe.

The author would like to express his thanks to Dr E. A. Baker and Dr M. A. Ellison for their advice and encouragement, and also his gratitude to Professor W. M. H. Greaves for the very many suggestions and criticisms he made throughout the whole of this investigation.

Royal Observatory,
Edinburgh :
1951 *May* 28.

References

(1) G. Mie, *Ann. d. Physik*, **25,** 377, 1908.
(2) H. C. van de Hulst, *Recherches Astronomiques de l'Observatoire d'Utrecht*, XI, parts 1 and 2, 1946, 1949.
(3) H. G. Houghton and W. R. Chalker, *J. Opt. Soc. America*, **39,** 955, 1949.
(4) J. L. Greenstein, *Harvard Circular*, No. 422, 1937.
(5) F. W. P. Götz, *Astr. Nachr.*, **255,** 63, 1935.
(6) W. M. H. Greaves, C. Davidson and E. Martin, *M.N.*, **94,** 488, 1934.
(7) G. A. Bull, *Met. Mag.*, **80,** 1, 1951.
(8) E. M. Elsley, *Weather*, **6,** 22, 1951.
(9) H. S. Hogg, *J.R.A.S. Canada*, **44,** 241, 1950.

Section Ten
Visibility

Theorie der horizontalen Sichtweite[1]).

Von Harald Koschmieder, Frankfurt a. M.

Dem Physikalischen Verein zur Jahrhundertfeier gewidmet.

Inhalt: § 1. Definitionen, Problem und Geschichtliches.
  § 2. Analytische Darstellung der Helligkeit H_s des schwarzen Körpers.
  § 3. Auswertung des Helligkeitsintegrals mit Hilfe der beobachteten Himmelshelligkeit i_g.
  § 4. Sichtweite eines schwarzen, weißen, farbigen Schirmes vor dem Horizont.
  § 5. Gültigkeitsgrenzen der entwickelten Beziehungen.
  Zusammenfassung.

§ 1. Definitionen, Problem und Geschichtliches.

Vorausgeschickt seien folgende Definitionen:

A) Die Sichtweite s sei diejenige Entfernung, in der der Kontrast K eines bestimmten Zieles Z gegen seine Umgebung U unter einen kleinen Wert ε (Reizschwelle) sinkt.

B) Die „in Richtung auf ein bestimmtes Ziel meßbare Helligkeit" H sei diejenige Licht-(Energie-)menge, die auf die zur Richtung Ziel—Auge normale Flächeneinheit am Beobachtungsort senkrecht zur Flächeneinheit einfällt.

C) Der Helligkeitskontrast K eines bestimmten Zieles gegen seine Umgebung sei gegeben durch

$$K(Z, U) = \frac{H_u - H_z}{H_u}.$$

Der Kontrast hängt aber auch von dem Farbunterschied von Ziel und Umgebung ab. Demzufolge werden später drei wesentliche Fälle behandelt werden: die Sichtweite a) des schwarzen Körpers, b) der homogen und diffus reflektierenden Lambert'schen Fläche von der Albedo $\mathfrak{A} = 1$, c) der selektiv und diffus reflektierenden Lambert'schen Fläche von der Albedo $\mathfrak{A} = \mathfrak{A}(\lambda)$. Diese drei Spezialfälle erschöpfen wesentlich das Problem. Dabei betrachten wir als Umgebung stets den Horizont, weil dieser Fall in Praxi zumeist der gegebene ist.

Am einfachsten liegen die Verhältnisse bei dem schwarzen Körper. Diese führte ich als Grundlage meiner Untersuchung ein, ohne Kenntnis einer später zu zitierenden Abhandlung von L. Weber. In der Sehstrahlenpyramide, die

---

1) Habilitationsschrift Frankfurt a. M., Oktober 1924.
Beiträge zur Physik der freien Atmosphäre. XII.

von dem Auge und der Zielfläche gebildet wird, werden Lichtmengen, die in einer beliebigen Richtung in diese Pyramide einfallen, partiell in das Auge gestreut und zwar aus drei Gründen: a) infolge Reflexion, Beugung und Brechung an Wassertröpfchen, b) infolge Reflexion und Beugung an Staubteilchen, c) infolge der Rayleigh'schen Zerstreuung an den Luftmolekeln selbst. Die Größe der Streuung wird abhängen von dem Winkel φ_1, den der einfallende Strahl mit der Richtung Ziel—Auge bildet. Diese Abhängigkeit, die für mittlere Verhältnisse Chr. W i e n e r [1] zu bestimmen versucht hat, sei durch eine Zerstreuungsfunktion β gegeben. E s w i r d j e d o c h e i n e M e t h o d e e n t w i c k e l t w e r d e n, d i e d i e o b e n g e g e b e n e n P r o b l e m e o h n e K e n n t n i s d i e s e r Z e r s t r e u u n g s f u n k t i o n z u l ö s e n g e s t a t t e t. Infolge der Existenz dieser Zerstreuungsfunktion, die schon aus der Rayleigh'schen Theorie folgt, wird die Entfernung s, in welcher die aus der Sehstrahlpyramide ins Auge gestreute Lichtmenge eine der Horizonthelligkeit hinreichend gleiche Helligkeit erzeugt, d. h. die Sichtweite, eine Funktion des Winkels A sein, den die Richtung Auge—Ziel mit dem Sonnenmeridian bildet. Es wird daher zu den obengestellten Problemen noch die Aufgabe hinzutreten, s als $s(A)$ darzustellen.

Mit dieser Auffassung des Sichtproblems stehen in Übereinstimmung die Arbeiten von L. Webers Schüler G. H a e c k e r [2] und von L. W e b e r [3] selbst. G. Haecker definierte die Sichtweite als diejenige Entfernung, in der der Kontrast zwischen einer schwarzen und weißen Fläche verschwindet, und L. Weber als diejenige Entfernung, in der die in Richtung auf den schwarzen Körper

$$\text{meßbare Helligkeit gleich der räumlichen Beleuchtungsstärke } (B) = \int_0^{4\pi} H d\omega$$

wird, wo H die Himmelshelligkeit bzw. das Unterlicht ist. Durch diese Definition entfällt aber der interessante Teil des Problemes, $s = s(A)$ darzustellen, der auch für die Praxis von Bedeutung ist. Es ist eben notwendig, den Horizont als Umgebung einzuführen. Doch hat L. Weber den Kernpunkt des Problems mit aller Schärfe erkannt und die Aufhellung des Zwischenmediums auch photometrisch gemessen. Hieraus hat er dann, wie er selbst angibt, unter stark vereinfachenden Annahmen die Sichtweite berechnet, wobei er zu zufriedenstellender Übereinstimmung zwischen Rechnung und Beobachtung gekommen ist. Seine Beobachtungen beziehen sich aber nur auf einen gleichmäßigen grauen Himmel. Ebenso wie von L. Weber ist auch neuerdings von A. S c h m a u ß [4] und von R. D i e t z i u s [5] wieder betont worden, daß die Sichtweite im wesentlichen n i c h t bedingt ist durch die Extinktion der vom Ziel nach dem Auge gerichteten Lichtstrahlen, sondern durch die Aufhellung der zwischen Ziel und Auge befindlichen Luftmassen.

Einen anderen Weg, dem ich jedoch nicht zu folgen vermag, gehen H. D e m b e r und M. U i b e [6]. Bei dem Versuch, die experimentell gefundene Proportionalität

---

1) Nova Acta der Kaiserl. Leop. Carol. Akad. d. Naturf., LXXIII, Nr. 1, Halle 1900.
2) Diss. Kiel 1905.
3) Ann. d. Physik, **51**, S. 427—449, 1916.
4) Met. Zeitschr., **35**, S, 185, 1918; ebd. **38**, S. 225, 1921.
5) Beitr. z. Physik d. fr. Atmosphäre X, S. 202, 1922.
6) Sitzungsberichte der Math.-Phys. Kl. der Kgl. Sächs. Ges. d. Wiss. zu Leipzig, Bd. LXIX, S. 391—411, 1917.

zwischen der Entfernung Beobachter—Himmelsgewölbe und $\sqrt{H}$ ($H =$ Himmelshelligkeit, l. c. § 3) auch theoretisch zu belegen (§ 2), gehen H. Dember und M. Uibe von der Annahme aus, daß die Luftmasse für jede beliebige Zenitdistanz der Sehrichtung „gleichmäßig hell" sei, d. h. daß der auf einer Strecke s durch Zerstreuung verlorene Anteil durch den infolge Zerstreuung gewonnenen gerade kompensiert wird, sodaß einfach gilt $J_s = s^{-2} \cdot J_1$, wenn J_1 die Intensität des von einer Molekel zerstreuten Lichtes bedeutet. Wird s so groß, daß J_s gerade noch hinreicht, um im Auge einen Lichteindruck hervorzurufen, so nennen die genannten Autoren s die „maximale Sichtweite". Mit dieser „maximalen Sichtweite", die man vielleicht als maximale „Empfindungsweite" bezeichnen könnte, identifizieren sie dann (§ 6) die Entfernung, unter der z. B. ein Pilotballon verschwindet! Selbst unter der Annahme, daß geometrisch betrachtet bei den l. c. angegebenen großen Entfernungen des Piloten (fast 40 km) der Pilot als punktförmig betrachtet werden kann, so ist das von dem Piloten nach dem Auge gelangende Licht doch durch selektive Reflexion entstanden, also in ganz anderer Weise von der Beleuchtungsrichtung und Beleuchtungsstärke abhängig, als die durch Rayleigh'sche Zerstreuung entstehenden Intensitäten.

§ 2. Analytische Darstellung der Helligkeit H_s des schwarzen Körpers.

Es läge nahe, hier einen Ansatz von R. Mecke[1] zu verfolgen. Doch führt er nicht zu einer Verknüpfung von Himmelshelligkeit und Helligkeit vor dem schwarzen Körper, eine Beziehung, die für das Problem der Sichtweite gerade fruchtbar ist, und zu einfachen Resultaten führen wird.

Die Rechnung wird vielmehr im engen Anschluß an die grundlegende Methode von Chr. Wiener[2] durchgeführt werden, die Wiener benutzte, um die Himmelshelligkeit zu berechnen. Im vorliegenden Problem sind jedoch noch zwei Integrationen mehr auszuführen, eine über die Einheitskugel und eine über den Raum der Sehstrahlpyramide. Die analytische Darstellung wird gleichzeitig auch die oben skizzierten Strahlenwege am klarsten beschreiben.

Zur Vereinfachung des Problems machen wir eine Reihe von Annahmen, die stets gemacht werden und die P. Gruner[3] klargelegt hat.

a) Die Atmosphäre sei als trübes Mittel aufzufassen.

b) Jedes Raumelement enthalte eine überaus große Zahl von Körperchen, die von höherer Ordnung kleiner sind als das betrachtete Raumelement.

c) Jedes Körperchen übe auf die darauffallende Strahlung eine zerstreuende Wirkung aus, als ob es allein in dem Raum vorhanden wäre.

d) Das von einem Volumelement ausgehende zerstreute Licht werde als von einer punktförmigen Lichtquelle ausgehend betrachtet, deren Intensität nach c) der Zahl der trübenden Teilchen proportional ist.

e) Die Lichtstrahlen seien stets als geradlinig angenommen.

f) Es sei ein wolkenloser Himmel vorausgesetzt.

Über Gruner hinausgehend machen wir noch drei weitere Annahmen:

---

1) Annalen der Physik, **65**, S. 257—273, 1921.

2) l. c. XCI, Nr. 2, Halle 1909. Den ersten Hinweis auf die „Knickmethode" verdanke ich jedoch Herrn Prof. Madelung.

3) Beitr. z. Physik d. fr. Atmosphäre VIII, S. 122—155. 1919.

g) In Richtung der Horizontalen sei der Zerstreuungskoeffizient a konstant, insbesondere längs der Erdoberfläche, wo er den Wert a_0 habe.

h) Die Erdoberfläche sei eben in der Horizontalen und diffus reflektierend.

i) Die lineare Abmessung des gesamten Zieles sei klein gegen die Entfernung Auge—Ziel.

Die übrigen Annahmen von Gruner führen wir nicht ein, insbesondere nicht die Annahme einer nur primären Zerstreuung, die auch schon in der Theorie der Himmelspolarisation aufgegeben werden mußte. Der Bequemlichkeit wegen führen wir noch die homogene Atmosphäre von der Dicke $d = 7.991$ km ein.

Da alle überhaupt ins Auge gelangenden Lichtstrahlen von der Sonne ausgehen, zerlegen wir sie in drei Gruppen: Gruppe I nimmt ihren Weg nur durch die Atmosphäre, Gruppe II gelangt nach Reflexion an der Erdoberfläche erst mittelbar infolge weiterer Zerstreuungen in die Sehstrahlpyramide. Gruppe III gelangt dagegen unmittelbar nach Reflexion der Erdoberfläche in die Sehstrahlpyramide. Vergl. hierzu Fig. 1.

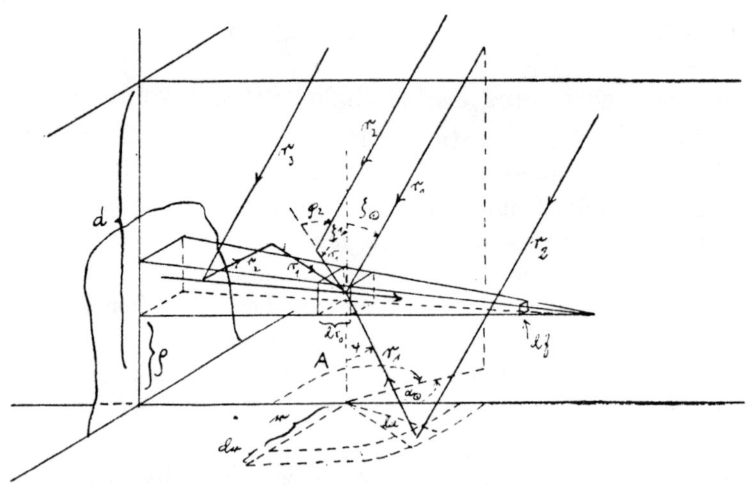

Fig. 1

I. Wir betrachten zunächst die Strahlen der Gruppe I und zerlegen sie in unendlich viele Klassen je nach der Anzahl ν der Knicke, die sie auf ihrem Weg vom Eintritt in die Atmosphäre bis zum Auge aufweisen. Mindestens der letzte Knick liegt immer in der Sehstrahlpyramide.

$\nu = 1$.) Diese Strahlen haben nur einen Knick und zwar ist dieser Knick in der Sehstrahlpyramide gelegen. (Vgl. Fig. 1) Ist J_0 die am Rande der Atmosphäre durch die Sonne erzeugte Beleuchtungsstärke, $\alpha_\odot$ und $\zeta_\odot$ Azimut und Zenitdistanz der Sonne, a der Zerstreuungskoeffizient der Luft pro km, so ist die bei einem Volumelement $d\tau_1$ der Sehstrahlpyramide wirksame Beleuchtungsstärke gegeben durch

$$1) \qquad (\alpha) = J_0 . e^{-\int_0^{d \sec \zeta_\odot} a \, dr_1}$$

Der Übersichtlichkeit halber wollen wir stets schreiben

$$2) \qquad \int_0^{r_\nu} a\,dr_\nu = ar \ .$$

Das Volumelement sei durch seine Entfernung r_0 vom Auge charakterisiert. Dann ist

$$3) \qquad d\tau_1 = df \cdot r_0^2\, dr_0,$$

wenn df der der Sehstrahlpyramide zugehörige räumliche Winkel ist, der klein sei gegenüber dem Öffnungswinkel des Auges und klein gegenüber dem dem Ziel zugehörigen räumlichen Winkel. Von $d\tau_1$ wird jetzt eine gewisse Lichtmenge (β) in die Richtung Ziel—Auge gestreut, die $d\tau_1$ selbst proportional ist. (β) ist außerdem proportional einer Zerstreuungsfunktion, die a als Faktor enthalten muß, also etwa $a \cdot \mathfrak{Z}$, wobei $\mathfrak{Z}$ wieder eine Funktion des Winkels φ_1 ist, den der in $d\tau_1$ einfallende Strahl mit dem austretenden bildet. $\mathfrak{Z}$ genügt also der Bedingung

$$4) \qquad \int_0^{2\pi} d\psi \int_0^{\pi} d\chi \sin\chi \cdot \mathfrak{Z}(\psi, \chi) = 1.$$

Es ist $\mathfrak{Z}(\varphi_1)$ diejenige Lichtmenge, die in den räumlichen Winkel 1 gestreut würde, der um φ_1 gegen den einfallenden Strahl geneigt ist, wenn auf jedes räumliche Winkelelement derselbe Betrag entfiele, wie auf das durch φ_1 charakterisierte. Es kann also $\mathfrak{Z} > 1$ werden. Damit wird schließlich

$$5) \qquad (\beta) = (\alpha) \cdot d\tau_1 \cdot a_0 \,\mathfrak{Z}(\varphi_1).$$

Es sei noch bemerkt, daß es natürlich auf dasselbe hinauskommt, ob man die Zerstreuung unmittelbar $d\tau_1$ proportional setzt oder dem Wege des Lichtstrahls innerhalb der Sehstrahlpyramide in Analogie zum Bouguer'schen Ansatz. Der Weg in der Sehstrahlpyramide erhält zwar den Faktor $\sec\varphi_1$, dafür aber das zu dr_0 gehörige zur Richtung des einfallenden Strahles normale Flächenstück den Faktor $\cos\varphi_1$, d. h. die Zerstreuung ist unabhängig von der Form des Raumelementes, nur abhängig von dessen Größe, bzw. der darin enthaltenen Massen.

Nach Voraussetzung $d)$ ist das Volumelement als punktförmige Lichtquelle zu betrachten, infolgedessen wird die auf die Flächeneinheit bei dem Auge einfallende Lichtmenge gegeben durch

$$6) \qquad (\gamma) = (\beta) \cdot \frac{e^{-a_0 r_0}}{r_0^2}.$$

Diese Beleuchtung ist die durch einen Strahl bewirkte Helligkeit dh_1, wobei der Index die Anzahl der Knicke andeuten soll. Mit 1), 3) und 5) erhält man endgültig aus 6)

$$7) \qquad dh_1 = J_0\, df \cdot e^{-a\,d\sec\zeta_{\odot}} \cdot a_0\, \mathfrak{Z}(\varphi_1) \cdot e^{-a_0 r_0}\, dr_0.$$

$\nu = 2$.) Der eine Knick ist wieder in der Sehstrahlpyramide gelegen, der andere an einem Punkte P_2 innerhalb der homogenen Atmosphäre, der durch α_1, ζ_1 und r_1 in Bezug auf $d\tau_1$ gegeben sei. Bei P_2 fällt dann auf die Flächeneinheit die Lichtmenge

8)
$$(\alpha_2) = J_0 \cdot e^{-a(d - r_1 \sin \zeta_1)\sec \zeta_\odot}.$$

Von dem Volumelement $d\tau_2$ bei P_2 wird, wenn $d\tau_2 = r_1^2 \sin \zeta_1 \, d\zeta_1 \, d\alpha_1 \, dr_1$, in die Richtung auf $d\tau_1$ gestreut der Betrag

9)
$$(\beta_2) = (\alpha_2) \cdot d\tau_2 \cdot a_2 \, \mathfrak{Z}(\varphi_2).$$

Die bei $d\tau_1$ von $d\tau_2$ bewirkte Beleuchtung ist gegeben durch

10)
$$(\gamma_2) = (\beta_2) \cdot \frac{e^{-ar_1}}{r_1^2}.$$

Von $d\tau_1$ wird nach dem Auge gestreut

11)
$$(\delta_2) = (\gamma_2) \cdot d\tau_1 \cdot a_0 \, \mathfrak{Z}(\varphi_1),$$

wovon auf die Flächeneinheit bei dem Auge der Betrag einfällt

12)
$$(\varepsilon_2) = (\delta_2) \cdot \frac{e^{-a_0 r_0}}{r_1^2},$$

sodaß mit 8) bis 12) schließlich folgt

13)
$$dh_2 = J_0 \, df \cdot a_2 \, a_0 \cdot \mathfrak{Z}(\varphi_2)\,\mathfrak{Z}(\varphi_1) \cdot e^{-a(r_2 + r_1 + r_0)} \cdot \sin \zeta_1 \, d\zeta_1 \, d\alpha_1 \, dr_1 \cdot dr_0.$$

Es muß hierbei hervorgehoben werden, daß auch P_2 innerhalb der Sehstrahlpyramide gelegen sein kann.

$\nu = n$). In derselben Weise erhält man für $\nu = n$, d. h. also für einen Lichtstrahl, der n Knicke aufweist, von denen mindestens der letzte in der Sehstrahlpyramide gelegen ist,

14)
$$dh_n = J_0 df \cdot \prod_{\mu}^{1,n} a_\mu \, \mathfrak{Z}(\varphi_\mu) \cdot e^{-ar_\mu} \cdot \prod_{\lambda}^{1,n-1} \sin \zeta_\lambda \, d\zeta_\lambda \, d\alpha_\lambda \, dr_\lambda \cdot e^{-ar_0} \, dr_0, \quad \text{wobei } a_1 \equiv a_0.$$

Hierbei muß hervorgehoben werden, daß dh_n nur solche Strahlen betrifft, die n Knicke aufweisen. Das gilt wegen des Auftretens von n Faktoren a, auch dann, wenn r_μ verschwindet. Das Verschwinden von r_μ besagt dann, daß bereits innerhalb von $d\tau_\mu$ eine doppelte Zerstreuung erfolgt. Keinesfalls hat das Verschwinden von r_μ den Übergang von dh_n in dh_{n-1} zur Folge.

Aus den Werten dh_ν, die immer nur für ei n e n Lichtstrahl gelten, sind jetzt durch Integration die H e l l i g k e i t e n h_ν zu bilden. die von allen nur möglichen Lichtstrahlen erzeugt werden, die ν Knicke aufweisen.

$\nu = 1$). In 7) ist für $\nu = 1$ φ_1 eine Konstante und zwar bestimmt sich φ_1 aus dem rechtwinkligen sphärischen Dreieck Sonne, Auge und Fußpunkt des Sonnenvertikals zu

$$\cos \varphi_1 = \sin \zeta_\odot \cos \alpha_\odot.$$

Dagegen ist r_0 eine Veränderliche, es fungieren alle Raumelemente der Sehstrahlpyramide als $d\tau_1$, sodaß über r_0 zu integrieren ist. Ist l die Länge der Sehstrahlpyramide, so folgt

15)
$$h_1 = \left(1 - e^{-a_0 l}\right) J_0 \, df \cdot \mathfrak{Z}(\varphi_1) \cdot e^{-ar_1}.$$

$\nu = 2$.) Wie für $\nu = 1$ fungieren auch hier alle Volumelemente der Sehstrahlpyramide als $d\tau_1$. Als $d\tau_2$ fungieren alle Raumelemente der homogenen Atmosphäre, gemäß unserer Voraussetzung (i). Dann ist φ_1 keine Konstante mehr, vielmehr wird

$$\varphi_1 = \sin \zeta_1 \cos \alpha_1,$$
$$\varphi_2 = \cos \zeta_\odot \cos \zeta_1 + \sin \zeta_\odot \sin \zeta_1 \cos (\alpha_\odot - \alpha_1),$$

aber beide bleiben unabhängig von r_0. Die Integration über r_0 läßt sich also sofort ausführen und es folgt

$$16)\ h_2 = \left(1 - e^{-a_0 l}\right) J_0\, df \cdot \int_0^{2\pi} d\alpha_1 \int_0^{\pi/2} d\zeta_1 \int_0^{d \sec \zeta_1} dr_1\, a_2\, \mathfrak{B}(\varphi_2)\, \mathfrak{B}(\varphi_1) \sin \zeta_1\, e^{-a(r_2 + r_1)}$$

Man erhält noch konstante Grenzen durch die Substitution

$$z_1 = r_1 \cos \zeta_1.$$

Damit wird

$$17)\ h_2 = \left(1 - e^{-a_0 l}\right) J_0\, df \cdot \int_0^{2\pi} d\alpha_1 \int_0^{\pi/2} d\zeta_1 \int_0^{d} dz_1\, a_2\, \mathfrak{B}(\varphi_2)\, \mathfrak{B}(\varphi_1) \sin \zeta_1 \cos \zeta_1\, e^{-a(r_2 + r_1)}.$$

$\nu > 2$.) Hierfür werden die Integrale verwickelter, da $\zeta_2, \zeta_3 \ldots$ auch Werte $> \pi/2$ annehmen. So folgt für $\nu = 3$

$$18) \qquad\qquad h_3 = \left(1 - e^{-a_0 l}\right) J_0\, df \cdot$$

$$\int_0^{2\pi} d\alpha_1 \int_0^{\pi/2} d\zeta_1 \left\{ \int_0^{d \sec \zeta_1} dr_1 \int_0^{2\pi} d\alpha_2 \int_0^{\pi/2} d\zeta_2 \int_0^{(d - r_1 \cos \zeta_1) \sec \zeta_2} dr_2\, a_3\, a_2\, \mathfrak{B}_3\, \mathfrak{B}_2\, \mathfrak{B}_1 \sin \zeta_2 \sin \zeta_1\, e^{-a(r_3 + r_2 + r_1)} \right.$$

$$\left. + \int_0^{d \sec \zeta_1} dr_1 \int_0^{2\pi} d\alpha_2 \int_{\pi/2}^{\pi} d\zeta_2 \int_0^{r_1 \cos \zeta_1 \sec \zeta_2} dr_2\, a_3\, a_2\, \mathfrak{B}_3\, \mathfrak{B}_2\, \mathfrak{B}_1 \sin \zeta_2 \sin \zeta_1\, e^{-a(r_3 + r_2 + r_1)} \right\}.$$

Der zweite Summand rührt eben davon her, daß $d\tau_2$ für $\nu > 2$ auch von unten beleuchtet wird. h_4 wird dann ein dreifaches Raumintegral, es treten vier Summanden auf, h_ν wird ein $(\nu - 1)$faches Raumintegral mit $2^{\nu-2}$ Summanden.

Die Gesamthelligkeit, unter der der schwarze Körper erscheint, ist gegeben durch

$$19) \qquad\qquad H_s = \sum_\nu h_\nu.$$

Nach dem vorliegenden physikalischen Problem muß diese Reihe konvergieren. Daß sie auch formal konvergiert, ergibt sich aus folgender Überlegung:[1] Chr. Wiener hat l. c. die in 16) und 18) enthaltenen Ausdrücke

$$20) \qquad\qquad h_1' = J_0\, df \cdot \int_0^{d \sec \zeta_1} dr_1\, a_2\, \mathfrak{B}(\varphi_2) e^{-a(r_2 + r_1)}$$

---

[1] Ein strenger Konvergenzbeweis ist mir nicht geglückt.

21)
$$h'_2 = J_0\, df .$$

$$\int_0^{d\sec\zeta_1} dr_1 \int_0^{2\pi} d\alpha_2 \left\{ \int_0^{\pi/2} d\zeta_2 \int_0^{(d-r_1\cos\zeta_1)\sec\zeta_2} dr_2\, a_3\, a_2\, \mathfrak{B}_3\, \mathfrak{B}_2\, \sin\zeta_2\, e^{-a(r_3+r_2+r_1)} \right.$$

$$\left. + \int_{\pi/2}^{\pi} d\zeta_2 \int_0^{r_1\cos\zeta_1\sec\zeta_2} dr_2\, a_3\, a_2\, \mathfrak{B}_3\, \mathfrak{B}_2\, \sin\zeta_2\, e^{-a(r_3+r_2+r_1)} \right\}$$

durch mechanische Quadratur berechnet, und zwar für verschiedene α_1 und ζ_1. Beide Ausdrücke und die weiteren in derselben Weise zu entwickelnden haben eine einfache physikalische Bedeutung: $h'_1(\alpha_1, \zeta_1)$ ist diejenige Helligkeit des Himmelselementes (α_1, ζ_1), die nur durch solche Strahlen hervorgerufen wird, die auf ihrem Weg durch die Atmosphäre einen einzigen Knick erleiden. Die Himmelshelligkeit h'_2 rührt nur von solchen Strahlen her, die einen zweifachen Knick erleiden u. s. f. Darauf fand Wiener durch nochmalige mechanische Quadratur, daß

22)
$$\int_0^{\pi/2} \int_0^{2\pi} h'_2(\alpha_1\,\zeta_1)\sin\zeta_1\, d\zeta_1\, d\alpha_1 : \int_0^{\pi/2} \int_0^{2\pi} h'_1(\alpha_1\,\zeta_1)\sin\zeta_1\, d\zeta_1\, d\alpha_1 = 0.2938.$$

Die Integranden in 22) unterscheiden sich von denen in 16) und 18) nur durch den Faktor $\mathfrak{B}(\varphi_1)$. Wendet man auf 16) und 18) den Mittelwertsatz an und bildet das der Gleichung 22) entsprechende Verhältnis

23)
$$\mathfrak{B}(\bar{\bar{\varphi}}_1)\cdot\int_0^{\pi/2}\int_0^{2\pi} h'_2 \sin\zeta_1\, d\zeta_1\, d\alpha_1 : \mathfrak{B}(\bar{\varphi}_1)\cdot\int_0^{\pi/2}\int_0^{2\pi} h'_1 \sin\zeta_1\, d\zeta_1\, d\alpha_1 = h_3 : h_2,$$

so ist im allgemeinen

24)
$$\mathfrak{B}(\bar{\bar{\varphi}}_1) : \mathfrak{B}(\bar{\varphi}_1) \neq 1,$$

aber zweifellos ist bei den kleinen Schwankungen von $\mathfrak{B}(\varphi_1)$

25)
$$\mathfrak{B}(\bar{\bar{\varphi}}_1) : \mathfrak{B}(\bar{\varphi}_1) < \frac{1}{0.2938},$$

d. h. auch die h_ν in 19) nehmen mit wachsendem Index ab.

II. Für die Strahlen der Gruppe II ist zu bemerken, daß $\mathfrak{B}\cdot a$ für den Knick an der Erdoberfläche den Wert $\dfrac{\mathfrak{A}}{\pi}$ annimmt, wenn die Erde als diffus reflektierend mit der Albedo $\mathfrak{A}$ angesehen wird. Wir können uns also die durch Strahlen der Gruppe II vor dem schwarzen Körper erzeugten Helligkeiten in 17) bis 19) enthalten denken, dadurch, daß in einer dünnen Oberflächenschicht an der unteren Grenze der homogenen Atmosphäre $a\mathfrak{B}$ den Wert $\dfrac{\mathfrak{A}}{\pi}$ annimmt.

III. Die Beleuchtung von $d\tau_1$ durch die Strahlen der Gruppe III läßt sich leicht berechnen. Bezeichnet man mit ϱ den Abstand der horizontalen Sehstrahlpyramide von der Erdoberfläche, mit ψ den Winkel, den das Lot von $d\tau_1$ auf die Erdoberfläche mit der Linie bildet, die $d\tau_1$ mit einem Flächenelement $r\, dr\, d\alpha$ der Erdoberfläche verbindet, so erhält man für die Helligkeit, die in Richtung auf den schwarzen Körper durch einen Lichtstrahl auf dem Wege über

$$\mathfrak{r}\,d\mathfrak{r}\,d\alpha \;=\; \frac{\varrho^2}{\cos^2\psi}\,\operatorname{tg}\psi\,d\psi\,d\alpha$$

erzeugt wird, den Ausdruck

$$dh_1^u \;=\; J_0\,e^{-ar_2}\cdot\left(\frac{\varrho^2}{\cos^2\psi}\,\operatorname{tg}\psi\,d\psi\,d\alpha\right)\cos\xi_\odot\cdot\cos\psi\cdot\frac{\mathfrak{A}}{\pi}\cdot\frac{e^{-a\varrho\sec\psi}}{\varrho^2\sec^2\psi}\cdot df\,r_0^2\,dr_0\cdot a_0\,\mathfrak{Z}(\varphi_1)$$

$$\cdot\frac{e^{-a_0 r_0}}{r_0^2}\cdot$$

Hierüber ist in den Grenzen $0\leqq\alpha\leqq 2\pi$, $0\leqq\psi\leqq\pi/2$, $0\leqq r_0\leqq l$ zu integrieren. Dann folgt

$$15\mathrm{u})\quad h_1^u \;=\; \left(1-e^{-a_0 l}\right)J_0\,df\left\{\frac{\mathfrak{A}}{\pi}\cos\zeta_\odot\,e^{-ar_2}\cdot\int_0^{\pi/2}d\psi\int_0^{2\pi}d\alpha\,e^{-a\varrho\sec\psi}\cdot\sin\psi\cdot\mathfrak{Z}(\alpha_1\psi)\right\}.$$

Die weiteren Integrale h_ν^u sind ähnlich denen der Gruppe I zu bilden. Die Strahlen sind sämtlich dadurch definiert, daß der letzte Knick in der Sehstrahlpyramide und mindestens der vorletzte an der Erdoberfläche gelegen ist. Wir schreiben in Analogie zu 19)

$$19\mathrm{u})\qquad\qquad H^u \;=\; \sum_\nu h_\nu^u,$$

und dafür erhalten wir den aus dem folgenden § ersichtlichen Wert

$$27\mathrm{u})\qquad H^u \;=\; \left(1-c^{-a_0 l}\right)\cdot\left\{\frac{\mathfrak{A}}{\pi}\cdot\int_0^{\pi/2}d\zeta_1\int_0^{2\pi}d\alpha_1\,\sin\zeta_1\cos\zeta_1\cdot i_g(\alpha_1,\zeta_1)\right.$$

$$\left.\int_0^{\pi/2}d\psi\int_0^{2\pi}d\alpha\,e^{-a\varrho\sec\psi}\sin\psi\cdot\mathfrak{Z}(\alpha,\psi)\right\}{}^1).$$

Um diesen Ausdruck nicht immer mitschleppen zu müssen, wollen wir uns über seine Größenordnung orientieren. Die Albedo $\mathfrak{A}$ der Erde ist für Ackererde

---

1) Die Gleichung 27u) führt zu einem Resultat, das für die Theorie der Messung der Erd-Albedo vom Ballon aus wichtig ist. Bestimmt man die Beleuchtung, die ein horizontales kleines Flächenstück df', dessen Schatten vernachlässigt werden kann, in der Höhe ϱ über der diffus reflektierenden Erdoberfläche von dieser erfährt, so erhält man dafür unter Vernachlässigung der Zerstreuung

$$dB^u \;=\; J_0\,(\varrho^2\sec^2\psi\,\operatorname{tg}\psi\,d\psi\,d\alpha)\cos\zeta_\odot\cdot\cos\psi\cdot\frac{\mathfrak{A}}{\pi}\cdot\frac{1}{\varrho^2\sec^2\psi}\cdot df'\cos\psi$$

und daraus durch Integration

$$B^u \;=\; \left(J_0\cos\zeta_\odot\cdot\frac{\mathfrak{A}}{\pi}\right)\cdot\pi\cdot df'.$$

D. h. die Beleuchtung von df ist unabhängig von seiner Höhe über dem Erdboden. — Bei genaueren Messungen ist aber sehr wohl die Zerstreuung zu berücksichtigen, da bei größeren Höhen (etwa bei Messung vom Freiballon aus) auch ein von der Erdoberfläche ganz unabhängiges Unterlicht erzeugt wird. Doch ist hier nicht der Ort, näher darauf einzugehen.

oder Wiese zu etwa 0.1 bestimmt worden [1]). Aus später ersichtlichen Gründen (§ 3) sind die Schwankungen von β als klein anzusehen, daher wird mit Berücksichtigung von 4) ein mittleres $\overline{\beta}$ sicher kleiner als 1. Ebenso ist $e^{-a\varrho\sec\psi} < 1$. Wir nehmen weiter

$$\int_0^{\pi/2} d\zeta_1 \int_0^{2\pi} d\alpha_1 \sin\zeta_1 \cos\zeta_1 . i_g(\alpha_1, \zeta_1) \text{ sicher kleiner } \frac{1}{10} \int_0^{\pi/2} d\zeta_1 \int_0^{2\pi} d\alpha_1 \sin\zeta_1 i_g(\alpha_1, \zeta_1),$$

indem wir berücksichtigen, daß die $i_g(\alpha_1, \zeta_1)$ ihrer Größe nach gerade für die Horizontgebiete den wesentlichen Beitrag zur Summe liefern, wofür aber $\cos\zeta_1$ klein wird. ($\cos 80^0 = 0.13$). Da weiterhin $\int_0^{\pi/2} \int_0^{2\pi} d\psi \, d\alpha \sin\psi = 2\pi$, so erhalten wir aus 27) und 27 u)

$$H^u < \frac{\mathfrak{A}}{\pi} \cdot \frac{1}{10} H_s . 2\pi, \; dh . \; H^u < 0.02 \, H_s,$$

d. h. wir können H^u gegenüber H_s vernachlässigen.

§ 3. Die Auswertung des Helligkeitsintegrals mit Hilfe der beobachteten Himmelshelligkeit i_g.

Die uns interessierenden Helligkeiten h_ν lassen sich nach den Beziehungen 15), 16) und 18) usw. zahlenmäßig nicht berechnen, einmal infolge der Unmöglichkeit, die Integrationen 16), 18) u. s. f. unbestimmt auszuführen, und weiter infolge der z. Z. noch bestehenden Unkenntnis des Zerstreuungskoeffizienten und der Zerstreuungsfunktion als Funktion der Höhe des Raumelementes über der Erdoberfläche. Man käme in jeder Beziehung auf dieselben Schwierigkeiten zurück, die der Theorie der Himmelshelligkeit entgegenstehen. Wie eng diese mit dem Sichtproblem verknüpft ist, geht bereits aus den Erörterungen hervor, die sich an 20) und 21) anschlossen. Wir wollen den dort gegebenen Hinweis auf die Bedeutung der h_ν' konsequent verfolgen und nehmen die Himmelshelligkeit als gegeben an. Sie ist durch Beobachtungen als Funktion von α_1 u. ζ_1 bekannt, und sei im Anschluß an C. Dorno [2]), dem wir die umfangreichsten und zuverlässigsten Messungen verdanken, mit i_g bezeichnet. Mit unseren früheren Bezeichnungen besteht damit die Beziehung

$$26) \qquad\qquad i_g = \underset{\nu}{\Sigma} h_\nu'.$$

Damit folgt aus 20) u. s. f., 15), 16), 18) u. s. f. und 19) die in Richtung auf den

---

1) C. Dorno, Veröffentl. des Preuß. Meteor. Inst., Abhandl. Bd. VI, S. 214; Wiener, l. c. Nr. 2, S. 158; K. Stuchtey u. A. Wegener, Götting. Nachr. 19, S. 21.

2) l. c. Weitere Messungen liegen vor von L. Weber, Handbuch d. Hygiene, Bd. IV; W. Schramm, Diss. Kiel 1901; M. Übe, Abhandl. der Math. Phys. Kl. d. Sächs. Ges. d. Wiss. XXXV, No. 6. Es ist bemerkenswert, daß sämtliche Beobachtungen ein sekundäres Maximum der Horizonthelligkeit in 180^0 Entfernung von der Sonne liefern, wie dies durch die Theorie von Wiener gegeben wird

schwarzen Körper meßbare Helligkeit H_s

$$27) \qquad H_s = \left(1 - e^{-a_0 l}\right) \cdot \int_0^{2\pi} d\alpha_1 \int_0^{\pi/2} d\zeta_1 \cdot \sin \zeta_1 \cdot \mathfrak{B}(\varphi_1) \cdot i_g(\alpha_1, \zeta_1).$$

Es sei hier nochmals darauf hingewiesen, daß in 27) alle nur denkbaren Strahlenwege berücksichtigt sind, wie schon im § 2 immer wieder betont wurde, daß je nach der Anzahl ν der betrachteten Knicke **mehr als ein Knick** in der Sehstrahlpyramide gelegen sein kann, d. h. also, daß in 27) auch diejenigen Strahlen berücksichtigt sind, die aus der Sehstrahlpyramide herausgestreut werden und auf Umwegen doch wieder das Auge erreichen. **Es erschien mir als eine Hauptaufgabe des § 2, diese Vorgänge exakt zu beschreiben und** so eine strenge Herleitung der Gleichung 27) zu ermöglichen.

In i_g, also der gemessenen Himmelshelligkeit, sind auch alle Strahlen der Gruppe II enthalten, das sind diejenigen Strahlen, die nach Reflexion am Erdboden mindestens einen Knick erleiden, bevor sie in der Sehstrahlpyramide ins Auge gelenkt werden. Nicht enthalten in i_g sind dagegen die Strahlen der Gruppe III, in unserem Fall das sogenannte Unterlicht, da wir die Sehstrahlpyramide auf der Erdoberfläche liegend annehmen. Wir können es vernachlässigen.

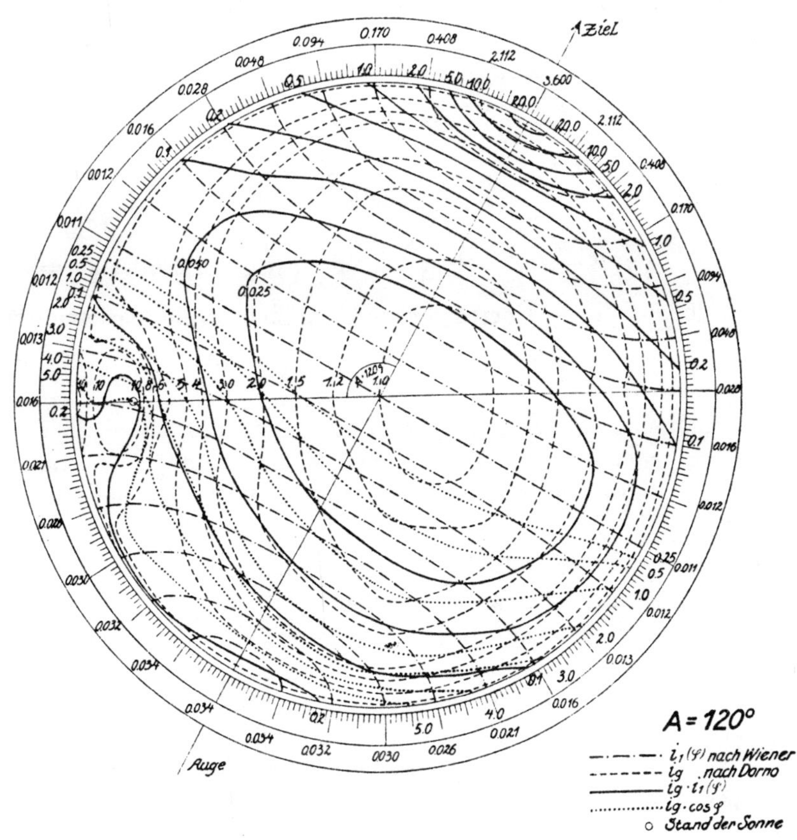

Fig. 2

Der Ausdruck 27) läßt sich nun graphisch auswerten. Ich habe das getan für $\zeta_\odot = 70^0$ und einen wolkenlosen Himmel, sowie für $A = 0$, 30, 60, 90, 120, 150 und 180°. In eine flächentreue Projektion der Himmelshalbkugel wurden die Linien gleicher i_g, die Isophoten, eingetragen, nach der Tafel III b nach Dorno l. c. Außerdem wurden eingetragen die Linien gleicher $\mathcal{B}(\varphi_1)$, die zahlenmäßig unter bestimmten, aber wohl anfechtbaren Voraussetzungen von Chr. Wiener berechnet worden sind (i_1 S. 236, l. c. Nr. 1), Diese Kurven wurden graphisch miteinander multipliziert, sodaß in den Darstellungen die dritte Kurvenschar $i_g \cdot i_1$ erscheint. Um das Verfahren anschaulich zu machen, sei wenigstens eine Tafel ($A = 120^0$) reproduziert vgl. Fig. 2. Darauf wurden die Flächen zwischen zwei Kurven $i_g \cdot i_1$ ausplanimetriert und diese Flächeninhalte mit einem mittleren $\overline{i_g i_1}$ multipliziert, das auf folgende Weise zu erhalten war:

Da die von zwei $i_g i_1$ - Kurven begrenzten Himmelszonen durch die Linie Auge—Ziel in zwei Flächen zerlegt werden, die sich annähernd als ein Rechteck mit aufgesetztem Dreieck darstellen lassen (A B C D, Fig. 3), so gilt

$$y = a + bx$$

und für das Intervall AB auch

$$i_g i_1 = \alpha + \beta x.$$

Dann bestimmt sich das gesuchte $\overline{i_g i_1}$ aus

$$\overline{i_g i_1} = \frac{\int i_g i_1 \cdot y dx}{\int y dx}$$

zu

$$\overline{i_g i_1} = (i_g i_1)_1 + \frac{\varDelta (i_g i_1)}{2} + \varDelta (i_g i_1) \cdot \frac{\varDelta s}{6 (2s + \varDelta s)}.$$

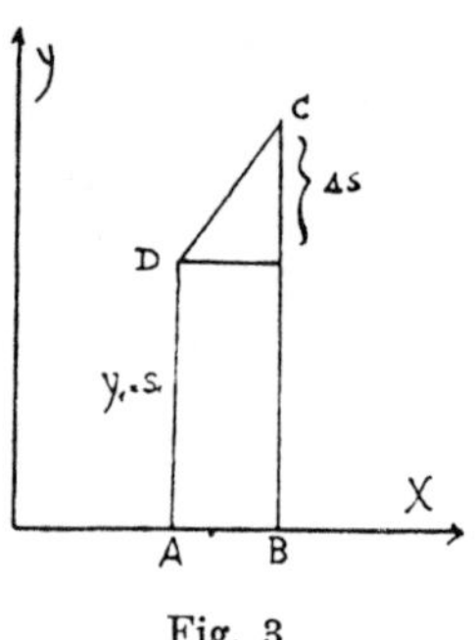

Fig. 3

Da der Faktor $\dfrac{\varDelta s}{6} \cdot (2s + \varDelta s)^{-1}$ selbst im ungünstigsten Falle ($s_1 = 0$) erst $\dfrac{1}{6}$ wird, so wurde auf dieses Korrektionsglied in anbetracht der anderen vorhandenen Ungenauigkeiten verzichtet und für $\overline{i_g i_1}$ das Mittel aus den beiden Werten $(i_g i_1)_A$ und $(i_g i_1)_B$ genommen.

Die so erhaltenen Produkte Π aus $\overline{i_g i_1}$ und dem Flächeninhalt F der Zonen wurden addiert und lieferten die Summen σ_1 und σ_2 für die Himmelshälften Ziel und Auge. Dabei wurden die F so ausgedrückt, daß ihre Gesamtsumme 2π ergibt.

Hierbei wurde auf den Helligkeitsanstieg auf die Sonne hin jenseits der 10.0 Isophote keine Rücksicht genommen, sondern der von der 10.0 Isophote umschlossenen Fläche das konstante Gewicht 10.0 erteilt. Der Anstieg der Helligkeit auf die Sonne hin wurde vielmehr für einen Ring von 3 Grad Breite, vom Sonnenrande ab gezählt, rechnerisch ermittelt, da hierfür die graphische Methode zu ungenau wird. Es wurden also die Kreisringe berechnet, bis 1 Grad Sonnenabstand von 0.1 Grad Breite, bis 3.0 Grad Sonnenabstand von 0.5 Grad Breite, und diese Flächeninhalte mit ($i_g - 10.0$) multipliziert, wobei für i_g die Werte von Dorno, l. c. Tab. 17., eingesetzt wurden. Für die Sonnenabstände über 1.0^0 mußten jedoch kleinere Werte für i_g eingesetzt werden, um Anschluß an die

Tafel III b bei Dorno zu erhalten. Natürlich wurde hier (wie auch später für die Sonne selbst) die Himmelshalbkugel $= 2\pi$ gesetzt. Diese Werte $F \cdot (i_g - 10.0)$ wurden dann mit demjenigen Werte $i_1 = \beta(\varphi_1)$ multipliziert, der für den Sonnenmittelpunkt gültig ist. Das konnte getan werden, da i_1 sich in dem engen Intervall hinreichend linear ändert und die Kreisringe von dieser i_1-Linie hinreichend halbiert werden. Die Summe dieser Werte $i_1 \cdot F \cdot (i_g - 10.0)$ ist mit R_3 bezeichnet.

Mit der Sonne selbst wurde in derselben Weise wie mit den Kreisringen um die Sonne verfahren. Die Flächenhelle der Sonne selbst ist gleichfalls bei Dorno, l. c. Tab. 17[4], angegeben. Der Wert $i_g i_1 F$ wird mit $\odot$ bezeichnet.

Da φ_1, also auch $\beta(\varphi_1)$ sich ändert für jeden bestimmten Ort des Himmels, wenn die Blickrichtung sich ändert, d. h. der von der Blickrichtung und dem Sonnenvertikal gebildeten Winkel A, so ist sicher

$$H_s = H_s(A).$$

Es wurden daher die eben beschriebenen Berechnungen für verschiedene A durchgeführt. Damit ergaben sich die Werte der Tab. 1 für

$$(H) = \int_0^{2\pi} d\alpha_1 \int_0^{\pi/2} d\xi_1 \cdot \sin\xi_1 \cdot \beta(\varphi_1) \cdot i_g(\alpha_1, \xi_1).$$

Tabelle 1.

A	0	30	60	90	120	150	180
σ_1	3.89	2.82	2.37	2.04	1.16	1.98	2.15
σ_2	0.18	0.18	0.18	0.21	0.17	0.25	0.29
R_3	0.00₅	0.00₂	—	—	—	—	—
$\odot$	30.76	9.80	1.92	0.84	1.15	2.14	2.45
(H)	34.83	12.80	4.47	3.09	3.18	4.37	4.89

$$(H) = \sigma_1 + \sigma_2 + R_3 + \odot.$$

Die Helligkeit H_s, die (H) proportional ist, schwankt zwischen den Grenzen 1 und 11. Dieses Resultat ist aus später ersichtlichen Gründen unbrauchbar. Die Schwankung dürfte über die Grenzen 1 und 2 nicht hinausgehen. Dieses unbefriedigende Ergebnis hat seinen Grund in der mindestens für das Hochgebirge viel zu großen Amplitude der Wienerschen Zerstreuungsfunktion i_1. Ihre Maximalwerte sind $i_1|_{\varphi_1 = 0} = 3{,}600$ und $i_1|_{\varphi = 90^0} = 0{,}011$, sie schwanken also zwischen 1 und 350.

Man kann sich aber von der Zerstreuungsfunktion vollständig freimachen, indem man den Weg konsequent weitergeht, der zu der Einführung von i_g führte. Geht man nämlich in 27) zu $l = \infty$ über, so erhält man

$$28) \qquad H_s|_{l = \infty} = \int_0^{2\pi} da_1 \int_0^{\pi/2} d\xi_1 \sin\xi_1 \, \beta(\varphi_1) \cdot i_g(\alpha_1, \xi_1).$$

Setzt man diesen Wert auf der rechten Seite von 27) ein, so folgt

$$29) \qquad H_s = \left(1 - e^{-a_0 l}\right) \cdot H_s|_{l = \infty}.$$

Nun ist aber $H_s\big|_{l=\infty}$ nichts anderes als die Horizonthelligkeit, die wir zum Zeichen dafür, daß sie nur noch Funktion von A ist, mit $i_g(A)$ bezeichnen wollen. Damit erhalten wir also

$$30) \qquad \boxed{H_s = \left(1 - e^{-a_0 l}\right) \cdot i_g(A).}$$

In Worten: **Die in Richtung auf den schwarzen Körper meßbare Helligkeit ist ein nur von der Entfernung Auge—schwarzer Körper abhängiger Prozentsatz der Horizonthelligkeit in demselben Azimut A.**

Dieses Resultat setzt uns in den Stand, sofort die gesuchten Sichtweiten zu berechnen. Bevor wir dazu übergehen, merken wir noch folgende Beziehung an: Schreibt man in 28) links für $H_s\big|_{l=\infty}$ $i_g(A)$, so folgt

$$31) \qquad i_g(A) = \int_0^{2\pi} d\alpha_1 \int_0^{\pi/2} d\xi_1 \sin\xi_1 \cdot \mathfrak{Z}(\varphi_1) \cdot i_g(\alpha_1, \xi_1),$$

eine Beziehung, die wir als Bestimmungsgleichung für die Zerstreuungsfunktion $\mathfrak{Z}$ auffassen können.

Aus 30) erkennen wir auch den Grund, warum die Schwankungen von (H) in Tab. 1 zu groß sind. (H) ist identisch mit $H_s\big|_{l=\infty} = i_g(A)$, und für $i_g(A)$ finden wir bei Dorno, l. c. Taf. III b, die Werte:

$$
\begin{array}{lcccccccc}
A = & 0 & 10 & 30 & 60 & 90 & 120 & 150 & 180 \\
i_g(A) = & 12.0 & 10.0 & 8.0 & 7.0 & 6.0 & 6.0 & 6.0 & 7.0
\end{array}
$$

d. h. $i_g(A)$ schwankt nur zwischen 1 und 2, also auch (H).

§ 4. Die Sichtweite eines schwarzen, weißen, farbigen Schirmes vor dem Horizont.

Bei der Berechnung der Sichtweiten behalten wir die drei eingangs erwähnten Spezialfälle bei.

1. $\mathfrak{A} = 0$. Bedeutet s_s die Sichtweite des schwarzen Körpers gegen den Horizont, so lautet gemäß der Definition (A) die Bestimmungsgleichung für s_s

$$32) \qquad \frac{H_{\mathrm{hor}}}{H_s} = \frac{i_g(A)}{i_g(A)\left(1 - e^{-a_0 s_s}\right)} = 1 + \varepsilon,$$

woraus folgt

$$33) \qquad \boxed{s_s = \frac{1}{a_0} \ln \frac{1 + \varepsilon}{\varepsilon}.}$$

In Worten: **Die Sichtweite des schwarzen Körpers gegen den Horizont ist umgekehrt proportional dem Zerstreuungskoeffizienten**

a_0 und unabhängig von dem Winkel A, der durch den Sonnenvertikal und die Blickrichtung gebildet wird[1]).

Tabelle 2.

s_s in km	a_0 pro km	p pro km	p pro m
0.1	39.12	1.00×10^{-17}	0.9616
0.5	7.824	3.98×10^{-4}	0.9922
1.0	3.918	2.00×10^{-2}	0.9961
5.0	0.7824	0.457×10^0	0.9992
10.0	0.3912	0.676	0.9996
15.0	0.2609	0.7706	
20.0	0.1956	0.8224	
25.0	0.1566	0.8551	
30.0	0.1304	0.8778	
35.0	0.1118	0.8943	
40.0	0.0978	0.9062	$\varepsilon = 2\%$
45.0	0.0870	0.9166	
50.0	0.0783	0.9247	$\dfrac{1+\varepsilon}{\varepsilon} = 50$
75.0	0.0522	0.9491	
100.0	0.0391	0.9616	
125.0	0.0313	0.9692	
150.0	0.0261	0.9743	
175.0	0.0227	0.9776	
200.0	0.0196	0.9806	

---

1) Der erste Teil dieses Resultates ist bereits von Haecker l. c auf anderem Wege abgeleitet worden. H. definierte die Sichtweite $\bar{s}$ als diejenige Entfernung, in der der Kontrast zwischen einer schwarzen und weißen Fläche bis auf ε abgefallen ist. Er erhielt in unserer Schreibweise

$$\bar{s} = \frac{1}{a_0}\ln\frac{1}{2\varepsilon} \simeq \frac{1}{a_0}\ln\frac{1+2\varepsilon}{2\varepsilon},$$

für einen gleichmäßig hellen Himmel. Wir kommen für $\bar{s}$ zu demselben Resultat mit Hülfe der Formel 36). Es wird nämlich hier

$$H_w = H_w^1 \cdot e^{-a_0\bar{s}} + i_g(A)\cdot\left(1 - e^{-a_0\vec{s}}\right),$$

$$H_s = \qquad\qquad i_g(A)\cdot\left(1 - e^{-a_0\vec{s}}\right),$$

wobei für $H = \text{const} = i_g(A) = i_g$

$$H_w^1 = \frac{i_g(A)}{\pi}\cdot\int_0^\pi da_1 \int_0^{\pi/2} d\zeta_1 \sin\zeta_1\cdot\cos\zeta_1 = \frac{i_g(A)}{2}.$$

Damit folgt aus $H_w\cdot H_s^{-1} = 1 + \varepsilon$ schließlich $\bar{s} = \frac{1}{a_0}\cdot\ln\frac{1+2\varepsilon}{2\varepsilon}$. D. h. die durch den Kontrast schwarz-weiß definierte Sichtweite $\bar{s}$ ist kleiner als die durch den Kontrast schwarz-Horizont definierte Sichtweite s_s. Das ist auch sofort einzu-

Weder der erste noch der zweite Teil unserer Schlußfolgerung läßt sich aus der Erfahrung bestätigen. Die Messungen von Weber l. c. sind die einzigen, die den hiesigen Bedingungen entsprechen. In dem einen verwertbaren Falle ergab sich eine berechnete Sichtweite von 3,3 km, während die beobachtete 4,7 betrug. Auch liegen bisher keine einwandfreien Beobachtungen von a_o vor in Richtung der Horizontalen. Für $p = e^{-a_o}$ sind Werte von Oddone Schlagintweit und Wild[1]) angegeben, doch sind die Messungen unvollständig definiert. Wir geben daher umgekehrt zu s_o die zugehörigen a_o Werte in Tab. 2 an nach 33), indem wir $\varepsilon = 2^0/_0$, $\frac{1+\varepsilon}{\varepsilon} = 50$ setzen und auch für späteren praktischen Gebrauch die p Werte beifügen.

Die Werte der Tabelle 2 beantworten zugleich eine Frage, die für die Scheinwerfertechnik von Bedeutung ist: Mit welchen Extinktionskoeffizienten hat man nachts zu rechnen, wenn am Abend z. B. die Sichtweite 5 km betrug und die Atmosphäre ihren Zustand in der Zwischenzeit nicht wesentlich geändert hat. Die a_o haben dann für Scheinwerferlicht als „Extinktions"koeffizienten zu gelten.

2. $\mathfrak{A} = 1$. Wir betrachten jetzt eine Lambert'sche Fläche von der Albedo 1. Wir bestimmen zunächst deren Flächenhelle H_w^1, die bekanntlich gegeben ist durch

$$34) \qquad H_w^1 = \frac{1}{\pi} \int_0^\pi d\alpha_1 \int_0^{\pi/2} d\zeta_1 \sin \zeta_1\, i_g(\alpha_1,\ \zeta_1) \cdot \cos \varphi_1.$$

Da eine vertikal aufgestellte Fläche von einer Himmelshälfte beleuchtet wird, so ist $0 \leqq \alpha_1 \leqq \pi$. Diese Integrale 34) sind in derselben Weise auszuwerten wie 27). Es wurde das getan wieder für $A = 0$, 30, 60, 90, 120, 150 und 180 Grad. In derselben Weise wie früher wurden R_s und $\odot$ berechnet. Überall tritt an Stelle von i_1 hier $\cos \varphi_1$, vgl. Fig. 2. Die Ergebnisse sind in Tabelle 3 zusammengestellt. H_w^1 in der Tab. 3 ist in denselben Einheiten ausgedrückt wie die $i_g(\alpha_1 \zeta_1)$, d. h. in Zenithelligkeiten.

Tabelle 3.

A	0	30	60	90	120	150	180
$\frac{1}{\pi}\sigma$	1.48	1.42	1.46	1.63	1.97	2.27	2.51
$\frac{1}{\pi}R_s$	—	—	—	—	0.16	0.27	0.31
$\frac{1}{\pi}\odot$	—	—	—	—	11.45	20.23	22.92
$H_w^1(A)$	1.48	1.42	1.46	1.63	13.58	22.77	25.74

$$H_w^1(A) = \frac{1}{\pi}\{\sigma + R_s + \odot\}.$$

sehen: H_w wächst mit l von $\frac{1}{2}i_g(A)$ bis auf $i_g(A)$ für s_w, ebenso wächst H_s mit l von 0 bis auf $i_g(A)$. Die Sichtweite $\bar{s}$ ist dann erreicht, wenn H_w annähernd gleich H_s ist, die dann aber beide $< i_g(A)$ sind.

1) Sämtlich zitiert in Perntner-Exner, Met. Optik, 2. Aufl. S. 730. Den Beobachtungen von Oddone haftet ein prinzipieller Mangel an, indem nämlich bei den Messungen Bergspitzen anvisiert wurden, sodaß die Linie Auge—Ziel ganz verschiedenartig getrübte Luftmassen durchschnitt.

Die Bestimmungsgleichung für die Sichtweite s_w des weißen Schirmes vor dem Horizont erhalten wir wieder aus der Definition (A). Entsprechend 32) setzen wir an

$$34) \qquad \frac{H_{hor}}{H_w} = 1 + \varepsilon.$$

Infolge der Zerstreuung wird H_w^1 in der Entfernung s_w auf $H_w^1 e^{-a_0 s_w}$ herabgedrückt. Zu diesem Betrag tritt aber noch die Lichtmenge hinzu, die aus der durch Auge und Ziel gebildeten Sehstrahlpyramide ins Auge gestreut wird. Diese ist gegeben durch 30), sodaß folgt

$$35) \qquad H_w = H_w^1 e^{-a_0 s_w} + \left(1 - e^{-a_0 s_w}\right) \cdot i_g(A).$$

Aus 34) und 35) erhalten wir dann schließlich

$$36) \qquad \boxed{s_w = \frac{1}{a_0} \ln\left[\frac{1+\varepsilon}{\varepsilon}\left(1 - \frac{H_w^1(A)}{i_g(A)}\right)\right].}$$

Das Vorzeichen von ε wird schon durch das Verhältnis $H_w^1 \cdot i_g(A)^{-1} = q(A)$ bestimmt. Man erkennt das daraus, daß man den reziproken Wert von 34) bildet und für $\frac{1}{1+\varepsilon}$ die geometrische Reihe schreibt. Die höheren Potenzen von ε haben keinen Einfluß mehr auf das Vorzeichen der rechten Seite. Man sieht, daß

$$\varepsilon \lessgtr 0 \text{ für } q - 1 \gtrless 0.$$

Für $H_w^1 = 0$ geht 36) in 33) über, wie das Problem es verlangt. Mit 36) ist der erste Teil unserer Aufgabe erledigt.

Um s_w als $s_w(A)$ darzustellen, benutzen wir 36) und Tabelle 3. Als Beispiel rechnen wir den Fall durch, daß $s_s = 15.0$ km, also nach Tab. 2 $a_0 = 0.2609$ ist. Dann erhalten wir

Tabelle 4.

A in Grad	0	30	60	90	120	150	180
s_w in km	14.5	14.25	14.1	13.8	15.9	19.0	18.6

In Worten: Steht, vom Beschauer aus geurteilt, der Schirm unter der Sonne, so ist seine Sichtweite ein sehr schwaches sekundäres Maximum. Die Sichtweite nimmt mit der Entfernung des Schirmes vom Sonnenfußpunkt langsam und gleichmäßig ab und erreicht für einen Abstand von 90^0 ihr Minimum. Sobald der Schirm von der Sonne beleuchtet wird, d.h. für einen größeren Abstand als 90^0, nimmt die Sichtweite rasch zu und erreicht etwa bei $A = 150^0$ ihr Maximum, während sie ein sehr schwaches sekundäres Minimum wird, wenn der Schirm der Sonne gegenüber steht.

Diese überraschende Abhängigkeit der Sichtweite von A wird wenigstens

qualitativ durch die Beobachtungen von A. Wigand[1]) bestätigt. In Figur 4
sind die von Wigand beobachteten und die von mir berechneten Werte nach
Tabelle 4 eingetragen. Wenn man berücksichtigt, daß meinen Berechnungen der
wolkenlose Himmel des Hochgebirges zugrunde liegt, während Wigand bei verschiedenen Wetterlagen und Tageszeiten seine Beobachtungen in Mitteldeutsch-
land ausgeführt hat, muß man mit der Übereinstimmung zufrieden sein, die im
wesentlichen darin besteht, daß die Sichtweite ihr Maximum nicht bei $A = 180^0$
und ihr Minimum nicht bei $A = 0^0$ erreicht.

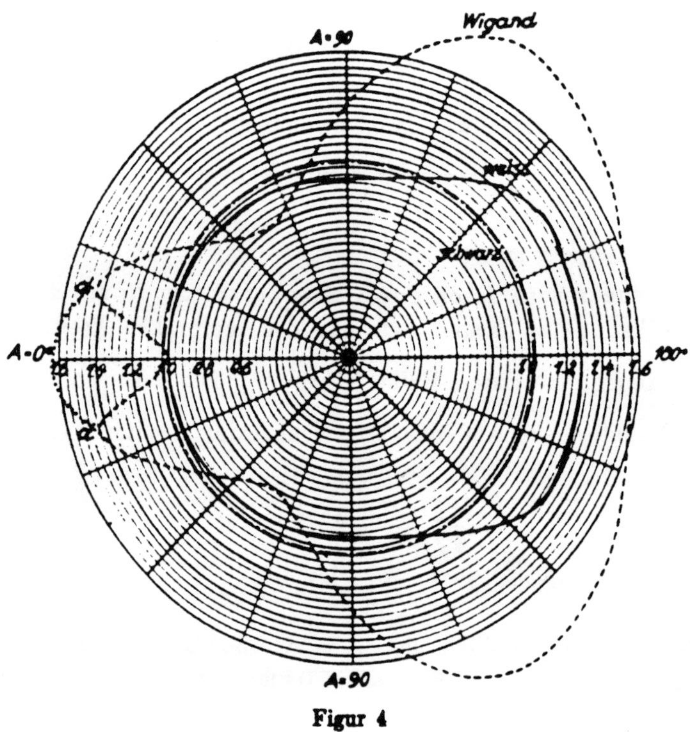

Figur 4

Ein Wert der Wigand'schen Messung jedoch, der mit dem von mir berechneten Verlauf gar nicht übereinzustimmen scheint, ist das Maximum d. Und
doch erscheint mir dieser Wert durchaus möglich, dagegen erscheint mir das
Minimum für $A = 0$ noch nicht hinreichend gesichert. Durch Bestrahlung der
Milchglasplatte des Wigand'schen Sichtmessers durch direktes Sonnenlicht tritt
nämlich eine Zusatzhelligkeit H^+ der Milchglasplatte zu den Helligkeiten des
Zieles H_s und des Horizontes H_{hor} hinzu. Statt mit dem wahren Kontrast

$$\varepsilon = \frac{H_{hor} - H_s}{H_s}$$

mißt man dann die Sichtweite mit Hilfe des Wertes

---

1) Phys. Zeitschr. 23, 1922, S. 288. Wie mir Herr Prof. Wigand freundlichst mitteilt, wurde
bei den in Fig. 4 wiedergegebenen Beobachtungsergebnissen ebenfalls der Kontrast des Zieles gegen
den Horizont benutzt.

$$(\varepsilon) = \frac{(H_{\text{hor}_s} + H^+) - (H_s + H^+)}{H_s + H^+},$$

sodaß also stets $(\varepsilon) < \varepsilon$.

H^+ wird nun wegen der Zerstreuungsfunktion der Milchglasplatte (vgl. Wigand a. a. O.) für kleine Abstände des Zieles von der Sonne besonders groß, der wahrnehmbare Kontrast (ε) und damit die „gemessene" Sichtweite erheblich zu klein. Es erscheint mir daher empfehlenswert, die Milchglasplatte des Wigandschen Sichtmessers durch einen geschwärzten Tubus vor direktem Sonnenlicht zu schützen. Dann wird sich vermutlich für die gemessene $s(A)-$Kurve der in Fig. 4 punktierte Verlauf dmd ergeben.

Es ist noch zu beachten, daß in unserer Theorie 1. die Erdoberfläche als diffus reflektierend angenommen wurde (h), zweitens die Strahlen der Gruppe III vernachlässigt wurden. Wenn die Voraussetzung (h) nicht zutrifft, sondern doch eine Zerstreuungsfunktion für die Zerstreuung an der Erdoberfläche besteht[1]), so sind vor allem für $A = 150^0$ und 180^0 die Strahlen der Gruppe III nicht mehr zu vernachlässigen, was in diesem Falle eine erhebliche Vergrößerung von H_r^1 für $A = 150^0$ und $A = 180^0$ und damit eine Schwankung von s_w mit A zur Folge hätte, die die von Wigand gemessene vielleicht erreichen könnte.

Schließlich geben wir hier noch die Gangfaktoren der Theorie und der Wigand'schen Messungen in Tabelle 5 zahlenmäßig an.

Tabelle 5.

A	0	15	30	60	90	120	150	180
$C \ln\left[\dfrac{1+\varepsilon}{\varepsilon}\left(1 - q(A)\right)\right]$	1.000	(0.991)	0.983	0.972	0.950	1.083	1.307	1.290
Wigand	1.000	1.41	1.12	0.81	1.31	1.88	1.69	1.56

3. $\mathfrak{A} = \mathfrak{A}(\lambda)$. Hier treten zwei Kontraste auf, die zunächst voneinander unabhängig wirken: Helligkeits- und Farbkontraste. Diese sind also getrennt zu berechnen: Von den Sichtweiten, die sich hieraus ergeben, ist stets die größere zu nehmen, da bei gleicher Helligkeit Ziel und Umgebung sich noch durch die Farbe unterscheiden könnten und umgekehrt.

a) Der Farbkontrast verschwindet, wenn das von dem Schirm reflektierte, durch λ charakterisierte Licht auf einen bestimmten Bruchteil ε' der Lichtmenge abgeklungen ist, die aus der Sehstrahlpyramide ins Auge gestreut wird. Damit ergibt sich als Bestimmungsgleichung

$$37) \qquad \frac{H_s}{H_\lambda} = \frac{i_g(A)\left(1 - e^{-a_0 s_\lambda}\right)}{H_\lambda^1 e^{-a_\lambda s_\lambda}} = \frac{1}{\varepsilon'}.$$

Hierbei ist zu beachten, daß nun zwei verschiedene Zerstreuungskoeffizienten auftreten, die miteinander in der Beziehung stehen

$$e^{-a_0 x} \Sigma J_0(\lambda)\varDelta\lambda = \Sigma e^{-a_\lambda x} \cdot J_0(\lambda)\varDelta\lambda.$$

---

1) Vgl z. B. die von Thaler, Diss. Kiel 1903, sogar für Magnesiumoxyd usw. experimentell ermittelte Zerstreuungsfunktion.

Es ist nicht wahrscheinlich, daß für die bodennahen Schichten, die für unser Problem nur in Frage kommen, $a_\lambda = C . \lambda^{-4}$ gemäß der Rayleigh'schen Theorie gegeben ist. Die Beobachtung der spektralen Extinktion bereits des Sternenlichtes ergeben eine stärkere Annäherung an $a_\lambda = C' . \lambda^{-2}$, je geringere Meereshöhe die beobachtende Station hatte. Jedenfalls ist $(a_0 - a_\lambda)$ eine kleine Größe, sodaß man $e^{-a_0 s_\lambda} = e^{-[a_\lambda s_\lambda + (a_0 - a_\lambda) s_\lambda]}$ nach Potenzen von $(a_0 - a_\lambda) = \varDelta a$ entwickeln kann. Dann erhält man

$$38) \qquad s_\lambda = \frac{1}{a_\lambda} \cdot \ln\left(\frac{1}{\varepsilon'} \cdot \frac{H_\lambda^1}{i_s(A)} + 1 - a_\lambda \varDelta a \right).$$

Wir wollen hier wieder das Beispiel rechnen $\zeta_\odot = 70^0$, $s_s = 15\,\mathrm{km}$. Wir wählen $\lambda = 650\,\mu\mu$ und nehmen den Spektralbereich $600-700\,\mu\mu$. Für diesen Bereich sei $\mathfrak{A}(\lambda) = \mathfrak{A}_{600-700} = 1$, sonst $= 0$. Das Himmelslicht selbst besitze noch einen kleinen Rotgehalt, den es auch besitzen muß, da in diffusem Tageslicht ein roter Schirm nicht schwarz erscheint. Spektrale Beobachtungen des Himmelslichtes liegen meines Wissens noch nicht vor, der Rotgehalt sei zu $10\,{}^0/_0$ angenommen. Das Sonnenlicht hat nach den Berechnungen von F. Linke[1]) in unserem Fall und für den Trübungsfaktor[2]) $T = 2$ einen Rotgehalt von $18.1\,{}^0/_0$. Für ε' setzen wir zwei extreme Werte ein, $\varepsilon' = 5\,{}^0/_0$ und $\varepsilon' = 1\,{}^0/_0$ und vernachlässigen $a_\lambda \varDelta a$. Aus $a_1 : a_2 = \lambda_2^2 : \lambda_1^2$ erhält man, wenn man a_0 mit dem a_λ des optischen Schwerpunktes im Horizontlicht (etwa $450\,\mu\mu$) identifiziert, $a_{650} = \frac{a_0}{2.09}$, wo für a_0 entsprechend $s_s = 15\,\mathrm{km}$ der Wert 0.2609 einzusetzen ist. Damit und aus Tabelle 3 erhält man Tab. 6.

Tabelle 6.

A	0	30	60	90	120	150	180 Grad
$\sigma \cdot \dfrac{1}{\pi}$	0.5	0.14	0.15	0.16	0.20	0.23	0.25
$R_s \dfrac{1}{\pi}$	—	—	—	—	0.02	0.02	0.03
$\odot \cdot \dfrac{1}{\pi}$	—	—	—	—	2.07	3.61	4.15
H_λ^1	0.15	0.14	0.15	0.16	2.29	3.86	4.43
s_λ für $\varepsilon' = 5\,{}^0/_0$	1.79	2.41	2.86	3.43	4.55	6.57	6.56
s_λ „ $\varepsilon' = 1\,{}^0/_0$	6.50	8.12	9.17	10.43	12.62	16.09	15.98

b) Der Helligkeitskontrast verschwindet, wenn entsprechend der Bedingung 37)

---

1) Noch nicht veröffentlicht. Herr Prof. Linke stellte mir freundlichst das Manuskript zur Verfügung.

2) T ist definiert durch $J_{m,r} = J_0 e^{-a_r \cdot T \cdot m}$, wenn $m = \sec \zeta_\odot$. T ist praktisch von der Sonnenhöhe unabhängig, also ein Maß der Lufttrübung. Vgl. F. Linke. Transmissionskoeffizient und Trübungsfaktor. Beitr. z. Phys. d. fr. Atm., X, S. 91—103.

39)
$$\frac{H_{hor}}{\overline{\jmath\jmath}\;e^{-a_\lambda s_\lambda} + H_s} = \frac{1}{\varepsilon' + \left(1 - e^{-a_0 s_\lambda}\right)} = 1 + \varepsilon.$$

Hieraus folgt

40)
$$s_\lambda = \frac{1}{a_0}\ln\frac{1+\varepsilon}{\varepsilon+\varepsilon'}.$$

Die Sichtweite s_λ wird also auch hier wieder von A unabhängig. Dies rührt davon her, daß die durch Reflexion erzeugte Helligkeit des Schirmes infolge der Bedingung 38) eben verschwindend klein sein muß, d. h. wir kommen hier im wesentlichen auf den schwarzen Körper zurück.

Zahlenmäßig ergibt sich aus 40)

für $s_s = 15.0$ km: $s_\lambda = 10.2$ km für $\varepsilon' = 5\%$ u. s_λ 13.4 km für $\varepsilon' = 1\%$.

Die Ergebnisse nach 38) und 40) können wir so in Worte fassen:
Die Sichtweite $s_{\lambda,}$ infolge des Farbenkontrastes hat ein Minimum für $A = 0$ und wächst bis $A = 180$ in einem Verhältnis, das nur abhängig ist von der Farbempfindlichkeit (ε'). Die Sichtweite $s_{\lambda,}$ infolge des Helligkeitskontrastes gegen den Horizont ist unabhängig von A ebenso wie s_s, aber kleiner als s_s. Dagegen ist $s_{\lambda,}$ größer als $s_{\lambda,}$ (die Sichtweite infolge des Helligkeitskontrastes wird von der Sichtweite infolge des Farbkontrastes nur für äußerste Farbenempfindlichkeit ($\varepsilon' = 1\%$) übertroffen). Somit ist die Sichtweite eines selektiv reflektierenden Schirmes praktisch vom Azimut unabhängig und ist dem mittleren Zerstreuungskoeffizienten a_0 umgekehrt proportional.

§ 5. Die Gültigkeitsgrenzen der hier entwickelten Beziehungen.

Wie betont, ist die Abhängigkeit der Sichtweite vom Azimut eine Funktion von $\zeta_\odot$. Sie verschwindet natürlich für Zenithsonne und erreicht ihr Maximum für Horizontsonne. Der letzte Fall konnte aber nicht berechnet werden, da hierfür noch die Daten der Flächenhelligkeit der Sonne und des 3^0-Ringes fehlen. Als Beispiel wurde stets der Fall $\zeta_\odot = 70^0$ durchgerechnet.

Über die einzelnen Komponenten des Faktors a läßt sich nichts aussagen, bis nicht die Frage geklärt ist, was eigentlich der die Atmosphäre trübende Bestandteil, der „Dunst" ist. Ob Reflexion, Brechung, Beugung oder Rayleighsche Zerstreuung wirksam ist, spielt bei unseren Überlegungen gar keine Rolle, da alle diese Komponenten auf jeden Fall lichtzerstreuend wirken. a ist ein Zerstreuungsfaktor, der durch $dJ = -a J(x)\,dx$ scharf definiert ist. Man kann a mit großer Annäherung als den Bruchteil definieren, um den die Intensität J eines sehr schmalen, parallelen Lichtbündels auf dem Wege 1 in einem sehr engen, absolut schwarzen, mit Luft gefüllten Rohre geschwächt wird.

Für unsere Entwickelungen ist es ebenso ganz gleichgültig, ob eine sekundäre Zerstreuung sichtbaren Lichtes tatsächlich erfolgt oder nicht. Durch Einführung der Himmelshelligkeit als gemessener Größe werden wir davon gänzlich unabhängig (vgl. Gleichung 27)).

Dagegen ist die Absorption vernachlässigt worden. Es ist garnicht undenkbar — das hängt wieder ganz von der Natur des „Dunstes" ab —, daß gerade in den untersten Schichten die Absorption doch nicht vernachlässigt werden darf. Das ändert aber an unseren Herleitungen wenig: In § 2 und 3 tritt in dem Exponenten von e stets anstelle von a die Summe $(a+b)$, wenn b der Absorptionskoeffizient ist, sodaß schließlich, wie eine Nachrechnung zeigt, in den Sichtweiten des § 4 antelle von $\frac{1}{a_0}$ der Faktor $\frac{1}{a_0+b}$ tritt.

Die Beziehungen gelten sämtlich nur für kleine Schirme. Als klein ist schon ein Schirm anzusehen, wenn er vom Mittelpunkt von l aus gesehen unter einem Winkel von weniger als 5^0 erscheint, weil dann die zuviel mitgenommenen Himmelsflächenstücke gegenüber der Himmelshalbkugel unbedenklich vernachlässigt werden dürfen. Für große Schirme ließe sich die Aufgabe durch mehrfache Anwendung desselben Integrationsverfahrens lösen, indem man für einzelne Punkte auf der Strecke Auge—Ziel immer nur über die von den Schirmen unverdeckten Himmelsteile F integriert und so

$$\iint\limits_F d\xi_1\, d\alpha_1 \sin\xi_1\, i_g(\alpha_1\,\xi_1)\, \mathfrak{B}(\alpha_1,\,\xi_1) \,.\, df\, dr_0 \,.\, e^{-a_0 r_0}$$

als Funktion von r_0 darstellt, wofür allerdings $\mathfrak{B}$ bekannt sein müßte. Doch ist dieser Fall für die Sichttheorie praktisch bedeutungslos.

Weiter gelten die Beziehungen zunächst nur für objektive Messung, etwa mit der lichtelektrischen Zelle. Für das Auge liegen die Verhältnisse etwas komplizierter. Da sich unsere Darstellung nur auf den Tag bezieht — bei Nacht wird ja die Sichtweite meist ein reines Extinktionsproblem sein —, so können wir für das helladaptierte Auge und für einen konstanten Sehwinkel auch ε als Konstante ansehen, und zwar für die ganze Tageszeit und für die in Betracht kommenden Helligkeiten [1]. (Für Helligkeiten, die kleiner oder größer als die hier genannten sind, wird jedoch die Empfindlichkeit kleiner, also ε größer, die dazu gehörigen Sichtweiten ceteris paribus kleiner). Dagegen ist der Wert von ε in starkem Maße abhängig von dem Winkel, unter dem das Ziel dem Auge erscheint. Setzen wir ε für den Gesichtswinkel 1^0 gleich der Einheit, so ist (Helmholtz, Physiolog. Optik, Bd. II, S. 286)

Gesichtswinkel	0.25^0	0.5^0	1.0^0	2.0^0	3.0^0	5.0^0	10^0	15^0	20^0	30^0
ε	4.3	2.0	1.0	0.6	0.5	0.5	0.5	0.5	0.5	0.5

Mit Hilfe dieser Tabelle wären die Sichtweiten verschieden großer Ziele aufeinander zu reduzieren. Für die praktische Sichtmessung selbst folgt daraus, daß man nach Möglichkeit solche Ziele benutzen wird, die unter demselben räumlichen Winkel erscheinen.

Es bietet keine Schwierigkeit mehr, die analytischen Beziehungen des § 2 auch auf das Problem der vertikalen oder noch allgemeiner der nichthorizontalen Sichtweite zu übertragen. Da aber Himmelshelligkeitsmessungen vom Ballon aus bisher leider noch fehlen, so wurde auf diese Übertragung verzichtet. da sie zu keinen zahlenmäßigen Angaben über die Sichtweite führt.

---

1) Vgl. z. B. Helmholtz, Physiologische Optik, Bd. II, S. 146.

Zusammenfassung. Der Ausgangspunkt ist die Aufhellung der Luft vor
dem schwarzen Körper (wie bei Weber). Die Knickmethode (Wiener, Madelung)
führt zu einem Ausdruck für die Helligkeit in Richtung auf den schwarzen
Körper, der sich bei bekannter Zerstreuungsfunktion in der oben angegebenen
Weise graphisch auswerten läßt. Dazu muß die Himmelshelligkeit als gemessene
Größe eingeführt werden. Ein Grenzübergang macht aber von der Zerstreuungsfunktion unabhängig. Die Sichtweiten selbst ergeben sich dann in einfacher
Weise, nur die des weißen Schirmes erfordert eine nochmalige Anwendung der
angegebenen Integrationsmethode. Die Ergebnisse gelten ganz unabhängig davon, welche Anschauung man sich über die Einzelvorgänge der Lichtzerstreuung,
über die Existenz mehrfacher Zerstreuungen und über die Form der Zerstreuungsfunktion macht.

Meteorolog. Geophysik. Institut der Universität, März 1924.

Theorie der horizontalen Sichtweite II: Kontrast und Sichtweite.

Von Harald Koschmieder, Frankfurt a. Main.

In den Vordergrund der folgenden Untersuchungen wird der Kontrast gestellt werden. Er wird in einer etwas anderen Weise definiert werden, als ich es früher in meiner „Theorie der horizontalen Sichtweite“ [1]) getan habe. Der Kontrast besitzt den Vorteil, sich feldmäßig darstellen zu lassen und führt so zu einem viel tieferen Verständnis der für die Sichtweite maßgebenden Verhältnisse. Durch seine Darstellung wird für die Sichtweite einer weißen Lambertschen Fläche eine wesentliche Vervollständigung gewonnen werden und für einen farbigen Schirm werden etwas überraschende Ergebnisse erzielt. Hierbei zeigt sich, daß die früher gegebene Formel für die Sichtweite eines farbigen Schirmes I 39) und I 40) mit einer unerlaubten Näherung rechnete, die durch einen Schreibfehler in I 38) zulässig erschien. Alle anderen Beziehungen bleiben erhalten, die durch die neue Kontrastdefinition eintretenden Änderungen haben auf die Sichtweiten keinen Einfluß.

Eingefügt sei noch eine einfache Ableitung der in Richtung auf die schwarzen Körper meßbaren Helligkeit, die dieselbe Strenge besitzt, wie das Wiener-Madelungsche Knickverfahren.

Es erübrigt noch zu bemerken, daß sich alle Untersuchungen wieder auf die horizontale Sichtweite beziehen. Als Umgebung wird stets der Horizont (h) benutzt. Hier wird ebenso wie in I nicht des Umstandes Rechnung getragen, daß nach Einbringen eines Zieles die Helligkeit der unmittelbar benachbarten Horizontteile eine Änderung erfährt. Wie weit diese Voraussetzung erfüllt ist, muß die Erfahrung lehren.

§ 1. **Definitionen.** Die Definitionen I A) und I B) bleiben in Kraft. Dagegen benutzen wir nun folgende Definition:

II C) Der Helligkeitskontrast eines bestimmten Zieles Z gegen den Horizont h sei gegeben durch

$$K(Z, h) = \frac{H_z - H_h}{H_h}.$$

Diese Definition entspricht gut dem Sprachgebrauch. Ist die in Richtung auf das Ziel meßbare Helligkeit kleiner als die Horizonthelligkeit, oder kurz gesagt,

---

1) Beitr. z. Phys. d. fr. Atmosph. XII, S. 33—53, 1925, im folgenden der Kürze halber mit I bezeichnet.

erscheint das Ziel dunkler als der Horizont, so ist der Kontrast negativ und zwar ist $K = -1$ für $H_z = 0$. Mit zunehmendem H_z wächst der Kontrast linear und geht für $H_z = H_h$ stetig durch die Null hindurch. Es wird $K = +\infty$ für $H_z = \infty$. Es ist wichtig festzuhalten, daß der Kontrast eine mit Vorzeichen behaftete Größe ist und daß sein Durchgang durch die Null stetig erfolgt.

§ 2. **Die Helligkeit in Richtung auf den schwarzen Körper.** (Das Luftlicht.) Wir betrachten ein bestimmtes Raumelement einer Sehstrahl-Pyramide, die durch Auge und schwarzen Körper gegeben ist. Die Größe dieses Raumelementes ist nach I 3) $dfr_0^2 . dr_0$. Die hieraus in Richtung auf das Auge gestreute Lichtmenge soll unabhängig sein von der Entfernung Volumenelement — schwarzer Körper. Das setzt eine lineare Abmessung des schwarzen Körpers voraus, die klein ist gegenüber seiner Entfernung vom Auge. In den allermeisten Fällen wird diese Voraussetzung erfüllt sein. Doch ist sie es oft nicht mehr bei sehr dichtem Nebel. Darauf sei hingewiesen. Unter dieser einschränkenden Voraussetzung, die übrigens auch in I ausdrücklich gemacht wurde, ist die in Richtung auf das Auge gestreute Lichtmenge gegeben durch $C . dfr_0^2 . dr_0$, wo C hinsichtlich r_0 eine Konstante ist. In C muß der Zerstreuungskoeffizient a_0 als Faktor enthalten sein, wir schreiben dafür $a_0 . c$, wo c aus den Randbedingungen zu bestimmen ist. Die von dem einen, herausgegriffenen Raumelement schließlich ins Auge gelangende Lichtmenge ist

$$1) \qquad dI = a_0 \cdot c \cdot dfr_0^2 dr_0 \cdot \frac{e^{-a_0 r_0}}{r_0^2} = \bar{c} \cdot a_0 \cdot dr_0\, e^{-a_0 r_0},$$

wenn wir $df.c$ in eine neue Konstante $\bar{c}$ zusammenfassen. Die von allen Raumelementen in das Auge gelangende Lichtmenge, d. h. das Luftlicht, ist dann einfach

$$2) \qquad H_s = \int_0^l dI = \bar{c}e^{-a_0 l} + C'.$$

Die Randbedingungen sind folgende:

a) Ist der schwarze Körper unendlich weit vom Auge, so ist das Luftlicht gleich der Horizonthelligkeit in demselben azimutalen Sonnenbestande, d. h.

$$3) \qquad H_{s|_{l=\infty}} = i_g \text{ liefert } C' = i_g.$$

b) Ist der schwarze Körper unmittelbar vor dem Auge, so ist das Luftlicht gleich Null, d. h.

$$4) \qquad H_{s|_{l=0}} = 0 \text{ liefert } \bar{c} = -C'.$$

Damit folgt entsprechend I 30):

$$5) \qquad H_s = i_g\left(1 - e^{-a_0 l}\right).$$

§ 3. **Der schwarze Körper.** Nach II C) ist der Kontrast

$$6) \qquad K(s, h) = \frac{H_s}{H_h} - 1 = -e^{-a_0 l}$$

Der Kontrast ist also stets negativ und für jede bestimmte Entfernung in allen Azimuthen der gleiche. Wird $K = -\varepsilon$, so folgt für die Sichtweite

$$7) \qquad s_s = \frac{1}{a_0} l_n \frac{1}{+\varepsilon}.$$

Der Unterschied dieser Beziehungen gegen I 33) rührt von der neuen Kontrastdefinition her, auf die Sichtweite selbst hat diese Veränderung keinen Einfluß, sodaß also I Tab. 2 gültig bleibt.

Es hat mehrfach Erstaunen hervorgerufen, daß sich die Sichtweite des schwarzen Körpers als unabhängig vom Azimut erweist. Die Ursache liegt eigentlich auf der Hand: Das Luftlicht ist für jede bestimmte Entfernung bei allen Azimuten ein und derselbe Prozentsatz des Horizontlichtes. Die Zerstreuungsfunktion, die das Horizontlicht für verschiedene Azimute größer oder kleiner macht, ändert das Luftlicht in demselben Verhältnis. Der schwarze Körper hat ja keine andere Funktion, als die Sehstrahlpyramide zu begrenzen.

§ 4. Die weiße Lambertsche Fläche. Wieder nach II C) ist der Kontrast gegeben durch

$$8) \qquad K(w, h) = \frac{H_w}{H_h} - 1 = \left(\frac{H_w^1}{i_g} - 1\right) \cdot e^{-a_0 l}.$$

Da sowohl H_w^1 wie i_g längs des Horizontes vom azimutalen Sonnenabstande A abhängen, so ändert sich der Faktor $\dfrac{H_w^1}{i_g} - 1$ längs des Horizontes. Da dort $H_w^1 \gtreqless i_g$ wird, so ist $K \gtreqless 0$. Um über das Verhalten des Kontrastes Klarheit zu schaffen, wurde er feldmäßig dargestellt. Dazu wurde H_w^1 nach meinen Berechnungen I Tab. 3 graphisch dargestellt (vgl. Fig. 1; die Abszisse ist der azimutale Sonnenabstand A, die Ordinate die Helligkeit, Zenit $= 1$ gesetzt). $\dfrac{dH_w^1}{dA}$ ist für $A = 90$ unstetig. Für i_g wurden wieder die Dornoschen Werte I S. 46 benutzt, aber ebenfalls graphisch ausgeglichen. Die den Dornoschen Werten entsprechende Kurve in Fig. 1 ist mit I bezeichnet. Man sieht, daß die beiden Kurven sich schneiden, und zwar bei $A = 101^\circ$. Dort also ist der Kontrast $= 0$! Für kleinere A ist er negativ, für größere ist er positiv. Seine Werte wurden mit Hilfe der Fig. 1 berechnet für $A = 0, 30, 60, 90, 92, 96, 106, 110, 120, 160$ und 180. Für a_0 wurde der Wert für $s_s = 15$ km eingesetzt nach I Tab. 2. In Figur 2 ist das so berechnete Kontrastfeld dargestellt. (Polarkoordinaten: azimutaler Sonnenabstand A in Winkelgraden und Entfernung l in km.) Die den Linien gleichen Kontrastes beigeschriebenen Zahlen sind Prozente. Die Fig. 2 bringt einen ganz einfachen Tatbestand zum Ausdruck: Unter der Sonne ist ein weißer

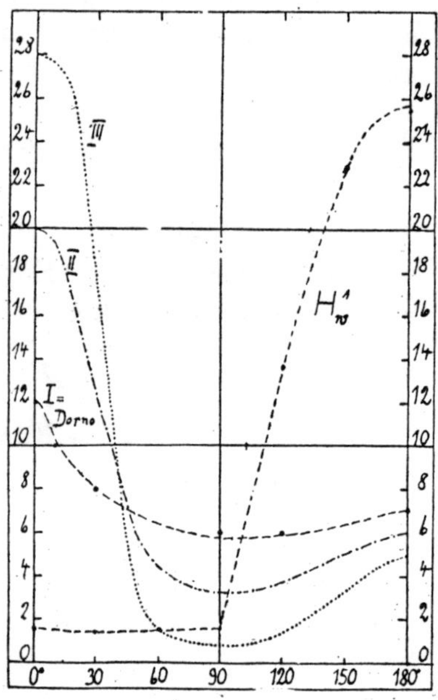

Figur 1.

Schirm dunkler als der Horizont und der Sonne gegenüber ist er heller als der
Horizont. Dann muß es auch ein Azimut geben, wo er eben so hell ist wie der
Horizont.

Für die Sichtweite ergibt sich nach (8)

$$9) \qquad s_w = \frac{1}{a_0}\, ln\left[\frac{1}{\varepsilon}\left(\frac{H_w^1}{i_g}-1\right)\right].$$

Der Unterschied gegen I 36) rührt wieder von der neuen Kontrastdefinition
her, auf die Sichtweite selbst hat diese Änderung keinen Einfluß, insbesondere
behalten die Tabellen I 4 und I 5 ihre Gültigkeit. Aber diese beiden Tabellen
sind um einen wesentlichen Bestandteil zu vervollständigen! Offensichtlich muß
die Sichtweite für das Azimut, in dem der Kontrast verschwindet, ebenfalls verschwinden. Für eine vorgegebene Reizschwelle $\pm\,\varepsilon$ verschwindet die Sichtweite

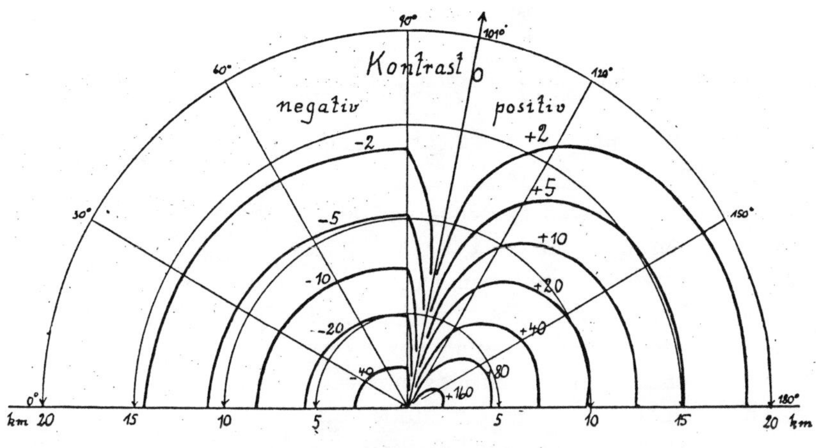

Figur 2.

bereits für $\dfrac{H_w^1}{i_g}-1 = \pm\,\varepsilon$. Dieses Ergebnis ist in Fig. 3 dargestellt, wo $\dfrac{s_w(A)}{s_w(0)}$ als
Funktion von A aufgetragen ist. Die Werte wurden mit Hilfe von Fig. 1 für
dieselben Azimute berechnet. Das rechtwinklige Koordinatensystem wurde gewählt, um die Verhältnisse in der Nähe der Nullstelle deutlich zum Ausdruck
zu bringen. Eine Darstellung der Sichtweite in Polarkoordinaten sind ja die
Linien $\varepsilon = \pm\,2\,^0/_0$ in Fig. 2.

Nun ist noch ein Moment zu berücksichtigen: H_w^1 und i_g sind nicht nur
Funktionen von A, sondern auch von a_0. Die Beziehung (9) lautet daher vollständig

$$10) \qquad s_w = \frac{1}{a_0}\, ln\left[\frac{1}{\varepsilon}\left(\frac{H_w^1(A,\,a_0)}{i_g(A,\,a_0)}-1\right)\right].$$

Je größer die Trübung, je größer a_0, umso größer sind auch die Schwankungen
der Horizonthelligkeit im Azimut. Man hat Gelegenheit, diese Tatsache besonders gut im Herbst zu beobachten: Ist die Dunstschicht nicht übermäßig hoch
(100 bis 200 m), so ist unter der Sonne der Horizont blendend weiß, der Sonne

gegenüber grauschwarz. M. Uibe [1]) findet in Güimar (300 m Höhe) $\dfrac{i_g(0)}{i_g(180)} = 4:1$,

auf dem Pik dagegen (3700 m) $\dfrac{i_g(0)}{i_g(180)} = 2:1$. In der (trüben) Ebene sind also die Schwankungen der Horizonthelligkeit zweifellos größer als im (klaren) Hochgebirge, insofern sind die Dornoschen Werte für Davos wohl für die Ebene nicht mehr übertragbar. Mir sind leider keinerlei Messungen der Horizonthelligkeit in der Ebene bzw. bei starker Trübung bekannt geworden, sodaß

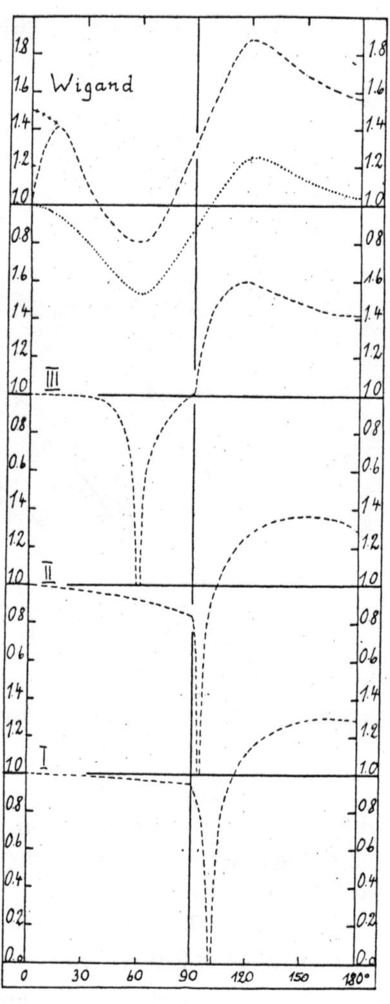

Figur 3.

nichts anderes übrig bleibt, als sich eine Horizonthelligkeit zu konstruieren, wobei selbst bei plausiblen Annahmen große Willkürlichkeiten unvermeidlich sind. Das schadet weiter nichts, die Ergebnisse sind nur qualitativ zu werten. Solche willkürlich angenommenen Horizonthelligkeiten sind in Fig. I, II und III dargestellt. Verlangt wurde von den Kurven lediglich, daß $\dfrac{\partial i_g}{\partial A} = 0$ für $A = 0$, 90, 180° und $\dfrac{\partial}{\partial a_0} \displaystyle\int_0^{180} i_g\, dA = 0$. Da es sich nur um eine dünne Dunstschicht handeln soll, wurde H_w^1 konstant gehalten.

Die mit diesen i_g-Kurven berechneten Gangfaktoren sind in Fig. 3, II und III in derselben Weise dargestellt wie die mit den Dornoschen Werten berechneten Gangfaktoren. Man sieht folgendes: Je größer die Schwankung von $i_g(A)$, um so näher rücken Nullstelle und Maximum der Sichtweite der Sonne. Ob die Nullstelle tatsächlich einen geringeren Abstand als 90° von der Sonne erhalten kann, muß die Erfahrung lehren: je dunkler der Horizont der Sonne gegenüber, umso kleiner wird ja auch die Flächenhelle des Schirmes unter der Sonne. Dafür sprechen lediglich die Wigandschen Messungen mit ihrem ausgesprochenen Minimum für $A = 60°$. Man sieht, daß man durch geeignete Veränderung der Horizonthelligkeit und der Flächenhelle des weißen Schirmes jede Abhängigkeit der Sichtweite vom Azimut (sogar das Wigandsche Minimum für $A = 0$) erzwingen kann: Hier helfen also nur Messungen weiter. — Sicher aber steht das Ergebnis: Solange eine Zunahme der Trübung mit einer Zunahme der Schwankung in der Horizonthelligkeit verbunden ist, solange nimmt die Schwankung der Sichtweite im Azimut mit der Trübung zu.

Noch ein Wort zu den Wigandschen Messungen. Diese sind ebenfalls in

---

1) Abhandl. der math.-phys. Kl. der Kgl. Sächs. Ges. d. Wiss., XXXV, 1918; vgl. besonders Tafel 17 u. 20.

der Fig. 3 dargestellt, gestrichelt in der von Wigand gegebenen Weise. Nach meinen Ausführungen I S. 50 ist für $A = 0$ bis $A = 15$ der gekreuzelte Verlauf wahrscheinlich. Bildet man aus der so abgeänderten Kurve $\dfrac{s_w(A)}{s_w(0)}$, so erhält man den punktierten Verlauf. Der Gang mit dem Azimut wird wesentlich geringer und steht in befriedigender Übereinstimmung mit der Kurve III.

§ 5. Der rote Schirm. Wir betrachten hier zunächst nur den Helligkeitskontrast. Dieser ist nach IIC

$$11) \qquad K(\lambda, h) = \frac{H_\lambda}{H_h} - 1 = \frac{H_\lambda^l}{i_g}\, e^{-a_\lambda l} - e^{-a_0 l}.$$

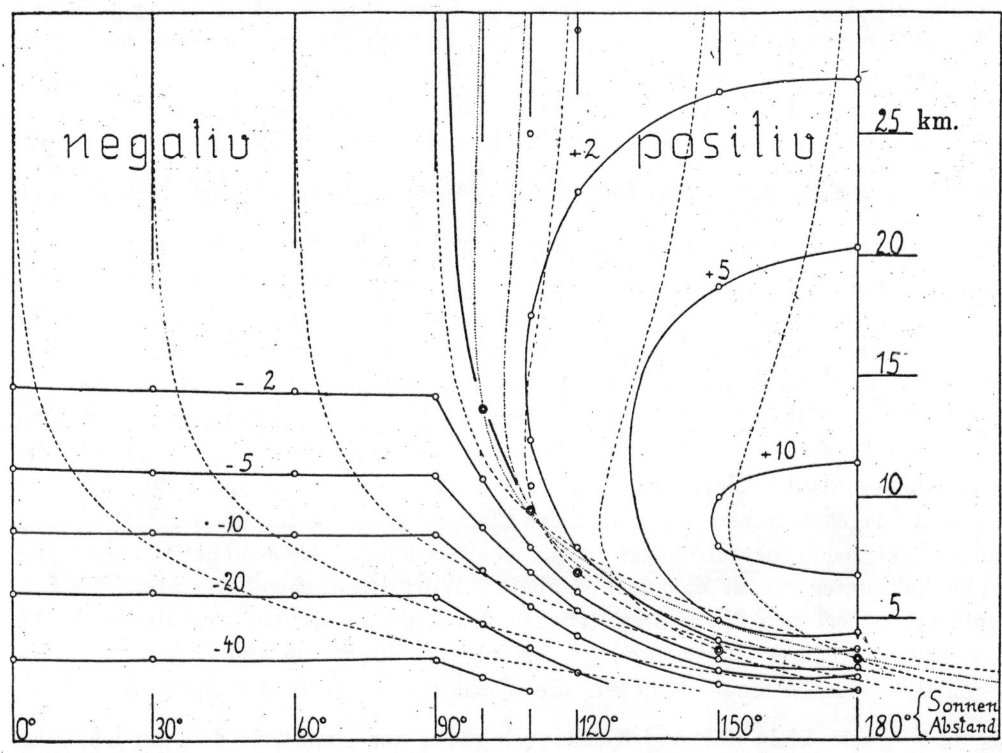

Figur 4.

Der Berechnung des Kontrastes wurden zugrunde gelegt die H_λ^l-Werte nach I Tab. 6 und die i_g-Werte von Dorno. Beide wurden graphisch ausgeglichen, wodurch die hier benutzten i_g-Werte etwas von denen abweichen, die für den weißen Schirm benutzt wurden:

Benutzte i_g-Werte

A	0	30	60	90	92	96	100	106	110	120	150	160	170	180
für weißen Schirm	12.0	8.0	6.4	5.70	5.73	5.76	—	5.82	5.85	5.9	6.4	6.6	6.9	7.0
„ roten „	12.0	8.6	6.6	5.66	5.68	5.70	5.72	—	5.76	5.8	6.4	—	—	7.0

Die quantitativen Abweichungen sind für die Ergebnisse unwesentlich, die Lage der Extrema bleibt erhalten.

Nach I Tab. 6 ist $\frac{H_\lambda^1}{i_g} < 1$, aber es ist $a_\lambda < a_0$. Infolgedessen nimmt $e^{-a_0 l}$ sehr viel schneller ab als $\frac{H_\lambda^1}{i_g} e^{-a_\lambda l}$, sodaß für jeden bestimmten Wert $\frac{H_\lambda^1}{i_g}$, d. h. für jedes Azimut, mit wachsender Entfernung l der Kontrast einmal sein Vorzeichen wechselt, weil ja die Gleichung $\frac{H_\lambda^1}{i_g} e^{-a_\lambda l} = e^{-a_0 l}$ außer $l = \infty$ stets noch eine Wurzel hat. Je größer $\frac{H_\lambda^1}{i_g}$, umso näher wird diese Umkehrstelle dem Auge liegen. Für $l = \infty$ wird $K = 0$.

Es wurde nun der Kontrast berechnet für die Azimute 0, 30, 60, 90, 100, 110, 120, 150, 180°. In Fig. 4 sind diese berechneten Werte gestrichelt eingetragen, als Abszisse dient die zu dem jeweiligen Sonnenabstand gehörige vertikale Entfernungsgerade, negativer Kontrast ist nach rechts, positiver nach links aufgetragen, derart, daß der Abstand von 30° gleich 12% ist. — Betrachten wir zunächst die Kontrastkurve für $A = 180$. Abszisse für $A = 180$ ist die in der Figur rechtstehende, vertikale Entfernungsgerade. Der Sonne gegenüber besitzt der Schirm die größte Helligkeit, $\frac{H_\lambda^1}{i_g} = 0.633$, aber er ist doch dunkler als der Horizont, d. h. $K < 0$. Schieben wir jetzt den Schirm hinaus, so nimmt die scheinbare Flächenhelligkeit infolge der Zerstreuung ab und · zwar mit $e^{-a_\lambda l}$. Das Luftlicht nimmt aber schneller zu $\left(\text{mit } 1 - e^{-a_0 l}\right)$, als die scheinbare Flächenhelligkeit abnimmt, d. h. die Summe beider wächst mit der Entfernung, und zwar am schnellsten zuerst. Sie geht dann durch die Horizonthelligkeit hindurch, d. h. $K = 0$, wächst noch weiter, aber langsamer, der Schirm erscheint infolge seiner eigenen Flächenhelligkeit und des dazugetretenen Luftlichtes heller als der Horizont, d. h. $K > 0$. Dann wird das Luftlicht praktisch konstant, dagegen nimmt die scheinbare Flächenhelligkeit infolge der Zerstreuung weiter ab, sodaß der Kontrast zwar positiv bleibt, aber kleiner wird, und schließlich unter einem vernachlässigbaren Wert (Reizschwelle) sinkt. Im Unendlichen nähert sich die Summe Reflektiertes Licht + Luftlicht dem Horizontlicht von einem größeren Wert, und der Kontrast sinkt von der positiven Seite gegen 0. — Im Azimut 0 liegen die Verhältnisse insofern wesentlich anders, als da $\frac{H_\lambda^1}{i_g}$ sehr klein (0.125) ist. Dadurch rückt der Punkt, in dem der Kontrast verschwindet, sehr weit hinaus, und der positive Kontrast bleibt immer unter der Reizschwelle. Hier, für $A = 0$, ist infolge der geringen Flächenhelle des Schirmes der Kontrast beinahe allein gegeben durch den zweiten Summanden.

Bestimmt man jetzt graphisch alle die Punkte, die denselben Kontrastwert besitzen, — die Punkte liegen sämtlich auf den vertikalen Entfernungsgeraden für $A = 0$, 30, 60, 90, 100, 110, 120, 150, 180 und sind durch einen kleinen Kreis gekennzeichnet — so kann man Linien gleichen Kontrastes ziehen. Diese sind in Figur 4 durchlaufend eingezeichnet. Damit ist das Kontrastfeld gegeben. Es zerfällt, wie das des weißen Schirmes, in zwei Teile, positiven und negativen Kontrastes, doch ist hier die 0-Linie nicht eine gerade Linie, die durch den Ursprung geht, wie beim weißen Schirm, sondern eine gekrümmte Kurve. Sie ist in Figur 4 dick ausgezogen, aber nur ein Stück, um nicht die Darstellung

zu überlasten. Ihre Fortsetzung ist durch die doppelt eingekreisten Punkte
gegeben.

Die Sichtweite infolge des Helligkeitskontrastes ist gegeben
durch die Linien $K = \pm 2\,{}^0\!/_0$. Diese sind in Figur 5 eingetragen (wieder in

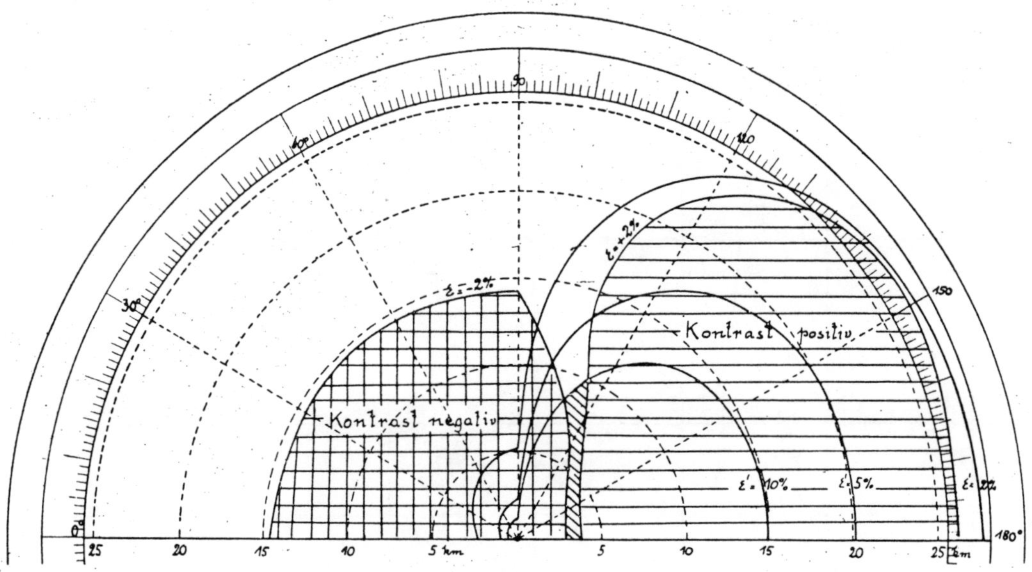

Figur 5.

Polarkoordinaten), die Gebiete, in denen der Helligkeitskontrast $\geqq |2|\,{}^0\!/_0$, d. h.
die Sichtgebiete, sind schraffiert, und zwar das Gebiet positiven Kontrastes horizontal, das negativen Kontrastes horizontal und vertikal. Beide sind durch
eine Unsichtigkeitszone $(K < |2|\,{}^0\!/_0)$ getrennt, die natürlich die oben be-
sprochene Nullinie des Kontrastes enthält. Es könnte auffallen, daß die Sichtweite des farbigen Schirmes über die des weißen (vergl. Figur 2) bei gleichen a
hinausgeht. Dies ist eine Folge davon, daß $a_\lambda < a_0$.

Es erübrigt noch eine Untersuchung, wie weit Farbunterschiede auf die
Sichtweite von Einfluß sind. Wir haben also zu fragen: In welcher Entfernung
ist das selektiv reflektierte, durch λ gekennzeichnete Licht auf einen vorgegebenen Bruchteil ε' des Luftlichtes gesunken. Diese Sichtweite s_λ' wird wie in
I 37) gegeben durch

$$12) \qquad\qquad \frac{H_\lambda}{H_s} = \varepsilon'$$

zu

$$13) \qquad\qquad s_\lambda' = \frac{1}{a_\lambda}\ln\left[1 + \frac{1}{\varepsilon'}\,\frac{H_\lambda^1}{i_g} - s_\lambda'\,\varDelta a\right]$$

(nicht $a_\lambda\,\varDelta a$ wie in I 38) irrtümlich steht). Am besten löst man die Gleichung
(12) graphisch, indem man sie in der Form

$$14) \qquad\qquad H_\lambda^1 e^{-a_\lambda s_\lambda'} = \varepsilon'\,i_g\bigl(1 - e^{-a_0 s_\lambda'}\bigr)$$

591

schreibt, die linke Seite gleich $\varphi(l)$, die Rechte gleich $\psi(l)$ setzt und diese Kurven aufträgt (vergl. Fig. 6, Abszisse l in km, Ordinaten φ und ψ). Die Schnittpunkte von φ und ψ geben dann die gesuchten s_λ'. Dieses ist durchgeführt für $\varepsilon' = 2,\ 5,\ 10\%$ und $A = 0,\ 30,\ 60,\ 90,\ 92,\ 96,\ 100,\ 110,\ 120,\ 150,\ 180$. Verbindet man die so erhaltenen Schnittpunkte, so erhält man Linien, längs deren zu jedem Azimut das dazugehörige s_λ' ermittelt werden kann. (In ähnlicher Weise läßt sich auch der Helligkeitskontrast bestimmen, für den Fall $K = -2\%$ ist das auch in Fig. 6 durchgeführt.)

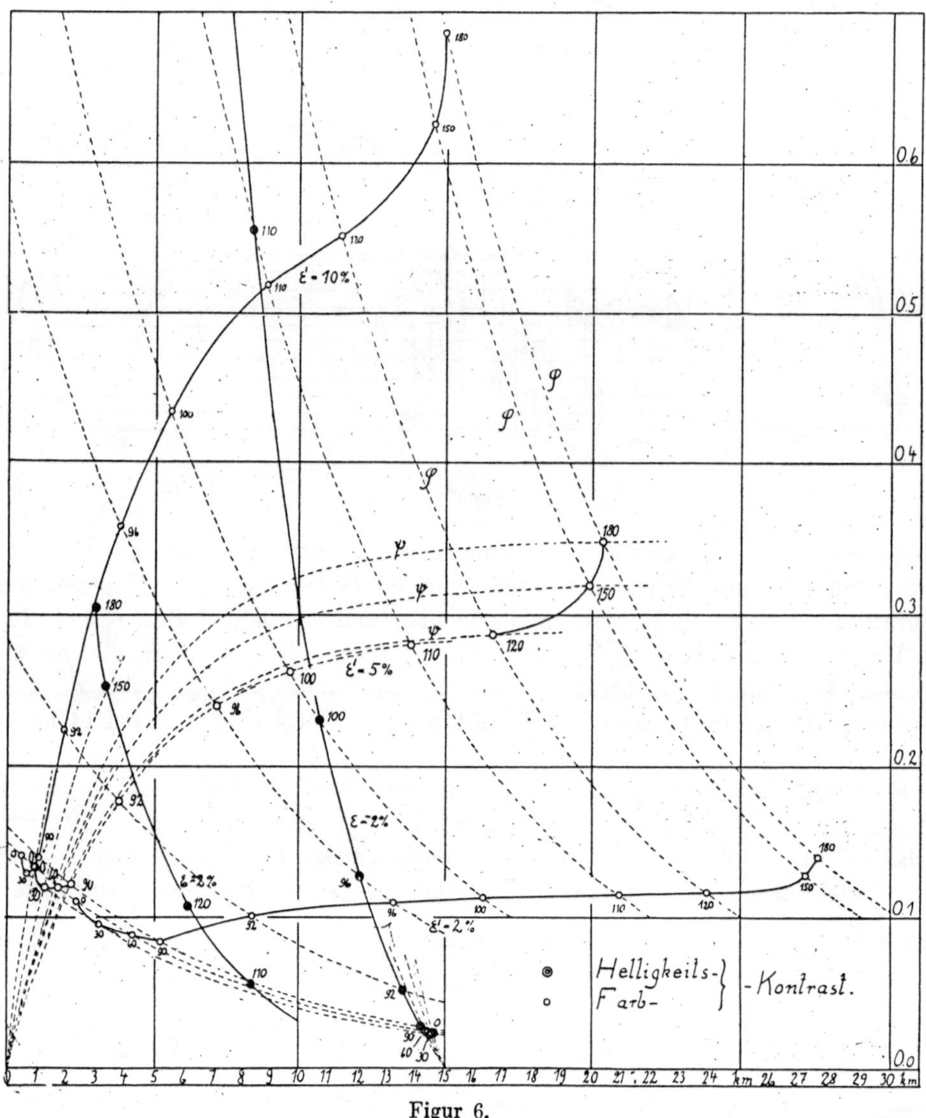

Figur 6.

Die s_λ' Werte sind für $\varepsilon' = 2,\ 5,\ 10\%$ in Figur 5 eingetragen. Man sieht, daß die s_λ'-Kurve die s_1-Kurve — natürlich nur für $\varepsilon' = \varepsilon$ — umfaßt, wie es ja sein muß. Nimmt man $\varepsilon' = 10\%$, so erhält man endgültig für den roten Schirm ein Sichtgebiet, daß sich aus den 3 schraffierten Einzelfällen zusammensetzt. In dem schrägschraffierten Teil ist der Schirm nur durch seine Farbe, nicht aber durch seine Helligkeit erkennbar. In einem Winkelbereich um $A = 110$ ist eine absolute Unsichtigkeitszone vorhanden.

§ 6. **Gültigkeitsgrenzen.** Die Ergebnisse des § 4 gelten streng für monochromatisches Licht und für den Fall, daß das vom Schirm reflektierte Licht und das Luftlicht gleiche Wellenlänge besitzen. Die Voraussetzung gleicher Farbe von Schirm und Umgebung wird in Praxi zumeist erfüllt sein, da der Horizont weiß erscheint, wenigstens im Gegensatz zu den zenitalen Gegenden des Himmels. Da Weiß aber eine Mischfarbe ist, so können Aussagen über das Spektrum des Luftlichtes von vornherein nicht gemacht werden. Es muß daher die Erfahrung lehren, wie weit man berechtigt ist, mit einem monochromatischen Zerstreuungskoeffizienten zu rechnen. Die Aussage, daß ein weißer Schirm mit zunehmender Entfernung vom Auge seine Farbe nicht wesentlich ändert, läßt ebenfalls keine eindeutigen Schlüsse zu.

Die Ergebnisse des § 5 gelten strenge für monochromatisches Licht und für den Fall, daß der Zerstreuungskoeffizient für das Luftlicht gleich a_0 und für das von dem Schirm reflektierte Licht gleich $\dfrac{a_0}{2.09}$ ist. Diese Zerstreuungskoeffizienten wurden gewählt, einmal um Anschluß an meine Berechnungen I Tab. 6 zu gewinnen und zweitens einen extremen Fall darzustellen. Welche Wellenlängen diesen Zerstreuungskoeffizienten zuzuordnen sind, kann zunächst nicht entschieden werden: die Funktion $a_\lambda = f(\lambda)$ hat wohl die Form $a_\lambda = C \cdot \lambda^{-n}$ aber n ist unbekannt. ($n = 4$ nach der Rayleigh'schen Theorie, gut bestätigt durch Abney's Messungen; $n = 2$ nach Müller's Messungen; vergl. Exner-Perntner, Met. Optik, 2. Aufl., S. 754.)

Aus den angeführten Gültigkeitsgrenzen ist ersichtlich, daß ein zahlenmäßiger Vergleich der Ergebnisse der § 4 und 5 hinsichtlich der Sichtweite nicht zulässig ist.

§ 7. **Bemerkungen über Sichtmessungen.** In den vorstehenden Untersuchungen ist wiederholt darauf hingewiesen worden, daß eine weitergehende Darstellung der Sichtverhältnisse experimentelle Untersuchungen erfordert. Diese seien hier zusammengestellt.

a) Der Beobachtungsplatz ist in die Ebene zu legen, damit eine Konstanz des Zerstreuungskoeffizienten in der Horizontalen vorausgesetzt werden darf.

b) Das Ziel muß von einer physikalisch streng definierten Beschaffenheit sein, die leicht reproduzierbar ist. In beliebigen Entfernungen soll das Ziel unter dem gleichen räumlichen Winkel erscheinen.

c) Außer der Helligkeit des schwarzen und des weißen Körpers ist stets die Horizonthelligkeit im selben Azimut zu messen und mit der Zenithelligkeit zu vergleichen.

Eine weitere große Förderung der hier vorliegenden Aufgabe würden folgende Untersuchungen gewähren:

d) Spektrale Zusammensetzung des Himmels- und Horizontlichtes.

e) Spektrale Zusammensetzung des in Richtung auf den Schirm wahrnehmbaren Lichtes und deren Änderung mit zunehmender Entfernung Auge—Schirm.

f) Schließlich (nach dem Vorgang von F. Linke und A. Wigand) die Änderung der Trübung mit der Höhe.

Wenn diese Forderungen zunächst kaum erfüllbar erscheinen, so müssen wenigstens die Forderungen b) und c) erfüllt werden, wenn eine quantitative Analyse durchgeführt werden soll.

Mit welchen Mitteln am besten gemessen wird, muß die Erfahrung lehren. Am meisten geeignet erscheint mir die Anwendung der energetischen Methode, d. h. der Photozelle. Leider waren mir die Arbeiten von E. Goldberg, der allerdings Vergleichsphotometer benutzte, nicht zugänglich. Soweit ich aus den knappen Referaten ersehen kann, erfüllen sie aber die Bedingung b) und c), sodaß ihre ausführliche Veröffentlichung sehr zu begrüßen wäre.

§ 8. Ergebnisse. Für monochromatisches Licht mit gleicher Wellenlänge für Schirm und Umgebung gibt es im allgemeinen eine Nullinie des Kontrastes, die als gerade Linie unter bestimmten, zum Sonnenvertikal symmetrischen Azimuten vom Beobachtungsort ausläuft und sich erst im Unendlichen schließt.

Für monochromatisches Licht mit ungleicher Wellenlänge für Schirm und Umgebung gibt es ebenfalls eine Nullinie des Kontrastes, die aber im allgemeinen nicht vom Beobachtungsorte ausläuft, sich aber bereits im Endlichen schließt.

Met. Geophys. Inst. d. Universität, März 1925.

The Reduction of Apparent Contrast by the Atmosphere

SEIBERT Q. DUNTLEY*

Massachusetts Institute of Technology, Cambridge, Massachusetts

(Received August 11, 1947)

A veil of atmospheric haze reduces the visibility of all distant objects by decreasing their apparent contrast. In this paper equations are derived which describe the manner in which the apparent contrast of any object depends upon the distance of the observer. The treatment is not limited to horizontal paths of sight, but applies also to the apparent contrast of objects on the ground as seen from the air, and to the apparent contrast of objects aloft as viewed from the ground. The equations are not limited to the case of a homogeneous standard atmosphere; they may be applied to many kinds of non-standard atmospheric conditions. For every path of sight there exists a luminance level which will be transmitted unchanged. The apparent luminance of any receding object approaches this equilibrium level. For many paths of sight the equilibrium luminance is matched by the luminance of some portion of the horizon sky.

THIS paper is an outgrowth of research begun during the war as part of a study of the visibility of distant objects;[1] it is presented in the belief that the theory may find peacetime uses. The limiting range at which a given object will be visible can be predicted[2] from data concerning contrast thresholds for the human eye[3] if a proper allowance is made for the reduction in apparent contrast caused by the atmosphere.[4,5] It is the purpose of this paper to set forth the laws which govern such a reduction of contrast along paths of sight which may be horizontal or may be inclined upward or downward. The simplest case is represented by an atmosphere uniform in lighting and in composition but decreasing regularly in density with increasing altitude. Such an idealized atmosphere will be treated first, but a method for allowing for departures from the idealized conditions will be presented near the end of the paper.

1. FUNDAMENTALS

The apparent luminance of any distant object is governed by two processes which operate concurrently: (1) light emanating from the object is gradually attenuated by scattering and by absorption; (2) daylight is scattered toward the observer all along the path of sight. The balance between that portion of the light emanating from the object which reaches the observer and the space light contributed by the intervening air determines the apparent luminance of the object. For example, the apparent luminance of a dark mountain increases with distance because of the addition of space light until the mountain finally is invisible because its apparent luminance matches that of the sky background. In like manner, the apparent luminance of a sunlit, snow-clad peak decreases with distance until, at a sufficient range, it too matches its sky background. In this case the reduction in apparent luminance is due to attenuation of light from the mountain by scattering and by absorption.

An insight into the relation between these two processes can be gained by regarding the atmosphere as a diffusing material, lighted throughout by natural sources such as the sun or moon, the sky, and the ground.** A useful method for describing the behavior of diffusing materials wherein both primary and secondary scattering play part was proposed by Schuster[6] in 1903, and it has subsequently been employed to describe the optical properties of most of the diffusing materials of interest to industry.[7] In an earlier paper,[8] the Schuster-type treatments by

* Formerly Technical Aide, Section 16.3, National Defense Research Committee.

[1] Summary Technical Report of N.D.R.C. Division 16, Vol. 2.

[2] S. Q. Duntley, J. Opt. Soc. Am. **36**, 359A (1946).

[3] H. R. Blackwell, J. Opt. Soc. Am. **36**, 624 (1946).

[4] S. Q. Duntley, J. Opt. Soc. Am. **36**, 359A (1946).

[5] S. Q. Duntley, J. Opt. Soc. Am. **36**, 713A (1946).

** An approach to the problem by the integral equation method has been made by E. S. Kousnetsov, whose paper entitled "Theory of non-horizontal visibility" appears in the Bulletin de L'Academe des Sciences de l'Union des Republiques Sovietiques Socialistes, No. 5 (1943).

[6] A. Schuster, Phil. Mag. **5**, 243 (1903).

[7] S. Q. Duntley, J. Opt. Soc. Am. **33**, 252 (1943).

[8] S. Q. Duntley, J. Opt. Soc. Am. **32**, 61 (1942).

various authors were classified in accordance with the number of constants which they employed to characterize the particular diffusing material with which they were concerned. In the case of the atmosphere it has been found that two specially defined constants suffice under most of the commonly encountered seeing conditions.

1.1 The Differential Equations

Schuster conceived of two fluxes moving in opposite directions through the diffusing material, each contributing back-scattered flux to the other. Simultaneous differential equations can be used to describe the changes in these fluxes as they pass through a lamina of path. In Fig. 1 an extended emitting object has a spectral radiant emittance, W_0, in the direction of the observer, who is assumed to view the surface of the object normally along a path of sight inclined at an angle θ with the horizontal. At a distance r is a parallel-sided flat lamina of atmosphere having a thickness dr. The surfaces of the lamina are perpendicular to the line of sight. Let the diffused spectral irradiance on the lower side of the lamina be denoted by t, and the diffused spectral irradiance on the upper side of the lamina be denoted by s.

In passing through the lamina, the upward-moving radiant flux, t, will be diminished by absorption and by scattering, the magnitude of the change (dt/dr) being proportional to the flux t. Let the constant of proportionality be written $\mu_r + B_r$, where μ_r will be called the absorption coefficient and B_r the back-scattering coefficient, respectively. The subscript r denotes that, in general, these quantities vary along the path

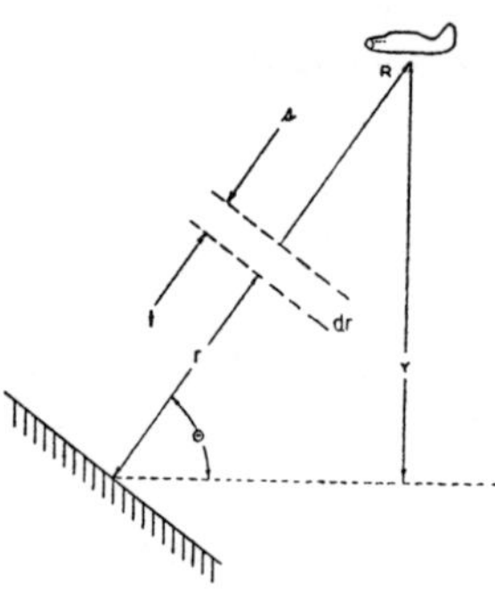

FIG. 1.

of sight. The flux t will also be augmented within the lamina by space light (l_t), and by back-scattered light from the oppositely moving flux body $(B_r s)$. In addition to its diffuse emittance, the object may also produce a directed beam. The diffused flux t will then be augmented by scattering from the beam $(F_r' I_r')$, where I_r' annotes the spectral irradiance on the lamina caused by the directed beam, and the proportionality coefficient F_r' is the forward-scattering coefficient for directed flux. The primes are used to indicate that these quantities refer to light which is not diffused. The notation adopted here conforms with the recommendations of the International Commission on Illumination for the specification of opal glass.[9] All of the above-mentioned changes in t are concurrent, as indicated by the following differential equation:

$$dt/dr = -\mu_r t - B_r t + l_t + B_r s + F_r' I_r'. \tag{1}$$

Corresponding concurrent changes take place in the downward moving diffused flux s when it traverses the lamina. These changes are represented by the following differential equation wherein the proportionality coefficient B_r' is the backward-scattering coefficient for directed flux.

$$-ds/dr = -\mu_r s - B_r s + l_s + B_r t + B_r' I_r'. \tag{2}$$

Differential equations (1) and (2) may be thought of as a steady-state equation of continuity for flux permeating the atmosphere. They involve the assumption that the radiation is monochromatic and the assumption that the medium may be divided laminarwise without destroying its over-all homogeneity. The treatment is thermodynamic in nature, and attempts no description of the behavior of light in passing over the airborne particles. For this reason, effects such as those discussed by Rayleigh will appear as spectral variations in the absorption and scattering coefficients.

The contributions arising from space light are some fractions of the radiant density in the lamina. Therefore, the terms l_t and l_s can be replaced by $\tau_r q_r$ and $\sigma_r q_r$, where q_r is the spectral radiant density of space light at the lamina and τ_r and σ_r are *scattering-rate coefficients*.

If the observer's path of sight is horizontal,

<hr>

[9] International Commission on Illumination, Neuvième Session, p. 3 (1937).

and if the atmosphere is homogeneous in its lighting and in its composition, the spectral radiant density q is the same at all points. Along inclined paths of sight, however, a variation of radiant density with altitude might be expected. Experiments*** conducted during the recent war by the Louis Comfort Tiffany Foundation, under a contract with the Office of Scientific Research and Development, sought to measure this variation, but none was discovered up to the highest altitude tested, 18,000 feet. In all cases the experiments were conducted with solar elevation in the neighborhood of 50 degrees. There seems, however, reason to expect that variations of q with altitude would have been found for low values of solar elevation. For example, just after sunset the radiant density is much greater at high altitudes than it is near the ground. The range of solar elevation within which the variation of q with altitude may be neglected is not known. A further experimental investigation of this matter should be undertaken. In the discussion which follows, q will be assumed to the same at all points along the path of sight.

The terms $B_r s$ and $B_r t$ which represent the contributions arising from back-scattering from the oppositely moving flux body are, to a first approximation, constant along the path of sight, except in the case of objects whose inherent spectral radiance is much greater than that of their background. If $B_r s$ and $B_r t$ are constant, they may be regarded as equivalent to an increase in the radiant density q. Hence, in Eq. (1) the terms $l_t + B_r s$ can be replaced by $\tau_r q$, and in Eq. (2) the terms $l_s + B_r t$ by $\sigma_r q$.

Inasmuch as the flux is attenuated both by absorption and by scattering, the terms $\mu_r t + B_r t$ in Eq. (1) can be replaced by $\beta_r t$, and the terms $\mu_r s + B_r s$ in Eq. (2) can be replaced by $\beta_r s$, where β_r is referred to as the *atmospheric attenuation coefficient*.

In the interest of simplicity let the discussion be restricted to cases in which the terms $F_r' I_r'$ and $B_r' I_r'$ are negligible; the special case of an intense, highly collimated source, such as a searchlight, being treated separately in Section

2.4 of this paper. As a further simplification, let the discussion be limited to those cases in which the effect of the object upon q can be neglected throughout all but a negligible portion of the path of sight; by means of this restriction, the 1-dimensional differential equations (1) and (2) can be applied to situations in which the object subtends but a small solid angle at the eye of the observer. The discussion of the reduction by the atmosphere of the apparent contrast of objects which subtend a large angle at the eye of the observer will be reserved for a subsequent paper. Equations (1) and (2) may now be written

$$dt/dr = -\beta_r t + \tau_r q, \tag{3}$$

$$-ds/dr = -\beta_r s + \sigma_r q. \tag{4}$$

Because of the assumptions made in passing from Eqs. (1) and (2) to Eqs. (3) and (4), the latter are not intended to be valid for objects possessing inherent radiance greatly in excess of their background, or along steeply inclined paths of sight when the sun is low. However, Eqs. (3) and (4) will be discussed fully because they treat correctly most of the commonly encountered problems in visibility, and because many of the principles of atmospheric optics can be derived simply from them. A more general discussion based upon Eqs. (1) and (2) will be reserved for a subsequent paper.

1.2 The Solutions

Since $B_r s$ and $B_r t$ have been assumed constant along the path of sight, the feedback of flux is no longer a variable, and Eqs. (3) and (4) need not be solved simultaneously. They may be rearranged in a form permitting integration by replacing β_r by $\beta_0 f(r)$, τ_r by $\tau_0 f(r)$, and σ_r by $\sigma_0 f(r)$, where β_0, τ_0, and σ_0 are the values assumed by the respective variables at the lower end of the path of sight. Equations (3) and (4) may now be written:

$$\int_{W_o}^{W_R} \frac{dt}{-\beta_0 t + \tau_0 q} = \int_o^R f(r)dr, \tag{5}$$

$$\int_{W_o}^{W_R} \frac{ds}{-\beta_0 s + \sigma_0 q} = -\int_R^o f(r)dr. \tag{6}$$

The limits of integration in Eq. (5) are arranged to correspond with the case of an observer looking

*** The experiments referred to in this paper were conducted under the author's supervision by the Louis Comfort Tiffany Foundation under contract OEMsr 597 with the Office of Scientific Research and Development.

downward along a slant path of length R at an object whose *apparent spectral radiant emittance* is W_R, but whose *inherent spectral radiant emittance* is W_0. The limits of Eq. (6) are arranged to represent the corresponding case of an observer looking upward along the slant path. It will be noted that the equations are identical except for the scattering-rate coefficients τ_0 and σ_0.

In the case of a horizontal path of sight through a homogeneous atmosphere, $f(r)=1$ since β_0, τ_0, and σ_0 are the same at all points. The integrals on the right then reduce to R, the range of the object. In the case of an inclined path of sight, $f(r) \neq 1$ because the air density decreases with increasing altitude (see Section 3.1). Let the integral of $f(r)$ along the path of sight be called the *optical slant range* and denoted by $\bar{R}$. Thus

$$\bar{R} = \int_o^R f(r)dr. \qquad (7)$$

Physically, $\bar{R}$ represents that horizontal distance in a homogeneous atmosphere for which the attenuation is the same as that actually encountered along the true path of length R.†

The integrated forms of Eqs. (5) and (6) may now be written:

$$W_R = \frac{\tau_0 q}{\beta_0}(1 - e^{-\beta_0 \bar{R}}) + W_o e^{-\beta_0 \bar{R}}, \qquad (8)$$

$$W_R = \frac{\sigma_0 q}{\beta_0}(1 - e^{-\beta_0 \bar{R}}) + W_o e^{-\beta_0 \bar{R}}. \qquad (9)$$

Equation (8) applies to an observer looking downward along a slant path; Eq. (9) to an observer looking upward. Both equations are seen to be comprised of two terms, the first representing the space light and the second representing flux from the object which reaches the observer.

<hr>

† An analogous procedure was adopted in a treatment of this subject by the Eastman Kodak Company in a monograph on aerial haze published in 1923 (reference 10), and more recently in a paper by Tousey and Hulburt of the Naval Research Laboratory (reference 11).

[10] "Aerial haze and its effect on photography from the air," Research Laboratory, Eastman Kodak Company, D. Van Nostrand Company, Inc. (1923).

[11] R. Tousey and E. O. Hulburt, J. Opt. Soc. Am. 37, 78 (1947).

If the quantities in Eqs. (8) and (9) showed no wave-length dependence, these relations would hold for the apparent luminance of distant objects. The field experiments described in Section 2.4 indicate that for achromatic objects the attenuation of luminance follows the same law. Hence, for practical purposes, Eqs. (8) and (9) can be written in terms of the *inherent luminance* of the object B_0 and its *apparent luminance* B_R as seen from a distance R:

$$B_R = \frac{\tau_0 q}{\beta_0}(1 - e^{-\beta_0 \bar{R}}) + B_o e^{-\beta_0 \bar{R}}, \qquad (10)$$

$$B_R = \frac{\sigma_0 q}{\beta_0}(1 - e^{-\beta_0 \bar{R}}) + B_o e^{-\beta_0 \bar{R}}, \qquad (11)$$

where the symbols σ_0, τ_0, q, and β_0 now refer to luminous quantities.

An equation of this type for horizontal paths of sight was published in 1924 by Koschmieder.[12] His derivation is given by Middleton,[13] whose standard text on visibility includes an extensive survey of the literature.

2. VISIBILITY

From the standpoint of visibility the most important consequence of Eqs. (10) and (11) is the result that luminance differences are exponentially attenuated:

$$(\mathbf{B}_R - B_R) = (\mathbf{B}_o - B_o)e^{-\beta_0 \bar{R}}, \qquad (12)$$

where $\mathbf{B}$ refers to the luminance of the object and B refers to the luminance of the background. Since the perceptual capacity of the human eye is best described in terms of contrast thresholds, a relation similar to Eq. (12) but involving the contrast of the object against its background is required. Let the *inherent contrast* of the object C_0 and the *apparent contrast* C_R be defined as follows:

$$C_o = \frac{\mathbf{B}_o - B_o}{B_o}, \qquad (13)$$

$$C_R = \frac{\mathbf{B}_R - B_R}{B_R}. \qquad (14)$$

<hr>

[12] H. Koschmieder, Beitr. z. Phys. d. freien Atm. 12, 33 (1924), and 12, 171 (1924).

[13] W. E. K. Middleton, *Visibility in Meteorology* (University of Toronto Press, Toronto, 1941).

Then Eq. (12) may be written

$$C_R = \frac{B_o}{B_R} C_o e^{-\beta_o \bar{R}}.$$ (15)

Equation (15) is the law of contrast reduction by the atmosphere expressed in its most general form.

2.1 Visibility Upward

In the case of an observer looking upward, B_R in Eq. (15) is the apparent luminance of the sky in the direction of the object, while B_o is the apparent luminance of the sky in the direction of sight as seen by an observer at the object. From Eq. (11) the luminance of the sky as seen from the ground is

$$B_R = \frac{\sigma_o q}{\beta_o}(1 - e^{-\beta_o \bar{R}_{0,\infty}}),$$ (16)

where subscripts $0,\infty$ indicate that Eq. (7) is to be integrated from zero to infinity. Similarly,

$$B_o = \frac{\sigma_o q}{\beta_o}(1 - e^{-\beta_o \bar{R}_{R,\infty}}).$$ (17)

Substituting (16) and (17) in (15) yields the law of atmospheric contrast reduction for an observer looking upward along an inclined path of sight:

$$C_R = C_o e^{-\beta_o \bar{R}}\left[\frac{1 - e^{-\beta_o \bar{R}_{R,\infty}}}{1 - e^{-\beta_o \bar{R}_{0,\infty}}}\right].$$ (18)

The factor $\sigma_o q/\beta_o$ has vanished, leaving the contrast attenuation affected only by the non-directional quantity β_o, and $\bar{R}$ which, in a homogeneous atmosphere, depends only on the inclination θ of the path of sight and R. Thus (see Section 3.1) the contrast attenuation for an observer looking upward along an inclined path is independent of the direction of the sun. For example, an object of fixed inherent contrast moving at constant altitude in a circular path centered on the observer will present the same apparent contrast at all points of its path. For the special case of a horizontal path of sight, this theorem is equivalent to the common statement that the daylight visual range (see Section

2.4) is the same in all directions. When objects are viewed against the sky along horizontal paths of sight, Eqs. (15) and (18) reduce to

$$C_R = C_o e^{-\beta_o R}.$$ (19)

2.2 Optical Equilibrium

Equation (3) shows that $dt/dr = 0$ when $t = \tau_r q/\beta_r = \tau_o q/\beta_o$. Similarly, Eq. (4) shows that $ds/dt = 0$ when $s = \sigma_r q/\beta_r = \sigma_o q/\beta_o$. Thus certain luminance levels are transmitted by the atmosphere without attenuation. These equilibrium levels are such that within each lamina of path the added space light exactly compensates for the attenuation. This condition has been called *optical equilibrium*.

It is a matter of common experience that the apparent luminance of the horizon does not change when an observer moves toward it. This indicates that in a homogeneous atmosphere, wherein q, σ_o, τ_o, and β_o do not vary along the path of sight, the luminance of the horizon is the equilibrium value. Since the luminance of the horizon depends upon direction relative to the sun, and since β_o and q do not vary with direction, there must be a compensating directional variation in the scattering-rate coefficients τ_o or σ_o.

It often happens that some horizontal path of sight bears the same angle with respect to the sun as does the inclined path of sight. This is illustrated by Fig. 2. Usually there are two horizontal directions, n and n', from which sunlight is scattered at the same angle as the sunlight which is scattered downward along the slant path, and two opposite directions, m and m', from which sunlight is scattered at the same angle as the sunlight which is scattered upward along the slant path. Thus the value of $\tau_o q/\beta_o$ for the slant path equals the luminance of the horizon sky B_m as seen by an observer at the object looking in the m or m' direction. Similarly, B_n, the luminance of the horizon sky in the n or n' directions, equals $\sigma_o q/\beta_o$.

As a consequence of the phenomenon of optical equilibrium an observer looking upward sees the apparent luminance of any receding object approach the luminance of the horizon sky in the n directions. As seen by an observer looking horizontally, the apparent luminance of a receding

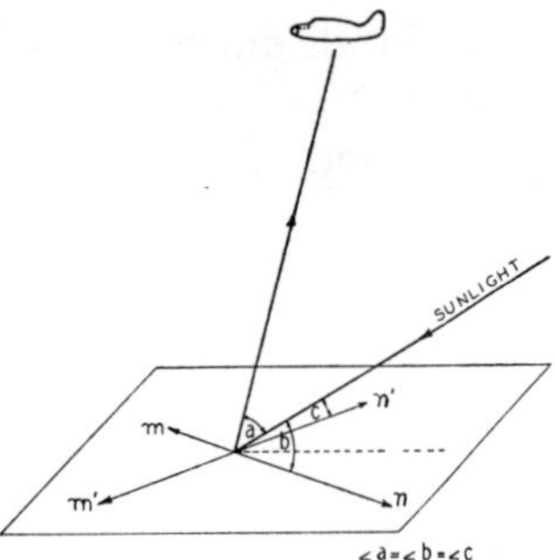

FIG. 2. Arrows n and n' indicate the directions in which the luminance of the horizon sky is determined by light scattered from the rays of the sun at the same angle as light scattered downward along the path of sight. Arrows m and m' indicate the directions in which the luminance of the horizon sky is determined by light scattered from the rays of the sun at the same angle as light scattered upward along the path of sight.

object changes with distance in such a manner as to approach the luminance of the horizon sky in the direction of the object. An ascending observer sees the apparent luminance of all objects on the ground approach the apparent luminance seen by an observer on the ground who looks at the horizon sky in the m directions.

2.3 Visibility Downward

Aviators commonly report a "circle of visibility" beyond which nothing can be distinguished on the ground. In a sense this is a misnomer, because the curve often departs markedly from circular shape. An insight into this phenomena can be gained by combining Eqs. (10) and (15) as follows:

$$C_R = C_o \left[1 - \frac{\tau_o q}{\beta_o B_o}(1 - e^{\beta_o \bar{R}}) \right]^{-1}. \qquad (20)$$

When the relation between the direction of the sun and the path of sight is such that the m and n horizontal directions exist (see Fig. 2), $B_m = \tau_o q/\beta_o$ and the law of contrast attenuation

TABLE I.

Sky condition	Ground condition	Sky-ground ratio
Overcast	Fresh snow	1
Overcast	Desert	7
Overcast	Forest	25
Clear	Fresh snow	0.2
Clear	Desert	1.4
Clear	Forest	5

may be written:

$$C_R = C_o \left[1 - \frac{B_m}{B_o}(1 - e^{\beta_o \bar{R}}) \right]^{-1}. \qquad (21)$$

The quantity B_m/B_o has been called the *sky-ground ratio*. Typical values of this quantity are given in Table I. Both B_m and B_o can be measured with a conventional luminance photometer, but the evaluation of $\tau_o q/\beta_o$ by three separate photometric measurements requires uncommon apparatus. Such measurements can be made, however, when because of geometry the m and m' directions do not exist, or when obstacles or non-uniform weather conditions make direct measurement of B_m impossible or meaningless. Henceforth, the term *sky-ground ratio* will be used to denote $\tau_o q/\beta_o B_o$ under all circumstances.

The shape of the "circle of visibility" cannot be expressed analytically, even for an idealized homogeneous atmosphere, partly because the inherent contrast of natural terrains usually depends upon the bearing of the sun relative to the line of sight, and partly because the magnitude of the just-visible apparent contrast depends upon both the angular size of the object and the state of adaptation of the observer's eye. A nomographic method for taking account of these variables has been described,[2] but that treatment cannot yield an analytic expression for the "circle of visibility." Under circumstances such that the "circle" is defined by the condition that the ratio C_R/C_o is constant, the optical slant range $\bar{R}$ is seen to depend upon τ_o, q, β_o, and B_o. Of these, only τ_o depends upon the direction of the line of sight, and hence it might at first be supposed that the "circle of visibility" is a functional polar plot of τ_o or of the brightness B_m of the horizon sky as seen by an observer on the ground. This is not strictly true, however, because the relation between $\bar{R}$ and R depends upon the angular elevation θ of the path of sight.

2.4 Horizontal Visibility

When an object is viewed along a horizontal path of sight, $\tau_o q/\beta_o = B_H$, the luminance of the horizon sky in the direction of the object. The law of contrast attenuation then becomes

$$C_R = C_0 \left[1 - \frac{B_H}{B_0}(1 - e^{\beta_0 R}) \right]^{-1}. \qquad (22)$$

In the case of an object viewed along a horizontal path of sight against the horizon sky as a background, $B_0 = B_H$ and Eq. (22) reduces to the form of Eq. (19).

An experiment designed to test the validity of Eq. (19) for dark targets was conducted by the Louis Comfort Tiffany Foundation under their OSRD contract. Large, glossy black, vertical panels were erected at distances of approximately 1500, 3000, and 6000 yards, and their apparent luminances and that of the horizon sky were measured by means of both photographic and photoelectric telephotometers. Figures 3 and 4 show the results of two such tests.

In discussing Eqs. (8) and (9) it was pointed out that the second term on the right represents the fraction of the flux emanating from the object which reaches the observer. Let the *contrast transmittance* T_R of a horizontal path of length R be defined from Eq. (19) as

$$T_R = C_R/C_o = e^{-\beta_o R}. \qquad (23)$$

The contrast transmittance T of a unit distance (mile, yard, etc.) of atmosphere is

$$T = e^{-\beta_o}. \qquad (24)$$

Equation (24) can, if desired, be combined with Eqs. (8) through (22). For example, Eqs. (10) and (19) may be written:

$$B_R = B_H(1 - T^{\overline{R}}) - B_o T^{\overline{R}}, \qquad (25)$$

$$C_R = C_o T^R. \qquad (26)$$

When the light emitted by the object is intense and highly collimated, Eqs. (3) and (4) and relations derived therefrom should not be used directly, for the terms of Eqs. (1) and (2) which involve I_r' must be considered. Although a completely general solution for the searchlight problem appears very involved, advantage may be taken of the limited angular opening of the searchlight beam and the problem may be divided into two independent limiting cases. Consider first the unlighted searchlight as an object. Its inherent luminance then differs from that of its background by no more than does that of other non-self-luminous objects, and therefore its apparent luminance is given by Eqs. (8) through (22). When the searchlight is lit, a highly collimated narrow beam is superimposed upon the path of sight, causing an irradiance I_R' at the observer. Because the collimated beam is narrow, the secondary scattering terms are negligible,[14] and the differential equation for the collimated beam is

$$dI_r'/dr = -\mu_r' I_r' - B_r' I_r' - F_r' I_r'$$
$$= -\beta_o' f(r) I_r', \qquad (27)$$

where $\beta_o' = \mu_o' + B_o' + F_o'$.

From the solution of Eq. (27), the transmittance of the atmosphere for a collimated beam (the *beam transmittance*) is

$$T_R' = e^{-\beta_o' \overline{R}}. \qquad (28)$$

A photoelectric *transmissometer* for measuring T_R' has been developed by members of the staff of the National Bureau of Standards.[15] Comparison of β_0 as computed from the slope of the lines in Figs. 3 and 4 with corresponding values of β_0' computed by means of Eq. (28) from data secured with a transmissometer show agreement to within 10 percent. Although this agreement suggests that measured values of T_R' may sometimes be a useful substitute for data on T_R, differences should be expected, and an extensive

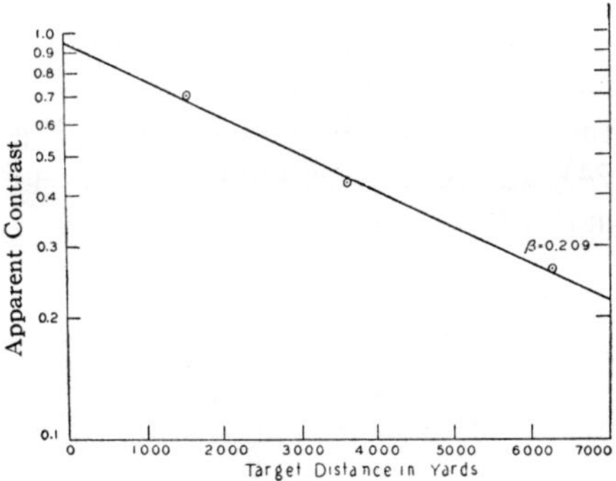

Fig. 3. Apparent contrast of distant black panels as measured with a photographic telephotometer. From the slope of the line: $\beta_o = 0.209$ per thousand yards; $v = 18,700$ yards

Time: 11 A.M.
Sky: 7/10 cirrus H_6
Ceiling: 25,000 feet
Estimated visibility: between 3 and 12 miles
Atmospheric pressure: 1011.9 millibars
Temperature: 74 degrees F
Dew point: 65 degrees F
Relative humidity: 74 percent
Wind: NW light.

[14] F. Benford, J. Opt. Soc. Am. **36**, 524 (1946).
[15] C. A. Douglas and L. L. Young, Technical Development Report No. 47, Civil Aeronautics Administration.

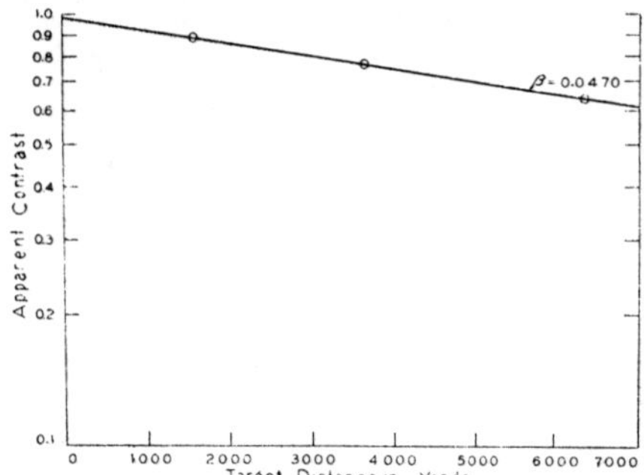

FIG. 4. Apparent contrast of distant black panels as measured with a photographic telephotometer. From the slope of the line: $\beta_0 = 0.047$ per thousand yards: $v = 83,200$ yards

Time: noon
Sky: clear
Estimated visibility: 50 miles
Atmospheric pressure: 1014.6 millibars
Temperature: 57 degrees F
Dew point: 47 degrees F
Relative humidity: 69 percent
Wind: NW 6 miles per hour.

experimental comparison of these two quantities will be required before the circumstances under which data on T_R' can be used in visibility problems will be known.

The most commonly used quantities for expressing the attenuating properties of the atmosphere are β_0, T, and the *meteorological range v*. Meteorological range is defined as that horizontal distance for which the contrast transmittance of the atmosphere is two percent. From Eqs. (23) and (24), v is seen to be related to β_0 and T by the relation††

$$v = \frac{3.912}{\beta_0} = \frac{1.699}{\log_{10}\frac{1}{T}}. \tag{29}$$

Meteorologists customarily make a visual judgment of the optical effect of the atmosphere by inspecting distant black objects seen against the horizon sky. By international agreement[13,16] the *daylight visual range* is the distance at which a large dark object on the horizon is just recognizable. By a "large dark object" is meant an object that subtends so large an angle at the eye

of the observer that if the subtended angle were greater the reported value of daylight visual range would not be changed. Because the distant dark objects available to meteorologists are rarely of sufficient angular size to fulfill this requirement, the reported *"visibility"* is usually somewhat less than the true daylight visual range. A comparison between "visibility" as reported by the staff meteorologist of the Tiffany Foundation and meteorological range as calculated from measurements of beam transmittance showed that, on the average, the reported "visibility" was three-fourths of the meteorological range. This conclusion is in agreement with the results of a similar experiment conducted before the war by personnel of the National Bureau of Standards.[15]

2.5 Allegedly Complicating Effects

Two allegedly complicating effects are referred to in the literature[13] as the *ground-glass plate effect* and the *edge effect*. The former refers to loss of sharp detail by low angle scattering, and the latter refers to an addition to the apparent luminance of an object because of light diffused around its edges. Since neither effect appeared to be supported by data of certain validity or by sound theoretical reasoning, the Louis Comfort Tiffany Foundation undertook, under its OSRD contract, to search for the effects and to determine their magnitudes under typical conditions.

The ground-glass plate effect was explored photographically. Distant resolving power targets were photographed in both clear and hazy weather with a camera having a focal length of 10 feet. When the contrast (gamma) of the photographic process was made equal to $1/T_R'$ determined from measurements of beam transmittance, the targets were resolved equally well in all photographs. The experiment was repeated using natural objects, and the conclusion was reached that no fine details were obliterated by haze. This conclusion concurs with the theoretical prediction of Middleton.[17]

The edge effect was explored with a high precision photoelectric telephotometer. This instrument was used to compare the apparent lumi-

---

†† It will be noted from Eqs. (24) and (29) that $\beta_0 = 2.303D$, where D is the optical density of a horizontal path of unit length.

[16] Organization Météorologique Internationale, "Conférence des directeurs à Varsovie 1935," Vol. 1, No. 29 Leyden 1936.

[17] W. E. K. Middleton, J. Opt. Soc. Am. **32**, 139 (1942).

Table II.

Altitude in feet	Relative number of molecules per unit volume
0	1.000
1,000	0.956
2,000	0.918
3,000	0.878
4,000	0.841
5,000	0.804
6,000	0.770
7,000	0.736
8,000	0.703
9,000	0.672
10,000	0.642
12,000	0.586
14,000	0.534
16,000	0.485
18,000	0.440
20,000	0.399
22,000	0.361
24,000	0.326
26,000	0.295
28,000	0.266
30,000	0.239

nance of several equidistant black objects visible against a background of horizon sky. The angular size of the objects ranged from 0.8 minute to more than one degree. No difference in the apparent luminance of the objects was found.

3. OPTICAL SLANT RANGE

Throughout the foregoing discussion the optical slant range $\bar{R}$ has been left as a completely general function of the actual slant range R (see Eq. (7)). All the equations involving $\bar{R}$ thus far derived are, therefore, valid regardless of the nature of the variation of the coefficients β_r, σ_r, and τ_r with altitude. However, means for evaluating $\bar{R}$ must be provided before most of the preceding equations become useful.

3.1 The Optical Standard Atmosphere

Meteorologists sometimes refer to a "standard atmosphere" in which pressure varies with altitude in a prescribed manner and in which there is a constant lapse rate.[18] Let an *optical standard atmosphere* be defined as one in which the scattering particles are the same at all altitudes, but in which the relative number of particles per unit of volume (N) decreases regularly with altitude in the manner shown in Table II. The figures in Table II have been obtained by applying the

<hr>

[18] *Encyclopaedia Britannica*, **3**, 129 (1945).

equation of state of a perfect gas to the "standard atmosphere."

An analytic expression for the data in Table II, would permit the integration indicated by Eq. (7) to be performed. Figure 5 shows a semilogarithmic plot of N as a function of altitude. The points represent the optical standard atmosphere as defined by Table II. The straight line has the equation

$$N/N_o = e^{-y/21,700}, \qquad (30)$$

where y is the altitude above sea level expressed in feet. Along any slant path $y = r\,\sin\theta$, and thus, if $f(r)$ is assumed equal to N/N_o, Eq. (7) becomes

$$\bar{R} = 21,700\ \csc\theta\left[e^{-R_1\,\sin\theta/21,700}\right.$$
$$\left. - e^{-R_2\,\sin\theta/21,700}\right]. \qquad (31)$$

In the special case of an observer at sea level looking upward, or of an observer aloft looking at an object at sea level, Eq. (31) becomes

$$\bar{R} = 21,700\ \csc\theta\left[1 - e^{-R\,\sin\theta/21,700}\right]. \qquad (32)$$

In Eqs. (31) and (32) all distances are to be expressed in feet.

Figure 6 shows a plot of Eq. (32) for values of R from zero to 425,000 feet. The solid lines correspond with various values of θ from zero to 90 degrees. The broken curves indicate loci of equal altitude. Such a plot has been called an *optical slant-range diagram*. Figure 7 is an expansion of Fig. 6 near the origin.

3.2 Non-Standard Conditions

Non-standard atmospheric conditions are of common occurrence; often the path of sight

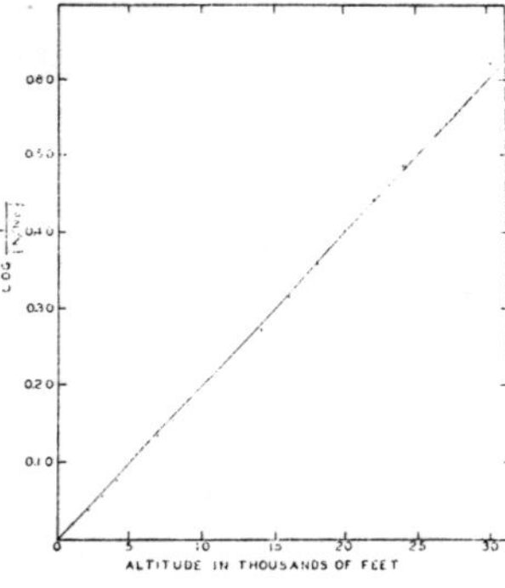

FIG. 5. The points represent the relative number of scattering particles per unit of volume in an optical standard atmosphere.

603

traverses strata, at the boundaries of which the optical properties of the atmosphere may change abruptly. Between stratum boundaries, however, the coefficients may vary in the manner of the optical standard atmosphere. This type of departure from standard conditions is readily allowed for on the optical slant-range diagram by a procedure illustrated in the following example.

Suppose that on a clear summer day the layer of air near the ground has the typical murky appearance caused by large suspended particles of water. Such ground haze often has a sharply defined upper boundary, above which the atmosphere contains very little condensed water. Above such a boundary the meteorological range is usually several times as great as below it. Figure 8 illustrates how the discontinuity can be represented on the optical slant-range diagram. For simplicity, only the curve for $\theta = 25$ degrees is shown. Let it be assumed that the boundary occurs at an altitude of 5000 feet, and that the

meteorological range is five times greater above the boundary than below it. Beginning at the point corresponding with 5000 feet, a new curve has been drawn having five times the slope of the original curve. The relation between $\bar{R}$ and R is then represented by the accentuated line which follows the normal curve up to altitude 5000 feet and the steeper curve thereafter. If the boundary is diffuse rather than sharp, the accentuated line can be rounded off to avoid the abrupt change in slope.

The character and altitude of stratum boundaries can be observed easily from a plane ascending or descending through them. In many cases the pilot can also make an estimate of the ratio of the meteorological range above and below the boundary. Proficiency in describing the stratification of the atmosphere is acquired very quickly by any flyer, once he understands what to look for. Moreover, stratification should correlate with other meteorological conditions, and ex-

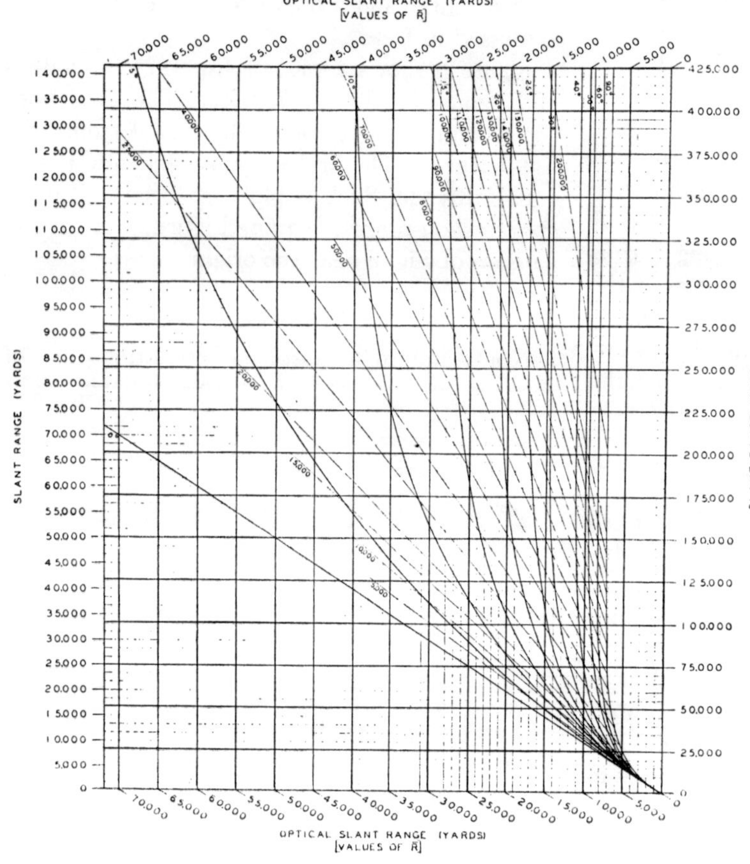

Fig. 6. Optical slant-range diagram for the optical standard atmosphere. Solid curves represent the relation between $\bar{R}$ and R for various sight-path elevation angles θ. Broken lines represent loci of equal altitude expressed in feet.

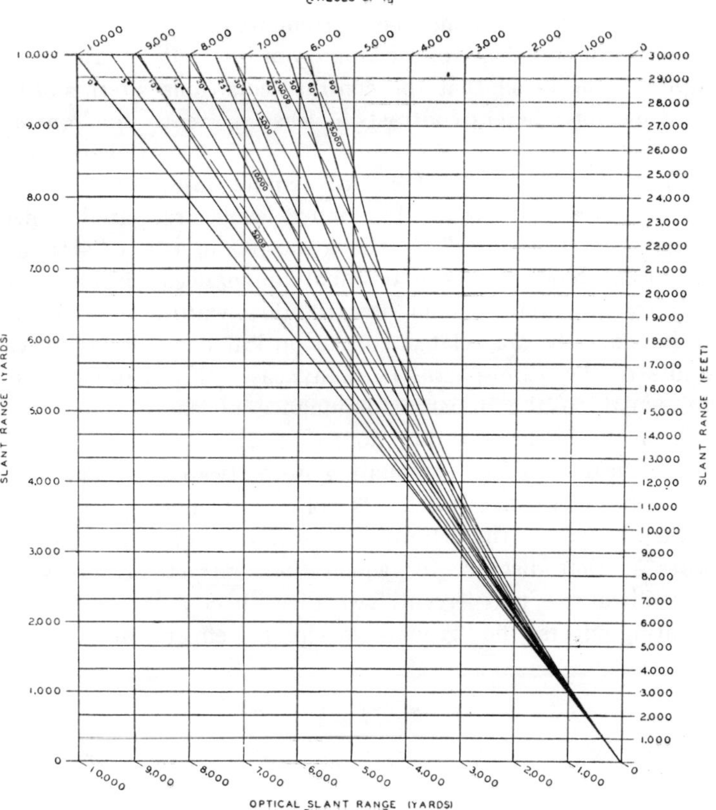

FIG. 7. Optical slant-range diagram similar to Fig. 4, but adapted to the solution to problems involving short slant ranges.

perience may enable very intelligent guesses to be made by an observer on the ground. Statistical information concerning the frequency of occurrence of common stratification conditions in a given locality can be accumulated in the same manner as other meteorological data.

Experience in drawing curves of modified slope on the optical slant-range diagram is quickly acquired with practice. Freehand curves are usually as precise as are warranted by estimates of the ratio of meteorological ranges within the strata. Often straight lines are sufficient approximations for curves of very great or very small slopes. An example of the latter occurs in the case of an opaque cloud deck within which the meteorological range is very short. Usually the lower boundary of the cloud layer is sharply defined; it can be represented on the optical slant-range diagram by a horizontal straight line passing through the point on the standard curve which corresponds to the altitude of the ceiling.

Along horizontal paths of sight local variations in q may be caused by cloud shadows or by variations in the reflectivity of the natural terrains. Moreover, local wind or temperature conditions may cause β_0, τ_0, and σ_0 to vary along the path of sight. If information concerning the variations is available or if they can be estimated, $\bar{R}$ can be found by altering the shape of the curve for $\theta = 0$ degrees on the optical slant-range diagram.

4. SUMMARY

A general expression (15) for the reduction of apparent contrast by the atmosphere has been obtained for the most commonly encountered outdoor visual tasks, allowance having been made for absorption and for both primary and secondary scattering. Special relations have been derived for the case of visibility upward (18), visibility downward (20), and horizontal visibility (22). In the special case of an object seen against a background of horizon sky, the apparent contrast is exponentially attenuated (19). Experimental evidence of the validity of Eq. (19) supports the belief that the contrast equations are valid for inherent contrasts at least as great as unity.

From the principle of optical equilibrium, which states that certain luminance levels are transmitted unchanged by the atmosphere, it has been shown how the terminal luminances seen by observers looking upward, downward, or horizontally are related to the optical constants of the atmosphere. In many cases they are equal to the apparent luminance of the horizon sky in specified directions (Fig. 2).

Specification of the attenuating properties of the atmosphere in terms of its contrast transmittance leads to the definition of *meteorological range* as that distance for which the contrast transmittance is two percent. Two allegedly complicating effects (the "ground-glass plate" effect and the "edge" effect) were searched for but not found.

Because all distances have been expressed in terms of the *optical slant range* $\bar{R}$, the equivalent horizontal path through a homogeneous atmosphere, the contrast reduction equations can be used along any path of sight if $\bar{R}$ can be evaluated. Whenever atmospheric conditions simulate a "standard atmosphere," this can be done by means of an optical slant-range diagram (Figs. 4 and 5). When non-standard atmospheric conditions prevail the diagram often can be modified to conform with the state of the atmosphere.

5. ACKNOWLEDGMENTS

The sustained and stimulating interest of all the personnel of NDRC Section 16.3, its Army and Navy liason officers, and many of its contractors is gratefully acknowledged. The author is indebted to the United States Weather Bureau and to the National Bureau of Standards for

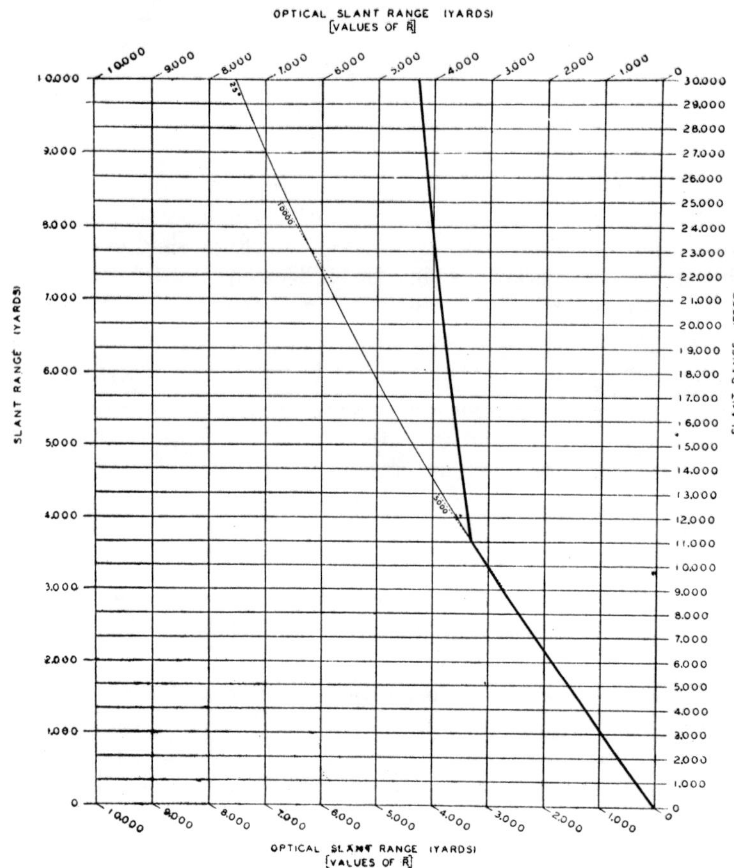

FIG. 8. Optical slant-range diagram for $\theta = 25$ degrees. Accentuated curve shows relation between $\bar{R}$ and R when the ground haze has a sharp upper boundary at 5000 feet, above which the meteorological range is five times greater than it is below the boundary.

their interest and assistance in the experimental aspects of the work. Prominent individual contributions to the experiments were made by Mr. C. A. Elford of the U. S. Weather Bureau, Mr. C. A. Douglas of the National Bureau of Standards, Dr. D. L. MacAdam of the Eastman Kodak Company, and Messrs. W. P. Greenwood and H. H. Lane of the Louis Comfort Tiffany Foundation. Flight facilities were provided by the Army Air Forces.

Optics of sunbeams

David K. Lynch

Thule Scientific, 22914 Portage Circle Drive, Topanga, California 90290

Received November 17, 1986; accepted November 28, 1986

Crepuscular and solar rays are visible by means of the contrast between sunlit and (usually) cloud-shaded portions of the atmosphere. Their visibility depends on (1) the volume angular-scattering coefficient $\beta(\phi)$, (2) the brightness of the sunlit sky, and (3) the integrated optical path through the shadowed regions of the atmosphere. We show that the geometry of path length through the umbra (3) is sufficiently important that it can account for most of the observed properties of rays, except when the volume angular-scattering coefficient $\beta(\phi)$ is sharply peaked in the forward direction.

1. INTRODUCTION

"Sunbeams" is the colloquial name for rays formed by the shadows of scattered or broken clouds seen in the Earth's atmosphere. Other common names are "crepuscular rays," "Buddha's fingers," "ropes of Maui," "sun drawing water," and "backstays of the Sun." If the ray is formed when the Sun shines through a gap in the clouds it is bright. When a relatively isolated cloud casts its shadow on the sky the ray appears dark. In both instances, scattering particles are necessary to render the ray visible. The scatterers may be dust, snow, rain, or simply air molecules acting as Rayleigh scatterers. They are visible by means of the contrast between sunlight scattered by the atmosphere and skylight scattered in the umbra of a cloud. The usual appearance of rays is of one or more straight shafts of light that appear to diverge from the Sun and that are sometimes seen opposite the Sun as rays converging to the antisolar point. All the shadows are parallel, but, owing to the small distance of the observer to the rays (compared with his distance from the Sun), perspective makes the rays appear to diverge.

There is surprisingly little quantitative information about rays. Indeed, the only scientific literature known to this author is among popular scientific treatises (e.g., papers by Minnaert,[1] Meinel and Meinel,[2] Greenler,[3] and Montheith,[4] although some of their geometrical properties are similar to those of mountain shadows.[5,6] In this paper we present a first-order analysis of the contrast of rays and show that many aspects of their visibility are independent of the scattering mechanism and can be explained purely by the geometry of shadows.

2. OBSERVATIONS

Rays are neither uniformly visible over the entire sky nor as likely to be seen at any time throughout the day. Careful study of rays reveals the following properties:

(1) Rays are most common when the Sun is low.

(2) Rays are most common when the air is clear.

(3) The rays' greatest visibility (contrast) occurs within 15° of the Sun and the antisolar point.

(4) Rays near the antisolar point reach their maximum visibility within a few degrees of the antisolar point.

(5) Antisolar rays are usually associated with large clouds overhead or near the antisolar point and are seldom observed from small clouds, unless the cloud is near the antisolar point.

(6) Maximum intensity contrast occurs near the Sun, and maximum color contrast occurs near the antisolar point.

In Section 3 we shall attempt to explain these observed characteristics of rays in terms of the factors that contribute to the ray's contrast.

3. THEORY

We shall assume throughout the following discussion that the line of sight is optically thin, i.e., that the brightness is directly proportional to the number of scatterers. Referring to Fig. 1, consider a unit volume, whose volume angular-scattering coefficient is $\beta(\phi)$, located a distance r from observers positioned at points A through G. The Sun is on the horizon or slightly above it (elevation, ~0.0). The observer looking through unshadowed air receives an amount of light E_b:

$$E_b = \int_0^\infty E_\odot \beta(\phi) \mathrm{d}r = \int_0^{r_1} E_\odot \beta(\phi) \mathrm{d}r$$

$$+ \int_{r_1}^{r_2} E_\odot \beta(\phi) \mathrm{d}r + \int_{r_2}^\infty E_\odot \beta(\phi) \mathrm{d}r, \qquad (1)$$

where $E_\odot$ is the solar illuminance and ϕ is the scattering angle. In this case ϕ is also the elevation angle of the observer's line of sight. The integral is calculated over the entire path through the atmosphere. E_b is a strong function of solar elevation ϕ, especially near the horizon where the air mass increases rapidly.[7]

Some first-order insights into ray contrast can be obtained by analyzing a ray formed by a cloud whose vertical thickness is T_c and whose cloud base is y_c. Let the Sun illuminate the cloud located at a horizontal distance x_c from the observer (at points A through G). We shall chose a cloud whose base height is 2 km and whose vertical thickness is 1 km, both of which are sufficiently small compared with the density scale height of the atmosphere (~8.5 km) that vertical variations in the scattering parameters can be safely ignored.

The locations of the observers at points A through G are x_c = $1.25U, 1.0U, 0.75U, 0.50U, 0.25U, 0.0U$, and $-0.25U$, respectively. The length of the umbra U is

$$U = T_c \times 57.3/0.5,$$

where 0.5 in the above expression represents the half-degree diameter of the Sun. In this example $U = 115$ km. The brightness of the sky along a line of sight through the umbra is the ray brightness E_r:

$$E_r = \int_0^{r_1} E_\odot \beta(\phi)dr + \int_{r_1}^{r_2} S\, dr + \int_{r_2}^{\infty} E_\odot \beta(\phi)dr. \qquad (2)$$

The second term represents skylight scattered in the umbra to the observer. Obviously, S is itself the integral

$$S = \int_0^{2\pi} \int_0^{\pi/2} E(\theta,\varphi)\beta[\phi'(\theta,\varphi)]\sin\theta\,d\theta d\varphi,$$

where $E(\theta, \varphi)$ is the brightness of the sky illuminating the shadow, as a function of the polar angles θ and φ, and $\beta(\phi')$ is the same function as $\beta(\phi)$, the primes indicating that the scattering angle is not the elevation of the observer's line of sight. The contrast C of the ray is defined as

$$C = (E_r - E_b)/E_b.$$

The contrast of the ray is determined by the brightness of the sky viewed through the shadow as compared with the brightness of the fully illuminated sky. To first order, let us consider the contribution to the contrast of the ray as being due only to those regions of space occupied by the umbra of the shadow and compare it with the brightness of the sky in a completely unshadowed region. The expression for the contrast can be simplified by noting that the first and third terms in Eqs. (1) and (2) are identical and thus

$$C = \frac{\int_{r_1}^{r_2}(S - E_\odot\beta)dr}{\int_0^{\infty} E_\odot\beta\, dr} \simeq \frac{(S - E_\odot\beta)(r_2 - r_1)}{\int_0^{\infty} E_\odot\beta\, dr}.$$

For an optically thin atmosphere $r_2 - r_1 = D$. Clearly, C is proportional to $(S - \beta E_\odot)$ and D, the path through the umbra. In general, all three quantities will depend on ϕ. A simple analysis reveals that, for positive (i.e., realistic) values of D,

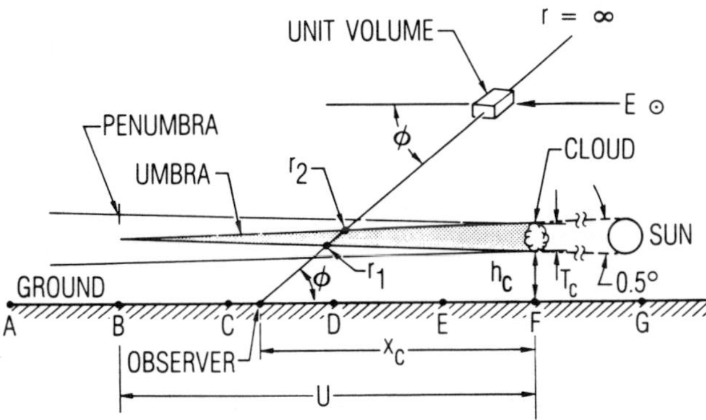

Fig. 1. Geometry of ray formation when the Sun is on the horizon. Observers are located at points A through G, and the Sun is taken as being very near the horizon.

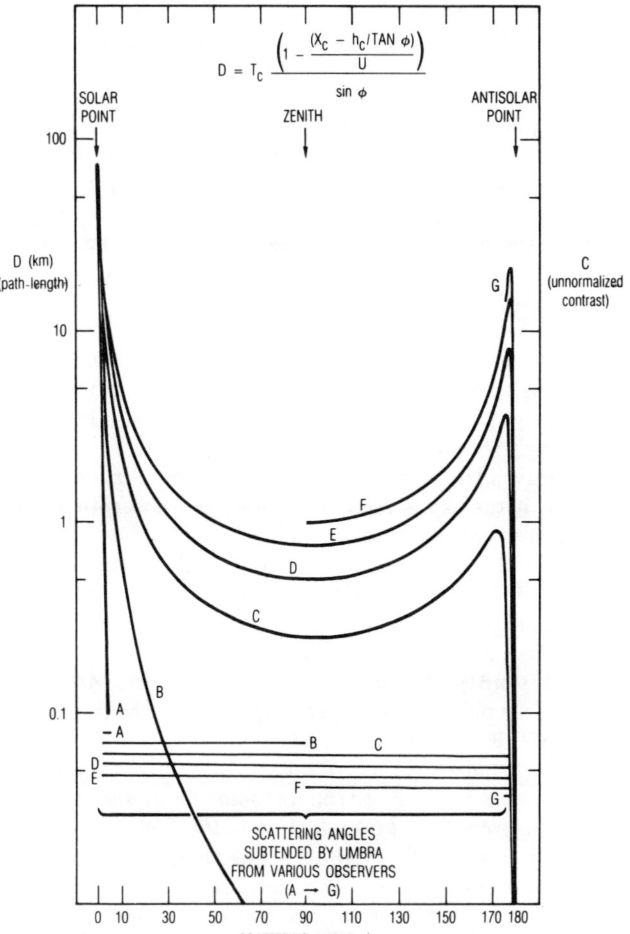

Fig. 2. Path length D (in kilometers) through the umbra (stippled region in Fig. 1) as a function of scattering angle ϕ (in degrees) for various observers located at positions A through G. Parameters T_c = 1 km, h_c = 2 km, Sun elevation = 0°, horizontal cloud distance x_c = $1.25U, 1.00U, 0.75U, 0.50U, 0.25U, 0.0U$, and $-0.25U$, where U is the length of the umbra, in this case 115 km.

$$D(\phi) = T_c\left[-\frac{(x_c - b_c/\tan\phi)}{u}\right]\csc\phi, \qquad (3)$$

where the first term is the vertical thickness of the umbra and the second term $[\csc(\phi)]$ scales the vertical thickness by the slant path through the umbra at a line-of-sight elevation of ϕ.

If β is not a strong function of ϕ, which it is not for Rayleigh scattering, then C is proportional to D, the geometrical line-of-sight distance through the umbra. When $\beta(\phi)$ displays strong forward and backward scattering, as in the case of dust particles, C is further enhanced in these directions. $D(\phi)$ is shown in Fig. 2 for each of the observers located at points A through G in Fig. 1. $D(\phi)$ represents the unnormalized contrast of the ray.

4. DISCUSSION

A number of features are quickly deduced from Fig. 2. Ray contrast is higher near the solar and antisolar points because the line of sight through the umbra, which scales roughly as csc(ϕ), is greatest near 0° and 180°. As defined in Eq. (3), D would go to infinity at the extreme scattering angles, but in practice it does not; $\phi = 0°$ is never reached because the cloud is not infinitely far away, and $\phi = 180°$ occurs beyond the location where the umbra ceases to exist because U is finite. The sharp drop in contrast that reaches a broad minimum near the zenith ($\phi = 90°$) is also due primarily to the csc ϕ term in Eq. (3).

With the calculated properties of $D(\phi)$ in mind, let us recall the observed characteristics of rays listed in Section 2 and examine them with respect to $D(\phi)$.

1. Rays Are Most Common When the Sun Is Low. If the Sun is high, the umbra (length U) intersects the ground and effectively shortens it to the point that D will also be correspondingly shortened.

2. Rays Are Most Common When the Air Is Clear. Clearly, if the sky is hazy, $E_\odot$ is low, and absorption may diminish the ray's contrast.

3. The Rays' Greatest Visibility (Contrast) Occurs Within 15° of the Sun and the Antisolar Point. This point has been explained in the first paragraph of this section.

4. Rays Near the Antisolar Point Reach Their Maximum Visibility Within a Few Degrees of the Antisolar Point. This point has also been explained in the first paragraph of this section.

5. Antisolar Rays Are Usually Associated with Large Clouds Overhead or Near the Antisolar Point and Are Seldom Seen from Small Clouds, Unless the Clouds Are Near the Antisolar Point. Figure 2 shows that, if the umbra of the cloud does not reach very near the antisolar point, as with a small cloud, there will be no ray seen here. A large cloud may have a long umbra that may reach to the antisolar point. Owing to the length of the umbra, an intermediate cloud may have an antisolar ray if it is overhead. Note that maximum contrast for a ray overhead occurs when the cloud is nearly overhead (observers located at points E and F in Fig. 2); this is because the umbra is thickest here. Owing to the high contrast of rays near the antisolar point (observer at point G in Fig. 1), even a tiny cloud can cause a short, high-contrast ray.

6. Maximum Intensity Contrast (C) Occurs Near the Sun, and Maximum Color Contrast Occurs Near the Antisolar Point. The color contrast can not be explained solely in terms of $D(\phi)$; it is caused by the scattering mechanism and the illumination of the umbra. Near the antisolar point the ray may be seen against the pinkish Belts of Venus. Since the umbra is illuminated by the blue sky and therefore appears bluish, the umbra and background sky are near opposite ends of the spectrum and thus display moderately strong color contrast. In the direction of the Sun, the sky tends to be whitish or whatever color that the Sun appears to be, owing to scattered light from the aureole. The rays, especially when they are due to dust particles, strongly forward scatter light from the whitish skylight; thus the color contrast is low.

5. CONCLUSION

We have compared the observed properties of rays with their theoretical contrast based on a line-of-sight calculation. To first order, the contrast of rays is dominated by the line-of-sight path length through the umbra. Furthermore, their contrast is often independent of the scattering mechanism producing them.

ACKNOWLEDGMENTS

I gratefully acknowledge many useful discussions with W. C. Livingston.

REFERENCES

1. M. Minnaert, *Light and Colour in the Open Air* (Dover, New York, 1954).
2. A. Meinel and M. Meinel, *Sunsets, Twilights and Evening Skies* (Cambridge U. Press, Cambridge, 1983).
3. R. G. Greenler, *Rainbows, Haloes, and Glories* (Cambridge U. Press, Cambridge, 1980).
4. J. L. Montheith, "Crepuscular rays formed by the Western Ghats," Weather **41**, 292–299 (1986).
5. W. Livingston and D. Lynch, "Mountain shadow phenomena," Appl. Opt. **18**, 265–269 (1979).
6. D. K. Lynch, "Mountain shadow phenomena 2. The spike seen by the off-summit observer," Appl. Opt. **19**, 1585–1589 (1980).
7. W. C. Allen, *Astrophysical Quantities* (Athlone, London, 1973).

Section Eleven
Backscattering

ERRORS INHERENT IN THE RADAR MEASUREMENT OF RAINFALL AT ATTENUATING WAVELENGTHS

By Walter Hitschfeld

McGill University [1]

and Jack Bordan

Sir George Williams College [2]

(Manuscript received 22 June 1953)

ABSTRACT

An equation for the rate of rainfall at a given range from the radar is derived. This is expressed in terms of the power level of the received signal (corrected for attenuation by intervening cloud and atmospheric gases) and takes account of radar attenuation due to intervening rain. The equation includes a constant which measures the performance of the radar and is determined by direct calibration.

At attenuating wavelengths (at 3 cm; to some extent at 5.6 cm) a small error in the calibration constant causes a large error in the measured rainfall. This error, which varies with range and may thus cause serious distortion, is, in fact, liable to be more serious than that caused if the attenuation were neglected entirely. Correcting for attenuation is therefore not recommended, unless the calibration error may be held within extremely narrow limits.

Very small calibration errors may be achieved by calibrating the radar by means of a rain gauge located at a point where the attenuation is appreciable. At points of smaller attenuation, a satisfactory degree of accuracy in the calculated rate of rainfall then results.

At wavelengths such as 10 cm, where the attenuation is negligible, errors in the constant still affect the measured rain, but neither so seriously, nor in a manner involving the range, thus causing no distortion.

An examination of the relative importance of the attenuation by gases and cloud at three wavelengths similarly emphasizes the difficulties associated with quantitative work at the shorter wavelengths.

---

1. Introduction

Most early work on weather radar was done by means of 10-cm equipment. But since the power returned by atmospheric scatterers is inversely proportional to the fourth power of the wavelength, there has been an increasing tendency to turn to shorter wavelengths. The resulting gain in sensitivity and resolution has partly obscured an essential disadvantage of all wavelengths less than 10 cm, namely attenuation by water. Atlas and Banks (1951) have pointed this out forcefully, and for a number of hypothetical rainfalls have calculated the echo pattern to be expected on the radar screen for wavelengths of 3.2 and 10 cm. They recognized two factors distorting the pattern as reported by the radar, and therefore requiring correction. One, applying at all wavelengths, is the decrease of the echo power with range (an inverse-square or an inverse-fourth power law, depending on whether the cross-section of the beam is, or is not, filled with target); the other is attenuation by rain and cloud. The effect of attenuation at 10 cm is practically negligible, but becomes increasingly more serious as the wavelength is reduced.

In this paper the opposite, and possibly the more practical, procedure is attempted. With the radar-power observations as the starting point, it is shown how the actual rain pattern may be deduced. To accomplish this, an equation giving the actual rain as an explicit function of the observed signal power is derived, taking care of both the above causes of distortion. This equation is relatively simple, and might possibly point the way to methods of automatic correction, if it were not for the unexpectedly great importance of the calibration error which, in the case of attenuating wavelengths, renders correcting almost useless. Only in special cases, when the calibration error can be kept within really narrow limits, is correcting possible. Attenuating wavelengths thus appear to be of limited applicability in rain measurement; in addition, attenuation reduces very substantially the sensitivity for the detection of precipitation viewed through intervening rain (Hitschfeld and Marshall, 1954).

In this work, the uncertainties and errors introduced by the random nature of the weather echo [as described by Marshall and Hitschfeld (1953), and Wallace (1953)] will be disregarded. Thus it will be assumed throughout that the average value of the echo power corresponding to any point in space is precisely known. Similarly, little account will be taken

[1] The research reported in this paper has been sponsored by the Geophysics Research Directorate, Air Force Cambridge Research Center, under Contract AF 19(122)–217.

[2] Part-time research assistant at McGill University.

of the errors inherent in the relationships between radar reflectivity and attenuation, and the rate of rainfall,

$$\text{reflectivity} = S_1 \times (\text{rate of rainfall})^{1.6},$$
$$\text{attenuation} = S_2 \times (\text{rate of rainfall})^{\alpha}, \tag{1}$$

where S_1, S_2 and α are functions of the wavelength only. These equations are useful working relations, although they involve the assumption of an idealized drop-size distribution which may cause errors in the rainfall of as much as 40 per cent (Wexler, 1948). Equations (1) are also based on approximations in the theory of microwave scattering from raindrops which are valid at wavelengths 3 cm or longer (Ryde, 1946; Marshall *et al*, 1952).

2. Theory

The radar equation.—The average value of the power returned from precipitation is, in the absence of attenuation, given by

$$P_{r0} = \frac{P_0 h A_e \Sigma\sigma}{8\pi r^2} F, \tag{2}$$

where P_0 is the power transmitted, A_e the effective antenna aperture for extended targets, h the pulse length, r the range, $\Sigma\sigma$ the sum of the back-scatter cross-sections of all the particles in unit volume, and F is a factor to allow for discrepancies between existing theory and experiment. [Austin and Williams (1951), for instance, found F to be about 0.2 in their careful rain measurements by means of a SCR-615-B (10 cm) radar.]

The quantity $\Sigma\sigma$ is, according to Rayleigh's treatment of the scattering from spheres, proportional to $Z = \Sigma D^6$, the sum of the sixth powers of the drop diameters in unit volume. The quantity Z, in turn, has been correlated with R, the rate of rainfall, by a number of workers, including Wexler (1948), who also reviewed much of the earlier work. Marshall *et al* (1952) recommend

$$Z = 200\, R^{1.6}, \tag{3a}$$

with Z in mm^6/m^3 and R in mm/hr as the best present correlation, and deduce

$$\Sigma\sigma = \frac{2.0\pi^5\, |\kappa|^2\, 10^{-10}}{\lambda^4}\, R^{1.6}. \tag{3b}$$

Here, $\kappa = (m^2 - 1)/(m^2 + 1)$, where m is the complex index of refraction of the scattering particle, and λ is the wavelength.

Combination of (2) and (3b) leads to

$$P_{r0} = \left(\frac{1}{a_0}\right)^{1.6} \frac{R^{1.6}}{r^2} \quad \text{or} \quad R = a_0 P_{r0}^{1/1.6} r^{2/1.6}, \tag{4}$$

where

$$\left(\frac{1}{a_0}\right)^{1.6} = \frac{2.5\,\pi^4 (10^{-11}) P_0 A_e h\, |\kappa|^2\, F}{\lambda^4}. \tag{5}$$

In practice, A_e, P_0 and F are difficult to determine absolutely; for this reason, the constant a_0 is generally obtained by direct calibration.

The preceding theory applies as long as the beam is filled with precipitation; if it is not, the formula to be used instead of (4) may be shown to be

$$P_{r0}' = \left(\frac{1}{a_0}\right)^{1.6} (r')^2 \frac{R^{1.6}}{r^4} \quad \text{or} \quad R = \frac{a_0 P_{r0}^{1/1.6} r^{4/1.6}}{(r')^{2/1.6}}, \tag{6}$$

where r' is the range at which the storm under observation would just fill the beam.

Attenuation.—In (4) and (6) no account has been taken of the attenuation, which causes a drop in the received power. If the thickness of the absorbing medium to be traversed is dr, the drop in the power received, dP_r, may be written

$$dP_r = -2\Lambda' P_r\, dr, \tag{7}$$

where Λ' is the attenuation constant of the medium in units of $(\text{length})^{-1}$. The factor 2 arises from the fact that, in the case of radar, each element of the path is traversed twice. In general, Λ' is a function of the range, so that

$$P_r = P_{r0} \exp\left(-2\int_0^r \Lambda'\, dr\right) \tag{8}$$

is the expression for the power received from the target at range r, where P_{r0} is the power that would be received in the absence of intervening attenuation. In practice, it is more convenient to write (8) in the form

drop in power level due to attenuation, in db,

$$= 10 \log\,(P_r/P_{r0}) = -2\int_0^r \Lambda\, dr, \tag{9}$$

where Λ is the attenuation caused by the medium in db/mi, r being measured in miles. Using (4) for P_{r0}, one obtains

$$10 \log P_r = 16 \log R$$
$$- 2\int_0^r \Lambda\, dr - 20 \log r - 16 \log a_0. \tag{10}$$

This, with some modifications, is equation (7) of Atlas and Banks (1951).

The attenuation constant Λ as used here includes attenuation by the atmospheric gases (including water vapor), by cloud and by rain. Formulas for the attenuation by the gases, k_1 (db/km or db/mi) are shown in table 1; probable maximum and minimum values of k_1 are listed in table 2. Table 3 lists values of the attenuation constant for water cloud, k_2, in (db/mi) $(\text{gm}/\text{m}^3)^{-1}$.

TABLE 1. k_1 (db/km). P is pressure in atmospheres; W is absolute humidity in g/m³.
(Source: K. L. S. Gunn and T. W. R. East, unpublished manuscript.)

Temperature (deg C)	Wavelength (cm)		
	3.2	5.6	10.0
20	$(7 \times 10^{-3})P^2 + (7 \times 10^{-4})PW$	$(7 \times 10^{-3})P^2 + (2 \times 10^{-4})PW$	$(7 \times 10^{-3})P^2 + 10^{-4}PW$
0	$(9 \times 10^{-3})P^2 + (8 \times 10^{-4})PW$	$(8 \times 10^{-3})P^2 + (3 \times 10^{-4})PW$	

TABLE 2. Range of k_1 (db/mi) of interest in this paper. Upper limit: $T = 10C$, $P = 1$ atm, $R.H. = 100$ per cent; Lower limit: $T = 0C$, $P = 0.5$ atm, $R.H. = 50$ per cent.

	Wavelength (cm)		
	3.2	5.6	10.0
Upper	24×10^{-3}	18×10^{-3}	13×10^{-3}
Lower	66×10^{-4}	48×10^{-4}	32×10^{-4}

TABLE 4. Values of K and α in (12), when R is in mm/hr and dr in mi.

	Wavelength (cm)		
	3.2	5.6	10.0
K:	144×10^{-4}	47×10^{-4}	4.9×10^{-4}
α:	1.3	1.1	1.0

The variation of the rain attenuation with rain intensity is shown in fig. 1, where the quantity K', defined by the equation

$$\text{rain attenuation, in db,} = \int K'R\,dr, \qquad (11)$$

with R in mm/hr and r in miles, is plotted. Data for this curve, based on the same drop-size distributions as (3a), are taken from Marshall *et al* (1952). Using the equations of the best straight lines for the loci of fig. 1, one may rewrite (11) as

$$\text{rain attenuation, in db,} = \int KR^\alpha\,dr, \qquad (12)$$

where $KR^{1-\alpha} = K'$. Values of K and α are listed in table 4. Attenuation by snow, ice and ice cloud may be neglected at all wavelengths.

In light of the above, the term $\int \Lambda\,dr$ of (10) should be replaced by

$$\int (k_1\,dr + k_2M\,dr + KR^\alpha\,dr), \qquad (13)$$

where M is the cloud density in gm/m³.

This paper is primarily concerned with the last and most important term, attenuation by rain. It should be noted, however, that the other two terms are not always negligible. Attenuation by the gases is rarely a problem. It may be corrected for relatively easily, as k_1 depends only on the pressure and relative humidity, approximate data for which are generally available.

The effect of cloud is usually more difficult to assess; clouds attenuate appreciably, but produce no observable echo on wavelengths 3 cm or longer, except possibly on the highest-powered radars. It is therefore not possible to devise a method for correcting this

TABLE 3. k_2 $[(db/mi)(g/m^3)^{-1}]$. Values for 0C and calculated on basis of scattering theory (Ryde, 1946).

	Wavelength (cm)		
	3.2	5.6	10.0
k_2:	137×10^{-3}	41×10^{-3}	14×10^{-3}

attenuation, as will be attempted in this paper for rain.

The determination of the rate of rainfall.—Equation (10) and expression (13) may be combined to read

$$10 \log P_r = 16 \log R$$
$$- 2 \int_0^r (k_1\,dr + k_2M\,dr + KR^\alpha\,dr)$$
$$- 20 \log r - 16 \log a_0. \qquad (14)$$

This relation expresses the received power as a function of R; it is practically more important, however, to have R as a function of P_r. A solution of (14) for R will now be obtained. The terms P_r and $\int_0^r (k_1 + k_2M)\,dr$ are functions of the range r; grouping them together, we may write (14) in the form

$$y = 16 \log R - 20 \log r$$
$$- 2K \int_0^r R^\alpha\,dr - 16 \log a_0, \qquad (15)$$

where

$$y = 10 \log P_r + 2 \int_0^r (k_1 + k_2M)\,dr. \qquad (16)$$

The quantity y is the power level of the received power, corrected for attenuation by intervening gases and cloud. Equation (15) may be written in a mathematically more convenient form,

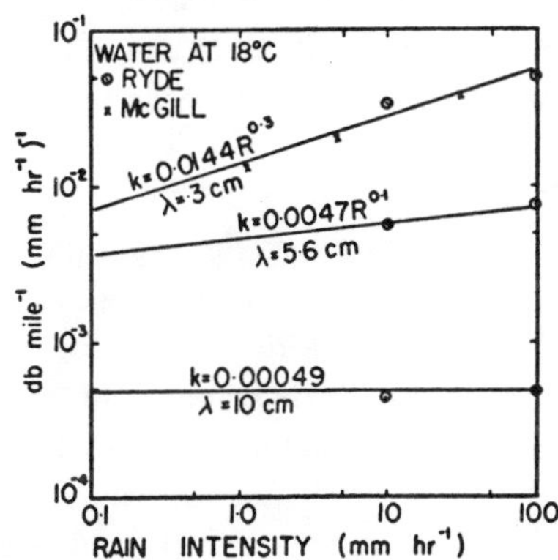

Fig. 1. (Attenuation by rain/rate of rainfall) as function of rate of rainfall at three wavelengths.

$$y = A \ln R - B \ln r - C \int_0^r R^\alpha \, dr - A \ln a_0, \quad (17)$$

where

$$A = 16 \times 0.4343 = 6.944,$$
$$B = 20 \times 0.4343 = 8.680, \quad (17a)$$
$$C = 2K.$$

Differentiation of (17) results in

$$\frac{dy}{dr} = \frac{A}{R} \frac{dR}{dr} - \frac{B}{r} - CR^\alpha.$$

By means of the substitution $U = R^{-\alpha}$, a linear differential equation is obtained which may then be solved directly. The result is

$$R = \frac{r^{B/A} e^{y/A}}{\left[\left(\dfrac{1}{a} \right)^\alpha - \dfrac{\alpha C}{A} \displaystyle\int_0^r r^{B\alpha/A} e^{y\alpha/A} \, dr \right]^{1/\alpha}}, \quad (18)$$

where $(1/a)^\alpha$ is the constant of integration. In principle, a may be determined by combining (15) and (18). This is difficult algebraically, and in any case would give a in terms of the constant a_0, introduced in (4) as a calibration constant. It is therefore simpler to regard a as another form of the calibration constant of the radar. As such, it may be obtained if the intensity of rainfall R_1 at range r_1 is known, while $y(r)$ is the known distribution of power received at the radar from all points between the radar and r_1 [corrected according to equation (16), if necessary]. It follows then from (18) that

$$\left(\frac{1}{a} \right)^\alpha = \frac{r_1^{B\alpha/A} e^{y_1\alpha/A}}{(R_1)^\alpha} + \frac{\alpha C}{A} \int_0^{r_1} r^{B\alpha/A} e^{y\alpha/A} \, dr. \quad (19)$$

The quantity a may be interpreted as the minimum detectable value of the rainfall at range one mile, when the received power has the minimum value that can be received, and when the attenuation due to intervening rain is zero or negligible.

General discussion of the equation for R.—Equation (18) expresses the rate of rainfall at any range in terms of the range and the power level of the signal received from that range and from all points closer in. As this equation makes full allowance for the attenuation, it suggests that attenuating wavelengths may be used for the measurement of R.

There are, however, some restrictions to this statement. Clearly R cannot be determined for values of r for which y is less than the minimum detectable value. This maximum range will depend on the performance of the set, and on the amount of intervening attenuation. An inevitable, though always small, error is also caused by the attenuation at the near edge of the

storm by regions returning an undetectably small signal. This means that in practice the integral $\int_0^r r^{B\alpha/A} e^{y\alpha/A} \, dr$ cannot be worked out from the required lower limit (zero), but only from the point where y becomes detectable.

By far the most serious practical limitation to the usefulness of (18) arises from the smallness of the denominator when the attenuation is appreciable. This is particularly the case when a is in error. This may arise from faulty calibration, and/or from the errors inherent in the relationships between radar reflectivity and attenuation and the rate of rainfall [equations (3) and (12)]. As such errors in a must be anticipated, the remaining sections of this paper will be devoted chiefly to an examination of their consequences.

Only two further points will be made here. For non-attenuating wavelengths $C = 0$, and hence the denominator in (18) is $1/a$, a constant. Any error in a therefore affects all calculated values of R equally. Calculated cross-sections through rain storms will then be everywhere in error to this extent, but the general shape of the computed profile will be correct.

Secondly, (18) and (19) suggest that it is advantageous to determine a (to calibrate the radar) by means of a rain gauge deep in the storm, *i.e.*, at a range, say r_n, where the attenuation is appreciable. Any error that is then made in measuring R enters into the calculation of $(1/a)^\alpha$ in only one of two positive terms [equation (19)]. Moreover, when the value of $(1/a)^\alpha$ thus determined is used in connection with (18), the error in the denominator and thus the error in R will, for all values of $r < r_n$, be less than the error made in the direct measurement of R at range r_n. This effect is discussed below in more detail.

3. Effect of the calibration error

Error analysis.—If the true value a of the calibration constant is known, and used in (18), the denominator will always be positive, and R can be calculated for all ranges for which the power returned (P_r) has a detectable value.

If the calibration constant is in error, two cases may be distinguished. The performance of the radar may be overestimated, *i.e.*, the calibration constant is taken to be a', which is smaller than the proper value a. This means that the rainfall deduced from a given received power is less than the true rainfall. In this case, the term $1/a'$, replacing the true $1/a$ in (18), is bigger than it ought to be, and, as the range increases, progressively less and less adequate allowance is made for the rain attenuation.

If, on the other hand, the performance of the radar is underestimated, the rainfall corresponding to a given received power is overestimated. Progressively more than necessary allowance will then be made for the attenuation, as the range increases. The quantity a' is now greater than a, making the denominator in (18) smaller than it ought to be. In fact, as the range increases, it is quite possible that the denominator vanishes, thus leading to an infinite value of the radar rainfall. As the subsequent analysis will show, this absurd result can be obtained at quite modest values of the attenuation and the calibration error. This was first noted (Bordan, 1952) in attempts to correct actual 3-cm radar photographs for attenuation by a graphical technique based on (14).

The effect of an error in the calibration constant a will now be studied quantitatively.

Let R be the true rate of rainfall at range r. This is the value which (18) would give if the calibration constant a were free from error. If the error in a is Δa, the computed rate of rainfall will be R' instead of R. The relation between the error factor R'/R and Δa, the calibration error, may be obtained from (18). The algebra is fairly involved, but elementary, and is omitted here. The result may be expressed in two ways,

$$\frac{R'}{R} = \left[1 + \left\{ \left(1 - \frac{\Delta a}{a'} \right)^{\alpha} - 1 \right\} \right.$$
$$\left. \times \exp \left\{ \frac{C\alpha}{A} \int_0^r R^{\alpha} \, dr \right\} \right]^{-1/\alpha}, \quad (20)$$

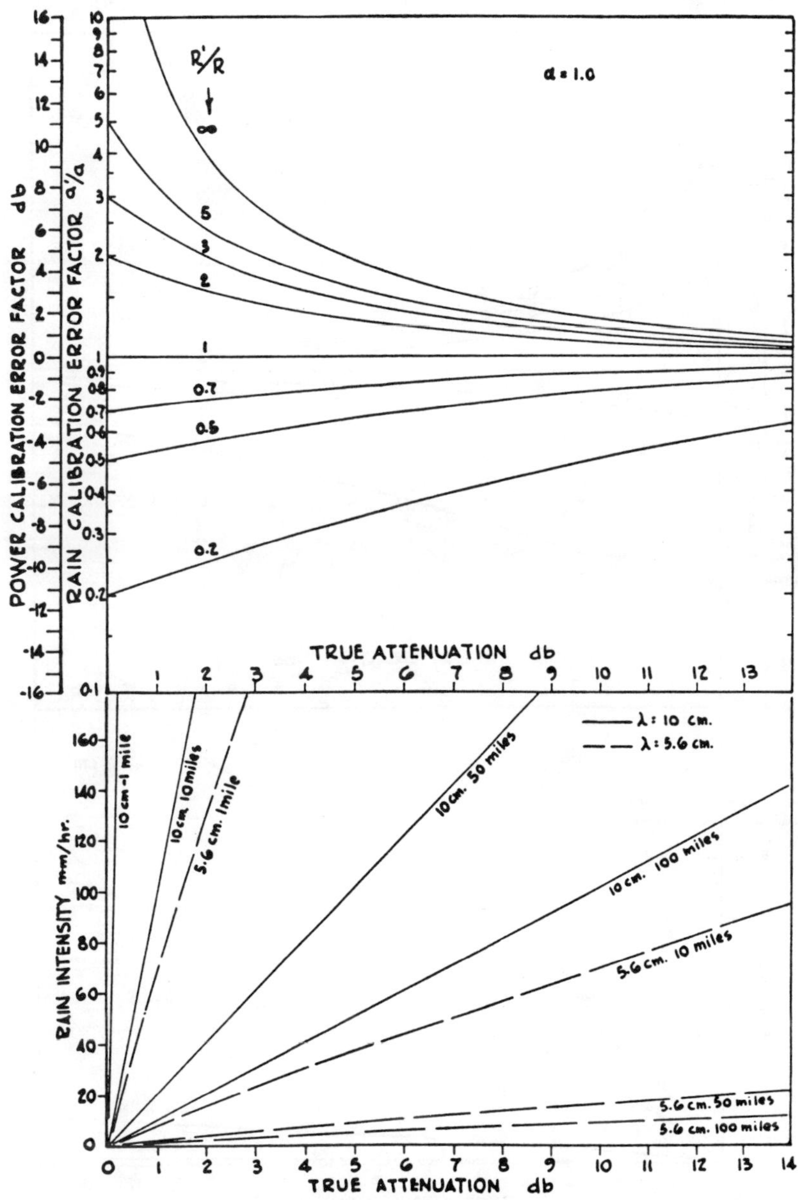

FIG. 2. Top: calibration-error factor *vs.* true attenuation, plotted for various values of error factor in measured rain; curves are drawn for $\alpha = 1.0$, and so apply at $\lambda = 10$ cm, and approximately at $\lambda = 5.6$ cm. Bottom: curves showing attenuation caused on wavelengths 10 and 5.6 cm as function of rain intensity for given thickness of storm.

and

$$\frac{R'}{R} = \left[1 + \left\{ \left(1 - \frac{\Delta a}{a'} \right)^{-\alpha} - 1 \right\} \right.$$

$$\left. \times \exp\left\{ \frac{C\alpha}{A} \int_0^r \left(R' \right)^\alpha dr \right\} \right]^{1/\alpha}, \quad (20')$$

where $a' = a + \Delta a$.

Equations (20) and (20′), despite their complexity, express the error factors R'/R in the radar rainfall in the most useful terms. Equation (20) gives the error factor to be expected in terms of the true attenuation $C\int R^\alpha dr$. This is the equation to be studied when the suitability of a radar set is to be assessed for quantitative measurements of hypothetical weather situations. Equation (20′), on the other hand, since it gives R'/R in terms of $C\int (R')^\alpha dr$, the apparent or measured

attenuation, allows the operator of a given equipment to estimate the possible or probable error in his rainfall measurements, provided he knows the possible or probable value of the calibration error. The meaning of these results will become clear when figs. 2 and 3 are examined. [To facilitate the calculations for these figures, and for subsequent work, it was found convenient to re-cast (20) and (20′) in "universal form" by means of the substitutions

$$\phi = A/\alpha \ln \left| (a'/a)^{-\alpha} - 1 \right| \quad \text{and} \quad \rho = C \int_0^r R^\alpha dr$$

for (20), and

$$\phi' = A/\alpha \ln \left| (a'/a)^\alpha - 1 \right| \quad \text{and} \quad \rho' = C \int_0^r (R')^\alpha dr$$

for (20′).]

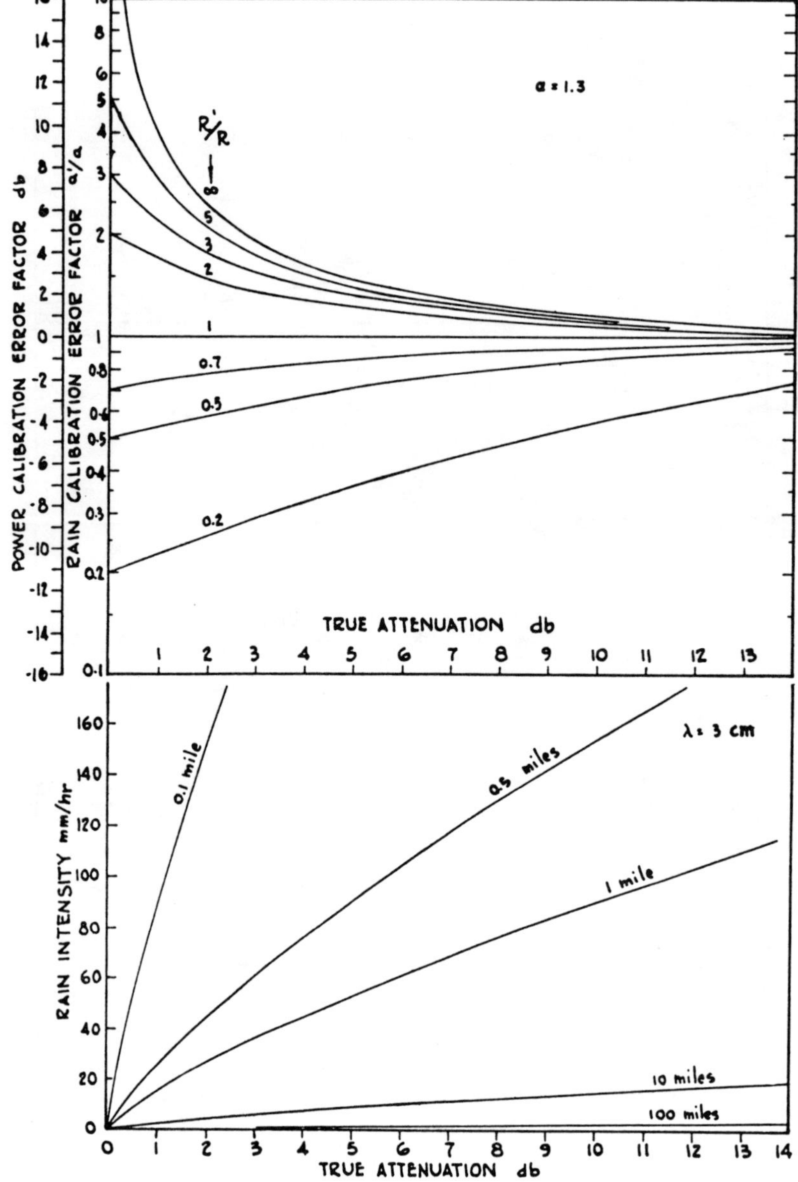

Fig. 3. Same as fig. 2, except that curves are drawn for $\alpha = 1.3$, and so apply at 3-cm wavelength. A rain storm of intensity 10 mm/hr, extending through 10 mi, causes about 5 db of attenuation on 3-cm radar; if radar calibration constant is overestimated by 20 per cent ($a'/a = 1.2$), corrected radar rainfall will be twice true rainfall.

Figs. 2 and 3 are based on (20); their upper parts are plots of the calibration-error factor against the true attenuation (in db) for a number of values of R'/R. Fig. 2 is drawn for $\alpha = 1.0$, and thus applies at wavelength 10 cm and approximately at 5.6 cm (see table 1 or fig. 1). It also applies more or less exactly at all wavelengths between 5.6 and 10 cm. Fig. 3 is drawn for $\alpha = 1.3$, thus applying at a wavelength of 3 cm. The calibration error (along the ordinates) is expressed in two ways. If the calibration error is considered to be due primarily to a faulty rain-gauge reading, it is best thought of as the "rain calibration-error factor a'/a." Note that a and R are measured in the same units. One may equally well consider the error to be associated with radar-power reading, in which case it should be expressed in db, and on account of the relation

$$R \propto P_r^{1.6} \quad \text{or} \quad a \propto P_r^{1.6}$$
$$\text{(at constant } a) \qquad \text{(at constant } R)$$

a factor 10 in a corresponds to a 16-db change in P_r. The "power calibration-error factor (db)" is thus $10 \log (P_r'/P_r)$ or $16 \log (a'/a)$. The upper parts of figs. 2 and 3 are very similar; it is on account of the apparent insensitivity of the data to the exact value of α that no separate curves for $\alpha = 1.1$ (*i.e.*, for 5.6-cm wavelength) are presented. To illustrate the meaning of the curves, consider, for instance, that when $\alpha = 1.3$, an attenuation of 5 db, combined with an error in a of only $+50$ per cent ($a'/a = 1.50$), will lead to an infinite value of R'. To obtain any reasonable

accuracy in R (say, R'/R between 0.5 and 2) the uncertainty in a would have to be held to within less than 30 per cent (actually, $0.73 < a'/a < 1.2$).

The great difference between the various wavelengths comes in when it is asked what actual weather situations might give rise to the above 5 db of attenuation. Reference may then be made to the lower parts of the figures. From them it appears that on 3 cm, a storm of intensity 52 mm/hr and effective thickness 1 mi, or 8 mm/hr extending through 10 mi (*i.e.*, less than 80 mm hr⁻¹ mi) would just cause this attenuation. On 10 cm, on the other hand, it is virtually impossible to obtain an attenuation of 5 db, as a storm of intensity 100 mm/hr extending over 50 mi (5000 mm hr⁻¹ mi) would be required. On 5.6 cm, the amount of intervening rainfall is about 300 mm hr⁻¹ mi, corresponding to 30 mm/hr extending over 10 mi, or 6 mm/hr extending over 50 mi. Thus, it appears that the 5 db of attenuation (which may cause such serious errors in any radar measurement of R) is never encountered on 10 cm, probably rarely on 5.6, and frequently on 3 cm.

Figs. 2 and 3, having the true attenuation as abscissas, are [like equation (20)] designed for use when the usefulness of various-wavelengths equipment is to be assessed. Fig. 4 [based on (20')] is designed for the operator of 3-cm equipment. (Similar curves apply at other wavelengths.) In contrast to the preceding figures, the abscissa here is the *measured* attenuation. The same scales were used as before, though when the attenuation is measured by means of a radar of under-

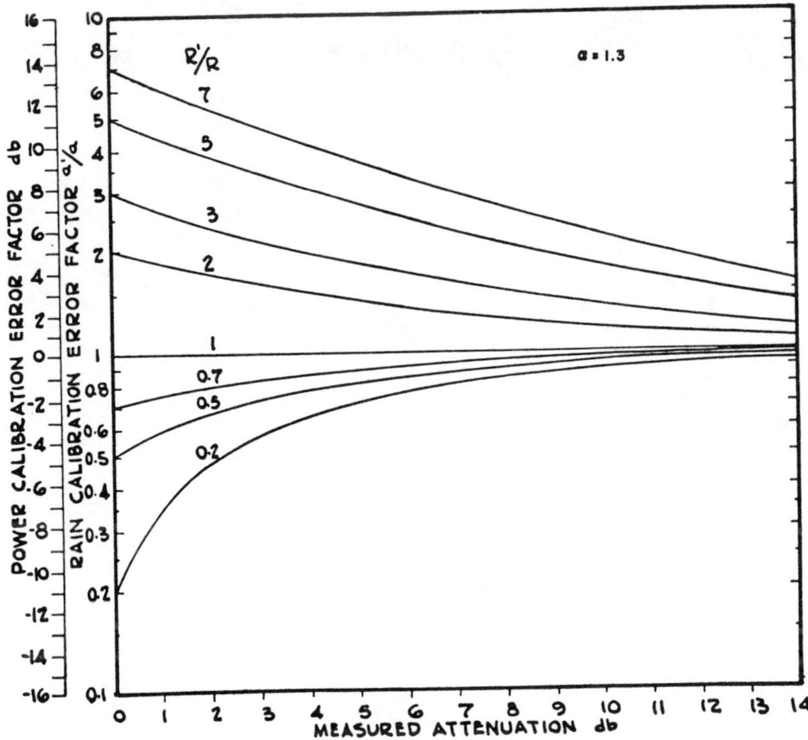

FIG. 4. Meaning of curves is much the same as that of fig. 2, except that abscissa here is attenuation as measured by radar. $\alpha = 1.3$, hence figure applies primarily for 3 cm; but corresponding $\alpha = 1$ curves are very similar.

estimated performance (a' greater than a) there are no real upper limits, and it is entirely arbitrary to stop the abscissa at 14 db.

The futility of correcting for attenuation.—It has become apparent that, unless the calibration error can be kept within extremely narrow limits, the errors in the value of R measured on attenuating wavelengths may easily become so great as to render the correcting procedure worse than useless. It is therefore worth comparing these errors due to a faulty attenuation correction with the errors due to the attenuation itself, when no correction for attenuation is attempted.

Such comparison for $\alpha = 1.3$ (*i.e.*, for 3-cm wavelength, the most important case) is made in fig. 5. This is a plot, against the true attenuation, of the calibration error expressed as $10 \log a'/a$. "Rdb" has been used here as the symbol for unit value of this quantity to distinguish it from the corresponding error in the power level of the received signal, which it will

be recalled would be $16 \log a'/a$.[3] In fig. 5, the set of parallel straight lines are lines of constant error in measured rain intensity when no correction for rain attenuation is made. This error, too, is measured in Rdb, *i.e.*, the numbers in the circles are the values of $10 \log R'/R$ resulting from the calibration error and the attenuation which is not corrected. The set of numbers (not in circles) throughout the diagram are the errors in R (values of $10 \log R'/R$) resulting from the calibration error when the attenuation is corrected for according to (18).

Comparison of the errors in R incurred with and without attenuation correction is made by comparing the uncircled numbers in the diagram and the circled numbers referring to the straight loci. If the occasional difference in sign of the two errors is disregarded, it

---

[3] While this term "Rdb" is introduced here with some hesitation, it was considered desirable to have a logarithmic unit for the calibration error, which would not permit confusion with db-intervals in the level of the received power.

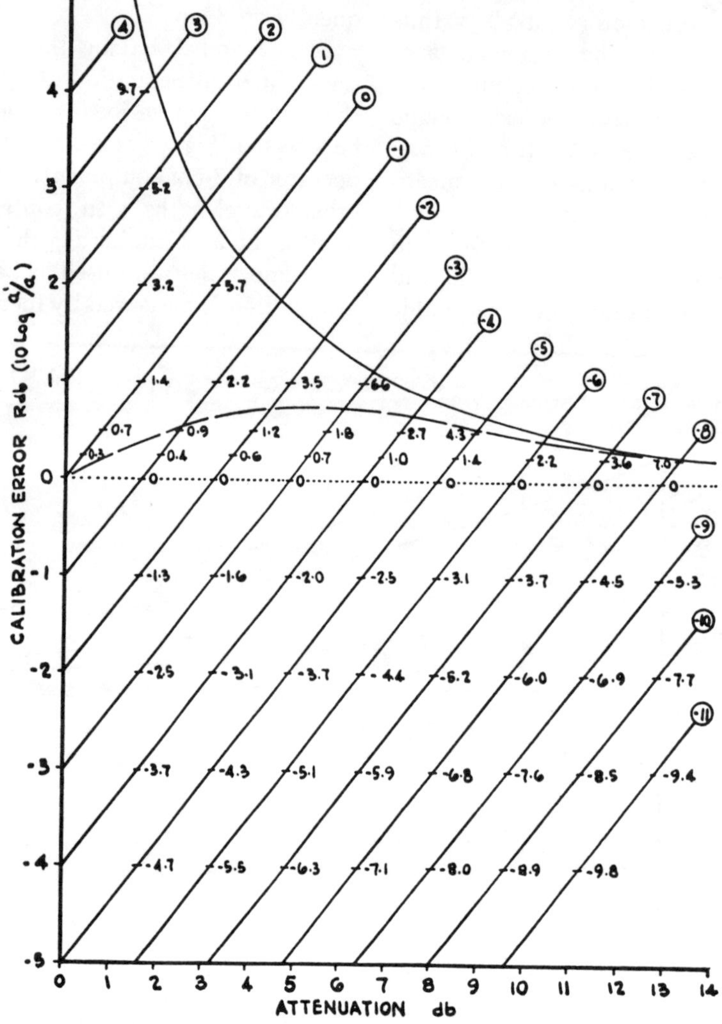

FIG. 5. Comparison of rainfall as measured by 3-cm radar with and without correcting for attenuation. *Uncircled numbers* show errors in corrected radar rainfall, in Rdb; *solid oblique lines* are lines of constant error in uncorrected radar rainfall (circled numbers show error in uncorrected radar rainfall, in Rdb); *solid curve:* line of infinite error in corrected radar rainfall; *dashed curve:* locus where errors in corrected and uncorrected radar rainfalls are equal; *dotted line:* locus of zero error in corrected radar rainfall.

appears that the error in the "corrected" radar rainfall is smaller than that in the uncorrected rainfall in the lower regions of the diagram, but the errors in the corrected radar rainfall become excessive near the top. Two (curved) loci have been drawn to emphasize this point. The solid curve is the locus at, or above, which the "corrected" radar rainfall is infinite. The broken curve is the locus along which the errors with and without attenuation correction are equal. Below this line, correcting is always better than not correcting. Above it, correcting rapidly leads to much greater errors than not correcting.

On its face, fig. 5 might suggest that, in more than half the cases, correcting is of advantage. While this is true, correcting is probably not, on the whole, worthwhile; the advantage to be gained by it below the horizontal dotted line drawn for zero calibration error is only small, whereas the errors due to correcting rapidly become enormous above the dashed curve. Consider, for instance, an attenuation of 5 db. When $a'/a = 1.4$ (an error of $+1.5$ Rdb in a), the "corrected" radar rainfall will be infinite, the uncorrected radar rainfall will be in error by only -1.6 Rdb. When $a'/a = 0.71$ (an error of -1.5 Rdb in a), the corrected

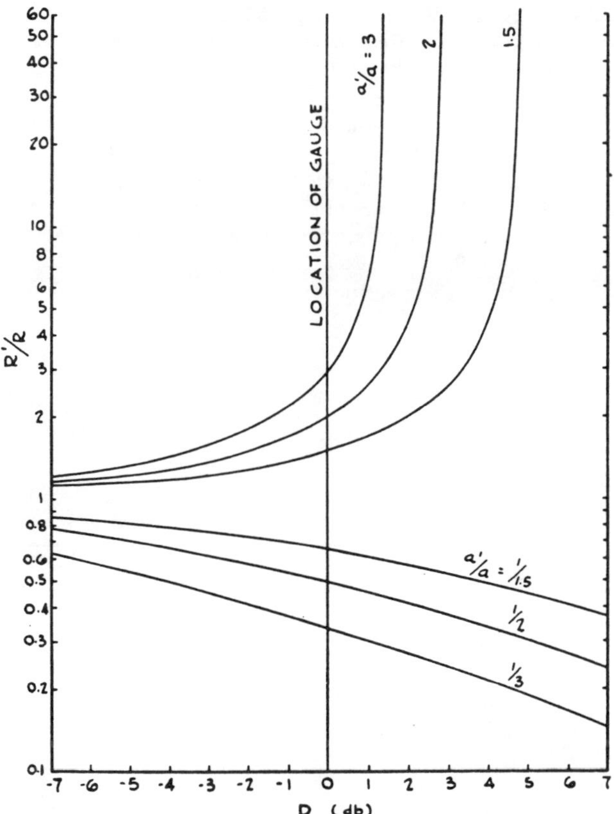

FIG. 6. Shows how errors in corrected radar rainfall vary with respect to gauge position. Distances (along abscissa) are measured in db of attenuation. Curves are valid for $\alpha = 1.3$, and thus apply at $\lambda = 3$ cm, but approximately indicate situation at other wavelengths. When calibration error factor a'/a is 3, then, at gauge, R'/R equals 3 also; at a point of 5 db less attenuation, R'/R is only 1.32, while at less than 1.5 db beyond the gauge, error in corrected radar rainfall is "infinite."

radar rainfall is in error by -2.6 Rdb, the uncorrected by -4.5 Rdb.

In the face of the generally unavoidable calibration errors, considerable doubt exists therefore as to the advisability of using attenuating wavelengths, with or without correcting for the attenuation, for rain measurements. The only important exception to this conclusion arises from the possibility of confining the calibration error to very narrow limits (rather better than ± 1 Rdb), as discussed in the following subsection.

Calibration of the radar by a rain gauge deep in the storm.—A suggestion made earlier for reducing the calibration error will now be evaluated. An inspection of (18) revealed that, at attenuating wavelengths, it is best to calibrate the radar by means of a rain gauge located in the storm at a point where the attenuation is appreciable. Any one of the figures 2 to 4 may be used to illustrate this point, but fig. 6 shows the general case for 3-cm radar. Here the ratio R'/R (the error factor in the corrected radar rainfall) is plotted against D, the distance between the points where measurement and calibration are made, expressed in decibels of attenuation. Each curve applies to a given value of a'/a, the calibration-error factor. When D is negative, measurement of R is made where the attenuation is smaller than at the calibrating gauge. A positive D (right half of fig. 6) means that the gauge is closer to the radar (attenuation-wise) than the measuring point. The graphs shown are horizontal displacements of each other; hence, the reader may add curves for any desired value of a'/a by moving an existing curve until it crosses the $D = 0$ axis at the point where R'/R equals the desired value of a'/a.

The negative conclusions of the previous subsection about rain measurements on short wavelengths should be qualified in light of the above finding. Rain measurements by short-wavelength radar appear, in fact, to be possible with a degree of accuracy better than that of the rain gauge at all points of attenuation smaller than that of the calibrating point. Attenuation causes serious errors only where it is greater than at the calibrating point. Where the attenuation is smaller, the radar yields rain values more precise than the gauge itself.

It should be recognized that the above analysis and fig. 6 apply only when y (the level of the received power corrected for gas and cloud attenuation) is accurately known everywhere between the gauge and the radar. In practice, this can be achieved only if the signal power to be measured is large compared to the noise level. It is therefore desirable that the attenuation intervening between gauge and radar be due to an intense but small storm, rather than be caused by light and extended rain.

Difficulties may also arise in connection with a. Throughout this paper, and notably in this section, it has been assumed that a is a constant, determinable by a single calibration, and is thus afflicted only with the calibration error as discussed above. There may, however, be other less easily defined errors in a, such as variations in the numerical factors ("2.0" and "1.6") or in $|\kappa|^2$ in (3a) and (3b), which, according to (5), is involved in a_0 and so in a. These constants may, to some extent, depend on the nature of the storm under study. While a may thus be known to some degree of precision at one place, its value may be in error more seriously at another due to a change in the nature of the storm. Such hard-to-specify variations in a would affect the radar-measured rainfall in a manner similar to the calibration error. They too would have a much more limited effect when non-attenuating wavelengths are used.

4. Conclusions

A new equation (18) was derived, giving the rate of rainfall explicitly in terms of the power received at the radar, with allowance for the attenuation by distance and by intervening atmospheric gases and cloud, and by rain. While this equation is relatively simple in principle, error analysis showed that the performance of the radar has to be known very accurately, if serious errors in the rainfall measured on attenuating wavelengths are to be avoided (figs. 2 to 5, notably 5). Calibration of the requisite precision is probably not usually possible, and hence radars using attenuating wavelengths are not generally recommended for quantitative rain measurements. Satisfactory rain measurements at such wavelengths are shown to be possible, however, at points at which the attenuation due to intervening rain is smaller than at the calibrating point (fig. 6). The advantage of calibrating by a rain gauge deep in the storm, as well as the difficulties attendant on the factors A_e and F in the fundamental radar equation (2), strongly underline a suggestion implicit in this paper: a radar used for rain measurement should be calibrated against a rain gauge, rather than by any other means.

Acknowledgment.—The writers wish to acknowledge Professor J. S. Marshall's interest and helpful criticism.

REFERENCES

Atlas, D., and H. C. Banks, 1951: The interpretation of microwave reflections from rainfall. *J. Meteor.*, **8**, 271–282.

Austin, P. M., and E. L. Williams, 1951: *Comparison of radar signal intensity with precipitation rate.* Cambridge, Mass. Inst. Tech. (Tech. Rep. No. 14 to Signal Corps Engr. Labs.), 43 pp.

Bordan, J., 1952: *Radar measurement of rainfall at attenuating wavelengths.* Montreal, McGill Univ. (M.Sc. thesis).

Hitschfeld, W., and J. S. Marshall, 1954: Effect of attenuation on a choice of wavelength for weather detection by radar. To be published in *Proc. Inst. Radio Engrs.*

Marshall, J. S., K. L. S. Gunn, and T. W. R. East, 1952: *The microwave properties of precipitation particles.* Montreal, McGill Univ., (Sci. Rep. MW-7, Stormy Weather Group), 40 pp.

Marshall, J. S., and W. Hitschfeld, 1953: Interpretation of the fluctuating echo from randomly distributed scatterers: part 1. *Canad. J. Phys.*, **31**, 962–994.

Ryde, J. W., 1946: The attenuation and radar echoes produced at centimetre wavelengths by various meteorological phenomena. In *Meteorological factors in radio wave propagation.* London, Physical Society, 169–189.

Wallace, P. R., 1953: Interpretation of the fluctuating echo from randomly distributed scatterers: part 2. *Canad. J. Phys.*, **31**, 995–1009.

Wexler, R., 1948: Rain intensities by radar. *J. Meteor.*, **5**, 171–173.

Determination of Aerosol Height Distributions by Lidar[1]

Frederick G. Fernald[2], Benjamin M. Herman[2] and John A. Reagan[3]

The University of Arizona, Tucson 85721

(Manuscript received 9 July 1971, in revised form 30 November 1971)

ABSTRACT

A new analytic solution to the lidar equation is presented, which realistically considers the scattering properties of the aerosols and the molecular atmosphere individually. With this solution, it is shown, in turbid atmospheres where the aerosols dominate the scattering properties, that accurate vertical profiles of the volume extinction cross section can be obtained with an uncalibrated lidar, provided that the total transmittance of the atmospheric layer being investigated is known. This solution is applied to data samples collected under very clear and under very dusty conditions.

1. Introduction

An obvious way to measure aerosols in the atmosphere is with lidar, for it can sample aerosols within the lowest 30 km of the atmosphere in 200 μsec, while airborne particle counters require 15 min to 1 hr at a single level before an adequate sample can be collected (Blifford, 1970). Lidar is considerably easier to use and provides much better spatial and temporal resolution. It is no simple matter, though, to obtain a reasonable description of the scattering properties of the aerosols from lidar data. Without any data from supporting experiments, the system must be well calibrated, and the physical makeup of the scatterers themselves should be known. It will be shown how these difficulties may be avoided by supplementing the lidar observations with one additional piece of information, namely, the transmittance of the layer in question.

The two scattering properties of the polydispersion of particulate scatterers which affect the amplitude of the lidar signal returned from the atmosphere are 1) the total volume extinction cross section σ which determines the attenuation of the laser pulse as it traverses its path, and 2) the backscattering cross section β which is a measure of the effectiveness of the scatterers as a target. Both parameters can vary with range. Usually, though, the composition and relative size distribution of the scatterers is assumed to remain constant so that $\sigma(Z)$ and $\beta(Z)$ at any height Z along the beam are just proportional to the total number density of the scatters at the height; and their ratio, $S = \sigma(Z)/\beta(Z)$, remains constant. Blifford's (1970) and Blifford and

Ringer's (1969) measurements with airborne particle counters show that, while the aerosol size distribution will change from day to day, on any one day the relative size distribution remains reasonably constant with height. Further, the aerosol optical depth will generally be almost entirely due to scatterers located within a well-mixed surface layer; and it will be quite valid to assume throughout this layer that the aerosols can be characterized by a single value of S.

One of the initial tasks, therefore, is the determination of the ratio S. This ratio can be predicted from Mie theory for polydispersions of spherical scatterers (e.g., Deirmendjian, 1969, or McCormick *et al.*, 1968). As Deirmendjian noted, though, the calculated scattering parameters are sensitive to both the choice of refractive index and the relative size distribution selected. Further, Holland and Gagne's (1970) experimental measurements on a system of irregularly shaped particles indicate that, while the total scattering cross sections agree with those predicted for an equivalent collection of spherical particles, the backscattering cross sections can be considerably below those predicted theoretically. It is true that these were highly irregular scatterers representing an extreme unlikely to occur in nature, but these results emphasize the fact that an assumption that these particles are spherical could possibly lead to considerable error.

It will be shown below, if the lidar observations are supplemented with simultaneous measurements of the total optical depth of the atmosphere at the ruby wavelength, that the ratio S becomes one of the results of the solution of the lidar equation. In addition, the vertical distribution of the extinction and backscattering cross sections of the aerosols can be obtained without any *a priori* assumptions as to their shape, index of refraction or size distribution, other than that their

[1] The research reported in this article has been supported by the Atmospheric Sciences Section of the National Science Foundation under Grants GA-16764, GA-1431 and GA-24134, and the Office of Naval Research under Grant N00014-67-A-0209-0009.

[2] Institute of Atmospheric Physics.

[3] Department of Electrical Engineering.

relative size distribution remains constant with height. The solution of the lidar equation is developed specifically for this problem, but can be readily applied to the case where S is one of the predetermined inputs to the problem.

One further complication is included. To date, published analytical solutions to the lidar equation have considered only one class of scatterers (Barrett and Ben-Dov, 1967, and Davis, 1969). In mildly turbid atmospheres, certainly typical of those prevailing over Tucson, scatter from both molecules and aerosols will contribute significantly to the problem and must be considered in any solution to the lidar equation. In developing this more general solution, the simpler case with one class of scatterers is considered first. These results are identical to those of Barrett and Ben-Dov (1967), and Davis (1969), but are obtained by using a transformation variable which has direct physical interpretation in the problem. Barrett and Ben-Dov have gone through an excessive number of variable transformations before arriving at their solution. Davis, in preserving a notation that presents the return signal strengths as the number of decibels above or below some reference signal, offers a nice solution, but ends with a notation that is awkward to interpret. Actually, the solution to be developed below was suggested to the authors after a careful review of the solution presented in the Appendix to Barrett and Ben-Dov's article. In their solution, they worked through a number of transformation variables, one of which (their w) is almost twice the optical depth from the lidar to the level in question. This led to the selection of the two-way transmittance as the transformation variable, and, as noted above, this is not just a convenient collection of mathematical terms, but is a variable significant to the problem.

The primary purpose here will be to present a complete analytic solution to the lidar equation for conditions where both molecules and aerosols contribute significantly to the problem. This solution is then applied to lidar observations collected at Tucson, Ariz. A rather extensive error analysis has been incorporated into the final computer program, and these results will also be discussed.

2. Solution to the lidar equation

The basic equation for a vertically looking, monostatic lidar is

$$P(Z) = ECZ^{-2}\beta(Z)T^2(Z), \tag{1}$$

where $P(Z)$ is the signal whose amplitude is proportional to the power received from a scattering volume at height Z; E is an output monitor pulse whose amplitude is proportional to the transmitted energy; C is a calibration constant which includes the losses in the transmitting and receiving optics, the effective receiver aperture, plus any coefficients required to make the

results dimensionally correct; $\beta(Z)$ is the backscattering cross section at height Z; and $T(Z)$ is the transmittance from the lidar to height Z. If the ratio of the extinction cross section to the backscattering cross section, $S = \sigma(Z)/\beta(Z)$, remains constant over the range interval being investigated, then the two-way transmittance, $T^2(Z)$, in Eq. (1) is

$$T^2(Z) = \exp\left[-2\int_0^Z \sigma(z)dz\right], \tag{2a}$$

$$= \exp\left[-2S\int_0^Z \beta(z)dz\right]. \tag{2b}$$

A solution to (1) can be obtained very easily if the unknown variables are expressed in terms of $T^2(Z)$. If (2b) is differentiated with respect to range, and then solved for $\beta(Z)$, one obtains

$$\beta(Z) = \frac{-1}{2ST^2(Z)}\frac{dT^2(Z)}{dZ}. \tag{3}$$

The lidar equation then becomes

$$P(Z) = -\frac{EC}{2SZ^2}\frac{dT^2(Z)}{dZ}. \tag{4}$$

Separating variables and integrating from height 0 to Z then gives

$$T^2(Z) = 1 - \frac{2S}{C}\int_0^Z \frac{P(z)z^2}{E}dz. \tag{5}$$

Substituting this back into (1) and solving for $\beta(Z)$ results in

$$\beta(Z) = \frac{P(Z)Z^2}{CE}\left[1 - \frac{2S}{C}\int_0^Z \frac{P(z)z^2}{E}dz\right]^{-1}. \tag{6}$$

[Eq. (6) is equivalent to Eq. (A10) of Barrett and Ben-Dov (1967) and Eq. (9) of Davis (1969).]

Eq. (5) has been skipped over in the past and is especially important to problems such as that presented by Davis (1969). For this type of problem, S is not known initially, but the total two-way transmittance, $T^2(Z^*)$, over some definite range increment 0 to Z^*, is known. The integral term can be obtained by numerically integrating the data obtained from the lidar returns; and, if the calibration constant C is known, S can be determined directly from (5) expressed as

$$S = C[1 - T^2(Z^*)]\left[2\int_0^{Z^*} \frac{P(z)z^2}{E}dz\right]^{-1}. \tag{7}$$

Thus, Davis did not have to iterate through successive guesses of S (his b), but could have obtained it directly from Eq. (7). Note also, that even if the lidar is uncalibrated, the ratio S/C can be determined from (7), and the value of $T^2(Z)$ at any intermediate range Z between 0 and Z^* can be calculated from (5).

An analogous solution can be formulated for the problem where two distinct classes of scatterers must be considered; for example, Rayleigh scatter from the molecular atmosphere (subscript R) and Mie scatter from particulates suspended in this atmosphere (subscript P). The lidar equation now becomes

$$P(Z) = ECZ^{-2}[\beta_R(Z)+\beta_P(Z)]T_R^2(Z)T_P^2(Z), \quad (8a)$$

where

$$T_R^2(Z) = \exp\left[-2\left(\frac{8\pi}{3}\right)\int_0^Z \beta_R(z)dz\right], \quad (8b)$$

$$T_P^2(Z) = \exp\left[-2S\int_0^Z \beta_P(z)dz\right]. \quad (8c)$$

In the above, S is the ratio of the extinction cross section to the backscattering cross section as defined earlier, but applies only to the particulate scatterers; the equivalent ratio for the molecular scatterers is just $8\pi/3$ if β_R has units [length steradians]$^{-1}$ and σ_R has units [length]$^{-1}$; $T_R^2(Z)$ and $\beta_R(Z)$ can be calculated within a few percent error from standard atmosphere data, or if more accuracy is required, from local radiosonde measurements; and $P(Z)$ is obtained from the lidar returns. Therefore, $\beta_P(Z)$ and $T_P^2(Z)$ are the only unknowns in (8a). If $\beta_P(Z)$ is expressed in terms of the two-way transmittance by applying Eq. (3) to the particulate scattering properties, (8a) then becomes

$$P(Z) = \frac{EC}{Z^2}\left[\beta_R(Z)-\frac{1}{2ST_P^2(Z)}\frac{dT_P^2(Z)}{dZ}\right]$$
$$\times T_R^2(Z)T_P^2(Z). \quad (9)$$

This can be expressed in the form

$$\frac{dT_P^2(Z)}{dZ}-2S\beta_R(Z)T_P^2(Z) = -\frac{2SP(Z)Z^2}{CET_R^2(Z)}, \quad (10)$$

and when solved for $T_P^2(Z)$ gives

$$T_P^2(Z) = \exp\left[2S\int_0^Z \beta_R(z)dz\right]$$
$$\times\left\{1-\frac{2S}{C}\int_0^Z \frac{P(z)z^2}{ET_R^2(z)}\exp\left[-2S\int_0^Z \beta_R(z')dz'\right]dz\right\}.$$
$$(11)$$

Note, however, that since

$$\exp\left[2S\int_0^Z \beta_R(z)dz\right] = T_R(Z)^{-3S/(4\pi)}. \quad (12)$$

Eq. (11) becomes

$$T_P^2(Z) = T_R(Z)^{-3S/(4\pi)}$$
$$\times\left[1-\frac{2S}{C}\int_0^Z \frac{P(z)z^2}{E}T_R(z)^{(3S/4\pi)-2}dz\right]. \quad (13)$$

Substituting this back into (8a) and solving for $\beta_P(Z)$ then yields

$$\beta_P(Z) = \frac{P(Z)Z^2}{CE}T_R(Z)^{(3S/4\pi)-2}$$
$$\times\left[1-\frac{2S}{C}\int_0^Z \frac{P(z)z^2}{E}T_R(z)^{(3S/4\pi)-2}dz\right]^{-1}-\beta_R(Z). \quad (14)$$

If the ratio S is known, then $T_P^2(Z)$ and $\beta_P(Z)$ can be determined directly from (13) and (14), respectively. On the other hand, in problems where S is not known initially, if the system is calibrated and if the two-way transmittance of the aerosols from the surface to height Z^*, $T_P^2(Z^*)$, can be estimated independently from solar radiometer measurements, S can be calculated from (13) put in the form

$$S = C[1-T_P^2(Z^*)T_R(Z^*)^{3S/(4\pi)}]$$
$$\times\left[2\int_0^{Z^*} \frac{P(z)z^2}{E}T_R(z)^{(3S/4\pi)-2}dz\right]^{-1}. \quad (15)$$

The height Z^* is the maximum height at which the signal can be easily read from the oscilloscope trace of the lidar return recorded on 35-mm film; and, for the data analyzed here, standard errors in the individual observations at this height have not exceeded 10%.

In general, an iterative solution of this transcendental equation will converge rapidly, as $T_R(Z^*)$ will be quite close to 1 and $T_R(Z^*)$ raised to the powers $3S/(4\pi)$ and $3S/(4\pi)-2$ will be relatively insensitive to variations in S. In very clean atmospheres, though, as the value of $T_P^2(Z^*)$ approaches 1, the term $[1-T_P^2(Z^*)T_R(Z^*)^{3S/(4\pi)}]$ becomes more sensitive to any variations in S, and the solution converges more slowly. Complications arise, though, with the analysis of real data, for both the integral expression and $T_P^2(Z^*)$ appearing in (15) can not be evaluated precisely at the outset of the iterative solution. In fact, their evaluation becomes incorporated into the iterative solution, as will be shown below.

The basic equations have been developed for a vertically looking lidar (zenith angle of zero degrees). Ideally, a lidar so positioned could collect usable data all the way from the surface to its maximum range. The lidar employed here, though, has individual transmitting and receiving optics parallel to each other and approximately 1 m apart. The beam width of the receiver ($\sim$4 mrad) is larger than that of the transmitter ($\sim$0.75 mrad), so that, at ranges beyond roughly 600 m, the entire transmitted pulse is in the field of view of the receiver, and only these data can be easily interpreted. Therefore, in order to extend the data down close to the surface, observations at larger zenith angles are required.

Before data collected at these larger zenith angles can be used in the basic equations, though, they must be corrected to approximate the returns a vertically looking lidar would have recorded. If the scattering properties of the atmosphere are assumed to be horizontally stratified, then the scattering properties away from the lidar at slant range R can be assumed identical to those directly overhead at the same height $Z = R/\sec\theta$. The range-corrected and output-normalized signal, $P(Z)Z^2/E$, that would have been returned from height Z had the lidar been looking vertically, is from simple geometric considerations

$$\frac{P(Z)Z^2}{E} = \frac{P(R)R^2}{E}[T_R^2(Z)T_P^2(Z)]^{1-\sec\theta}, \qquad (16)$$

where $P(R)$ is the signal returned from slant range R with the lidar positioned at zenith angle θ. The expression $[T_R^2(Z)T_P^2(Z)]^{1-\sec\theta}$ corrects for the additional attenuation experienced by the laser pulse in traversing a slant path to height Z instead of a vertical path. Note, though, that this includes $T_P^2(Z)$, which is one of the results to be calculated. On the initial run through the data, $T_P^2(Z)$ is set equal to one at all heights except Z^*, so that the resulting values of $P(Z)Z^2/E$ will be overestimated, tending to cause the initial computations of S to be underestimated. This first estimate of S is then used in (13) to calculate new values of $T_P^2(Z)$, which in turn are used to correct the data for the second run, etc., until the solution converges.

As noted above, $T_P^2(Z^*)$ must also be continually updated. It depends on $\tau_P(0,Z^*)$, the aerosol optical depth from the surface to height Z^*, as given by

$$T_P^2(Z^*) = \exp[-2\tau_P(0,Z^*)]. \qquad (17)$$

The solar radiometer, though, only measures the total optical depth of the atmosphere, τ_{tot}. The quantity $\tau_P(0,Z^*)$ is just τ_{tot} less the effects of ozone absorption, 0.007, the optical depth of the entire molecular atmosphere, 0.036, and the optical depth of the aerosols above Z^*, $\tau_P(Z^*,\infty)$. However, $\tau_P(Z^*,\infty)$ can not be measured directly. It was estimated for the initial run through the iterative solution. On subsequent runs, it was approximated by modeling the backscattering cross section of the aerosols above Z^* and then determining the optical depth of these aerosols from

$$\tau_P(Z^*,\infty) = S\int_{Z^*}^{\infty} \beta_P(z)dz. \qquad (18)$$

Estimates of S and $\beta(Z^*)$ were obtained from (15) and (14), respectively. The values for $\beta_P(Z)$ at heights above Z^* were based on previously published lidar results from the upper troposphere and stratosphere, as follows.

Schuster (1970) and Goyer and Watson (1968) show the ratio of the total backscattering cross section to the molecular backscattering cross section, $(\beta_P + \beta_R)/\beta_R$, rising steadily from 15 km to a peak at approximately 18 km. At 15 km, the aerosols have a backscattering cross section, $\beta_P(15) \approx 0.3\,\beta_R(15)$. At 18 km, the aerosol backscattering cross section is shown to be anywhere from 0.5–2.5 times the backscattering cross section of the molecular atmosphere at that level. Goyer and Watson's value of $\beta_P(18) \approx 0.8\,\beta_R(18)$, collected over New Mexico, was selected as the best representation of the conditions over Tucson.

The profile of the backscattering cross section of the aerosols above Z^*, therefore, was modeled by allowing $\beta_P(Z)$ to fall exponentially from its value at Z^* to the estimated value at 15 km, then letting it rise exponentially from 15 to 18 km, and finally having it fall off exponentially to infinity with a scale height of 3.75 km (see Elterman, 1968). The value of $\tau_P(Z^*,\infty)$ was obtained by integrating under this model curve according to (18). This was subtracted from $\tau_P(0,\infty)$, obtained with the solar radiometer, to give $\tau_P(0,Z^*)$, which led to $T_P^2(Z^*)$ via (17).

Attempts were made to obtain the calibration constant C using a slightly modified version of the method outlined by Hall and Ageno (1970), in which all lidar returns from the atmosphere are referenced against returns from a standard target consisting of a flat aluminum plate coated with Eastman White Reflectance Paint. This should approximate a Lambert reflector with a reflectivity of 0.992 at the ruby wavelength. The calibration constants obtained by this scheme, however, were quite variable and appeared to be approximately 40% too high.

An alternate scheme was therefore employed, making use of the fact that Tucson experiences some exceptionally clear days on which the atmosphere in the region from 3–6 km can be assumed to be reasonably free of aerosols. On these days, the system can be calibrated against the known molecular backscattering properties of this region.

This calibration fits readily into the iterative solution. If Eq. (14) is solved for C after $\beta_P(Z)$ is set equal to zero at height Z, and after S/C is replaced by the expression obtained when Eq. (15) is solved for S/C, the following equation results:

$$C = \frac{P(Z)Z^2}{E}\frac{T_R(Z)^{(3S/4\pi)-2}}{\beta_R(Z)}$$

$$\times \left[1 - \left(1 - T_P^2(Z^*)T_R(Z^*)^{3S/(4\pi)}\right.\right. \Bigg/$$

$$\left.\int_0^{Z^*} \frac{P(z)z^2}{E}T_R(z)^{(3S/4\pi)-2}dz\right)$$

$$\times \int_0^{Z} \frac{P(z)z^2}{E}T_R(z)^{(3S/4\pi)-2}dz\Bigg]^{-1}, \qquad (19)$$

where C can be recalculated at the end of each complete iterative cycle, providing an updated value for the

TABLE 1. Measured and computed parameters from lidar data. See text for explanation.

Date	τ_{tot}	$\tau_P(0,\infty)$	$\tau_P(0,Z^*)$	Z^* (−km)	S (sr)	Error in S (per cent)
5 April	0.206	0.163	0.141	6.75	12.7	30
6 April	0.336	0.293	0.267	4.20	14.5	11
10 May	0.050	0.007	0.0055	6.00	2.5	180
18 May	0.046	0.003	0.0020	6.00	1.6	630

TABLE 2. Theoretically determined values of S vs the shape ν of the aerosol distribution function.

ν	2.0	2.2	2.4	2.6	2.8	3.0	3.2	3.4	3.6	3.8	4.0
S	12.5	13.3	14.5	16.0	17.7	19.6	21.3	22.7	24.5	24.5	22.6

next run through the data. This calculation can be performed at a single level where it is reasonable to suppose that only molecular scatter predominates; or a search can be made by performing the above calculation over all levels, and the level at which the minimum value is obtained will correspond to the level at which the aerosols are contributing least to the net lidar return.

3. Data collection and analysis

Data collected at Tucson, Ariz., on 5 April, 6 April, 10 May and 18 May 1971 have been analyzed. Vertical profiles of scattering parameters were derived for each day from 16 lidar observations (four observations collected at each of four elevation angles). The individual lidar observations were reduced to some 30–40 points selected at 150 m intervals.

Eq. (19) was used to compute the calibration constant on 10 and 18 May, as the atmosphere over Tucson was very clear. Solar radiometer data collected at the times of the lidar observations showed total aerosol optical depths of 0.007 and 0.003, respectively. The value of C obtained on 10 May was slightly below that of 18 May and has been used in the analysis of the lidar data presented below. Meteorologically, these two days were somewhat similar. In both cases, dry cold fronts had

moved eastward through Arizona on the previous day and very clean air followed in the northwest flow behind these fronts.

On 5 and 6 April, the atmosphere over Tucson was very dusty, with total aerosol optical depths at the time of the lidar observations of 0.16 and 0.29, and with surface visibilities on the order of 50 and 20 km, respectively. Early on 4 April, a dry cold front moved down from the north through Arizona, New Mexico and western Texas, stirring up considerable dust, which was then blown toward Tucson by the easterly low-level anticyclonic flow which set in over the Southwest on 5 and 6 April.

Table 1 lists some of the data pertinent to the problem: these include the total optical depth of the atmosphere (τ_{tot}) at 690 nm determined from the solar radiometer located at the University of Arizona 6 mi south of the lidar site; the total aerosol optical depth $[\tau_P(0,\infty)]$; the estimated aerosol optical depth $[\tau_P(0,Z^*)]$ of the layer from the surface to Z^*; and $S = \sigma_P(Z)/\beta_P(Z)$, one of the major results of the calculations.

Theoretically determined values of S are presented in Table 2. The aerosols have been assumed spherical with a size distribution function that follows a Junge law given by

$$\frac{dn}{d\log r} = cr^{-\nu},$$

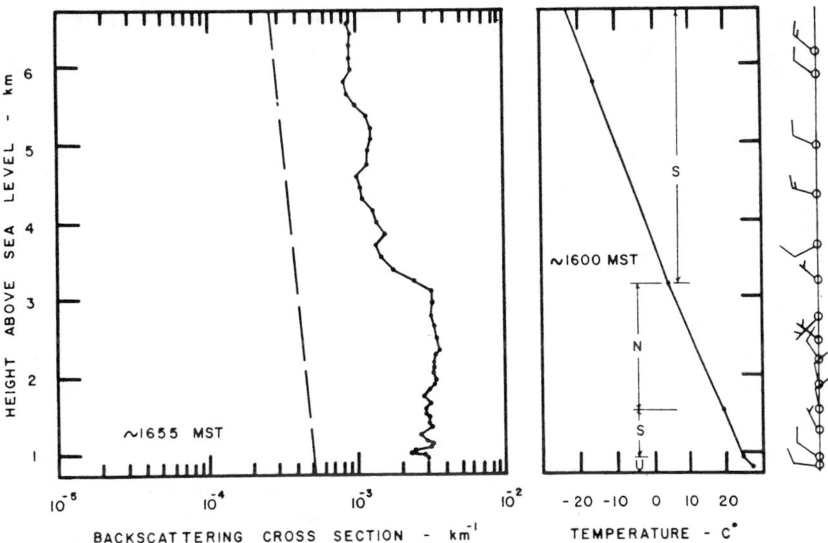

FIG. 1. Vertical profiles of the backscattering cross section, temperature and wind for 5 April 1971. (Units of backscattering cross section should be km⁻¹ sr⁻¹.)

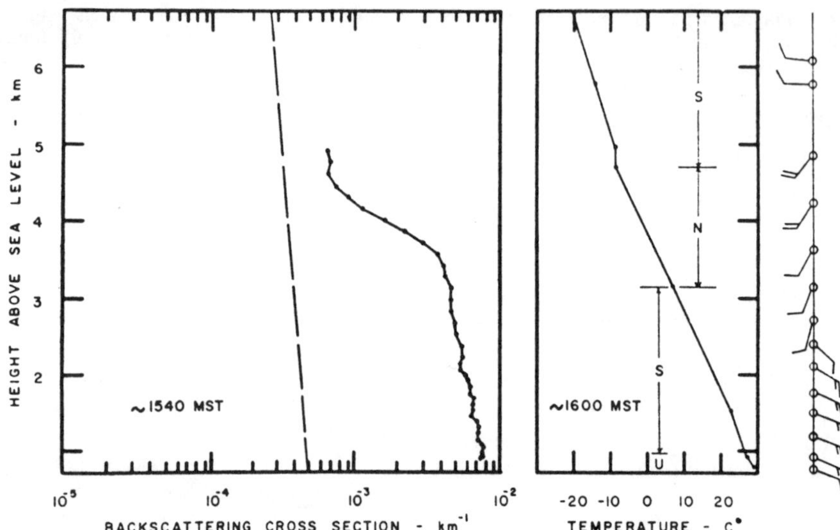

FIG. 2. Vertical profiles for the backscattering cross section, temperature and wind for
6 April 1971. (Units of backscattering cross section should be km^{-1} sr^{-1}.)

where n is the number of particles per unit volume of
radius r per unit interval of r, and c and ν are constants.
The value of c determines the total number density of
the particles, while the value of ν determines the shape
of the distribution function. Computations were carried
out at the ruby wavelength for aerosols with a real
index of refraction of 1.54 and for values of ν between
2 and 4 (typical for atmospheric aerosols). Note that
experimentally determined values of S for the dusty
skies of 5 and 6 April correspond to the smaller values
of ν in Table 2 and that small values of ν indicate a
preponderance of large-sized particles. Possible reasons
for the large discrepancies between the theoretical and
experimental values of S for 10 and 18 May are noted
in the discussion of errors presented below.

Figs. 1–4 show the calculated backscattering cross
sections of the aerosols, along with the temperature and
wind profiles obtained from the Tucson radiosonde
ascent closest to the time of lidar observations. The
backscattering cross sections of the aerosols can be
compared with the dashed line depicting the molecular
backscattering cross section. The stability of the atmosphere is noted along the temperature profile, with N
indicating a dry adiabatic lapse rate or neutral stability,
S a stable lapse rate which (with the very dry atmosphere experienced here) includes any lapse rate that
is less than dry adiabatic, and U a superadiabatic,
absolutely unstable condition. In general, marked
variations in the scattering properties of the aerosols
are closely aligned with the thermal structure. Such

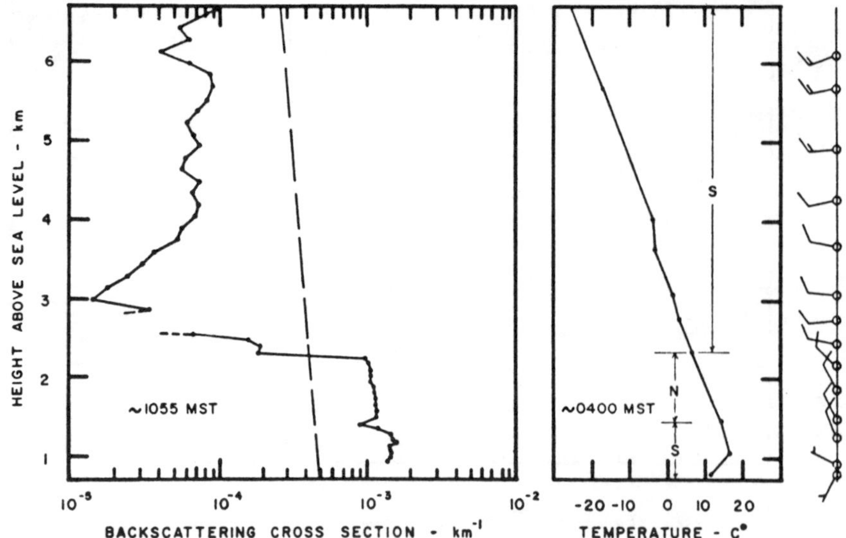

FIG. 3. Vertical profiles of the backscattering cross section, temperature and wind for
10 May 1971. (Units of backscattering cross section should be km^{-1} sr^{-1}.)

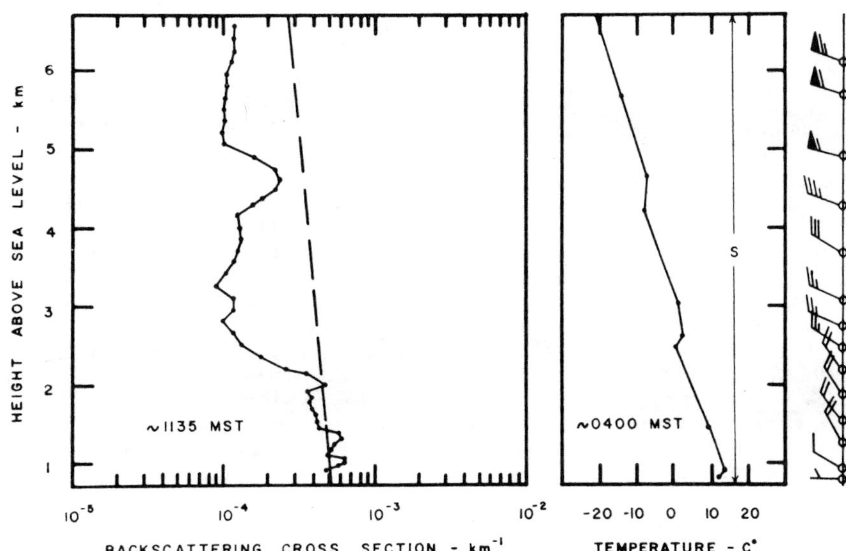

FIG. 4. Vertical profiles of the backscattering cross section, temperature and wind for 18 May 1971. (Units of backscattering cross section should be km⁻¹ sr⁻¹.)

stratification of aerosols has been noted in previous investigations (e.g., Collis *et al.*, 1964, and Viezee and Oblanas, 1969).

A complete error analysis has been performed, but only the general conclusions will be discussed here. This analysis has shown that errors in $T_P{}^2(Z^*)$ and C will contribute greatest to the final errors in S, $\beta_P(Z)$ and $\sigma_P(Z)$, and that errors in data points, $P(Z)$, taken from the lidar returns and the output monitor signal E, can essentially be ignored.

Errors in $T_P{}^2(Z^*)$ will be the prime source of errors in the calculated values of $\sigma_P(Z)$ and S, but their effect will diminish as the atmosphere becomes more turbid. Therefore, on the relatively turbid days of 5 and 6 April, errors of 7 and 6% in $T_P{}^2(Z^*)$ resulted in $\sigma_P(Z)$ profiles and values for S with standard errors of roughly 30 and 11%, respectively. On the other hand, errors in C (estimated to be 6%) were the primary source of the resulting errors (approximately 12 and 8%) in the calculated values of $\beta_P(Z)$.

On very clear days, estimates of $T_P{}^2(Z^*)$ are quite accurate (standard errors of 2.2% were estimated for both the 10th and 18th of May). Errors in $T_P{}^2(Z^*)$, though, become greatly amplified, so that on these days the computed errors in S and $\sigma_P(Z)$ were 180 and 630%, respectively. Thus, for these very clear days, it was essentially impossible to determine S or $\sigma_P(Z)$. Further, when the aerosols are the primary source of backscattered energy, the errors in $\beta_P(Z)$ will be relatively small compared to those resulting when the net return is overwhelmed by the signal from the molecular atmosphere. Thus, on 10 May below 2.2 km, where $\beta_P(Z) \approx 2 \times \beta_R(Z)$, the values of $P_P(Z)$ were accurate to within 11%. Above 2.2 km, however, $\beta_P(Z) \ll \beta_R(Z)$, and the errors ranged from 40 to 150%. On 18 May

below 1.5 km, the backscattering cross section of the aerosols was only slightly greater than that of the molecular atmosphere, and the values of $\beta_P(Z)$ were accurate to within 15%. Above this height, $\beta_P(Z)$ fell off rapidly, and its error increased to over 100%. Thus, while the very clear days offered the best means of calibration, the lidar calculations of S and $\sigma_P(Z)$, and $\beta_P(Z)$ when $\beta_P(Z) \lesssim \beta_R(Z)$, can not be relied upon, as they were considerably more sensitive to errors in the input data than on the more turbid days.

4. Conclusions

To date, no accurate description of atmospheric aerosols has been obtained from lidar measurements, primarily because 1) lidars are inherently difficult to calibrate, and 2) because the analytic tools available for the analysis of the data have not been developed sufficiently. This paper has attacked the second problem listed above, by introducing an advanced analytic solution to the lidar equation that considers the realistic situation of two classes of scatterers, aerosols and molecules. It has also emphasized the importance of supplementing lidar experiments with additional data, such as solar radiometer measurements. When this is done, a true profile of the total extinction cross section can be obtained in turbid atmospheres, even though the lidar is uncalibrated. Under very clear skies, as might be anticipated, the aerosol scattering properties (being comparable to or less than the molecular scattering properties) tend to become submerged in the observational errors. Precise input data become especially important here. In any case, for these analytic procedures to be fully implemented, the lidar must be accurately calibrated.

629

REFERENCES

Barrett, Earl W., and Oden Ben-Dov, 1967: Application of the lidar to air pollution measurements. *J. Appl. Meteor.*, **6**, 500–515.

Blifford, Irving H., Jr., 1970: Tropospheric aerosols. *J. Geophys. Res.*, **75**, 3099–3103.

——, and Lynn D. Ringer, 1969: The size and number distribution of aerosols in the continental troposphere. *J. Atmos. Sci.*, **26**, 716–726.

Collis, R. T. H., F. G. Fernald and M. G. H. Ligda, 1964: Laser radar echoes from a stratified clear atmosphere. *Nature*, **203**, 1274–1275.

Davis, P. A., 1969: The analysis of lidar signatures of cirrus clouds. *Appl. Opt.*, **8**, 2044–2102.

Deirmendjian, D., 1969: *Electromagnetic Scattering on Spherical Polydispersions*. New York, Elsevier Publ. Co., 290 pp.

Elterman, L., 1968: UV, visible, and IR attenuation for altitudes to 50 km. Environ. Res. Papers No. 285, Air Force Cambridge Research Laboratories, Bedford, Mass., 49 pp.

Goyer, Guy G., and Robert D. Watson, 1968: Laser techniques for observing the upper atmosphere. *Bull. Amer. Meteor. Soc.*, **49**, 890–895.

Hall, F. F., and H. Y. Ageno, 1970: Absolute calibration of a laser system for atmospheric probing. *Appl. Opt.*, **9**, 1820–1824.

Holland, A. C., and G. Gagne, 1970: The scattering of polarized light by polydisperse systems of irregular particles. *Appl. Opt.*, **9**, 1113–1121.

McCormick, M. P., J. D. Lawrence and F. R. Crownfield, Jr., 1968: Mie total and differential backscattering cross sections at laser wavelengths for Junge aerosol models. *Appl. Opt.*, **7**, 2424–2425.

Schuster, Burton, 1970: Detection of tropospheric and stratospheric aerosol layers by optical radar (lidar). *J. Geophys. Res.*, **75**, 3123–3132.

Viezee, William, and John Oblanas, 1969: Lidar observed haze layers associated with thermal structure in the lower atmosphere. *J. Atmos. Sci.*, **8**, 369–375.

Analysis of atmospheric lidar observations: some comments

Frederick G. Fernald

University of Denver, Physics Department, Denver, Colorado 80208.
Received 1 October 1983.
0003-6935/84/040652-02$02.00/0.

There have been many discussions of solutions to the lidar equation for elastic scattering (e.g., Fernald *et al.*,[1] Klett,[2] Davis,[3] and Collis and Russell[4]). Most of these are simply variations on Hitschfeld and Bordan's[5] solution for meteorological radars. Klett[2] recently restated this solution in a very convenient form for the analysis of lidar observations collected in very turbid atmospheres. His paper has prompted a restatement of the more general solution of Fernald *et al.*[1] which is also applicable to mildly turbid atmospheres where both aerosol and molecular scatterers must be considered in the analysis. This has led to a simple numerical scheme for the computer analysis of lidar measurements.

The lidar equation for two distinct classes of scatters (Fernald *et al.*[1]) is

$$P(Z) = ECZ^{-2}[\beta_1(Z) + \beta_2(Z)]T_1^2(Z)T_2^2(Z), \qquad (1)$$

where

$P(Z)$ = the return signal that is proportional to the received power from a scattering volume at slant range Z,

E = an output energy monitor pulse which is proportional to the transmitted energy,

C = the calibration constant of the instrument which includes losses in the transmitting and receiving optics and the effective receiver aperture,

$\beta_1(Z)$ and $\beta_2(Z)$ = respectively, the backscattering cross sections of the aerosols and molecules at slant range Z,

$T_1(Z) = \exp[-\int_0^Z \sigma_1(z)dz]$ = the aerosol transmittance,

$T_2(Z) = \exp[-\int_0^Z \sigma_2(z)dz]$ = the molecular atmosphere transmittance, and

where $\sigma_1(Z)$ and $\sigma_2(Z)$ = respectively, the extinction cross sections of the aerosols and molecules at range Z.

The molecular atmosphere scattering properties, $\beta_2(Z)$ and $\sigma_2(Z)$, can be determined from the best available meteorological data or approximated from appropriate standard atmospheres; so that only the aerosol scattering properties, $\beta_1(Z)$ and $\sigma_1(Z)$, remain to be determined. One further simplifying assumption is that the extinction-to-backscattering ratio for aerosols, $S_1 = \sigma_1(Z)/\beta_1(Z)$, remains constant with range. It essentially states that the size distribution and composition of the aerosol scatterers are not changing with range from the lidar, and that variations in backscattering from aerosols are due to changes in their number density. This is not exceedingly restrictive. In the numerical analysis of lidar data, the atmosphere can be divided into layers, with S_1 allowed to vary among the layers. Collis and Russell,[4] Pinnick *et al.*,[6] and Russell *et al.*[7] can be referenced for values for this ratio. The corresponding ratio for the molecular scatterers is the constant $S_2 = \sigma_2(Z)/\beta_2(Z) = 8\pi/3$.

The solution to Eq. (1) for the aerosol backscattering cross sections (Fernald *et al.*[1]) then becomes

$\beta_1(Z) =$

$$\frac{P(Z)Z^2 \exp[-2(S_1 - S_2)\int_0^Z \beta_2(z)dz]}{CE - 2S_1 \int_0^Z P(z)z^2 \exp[-2(S_1 - S_2)\int_0^z \beta_2(z')dz']dz}$$
$$- \beta_2(Z). \qquad (2)$$

If *a priori* information can be used to specify the value of the aerosol and molecular scattering cross sections at a specific range Z_c, the lidar can be calibrated by solving Eq. (2) for CE in terms of these scattering properties and

$\beta_1(Z) + \beta_2(Z) =$

$$\frac{X(Z) \exp[-2(S_1 - S_2)\int_{Z_c}^Z \beta_2(z)dz]}{\dfrac{X(Z_c)}{\beta_1(Z_c) + \beta_2(Z_c)} - 2S_1 \int_{Z_c}^Z X(z) \exp[-2(S_1 - S_2)\int_{Z_c}^z \beta_2(z')d']dz},$$
$$\qquad (3)$$

where $X(Z)$ is the range normalized signal $P(Z)Z^2$.[2] The total backscattering cross section at range Z is now expressed as a function of the scattering properties at the calibration range and those of the intervening atmosphere between the ranges Z_c and Z.

Equation (3) leads to a simple numerical integration scheme. If

$$A(I,I + 1) = (S_1 - S_2)[\beta_2(I) + \beta_2(I + 1)]\Delta Z \qquad (4)$$

is used to replace the exponential terms that incorporate the effects of aerosol extinction between adjacent data points range ΔZ apart, the total backscattering cross section at range $Z(I + 1)$, one data step beyond the calibration range $Z(I)$, becomes

$\beta_1(I + 1) + \beta_2(I + 1) =$

$$\frac{X(I + 1) \exp[-A(I,I + 1)]}{\dfrac{X(I)}{\beta_1(I) + \beta_2(I)} - S_1\{X(I) + X(I + 1) \exp[-A(I,I + 1)]\}\Delta Z}. \qquad (5)$$

Similarly, the total backscattering cross section at $Z(I - 1)$, one step before the calibration range $Z(I)$, becomes

$\beta_1(I - 1) + \beta_2(I - 1) =$

$$\frac{X(I - 1) \exp[+A(I - 1,I)]}{\dfrac{X(I)}{\beta_1(I) + \beta_2(I)} + S_1\{X(I) + X(I - 1) \exp[+A(I - 1,I)]\}\Delta Z}. \qquad (6)$$

Solutions in terms of aerosol extinction are correspondingly

$\sigma_1(I + 1) + \dfrac{S_1}{S_2}\sigma_2(I + 1) =$

$$\frac{X(I + 1) \exp[-A(I,I + 1)]}{\dfrac{X(I)}{\sigma_1(I) + S_1/S_2\sigma_2(I)} - \{X(I) + X(I + 1) \exp[-A(I,I + 1)]\}\Delta Z}, \qquad (7)$$

$\sigma_1(I - 1) + \dfrac{S_1}{S_2}\sigma_2(I - 1) =$

$$\frac{X(I - 1) \exp[+A(I - 1,I)]}{\dfrac{X(I)}{\sigma_1(I) + S_1/S_2\sigma_2(I)} + \{X(I) + X(I - 1) \exp[+A(I - 1,I)]\}\Delta Z}. \qquad (8)$$

The lidar data can, therefore, be analyzed in successive steps that can move either out or in from the assigned calibration range.

Some general comments can now be made concerning the application of Eqs. (5)–(8) to different atmospheric conditions. They are dependent on the laser wavelength, the extent to which multiple scattering can be ignored, and the data sampling interval ΔZ of the specific lidar system being used. The conclusions concerning highly turbid atmosphere are basically a reiteration of those of Klett.[2]

For highly turbid atmospheres ($\sigma_1 \gg \sigma_2$), the molecular scatterers can be ignored and Eqs. (6) and (8) reduce to Klett's.[2] In these atmospheric conditions the two terms in the denominators will be of comparable magnitude so that outward stepwise integration, Eqs. (5) and (7), can become very unstable. On the other hand, inward stepwise integration is very stable and rapidly loses its dependence on the initial guess of the scattering cross sections attributed to the calibration range. In this sense, uncalibrated lidars can yield the extinction properties of highly turbid atmospheres. Equations (6) and (8) for highly turbid atmospheres become

$$\beta_1(I-1) = \frac{X(I-1)}{\dfrac{X(I)}{\beta_1(I)} + S_1[X(I) + X(I-1)]\Delta Z}, \qquad (9)$$

$$\sigma_1(I-1) = \frac{X(I-1)}{\dfrac{X(I)}{\sigma_1(I)} + [X(I) + X(I-1)]\Delta Z}. \qquad (10)$$

When the aerosol and molecular scattering cross sections are of a comparable magnitude (during light to moderate air pollution events or in stratospheric studies), the second terms in the denominators of Eqs. (5)–(8) will be considerably smaller than the first terms. Numerically stable solutions are, therefore, possible when stepping in either direction from the calibration level. In these atmospheric conditions, the analyses will be dependent on the aerosol and molecular backscattering cross sections assigned to the calibration level. Net aerosol extinction will be small. It will be tied to the values selected for S_1, the aerosol extinction-to-backscattering ratio which can vary over a relatively wide range without greatly affecting the backscattering cross sections computed for Eqs. (5) and (6).

The analyses developed above lend themselves readily to qualitative statements collected in highly turbid, moderately turbid, and relatively clean atmospheres. The precise definition of these atmospheres will vary among lidar systems, primarily with the laser wavelength and data sampling interval.

In highly turbid atmospheres, aerosols dominate the scattering process to the extent that molecular scattering can be ignored. From Eq. (10) it can be demonstrated that an uncalibrated lidar can readily yield aerosol extinction profiles. On the other hand, backscattering profiles, Eq. (9), are directly dependent on an accurate knowledge of the extinction-to-backscattering ratio S_1.

In relatively clean atmospheres, the basic result of the analysis is the aerosol backscattering cross section, and the aerosol extinction now becomes dependent on an accurate knowledge of the extinction-to-backscattering ratio.

For moderately turbid atmospheres, lying in some ill-defined region between the two cases discussed above, the analyses will be sensitive to both the extinction and backscattering properties of the aerosols. The lidar system must be accurately calibrated, and the extinction-to-backscattering ratio must be reasonably well established. Equations (5)–(8) lend themselves to a very compact sequence of FORTRAN statements for the computer analysis of digitized lidar observations.

References

1. F. G. Fernald, B. M. Herman, and J. A. Reagan, J. Appl. Meteorol. 11, 482 (1972).
2. J. D. Klett, Appl. Opt. 20, 211 (1981).
3. P. A. Davis, Appl. Opt. 8, 2099 (1969).
4. R. T. H. Collis and P. B. Russell, in *Laser Monitoring of the Atmosphere*, E.D. Hinkley, Ed. (Springer, New York, 1976), p. 117.
5. W. Hitschfeld and J. Bordan, J. Meteorol. 11, 58 (1954).
6. R. G. Pinnick, J. M. Rosen, and D. J. Hofman, J. Atmos. Sci. 33, 304 (1976).
7. P. B. Russell, T. J. Swissler, M. P. McCormick, W. P. Chu, J. M. Livingston, and T. J. Pepin, J. Atmos. Sci. 38, 1279 (1981).

Lidar inversion with variable backscatter/extinction ratios

James D. Klett

The conventional approach to solving the single-scattering lidar equation makes use of the assumption of a power law relation between backscatter and extinction with a fixed exponent and constant of proportionality. An alternative formulation is given herein which assumes the proportionality factor in the power law relationship is itself a function of range or extinction. The resulting lidar equation is solvable as before, and examples are given to show how even an approximate description of deviations from the power law form can yield an improved inversion solution for the extinction. A further generalization is given which includes the effects of a background of Rayleigh scatterers.

I. Introduction

The volume extinction $[\sigma(k)]$ and backscatter $[\beta(k)]$ coefficients at wave number k due to particulate scattering and absorption are given by integrals over the size distribution $n(r)$ of (assumed spherical) particles of radius r:

$$\sigma(k) = \pi \int r^2 Q(\tilde{m},kr)n(r)dr, \tag{1}$$

$$\beta(k) = \pi \int r^2 Q_\pi(\tilde{m},kr)n(r)dr, \tag{2}$$

where Q and Q_π are, respectively, the extinction and backscatter efficiencies, as determined by the Mie theory of scattering; both are functions of the complex index of refraction $\tilde{m}$ and size parameter kr. (Gaseous absorption contributions to σ are not included in this paper.) It is clear from their definition as weighted integrals over the size distribution that the ratio of backscatter to extinction will be, for a given wavelength and complex index, only a function of the shape of the distribution. (Shape here simply refers to the size distribution known to within an arbitrary scale factor.) Therefore, if the spectral shape and composition of the scattering aerosol are spatially invariant, so too will be the proportionality between β and σ. On the other hand, if the aerosol spectral form or composition does vary with location, β and σ can be regarded as having a proportionality which is range dependent.

The author is with PAR Associates, 1077 La Quinta, Las Cruces, New Mexico 88005.

Received 10 September 1984.

0003-6935/85/111638-06$02.00/0.

The traditional way to account for any dependence between β and σ is to assume a power law relationship of the form

$$\beta = Bo\sigma^k, \tag{3}$$

where Bo and k are constants, but k is no longer unity. Then the single-backscattering lidar equation may, with the use of Eq. (3), be reduced to the form

$$\frac{dS}{dr} = \frac{1}{\beta}\frac{d\beta}{dr} - 2\sigma, \tag{4}$$

$$= \frac{k}{\sigma}\frac{d\sigma}{dr} - 2\sigma, \tag{5}$$

with $S(r) = \ln[r^2 P(r)]$, and where $P(r)$ is the power received from range r. The stable solution to Eq. (5) is

$$\sigma(r) = \frac{\exp[(S - Sm)/k]}{\left\{\sigma_m^{-1} + \dfrac{2}{k}\displaystyle\int_r^{r_m} \exp[(S - Sm)/k]dr'\right\}}, \tag{6}$$

where $Sm = S(r_m)$, $\sigma_m = \sigma(r_m)$, and $r \leq r_m$.[1] In addition to requiring the validity of Eq. (3) and the absence of multiple scattering, Eq. (6) also assumes that gaseous absorption effects are negligible.

From the discussion preceding Eq. (3) it is clear that it would be more realistic to regard β and σ as linearly related but with a proportionality which is itself a function of range or extinction. An approximate and simple way of formulating the inversion problem in this fashion is given in this paper. Any available information concerning deviations from a strict linear relationship can be easily incorporated into the solution implementation. An example of the backscatter/extinction variation for water clouds is given, and it is shown how even a very approximate description of this variation can yield an improved inversion solution for the extinction. Finally, a further generalization is carried out which includes the effects of a background of Rayleigh scatterers.

II. Modified Inversion Solution for Variable *B*

Suppose now that β and σ are related according to

$$\beta = B\sigma^k, \qquad (7)$$

where B is a function of r, and k is a constant. (From the discussion above one would generally set $k = 1$, but it is retained here as a fixed parameter primarily to facilitate direct comparison below between new and old expressions.) Then on substitution of this expression into Eq. (4) one finds

$$\frac{dS}{dr} = \frac{1}{B}\frac{dB}{dr} + \frac{k}{\sigma}\frac{d\sigma}{dr} - 2\sigma, \qquad (8)$$

or

$$\frac{d(S - \ln B)}{dr} = \frac{k}{\sigma}\frac{d\sigma}{dr} - 2\sigma. \qquad (9)$$

But this is the same as Eq. (5) except that S is now replaced by $S - \ln B$. Hence one can write down the solution immediately in the form of Eq. (6):

$$\sigma(r) = \frac{\exp[(S - Sm - \ln B + \ln Bm)/k]}{\left\{\sigma_m^{-1} + \dfrac{2}{k}\displaystyle\int_r^{r_m} \exp[(S - Sm - \ln B + \ln Bm)/k]dr'\right\}} \qquad (10)$$

or

$$\sigma(r) = \frac{(Bm/B)^{1/k}\exp[(S - Sm)/k]}{\left\{\sigma_m^{-1} + \dfrac{2}{k}\displaystyle\int_r^{r_m}(Bm/B)^{1/k}\exp[(S - Sm)/k]dr'\right\}}, \qquad (11)$$

where $Bm = B(r_m)$.

An apparent difficulty with Eq. (11) is that in general we cannot expect to know $B(r)$ precisely until we also know $\sigma(r)$; i.e., it is really part of the solution. However, an approximate solution to Eq. (11), still more accurate than the default version with $Bm/B = 1$ [i.e., Eq. (6)], can be obtained if there is available at least some information on the dependence of B on either r or σ. Suppose, for example, that for the aerosol in question there has been established an approximate empirical correlation of the form $B = f(\sigma)$. Then an approximate solution can be obtained by a two (or more if needed)-stage iteration: First, let $Bm/B = 1$ and solve Eq. (11) or (6) to obtain the first-order solution σ_1. Second, solve Eq. (11) for the second-order solution σ_2, where now we use

$$B(r) = f[\sigma_1(r)]. \qquad (12)$$

We shall now turn to an example of this approach and compare the outcome with the familiar method based on Eqs. (3) and (6).

III. Inversion Examples for Conditions of Fog and Low Clouds

An example of the variability of B with range for conditions of fog and low clouds is shown in Fig. 1. The data displayed (supplied by E. Measures, Atmospheric Sciences Laboratory, White Sands Missile Range, N.M. 88002, personal communication) derives from *in situ* measurements made by balloon-borne particle counters in Meppen, Germany, in the fall of 1980.[2,3] Mie theory was applied at a wavelength of 1.06 μm to obtain profiles with height of β and σ under the assumption that the

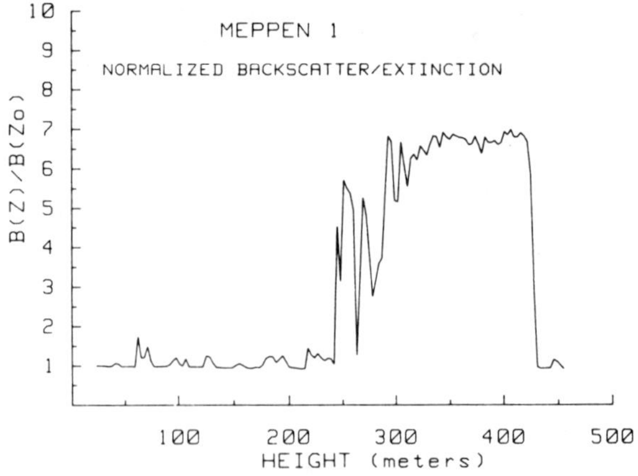

Fig. 1. Dependence on range of backscatter-to-extinction ratio for low stratus cloud in Meppen, Germany, 1980. Ratio normalized by value at near range point Zo. Data based on *in situ* measurements of drop size distributions. Light wavelength = 1.06 μm (adapted from calculations of E. Measures, ASL/WSMR; personal communication).

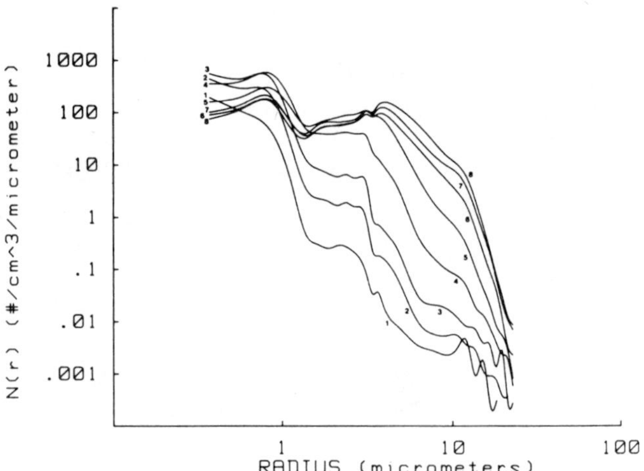

Fig. 2. Drop size distributions from in-cloud measurements in Meppen, Germany, 1980. The eight distributions shown are averages for the following labeled ranges of extinction (km): 1, (0.1,0.6); 2, (0.6,1.5); 3, (1.5,3); 4, (3,15); 5, (15,40); 6, (40,65); 7, (65,100); 8, (100,200) (from Ref. 3).

particles were spherical drops of pure water. A set of characteristic aerosol spectra obtained by these measurements for various extinction intervals is shown in Fig. 2.[3]

It is evident in Fig. 2 that the data may be somewhat incomplete, as the distributions are truncated at small sizes where the particle concentrations are still high. This spectral truncation is due to limited instrumental resolution. In addition, one should perhaps take into account the variation of $\tilde{m}$ with size; for example, the presence of soot particles in the smaller droplets may significantly affect their optical properties.[4,5] But in spite of these limitations, the data shown serve well to indicate the kind of behavior that is expected to occur typically. The trends shown in Fig. 2 are, for example, quite consistent with what one expects for water clouds,

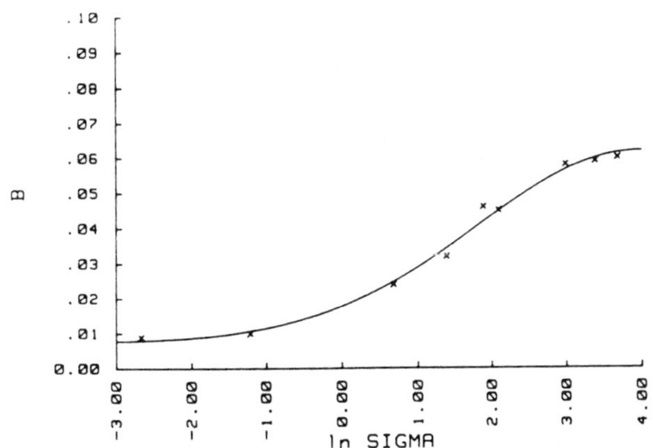

Fig. 3.   Log-Gaussian fit [Eq. (13)] of backscatter to extinction ratio vs extinction for Meppen data.

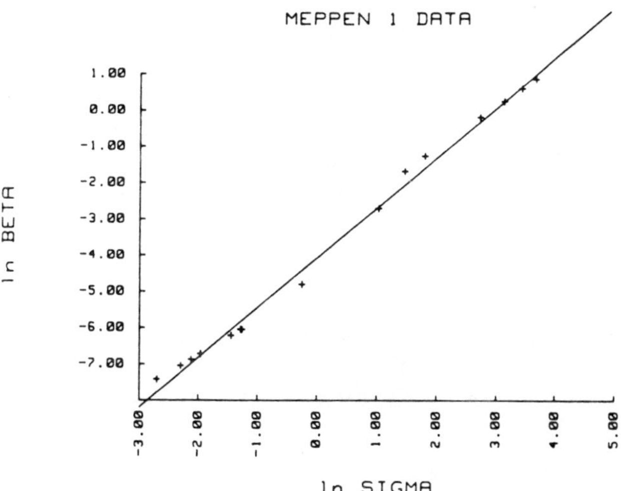

Fig. 4.   Dependence of backscatter-to-extinction ratio on extinction for Meppen data.

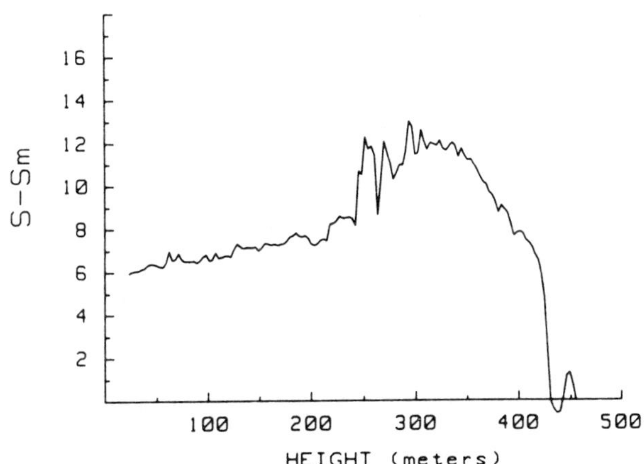

Fig. 5.   Simulated relative range-corrected logarithmic signal, $S - Sm$ [Eq. (14)], for Meppen data.

namely, that the drop spectrum will evolve toward a bimodal shape and a larger average drop size with increasing liquid water content.[6] The continuing flow of water mass up the size spectrum to larger drops is also expected to be positively correlated with increasing extinction, especially if condensation continues to replenish the spectrum at small drop sizes. Such continued production of small drops by condensation and diffusional growth is to be expected for actively growing clouds with an evolving bimodal drop distribution as long as an upward flux of moist air continues.

From the profiles of β and σ described above one can also obtain an empirical correlation between B and σ. This is shown in Fig. 3, where a lognormal curve has been fitted to a total of nine data pairs; the equation of the curve is

$$f(\sigma) = 0.0074 + 0.055 \exp\{-[(\ln\sigma - 4)/3.1]^2\}. \qquad (13)$$

One may alternatively use linear regression applied to a plot of β and σ to find values for Bo and k in the power law expression of Eq. (3). The best fit line is shown in Fig. 4 and has the parameters $Bo = 0.017$ and $k = 1.34$. The basis for the departure from $k = 1$ can be discerned by comparing Figs. 2 and 4. The plotted data in the latter figure tend to trace out a slightly undulating curve, the ends of which have a smaller slope (k near unity) than the middle section. This corresponds qualitatively to the similar unimodal shapes of spectra 1–3 and the similar bimodal shapes of spectra 6–8 in Fig. 2. Such behavior is consistent with the expectation that $k = 1$ over any interval where the spectral shape is invariant, as discussed above. In general, it is the changing proportionality between β and σ with changing spectral composition or shape that causes any overall power law description of $\beta(\sigma)$ to produce $k \neq 1$.

By integration of Eq. (4) the relative signal, $S - Sm$, appearing in the inversion solution may be obtained as a function of β and σ:

$$S - Sm = \ln\frac{\beta}{\beta_m} + 2 \int_r^{r_m} \sigma dr', \qquad (14)$$

where $\beta_m = \beta(r_m)$. Then by substituting the calculated profiles of $\beta(r)$ and $\sigma(r)$ for the Meppen data, a simulation of the relative signal is available (Fig. 5) from which one may in turn use Eq. (6) or (11) to attempt to retrieve $\sigma(r)$ for various choices of k or B. For the purpose of this paper the correct value for σ_m will be assumed as a given quantity. (The sometimes formidable but separate problem of estimating σ_m lies outside the scope of this paper.)

Let us first consider use of Eq. (6) with $k = 1$ or 1.3. The inversion results shown in Figs. 6 and 7 demonstrate that the better choice is $k = 1.3$, as expected from the previous fit of the data to the power law form. (Results like those shown in Figs. 6 and 7 have been obtained previously and independently from the Meppen data by E. Measures, ASL/WSMR; private communication.) The solution for $k = 1$ is seen to overestimate the extinction by nearly an order of magnitude in the first 250 m, where the visibility is high.

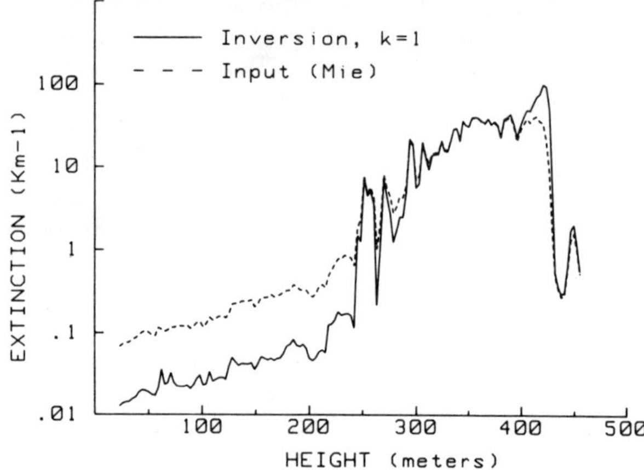

Fig. 6. For Meppen data, comparison of input profile of extinction, based on Mie theory computations over measured size distributions to inversion solution [Eq. (6)] with $k = 1$.

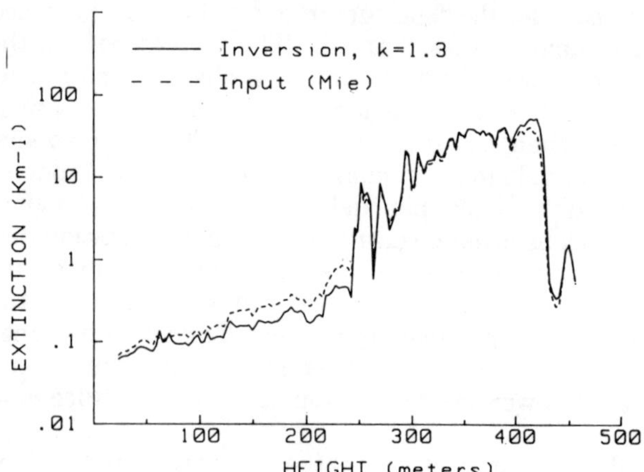

Fig. 7. Same as Fig. 6 but with $k = 1.3$.

The inversion using Eq. (11) with $k = 1$ and B given by Eq. (13) is shown in Fig. 8. The agreement with the actual extinction profile is seen to be considerably better than was possible with either of the inversions based on Eq. (6). The corresponding estimate of $B(r)$ from Eq. (13) also agrees well with the actual profile, as shown in Fig. 9. The discrepancies that do occur can be attributed in part to the limited information content of the nine data pairs on which Eq. (13) is based and also in part to the fact that the relationship between B and σ is not quite unique in principle: The procedure of representing $B(r)$ by $f[\sigma(r)]$ would be mathematically proper if there were a one-to-one transformation between B and σ. But in general it is in fact many-one, as more than one value of B may occur, at different locations for the same value of σ. However, in practice such sources of error would likely prove less important than a host of other uncertainties which arise due to factors such as noise, multiple scattering, gaseous absorption, instrumental problems, and especially errors in σ_m.

One might also attempt to improve the inversion solutions by letting the exponent k become a function of range or extinction instead of a fixed constant. For example, one might choose $k = 1$ over intervals where the drop spectral form is thought to be relatively fixed and $k = 1.3$ over the transition region from one characteristic spectral shape to another (e.g., unimodal to bimodal). However, once k ceases to be a constant, the inversion solutions Eq. (6) or (11) no longer are valid, as the passage from Eq. (4) to (5) or (8) can no longer be made unless terms describing gradients in k are also included. Numerical experiments show that these extra terms are significant, as the ploy of merely letting k become a variable in Eq. (6) or (11) produces very poor results.

IV. Inclusion of Effects of Rayleigh Scattering

An extension of the inversion solution Eq. (6) to include the effects of a background of known molecular or Rayleigh scatterers has been carried out recently by

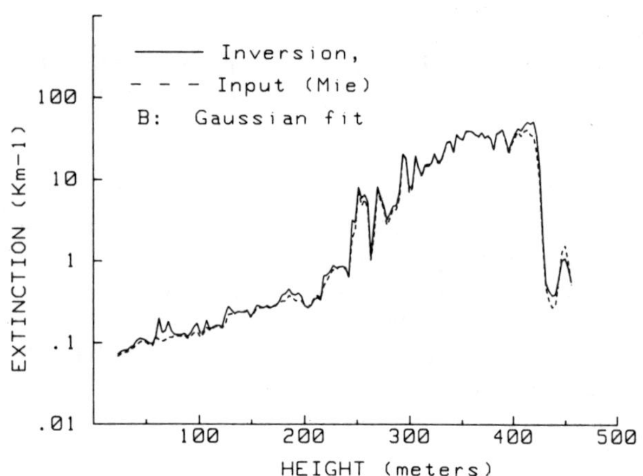

Fig. 8. For Meppen data, comparison of input and inversion profiles of extinction. Inversion via Eq. (11) with B from Eq. (13) and $k = 1$.

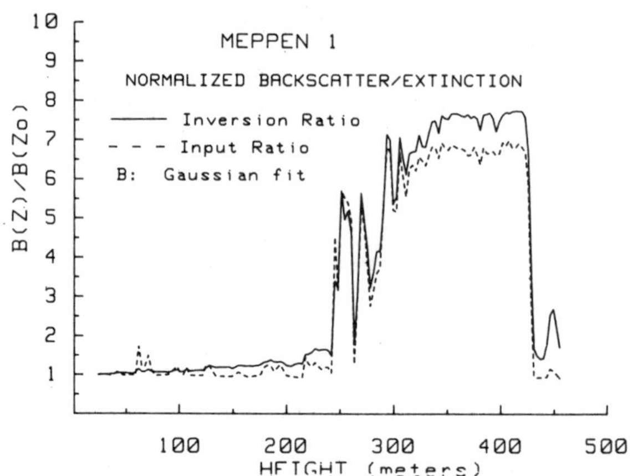

Fig. 9. For Meppen data, comparison of input and inversion profiles of backscatter-to-extinction ratio. Inversion via Eq. (11) with B from Eq. (13) and $k = 1$.

Fernald.[7] In this section a similar extension of the modified inversion solution Eq. (11) will be developed.

Let the total backscatter coefficient be expressed as

$$\beta(r) = B_P(r)\sigma_P(r) + B_R\sigma_R(z), \tag{15}$$

with

$$B_R = 3/(8\pi), \tag{16}$$

$$B_P(r) \approx f[\sigma_P(r)]. \tag{17}$$

Subscripts P and R refer to the aerosol particle and Rayleigh contributions, respectively. Thus Eq. (16) describes the fixed backscatter-to-extinction ratio for Rayleigh scatterers, while Eq. (17) describes the ratio for the particles according to formulation of Sec. II, with the function $f(\sigma_P)$ being the empirical connection between B_P and σ_P. Also the Rayleigh extinction coefficient $\sigma_R(z)$ is assumed to be a known function of height z.

On substituting Eq. (15) into the basic lidar differential equation, Eq. (4), and noting also that now $\sigma = \sigma_P + \sigma_R$, we obtain

$$\frac{dS}{dr} = \frac{1}{\beta}\frac{d\beta}{dr} - 2(B_P^{-1}\beta_P + B_R^{-1}\beta_R), \tag{18}$$

$$= \frac{1}{\beta}\frac{d\beta}{dr} - 2B_P^{-1}\beta + 2(B_P^{-1} - B_R^{-1})\beta_R. \tag{19}$$

Therefore, if one defines a new signal variable S' as

$$S' - Sm' = S - Sm + \frac{2}{B_R}\int_r^{rm}\beta_R dr' - 2\int_r^{rm}\frac{\beta_R dr'}{B_P}, \tag{20}$$

Eq. (19) can be rearranged to read as follows:

$$\frac{dS'}{dr} = \frac{1}{\beta}\frac{d\beta}{dr} - \frac{2\beta}{B_P}. \tag{21}$$

But this has the same form as Eq. (5), except for the presence of the function $B_P(r)$. The solution to Eq. (21) can, therefore, be constructed in the same fashion as Eq. (6) to obtain

$$\beta(r) = \frac{\exp(S' - Sm')}{\left[\beta_m^{-1} + 2\int_r^{rm}\frac{\exp(S' - Sm')dr'}{B_P}\right]}, \tag{22}$$

Eqs. (20) and (22) comprise the solution to the problem stated at the beginning of the section. Note that if $\beta_R \to 0$, Eq. (11) is recovered (with $k = 1$). Similarly, if B_P is a constant, Eq. (22) becomes equivalent to the solution obtained by Fernald.[7]

Implementation of Eqs. (20) and (22) would parallel the procedures already described. Examples of the application of this model will be presented elsewhere; here the goal has been merely to present formally this particular generalization to show how a known background level of scattering could be included in the formulation of Sec. II.

V. Discussion

A new theoretical framework has been presented for dealing with the problem of variations in the backscatter-to-extinction ratio that occur because of varia-tions in aerosol particle composition or spectral shape. It has been shown that if this ratio can be empirically correlated with extinction magnitude, this information can then be incorporated into a modified stable inversion solution to improve the accuracy of retrieved extinction profiles. With this physically direct and simple approach the backscatter/extinction ratio becomes a modulating factor in an otherwise linear relationship between backscatter and extinction.

The familiar alternative approach, whereby information on backscatter and extinction is cast in the form of an empirical power law relationship, has a less direct physical justification but nevertheless also has practical value with respect to the accuracy of retrieved profiles of extinction so long as the exponent is a constant representative of all the empirical data. However, attempts to refine the inversion solutions by letting the exponent become a variable parameter are ill-advised, as the available solution forms no longer apply, and serious errors may result.

An example has been given of an aerosol for which the backscatter/extinction ratio could be described approximately as a function of extinction [Eq. (13) and Fig. 3]. This was provided as an illustration of the technique of incorporating information on departures from a linear relationship between β and σ to invoke the new inversion solution. The functional form of the particular example chosen was, of course, not intended to be viewed as a universally applicable relationship. Nevertheless, some of its shape characteristics were found to be qualitatively as expected, and similar reasoning based on such factors as meteorological conditions, or the source and age of the aerosol, might similarly provide some insight into the approximate deviations from a linear relationship between β and σ in various real applications.

It should also be noted that great precision in the functional description for β/σ is not required to obtain useful results. As many numerical simulations have shown, even if nothing is known about the aerosol, the standard default assumption that β and σ are proportional may be used to obtain extinction profiles that often reflect surprisingly well the qualitative and quantitative trends of the actual distributions. This is fundamentally a consequence of the stability of the solution form given by Eq. (6). (The contrary assertion that β/σ must be very accurately known to get any useful information on σ is still made occasionally, but this is an erroneous notion that first came about from studies based on the unstable form of the lidar inversion solution. The hazards of that approach in general and in the context of the relationship between backscatter and extinction have been discussed elsewhere.[1])

Mathematically speaking, the present formulation relaxes the previous requirement that any account, theoretical or experimental, of the relationship between backscatter and extinction must be cast into an approximate power law form to make use of the standard inversion solution. With Eq. (11) any functional form, $\beta = f'(\sigma)$ can be used to obtain quickly an approximate solution for σ by substituting $B = f'(\sigma)/\sigma$ in Eq. (11) and iterating a few times as described in Sec. II. As a simple example, the usual solution based on the power law form of Eq. (3) may be recovered by substituting $B = B_0 \sigma^{k-1}$ into Eq. (11) and iterating.

Finally, the numerical examples given in this paper suggest that for the same amount of aerosol scattering information, the modulated backscatter/extinction ratio approach gives better results than the power law formulation. Further study of the new modified inversion solution will of course be required to assess more fully its practical utility.

This work was performed under contract to the U.S. Army Atmospheric Sciences Laboratory, White Sands Missile Range, N.M. 88002.

References

1. J. D. Klett, "Stable Analytical Inversion Solution for Processing Lidar Returns," Appl. Opt. **20,** 211 (1981).

2. J. D. Lindberg, Ed., "Early Wintertime European Fog and Haze: Report on Project Meppen 80," ASL-TR-0108, U.S. Army Atmospheric Sciences Laboratory, White Sands Missile Range, N.M. 88002 (1982).

3. J. D. Lindberg, R. B. Loveland, L. D. Duncan, M. B. Richardson, and J. Esparza, "Vertical Profiles of Extinction and Particle Size Distribution Measurements Made in European Wintertime Fog and Haze," ASL-TR-0151, U.S. Army Atmospheric Sciences Laboratory, White Sands Missile Range, N.M. 88002 (1984).

4. J. W. Fitzgerald, "Effect of Relative Humidity on the Aerosol Backscattering Coefficient at 0.694- and 10.6-m Wavelengths," Appl. Opt. **23,** 411 (1984).

5. C. F. Bohren, and D. R. Huffman, *Absorption and Scattering of Light by Small Particles* (Wiley, New York, 1983), 530 pp.

6. H. R. Pruppacher and J. D. Klett, *Microphysics of Clouds and Precipitation* (Reidel, Dordrecht, Holland, 1978), 714 pp.

7. F. G. Fernald, "Analysis of Atmospheric Lidar Observations: Some Comments," Appl. Opt. **23,** 652 (1984).

Author Index

Airy, G. B., On the intensity of light in the neighbourhood of a caustic *and* supplement, 329

Ångström, A., On the atmospheric transmission of sun radiation and on dust in the air, 505

———, On the atmospheric transmission of sun radiation II, 516

Bohren, C. F., Multiple scattering of light and some of its observable consequences, 165

Bordan, J., Errors inherent in the radar measurement of rainfall at attenuating wavelengths, 613

Bryant, H. C., Mie theory and the glory, 374

Chandrasekhar, S., The illumination and polarization of the sunlit sky on Rayleigh scattering, 261

Cox, A. J., Mie theory and the glory, 374

Deirmendjian, D., Scattering and polarization properties of water clouds and hazes in the visible and infrared, 210

Duntley, S. Q., The reduction of apparent contrast by the atmosphere, 595

Elbert, D. D., The illumination and polarization of the sunlit sky on Rayleigh scattering, 261

Fernald, F. G., Analysis of atmospheric lidar observations: some comments, 631

———, Determination of aerosol height distributions by lidar, 623

Fraser, A. B., The green flash and clear air turbulence, 494

———, What size of ice crystals causes the halos?, 420

———, Why can the supernumerary bows be seen in a rain shower?, 393

Greenler, R. G., Circumscribed halos, 408

———, Infrared rainbow, 378

———, Multiple-scattering effects in halo phenomena, 433

Hansen, J. E., Interpretation of the polarization of Venus, 288

Hastings, C. S., A general theory of halos, 399

Herman, B. M., Determination of aerosol height distributions by lidar, 623

Hitschfeld, W., Errors inherent in the radar measurement of rainfall at attenuating wavelengths, 613

Hovenier, J. W., Interpretation of the polarization of Venus, 288

Howell, H. B., Matrix methods for multiple-scattering problems, 149

Hulburt, E. O., Brightness and polarization of the daylight sky at various altitudes above sea level, 273

———, Explanation of the brightness and color of the sky, particularly the twilight sky, 75

Jacobowitz, H., Matrix methods for multiple-scattering problems, 149

Jayaweera, K., Numerically stable algorithm for discrete-ordinate-method radiative transfer in multiple scattering and emitting layered media, 175

Joseph, J. H., The delta-Eddington approximation for radiative flux transfer, 157

King, L. V., On the complex anisotropic molecule in relation to the dispersion and scattering of light, 50

Klett, J. D., Lidar inversion with variable backscatter/extinction ratios, 633

Können, G. P., Skylight polarization during a total solar eclipse: a quantitative model, 319

Koschmieder, H., Theorie der horizontalen sichtweite, 561

———, Theorie der horizontalen sichtweite II: kontrast und sichtweite, 584

Liou, K.-N., Solar radiative transfer in cirrus clouds. part I: single-scattering and optical properties of hexagonal ice crystals, 224

———, Solar radiative transfer in cirrus clouds. part II: theory and computation of multiple scattering in an anisotropic medium, 241

Ludlam, F. H., Noctilucent clouds, 185

Lynch, D. K., Optics of sunbeams, 608

Mallmann, A. J., Circumscribed halos, 408

Middleton, W. E. K., Random reflections on the history of atmospheric optics, 3

Nichols, E. L., Theories of the color of the sky, 35

Nussenzveig, H. M., Complex angular momentum theory of the rainbow and the glory, 380

Penndorf, R., Tables of the refractive index for standard air and the Rayleigh scattering coefficients for the spectral region between 0.2 and $20.0\,\mu$ and their application to atmospheric optics, 81

Lord Rayleigh (J. W. Strutt), On the light from the sky, its polarization and colour, 9

————, On the transmission of light through an atmosphere containing small particles in suspension, and on the origin of the blue of the sky, 26

Lord Rayleigh (R. J. Strutt), Normal atmospheric dispersion as the cause of the "green flash" at sunset, with illustrative experiments, 473

Reagan, J. A., Determination of aerosol height distributions by lidar, 623

Rozenberg, G. V., Light scattering in the earth's atmosphere, 88

Schuster, A., Radiation through a foggy atmosphere, 127

Sekera, Z., Light scattering in the atmosphere and the polarization of sky light, 312

Shaw, G. E., Observations and theoretical reconstruction of the green flash, 481

Simpson, G. C., Coronae and iridescent clouds, 358

Stamnes, K., Numerically stable algorithm for discrete-ordinate-method radiative transfer in multiple scattering and emitting layered media, 175

Takano, Y., Solar radiative transfer in cirrus clouds. part I: single-scattering and optical properties of hexagonal ice crystals, 224

————, Solar radiative transfer in cirrus clouds. part II: theory and computation of multiple scattering in an anisotropic medium, 241

Tape, W., Some ice crystals that made halos, 427

Tousey, R., Brightness and polarization of the daylight sky at various altitudes above sea level, 273

Tränkle, E., Multiple-scattering effects in halo phenomena, 433

Tricker, R. A. R., A simple theory of certain heliacal and anthelic halo arcs: the long hexagonal ice prism as a kaleidoscope, 412

Tsay, S.-C., Numerically stable algorithm for discrete-ordinate-method radiative transfer in multiple scattering and emitting layered media, 175

Twomey, S., Matrix methods for multiple-scattering problems, 149

————, The influence of pollution on the shortwave albedo of clouds, 220

van de Hulst, H. C., A theory of the anti-coronae, 367

Wegener, A., Elementare theorie der atmosphärischen spiegelungen, 445

Weinman, J. A., The delta-Eddington approximation for radiative flux transfer, 157

Wilson, R., The blue sun of 1950 September, 546

Wiscombe, W. J., Numerically stable algorithm for discrete-ordinate-method radiative transfer in multiple scattering and emitting layered media, 175

————, The delta-Eddington approximation for radiative flux transfer, 157

Young, A. T., On the Rayleigh-scattering optical depth of the atmosphere, 115

————, Rayleigh scattering, 118

Subject Index

absorption, by atmospheric aerosols, 220
aerosol height distributions, 623
Airy theory, 329, 380, 393
anti-corona, 367, 374, 380
arcs and halos, 224, 241, 399, 412, 427, 433
atmospheric aerosols
    absorption by, 220
    scattering by, 35, 481, 505, 516, 623
atmospheric contrast, 608
atmospheric dispersion, 50, 81, 473, 481
atmospheric optics, history of, 3, 35, 88, 118
atmospheric refraction, 445, 473, 494
atmospheric transmission, 26, 88
    of sunlight, 505, 516, 546

backscattering, 367, 374, 613, 623, 631, 633
blue moon, 546
blue sun, 546
bows
    cloud, 210
    supernumerary, 393

cloud bow, 210
clouds
    albedo of, 220
    cirrus, 224, 241
    noctilucent, 185
    of Venus, 288
complex angular momentum theory, 380
contrast reduction, 584, 595
coronas, 358
crepuscular rays, 608

delta-Eddington theory, 157
discrete ordinate method, 175
dispersion, atmospheric, 50, 81, 473, 481
distortion, solar, 445, 494

extinction, 50, 88, 115, 505, 516, 546, 633

glory, 210, 367, 374, 380
green flash, 473, 481, 494

halos, 408, 420
    and arcs, 224, 241, 399, 412, 427, 433

ice crystals
    orientation of, 420
    scattering by, 224, 241, 399, 408, 412, 420

iridescence, 358

lidar equation, 613, 623, 631, 633

Mie theory, 210, 367, 374, 380
mirages, 445, 494
molecular anisotropy, 50, 115, 118

neutral points, 261

particle size distribution, 210

radiative transfer, 127, 149, 157, 165, 175, 241, 312
rainbow, 3, 329, 378, 380, 393
Rayleigh scattering, 9, 75, 115, 118, 261, 312, 633
refraction, atmospheric, 445, 473, 494
refractive index of air, 81

scattered light, polarization of, 241, 288, 367
scattering
    by atmospheric aerosols, 35, 481, 505, 516, 623
    by clouds, 165, 185, 210, 241
    by droplets, 210, 220, 288, 329, 358, 367, 546
    by drops, 374, 378, 393
    by ice crystals, 224, 241, 399, 408, 412, 420
    infrared, 378
    by molecules, 26, 50, 81, 118
    multiple, 127, 149, 157, 165, 175, 241, 261, 312, 433
        in anisotropic media, 241
    Rayleigh, 9, 75, 115, 118, 261, 312, 633
scattering coefficient of air, 81
sky brightness, 75, 88, 273
sky color, 3, 9, 26, 35, 75, 88, 118, 165
sky polarization, 9, 88, 261, 273, 312, 319
solar distortion, 445, 494
Stokes parameter, 261, 312
sun pillars, 427
sunlight, atmospheric transmission of, 505, 516, 546

transmission
    atmospheric, 26, 88
        of sunlight, 505, 516, 546
    by drops, 613
twilight color, 75
two-stream theory, 127, 165

visibility, 3, 561, 584, 595
visual range, 3, 561, 584, 595